T0111869

ARTIFICIAL INTELLIGENCE, BLOCKCHAIN, COMPUTING AND SECURITY, VOLUME 2

This book contains the conference proceedings of ICABCS 2023, a non-profit conference with the objective to provide a platform that allows academicians, researchers, scholars and students from various institutions, universities and industries in India and abroad to exchange their research and innovative ideas in the field of Artificial Intelligence, Blockchain, Computing and Security.

It explores the recent advancements in the field of Artificial Intelligence, Blockchain, Communication and Security in this digital era for novice to profound knowledge about cutting edges in Artificial Intelligence, financial, secure transaction, monitoring, real time assistance and security for advanced stage learners/ researchers/ academicians. The key features of this book are:

- Broad knowledge and research trends in AI and Blockchain with security and their role in smart living assistance
- Depiction of system model and architecture for clear picture of AI in real life
- Discussion on the role of AI and Blockchain in various real-life problems across sectors including banking, healthcare, navigation, communication, security
- Explanation of the challenges and opportunities in AI and Blockchain based healthcare, education, banking, and related industries

This book will be of great interest to researchers, academicians, undergraduate students, postgraduate students, research scholars, industry professionals, technologists and entrepreneurs.

PROCEEDINGS OF THE INTERNATIONAL CONFERENCE ON ARTIFICIAL INTELLIGENCE, BLOCKCHAIN, COMPUTING AND SECURITY (ICABCS 2023), GR. NOIDA, UP, INDIA, 24–25 FEBRUARY 2023

Artificial Intelligence, Blockchain, Computing and Security

Volume 2

Edited by

Arvind Dagur
School of Computing Science and Engineering, Galgotias University, Gr. Noida

Karan Singh
School of Computer & Systems Sciences, JNU New Delhi

Pawan Singh Mehra
Department of Computer Science and Engineering, Delhi Technological University, New Delhi

Dhirendra Kumar Shukla
School of Computing Science and Engineering, Galgotias University, Gr. Noida

CRC Press is an imprint of the
Taylor & Francis Group, an **informa** business
A BALKEMA BOOK

First published 2023
by CRC Press/Balkema
4 Park Square, Milton Park, Abingdon, Oxon, OX14 4RN
and by CRC Press/Balkema
2385 NW Executive Center Drive, Suite 320, Boca Raton FL 33431

CRC Press/Balkema is an imprint of the Taylor & Francis Group, an informa business

British Library Cataloguing-in-Publication Data

A catalogue record for this book is available from the British Library

Library of Congress Cataloging-in-Publication Data
A catalog record has been requested for this book

SET
ISBN: 978-1-032-66966-3 (hbk)
ISBN: 978-1-032-68590-8 (pbk)

Volume 1
ISBN: 978-1-032-49393-0 (hbk)
ISBN: 978-1-032-49397-8 (pbk)
ISBN: 978-1-003-39358-0 (ebk)
DOI: 10.1201/9781003393580

Volume 2
ISBN: 978-1-032-67841-2 (hbk)
ISBN: 978-1-032-68498-7 (pbk)
ISBN: 978-1-032-68499-4 (ebk)
DOI: 10.1201/9781032684994

Typeset in Times New Roman
by MPS Limited, Chennai, India

Table of Contents

Smart Intelligent Computing in Advanced Sectors (SICAS)

Sharing multimedia data using deep learning approaches

Deep learning, AI and IOT for computer vision and blockchain

Artificial Intelligence, Blockchain, Computing and Security – Dagur et al. (Eds)
© 2024 The Editor(s), ISBN: 978-1-032-67841-2

Preface

On the behalf of organising committee, I would like to extend my heartiest welcome to the first international conference on Artificial Intelligence, Blockchain, Computing and Security (ICABCS 2023).

ICABCS 2023 is a non-profit conference and the objective is to provide a platform for academicians, researchers, scholars and students from various institutions, universities and industries in India and abroad, to exchange their research and innovative ideas in the field of Artificial Intelligence, Blockchain, Computing and Security. We invited all students, research scholars, academicians, engineers, scientists and industrialists working in the field of Artificial Intelligence, Blockchain, Computing and Security from all over the world. We warmly welcomed all the authors to submit their research in conference ICABCS 2023 to share their knowledge and experience among each other.

This two-day international conference (ICABCS 2023) was organized at Galgotias University on 24th and 25th February 2023. The inauguration was done on 24th February 2023 at Swami Vivekananda Auditorium of Galgotias University. In the inauguration ceremony, Professor Shri Niwas Singh, Director, Atal Bihari Bajpai Indian Institute of Information Technology and Management, Gwalior attended as Chief Guest. Professor Rajeev Tripathi, former Director Motilal Nehru National Institute of Technology Allahabad and Professor D.K. Lobiyal, Jawaharlal Nehru University attended as Guests of Honour. In the inauguration ceremony of the program, the Vice-Chancellor of the University, Professor K. Mallikarjuna Babu, Advisor to Chancellor, Professor Renu Luthra, Dean SCSE, Professor Munish Sabharwal welcome the guests with welcome address. The Registrar, COE and Deans of all the Schools were present. Conference Chair Professor Arvind Dagur told that in this conference more than 1000 research papers were received from more than ten countries, on the basis of blind review of two reviewers, more than 272 research papers were accepted and invited for presentation in the conference. The Chief Guest, Honorable Guests and Experts delivered lectures on Artificial Intelligence, Block Chain and Computing Security and motivated the participants for quality research. The Pro Vice-Chancellor, Professor Avadhesh Kumar, delivered the vote of thanks to conclude the inauguration ceremony. During the two-day conference, more than 272 research papers were presented in 22 technical sessions. The closing ceremony was presided over by Prof. Awadhesh Kumar, Pro-VC of the University and Conference Chair Professor Arvind Dagur, on behalf of the Organizing Committee. Conference Chair, Professor Arvind Dagur thanked Chancellor Mr. Sunil Galgotia, CEO Mr. Dhruv Galgotia, Director Operation Ms. Aradhana Galgotia, Vice Chancellor, Pro Vice-Chancellor, Registrar, Dean SCSE, Dean Engineering and university family for their co-operation and support.

Finally, once again I would like to thank to all participants for their contribution to the conference and all the organising committee members for their valuable support to organise the conference successfully. I highly believed that this conference was a captivating and fascinating platform for every participant.

On the behalf of editors
Dr. Arvind Dagur

Acknowledgements

It gives me immense pleasure to note that Galgotias University, Greater Noida, India is organizing the International Conference on Artificial Intelligence, Blockchain, Computing and Security (ICABCS 2023) on 24th and 25th February 2023. On behalf of the organizing committee, I would like to convey my sincere thanks to our Chief Patron, Honorable Shri Sunil Galgotia, Chancellor, GU and Hon'ble Shri Dhruv Galgotia CEO, GU for providing all the necessary support and facilities required to make ICABCS-2023 a successful conference. I convey my thanks to Prof. (Dr) K. Mallikharjuna Babu, Vice Chancellor and Prof. (Dr) Renu Luthra advisor to the chancellor for their continuous support and encouragement, without which it was not possible to achieve. I want to convey my sincere thanks to them for providing technical sponsorship and for showing their confidence in Galgotias University to provide us the opportunity to organize ICABCS-2023 and personally thank to all the participants of ICABCS 2023. I heartily welcome all the distinguished keynote speakers, guest, session chairs and all the authors presenting papers. In the end, I would convey my thanks to all the reviewers, organizing committee members, faculty and student volunteers for putting their effort into making the conference ICABCS 2023 a grand success.

Thank you,
Prof.(Dr.) Arvind Dagur
Organizing Chair ICABCS 2023,
Galgotias University

Artificial Intelligence, Blockchain, Computing and Security – Dagur et al. (Eds)
© 2024 The Editor(s), ISBN: 978-1-032-67841-2

Committee Members

Scientific committee

Prof. Valentina Emilia Balas
Aurel Vlaicu University of Arad, Romania

Prof. Toshio Fukuda
Nagoya University, Japan

Dr. Vincenzo Piuri
University of Milan, Italy

Dr. Ahmad Elngar
Beni-Suef University, Egypt

Dr. Malik Alazzam
Lone Star College – Victory Center. Houston,TX, United States

Dr. Osamah Ibrahim Khalaf
Profesor, Al-Nahrain University, College of Information Engineering, Baghdad, Iraq

Dr. TheyaznHassnHadi
King Faisal University, Saudi Arabia

Md Atiqur Rahman Ahad
Osaka University, Japan, University of Dhaka, Bangladesh

Prof. (Dr) Sanjay Nadkarni
Director of Innovation and Research, The Emirates Academy of Hospitality Management, Dubai, UAE

Dr. Ghaida Muttashar Abdulsahib
Department of Computer Engineering, University of Technology, Baghdad, Iraq

Dr. R. John Martin
Assistant Professor, School of Computer Science and Information Technology, Jazan University

Dr. Mohit Vij
Associate Professor, Liwa College of Technology, Abu Dhabi, United Arab Emirates

Dr. Syed MD Faisal Ali khan
Lecture & Head – DSU, CBA, Jazan University

Dr. Dilbag Singh
Research Professor, School of Electrical Engineering and Computer Science, Gwangju Institute of Science and Technology, South Korea

Dr. S B Goyal
Dean & Director, Faculty of Information Technology, City University, Malaysia

Dr. Shakhzod Suvanov
Faculty of Digital Technologies, Department of Mathematical Modeling, Samarkand State University, Samarkand Uzbekistan

National Advisory Committee

Organizing committee

Chief Patron
Shri Suneel Galgotia,
Chancellor, Galgotias University, Greater Noida, India

Patrons
Shri Dhruv Galgotia,
CEO, Galgotias University, Greater Noida, India
Prof.(Dr.) Mallikharjuna Babu Kayala,
Vice-Chancellor, Galgotias University, Greater Noida, India
Ms. Aradhna Galgotia,
Director Operations, Galgotias University, Greater Noida, India

General Chairs
Prof. (Dr.) Avadhesh Kumar,
Pro-VC, Galgotias University, Greater Noida, India
Prof. (Dr.) Munish Sabharwal,
Dean, SCSE, Galgotias University, Greater Noida, India

Conference Chairs
Prof. (Dr.) Arvind Dagur,
Professor, Galgotias University, Greater Noida, India
Dr. Karan Singh,
Professor, JNU New Delhi, India
Dr. Pawan Singh Mehra, DTU, New Delhi

Conference Co-Chairs
Prof. (Dr.) Dr. Amit Kumar Goel,
HOD (CSE) and Professor, Galgotias University, Greater Noida, India
Prof. (Dr.) Krishan Kant Agarwal,
Professor, Galgotias University, Greater Noida, India
Dr. Dhirendra Kumar Shukla,
Associate Professor, Galgotias University, Greater Noida, India

Organizing Chairs
Dr. Abdul Aleem,
Associate Professor, Galgotias University, Greater Noida, India
Dr. Vikash Kumar Mishra,
Assistant Professor, Galgotias University, Greater Noida, India

Technical Program Chairs
Dr. Shiv Kumar Verma, Professor, SCSE, Galgotias University
Dr. SPS Chauhan, Professor, SCSE, Galgotias University
Dr. Ganga Sharma, Professor, SCSE, Galgotias, University
Dr. Anshu Kumar Dwivedi, Professor, BIT, Gorakhpur

Finance Chair
Dr. Aanjey Mani Tripathi, Associate Professor, Galgotias University
Dr. Dhirendra Kumar Shukla, Associate Professor, Galgotias University

Conference Organizing Committee
Dr. Gambhir Singh, Professor, Galgotias University
Dr. Arvinda Kushwaha, Professor, ABESIT, Ghaziabad
Dr. Sanjeev Kumar Prasad, Professor, SCSE, Galgotias University
Dr. Sampath Kumar K, Professor, Galgotias University
Dr. Vimal Kumar, Associate Professor, Galgotias University
Dr. T. Ganesh Kumar, Associate Professor, Galgotias University
Dr. Atul Kumar Singh, Assistant Professor, Galgotias University
Dr. Anuj Kumar Singh, Assistant Professor, Galgotias University

Media and Publicity Chairs
Dr. Ajay Shanker Singh, Professor, Galgotias University
Dr. Ajeet Kumar, Professor, Galgotias University
Dr. Santosh Srivastava, Professor, Galgotias University

Cultural Program Chairs
Ms. Garima Pandey, Assistant Professor, Galgotias University
Ms. Heena Khera, Assistant Professor, Galgotias University
Ms. Ambika Gupta, Assistant Professor, Galgotias University
Ms Kimmi Gupta, Assistant Professor, Galgotias University

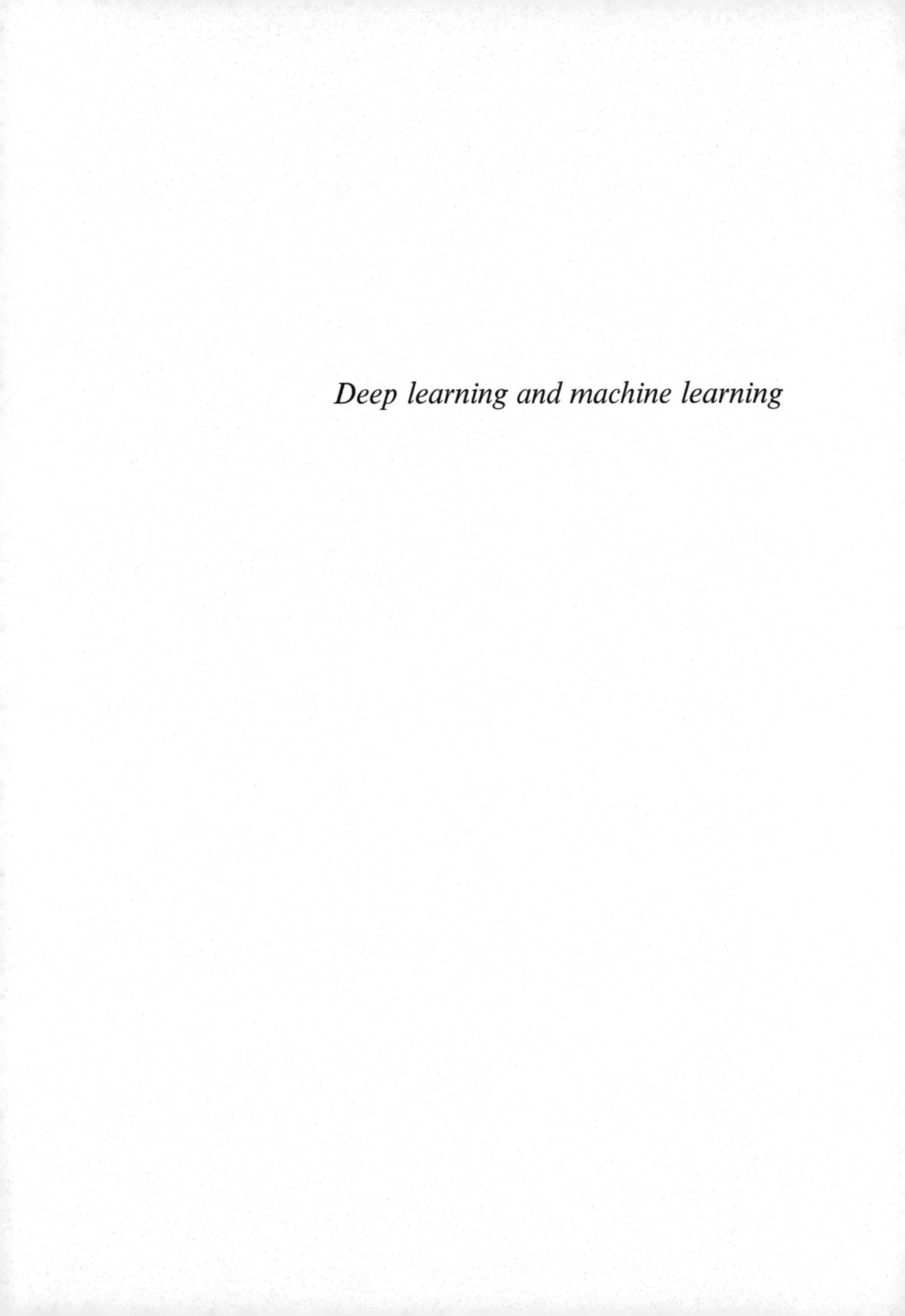

Deep learning and machine learning

Artificial Intelligence, Blockchain, Computing and Security – Dagur et al. (Eds)
© 2024 The Author(s), ISBN: 978-1-032-67841-2

Machine learning model for cardiovascular disease prediction

Arundhati Uplopwar & Sunil Kumar
Department of Computer Science & Engineering, Amity University, Noida, India

Arvinda Kushwaha
Department of Computer Science & Engineering, GCET, Greater Noida, India

ABSTRACT: Cardiovascular Disease is a major global health issue specially post COVID era and could increase to 8 million by 2030. The major cause of chronic heart disease in younger population nowadays is due to stressful lifestyle, obesity, eating habits, lack of physical moments, post COVID symptoms. Moreover major population in our country is least aware about the regular routine health checkup and hence Cardiovascular Disease comes out to be the highest cause of life threatening disease. Machine learning has become important tool to predict early occurrence of the Heart diseases so that the mortality rate could be reduced to quite extent. Auto Prognosis optimizes the pipeline configuration efficiently that automatically selects best possible ML Model on any dataset. Auto sklearn in AutoMl is used on Kaggle Heart disease dataset in study by building a classifier predicting 86% accuracy.

1 INTRODUCTION

According to WHO Heart disease also known as Cardiovascular Diseases are major cause to death today. After the post COVID period there is 17% raise the CVD in quite younger population. The major attribute responsible for CVD is Heart Rate (HR).Normal Heart Rate of a person is 60 to 100 beats per Minute. However the other factors which may cause variations in the HR or abnormal heart bit is due to unhealthy life style like Lack of exercise, Smoking habits, High Blood pressure, Diabetes, increased stress level, Fast food, Regular use of alcohol, obesity etc doubles the risk of Various Cardiovascular diseases, so it becomes very important to detect the CVD as early as possible.

Recent advancement in the field of Artificial Intelligence and Machine learning is creating a revolution in health care industry to predict and diagnose the early detection of heart disease so that the precautionary measures and treatment can be taken earlier. Big data in Medical industry is used to analyze, preprocessed for intelligent disease prediction by building a model based upon the large dataset. Machine learning focus on the development of computer programs which access a large amount of data and learn from themselves by building the relationship between the variables. Major Cardiovascular diseases are Coronary Artery disease, Congestive heart failure, Arrhythmia, stroke, Peripheral artery disease, congenital heart disease, and Myocarditis. Heart disease can be diagnosed by chest X-ray, blood test, Electrocardiogram or EKG quickly records the electrical signals of heart. In this paper section II shows recent related work done, Section III represents the general classification Model for Machine Learning Algorithms, Section IV represents a general Prediction model for Heart Disease prediction using ML, Section V Represents a Proposed Model for CVD prediction using Auto Prognosis, section VI represents the Results and discussion followed by conclusion.

DOI: 10.1201/9781032684994-1

2 RELATED WORK

In one of the recent study being done on Cardiovascular Disease Prediction various Machine Learning algorithms were compared in which Random Tree model performed well with very less prediction time of 0.01 sec [1]. In another study carried out for CVD prediction XGBoost was used to test alternative Decision Tree classification algorithm to improve the accuracy of model and two SVM classifier were used in the framework where XGBoost gave 95% accuracy [2].

ECG is a diagnostic technique which records heart's electrical signal, in a study conducted recently Radial Basis Function Neural Network (RBFNN) is used to measure local minima in the signal to identify normal and abnormal ECG signals [3]. In another study carried out four classification models MLP,SVM,RF and NB were evaluated on the dataset where in SVM performed best with 91.67% accuracy [4]. In another recent study carried out to detect heart disease in two step process on dataset taken from UCI Repository, various Machine Learning algorithms LR, SVM, KNN, RF, Gradient Boosting classifiers were used on dataset in which Logistic regression gave high accuracy rate of 95% [5]. Coronary Heart disease was predicted in one of the study using 11 ML classifiers wherein 95% accuracy was achieved by Gradient Boosted Tree and RF gave 96% accuracy [6]. In one of the study carried out on Coronary Artery classification score 5 ML models were performed on 31 variables were RF, RBFNN, Kernel ride regression(KRR), SVM and KNN in which RF achieved the best performance accuracy [7]. In one of the trending study carried out on Heart disease prediction, performance of ARIMA, Linear Regression, SVR ,KNN, Decision Tree Regressor, Random Forest Regressor and LSTM was performed in which ARIMA model and LR were effective in predicting Heart rate [8]. In one of the study carried out SMOTE based technique is used on unbalanced dataset using XGBoost algorithm and found that XGBoost algorithm performed best [9]. CVD had significant rise after post acute COVID-19 period. Various Machine learning algorithms Logistic Regression, Random Forest Classifier and XGBoost classifier to study development of heart failure during the post COVID period and prominent results were seen [10]. In a recent study Automated Machine learning performance is used to develop optimized machine learning pipeline without requiring significantly technical expertise using Auto prognosis [11]. In one of the study conducted on heart disease prediction AutoML acts as best tool for imbalanced healthcare datasets [13].

3 CLASSIFICATION OF MACHINE LEARNING ALGORITHMS FOR HEART DISEASE PREDICTION

Machine Learning methods are emerging tools used to predict and diagnose the heart disease at an early stage so that some preventive measures can be taken. Various categories of machine Learning Algorithms are shown in Figure 1 [12].

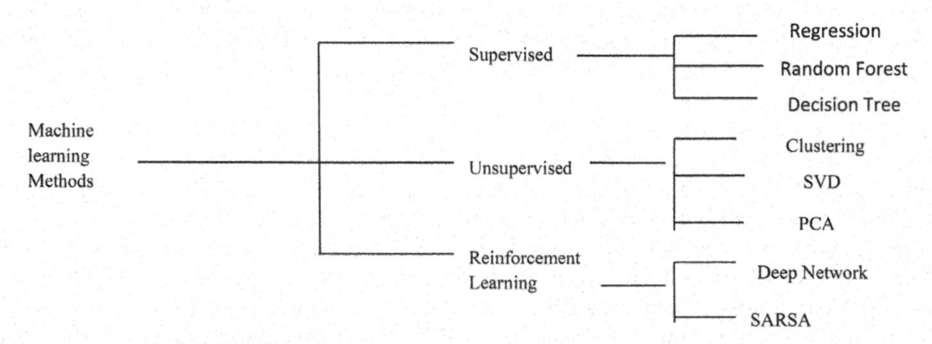

Figure 1. Machine learning algorithms.

4 GENERAL PREDICTION MODEL FOR HEART DISEASE DETECTION USING ML

The dataset for heart disease present in various repository based upon different attributes are taken as a raw information which is then cleaned, preprocessed, Machine learning classification techniques are applied on the trained dataset and then disease prediction model is build as shown in Figure 2.

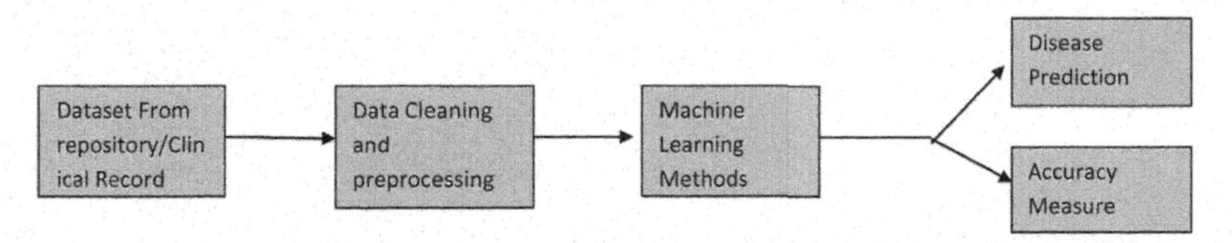

Figure 2. Disease prediction model using ML.

4.1 *Challenges faced by machine learning model*

1. Data Size-Machine Learning Algorithm generally shows Low performance on large data size, hence data sample size is key factor to estimate the accuracy level of Model.
2. There is no specific or standard representation of data splitting for Training and Testing, Some study splits the data into 80% Training and 20% Testing or 70% Training and 30% Testing.
3. Machine Learning Model does not update itself with respect to the additional or new predictive variables or dataset thus making interpretation more challenging.

4.2 *Pipeline machine learning*

Machine Learning pipeline is a way to automate the work flow to produce Machine Learning model.ML pipeline consist of sequential steps from data extraction, data cleaning, data preprocessing to Model training and deployment.

Auto Prognosis 2.0 is a framework and software package that allows Healthcare professional to use Machine Learning to develop diagnostic Model. Auto prognosis uses the features of AutoML. Automated Machine learning improves the efficiency of machine learning and act as important resource to Non machine learning experts. AutoML preprocess the data, selects appropriate Model, optimize hyper parameters and analyze the results obtained. Auto Prognosis tool empowers health care professionals with the following features

- Automatically Builds and optimize Machine Learning pipelines.
- Allows the professionals to understand the value of information by enabling proper selection of variable thus streamlining the complex phases in Machine learning workflow.
- Provides interpretability by debugging and automatically updates the system with the variations in the dataset.
- Tune model hyper parameters automatically to identify the best performing Model.

5 PROPOSED MODEL FOR HEART DISEASE PREDICTION USING AUTO PROGNOSIS 2.0

Auto Prognosis takes Raw Clinical dataset and handles missing data imputation, future processing, Model selection, interpretation by creating a ML pipeline optimization and configuration thus reducing the technical expertise to build clinical Model as shown in Figure 3.

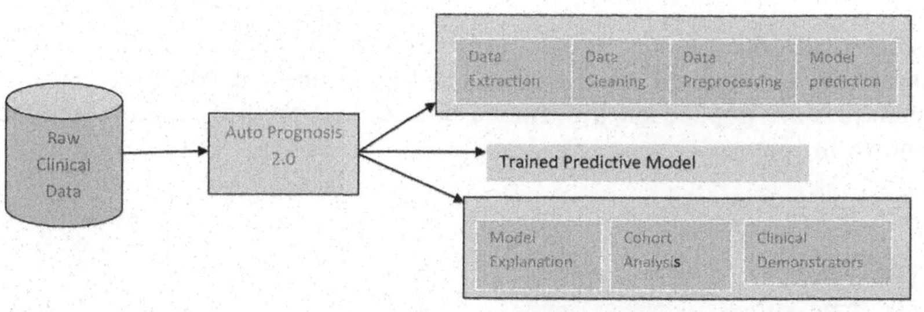

Figure 3. Disease prediction model using auto prognosis.

6 RESULTS AND DISCUSSION

The dataset was collected from kaggle Repository based upon the attributes age, sex, chest pain, cholesterol level, treetops, fbs ect, autos learn tool is used in AutoML.Preprocessed data set on heart disease is then visualized using visualization tool pipeline profiler on Google colaboratory platform using python. The accuracy check was performed with target attribute as age. The accuracy came out to be 86% . The dataset was split on x and y values and 20% data was used for training purpose. Overall Model is already sorted in descending order of their accuries, linear support vector Machine showed highest accuracy and balancing activity carried out while creating the classifier. It is seen that Autos learn using AutoML predicted higher accuracy on the following dataset. The dataset consist of 14 attributes (columns) and 1025 rows. The statistical outline of subset attributes are shown in Table 1.

Table 1. Statistical outline of subset attributes.

age	sex	cp	trestbps	chol	fbs	restecg	thalach	exang	oldpeak	slope	ca	thal	target	
0	52	1	0	125	212	0	1	168	0	1.0	2	2	3	0
1	53	1	0	140	203	1	0	155	1	3.1	0	0	3	0
2	70	1	0	145	174	0	1	125	1	2.6	0	0	3	0
3	61	1	0	148	203	0	1	161	0	0.0	2	1	3	0
4	62	0	0	138	294	1	1	106	0	1.9	1	3	2	0

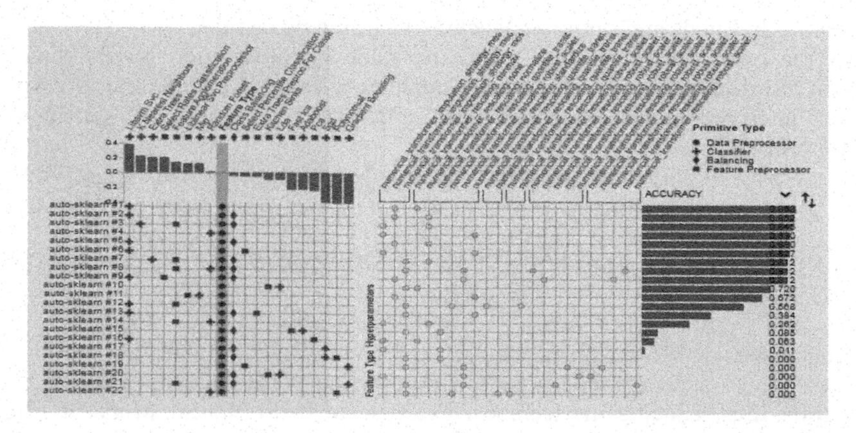

Figure 4. Result analysis.

7 CONCLUSION

Machine Learning acts as popular and most widely used techniques for predicting cardio-vascular diseases.AutoPrognosis uses AutoML and pipelining in machine learning to improve the efficiency by reducing complex phases of Machine learning algorithms for developing and deploying models accurately. This paper shows the use of Automated Machine Learning applied on the Heart disease dataset to predict the accuracy.

REFERENCES

[1] *Clinical Data Analysis for Prediction of Cardiovascular Disease using Machine Learning Techniques*, Rajkumar Gangappa Nadakinamani *et al.*, National Library of Medicine, Jan 2022, PMID 35069715.

[2] *Machine Learning Technology Based Heart Disease Detection Models*, Umarani Nagavelli *et al.*, Hindawi, JOHAE, Volume 2022, Article ID 7351061, 9 pages.

[3] *An Innovative Machine Learning Approach for Classifying ECG Signal in Healthcare Devices*, Kishore B *et al.*, Hindawi, JOHAE, Volume 2022, Article ID 7194419, 16 pages.

[4] *Heart Disease Prediction Using Machine Learning,* Chaimaa *et al.*, *ASET*, IEEE Xplore, 2022, Accession Number 21668689.

[5] *Heart Disease Prediction using Machine Learning Techniques*, Mohammed Khalid Hossen, AJOCST, 2022, 5(3):146–154.

[6] *Effectively Predicting the Presence of Coronary Heart Disease Using Machine Learning Classifiers*, Ch Anwar UL Hassan *et al.*, Sensors, 2022, 22, 7227, MDPI.

[7] Yue Huang *et al.*, *Computers in Biology and Medicine*, 151(2022),106297.

[8] *Predictive Analysis of Heart Rate using Machine Learning* Techniques, Matthew Oyeleye *et al.*, NLM, 2022 Feb, 19;19(4):2417, PMID 35206603.

[9] *A Heart Disease Prediction Model Based on Future Optimization and Smote-XGboost Algorithm*, Jian Yang and Jinan Guan, Information 2022, 13, 475.

[10] *Using Machine Learning to Predict Development of Heart Failure, During Post-Accute COVID-19,* by Race and Ethnicity, Emily Cathay *et al.*, 2022, SIEDS, IEEE.

[11] Auto Prognosis 2.0: Democratizing Diagnostic and Prognostic Modeling in Healthcare with Automated Machine Learning, Fergus Imrie *et al.*, *arXiv*: 2210.12090v1[cs.LG], 2022.

[12] *Machine Learning Algorithms-A Review*, Atta Mahesh, IJSR, Volume 9, Issue 1, January 2020, pp. 381–386.

[13] Benchmarking AutoML Frameworks for Disease Prediction using Medical Claims, Roland Albert. A. Romero *et al. BioData Mining*, (2022) 15:15, pp. 1–13.

Artificial Intelligence, Blockchain, Computing and Security – Dagur et al. (Eds)
© 2024 The Author(s), ISBN: 978-1-032-67841-2

A comprehensive review on breast cancer prediction using deep neural network

Aditee Singh, Anuraj Jain, Anmol Mishra, Sur Singh Rawat & G. Mahesh
Department of Computer Science and Engineering JSS Academy of Technical Education, Noida

Gambhir Singh
School of Computing Sciences and Engineering, Galgotias University, Greater Noida, India

ABSTRACT: Breast cancer is the largest and most prevalent cancer type among women, according to the World Health Organization (WHO). Breast cancer affects millions of women annually and is the most common cancer disease in women. It also accounts for the majority of women's cancer-related deaths. Lung cancer is the primary cause of cancer-related death, followed by breast cancer. Early detection is essential for survival and the elimination of this issue. There are two methods one is Early diagnosis and other is screening which are available for earlier detection of breast cancer. Early diagnostic initiatives based on knowledge of early signs and symptoms and timely referral to diagnosis Based on the attributes that the data provides, several machine learning algorithms are utilised to determine whether a tumour is benign or malignant. The proposed work can be used to forecast the results of many techniques, and based on the need, the appropriate technique can be applied.

Keywords: Breast cancer, Prediction, Diagnosis, Screening, Machine learning, Benign, Malignant, Analysis

1 INTRODUCTION

Cancer is a terrible disease that develops when our bodies' normal cells begin to divide abnormally and come into contact with other normal cells, turning them malignant. These cells have the potential to disperse throughout the body, obliterating healthy body cells, and forming a mass known as a tumour [1]. Breast cancer develops from abnormal cell growth in the breast, and the most common symptoms are changes in the breast's form and skin, as well as lumps. Breast cancer can also affect men, despite the fact that it is an invasive malignancy in women [2].

Lung cancer is the leading cause of death for women, with breast cancer coming in second. According to history, breast cancer is the most common and dangerous disease in the world. Either invasive or non-invasive treatment options are possible. While non-invasive cancers are pre-cancerous and remain in the original organ, invasive cancers are cancerous, malignant, and spread to other organs. They do, however, eventually progress to aggressive breast cancer.

2 TECHNOLOGY

We will employ the process of transfer learning by enhancing the images from the provided data set in order to detect the type of breast cancer known as infiltrating ductal carcinoma, which begins in the milk ducts and spreads to adjacent tissues.

 DOI: 10.1201/9781032684994-2

2.1 *Under sampling and data preparation*

It is a regular occurrence for there to be an imbalance between the quantity of photographs that reveal positive occurrence of the disease and those that do not while images are being captured and collected for medical analysis.

The development of this imbalance could result in improper model training. One type of result may be trained into the model to be more or less reliably detected than the other category.

In order to get around this restriction, we resample our dataset.

2.2 *Augmentation of image*

A method used to increase the size of the dataset accessible is image augmentation. It is used to get over the data shortage and to train the model to work with a variety of varied images. The efficiency of the model may be constrained as a result of training it on a certain sort of image. Making minor changes to the photos, such as zooming, mirror imaging, or rotating the image in either direction, can result in a brand-new, unexpected image that extends the images from the current data set and improves the model's effectiveness.

2.3 *Machine learning*

In the discipline of machine learning, machines are taught to carry out activities without human assistance. The computer then learns how to carry out the activity as a human would and finds ways to do so. In machine learning, three different sorts of approaches are employed.

2.3.1 *Supervised learning*

This method tries to train the computer based on the findings after providing it with both the label and the data. Naive bays, random forests, support vector machines, etc. are a few examples.

2.3.2 *Unsupervised learning*

With this method, all that is available is the dataset—no labels. One example of this method is clustering.

2.3.3 *Reinforcement learning*

Reinforcement learning is a method where a computer learns from its errors and keeps learning until the output.

2.4 *Deep learning*

Deep Learning is a branch of machine learning that deals with neural networks and draws inspiration from the human brain. We have investigated the following Deep Learning models:

2.4.1 *VGG16 and VGG19*

The VGG16 is a deep convolutional neural network with 16 layers. It accepts as input a photograph with a dimension of (224, 224, 3). The network's first two layers have a max pool layer with a 2x2 stride and 64 channels each with a 3x3 filter size. Convolutional neural network with 19 layers is called VGG19.

2.4.2 *ResNet 50*

ResNet 50 features a 50-layer design. There are leftover blocks there, placed one above the other. It is used for picture classification, object recognition, and object identification.

2.4.3 *DenseNet 121*

DenseNet is a straightforward design with a highly specific filter. Once more, it is employed for picture recognition and categorization.

3 LITERATURE REVIEW

The information about existing research projects that are linked to this section is provided. There are currently only two methods for finding breast cancer. Machine learning comes first, followed by deep learning. There has been a lot of study on machine learning; here, we review the majority of it and discuss the tools utilised, accuracy, and benefits.

Megha Rathi *et al.* proposed a machine learning model based on a hybrid method from [1]. This method was put into practise to determine the optimal outcomes using MRMR feature selection with four classifiers. The four classifiers SVM, Naive Bays, End Meta, and Function Tree were utilised by the author, and they were all compared. SVM was discovered to be an effective classifier.

To discover the best results, M. Tahmooresi *et al.* suggested an additional machine learning-based strategy based on [2]. Nevertheless, SVM was a solid classifier that had the best overall accuracy. Decision trees, SVM, KNN, and ANN have all been compared. The datasets for blood and images were subjected to it. The machine learning model provided by Anusha Bharat *et al.* is based on [4]. It used four distinct classifiers: SVM, decision tree (CART), KNN, and Nave Bayes.The author claims that KNN provided the superior accuracy. SVM had its limitations.

Table 1. Existing related work.

Author & ref	*Dataset*	*Technique Used*	*Advantage*	*Accuracy*
1. Wang *et al.* [18]	Electronic health records	Logistic regression	Using logistic regression, predict 5-year survivorship	96.4%
2. Akbugday [19]	Breast Cancer Wisconsin dataset	KNN and SVM	A k-NN classifier's ideal k-Value, g KNN is a simple, lazy learning algorithm that takes very little time to create.	KNN96.85% NAÏVE BAYES – 95.99% SVM – 96.85%
3. V Chaurisya & S Paul [21]	Wisconsin breast cancer	Statistical Feature Selection	Patient features sortedout from data materials are statisticaly tested based on the type of individual feature. Then 51 attributes or featuresare selected out, a feature's importance score is calculated. XG Boost algo is done by repaeting 10-fold cross validations	92.3 %
4. Keles, M. Kaya, [20]	Wisconsin Diagnostic Breast Cancer dataset	SVM vs KNN, decision trees and Naives bayes	SVMs locate the hyperplane that divides the data points into two classes by mapping the input vector into a higher-dimensional feature space. The instances that are closest to the border have the greatest marginal distance from the decision hyperplane.	96.91%
5. R. Preetha *et al.* [22]	Wisconsin breast cancer dataset.	Data Mining techniques	Find the cancer that is concealed and related with it.	97.34%
6. Naresh Khuriwal *et al.* [23]	Mammogram MIAS database.	Deep learning	Using CNN, it had a 98% accuracy rate.	98%
7. Delen *et al.* [23]	Cancer Society	ADABOOST	Low mistake rate; good performance in the set of low noise data. This algorithm's benefit is that it needs fewer input parameters and minimal prior information on the weak learner.	97.5 %

(*continued*)

Table 1. Continued

Author & ref	Dataset	Technique Used	Advantage	Accuracy
8. Ajay kumar *et al.* [8]	BCDW11 and WBCD32 dataset from UCI Repository.	Classification techniques like SVM, KNN, Naïve Bayes and Decision Tree	SVM provided 97.89% accuracy when using WBCD32 and 97.13% accuracy while using BCDW11.	97.13%
9. Khourdifi *et al.* [6]	Wisconin breast cancer dataset	Fast CorrelationBased Filter with SVM, Random Forest, Naive Bayes	By eliminating redundant and useless features that serve no purpose in the classification task procedures, attributes are minimised.	96.1%
10. R. Chtihrakkannan, P. Kavitha *et al.* [24]	Mammogram images.	Machine learning techniques	Using DNN, it has a 96% accuracy rate.	96%

Ebru Ayndindag Bayrak *et al.* compared machine learning methods using [5]. The Wisconsin breast cancer dataset and WEKA were used in the comparison. The author claims that SVM produced superior performance matrix findings. Following that, deep learning techniques were developed to solve the machine learning problem

Based on [6], Shewtha K *et al.* presented the deep learning convolution neural network model. CNN used a variety of models, the most notable of which were Mobile Net and Inception V3. When the author examined the two models, he noticed that Inception V3 delivered greater Accuracy.

Source: [7], Ch. The model for supervised machine learning was proposed by Shravya *et al.* This study used classifiers including KNN, SVM, and Logistic Regression. This suggests that SVM was an effective classifier with around 88 percent accuracy on the Python platform.From [10], Kalyani Wadkar *et al.* suggested the model based on ANN in this study, and SVM classifier evaluated its performance. The author claims that SVM had a 91% accuracy rate and ANN had 97%.

Model-based SVM and Grid search were recommended by Vishal Deshwal *et al.* based on [11]. The author initially applied the SVM study before utilising it with Grid search. The author conducted comparisons to determine which was best. Comparatively speaking, the new model was constructed. Grid search was used to attain the higher accuracy

S. Shamy *et al.* suggested the model based on CNN and k-mean GMM from [11]. The author used the texture feature extraction technique after first determining ROI. The author's accuracy rate was 95.8%. The author's choice of MIAS data. V Sansya Vijayam *et al.* proposed the deep learning-based model from [13]. The author concentrated on CNN for classification and Lloyd's technique for clustering. The suggested strategies succeeded in achieving the 96% accuracy.

Puspanjali Mohapatra *et al.* suggested a technique based on deep learning to improve histopathology images, which was taken from [14]. Many techniques, including PCA and LDA, were employed in this research for feature extraction.Deep learning reached 81% accuracy by using CNN. However, accuracy increased to 89% when the photos were trained on a GPU.Chandra Church Chatterjee *et al.* introduced the deep residual neural network-based technique for IDC prediction based on [15]. The author used histopathology image data as her dataset. With an AUROC score of 0.9996, the author's accuracy was 99.29%.

The deep learning method from [16], which required enlarging the dataset to obtain the optimal accuracy, was proposed by Canh Phong Nguyen *et al.* According to S. Gokhale *et al.* proposed's approach from [16], clinicians are aware of and have firsthand experience with the fact that breast cancer develops when some breast cells start to grow abnormally. The Naive Bayesian Classifier, k-Nearest Neighbor, Support Vector Machine, Artificial Neural Network, and Random Forest are the four machine learning classifiers used in this study.

The ability to anticipate breast cancer is an active area of study, according to Pragya Chauhan and Amit Swami *et al.* proposed's approach, which was based on the research from [17]. For the purpose of detecting and predicting breast cancer, they employed divergent machine learning techniques. Decision trees, random forests, support vector machines, neural networks, linear models, and naive bayes methods are examples of prediction techniques.

Priyanka Gandhi and Prof. Shalini L of VIT University, Vellore examined machine learning strategies for breast cancer prediction using data from [19]. To increase diagnosis accuracy, ML techniques are looked into in this work. Among the techniques compared are CART, Random Forest, and K-Nearest Neighbors. The UC Irvine Machine Learning Repository provided the dataset. It is found that, in comparison, the KNN algorithm performs better than the other methods. The most precise model was K-Nearest Neighbor.

In their work Performance Evaluation of Machine Learning Methods for Breast Cancer Prediction, Yixuan Li and Zixuan Chen used two datasets from [21]. The study begins with data from the WBCD dataset, which contains 699 volunteers and eleven qualities, as well as data from the BCCD dataset, which includes 116 participants and nine features. Using preprocessed data from the WBCD dataset, we then selected 683 people, each of whom had nine features and an index indicating whether or not they had a malignant tumour.The results showed that RF was selected as the main classification model in this investigation after analysing the accuracy, F-measure metric, and ROC curve of 5 classification models. Therefore, the findings of this study serve as a reference for professionals to identify the traits of cancer. There are still several issues with this study that need to be resolved in follow-up research. For instance, despite the fact that more indices have not yet been discovered, this study only gathers data on 10 qualities during this experiment.

4 DISCUSSION

Based on Table 1, it can be concluded that deep learning outperforms machine learning. Several datasets with varying outcomes are used to compute the results. The enhanced performance is also due to some dataset augmentation. This section's study makes use of dataset augmentation and enhancement. When compared to hybrid techniques, [3] determined that SVM was an excellent classifier. According to [5], an extreme learning machine outperformed several other methods. The author compared the two CNN models from [8–11] and discovered that Inception 3 outperformed Mobile Net. SVM and SVM with grid search were studied in paper [12], and the grid search method generated better results. In that paper [13], the author proposed a CNN-related model and applied the feature extraction method to achieve a superior result. The study's author [15] proposed a CNN model, compared it to machine learning approaches, and discovered that CNN performed better. In the paper [16], the author increased performance by working on the deep residual network.

5 MACHINE LEARNING TECHNIQUES

5.1 *Discreate fourier transform algorithm*

The Discrete Fourier Transform method analyses the starting sequence of inputs in the frequency domain. If the original sequence includes all non-0s values of a function, the function's DTFT is contiguous (and periodic), and the DFT delivers discrete samples of one cycle. If the original sequence is one cycle of a periodic function, the DFT returns all non-zero values of one DTFT cycle. As a result, the DFT is said to reflect the original input sequence in frequency domain.

5.2 *Naive Bayes*

Naive Bayes is a classification strategy that assumes predictor independence in order to build on Bayes' Theorem. Simply expressed, a Naive Bayes classifierbelieves that the proximity of one element within a class has no influence on the proximity of another element within the same class. The term "Naive" refers to a set of traits that freely raise the likelihood of a class despite the fact that they aredependent on one another or on the absence of the opposing features.

5.3 *Logistic Regression(LR)*

Logistic regression belongs to the linear classifiers and is practically equivalent to statistical and polynomial regression. It is quick and simple, you can grasp the findings more readily and quickly. Logistic regression models the chance that a reaction falls into a specific categorization. A logistic regression model uses the Sigmoid function.

5.4 *K-Nearest Neighbors(KNN)*

The supervised machine learning approach k-nearest neighbours may be used to address classification and regression issues (KNN). KNN evaluates the distances betwen a each example in the data and query , picks the nearest K instances to the query, and then either averages the label or selects the label with thehighest frequency for categorising data (incase of regreession).

6 CONCLUSION

In order to reduce the death rates cause by breast cancer, breast cancer prediction technologies as well as algorithms may be used to identify and counter the effects of this disease in the initial stages. For more efficient prediction of the disease and for high accuracy of the system, inspiration can be taken from various other proposed models and algorithms, considering broadening the dataset by image augmentation and collecting the important features from these samples. Breast cancer is diagnosed using machine learning and deep learning approaches. Previous research has proven that machine leanring approaches outperform in their respective industries. Machine leanring, on the other hand, produces higher results on linear data.

REFERENCES

[1] *Cancer Treatment Centers for America* https://www.cancercenter.com/

[2] *MayoClinic* https://www.mayoclinic.org/diseases-conditions/breastcancer/symptomscauses/syc-2035247

[3] Megha Rathi, Vikas Pareek, "Hybrid Approach to Predict Breast Cancer Using Machine Learning Techniques," *International Journal of Computer Science Engineering*, vol. 5, no. 3, pp. 125–136, 2016.

[4] Tahmooresi M., Afshar A., Bashari B. Rad, Nowshath K. B., Bamiah M. A., "Early Detection of Breast Cancer Using Machine Learning Techniques," *Journal of Telecommunication, Electronic and Computer Engineering*, vol. 10, no. 3–2, pp. 21–27, 2018.

[5] Muhammet Fatih Aslam, Yunus Celik, Kadir Sabanci, Akif Durdu, "Breast Cancer Diagnosis by Different Machine Learning Method Using Blood Analysis Data," *International Journal of Intelligent System and Applications in Engineering*, vol. 6, no. 4, pp. 289–293, 2018.

[6] Khourdifi Y. and Bahaj M., "Feature Selection with Fast Correlation-Based Filter for Breast Cancer Prediction and Classification Using Machine Learning Algorithms," *2018 International Symposium on Advanced Electrical and Communication Technologies (ISAECT)*, Rabat, Morocco, 2018, pp. 1–6.

[7] Shwetha K, Spoorthi M, Sindhu S S, Chaithra D, "Breast Cancer Detection Using Deep Learning Technique," *International Journal of Engineering Research & Technology*, vol. 6, no. 13, pp. 1–4, 2018.

[8] Ajay Kumar, R. Sushil, A. K. Tiwari, "Comparative Study of Classification Techniques for Breast Cancer Diagnosis," *International Journal of Computer Science and Engineering*, vol. 7, no. 1, pp. 234–240, 2019.

[9] Sivapriya J, Aravind Kumar V, Siddarth Sai S, Sriram S, "Breast Cancer Prediction Using Machine Learning," *International Journal of Recent Technology and Engineering*, vol. 8, no. 4, pp. 4879–4881, 2019.

[10] Kalyani Wadkar, Prashant Pathak, Nikhil Wagh, "Breast Cancer Detection Using ANN Network and Performance Analysis with SVM," *International Journal of Computer Engineering and Technology*, vol. 10, no. 3, pp. 75–86, 2019.

[11] Vishal Deshwal, Mukta Sharma, "Breast Cancer Detection Using SVM Classifier with Grid Search Techniques," *International Journal of Computer Application*, vol. 178, no. 31, pp. 18–23, 2019.

[12] Shamy S., Dheeba J., "A Research on Detection and Classification of Breast Cancer Using kmeans GMM & CNN Algorithms," *International Journal of Engineering and Advanced Technology*, vol. 8, no. 6S, pp. 501–505, 2019.

[13] Sansya Vijayan V, Lekshmy P L, "Deep Learning Based Prediction of Breast Cancer in Histopathological Images," *International Journal of Engineering Research & Technology*, vol. 8, no. 07, pp. 148–152, 2019.

[14] Puspanjali Mohapatra, Baldev Panda, Samikshya Swain, "Enhancing Histopathological Breast Cancer Image Classification Using Deep Learning," *International Journal of Innovative technology and Exploring Engineering*, vol. 8, no. 7, pp. 2024–2032, 2019.

[15] Chandra Churh Chatterjee, Gopal Krishan, "A Noval Method for IDC Prediction in Breast Cancer Histopathology Images Using Deep Residual Neural Networks," *2nd International Conference on Intelligent Communication and Computational techniques(ICCT)*, pp. 95–100, 2019.

[16] Canh Phong Nguyen, Anh Hoang Vo, BaoThien Nguyen, "Breast Cancer Histology Image Classification Using Deep Learning," *19th International Symposium on Communication and Information Technologies(ISCIT)*, pp. 366–370, 2019.

[17] Wang, Zhang D. and Huang Y. H. "*Breast Cancer Prediction Using Machine Learning*" (2018), Vol. 66, NO. 7.

[18] Akbugday B., "Classification of Breast Cancer Data Using Machine Learning Algorithms," *2019 Medical Technologies Congress (TIPTEKNO)*, Izmir, Turkey, 2019, pp. 1–4.

[19] Keles, M. Kaya, "Breast Cancer Prediction and Detection Using Data Mining Classification Algorithms: A Comparative Study." *Tehnicki Vjesnik – Technical Gazette*, vol. 26, no. 1, 2019, p. 149+.

[20] Chaurasia V. and Pal S., "Data Mining Techniques: To Predict and Resolve Breast Cancer Survivability", *IJCSMC*, Vol. 3, Issue. 1, January 2014, pg. 10–22

[21] reetha R., Vinila Jinny S., "A Research on Breast Cancer Prediction Using Data Mining techniques," *International Journal of Innovative Technology and Exploring Engineering*, vol. 8, no. 11S2, pp. 362–370, 2019.

[22] Mehra, P. S., Jain, K., Chawla, D., Dagur, A., Singh, S., & Sharma, J. (2022). *GWO-EFUCA: Grey Wolf Optimisation and Fuzzy Logic Based Unequal Clustering and Routing Protocol for Sustainable WSN-based Internet of Things.*

[23] Mehra, P. S., Mehra, Y. B., Dagur, A., Dwivedi, A. K., Doja, M. N., & Jamshed, A. (2021). COVID-19 Suspected Person Detection and Identification Using Thermal Imaging-based Closed Circuit Television Camera and Tracking Using Drone in Internet of Things. *International Journal of Computer Applications in Technology*, 66(3–4), 340–349.

[24] Sri Hari Nallamala, Pragnyaban Mishra, Suvarna Vani Koneru, "Breast Cancer Detection Using Machine Learning Way," *International Journal of Recent Technology and Engineering*, vol. 8, no. 2S3, pp. 1402–1405, 2019.

Artificial Intelligence, Blockchain, Computing and Security – Dagur et al. (Eds)

A review paper on the diagnosis of lung cancer using machine learning

Sonia Kukreja
Assistant Professor, Galgotias University

Anish Kumar & Ghufran Ahmad Khan
Galgotias University

ABSTRACT: Lung cancer could be caused by lung cells that grow out of control. To beat lung cancer, you need to find it early. Lung cancer has been the leading cause of cancer-related deaths this century. Many reliable systems for the treatment of lung cancer that are simple to use and lower in cost have been developed using the data science technology. This research compares a number of machine learning-based methods for accurately detecting lung cancer from 2019 to 2022. There are too many ways to find lung cancer now, and most of them rely on CT scans or x-rays. Almost all of them use image classification methods to find lung cancer nodules. This is combined with several segmentation methods and a wide range of classifier algorithms to get a more accurate result. Based on the results of this study, it seems that CT scan images are more accurate. Because of this, CT scan pictures are often used to find cancer [6]. In terms of accuracy, a random forest-based model does better than other machine learning techniques.

Keywords: Random Forest, Image classification, Deep learning, CT scan, CNN

1 INTRODUCTION

When a few of the body's cells grow out of control and spread to other internal organs, it becomes cancer. Cancer may appear almost anywhere in the trillions of cells that make up the human body. If not treated properly, tumors may cause several illnesses. Tumor cells multiply uncontrolled, unlike normal cells that replace old or damaged ones [7]. They sometimes suffer modifications (mutations). Damaged cells that grow uncontrolled and expand into enormous amounts of tissue become tumors. Tumors may be either benign (noncancerous) or malignant (cancerous). The growth and spread of benign tumors are often sluggish. Malignant tumors have the ability to spread throughout the body, develop quickly, infect neighboring normal tissues, and do great damage as show in the Figure 1. Modern technologies like low-dose computed tomography and

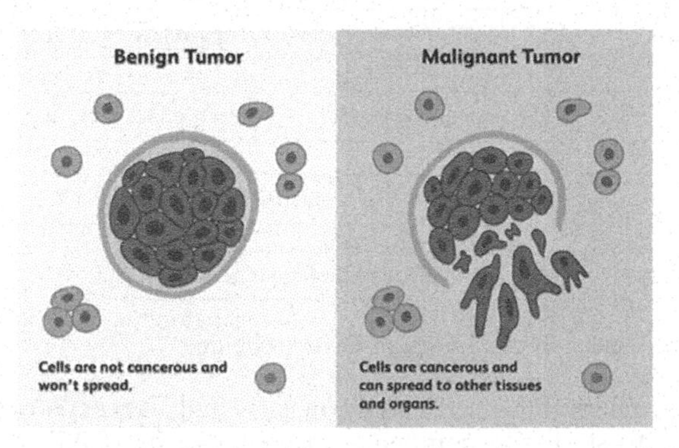

Figure 1. Benign tumor and malignant tumor.

DOI: 10.1201/9781032684994-3

other methods for early lung cancer identification make it feasible to treat lung cancer sooner [12]. Lung cancer is caused by an abnormal tissue development termed a nodule, which arises from cells in the airways of the respiratory system. These cells look spherical on chest X-rays and are always in contrast. If lung nodules are detected early, patients may have a better chance of survival. Evaluation of raw chest X-ray pictures is laborious and difficult because lung nodules are hard to see [10]. Computer-aided diagnostics must find a small nodule in a big 3D lung CT image [5]. The CT image is overloaded with background noise from air and bone and other nearby muscles, fat, organs, and blood vessels; therefore, this noise must first be reduced in order for the CAD systems to search efficiently. Our classification pipelines include nodule candidate, image pre-processing, and malignancy classification detection. Machine learning may help identify and treat lung nodules in AI-generated CT images. There are multiple lung cancer detection methods available today. A general block diagram of the machine learning approach for lung cancer detection is shown in Figure 2. We evaluated new cancer detection methods to choose the best [10].

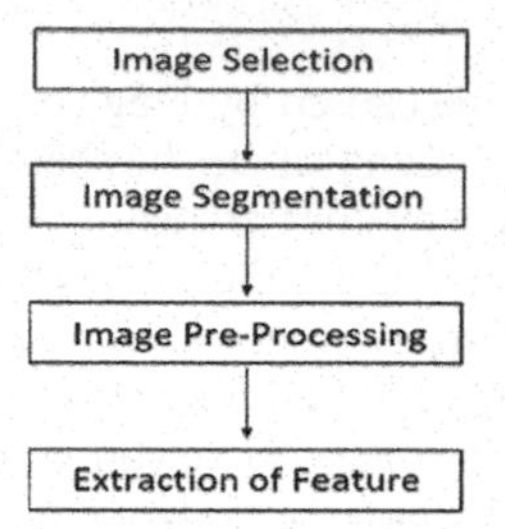

Figure 2. General block diagram for lung cancer detection using machine learning approach.

2 LITERATURE REVIEW

Statistical and machine learning methods have improved labor, business, etc. Therefore, machine learning methods are also being applied in the area of medicine to identify diseases and treat them appropriately [9]. To better predict lung cancer, researchers have used deep learning and neural networks. Several researchers published a study in October 2019 based on a supervised machine technique called "random forest" that may aid in the early identification and diagnosis of lung cancer. Lung cancer data is categorized using the Random Forest Algorithm in this study. Healthcare research uses Random Forest, the most precise learning algorithm. The suggested solution, based on the random forest algorithm, has an accuracy of 100% for dataset 1 and 96.31% for dataset 2. In comparison to previous techniques, the Random Forest Algorithm demonstrated greater accuracy in the prediction of lung cancer [3].

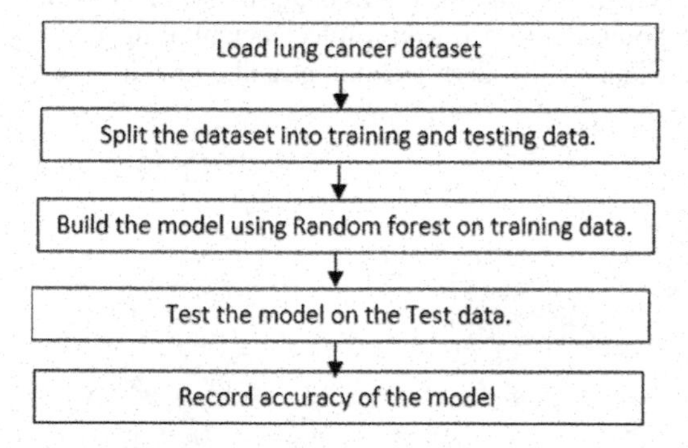

Figure 3. Lung cancer detection using random forest technique.

In 2020, some researchers published a study on early lung cancer detection using statistical methods. According to studies, a computer-aided diagnostic (CAD) system's capacity to identify lung nodule cancer helps clinicians and reduces deaths In this research, a CAD system detects lung nodule cancer early. A picture's contrast is increased using Contrast Limited Adaptive Histogram Equalization (CLAHE). Morphological filters remove background and geometrical objects from lung tumours segregated by Otsu thresholding. A discrete wavelet transform denoises the image. The Gray Level Co-occurrence Matrix (GLCM) is used to extract variables including correlation, energy, and others. Features are chosen using principal component analysis (PCA). A picture's malignant or benign status is determined via SVM. It uses CT scan images to reliably identify lung nodules. This model detected 97.34% and 96.55% of nodules in the LIDC and ELCAP datasets, respectively [11].

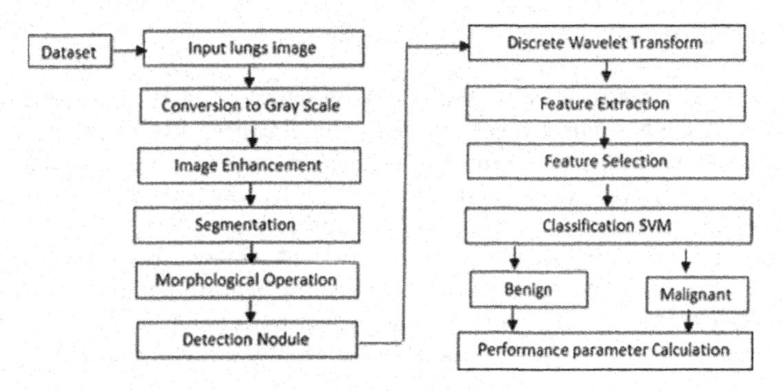

Figure 4. Flow diagram of lung cancer detection using SVM algorithm.

In 2021, Gopichand Shelke, Shraddha Patil, and Smita Raut presented "Lung Cancer Detection Using a Machine Learning Approach." Their technique finds tumors in photographs using machine learning and digital image processing. Digital image processing and machine learning rules comprise the model. Digital image processing includes image acquisition, grayscale conversion, noise reduction, picture binarization, segmentation, characteristic extraction, machine learning, and most cancers' movement identification. Machine learning algorithmic decision trees forecast outcomes in the second step. The design's accuracy was 78% [8].

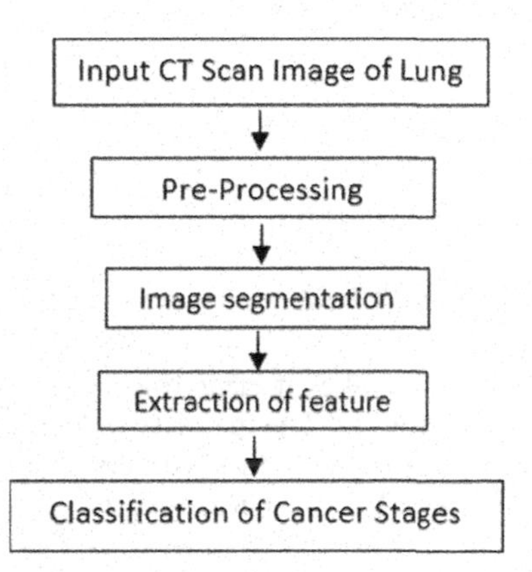

Figure 5. Steps invoved in data processsing.

Researchers published a machine-learning paper at the inaugural international conference on artificial intelligence and data analytics in 2021. According to their study report, their main purpose was to detect and categorise lung cancers including adenocarcinoma, big cell carcinoma, and squamous-cell carcinoma. Machine learning techniques were used to create a hybrid lung cancer detection system. Three-patched local binary pattern descriptor-based discrete cosine transform extraction was used to detect and diagnose lung cancer. This study uses an advanced machine learning system to identify lung cancer more accurately and assist diagnose and cure it. According to the paper, the proposed lung cancer detection method worked best when using chest CT scan image datasets. The proposed method beats support vector techniques and the K-nearest neighbours algorithm with 93% and 91% accuracy, respectively [2].

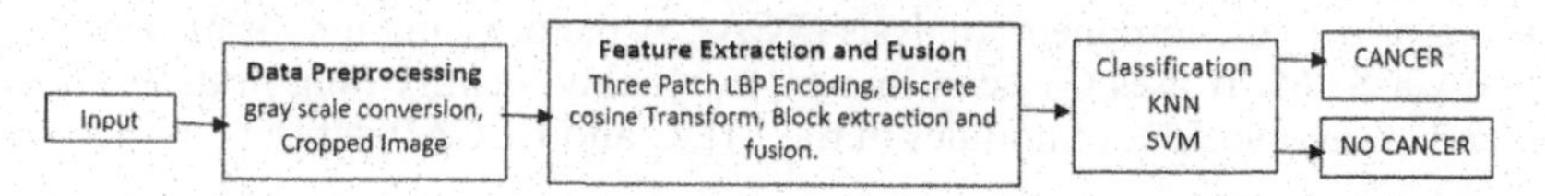

Figure 6. Block diagram of lungs cancer detection using KNN.

A correlation study was released in May 2021 by many scholars. This research examines SVM, KNN, and CNN's effectiveness in early lung cancer identification to save lives. Research proposes predicting and diagnosing lung tumor stages. As illustrated in Figure 7, the suggested model starts with data preparation, then feature selection, classification, and assessment. Weka algorithms were commonly employed to characterize this study's lung cancer datasets. New Zealand scientists created WEKA. This publication selected features using the correlation attribute approach. Attributes are feature extraction methods. SVM, KNN, and CNN were used in this study, each with varied accuracy. SVM has 95.56 accuracy, KNN 88.40, and CNN 92.11. The proposed model's KNN classifier reduces accuracy [4].

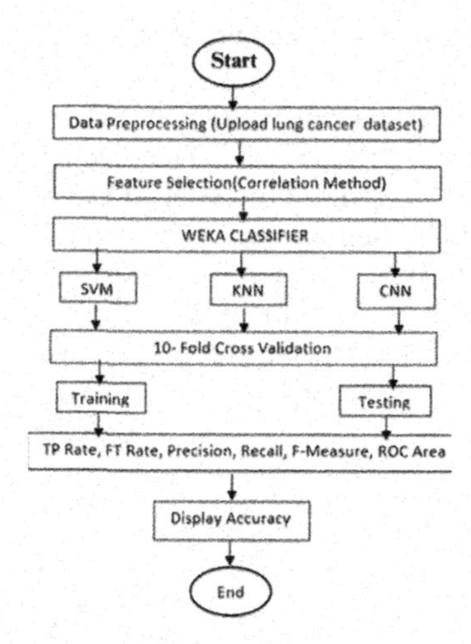

Figure 7. Lungs cancer detection using WEKA Technique and correlation method.

The most recent literature review makes it clear that lung cancer has received significant recognition from the scientific community. The majority of solutions have relied on traditional machine learning and neural network techniques. Others used deep learning techniques to analyze both types of photos (CT and X-ray). Therefore, the goal of this research is to examine different techniques and identify the most effective way for identifying lung cancer [1].

Table 1. A table comparing the five research papers published between 2019 and 2022.

References	Year	Methods	Results
1	2019	Random Forest Algorithm to categorize data	The recommended approach achieved 100% accuracy for dataset 1 and 96.31% for dataset 2.
6	2020	Support Vector Machine (SVM)	This model's accuracy for LIDC and ELCAP was 97.34% and 96.55%, respectively.

(continued)

Table 1. Continued

References	Year	Methods	Results
8	2021	Decision tree Machine learning algorithm	The accuracy of the model was 78%.
9	2021	Support Vector Machine (SVM) and K-nearest neighbor's	SVM – 93% K-NN – 91%
10	2021	Correlation Method and WEKA Technique.	SVM – 95.56% K-NN – 88.40% CNN – 92.11%

3 CONCLUSION

If lung cancer is identified at an early stage, therapy may begin to prevent the illness from progressing to a dangerous level. Thus, this paper investigated many machine learning algorithms for lung cancer diagnosis utilizing CT scans or X-rays. SVM, K-NN, CNN, deep learning techniques, random forests, decision trees, and others have been used by researchers in the past. Thus, this study's extensive evaluation suggests that random forest approaches were more accurate than other machine learning methods. The suggested method gave 100% accuracy for dataset 1 and 96.31% for dataset 2.

REFERENCES

[1] Abhir Bhandary, G. Ananth Prabhu, V. Rajinikanth, K. Palani Thanaraj, Suresh Chandra Satapathy, David E. Robbins Charles Shasky Yu-Dong Zhang, João Manuel R.S. Tavares, N. Sri Madhava Raja, "*Deep Learning Framework to Detect Lung Abnormality – A Study with Chest X-Ray and Lung CT Scan Images*" In 2020.

[2] Amjad Rehman, Muhammad Kashif, Ibrahim Abunadi, and Noor Ayesha "Lung Cancer Detection and Classification from Chest CT Scans Using Machine Learning Techniques" *In 2021 1st International Conference on Artificial Intelligence and Data Analytics.*

[3] A. Rajini and M.A. Jabbar "*Lung Cancer Prediction Using Random Forest*" In oct 2019.

[4] Dakhaz Mustafa Abdullah, Adnan Mohsin Abdulazeez, & Amira Bibo Sallow "Lung Cancer Prediction and Classification based on Correlation Selection Method Using Machine Learning Techniques" In *May 2021 in Qubahan Academic Journal.*

[5] Haron, H., Zeebaree, D. Q., Zebari, and Abdulazeez, "Trainable Model Based on New Uniform LBP Feature to Identify the Risk of the Breast Cancer". *In 2019 International Conference on Advanced Science and Engineering (ICOASE).*

[6] Jothilakshmi, R., and SV, R. G. "Early Lung Cancer Detection Using Machine Learning and Image Processing" In *Journal of Engineering Sciences*, in 2020.

[7] Sherawat, D., Sonia & Rawat, "A. Brain Tumor Detection Using Machine Learning in GUI". In *Proceedings of Integrated Intelligence Enable Networks and Computing* (pp. 9–17). Springer, Singapore in 2021.

[8] Smita Raut, Shraddha Patil, Gopichand Shelke, "Lung Cancer Detection using Machine Learning Approach", *International Journal of Advance Scientific Research and Engineering Trends (IJASRET)* in 2021.

[9] Sonia Kukreja, Munish Sabharwal, D.S. Gill, *"A Survey of Machine Learning Algorithms for Lung Cancer Detection"* In *4th International Conference on Advances in Computing, Communication Control and Networking* 2022.

[10] Tanzila Saba "Recent Advancement in Cancer Detection Using Machine Learning: Systematic Survey of Decades, Comparisons and Challenges" In *Journal of Infection and Public Health*, in 2020.

[11] Waseem Abbas, Khan Bahadar Khan, Muhammad Aqeel, Muhammad Adeel Azam Muhammad Hamza Ghouri, Fawwad Hassan Jaskani, *"Lungs Nodule Cancer Detection Using Statistical Techniques" IEEE 23rd International Multitopic Conference (INMIC)* 2020.

[12] Yu, K. H., Lee, T. L. M., Yen, M. H., Kou, S. C., Rosen, B., Chiang, J. H., and Kohane, I. S. "Reproducible Machine Learning Methods for Lung Cancer Detection Using Computed Tomography Images: Algorithm Development and Validation" *Journal of Medical Internet Research*, In 2020.

Loan eligibility prediction model using machine learning algorithms

Muskan Gupta, Prakarti Singh & Vijay Kumar Sharma
Meerut Institute of Engineering and Technology, Meerut, U.P., India

ABSTRACT: Nowadays, anything can be bought on a loan. The lender provides loans to borrowers with the guarantee of repayment. Previously there were manual processes to check whether it is safe to lend money to a certain borrower or not; which can result in a lot of misconceptions in selecting a genuine applicant. Henceforth, via machine learning algorithms we tried to develop a loan approval prediction system that automatically predicts whether a loan should be approved or rejected for a certain individual based on some details like marital status, income, credit history, etc. filled by the applicant. Naive Bayes classification algorithm, logistic regression, random forest algorithm, SVM algorithm, and decision tree algorithms are all supervised learning algorithms that are implemented and compared for accuracy in this paper, and amongst them, it is observed that the random forest algorithm provides much higher accuracy as compared logistic regression and other classification algorithms.

Keywords: Loan, SVM, Naive Bayes, Supervised learning, Decision tree, Prediction, Machine learning, Logistic Regression, Random Forest, classification, repayment

1 INTRODUCTION

A loan is a vital aspect of any banking organization, they provide various types of loans like home loans, personal loans, business loans, etc., and make a profit through the interest from those loans. Also, a borrower enjoys a lot of benefits from a loan, one can repay the bank at one's convenience as long as the installments are regular and on time. When a lender lends money to a borrower, they are expected to repay the loan and the interest owed. The interest payments from the lender act like a cash flow for the lender. But the lack of fulfillment of the commitment by the borrowers results in resistance to the lender's cash flow. So, in the financial sector, to protect the cash flow of the lender and reduce the chances of losses, one needs to consider various factors before approving a loan to any applicant. Previously all the verification and validation were done manually by bank employees; bank employees checked the details and the loan was given to eligible applicants. However, manually checking each application for loan approval can be a tedious process for bank employees and can result in errors which in turn make way for losses to an organization. Also, there are high chances of credit risks which are caused due to the lack of repayment by loan defaults that contributes to the major source of risk the banking industry encounters. The capacity to repay and income are significant factors on which the eligibility for loan approval majorly depends; But keeping in mind, the risk that lender faces we wanted to build a model to predict, check, and validate customers who stand eligible for a loan. So, we build an automated loan prediction model that can be significantly reduced. Human errors can also be avoided using this automated system. Henceforth, potential losses can be prevented, time can be saved, and the company can focus more on eligible customers.

In this paper, we have attempted to automate the loan prediction process to save time and reduce errors that happen due to applicants' vast amounts of data. We have considered various

DOI: 10.1201/9781032684994-4

parameters or attributes that can affect the eligibility for loan eligibility prediction, including employment, education, marital status, etc. For the same, we have trained our model on previous records of customers acquiring loans from the bank and tested for its accuracy. Then our trained model is fed with the test data on which the prediction has to be made. We have used two machine learning classification models to develop the same. During the implementation experiments, we learned that the Random Forest algorithm is better at precisely classifying.

2 LITERATURE REVIEW

Kumar and others gave "Customer Loan Eligibility Prediction using Machine Learning Algorithms in Banking Sector" [1]. Shaik and others gave "Customer Loan Eligibility Prediction using Machine Learning" [2]. Reddy and others gave "Machine Learning based Loan Eligibility Prediction using Random Forest Model" [3]. Sarkar, A. gave Machine learning techniques for recognizing loan eligibility [5]. Ashlesha Vaidya used logistic regression for loan approval prediction as a probabilistic and predictive approach [8]. Kadam and others gave "Prediction for Loan Approval using Machine Learning Algorithm" [4]. T. Sunitha and colleagues used Logistic Regression and a Binary Tree approach to predict loan Status. Malakauskas and others in their paper gave financial distress predictions for small and medium enterprises using machine learning techniques [6]. In T. Mohana Kavya and J. Tejaswini's paper, they talked about a loan prediction system that helps each feature assign an associated weight for the processing to be done on the same features; After implementing different classifiers, they concluded that amongst various classifiers decision tree performs better on loan eligibility prediction [7]. Ranpreet Kaur and Anchal Goyal combine two or more algorithms for better predictive performance termed an ensemble algorithm. In their research paper, they provided a structured literature review in which various stand-alone machine-learning algorithms were compared with ensemble models. Finally, they concluded that combining two or more algorithms contribute to higher performance accuracy [9]. Ghatasheh N in his paper stated that the best classification algorithm for predicting the eligibility of loan approval is the random forest algorithm. Hassani B, Guegan D, and Addo PM in their paper selected the multilayer neural network models (deep learning), gradient boosting algorithm, random forest algorithm, and logistic regression model to illustrate that the quality of data,i.e, cleaning, and analysis before modeling is significantly beneficial, also the selection of attributes and algorithms are major aspects of deciding for loan approval.

Logistic regression is also a supervised learning method, that deals with classification and regression problems. A logistic regression model predicts the probability estimate for our target variable which ranges between 0 and 1, by drawing a relationship graph between the dependent and one or more independent variables. It uses a logit function that measures the log odds for the Favor of the event; the logit function is an inverse of the sigmoid function It provides binary outcomes, such as yes or no, to a class label. Logistic regression generates an S-shaped curve [11]

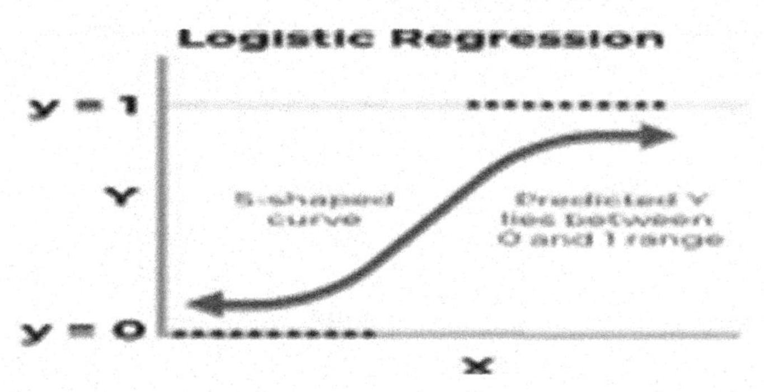

Figure 1. The S-shaped curve in logistic regression.

Random Decision Forest or simply random forest falls under the category of a supervised learning algorithm in machine learning that is like a decision tree with classification and regression problems. It is widely used due to its accuracy, simplicity, and flexibility. Tim kam ho. first proposed the term "random decision forest" in the year 1995; Later by Leo Breiman and Adele Cutler, the random forest was created by extending the algorithm, in 2006. A random forest randomly selects features to make a forest of decision trees, finally the mode or average of all the decision trees will be our final result. According to the theories, using the Random Forest classifier, we gain immunity from overfitting with a higher rate of accuracy and, better performance on large datasets [12].

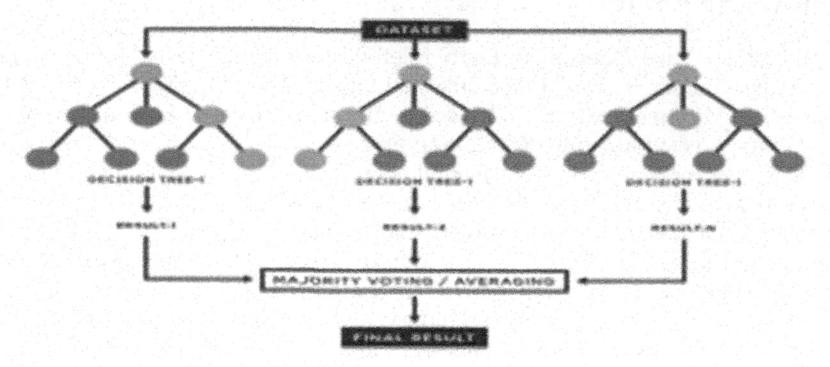

Figure 2. Random forest diagram.

3 PROPOSED METHODOLOGY

The major objective of the model is to automate the loan approval process which is made possible by training the same with previous datasets that precisely predicts the class label for loan status, i.e., whether or not the loan should be approved. Loans play an essential role for both the borrower and the granter. The granter grants loans to borrowers with the guarantee of repayment. When various people apply for a loan, it's hard to know if it is safe to give an individual a loan. Or is there a guarantee of repayment? Henceforth, to avoid all the misconceptions generated using manual processes for loan approval we have generated a loan prediction model that automates the whole process for loan approval to ensure complete repayment, it reduces the delays, saves time, and focuses more on eligible customers. For the same, we have trained our model on records of customers, that had acquired loans from the bank in the past and tested for its accuracy. Then our trained model is fed with the test data and the prediction has to be made.

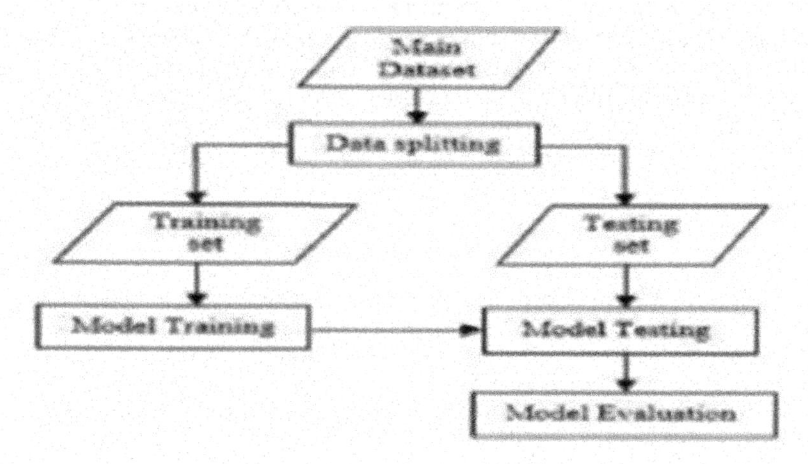

Figure 3. The general architecture of the model.

3.1 *Data collection*

The very first step is to load the dataset on which we can train our model and further check its accuracy. The dataset is in CSV format and comprises 13 features in total. Dataset: – The data set for training and testing the model is taken from Geeks for geeks https://www.geeksforgeeks.org/loan-approval-prediction-usingmachine-learning/ [10].

Table 1. Specifications of the dataset.

Field Name	Value
Number of Rows	614
Number of Columns	13
Number of Categorical Values	7
Number of numeric values	6
Number of target variables	1

	Loan_ID	Gender	Married	Dependents	Education	Self_Employed	ApplicantIncome	CoapplicantIncome	LoanAmount	Loan_Amount_Term	Credit_History	Property_Area	Loan_Status
0	LP001002	Male	No	0.0	Graduate	No	5849	0.0	NaN	360.0	1.0	Urban	Y
1	LP001003	Male	Yes	1.0	Graduate	No	4583	1508.0	128.0	360.0	1.0	Rural	N
2	LP001005	Male	Yes	0.0	Graduate	Yes	3000	0.0	66.0	360.0	1.0	Urban	Y
3	LP001006	Male	Yes	0.0	Not Graduate	No	2583	2358.0	120.0	360.0	1.0	Urban	Y
4	LP001008	Male	No	0.0	Graduate	No	6000	0.0	141.0	360.0	1.0	Urban	Y

Figure 4. Representation of datasets pre-processing.

The dataset comprises 7 categorical variables: –

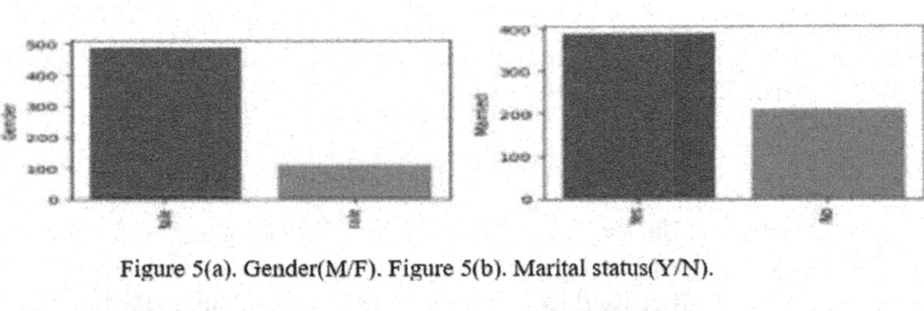

Figure 5(a). Gender(M/F). Figure 5(b). Marital status(Y/N).

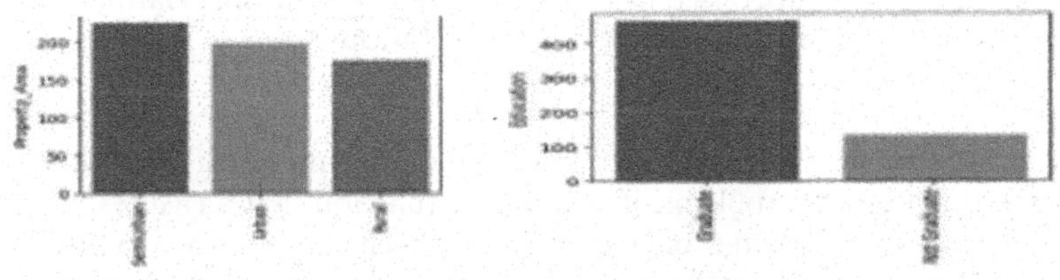

Figure 5(c). Property area. Figure 5(d). Education status.

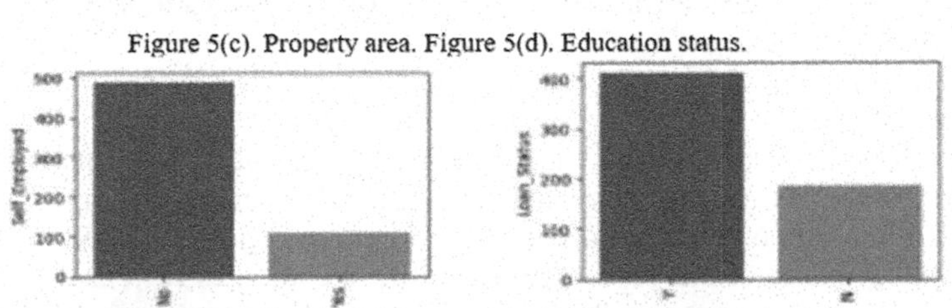

Figure 5(e). Employment status(Y/N). Figure 5(f). Loan status(Y/N).

Figure 5. Bar graph representation of categorical data from dataset.

3.2 *Data pre-processing*

After data gathering, the next step is to get rid of any kind of outliers, and unnecessary data from the dataset and to fill in any sort of missing values that can affect the precision of our model. All these pre-processing is important and helps in optimizing the dataset for further steps. Herein, unnecessary or impropriate data values are removed, and missing values are filled. Many attributes in our dataset are comprised of various null values that need to be taken care of. We either drop certain columns that seem less significant or fill the missing values with mean, median, or mode. After all the cleaning, the count of null values is.

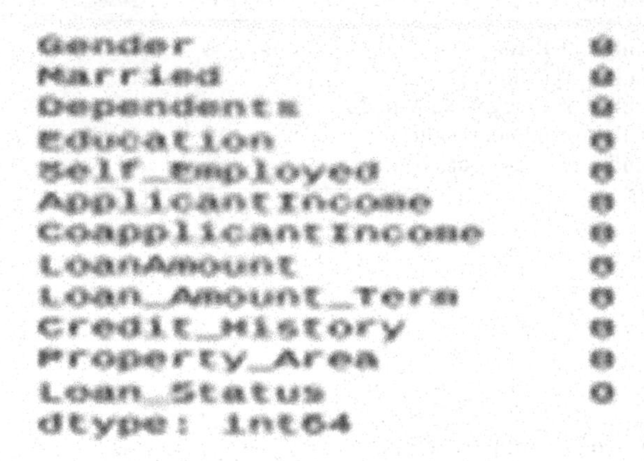

Figure 6. Count of null values after cleaning the dataset.

3.3 *Data splitting*

The test and train data were split from the training dataset itself. The model is trained on the training data after selecting a specific machine learning algorithm and then the accuracy of our trained data model is checked on the test data.

3.4 *Selecting a model*

Next step is to train the model on a selected machine learning algorithm. Here we have applied random forest and logistic regression algorithms to our model. Random forest and logistic regression are both supervised learning techniques of machine learning. In this paper, we have compared the accuracy achieved by both algorithms.

3.4.1 *Logistic regression*

Logistic regression comes under the supervised learning technique of ML, which deals with classification and regression problems, it uses a logit function that measures the log odds for the Favor of the event. It provides binary outcomes, such as yes or no, to a class label. We achieved 80.03% accuracy while training the model and 80.8% accuracy while testing the model using logistic regression.

3.4.2 *Random forest*

Random forests are also a supervised learning technique but are an ensemble learning method for problems like classification and regression. An ensemble learning technique refers to the combination of two or more machine learning algorithms used to increase the accuracy and performance of our model. Random forest combines various decision trees to achieve the result, i.e., the average value from various decision trees is the final result given by the random forest. We achieved 98.01% accuracy for the training dataset and 82.5% accuracy for the test dataset using a random forest classifier.

3.5 *Predicting the outcomes*

The Classifier is fed with the required parameters or attributes and then the classifier with a higher accuracy rate is then used to predict the outcomes on the new dataset. We have more accuracy using a random forest classifier, so made predictions using the same. Here, 1 represents that the person is eligible for a loan, and 0 represents that person is not eligible.

```
array([1, 1, 0, 1, 1, 1, 1, 1, 1, 0, 1, 1, 1, 1, 1, 0, 0, 1, 1, 1, 1, 1,
       0, 1, 1, 1, 1, 0, 0, 1, 1, 1, 0, 1, 1, 1, 0, 1, 0, 1, 1, 1, 1, 1,
       1, 1, 1, 0, 1, 1, 1, 1, 1, 1, 1, 0, 1, 0, 0, 1, 1, 1, 1, 1, 1, 1,
       0, 1, 1, 0, 1, 1, 1, 1, 0, 1, 1, 1, 1, 0, 1, 0, 1, 1, 1, 1, 1, 1,
       1, 1, 1, 0, 1, 1, 1, 0, 1, 1, 1, 0, 0, 1, 1, 1, 1, 1, 1, 1, 1, 1,
       1, 1, 0, 1, 1, 1, 1, 0, 1, 1, 1, 1, 1, 1, 1, 1, 1, 1, 0, 1, 1, 1,
       0, 1, 1, 0, 1, 1, 1, 1, 1, 0, 1, 1, 0, 0, 0, 1, 1, 0, 0, 0, 1, 1,
       1, 1, 1, 1, 1, 1, 1, 0, 0, 1, 1, 1, 0, 1, 1, 1, 1, 0, 1, 1, 1, 1,
       1, 0, 1, 1, 0, 0, 0, 0, 1, 1, 1, 1, 1, 1, 0, 0, 1, 1, 1, 0, 1, 1,
       0, 1, 1, 1, 1, 0, 0, 1, 1, 0, 0, 1, 0, 1, 1, 1, 1, 1, 1, 1, 0, 0,
       1, 0, 1, 1, 0, 1, 1, 1, 1, 1, 1, 0, 1, 0, 1, 1, 1, 1, 1, 1])
```

Figure 7. Predictions on a test dataset.

4 RESULTS

After comparing various prediction algorithms, it is observed that the accuracy of the Random Forest algorithm is more than other algorithms. Although the accuracy of all the algorithms was much higher during the testing phase as we increased and changed the data for testing, the accuracy shown was comparatively lower. Below is the accuracy-based comparison of various prediction algorithms after testing.

Table 2. Accuracy-based comparison of various prediction algorithms.

PREDICTION ALGORITHM	ACCURACY
Logistic Regression	80.8
Random Forest	82.5
Decision Tree	67.9
Naive Bayes	81.6
SVM	69.1

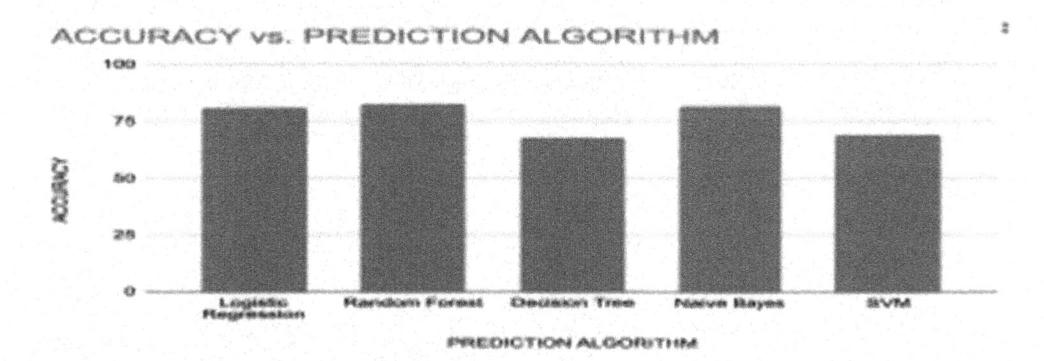

Figure 8. Graphical representation of accuracy-based comparisons.

5 CONCLUSION

This project helps in automating the loan approval process. Basically, instead of manually checking each applicant's details, we train the model on the previous dataset which is later applied to predict or classify whether the person is eligible for the loan to ensure complete repayment from the same. After all the analysis and experiments on the dataset, it can be concluded that the random forest classifier performs better in terms of accuracy and prediction than the logistic regression algorithm and hence we made our classifier model on the same basis. We have got 82.5% accurate prediction using our classifier model.

REFERENCES

[1] Kumar, Ch Naveen, D. Keerthana, M. Kavitha, and M. Kalyani. "Customer Loan Eligibility Prediction using Machine Learning Algorithms in Banking Sector." In *2022 7th International Conference on Communication and Electronics Systems (ICCES)*, pp. 1007–1012. IEEE, 2022.

[2] Shaik, Amjan, Kunduru Sai Asritha, Neelam Lahre, Bollu Joshua, and Velagapudi Sri Harsha. "Customer Loan Eligibility Prediction using Machine Learning." *Journal of Algebraic Statistics* 13, no. 3 (2022): 2053–2062.

[3] Reddy, Chenchireddygari Sudharshan, Adoni Salauddin Siddiq, and N. Jayapandian. "Machine Learning based Loan Eligibility Prediction using Random Forest Model." In *2022 7th International Conference on Communication and Electronics Systems (ICCES)*, pp. 1073–1079. IEEE, 2022.

[4] Kadam, A., S. Nikam, A. Aher, G. Shelke, and A. Chandgude. "Prediction for Loan Approval Using Machine Learning Algorithm." *International Research Journal of Engineering and Technology* 8, no. 04 (2021): 4089–4092.

[5] Sarkar, A. "Machine Learning Techniques for Recognizing the Loan Eligibility." *International Research Journal of Modernization in Engineering Technology and Science* 3, no. 12 (2021).

[6] Malakauskas, Aidas, and Aušrinė Lakštutienė. "Financial Distress Prediction for Small and Medium Enterprises using Machine Learning Techniques." *Engineering Economics* 32, no. 1 (2021): 4–14.

[7] Tejaswini, J., T. Mohana Kavya, R. Devi Naga Ramya, P. Sai Triveni, and Venkata Rao Maddumala. "Accurate Loan Approval Prediction based on Machine Learning Approach." *Journal of Engineering Science 11*, no. 4 (2020): 523–532.

[8] Vaidya, Ashlesha. "Predictive and Probabilistic Approach using Logistic Regression: Application to Prediction of Loan Approval." In *2017 8th International Conference on Computing, Communication and Networking Technologies (ICCCNT)*, pp. 1–6. IEEE, 2017.

[9] Goyal, Anchal, and Ranpreet Kaur. "A Survey on Ensemble Model for Loan Prediction." *International Journal of Engineering Trends and Applications (IJETA) 3*, no. 1 (2016): 32–37.

[10] Mehra, P. S., Jain, K., Chawla, D., Dagur, A., Singh, S., & Sharma, J. (2022). *GWO-EFUCA: Grey Wolf Optimisation and Fuzzy Logic based Unequal Clustering and Routing Protocol for Sustainable WSN-based Internet of Things.*

[11] Mehra, P. S., Mehra, Y. B., Dagur, A., Dwivedi, A. K., Doja, M. N., & Jamshed, A. (2021). COVID-19 Suspected Person Detection and Identification Using Thermal Imaging-based Closed Circuit Television Camera and Tracking Using Drone in Internet of Things. *International Journal of Computer Applications in Technology*, 66(3–4), 340–349.

[12] Kumar, A., & Alam, B. (2016). Real-Time Fault Tolerance Task Scheduling Algorithm with Minimum Energy Consumption. In *Proceedings of the Second International Conference on Computer and Communication Technologies: IC3T* 2015, Volume 2 (pp. 441–448). Springer India.

Artificial Intelligence, Blockchain, Computing and Security – Dagur et al. (Eds)
© 2024 The Author(s), ISBN: 978-1-032-67841-2

Generative Adversarial Networks (GANs): Introduction and vista

Jyoti Kesarwani & Himanshu Rai
United University, Prayagraj

ABSTRACT: A particular type of deep learning model [11] used for unsupervised learning is called a generative adversarial network (GAN), which was initially propound by Ian Goodfellow and others in a 2014 [22]. GANs have been successful in generating an extensive data, including images, audio, and text, and have many potential applications, such as generating synthetic training data, improving data augmentation, and creating new content in fields such as art and music. Despite their success, GANs have also faced challenges, including the difficulty of training them and the potential ethical implications of generating synthetic data. Nevertheless, GANs have remained an active area of research and have continued to evolve and improve over time. The goal of this research is to offer a thorough review of GANs. The theoretical foundations of GANs and a number of GAN variants are covered in this paper. After that, we explore the standards for evaluating GANs.

Keywords: generative adversarial networks, zero-sum games, computer vision

1 INTRODUCTION

Game theory serves as the foundation for the GAN principle, which was developed by Ian Goodfellow *et al.* [2]. GAN consists a generator and a discriminator model. By using GAN, an actual image or video of a person can be converted into a fake image or video [3]. Hence, this has posed a significant threat to fake news, so the challenge is to differentiate fake images/videos from the original image. The field of image processing needs to identify between fake and original images or videos.

As previously mentioned, GAN uses two models which work simultaneously: a generator and a discriminator. The generator's main objective is to provide data that appears actual in an attempt to fool the discriminator. It tries to generate a model distribution that looks like real samples. The discriminator model acts as a binary classifier and makes the most precise distinction between genuine and fake samples produced by the generator. The discriminator and generator serve as adversaries to one another. The basic architecture of GANs is shown in Figure 1.

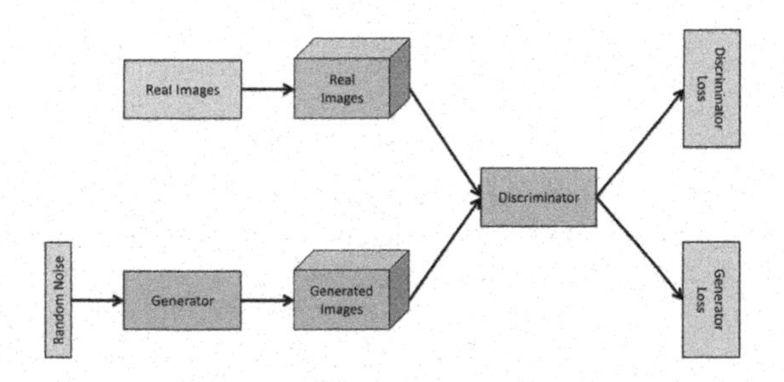

Figure 1. Architecture of basic GAN's.

DOI: 10.1201/9781032684994-5

Generative models act as an important role in Artificial Intelligence. There are two branches of Machine learning one is generative methods and other is discriminative methods. Parameter estimation, distribution hypothesis, and sample new data from the estimated models all methods are used in Generative methods. Research point of view, there are two outlooks: data understood by humans and data machines understood by machines.

In the last some years GAN has been popular and mostly used in various areas. GANs are used in data generation (means to generate a huge amount of data such as chatting records, small blog messages, and search records every day) and augmentation [4] [5], image-to-image translation, etc. Image processing, natural language processing, cross-model retrieval, computer vision, healthcare, biology, fashion, sports, and music are applications of GANs.

GAN optimization depends on the minimax game. This game used a decision rule applied in many fields such as artificial intelligence, decision theory, game theory, and statistics to minimize the loss and maximize the gain.

Generative Adversarial networks (GANs) are used to develop Deepfakes. Deepfake means to interchange faces of famous celebrities, persons, and politicians over porn images and videos [10]. Deepfakes produce ambiguous knowledge and rumors about famous politicians and misleading information on social media, which is harmful to our society.

2 RELATED WORK

2.1 *GAN basics*

The generative model known as Generative Adversarial Networks (GANs) was created by Goodfellow *et al.* in 2014 and employs a min-max, zero-sum game theory method involving two neural networks: generator and discriminator [6] . The generator creates fake sample distributions p_{data} and image represented as $G(z)$ whereas the discriminator's objective is to tell the differentiate original and forged distributions. The architecture of a GAN consists of a generator(G) that takes a random noise vector as input and generates an image and a discriminator(D) that receives both real and fake samples [18] and acts as a binary classifier to determine which is which. During training, the parameters are updated the back-propagation approach [12] for the generator and discriminator, and the minimax game is used to optimise the GAN.

One of the key application areas of GANs is of image applications such as image synthesis, image augmentation and image-to-image translation [7]. Several researchers proposed various GAN models [23] for the same such as Pix2Pix GAN , Cycle GAN [1] [8], Dual GAN [21], LS-GAN [24].

First time Goodfellow introduced GANs and it used the Minimax loss function. The following function tries to minimize by the Generators and to maximizes by the Discriminators. d(x) represents the probability of sample x generated from real data. If the input is real, the discriminator will make an effort to approach d(x) as 1. The discriminator work to approach d(g(z)) as 0, if the input data from g(z). GAN is based on a zero-sum game between generator G and discriminator D. The following Minimax loss is as follows

$$\max_{d} \min_{g} V(g, d) = nEx \sim pdata(x)[\log d(x)] + E_{z \sim pz}[\log (1 - d(g(x)))] \tag{1}$$

In GAN, we have to teach the discriminator d, which accepts input from generated data g (z) or actual data x and requires maximum efficiency as 1. We must train generator g, which aims to minimize as 0 (1–d(g(z))).

$$d^*_{g(x)} = \frac{p_{data(x)}}{p_{data(x)} + p_{g(x)}}$$

2.2 *Deep convolutional GANs*

A Deep convolutional GAN [5] is an extension of the fully connected network. For image data, CNNs are hugely well-suited. First, research using the CIFAR–10 dataset suggested that utilising CNNs would make training the generator and discriminator more challenging since CNNs needed to be applied to supervised learning with an equivalent amount of ability and graphical power. Deep convolutional GAN used a family of network architectures [15] to accomplish during the training. The translation of image space to a latent space that has smaller dimensions, accounting for differences in location and sampling rates, as well as mapping to the discriminator from image space is managed by both up-sampling and down-sampling operators.

2.3 *Conditional GANs*

Conditional GAN (CGAN) [1] established variable c which is known as conditional variable, the variable can be text, label, or extra data. In order to add certain conditions to the framework, the conditional variable c is implemented in the two components of the generator and the discriminator and data generation process influenced by that conditional variable. The conditional GAN extended from 2D GAN by developing the generator and discriminator networks in class conditional by Mirza *et al.* Multi-modal data representations are preferably used in Conditional GAN, this is the advantage of Conditional GAN. The basic architecture of the conditional is shown in Figure 2. To detect similarities, we can relate Conditional GAN to InfoGAN [16], which divided the source of noise into a "latent code" and an uncompressed source vector. Unsupervised learning is used by latent code, which is used to search object classes. InfoGAN handles complex intertangled factors in image representation, as well as differences in emotional expressions, face pose, and lighting effect of a facial image [16].

$$\max_{d} \min_{g} V(g, d) = nEx(x)[\log d(x|c)] + E_{z\sim pz}[\log (1 - d(g(z|c)))] \tag{2}$$

2.4 *Adversarial Autoencoders*

Makhzani *et al.* [17] developed Adversarial Autoencoder (AAE) in 2015. An encoder and a decoder compose the autoencoder networks. In order to impose a certain probability on the latent variable that in its coding layer, adversarial learning is performed to the AAE.

As one can see from Figure 3 the basic structure of AAE contains three blocks namely B1, B2, and B3. Each block has several layers with an activation function taken for non-linearity. The encoder block in the figure is B1 with real data point $x \in R_d$ with B2 as the decoder block containing reconstructed data points $\sim x \in R_d$ as its input. The latent layer is represented by $z \in R_p$ with $z \leq p$ act as the low dimensional layers. The conditional distributions model in both encoder and decoder as $p(z|x)$ and $p(x|z)$, consequently. Consider q(z) as showed latent variable distribution inside the autoencoder. The latent variable employed posterior distribution. B1 and B3 blocks are accordingly represented as generator and discriminator in the adversarial network. We are given the earlier distribution for the latent variable, that is shown by the symbol p(z). The p-dimensional normal distribution over $N(0,I)$ is referred to as the previously mentioned distribution. Block B1 as an encoder for autoencoder acts as the generator G, which gives output the latent variable by using posterior distribution as $g(x) = z \sim q(z)$ [17]. Block B3 as discriminator D generates one output neuron from the activation function sigmoid. In order to determine if latent variable z is an accurate latent variable, the discriminator uses either the previous distribution $p(z)$ or the encoder generated latent variable.

As discussed above, the adversarial network and the autoencoder shared block B1. Both the adversarial network and autoencoder become step by step stronger with adversarial training

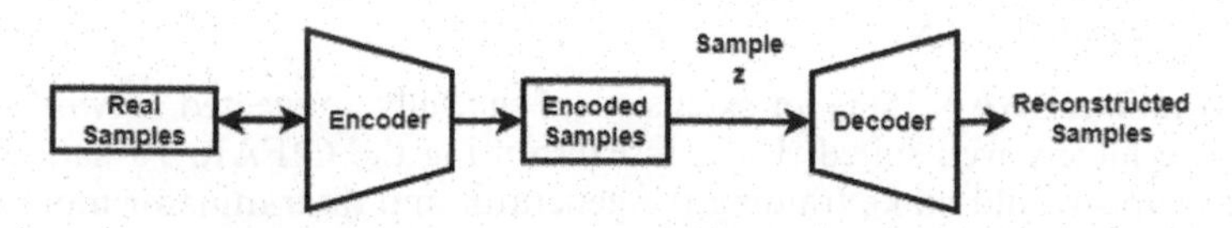

Figure 2. Architecture of adversarial autoencoder GAN.

because the latent variable provided by the autoencoder, which has a prior distribution and reflects a real latent variable. The autoencoder's goal is to deceive the discriminator in this way. However, the discriminator wants to evolve and produce exact outcomes.

2.5 *Wasserstein GAN (WGAN)*

Wasserstein GAN (WGAN) [18]is a new GAN algorithm developed to replace traditional GAN and improve model stability during learning and prevent problems like mode collapse. The objective function in GAN training, Jensen-Shannon divergence, was found to cause instability, to train the GAN, the WGAN employs the Wasserstein distance. The Wasserstein distance is appropriate for transformations and meets the 1–Lipschitz condition. Weight clipping is used by WGAN to preserve weight values within specified ranges when evaluating the model. The Wasserstein distance is also applied in various disciplines such as medical and remote sensing using the Earth Mover's distance [2]. The Wasserstein GAN's fundamental framework is as in Figure 3.

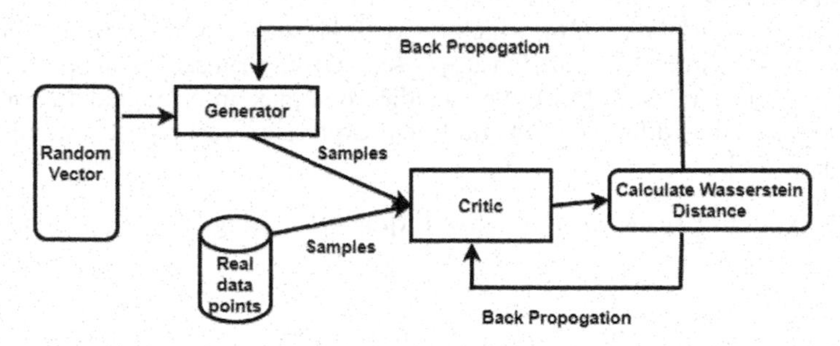

Figure 3. Architecture of wasserstein GAN.

2.6 *InfoGAN*

InfoGAN [21] represent the information between observations and a compact subsets of latent variables. The model separates the noise vector, G, into two parts: an incompressible noise z, and a latent code c, which is used to represent the semantic data distribution in a structured form. The InfoGAN objective function is to minimize G while maximizing D, using the equation

$$\max_{d} \min_{g} f_I(g, d) = f(g, d) - \lambda I(c; g(z, c)) \tag{3}$$

The output of generator g is g(z,c). I stands for mutual information, and lambda for regularisation parameter. I(c,g(z,c)) is known asÂ mutual information,Â used by InfoGAN, and it also indicates the correlation between c and g(z,c). Figure 4 depicts the fundamental architecture of infoGAN.

2.7 *CycleGAN*

The primary goal of image-to-image translation is to map the input and output visuals. If training paired data is available, image-to-image translation is applied. CycleGAN [19] is an image-to-image translation technique that uses automatic training without paired examples.

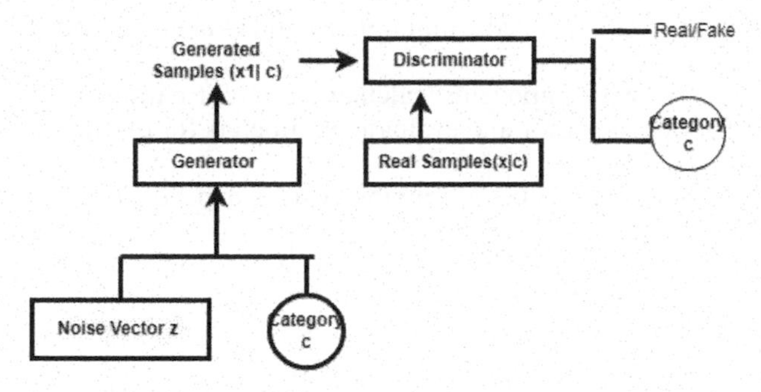

Figure 4. Architecture of infoGAN.

i.e changing sky color from sunny to cloudy, translating photographs of horses to zebra in vice versa. For unpaired data, CycleGAN is very important.

CycleGAN is implemented by using pairs of Generators and Discriminators. The First Generator G is used to transform images from the domain X to the domain Y. Images are transformed using Generator F from Y to X. Both the discriminator DY and the discriminator Dx distinguish between x and F(y). Figure 5 displays the basic architecture of Cycle GAN.

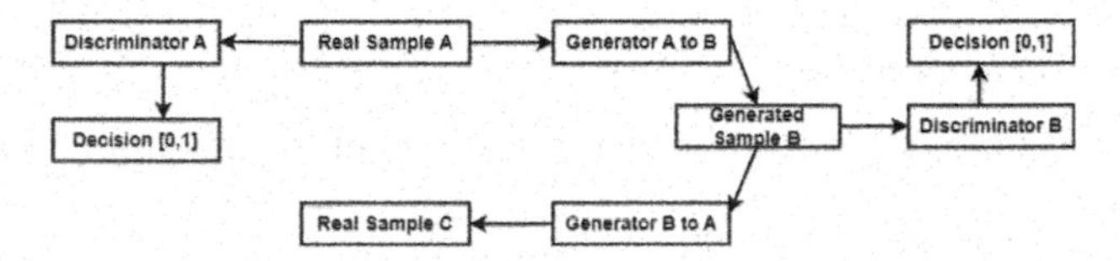

Figure 5. Architecture of cycle GAN's.

2.8 *Loss sensitive GAN(LS-GAN)*

The similar Lipschitz limitations was used by LS-GAN [24] and WGAN. LS-GAN aims for the objective function to satisfy the Lipschitz constraints. GAN structure does not change in WGAN and LS-GAN but it improves the optimization method and learning parameters. In LS-GAN, the loss function is determined by a parameter called theta. The objective of LS-GAN is to create samples whose loss value is greater than that of real data.

2.9 *StackGAN*

StackGAN [9] proposed to convert text description into high realistic photosynthesis. Stage I of StackGAN takes text description as input and generates primary colors of objects, sketch primitive shape objects and images with lower resolution . Stage I outcomes of StackGAN are fed into Stage II outcomes of StackGAN, which likewise accepts textual descriptions as input and generates high-quality photos with realism descriptions as output. Using text descriptions as input, StackGAN suggested to produce 256×256 photo-realistic graphics.

3 GAN ALGORITHM

When the models are neural network then GANs framework used for implementation. Generators distribution pg on data x and noise vector taken as input, denoted as pz(z) where noise variable is z. A GAN-represented mapping $g(z, \theta g)$ from noise space to data space. A parameterized neural network θg represented by a differentiable function which is denoted by G. The another neural network $d(x, \theta d)$ with parameters Ó·d, we get a single scalar D(x) as output. We get probability that x from the data represented by $d(x)$ in place of the

Generator G. Generator G is trained to minimise $log(1 - d(g(z))$, and discriminator D is trained to maximise together.

A training consists of two components: calculate different parameters for generator that maximum foolish the discriminator and evaluate parameters for discriminator that generate accurate classification.

The function V(g, d) applied in GAN training costs is dependent on the both models

$$\max_{d} \min_{g} V(g, d)$$

$$V(g, d) = E_{pdata}(x) \log d(x) + E_{pg(x)} \log(1 - d(x)) \quad (4)$$

At the GAN training, one model parameter is updated, while other model parameters are fixed. A fixed generator gives an individual optimum discriminator $d(x) = p_data(x)$ $p_data(x) + p_g(x)$ proved by Goodfellow *et al.* [2]. The optimum generator G gets when $p_{data(x)} = p_{g(x)}$, identical to the optimum discriminator that predicts a value 0.5 drawn for every sample x. When the discriminator D maximum confused the generator and can not differentiate real samples from fake samples, then the generator is optimum.

Hence, the discriminator is trained continuously for optimal regarding the generator and then again updated the generator. In other words, the generator and discriminator are being updated concurrently. In GAN training iteration we repeat the following steps.

3.1 *Generator training algorithm*

Step 1: Take a z vector of random noise and, applying on the generator and generate a fake example x^*

Step 2: Fake example x^* as input in the Discriminator

Step 3: Calculate the backpropogate error and classification error, for updation of the Disciminator's training parameters and trying to maximize the Disciminators error.

3.2 *Discriminator training algorithm*

Step 1: Training dataset used to get a random real example x

Step 2: Take a z vector of random noise and, applying on the generator and generate a fake example x^*

Step 3: Discriminator used for classification between real example x and fake example x^*

Step 4: To update the generator's training configurations and try to reduce the classification error, estimate the backpropagate error and classification error.

4 GAN EVALUATION METRICS

4.1 *Inception Scores(IS)*

The idea of Inception Scores originated in [25]. GAN used this metric very widely. The main goal of Inception Scores(IS) is to determine two features of generated images: image diverse and image quality. If the GAN model produced high quality samples and samples are diverse then it gives a higher IS value. In this generated image sample measure as the conditional label distribution $P(y|x)$.

If mode collapse happened in generative models then Inception Scores(IS) might be given higher values this is the drawback of Inception Scores(IS).

4.2 *Frechet Inception Distance (FID)*

The distinction between feature vectors for actual images and produced images is determined by the Frechet Inception Distance (FID) [13]. For both actual and synthetic images, the mean and covariance are computed using the multivariate Gaussian. Because it provides constancy in human traits and is sensitive to little changes in real distribution, FID has mostly been employed.

4.3 *Mode Score (MS)*

In comparison to Inception Scores (IS), Mode Score (MS) [25] is a better statistic. Dissimilarity can measure between real samples and generated samples by Mode Score(MS).

5 GAN APPLICATIONS

Gan applications in image processing are image editing such as the color of hair, adding of smile, image de-raining, image super-resolution, and texture synthesis. similarly,GANs synthesizing videos is applied into 3 types UnConditional, Conditional and prediction video generation. Generative models are used in biology for jobs involves such as protein sequence modeling and design [20], and imputation and biological generation of images [18]. GAN is used in medical imagining as image super-resolution , synthetic data synthesis and image augmentation.

6 CONCLUSIONS

In conclusion, GANs (Generative Adversarial Networks) are a specific type of model that are used to produce fake data that looks like a training set. They have been utilized for various purposes and have the ability to generate different types of data such as audio, images and text. Despite challenges such as instability in the training process and ethical concerns, GANs are one of several generative models that include VAEs, normalizing flow models, and autoregressive models. Each model has its own strengths and limitations. Generative models have the potential to change the machine learning and AI fields, and GANs have already had a major impact on many domains.This study aims to offer an extensive overview of generative adversarial networks (GANs). The paper discussed the theory, variants, algorithm, and applications of GANs. GANs are popular because they can learn nonlinear correlations between data and latent spaces and can use huge volume of unlabeled data, making it similar to supervised learning.

REFERENCES

[1] Mirza, M., Osindero, S. (2014). Conditional Generative Adversarial Nets. *ArXiv Preprint ArXiv*:1411.1784.

[2] Creswell, A., White, T., Dumoulin, V., Arulkumaran, K., Sengupta, B., Bharath, A. A. (2018). Generative Adversarial Networks: An Overview. *IEEE Signal Processing Magazine*, 35(1), 53–65.

[3] Nataraj, L., Mohammed, T. M., Chandrasekaran, S., Flenner, A., Bappy, J. H., Roy-Chowdhury, A. K., Manjunath, B. S. (2019). Detecting GAN Generated Fake Images Using Co-occurrence Matrices. *ArXiv Preprint ArXiv*:1903.06836.

[4] Karras, T., Aila, T., Laine, S., Lehtinen, J. (2017). Progressive Growing of Gans for Improved Quality, Stability, and Variation. *ArXiv Preprint ArXiv*:1710.10196.

[5] Radford, A., Metz, L., Chintala, S. (2015). Unsupervised Representation Learning with Deep Convolutional Generative Adversarial Networks. *ArXiv Preprint ArXiv*:1511.06434.

[6] Gui, J., Sun, Z., Wen, Y., Tao, D., Ye, J. (2021). A Review on Generative Adversarial Networks: Algorithms, Theory, and Applications. *IEEE Transactions on Knowledge and Data Engineering*.

[7] Isola, Phillip, Jun-Yan Zhu, Tinghui Zhou, and Alexei A. Efros. "Image-to-image Translation with Conditional Adversarial Networks." In *Proceedings of the IEEE Conference on Computer Vision and Pattern Recognition*, pp. 1125–1134. 2017.

[8] Almahairi, Amjad, Sai Rajeshwar, Alessandro Sordoni, Philip Bachman, and Aaron Courville. "Augmented Cyclegan: Learning many-to-many Mappings from Unpaired Data." In *International Conference on Machine Learning*, pp. 195–204. PMLR, 2018.

[9] Zhang, Han, Tao Xu, Hongsheng Li, Shaoting Zhang, Xiaogang Wang, Xiaolei Huang, and Dimitris N. Metaxas. "Stackgan: Text to Photo-realistic Image Synthesis with Stacked Generative Adversarial Networks." In *Proceedings of the IEEE InternationalConference on Computer Vision*, pp. 5907–5915. 2017.

[10] Almars, A. M. (2021). Deepfakes Detection Techniques Using Deep Learning: A Survey. *Journal of Computer and Communications*, 9(5), 20–35.

[11] LeCun, Y., Bengio, Y., Hinton, G. (2015). *Deep Learning. Nature*, 521(7553), 436–444.

[12] Creswell, A., White, T., Dumoulin, V., Arulkumaran, K., Sengupta, B., Bharath, A. A. (2018). *Generative Adversarial Networks: An Overview. IEEE Signal Processing Magazine*, 35(1), 53–65.

[13] Heusel, Martin, Hubert Ramsauer, Thomas Unterthiner, Bernhard Nessler, and Sepp Hochreiter. "*Gans Trained by a Two Time-scale Update Rule Converge to a Local Nash Equilibrium.*"

[14] Harshvardhan, G. M., Gourisaria, M. K., Pandey, M., Rautaray, S. S. (2020). A Comprehensive Survey and Analysis of Generative Models in *Machine Learning. Computer Science Review*, 38, 100285.

[15] Dash, A., Ye, J., Wang, G. (2021). A Review of Generative Adversarial Networks (GANs) and Its Applications in a Wide Variety of Disciplines–From Medical to Remote Sensing. *ArXiv Preprint ArXiv*:2110.01442.

[16] Zhang, H., Sindagi, V., Patel, V. M. (2019). Image de-raining Using a Conditional Generative Adversarial Network. *IEEE Transactions on Circuits and Systems for Video Technology*, 30(11), 3943–3956.

[17] Makhzani, A., Shlens, J., Jaitly, N., Goodfellow, I., Frey, B. (2015). Adversarial Autoencoders. *ArXiv Preprint ArXiv*:1511.05644.

[18] Arjovsky, M., Chintala, S., Bottou, L. (2017, July). Wasserstein Generative Adversarial Networks. In *International Conference on Machine Learning* (pp. 214–223). PMLR.

[19] Chu, Casey, Andrey Zhmoginov, and Mark Sandler. "Cyclegan, a Master of Steganography." *ArXiv Preprint ArXiv*:1712.02950 (2017).

[20] Isola, P., Zhu, J. Y., Zhou, T., Efros, A. A. (2017). Image-to-image Translation with Conditional Adversarial Networks. In *Proceedings of the IEEE Conference on Computer Vision and Pattern Recognition* (pp. 1125–1134).

[21] Chen, Xi, Yan Duan, Rein Houthooft, John Schulman, Ilya Sutskever, and Pieter Abbeel. "Infogan: Interpretable Representation Learning by Information Maximizing Generative Adversarial Nets." *Advances in Neural Information Processing Systems* 29 (2016).

[22] Wang, K., Gou, C., Duan, Y., Lin, Y., Zheng, X., Wang, F. Y. (2017). Generative Adversarial Networks: Introduction and Outlook. *IEEE/CAA Journal of Automatica Sinica*, 4(4), 588–598.

[23] Pan, Z., Yu, W., Yi, X., Khan, A., Yuan, F., Zheng, Y. (2019). Recent Progress on Generative Adversarial Networks (GANs): *A Survey. IEEE Access*, 7, 36322–36333.

[24] Qi, G. J. (2020). Loss-sensitive Generative Adversarial Networks on Lipschitz Densities. *International Journal of Computer Vision*, 128(5), 1118–1140.

[25] Salimans, T., Goodfellow, I., Zaremba, W., Cheung, V., Radford, A., Chen, X. (2016). Improved Techniques for Training gans. *Advances in Neural Information Processing Systems*, 29.

Artificial Intelligence, Blockchain, Computing and Security – Dagur et al. (Eds)

Traffic flux prediction on scarce dataset using transfer learning

Rishabh Jain, Sunita Dhinra & Kamaldeep Joshi
CSE Department, UIET, MDU, Rohtak, Haryana, India

ABSTRACT: It is vital for everyone to reach their destination on time, which is only possible by checking the various routes to the destination and choosing the one which takes minimum time. So, to ensure that path chosen is the efficient one we need a traffic flux prediction model, which provides the path with minimum traffic accurately and precisely. But sometimes, the traffic data contains some voids due to various factors like faulty input devices, drained battery, etc., which causes the scarcity of the data. To ensure that the traffic data remains complete to provide optimal results we need to impute those missing data values. In order to do so, we have proposed an approach which can complete those missing values and then can use the new data in the prediction model to achieve our objective.

Keywords: Missing Data, Transfer Learning, Data Imputation, Traffic Management System.

1 INTRODUCTION

In today's modern era, commuting from one place to another through various means of road transport is a common manifestation. With so many travelers commuting every now and then, the chances of roadways getting swarmed are highly likely. A uniform flow of traffic is necessary for the smooth working of a society which can be achieved by managing the increasing traffic flow. The prediction of traffic flux is expected to help authorities to come up with appropriate counter measures such as in determining whether to choose a particular route or not, or by displaying the time delay to reach the specific destination accurately. So, if the prediction of traffic flux is calculated accurately and precisely, the information generated, if sourced to the travelers in real time, then this would eventually help them reach their destination in time and avoid heavy traffic.

For traffic flux prediction data is gathered through numerous input sensors that have been setup across many road networks containing various types of input sensors such as loop detectors, sensors, etc. to serve the purpose of constant traffic state observation. These sensors generate an ample amount of data during traffic flow monitoring in real time. But unfavorable scenarios caused by natural or man-made events, problems such as insufficient readings, faulty devices resulting into power loss, problem in data exchange and loosing of sensor observations result into development of data voids in the dataset which consequently weakens the accuracy of the model and quality of the dataset.

Various studies were proposed to resolve the particular issue of incomplete dataset, but still they lacked in few areas and left some room for improvement. In paper [6], two machine learning techniques were used for updating missing data without considering the length of the gap. Their spatial context sensing method has demonstrated that when there is enough environmental information, missing values can be assumed even in the situations where information lacks. Estimation of unavailable data is a crucial component of data processing for intelligent traffic prediction systems. Using a modified k-nearest neighbor method,

DOI: 10.1201/9781032684994-6

researches suggested a data-driven estimation technique for road segments according to their geographical and spatial relation. The authors tried to simultaneously estimate missing data from a segment with numerous sensors that are correlated with one another which indicated that this approach performed better for all kind of missing data.

One of the approaches to deal with the complication of data voids is by recognizing patterns from numerous traffic datasets and implementing it on our dataset to make up for any voids in the data. Problem of missing data was also considered in various large road networks and several grid and tensor-based approaches were proposed to estimate these lost values and worldwide traffic patterns can be extracted from insufficient information. Tensors are mathematical constructs that expands scalars, vectors and grids to upper dimension. They analyzed the effectiveness of several methods for various variables, including pavement markings, weekdays, and the influence of matrix factorization selection on the prediction accuracy of recovered data traffic. KNN-based approach was used in order to achieve multistep forecasting and improve its accuracy based on spatiotemporal correlation. Determined equivalent distances through static and dynamic data, were used to substitute the precise separations between road segments. Instead of just a time series, a space-time state grid was formed to describe the traffic status of a road segment. On the basis of the shifting patterns of the traffic during a day, free and busy traffic hours were established. For imputation of the traffic data, suggested de-noising stacked auto encoders.

Another method to interact with the data void is to take the improper data as it is and create a traffic prediction model. According to the research conducted in the paper, when a sensor malfunctions, data recorded before the malfunction and/or after it, are used to reconstruct the missing information. A. Garnier *et al.* created an estimation tool build on the idea of time series using a feedforward multi-layer Perceptron. Missing information was estimated using an artificial neural network. Even in cases of prolonged sensor failures, the average relative error that was measured did not surpass 6%. These estimating techniques are now being used in a commercial embedded system developed by Pyrescom with reduced energy requirements. In this paper, we have introduced an approach, which will be discussed in following sections, that fulfill this requirement and present better predicting system even when the dataset is lacking some values.

2 RELATED WORK

Table 1. Literature review.

Paper Name	Author	Algorithms	Merit
"*A neural network approach for traffic prediction and routing with missing data imputation for intelligent transportation system.*"	Robin Kuok Cheong Chan *et al.*	WEMDI (Weighted missing data imputation), SUMO, MVR (Multifactor vehicle rerouting)	The current MVR system was enhanced by the introduction of additional 6.5% to 19.4%
"*Missing data imputation for traffic flow based on Combination of fuzzy neural network and rough set theory.*"	Tang, JinjunZhang *et al.*	FRS (Fuzzy Rough Set), FNN (Fuzzy Neural Network), KNN (K Nearest Neighbor)	The suggested LSTM network technique for forecasting traffic volume is reliable
"*Missing Value Imputation for Traffic-Related Time Series Data Based On a Multi-View Learning Method.*"	Linchao Li, Jian Zhang, Yonggang Wang, and Bin Ran	LSTM (Long-Short Term Memory) SVR (Support Vector Regression) CF (Collaborative Filtering), MVLM (Multi View Learning Method)	Even with a high missing ratio, MVLM can handle missing patterns.

(continued)

Table 1. Continued

Paper Name	Author	Algorithms	Merit
"LSTM-based Traffic Flow Prediction with Missing Data."	Tian, Yan & Zhang, Kaili & Li, Jianyuan & Lin, Xianxuan	LSTM (long-short term memory), Multiscale temporal smoothing, LSTM- M	The accuracy of the proposed LSTM-M approach surpasses that of numerous cutting
"Recurrent neural networks for multivariate time series with missing values."	Z. Che, S. Purushotham, K. Cho, D. Sontag, Y. Liu	GRU-D (Gated recurrent unit), GRU	To improve prediction outcomes, the method identifies long-term temporal relationships in time series
"On the imputation of missing data forroad traffic forecasting: New insights and novel techniques."	Ibai *et al.*	Facial context sensing, automated clustering analysis	Displays the possibility in the absence of data, replace missing values.
"Diffusion convolutional recurrent neural network: Data-driven traffic forecasting."	Y. Li *et al.*	DCRNN (Diffusion Convolutional Recurrent Neural Network)	Neural network that captures the spatiotem-poral dependencies that are dynamic in nature.
"Deep forecast: Deep learning based spatio temporal forecasting."	A. Ghaderi *et al.*	Spatio-temporal wind speed forecasting algorithm, RNN (Recurrent neural Network)	In the suggested the space-time information is represented by a graph
"LSTM network:a deep learning approach for short-term traffic forecast."	Z. Zhao, W. Chen, X. Wu, P. C. Chen, J.Liu	LSTM (Long-Short Term Memory)	The suggested LSTM network technique for forecasting traffic volume is reliable.
"Deep spatio-temporal residual networks for citywide crowd flows prediction."	J. Zhang, Y. Zheng, D. Qi	ST-ResNet (Spacio-Temporal Residual Neural Network)	Generated results that greatly outperformed six benchmark HA, ARIMA.

3 METHODOLOGY

In Figure 1 LSTM is the core idea, understanding it is essential. Even though Sequence Prediction was an easy task for RNNs to perform, learning long-term reliance can still be challenging due to its modest gradient, which makes adjusting the weights and biases of existing layers utterly worthless even after numerous iterative training sessions. This problem of ineffective amendment is resolved by the LTSM through its incorporation of memory units which guides the network during memory resetting and memory updating.

Time-series analysis frequently uses LSTM networks. The historical traffic information may be thought of as a priori knowledge for a particular monitoring location in the roadway. In contrast to the traditional LSTM network, the proposed model incorporates ODC matrices, and both analysis and data training reveal temporal-spatial connection. The constant traffic forecasting can also be divided using the pipelined LSTM network into some short-term forecasting processes, and it can produce findings for several traffic flow forecasts

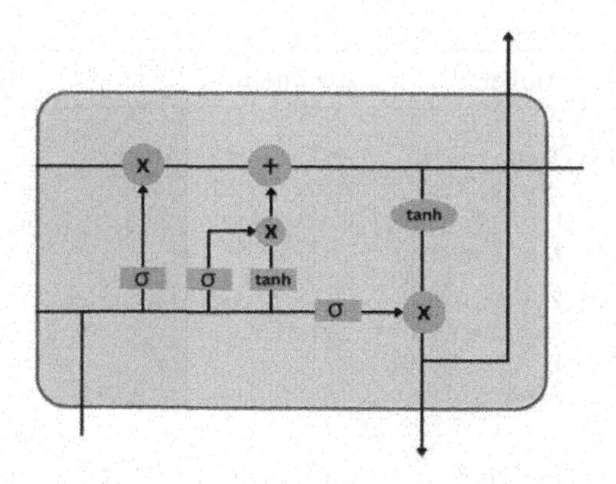

Figure 1. LSTM.

in future rather than over an extended period of time. The data from each observation site is organized into a temporal sequence in the proposed method for quick traffic prediction.

In [10] the 2D network, horizontal axis denotes time variation and vertical axis denotes indexes of different monitored points, which are arranged in increasing order. First these indexes are assigned, then we compute the distance in space axis.Time difference of these multiple layers are represented as Δt1, Δt2, . . . , Δtm, which satisfies the constraints

$$T_f = \sum_{i=1}^{m} \Delta t_i$$

here T_f = the prediction time,
m = number of layers (<=8)
Δt1, Δt2, . . . , Δtm are then calibrated by minimizing the sum of square errors.
In time t, complete connection layer is supplied to link the result of former moment t-1.
Assume S_{t-1} as the data of traffic of roadways at an instant t-1, indicates by

$$S_{t-1} = \left[X_{1,t-1}, X_{2,t-1,\ldots,} X_{k,t-1}\right]^T$$

$$I_t = \left[X_{1,t,} X_{2,t,\ldots,} X_{k,t}\right]^T$$

and memory unit's input at an instant t is denoted by
The relation of the S_{t-1} and It is
here M(t, Δt) = ODC matrix, and

$$I_t = M(t, \Delta t) * \mathrm{repmat}(S_{t-1,m})$$

'*' = the product,
repmat(S_{t-1}, m) = newly built matrix
Size of repmat(S_{t-1}, m) = size of M(t, Δt)

It's ith column the respective multiplication of elements in S_{t-1} and the elements in ith column of M(t, Δt). The kth memory unit will take the vector Xk,t and outcome is relies on memory unit evaluation. The time–space relation is embedded in the LSTM network of two-dimension. The prediction outputs are close with the traffic's data and communication among various monitored channels.

4 RESULT

In LSTM (LSTM-DO-TR) made use of dual memory cell layers. Target replication improves performance across the board by reducing overfitting and accelerating learning.

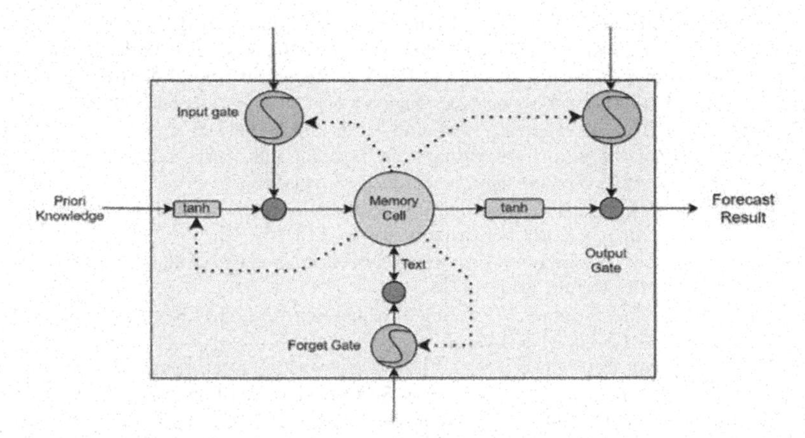

Figure 2. LSTM memory unit.

Additionally, we discovered that the LSTM equipped with target replication gains the ability to produce accurate verification.

Particularly when the prediction duration is lengthy, the performance our model outperforms that of the SAE, RBF, SVM, and ARIMA models, while other deep learning techniques like RBF network, RNN, or SAE, SVM exhibited weakness. LSTM RNNs outperformed baselines with fixed windows just using the first and latest six hours.

Although our data collection is substantial according to clinical benchmarks, it is modest when compared to datasets used in deep learning applications like voice and visual recognition. Regularization is crucial at this size. Our research shows that dropout, target replication, and auxiliary outputs all help close the generalization gap. This might be due to the fact that target replication offers greater benefits than regularization and lessens the challenge of learning long-range interdependence by supplying local targets.

5 CONCLUSION

We proposed a linear-based model for traffic flux prediction Figure 2 even when there is data scarcity. The tested results demonstrated that our approach outperforms several other methods in terms of accuracy by first imputing the missing values in the original dataset and then using raw features, which reduces the steps like feature engineering and preprocessing. Moreover, our data imputation process improves the data, by filling up the voids present in them, which ensures that the model generates the output precisely.

By fusing the activity among the roadways in time-space bound, a successive LSTM network is built. Tests deduced that the approach for traffic flux prediction is resilient.

REFERENCES

[1] Robin Kuok Cheong Chan, Joanne Mun-YeeLim, Rajendran Parthiban, "A Neural Network Approach for Traffic Prediction and Routing with Missing Data Imputation for Intelligent Transportation System", *Expert Systems with Applications: Elsevier*, 1 June 2021, vol. 171.

[2] Tang, Jinjun & Zhang, Xinshao & Yin, Weiqi &Zou, Yajie & Wang, Yinhai (2020). "Missing Data Imputation for Traffic Flow Based on Combination of Fuzzy Neural Network and Rough Set Theory". *Journal of Intelligent Transportation Systems*. 25.1–16.10.1080/15472450.2020.1713772.

[3] Linchao Li, Jian Zhang, Yonggang Wang, and Bin Ran. "Missing Value Imputation for Traffic-Related Time Series Data Based on a Multi-View Learning Method", *IEEE Transactions on Intelligent Transportation Systems*, pp. 1524–9050, 8 August 2019.

[4] Tian, Yan Zhang, Kaili Li, Jianyuan Lin, Xianxuan Yang, Bailin. (2018). "LSTM-based Traffic Flow Prediction with Missing Data". *Neurocomputing*. 318. 10.1016/j.neucom.2018.08.067.

[5] Che Z., Purushotham S., Cho K., Sontag D., Liu Y., "Recurrent Neural Networks for Multivariate Time Series with Missing Values", *Scientific Reports* 8 (1) (17 April 2018) 6085.

[6] Ibai Lañaa, Ignacio (Iñaki) Olabarrietaa, Manuel Vélez and Javier Del Sera, "On The Imputation of Missing Data for Road Traffic Forecasting: New Insights and Novel Techniques", *Transportation Research Part C: Emerging Technologies*, 9 March 2018, vol. 90.

[7] Li Y., Yu R., Shahabi C., Liu Y., "Diffusion Convolutional Recurrent Neural Network: Data- Driven Traffic Forecasting", *in: ICLR*, (22 February 2018), pp. 147–155.

[8] Ghaderi A., Sanandaji B., Ghaderi F., "Deep Forecast: Deep Learning-based Spatio Temporal Forecasting", *in: ICML*, 24 July 2017, pp. 264–271.

[9] Lipton Z., Kale D., Elkan C., et al., "Learning to Diagnose with LSTM Recurrent Neural Networks", *in: ICLR*, 21 March 2017, pp. 1456–1463 Z.

[10] Zhao, W. Chen, X. Wu, P. C. Chen, J. Liu, "LSTM Network: A Deep Learning Approach for Short-term Traffic Forecast", *IET Intelligent Transport Systems* 11 (2) (24 Feb 2017) 68–75.

[11] Zhang J., Zheng Y., Qi D., "Deep Spatio-temporal Residual Networks for Citywide Crowd flows Prediction", *in: AAAI*, 10 Jan 2017, pp. 1655–1661.

[12] Duan Y., Lv Y., Liu Y.-L., and Wang F.-Y., "An Efficient Realization of Deep Learning for Traffic Data Imputation", *Transportation Research Part C Emerging Technologies*, vol. 72, pp. 168–181, Nov.2016.

Artificial Intelligence, Blockchain, Computing and Security – Dagur et al. (Eds)
© 2024 The Author(s), ISBN: 978-1-032-67841-2

Intelligent assessment of MCQs using machine learning

Shoumik Mukherjee, Harsh Goel & Abdul Aleem
SCSE, Galgotias University

ABSTRACT: Online tests are an important way to test student's ability to succeed. This research proposes smart evaluation through an online education system that combines exam processing and assessment techniques through the use of association rule mining. Association rule mining is one of the various machine learning techniques that are used for data modelling. The model proposed is of a web-based placement test system, which addresses research obstacles, and benefits end users through software development. The people considered for research are students, who enrolled in computer course department of Galgotias University. The trainer constructs lesson-based questions that are available online to students. Users registered on the forum can access the electronic information provided and achieve a range of tasks through the online education system including participation in online tests. Users can receive online tests, with multimedia, tutorial content, and can provide electronic test responses. After completion of exam, students are given a grade based on smart evaluation methodology, which considers the pattern of wrongly marked answers and deduct marks accordingly. The patterns are checked through association rule mining.

1 INTRODUCTION

Online tests content providers focus on creating effective test questions and delivering test answers to students [1]. This article present strategies related to the fundamentals of the examination system, smart assessment and post-submission advice after compliance. Existing examination systems are automated and computer are compliant enough to provide faster and effective mechanism for assessment of such online examinations. Supervisors, teachers, students who attend online tests can interact with the program through these projects, thus enabling effective execution and monitoring of various online test tasks such as conducting the tests on a programmed base and deliver results to that application or student. Details of student who have tried the Internet Test are kept in the administrator. This article proposes a web-based examination system, which evaluates MCQs through smart evaluation and deduct marks for cheated questions.

Students who appear in the examination system have to answer MCQs, which have just one correct option. However, when students cheat for the answers, they marked the answers same as their friends. There is one unique right answer, which is to be same for all the correct answers. However, wrong answers can be marked with three remaining options in MCQs. If two students have similar pattern for wrongly marked answers, then it is a case of cheating. To counter this, assessment can be done considering the pattern of wrongly marked answers. In machine learning, we can make use of association rule mining to determine the pattern of wrongly marked answers.

2 LITERATURE SURVEY

Online examination gadget is one of the techniques of attractive assessments which doesn't necessitate any form of a bit of term paper or a marker. It's miles the rapid developing

DOI: 10.1201/9781032684994-7

technique to obtain assessments over online velocity and exactness is the motive at the back of the well-known of this method due to the fact speed and accuracy is the backbone of this device. Many researchers have already research approximately online examination gadget. This article briefly reviews the research on online exam tools Zhenming *et al.* [2] urbanized a web exam device based totally on net browser/server skeleton. Zhenming *et al.* advanced a web exam gadget based totally on net browser/server framework, which supports some premium fundamental features, vendors out the exam and provide the car grading machine for goal questions and operating questions like programming, edit MS word, electricity point, MS windows, Excel etc.

Aleem and Gore [3] has proposed a new marking scheme for evaluation of MCQs, which is a step further towards smarter evaluation. The authors gave the examinees option to select more than one option for answering MCQs. However, the marks awarded/deducted were in proportion of the number of options selected. Similarly, Guzman *et al.* [4] evolved an online examination device called as SIETTE, which is a system for smart assessment through the usage of distant learning.

Ayo *et al.* [5] planned a mock-up of e-examination. The software turned into evolved in non-public college in Nigeria. The motive in the back of the evolved which include software is to behavior the access exam for all Nigeria universities referred to as JAMB (Joint Admission Matriculation Board). This software turned into designed and examined in Covenant college they had been the personal university in Nigeria. They determined the software program certainly beneficial for undertaking neat and smooth with accuracy front exam. It's miles eliminates the issues which might be related to the traditional strategies of entrance examination.

Jim and Sean [6] proposed that e-assessment may be taken in distinct approaches. The authors proposed that the e-evaluation can be taken in diverse conduct. to start with they made a e-evaluation via net after which the mind-set has been accomplished that we can beautify the e-assessment to a on line examination system. and that they notion that there will be a many one-of-a-kind ways to take the evaluation and e-examination. They constantly introduced the satisfied material and that they take the form of e-examination portal eventually. Due to the fact it's far the ways away one of a kind and correct than of the conventional process to obtain any of the access exam or estimation computer based absolutely exam is one of the finest and happy approach of taking.

These are the really necessary keywords of any developed system. Guzman and Cenejo [7] enhanced the online examination system SIETTE (System of intelligent Evaluation using Tests for Tele education). The above developed system supports the login and some basic features but doesn't supports the premium features such as random questions selection, random choices distribution, resumption capabilities, random questions distribution.

The application of association rule mining on educational systems is an evolving concept. Association rule mining has been able to track manuscripts evolution to perfection [8]. But its application for smarter assessment has not been done before. Many online assessment systems [9,10] exist, which may be enhanced for smart assessment of MCQs. This article aims to use association rule mining for smarter assessment of MCQs, which could deduct marks for the cheated wrong answer.

3 PROPOSED WORK

3.1 *Existing system*

The present system is a manual in which users store archives such as scholar information, teacher details, schedule information and answers regarding student who have tried to test as per schedule. It is very hard to store old data. The subsequent problems of the accessible arrangement give emphasis to the necessities required for creating a web-based examination system, which are listed as follows:

1. Many copy of the questions should be made
2. Too much repair work which is why you are delayed in delivering results
3. Multiple table work results for each lesson

3.2 *Proposal*

This function is used to perform online tests. student can take a seat in personality terminal and sign in to engrave exams in the allotted time. Questions should be given to readers. This proposal gives the administrator space to add new tests. Moreover, a smart assessment can be done using Associtaiton Rule Mining (ARM). Association rule mining finds interesting institutions and relationships among massive units of data objects. This rule shows how often a itemset takes place in a transaction. a regular example is market primarily based evaluation.

Market based totally evaluation is one of the key techniques used by big members of the family to expose institutions among items. It lets in retailers to pick out relationships between the gadgets that people buy collectively frequently. Given a set of transactions, we will discover rules so one can expect the incidence of an item based totally at the occurrences of other objects in the transaction. Table 1 shows the data for conducting a demo exam of 10 students. The examination question paper has 6 questions. The objective of the research is to find pattern of wrong answer keys having high similarity, so that cheaters could be identified. It was found that out of 10 students, 3 patterns were 2A4B5C and 3 pattern were 3A4C. So total cheater found was, [(3+3)/10] *100 i.e, 60%.

Table 1. An Illustration of deriving patterns from the MCQs assessment.

SN.	1	2	3	4	5	6	Pattern
1	R	R	R	R	R	C	6C
2	R	A	R	B	C	R	2A4B5C
3	R	A	R	B	C	R	2A4B5C
4	R	A	R	B	C	R	2A4B5C
5	R	R	R	C	C	R	4C5C
6	C	B	A	R	R	R	1C2B3A
7	R	R	R	R	R	R	NIL
8	R	R	A	C	R	R	3A4C
9	R	R	A	C	R	R	3A4C
10	R	R	A	C	R	R	3A4C

4 IMPLEMENTATION

The nearly all-important step in achieve a successful innovative arrangement and in generous assurance to an original arrangement for user is that it will employment proficiently and successfully.

Figure 1 shows the sequence diagram of proposed smart assessment system. The proposed system has been executed on a Pentium-IV CPU with 256 MB RAM, 512 KB Archive Memory 10 GB CD Microsoft Companionable 101 or higher Keyboard. The software essentials for this tool are a Windows operating system. The tool has been developed using PHP as base language. HTML, CSS, and JAVASCRIPT have been utilized as Front-End, where MySQL has been employed as Back-End. The web server utilized for hosting the application is Apache server.

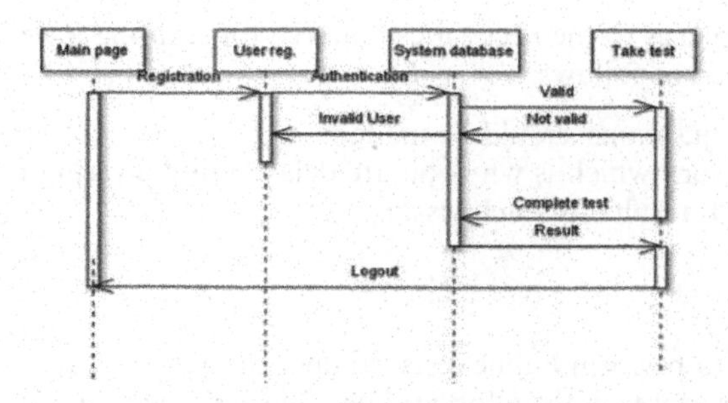

Figure 1. Sequence diagram of proposed smart assessment system.

The organization can only be used behind a careful examination and if it is created to be working in accordance with the definition. It includes careful planning, the investigation of the current system and its implications for implementation, the design of ways to achieve change and the assessment of changes in roads are part of planning. Two main farm duties for the accomplishment of user learning and guidance and program evaluation. Figure 2 shows the collaboration diagram of the proposed system.

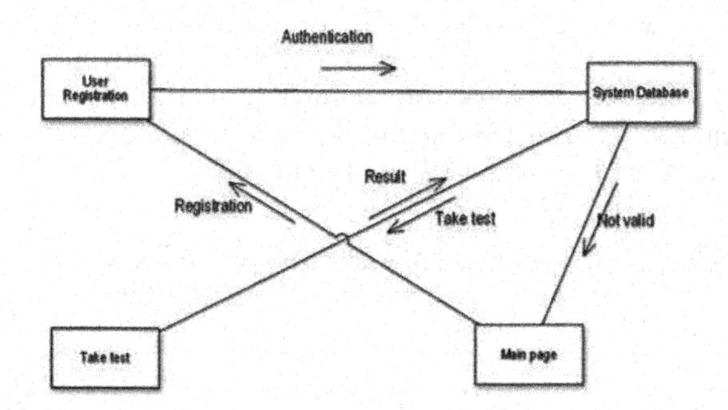

Figure 2. Collaboration diagram of smart assessment system.

If the system is working hard, it will be more involved to analyze the plans and the design effort required to implement it. The implementation phase includes several functions, compulsory hardware and software attainment is done. The arrangement may well need a few software to advance. In this case, the plans are documented and evaluated. The customer next switches to his original entirely experienced organization and the older arrangement are discontinue. Figures 3 and 4 shows the homepage and dashboard of the proposed smart assessment cum examination system.

Figure 3. Home page of the proposed smart assessment system.

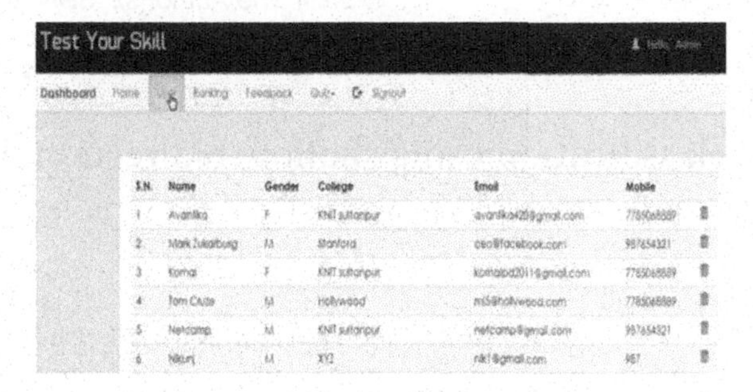

Figure 4. Dashboard of the proposed smart assessment system.

5 CONCLUSION AND FUTURE WORK

This article presented the automation of the complete examination system, which improves effectiveness and provide a user-friendly visual system that prove to be better as compared to the existing system. The proposed system provides relevant access to authorized user depending on their permissions. The system successfully overcome communication delay, making update of information much easier. System security, data security and consistency are remarkable features.The smarter assessment of MCQs is a novel step which discourage the examinees to cheat in the examination. The novelty lies in the application of association rule mining for detecting patterns in the wrongly marked responses which have been cheated from other examinees. The system has plentiful space for future change if required. We have found that the project can be done in a improved way. In particular, when we ask for details about a specific plan it simply indicates the date of the test and the forum. Therefore, after receiving the information we can access the information in a more user-friendly way.

REFERENCES

[1] A. Aleem, and M. M. Gore, "C-BEAM: A Confidence-based Evaluation of MCQs for Providing Feedback to Instructors", *Computer Applications in Engineering Education*, volume 27, issue 1, p. 112–127, 2019.

[2] Y. Zhenming, Z. Liang, and Z. Guohua. A Novel Web-based Online Examination System for Computer Science Education. *In 33rd ASEE/IEEE Frontiers in Education Conference* (pp. 5–8), 2003.

[3] A. Aleem and M. M. Gore, "The Choice is Yours: The Effects of Optional Questions in Engineering Examinations", *Computer Applications in Engineering Education*, vol. 27, issue no.5, pp:1087–1102, 2019.

[4] R. Conejo, E. Guzmán, and M. Trella. The SIETTE Automatic Assessment Environment. *International Journal of Artificial Intelligence in Education*, 26(1), 270–292, 2016.

[5] Ayo, C.K., Akinyemi, I.O., Adebiyi, A.A., Ekong, U.O.: The Prospects of E-examination Implementation in Nigeria. *Turk. Online J. Distance Educ. – TOJDE* 8(4), 125–135 (2007). ISSN 1302–6488, Article No. 10.

[6] R. Jim and M. Sean, *L*iterature Reviews of E-assessment, Future Lab Series, Report 10: ISBN: 0-9544695-8-5. Retrieved on October 4th 2013 from http://hal.archives-ouvertes.fr/docs/00/19/04/ 40/PDF/ridgway-j-2004-r10.pdf

[7] Guzmán, E., and Conejo, R. (2002). Simultaneous Evaluation of Multiple Topics in Siette. In S. Cerri, G. Gouardères, & F. Paraguaçu (Eds.), Intelligent Tutorial Systems, *6th International Conference*. Lecture Notes in Computer Science 2363 (pp. 739–748). Berlin: Springer.

[8] A. Aleem, and A. Kumar, and M. M. Gore, "A Study of Manuscripts Evolution to Perfection". In *Proceedings of 2nd International Conference on Advanced Computing and Software Engineering* (ICACSE-2019), pp:278–282, 2019.

[9] Z. M. Yuan, L. Zhang, G. H. Zhan, A Novel Web-based Online Examination System for Computer Science Education, In *Proceeding of the 33rd Annual Frontiers in Education*, 2013, S3F7-10.

[10] Zhang, L., Zhuang, Y.T., Yuan, Z.M., Zhan, G.H.: A Web-based Examination and Evaluation System for Computer Education. In: *Sixth IEEE International Conference on Advanced Learning Technologies (ICALT 2006)*, pp. 120–124. IEEE (2006)

Artificial Intelligence, Blockchain, Computing and Security – Dagur et al. (Eds)
 ISBN: 978-1-032-67841-2

Disaster tweet classification using shallow bidirectional LSTM with attention layer

Sipra Singh* & Jagrati Singh*
Centre for Advanced Studies, Dr. A.P.J. Abdul Kalam Technical University, Lucknow, U.P., India

ABSTRACT: During the time of crisis, disaster-related tweets on Twitter can provide various details about people who have been hurt or killed, who are missing or have been found, and who have had damage to utilities and infrastructure, which can assist governmental also nonprofit organizations in prioritizing their aid and rescue efforts. Due to the massive amount of these tweets, it is crucial to construct a model that will divide them into distinct classifications. In this paper an attention-based shallow Bi-directional Long Short-Term Memory (BLSTM) model is proposed to categorize disaster and non-disaster tweets. To extract semantic meaning from tweets, Global Vectors for word representation (GLoVe) is used. From the result, the suggested attention-based BLSTM model performs better when compared to several machine learning models. The proposed shallow BLSTM with attention layer model achieved an accuracy of 88.79 percent.

Keywords: Bidirectional LSTM (BLSTM), Disaster, GloVe, Long Short-Term Memory (LSTM).

1 INTRODUCTION

Over the past 10 years, social media platforms like Twitter, Instagram, Facebook, and Snapchat have become incredibly popular and have produced unheard amounts of unstructured data. As social media generates huge amounts of data as well as a wide range of data types, including text, photos, and videos. Globally, 3.5 billion people used social media in 2020, which is roughly 45% of the world's population. Every day, Twitter users send a total of 500 million tweets, or 5787 tweets every second. Users can express ideas on a variety of issues and their concerns on these platforms. As a result, numerous firms have the chance to use data analytics tools and algorithms to recognise people's emotions and provoke an emotional response. People who are in close proximity to actual disasters may rapidly post about them on their social media accounts, making sentiment analysis used by disaster rescue and response teams one of its most important applications. The objective is to assess the material for any mentions of a disaster (such as a wildfire, an earthquake, etc.) that could help promptly mobilise response teams. Using disaster-keyword filtering is a simple way to gather tweets about disasters. For example, tweets can be filtered using a dictionary with relevant keywords (e.g., "flood", "earthquake") or specific hashtags (e.g., "#NepalQuake", "#boulderflood", "#coloradoflood"). These descriptive phrases are various and vague and individual users frequently change their hashtags over time. As a result, a sizable fraction of the gathered tweets might be pointless. To recognize disaster-related tweets, machine learning (ML) techniques and more recently, deep learning (DL) methods are frequently modelled as an automatic categorization problem. Tweets can only be 280 characters long

*Corresponding Authors: 21mcs10@cas.res.in and jagrati@cas.res.in

 DOI: 10.1201/9781032684994-8

and frequently contain uncommon acronyms and spelling errors; it is quite difficult to categorize them.

In this study, a contrast examination of traditional machine learning and deep learning-based classifiers for categorizing tweets about disasters has been conducted. For traditional machine learning, TF-IDF feature combinations in the unigram, bigram, trigram weight ranges, and GloVe [1] embedding words in deep neural networks were applied. This paper's contribution can be summed up as follows:

(i) The disaster-related dataset is trained and tested on seven different conventional machine learning algorithms as well as the deep learning algorithm.
(ii) Examining how the classification job is affected by the TF-IDF feature as well as the word embedding vector.
(iii) Evaluating the effectiveness of the proposed model i.e., BLSTM with attention layer versus traditional machine learning models.

2 RELATED WORK

The issue of automatic classification of social media messages related to crises has been tackled using a variety of computational techniques. Three categories can be used to broadly classify these techniques. The conventional Machine learning (ML) methods are noted for their effective and efficient explainable feature-based predictions. For instance, [3] and [4] used a modification technique along with Logistic Regression and Naive Bayes to classify crisis tweets. Numerous issues relating to the processing of brief messages have been successfully solved using deep learning techniques based on neural networks [5,6,7].

A group of attention-based neural network models called Transformers was inspired by the original transformer paper [8]. In [9], a multitask learning model utilizing the neural network is used in the concept of the deep neural network for analyzing disaster-related tweets, which was introduced in [10,11] suggested a deep neural network with 29 layers also for examining disaster-related tweets. [2] demonstrated that a 2 neural network could be trained for catastrophe datasets more effectively when combined with external pre-trained word vectors i.e., GloVe. These earlier works served as an inspiration for the deep learning approach. Due to the character limit on tweets, attention layers with tailored auxiliary features that are specific to a given domain can have a significant impact. An attention-based deep neural network is used in this study to divide tweets into two categories based on whether they refer to actual disasters or not.

3 PROPOSED APPROACH

The methodology used to categorize tweets about disasters is covered in detail in this section. For the creation of our system, we used a deep neural network-based model, Shallow Bidirectional Long-Short-Term-Memory (BLSTM) with an attention layer compared with different seven machine learning classifiers. The different categories are used in machine learning: (i) Extreme Gradient Boosting Classifier (XGB), (ii) Naive Bayes (NB), (iii) Random Forest (RF), (iv) CatBoost Classifier, (CBC) (v) Logistic Regression (LR), (vi) Support Vector Classifier (SVC), (vii) Voting Classifier (VC). The expression "Frequency-Inverse Document Frequency" (TF-IDF) vector was applied as an input to classifiers in traditional machine learning techniques. To test every classifier, every combination of Features from the 1-, 2-, and 3-gram TF-IDF were retrieved.

3.1 *Dataset*

The data set consists of two datasets with a combined total of 10,848 tweets: a training dataset (7,593 rows), and a testing dataset (3,255 rows). The training dataset differs from the testing dataset because it contains a target attribute, whereas the testing dataset does not. The data set was acquired from the Kaggle website, a popular online forum and network for data scientists and machine learning (ML) practitioners.

3.2 *Data analysis*

3.2.1 *Data understanding*

The working group of data needs to be comprehended and analyzed; this is an essential first step in attempting to highlight some of the features and traits of the collecting data. For instance, the structure, type, occurrence, empty values, and other aspects of each of the five characteristics must be considered. The creation of word clouds, calculating sentiments, and graphing attributes to learn much more about them, such as location features and keyword features, are additional components of exploratory data analysis (EDA).

3.2.2 *Data set exploration*

We will first load the working data set and all the packages needed to do exploratory data analysis (EDA), data preprocessing, visualization, evaluation and data modelling. We have 7,593 observations across 5 variables in our data collection. Each record's individual identification number (integer) is contained in the first variable, "id." Each tweet has a keyword, which is represented by the character variable "Keyword." The term "location" is similar to the character type and relates to the place where the tweets were published. The primary attribute in this dataset is "text" which is the body of the tweet and also a character type. The last variable, "target," is an integer with a value either of "1" or "0". The target indicates each record's output and label, where "1" represents the tweet related to a real disaster on the other hand "0" denotes fake catastrophe tweets. Additionally, by looking at null values, we observed "id", "target" and "text" doesn't contain null values, whereas "Keyword" contains 87 missing data, and "location" contains 3638 null values, as shown in Figure 1.

Next, the "target" variable was visualized, as shown in Figure 2. It has been highlighted that the information gathering has more entries with "0" label, which denotes fictitious tweets about disasters, than those with "1" label. A potential option is to sample an equal number of tweets from each label in turn to have a balancing subset with 50 per cent of each desired variable in it.

Then, analyzing the top 20 disaster and non-disaster keywords the same procedures were followed for the "location" variable.

3.2.3 *Data pre-processing*

Following visualization and data exploration the pre-processing stage for the functioning data set comes in this section. Data preprocessing is as crucial as model creation, and it typically takes the effort and time than any steps. Preparing and cleaning text data for use in the model is known as text preprocessing. Because capital letters and lower-case letters are

	number of nulls	percentage of nulls
id	0.0	0.000000
keyword	87.0	0.007999
location	3638.0	0.334498
text	0.0	0.000000
target	0.0	0.000000

Figure 1. Dataset null values.

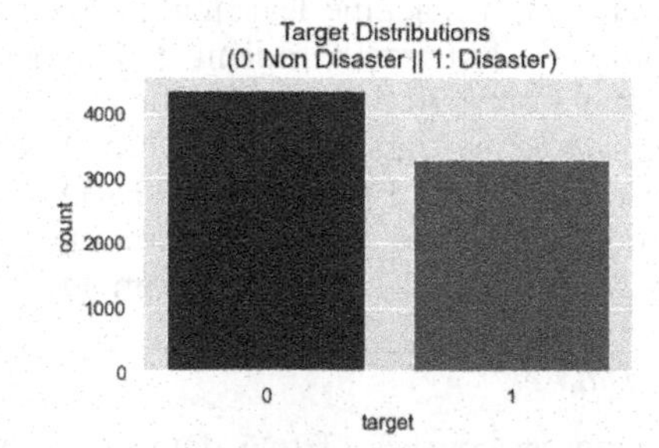

Figure 2. Distribution of target column.

read differently by computers, all text letters will be changed to the lower case. Combining digits and words in a text makes it difficult for a machine to understand. Consequently, it is better to avoid combining words with numbers. Because this kind of statement can be confusing, it is best to eliminate it or substitute an empty string in its place.

The most frequent keywords are in stop words, which provide no useful information. Phrases like "are," "this," "there," and others are examples of stop words. Using an online list of prohibited terms, the stop words were eliminated. Special characters like @, $, #, /, $, and emojis were removed. Utilizing stemming often entails removing a word's suffix and reaching the word's root. For instance, the base word or root word "Fly" remains after the suffix "ing" is removed from the term "Flying." These suffixes are used to transform the stem word into a new term. Finally, after completing pre-processing stages of all the previous data, then training subset, and a testing subset were prepared for usage in the modelling phase.

3.3 *BLSTM model with attention layer*

An input layer, an embedding layer, a BLSTM layer, an attention layer, and an output layer are the five main parts of the suggested deep learning model. The essential system graph is shown in Figure 3.

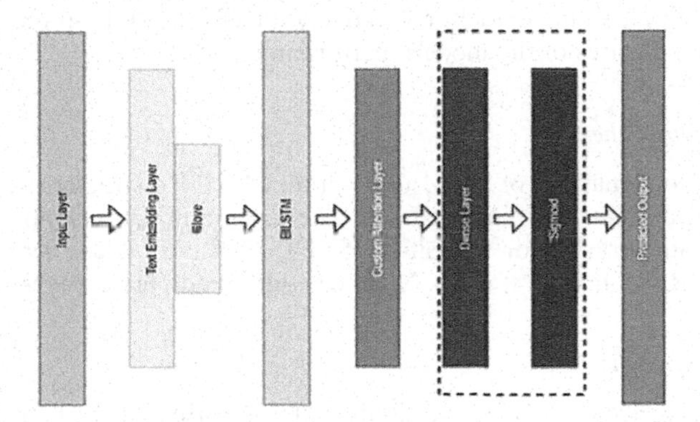

Figure 3. System architecture of proposed model.

3.3.1 *Input layer*

Tweets that have already been analyzed are supplied into the input layer, which will have linked to an embedding layer.

3.3.2 *Embedding layer*

Using lookup tables, this layer transforms an input into vectors with real values. Word embedding with prior training is useful in determining the semantic definition of words and enhancing the classification model. In this study, a pre-trained word vector called GloVe [1] was used to create feature word vectors using a statistical model based on co-occurrences. Every tokenized word in every tweet is mapped to its appropriate word vector table with the use of embedding. To converge the feature into the vector matrix, the necessary padding is applied.

3.3.3 *BLSTM layer*

A specific Recurrent Neural Network (RNN) which learns long-term dependencies is the Long-Short Term Memory (LSTM). Input is transmitted through a bidirectional LSTM (BLSTM) in both forward and backward directions, unlike LSTM, which could only observe and learn from the past input data. For the different applications involved in understanding complicated language, the BLSTM's bidirectional feature is essential [12] because it can record both the present and the past context of the input sequence.

3.3.4 *Attention layer*

The fundamental idea of attention [13] was derived from the observation that not every word in a phrase equally contributes to the semantic meaning. To locate the words in a tweet that have a closer semantic relationship, there employed a word-level, differentiable, deterministic attention strategy.

3.3.5 *Output layer*

After performing binary classification, the dense layer employs the sigmoid activation function. For the desired output, the model generates binary values. Further, describe comprehensive details on model hyperparameters and evaluation results.

4 EXPERIMENTAL STUDIES

The models are implemented at Google Collaboratory. Scikit-learn, keras, Tensorflow as the back-end with Python libraries, are used to implement the deep learning models and the conventional machine learning models, respectively. Precision (P), F1-score (F1), Recall (R), and Accuracy are measures of the models' performance.

4.1 *Compared approaches*

Many alternative combinations of 1-, 2-, and 3-gram TF-IDF characteristics were employed for the traditional machine learning methods. The seven different classifiers are as: Voting Classifier (VC), Support Vector Classifier (SVC), CatBoost Classifier (CBC), Logistic Regression (LR), Random Forest (RF), XGB Classifier and Naive Bayes (NB).

4.2 *Experimental results*

Word embedding vectors (GoVe) were used in the experiments with the proposed deep neural network. The F1-score, accuracy, and other well-known evaluation metrics are utilised to compare and validate the experimental results of the models of all the traditional machine learning and deep learning classifiers for disaster tweets. It is essential to have a set of ideal parameters to get the performance outcomes you want. There, precise parameter tuning is done, and an optimal set is chosen for the experiment. For improved evaluation and model reproducibility, the same parameter was employed. The model's input parameters are shown in Table 1.

Table 1. Hyper-parameter values.

Hyperparameter	Value/Description
Text embedding	Dimension: 100
BLSTM Layer	1 layer; 48 hidden units (Forward and Backward)
Drop-out rate	Dense layer: 0.2
Attention Layer	1 layer; 48 hidden units
Activation function	Output Dense layer: Sigmoid
Adam optimizer	Learning rate = 0.001
Loss	Binary cross entropy
Epochs and batch	Epochs = 100; batch size = 64

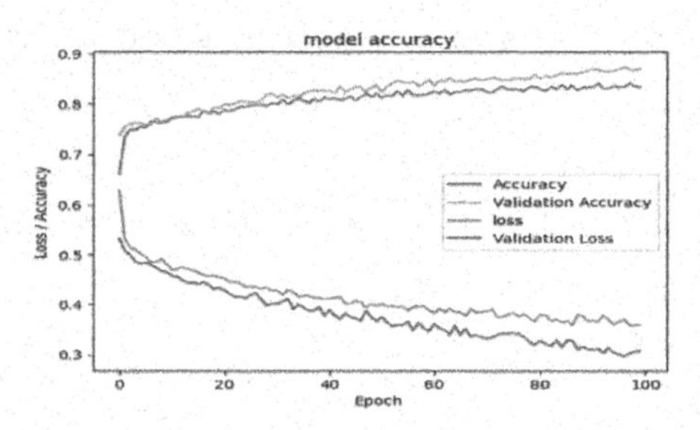

Figure 4. Accuracy and loss curve.

Table 2. Classifier evaluation.

Model	P	R	F1	Accuracy
LR BoW3	0.77	0.70	0.73	0.70
LR BoW2	0.79	0.71	0.70	0.74
CBC Tf-Idf1	0.80	0.79	0.79	0.79
XGB Tf-Idf1	0.80	0.80	0.79	0.79
RF Tf-Idf1	0.80	0.80	0.80	0.80
NB Tf-Idf1	0.82	0.81	0.81	0.81
LR BoW1	0.82	0.82	0.81	0.81
SVC Tf-Idf1	0.82	0.82	0.82	0.82
LR Tf-Idf1	0.82	0.82	0.82	0.82
VC Tf-Idf1	0.84	0.83	0.83	0.83
BLSTM	0.85	0.85	0.83	0.84
BLSTM with Attention Layer	**0.86**	**0.85**	**0.85**	**0.88**

The training data is divided into 64 batches. The optimizer in use is called the Adam optimizer. 100 training epochs are applied to the data. The metric that has been optimized is accuracy, and the loss is a binary cross-entropy loss. The accuracy was 86.9 per cent at the conclusion of the training data, and the loss was almost 30 percent, as shown in Figure 4. On test data, the model has an accuracy of 88.79%. In comparison to conventional machine learning algorithms, this is higher. Therefore, compared to other algorithms, BLSTM with an attention layer operates with more accuracy.

As shown in Table 2, a deep neural network-based model, specifically a bi-directional LSTM with an attention layer, outperformed traditional machine learning techniques throughout the catastrophic events.

5 CONCLUSIONS

The proposed model is more effective than conventional machine learning classification techniques, providing higher accuracy for identification and quicker reaction to tweets about disasters. By running in both directions efficiently, bidirectional LSTM with an attention layer has consistently been shown to perform significantly better than other neural network methods for text analysis or sentiment analysis. The combination of qualitative and quantitative analyses can aid in the officials' understanding of the urgency. Therefore, it's important to motivate the public to share relevant information on social media with the appropriate hashtags. As a result, this approach for disaster management functions better than older mass-media communication methods. The model can be integrated with sentiment analysis, and the output from both can be merged to generate a hybrid model with higher accuracy. Because of this, there is a great need and scope for further research on this issue.

REFERENCES

[1] Kumar A., Singh J. P., and Saumya S., "A Comparative Analysis of Machine Learning Techniques for Disaster-related Tweet Classification," in *IEEE Region 10 Humanitarian Technology Conference, R10-HTC*, Nov. 2019, vol. 2019-November. doi: 10.1109/R10-HTC47129.2019.9042443.

[2] Karnati A. and Reddy Boyapally S., *"Natural Language Processing with Disaster Tweets Using Bidirectional LSTM."*

[3] Li H., Li X., Caragea D., and Caragea C., *"Comparison of Word Embeddings and Sentence Encodings as Generalized Representations for Crisis Tweet Classification Tasks." [Online]. Available:* http://aidr.qcri.org/

[4] Li H., Caragea D., Caragea C., and Herndon N., "Disaster Response Aided by Tweet Classification with a Domain Adaptation Approach," *Journal of Contingencies and Crisis Management*, vol. 26, no. 1, pp. 16–27, Mar. 2018, doi: 10.1111/1468-5973.12194.

[5] Kumar A., Singh J. P., and Saumya S., "A Comparative Analysis of Machine Learning Techniques for Disaster-related Tweet Classification," in *IEEE Region 10 Humanitarian Technology Conference, R10-HTC*, Nov. 2019, vol. 2019-November. doi: 10.1109/R10-HTC47129.2019.9042443.

[6] Kumar A., Singh J. P., Dwivedi Y. K., and Rana N. P., "A Deep Multi-modal Neural Network for Informative Twitter Content Classification During Emergencies," *Ann Oper Res*, 2020, doi: 10.1007/s10479-020-03514-x.

[7] Chowdhury J. R., Caragea C., and Caragea D., *"On Identifying Hashtags in Disaster Twitter Data."* [Online]. Available: www.aaai.org

[8] Wang C., Nulty P., and Lillis D., "Congcong Wang *et al. Transformer-based Multi-task Learning for Disaster Tweet Categorisation Transformer-based Multi-task Learning for Disaster Tweet Categorisation."* [Online]. Available: https://github.com/wangcongcong123/crisis-mtl

[9] Li H., Caragea D., and Caragea C., *"Combining Self-training with Deep Learning for Disaster Tweet Classification."*

[10] Adwaith D., Abishake A. K., Raghul S. V., and Sivasankar E., "Enhancing Multimodal Disaster Tweet Classification Using State-of-the-art Deep Learning Networks," *Multimed Tools Appl*, vol. 81, no. 13, pp. 18483–18501, May 2022, doi: 10.1007/s11042-022-12217-3.

[11] Algiriyage N., Sampath R., Prasanna R., Doyle E. E. H., Stock K., and Johnston D., "Identifying Disaster-related Tweets: *A Large-Scale Detection Model Comparison.*"

[12] Wang S. and Jiang J., *"Learning Natural Language Inference with LSTM,"* Dec. 2015, [Online]. Available: http://arxiv.org/abs/1512.08849

[13] Calixto I., Liu Q., and Campbell N., *"Incorporating Global Visual Features into Attention-Based Neural Machine Translation,"* Jan. 2017, [Online]. Available: http://arxiv.org/abs/1701.06521

Artificial Intelligence, Blockchain, Computing and Security – Dagur et al. (Eds)

Performance enhancement of deep learning-based face detection system

Vinod Motiram Rathod*
Research Scholar, Suresh Gyan Vihar University, Jaipur, Rajasthan, India

Om Prakash Sharma*
Professor, Suresh Gyan Vihar University, Jaipur, Rajasthan, India

ABSTRACT: In the last couple of years, many industries and organizations have made masks compulsory for their employee. As a result, the detection of the face becomes very difficult. According to the National Institute of Standard Technology 2020 report, state-of-the-art methods have a 20% to 50% error rate while recognizing the face. Due to the mask covering the front, many scholars have lately sought to address this problem using a variety of methodologies, including Convolutional Neural Networks. In many cases, researchers used traits from the occluded (mask-covered) and non-occluded areas of the face to identify individuals. Due to the high error rate, these two traits or situations could be more sustainable for recognizing the face accurately. The Principal Component Analysis (PCA) and other methods can be used to overcome more issues of face detection. The survey discussed in this paper will help the researchers get better insight into the techniques for face detection.

Keywords: Face recognition, Principal Component Analysis (PCA), and Conventional Neural Network (CNN)

1 INTRODUCTION

Face detection is an enhanced way of human identification with security and privacy applications. Thumb impressions and retinal scans are replacing conventional security measures. Security, privacy, and data protection have been transformed by technology. Numerous more identification techniques were adopted for improved safety, such as speech recognition [1], eye impression, and face recognition [2,3]. Airports utilize face recognition for security and criminal identification [4]. The identification systems [5,6] employ embedded feature extractors as building blocks. Two-dimensional PCA face rearrangement is used in many applications [7,8,15,16,18]. CNN [2,9,11,12,17,20] and SVM-KNN [10,19] have improved face rearrangement.

Natural disasters have prompted humans to push technology to its limits. During the last decades, hurricanes, floods, earthquakes, etc., have led to system improvements. COVID-19 has disrupted ordinary life [21–23]. Business, travel, and schools were initially shut down, which are needed for economic growth.

With COVID-19 safeguards and SOP, industries, schools, offices, and markets were opened. Masking and social distance were made compulsory [14].

At present, we reviewed different algorithms used in facial recognition. We found that there are generally two methods for Facial Expression Recognition. The first method includes the whole functional engineering, and the second is based on the construction of the model. This paper consists of 5 sections. Section 1 deals with a brief introduction. The need for face recognition and

*Corresponding Author: avinodrathod.becomps@gmail.com and om.sharma@mygyanvihar.com

DOI: 10.1201/9781032684994-9

the research approach is discussed in paragraphs two and section 3, and respectively section 4 defines the Review and discussion. Finally, section 5 concludes the work related to the field.

2 THE NEED FOR FACE RECOGNITION

Facial recognition algorithms may be used to find people in real-time, on-screen, or motion pictures. Such as fingerprints (preventing ID fraud and identity theft).

3 RESEARCH APPROACH

The prime objective of survey is to compile data that supports face detection and prediction using machine learning models. The following categories are used as the basis for the analysis in the current reviews.

1. Face recognition is simulated using a machine learning algorithm.
2. Tools used in the face detection
3. Dataset used in face detection.

4 REVIEW AND DISCUSSION

We carefully selected various research publications based on the categories presented in the sections.

4.1 *Face recognition employing a machine learning algorithm*

The details of the most popular machine learning algorithms are mentioned below.

4.1.1 *Support Vector Machine (SVM)*

Most commonly used method that divides into two main types Supervised and Unsupervised. For prediction purposes, SVM is a handy tool.

4.1.2 *Artificial Neural Network (ANN)*

An artificial neural network is a computer network that is designed especially after the bayesian neurons networks that make up the architecture of the human brain.

4.1.3 *Linear regression*

Linear regression is one of the most basic and extensively used Machine Learning algorithms. It is a statistical method used to do predictive analysis.

4.1.4 *Recurrent Neural Network (RNN)*

A recurrent neural network (RNN) is a type of artificial neural network that employs statistical model or periodic input. These deep learning algorithms are frequently employed to solve categorical or temporal issues in applications like as speech recognition, pictorial tagging, language translation, and natural language processing (NLP).

4.1.5 *Long Short-Term Memory (LSTM)*

An advanced, recurrent neural network (RNN) used to capture facial expressions' temporal and contextual information using long short-term memory.

4.1.6 *K-means*

The K-mean can be used to cluster face features.

From Figure 1, it is found that most of the researchers are using the convolutional neural network, Long Short-Term Memory, and Recurrent Neural Network algorithm for face detection.

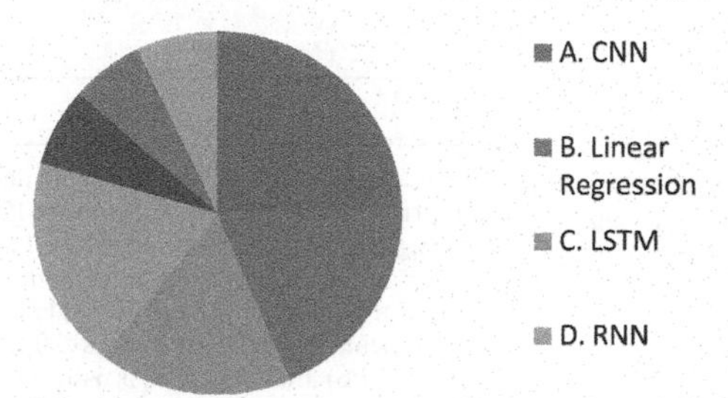

Figure 1. Algorithm used for the face detection.

4.2 *Tools used in the face detection*

We focused more on the statistical techniques used in face detection studies. Different statistical tools are used in the analysis. A description of some of the methods is provided below.

4.2.1 *Deep face*

One of the most popular analytic tools is Deep face. It briefly examines relevant *aspects like age, gender, and emotion. Additionally, it has a hybrid face recognition system.*

4.2.2 *FaceNet*

FaceNet Based on the CVPR 2015 paper FaceNet, OpenFace implements facial recognition using deep neural networks in Python and Torch.

4.2.3 *Kairos API*

Another mainly used tool is the Kairos API. Kairos API is a framework for web developers that enable them to create face recognition-enabled applications.

4.3 *Dataset used in face detection*

Different datasets were used in the face detection system, but only some were available. The majority of the participants used available datasets to detect the face. These information sets are utilized for categorization and prediction—Table 1 lists various types of datasets available and used in different studies.

Table 1. Dataset used by selected studies.

Data Set (Year)	Function/Uses	Size	Projects
JAFFE (POSED) (2022)	This dataset is used to find the different expressions of the face.	The JAFFE dataset contains 213 photos of various face expressions on ten distinct Japanese girls. Each user was instructed to generate seven facial gestures, six out of which were basic and one of which was neutral, and the pictures were evaluated by 60 observers who produced average semantic scores on each face emotion.	Facial expression recognition.
OULUCASIA VIS (2022)	Facial expression Recognition	The Oulu-CASIA NIR&VIS facial expression database involves 80 people ranging in age from 23 to 58, which each have six reactions (surprise, happiness, sorrow, rage, fear, and disgust). Males think up 73.8% of the community.	Facial expression recognition

(*continued*)

Table 1. Continued

Data Set (Year)	Function/Uses	Size	Projects
CK+(POSED) (2021)	A complete dataset for action unit and emotion-specified expression.	The Extended Cohn-Kanade (CK+) database comprises 593 video sequences from 123 men ranging in age from 18 to 50, as well as gender and ethnicity. Each documentary showed a frontal shift from neutral to a certain peak emotion, captured at 30 frames a second (FPS) and 640×490 or 640×480 pixels in resolution.	Emotion specified expression
Yale Face Database (2020)	This data is quite valuable for getting started with facial recognition tests.	The database contains 165 GIF images of 15 subjects	Image detection.
Tufts-Face-Database (2019)	This database gives researchers working on innovative recognition algorithms a platform to assess AI algorithm performance, Accuracy, and fairness.	The collection contains over 10,000 photographs of 74 women and 38 men from some of the more than 15 countries, with ages ranging between 4 to 70.	3D face recognition, heterogamous face recognition etc.
Detection of Real and Fake Faces (2019)	This database helps us to recognize the Real and the Fake faces	The size of the dataset is 215MB.	Fake face detection. Women Safety Application.
Dataset for Analysing Facial Emotions from Google (2018)	This Google database is a vast facial expression database consisting of face photo triplets with human annotations showing which two faces in each triplet create the most similar facial expression pair.	The dataset is 200MB in size and contains 500K triplets and 156K face pictures.	Image retrieval using expressions. Photo album summarization based on an expression. Classification of emotions.
UTKFace is a large-scale face data set. (2017)	The UTKFace dataset contains many faces ranging from 0 to 116 years old. Poses, facial expressions, lighting, Occlusion, resolution, and other elements are present in the images.	The heterogeneous nature around 20K snapshots annotated with time of life, gender, and region.	Facial recognition. Age estimate, progression, regression, land-mark localization, and so on.
CelebFaces Attributes (Celebi) Dataset on a Large Scale (2015)	The photos in this database include a variety of position variations as well as complex backgrounds.	There are 10,177 identities in the database, 202,599 sample images, five iconic spots, and 40 digital descriptor labels per image.	Face attribute identification, Face detection and face editing.

After studying face identification methods such as Convolutional Neural Networks, Support Vector Machines, Visual Geometry Groups, and Residual Networks, We discovered that there is a reason to increase face identification accuracy since the existing state-of-the-art approach needs more rigorous work investigation. As a result, we need a mechanism to improve Accuracy.

Table 2. Summary of different methods.

Reference No.	Author	Method Used	Advantages	Limitations
[1,5,11,14]	Susanta Malakar *et al.* Shubham Gupta *et al.* Warinthorn Naultim *et al.* Zijian Zhang *et al.*	Principal Component Analysis (PCA)	Occluded part accuracy is up to 15% Partial face recognition	Cannot Construct the Occluded part accurately. Needs to better recognize the partial face with more Accuracy or faces with Occlusion. More security can be added, such as the retina. The recognition rate can be improved.
[2,4,12,14,19–21]	Muhammad Ihtisham Amin *et al.*, Jinu Lilly Joseph *et al.*, Paras Jain *et al.*, Meijin Lin *et al.*, Yelanati Ayyappa *et al.*, Renjith Thomas	Convolutional Neural Network (CNN)	Achieved an accuracy of 97.67% in masked face recognition. Face recognition accuracy of 67.18%. Identify seven different emotions on the face. Automatic analysis of Facial expressions.	Complex systems increase the time required for real-time implementation. Application deployed on the android device only. More emotions to be recognized. If the image is in motion, there is no alternative to capture it.
[3,13]	Nitendra Mishra *et al.*	Local Binary Pattern(LBP), Gabor filter	Images can be captured easily.	You are minimizing the quantity of time spent on video preparation—detection, and recognition to above 90%.
[6,22]	R. Satheesh Kumar *et al.*	Deep Learning Methods, Deep Perceptual Mapping	Faces can be identified if people wear or don't wear a mask	Difficult to identify the face in a crowded place such as Seminar Halls etc.
[7,23]	Ali Elmahmudi *et al.* Zheng Chen *et al.*	Deep Learning Methods, VGG – Face Model	Faces can be recognized easily. Students' expression and behavior recognition	Difficult to recognize the student's expression whenever there are student movements occur.
[8,15]	Chandra Prabha K *et al.*, Kolipaka preethi *et al.*	Neural Network, LBHP(Local Binary Histogram Pattern) Algorithm	Images can be captured when they are static. Recognizes the images and marks their attendance automatically	The performance of the face recognition security framework is improved. To update the attendance of multiple people.
[9]	Mohammad Abuzneid *et al.*	BPNN(Back Propagation Neural Network)	It will lead to a robust face recognition system.	Converge faster and more accurately.

(*continued*)

Table 2. Continued

Reference No.	Author	Method Used	Advantages	Limitations
[10,19]	Meijin Lin *et al.* Radha Guha *et al.*	Support Vector Machine(SVM)	Has better recognition accuracy for small sample space	It may overcome the over-fitting problem. Needs more Accuracy in automatic face recognition.
[16–18]	Muhammad Haziq Rusli *et al.* Asep Hadian Sudrajat Ganidiastra *et al.* Sanika Tanmay Ratnaparkhi *et al.*	Multi-Task cascaded Neural Network(MTCN)	Has an Accuracy of 98.61%.	It works for small datasets. Depending on the image, the amount of a database used for training will decrease. The system can be extended to identify multiple faces at once.

5 CONCLUSION

The objective of this survey is to identify the directions for future face recognition using deep learning based on the current literature review. Machine learning can help us to recognize faces, but deep learning overcomes the limitations of previously used machine learning algorithms.

REFERENCES

[1] Susanta Malakar *et al.* Masked Face Recognition Using Principal Component Analysis and Deep Learning 18th International Conference on Electrical Engineering 2021 DOI:10.1109/ecticon51831.2021.9454857 978-1-6654-0382-5/20/$31.00.

[2] Muhammad Ihtisham Amin, Muhammad Adeel Hafeez, Rana Touseef, and Qasim Awais Person Identification with Masked Face and Thumb Images under Pandemic of COVID-19 7th International Conference on Control, Instrumentation and Automation (ICCIA)|DOI: 10.1109/ICCIA52082.2021.94035772021|978-1-6654-0350-4/20/$31.00 ©2021 IEEE.

[3] Nitendra Mishra, Aruna Bhatt Feature Extraction Techniques in Facial Expression Recognition 5th International Conference on Intelligent Computing and Control Systems (ICICCS) 2021 DOI: 10.1109/ICICCS51141.2021.9432192 | 978-1-6654-1272-8/21/$31.00 ©2021 IEEE.

[4] Jinu Lilly Joseph, Santhosh P. Mathew Facial Expression Recognition for the Blind Using Deep Learning 4th International Conference on Computing, Power and Communication Technologies (GUCON) DOI: 10.1109/GUCON50781.2021.9574035 | 978 978-1-7281-9951-1/21/$31.00 ©2021 IEEE.

[5] Paras Jain, M Murali, and Amaan Ali Face Emotion Detection Using Deep Learning Fifth International Conference on I-SMAC (IoT in Social, Mobile, Analytics, and Cloud) (I-SMAC) DOI: 10.1109/I-SMAC52330.2021.9641053 ".978-1-6654-2642-8/21/$31.00 ©2021 IEEE.

[6] Andres Espinel, Noel Perez, Daniel Riofrio Diego Benitez, and Ricardo Flores Moyano Face Gesture Recognition Using Deep-Learning Models IEEE Colombian Conference on Applications of Computational Intelligence (ColCACI) | DOI: 10.1109/ColCACI52978.2021.9469528.| 978-1-6654-3534-5/20/$31.00 ©2021 IEEE.

[7] R. Satheesh Kumar, Anagha Rajendran, Amrutha . V, Gopika .T. Raghu Deep Learning Model for Face Mask Based Attendance System in the Era of the Covid-19 Pandemic 7th International Conference on Advanced Computing and Communication Systems (ICACCS).| DOI: 10.1109/ICACCS51430.2021.9441735. 978-1-6654-0521-8/20/$31.00 ©2021 IEEE.

[8] Ali Elmahmudi, Hassan Ugail Experiments on Deep Face Recognition using Partial Faces Sixth International Conference on Inventive Computation Technologies [ICICT 2021] DOI 10.1109/CW.2018.00071 | 978-1-5386-7315-7/18/$31.00 ©2018 IEEE Part Number: CFP21F70:ART; ISBN: 978-1-7281-8501-9.

[9] Chandra Prabha K, Prabha Selavaraj, Vijaya Kumar Burugari, Kanmani P Image extraction for vehicle theft detection using Neural Network International Conference on Computer Communication

and Informatics (ICCCI -2021), Jan. 27–29, 2021, Coimbatore, INDIA DOI: 10.1109/ ICCCI50826.2021.9457023.| 978-1-7281-5875-4/20/$31.00 ©2021 IEEE.

[10] Abuzneid, Mohannad; Mahmood, Ausif. Improving Human Face Recognition using Deep Learning based Image Registration and MultiClassifier Approaches [IEEE 2018 IEEE/ACS 15th International Conference on Computer Systems and Applications(AICCSA)1–2 DOI: 10.1109 / AICCSA. 2018.8612896 | 978-1-5386-9120-5/18/$31.00 ©2018 IEEE.

[11] Meijin Lin; Zihan Zhang; Weijia Zheng A Small Sample Face Recognition Method Based on Deep Learning. 2020 IEEE 20th International Conference on Communication Technology (ICCT), DOI:10.1109/icct50939.2020.9295707.

[12] Yalanati Ayyappa, P. Neelakanteswara, P. Neelakanteswara, Yamini Tondeti, CMAK Zeelan Basha Automatic Face Mask Recognition System With FCM AND BPNN 5th International Conference onComputing Methodologies and Communication (ICCMC 2021) DOI 10.1109// ICCMC51019.2021.9418243 ©2021 IEEE |978-1-6654-0360-3/20/$31.00 ©2021 IEEE.

[13] Harikrishnan Sudarsan, Arya Sudarshan, Sadashiv , Aravind, S. Remya Ajai A. Vision-Face Recognition Attendance Monitoring System for Surveillance using Deep Learning Technology andComputer Vision IEEE 2019 International Conference on Vision Towards Emerging Trends inCommunication and Networking (ViTECoN) - Vellore, India (2019.3.30-2019.3.31)], 1–5. DOI:10.1109/ViTECoN.2019.8899418 | 978-1-5386-9353-7/19/$31.00 ©2019 IEEE.

[14] Xiujie Qu, Tianbo Wei, Cheng Peng, Peng Du A Fast Face Recognition System Based On Deep Learning 2018 11th International Symposium on Computational Intelligence and Design (ISCID)DOI 10.1109/ISCID.2018.00072 | 2473-3547/18/$31.00 ©2018 IEEE.

[15] Kolipaka Preethi, Swathy Vodithala; (2021). Automated Smart Attendance System Using Face Recognition 2021 5th International Conference on Intelligent Computing and Control Systems (ICICCS), DOI:10.1109/ICICCS51141.2021.9432140 |978-1-6654-1272-8/21/$31.00 ©2021 IEEE.

[16] Shubham Gupta, Divanshu Jain, Milind Thomas Themalil, Electronic Voting Mechanism using Microcontroller ATmega328P with Face Recognition. 2021 5th International Conference on Computing Methodologies and Communication (ICCMC), –.DOI:10.1109/iccmc51019.2021.9418372 | | 978-1-6654-0360- 3/20/$31.00 ©2021 IEEE.

[17] Asep Hadian Sudrajat Ganidisastra ;Yoanes Bandung. An Incremental Training on Deep Learning Face Recognition for M-Learning Online Exam Proctoring. 2021 IEEE Asia Pacific Conference onWireless and Mobile (APWiMob). DOI:10.1109/APWiMob51111.2021.9435232 | | 978-1-7281-9475-2/21/$31.00 ©2021

[18] Radha Guha. A Report on Automatic Face Recognition: Traditional to Modern Deep Learning Techniques 2021 6th International Conference for Convergence in Technology (I2CT),–.DOI:10.1109/ i2ct51068.2021.9418068 | 978-1-7281-8876-8/21/$31.00 ©2021 IEEE

[19] Vallabhaneni Sri Harsha Sai; Image classification for user feedback using Deep Learning Techniques. 5th International Conference on Computing Methodologies and Communication (ICCMC) –. DOI: 10.1109/ICCMC51019.2021.9418419.| 978-1-6654-0360-3/20/$31.00 ©2021 IEEE.

[20] Dane Brown Mobile Attendance Based on Face Detection and Recognition using OpenVINO. 2021 International Conference on Artificial Intelligence and Smart Systems (ICAIS),–DOI:10.1109/ icais50930.2021.9395836 | 978-1-7281-9537-7/20/$31.00 ©2021 IEEE.

[21] Cho, MyeongAh; Kim, Taeoh; Kim, Ig-Jae; Lee, Kyungjae; Lee, Sangyoun. Relational Deep Feature Learning for Heterogeneous Face Recognition. *IEEE Transactions on Information Forensics and Security*, 16(), 376–388. DOI:10.1109/TIFS.2020.3013186.

[22] Harshada Badave, Madhav Kuber; Head Pose Estimation Based Robust Multicamera Face Recognition. 2021 International Conference on Artificial Intelligence and Smart Systems (ICAIS) DOI:10.1109/icais50930.2021.9395954 | 978-1-7281-9537-7/20/$31.00 ©2021 IEEE.

[23] Chethana H.T, Trisiladevi C. Nagavi. A Heterogeneous Face Recognition Approach for Matching Composite Sketches with Age Variation Digital Images. 2021 Sixth International Conference on Wireless Communications, Signal Processing and Networking (WiSPNET)–. DOI:10.1109/wispnet51692.2021.9419436 | 978-1-6654-4086-8/21/$31.00 ©2021 IEEE

Artificial Intelligence, Blockchain, Computing and Security – Dagur et al. (Eds)
 ISBN: 978-1-032-67841-2

Text analytics using natural language processing: A survey

Sherish Johri
Research Scholar, Department of Computer Science Engineering, SRMIST Delhi-NCR Campus, Ghaziabad

Dhowmya Bhatt
Associate Professor, Department of Computer Science Engineering SRMIST Delhi-NCR Campus, Ghaziabad

Amit Singhal
Professor, Department of Computer Science Engineering RKGIT, Ghaziabad

ABSTRACT: Text mining classifies, clusters, extract useful information, searches, and analyses natural language texts to uncover patterns. Text mining extracts and Natural language processing (NLP) may create organized data from unstructured documents. It converts unstructured phrases and words into quantitative data that can be linked to database information and analyzed using data mining techniques. In this review, we examine a variety of text mining methods and analyses different datasets. In everyday conversations, people neglect spelling and grammar, which can lead to lexical, syntactic, and semantic issues. Consequently, data analysis and pattern extraction are more challenging. The primary purpose of this research a paper is to review diverse datasets, approaches, and methodologies over the past decade. This paper asserts that text analytics may provide insight into textual data, discusses text analytics research, and evaluates the efficacy of text analytics tools.

Keywords: Text Analytics, Knowledge extraction, NLP, Big data, Tools

1 INTRODUCTION

A form of text classification known as sentiment analysis organizes texts into categories according to the emotional tenor of the thoughts that are expressed within the text [1]. As a consequence of this, it constitutes an important contribution to the discipline of Natural Language Processing. The interplay between human and computer language constitutes the fundamental basis for natural language processing, which is a subdomain of computer science and artificial intelligence that has garnered significant attention in recent times [3]. The electoral officials, stock dealers, and merchants can all benefit greatly from having access to this field.

Sentiment analysis detects text polarity. It determines whether it's favorable or adverse text [5]. It's also termed opinion mining because it determines the speaker's opinion. User opinions are collected for this analysis. Sentiment analysis is defined as "the process of obtaining, transformation, and interpreting opinions from a text in order to categorize them as either favorable, adverse, or natural sentiments" [2]. This is done in order to categorize the opinions as either positive, negative, or natural. Natural Language Processing (also known as NLP) is implemented in this method. NLP uses tools, techniques, and algorithms to analyses and comprehend unstructured text, voice, etc. In addition to Intelligent Tutor

 DOI: 10.1201/9781032684994-10

Systems, Natural Language Processing also makes significant use of Speech Processing, Data Mining, Text Mining, and Text Analytics. Big data refers to big or voluminous data sets that have a more diversified, complicated and vast structure, creating challenges for storing, processing, and visually representing the data for subsequent processes or outcomes [6]. Tweets, retweets, status updates, Facebook updates, memes, magazines, websites, and blogs provide massive volumes of data in the digital era. Massive volumes of user-provided information are collected by social platforms. Twitter, a microblogging platform, generates over 500 million tweets every day, with users sending approximately 6,000 tweets per second. Sometimes it has significance, and sometimes it has less. Even brick-and-mortar and online retailers create a lot of textual data, including product info, metadata, and customer reviews. Some of this textual data may also be employed for marketing reasons. There are several challenges associated with textual data. The first issue is to effectively store and manage this data, while the second challenge is to analyses this data and attempt to derive significant patterns and actionable insights from it. Big data text analytics work as a knowledge management enabler (KM). By visualizing and analyzing unstructured data, big data text analytics may help organizations enhance their KM. Text mining is a method of information discovery that involves the extraction of trends from natural text for example social media that are both intriguing and non-trivial.

2 BIG DATA AND TEXT ANALYTICS

Text Analytics, also known as Intelligent Text Analysis, Text Data Mining, or Knowledge-Discovery, is a form of data mining that seeks to identify patterns and trends within large, unstructured textual datasets. The term natural language processing (NLP) refers to a language that is the native voice of a people or is known as unstructured data. However, NLP often belongs to a single language and follows the syntax and semantics of that language. It was developed by humans via their utilization of natural resources. Text Analytics or Keyword extraction technologies help firms make better decisions by content can lead, entities, emotion, subject, and purpose.

3 PAST THEORY OF NLP AND TEXT ANALYTICS

There are various terms that need to be understand while talking about NLP. Phonetics, Phonology, Syntax, Semantics (Lexical semantics and Compositional semantics), Morphology, Lexicon, Pragmatics, Discourse analysis, Stylistics, Semiotics comes under the linguistics. Another type of language contains Words Most of the time, words may be placed into one of the following broad categories such as noun, verb, adjectives and adverb under the sub division of the categories like singular noun, proper noun etc. symbols followed by clauses and clauses followed by the phrases and phrases followed by the words plays important role in NLP. In addition, phrases having five major categories noun, verb, adjective, adverb and prepositional phases. Several rules can construct these five syntactic phrases from words. Text Analytics is the newest term for Natural Language. Text classification, grouping, summarization, sentiment analysis, feature extraction, similarity analyzation, and link modelling are among its capabilities.

4 STATE OF AN ART

In this research author proposes an information extraction system pipeline that uses deep learning to categorise key-value pairs and then link them. The pipeline develops two merging and two pairing rules. Experiments have shown the system's performance (Tan *et al.* 2023).

The authors created a text pre-handling process for authority reports to remove unwanted content and recommend pre-processing legal papers for content classification, text layout, and specialised commitments to locate the important terms in legal documents. Natural language processing techniques like bag of terms, measure vectorization, and study related to NLP is plays significant role to extract the information (Mandal *et al.* 2023). The author discusses weakly supervised methods that input to describing causal knowledge needed in the text document (Manik Bhandari *et al.* 2020). In this paper the author proposed that legal document analysis employs semi-structured and unstructured text. Natural languages preserve legal text. Text mining, a machine learning subfield is used to analyse such documents (Wagh & Rupali 2013).

The information in the following table is presented in tabular format and span the years 2023 to 2013 and divided into the categories. First section discusses about the dataset and second section describe about the technologies and methodologies used in the previous research. This research seeks to identify the most recent approaches and findings in the fields of text mining and knowledge extraction. This research paper also examines text analytics-related tools. In addition, future research or work gaps are analyzed.

4.1 *Dataset*

The preceding study made use of a variety of datasets, which are presented in this section.

Table 1. Dataset.

No.	Reference number	Autor	Year	Dataset
1	1	QiuXing Michelle Tan, Qi Cao, Chee Kiat Seow,	2023	1. FUNSD dataset 2. Cargo Invoices dataset
2	3	Jose Ramon Sauraa, Daniel Palacios Marquésb, Domingo Ribeiro-Sorianoc	2023	Social Media (Twitter)
3	2	Souraneel Mandal, Sajib Saha & Tanaya Das	2022	1. Authoritative Records 2. Legal documents
4	7	Md Tarique Jamal Ansari 1*, Naseem Ahmad Khan	2021	Social Media (Tweets related to COVID-19 vaccine)
5	5	Pranav Shetty, Rampi Ramprasad	2021	Polymer Papers
6	6	N Chilman, X Song, A Roberts, E Tolani, R Stewart	2021	Electronic health records from a large secondary mental healthcare provider in south London
7	4	Manik Bhandari, Mark Feblowit, Oktie Hassanzadeh, Kavitha Srinivas, Shirin Sohrabi	2020	BECauSE 2.0 corpus for cause and effects and the SemEval, commonly used in causal extraction
8	8	Xieling Chen, Haoran Xie, Gary Cheng, Leonard K.M. Poon,Mingming Leng and Fu Lee Wang	2020	Bibliographical data collected from educational websites
9	9	Caitlin Dreisbach, Theresa A. Koleck	2019	Electronic patient-authored text (ePAT)
10	10	Joo-Chang Kim & Kyungyong Chung	2018	Web-generated health big data
11	11	SA. Salloum, Mostafa Al-Emra, AA Monem, Shaalan	2017	Social Media: Facebook and Twitter
12	12	Dr. S.Vijayarani1 and Ms. R. Janani2	2016	Tools Information
13	13	Rupali Sunil Wagh	2013	Legal Documents

4.2 *Technologies, methodology and findings*

In the following section, many techniques and approaches that have been employed throughout the previous ten years is discussed. In addition, a discussion of the many results that have been compiled up to this point is included in this part.

Table 2. Technologies, methodology and findings.

No.	Reference no	Techniques	Methodology	Findings
1	2	Used NLP techniques like bag of terms, measure vectorization	1. Pre-processing of data set for content classification 2. Text layout, and specialised obligations to discover the key terms	This case study used text mining to do the analysis on legal documents.
2	1	Used Regular Expression, NLP and Layout Detection	1. System Architecture Pipeline for cargo invoices 2. Key-value label categorization and linking utilize deep learning methods	It can only locate nearby bounding boxes horizontally. Thus, the system requires vertically close key-value pairs to improve.
3	3	Python, Textblob sentiment analysis	Three phases of social media-mining supported by the CATA theoretical framework	Twitter-based open innovation and theoretical factors associated to open innovation with PLS-SEM, AMOS, or LISREL
4	7	Naïve bayes classification	Sentiment analysis, Deep learning	NLP skills would enable user-customized interactions.
5	5	word vectors, Tokenization	NLP Methods	Future work will normalise PNEs to list distinct polymers.
6	6	e CRIS platform using SQL queries	ML and Rule-based approaches	As NLP advances, this application can identify occupation temporality and improve connection categorization for health and social care vocations.
7	4	Pattern Matching and Phrase Extraction	CaKNowLI uses just the text corpus to automatically extract high-quality cause-effect pair-ings.	Further enhance in decision support and event forecasting
8	8	Topical trend visualization and test analysis, term frequency-inverse document frequencies	Structural topic modelling	Clinical trials research needs NLP to improve quality of life.
9	9	Natural language processing (NLP) and text mining techniques	Precision, Recall, and F-measure	Future study should incorporate ePAT patient requirements and symptom science.
10	10	Apriori mining algorithm, the association rules, Morphological analysis	1. TF-C-IDF term frequency-inverse document frequencies evaluates words in the candidate corpus. 2. F-measure	
11	11	Text clustering, and categorization, Association rule extraction & trend ana-lysis	Review paper	Future studies should include Arabic sentiment analysis. Standard Arabic avoids short vowels, has flexible word order, and compli-cated morphology.
12	12	Comparisons of 7 tools	Tokenization with Python NLTK	Common Tokenization tool for all the languages need to be generated
13	13	Machine Learning	Clustering	

5 CONCLUSION

Within the scope of this paper, we investigate a wide range of text mining approaches and conduct analysis on several data sets. People tend to be careless with their spelling and grammar in ordinary discussions, which can result in lexical, syntactic, and semantic errors. As a direct result of this, data analysis and the identification of patterns are more difficult. The fundamental objective of this research article is to examine various datasets, methods, and approaches that have been developed over the course of the previous ten years. This article makes the claim that text analytics may be able to give insight into textual data, covers research on text analytics, and examines the effectiveness of technologies that are used for text analytics.

REFERENCES

[1] Tan Q. M., Cao Q., Seow C. K., and Yau P. C., "*Information Extraction System for Cargo Invoices,*" 2023.

[2] Mandal S., Saha S., and Das T., 2023, "*An Approach to Extract Major Parameters of Legal Documents Using Text Analytics BT – ICT Analysis and Applications*" pp. 331–338.

[3] Saura J. R. *et al.*, "Exploring the Boundaries of Open Innovation: Evidence from Social Media Mining," *Technovation*, vol. 119, 2023,

[4] Bhandari M., Feblowitz M., Hassanzadeh O., Srinivas K., and Sohrabi S., "*Causal Knowledge Extraction from Text using Natural Language Inference (Student Abstract)*," 2021.

[5] Shetty P. & Ramprasad, 2021 "Automated Knowledge Extraction from Polymer Literature Using Natural Language Processing," *iScience*, vol. 24, no. 1, p. 101922, 2021

[6] Chilman N. et al., 2021 "Text Mining Occupations from the Mental Health Electronic Health Record: A NLP Approach Using Records from the Clinical Record Interactive Search (CRIS) platform in South London, UK," *BMJ Open*, vol. 11, no. 3, 2021.

[7] Ansari M. T. J. & Khan, "Worldwide COVID-19 Vaccines Sentiment Analysis Through Twitter Content.," *Electron. J. Gen. Med.*, vol. 18, no. 6, 2021.

[8] Chen X. *et al.* "Trends and Features of the Applications of Natural Language Processing Techniques for Clinical Trials Text Analysis," *Applied Sciences*, vol. 10, no. 6. 2020

[9] Dreisbach C. *et al* "A Systematic Review of Natural Language Processing and Text Mining of Symptoms from Electronic Patient-authored Text Data," *Int. J. Med. Inform.*, vol. 125, pp. 37–46, 2019

[10] Kim and Chung, 2019 "Associative Feature Information Extraction Using Text Mining from Health Big Data," *Wirel. Pers. Commun.*, vol. 105, no. 2, pp. 691–707, 2019

[11] Salloum S. *et al.*, 2017 "A Survey of Text Mining in Social Media: Facebook and Twitter Perspectives," *Adv. Sci. Technol. Eng. Syst. J.*, vol. 2, pp. 127–133, Jan.

[12] Mohan V., 2016, "*Text Mining: Open Source Tokenization Tools: An Analysis,*" vol. 3

[13] Wagh R., 2013 "Knowledge Discovery from Legal Documents Dataset using Text Mining Techniques," *IJCA*, vol. 66 doi: 10.5120/11258-6501.

[14] Aggarwal C. C. 2011, "An Introduction to Social Network Data Analytics," in *Social Network Data analytics*, Springer

[15] Kano Y. *et al.* 2009, "Data Mining: Concept and Techniques," *Oxford J. Bioinforma.*, vol. 25, no. 15

[16] Ben-Dov & Feldman 2005, *"Text Mining and Information Extraction BT – Data Mining and Knowledge Discovery Handbook,"* O. Maimon and L. Rokach, Eds. Boston, MA: Springer US,

[17] Singhal S and Jena M, 2013 "A Study on WEKA Tool for Data Preprocessing, Classification and Clustering", *International Journal of Innovative Technology and Exploring Engineering (IJITEE)*,

[18] Singh B and Singhal S, 2020, Automated Personality Classification Using Data Mining Techniques (May 16, 2020). *Proceedings of the International Conference on Innovative Computing & Communications (ICICC)*

[19] Singhal S. *et al.*, "State of The Art of Machine Learning for Product Sustainability," *2020 2nd International Conference on Advances in Computing, Communication Control and Networking (ICACCCN)*, Greater Noida, India, 2020, pp. 197–202,

Artificial Intelligence, Blockchain, Computing and Security – Dagur et al. (Eds)
© 2024 The Author(s), ISBN: 978-1-032-67841-2

Detection of diabetic retinopathy using deep learning

Shruti Saxena*, Ajay Kumar* & Aditi Goel*
Department of Information Technology, Meerut Institute of Engineering and Technology, Meerut (U.P.) India

ABSTRACT: When untreated for a given amount of time, the eye condition known as diabetic retinopathy (DR), affects people with diabetes, results in blindness. In diabetic patients, it results in vascular abnormalities that cause blindness. If treated in the early stages, it can prevent vision issues such as glaucoma, vitreous hemorrhage, and retinal detachment. In this study, non-proliferative diabetic retinopathy is automatically detected and graded from retinal fundus pictures using CNN.This model has the potential to perform amazingly well in small real-time applications because of its extraordinary durability and light weight that only require a modest amount of computer resources, hence accelerating the screening process. Google colab was used to train the dataset. In most models, diabetic retinopathy can only be divided into 3 categories, with mild and moderate NPDR being denoted as a single group. In this paper, we will categories mild and moderate as two different categories.

Keywords: Conventional Neural Network, Diabetic Retinopathy, K-fold Cross Validation, Fundus images, Gaussian Blur

1 INTRODUCTION

One of the most widespread diseases is diabetes, and among people of working age, diabetic retinopathy is thought to be the main reason of blindness. This eye ailment is resulting from the blood vessels in the back of the eye being damaged by high blood sugar and blood pressure. We attempted to solve this issue by employing a deep learning model that is capable of determining whether or not an eye has DR to diagnose the condition in the early stages and effectively. We suggest a technique that will result in a straightforward statement expressing the information in the retinal fundus image.The severity levels of DR can be classified into many classes depending on the irregularity, such as class 0, 1, 2, 3 and 4, which are known successively as normal, mild, moderate, severe, and proliferative DR.We suggest a simple CNN model that can achieve 90% accuracy when trained on fundus pictures.

2 LITERATURE REVIEW

Chakrabarty (2018) published a deep learning method for the detection of DR. It first converts photos to grayscale, resizes them to 1000*1000, scales the pixel values to a range of 0 to 1, and then feeds the preprocessed image into CNN [8] in order to predict the category to which a picture belongs. Qomariah *et al.* (2019) suggested utilising the support vector machine (SVM) method to extract features from CNN using transfer learning [15]. Particle swarm optimization (PSO) was used by Herliana *et al.* (2019) to choose the best DR features, and then neural network classification was used to further categorise the features [26]. Kar & Maity (2019) constructed an aDR identification utilising an automated lesion detection technique employing a variety of publicly

*Corresponding Authors: shruti.saxena.it.2019@miet.ac.in, ajay.kumar@miet.ac.in and aditi.goel.it.2019@miet.ac.in

DOI: 10.1201/9781032684994-11

accessible datasets. The outcomes revealed 96.45% sensitivity and 97.71% accuracy [9]. Red lesions are detected more slowly by this type. Numerous machine learning techniques have been employed to automatically predict diabetic retinopathy in patients, including Decision Trees, Logistic Regression, K-nearest Neighbor (K-NN), Random Forest, Support Vector Machines (SVM), etc. Blood vessels, micro-aneurysms, and hard exudates were separated by Carrera *et al.* in order to extract the characteristics that were given to the support vector machine (SVM) to assess DR [5]. Qomariah *et al.* used the transfer learning technique to take advantage of the key properties of the Convolution Neural Network (CNNfully)'s connected layer as the input features for the classification process utilising support vector machines. They primarily classified NPDR classifications as being either typical or severe, which was a problem [16]. Kobat, S.G., the input DR The final fully linked and global average pooling layers of the DenseNet201 architecture were used to extract deep features from images as well as each of the eight vertically and horizontally divided spots [11]. The OD (optical disc) and BV are initially segmented using two separate U-Net models (blood vessel). Second, the symmetric hybrid CNN-SVD model was created, which locates retinal biomarkers such exudates, haemorrhages, and micro aneurysms to detect DR. After OD and BV extraction, this model was preprocessed to apply Inception-V3 based on transfer learning to choose and extract the most distinctive traits [4].

3 METHODOLOGY

Building a reliable technique for the identify the result of diabetic retinopathy is the main goal of this project.

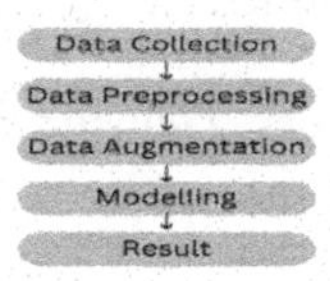

Figure 1. Methodology.

The Dataset: We used the dataset from the "APTOS 2019 Blindness Detection Kaggle" competition in order to train and then test the deep learning model.

Table 1. Different stages of DR.

S.N.	Severity Level	Observable Findings
1	No Diabetic Retinopathy	N DR
2	Mild Diabetic Retinopathy	Mild DR
3	Moderate Diabetic Retinopathy	Moderate
4	Severe Diabetic Retinopathy	Severe
5	Proliferative Diabetic Retinopathy	Proliferative DR

*Total images: 3661 files.

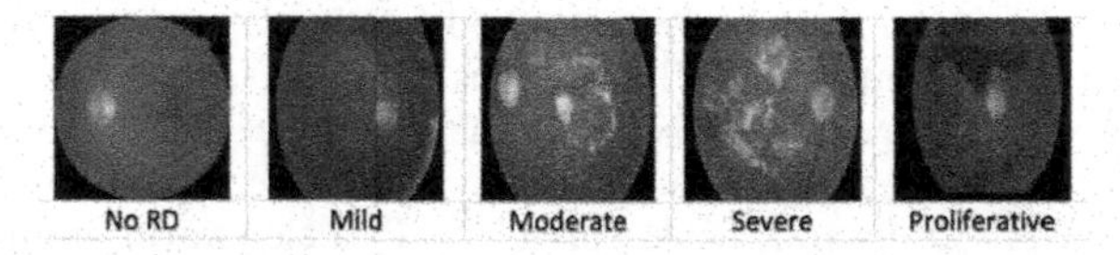

Figure 2. Images of various diabetic retinopathy severity levels in eye scans.

3.1 *Data preprocessing*

Preprocessing is necessary to extract the images in the standard format. The following activities are part of the preprocessing phase:

1. Removing the black border: In the dark backdrop surrounding the photos is omitted because it serves no use and does not offer any information to that same fundus image.
2. Remove the black corner: Due to the fundus image's rounded shape, even after the black border was removed, some black corners remained. In this stag, the image's black corners are selected.
3. Resizing image: A 256*256 (width*height) resizing has been done to the images.
4. Applying the Gaussian Blur: By setting the kernel size to 256/6, the images are blurred with a Gaussian filter. This technique aids in eliminating Gaussian noise.

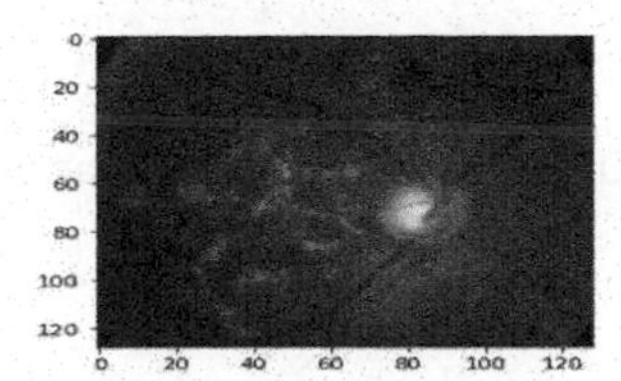

Figure 3. Before applying gaussian blur.

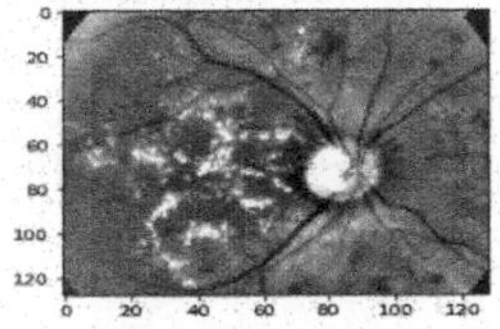

Figure 4. After applying gaussian blur.

3.2 *Data augmentation*

Data augmentation techniques apply a variety of operations to photos, such as picture scaling, geometric manipulation, noise addition, alterations to the illumination, and image flipping.

3.3 *Proposed model description*

The convolutional layer's activation map or input image can be processed with a series of linear filters to extract a variety of characteristics at either a low or high level[4], including edges, curves, blood vessels, etc. A 3x3 convolution produces the following result:

$$\mathrm{y(l,m,n) = (k=1)3\ (i=1)3\ (j=1)3w(l,i,j,k)x(i+m-1,j+n-1,k)+b(1)} \tag{1}$$

where the values of the convolutional layer's weights and biases are denoted by the w(l,i,j,k) and b (l), respectively, while the image's grey level is represented by the letters x(i,j,k). Strong low dimensional features with the definition of z (l, m, n) are generated by a 2×2 max pooling layer.

$$\mathrm{z(l,m+1,n+1)} = Max\begin{pmatrix} y(l,2m+1,2n+1) & y(l,2m+1,2n+2) \\ y(l,2m+2,2n+1) & y(l,2m+2,2n+2) \end{pmatrix} \tag{2}$$

A collection of neurons in a completely linked layer are those that have all of their activation maps from neurons in earlier levels connected to them. Convolutional and initial fully connected layer outputs are often handled by a Rectified Linear Unit (ReLU), which is explained as:

$$a_i = \begin{cases} b_i & b_i > 0 \\ 0 & b_i < 0 \end{cases} \tag{3}$$

where a_i is equivalent activation produced by the ReLU and b_i is indeed an source to the ReLU. However, at the network's conclusion, the soft max activation function is utilised to ascertain the probability distribution for each output from the last fully connected layer, is explained as:

$$a_i = \frac{e^{-c_i}}{\sum_{j=0}^{L} e^{-c_j}} \tag{4}$$

where L is the class count, and a_i is the matching SoftMax activation, and c_i is the ith output of the final completely connected layer.

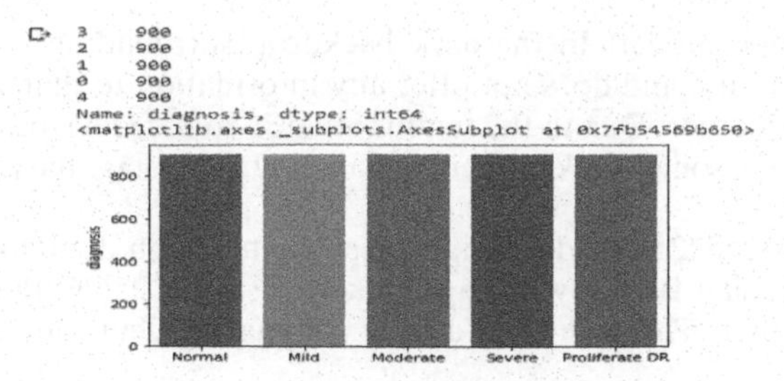

Figure 5. Visualizing balanced data.

The gap between the SoftMax's anticipated outputs and the desired outputs is known as the cross entropy loss e. And it is defined as follows:

$$e = -\sum_{j=0}^{L} \widehat{a_j} \log(a_j) \tag{5}$$

where the real probability is a_j (i.e. the anticipated desired probability for the last completely connected layer jth output for a specific fundus image corresponding to a specific class.). The Stochastic Gradient Descent (SGD) is then used to improve the parameters of the model that will alloweffective image categorization by minimizing the cross entropy loss.

The suggested CNN design is a VGG-19 modified in some way, with two convolution layers and middle two phases are supplemented with rectified linear units and a three-neuron layer in place of the ending fully connected layer with 1000 neurons.

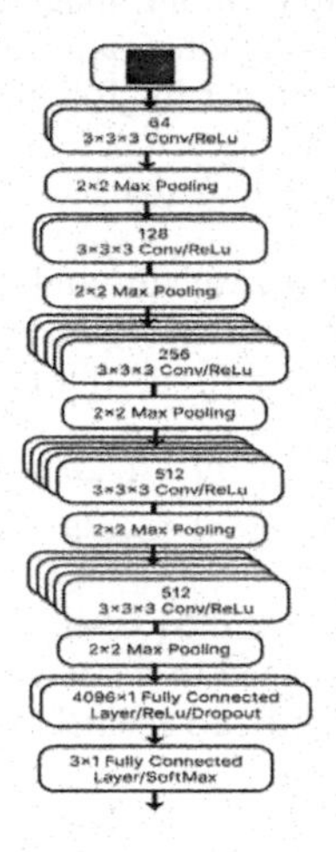

Figure 6. Proposed CNN architecture.

4 RESULT AND DISCUSSION

3661 fundus pictures were used to validate the model. The validation procedure moved quickly. We achieved a 77% validation accuracy and the mean evaluated the model's accuracy is 66%. Data used for training was divided into sets for validation and training using the stratified K-Fold cross validation technique. For each division, collectandtrainthemodel. Our suggested CNN model is run in Google Colab using the Tensorflow Python package, which supports GPU support. Following the preservation and processing of the original fundus pictures, Utilizing stratified 5-fold cross-validation, the suggested model was trained and validated, with the remaining folds

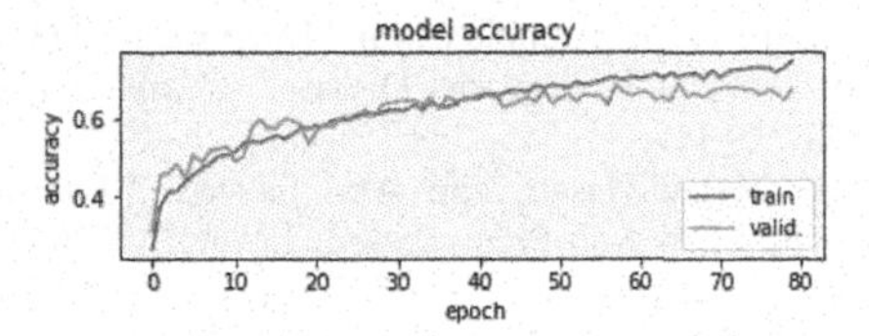

Figure 7. Proposed model accuracy.

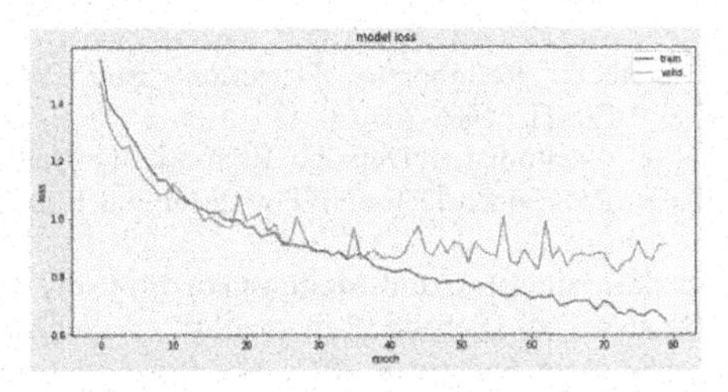

Figure 8. Training loss and validation oss.

serving as training folds and the first fold serving as validation fold. A lack of CPU resources led to the batch size being set at 40.

The highest observed validation and training accuracy were 0.68 and 0.77, respectively.

5 CONCLUSION

There weren't many works where CNN was employed and great results were obtained, according to a thorough review of the existing works. High-end equipment had to be employed because the method was exceedingly computationally intensive, even among those who had good results.The suggested approach performed well in both detecting and classifying non-proliferative diabetic retinopathy. Since the categorization greatly depends on the quality of the images fed. In this study, we applied the transfer learning strategy from the VGG16 model that had been trained beforehand. But first, the fundus images had to be processed before features could be extracted and categorized. We have observed that the accuracy rate rises as the number of epochs grows before decreasing to a local minimum. However, the average validation accuracy for our suggested model was 0.68. We will address the local minimum problem in our upcoming work.Additionally, we will significantly increase accuracy. The suggested method still has room for development.

REFERENCES

[1] Al-Smadi, Mohammed, *et al.* "A Transfer Learning with Deep Neural Network Approach for Diabetic Retinopathy Classification." *International Journal of Electrical and Computer Engineering 11.4* (2021): 3492.

[2] Amalia, R., A. Bustamam, and D. Sarwinda. "Detection and Description Generation of Diabetic Retinopathy Using Convolutional Neural Network and Long Short-term Memory." *Journal of Physics: Conference Series.* Vol. 1722. No. 1. IOP Publishing, 2021.

[3] Bhardwaj, Charu, Shruti Jain, and Meenakshi Sood. "Transfer Learning Based Robust Automatic Detection System for Diabetic Retinopathy Grading." *Neural Computing and Applications 33.20* (2021): 13999–14019.

[4] Bilal, Anas, *et al.* "AI-Based Automatic Detection and Classification of Diabetic Retinopathy Using U-Net and Deep Learning." *Symmetry 14.7* (2022): 1427.

[5] Carrera, Enrique V., Andrés González, and Ricardo Carrera. "Automated Detection of Diabetic Retinopathy Using SVM." 2017 *IEEE XXIV International Conference on Electronics, Electrical Engineering and Computing (INTERCON).* IEEE, 2017.

[6] Indumathi, N., B. Kalanjiyam, and R. Ramalakshmi. "Deep Learning Classification of Retinal Images for the Early Detection of Diabetic Retinopathy Disease." *Computational Intelligence for Information Retrieval*. CRC Press, 2021. 93–113.

[7] Islam, MdRobiul, Md Al MehediHasan, and Abu Sayeed. "Transfer Learning Based Diabetic Retinopathy Detection with a Novel Preprocessed Layer." 2020 *IEEE Region 10 Symposium (TENSYMP)*. IEEE, 2020.

[8] Jaggi, ArshdeepKaur, *et al.* "Diabetes Prediction Using Machine Learning." *Intelligent Systems*. Springer, Singapore, 2021. 383–392.

[9] Kanimozhi, J., P. Vasuki, and S. MdRoomi. "Fundus Image Lesion Detection Algorithm for Diabetic Retinopathy Screening." *Journal of Ambient Intelligence and Humanized Computing 12.7* (2021): 7407–7416.

[10] Khan, Zubair, *et al.* "Diabetic Retinopathy Detection using VGG-NIN a Deep Learning Architecture." *IEEE Access* 9 (2021): 61408–61416.

[11] Kobat, SabihaGungor, *et al.* "Automated Diabetic Retinopathy Detection Using Horizontal and Vertical Patch Division-based Pre-trained DenseNET with Digital Fundus Images." *Diagnostics* 12.8 (2022): 1975.

[12] Mohanty, Cheena, Sakuntala Mahapatra, and Madhusmita Mohanty. "A Study on Digital Fundus Images of Retina for Analysis of Diabetic Retinopathy." *Advances in Machine Learning and Computational Intelligence*. Springer, Singapore, 2021. 445–455.

[13] Nahiduzzaman, Md, *et al.* "Diabetic Retinopathy Identification Using Parallel Convolutional Neural Network Based Feature Extractor and ELM Classifier." *Expert Systems with Applications* (2023): 119557.

[14] Nguyen, Quang H., *et al.* "Diabetic Retinopathy Detection Using Deep Learning." *Proceedings of the 4th International Conference on Machine Learning and Soft Computing.* 2020.

[15] Noor, Farhan Nabil Mohd, *et al.* "The Diagnosis of Diabetic Retinopathy: A Transfer Learning with Support Vector Machine Approach." *International Conference on Innovative Technology, Engineering and Science*. Springer, Cham, 2021.

[16] Qomariah, DinialUtamiNurul, HandayaniTjandrasa, and ChastineFatichah. "Classification of Diabetic Retinopathy and Normal Retinal Images Using CNN and SVM." 2019 *12th International Conference on Information & Communication Technology and System (ICTS)*. IEEE, 2019.

[17] Raja Kumar, R., *et al.* "Detection of Diabetic Retinopathy Using Deep Convolutional Neural Networks." *Computational Vision and Bio-Inspired Computing*. Springer, Singapore, 2021. 415–430.

[18] Samanta, Abhishek, *et al.* "Automated Detection of Diabetic Retinopathy Using Convolutional Neural Networks on a Small Dataset." *Pattern Recognition Letters* 135 (2020): 293–298.

[19] Saranya, P., and S. Prabakaran. "Automatic Detection of Non-proliferative Diabetic Retinopathy in Retinal Fundus Images Using Convolution Neural Network." *Journal of Ambient Intelligence and Humanized Computing* (2020): 1–10.

[20] Shaban, Mohamed, *et al.* "A Convolutional Neural Network for the Screening and Staging of Diabetic Retinopathy." *PLoS One* 15.6 (2020): e0233514.

[21] Shanthini, A., *et al.* "Threshold Segmentation Based Multi-layer Analysis for Detecting Diabetic Retinopathy Using Convolution Neural Network." *Journal of Ambient Intelligence and Humanized Computing* (2021): 1–15.

[22] Tufail, Ahsan Bin, *et al.* "Diagnosis of Diabetic Retinopathy Through Retinal Fundus Images and 3D Convolutional Neural Networks with the Limited Number of Samples." *Wireless Communications and Mobile Computing 2021* (2021).

[23] Tymchenko, Borys, Philip Marchenko, and Dmitry Spodarets. "Deep Learning Approach to Diabetic Retinopathy Detection." arXiv *preprint arXiv*:2003.02261 (2020).

[24] Vives-Boix, Víctor, and Daniel Ruiz-Fernández. "Diabetic Retinopathy Detection Through Convolutional Neural Networks with Synaptic Metaplasticity." *Computer Methods and Programs in Biomedicine* 206 (2021): 106094.

[25] V. Raman, P. Then and P. Sumari, "Proposed Retinal Abnormality Detection and Classification Approach: Computer Aided Detection for Diabetic Retinopathy by Machine Learning Approaches," *2016 8th IEEE International Conference on Communication Software and Networks (ICCSN)*, Beijing, 2016, pp. 636–641.

[26] Zago, Gabriel Tozatto, *et al.* "Diabetic Retinopathy Detection Using Red Lesion Localization and Convolutional Neural Networks." *Computers in Biology and Medicine* 116 (2020): 103537.

Artificial Intelligence, Blockchain, Computing and Security – Dagur et al. (Eds)
 ISBN: 978-1-032-67841-2

Detection of diabetic retinopathy from iris image using deep neural network methodologies: A survey

Ajay Kumar
Research Scholar, Department of Computer Science Engineering, SRMIST Delhi-NCR Campus, Ghaziabad

Dhowmya Bhatt
Associate Professor, Department of Computer Science Engineering, SRMIST Delhi-NCR Campus, Ghaziabad

Vimal Kumar
Associate Professor, Department of Computer Science Engineering, Galgotias University, Greater Noida

ABSTRACT: A key contributing factor to diabetic retinopathy, which damages the human retina and causes visual issues and eventual blindness, is diabetes mellitus. According to statistics, 80% of diabetics battle chronic diabetes for fifteen to twenty years before developing diabetic retinopathy. Although manual diagnosis is a viable option for treating this illness, it is also cumbersome and excessive, necessitating the development of a cutting-edge method instead. Therefore, such health conditions need early detection and diagnosis in order to avoid diabetic retinopathy from causing blindness. Deep learning is more effective for small datasets and produces better results in image classification and feature extraction with large datasets, which is why it is advocated together with machine learning models for this task. The reasons, several ML/DL models, an evaluation of these modeling techniques, and opportunities for future early detection of diabetic retinopathy are all outlined in this work.

Keywords: Conventional Neural Network (CNN), Diabetic Retinopathy, Deep Learning, Diabetes, Medical image analysis

1 INTRODUCTION

A dangerous side effect of diabetes mellitus called diabetic retinopathy (DR) damages the blood vessels located in the retina, which can result in blindness. After fifteen to twenty years with the condition, it affects 80% of diabetic people (Saeedi *et al.* 2019) Early identification and treatment of DR are essential given the growing worldwide diabetes epidemic, which will likely impact 360 million population by 2030(Khatri *et al.* 2022).The need for ophthalmologists is growing, particularly in India (Wong *et al.* 2019), which necessitates the creation of Automated Intelligent Detection systems that use deep learning methods like Convolutional Networks and Deep Neural Networks to accurately diagnose DR in its early stages (Thomas *et al.* 2019). There are various approaches for diagnosing DR under supervision.

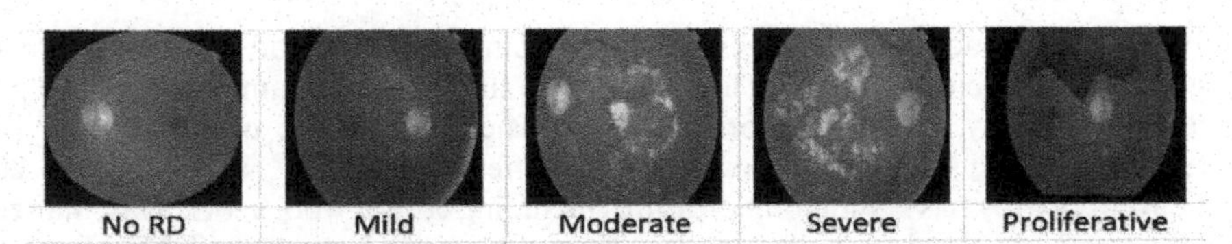

Figure 1. Variety in diabetic retinopathy phases: (a) No DR, (b) Mild, (c) Moderate, (d) Severe (e) Pdr.

DOI: 10.1201/9781032684994-12

Deep learning is a form of machine learning that is capable of handling complex and automated tasks with excellent results. It uses deep neural network architectures like CNN, DCNN, and DNN to analyze multidimensional data, making accurate decisions and applying domain knowledge (Pratt *et al.* 2016). This surpasses traditional machine learning models and is being used for image classification and feature extraction. Examples of deep learning models include VGG-16, AlexNet, ResNet, GoogleNet, LSTM, and Generative-Adversarial Network (Gadekallu *et al.* 2020). Supervised methods for diagnosing diabetic retinopathy are also mentioned.

The following are the key objectives of this review:

1. To discuss the effectiveness of databases that are available to the public and to draw focus on the issues with DR.
2. An extensive analysis of popular ML and DL techniques for DR detection.
3. Will talk about some potential future areas of research and difficulties that DR diagnostic researchers will need to take into account in the future.

2 FEATURES OF DIABETIC RETINOPATHY

To detect diabetes mellitus in its early stages and prevent blindness, various characteristics of Diabetic Retinopathy can be used to identify and classify the disease. Correctly identifying these features is crucial for an effective detection system, as shown in Figure 2, which displays multiple Diabetic Retinopathy features for disease identification.

2.1 *Micro aneurysms*

Microaneurysms, which are tiny, localized dilations of arteries that are red and present in clusters or alone, are the target illness of diabetic retinopathy. Their size ranges from one to three pixels (10-100). For the intent of identifying potential microaneurysms, a machine learning approach called Directional Local Contrast (DLC) is applied (Samanta *et al.* 2020). An improved improvement procedure based on a review of the Eigen values of the Hessian matrix is used to first enhance and segment blood vessels. After blood vessels have been removed, areas with microaneurysms are found by looking at linked components and form characteristics (Long *et al.* 2020).

2.2 *Exudates*

Exudates are yellowish-white in color and are made up of extra-cellular lipids and proteins from blood leakage from abnormal retinal capillaries. They can start as small specks but can grow into large patches and cause a circular shape with a diameter of 1 to 6 pixels. There are two types of exudates: soft exudates with soft boundaries and cloudy structure, and hard exudates with distinct boundaries and bright structure. These can lead to thickening of the retina and dysfunction of the macula (Patwari *et al.* 2018).

2.3 *Hemorrhages*

Hemorrhages are blood vessel damages caused by changes in the blood vessel wall structure leading to the risk of blood leakage. These vessels are small, with a diameter of 3 to 10 pixels and can break easily. When Hemorrhages and Micro-aneurysms occur together, they are referred to as red lesions and can appear in different shapes. (N. Nasir *et al.* 2022) Diabetic Retinopathy has several other features including cotton wool spots, avascular zone, optic disc, retinal blood vessels, neovascularization and Intra-retinal Micro-vascular abnormalities.

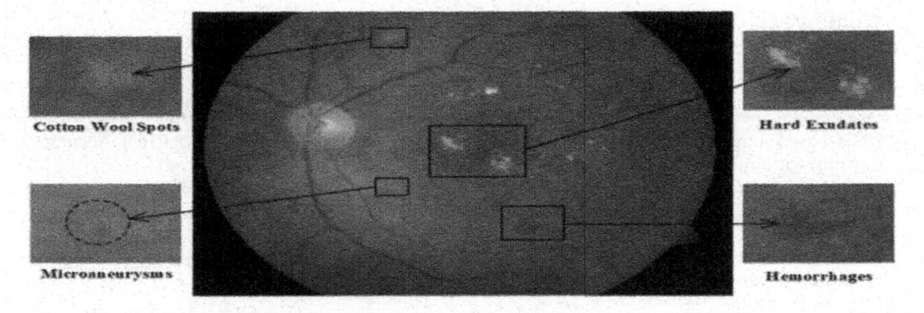

Figure 2. Features of diabetic retinopathy.

2.4 *Blood vessel segmentation*

For diagnosis, treatment, and clinical impact assessment in a variety of medical disciplines including neurosurgery, ophthalmology, and laryngology, it is crucial to segment blood arteries in medical pictures (Qiao *et al.* 2020). Various medical imaging technologies are used with automatic or semi-automatic vascular feature extraction to assist doctors. In the fundus picture, MAs, HEs, and EXs are depicted in Figure 2.

3 REVIEW OF THE LITERATURE ON DEEP LEARNING MODELS FOR EARLY DIABETIC RETINOPATHY DETECTION

The most significant studies on the deep learning CNN methodology for DR detection were carried by a number of researchers, and the results were automatically delivered using software for analyzing retinal blood vessels.

The table shows how various studies used CNN to detect DR using different techniques and datasets. Most studies used open-source datasets such as e-Ophtha, DIARETDB1, etc. Different CNN architectures like GoogleNet, ResNet, etc. were used, resulting in varying

Table 1. Summary of literature review papers.

Paper Name	Methodology	Image/Datasets	Classifiers	Results/Outcome
Y. Sun [15]	Utilize the CNN technique to analyses unconnected, one-dimensional data sets.	A total of 301 hospitalized patients' 3500 fundus photographs	Redesigns the network architecture of the conventional LeNet model and creates a new model by including BN layer. BNCNN	Training accuracy = 99.85% Testing accuracy = 97.56%
M. Hamzah Abed *et al.* [16]	Pre-processing visual improvement, categorization of photos into both healthy and ill patients.	DiaretDB0, DiaretDB1 and DrimDB	CNN	DiaretDB0 accuracy = 100%, DiaretDB1 accuracy = 99.495% DrimDB accuracy = 97.55%
Kadam et al [17]	utilising the thermal pictures in the CNN algorithm	Images thermal	CNN algorithm, random forest	The relevant characteristics are retrieved from these photographs after they have been pre-processed by converting them to GRAY from RGB. in order to identify diabetic retinopath

(*continued*)

Table 1. Continued

Paper Name	Methodology	Image/Datasets	Classifiers	Results/Outcome
SaManta *et al.* [18]	CNN-based transfer learning with a tiny dataset	smaller dataset of skewed classes, consisting of 419 validation photos and 3050 training images	Cohen's Kappa Dense Net	84.10% Accuracy
Xu *et al.* [19]	The studied neural network's depth runs between 9 and 18, while the convolution kernel size is between 1 and 5.Different CNN designs.	Kaggle Community	CNN	94.5% accuracy
Gulshan *et al.* [20]	a deep machine learning-based algorithm	Data from EyePACS-1 (9963 images from 4997 patients) Set of Messidor-2 data (1748 images from 874 patients)	The CNN algorithm's Inception-v3 architecture	Operating curve for EyePACS-1 = 0.991 90.3% sensitivity, 98.1% specificity Messidor-2 running curve = 0.990 Sensitivity=87.0% specificity=98.5%
Carson *et al.* [21]	CNN uses a variety of categorization algorithms, including deep learning (2,3,4-ary)	KaggleEyePACS Dataset, Messidor-1 dataset,	Pre-trained model (AlexNet) and (GoogLeNet)	Sensitivity=94% accuracy 2-ary = 75.5%, 3-ary=69.7%, 4-ary = 51.25%
Mateen *et al.* [22]	CNN-based training framework	Database of e-Ophtha and DIARETDB1	Inception-V3, VGG-19, ResNet-50.	Accuracy= 98.91% Accuracy =98.43%
Kajan *et al.* [23]	Applying transfer learning, CNN	DR EyePacs Database: 25 790 different retinal pictures	Inception-v3 network	Accuracy Training Data=90.97% Testing data = 70.29%
Rakhlin *et al.* [24]	Utilize the CNN technique	KaggleEyePACS Dataset, Messidor-2 dataset	Redesigned of VGG architecture	MESSIDOR-2 AUC = 0.97. Sensitivity = 99% specificity = 71% Kaggle dataset AUC = 0.923 Sensitivity=87% specificity = 92%
Mansour *et al.* [25]	the technique incorporates multilayer optimization	KaggleEyePACS Dataset, Messidor-2 dataset	DNN model based on CNN, AlexNet	Sensitivity = 99%, Specificity=71%, accuracy = 97.93% area under the ROC curve=s 0.97.
Lam *et al.*[26]	CNN-based training framework	243 Medical Images	CNN, Alex Net, ResNet, VGG16, Inception-V3, GoogleNet	ResNet accuracy =92% & 95% Inception-v3 accuracy= 96% &98% AlexNet accuracy =74% & 79%
Dutta *et al.*[27]	VGG-16 and DNN	Kaggle Community	VGG-16 architecture	Accuracy Testing Data(600 Picture) =78.3% Testing Data(300 Picture)= 72.49%
Rakhlin [28]	DCNN	KaggleEyePACS Dataset, Messidor-2 dataset	VGG-16 architecture	MESSIDOR-2 ROC= 0.97 Kaggle dataset AUC = 0.93

accuracy (Sesikala *et al.* 2022). Studies that used fewer datasets had better accuracy and faster results compared to studies using multiple datasets.

4 CONCLUSION

The study addresses the problem of diabetic retinopathy, a serious illness that can result in blindness. The disease's symptoms, causes, and effects on eyesight are all addressed. The report also discusses a variety of methods for deep learning and machine learning that are essential for the early detection and diagnosis of diabetic retinopathy. It exemplifies the advantages of deep learning methods over traditional methods for machine learning, particularly in managing enormous volumes of data and providing effective outcomes. This research gives a general review of several deep learning models, their architectures, and how they might be used with hybrid approaches to enhance the precision of retinal fundus images diagnosis. This survey will be helpful for researchers interested in diabetic retinopathy, deep learning, and medical imaging and will encourage the use of innovative technologies for better outcomes.

REFERENCES

[1] Khatri, M. *Diabetes Complications.* Available online: https://www.webmd.com/diabetes/diabetes-complications.

[2] Thomas R. L. *et al.*, "IDF Diabetes Atlas: A Review of Studies Utilising Retinal Photography on the Global Prevalence of Diabetes Related Retinopathy Between 2015 and 2018," *Diabetes Res. Clin. Pract.* 157, 107840 (2019).

[3] Saeedi P. *et al.*, "Global and Regional Diabetes Prevalence Estimates for 2019 and Projections for 2030 and 2045: Results from the International Diabetes Federation Diabetes Atlas," *Diabetes Res. Clin. Pract.* 157, 107843 (2019).

[4] Pratt H, Coenen F, Broadbent DM, Harding SP, Zheng Y. Convolutional Neural Networks for Diabetic Retinopathy. *Procedia Computer Science.* 2016;90:200–205.

[5] Wong T. Y. and Sabanayagam C., "Strategies to Tackle the Global Burden of Diabetic Retinopathy: from Epidemiology to Artificial Intelligence," *Ophthalmologica* 243(1), 9–20 (2020).

[6] Early Treatment Diabetic Retinopathy Study Research Group. Grading Diabetic Retinopathy from Stereoscopic Color Fundus Photographs- an Extension of the Modified Airlie House Classification. *Ophthalmology* 2020, 127, S99–S119. [CrossRef] [PubMed]

[7] Malik, U. *Most Common Eye Problems—Signs, Symptoms and Treatment Options.* 2021. Available online: https://irisvision.com/most-common-eye-problems-signs-symptoms-and-treatment

[8] Centers for Disease Control and Prevention. *Common Eye Disorders and Diseases.* 2020. Available online: https://www.cdc.gov/visionhealth/basics/ced/index.html

[9] Gadekallu TR, Khare N, Bhattacharya S, Singh S, Maddikunta PKR, Srivastava G. Deep Neural Networks to Predict Diabetic Retinopathy. J. Ambient Intell. Humaniz. Comput; 2020.

[10] Samanta A, Saha A, Satapathy SC, Fernandes SL, Zhang YD. Automated Detection of Diabetic Retinopathy Using Convolutional Neural Networks on a Small Dataset. *Pattern Recognition Letters.*2020

[11] Ian Goodfellow, Yoshua Bengio, and Aaron Courville. (2016). *Deep Learning.* MIT Press; 2020. Available:http://www.deeplearningbook.org

[12] Long, S., Chen, J., Hu, A. *et al.* Microaneurysms Detection in Color Fundus Images Using Machine Learning Based on Directional Local Contrast. *BioMed Eng OnLine* 19, 21 (2020).

[13] Patwari, M.B., Manza, R.R., Rajput, Y.M., Saswade, M., & Deshpande, N.A. (2013). Review on Detection and Classification of Diabetic Retinopathy Lesions Using Image Processing Techniques. *International Journal of Engineering Research and Technology.*

[14] L. Qiao, Y. Zhu, and H. Zhou, "Diabetic Retinopathy Detection Using Prognosis of Microaneurysm and Early Diagnosis System for Non-proliferative Diabetic Retinopathy Based on Deep Learning Slgorithms," *IEEE Access* 8, 104292–104302 (2020).

[15] Sun Y. "The Neural Network of One-Dimensional Convolution-An Example of the Diagnosis of Diabetic Retinopathy," *in IEEE Access*. 2019;7:69657-69666.

[16] hamzah abed M, Muhammed LAN, S. J. a. e.-p. Hussein Toman, "*Diabetic Retinopathy Diagnosis based on Convolutional Neural Network.*" 2008;00148 Accessed on: July 01, 2020. Available:https://ui.adsabs.harvard.edu/ab s/2020arXiv200800148H

[17] Kadam S, Pradhan V, Kate K. *Convolution Neural Network-based Method for Automatic Detection of Diabetic Eye Disease Using Thermal Images*. 2020;07(05):5.

[18] Samanta A, Saha A, Satapathy SC, Fernandes SL, Zhang YD. Automated Detection of Diabetic Retinopathy Using Convolutional Neural Networks on a Small Dataset. *Pattern Recognition Letters*.

[19] Xu K, Feng D, Mi H. Deep Convolutional Neural Network-based Early Automated Detection of Diabetic Retinopathy Using Fundus Image. *Molecules*. 2017;22(12):2054.

[20] Gulshan V, Peng L, Coram M, Stumpe MC, Wu D, Narayanaswamy A, Webster DR. Development and Validation of a Deep Learning Algorithm for Detection of Diabetic Retinopathy in Retinal Fundus Photographs. *JAMA*. 2016;316(22):2402-2410.

[21] Lam C, Yi D, Guo M, Lindsey T. Automated Detection of Diabetic Retinopathy Using Deep Learning. *AMIA Summits on Translational Science Proceedings*. 2018;147.

[22] Mateen M, Wen J, Nasrullah N, Sun S. Hayat S. Exudate Detection for Diabetic Retinopathy Using Pretrained Convolutional Neural Networks. *Complexity*; 2020.

[23] Graham B. *Kaggle Diabetic Retinopathy Detection Competition Report. The University of Warwick*; 2015.

[24] Pratt H, Coenen F, Broadbent DM, Harding SP, Zheng Y. Convolutional Neural Networks for Diabetic Retinopathy. *Procedia Computer Science*. 2016;90:200–205.

[25] Das D., Biswas S. K., Bandyopadhyay S. and Laskar R. H., "Deep Learning Techniques for Early Detection of Diabetic Retinopathy: Recent Developments and Techniques," *2020 5th International Conference on Computing, Communication and Security (ICCCS), Patna, India*, 2020, pp. 1–7, doi: 10.1109/ICCCS49678.2020.9276781.

[26] Lam C, Yu C, Huang L, Rubin D (2018) Retinal Lesion Detection with Deep Learning Using Image Patches. *Multidisciplinary Ophthalmic Imaging* 59(1):590–596. https://doi.org/10.1167/iovs.17-22721

[27] Dutta S, Manideep BCS, Basha SM, Caytiles RD, Iyengar NCSN (2018) Classification of Diabetic Retinopathy Images by Using Deep Learning Models. *Int J Grid Distrib Comput* 11(1):89–106. https://doi.org/10.14257/ijgdc.2018.11.1.09.

[28] Rakhlin A (2017) Diabetic Retinopathy Detection Through Integration of Deep Learning Classification Framework. bioRxiv, p.225508. https://doi.org/10.1101/225508Mateen M, Wen J, Nasrullah N, Sun S. Hayat S. Exudate detection for diabetic retinopathy using pretrained convolutional neural networks. Complexity; 2020.

[29] Nasir N., Oswald P., Alshaltone O., Barneih F., Al Shabi M. and Al-Shammaa A., "Deep DR: Detection of Diabetic Retinopathy Using a Convolutional Neural Network," *2022 Advances in Science and Engineering Technology International Conferences (ASET), Dubai, United Arab Emirates,* 2022, pp. 1–5, doi: 10.1109/ASET53988.2022.9734314.

[30] Sesikala B., Harikiran J. and SaiChandana B., "A Study on Diabetic Retinopathy Detection, Segmentation and Classification using Deep and Machine Learning Techniques," *2022 6th International Conference on Trends in Electronics and Informatics (ICOEI)*, Tirunelveli, India, 2022, pp. 1419–1424, doi: 10.1109/ICOEI53556.2022.9776690.

[31] Gargeya R, Leng T. Automated Identification of Diabetic Retinopathy Using Deep Learning. *Ophthalmology*. 2017;124(7):962–969.

[32] Kajan S, Goga J, Lacko K, Pavlovičová J. Detection of Diabetic Retinopathy Using Pretrained Deep Neural Networks. In 2020 Cybernetics & Informatics (K&I). IEEE. 2020;1–5.

Artificial Intelligence, Blockchain, Computing and Security – Dagur et al. (Eds)
© 2024 The Author(s), ISBN: 978-1-032-67841-2

Implementation and deployment of the integration of IoT devices in the development of GIS in the cloud platform

R. Krishna Kumari*
Department of Mathematics, Panimalar Engineering College, Chennai, Tamil Nadu, India

N. Gireesh*
Department of Electronics and Communication Engineering, School of Engineering, Mohan Babu University, Tirupati, Andhra Pradesh, India

R. Bhavani*
Institute of Computer Science and Engineering, Saveetha School of Engineering, Saveetha Institute of Medical and Technical Sciences, Chennai, Tamil Nadu, India

K. Sathish*
Department of Computer Science and Engineering, Madanapalle Institute of Technology and Science, Madanapalle, Andhra Pradesh, India

G.A. Senthil*
Department of Information Technology, Agni College of Technology, Chennai, Tamil Nadu, India

S. Praveena*
Department of Electronics and Communication Engineering, Mahatma Gandhi Institute of Technology, Hyderabad, Telangana, India

ABSTRACT: The technological infrastructure provided by cloud providers, and used as a deployment platform, allows GIS to take advantage of its high storage and processing capacities; while the use of IoT devices allows the automation of the data collection process. However, the current implementations are carried out in an ad hoc manner, without providing solutions that facilitate their design, implementation and reuse. In this work, we propose a development approach that, based on models that describe at a high level of abstraction, both the system architecture and the interaction between its services and IoT devices, allows guiding the implementation and deployment activities in cloud environments. This has been illustrated with the design and implementation of a GIS application analyse the collected spatial data with air quality sensors and were deployed on the Google Cloud platform.

Keywords: GIS, cloud computing, SoaML, sensors, IoT

1 INTRODUCTION

A GIS application (Geographic Information System) is made up of services that allow collecting, storing, analyzing, managing and visualizing geographic information; being able to

*Corresponding Authors: krishrengan@gmail.com, naminenigireesh@gmail.com, srbhavani2016@gmail.com, sathish1234u@gmail.com, senthilga@gmail.com and spraveena_ece@mgit.ac.in

DOI: 10.1201/9781032684994-13

be deployed in Web environments, thus acquiring the ability to share geographic information via the Internet (Yue et al. 2012). GIS require great storage and processing capacity, so the current trend is to deploy them on platforms provided by cloud providers (Bhat et al. 2011). These platforms on the ground provide high processing and storage capacity, but which facilitate the dynamic use of resources according to the application demand and payment based on metrics consumption (pay-per-use) (Liu et al. 2013). However, none of these proposals supports the design of architecture and interaction between GIS services, nor does it suggest how to approach implementation and deployment. The architectural style Service Oriented Architecture (SOA) provides the ability to integrate services considering the heterogeneity of IoT technologies, Web services for processing and visualizing information (Chen et al. 2016), of geographic information in this case.

2 LITERATURE REVIEW

There are several studies that propose the integration of cloud technologies and IoT devices. (Ara et al. 2016) propose a case study on the integration of sensors with services deployed on cloud platforms. However, despite the fact that they implement a studio case, they do not provide details of their implementation or propose mechanisms that facilitate the integration between application services and IoT devices. On the other hand, (Gubbi et al. 2013) propose an application architecture that facilitates the interaction between IoT devices and information processing and presentation services deployed in cloud environments. In the case of GIS applications, (Sagl et al. 2011) propose and implement an application architecture whose data acquisition process is carried out by sensors, whereas other processes are supported by specialized Web services. (Bröring et al. 2011) propose a Space Data Infrastructure (IDE) where measurements taken by sensors are collected to later make them available through Web services under the SWE (Sensor Web Enablement) standard of the OGC. (Chen et al. 2016) propose the integration of Web services and IoT devices using the SOA architectural style to design the architecture of the application. (Zuñiga-Prieto et al. 2009) and (Zúñiga-Prieto et al. 2016), propose a process for the incremental integration of web services through service-oriented Architecture Description Languages.

3 IMPLEMENTATIONS OF THE APPLICATION DESIGN

Once the software architect has designed the architecture of the application (Application Architecture Model), the personnel in charge of the development phase proceed to the implementation phase of the design. First, developers create a project for each Service Contract; this project includes software artifacts such as classes that define the Interfaces, Messages, and Data Types described in the Service Agreement. Additionally, it includes a software artifact with the implementation of the Interaction Protocol described as a sequence diagram within each Service Contract. The Service Contracts are implemented as Web services that will orchestrate the interaction between the services offered by the Participants involved.

Subsequently, for each Participant, the developers create software artifacts whose source code is the implementation of the Interfaces that correspond to them according to their Role, according to the respective Service Contract. This artifact implements the business logic of the service that the participant will offer. In the event that the Participant is the one who initiates the interaction, an operation will be included that invokes the web service that implements the Service Contract of which it participates (orchestration service). The business logic of the participants will be implemented as web services, except for the one corresponding to the service that initiates the interaction, which could be part of a client application.

The implementation code of the Participant that initiates the interaction must take into account that the first method called must supply all the initialization parameters required in the operations of the other participants. For example, if the insertion of an observation is intended, the Participant that initiates the interaction must provide data such as type of measurement, units of measurement, etc., so that the Participant of type IoT Device can operate.

Once the implementation code is developed, the next step is the configuration of services. The access points must be assigned so that the Service Contract orchestrates the interaction between the services offered by the Participants. Participants initiating interactions in the Service Agreement must also be able to access the corresponding access point. For this, it is necessary to consider technical aspects of connection and visibility. For example, when the Service Contract connects to the access point of an IoT Device participant, access mechanisms such as VPN (Virtual Private Network), public IP or domain name must be provided.

4 PRACTICAL EXAMPLE: COLLECTION OF POLLUTION MEASUREMENTS

To illustrate the proposal presented, the following practical example was raised: An environmental control company needs to implement a spatial data analysis system for air quality, for this, it wants to automate the data collection process of CO2 pollution levels through the use of geo-positioned sensors. Additionally, it requires an infrastructure that provides high storage and processing capacities, and in turn allows you to manage your data through the Internet. With this purpose, the company has raised as a requirement that the system to be built satisfies the logic shown in Figure 1; where, the services that make up the application will be deployed using infrastructures provided by cloud providers.

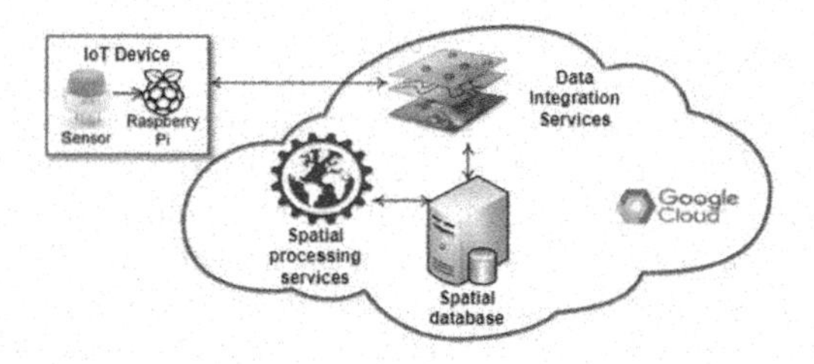

Figure 1. Practical example.

4.1 *Specification of the application architecture*

For the architectural design of the application, the standards proposed by the OGC were followed. The Application Architecture Model (see Figure 2) was created using the UML Profile with the Eclipse IDE Papyrus plug-in.

In the first place, the Participants, or services, required by the proposed practical example, were categorized by type, and the OGC standard that they implement was defined: i) Business Service, services under the SOS standard to manage the observations collected of sensors (sosServices), services under the WPS standard for spatial processing (wpsServices) and services under the WMS standard for visualization of the resulting spatial information, ii) Cloud Resource services, such as Spatial Database Services (spatialDB), and iii) IoT Devices for the collection of information through sensors (IoT Device).

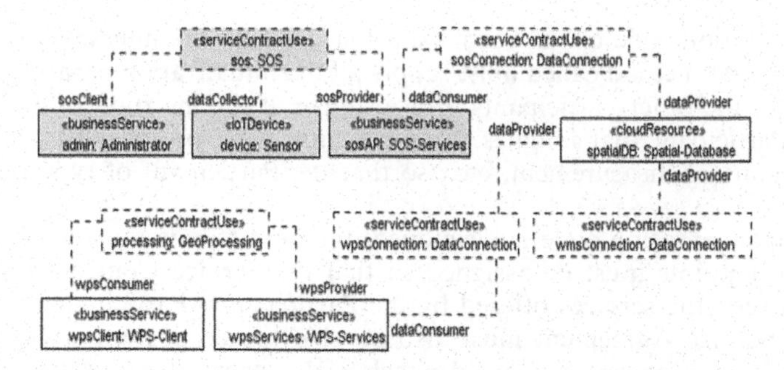

Figure 2. Model of the architecture of the application.

4.2 *Creating the implementation code*

For each Participant of non-external Business Service type (attribute is External = false) the respective implementation code was created. For Business Service type participants that implement third-party software, the respective deployment artifacts provided by third parties were used, for example, 52 North for sensor information integration services, Geoserver for spatial information processing services. In the case of Service Contracts, their internal architectural elements were implemented and deployed as orchestration Web services on the Google Cloud platform.

This practical example has allowed illustrating the applicability of the description language proposed in this work (based on an extension of SoaML through profiles), greatly contributing to the design, implementation and integration tasks of both proprietary software artifacts and third-party software artifacts.

5 CONCLUSIONS

For the application architecture specification, a UML Profile was created that extends a language created specifically for the description of service-oriented architectures – SoaML. In addition, the application of Sensor Web Enablement standards formulated by the Open Geospatial Consortium for the design of the interaction with IoT devices is proposed. The applicability of the proposed process was illustrated with a practical example of a GIS where the services that make up its architecture were deployed on the Google Cloud platform and its data collection process was carried out by air quality sensors under the IoT paradigm.

REFERENCES

Ara, T., Gajkumar Shah, P., Prabhakar, M. (2016). Internet of Things Architecture and Applications: A Survey. *Indian Journal of Science & Technology*, 9(45), 1–7. https://doi.org/10.17485/ijst/2016/v9i45/106507

Bhat, M.A., Ahmad, B. (2011). Cloud Computing: A Solution to Geographical Information Systems (GIS). *International Journal on Computer Science and Engineering*, 3(2), 594–600.

Broering A, Echterhoff J, Jirka S, Simonis I, Everding T, Stasch C, Liang S, Lemmens R (2011). New Generation Sensor Web Enablement. *Sensors (Basel)*, 11(3), 2652–99. https://doi.org/10.3390/s110302652

Chen, I.R., Guo, J., Bao, F. (2016). Trust Management for SOA-based IoT and its Application to Service Composition. *IEEE Transactions on Services Computing*, 9(3), 482–495. https://doi.org/10.1109/TSC.2014.2365797

Gubbi, J., Buyya, R., Marusic, S., Palaniswami, M. (2013). Internet of Things (IoT): A Vision, Architectural Elements, andFuture Directions. *Future Generation Computer Systems*, (7), 1645–1660. https://doi.org/10.1016/j.future.2013.01.010

Liu, Z. (2013). Typical Characteristics of Cloud GIS and Several Key Issues of Cloud Spatial Decision Support System. *4th International Conference on Software Engineering and Service Science (ICSESS)*, pp. 668–671. https://doi.org/10.1109/ICSESS.2013.6615395

OMG (2012). Service Oriented Architecture Modeling Language (SoaML). *Object Management Group (OMG)*. Available at https://www.omg.org/spec/SoaML/About-SoaML/

Sagl G, Lippautz M, Resch B, Mittleboeck M (2011). Near Real-time Geo-analyses for Emergency Support: An Radiation Safety Exercise. *14th AGILE International Conference on Geographic Information Science*, Utrecht, The Netherlands, pp. 1–8.

Yue P, Zhou H, Gong J, Hu L (2012). Geoprocessing in Cloud Computing Platforms – a Comparative Analysis. *International Journal of Digital Earth*, 6(4), 404–425. https://doi.org/10.1080/17538947.2012.748847

Zuñiga-Prieto, M., Gonzalez-Huerta, J., Abrahão, S., Insfran, E. (2009). Dynamic Reconfiguration of Cloud Application Architectures. *Journal of Software: Practice and Experience*, 48(2), 327–344. https://doi.org/10.1002/spe.2457

Zúñiga-Prieto, M., Insfran, E., Abrahão, S. (2016). Architecture Description Language for Incremental Integration of Cloud Services Architectures. *10th International Symposium on Maintenance and Evolution of Service-Oriented and Cloud-Based Environments (MESOCA)*, pp. 16–23. https://doi.org/10.1109/MESOCA.2016.10.

Artificial Intelligence, Blockchain, Computing and Security – Dagur et al. (Eds)

GA-ESE: A high-performance heart disease classification using hybrid machine learning approach

Pradeep Kumar Kushwaha, Arvind Dagur & Dhirendra Shukla
SCSE, Galgotias University, Greater Noida, Uttar Pradesh, India

ABSTRACT: The primary cause of death worldwide is heart disease, which claims more lives than cancer. Furthermore, the prevalence of individuals encountering progressive heart failure is projected to surge by 30% by the year 2030, exacerbating the situation. In our research, we applied genetic algorithm-based feature selection approach and conducted a comparative analysis of several machine learning algorithms, namely Random Forest, Multilayer Perceptron, K-Nearest Neighbor, Extra Tree, Extreme Gradient Boosting, Support Vector Classifier, Stochastic Gradient Descent, AdaBoost, Decision Tree, and Gradient Boosting with both before and after feature selection. Further an enhanced stacked ensemble using genetic algorithm (GA-ESE) has been introduced to evaluate the performance of metaheuristic classifiers. Various table and plot were used to present the results of the proposed classifiers along with metaheuristic classifiers. With a prediction accuracy of 98.76%, the proposed GA-ESE approach demonstrated superior performance compared to all other machine learning classifiers in our study.

1 INTRODUCTION

In the modern era, medical diagnosis and treatment have greatly improved the lives of patients. Preventative measures can be taken to avoid heart disease, but if one already has the condition, there are treatments available to alleviate symptoms. Medication is often used to treat heart disease, but advancements in technology have led to alternative treatments. Moreover, this technology is currently being employed to enable early disease detection, assisting physicians in making well-informed decisions regarding their patients. Advance Machine learning has the capability to significantly benefit the medical field, enabling patients to understand their medical information and aiding physicians in making more effective diagnoses and treatment plans. Animal welfare has also been improved using machine learning. As per the report by WHO [11], heart disease affects a vast number of individuals globally, with critical hazard factors such as smoking, bad cholesterol and irregular blood pressure.

The end-stage prediction of these illnesses is typically associated with a significant increase in mortality among heart patients. This lack of accuracy highlights the need to develop proficient algorithms for disease prediction [7].

The present study aims to achieve the following objectives:

1. After preprocessing the dataset, apply genetic algorithms to perform feature selection.
2. Utilize 10-fold cross-validation to apply a range of classification techniques to the dataset.
3. Assess the efficiency of the different techniques both before and after applying feature selection techniques.
4. Assess the performance of the proposed classification approach, which incorporates Genetic Algorithms for feature selection (GA-ESE).

DOI: 10.1201/9781032684994-14

4) By critically analyzing the proposed GA-ESE algorithm in comparison to other techniques used in the study, a comprehensive assessment can be made

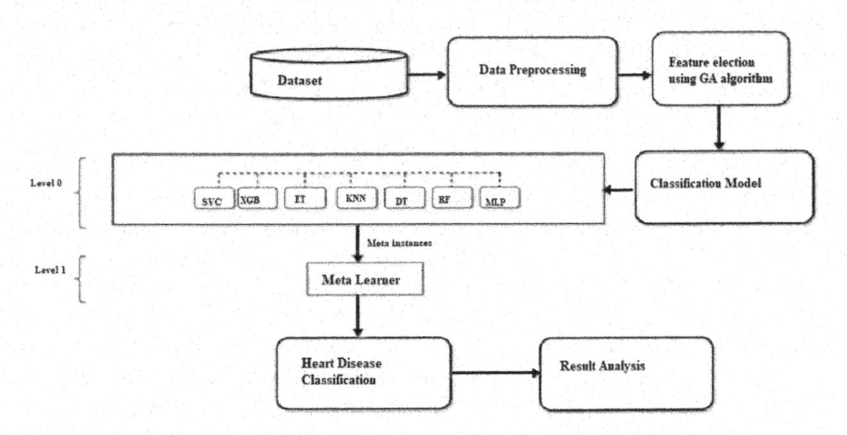

Figure 1. Proposed framework of architecture.

Advancement in Machine learning techniques have the capability to revolutionize the medical industry by implementing the fast and diagnosis of disease in less amount of time, adapted treatment plans, and improved patient improvement for disease. By analyzing large amounts of medical data, machine learning can identify risk factors, predict the likelihood of developing heart disease, and develop targeted prevention strategies. Additionally, machine learning can reduce the occurrence of medical errors by identifying patterns and trends in medical data. As various recent research in this field endures to development, we can expect to see even more advanced applications of machine learning technology in the upcoming future.

We have organized this research article in following way: In the section 2 we outline the related literature, Section 3 despite the proposed methodology, practical simulation work has been discussed in section 4, Results analysis using proposed algorithm is presented with section 5 and the future work of proposed algorithm is defined in final section of research article.

2 LITERATURE REVIEW

Several studies have showcased the successful utilization of machine learning (ML) models in detecting heart diseases. In their study, Tasnim and Habiba (2021) achieved a remarkable milestone by developing an automated method for predicting the risk of heart disease (HD). Among the classification methods, the RF-PCA algorithm demonstrated the highest accuracy, reaching 92.84% [18]. Saqlain *et al.* (2019) conducted a study where they developed a prediction model for heart diseases. The model employed the FS algorithm for extract the high value feature and utilized and therefore SVM techniques. The results demonstrated the efficiency of 81.90% accuracy [15].

In their research, Lakshmanarao *et al.* (2021) proposed a novel machine learning approach for heart disease prediction [10]. Kondababu *et al.* (2021) introduced various classification methods and features into their predictive proposed architecture. They also incorporated several combined features within system. Among these methods, HRFLM demonstrated the highest accuracy, reaching an accuracy level of 88.7% [9]. In their research, Raja *et al.* (2021) discussed a reliable system using the Random Forest (RF) methods, which is a prevailing machine learning adaptive method. The researchers applied this system to effectively predict cardiac diseases [14]. In another research S. Mahapatra *et al.* (2022) proposed the concept of various stacking ensemble for predicting emergency readmission for the heart disease patients [12]. In the research article proposed by Farman Ali *et al.* [2], authors used the two set of selected features and further employed back propagation based fussy neural network to detect the efficiency of model. The accuracy in term of predicting the mortality was proposed by Zhou *et al.* [19]. The research described in [17] focuses on the implementation of an ensemble learning model that utilizes bagging technique for the detection of Ischemia or coronary artery heart disease.

In [3,5], the authors put forth a cardiovascular disease diagnosis model which uses the ensemble learning based hybrid disease prediction model. They conducted a comparison of various existing the same techniques, highlighting with proposed system which surpasses other existing models. According to the research article introduced by Al-Sayyed et al, their approach combines an ensemble filter with an enhanced version of the Intelligent Water Drop (IWD) wrapper [1]. Another authors, A. Hashemi *et al.* [8] introduces a pioneering approach where advance method for feature selection is formulated, marking the first instance of such modelling. In 2023, P. Srivastava *et al.* [16] proposed hybrid model that utilizes deep learning techniques, specifically CNN and Bi-LSTM, to predict the disease in individuals. Furthermore, the model aims to raise awareness or provide diagnoses based on the generated predictions.

Various authors developed the concept of hybrid machine learning algorithm to improve the prediction capability [1,4,6,13].

3 METHODOLOGY USED

In this study, the proposed methodology involves taking real-life datasets as input and subjecting them to a preprocessing phase. The data collection approach is initially outlined in this section of the study. Subsequently, a data preprocessing technique is proposed to enhance the dataset's quality. Moreover, the section proceeds to describe a feature selection model, followed by an extensive analysis of performance evaluation metrics for numerous techniques. Algorithm 1 despite the details workflow of proposed system.

Algorithm 1 Structure of used methodology

Step 1: Dataset form Kaggle.
Step 2: Data cleaning the dataset.
Step 3: Data Normalization and scaling.
Step 4: Removal of outfitted data.
Step 5: Retrieval of best feature using Genetic algorithm.
Step 6: Performance evaluation of defined classifiers with both before and after feature selection algorithm.
Step 7: Performance analysis and comparison of proposed enhanced stacked ensemble classifier.

3.1 *Data collection*

The aim of the proposed approach in this study is to examine and anticipate the course of treatment for CHD based on the patient's medical antiquity and current disorder. In the research article we have used heart disease dataset from the Kaggle. This dataset contains the records of 1193 people of US, UK, Switzerland and Hungary effected from major heart disease. To enhance the dataset's quality, feature engineering techniques are employed. These techniques involve renaming columns with unconventional naming patterns and encoding features into categorical variables. As a result, the dataset comprises 17 features and one target feature, which are utilized for data analysis purposes.

Table 1. Data set description.

Used Dataset	Features	Instances	Level
Cleveland Dataset	12	1190	2

3.1.1 *Data preprocessing techniques*

The various techniques used for cleaning and organizing the data is known as data pre-processing. Raw data is often riddled with errors and noise in real-world scenarios, which makes this phase crucial. Data pre-processing uses different techniques to enhance the quality of the data. Our study followed predetermined procedures for data pre-processing after performing exploratory data analysis (EDA) to ensure that we have preprocessed the data in effective manner, This process includes a chain of steps, such as data cleaning, normalization, dimensionality reduction, and transformation, to transform raw data into a format that is suitable for ML algorithms. The superiority of the preprocessed data directly affects the performance of the machine learning models. Without proper data pre-processing, the models may generate inaccurate results or produce low-quality outputs. Therefore, it is important to conduct an exploratory data analysis (EDA) to understand the data and apply various techniques to pre-process it effectively.

3.2 *Feature selection using genetic algorithm*

The feature selection process for the dataset is conducted using a genetic algorithm, which is inspired by Charles Darwin's concept of natural selection. Genetic algorithms utilize a fitness function to determine the suitability of individuals in a population. Genetic algorithms are particularly useful for solving combinatorial and optimization problems involved in identifying the most relevant subset of variables. One advantage of this technique is its ability to leverage the success of past solutions to drive the search for an optimal solution. By iteratively evolving and selecting individuals based on their fitness, genetic algorithms strive to find the best possible solution to the problem at hand.

Algorithm 2: Modified genetic algorithm

---

1. Reset the genetic operation (GP) on population P.
2. Evaluate the fitness function F for each individual in P.
3. Set iteration counter i to 0.
4. Set Best_FF to the maximum fitness value encountered in P.
5. While (i < Max_Itr and Best_F < Max_F) do:
   a. Increment i by 1.
   b. Select individuals for crossover and mutation (c, m) from P.
   c. Perform crossover and mutation operations on c and m to generate new offspring.
   d. Evaluate the fitness score (Fc) for the offspring.
   e. Replace individuals in P with the offspring if Fc is better than their fitness.
   f. Update Best_FF if a new maximum fitness value is found.
6. End while.

Table 2 describes the outcome of applied Algorithm 2, it has been observed that the efficiency of proposed algorithm secures the accuracy of 89%, which shows the major cause to adopt in this research article.

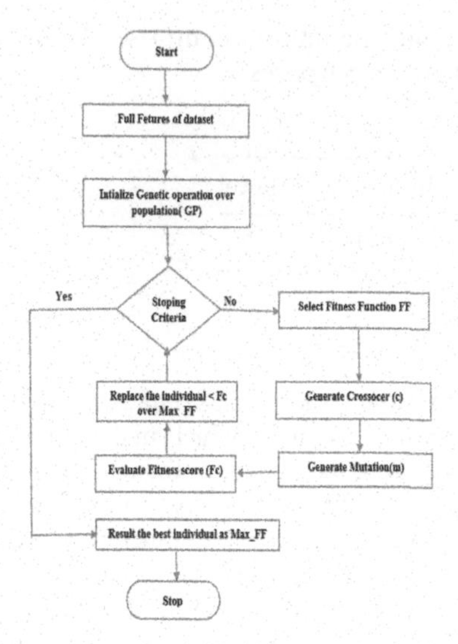

Figure 2. Proposed flowchart of enhanced genetic algorithm.

Table 2. Details of selected features using genetic algorithm.

S. No	Feature	Outcome	Remark	Accuracy
1	Age	1	Selected	89.45%
2	Sex	1	Selected	
3	Chest Pain	1	Selected	
4	resting bp s	1	Selected	
5	cholesterol	0	Rejected	
6	fasting blood sugar	1	Selected	
7	resting ecg	0	Rejected	
8	max heart rate	0	Rejected	
9	exercise	1	Selected	
10	oldpeak	1	Selected	
11	ST slope	0	Rejected	

3.3 *Enhanced stacked ensemble*

Anticipating the best-performing learner for prediction on a given dataset is often challenging. The enhanced stacked learner involves selecting a set of significant algorithms for the classification problem and evaluating their individual efficiency of prescribed model on the data using advance sampling method. Figure 3 shows the details flowchart of proposed enhanced stacked ensemble.

4 EXPERIMENTAL RESULT ANALYSIS

The experiments were conducted in Python language on the Google Colab platform. In the proposed research article we imposed SVC, XGB, ET, KNN, DT, RF and MLP machine learning classifiers on both before feature selection and after features selection algorithm. Perfomance analysis has been evaluated with various attributes. Table 2 shows the comparison analysis of performance metrices.

RF classifier outperform with genetic algorithm feature selection approach with accuracy of 94% and ET, DT and MLP classifier improve their prediction accuracy while using the proposed feature selection approach 93%, 88% and 87% respectively. Figure 3 illustrate the

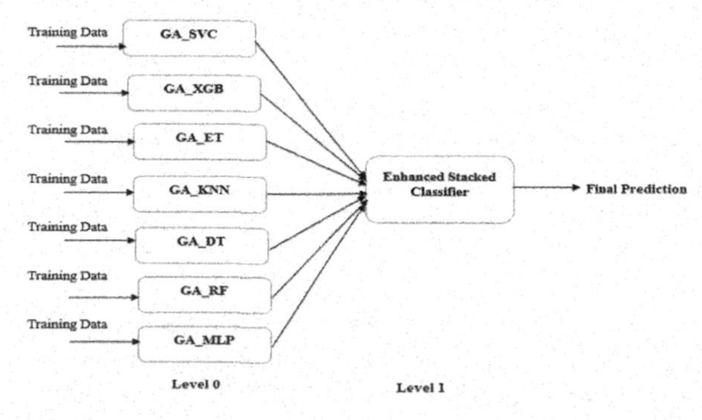

Figure 3. Workflow of enhanced stacked classifier.

comparison of accuracy. Prediction capability of various meta classifiers are discussed in Table 4. SVC classifiers outperform with 98% prediction accuracy. Accuracy comparison of various meta classifiers have been discussed in Figure 5. It has been observed that proposed GA_ESE classifiers outperform with an accuracy of 98.89%. GA_SVC and GA_RF also sustain the accuracy of 98.35% and 97.37% respectively. Another performance evaluation has been described in the form of precision through figure.

Table 3. Performance analysis.

	Without feature selection				With feature selection			
Model	Accuracy	Precision	F1 Score	ROC	Accuracy	Precision	F1 Score	ROC
GA_SVC	0.84	0.9	0.85	0.84	0.84	0.86	0.85	0.84
GA_XGB	0.88	0.88	0.89	0.88	0.88	0.88	0.89	0.88
GA_ET	0.91	0.9	0.92	0.91	0.93	0.91	0.94	0.91
GA_KNN	0.85	0.87	0.86	0.85	0.85	0.87	0.86	0.85
GA_DT	0.85	0.84	0.86	0.85	0.88	0.86	0.88	0.87
GA_RF	0.92	0.9	0.93	0.92	0.94	0.92	0.94	0.94
GA_MLP	0.85	0.87	0.85	0.85	0.87	0.87	0.85	0.85

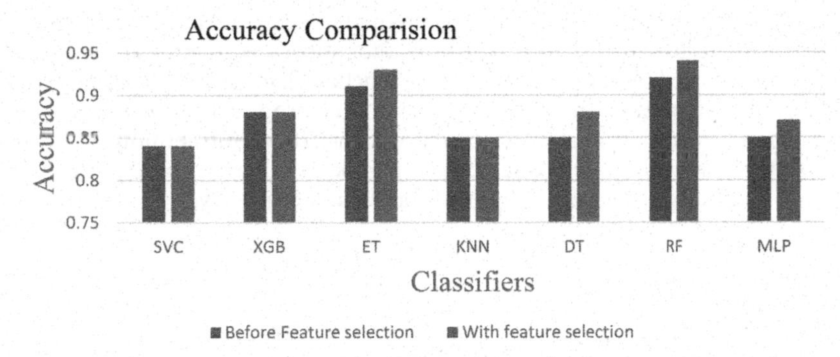

Figure 4. Accuracy comparison of various classifier with and without feature selection.

Table 4. Performance of various meta classifiers.

	Accuracy	Precision	F1 Score	ROC
GA_SVC	0.9835	0.92	0.87	0.9
GA_XGB	0.9265	0.94	0.92	0.93
GA_ET	0.9145	0.9	0.92	0.94
GA_KNN	0.9737	0.92	0.9	0.87
GA_DT	0.9265	0.89	0.93	0.89
GA_RF	0.9758	0.93	0.95	0.94
GA_MLP	0.8948	0.91	0.87	0.89
Proposed GA_ESE Classifier	0.9876	0.96	0.92	0.97

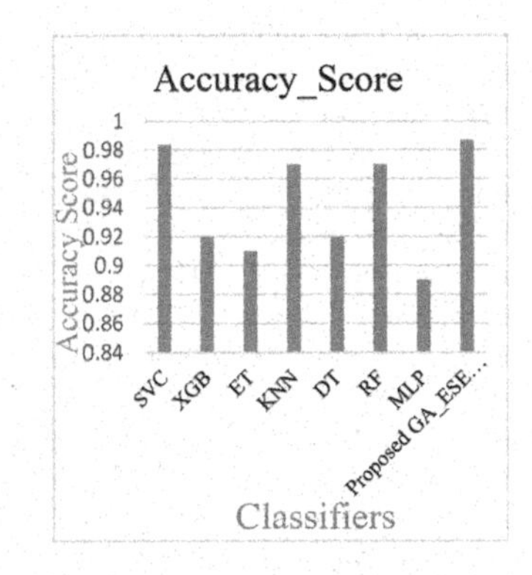

Figure 5. Accuracy score of various classifiers.

Precision_Score
Precision Score
0.98
0.96
0.94
0.92
0.9
0.88
0.86
0.84
SVC
XGB
ET
KNN
DT
RF
MLP
Proposed GA_ESE Classifier
Classifires

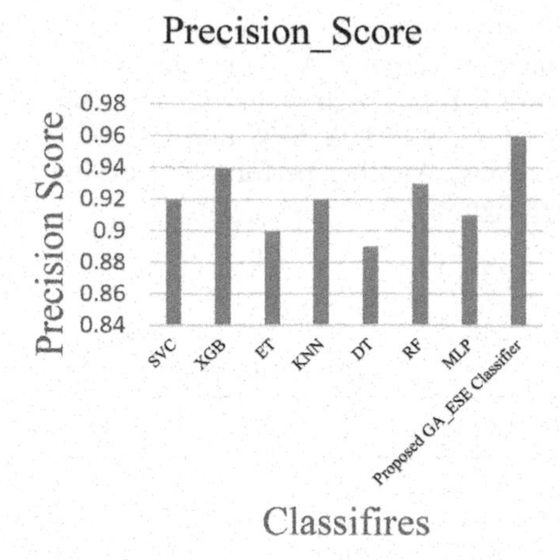

Figure 6. Precision score of various classifiers.

F1_Score
F1 Score
0.98
0.96
0.94
0.92
0.9
0.88
0.86
0.84
0.82
SVC
XGB
ET
KNN
DT
RF
MLP
Proposed GA_ESE...
Classifier

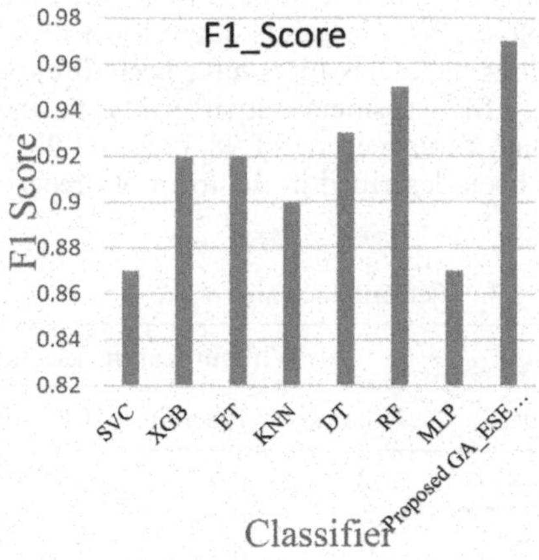

Figure 7. Precision score of various classifiers.

5 CONCLUSION

Heart related diseases are the leading cause of mortality worldwide, and timely recognition can assistance in prevention of its progression. Within our research, we presented an innovative hybrid machine learning approach for predicting heart disease. This approach integrates genetic algorithms for feature selection and utilizes a super learner for accurate predictions. We deployed our approach on seven benchmark datasets as presented in Table 4 and compared it with existing methods in Table 3. Our proposed algorithm outperformed other methods, achieving an impressive accuracy of 98.78%. Furthermore, this algorithm can be utilized for predicting other diseases and can be evaluated using other feature selection techniques and larger datasets to improve prediction efficiency.

REFERENCES

[1] Alhenawi E.,Alazzam H., Al-Sayyed R., Abualghanam O., and Adwan O. 2022. Hybrid Feature Selection Method for Intrusion Detection Systems Based on an Improved Intelligent Water Drop Algorithm. *Cybern. Inf. Technol.*, vol. 22, no. 4, pp. 73–90, doi: 10.2478/cait-2022-0040.

[2] Ali Farman, El-Sappagh Shaker, Riazul SM. Islam, Kwak Daehan, Ali Amjad, Imran Muhammad, Kwak Kyung-Sup. 2020. A Smart Healthcare Monitoring System for Heart Disease Prediction based on deep ensemble Learning and Feature Fusion, *Inf. Fusion* 63, 208–222.

[3] Asif M., Wajid U. 2020, Heart Disease Prediction using Ensemble Learning and Feature Selection *Techniques,* Healthc. Inform. *Res.* 26 (4), pp: 279–289.

[4] Bataineh A Ali, and Manacek S. 2022. MLP-PSO Hybrid Algorithm for Heart Disease Prediction. *J. Pers. Med.*, vol. 12, no. 8, p. 1208, doi: 10.3390/jpm12081208.

[5] Chadaga K., Prabhu S., Sampathila N., Chadaga R., KS S., Sengupta S. 2022. Predicting Cervical Cancer Biopsy Results Using Demographic and Epidemiological Parameters: A Custom Stacked Ensemble Machine Learning Approach, *Cogent Eng.* 9 (1), 2143040.

[6] Doppala B.P., Bhattacharyya D., Chakkravarthy M., and hoon Kim T. 2021. A Hybrid Machine Learning Approach to Identify Coronary Diseases Using Feature Selection Mechanism on Heart Disease Dataset. *Distrib. Parallel Databases*, no. 0123456789, 2021, doi: 10.1007/s10619-021-07329-y.

[7] Goel R. 2021. Heart Disease Prediction Using Various Algorithms of Machine Learning, *Ssrn Electron. J.*, Https://Doi.Org/10.2139/Ssrn.3884968

[8] Hashemi, Dowlatshahi B. D., and Nezamabadi-pour H. 2022. Ensemble of Feature Selection Algorithms: A Multi-criteria Decision-making Approach, Int. J. Mach. Learn. Cybern., vol. 13, no. 1, pp. 49–69, 2022, doi:10.1007/s13042-021-01347-z.

[9] Kondababu A., Siddhartha B., Kumar B.B., Penumutchi B. 2021. A Comparative Study on Machine Learning Based Heart Disease Prediction, *Mater.* Today Proc., Https://Doi.Org/10.1016/J. Matpr.2021.01.475.

[10] Lakshmanarao A., Srisaila A., T.S.R. Kiran, *Heart Disease Prediction Using Feature Selection And Ensemble Learning Techniques*, 2021, Https://Doi.Org/10.1109/ Icicv50876.2021.9388482.

[11] Louridi N., Douzi S., Ouahidi B. El. 2021. Machine Learning Based Identification of Patients with a cardiovascular defect. *J. Big Data*, Https://Doi.Org/10.1186/ S40537-021-00524-9.

[12] Mohapatra S., Maneesha S., Patra P.K., and Mohanty S. 2022. Heart Diseases Prediction Based On Stacking Classifiers Model. *Procedia Comput. Sci.*, Vol. 218, No.2022, Pp. 1621–1630, Doi: 10.1016/J. Procs.2023.01.140.

[13] Nouri-moghaddam J.A.B. 2022. A Hybrid Method for Heart Disease Diagnosis Utilizing Feature Selection Based Ensemble Classifier Model Generation. *Iran J. Comput. Sci.*, vol. 5, no. 3, pp. 229–246, doi: 10.1007/s42044-022-00104-x.

[14] Raja M.S., Anurag M., Reddy C.P.2021. *Machine Learning Based Heart Disease Prediction System*, Https://Doi.Org/10.1109/ Iccci50826.2021.9402653.

[15] Shah, S. M. S., Shah, F. Hussain A., S. A., & Batool, S. 2020. Support Vector Machines-based Heart Disease Diagnosis using Feature Subset, Wrapping Selection and Extraction Methods. *Computers & Electrical Engineering*, 84, 106628.

[16] Shrivastava P.K., Sharma M., Sharma S., and Kumar A. 2022. HCBiLSTM: A Hybrid Model for Predicting Heart Disease using CNN and BiLSTM Algorithms. *Meas. Sensors*, vol. 25, pp: 100657, 2023, doi: 10.1016/j.measen.2022.100657.

[17] Tao Rong, Zhang Shulin, Huang Xiao, Tao Minfang, Ma Jian, Ma Shixin, Zhang Chaoxiang. 2018. Magnetocardiography-based Ischemic Heart Disease Detection and Localization Using Machine Learning Methods, *IEEE Trans. Biomed. Eng.* 66 (6), pp: 1658–1667.

[18] Tasnim F., Habiba S.U. 2021. *A Comparative study on Heart Disease Prediction Using Data Mining Techniques And Feature Selection,* Https://Doi.Org/10.1109/ Icrest51555.2021.9331158

[19] Zhou Y., Liao Y., Wu Z. 2020. A Novel Machine Learning-based Model for Predicting All-cause Mortality in Patients with Acute Myocardial Infarction, *Int. J. Med. Inform.* 143, 104260.

Artificial Intelligence, Blockchain, Computing and Security – Dagur et al. (Eds)
© 2024 The Author(s), ISBN: 978-1-032-67841-2

Collaborative logistics, tools of machine and supply chain services in the world wide industry 4.0 framework

A. Suresh Kumar*
Department of CSE, Graphic Era Deemed to be University, Dehradun, India

Anil B. Desi
Department of CSE, Graphic Era Hill University, Dehradun, India

ABSTRACT: This study describes the importance of the usage of collaborative logistics and machine tools in the global competitive market. Moreover, accurate strategies adoption with the technology such as industry 4.0 is specifically provided to highlight its importance. Globalization in the field of online services or technology is making every competitive industry have changes in it. Supply chain related to supply in the field of different technological advancements, accuracy is required to have sustenance. Collaborative logistics is the term, which helps to have operation between the market competitors to serve the common purpose of having a strong market position. Industry 4.0 is serving the purpose of making the operations more automated and even applying to have overall development making the company more appropriate.

Keywords: Collaborative Logistics, Machine tools, Supply chain services, Industry 4.0, Globalization, and Tough competition.

1 INTRODUCTION

Globalization is increasing at a faster pace and every industry is being adhered to it, which requires further technological development. The objectives mentioned below state the requirement of this different technology, which is getting involved in this global market daily. Different types of industrial revolutions are taking place according to market conditions and therefore proper application terms are required to be enhanced [one]. The supply chain services included different steps, which are to be accordingly followed in order to overall better performance. The use of the Industry 4.0 framework helps to create more scope in resource optimization, which supports gathering the trust of their stakeholders. The use of supply chain services makes the company have proper cooperation between suppliers and their manufacturers. The use of peer-reviewed sources such as scholars used in the methodology section has helped to have more informative data available.

The use of different data available has helped to create more scope to include different types of arithmetical intelligence in the development of an organization. Different types of measures or techniques were undertaken in this study to have a more exposable approach to use in a global world [2]. The framework related to and the challenges faced in its implementation due to the complex term included in it. Collaborative logistics helps to create the collaborative approach in an appropriate manner. Machine tools include different types of strategies, which is to be implemented accordingly.

2 OBJECTIVE

The use of objectives helps to create the purpose of the study to be described further.

- To know the importance of supply chain services
- To examine the usage of different machine tools helps to create more global scope

*Corresponding Author: Suresh79.slm@gmail.com

 DOI: 10.1201/9781032684994-15

- To investigate the challenges in developing different strategies related to advanced technology implementation
- To highlight the need for applying collaborative logistics
- To implement different approaches in order to explore the advantages of the framework related to industry 4.0.
- To find out different types of measures in securing the usage of advanced technological development.

3 METHODOLOGIES

The global world is facing tough competition due to enhanced knowledge advancement and the facilities being developed with the ongoing days. Therefore, different types of changes are required in any field whether education or in the business field to develop it accordingly. Collaborative Logistics is the concept in which all the competitors come together to serve one common purpose is increasing in the current market situation [3]. The use of different company data collected helps to know a more in-depth understanding of it. An industry 4.0 framework creates an influenced end-to-end digitization in the case of handling physical resources. The use of different peer-reviewed sources has helped to collect much information, which explores the study's context in the global scenario of technology.

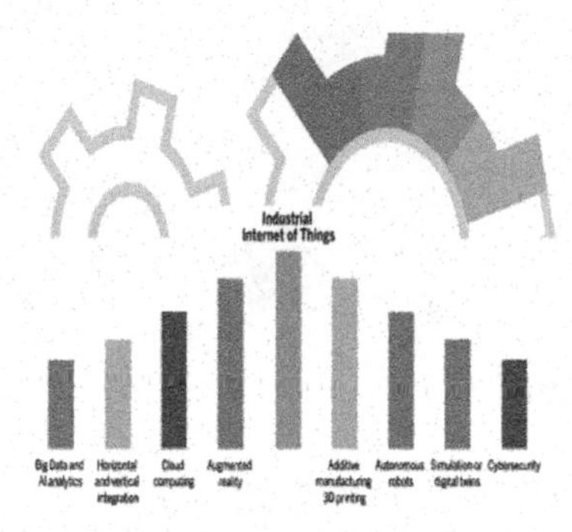

Figure 1. Industry 4.0 technologies.

4 COLLABORATIVE LOGISTICS

Competitors are coming in one place to serve the common purpose of customer satisfaction and others are the important concept related to it. Customer satisfaction importance is given here high importance, which serves the purpose of having overall development due to market position improvement [4]. There are four types of logistics, which include production logistics, recycling logistics, recovery logistics, and sales logistics. This collaboration is implemented by taking the approach each member taking part in this approach adoption and therefore, components recognition is important to serve [16]. The various principles, which are required to maintain the authenticity of this collaboration, are positive interdependence, Individual accountability, Group Processing, and Interpersonal skills.

5 SUPPLY CHAIN SERVICES

The chain including suppliers and customers in the two ends of the process is known as supply chain services, which helps to make the process of operation more successful.

5.1 *Integration*

Proper communication and integration are required from the start to the execution process to have successful operation conditions and in turn better customer satisfaction [5]. The technology-related supply chain is a much more complex term and strict integration, which is required here.

5.2 *Operations*

Information related to data analysis and sharing activities supports to have reduced resource wastage. It gives the process a smoother, path fulfillment, and less expensive manner.

5.3 *Purchasing*

Demand forecasting is one of the advantages, which is added in the context of having proper supply chain services to its customers [11]. Supply tracking, demand cycles, competing producers, and right quantity are the terms related to it.

5.4 *Distribution*

Centralization of the return process is given importance in the term of Distribution, which helps to evaluate the real-time inventory. Order status and location of the stock have to be regularly checked in this process of distribution.

Figure 2. Industry 4.0's components.

6 SUPPLY CHAIN SERVICES V/S LOGISTICS

Table 1. Here specifies that there is a link between the services of the supply chain and logistics [6].

	Supply Chain	Logistics
Management process	There are many activities such as sourcing materials, labor, planning, and facilities management, which are part of the supply chain.	It is one of the activities of logistics with advanced process inclusion
Cost-effectiveness	Operational performance is high in the process of supply chain management in the process of competitive advantage.	Effective delivery is possible due to the application of logistics.
Background	Keith Oliver and 1st implemented by the Ford motor company started this type of chain in the 20th century.	It started in the days of historian rule in 356 BC as master of logistics.
Importance	It oversees the development of the raw materials to final goods with proper allocation of all the strategies specified.	This term is more centered on the transport of services or goods in the internal part of the company.

7 INDUSTRY 4.0 FRAMEWORK

This framework of Industrial 4.0 helps to make the manufacturing and distribution process more accurate and efficient. It includes various types of technologies which makes some of the process of operation automated (7). Moreover, automation or machine use makes changes of mistakes in the process of operation examination.

7.1 *Augmented reality*

The combination of the real world and the content of computer generation with that interactive experience is the part included in it [10]. The elements of AR are Virtual words, accurate 3D, and real-time interaction which helps to have an interlink between the virtual world and real objects.

7.2 *Computer-based algorithms*

It is one of the problem-solving techniques included in this augmented reality, which has a set of instructions [11]. Additionally, which are done on the understanding of alternatives available help to solve hurdles created to perform tasks.

7.3 *Vertical and horizontal integration*

This is considered the backbone of this Industry 4. Zero, which helps with the application of complex technology and operation based on terms, specified [7]. Horizontal integration helps in having multiple facilities related to production across the whole chain of supply. Layers organization effectively is included in Vertical integration.

8 CYBER SECURITY

The enhanced level of competition and globalization is making the completely industrial sector transfer their methods of traditional approaches to technological approaches. However, the issue of cyber threats is increasing in an appropriate manner, which in turn can lapse the operations of an organization [8]. The Increase connectivity and big data usage are making more chances of cyber security being in the paramount state. Moreover, the use of different tools and technologies such as Artificial intelligence helps to create more security development in case adhered with proper implementation [9]. Blockchain and machine learning are the effective steps to be undertaken in order to have more security in case of cybercrime.

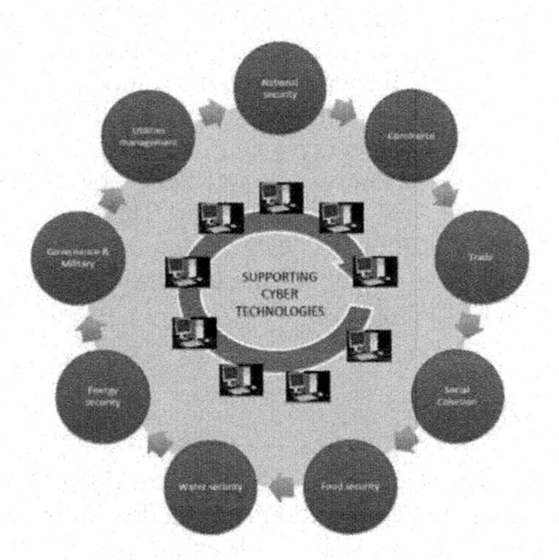

Figure 3. Factors link with cyber security.

8.1 *Machine tools usage*

Advanced technologies such as various machine tools in the implementation of services related to the supply chain have been considered to know its output results [9]. It helps to have proper components of product delivery and more specifications in applying complex strategies. It includes the functions such as securing, controlling and providing motion of relativity.

8.2 *Cyber security principles*

The use of four principles related to cyber security supports the proper management system with accuracy and security (11). They are Governing, Protecting, Detecting, and Responding, which helps to make the process of securing more appropriate and defined.

9 ADVANTAGES AND CHALLENGES REGARDING SUPPLY CHAIN IMPLEMENTATION

Table 2. Is describing the usage with its importance in the supply chain application [11]. Moreover, which is linked with different types of challenges faced in this process of implementation?

Advantages	Challenges
It provides with better collaboration between suppliers to customers.	The complexities adhered with this concept are making the overall internal management weak.
The visibility is better due to the implementation of different data analytics	Effective management of risk with the help of tools is to be applied and in case of some discrepancies, the cost is exaggerated.
Risk mitigation factors are improved by adopting this method of supply chain implementation.	The global chain of supply includes the risk of failure due to market fluctuations.
It supports the minimization of overhead costs and inventory management.	Cost management is one of the challenges in an effective manner that is mandatory in this process of implementation.

10 INDIVIDUAL ACCOUNTABILITY IN COLLABORATIVE LOGISTICS

Team collaboration is one of the approaches to be applied in order to have proper maintenance of the operations taking place under any individual or an organization [11]. Cooperation learning in the process of integration helps to have individual accountability. Collaborative logistics are the process of having collaboration within the market and the competitors come to serve the common purpose of customer satisfaction. The term individual accountability helps an individual to take responsibility for their work completion process and according is needed to acquire the information. It even helps to create more motivation and knowledge related to the complex term engage in the process of technological implementation. There are four steps to establishing accountability Defining, Communication, Assessing, and Monitoring. Integration related to digital environments in value helps to make the data collection more improvised and the overall motivation factor gets sustained to a high level.

11 PROBLEM STATEMENT

The trends of the market whether globally or nationally are fluctuating at a faster pace due to various reasons such as inflation and competition [12]. Additionally, crises due to overdue debt and lack or avoidance of recognizing the change is been recognized. Proper use of resources and data has been used to clarify the importance of technological development. This study has used approaches, which have given more appropriate information collection. The complexities adhered with the use of technological development are the main reasons to have challenges in this process [13]. The applying capacities are related to internal improvement, which requires proper guidance and collaboration. Collaborative logistics in an interactive approach, however, the cooperation in this process is much more difficult to adopt and specific regulatory rules implication is required in this process. Machine tools adoption with the help of AI implementation is the better approach to be adopted with efficient technological development.

12 CONCLUSION

Thus, it can be concluded that the use of technological changes and improvisation is required in any field to have more skill improvement and more output production. The different kinds of approaches such as collaboration logistics tend to bring different individuals into one place in order to serve one purpose. This study has collected information from peer-reviewed sources, which helps to analyze the study's relevance in an appropriate manner. Individual accountability is an important concept of making responsibilities development and is one of the approaches to having an accurate technological application.

REFERENCES

[1] Aceto, G., Persico, V., & Pescapé, A. (2019). A Survey on Information and Communication Technologies for Industry 4.0: State-of-the-art, Taxonomies, Perspectives, and Challenges. *IEEE Communications Surveys & Tutorials*, 21(4), 3467–3501.

[2] Angrisani, L., Arpaia, P., Esposito, A., & Moccaldi, N. (2019). A Wearable Brain–computer Interface Instrument for Augmented Reality-based Inspection in Industry 4.0. IEEE Transactions on Instrumentation and Measurement, 69(4), 1530–1539.

[3] Arachchige, P. C. M., Bertok, P., Khalil, I., Liu, D., Camtepe, S., & Atiquzzaman, M. (2020). A Trustworthy Privacy Preserving Framework for Machine Learning in Industrial IoT Systems. *IEEE Transactions on Industrial Informatics*, 16(9), 6092–6102.

[4] Bajic, B., Rikalovic, A., Suzic, N., & Piuri, V. (2020). Industry 4.0 Implementation Challenges and Opportunities: A Managerial Perspective. *IEEE Systems Journal*, 15(1), 546–559.

[5] Cao, Z., Zhou, P., Li, R., Huang, S., & Wu, D. (2020). Multiagent Deep Reinforcement Learning for Joint Multichannel Access and Task Offloading of Mobile-edge Computing in Industry 4.0. *IEEE Internet of Things Journal*, 7(7), 6201–6213.

[6] Chang, S. E., & Chen, Y. (2020). When Blockchain Meets Supply Chain: A Systematic Literature Review on Current Development and Potential Applications. *IEEE ACCESS*, 8, 62478–62494.

[7] Compare, M., Baraldi, P., & Zio, E. (2019). Challenges to IoT-enabled Predictive Maintenance for Industry 4.0. *IEEE Internet of Things Journal*, 7(5), 4585–4597.

[8] Jiang, D., Wang, Y., Lv, Z., Qi, S., & Singh, S. (2019). Big Data Analysis Based Network Behavior Insight of Cellular Networks for Industry 4.0 Applications. *IEEE Transactions on Industrial Informatics*, 16(2), 1310–1320.

[9] Jiang, W. (2019). An Intelligent Supply Chain Information Collaboration Model based on Internet of Things and Big Data. *IEEE access*, 7, 58324–58335

[10] Peres, R. S., Jia, X., Lee, J., Sun, K., Colombo, A. W., & Barata, J. (2020). Industrial Artificial Intelligence in Industry 4.0-Systematic Review, Challenges and Outlook. *IEEE Access*, 8, 220121–220139.

[11] Qu, Y., Pokhrel, S. R., Garg, S., Gao, L., & Xiang, Y. (2020). A Blockchained Federated Learning Framework for Cognitive Computing in Industry 4.0 Networks. *IEEE Transactions on Industrial Informatics*, 17(4), 2964–2973.

[12] Quayson, M., Bai, C., & Osei, V. (2020). Digital Inclusion for Resilient Post-COVID-19 Supply Chains: Smallholder Farmer Perspectives. *IEEE Engineering Management Review*, 48(3), 104–110.

[13] Sverko, M., Grbac, T. G., & Mikuc, M. (2022). Scada Systems with Focus on Continuous Manufacturing and Steel Industry: A Survey on Architectures, Standards, Challenges and Industry 5.0. *IEEE Access*, 10, 109395–109430.

Artificial Intelligence, Blockchain, Computing and Security – Dagur et al. (Eds)
© 2024 The Author(s), ISBN: 978-1-032-67841-2

Feature selection based hybrid machine learning classification model for diabetes mellitus type-II

Bhanu Prakash Lohani, Arvind Dagur & Dhirendra Shukla
SCSE, Galgotias University, Greater Noida, U.P. India

ABSTRACT: Diabetes is a long-term metabolic condition impacting a substantial number of individuals globally. Precise categorization of diabetes types plays a vital role in ensuring efficient treatment and control. Over the past years, the fusion of feature selection techniques and machine learning algorithms has demonstrated potential in enhancing classification accuracy and diminishing the complexity of extensive diabetes datasets. The primary objective of this research paper is to investigate the utilization of feature selection methods along with machine learning algorithms for the purpose of classifying diabetes. In this paper we have applied various Machine learning algorithm ie. Support Vector Machine (SVM), Random Forest (RF), KNN and Naive Bayes classifier after feature selection using Genetic algorithm and the performance is measured in the form of classification accuracy. Then we have proposed an ensemble model of the chosen Machine learning algorithm and achieved an accuracy of 93.82%.

Keywords: Diabetes, Feature Selection, Classification, SVM, Random Forest, Ensemble Model

1 INTRODUCTION

A common and growing global health concern that affects millions of people worldwide is diabetes, a chronic metabolic condition. It is characterized by high blood sugar levels brought on by the body's inability to make or use insulin properly. This hormonal imbalance throws off the way glucose is regulated, which has a number of negative effects on long-term health. Diabetes diagnostics and control are critical for efficient treatment and averting complications [2]. Machine learning techniques mixed with feature selection methods have emerged as potential diabetes classification methodologies. These hybrid models try to extract significant elements from massive datasets and enhance diabetes diagnosis accuracy. Creating reliable and effective categorization models has become crucial as large and complicated healthcare datasets are becoming more and more readily available. By locating the most pertinent and instructive characteristics from a given dataset, feature selection, a crucial stage in the machine learning process, significantly improves the performance of classification models [13]. By choosing a subset of features that contribute the most to the classification task, feature selection primarily aims to lower the dimensionality of the dataset. Feature selection enhances the efficiency, interpretability, and generalization abilities of the model by removing irrelevant, redundant, or noisy features. In many different sectors, traditional feature selection techniques like filter, wrapper, and embedding approaches have been widely used. When working with high-dimensional data, these approaches, however, frequently run into problems such as overfitting or poor generalization performance. We present a novel hybrid classification model for diabetes that uses feature selection methods to

 DOI: 10.1201/9781032684994-16

enhance classification accuracy in order to overcome these issues. Our strategy intends to leverage the complimentary nature of feature selection and classification algorithms to combine their strengths, improving the overall functionality of the diabetes classification system. Our model decreases computing complexity while also making the classification process easier to understand and comprehend by using a smaller group of discriminative features.

2 RELATED WORK

Prabha *et al.* (2021) [8] used a dataset of 217 participants to develop a classification model that can detect diabetes based on the wristband photoplethysmography (PPG) signal and basic physiological parameters (PhyP). Their model achieved high accuracy and hence, they concluded that detection through wearable sensors is a more convenient and simpler alternative in routine applications. Saxena *et al.* (2022) [11]in their comparative study to classify diabetes mellitus using different classifiers like multilayer perceptron, decision trees etc. published accuracies in the range of 76 to 80. M.K Hasan *et al.* (2020) [6] proposed a ensembled model for diabetes classification, in their work they have used the concept of pre-processing, outlier handling, feature selection and k fold cross validation and then presented the performance analysis with respect to sensitivity, false omission rate. Umair Muneer Butt *et al.* [4] worked on a Machine learning embedded classification and prediction model where the author used LSTM, MLP, Logistic regression and Random Forest for experimental work and concluded that MLP outperform with an accuracy of 87.26% Yogita *et al.* [5] worked on dataset collected from a hospital & research centre of Nagpur and applied Logistic regression, SVM, Naïve Bayes and Random Forest ML algorithm to find the various quantitative measure. The main aim of author was early detection of diabetes. Pooja *et al.* [9] worked on classification algorithm SVM, Decision Tree, Extreme gradient boosting and Random forest in their research work and find the classification accuracy of 77.40% using SVM, for Decision tree achieved 74.69%, for Extreme gradient boosting achieved accuracy of 78.28% and using Random forest achieved accuracy of 82.70%. R. Vaishali *et al.* [14] worked on PIMA dataset and used MOE fuzzy classifier after feature selection using genetic algorithm and achieved accuracy of 83.043%. Rukhsar *et al.* [12] proposed an algorithm which contain pre-processing by Sampling SMORT for pruning and done the comparative study with conventional algorithm like RF, J48 and RT. And author come to the conclusion that the proposed system outperforms over the conventional one.

3 PROPOSED METHODOLOGY & MATERIAL

In this section, we will describe the data set we have used and describe the proposed methodology for the research work. We have used the PIMA dataset for the classification purpose. The size of the dataset is 768,9 where 768 rows represent the data and 9 columns represent the feature. Figure 1 describes the proposed methodology. The methodology steps are defied below.

Step 1: Data Collection—PIMA data set
Step 2: Pre-processing (Data Cleaning, Transformation, Outlier Analysis)
Step 3: Train Test split with ratio 80:20
Step 4: Applying Classifiers (SVM, RF, KNN, Naive Bayes) on the chosen dataset to find Accuracy
Step 5: To find Accuracy with Ensemble Model
Step 6: Result Analysis

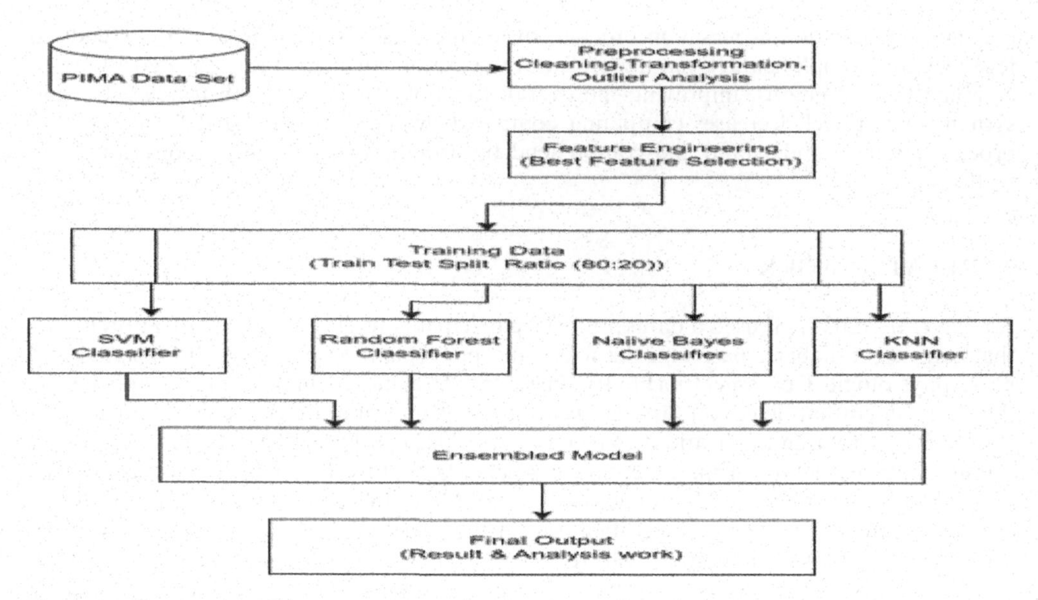

Figure 1. Proposed model.

4 MACHINE LEARNING TECHNIQUES

In this section we have defined the existing machine learning methods used for classification purpose.

4.1 *Decision trees and Support Vector Machines (SVM)*

In diabetes classification, the combination of SVM and decision trees has been widely used. SVM is a strong method for categorizing data, whereas decision trees provide interpretable rules for classification [1]. The hybrid model takes advantage of the SVM's capacity to deal with complex boundaries as well as the decision tree's interpretability [3]. This combination, however, may suffer from overfitting and problems with high-dimensional datasets.

4.2 *k-Nearest Neighbors Neural Networks (KNN)*

For diabetes categorization, the merging of neural networks with KNN has been investigated. KNN considers the similarity between cases for categorization, but neural networks excel at capturing complicated relationships in data. The neural network's capacity to learn detailed patterns and the KNN's ability to manage local data features enhance the hybrid model [10]. This combination, however, may be plagued by the curse of dimensionality and computationally costly training.

4.3 *Fuzzy logic with random forests*

For diabetes categorization, hybrid models integrating fuzzy logic and random forests have been used. Fuzzy logic can handle data ambiguity and imprecision, but random forests can give robust categorization by aggregating numerous decision trees [15]. The hybrid model benefits from the ability of fuzzy logic manage ambiguity in diabetes data and the ability of random forests to capture complicated interactions. However, this combination may result in greater computational complexity and interpretability issues.

5 FEATURE SELECTION

Feature selection is critical in constructing successful diabetes classification models. Several feature selection strategies have been investigated in prior research. Filter techniques such as chi-square, data gain, and correlation-based feature selection are examples. Wrapper techniques such as genetic algorithms, simulated annealing, and particle swarm optimization were also used. Furthermore, embedding approaches such as LASSO, decision trees, and random forests have been used to determine the most useful diabetic categorization characteristics. In our research work we have chosen genetic algorithm for feature selection.

5.1 *Genetic algorithm*

Genetic algorithms simulate the natural selection process in order to find the best feature subset. They start with a population of potential feature subsets and evolve them through selection, crossover, and mutation. Although genetic algorithms are good at dealing with vast feature spaces, they can be computationally expensive [7]. When we have performed feature selection on the selected dataset using genetic algorithm, we have selected 4 features out of 8 features from the dataset.

6 ENSEMBLE MODEL

Diabetes classification has made extensive use of ensemble models such as AdaBoost, bagging, and stacking. To boost overall performance, these models incorporate many base classifiers. AdaBoost gives more weight to misclassified occurrences, with a focus on tough samples. Using bootstrap samples, Bagging creates numerous classifiers and aggregates their predictions. Stacking uses a meta-classifier to combine the predictions of numerous classifiers. Ensemble models are good at reducing bias and variation, resulting in increased accuracy. They may, however, suffer from increasing computing complexity and model interpretation challenges. In our research work we have ensembled 4 classifiers SVM, Random Forest, Naïve Bayes and KNN classifiers and then find the final output in terms of accuracy achieved.

7 RESULT & DISCUSSION

We have performed the experiment in python language with the help of google colab. After the pre-processing step for selecting the best feature, we have applied genetic algorithm and we found 4best features from the dataset. Then we have applied various classifier for getting the classification accuracy before feature selection and after feature selection. Table 1. depicts the accuracy, precision,F1 Score and ROC achieved by various classifiers used for experimental work before feature selection. Table 2. Depicts the various performance measures after feature selection. From both the Tables 1 and 2 we can say that when we have

Table 1. Performance of classifiers after before selection.

Classifier	SVM	RF	Naïve Bayes	KNN	Proposed - Ensembled Model
Accuracy Achieved (%)	72.89	71.98	69.68	73.45	87.87
Precision	65.12	61.24	57.42	61.25	76.45
F1 Score	51.65	55.34	47.87	58.54	82.32
ROC	64.67	66.24	61.46	68.34	78.83

used ensemble model the performance accuracy enhanced. So, the proposed Hybrid concept of Feature selection and ensembeling of model outperform with an accuracy of 93.82%.

Table 2. Performance of classifier after feature selection.

Classifier	SVM	RF	Naïve Bayes	KNN	Proposed - Ensembled Model
Accuracy Achieved (%)	89.7	83.52	87.85	86.66	93.82
Precision	86.31	80.14	84.81	82.36	88.25
F1 Score	78.13	72.43	76.25	74.54	86.54
ROC	75.23	69.64	73.75	69.82	81.47

8 CONCLUSION

In this paper we have collected the data set of diabetes and applied feature selection algorithm to obtained best feature from the dataset. The Comparative study in terms of accuracy is done for various classifiers i.e., SVM, RF, KNN and Naïve Bayes. Then we have done the ensembeling of the said classifier and obtained the accuracy 0t 93.82% for the selected dataset after feature selection using genetic algorithm which is the best result. It means ensemble model can provide better accuracy rather than single one classifier. We have used the ensemble model for diabetes classification in this paper, further in future we can apply ensemble model for other disease classification like heart disease, Parkinson's disease etc.

REFERENCES

[1] Ajay V, Metun, Reddy B. P. K and Tarale P. R. 2022. An Aquila-optimized SVM Classifier for Diabetes Prediction, "*International Conference on Artificial Intelligence and Data Engineering (AIDE)*, Karkala, India, pp. 29–34, doi: 10.1109/AIDE57180.2022.10060334.

[2] Alehegn M, Joshi R, and Mulay P. 2018. Analysis and Prediction of Diabetes Mellitus Using Machine Learning Algorithm. *Int. J. Pure Appl. Math.*, vol. 118, no. Special Issue 9, pp. 871–878.

[3] Azad C, Bhushan B, Sharma R, Shankar A, Singh K. K, and Khamparia A. 2022. Prediction Model Using SMOTE, Genetic Algorithm and Decision Tree (PMSGD) for Classification of Diabetes Mellitus. *Multimed. Syst.*, vol. 28, no. 4, pp. 1289–1307, 2022, doi: 10.1007/s00530-021-00817-2.

[4] Butt U. M. *et al.* 2021. *Machine Learning Based Diabetes Classification and Prediction for Healthcare Applications.*

[5] Dubey Y, Wankhede P, Borkar T, Borkar A and Mitra K. 2021. Diabetes Prediction and Classification using Machine Learning Algorithms. *IEEE International Conference on Biomedical Engineering, Computer and Information Technology for Health (BECITHCON)*, Dhaka, Bangladesh, pp. 60–63, doi: 10.1109/BECITHCON54710.2021.9893653.

[6] Hasan M.K., Alam M.A., Das D., E. Hossain and Hasan M. 2020. Diabetes Prediction Using Ensembling of Different Machine Learning Classifiers. *IEEE Access*, vol. 8, pp. 76516–76531, doi: 10.1109/ACCESS.2020.2989857.

[7] Khanse S, Bhandari P, Singru R, Runwal N and Dharane A, 2020. Comparative Study of Genetic Algorithm and Artificial Neural Network for Multi-class Classification based on Type-2 Diabetes Treatment Recommendation model. *Sixth International Conference on Parallel, Distributed and Grid Computing (PDGC)*, Waknaghat, India, pp. 538–543, doi: 10.1109/PDGC50313.2020.9315837.

[8] Prabha, A., Yadav, J., Rani, A., & Singh. 2021. Design of Intelligent Diabetes Mellitus Detection System Using Hybrid Feature Selection based XGBoost Classifier. *Computers in Biology and Medicine*, 136, 104664. https://doi.org/10.1016/j.compbiomed.2021.104664

[9] Rani P, Lamba R, Sachdeva R. K, Bathla P and Aledaily A. N. 2023. Diabetes Prediction Using Machine Learning Classification Algorithms. *International Conference on Smart Computing and Application (ICSCA)*, Hail, Saudi Arabia, pp. 1–5, doi: 10.1109/ICSCA57840.2023.10087827.

[10] Salem H. Shams M. Y, Elzeki O. M., Elfattah M. A., Al-amri J. F., and Elnazer S. 2022. Fine-Tuning Fuzzy KNN Classifier Based on Uncertainty Membership for the Medical Diagnosis of Diabetes. *Appl. Sci.*, vol. 12, no. 3, pp. 1–26, doi: 10.3390/app12030950.

[11] Saxena, R., Sharma, S., Gupta, M., & Sampada, G. C. 2022. A Novel Approach for Feature Selection and Classification of Diabetes Mellitus: Machine Learning Methods. *Computational Intelligence and Neuroscience*, 2022, 1–11. https://doi.org/10.1155/2022/3820360

[12] Syed R, Gupta R. K and Pathik N. 2018. An Advance Tree Adaptive Data Classification for the Diabetes Disease Prediction. *International Conference on Recent Innovations in Electrical, Electronics & Communication Engineering (ICRIEECE)*, Bhubaneswar, India, pp. 1793–1798, doi: 10.1109/ICRIEECE44171.2018.9009180.

[13] Thakkar H, Shah V, Yagnik H, and Shah M. 2021. Comparative Anatomization of Data Mining and Fuzzy Logic Techniques used in Diabetes Prognosis. *Clin. eHealth*, vol. 4, pp. 12–23, doi: 10.1016/j.ceh.2020.11.001.

[14] Vaishali R, Sasikala R, Ramasubbareddy S, Remya S and Nalluri S. 2017.Genetic Algorithm based Feature Selection and MOE Fuzzy Classification Algorithm on Pima Indians Diabetes Dataset. *International Conference on Computing Networking and Informatics (ICCNI)*, Lagos, Nigeria, pp. 1–5, doi: 10.1109/ICCNI.2017.8123815.

[15] Wang X. *et al.* 2021.Exploratory Study on Classification of Diabetes Mellitus Through a Combined Random Forest Classifier. *BMC Med. Inform. Decis. Mak.*, vol. 21, no. 1, pp. 1–14, doi: 10.1186/s12911-021-01471-4.

Artificial Intelligence, Blockchain, Computing and Security – Dagur et al. (Eds)
© 2024 The Author(s), ISBN: 978-1-032-67841-2

Design and analysis of predictive model to detect fake news in online content

Ratre Sushila
Department of Engineering and Technology, Bharati Vidyapeeth Deemed University, Navi Mumbai, India

Rohatgi Divya & Bhise Rajesh
Amity University Maharashtra, India

ABSTRACT: Exponential growth of Internet and social media during the past few decades has helped the users to pass any information without even analyzing. Out of these many are fake information which has got no authenticity and relevance. The fake news can be misleading and can pose dangerous threat to public health especially in situations of pandemic like Covid-19. Incorrect or misleading information may include information regarding medical, medicine, doctors, presciprtions etc. Sometimes even an expert has to analyse various aspects simultaneously to classify the news as fake or not. To solve this problem we have proposed a prediction model for fake news detection using machine learning approach. Also a comparative analysis of other Machine Learning algorithms like Logistic Regression, Naive Bayes, SVM, Decision Tree, Random Forest and KNN is performed on the dataset to verify the efficiency of the methodology. Based on the experimental results, it can be inferred that SVM classifier gives better performance as compared to others with an accuracy of 82.85%.

Keywords: Fake news, Machine learning algorithm, SVM classifier, KNN

1 INTRODUCTION

1.1 *Growth of internet and social media*

The growth of Internet and social media has created a vast opportunity for the users to disseminate any kind of information with very low cost. This has paved the way for growth but in some ways, it has also lead to passing of fake news very instantly which can pose to threat, chaos and other unwanted issues to mankind. To detect whether a news is fake or not requires expertise from relevant domain of knowledge and sometimes the expert has to look on various facets of decision making to have a conclusion. The availability of expert and cost associated in expert judgement is also present. The better solution can be to have computer-based approach to do prediction. Machine learning algorithms are one of the best approaches for prediction in such scenarios. Any type of fake news has two major divisions regarding their authenticity and the intention behind the news. Fake news is not authentic as they have incorrect information that is difficult to validate. The intention means the purpose behind of creating fake news. It may be for publicity, deceiving the reader, spreading rumors to have undue advantage etc. Earlier when growth of internet and other social media was not there, generating and spreading fake news was difficult but with the expansion of new technologies and exponential growth of internet, it has become very easy to generate and spread fake news almost instantly. Consequently, the impact of fake news has also increased, so some measures are required to have a check and control on such news. Many accessible misleading-information discovery websites may well be utilized to look for pre-checked

 DOI: 10.1201/9781032684994-17

information. In any case, these websites are generally human-based, where the examination of information is carried out physically. This investigation is performed by master investigators who are personally commonplace with the subject setting. The manual approach is moderate, costly, profoundly subjective, one-sided, and has gotten to be unreasonable due to the gigantic volume of accessible information on social systems.

1.2 *Affects of fake news*

Fake news not only creates a bad impact on a person but in the long run it also affects the society in a negative way. Because of huge volume of false information spreading everywhere, fake news can disrupt the "news ecosystem's balance.". Instead of "most popular true mainstream news," the "most popular false mainstream news" was significantly more widely distributed on Facebook during the 2016 Presidential Election. This indicates how consumers are more likely to pay attention to modified data than to true information. This is very disturbing as such type of incorrect news not only "persuades consumers to accept biased or misleading views" so as to express a manipulator's goal, but it also alters how consumers react to actual news. People that manipulate information want to create confusion so that people's capacity to discriminate between right or wrong is hampered even more. This is also one of the major motives that why some unscrupulous people are engaged in creating and generating fake news. Subsequently, the method which can do automated classification of information is required. This paper includes ML (Machine Learning) techniques to make a model that can uncover records that are, with high probability, fake news stories and articles. By gathering occurrences of both veritable and fake news and planning a demonstrate, it ought to be conceivable to orchestrate fake news stories with a particular level of exactness.

1.3 *Main objective*

The main objective of this paper is to check for the accuracy of the system regarding fake news by using machine learning algorithms. The result of this extend ought to be to choose how much can be fulfilled in this errand by dismembering plans contained within the content and tie to the exterior information almost the world. This kind of arrangement isn't anticipated to be an end-to-end arrangement for fake news. Instead of being an end-to-end solution, this venture is expected to be one arrangement that may well be utilized to assist individuals who are endeavouring to classify fake news. On the other hand, it may be one apparatus that's utilized in future applications that intellectuals combine diverse gadgets to create an end-to-end arrangement for computerization of the strategy of fake news classification.

2 LITERATURE REVIEW

Thomas Felber *et al.* in their paper [1] apply classical machine learning algorithms together with several linguistic features, such as n-grams, readability, emotional tone and punctuation. They found that linear SVM performed best with a weighted average F1 score of 95.19Z Khanam *et al.* [2] explores the traditional machine learning models to choose the best, in order to create a model of a product with supervised machine learning algorithm, that can classify fake news as true or false, by using tools like scikit – learn, NLP for textual analysis. They used python scikit-learn library to perform tokenization and feature extraction of text data. Jamal Abdul Nasir *et al.* [3] proposed a novel hybrid deep learning model that combines convolutional and recurrent neural networks for fake news classification. This model used two fake news datasets (ISO and FA-KES) and it was verified with more accuracy than other non hybrid baseline methods. Deep learning model that combines CNN-RNN is used that achieves an accuracy of 95William Scott Paka *et al.* [4] propose Cross-SEAN, a cross-stitch based semi-supervised end-to-end neural attention model which leverages the large amount of unlabelled data. Cross-SEAN has demonstrated 0.95 F1 Score on CTF based on Twitter dataset of

COVID-19 which had properly labelled correct and fake or incorrect tweets. Md Shad Akhtar *et al.* [5] describes the details of shared tasks on COVID-19 Fake News Detection in English and Hostile Post Detection in Hindi. The most successful models were BERT or its variations. Machine Learning models like SVM and logistic regression were used as baseline models and achieve an F1-Score of 93.32Gahirwal Manishan *et al.* [6] proposes a system that classifies unreliable news into different categories after computing an F-score. This system aims to use various NLP and classification techniques, Machine learning, and Artificial Intelligence to help achieve maximum accuracy. Julio C. S. Reis *et al.* [7] present a new set of features and measures the prediction performance of current approaches and features for automatic detection of fake news. The results in the paper demonstrated on the usage and importance of features in order to find fake news. Jagrati Sahu *et al.* [8] proposes a model for recognizing forged news messages from twitter posts, by figuring out how to anticipate precision appraisals, in view of computerizing forged news identification in Twitter datasets. The paper also compared various machine learning algorithms to find the efficiency on the given dataset. Simon Lorent *et al.* [9] evaluate the performance of Attention Mechanism for Fake News detection on two datasets and consequently results were compared against LSTM and other traditional algorithms. Ifthikar Ahmed *et al.* [10] propose to use machine learning ensemble approach for automated classification of news articles. In the paper, authors gave various textual properties which may be utilized to differentiate between fake news with real news and subsequently we may use these textual properties for training to various ML algorithms using different ensemble methods. The authors used 4 real world data sets for their work. Alim Al Ayub Ahmed *et al.* [11] provides a systematic literature review about the use of supervised machine learning classifiers trained with three different models for feature extraction for detecting the fake news. Steni Mol T S *et al.* [12] presents a review and comprehensive analysis of the articles in recent literature which were about detecting fake news over social media. Uma Sharma *et al.* [13] aims to provide the user with ability to classify the news as fake or real and also checks the authenticity of the website publishing the news. Aswini Thota *et al.* [14] present neural network architecture to accurately predict the stance between a given pair of headline and article body. The model shows an accuracy of 94.21% on test data. Shubham Dalbhanjan *et al.* [15] demonstrate model and methodology for fake news detection with help of machine learning and tongue processing and determine whether news is real or fake using passive Aggressive classifier. Defining with correctness of result upto 93.6% of accuracy.

3 IMPLEMENTATION

The proposed system is depicted in Figure 1. The process consist of various steps which are needed from data scrapping to collection till classification.

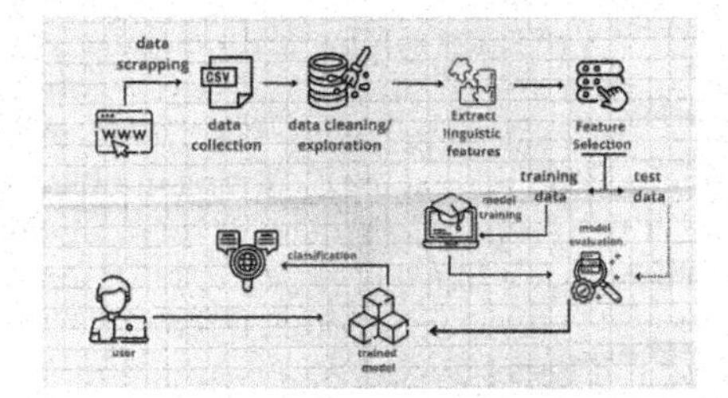

Figure 1. Proposed model for fake news detection.

3.1 *Data collection*

Online news can be collected from various points like websites, social media, search engines etc. The dataset which is collected is shown in Figure 2. The dataset contains 5 attributes which

```
df = pd.read_csv("fake_new_dataset.csv")
df.head()
```

	Unnamed: 0	title	text	subcategory	label
0	0	FACEBOOK DELETES MICHIGAN ANTI-LOCKDOWN GROUP ...	Facebook has shuttered a popular group for Mic...	false news	0
1	1	Other Viewpoints: COVID-19 is worse than the flu	We can now officially put to rest all comparis...	true	1
2	2	Bermuda's COVID-19 cases surpass 100	The Ministry of Health in Bermuda has confirme...	true	1
3	3	Purdue University says students face 'close to...	Purdue University President Mitch Daniels, the...	partially false	0
4	4	THE HIGH COST OF LOCKING DOWN AMERICA: "WE'VE ...	Locking down much of the country may have help...	false news	0

Figure 2. Data set.

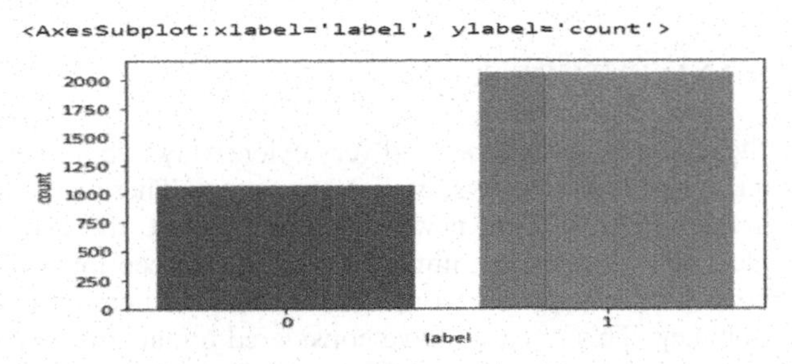

Figure 3. Count plot of label attribute.

are as follows: Index(Unamed:0) that is the number of rows. title which is the title of news article text which is the content of article subcategory which has value as fake news, true, partially false depending on the content of text. label which contains value as 0 for fake and 1 for real. Dataset has been taken from Kaggle for detection of fake news regarding COVID.

3.2 *Data cleaning*

Due to presence of many special type of characters and notations data cant be accessed directly in the form of text. Unfilterized usage of such raw will result in having hard time detecting structures in the data and will occasionally leads to errors. Hence we must filter the data. We can produce a number of features from text data, such as tor number of words, their frequencies, n-grams, and so on. By making a presenting the words so as get their implications, structural connections, and various settings they can be utilized in, ready to compute them and obtain their contents and generates Clusters and Classify them. Text preparation using the body can be termed as Word Vector Representation. The most challenging part is then creating the headline of the news article for modelling. Text analytics is done by changing over crude content to integral highlights. Experiments were perfomed using TF-IDF technique to filtere the raw textual information and extract feature.

3.3 *Feature selection*

Splitting the data is the most essential step in machine learning. We prepare our demonstrate on the trainset and test our information on the testing set. We part our information in prepare and test utilizing the train and test split work from Scikit learn. We part our 80% information for the preparing set and the remaining 20% information for the testing set.

4 RESULT ANALYSIS

We have used 6 models of machine Learning Algorithms. These are Logistic Regression (LR), Decision Tree (DT), Support Vector Machine (SVM), K-Nearest Neighbors, Random Forest Tree. Table 1. gives the comparison of all the algorithms used in this paper on the basis of their scores of accuracy, precision, recall, F1-Score.

Table 1. Comparison of accuracy, prediction, recall, F1-score of 6 algorithms.

Classifier	Accuracy (in%)	Precision	Recall	F1 Score
Logistic Regression	80.61	0.81	0.81	0.80
Decision Tree	72.92	0.73	0.73	0.73
SVM	82.85	0.83	0.83	0.82
NB	72.76	0.730.73	0.73	0.73
KNN	72.92	0.76	0.73	0.68

5 CONCLUSION AND FUTURE SCOPE

In current century the online factor will be used very extensively to complete the task online. Online programs and social media have started Newspapers. These were once favoured as tangible copies. The rising risk of fake news will only increase the complexity and sway people's opinion. Depending on the assumption, people's perception will change. As the belief goes way beyond the facts, people wont believe the news item accessible that contradicts a purportedly phoney. This research have considered 6 machine learning models on a Fake news dataset in terms of accuracy, precision, recall and F1 score. According to our experiment and research, some models like Support Vector Machine and Linear Regression (LR) had a much better performance on datasets. The experimental results of these two best models demonstrated that our proposed method achieved above 80% accuracy on the dataset was high as compared to other models. The results were compared with existing work and concluded that our SVM based model is much better with an accuracy of 82.85As far as future Future work is considered, different directions which results in expansions of modality set can be explored. Different deep learning models can be used for fake news detection and perform more experiments on other datasets in different languages.

REFERENCES

[1] Thomas, Felber; Constraint 2021: Machine Learning Models for COVID-19 Fake News detection Shared Task; *arXiv* preprint arXiv; 2101.0371.

[2] Z, Khanam: BN, Alwael; H, Sirafi; M, Rashid 2021: Fake News Detection Using Machine Learning Approaches: Z Khanam, *IOP Conf. Ser.; Mater. Sci. Eng.* 1099012040

[3] Jamal Abdul, Nasir; Osama, Subhani Khan; Iraklis Varlamis 2021: Fake News Detection: A Hybrid CNN-RNN Based Deep Learning Approach; *International Journal of Information Management Data Insights* 1 (1), 100007.

[4] William, Scott Paka; Rachit, Bansal; Abhay, Kaushik: Shubhashis, Sengupta: Tanmoy, Chakraborty. 2021. Cross-SEAN: A Cross-stitch Semi-supervised Neural Attention Model for COVID-19 Fake News Detection: *Applied Soft Computing* 107,107393.

[5] Md Shad, Akhtar; Tanmoy, Chakraborty 2021. Combating Online Hostile Posts in Regional Languages during Emergency Situation: *First International Workshop, Constraint*, 42.

[6] Gahirwal, Manishan; Tanvi, Kulkarnia; Sanjana, Moghe 2018; *International Journal of Advance Research, Ideas and Innovations in Technology*.

[7] Julio C. S. Reis; Andre, Correia; Fabricio, Murai; Adriano, Veloso; Fabricio, Benevenuto 2019; *Supervised Learning for Fake News Detection*, IEEE.

[8] Jagrati, Sahu; Uma, Sharma; Siddarth, Saran; Shankar, M. Patil 2016; Fake News Detection using Machine Learning Algorithms; *International Journal of Science and Research.*

[9] Simon, Lorent 2019; *Fake News Detection Using Machine Learning*; University Of Liege Faculty Of Applied Science Belgium.

[10] Iftikhar, Ahmad; Muhammad, Yousaf; Suhail, Yousaf; Muhammad, Ovais Ahmad 2020; *Fake News Detection Using Machine Learning Ensemble Method.*

[11] Alim Al Ayub Ahmed; Ayman, Aljarbouh; Praveen, Kumar Donepudi; Myung Suh Choi 2021; *Detecting Fake News using Machine Learning: A Systematic Literature Review.*

[12] Steni Mol T S, Sreeja P S 2020; *Fake News Detection on Social Media-a Review.*

[13] Uma, Sharma; Sidarth, Saran; Shankar, M. Patil 2021; *Fake News Detection using Machine Learning Algorithms.*

[14] Ashwini, Thota; Priyanka, Tilak; Simrat, Ahluwalia; nibrat, Lohia 2018; *Fake News Detection: A Deep Learning Approach.*

[15] Shubham, Dalbhanjan; Monika, Borkar; Sanjeevani, Pawar 2021; Fake News Detection Using Machine Learning.

Artificial Intelligence, Blockchain, Computing and Security – Dagur et al. (Eds)
 ISBN: 978-1-032-67841-2

Various methods to classify the polarity of text based customer reviews using sentiment analysis

Rahul Kumar Sharma & Arvind Dagur
School of Computing Science & Engineering, Galgotias University, Greater Noida, Uttar Pradesh

ABSTRACT: Sentiment analysis is the computational analysis of end users' attitudes, opinions, and feelings regarding a specific subject or item. Sentiment analysis groups the message based on its polarity whether it is neutral, negative, and positive. Sentiment analysis in NLP is one that is rapidly developing with new approaches and methods are being created to increase its efficacy and accuracy. The most recent technologies and research methodologies for sentiment analysis is briefly summarized in this review article. The recent updates in this field are thoroughly summarized in this survey article. The paper presents recent studies systematically in areas associated with sentiment analysis (business surveillance, polarity observation, social media monitoring). In addition, it also analyses the challenges and drawbacks of the present sentiment analysis methods and suggests possibilities for future research. Overall, this study is a helpful resource for academics and industry professionals who want to stay on top of the most recent developments in sentiment analysis.

Keywords: Sentiment Analysis, Emotional Recognition, Polarity, Sentiment Classification method, Feature Selection.

1 INTRODUCTION

Sentiment analysis is used to determine and extract emotion, views, attitudes, and sentiment conveyed in text data. It makes an advantage of NLP, machine learning, and data mining techniques to analyse and understand the subjective information conveyed by individuals or groups [5].

Today, our lives are significantly impacted by the Internet. The polarity of the product is the first thing a customer looks into before buying it from an online store. They also use these websites to see what other customers are saying about the product. Because of this, mining and sentiment extraction from these data have grown to be crucial areas of study. In the current digital era, there is wealth of textual information available from different sources including social networking sites, media platforms, customer reviews, surveys, news articles, and online forums. Extracting sentiment from this vast amount of data provides valuable insights into public opinion, customer feedback, brand perception, market trends, and more. The main objective of sentiment analysis is to categorized text into predefined sentiment categories, typically positive, negative, or neutral [1].

Polarity and subjectivity are two concepts that might be seen as elements of emotional analysis. All of the work completed at the sentence, document, and sub-sentence level is included in sentiment analysis.

1.1 *Subjectivity/objectivity*

Subjectivity and objectivity are concepts that relate to the presence or absence of personal opinions, emotions, and subjective judgments in text. In sentiment analysis, subjectivity refers to

DOI: 10.1201/9781032684994-18

the extent to which a piece of text expresses personal opinions or subjective viewpoints, while objectivity refers to the absence of personal opinions and a more factual or neutral tone.

Subjective text contains personal opinions, emotions, evaluations, or judgments. For example, in the sentence "I really enjoyed the movie," the phrase "enjoyed" indicates a positive sentiment.

Objective text presents facts, information, or descriptions without personal opinions or emotional language. For example, in the sentence "The temperature today is 25 degrees Celsius," the statement is objective and doesn't convey sentiment.

1.2 *Polarity*

Polarity in sentiment analysis refers to the classification of sentiment as positive, negative, or neutral. It is a key component of sentiment analysis; whose objective is to ascertain the sentiment or opinion expressed in a certain text.

Positive polarity is associated with expressions of positivity, satisfaction, happiness, approval, or favorable sentiment. Examples of positive sentiment include "I love this product," "The movie was fantastic," or "The service was excellent."

Negative polarity, on the other hand, indicates expressions of negativity, dissatisfaction, sadness, disapproval, or unfavorable sentiment. Examples of negative sentiment include "I hate this product," "The movie was terrible," or "The service was awful."

Neutral polarity is assigned when the text does not show a strong positive or negative sentiment. Neutral sentiment can represent a lack of opinion, a factual statement, or an emotionally neutral viewpoint [8]. Examples of neutral sentiment include "The weather is mild today" or "The book was published in 2005."

1.3 *Level of sentiment analysis*

The common levels of sentiment analysis include:

1.3.1 *Document-level sentiment analysis*

This stage of analysis focuses on the sentiment expressed in an entire document or piece of text as a whole. It aims to predict the overall opinion of the document into categories such as positive, negative, or neutral.

1.3.2 *Sentence-level sentiment analysis*

Sentence-level sentiment analysis allows for a more granular understanding of sentiment within a document, capturing variations and nuances in the sentiment expressed across different sentences [9].

1.3.3 *Aspect-based sentiment analysis*

This stage of analysis focuses on recognizing and analyzing the opinion expressed towards specific aspects or entities within a document. It involves identifying aspects or entities of interest (e.g., product features, services, or specific topics) and determining the sentiment associated with each aspect. Aspect-based sentiment analysis provides insights into how sentiment varies across different aspects and helps in understanding the strengths and weaknesses of specific entities or topics [9].

2 HISTORY OF SENTIMENT ANALYSIS

There are four significant phases in the development of sentiment analysis techniques.

2.1 *First stage (1950 till early 2000s)*

Lexicon-based methods were the primary techniques used during this period. Researchers developed sentiment lexicons, which are dictionaries containing words and phrases with their corresponding sentiment polarity, to analyse the sentiment of texts [10].

2.2 *Second stage (mid-2000s)*

Machine learning algorithms, such as SVM and Naïve Bayes, were introduced to sentiment analysis, which improved the accuracy. Researchers also explored the use of part-of-speech tagging and syntactic parsing to enhance the accuracy of sentiment analysis [11].

2.3 *Third stage (late 2000s to early 2010s)*

Deep learning techniques, such as CNN and RNN, were introduced to sentiment analysis, leading to significant improvements in accuracy [12].

2.4 *Fourth stage (2010s to present)*

Hybrid methods that combine multiple techniques, including lexicon-based, machine learning, and deep learning, have become the dominant approach in sentiment analysis. Researchers have also explored various techniques to enhance the robustness of sentiment analysis, such as domain adaptation and multi-task learning [13].

Overall, the history of sentiment analysis reflects the ongoing evolution of natural language processing as a field, with advances in machine learning and deep learning driving progress in sentiment analysis research.

3 LITERATURE SURVEY

Table 1. Article summary.

Ref # (Year)	Methodology Used	Dataset Used	Evaluation Parameter	Future Scope
15 (2016)	NLP Technique	Tweets from twitter	80.6% accuracy with NLP. To determine the polarity of tweets using the sentiment score.	By examining the tweets from various websites and sources and contrasting the results, the veracity of the analysis may be confirmed.
16 (2017)	Hybrid approach	Movie Reviews	95.43% accuracy with Hybrid Approach	Future research may focus on features that require intense learning methods to further improve emotion-based classification.
17 (2017)	Naive Bayes, SVM, Decision Tree	400000 reviews of the 4500 smartphones were obtained from Amazon	In all three models, SVM's accuracy reached a maximum of 81.75%.	We can use deep learning algorithms like LSTM, GRNN, and others to boost performance. Additionally, use the reviews to anticipate a product's rating.
18 (2018)	Hybrid approach	Reviews of eight films from Amazon	Accuracy = 82%	apply related procedures using LSTM or CNN to increase accuracy
19 (2018)	Naive Bayes classifier	Twitter Data Set	The maximum accuracy is obtained when tweet sentiments are extracted using the WSD sentiment analyzer (79%).	Based on Twitter, we can identify political party trends.

(*continued*)

Table 1. Continued

Ref # (Year)	Methodology Used	Dataset Used	Evaluation Parameter	Future Scope
20 (2019)	LSTM classification algorithm	Tourist review	With SVM, K-Nearest Neighbor, and LSTM, the accuracy was 84%, 90%, and 98%, respectively.	With deep learning, the proposed framework can be incorporated into numerous categories.
21 (2020)	Gated Recurrent Neural Network	Amazon product reviews	Find out accuracy values of 70.42 and 71.19 with the bidirectional LSTM and bidirectional gated recurrent unit respectively.	continue the work and develop a solid understanding of human interaction like irony and humor.
22 (2020)	SVM Classifier	Amazon (user reviews)	Got 97% accuracy with SVM.	Future efforts will concentrate on tackling more difficult issues including negativity and sarcasm, counterfeiting, and spam.

4 RESEARCH METHODOLOGY

The process of sentiment analysis involves several steps to analyses and extract sentiment from textual data.

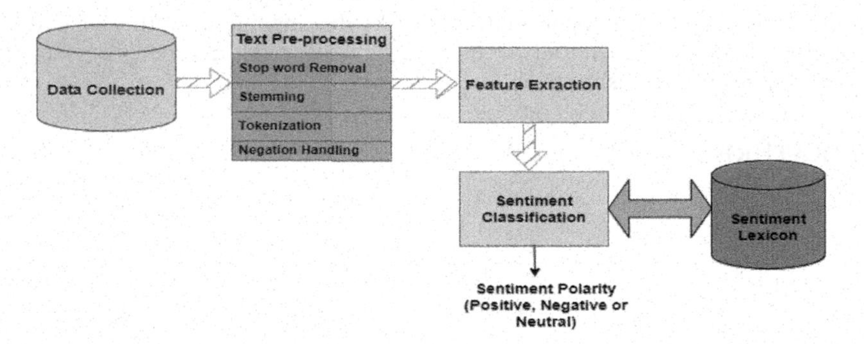

Figure 1. Sentiment analysis process.

4.1 *Data collection*

Gather the text data from relevant sources such as social media platforms, customer reviews, surveys, or other text-based sources. Ensure that the data is representative and covers the target domain or topic of interest.

4.2 *Text pre-processing*

Clean and pre-process the text data to remove noise and irrelevant information. This may involve steps such as removing special characters or punctuation, lower casing the text, handling contractions or abbreviations, and removing stop words [1].

4.3 *Feature extraction*

Take out relevant features from the text data to represent it in a numerical format that machine learning algorithms can process. This step may involve techniques such as bag-of-words representation, TF-IDF or word embedding to capture semantic information [2].

4.4 *Sentiment classification*

The text should be classified using a machine learning or deep learning algorithm into one of the established sentiment categories, such as positive, negative, or neutral [2].

4.5 *Model training*

Select an appropriate sentiment analysis algorithm or model (e.g., Naive Bayes, Support Vector Machines, LSTM, or transformer-based models) and train it on the labelled data. The model learns to associate textual features with sentiment labels during the training process [3].

4.6 *Model evaluation*

Evaluate the trained sentiment analysis model using a separate validation or test dataset. To evaluate the model's performance and determine whether it can accurately classify sentiment, use assessment metrics like accuracy, precision, recall, F1-score, or AUC-ROC [4].

A proper research methodology is crucial for several reasons:

4.6.1 *Rigor and validity*

A well-defined research methodology ensures that the study is conducted in a systematic and rigorous manner. It helps establish the validity of the research findings by providing a structured approach to collect data, analyze it, and draw meaningful conclusions. A proper methodology increases confidence in the results and ensures that the research is based on sound principles.

4.6.2 *Replicability and generalizability*

A clear research methodology enables other researchers to replicate the study, following the same steps and procedures. This replicability strengthens the validity of the research and allows for the generalizability of the findings to other contexts or populations. Without a proper methodology, it becomes difficult for others to reproduce the study, leading to uncertainty and limited applicability of the results.

4.6.3 *Ethical considerations*

A research methodology provides a framework to address ethical considerations in the study. It ensures that participants' rights and confidentiality are protected, informed consent is obtained, and any potential risks are minimized. A proper methodology helps researchers navigate ethical challenges and conduct research in an ethically responsible manner.

4.6.4 *Control of variables and bias*

A well-designed research methodology helps control for confounding variables and biases that could influence the results. It allows researchers to establish appropriate control groups, implement randomization techniques, and employ blinding procedures, depending on the study design. This control minimizes the impact of extraneous factors and enhances the internal validity of the research.

4.6.5 *Systematic literature review*

In the case of a literature review or review paper, a proper research methodology ensures a systematic and comprehensive approach to gather, analyze, and synthesizing relevant literature. It helps establish clear inclusion and exclusion criteria, conduct a thorough search, and critically evaluate the selected studies. A systematic methodology increases the reliability of the review and provides a comprehensive overview of the existing knowledge in the field.

4.6.6 *Research integrity and transparency*

A well-documented research methodology enhances the integrity and transparency of the research process. It allows other researchers to assess the quality and reliability of the study, promoting scientific accountability. Transparent reporting of the methodology also helps identify any limitations or potential biases in the research, contributing to the overall scientific integrity.

5 METHOD USED IN SENTIMENT ANALYSIS

The study covers the most widely used techniques based on historical evidence.

5.1 *Supervised learning*

Supervised learning is a popular approach where sentiment analysis models are trained on l datasets. Human annotators assign sentiment labels (positive, negative, neutral) to a set of texts, which are then used as training data. Various machine learning algorithms and deep learning classifier like recurrent or convolutional neural networks (RNN/CNN) are trained on these labelled data to predict sentiment in unseen texts [8].

5.2 *Semi-Supervised learning*

A combination of labelled and unlabeled data is employed in semi-supervised learning. Initially, a small portion of the data is labelled by human annotators, and sentiment analysis models are trained on this labelled data. Then, the trained models are used to predict the sentiment of unlabelled data, and the high-confidence predictions are used to augment the labelled dataset. This iterative process continues until a sufficient amount of labelled data is available for training.

5.3 *Unsupervised learning*

Unsupervised learning methods do not rely on labelled data for training. Instead, they aim to discover patterns, clusters, or latent representations within the data. Techniques like clustering, topic modelling, or dimensionality reduction are employed to identify sentiment patterns or group texts based on their sentiment similarities. Unsupervised learning approaches can be useful when labelled data is scarce or when exploring new domains [8].

5.4 *Lexicon-based methods*

Lexicon-based approaches leverage sentiment lexicons or dictionaries containing pre-defined sentiment scores for words or phrases. Each word is assigned a sentiment polarity (positive, negative, neutral) or a sentiment intensity score. Sentiment analysis is performed by aggregating the sentiment scores of the words present in the text. Lexicon-based approaches are useful when sentiment expressions are strongly tied to specific words or when quick estimations of sentiment are required [14].

5.5 *Deep learning*

Deep learning techniques, such as recurrent neural networks (RNNs), convolutional neural networks (CNNs), or transformers, have gained popularity in sentiment analysis. These models can capture the contextual and semantic information of text more effectively, learning complex patterns and dependencies. Deep learning approaches often require large amounts of labelled data and computational resources for training [14].

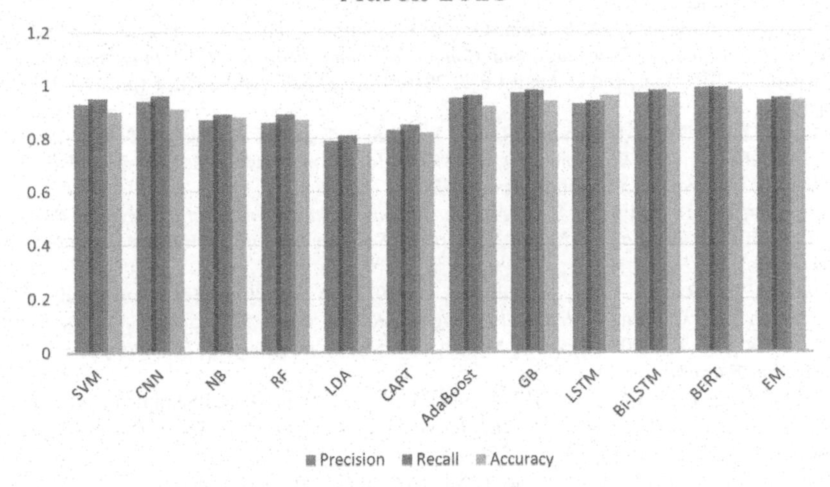

Figure 2. Comparison chart of various existing algorithms.

6 CONCLUSION

The study focuses on identifying significant themes for future sentiment analysis research since it aids in prioritizing research efforts and concentrating attention on regions with the most promise or potential impact. Such a publication can help direct future research toward solving significant issues and establishing new ways to increase the accuracy and robustness of sentiment analysis by summarizing the present state-of-the-art and highlighting gaps in the literature. Paper can also assist scholars, business professionals, and politicians alike in identifying new trends and directions in sentiment analysis research. The paper might encourage additional study and innovation in the subject by highlighting interesting research areas and examining their possible applications and ramifications.

REFERENCES

[1] Bird, S., Klein, E., & Loper, E. (2009). *Natural Language Processing with Python: Analyzing Text with the Natural Language Toolkit.* " O'Reilly Media, Inc.".

[2] Pang, B., & Lee, L. (2008). Opinion Mining and Sentiment Analysis. *Foundations and Trends® in Information Retrieval,* 2(1–2), 1–135.

[3] Socher, R., Perelygin, A., Wu, J., Chuang, J., Manning, C. D., Ng, A. Y., & Potts, C. (2013, October). Recursive Deep Models for Semantic Compositionality Over a Sentiment Treebank. In *Proceedings of the 2013 Conference on Empirical Methods in Natural Language Processing* (pp. 1631–1642).

[4] Feldman, R. (2013). Techniques and Applications for Sentiment Analysis. *Communications of the ACM,* 56(4), 82–89.

[5] Pang, B., & Lee, L. (2008). Opinion Mining and Sentiment Analysis. *Foundations and Trends® in Information Retrieval,* 2(1–2), 1–135.

[6] Liu, B. (2012). Sentiment Analysis and Opinion Mining. *Synthesis Lectures on Human Language Technologies,* 5(1), 1–167.

[7] Ravi, K., & Ravi, V. (2015). A Survey on Opinion Mining and Sentiment Analysis: Tasks, Approaches and Applications. *Knowledge-based Systems,* 89, 14–46.

[8] Kaur, H., & Mangat, V. (2017, February). A Survey of Sentiment Analysis Techniques. In 2017 *International Conference on I-SMAC (IoT in Social, Mobile,Aanalytics and Cloud)(I-SMAC)* (pp. 921–925). IEEE.

[9] Behdenna, S., Barigou, F., & Belalem, G. (2016). Sentiment Analysis at Document Level. In *Smart Trends in Information Technology and Computer Communications: First International Conference*, SmartCom 2016, Jaipur, India, August 6–7, 2016, Revised Selected Papers 1 (pp. 159–168). Springer Singapore.

[10] Winograd, T. (1972). Understanding Natural Language. *Cognitive Psychology*, 3(1), 1–191.

[11] SS971555RF11, L. R., & Rabiner, A. (1989). Tutorial on Hidden Markov Models and Selected Applications in Speech Recognition. *Proc. IEEE.*

[12] Christopher, D. M., & Hinrich, S. (1999). *Foundations of Statistical Natural Language Processing*.

[13] Chen, Y. (2015). *Convolutional Neural Network for Sentence Classification* (Master's thesis, University of Waterloo).

[14] Wankhade, M., Rao, A. C. S., & Kulkarni, C. (2022). A Survey on Sentiment Analysis Methods, Applications, and Challenges. *Artificial Intelligence Review*, 55(7), 5731–5780.

[15] Pang, B., & Lee, L. (2008). Opinion Mining and Sentiment Analysis. *Foundations and Trends® in Information Retrieval*, 2(1–2), 1–135.

[16] Sruthi, S., Sheik, R., & John, A. (2017, August). Reduced Feature Based Sentiment Analysis on Movie Reviews Using Key Terms. In *2017 IEEE International Conference on Signal Processing, Informatics, Communication and Energy Systems (SPICES)* (pp. 1–8). IEEE.

[17] Zeenia Singla, Sukhchandan Randhawa, Sushma Jain, "Sentiment Analysis of Customer Product Reviews Using Machine Learning", 2017 *International Conference on Intelligent Computing and Control (I2C2).*

[18] Kavousi, M., & Saadatmand, S. (2019). Estimating the Rating of the Reviews Based on the Text. In Data Analytics and Learning: *Proceedings of DAL 2018* (pp. 257–267). Springer Singapore.

[19] Hasan, A., Moin, S., Karim, A., & Shamshirband, S. (2018). Machine Learning-based Sentiment Analysis for Twitter Accounts. *Mathematical and Computational Applications*, 23(1), 11.

[20] Thakur, P., & Shrivastava, R. (2019). Sentiment Analysis of Tourist Review Using Supervised Long Short Term Memory Deep learning Approach. *Int. J. Innov. Res. Comput. Commun. Eng*, 7(2), 592–604.

[21] Sachin, S., Tripathi, A., Mahajan, N., Aggarwal, S., & Nagrath, P. (2020). Sentiment Analysis Using Gated Recurrent Neural Networks. SN Computer Science, 1, 1–13.

[22] Nandal, N., Tanwar, R., & Pruthi, J. (2020). Machine Learning Based Aspect Level Sentiment Analysis for Amazon Products. *Spatial Information Research*, 28, 601–607.

Artificial Intelligence, Blockchain, Computing and Security – Dagur et al. (Eds)
© 2024 The Author(s), ISBN: 978-1-032-67841-2

A systematic review on heart disease prediction using deep neural network for scalable datasets

Sh. Suvanov
Samarkand State University, Samarkand City, Uzbekistan

M. Sabharwal & V. Choudhary
Galgotias University, Uttar Pradesh, India

ABSTRACT: Deep Learning and Artificial Neural Networks (ANN) are technologies that simulate a human brain because it consists of several neurons. The input layer, hidden layers, and output layer of the network are thus manifested. The network attempts to learn from the data that is fed to it and then makes predictions accordingly. ANN is one of the most basic types of neural network. The Artificial Neural Network does not have any specific structure; it comprises multiple neural layers for prediction. The effective application of deep learning happens in the fields like finances, marketing, and retail which has inspired the other sectors. It can be effectively applied to healthcare and medical science, such as cardiology. Doctors have a considerable amount of data in hospitals and clinics; unfortunately, there is insufficient knowledge to use the data to make a persuasive heart disease prediction decision. This paper aims to bring an effective study of machine learning in Heart Disease Prediction for scalable datasets. The study also concludes that the significant factors affecting heart disease are classified into four factors: socio-demographic characteristics, family history, medical history, environmental risk factors, and dietary/lifestyle factors. It also sub-classifies them as adjustable and non-adjustable with a detailed list.

1 INTRODUCTION

Artificial Intelligence (AI) and Machine Learning (ML) are now frequently utilized in healthcare for the prediction of serious illnesses such as Type-2 Diabetes Risk Prediction (Kour 2021), Colorectal Cancer (Tanwar 2021), Alzheimer's (Savita & Sabharval 2021), and Fetal Brain Abnormality Detection (Durgadevi *et al.* 2021). You can get more information about machine Alzheimer's medical applications at (Anand *et al.* 2013), and the current study also used AI and ML for the prediction of heart disease. As per the World Health organization reports, in 2016, approximately 17.9 million people died from CVDs, 31% of total deaths worldwide, out of which 85% are due because of heart attack and stork. In 2015 approximately 17 million deaths under the age of 70 were non-communicable, i.e., 37% of all deaths globally. By the estimation of WHO, Heart attacks will cause over 23.6 million deaths worldwide by 2030. While keeping in mind such data, researchers had focused on designing a smart system to diagnose heart disease with higher accuracy and avoid inaccurate diagnoses. The healthcare sector is "information wealthy" but "intelligence deprived." A large amount of data is available in healthcare, but we required robust and efficient tools to analyze the hidden relationship between the data. Early prediction of heart disease will help to cure the disease and help the patient take preventive measures and get more accurate and efficient treatment. Even though health data is collected daily by the medical

DOI: 10.1201/9781032684994-19

practitioners and healthcare centres, machine learning and pattern matching on this data to retrieve the knowledge is relatively low, which can play a vital role in diagnosis and risk prediction. An intelligent AI-based prediction system for heart disease can predict the problem that may occur with the patient's heart sooner or later; it may also help to rescue the patient from death. That is one of the primary reasons why heart disease prediction is a substantial and essential problem. Since not all doctors can make an accurate diagnosis, that may result in incomplete and inaccurate analysis and treatment, leading to death and life loss. Also, hospitals generate a vast amount of patient data daily, saving many individuals' lives if utilized by an intelligent AI-based prediction system to predict the heart disease risk. Figure 1 shows the model diagram of an intelligent AI-based prediction system.

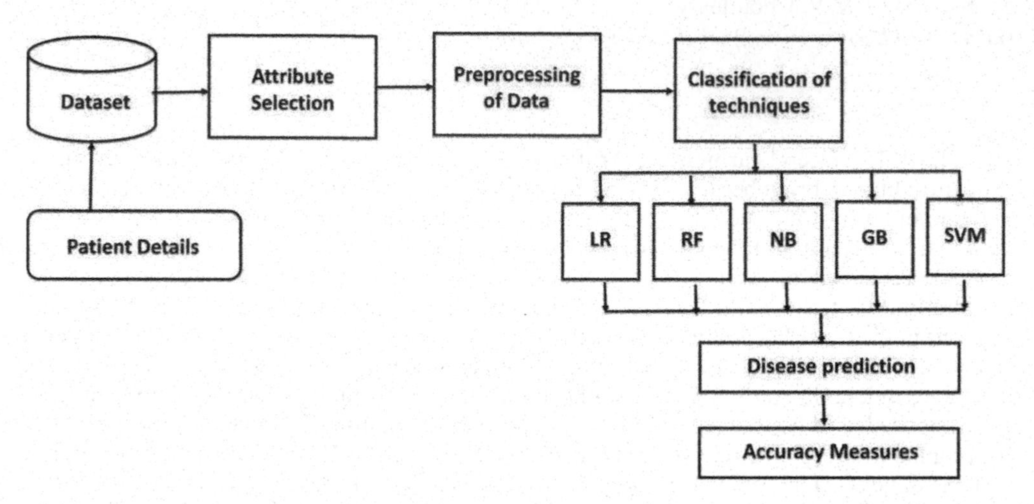

Figure 1. Block diagram of proposed system.

Machine learning methods for heart disease prediction provide more accurate diagnoses than other medical techniques. Artificial Neural Network (ANN) is the progressive branch of computer science; it makes computers think and make decisions like humans. We aim to review the current status of research on an artificial neural network to identify heart disease beforehand. The present study also discusses the risk factors for heart disease, various proposed techniques & artificial intelligence and machine learning-based systems, and findings in research papers on artificial neural networks for heart disease prediction between 2016 and 2020. Security is one of the key security services necessary for each application, whereas Authentication is one of the applications of information technology. (Anand & Khemchandani 2017; Anand 2019a, 2019b, 2019c). Some of those methods are beneficial to organizations and society. (Gaharana & Anand 2016; Heer & Anand 2020). With the high-speed network made possible by a 5G network, some of the (Dubey *et al.* 2018) computer science, communications, and distributed computing main applications produce improved results. (Anand *et al.* 2021). Home security uses a variety of IoT concepts and implementations. (Anand & Khemchandani 2020).

2 LITERATURE REVIEW

The detailed literature related to the problem can be classified into various classes as below:

- Studies describing risk factors for heart disease.
- Studies giving a comparative analysis of the heart disease prediction approaches using machine learning.

- Studies of heart disease prediction using ECG signal.
- Several studies use ANN for heart disease prediction.
- Studies that evaluate the application of an artificial neural network with a Multilayer perceptron neural network in the diagnosis and prediction of heart disease.
- Studies of Deep Neural Networks (DNN) for heart disease prediction.

2.1 *Studies describing risk factors for heart disease*

Pasalar and Sadegh Zadehare worked on the coronary disease prediction label dataset (Esfahani & Ghazanfari 2017; Jadon *et al.* 2017; Wu *et al.* 1993). Their study suggests categorising the heart disease process's risk factors into two classes: non-adjustable and adjustable. To them, Non-adjustable factors include age, gender, family history, and race. Modifiable factors include high obesity, smoking, diabetes, hyperlipidemia, high blood pressure, oral contraceptives, etc. (Munish & Sohan 2016; Sabharwal 2016, 2018). Some studies use data mining methods, classification prediction, and electrocardiogram pattern-based empirical model decomposition neural networks to forecast the disease by examining the factors and thereby results comparing the two implementations. (Abdou *et al.* 2018; Adhikari 2018; Hannun *et al.* 2019; Kaur & Arora 2018) The model is proposed that includes input variables related to four factors: personal information, medical history, diet, lifestyle applied ANN, heart blockage image processing BPNN (Asha *et al.* 2018; Kaur & Arora 2018). After calculating the necessary results, the person is classified as underweight, standard, and overweight. Further analysis is done using a person's medical history like diet, quantity, and frequency of consuming food, and these factors are again fed into the proposed model, in the last step, after collecting lifestyle information. The model is tested using medical data collected from different people and gives 70% accuracy (Atkov *et al.* 2012; Bhatla & Jyoti 2012). Various software tools were used along with soft computing and hypothesis testing (Sabharwal 2018). In the paper by Adhikari 2018, it is analyzed various patients' data who belong to India. Family history, smoking, hypertension, dyslipidemia, fasting glucose, obesity, lifestyle, and coronary artery bypass grafting (CABG) are analyzed risk factors. A model built using this data, trained and tested on new sample data, predicts the heart attack possibilities, preventing the various heart problems. (Adhikari 2018; Nikhar & Karandikar 2016). This model can help doctors in the decision-making, cure the patient well, and create transparency between the doctor and the patient (Mähringer-Kunz *et al.* 2020). Daniel La Freniere *et al.* 2016 find the critical risk factors using patients' current health situations, medical histories, and demographics. These factors can be employed to predict hypertension with higher accuracy. These risk factors are also used to identify whether a person will develop hypertension in the near future (Alam *et al.* 2020; Bhalerao & Gunjal 2013; Burse *et al.* 2019). (Rahim *et al.* 2016; Sabharwal M. 2018) used features collected from the healthcare unit to predict heart disease using ANN, which gives an accuracy of 84.47%. Machine learning algorithms are used to predict medical ailments like hypertension. Amin *et al.* 2013 created the system in which the model combines ANN and GA. The proposed system can be used for diagnosing hypertension in a patient. The study uses MATLAB to develop the model, and the results show an accuracy of almost a hundred percent for the PASCAL heart sound database. Vrbaski *et al.* 2019 presented an approach to making lipid profile predictions by studying the relationship between lipid profile and obesity. After training neural networks on the data-set gained from 1491 volunteers which uses measurements like triglycerides, high-density lipoprote in, low-density lipoprotein, level of total cholesterol, Etc. helps in determining the absence or presence of disease related to the heart. Menotti *et al.* 2015 predicted the heart attack risk data by questionnaire method. The risk of heart attack is predicted based on behavioural habits such as smoking, drinking alcohol, drinking soda(sugar), eating fruits, eating vegetables, doing exercises, and demographic variables like marital status, income level, and age. MBP algorithm is used for training and testing datasets (Karayılan & Kılıç 2017). It predicts whether there is a heart problem or not from the behavioural habits and the demographic variables. This enables the individuals to make lifestyle changes to promote their health. (Gligorijevis *et al.* 2017; Li *et al.* 2017) produce an ANN model to diagnose genetic heart disease threats in pregnant women based on the detailed epidemiological data. All individuals were interviewed face-to-face by expert obstetricians and gynaecologists and asked to fill a questionnaire of 36 variables from 5 categories: socio-demographic characteristics, pregnancy history, family history, environmental risk factors, and dietary/lifestyle behaviours during pregnancy (Guo 2020; Kalra 2018). The significant risk factors for heart disease are shown below in Table 1.

Table 1. Factors affecting coronary heart disease.

Characteristics	Factors
Socio-Demographic Characteristics	Race, Age, Gender, Marital Status, etc.
Environmental Risk Factors and Family History	Family History, Income Level, Environment Condition of Living and Working, etc
Dietary and Lifestyle Factors	Smoking, Drinking Alcohol, Drinking Soda(Sugar), Eating Fruits, Eating vegetables, etc. Sleep Duration per Day, doing exercises
Medical History	High Obesity, BMI, Diabetes, hypertension, Total Cholesterol Level, Triglycerides, PCG/ECG Signals, Oral Contraceptives, Hyperlipidemia, Dyslipidemia, High-Density Lipoprotein, Low-Density Lipoprotein, High Serum, Hemodynamics Factors, Aortic Pulse Wave Velocity Index Coronary Artery Bypass Grafting (CABG), etc

2.2 *Studies of comparative analysis of machine learning methods for the heart disease prediction*

A K Dwivedi uses six machine learning techniques to predict heart disease, including artificial neural networks based on numerous parameters (Dwivedi 2018). These approaches are validated using 10-fold cross-validation, and the research concludes that the effort can be prolonged by collecting the data from various clinics (Lee 2016). Figure 2 shows the prediction of heart disease using ANN-based models.

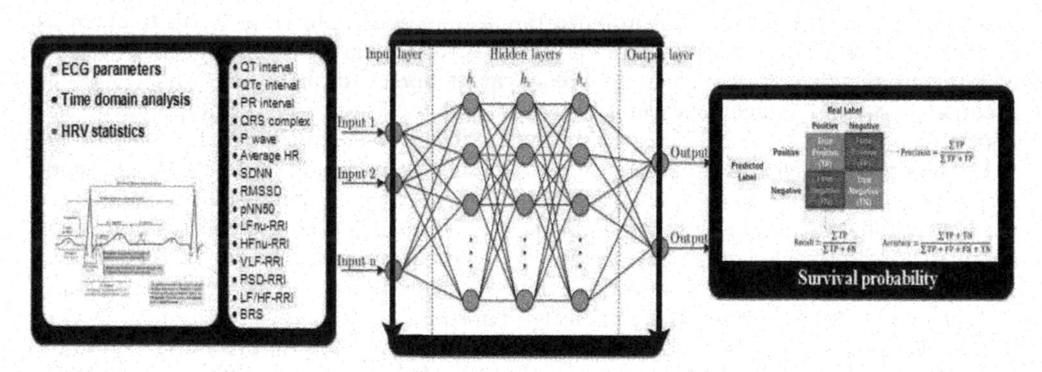

Figure 2. ANN-based model to predict the heart disease.

They collect the data from 393 subjects between 26–79 years old like sex, age, weight, being married or not, an individual's status in the family, participation in sport, sleep duration per day, smoking, sort of tobacco, BMI as shown in Table 1. Artificial neural network shows high accuracy in predicting CVD risk factors. The results obtained by this method can be used as an online case of ANN in solving problems that make it possible to predict the independent variables dependent non linearly. Costa *et al.* 2019 in their paper, used artificial neural networks for diagnosing people to recognize cardiovascular disease. A decision-making tool system is proposed as a helper for doctors that use the Amazon cloud server to analyze the hyper combination of parameters and achieve more than ninety percent accuracy (Costa *et al.* 2019; Desai *et al.* 2019).

In the paper, Hearn J *et al.* 2018 proposed the methods of examining cardiopulmonary exercise test data involving the simplification of complex time series data into simple precise indices used for prognostication. Feed-forward neural networks were used to predict the clinical deterioration in heart failure patients using cardiopulmonary exercise test- derived time-series data. Compared with the currently implemented prognostication methods, the proposed neural networks demonstrated tangible improvements in the discriminated ability and clinical utility (Mathan *et al.* 2018; Matis *et al.* 2017). In their research work, Krishnasree & Rao 2016 used two methods for heart disease prediction and investigated the outcomes of these methods and projected the model which had better accuracy (Husain *et al.* 2016; Zhang & Han 2017). Reddy *et al.* 2017 offered a system divided into two parts, such as prediction and performance models. The prediction model is used to diagnose a patient's condition after the evaluation based on the various heart parameters. The performance model is designed to assess the overall performance of the application (Sharma *et al.* 2017; Shinde *et al.* 2017). Several studies have used hybrid machine learning models for the heart disease perdition and compression among them. Costa *et al.* 2019 in their paper proposed a model that classifies the input layers as sick or healthy. It gains a recognition rate of more than ninety per cent by using the stochastic gradient descent for the error correction. In the paper, some metrics were also used: precision is used to not sort negative layers as positive ones, recall metric is to evaluate the model's ability to find all positive patterns, and f1 is to weigh an average accuracy. The developed system provides the diagnosis of cardiovascular diseases with the report of this decision. Poornima & Gladis 2018 proposed a heart disease prediction system by orthogonal Local Preserving projection and hybrid classifier. In the hybrid classifier, the Group Search Optimizer (GSO) algorithm was associated with the Levenberg-Marquardt (LM) Training algorithm in the neural network. The best weight acquired from the LM algorithm was associated with the GSO algorithm. The proposed LM algorithm-based neural network was prepared under n number of iterations, and another best weight was acquired, which improved the classification accuracy. Performances were assessed using metrics for accuracy, sensitivity, and specificity in order to attain the best result possible when using the GSO-LM algorithm in comparison to other methods. Desai *et al.* 2019 analyzed the advantages and disadvantages of models: BPNN and logistic regression. They developed a classification model to help in making a significant diagnostic decision. In their paper, Maheswari & Jasmine 2017 used neural network and logistic regression analysis. The logistic regression method is applied to choose the critical risk factors for heart disease. These gained risk factors are trained by using neural networks and tested whether they show any heart disease (Turabieh 2016). Nashif *et al.* 2018 in their research work, employed various data mining methods such as techniques like Naive Bayes, Support Vector Machine, Simple Logistic Regression, Random Forest, and ANN to build a Decision Support System for diagnosing heart disease. An android application was developed to be used by doctors, patients, medical students, etc. as a helpful instrument to predict heart disease. Mirian *et al.* 2017 proposed bi-variate logistic regression and neural network models to predict death and the heart block. The accuracies of both models are compared. MATLAB fits the models and compares the results. Esfahani & Ghazanfari 2017 created a hybrid model for heart disease prediction. Figure 3 shows the various phases of the Artificial Neural Network model based on the K-means for the prediction of heart disease.

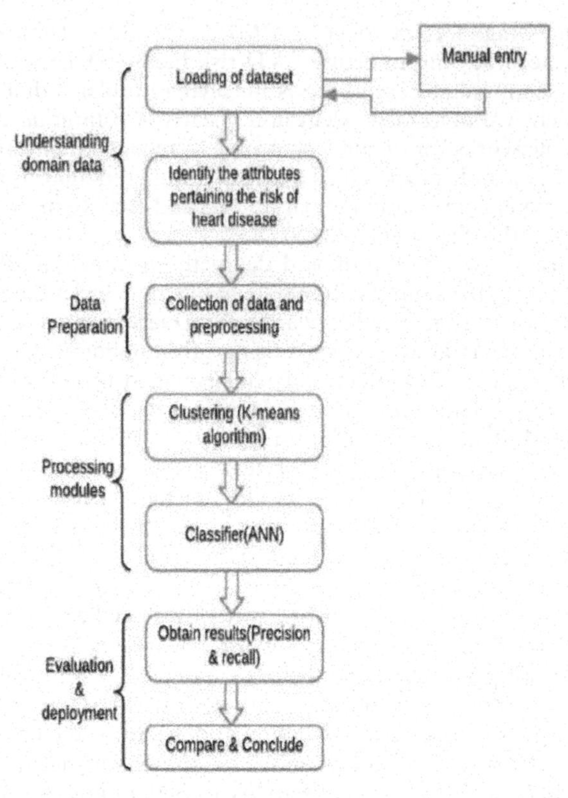

Figure 3. Phases of the K-means-based artificial neural network model for heart disease prediction.

The authors analyzed the various data mining techniques like decision tree, SVM, SVM with poly-function, neural networks, naive Bayes, and Rough Set, and compared their results based on accuracy. Among these algorithms, neural networks, Rough Set, and Naive Bayes achieve the best accuracy whereas the neural network algorithm shows the highest 86.9% accuracy. Therefore, the authors combined these three techniques and developed a hybrid prediction model. This ensemble classifier outperforms the result and shows 89% accuracy for the cardiovascular heart disease prediction (Malav *et al.* 2017). Nikhar & Karandika 2016 implemented three machine learning algorithms: neural networks, decision trees, and logistic regression to predict heart disease. It was shown that depending on the data set and models, the algorithms' accuracy may vary. Their results were compared. Bhalerao & Gunjal 2013 investigated a hybrid method of improved K- means and ANN to efficiently result in heart disease diagnosis with a high-level accuracy of 96.74%. Malav & Kadam 2018 carried out the predictive analysis on the heart disease data-set of the UCI repository using K-means and ANN techniques and proposed three algorithms, Naive Bayes, ANN, and Hybrid for heart disease prediction. Medical data contains a combination of fuzzy and crisp values, and it is classified according to their properties. It shows that the hybrid approach gives a higher accuracy rate of about 97%. Sharma *et al.* 2017 studied different neural network and decision tree algorithms and compared their results. They use neural networks, ID3, C4.5, CART, J48, and Naive Bayes algorithms, and the authors concluded that neural networks with 13 attributes give 99.25% accuracy, and ones with 15 attributes show 100%accuracy. In their study, Mohan *et al.* 2019 proposed a hybrid random forest linear model (HRFLM) to estimate heart disease with high accuracy. All experiments were done by using a standard UCI repository dataset with a Neural Network. The data is pre-processed, and then it is converted into necessary useful value. Before applying classifiers to the data, it is clustered based on the Decision Tree method. The proposed method used ANN with back propagation along with 13 input layers. Including ANN, several Machine Learning methods, such as DT, SVM, RF, NB, and k-NN, were used in all experiments, and the results were compared. The best models were chosen based on their rate of accuracy. The selection and modelling of objects continue to be repeated for various combinations of attributes. After gaining the results, the authors combined two models: the random forest method with a linear model because LM and RF become the models with the highest accuracy. The hybrid random forest linear model (HRFLM)prediction model for heart disease achieves 88.7% accuracy. Samadiani, Hassani & Frick 2021 proposed a system that applies three machine learning algorithms: ANNs, decision tree, and Adaboost algorithm to find coronary heart disease. As a result, an MLP neural network gives the best accuracy and predicts the disease with the accuracy of 94.53%, Decision Tree and Adaboost give an accuracy of 86.77% and 99.39%, respectively. Mehanovis *et al.* 2019 had created a heart disease prediction model utilising ANN, k-NN, and SVM. These algorithms are joined together and supplement each other. Various combinations of parameters were used for each of these methods. Besides, the majority voting method of ensemble learning is applied, and results are compared. Two classification methods were used. First of all, the problem is treated as multi-class classification, and then it is converted into a binary classification problem. The classification method reduces the number of outputs. In both situations, majority voting yields the maximum accuracy. Bhaskaru & Sree 2019 proposed the algorithm, which is a hybrid model using a differential evolution algorithm and Fuzzy Neural Network. Data were taken from the UCI repository and normalised to make them appropriate for the proposed algorithm. After this process, the k-fold cross-validation method was implemented to check the accuracy of the training data. Here, the input data is separated into ten parts where some are used for testing, and the other is for training. The classification accuracy is then calculated using a unique matrix called confusion and used to forecast the result. The proposed HDEFNN algorithm is compared with some algorithms like J48, Naive Bayes, and Random Forest. It is shown that the best accuracy is gained by the proposed algorithm (between Mohan *et al.* 2019 and Mehanovis *et al.* 2019) compared to the others. Besides, the execution time was also good.

2.3 *Studies of heart disease prediction using genetic algorithm*

In their research work (Mirian *et al.* 2016, 2017) An Artificial Neural Network and Genetic Algorithm (ANN-GA) hybrid model was proposed and put into practise using MATLAB to predict the heart block and deaths in myocardial infarction patients simultaneously. Bivariate Logistic Regression (BLR), ANN, and hybrid ANN-GA models give 77.7%, 83.69%, and 93.85% accuracy for the training and 78.48%, 84.81%, and 96.2% for the test data. Ahmed & Verma 2017 examine the neural network and the genetic algorithm to build a hybrid back propagation system. After building the heart's data, training and test data sets were created. The method contains 13 nodes on the input layer, seven nodes on the hidden layer, and only one node on the output layer. The result showed that the accuracy obtained is approximately 94.17%. Khourdifi & Bahaj 2019 used the Fast Correlation method to filter redundant features, and this improves the accuracy of the heart disease classification. The system architecture of the artificial neural network model for heart disease prediction is shown in Figure 4. They performed the different classification algorithms, including Artificial Neural Network optimised by Particle Swarm Optimization (PSO) combined with Ant Colony Optimization (ACO) approaches. This study also compared the results of different machine learning algorithms. Kaur & Arora 2018 analyzed the different types of data mining techniques, including ANN. The results showed that the KNN, Hybrid approach and ANN methods give more than 90% accuracy to predict heart disease. The studies that

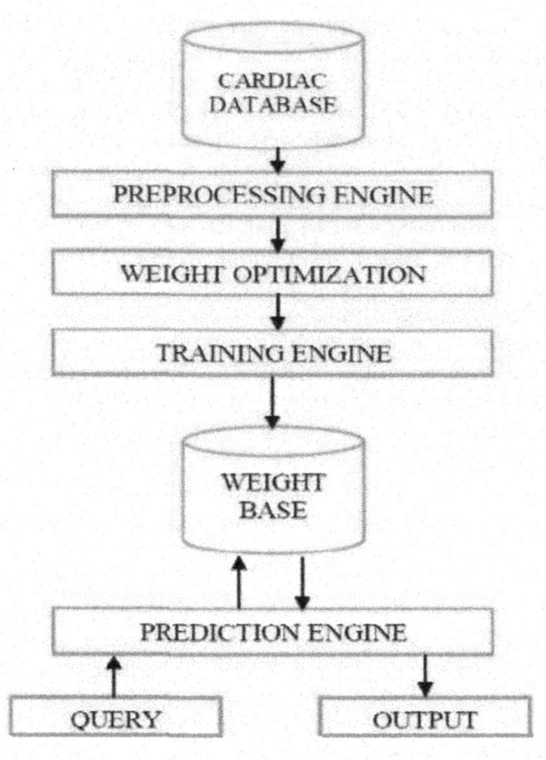

Figure 4. System architecture of the artificial neural network model for heart disease prediction network.

used genetic algorithms in predicting CVD risk developed a hybrid model using an artificial neural.

2.4 *Studies of heart disease prediction using ECG signal*

There have been several studies that use ECG Signals as input for heart disease prediction using ANN. Nabih *et al.* 2017 in their study, proposed an algorithm for heart disease prediction based on phono-cardiogram (PCG)signals. PCG signals are used for training and checking the proposed algorithm. Discrete Wavelet Transform (DWT)method eliminates the noise and extracts the features. Signal processing and neural networks toolboxes in MATLAB help to recognize the normal and abnormal ECG data. The PCG signals are classified using the neural network. Li *et al.* 2020 developed a system with an electrocardiogram that makes dynamic monitoring in real-time. The ECG wave form and characteristic parameter for the dynamic ECG signals are calculated using the wavelet transform method with an excellent filtering effect. The system detects a little change in ECG signals during specific symptoms for a short time and gives necessary instructions to doctors. Abdou *et al.* 2018 proposed a hybrid model to research arrhythmia very clearly. ECG signals are used as input layers; the neural network and linear regression predict heart disease. The proposed model results with higher accuracy in predicting heart diseases like tachycardia and bradycardia. Nanarkar & Chawan 2018 proposed a model to detect arrhythmia. Raw ECG signals that include noise were denoised using three different techniques: medians, moving average and notch filters, and by retrieving exactly nine features, it is fed as input. The database of gathered features is split into training and testing halves, and SVM and ANN classifiers are then applied to it. The authors concluded that the system can identify the type of arrhythmia disease by using a classification technique. In their paper, Wale *et al.* 2017 proposed a system that analysed ECG data to classify heart disease using ANN and GA. This study includes GA instead of the back-propagation algorithm because it has fast convergence, short time training, and predictive heart disease with high accuracy. In their research work, Gligorijevie *et al.* 2017 aimed to determine the risk level of death using ANN for patients suffering from acute myocardial infarction. The patients' following data were collected for three years; ECG monitoring of patients experienced 24 hours, short ECG tests, noninvasive beat-to- beat heart rate variability, and bar reflex sensitivity. The proposed classifier model gives the prediction with an accuracy of 88%, and it is concluded that artificial neural networks can indicate the patients with a higher risk of death. Mukherjee & Sharma 2019 strive to present their model for predicting cardiovascular diseases using ECG analysis and symptom-based detection. The researchers have built a multi-class deep neural network written in crystal, and the multi-layer model has ten input layer nodes, ten hidden layer nodes, and two output layers. The model has a promised accuracy of 97%. Rahim *et al.* 2016 analyzed cardio signals based on a radial base network using MATLAB. Algorithms were defined based on norm signals, bradycardia, and tachycardia. The radial neural network consists of 3 layers. First is the input layer: which transfers input signals, and the second, a hidden layer that consists of neuronal radial type, and the last is, as usual, the output layer performs a weighted summation result in the operation of the hidden layer. Savalia *et al.* 2017 implemented an algorithm using MATLAB, which is tested on ECG signals. The algorithm could distinguish 86% of Normal and Arrhythmia data successfully using neural networks to classify the signals. Also, the heart rate was calculated to detect diseases like Tachycardia and Bradycardia. The EGC signals of a patient, when combined with other risk factors as input to an ANN-based algorithm, provided high accuracy in predicting CVD risk.

2.5 *Several studies using ANN for heart disease prediction*

In their study, Yazdani & Ramakrishnan 2016 proposed the system based on different artificial neural network models. It uses standard heart disease data to find the best accuracy among the accuracy measures. The system provides an interface for doctors to predict heart diseases. Awan *et al.* 2018 discussed several different types of ANN, which give different accuracy rates. It is shown that before applying Principal Component Analysis (PCA), accuracy is 94.7%, and after applying it is up to 97.7%, it concludes that the accuracy can be further increased by changing the setting and making them more optimized according to each algorithm and based on the nature of the data. Mathan *et al.* 2018 used data mining methods to select the best predicting heart disease attributes. The paper showed that the neural networks and Gini index prediction model provide higher accuracy. The research paper comes to the conclusion that a decision tree built on classifiers for the neural system provides a notable degree of accuracy for predicting heart disease. Kim & Kang 2017 used neural network-feature correlation analysis to build a predictor of coronary heart disease risk. It works in two steps; the first can be called the features election stage in which the features are ranked after acceding to the importance in predicting the coronary heart disease, and in the second step, which is called the feature correlation analysis stage, the existence of correlations between the feature relations and the data of each neural network predictor output is determined. It is claimed that the proposed model will make better the coronary heart disease risk and decision support for the suitable treatment. Kuznetsov *et al.* 2018 aim to find the predictors of significant coronary lesions in patients with myocardial perfusion

abnormalities by the framework called SPECT using an ANN. An artificial neural network can diagnose heart disease in patients with abnormal SPECT with a good accuracy rate. The study constructs different neural network models by changing the hidden nodes and layers. Isma'eel *et al.* 2018 in their research article, compared the ANN model with two well-known predictive models the Diamond-Forrester and Morise Scores. ANN used stress imaging to predict ischemia. Research work compared an ANN model with the Diamond-Forrester and Moriserisk scores; the ANN model has a higher discriminatory power in forecasting ischemia with stress echocardiography and radionuclide stress testing. Suresh 2017 proposed to help doctors make a diagnosis and predict heart disease and prescribe the medicine based on predicted disease. Two methods are used for assessment, ANN, by testing the datasets, Case-Based Reasoning (CBR) image similarity search by drawing a comparison with stored images in the heart disease prediction database. The evaluation of CBR is also implemented for prescribing medicine from the history of old patients with Generalized Regression Neural Network and Radial basis function successfully. Kouser *et al.* 2018, including ANN, used integrated CBR techniques to increase their research accuracy. However, CBR is used to increase accuracy and forecast heart disease, compared with the RBF (Radial Basis Function). The studies using artificial neural networks and CBR helps to diagnose and predict CVD risk and prescribe medicines.

2.6 *Studies that evaluate the application of an artificial neural network with a multilayer perceptron neural network in the diagnosis and prediction of heart disease*

Atkov *et al.* 2012 tried to build a diagnostic model to diagnose the coronary heart disease risk using the data of clinical and hemodynamics factors and aortic pulse wave velocity index. Based on the ANNs topology of multi-layer perception, the best accuracy is gained with the three hidden layers. In their paper, Rufai *et al.* 2018 implied a Multilayer perceptron to predict the patient's heart disease status. To determine the optimal network parameters, several experiments were carried out. The proposed system achieved a diagnosing accuracy of 92.2%. Matis 2017 used the Multilayer Perceptron Classifier (MLP) to make heart disease predictions. Their research primarily works on determining the correct parameter settings for MLP, which are tested by using k-NN and LDA algorithms. Figure 5 shows a system for predicting heart disease using ANN, RBF, and CBR.

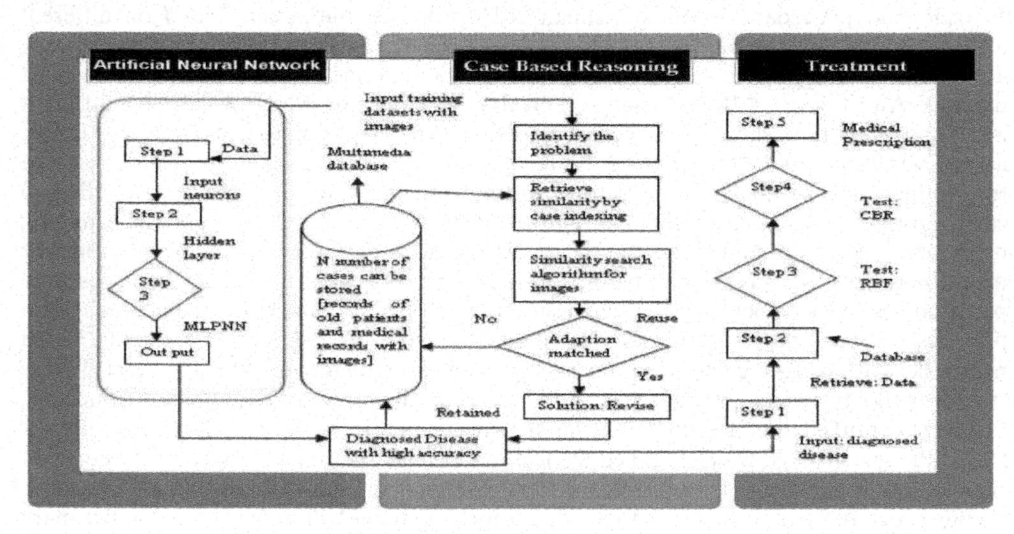

Figure 5. A system for predicting heart disease using ANN, RBF, and CBR.

The investigation shows that MLP gives almost 97% accuracy for heart disease prediction. Moghaddasi *et al.* 2017 in their research work, studied the Multilayer Perception by employing the data of Iran's heart centre. Kouser 2018 evaluated the application of artificial neural networks with a Multilayer perceptron to diagnose heart-disease and its type. Besides increasing the accuracy, the Case-Based Reasoning (CBR) technique is integrated with ANN. In addition to CBR's output, the medicine prescribed is used to know the medicine by comparing it with the original medicine and the RBF (Radial Basis Function). Kabirirada & Kardanmoghaddamb 2016 used a multi-layer perceptron and radial basis function. Authors compared their results with the other network models like SVM, PCA, and GFF gained by other authors. The Switzerland, Hungarian, Cleveland database is divided into three parts; 70% are used for training, 15% for testing, and the last 15% for validation purposes. There are 13 input layers and only one output layer in the proposed method. In the proposed method, the RBF neural network gives an accuracy of 95%, and MLP predicts coronary artery disease with an accuracy of 98%, which outperforms the other models. Several studies have evaluated applying Multi-layer perceptron with a back propagation algorithm to diagnose and predict heart disease. Kaur & Arora 2018 in their research work, present a system using data mining methods and ANN to predict heart disease. ANN is used to extend a multi-layer neural network with back propagation; this model shows the best early diagnosis results. The results show a neural network-based system diagnoses heart disease with high accuracy. Umasankar & Thiagarasu 2018 proposed a system called "TFIE- PSFS" to predict heart disease using a multi-layer perceptron and back propagation. The ANN model gives three outputs: "Yes" for a patient suffering from cardiovascular diseases, "No" for a patient not having heart diseases, and "Hesitant" for a patient, lies between the categories "Yes" and "No." Chaitanya *et al.* in their study, analyze parameters of Error-Back propagation that reach the accuracy of 88.5% for diagnosing heart disease. In their paper, Tarle & Jenal 2017 used the artificial neural network with back-propagation for predicting cardiovascular diseases and forecasting the probability of diseases like swine flu. Karayılan, & Kılıç 2017 also proposed the system that used ANN with back propagation to predict the heart disease while taking thirteen standard clinical features as input to the network; the network is trained using an algorithm to forecast the heart disease in patients with an accuracy of 95%. Burse *et al.* 2019 used different methods for ANN. They proposed a system for medical diagnosis using a standard back propagation algorithm. As an activation function, the bipolar sigmoid function is employed. Normalization, Principal Component Analysis, and LDA were employed for data pre-processing. In this work, a support vector machine model using LDA is also proposed. Kalra *et al.* 2018 in their research paper, proposes a system that uses the back propagation algorithm of an artificial neural network. After training 70% of the data, it gives an accuracy of 91% that claims to help doctors to detect heart disease. Turabieh 2016 in his research work, using the MATLAB tool, tries to increase the efficiency of the back propagation algorithm. For this, a new hybrid algorithm is suggested, which is made by combining ANN and GWO. In order to increase the convergence speed, GWO supports ANN to find the best initial weights and biases. In other words, GWO becomes a global search algorithm, as well as back propagation, which is a local search algorithm. Mhatre & Varma 2019 worked on a data set of the University of California at Irvine machine learning repository, trained using evolutionary-based back propagation neural network algorithm for disease prediction. Data is classified into several classes, and the last weights are saved in the weighted base. All attributes are used to forecast cardiovascular disease. The accuracy of the model is 78.763%. Poornima & Gladis 2018 proposed a heart disease prediction system that employed MLP with a back-propagation algorithm. The system has 15 input layers and works on are coded database of 303 records. The system generates 100% accuracy in heart disease prediction. Lee *et al.* 2016 in their study, developed a model to forecast ventricular tachycardia one hour before it starts. The researchers build an ANN model by using heart rate variability and respiratory rate variability parameters. First, an ANN model is created with 11 heart rate variability parameters, and second with three respiratory rate variability parameters. In both cases, an ANN model with five hidden neurons in one hidden layer gave accuracies of 73.5% and 82.4%, respectively. Finally, an ANN is built with one hidden layer having 13 hidden neurons with all heart and respiratory rate variability parameters, and the accuracy of the test shows 85.3% accuracy. Samuel *et al.* 2017 in their study, also offered a system in which two algorithms are used, such as the fuzzy process of the analytical hierarchy (Fuzzy AHP) and an artificial neural network. The first method, based on the attributes' contribution, is used to calculate the global weights. These scales are then used to train the ANN classifier to predict the failure risk in patients. In the proposed method, each of the attributes is considered a separate risk factor based on the medical data, and the result shows that it can forecast heart disease with an accuracy of 91.10%. The ANN classifiers in the study are trained and retrained until the networks' performance becomes relatively stable. In their paper, Shinde *et al.* 2017 used two neural network algorithms: back propagation and multilayered feed-forward to diagnose heart disease in several steps, and in the end, the algorithms show 92% accuracy.

2.7 *Several researchers have used MATLAB Neural Network Toolbox for analyzing data for predicting heart disease*

Mukherjee & Sharma 2019 exhibited a system to diagnose coronary illness by utilizing the significant dangerous elements. The method includes the two best data mining instruments, neural systems and hereditary calculations. The system was developed using MATLAB and they predicted the severity of the coronary disease with higher accuracy. The studies using MLP artificial neural networks have a very high degree of accuracy.

2.8 *Studies of Deep Neural Networks (DNN) for heart disease prediction*

Tomov & Tomov 2018 in their research work, investigated and used Deep Neural Networks (DNN)-based data analysis using routine clinical data in heart disease detection with an accuracy of 99% and developed a software called HEARO-5. Vijayashree & Sultana 2020 presented a trained deep neural network to investigate heart disease using Hybridized Ruzzo-Tompamemetic. After deleting the noise and reducing the overfitting of data, necessary features are elected to build and train a deep neural network, after which trained features are implemented to predict heart disease with higher accuracy. In their research work, Miao & Miao 2018 offers two models, a classification model and a diagnostic model consisting of two parts: a learning model based on learning in a deep neural network and a predictive model for the presence of heart disease. Based on the deep learning algorithms, the diagnosis model is used for predicting coronary heart disease. Asha *et al.* 2018 used neural networks and clustering to predict heart disease by taking the input as heart images and comparing them with the images in the data set. Noise is removed from the image using a median filter. After training, the prediction of heart disease with high accuracy is gained. Zhang & Han 2017 in their study, developed the system for the prediction of cardiovascular disease using the convolution neural network. Heart cycles are received from the heart signals at their various positions and spectrograms of these cycles are represented in proportional dimensions as input to CNN. Convolution neural networks can analyze different heart positions; the proposed system does not need heart signal segmentation. In the end, SVM makes the last decision for classifying the disease.

3 DISCUSSION AND ANALYSIS

Some effective techniques are elaborated on in this section for analysis.

3.1 *Kim, Changgyun, Youngdoo Son, and Sekyoung Youm. (2019) Chronic Disease Prediction Using Character-Recurrent Neural Network in The Presence of Missing Information. Applied Sciences 9, no. 10: 2170. https://doi.org/10.3390/app9102170*

In this work, the authors selected five widespread chronic diseases and related variables, including osteoarthritis, rheumatoid arthritis, osteoporosis, tuberculosis, asthma, thyroid disease, cancer, inflammation, and hepatitis. Pre-treatment splits the diseases and variables as critical variables. Then, regression analysis is used to determine the disease's variables and a prepossessed dataset is used to train Char-RNN.

3.2 *Kompella, Subhadra & Boddu, Vikas. (2019). Neural Network Based Intelligent System for Predicting Heart Disease. International Journal of Innovative Technology and Exploring Engineering. 8. 484–487*

Heart disease is predicted using a multilayered neural network- based method in two stages. Fourteen clinical attributes were fed as input, the network is trained on a training data-set using backpropagation learning algorithm. Figure 6 demonstrates the working of this schema.

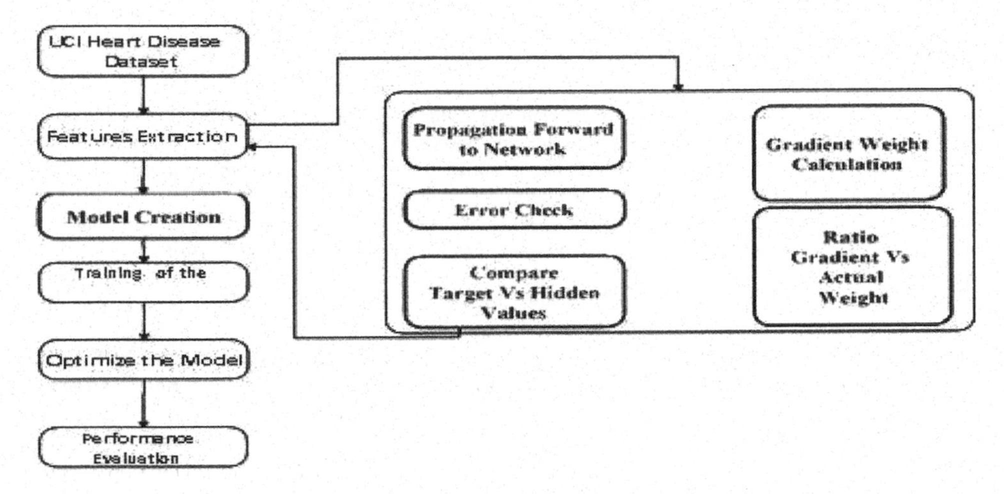

Figure 6. Heart disease prediction architectural block of K. Subhadra and Vikas B scheme.

4 CONCLUSION AND FUTURE SCOPE

The analysis of papers in this research concludes that the significant factors affecting heart disease are classified into four socio-demographic characteristics: family history, medical history, environmental risk factors, and dietary/lifestyle factors. Such factors can be further subclassified as non-adjustable factors like race, family history, age, gender, marital status, income level, etc. and modifiable factors like weight, high obesity, sleep duration per day, BMI, ECG Signals, smoking, drinking alcohol, drinking soda(sugar), eating fruits, eating vegetables, doing exercises, diabetes, hypertension, total cholesterol level, triglycerides, oral contraceptives, hyperlipidemia, dyslipidemia, high density lipoprotein, low-density lipoprotein, coronary artery bypass grafting (CABG), high serum, hemodynamics factors, and aortic pulse wave velocity index, etc. The literature review also concludes that an artificial neural network algorithm is robust in heart disease risk prediction. With unique models and relevant medical data, some researchers have achieved higher accuracy than ninety percent. It can also be concluded that most of the research works use the same data-set, taken from the centre of UCI, and it is hypothesised that the results might vary by using other available standard data-sets along with the data-set from the centre of UCI as well as a self-developed data set by collecting data from the varied small-sized population.The number of cases and causes of heart disease increases in the population; therefore, the goal is to develop a hybrid model using ANN that always gives high accuracy on the different data sets. The future study should also estimate the results using three or more data sets, two or more standard data sets, and one self-developed data set by collecting data from varied small-sized populations and estimating their cumulative accuracy. The selected data sets for the study should be such that it includes most of the factors affecting heart disease, as concluded above. Several studies for heart disease prediction using a hybrid classifier involving ANN use Phonocardiogram (PCG) and Electrocardiogram (ECG) signals, after eliminating the noise and extracting the features from the PCG/ECG signals using the wavelet transformation method. Therefore, the future study should also use features extracted from the PCG/ECG signals. The study shows that Artificial Neural Network (ANN) is one such progressive branch of computer science that gets computers to make decisions like humans and can be effectively applied to various medical science domains, especially cardiology. The future study should be easy to comprehend and implement with low computational and storage costs and the real world's deployment opportunities on a large population.

REFERENCES

Abdou AD, Ngom NF, Niang O. 2018. *Classification and Prediction of Arrhythmias from Electrocardiograms Patterns Based on Empirical Mode Decomposition and Neural Network*. In: Springer. 174–184.

Adhikari NCD. 2018. Prevention of Heart Problem Using Artificial Intelligence. *International Journal of Artificial Intelligence and Applications (IJAIA)* 9(2).

Ahmed & Verma, 2017 Prediction of Heart Diseases using Artificial Intelligence. *International Journal of Advanced Research in Computer and Communication Engineering* ISO 3297:2007 Certified, vol. 6, no. 6, June.

Alam KMR, Siddique N, Adeli H. 2020. A Dynamic Ensemble Learning Algorithm for Neural Networks. *Neural Computing and Applications*; 32(12): 8675–8690.

Albu A, Precup RE, Teban TA. 2019. Results and Challenges of Artificial Neural Networks Used for Decision-making and Control in Medical Applications. *Facta Universitatis, Series: Mechanical Engineering*; 17(3): 285–308.

Amin *et al.* 2013. Genetic Neural Network Based Data Mining in Prediction of Heart Disease Using Risk Factors. *IEEE Conference on Information & Communication Technologies*, Thuckalay, India, pp. 1227–1231, doi: 10.1109/CICT.2013.6558288.

Anand D, Khemchandani V, Sabharawal M, Cheikhrouhou O, Ben Fredj O. 2021. Lightweight Technical Implementation of Single Sign-On Authentication and Key Agreement Mechanism for Multiserver Architecture-Based Systems. *Security and Communication Networks*.

Anand D, Khemchandani V, Sharma RK. 2013. *Identity-based Cryptography Techniques and Applications (A Review)*. In: IEEE.; 343–348.

Anand D, Khemchandani V. 2017. An Analytical Method to Audit Indian E-governance System. *International Journal of Electronic Government Research (IJEGR)*; 13(3): 18–37.

Anand D, Khemchandani V. 2019a. Identity and Access Management Systems. *Security and Privacy of Electronic Healthcare Records: Concepts, Paradigms and Solutions*: 61.

Anand D, Khemchandani V. 2019b. Study of E-governance in India: A Survey. *International Journal of Electronic Security and Digital Forensics*; 11(2): 119–144.

Anand D, Khemchandani V. 2019c. Unified and Integrated Authentication and Key Agreement Scheme for E-governance System Without Verification Table. *Sādhanā*; 44(9): 1–14.

Anand D, Khemchandani V. 2020. *Data Security and Privacy Functions in Fog Computing for Healthcare 4.0*. In: Springer. (pp. 387–420).

Asha P, Sravani B, SatyaPriya P. 2018. *Heart Block Recognition Using Image Processing and Back Propagation Neural Networks*. In: Springer.; 2018 200–210.

Atkov OY, Gorokhova SG, Sboev AG, *et al.* 2012. Coronary Heart Disease Diagnosis by Artificial Neural Networks Including Genetic Polymorphisms and Clinical Parameters. *Journal of Cardiology*; 59(2): 190–194.

Awan *et al.* (2018) Prediction of Heart Disease Using Artificial Neural Network. *Vfast Transactions on Software Engineering*. http://vfast.org/journals/index.php/VTSE@2018 ISSN(e): 2309-3978; ISSN(p): 2411–6246. Volume 6, Number 1, January-December.

Bhalerao S, Gunjal B. 2013. Hybridization of Improved K-Means and Artificial Neural Network for Heart Disease Prediction. *Int.J. Comput. Sci. Trends Technol*; 4(3): 5461.

Bhaskaru O, Sree M. 2019. Accurate and Fast Diagnosis of Heart Disease Using Hybrid Differential Neural Network Algorithm. *Int.J. Eng. Adv. Technol*; 8(3S): 452–457.

Bhatla N, Jyoti K. 2012. An Analysis of Heart Disease Prediction Using Different Data Mining Techniques. *International Journal of Engineering*; 1(8): 1–4.

Burse K, Kirar VPS, Burse A, Burse R. 2019. *Various Preprocessing Methods for Neural Network Based Heart Disease Prediction*. In: Springer. (pp. 55–65).

Costa W, Figueiredo L, Alves E. 2019. *Application of an Artificial Neural Network for Heart Disease Diagnosis*. In: Springer.; 753–758.

Desai SD, Giraddi S, Narayankar P, Pudakalakatti NR, Sulegaon S. 2019. *Back-propagation Neural Network Versus Logistic Regression in Heart Disease Classification*. In: Springer. (pp. 133–144).

Dubey S, Anand D, Sharma J. 2018. Lfsr Based Block Cipher Technique for Text. *International Journal of Computer Sciences and Engineering*; 6(2): 53–60.

Durgadevi P, Vijayalakshmi S, Sabharwal M. 2021. Fetal Brain Abnormality Detection Through PSO (Particle Swarm Optimization) and Volume Estimation. *Annals of the Romanian Society for Cell Biology*: 2700–2714.

Dwivedi AK. 2018. Performance Evaluation of Different Machine Learning Techniques for Prediction of Heart Disease. *Neural Computing and Applications*; 29(10): 685–693.

Esfahani HA, Ghazanfari M. 2017. *Cardiovascular Disease Detection Using a New Ensemble Classifier*. In: IEEE.; 1011–1014.

Gaharana S, Anand D. 2016. A New Approach for Remote User Authentication in a Multi-server Environment Based on DYNAMIC-ID Using SMART-CARD. *International Journal of Computer Network and Information Security*; 8(10): 45.

Gligorijevis T, Ševarac Z, Milovanovis B, *et al.* 2017. Follow-up and Risk Assessment in Patients With Myocardial Infarction Using Artificial Neural Networks. *Complexity*.

Guo C, Zhang J, Liu Y, Xie Y, Han Z, Yu J. 2020. Recursion Enhanced Random Forest with an Improved Linear Model (RERF-ILM) for Heart Disease Detection on the Internet of Medical Things Platform. *IEEE Access*; 8: 59247–59256.

Hannun AY, Rajpurkar P, Haghpanahi M, *et al.* 2019. Cardiologist-level Arrhythmia Detection and Classification in Ambulatory Electrocardiograms Using a Deep Neural Network. *Nature Medicine* 25(1): 65–69.

Hearn J, Ross HJ, Mueller B, *et al.* 2018. Neural Networks for Prognostication of Patients With Heart Failure: Improving Performance Through the Incorporation of Breath-by-breath Data from Cardiopulmonary Exercise Testing. *Circulation: Heart Failure*; 11(8): e005193.

Heer S, Anand D. 2020. An Improved Hand Gesture Recognition System Based on Optimized msvm and Sift Feature Extraction Algorithm. *Tech. Rep., EasyChair*;

Husain K, Mohd Zahid MS, Ul Hassan S, Hasbullah S, Mandala S. 2016. Advances of ECG Sensors from Hardware, Software and Format Interoperability Perspectives. *Electronics* 2021; 10(2): 105.

Isma'eel HA, Sakr GE, Serhan M, *et al.* 2018. Artificial Neural Network-based Model Enhances Risk Stratification and Reduces Non-invasive Cardiac Stress Imaging Compared to Diamond–Forrester and Morise Risk Assessment Models: A Prospective Study. *Journal of Nuclear Cardiology*; 25(5): 1601–1609.

Jadon P, Anand D, Sharma J. 2017. CAPTCHA as Graphical Password: A Novel Approach to Enhance the Security in WWW. *International Journal of Latest Technology in Engineering, Management and Applied Science (IJLTEMAS)* VI(VIIIS): 62–66.

Kabirirada & Kardanmoghaddamb. 2016. Heart Disease Prediction by Using Artificial Neural Networks. *International Journal of Computer Science and Information Security*, vol. 14, no. 1, 2016.

Kalra A, Tomar R, Tomar U. 2018. *Heart Risk Prediction System Based on Supervised ANN*. In: Springer.: 405–413.

Karayılan T, Kılıç Ö. 2017. *Prediction of Heart Disease Using Neural Network*. In: IEEE.;: 719–723.

Kaur A, Arora J. 2018. Heart Disease Prediction Using Data Mining Techniques: A Survey. *International Journal of Advanced Research in Computer Science* 9(2).

Khourdifi, Youness and Mohamed Bahaj. 2019. Heart Disease Prediction and Classification Using Machine Learning Algorithms Optimized by Particle Swarm Optimization and Ant Colony Optimization. *International Journal of Intelligent Engineering and Systems* (2019): pp.242–252.

Kim JK, Kang S. 2017. Neural Network-based Coronary Heart Disease Risk Prediction Using Feature Correlation Analysis. *Journal of Healthcare Engineering*.

Kour H, Sabharwal M, Suvanov S, Anand D. *An Assessment of Type-2 Diabetes Risk Prediction Using Machine Learning Techniques*. In: Springer.; 2021: 113–122.

Kouser RR, Manikandan T, Kumar VV. 2018. Heart Disease Prediction System Using Artificial Neural Network, Radial Basis Function and Case Based Reasoning. *Journal of Computational and Theoretical Nanoscience*; 15(9–10): 2810–2817.

Krishnasree K, Rao MN. 2016. Diagnosis of Heart Disease Using Neural Networks-comparative Study of Bayesian Regularization with Multiple Regression Model. *Journal of Theoretical and Applied Information Technology*; 88(3): 638–643.

Kuznetsov V, Yaroslavskaya E, Krinochkin D, Teffenberg D, Dyachkov S. 2018. Artificial Neural Network for Prediction of Coronary Artery Disease in Patients with Abnormal Myocardial Perfusion. *European Journal of Heart Failure*; 20(S1): 484–484.

LaFreniere D, Zulkernine F, Barber D, Martin K. 2016. *Using Machine Learning to Predict Hypertension from a Clinical Dataset* In: IEEE.; 1–7.

Lee H, Shin SY, Seo M, Nam GB, Joo S. 2016. Prediction of Ventricular Tachycardia One Hour Before Occurrence Using Artificial Neural Networks. *Scientific Reports*; 6(1): 1–7.

Lee U, Han K, Cho H, *et al.* 2019. Intelligent Positive Computing with Mobile, Wearable, and IoT Devices: Literature Review and Research Directions. *Ad Hoc Networks*; 83: 8–24.

Li H, Luo M, Zheng J, *et al.* 2017. An Artificial Neural Network Prediction Model of Congenital Heart Disease Based on Risk Factors: A Hospital-based Case-control Study. *Medicine* 96(6).

Li Y, Yang M, Liu Z, *et al.* 2020. Detection and Diagnosis of Myocarditis in Young Patients Using ECG Analysis Based on Artificial Neural Networks. *Computing*; 102(1): 1–18.

Maheswari KU, Jasmine J. 2017. Neural Network Based Heart Disease Prediction. *International Journal of Engineering Research & Technology (IJERT)* 5(17): 1–4.

Mähringer-Kunz A, Wagner F, Hahn F, *et al.* 2020. Predicting Survival After Transarterial Chemoembolization for Hepatocellular Carcinoma Using a Neural Network: A Pilot Study. *Liver International*; 40(3): 694–703.

Malav A, Kadam K, Kamat P. 2017. Prediction of Heart Disease Using k-means and Artificial Neural Network as Hybrid Approach to Improve Accuracy. *International Journal of Engineering and Technology*; 9(4): 3081–3085.

Malav A, Kadam K. 2018. A Hybrid Approach for Heart Disease Prediction Using Artificial Neural Network and K-means. *International Journal of Pure and Applied Mathematics*; 118(8): 103–10.

Mathan K, Kumar PM, Panchatcharam P, Manogaran G, Varadharajan R. 2018. A Novel Gini Index Decision Tree Data Mining Method with Neural Network Classifiers for Prediction of Heart Disease. *Design Automation for Embedded Systems*; 22(3): 225–242.

Matis V, *et al.* 2017. *Effective Diagnosis of Heart Disease Presence Using Artificial Neural Networks*. In: Singidunum University.; 3–8.

Mehanovis D, Mašetis Z, Kešo D. 2019. *Prediction of Heart Diseases Using Majority Voting Ensemble Method.* In: Springer.; 491–498.

Menotti *et al.* 2015. Lifestyle Behavior and Lifetime Incidence of Heart Diseases. *International Journal of Cardiology*. Volume 201, 15 December, Pages 293–299.

Mhatre & Varma. 2019. Heart Disease Prediction using Evolutionary based Artificial Neural Network. *International Journal of Engineering Research & Technology (IJERT)* Vol. 8 Issue 08, August.

Miao KH, Miao JH. 2018. Coronary Heart Disease Diagnosis Using Deep Neural Networks. *Int. J. Adv. Comput. Sci. Appl*; 9(10): 1–8.

Mirbabaie M, Stieglitz S, Frick NR. 2021. Artificial Intelligence in Disease Diagnostics: A Critical Review and Classification on the Current State of Research Guiding Future Direction. *Health and Technology*: 1–39.

Mirian N, Sedehi M, Kheiri S, Ahmadi A. 2017. Joint Prediction of Occurrence of Heart Block and Death in Patient with Myocardial Infarction with Artificial Neural Network Model. *Koomesh*; 19(1): 241–247.

Mirian NS, Sedehi M, Kheiri S, Ahmadi A. 2016. A Hybrid ANN-GA Model to Prediction of Bivariate Binary Responses: Application to Joint Prediction of Occurrence of Heart Block and Death in Patients with Myocardial Infarction. *Journal of Research in Health Sciences*; 16(4): 190.

Moghaddasi H, Ahmadzadeh B, Rabiei R, Farahbakhsh M. 2017. Study on the Efficiency of a Multi-layer Perceptron Neural Network Based on the Number of Hidden Layers and Nodes for Diagnosing Coronary-Artery Disease. *Jentashapir Journal of Health Research*; 8(3).

Mohan S, Thirumalai C, Srivastava G. 2019. Effective Heart Disease Prediction Using Hybrid Machine Learning Techniques. *IEEE Sccess*; 7: 81542–81554.

Mukherjee S., Sharma A. 2019. Intelligent Heart Disease Prediction using Neural Network. *International Journal of Recent Technology and Engineering (IJRTE)* ISSN: 2277-3878, Volume-7 Issue-5, January.

Munish S, Sohan G. 2016. The Summation of Potential Biometric Types and Technologies for Authentication in E-banking. *Int. J. Sci. Rev. Res. Eng. Technol. (IJSRRET)* 1(2): 83–92.

Nabih-Ali M, El-Dahshan ESA, Yahia AS. 2017. A Review of Intelligent Systems for Heart Sound Signal Analysis. *Journal of Medical Engineering & Technology*; 41(7): 553–563.

Nanarkar HM, Chawan PM. 2018. A Survey on Classification and Identification of Arrhythmia Using Machine Learning Techniques. *International Research Journal of Engineering and Technology*; 5(10): 446–449.

Nashif S, Raihan MR, Islam MR, Imam MH. 2018. Heart Disease Detection by Using Machine Learning Algorithms and a Real-time Cardiovascular Health Monitoring System. *World Journal of Engineering and Technology*; 6(4): 854–873.

Nikhar S, Karandikar A. 2016. Prediction of Heart Disease Using Machine Learning Algorithms. *International Journal of Advanced Engineering, Management and Science*; 2(6): 239484.

Niveditha V, Ananthan T, Amudha S, Sam D, Srinidhi S. 2020. Detect and Classify Zero Day Malware Efficiently in Big Data Platform. *International Journal of Advanced Science and Technology*; 29(4s): 1947–1954.

Poornima V, Gladis D. 2018. A Novel Approach for Diagnosing Heart Disease with Hybrid Classifier. *Biomed Res*; 29(11): 2274–2280.

Prerana PS, Taneja K. 2015. Predictive Data Mining for Diagnosis of Thyroid Disease Using Neural Network. *International Journal of Research in Management, Science & Technology*; 3(2): 75–80.

Rahim M, Yelena R, Sevinc A. 2016. Identification and Prediction of Heart Disease Based on the Analysis Electrocardiosignals Using a Neural Network. *World Journal of Research and Review (WJRR)* ISSN: 2455-3956, Volume-3, Issue-1, July. Pages 24–27

Reddy MPSC, Palagi MP, Jaya S. 2017. Heart Disease Prediction Using Ann Algorithm in Data Mining. *International Journal of Computer Science and Mobile Computing*; 6(4): 168–172.

Rufai A. *et al.* (2018). Using Artificial Neural Networks to Diagnose Heart Disease. *International Journal of Computer Applications* (0975 – 8887) Volume 182 – No. 19, October.

Sabharwal M. 2016. *Contemporary Research: Intricacies and Aiding Software Tools Based on Expected Characteristics*. In: 28–29.

Sabharwal M. 2018. *The Use of Soft Computing Technique of Decision Tree in Selection of Appropriate Statistical Test for Hypothesis Testing*. In: Springer. (pp. 161–169).

Sajja TK, Kalluri HK. 2020. A Deep Learning Method for Prediction of Cardiovascular Disease Using Convolutional Neural Network. *Rev. d'Intelligence Artif.*; 34(5): 601–606.

Samuel OW, Asogbon GM, Sangaiah AK, Fang P, Li G. 2017. An Integrated Decision Support System Based on ANN and Fuzzy_AHP for Heart Failure Risk Prediction. *Expert Systems with Applications*; 68: 163 172.

Savalia S, Acosta E, Emamian V. 2017. Classification of Cardiovascular Disease Using Feature Extraction and Artificial Neural Networks. *Journal of Biosciences and Medicines*; 5(11): 64–79.

Savita SV, Sabharwal M. 2021. Alzheimer's Disease Detection Through Machine Learning. *Annals of the Romanian Society for Cell Biology*: 2782–2792.

Sharma M, Khan F, Ravichandran V. 2017. Comparing Data Mining Techniques Used For Heart Disease Prediction. *International Research Journal of Engineering and Technology*; 4: 1161–1167.

Shinde A, Kale S, Samant R, Naik A, Ghorpade S. 2017. Heart Disease Prediction System Using Multilayered Feed Forward Neural Network and Back Propagation Neural Network. *International Journal of Computer Applications*; 166(7): 32–36.

Sivaranjani R, Yuvaraj N. 2019. *Artificial Intelligence Model for Earlier Prediction of Cardiac Functionalities Using Multilayer Perceptron*. In:. 1362. IOP Publishing.; 012062.

Suresh A. 2017. Heart Disease Prediction System Using ANN, RBF and CBR. *International Journal of Pure and Applied Mathematics, (IJPAM)*; 117(21): 199–216.

Tanwar S, Vijayalakshmi S, Sabharwal M. 2021. Using Novel Method with Convolutional Neural Network for Colorectal Cancer Classification. *Annals of the Romanian Society for Cell Biology*: 2653–2671.

Tarle B. and Jena S. 2017. An Artificial Neural Network Based Pattern Classification Algorithm for Diagnosis of Heart Disease. *International Conference on Computing, Communication, Control and Automation (ICCUBEA)*, Pune, India. pp. 1–4, doi: 10.1109/ICCUBEA.2017.8463729.

Tomov NS, Tomov S. 2018. On Deep Neural Networks for Detecting Heart Disease. *arXiv* preprint arXiv:1808.07168.

Turabieh H. 2016. A Hybrid Ann-gwo Algorithm for Prediction of Heart Disease. *American Journal of Operations Research*; 6(2): 136–146.

Umasankar P, Thiagarasu V. 2018. A Novel Thrice Filtered Information Energy Based Particle Swarm Feature Selection for the Heart Disease Diagnosis. *International Journal of Pure and Applied Mathematics*; 119(15): 3485–3499.

Vijayashree J & Sultana HP. 2020. Heart Disease Classification Using Hybridized Ruzzo-Tompa Memetic Based Deep Trained. *Neocognitron Neural Network. Health and Technology*; 10(1): 207–216.

Vrbaški D, Vrbaški M, Kupusinac A, *et al.* 2019. Methods for Algorithmic Diagnosis of Metabolic Syndrome. *Artificial Intelligence in Medicine*; 101: 101708.

Wale AS, Sonawani SS, Karande SC. 2017. *ECG Signal Analysis and Prediction of Heart Attack with the Help of Optimized Neural Network Using Genetic Algorithm*. In:.

Willems SM, Abeln S, Feenstra KA, *et al.* 2019. The Potential Use of Big Data in Oncology. *Oral Oncology*; 98: 8–12.

Wu Y, Giger ML, Doi K, Vyborny CJ, Schmidt RA, Metz CE. 1993. Artificial Neural Networks in Mammography: Application to Decision Making in the Diagnosis of Breast Cancer. *Radiology*; 187(1): 81–87.

Yazdani, A., Ramakrishnan, K. (2016). Performance Evaluation of Artificial Neural Network Models for the Prediction of the Risk of Heart Disease. In: Ibrahim, F., Usman, J., Mohktar, M., Ahmad, M. (eds) *International Conference for Innovation in Biomedical Engineering and Life Sciences. ICIBEL 2015. IFMBE Proceedings*, vol 56. Springer, Singapore. https://doi.org/10.1007/978-981-10-0266-3_37

Zhang W, Han J. 2017. *Towards Heart Sound Classification Without Segmentation Using Convolutional Neural Network*. In: IEEE.; 1–4.

Artificial Intelligence, Blockchain, Computing and Security – Dagur et al. (Eds)
© 2024 The Author(s), ISBN: 978-1-032-67841-2

Breast cancer forecast and diagnosis using machine learning approaches: A comparative analysis

Ranjeet Kumar Dubey
Assistant Professor, Department of C.S.E., B.I.T., Gorakhpur

Rajesh Kumar Singh
Assistant Professor, Department of C.S.E., K.I.P.M., Gorakhpur

Shashank Srivastav & Anshu Kumar Dwivedi
Assistant Professor, Department of C.S.E., B.I.T., Gorakhpur

ABSTRACT: In recent times, breast cancer has been the disease that affects most women worldwide. Because people aren't aware of the first signs of cancer, the death rate from breast cancer keeps rising. It is already possible to identify breast cancer using various tools and environments with real-time algorithms. All training and classification fields have seen a rapid rise in terms of application of machine learning. In modern computer programming methods for categorizing breast cancer, the Deep Learning (DL) methodology is utilized to train a model employing a support vector machine and a Convolution Neural Network (CNN) for extracting the dominating features and detect breast cancer from test image samples. Using the dataset of the tissue cells utilized in the test samples, the automatic detection model divides the mammography images into 'Malignant' and 'Benign' breast cancers. This study compares the accuracy of the findings for identifying breast cancer for several kernels, such as sigmoid function, radial basis function, polynomial function, and linear function. With an accuracy of 95.78%, the linear kernel outperforms others. The training set of data contains a variety of tissue cell samples, and their recognition performance are tested until it exceeds future expectations.

Keywords: Machine Learning Algorithm, Breast Cancer, K-NN, Decision Tree, SVM

1 INTRODUCTION

One of the main reasons women aren't aware that they have cancer in its early stages is because of breast cancer. According to a survey, breast cancer is estimated to account for 39,350 women's deaths, and 245,650 new cases of the disease will be diagnosed in the United States in 2020. Breast cancer is a type of cancer that originates in the breast. Breast cancer develops when cells start to grow in an uncontrolled way. Breast cancer is currently surpassing lung cancer as the most common kind of cancer in women. Breast cancer can also start with creating a mass of cells known as a tumor. A malignant tumor has spread and is growing to other sections of the patient's body. The International Agency for Research on Cancer (ICRA) will publish statistics in December 2020. Breast cancer comes in numerous forms, but the most frequent are invasive carcinoma and ductal carcinoma in situ (DCIS). Breast cancer diagnoses have virtually quadrupled in the previous two decades [1]. The number of deaths from the disease has also climbed from 7.2 million in 2004 to 10.2 million in 2022. Invasive carcinoma is the most frequent type of breast cancer.

DOI: 10.1201/9781032684994-20

The successful application of enormously new Information and Communication Technologies (ICT) in the field of medical practice so that they can be used more particularly in cancer care is a vital stake in the new technology of use in the health system. There are numerous algorithms for classifying breast cancer and predicting its outcomes. Modelling and data mining is beneficial for predicting and classifying breast cancer. When applied to the healthcare industry, data mining plays a critical role in high performance, predicting breast cancer outcomes, cutting drug prices, improving patient diagnosis quality, and making real-time decisions to save lives. In this work, the performance of four classifiers is compared. The top five algorithms used in prediction tools include Random Forest (RF), Naïve Bayes (NB), k-Nearest Neighbours (KNN), Support Vector Machine (SVM) and Logistics Regression, which the scientific community considers to be the most useful data-mining algorithms [2]. Utilizing machine learning methods, our aim is to think about both the diagnosis and prognosis of breast cancer.

Section 2 presents the related methods and implementation of previous research and the findings of these algorithms are analyzed thoroughly. Our proposed work and research methodology are provided in Section 3. And the experiment results are shown in the second and final Section 4. Section 5 of this paper presents conclusion.

2 RELATED WORKS

Many studies on breast cancer have been published in recent years, and the majority of them include appropriate categorization for the best result. ML methods such as SVM, logistic regression, and KNN are prevalent methods compared to other methods used to predict breast cancer. By giving the proper treatment interventions at the right time, early identification of this condition could help slow the disease's progression and reduce fatality rates. Using various data sets, such as the SEER dataset program of the NCI, numerous researchers have gained insight into breast cancer, which provide information regarding female patients with lobular carcinoma and infiltrating duct breast cancer. The Wisconsin Dataset, as well as datasets from a variety of hospitals, comprise the images and labels or annotations for mammography scans. The authors complete the research by extracting and selecting the various features from datasets. There are significant research studies. Sudarshan Nayak [3] identified three-dimensional photos of breast cancer using several supervised learning techniques, and he concluded that SVM performed the best overall. Mohamed Bahaj and Youness Khoudfi [6] compared machine learning algorithms similarly and discovered that the SVM classifier performance is the best, surpassing KNN, RF, and NB with an accuracy of 97.49%. These methods are based on Multilayer Perceptron, which employs MLP to pass validation ten times and contains five layers. The author Latchoumiet TP [7] developed a Weighting Particle Swarm Optimization (WPSO) for the classification based on the SVM and discovered a KNN classification score of 96.8%. A. H. Osman proposes a proper explanation for the Wisconsin Breast-Cancer Diagnosis (WBCD) [8] by fusing a clustering technique with a powerful probabilistic vector support machine. The prediction that was made by the SVM method was 98.03% accurate. To determine the most efficient strategy for early detection and diagnosis of breast cancer, our research evaluates these using machine learning algorithms and methodologies.

3 METHODOLOGY

Modern technologies make predicting a woman's risk of acquiring breast cancer considerably more manageable and more accurate. Breast cancer prediction will be made via machine learning. We employed machine learning classifiers such as SVM, LR, and KNN on the WBCD to develop practical and predictive approaches for breast cancer screening. The architecture of the pre-processing work is detailed in Figure 1.

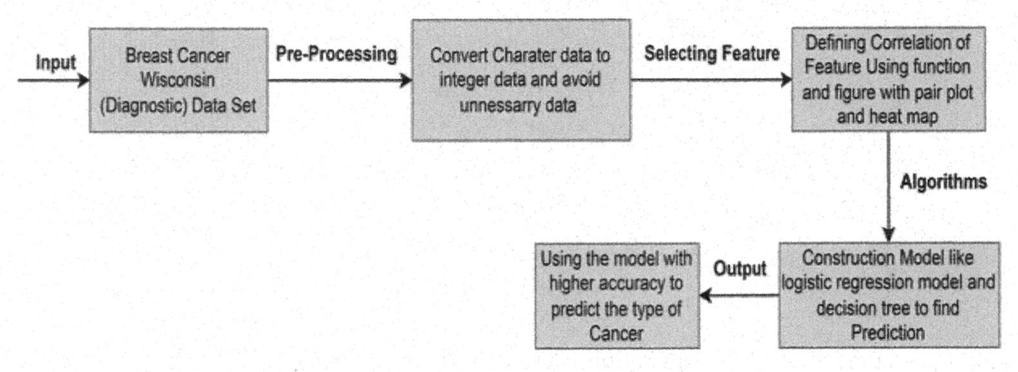

Figure 1. Block diagram of pre-processing work for breast cancer detection.

Breast cancer will be predicted using machine learning. In our dataset we have, 32 features (columns) and 570 items (rows). Our computer will be trained to recognize whether the found cancer is malignant or benign using the dataset's multivariate regression. Additionally, a logistic regression technique will be used to condense the attributes for speedier prediction by our computer. The remaining 550 data points will be utilized to train and test our computer, ensuring that it accurately predicts the kind of cancer.

3.1 *Machine learning methods*

We can use a machine learning method to perform efficient predictive analysis. The following are the ML algorithms used in our article:

a) **Support Vector Machine (SVM)** is a classifier that divides a dataset into multiple categories to locate the maximum marginal hyperplane (MMH) using the closest data points [9].
b) **The predictive modelling tool**, called decision trees, can be utilized in numerous fields. It can be built using an algorithm that can be divided into smaller datasets in various ways and based on multiple conditions because random decision forests are used to correct their training set's overfitting [10].
c) **K-Nearest Neighbors (K-NN)** is a supervised learning classification method. It teaches itself how to label new points by using a large number of labelled points. Before assigning a new tag, it looks at the labels on the issues [11].
d) **A Decision Tree** is a form of Classification which uses a discrete set of target variable values by building a decision tree model. Regression Trees are decision trees in which the goal variable has a constant value, usually a number [12]. Each node or leaf in these trees represents a class label, while the branches represent attribute combinations that lead to labels of classes.

3.2 *Dataset acquisition*

Our study uses the University of Madison (Wisconsin) Hospital's Breast Cancer Database's Breast cancer Wisconsin Diagnostic dataset [13,14]. The term "data acquisition" comprises two words: Data and Acquisition: Data is the unstructured or structured

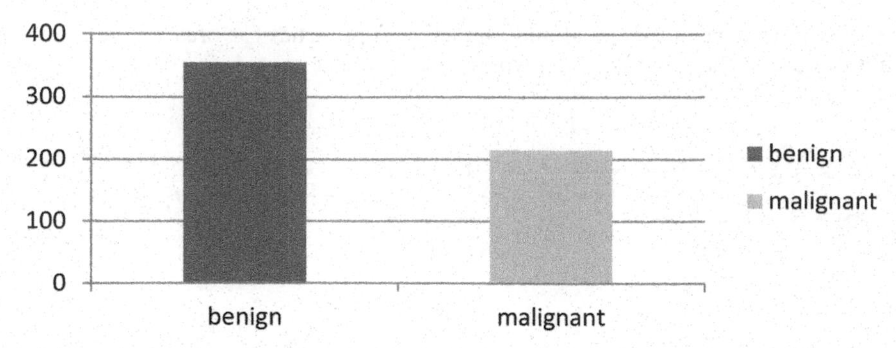

Figure 2. Breast cancer dataset acquisition result.

raw facts and figures, and acquisition is the process of acquiring data for a specific task. The dataset's features are derived from a digital breast cancer sample. There are 570 instances of the diagnostic (Benign: 355 Malignant: 215), data acquisition of two classes (62.28 percent benign and 37.11 percent malignant), utilized for additional mechanisms. The actual classification is depicted in Figure 2. It's the process of getting relevant organization data, putting it into the appropriate organization form, and loading it into the suitable system.

3.3 *Experimental setup*

All breast cancer studies on the application of machine learning methodologies addressed in this article used Scikitlearn and the Python computer language. Scikit-learn is a machine-learning library for the Python computer language, sometimes known as sklearn [15–17]. It includes clustering, regression and classification methods such as SVM, k-means, and DBSCAN. NumPy and SciPy are scientific and numerical libraries included with Python [18–20].

4 RESULT-DISCUSSION

Following the application of ML algorithms to WBCD Dataset. The Accuracy and the Confusion Matrix, Forecast, Sensitivity and AUC were used as a performance measure in machine learning to compare and determine the best breast cancer prediction model. A confusion matrix can be used to evaluate a classification issue in which the outcome might be one of the many classes. A table known as confusion matrix, contains True Negatives (TN), False Negatives (FN), True Positives (TP) and False Positives (FP) in each of the two dimensions. Based on the confusion matrix and other measure, following results are obtained and shown in Figure 3 and Table 1.

Table 1. Comparative analysis of breast cancer detection.

Metrics	SVM	KNN	Logistic Regression	Decision Tree
Accuracy	97.49	96.8	96.07	97.64
forecast	98.3	98.27	96.77	96.87
Sensitivity	91.06	91.47	94.23	98.47

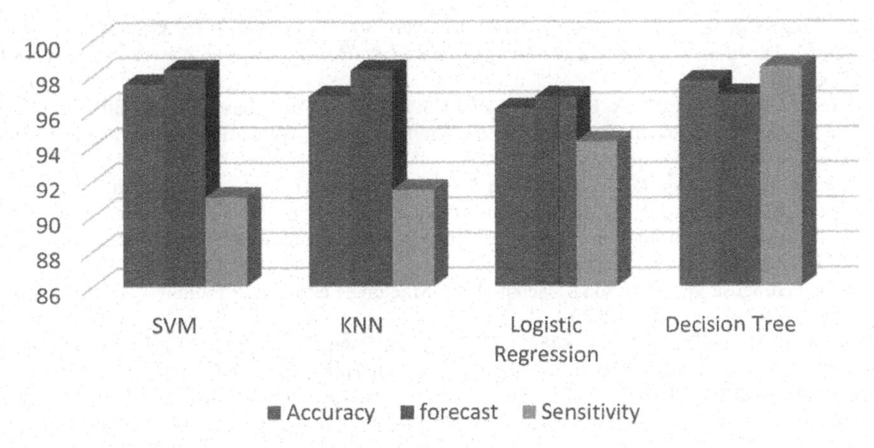

Figure 3. Tabular comparison of results with existing ML classification algorithms.

5 CONCLUSION

In this work, ML classification approaches like SVM, KNN, Decision Tree and Logistic Regression are utilized to predict the existence of breast cancer. To determine which method was best suited to the task, the accuracy of each was evaluated. In using "Breast Cancer (Diagnostic) through the Wisconsin Data Set," the SVM Classifier is the most accurate prediction method because it has the highest prediction accuracy. This article shows that Decision tree approach performed better compare to other approaches. Therefore, breast cancer can be recognized with almost absolute precision using this approach and its results are close to the SVM Classifier method in this dataset.

REFERENCES

[1] *'WHO | Breast Cancer'*, WHO. http://www.who.int/cancer/prevention/diagnosis-screening/breast-cancer/en/ (accessed Feb. 18, 2020).

[2] *Datafloq – Top 10 Data Mining Algorithms, Demystified.* https://datafloq.com/read/top-10-data-mining-algorithmsdemystified/1144. Accessed December 29, 2015.

[3] Nayak S. and Gope D., "Comparison of Supervised Learning Algorithms for RF-based Breast Cancer Detection," *2017 Computing and Electromagnetics International Workshop (CEM)*, Barcelona, 2017, pp.

[4] Gayathri B. M. and Sumathi C. P., "Comparative Study of Relevance Vector Machine with Various Machine Learning Techniques Used for Detecting Breast Cancer," *2016 IEEE International Conference on Computational Intelligence and Computing Research (ICCIC)*, Chennai, 2016, pp. 1–5.

[5] Asri H., Mousannif H., Moatassime H. A., and Noel T., 'Using Machine Learning Algorithms for Breast Cancer Risk Prediction and Diagnosis', *Procedia Computer Science*, vol. 83, pp. 1064–1069, 2016, doi: 10.1016/j.procs.2016.04.224.

[6] Khoudfi Y. and Bahaj M., *Applying Best Machine Learning Algorithms for Breast Cancer Prediction and Classification*, 978-1-5386-4225-2/18/$31.00 ©2018 IEEE.

[7] Latchoumi L., T. P., & Parthiban, "Abnormality Detection Using Weighed Particle Swarm Optimization and Smooth Support Vector Machine," *Biomed. Res.*, vol. 28, no. 11, pp. 4749–4751, 2017.

[8] Osman A. H., "An Enhanced Breast Cancer Diagnosis Scheme Based on Two-Step-SVM Technique," *Int. J. Adv. Comput. Sci. Appl.*, vol. 8, no. 4, pp. 158–165, 2017.

[9] Noble WS. What is a Support Vector Machine? *Nat Biotechnol.* 2006;24(12):1565–1567. doi:10.1038/nbt1206-1565.

[10] Larose DT. *Discovering Knowledge in Data*. Hoboken, NJ, USA: John Wiley & Sons, Inc.; 2004.

[11] Hastie T, Tibshirani R, Friedman J. *The Elements of Statistical Learning*. New York, NY: Springer-Verlag;2001.

[12] Dwivedi, A. K., & Sharma, A. K. (2021). I-FBECS: Improved Fuzzy Based Energy Efficient Clustering Using Biogeography-based Optimization in Wireless Sensor Network. *Transactions on Emerging Telecommunications Technologies*, 32(2), e4205.

[13] "*UCI Machine Learning Repository: Breast Cancer Wisconsin (Diagnostic) Data Set.*"

[14] Dwivedi, A. K., Mehra, P. S., Pal, O., Doja, M. N., & Alam, B. (2021). EETSP: Energy-efficient Two-stage Routing Protocol for Wireless Sensor Network-assisted Internet of Things. *International Journal of Communication Systems*, 34(17), e4965

[15] Fabian Pedregosa and all (2011). "Scikit-learn: Machine Learning in Python". *Journal of Machine Learning Research*. 12: 2825–2830.

[16] Dwivedi, A. K., & Sharma, A. K. (2021). EE-LEACH: Energy Enhancement in LEACH Using Fuzzy Logic for Homogeneous WSN. *Wireless Personal Communications*, 120(4), 3035–3055.

[17] Dwivedi, A. K., & Sharma, A. K. (2020). NEEF: A Novel Energy Efficient Fuzzy Logic-based Clustering Protocol for Wireless Sensor Network. *Scalable Computing: Practice and Experience*, 21(3), 555–568.

[18] Dwivedi, A. K., & Sharma, A. (2020). FEECA: Fuzzy Based Energy Efficient Clustering Approach in Wireless Sensor Network. *EAI Endorsed Transactions on Scalable Information Systems*, 7(27).

[19] Srivastav, S., & Singh, P. K. (2021). An Approach for Fast Compressed Text Matching and to Avoid False Matching Using WBTC and Wavelet tree. *EAI Endorsed Transactions on Scalable Information Systems*, 8(30), e6.

[20] Srivastav, S., Singh, P. K., & Yadav, D. (2021). A Method to Improve Exact Matching Results in Compressed Text using Parallel Wavelet Tree. *Scalable Computing: Practice and Experience*, 22(4), 387–400.

Smart Intelligent Computing in Advanced Sectors (SICAS)

Various possible attacks on Internet of Things (IoT)

Vidhu Kiran
Ch. Devi Lal State Engineering College, Sirsa, India

Susheela Hooda & Dr Tejinder Kaur
Chitkara University Institute of Engineering and Technology, Chitkara University, Punjab, India

ABSTRACT: Internet of Things (IoT) has emerged as a promising technology in a wide range of industries, including healthcare and information technology. Despite technological developments, there are still numerous security and integrity problems with the Internet of Things. The author evaluated literature review and discussed several potential attacks against the Internet of Things that can affect network traffic and resources. In this paper, author also compared various attacks with their effect on security which may occur in Internet of Things.

Keywords: IoT, Active attack, Passive attack, Confidentiality, Integrity.

1 INTRODUCTION

Due to the proliferation of smart internet-based devices, the Internet of Things (IoT) has gained significant popularity in the modern world. The IoT offers inter-connectivity and smart communication among digital devices such as Personal Computers (PCs), laptops, tablets, smart phones, Personal Digital Assistants (PDAs), and other handheld embedded devices (Atzori *et al.* 2010) (Jorge *et al.* 2015). The primary contribution of IoT is to promote connectivity among smart internet devices at anywhere, anytime, anyplace, with anything and anyone ideally using any path/network and any service. Thus, the IoT applications have primarily contributed to day-to-day human activities by enabling smart devices to manage routine life activities and chores. The IoT incorporates a vast amount of heterogeneous devices that are limited in memory, energy, and processing capabilities (Whitmore *et al.* 2015). Each device in IoT produces diverse data and it is challenging to transmit such huge volume heterogeneous data among the devices. The IoT routing protocols route the heterogeneous data using well pre-defined routing algorithms. However, the internet is the heart of the IoT network and hence, interconnectivity and security are two major issues of IoT. As a consequence, the quality of data transmissions for future IoT applications is mostly desired in the smart world (Maalel *et al.* 2013). Hence, the data is interrupted by hackers, malicious activities, and viruses due to the vulnerable characteristics of IoT, such as low computing capability, open network, and high volume data produced by heterogeneous devices.

Currently, the IoT is exploited in a lot of smart social life applications such as smart home security and intelligent transportation system (Ibarra *et al.* 2017). Although the IoT applications make human life more convenient, it lacks to assure complete security to the personal information of humans that may leak at exuded by attackers at any time. Consequently, attacker stoles or interrupts the signal of IoT once it will straightly damage the security level of the entire IoT system. With the vast exploitation of IoT, it offers a significant level of security according to the application type (Bandyopadhyay & Sen 2011). The IoT is a proliferated network, and there are no sufficient security solutions to the IoT, resulting in

DOI: 10.1201/9781032684994-21

considerable restrictions in the IoT development. Therefore, it is essential to propose meaningful security solutions to IoT routing protocols. A prominent routing mechanism, named as the Routing Protocol for Low Power and Lossy Networks (RPL) is a standardized routing mechanism proposed for IoT. However, the RPL is a fundamental routing solution primarily proposed for IPv6 over Low Power Wireless Personal Area Networks (6LoWPAN) is vulnerable to several types of security attacks (Wallgren *et al.* 2013). It suffers from offering high-quality security to the IoT and it is crucial to develop security solutions against various types of attacks against RPL. In addition to this, next section discussed various attack possible in Internet of Things.

2 ROUTING PROTOCOLS IN IOT

RPL is an effective routing protocol utilized for IoT (Mercy & Pravin 2014). The RPL routing protocol determines the routing paths between a source-destination pair as soon as possible. Currently exploited RPL for IoT is used key-based applications in pre-designed smart devices. However, the security level of RPL is weak, and it lacks to attain better performance under secure mission-critical applications. Moreover, the RPL protocol is susceptible to various types of routing attacks as the same as the sensor network, and also vulnerable to the attacks against the IoT. Most of the researchers define the security requirements of RPL over IoT, whereas there are no appropriate security solutions for such networks. Therefore, it is worth to analyze the routing attacks against RPL and it is essential to propose high RPL security against such attacks. Some of the RPL routing attacks such as route counterfeiting, message replay, version number falsification, IoT requires the security measures of traditional networks, as the IoT in the real-world is envisioned by connecting the heterogeneous devices and various technologies with the internet. Thus, it significantly increases the security demands associated with an IoT. The malicious device not only modifies the contents of messages but also takes control of an entire IoT system. The advanced technologies of IoT devices also pose several new security threats.

3 LITERATURE REVIEW

RPL is an effective routing protocol utilized for IoT (Mercy & Pravin 2014). The RPL routing protocol determines the routing paths between a source-destination pair as soon as possible. Currently exploited RPL for IoT is used key-based applications in pre-designed smart devices. However, the security level of RPL is weak, and it lacks to attain better performance under secure mission-critical applications. Moreover, the RPL protocol is susceptible to various types of routing attacks as the same as the sensor network, and also vulnerable to the attacks against the IoT. Most of the researchers define the security requirements of RPL over IoT, whereas there are no appropriate security solutions for such networks. Therefore, it is worth to analyze the routing attacks against RPL and it is essential to propose high RPL security against such attacks. Some of the RPL routing attacks such as route counterfeiting, message replay, version number falsification, IoT requires the security measures of traditional networks, as the IoT in the real-world is envisioned by connecting the heterogeneous devices and various technologies with the internet. Thus, it significantly increases the security demands associated with an IoT. The malicious device not only modifies the contents of messages but also takes control of an entire IoT system. The advanced technologies of IoT devices also pose several new security threats. In general, the routing path is established when information is transmitted to a destination node. Further, the roués are maintained or deleted according to the protocol process. In such routing, a misbehaving node may insert false information or dropping the messages for their benefits. For instance, a particular node transmits a vast amount of false amount to its neighboring node for creating the overflow in the routing table. Such malicious activities deny the real routed by occupying the routing table with spurious routing information. Such activity also drains the battery power of neighboring nodes quickly, resulting in reduced network performance.

4 VARIOUS AVAILABLE ROUTING ATTACKS IN IOT

- **Attacks on Network Resources:** long-term, this entails consuming handling, memory, and node energy. By restricting available connections, this could disrupt the network's accessibility and, as a consequence, its sustainability, which could be significantly impacted. We further categorize it into two types of attacks. The first kind of attack involves hostile nodes producing excess load on their own, corrupting the network. The second type of attack is an indirect one in which the attacker influence various nodes to generate a substantial amount of traffic.
- **Attacks on Network Topology:** These attacks are classified into one of two major categories: sub-optimization or isolation.
- **Attacks on Network Traffic:** The third category contains of attacks that focus on the RPL network traffic. It mostly consists of appropriation attacks and eavesdropping attacks.

4.1 *Security threats against RPL in IoT*

The heterogeneous IoT devices are restricted in battery power, memory, and computation capabilities. Due to the ubiquitous nature of IoT, Therefore, security plays a vital role in IoT. There are several security attacks in the RPL network, and each attack establishes its malicious activities based on the network topology, network traffic and resources (Mayzaud *et al.* 2016). Based on the topology, there are several types of security threats such as (DOS) attack. Based on the resources, the security threats are flooding, routing table overhead, increased rank attack and DAG inconsistency.

4.1.1 *Classification based on topology*

4.1.1.1 Selective forwarding

In this type, the attacker also selectively forwarding the packets to the destination, and drop the remaining packets during the transit. For instance, the attacker forwards the control messages of RPL correctly, whereas it mainly drops the data packets.

4.1.1.2 Sinkhole attacks

In this type, an attacker builds artificial routing paths and launching the attack by transmitting the data traffic through the established artificial routing paths. The sinkhole attack

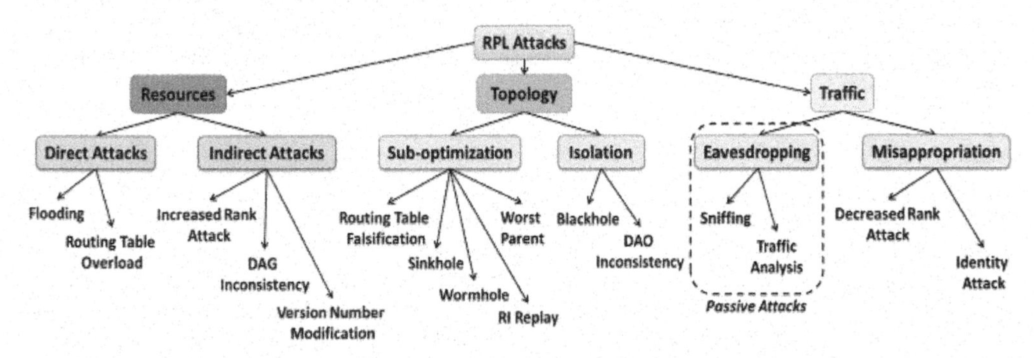

Figure 1. Classification of attacks.

does not create a more significant impact on RPL routing operations, as long as the attacker is separated. However, when the attacker is coupled with various types of attacks, it creates a severe impact on routing operations.

4.1.1.3 Hello flooding

Every node in IoT routing exploits hello packets for route establishment and data forwarding. In hello flooding attack, an attacker broadcasts unnecessary hello packets within the network in order to launch an attack by pretending as a neighboring device as many devices. The data packets are highly loosed in the presence of hello flooding attackers, as the source believes the flooding attacker is its neighbor for data forwarding, whereas the attacker is in out of communication range of source.

4.1.1.4 Wormhole attacks

The wormhole attack paths are rapidly forwards the data packets compared to typical routing paths. The wormhole attacker does not create strong influence until it is combined with other types of attackers in the network.

4.1.2 *Classification based on network traffic*

4.1.2.1 Clone ID and sybil attacks

In the Clone ID attack, an attacker obtains the real identities of other devices, and it launches an attack by pretending as multiple devices in the network. Likewise, the Sybil attacker reinforces a malicious event within the network and it motivates the devices to take wrong routing decisions. This type of attack creates a great influence on trust-based security solutions. The Sybil attacker can control a large part of the network by pretending as multiple devices and it reduces the overall system performance.

4.1.2.2 Sniffing attack

Sniffing attack launches the attack passively without disturbing the services of the network. These attacks aim to steal sensitive information, thereby compromising the confidentiality of the communication. The sniffing attack is performed through a compromised device or captured directly through the shared medium in the network.

4.1.2.3 Traffic analysis

If the presence of a traffic analysis attacker is near the root node, the attacker is able to obtain the routing information of almost all the edge nodes and launch other serious attacks in the network.

4.1.2.1 Decreased rank attack

In the RPL protocol, rank property depicts the location of each node within the DODAG. Rank plays a significant role in avoiding and detecting loop formation in the network. In rank attack, malicious nodes change the rank rule by posing themselves as the lowest rank for entering the parent selection. The neighboring node that receives the rank of the malicious node in the DIO packet elects the attacker as its respective parent node. These attacks are launched primarily before routing attacks such as sinkhole attacks for becoming part of the routing path.

4.1.3 *Classification based on resources*

4.1.3.1 Denial of service attacks

The DoS attack is one of the serious threats that disrupt the services in the network by continuously flooding fake data packets. In DDoS attacks, multiple attackers incur spurious data transmissions towards the targeted node or gateway node. These flooding attackers continuously send fake data packets to the gateway node for disabling it from performing its services. In case, a normal node sends packets to the gateway node, these packets are dropped, and the gateway node keeps on receiving the flooding packets.

4.1.3.2 DAG inconsistency attack

In a DAG inconsistency attack, the attacker node uses the datapath validation mechanism in RPL for launching the attack. The malicious node manipulates the RPL IPv6 header options to present a fake existence of DAG inconsistencies forcing the target to drop packets. This attack causes an RPL router to reset its DIO Trickle timer and thereby frequently transmitting DIO messages. It leads to a DoS attack, which increases the control overhead and energy consumption in the network.

Some of the RPL attacks and their properties are explained in the following table.

Table 1. Comparison of Various Types of RPL attacks over IoT.

Attacks	Description	Security Issues	Impact on Performance
Rank (Raza *et al.* 2013)	Aims to generate non-optimal paths and loops for routing	Confidentiality and Integrity	Reduces the packet delivery ratio and increases the delay
Sinkhole (Raza *et al.* 2013)	Compromises the nodes bypassing vast traffic via the attacker		High packet loss and delay in data delivery
Wormhole (Perazzo *et al.* 2018)	Disrupts the routing topology and the data traffic flow in the network		Inaccurate routing path discovery and high packet loss
Sybil and Clone ID (Wallgren 2013)	Perform node compromise to disrupt routing paths and prevents the traffic from reaching the destination		Reduced Routing Efficiency
Version number (Dvir *et al.* 2011)	Aims to change the version number and launching attacks		High control overhead, minimum packet delivery ratio, and a maximum end to end delay
Local repair Control overhead (Le *et al.* 2012)	Disrupts the control and data traffic flow		Routing performance degradation
Selective Forwarding (Wallgren 2013)	Aims to disrupt routing functionalities		Diminishes routing efficiency
Hello flooding (Wallgren 2013)	Aims to drain the battery power of devices quickly	Availability	High energy dissipation and poor network connectivity
Denial of Service (Kasinathan *et al.* 2013)	Denies the network services and makes the resources unavailable to nodes		Affecting data quality and unnecessary energy depletion at neighboring nodes
DODAG Information Solicitation (DIS) (Perrey *et al.* 2013)	An attacker aims to broadcast DIS messages continuously		Minimized packet delivery ratio, high packet delay, and High resource consumption
Neighbor attack (Perrey *et al.* 2013)	Erroneous route discovery and route disruption activities	Availability, Confidentiality & Integrity	High resource consumption at neighboring nodes
Blackhole (Jiang *et al.* 2018) (Ahmed & Ko 2016)	Aims to drop packets or increases the route traffic		Increases the control overhead and decreases the packet delivery ratio

5 CONCLUSION

With the exponential growth in the Internet of Things (IoT), its reach has extended to almost all application sectors such as healthcare, industrial manufacturing, smart homes, transportation, agriculture, and so on. Even though IoT aims to provide ubiquitous connectivity with effective solutions, the open deployment of IoT devices gives rise to security issues. The security threats specifically utilize the resource constraints and physical characteristics of IoT for launching security attacks to degrade network performance and disrupt its services. Author discussed various attack with compromised security features due to this.

REFERENCES

Ahmed, F., and Ko, Y. B. (2016). – Mitigation of Black hole Attacks in Routing Protocol for Low Power and Lossy Networks. *Security and Communication Networks*, 9(18).

Airehrour, D., Gutierrez, J., and Ray, S. K. (2016). – A Lightweight Trust Design for IoT Routing. International Dependable, Autonomic and Secure Computing, *14th Intl Conf on Pervasive Intelligence and Computing, 2nd IEEE Intl Conf on Big Data Intelligence and Computing and Cyber Science and Technology Congress*, pp. 552–557.

Airehrour, D., Jairo G., and Ray, S. (2016). – Secure Routing for Internet of Things: A Survey. *Journal of Network and Computer Applications*, pp.198–213.

Alaparthy, V. T., and Morgera, S. D. (2018). – A Multi-Level Intrusion Detection System for Wireless Sensor Networks based on Immune Theory. *IEEE Access*, 1(1).

Bandyopadhyay, D., and Sen, J. (2011). – Internet of Things: Applications and Challenges in Technology and Standardization. *Wireless Personal Communications*, 58(1), pp. 49–69.

Chang, K. (2014). - Bluetooth: A Viable Solution for IoT. *IEEE Wireless Communications*, 21(6), pp. 6–7.

Chen Y., Chanet, J. and Hou, K. (2012). – RPL Routing Protocol a Case Study: Precision Agriculture. *First China-France Workshop on Future Computing Technology (CF-WoFUCT)*, pp. 6–10.

Chen, C. M., Hsu, S. C., and Lai, G. H. (2016). - Defense Denial-of-Service Attacks on IPv6 Wireless Sensor Networks. *International Genetic and Evolutionary Computing, Springer International Publishing*, pp. 319–26.

Chze P. L. R. and Leong K. S. (2014). – A Secure Multi-Hop Routing for IoT Communication. *IEEE World Forum on Internet of Things (WF-IoT)*, pp. 428–32.

Danyang QIN (2016) "Research on Trust Sensing based Secure Routing Mechanism for Wireless Sensor Network" *IEEE Access*, VOL. XX, NO. Y, 2016

Dawans, S., and Bonaventure, O. (2012) – On link Estimation in Dense RPL Deployments. Local Computer Networks Workshops (LCN Workshops), *IEEE 37th Conference*, pp. 952–55.

Din, I. U., Guizani, M., Kim, B., Hassan, S., and Khan, M. K. (2018). - Trust Management Techniques for the Internet of Things: A Survey. *IEEE Access*, 1–1.

Ding, Y., Zhou, X. W., Cheng, Z. M., and Lin, F. H. (2013). – A Security Differential Game Model for Sensor Networks in Context of the Internet of Things. *Wireless Personal Communications*, 72(1), pp. 375–88.

Divya Sharma (2017) "A Detailed Classification of Routing Attacks against RPL in Internet of Things", *International Journal of Advance Research, Ideas and Innovations in Technology*, ISSN: 2454-132X Impact factor: 4.295 (Volume3, Issue1)

Djedjig, N., Tandjaoui, D., Medjek, F. (2015). – Trust-based RPL for the Internet of Things. *IEEE Symposium on Computer and Communication (ISCC)*.

Djedjig, N., Tandjaoui, D., Medjek, F., and Romdhani, I. (2017). – New Trust Metric for the RPL Routing Protocol. *8th International Conference on Information and Communication Systems (ICICS)*.

Duan, J., Gao, D., Yang, D., Foh, C. H., and Chen, H. H, (2014). - An Energy-Aware Trust Derivation Scheme with Game Theoretic Approach in Wireless Sensor Networks for IoT Applications. *IEEE Internet of Things Journal*, 1(1), pp.58–69.

Duan, J., Yang, D., Zhu, H., Zhang, S., and Zhao, J. (2014). – TSRF: A Trust-Aware Secure Routing Framework in Wireless Sensor Networks. *International Journal of Distributed Sensor Networks*, 10(1).

Jingpei Wang, Sun Bin (2013) – "Distributed Trust Management Mechanism for the Internet of Things", Proceedings of the 2nd International Conference on Computer Science and Electronics Engineering *(ICCSEE 2013)*

Miorandi D., Sicari S., De Pellegrini F., and Chlamtac I. (2012) – "Internet of Things: Vision, Applications and Research Challenges," *Ad Hoc Networks*, vol. 10, pp. 1497–1516, 2012.

Neeraj, Amitpal Singh (2016) – "Internet Of Things And Trust Management In IOT – Review", *International Research Journal of Engineering and Technology (IRJET)* e-ISSN: 2395 -0056 Volumemac_mac 03 Issue: 06 | June-2016 www.irjet.net p-ISSN: 2395-0072

Yosra Ben Saied (2013) – "Trust Management System Design for the Internet of Things: A Context-aware and Multiservice Approach", Computers & Security xxx (2013) 1–15

Zhao K. and Ge L., (2013) "A Survey on the Internet of Things Security," in *Computational Intelligence and Security (CIS), 2013 9th International Conference on*, 2013, pp. 663–667.

Artificial Intelligence, Blockchain, Computing and Security – Dagur et al. (Eds)
© 2024 The Author(s), ISBN: 978-1-032-67841-2

Analysis of machine learning approaches for predicting heart disease

Srikanta Kumar Mohapatra, Arpit Jain & Anshika
Chitkara University Institute of Engineering and Technology, Chitkara University, Rajpura, Punjab, India

ABSTRACT: Cardiovascular Disease (CVDs) is among the most the deadly diseases. So, it is important to tackle and diagnose the disease in its earlier stage such that it can be curable in its later stage. If researcher is successful in getting CVDs early it is very serene to minimize mortality produced by CVDs. ML algorithms are used to forecast the operation and performance for the disorder. Here 3 classification algorithms such as "Gradient Boosting Machine (GBM), Random Forest (RF), and Extra Tree Classifier (ETC)". Category dividers are trained, validated, and accredited by using different parameters. In conclusion, Random Forest (RF) is best system for predicting CVDs using Kaggle database with 98.5% accuracy.

Keywords: CVD, ML Algorithm, Classifier, GBM, RF, ETC

1 INTRODUCTION

A challenge for clinical experts is to get the correct detection of precise disorder like CVDs, etc.at an early stage. Cancer is considered as the no. 1 dreadful disease in the world [1] and CVDs is the 2nd most dreadful disease in the world. A large amount of available information related to this specific disease is found in websites, hospitals, or non-profit research institutes. It is usually very difficult to distinguish these data numbers and for choosing prime algorithm which has used for clinical experts. According to WHO, Heart Diseases are an important source of assassination, with approximately 17.9 million resides every year, representing 32% of all deaths globally. CVD is a category of cardiovascular disorders which incorporate heart disease. More than 75 percent of deaths by CVDs are occurred due to strokes and attacks on the heart. These deaths generally occur in old age. But it also occurs prematurely under the age of 70. Heart attacks and strokes are often the most serious and the main cause is blockages that prevent blood from flowing to the heart or brain [2]. In categorized division CVDs are divided into four types such as Coronary Artery Disease (CAD), Cerebro vascular, Peripheral Artery Disease (PAD), and Aortic Atherosclerosis. Among these four CAD is more dangerous as it decreases myocardial perfusion which causes heart failure due to myocardial infarction (MI) [3]. IoT thrives on health care systems such as monitoring of health progress and regulating fitness programs, etc. Much research has been done using health care system based on IoT methods to improve the effectiveness of monitoring [4]. Along with IoT, Machine Learning methods also contribute in the prediction of many diseases like Breast Cancer with accuracy of 98.7% [1], Brain Tumor with accuracy of 98.09% [5], etc.

DOI: 10.1201/9781032684994-22

Various related activities were introduced in the second section. After collecting data, implementing, and analyzing results, the third and fourth stages conclude with a conclusion.

2 RELATED WORK

In general, many researchers have researched related to CVDs that can be predicted using different machine learning methods. K. Ahmed *et al.* [6] have studied that if heart rate data has collected from Kaggle dataset which may be worn to make heart disease predictions that produces highest classification accuracy. M.K. Gourisaria *et al.* [7] used various machine learning approaches and SVM got best accuracy of approx. 85.25% using Regression model. H. Jindal [8] *et al.* have setup a cardio vascular prediction system so that we predict the heart disease, and concluded that Logistic Regression and KNN showed good accuracy than Naïve Bayes.

Y.S. Triana [9] *et al.* has diagnosed CVDs in Malaysia. Here they have taken datasets from UCI ML repository and Classification of Random Forest and Artificial Neural Network (ANN). To achieve maximum accuracy, they used a dataset splitting and K-Fold Cross-Validation. M.A. Butt *et al.* [10] used some ML techniques like SVM, ANN, Naïve Bayes, KNN, etc. and they got an average accuracy of 86.9%. S.K. Bharti *et al.* [11] have used a dataset containing 303 instances with 76 attributes, from which only 14 are chosen for testing, and compared various ML algorithms, out of which the output achieved the highest accuracy score using KNN. M. Pal *et al.* [12] has chosen a dataset containing 303 samples and 14 attributes and concluded that Random Forest gives out accuracy of 86.9% with a sensitivity value of 90.6%. M.M. Mijwil *et al.* [13] performed the prediction on the dataset containing 170 reports of the people using SVM and ANN and as the outcome; they got 89.1% as the best accuracy using SVM Model. Other research activities in the field of Healthcare are also carried in different fields [14,15]. From the above study, we can consider that the effective prediction of the CVDs using mentioned above machine learning methods are sufficient but also alternative large-scale datasets and various measurement features are also needed.

3 METHODOLOGY

To compare the behavior of different machine learning algorithms, we focus on efficiency and effective features of related algorithms. Some research questions are

(i) What are the types of method that have most effective effect?
(ii) Which is better for pondering efficiency?
(iii) Which method provides the highest level of accuracy?

3.1 *Cardio Vascular Diseases (CVDs) dataset*

The database has 1025 cases found in the Kaggle database. The total database has 12 attributes with no missing value [16]. The total number of classes is (type of chest pain) i.e., 497 general angina pain, 284 non-angina pain, 167 angina pain atypical, and 77 asymptomatic pain. Significant changes used include high heart rate, chest pain, cholesterol, and resting blood pressure. A roadmap of the implemented model (as shown in Figure 1) and attributes information given in Table 1.

3.2 *Training and validation*

The set of information is instructed as well as validated by diverse replicas. The instructing part is used to remove the features while the test method is worn to direct the performance of the database.

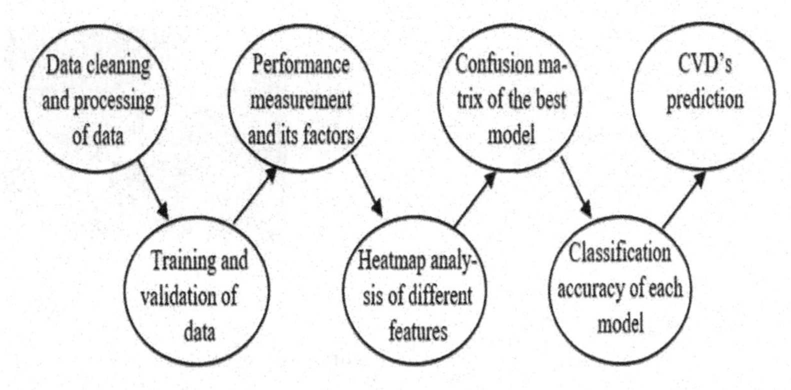

Figure 1. Roadmap of the implemented model.

Table 1. Attributes information.

Properties	Details
Age	Input Value As Years
M/F	M = 0, F = 1 (M: Male, F: Female)
BP(Resting)	In mm Hg while admitting
Cholesterol	Serum Cholesterol (mg/dl)
Blood Sugar(Fasting)	T = 1, F = 0
ECG(Resting)	Normal = 0, Left Ventricular Hypertrophy = 1, ST-T wave abnormality = 2
Heart Rate(At Max Level)	Max.
Anginal Value	Y = 1, No = 0
Old peak	ST depression acquired.
ST slope	Up_sloping = 0, Flat = 1, Down_sloping = 2

3.3 *Measuring performance*

The performance of prototype F1and precision parameters is assessed in each area using the algorithm. The confusion matrixes parameters are evaluated to determine the true and predicted values.

Accuracy: Rating is calculated using the ratio between the true Sample cases v/s Complete Samples.
Sensitivity: The ratio between the positive events predicted in all positive conditions.
Specificity: Compared to negative viewed values with all negative values.
Precision: Value of forecasted positive to rational positives.
F1Score: It is a real measure of performance by taking precision and memory.
Log-Loss: It is indication of how near prediction probability is to corresponding real/true value.
MCC: It is usually used to find binary division. So, +1 specifies superior achievement, and −1 specifies inferior.

Achievement evaluation of the replica has shown in Figure 2 and Table 2 respectively.

As some of the features related to CVDs are taken, so a heatmap analysis needed to be used to better visualize the number of occurrences in dataset are important parts in data visualizations in the form of graphs of data is discussed. Thus, the heatmap analysis of various factors is shown in Figure 3.

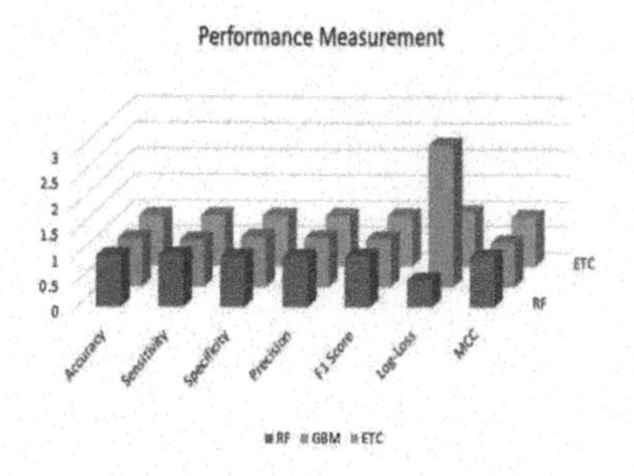

Figure 2. Performance measurement.

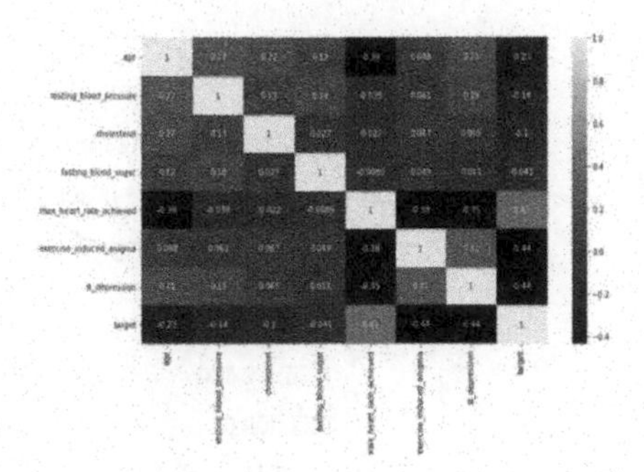

Figure 3. Heatmap analysis of various factors.

Table 2. Performance measurement.

Parameters	Accuracy	Sensitivity	Specificity	Precision	F1 Score	Log-Loss	MCC
Random Forest	0.985	1.000	0.969	0.971	0.985	0.515	0.970
Gradient Boosting	0.920	0.894	0.948	0.938	0.920	2.740	0.842
Extra Tree Classifier	0.970	0.970	0.969	0.971	0.971	1.030	0.940

4 RESULTS & DISCUSSIONS

The total number of instances is 1025. Parameters used for measuring performance are listed in Table 2. The table tells us that accuracy of the Random Forest (RF) is significantly better with respect to other strategies. The hyperparameters of our planned models and best hyperparameters used for performance tests are shown in Table 3. In the Random Forest (RF) accuracy is about 98.5%, while in GBM and ETC it is 92% and 97% respectively; and plotted the line graph as shown in Figure 5. For each model in the sequence, the confusion matrix is calculated, and the confusion matrix of Random Forest(RF) is shown in Figure 4.

Table 3. Hyperparameters and best hyperparameters used for each model.

Techniques	Hyperparameter	Best hyperparameter
RF	estimator-n, n-criterion	estimator-n = 5, n-criterion = 'entropy'
GB	estimator-n, max_feature-analysis	estimator-n = 100, max_feature-analysis = 'sqrt'
ETC	estimator-n	estimator-n = 100

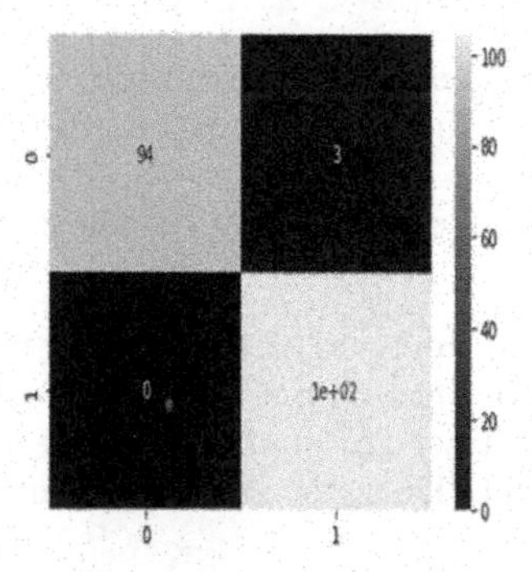

Figure 4. Confusion matrix of random forest.

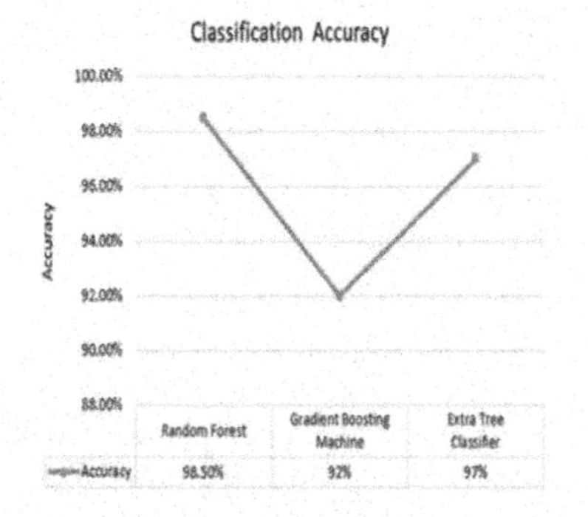

Figure 5. Classification accuracy of each model.

5 CONCLUSION

Cardiovascular diseases (CVDs) appear as a major cause of death in many countries in the region. In medical science, diagnosis is very costly. Therefore, new strategies shouldbe usedto find the appropriate solution. Therefore, this work indicates the ML methods have produced superior explanation. In this work, 3 ways to forecast CVDs "Gradient Boosting Machines (GBM), Random Forest (RF), and Extra Tree Classifier (ETC)". The basic methodof feature extraction is pertained to each method. Later than doing so with the right operational goal, it concludes that Random Forest (RF) has much better accuracy and performance than other algorithms. It shows an accuracy of about 98.5%. The lowest accuracy is generated by GBM, which is 92%.

REFERENCES

[1] Mohapatra, S.K., Jain, A. and Sahu, P., 2022, April. Comparative Approaches by Using Machine Learning Algorithms in Breast Cancer Prediction. *In 2022 2nd International Conference on Advance Computing and Innovative Technologies in Engineering (ICACITE)* (pp. 1874–1878). IEEE.

[2] "World Health Organization (WHO)" [Online]. Available: https://www.who.int/news-room/fact-sheets/detail/cardiovascular-diseases-(cvds) [Accessed 1 November 2022].

[3] Lopez E. O., Ballard B. D.and Jan A., "Cardiovascular Disease", *In StatPearls [Internet]*. StatPearls Publishing, 2021.

[4] Selvaraj S.and Sundaravaradhan S., "Challenges and Opportunities in IoT Healthcare Sys tems: A Systematic Review", *SN Applied Sciences*, Vol.2, no. 1, pp.1–8, 2020.

[5] Sahu, P., Sarangi, P.K., Mohapatra, S.K. and Sahoo, B.K., 2022. Detection and Classification of Encephalon Tumor Using Extreme Learning Machine Learning Algorithm Based on Deep Learning Method. In *Biologically Inspired Techniques in Many Criteria Decision Making* (pp. 285–295). Springer, Singapore.

[6] Ali M. M., Paul B. K., Ahmed K., Bui F. M., Quinn J. M. and Moni M. A., "Heart Disease Prediction Using Supervised Machine Learning Algorithms: Performance Analysis and Comparison", *Computers in Biology and Medicine*, 136, p.104672, 2021.

[7] Sarah S., Gourisaria M. K., Khare S. and Das H., "Heart Disease Prediction Using Core Machine Learning Techniques—A Comparative Study", *In Advances in Data and Information Sciences, Springer, Singapore*, pp. 247–260 2022.

[8] Jindal H., Agrawal S., Khera R., Jain R. and Nagrath P., "Heart Disease Prediction Using Machine Learning Algorithms", *In IOP Conference Series: Materials Science and Engineering, IOP Publishing*, Vol. 1022, no. 1, pp. 012072, 2021.

[9] Bakar W.A.W.A., Josdi N.L.N.B., Man M. B. and Triana Y. S., "An Evaluation of Artificial Neural Networks and Random Forests for Heart Disease Prediction", *Journal of Hunan University Natural Sciences*, Vol. 49, no. 2, 2022.

[10] Riyaz L., Butt M. A., Zaman M.and Ayob O., "Heart Disease Prediction Using Machine Learning Techniques: A Quantitative Review", *In International Conference on Innovative Computing and Communications Springer, Singapore*, pp. 81–94, 2022.

[11] Palvika, Shatakshi, Sharma, Y., Dagur, A., & Chaturvedi, R. (2019). Automated Bug Reporting System with Keyword-driven Framework. In *Soft Computing and Signal Processing: Proceedings of ICSCSP* 2018, Volume 2 (pp. 271–277). Springer Singapore.

[12] Kumar, A., & Alam, B. (2019). Energy Harvesting Earliest Deadline First Scheduling Algorithm for Increasing Lifetime of Real Time Systems. *International Journal of Electrical and Computer Engineering*, 9(1), 539.

[13] Kumar, A. & Alam, B. (2018). Task Scheduling in Real Time Systems with Energy Harvesting and Energy Minimization. *Journal of Computer Science*, 14(8), 1126–1133.

[14] Kumar, A., & Alam, B. (2014, February). Real Time Scheduling Algorithm for Fault Tolerant and Energy Minimization. In *2014 International Conference on Issues and Challenges in Intelligent Computing Techniques (ICICT)* (pp. 356–360). IEEE.

[15] Pradeep Ghantasala, G. S., Nageswara Rao, D., & Patan, R. (2022). Recognition of Dubious Tissue by Using Supervised Machine Learning Strategy. In *Applications of Computational Methods in Manufacturing and Product Design* (pp. 395–404). Springer, Singapore.

[16] *Kaggle Dataset [Online]* Available: https://www.kaggle.com/datasets/johnsmith88/heart-disease-dataset [Accessed 5 November 2022].

Artificial Intelligence, Blockchain, Computing and Security – Dagur et al. (Eds)
© 2024 The Author(s), ISBN: 978-1-032-67841-2

A study of Blockchain technology: Architecture, challenges and future trends

Suvarna Sharma* & Annu Priya*
Chitkara University Institute of Engineering and Technology, Chitkara University Punjab, India

ABSTRACT: The academic community, national governments, financial institutions, and energy supply businesses have all given blockchains, commonly referred to as distributed ledgers, a lot of attention. Blockchains are seen to have the potential to significantly improve lives and spur innovation, according to several individuals from different fields. Blockchains offer safe, transparent, and irreversible platforms that can accommodate innovative business ideas, particularly when integrated with smart contracts. This article provides an in-depth analysis of the fundamental ideas underlying blockchain technology, covering system topologies, challenges, characteristics, and a taxonomy of blockchain.

Keywords: Blockchain, public blockchain, private blockchain, consortium blockchain, hybrid blockchain, blockchain architecture.

1 INTRODUCTION

There are many distinct definitions of the blockchain by various writers, and as noted, there isn't a single, universal definition; it is crucial to comprehend the fundamental components of the blockchain (Fu *et al.* 2021). In 2016, Don & Alex Tapscott define the blockchain as a technology that immutable digital database of financial transactions can be set up to record not only financial transactions as well as virtually everything of value (Wenzheng *et al.* 2021). Now these days Internet of Value, or blockchain, become very demanding technology. The well-known blockchain technology is a chain of data-containing "blocks". It uses a distributed database system to support an ever-expanding list of immutable blocks. Without the use of intermediaries, blockchain enables direct connections between end users and suppliers for transparent communication and transactions.

In the IT industry, the blockchain has become one of the technologies that receives the most attention. Despite just being a notion a few years ago, this technology is already being deployed by many of businesses in significant industrial deployments. These blockchain technology's limitations don't make it any less revolutionary, but they have led some to question its dependability and effectiveness (Don *et al.* 2016).

1.1 *Architecture of Blockchain*

The Blockchain is preserved in a copy on each node in the blockchain network. Every network node has access to the transactions. A Blockchain's structure is seen in Figure 1

A list of blocks containing transactions listed in a certain sequence serves as a representation of the organizational structure of the blockchain technology (Sharma *et al.* 2021):

- Pointers
- Linked lists

*Corresponding Authors: suvarna.sharma@chitkara.edu.in and annu.priya@chitkara.edu.in

DOI: 10.1201/9781032684994-23

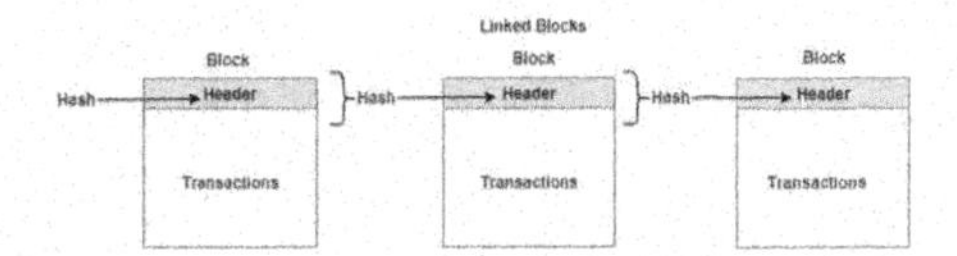

Figure 1. Blocks in the Blockchain architecture.

1.2 *Characteristics of Blockchain*

Blockchain has the following characteristics - decentralization, trustlessness, openness, immutability and anonymity. (Fu *et al.* 2021)

- Decentralization
- Detrusting
- Transparency
- Immutability
- Anonymity

The following are some uses of blockchain for businesses and organizations:

- Cost reduction
- Data history
- Data veracity, and data security

2 TYPES OF BLOCKCHAIN

Based on their applications and thresholds, blockchains are divided into these categories: public (permissionless) chains, private chains, hybrid chains, and consortium chains. (Zheng *et al.* 2017) (Andreev *et al.* 2018)

Table 1. Blockchain taxonomy.

Blockchain Type	Articles	Description	Application	Example
Public Blockchain	(Sai *et al.* 2021)	A permission-less system is one in which joining the network is not necessary for system participants who are public blockchain users. Every member may preserve a copy of the whole Blockchain since they can all publish, access, and validate new blocks here. Public blockchains are safe to create and use.	Government, Business	Bitcoin, Ethereum, Litecoin.
Private Blockchain	(Dinh *et al.* 2017)	For a specific business, private or permissioned blockchains are created. Users may join the network and participate in certain tasks that support the decentralized operation of the blockchain only upon invitation.	Supply Chain Management, Asset Ownership, Internal Voting	Multichain and Hyperledger projects (Fabric, Sawtooth), Corda, etc.
Hybrid Blockchain	(Ge *et al.* 2022)	The advantages of both public and private blockchains are combined using a hybrid blockchain. Both the private permission-based system and the public permission-less system of blockchain technology are utilized. On the blockchain, only a small percentage of the data or records may be made public; the rest must be kept hidden and private.	Retail market, Real Estate, Banking Sector	Dragonchain
Consortium Blockchain	(Dib *et al.* 2018)	Position the cursor Consortium blockchains are private blockchains even if they are designed for many businesses. Only invited and dependable individuals are allowed to join and manage the network.	Finance, Banking, Logistic, Healthcare and Insurance	Energy Web Foundation, R3, etc.

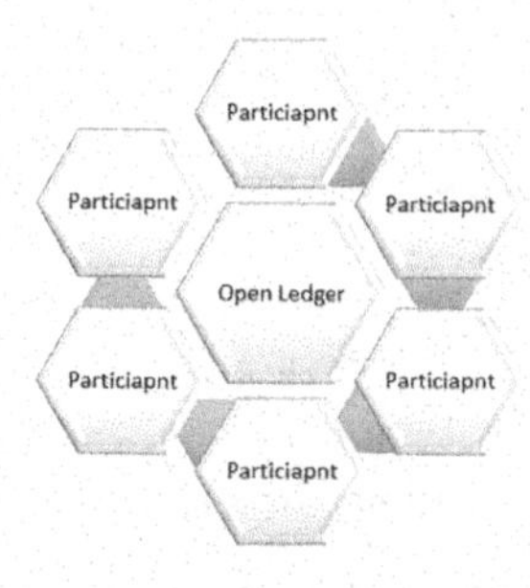

Figure 2. Public Blockchain.

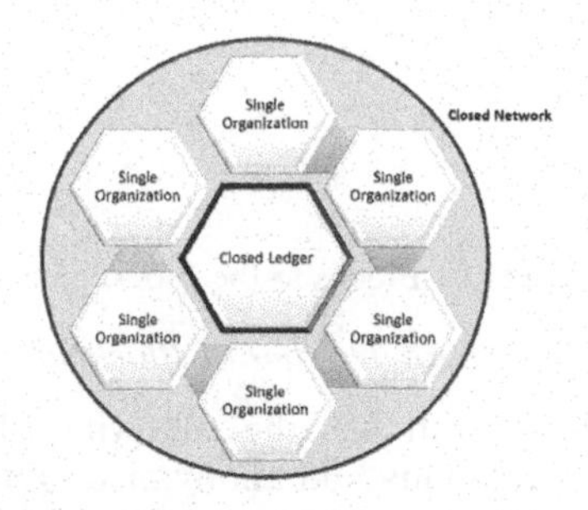

Figure 3. Private Blockchain.

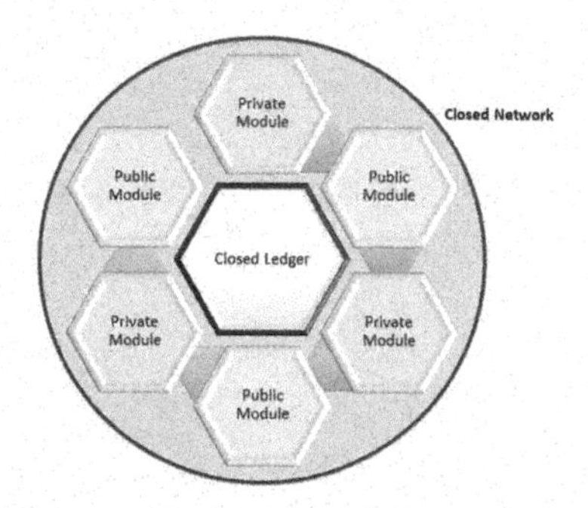

Figure 4. Hybrid Blockchain.

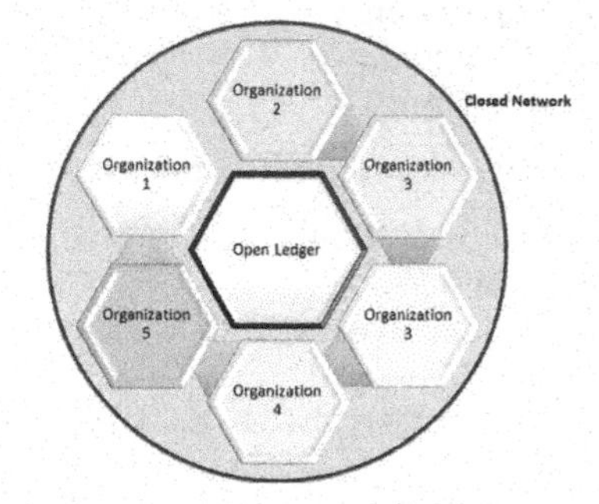

Figure 5. Consortium Blockchain.

3 CHALLENGES

Companies in asset management, retail, financial use cases, virtual currencies, e-contracts, decentralized exchanges, and other areas are already utilizing blockchain technology. However, blockchain isn't perfect, just like any other technology. The use of blockchain technology is constrained by a few operational, maintenance-related, and implementation-related obstacles (Wenzheng *et al.* 2021).

These existing problems amply demonstrate the blockchain's limitations:

- Poor efficiency
- Validation of signatures.
- Duplication.
- The use of resources and energy.
- Security hazard
- System malfunctions.
- Reaching an agreement.
- Issues with storage and limited expandability.
- Insufficient technical expertise.
- Human error.
- Extreme energy is needed.

4 FUTURE TRENDS

A chain of blocks is part of the emerging invention known as blockchain technology. The facts of each financial and non-financial transaction are principally stored in a distributed or

Table 2. Blockchain technology future scope.

Area	References	Scope
Health Care	(Ghosh *et al.* 2023) (Kushwaha *et al.* 2023)	Blockchain usage in healthcare has the ability to enhance patient outcomes as well as the entire healthcare system.
Digital Marketing	(Varma *et al.* 2022) (Rahman 2021)	Consumers may share and sell their information to advertising and advertisers directly thanks to blockchain technology. These days, data is what drives the market. Blockchain technology has the ability to give data owners targeted by digital advertising back control over their information.
Cyber Security	(Mahmood *et al.* 2022) (Wylde *et al.* 2022)	So many solutions to reduce new cyber security dangers can be provided by blockchain technology.
Finance	(Patel *et al.* 2022) (Treleaven *et al.* 2017)	Blockchain technology can streamline corporate operations while producing secure, reliable records of agreements and transactions in the banking and financial services sector.
Government	(Carter & Ubacht 2018) (Verma & Sheel 2022)	The value of blockchain in improving efficiency and transparency in public services run by the government.
IOT	(Reyna *et al.* 2018) (Abdelmaboud *et al.* 2022)	Innovations in blockchain technology and how they might be used in IoT applications to enhance quality of life are hot topics in today's research community.
supply chain	(Chang *et al.* 2022) (Van Nguyen *et al.* 2023)	The blockchain offers a great deal of potential to improve several parts of the Supply Chains.

digital ledger known as a "blockchain". A distributed database houses the unchanging and permanent data.

Following are some potential future applications of blockchain technology in various industries: (Monrat *et al.* 2019) (Rupa *et al.* 2022) (Pattanaik *et al.* 2022).

5 CONCLUSION

A transmitted database that records each transaction that has ever occurred in the system is what blockchain technology does. The main benefit of blockchain is that it eliminates the requirement for a trustworthy third party to mediate interactions between dubious parties. Blockchain technology offers a decentralized domain for exchanges, where each transaction is logged in an open record that is visible to all. Each of its consumers will receive privacy, security, security, and simplicity from Blockchain. As a result, it would appear that blockchain technology has the ability to improve both existing projects' employment opportunities and frameworks for tracking the historical development of artifacts through a far better, simpler record structure. In this study, we assess the situation and demonstrate how the qualities of this technology might support open science.

REFERENCES

Abdelmaboud, A., Ahmed, A. I. A., Abaker, M., Eisa, T. A. E., Albasheer, H., Ghorashi, S. A., & Karim, F. K. (2022). Blockchain for IoT Applications: Taxonomy, Platforms, Recent Advances, Challenges and Future Research Directions. *Electronics*, 11(4), 630.

Andreev, R. A., Andreeva, P. A., Krotov, L. N., & Krotova, E. L. (2018). Review of Blockchain Technology: Types of Blockchain and Their Application. *Intellekt. Sist. Proizv.*, 16(1), 11–14.

Carter, L., & Ubacht, J. (2018, May). Blockchain Applications in Government. In *Proceedings of the 19th Annual International Conference on Digital Government Research: Governance in the Data Age* (pp. 1–2).

Chang, A., El-Rayes, N., & Shi, J. (2022). Blockchain Technology for Supply Chain Management: A Comprehensive Review. *FinTech*, 1(2), 191–205.

Dagur, A., Kaushik, A., Rastogi, A., Singh, A., Kumar, A., & Chaturvedi, R. (2021). Optimization of Queries in Database of Cloud Computing. In Data Intelligence and Cognitive Informatics: *Proceedings of ICDICI 2020* (pp. 325–332). Springer Singapore

Dagur, A., Malik, N., Tyagi, P., Verma, R., Sharma, R., & Chaturvedi, R. (2021). Energy Enhancement of WSN Using Fuzzy C-means Clustering Algorithm. *In Data Intelligence and Cognitive Informatics: Proceedings of ICDICI 2020* (pp. 315–323). Springer Singapore.

Dib, O., Brousmiche, K. L., Durand, A., Thea, E., & Hamida, E. B. (2018). Consortium Blockchains: Overview, Applications and Challenges. *International Journal On Advances in Telecommunications*, 11 (1&2), 51–64.

Dinh, T. T. A., Wang, J., Chen, G., Liu, R., Ooi, B. C., & Tan, K. L. (2017, May). Blockbench: A Framework for Analyzing Private Blockchains. In *Proceedings of the 2017 ACM International Conference on Management of Data* (pp. 1085–1100).

Fu, X., Wang, H., & Shi, P. (2021). A Survey of Blockchain Consensus Algorithms: Mechanism, Design and Applications. *Science China Information Sciences*, 64(2), 1–15.

Ge, Z., Loghin, D., Ooi, B. C., Ruan, P., & Wang, T. (2022). Hybrid Blockchain Database Systems: Design and Performance. *Proceedings of the VLDB Endowment*, 15(5), 1092–1104.

Ghosh, P. K., Chakraborty, A., Hasan, M., Rashid, K., & Siddique, A. H. (2023). Blockchain Application in Healthcare Systems: A Review. *Systems*, 11(1), 38.

Kushwaha, S. S., Bairwa, A., Joshi, S., Chaurasia, S., Hemrajani, P., & Kumar, K. (2023). Blockchain in Healthcare, Supply-Chain Management, and Government Policies. In *Machine Learning, Blockchain, and Cyber Security in Smart Environments* (pp. 145–156). Chapman and Hall/CRC.

Li, W., He, M., & Haiquan, S. (2021, June). An Overview of Blockchain Ttechnology: Applications, Challenges and Future Trends. In *2021 IEEE 11th International Conference on Electronics Information and Emergency Communication (ICEIEC) 2021 IEEE 11th International Conference on Electronics Information and Emergency Communication (ICEIEC)* (pp. 31–39). IEEE.

Mahmood, S., Chadhar, M., & Firmin, S. (2022). Cybersecurity Challenges in Blockchain Technology: A Scoping Review. *Human Behavior and Emerging Technologies*, 2022.

Monrat, A. A., Schelén, O., & Andersson, K. (2019). A Survey of Blockchain from the Perspectives of Applications, Challenges, and Opportunities. *IEEE Access*, 7, 117134–117151.

Patel, R., Migliavacca, M., & Oriani, M. (2022). Blockchain in Banking and Finance: Is the Best Yet to Come? A Bibliometric Review. *Research in International Business and Finance*, 101718.

Pattanaik, R. K., Mohapatra, S. K., Mohanty, M. N., & Pattanayak, B. K. (2022). *System Identification Using Neuro Fuzzy Approach for IoT Application. Measurement: Sensors*, 24, 100485.

Rahman, K. T. (2021). Applications of Blockchain Technology for Digital Marketing: A Systematic Review. *Blockchain Technology and Applications for Digital Marketing*, 16–31.

Reyna, A., Martín, C., Chen, J., Soler, E., & Díaz, M. (2018). On Blockchain and Its Integration with IoT. Challenges and Opportunities. *Future Generation Computer Systems*, 88, 173–190.

Rupa, C., MidhunChakkarvarthy, D., Patan, R., Prakash, A. B., & Pradeep, G. G. (2022). Knowledge Engineering–based DApp Using Blockchain Technology for Protract Medical Certificates Privacy. *IET Communications*, 16(15), 1853–1864.

Tapscott, Don and Tapscott, Alex, "Blockchain Revolution: How the Technology Behind Bitcoin is Changing Money, Business, and the World," Penguin, 2016

Treleaven, P., Brown, R. G., & Yang, D. (2017). Blockchain Technology in Finance. *Computer*, 50(9), 14–17.

Van Nguyen, T., Cong Pham, H., Nhat Nguyen, M., Zhou, L., & Akbari, M. (2023). Data-driven Review of Blockchain Applications in Supply Chain Management: Key Research Themes and Future Directions. *International Journal of Production Research*, 1–23.

Varma, P., Nijjer, S., Kaur, B., & Sharma, S. (2022). Blockchain for Transformation in Digital Marketing. *In Handbook of Research on the Platform Economy and the Evolution of E-Commerce* (pp. 274–298). IGI Global.

Verma, S., & Sheel, A. (2022). Blockchain for Government Organizations: Past, Present and Future. *Journal of Global Operations and Strategic Sourcing*.

Wylde, V., Rawindaran, N., Lawrence, J., Balasubramanian, R., Prakash, E., Jayal, A., … & Platts, J. (2022). Cybersecurity, Data Privacy and Blockchain: A Review. *SN Computer Science*, 3(2), 1–12.

Zheng, Z., Xie, S., Dai, H., Chen, X., & Wang, H. (2017, June). An Overview of Blockchain Technology: Architecture, Consensus, and Future Trends. In *2017 IEEE International Congress on Big Data (BigData congress)* (pp. 557–564). IEEE.

Artificial Intelligence, Blockchain, Computing and Security – Dagur et al. (Eds)
© 2024 The Author(s), ISBN: 978-1-032-67841-2

Infrastructure as code: Mapping study

Abbas Mehdi* & Abhineet Anand*
Department of Computer Science and Engineering, Apex Institute of Technology (CSE), Chandigarh University, Gharuan, Mohali, Punjab, India

ABSTRACT: Framework as code (IaC) is the training to consequently arrange framework conditions and to arrange neighborhood and remote cases. Experts consider IaC as a central point of support to carry out DevOps rehearsals, which assists them with quickly conveying programming and administrations to end- clients. Data innovation (IT) associations, for example, GitHub, Mozilla, Facebook, Google and Netflix have taken on IaC. A precise planning concentrate on existing IaC examination can assist scientists with distinguishing potential exploration regions connected with IaC, for instance deformities and security blemishes that might happen in IaCscripts. Infrastructure as code (IaC) is a bunch of strategies which utilize "code (in lieu of than manual tasks) for getting arrangement with (virtual) machines as well as organizations, introducing conditions, and designing the turn of events and creation climate for the device or programming within reach. The framework constrained by this code incorporates both the actual machines ("exposed metal") and virtualized machines,docker holders, programming characterized virtual organizations. This code ought to be created and overseen utilizing a similar rendition control framework as some other vault, for outline, it ought to be built, tried, and warehoused in a version-controlled storehouse.

Keywords: IaC, Docker, REST APIs, Terraform, Cloud Computing

1 INTRODUCTION

IaC code can be utilized normally all through advancement, joining, and creation conditions. This improves climate equality and can eliminate situations where programming works in a single designer's current circumstance however not really for another engineer, or situations where programming works being developed yet not in the reconciliation or creation climate [9]. The foundation code utilized for IaC should be put away in a rendition-controlled storehouse. This empowers incredible forming of a sent framework. Any transformation of the foundation can be created utilizing the IaC code relating to the ideal release. Together, robotization and forming convey the possibility to reproduce an organization productively and reliably. This can be utilized to move back a switch made during improvement, reconciliation, or similarly as creation and to help inconvenience ticket recovery and investigating. IaC has the option to enable an IT activity set up as a regular occurrence called permanent foundation. In a conventional tasks approach, framework and application programming is associated with individual hubs. Additional time, every hub is exclusively fixed, programming is refreshed, and network and other setup boundaries are expanded depending on the situation. Setup floats might create, for instance, as the fix up level fluctuates across hubs [7]. In a couple of cutting-edge cases, hubs can be reproduced just from reinforcement, without any means to remake the design without any preparation. In an unchanging framework, patches, updates, and setup changes are never placed into the sent hubs. All things considered, another variant of the IaC

*Corresponding Authors: 786abbasmehdi@gmail.com and abhineet.e13847@cumail.in

DOI: 10.1201/9781032684994-24

code is worked with the adjustments that mirror the required changes to the sent foundation and applications. Climate equality permits the new adaptation to be evaluated being developed and joining conditions going before to creation, and climate forming make accessible the new changes to be moved back assuming there is an unforeseen issue subsequent to conveying to creation. Every one of the features about IaC become our motivation for this undertaking [6].

Code survey is generally pitched as a product "improvement" practice. Notwithstanding, as foundation including design of the executives, organization scripts, provisioning shows, bundles, and runbooks becomes code, those antiques become testable as code. Testability is a conspicuous excellence whose advantages are presently all around acknowledged by experts [10]. Similarly as framework code fills in as a typical language for both turn of events and tasks centered groups, survey is the normal gathering for code conversations and carries with it one more arrangement of virtues. There are multiple ways that designing associations survey the source code that they compose. By and large, the source code may be all printed out and afterward assessed by specialists before the code is run. Specialists could meet as a gathering irregularly to survey proposed changes at an undeniable level (maybe serving as an engineering item, or sending methodology synopsis). Designers could likewise review a whole code base for security or consistency reasons. These are fine practices. Yet, this article is about the kind of code survey where each change is inspected consistently as it is grown, for example, sending a fix to a mailing list or a GitHub pull request.

There are a few clear explanations behind doing code survey. Most straightforwardly, the cycle can find bugs before the product is running underway when they are decisively less expensive to address. Code survey additionally works on the "transport factor" for parts by guaranteeing that no less than two individuals comprehend how each change functions and all the more significantly why it was made. Code survey helps separate little territories of a specific style ("this is clearly grouping X code") and spread reliable prescribed procedures all through an association.

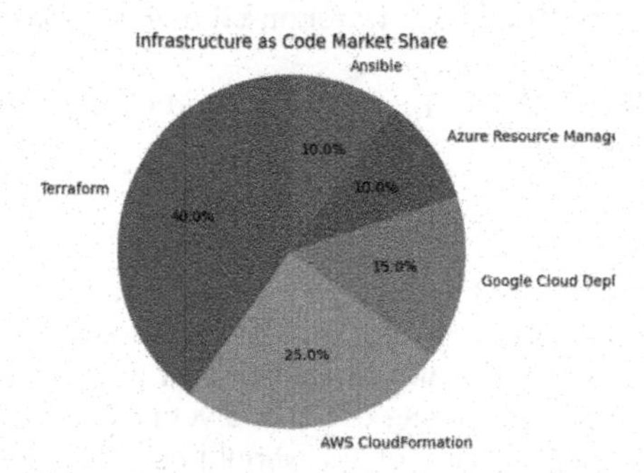

Figure 1. Infrastructure as code market share.

Foundation code can be hard to test, as it is seldom overwhelmed by unadulterated capabilities and nearly by definition collaborates broadly with the nearby operating system and distant frameworks. While testing and code survey never need to be restrictive practices, audit of foundation code enjoys the critical benefit of continuously being appropriate. Code might be hard to test or require a broad framework before the principal unit test could be composed, yet all changes can promptly be investigated. Code audit could try and be important to get that test framework working for foundation code. At AddThis we utilize a blend of chefspec (speedy running confirmation like "Was the right guidance passed to the design of the board framework?") and serverspec (make a VM and check: "Was the right catalog tree truly made?") for testing our setup the executive's recipes. When to utilize which one while fulfilling other helpful properties like limiting duplication takes significantly more subtlety than a basic unit test. Banter over the legitimate experiments is in many cases answerable for more editorial than the genuine code.

1.1 *Motivation*

IaC is firmly connected with DevOps. By mechanizing the production of execution/test conditions, IaC rehearses advanced lithe qualities. Decreasing the time that is between committing a shift to a framework and the shift being set into creation, while guaranteeing raised greatness incredibly improves programming improvement. The point of the review is to foster a code store for provisioning of virtualized asset occurrences for all machines associated with the item sending and incorporating them to the current engineer pipelines [15]. Added exertion of asset provisioning and arrangement of those assets can hinder advancement exertion. This could almost certainly make engineers skip coordination tests by and large. Requirement of such checks must be made doable assuming the whole situation is mechanized and all around incorporated into the testing pipeline. This ought to assist in authorizing greater quality checks and revelation of bugs and issues with willingness to be faster and dependable.

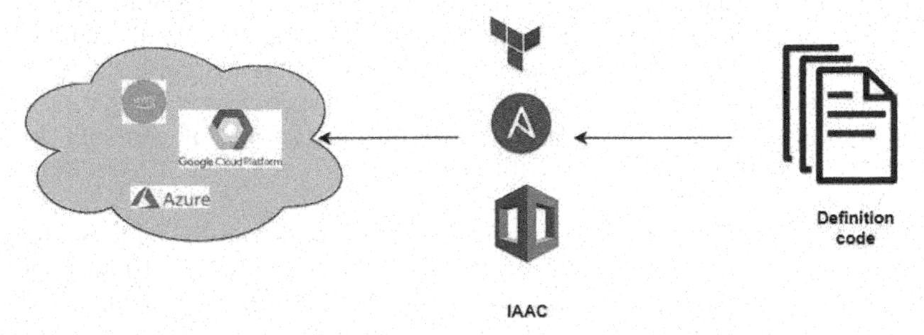

Figure 2. Infrastructure as code tools.

1.2 *Literature review*

Stefano Dalla Palma *et al.* [2022] Foundation as-code (IaC) is the DevOps work on empowering the executives and provisioning of framework through the meaning of machine-intelligible records, hereinafter alluded to as IaC scripts. Alsoto other source code curios, these documents might contain deserts that can block their right working. In this paper, we target surveying the job of item and cycle measurements while foreseeing flawed IaC scripts. We propose a completely incorporated AI structure for IaC Imperfection Expectation, that takes into consideration storehouse slithering, measurements assortment, model structure, and assessment. To assess it, we dissected 104 undertakings and utilized five AI classifiers to look at their exhibition in hailing dubious blemished IaC scripts. The vital consequences of the review report Irregular Woods as the best-performing model, with a middle AUC-PR of 0.93 and MCC of 0.80. Besides, basically for the gathered tasks, item measurements recognize damaged IaC scripts more precisely than process measurements. Our discoveries put a gauge for researching IaC Imperfection Expectation and the connection between the item and cycle measurements, and IaC contents' quality.[1]

Michael Howard *et al.* [2022] Fostering a product administration requires a severe programming improvement life cycle and interaction. This cycle requests controlling all application code through source control of the board as well as a thorough forming and spreading technique. Be that as it may, the stage and foundation likewise benefit from this meticulousness. Programming administrations should be conveyed to an objective runtime climate and provisioning that climate through manual client activities is dreary and mistake inclined. Provisioning physically additionally becomes restrictive as the quantity of assets develop and spread worldwide over various districts. The response is to apply a similar meticulousness to provisioning the framework as applied to fostering the application programming. Terraform gives a stage permitting framework assets to be characterized in code. This code not just permits the computerization of the framework provisioning yet in addition considers a severe turn of events and survey life cycle, same as the application software.[2]

Julio Sandobalín *et al.* [2019] Foundation as Code (IaC) is a methodology for framework robotization that depends on programming improvement rehearses. The IaC approach upholds code-driven apparatuses that utilize contents to determine the creation, refreshing and execution of cloud foundation assets. Since each cloud supplier offers an alternate kind of foundation, the meaning of a framework asset (e.g., a virtual machine) infers composing a few lines of code that enormously rely upon the objective cloud supplier. Model-driven apparatuses, in the meantime, dynamic the intricacy of utilizing IaC scripts through the undeniable level demonstrating of the cloud framework. In a past work, we introduced a framework displaying approach and device (Argon) for cloud provisioning that uses model-driven designing and supports the IaC approach. We utilized the Stomach muscle/BA hybrid plan to arrange the singular trials and the straight blended model to measurably break down the information gathered and consequently get exact discoveries. The aftereffects of the singular examinations and metaanalysis demonstrate that Argon is more viable as regards supporting the IaC approach with regards to characterizing the cloud foundation. The members likewise saw that Argon is simpler to utilize and more helpful for determining the foundation assets. Our discoveries recommend that Argon speeds up the provisioning system by demonstrating the cloud framework and robotizing the age of contents for various DevOps instruments when contrasted with Ansible, which is a code- driven device that is enormously utilized in practice.[3]

Shivam *et al.* [2020] Infrastructure as code (IaC) is a bunch of techniques which utilize "code (in lieu of than manual tasks) for getting arrangement with (virtual) machines as well as organizations, introducing conditions, and designing the turn of events and creation climate for the device or programming within reach. The foundation constrained by this code incorporates both the actual machines ("exposed metal") and virtualized machines, docker compartments, programming characterized virtual organizations. This code ought to be created and overseen utilizing a similar variant control framework as some other vault, for representation, it ought to be built, tried, and warehoused in a version controlled store. Despite the fact that IT administrators have long involved robotization by the utilization of impromptu pre arranging for errands, IaC innovation and practices arose with the presentation of distributed computing, and especially foundation as-a- administration (with enormous names like Google and Microsoft offering their own cloud framework administrations). While cloud-based specialist organizations empower the managerial control center that theoretical an intuitive application on top of REST APIs, it isn't practical to utilize an administration control center to make a computerized framework with additional perplexing hubs. For instance, making a new virtualized asset utilizing Microsoft Purplish blue requires an IT administrator venturing through 4 web-structures and filling nearly at least 20 fields. VMs are made and destroyed ordinarily during the day with arrangement and tests running ceaselessly, so playing out these assignments genuinely aren't fitting. [4]

Table 1. Basic comparison of related work.

Reference	Focus	Key Findings
[1]	Within-project defect prediction of Infrastructure-as-Code (IaC)	Developed a prediction model for IaC defects using process and product metrics
[2]	Terraform for automating Infrastructure	Introduces Terraform as a tool for automating IaaS and highlights its benefits and drawbacks
[3]	Comparison of model-driven and code-centric approaches to supporting IaC	Found that model-driven approaches have better support for correctness

2 PROPOSED METHOD

To foster the framework expected to computerize the job that needs to be done, the arrangement of virtual foundation with administrations pre-installed and in a prepared to convey stage, utilization of open-source device Terraform was picked. The work process around terraform was separated into three sections where the initial segment is Code [13]. During this stage, a decisive language is utilized to characterize out end express, this incorporates every one of the traits and properties of the ideal foundation. The subsequent stage is the arrangement stage, it utilizes a Terraform order and it is basically to approve our revelatory records and run static as well as unique keeps an eye on the code vaults. It likewise ascertains the delta between the ongoing assets accessible and the assets wanted, subsequently just provisioning those additional or adjusted. The third and last stage for our undertaking turns into the apply stage where our asset setup and Programming interface tokens are utilized to speak with true cloud supplier Programming interface end-focuses and work with those end focuses to convey us the assets we characterized. This stage involves specific result variables and is basically one more module in the Terraform parallel.

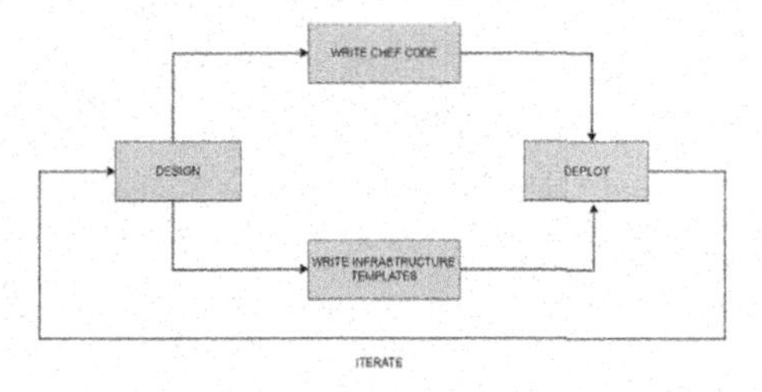

Figure 3. Infrastructure as code work flow.

2.1 *Results and analysis*

There are many devices utilized in CSD, for each stage (for example plan, execution, and testing), as well as each movement (for example drawing, teaming up, composing, and building). How much construction of the data is emphatically connected with the instrument? Some data is not difficult to catch and simple for human correspondence, for example, whiteboard representations or discussions intalks. Simultaneously, these kinds of data are hard for programmed handling. Source code then again, can be naturally handled.

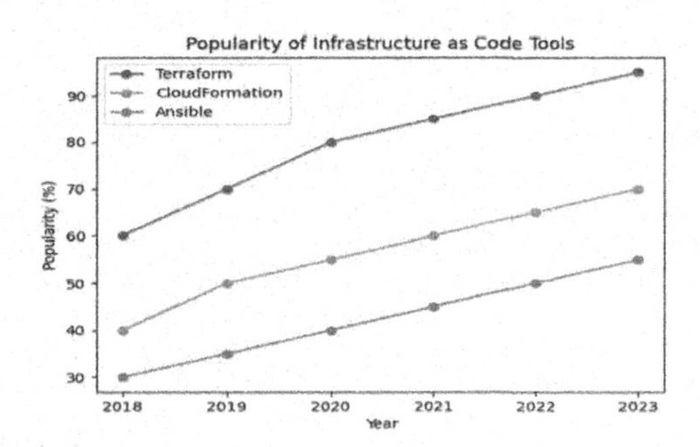

Figure 4. Infrastructure as code tools popularity.

3 CONCLUSION

Future exploration is expected to research in the event that barely enough forthright documentation can be restricted to significantly shaping considerations by utilizing casual whiteboard draws, along with the classified Programming Interface documentation. The forthright documentation ought to be joined by a plan when-done after a cycle is finished. Whatever can be created or figured out isn't expected to record since it is accessible whenever. Pertinent for activities, support, and information move are choices, contemplations about the product item and cycle, and group association.

REFERENCES

[1] Stefano Dalla Palma, Dario Di Nucci, Fabio Palomba, and Damian A. Tamburri "Within-Project Defect Prediction of Infrastructure-as-Code Using Product and Process Metrics" *IEEE Transactions on Software Engineering*, VOL. 48, NO. 6, JUNE 2022

[2] Michael Howard *Terraform — Automating Infrastructure as sService* 2022.

[3] Julio Sandobalín, Emilio Insfran, and Silvia Abrahão "On the Effectiveness of Tools to Support Infrastructure as Code: Model-Driven Versus Code- Centric "*IEEE Access*, date of current version January 28, 2020.

[4] Shivam, Poornima Kulkarni2 "Infrastructure as Code: A Devops First Approach" *International Research Journal of Engineering and Technology (IRJET) e-ISSN: 2395-0056* Volume: 07 Issue: 07 | July 2020

[5] MatejArtac, Tadej Dovsak, Elisabetta Di Nitto, Michele Guerriero, Damian Andrew Tamburri "*DevOps: Introducing Infrastructure-as-Code*" dec 2017

[6] Knauss E., Schneider K., Stapel K.A *Game for Taking Requirements Engineering More Seriously 2008 Third International Workshop on Multimedia and Enjoyable Requirements Engineering - Beyond Mere Descriptions and With More Fun and Games* (2008), pp. 22–26

[7] Kitchenham Barbara A, Brereton Pearl, Turner Mark, Niazi Mahmood K, Linkman Stephen, Pretorius Rialette, Budgen David *Refining the Systematic Literature Review Process– Two Participant-Observer Case Studies Empirical Softw. Eng.*, 15 (6) (2010), pp. 618–653

[8] Král Jaroslav, emlika Michael Requirements specification: What Strategy under What Conditions Proceedings - SERA 2007: *Fifth ACIS International Conference on Software Engineering Research, Management, and Applications* (2007), pp. 401–408.

[9] Lwakatare Lucy Ellen, Kuvaja Pasi, *Oivo Markku Relationship of DevOps to Agile, Lean and Continuous Deployment International Conference on Product-Focused\Software Process Improvement*, Springer (2016),

[10] Theunissen Theo, Hoppenbrouwers Stijn, Overbeek Siet se In Continuous Software Development, Tools are The Message for Documentation *Proceedings of the 23rd International Conference on Enterprise Information Systems, SCITEPRESS - Science and Technology Publications* (2021),

[11] Mussbacher G., Amyot D., Breu R., Bruel J.-M., Cheng B. H.C, Collet P., Combemale B., France R. B., Heldal R., Hill J., Kienzle J., Schöttle M., Steimann F., Stikkolorum D., and Whittle J., "The Relevance of Model Driven Engineering Thirty Years from Now," in *Model-Driven Engineering Languages and Systems*. Cham, Switzerland: Springer, 2014, pp. 183–200

[12] Agner L. T. W., Soares I. W., Stadzisz P. C., and Simão J. M., "A Brazilian Survey on UML and Model-driven Practices for Embedded Software Development," *J. Syst. Softw.*, vol. 86, no. 4, pp. 997–1005, Apr. 2013.

[13] Bernal A., Cambronero M. E., Valero V., Nunez A., and Canizares P. C., "A Framework for Modeling Cloud Infrastructures and User Interactions," *IEEE Access*, vol. 7, pp. 43269–43285, 2019.

[14] Bernal A., Cambronero M. E., Valero V., Nunez A., and Canizares P. C., "A Framework for Modeling Cloud Infrastructures and User Interactions," *IEEE Access*, vol. 7, pp. 43269–43285, 2019.

[15] Casola V., Benedictis A. D., Rak M., Villano U., Rios E., Rego A., and Capone G., "MUSA Deployer: Deployment of Multi-cloud Applications," in *Proc. IEEE 26th Int. Conf. Enabling Technol., Infrastruct. Collaborative Enterprises (WETICE)*, Jun. 2017, pp. 107–112.

[16] Ferry N., Chauvel F., Song H., Rossini A., Lushpenko M., and Solberg A., "CloudMF: Model-driven Management of multi-cloud Applications," *ACM Trans. Internet Technol.*, vol. 18, no. 2, pp. 1–24, Jan. 2018

[17] Parnin C., Helms E., Atlee C., Boughton H., Ghattas M., Glover A., Holman J., Micco J., Murphy B., Savor T., Stumm M., Whitaker S., and Williams L., "The Top10 adages in Continuous Deployment," *IEEE Softw.*, vol. 34, no. 3, pp. 86–95, May 2017.

[18] Geerling J., *Ansible for DevOps: Server and Configuration Management for Humans*. Victoria, BC, Canada: Leanpub, 2015.

[19] Di Nitto E., Matthews P., Petcu D., and Solberg A., *Model-Driven Development and Operation of Multi-Cloud Applications*. Cham, Switzerland: Springer, 2017.

[20] Steinberg D., Budinsky F., Merks E., and Paternostro M., *Eclipse Modeling Framework*. Reading, MA, USA: Addison- Wesley, 2008.

Artificial Intelligence, Blockchain, Computing and Security – Dagur et al. (Eds)
 ISBN: 978-1-032-67841-2

Virtualization as a cloud computing technology: Mapping study

Milandeep Kour Bali* & Abhineet Anand*
Department of Computer Science and Engineering Apex Institute of Technology (CSE), Chandigarh University Gharuan, Mohali, Punjab, India

ABSTRACT: The beginning of virtualization started up in the era of 1960s, which says that virtualization is the technology which is used as a splitting up and adding up of physical resources into logical resources as we want orin other words we can say that it is a technology that transfers resources from hardware into software for e.g. CPU, Memory, storage, network resources Virtualization In cloud used as scaling up and down of resources in the large amount of data that allow users to access when they need it. It allows providers to virtualize a server, storage or many physical hardware resources or data centers and they also provide various services such as infrastructure as a service (IAAS), software as a service (SAAS) and platform as a service (PAAS). Virtualization is used increasingly and performs countless benefits. As earlier, to manage our physical data centers traditional ways were used which results in more focus on physical hardware rather than on main performances. After the enhancement of virtualization, the traditional way gets improved and also reduces the cost, capex/opex i.e Capital Expenditure and Operational Expenditure.

Keywords: Virtualization, Xen server, Cloud Computing, Virtual Environment, Hyperviser

1 INTRODUCTION

In the general term virtualization refers as a single machine or instance of a physical object split into multiple virtual machines or instances of an object so in other word, we can say that we can run multiple virtual machines in a single physical hardware or physical server [1]. We can also simulate the interfaces of a physical resource by multiplexing, aggregation, emulation, multiplexing and emulation. Actually, with the help of virtualization multiple OS (operating system) and Applications can run on the same machine and its same physical hardware at the same time which is cost effective and increases the flexibility and utilization of physical hardware. The purpose of the cloud isto provide storage to users, applications and data which are stored on the cloud and they can access through any part of the world. Cloud Computing [2] is derived from two words cloud and computing where cloud refers to network or internet and computing is to compute storage, networking, server and other IT resources or services. This can be accessed through the internet from where anyone wants to access it. Many IT companies including small and large escorts the traditional way to build their interior infrastructure. The infrastructure means a specific room where all the resources including database, networking, mail, cables, routers, switches, firewalls, modems are stored and are managed by ourselves and that storage room is known as server room or physical data centers. In the traditional way we can manage our resources on premises or physical data centers and in cloud computing is basically on demand delivery of compute resources [11]. Whatever we can use the resources, we have to pay accordingly. In cloud computing, it provides us with services and deployment models.

*Corresponding Authors: 22mcc10008@cumail.in and abhineet.e13847@cumail.in

DOI: 10.1201/9781032684994-25

1.1 *Related work*

[1] He suggested safeguarding in the cloud to check and preserve both guest and client-server methodologies of software probity and completely obscure for both services. Client and service provider also proposed the development of cloud computing in terms of virtualization. Through a few implementation examples, the evaluation of server virtualization in the cloud platform is summed up.

[2] He proposed the development of virtualization and the architecture in his research and also analyzed the decrease in time and usage of energy. He also concluded the terminologies of virtualization and cloud computing. He suggested the methodologies of cloud computing and he also suggested the efficiency and cost effects of systems. He discussed the history of virtualization and examined the design of a commercial virtualization product. With the help of the VMware vCloud tools, we established the cloud platform and continued to apply virtualization technology to cloud computing.

[3] To protect our data in a traditional way and virtual cloud, strong techniques like encryption should be done in order to secure connections and policies. By doing these techniques, the rest of the resources as well as our data centers, interior structure will get secure and trustworthy, and transfer of data or resources can be performed. Although the deployment architecture offered by cloud computing has the capacity to address vulnerabilities found in conventional IS, the effectiveness of conventional countermeasures may be thwarted by the dynamic nature of the system. Depending on the specific requirements of the organization, many cloud models can be chosen

[4] He prominently researches virtualization and hypervisors including Type 1 and Type 2 hypervisors with their pros and cons. Virtualization also upgrades the working production of the applications of strong security. Different approaches can be used to administer hypervisors, and some hypervisors even support numerous methods. Each hypervisor administration interface, both locally and remotely accessible, needs to be secured. Typically, the virtualization management system allows users to enable or disable the functionality for remote administration. If a hypervisor supports remote administration, a firewall should be used to limit access to all remote administration interfaces. Additionally, connections between hypervisors should be secured. Havinga specialized management network that is isolated from all other networks and that only authorized administrators may access is one possibility. The virtualization solution itself, ora separate third-party solution like a virtual private network,must include FIPS-approved encryption techniques for management communications that are sent over untrusted networks.

Table 1. Basic comparison of related work.

Reference	Year	Techniques	Findings
[1]	2016	Direct execution	Provides a survey of virtualization technologies used in cloud computing
[2]	2015	Hypercals	Presents a comprehensive review of energy-efficient virtual machine placement algorithms in cloud data centers
[4]	2022	Exist to root on privileges instruction	Offers a survey of resource management techniques employed in virtualized cloud environments

1.2 *Analysis of virtualization*

Virtualization is basically a technology in which we can simulate our hardware into multiple virtual machines and we can distribute our resources respectively or equally. Each VM is isolated to one another and also, we can install different OS on hardware or single machine. Virtualization is of various types as below *1.2.1 A: Server virtualization*

As Figure 1. Server virtualization refers to the installation of the virtual machine management directly on the server system. In server virtualization, a single physical server is divided into many servers based on demand and for load balancing.[7]

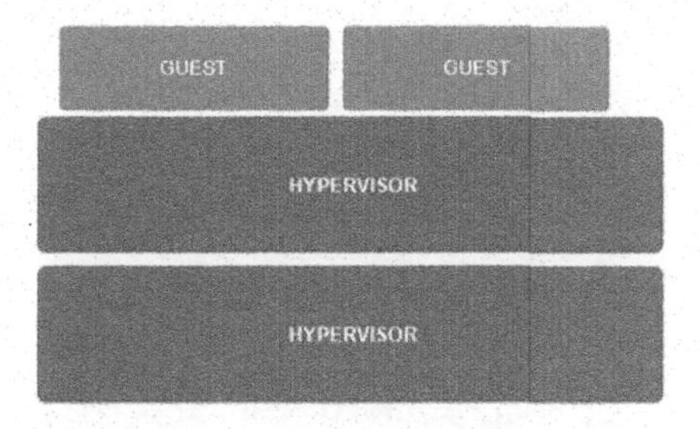

Figure 1. Server virtualization.

1.2.2 *B: Desktop virtualization*

As Figure 2. is based on a client and server model in which they respond to the request of the user. The user can remotely access their application from pc, laptop, Smartphone, tabs etc.[15]

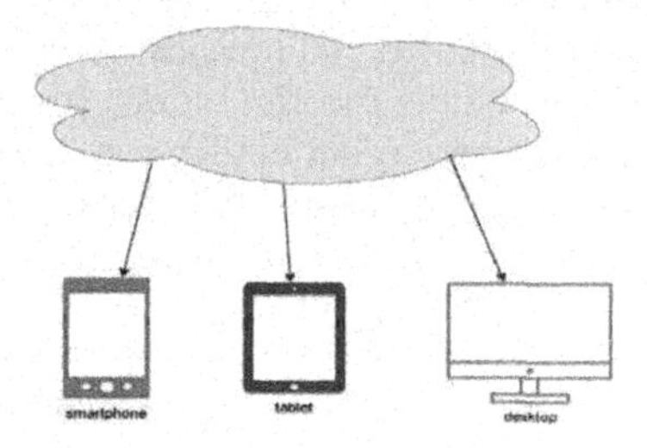

Figure 2. Desktop virtualization.

1.2.3 *C: Storage virtualization*

As Figure 3. In these storage is mostly used for backup and recovery purposes and multiple physical storage devices are grouped together, which appear as a single device.[18]

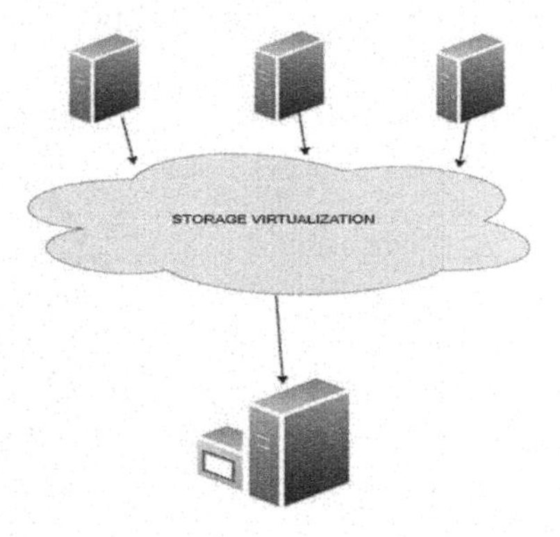

Figure 3. Storage virtualization.

1.2.4 *D: Hardware virtualization*

As illustrated in Figure 4, hardware virtualization is defined as the installation of virtual machine software or virtual machine manager (VMM) directly on the hardware system. Hardware virtualization is used to control and monitor the processor, memory, and other hardware resources.[13]

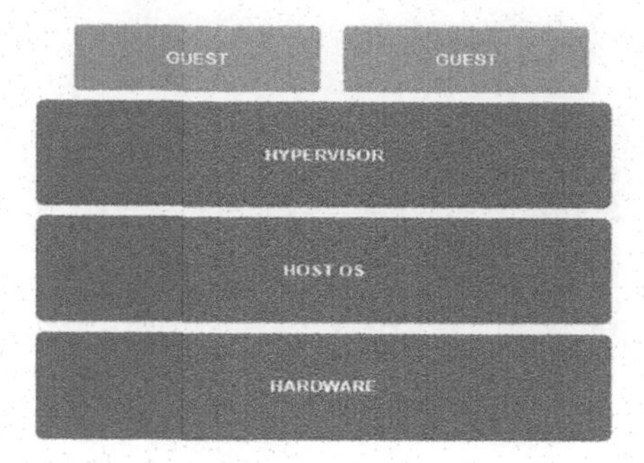

Figure 4. Hardware virtualization.

2 PROPOSED WORK

Here we had taken a case study of a company which is London based and has a data center of about 2000 sq feet, where it has installed 100 physical servers and large amount of server are needed to work because only one thing can run on one server and that one server can be AD server, web server or application server, but in this scenario 30 server are distributed into Infra and 70 for Applications [10]. The 30 servers include AD server, DNS server, financial/billing server, DHCP server, antivirus, firewalls and many more. These works are done physically before the virtualization, on an average each server contains 4GB RAM,60GB HDD and 2 core CPU. When these servers work in a huge data center, heating occurs and to control this the cooling system is also included for these 100 servers. Physical disk is attached with each of the 100 servers to store their database. For connectivity cabling and power whips are used which contains extra expenditure. To handle these data centers staff must be needed which increases our operational expenditure. The electricity and power backup also increased. In the end in the traditional way before virtualization the capital and operational expenditure also increased and the outcome of these are very low as per the expectations. So, for these huge issues virtualization are used as a solution.

Now in this case if virtualization is done then what will happen? Let us take one physical server and on this server, we will put virtualization software (also called hypervisor) and

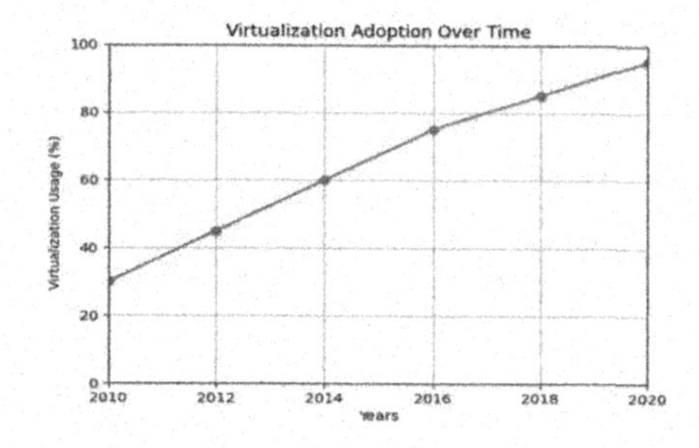

Figure 5. Analysis of virtualization.

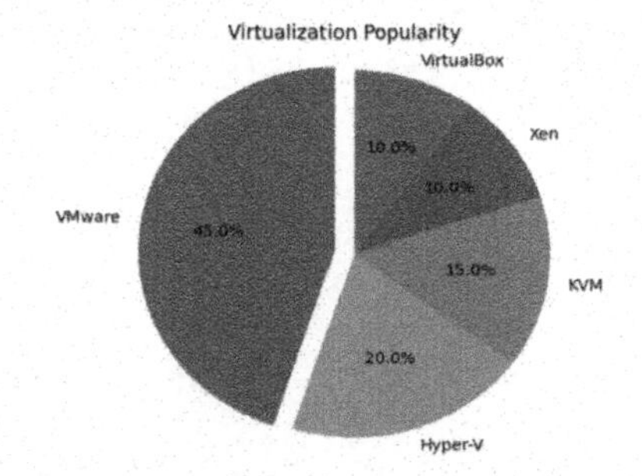

Figure 6. Virtualization popularity.

some virtual machine or virtual server. If our physical server contains 64GB RAM, 2TB HDD and 32 CPU,then these get distributed to each VM equally like 4GB RAM,100 GB HDD and 2 Core CPU on which any choice of server can be run. By this process after virtualization, we can access different VM on one server and these servers are isolated from each other and work virtually the same as in the traditional way. These VM behaves as individual identities which send the request to the physical host and then back from the main switches and cannot contact directly. After the virtualization the cost expenditure got reduced and physical amount of work also decreased. We can also say that one hardware can run a different operating server. In physical form 1 Os works only on one server i.e., ratio is 1:1 while in virtualization multiple work is performed on one server and the ratio is N: 1.

2.1 *Virtualization architecture*

The virtualization is basically based on the hypervisor and these hypervisors [9] are isolated to each and every OS (operating system) and applications are underlying the hardware so that a physical machine or host OS can run multiple virtual machines. Hypervisor is sometimes also called a virtual machine manager (VMM) and without a hypervisor we cannot simulate physical hardware. Hypervisor contains basically two types Type 1 Hypervisor And Type 2 Hypervisor

Type 1 hypervisor: As Figure 7. Is also called bare metal or native or enterprise hypervisor and is also known as firmware which is directly run on the system without an underlying OS [8]. Vmware ESXI or XEN is a Type 1 hypervisor that runs on the host server hardware without the underlying operating system.

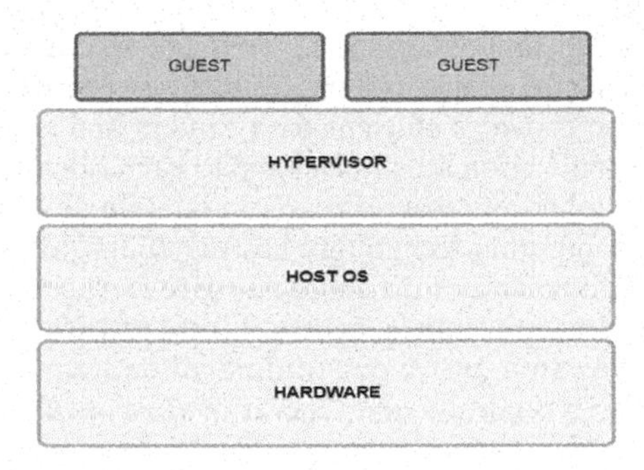

Figure 7. Type 1 hypervisor.

Type 2 hypervisor: As Figure 8. is also called a hosted hypervisor that runs with a conventional Operating system environment and the host os provided and Type 2 hypervisor

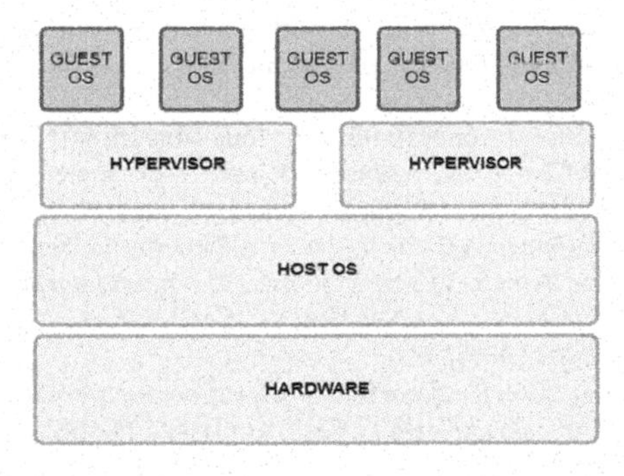

Figure 8. Type 2 hypervisor.

basically used for testing or learning purpose. Some examples of Type 2 hypervisors are Vmware Workstation and Virtualbox. Basically in Type 2 hypervisor we can run multiple virtual machines with underlying operating systems and Vmware workstation or Virtualbox software that helps to create multiple virtual machines in the same instance or single physical hardware.We can also run multiple applications on it.[6]

2.2 *Result analysis*

The proposed work we have given that the traditional method we were using gives low outcomes and high expenditures. But after virtualization the resources work efficiently and reduce the capital and operational expenditure which results in the profit in our business which implies that the traditional way gets improved. After the virtualization the cost expenditure gets reduced and physical amount of work also decreased. We can also say that one hardware can run a different operating server. In physical way 1 Os works only on one server i.e. ratio is 1:1 while in virtualization multiple work is performed on one server and the ratio is N:1.In the traditional way we can manage our resources on premises or physical data centers and in cloud computing is basically on demand delivery of compute resource. Whatever we can use the resources, we have to pay accordingly. In cloud computing, it provides us with services and deployment models.

3 CONCLUSION

Cloud computing is a rising technology in the IT area and will get assured only when vm is secured. Cloud computing enables various Services like Servers, networking, and databases over the web. Virtualization provides efficient data centers and low operation expenditure. Cloud Computing and Virtualization are interrelated to each other and cannot be separated. In virtualization various numbers of resources can run on a single server and are also isolated to each other. The cloud computing technology has the ability to provide consumer a wide range of services. The main objective of the cloud isto give storage to users, applications, and data that are saved in the cloud and can be accessed from anywhere in the world. After The Virtualization technique we can improve our traditional datacenter which also reduces the cost,capex/opex i.e Capital Expenditure and Operational Expenditure.

REFERENCES

[1] Goel N., Gupta A.and Singh S. N., "A Study Report on Virtualization Technique," *2016 International Conference on Computing, Communication and Automation (ICCCA)*, Greater Noida, India, 2016, pp. 1250–1255.

[2] Chen L., Patel S., Shen H. and Zhou Z., "Profiling and Understanding Virtualization Overhead in Cloud," *2015 44th International Conference on Parallel Processing*, Beijing, China, 2015, pp. 31–40, doi: 10.1109/ICPP.2015.12.

[3] Linthicum D. S., "Cloud-Native Applications and Cloud Migration: The Good, the Bad, and the Points Between," in *IEEE Cloud Computing*, vol. 4, no. 5, pp. 12–14, September/October 2017, doi: 10.1109/MCC.2017.4250932.

[4] Kapil D., Mittal V. and Gangodkar D. P., "Virtualization and Nested Virtualization Technology: Concept, Architecture and Attack Vector Model," *2022 International Conference on Computational Intelligence and Sustainable Engineering Solutions (CISES)*, Greater Noida, India, 2022, pp. 349–354, doi: 10.1109/CISES54857.2022.9844347.

[5] Benedikt J. *et al.*, "Virtualization Platform for Urban Infrastructure," *2022 22nd International Scientific Conference on Electric Power Engineering (EPE)*, Kouty nad Desnou, Czech Republic, 2022, pp. 1–5, doi: 10.1109/EPE54603.2022.9814159.

[6] Moravcik M., Kontsek M., Segec P. and Cymbalak D., "Kubernetes - Evolution of Virtualization," *2022 20th International Conference on Emerging eLearning Technologies and Applications (ICETA)*, Stary Smokovec, Slovakia, 2022, pp. 454–459, doi: 10.1109/ICETA57911.2022.9974681.
[7] Lv C., Zhang F., Gao X. and Zhu C., "LA-vIOMMU: An Efficient Hardware-Software Co-design of IOMMU Virtualization," *2022 IEEE Intl Conf on Parallel & Distributed Processing with Applications, Big Data & Cloud Computing, Sustainable Computing & Communications, Social Computing & Networking (ISPA/BDCloud/SocialCom/SustainCom)*, Melbourne, Australia, 2022, pp. 246–253, doi: 10.1109/ISPA-BDCloud-SocialCom-SustainCom57177.2022.00038.
[8] Rastogi R. and Aggarwal N., "A Review on Virtualization and Cloud Security," 2022 2nd *International Conference on Innovative Practices in Technology and Management (ICIPTM)*, Gautam Buddha Nagar, India, 2022, pp. 162–166, doi: 10.1109/ICIPTM54933.2022.9754172.
[9] Ahsan M. M., Gupta K. D., Nag A. K., Poudyal S., Kouzani A. Z.and Mahmud M. A. P., "Applications and Evaluations of Bio-Inspired Approaches in Cloud Security: A Review", *IEEE Access*, vol. 8, pp. 180799–180814, 2020
[10] Tim Mather, Subra Kumaraswamy and Shahed Latif, *Cloud Security and Privacy: An Enterprise Edition on Risks and Compliance (Theory in Practice) Ii*, O'Reilly Media, Sep. 2009, [online] Available: http://oreilly.com/catalog/9780596802776, ISBN 9780596802769.
[11] Chen L., Patel S., Shen H. and Zhou Z., "Profiling and Understanding Virtualization Overhead in Cloud", *44th International Conference on Parallel Processing*, 2015.
[12] Correa E. S., Fletscher L. A. and Botero J. F., "Virtual Data Center Embedding: A Survey", *IEEE Latin America Transactions*, 2015.
[13] Chen W., Xu L., Li G. and Xiang Y., "A Lightweight Virtualization Solution for Android Devices", *IEEE Transactions on Computers*, 2015.
[14] Viejo-Cortés J., Ruiz-De-Clavijo-Vázquez P., Ostúa-Arangüena E., Cano-Quiveu G. and Juan-Chico J., "Virtualization Environment for IT labs Development and Assessment," *2022 Congreso de Tecnología, Aprendizaje y Enseñanza de la Electrónica (XV Technologies Applied to Electronics Teaching Conference)*, Teruel, Spain, 2022, pp. 1–5, doi: 10.1109/TAEE54169.2022.9840655.
[15] Sun Y., "Research on Network Access Control System Model Based on Virtualization Technology," *2022 Fourth International Conference on Emerging Research in Electronics, Computer Science and Technology (ICERECT)*, Mandya, India, 2022, pp. 1–5, doi: 10.1109/ICERECT56837.2022.10059594.
[16] Shohei Mitani Taniya Singh Nakul Ghate *et al.* "Attribute-based low-complexity Network Access Control Policy with Optimal Grouping Algorithm" *IEICE Communications Express* vol. 11 pp. 10–14 2021.
[17] Nadiah M. Almutairy and Khalil H. A. Al-Shqeerat "A Survey on Security Challenges of Virtualization Technology in Cloud Computing" *International Journal of Computer Science and Information Technology* vol. 3 pp. 11–15 2019.
[18] Lambropoulos Georgios Mitropoulos Sarandis and Douligeris Christos "Improving Business Performance by Employing Virtualization Technology: A Case Study in the Financial Sector" *Computers* vol. 4 pp. 10–16 2021.
[19] Xiang X., Yu H.and Shu J., "Storage Virtualization Based Asynchronous Remote Mirror", 8th InternationalConference on Grid and Cooperative Computing, 2009.

Artificial Intelligence, Blockchain, Computing and Security – Dagur et al. (Eds)

Security challenges of IoT with its applications and architecture

Neha Bathla
Research Scholar, CSE Department of Computer Science and Engineering, Chitkara University Institute of Engineering & Technology, Chitkara University, Punjab, India

Amanpreet Kaur
Associate Professor, Department of Computer Science and Engineering, Chitkara University Institute of Engineering & Technology, Chitkara University, Punjab, India

ABSTRACT: IoT (Internet of Things) is the network of physical objects("things") in which embedded devices are used like sensors and software that are used for connecting the devices and objects over the internet. Currently, seven billion connected IoT devices today, by 2025 experts is expecting this number to grow to twenty-two billion. In this article we provide an introduction or overview of the IoT (Internet of Things), IoT architecture and functionality of each layer, its applications and security issues. The IoT are enabled by the newest developments in smart sensors and RFID. The basic principle of IoT is that it has smart sensors that work together directly without human participation to deliver the new applications. In the upcoming years, with the intelligent decision making IoT expected as bridge diverse technologies to enable new applications by connecting physical objects together. In this survey we also describe the importance of security and its challenges.

Keywords: IoT, RFID, Sensors, Actuators

1 INTRODUCTION

Internet of Thing is the arrangement or system of interconnected devices that transfers and exchange data over a wireless network without human involvement. The devices (like heart monitors, a remote or an automobile) with built-in sensors. The device that has an IP address interacts with each other to collect and share the data [2]. Through embedded technology these devices interact with the outer environment which assists these devices in taking the decisions. Since these devices can currently characterize themselves digitally. Here we can say that "This life change technology reduce the whole world to a small worldwide connected center by using a single key".

Internet: Means the Inter connectivity of the global network +

Things: Devices with embedded mechanism such as sensors, actuators, RFID (radio frequency identifier) tags, QR codes and many others.

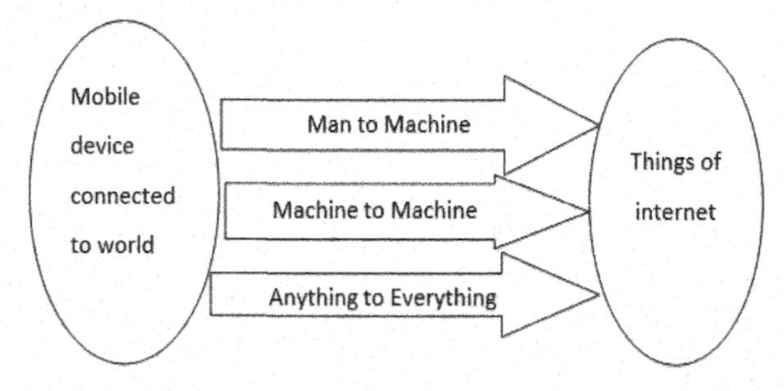

Figure 1. Internet of things.

DOI: 10.1201/9781032684994-26

2 SOME COMMUNICATION DEVICES IN IOT(INTERNET OF THINGS) ARE:

2.1 *Sensors*

Sensors/devices are used to collect data from the outer environment. Once the data gets software processes it. After processing it might decide to take an action like sending the alerts without participating (user) in it. Smart sensors are the essential enablers of IoT. Suppose, a situation of automatic monitoring of a farm so that it will just indicate the current condition of crops "let's take an example of 4 crops that needs water, After that I'm going to pouring water these crops so that this fulfils its need. That will possible because of the IoT technology works behind this [3–5].

1. The temperature sensor identifies the low temperature in the pot.
2. Then it activates the microprocessor such as Raspberry-Pi, Arduino boards.
3. Then it receives sensor signals over internet path like Wi-Fi, Bluetooth.
4. After that it will give alerts the user and motion sensor connected to the tap which turns on to pour this.

2.2 *Actuators*

[6] These are the devices which are using in comparison to sensors. It is a machine component that moves or controls the mechanism or the system. Sensors in the device sense the environment, then control signals are generated for the actuators according to the actions needed to perform. For example in electrical actuators it is actuated by a motor that converts electrical energy into the mechanical torque. So sensors and actuators are the very important part of IoT [7].

2.3 *RFID tags*

[8] It is a form of wireless communication that uses radio waves to identify & track objects (like books, vehicles, money etc). It uses electromagnetic fields to automatically identify and track tags attached to objects. RFID can also be used to track animals and birds by implanting RFID tags into them [2].

3 ARCHITECTURE OF IOT

The figure shown below shows the five-layer architecture of IoT. The different functions are performed by these five layers are discussedhere

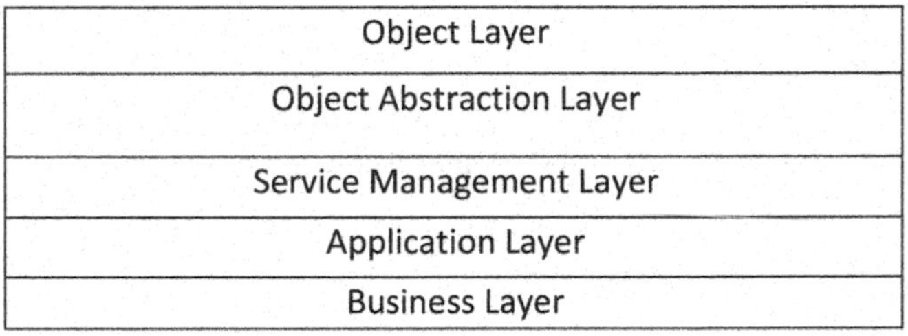

Object Layer
Object Abstraction Layer
Service Management Layer
Application Layer
Business Layer

Figure 2. Layers of IoT architecture.

3.1 *Object layer: [11]*

This layer is responsible for collecting and processing the data between the devices which has the capability for sensing and gathering the information as per the requirements or these physical objects collect the information(data)from the sensors, actuators. The data is transferred through a secure manner by this object layer. Presently, smart devices like sensors and actuators are compact, multipurpose, and efficient.

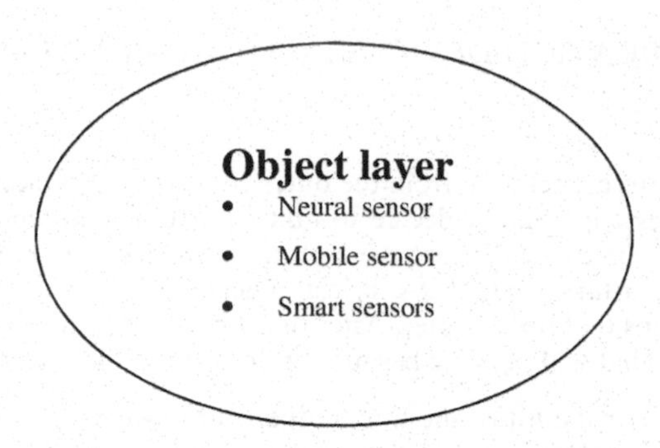

Figure 3. Object layer.

Many platform and services are used for the combination of these devices and the app developers were provided an access through these platforms, which is known as interoperability. The interoperability is implemented device to device communication between smart phones, watches, and wearable devices.

3.2 *Object abstraction layer: [11]*

From the object layer data is collected then it needs to be transmitted & processed. This layer connects the devices to the smart objects, server, cloud and network devices through various technologies, i.e., Wi-Fi, GSM, 3G, Bluetooth, NFC, etc. This layer also responsible for transferring the data between the two layers as shown in Figure 4. Many other responsibilities of this layer cloud computing and data management. Some applications are health monitoring, fitness tracking, and emergency services that are used for patient information through smart watches, mobiles, and body sensors.

3.3 *Service management layer: [10]*

This layer is responsible for the managing the IoT services. The function of this layer is to secure the IoT device analysis. It is also used to get the necessary information from the huge amount of data gathered by the sensors and to convert it into valuable result. In management service layer OSS(operational support system)is used for device management and configuration and also BSS (Billing support system) that is used for billing and reporting purposes.

3.4 *Application layer: [11]*

The function of this layer is to provide the data as per user command. This layer gives services to the healthcare sector, transportation sector and many more. Moreover, the IoT with the

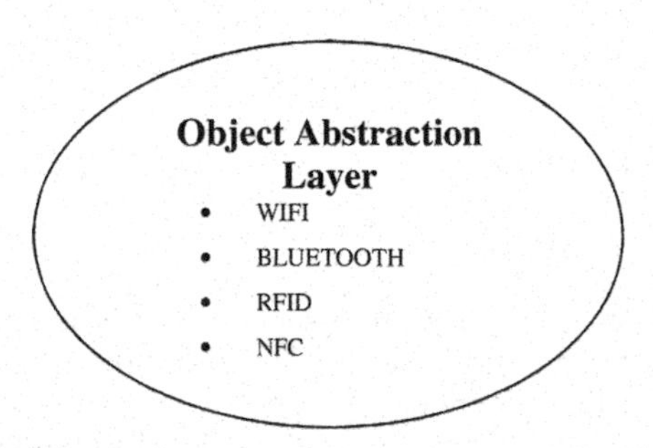

Figure 4. Object abstraction layer.

application layer improves the services by providing quality services, according to the user demand, which improves the social, financial, and technical aspects for the application layer. The responsibility of application layer is data formatting and the presentation in conventional network. In traditional networking Hypertext transfer protocol is used but this protocol is not appropriate because of the large overhead with limited resources. Other protocols are used with this IoT like MQTT and CoAP)

(d) Business Layer: [11] This function or responsibility of this layer is to manage overall/whole IoT system. This layer used many technologies like cloud computing, big data and databases etc. Based upon the data received from the upper layer i.e. application layer it makes the flowcharts and graphs.

4 LITERATURE SURVEY

This survey covers the related work that has been done in this field in last few years. Miorandi [2012] represents the next major leap onwards in the ICT sector of IoT. This survey concludes that with the heavy distribution of the embedded devices chances of merging the real and virtual world are consistently opens up the new scope in the business and research field [12]. Suo *et al.* [2012] In this article author summarized four layer architecture security in IoT by using the encryption algorithm and also represents that how to protect the data sensed by the sensors [13]. JainDeeksha *et al.* [2012] concluded that all the devices are identified uniquely on the network that takes decisions independently. Several issues of security and privacy and many applications are addressed in this article [14]. MadakamSomayya *et al.* [2015] proposed that the definition is not standard in worldwide and in the architecture the requirement of universal standardization is needed because the technologies which are using in the IoT are varying when it was used by many vendors, so in this case there is need of interoperability and there is also need of standard protocol for the better worldwide governance [15]. Santhi Sri *et al.* [2016] this paper summarized the latest trends in IoT architecture, protocols used for communication, design challenges, security issues and applications which have lot of scope in iot growth and gives a clear view of IoT thing [16]. Sathish Kumar *et al.* [2014] this survey presented the Internet of Things with its architecture, design goals and its applications in real life. They discussed many security and privacy issues at each layer of IoTs. They also identify many issues that occurred during the security and privacy that need to be identified by researchers to make a secure or trusted platform for the future IoT [17]. Chopra A. In [2020] This survey focused on security personnel of connectivity of old systems and operational Technology used at that time. By using of the right policies the connectivity of the devices across networks can be secured. There is a lot of security requirements needed to protect the information and lots of work has to done in each layer of IoT. By selecting the right policies personnel can get better operational technology across the world [18]. Alavi *et al.* [2018] summarized the issues that are generated in cities. The population has been increased in urban area because the population are moving from the rural to the urban area. So, there is requirement to give solution in different fields like healthcare, medical and energy sector. Also for the development the maintainable smart cities IoT is very important [19]. Li *et al.* [2019] proposed a dynamic approach for data centric IoT applications with respect to cloud platforms. The need of an proper device, software configuration and infrastructure needs well organised solutions to keep up the huge amount of IoT applications which are running on cloud platforms. IoT developers and researchers are actively engaged in developing solutions considering both massive platforms and heterogeneous nature of IoT objects and devices [20]. Noura *et al.* [2019] it proposed significance of interoperability interface in IoT. To get the efficient and reliable service it permit the integration of devices and services from various platform [21]. Bharath Kumar *et al.* [2018] stated that with the use of cloud computing and IoT technology the family members can monitoring the condition of patient in vegetative state remotely [22]. Sushan *et al.* [2019] stated the blueprint for the improvement in the protection of patient and its observation monitoring. This system was complex in the clinical settings and also the measurements of data sensors were not pertinent [23]. Durgadevi S., Anbazhagan *et al.* [2020] this article summarized IoT health monitoring system for wellness. It is cheap &generates congruous results. In the future IoT provides better results in this area [24]. Thilakarathne *et al.* [2022]: In the analysis, we noticed the security challenges of layered architecture in IOT. This paper explained different security challenges related to layered architecture -># application layer# data processing layer# network layer# physical layer. After discussing about these challenges, this also discussed on about how we can counter it. This research to be worthwhile resource to strengthen the security for future IOT application [37].

5 APPLICATIONS OF IOT

IoT is making our life a lot easier almost all appliances can now be monitored and in some cases operated remotely with the help of simple use case. There are many fields in which IoT plays an important role such as production, transportation, agriculture, medical and healthcare etc. This Figure 5 shows the various applications of IoT in different fields.

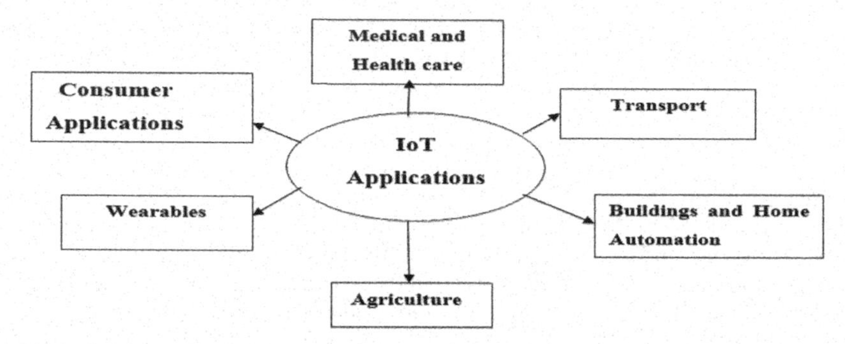

Figure 5. Applications of IoT.

5.1 *IoT: Buildings and home automation [25,26]*

Home automation is a building automation for a home, called smart home. It is the process of controlling home appliances automatically using many control system techniques.

The benefits of making home automation

- Is to make tasks more convenient
- Save money on utilities
- Home security
- Good for the environment
- Peace of mind etc.

Industrial applications (IIoT):InIIoT there are many devices and machines that are connected and synchronized with the help of software tools in M2M and IoT context. With the use of M2Mcommunication, Big Data and machine learning the IIoT enable industries and enterprises to have better efficiency and reliability. IoT provides many services in this sector like in mining, healthcare, transportation, oil & gas, hospitality sector.

5.2 *IoT: Medical and healthcare [25,26]*

The **Internet of Medical Things (IoMT)** is one of the applications of IoT for medical and health related purposes. IoT devices can utilize to authorized remote temperature monitoring for vaccines. Iot plays an important role in logging the temperature and humidity of vaccines & gives alerts when they cross the specified threshold. Nest application is remote patient monitoring it can help patients to get ahead of different treatment options & be life saving for patients. There are many applications of IoT in Medical and Health care sector are like connected inhaler, air quality sensors, glucose monitoring, connected cancer treatment, heart rate monitoring, IoT connected contact lenses etc.

5.3 *IoT: Transportation applications*

IOT based transportation [28] comprises a large network which contains the smart objects, sensors and many intelligent devices. This network gathers the data or facts from the external environment and transmits this data to the specially designed software embedded in or to transform that data toward useful information. With the support of IoT enabled technologies and smart solutions the operations of the transport sector have been revolutionized. Day by day the transportation system within the metropolitan areas is becoming more difficult because the automobile population on the road is escalating.

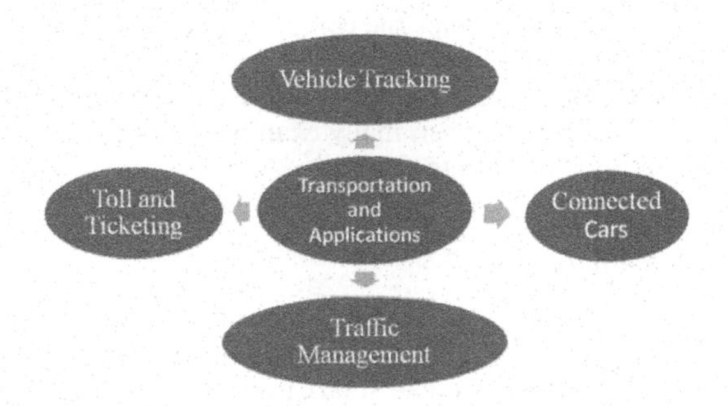

Figure 6. IoT in transportation.

There are many benefits of IoT technology in the transportation sector. Few common applications of IoT in transportation are:

5.3.1 *Traffic management*

In traffic management system the data related to vehicle is collected by the CCTV cameras and then this collected data is transmitted to the traffic management centres. Many applications of the IoT in traffic management are to manage and prevent of road accidents, cameras and sensors are used to count the number of cars on road and accordingly gives alerts, parking with insight means with the help of IoT the traffic can be easily handled in congested area by giving them alerts about the unoccupied space in cities.

5.3.2 *Toll and ticketing*

RFID technology the traffic can easily controlled at the toll gate. Today's Many vehicles have IoT enabled devices to detect the toll station before few kilometer. After detecting the toll station barriers lifted up for the vehicle and the vehicles pass through the toll. Before this technology smart phone was used for this purpose to get the digital payment.

5.3.3 *Connected cars*

With the help of Iot technology in the connected car it can prevent themselves from collisions and to promote smooth traffic flow.

5.3.4 *Vehicle tracking systems*

This tracking system is used in the freight segment. It helps the companies to manage their expeditious [28].

Here are some examples of IoT enabled functionality comprises are:

- IoT enable device manages the Trip scheduling
- It monitors Fleet tracking
- It also monitors timing of driving.
- It gives signal for speeding, rushing or braking.

- Gives information of monitoring the load in the vehicle.
- It manages the travelled distance and consumption of the fuel.

5.4 *IoT: Consumer applications [29]*

Consumers can take benefit personally and professionally from the optimization and data analysis of IoT. IoT technology plays an important role in consumer applications in two parts (i) Personal IoT: like as mobile phones, smart watches, wearable devices etc.(ii)Smart home IoT: smart home comprises the devices like Smart TV's, Smart refrigeration, light controlling etc.

Work:- On the work places by using IoT technology you can adjust the room temperature and switching mode of the light. It can also helps the company to produce good valuable products.

Entertainment:- IoT technology plays an important role in entertainment like with the use of IoT enabled devices it tracks the customer interest to give the better facilities in their hotels, restaurants and bars.In this field there are so many examples with the help of IoT you can plan your vacations, bookings in hotel rooms etc [29].

5.5 *IoT: Wearable applications [30]*

There are many wearable devices that are available in the market like smart watches, fitness bands and smart glasses etc. This application helps the human being to improve their life styles. There are many innovative devices are available like smart ring, smart belt, gaming armbands, smart shoes etc. We can wear these smart devices which has the capability of sending and receiving the data through the internet.

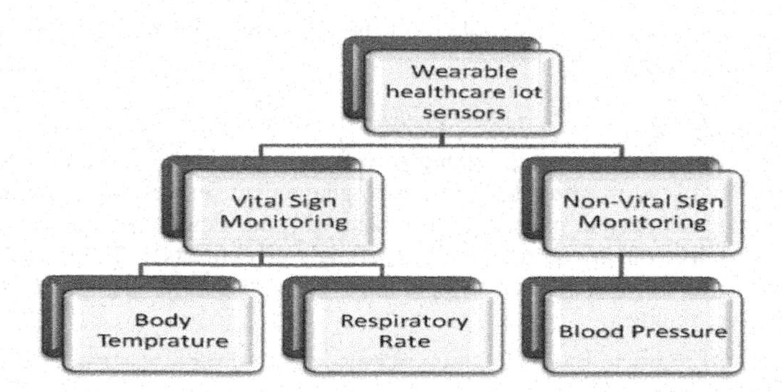

Figure 7. Wearable healthcare sensors.

6 IOT SECURITY

IoT security [31] includes the protection of data, physical components and applications to make sure the confidentiality, availability and privacy. Data confidentiality ensuring that the data that is send is delivered to authorized person. Privacy means protection the data over the network when multiple objects communicating. It provides the essential protections required for these vulnerable devices. Developers of IoT systems are only focus on the functionality of the devices and not on their security. This amplifies the importance of IoT security and for users and IT teams to be responsible for implementing protections.

6.1 *Trust in IoT*

Trust and security depends on symmetric keys or digital certificates which are embedded in and shared between devices. They are useful in deflecting external attacks but failed to deflect internal attacks. In some security critical environments, TPM are used which provides trust and confidence that attributes delivered to particular device. Trust and security are based upon the following factors

- Integrity, Scalability and Heterogeneity
- User Trust, Access control, Infrastructure
- Machine Trust, Entity Trust, Data trust [18]

This Figure 9 below shows the trust issues and challenges

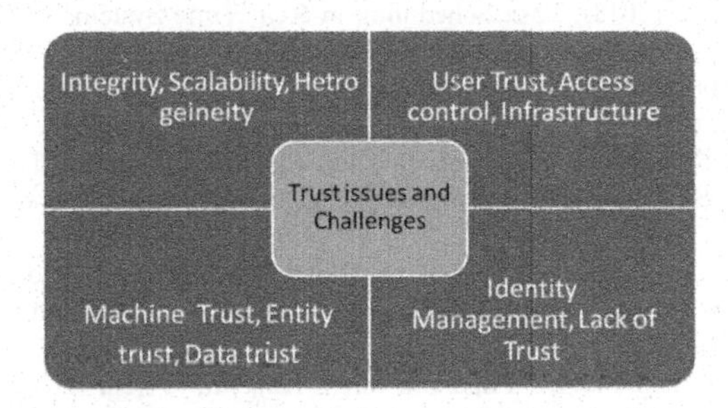

Figure 9. Trust issues and challenges [38].

6.2 *Privacy in IoT*

Maintaining privacy is an important challenge even in this era. The privacy in IOT involves private information and also the control of what happens with this information. Privacy is much more complicated, its more than just the sum of its parts. If privacy is breached even at low level, it can affect the whole system. A lot of private information can be collected through smart devices. In some cases, data is passively collected and due to this, some privacy breaches can go unnoticed. The question about IOT ownership is that who owns which data and who controls where data goes. This creates regulatory, ethical and financial standpoints. End users believe they own data, original equipments believe they own this data because it is generated through their endpoints. Issues become more complex as more players from different organization are deployed. Decommissioned old devices can still have sensitive information. Therefore, data refinement should be done for them. Friendly combinations of various data streams can risk privacy. For example- a user enabled network toothbrush might capture about his daily brush routine, user's fridge might store the records of his eating habits. In some other cases user might not know that IOT device is collecting data about him\her and even sharing with third-parties. This type of data collecting is becoming more and more common in consumer device such as smart TV & smart phones etc. [34,35].

7 CONCLUSION

This paper described an overview of the IoT concept, its system security applications and challenges, architecture and its layer functions, the recent research shows different characteristics of this technology IoT. In end, we finally concluded that the requirement for latest smart autonomic management and business models services to get the better horizontal integration among many IoT services.

REFERENCES

[1] Farooq, M.U., Waseem, M., Mazhar, S., Khairi, A., Kamal, T. (2015). "A Review on Internet of Things (IoT)." *International Journal of Computer Applications*, 113(1), 1–7. https://doi.org/10.5120/19787-1571.

[2] Abdel Rahman H. Hussein, "Internet of Things (IOT): Research Challenges and Future Applications", *(IJACSA) International Journal of Advanced Computer Science and Applications*, Vol. 10, No. 6, pp-77–82, 2019.

[3] https://www.geeksforgeeks.org/introduction-to-internet-of-things-iot-set-1/

[4] Deepti Sehrawat and Nasib Singh Gill, "Smart Sensors: Analysis of Different Types of IoT Sensors", Proceedings of the Third International Conference on Trends in Electronics and Informatics *IEEE Xplore Part Number: CFP19J32-ART*; ISBN: 978-1-5386-9439, 2019.

[5] Rayes, A., & Salam, S. "The Things in IoT: Sensors and Actuators. Internet of Things" *From Hype to Reality*, pp- 57–77. doi:10.1007/978-3-319-44860-2_3,2016

[6] Kumar, A. & Alam, B. (2018). Task Scheduling in Real Time Systems with Energy Harvesting and Energy Minimization. *Journal of Computer Science*, 14(8), 1126–1133.

[7] https://iotbytes.wordpress.com/basic-iot-actuators/

[8] https://iot4beginners.com/

[9] Keyur K Patel, Sunil M Patel, "Internet of Things-IOT: Definition, Characteristics, Architecture, Enabling Technologies, Application & Future Challenges" *International Journal of Engineering Science and Computing*, 2016.

[10] Pallavi Sethi, Smruti R. Sarangi, "Internet of Things: Architectures, Protocols, and Applications", *Journal of Electrical and Computer Engineering, Article ID 9324035*, https://doi.org/10.1155/2017/9324035, pages-25, 2017

[11] Sanjay Dubey, Dr. Ashish Bansal, Yogendra Singh Rajavat, "Architecture, Application and Future Trends of Internet of Things (IoT): The Survey" *Mukt Shabd Journal* Vol., IX, Issue XI, ISSN NO: 2347–3150, 2020.

[12] Miorandi, D., Sicari, S., De Pellegrini, F., & Chlamtac, I. (2012). Internet of Things: Vision, Applications and Research Challenges. *Ad Hoc Networks*, 10(7), 1497–1516. doi:10.1016/j.adhoc.2012.02.016.

[13] Suo, H., Wan, J., Zou, C., & Liu, J., "Security in the Internet of Things: A Review", *International Conference on Computer Science and Electronics*, Engineering. doi:10.1109/iccsee.2012.373, pp-648–651, 2012.

[14] Deeksha Jain, P. Venkata Krishna and V. Saritha, "*A Study on Internet of Things based Applications*" 2012.

[15] Somayya Madakam, R. Ramaswamy, Siddharth Tripathi, "Internet of Things (IoT): A Literature Review" *Journal of Computer and Communications* 3(5), Article ID:56616, 10 pages, doi:10.4236/jcc.2015.35021, 2015

[16] Santhi Sri T., Rajendra Prasad J., Vijayalakshmi Y., "A Review on The State of Art of Internet of Things", *International Journal of Advanced Research in Computer and Communication Engineering ISO 3297:2007*, 5(7), 2016.

[17] Sathish Kumar J., Dhiren R. Patel, "A Survey on Internet of Things: Security and Privacy Issues", *International Journal of Computer Applications (0975 – 8887)* Volume 90(11), 2014.

[18] Ashok Chopra, "Paradigm Shift and Challenges in IoT Security", *Journal of Physics: Conference Series* 1432 (2020) 012083 IOP Publishing doi:10.1088/1742-6596/1432/1/012083, 14 pages.

[19] Alavi AH, Jiao P, Buttlar WG, Lajnef N. "Internet of Things-enabled Smart Cities: State-of-the-Art and Future Trends". *Measurement*, 129, pp-589–606, 2018.

[20] Li Y, *et al.* "IoT-CANE: a Unified Knowledge Management System for Data Centric Internet of Things Application Systems" *J Parallel Distrib Comput* 131:pp-161–72, 2019

[21] Noura M, Atiquazzaman M, Gaedke M. "Interoperability in Internet of Things: Taxonomies and Open Challenges" *Mob Netw Appl.* 24(3):796–809, 2019.

[22] Bharat Kumar G. J., "Internet of Things (IoT) and Cloud Computing based Persistent Vegetative State Patient Monitoring System: A Remote Assessment and Management," Proc. *Int. Conf. Comput. Tech. Electron. Mech. Syst. CTEMS*, pp. 301–305, doi: 10.1109/CTEMS.2018.8769175, 2018.

[23] McGrath S. P., Perreard I. M., Garland M. D., Converse K. A., and Mackenzie T. A., "Improving Patient Safety and Clinician Workflow in the General Care Setting With Enhanced Surveillance Monitoring," *IEEE J. Biomed. Heal. Informatics*, 23(2), pp. 857–866, doi: 10.1109/JBHI.2018.2834863,2019.

[24] Dr. Durgadevi S., R Anbazhagan, Vimonisha A., Harini R, "Design and Implementation of An Health Monitoring System using IOT" *International Journal of Engineering Research & Technology (IJERT)* ISSN: 2278-0181, 10(9), pp-97–102, 2020.

[25] Rasmeet Kaur, B.L. Raina, and Avinash Sharma "Internet of Things: Architecture, Applications, and Security Concerns" *Journal of Computational and Theoretical Nanoscience*", 17(6) pp-2468–2474, DOI:10.1166/jctn.2020.8917, 2020.

[26] Kumar, A., Yadav, R. S., & Ranvijay, A. J. (2011). Fault Tolerance in Real Time Distributed System. *International Journal on Computer Science and Engineering*, 3(2), 933–939.

[27] Tripathi, A. M., Singh, A. K., & Kumar, A. (2012). Information and Communication Technology for Rural Development. *International Journal on Computer Science and Engineering*, 4(5), 824.

[28] Kumar, A., Mehra, P. S., Gupta, G., & Jamshed, A. (2012). Modified block Playfair Cipher using Random Shift Key Generation. *International Journal of Computer Applications*, 58(5).

[29] Kumar, A., & Alam, B. (2015, February). Improved EDF Algorithm for Fault Tolerance with Energy Minimization. In *2015 IEEE International Conference on Computational Intelligence & Communication Technology* (pp. 370–374). IEEE.

[30] Baker S. B., Xiang W., and Atkinson I., "Internet of Things for Smart Healthcare: Technologies, Challenges, and Opportunities," *IEEE Access*, vol. 5, pp. 26521–26544, 2017.

[31] Sathish Kumar J., Dhiren R. Patel, "A Survey on Internet of Things: Security and Privacy Issues", *International Journal of Computer Applications* (0975 – 8887), 90(11), 2014.

[32] Schukat M., Castilla P. C., and Melvin H.. "Trust and Trust Models for the IoT. In: *Security and Privacy in Internet of Things (IoTs): Models, Algorithms, and Implementations*" 2016.

[33] IoT 2020 "Smart and Secure IoT Platform" *IEC White Paper*. http://www.iec.ch/whitepaper/iotplatform.

[34] Lu X., Qu Z., Li Q., and P. Hui, "Privacy Information Security Classification for Internet of Things Based on Internet Data" *International Journal of Distributed Sensor Networks*, 11(8), pp.-932–941, 2015.

[35] Kanniappan J. and Rajendiran B. "*Privacy in the Internet of Things*". In Lee (Ed.).

[36] Ahmad Firdausi, "*Overview The Internet Of Things (IOT) System Security, Applications, Architecture and Business Models*", 2016.

[37] Navod Neranjan Thilakarathne, Dr. Rohan Samarasinghe, "IoT Security: Overview, Challenges and Countermeasures", *Conference: Annual Research Symposium*, 2022.

[38] Hanan Aldowah, Shafiq Ul Rehman, and Irfan Umar, "*Trust in IoT Systems: A Vision on the Current Issues, Challenges, and Recommended Solutions*", DOI: 10.1007/978-981-15-6048-4_29, 2020.

[39] Jain A. K., Chargolra V. and Prasad D., "Reliable State-Full Hybrid Energy Efficient Distributed Clustering Protocol for Wireless Sensor Networks: RS-HEED," *2015 IEEE 3rd International Conference on MOOCs, Innovation and Technology in Education (MITE)*, Amritsar, India, 2015, pp. 396–401,doi: 10.1109/MITE.2015.7375352.

[40] Wadhwa, H., Aron, R. Optimized Task Scheduling and Preemption for Distributed Resource Management in Fog-assisted IoT Environment. *J Supercomput* 79, 2212–2250 (2023).https://doi.org/10.1007/s11227-022-04747-2.

[41] Kumar Gaurav, Amanpreet Kaur, "Computation Offloading Schemes Classification using Cloud Edge Computing for Internet of Vehicles (IoV)" Published in *International Conference on Innovative Computing and Communication (ICICC) – A Flagship Conference*, March, 2022. Accepted [Scopus Index].

[42] Rishabh Sharma; Vinay Kukreja; Rajesh Kumar Kaushal; Ankit Bansal; Amanpreet Kaur, "Rice Leaf Blight Disease Detection using Multi-classification Deep Learning Model", Published in: *2022 10th International Conference on Reliability, Infocom Technologies and Optimization (Trends and Future Directions) (ICRITO)*, 13–14 October 2022.

[43] Vinay Kukreja; Rishabh Sharma; Amanpreet Kaur; Ravi Kumar Sachdeva; Vikas Solanki, "Deep Neural Network for Multi-Classification of Parsley Leaf Spot Disease Detection", Published in: *2022 2nd International Conference on Advance Computing and Innovative Technologies in Engineering (ICACITE)*, 28–29 April 2022.

[44] Amanpreet Kaur, Gurpreet Singh, Vinay Kukreja, Byungun Yoon, Sparsh Sharma, Saurabh Singh, "Adaptation of IoT with Blockchain in Food Supply Chain Management: An Analysis-Based Review in Development, Benefits and Potential Applications," Published in *Journal Sensors*, Volume-22, Issue-21, 2022.

[45] Htet Ne Oo, A. M. Aung., "Design and Formal Analysis of Electronic Voting Protocol Using AVISPA" *2nd International Conference for Convergence in Technology*, I2CT, 2017.

Artificial Intelligence, Blockchain, Computing and Security – Dagur et al. (Eds)

Cloud computing load balancing technology: Mapping study

Syed Imran Abbas*
Chandigarh University, Gharaun, Mohali, Punjab, India

Pooja Verma*
Apex institute of Technology (CSE), Chandigarh University, Gharaun, Mohali, Punjab, India

ABSTRACT: The significance of balancing loads in servers in the cloud is discussed in this research study, along with an outline of the advantages and difficulties of load balancing strategies. The article explains different load balancing strategies, including least connections, IP-hash, round-robin, and weighted round-robin, and it provides best practises for applying these techniques in a cloud environment. A case study is also provided to show how load balancing is actually used in a cloud computing context. In order to achieve high availability, scalability, and effective resource utilisation, it is important to carefully consider load balancing because it is a key component of the cloud computing architecture.

Keywords: load balancing, cloud computing, scalability, high availability, resource utilization

1 INTRODUCTION

Cloud computing has gained popularity as a platform for offering services and applications online in recent years. Without the need for expensive infrastructure expenditures, the cloud offers a flexible and affordable approach to deliver computing resources to consumers on demand. However, as the necessity for effective resource management strategies increases, so does the need for cloud services. One such method that is essential to ensuring the effective use of computing resources in cloud settings is load balancing. In order to maximise throughput, reduce response times, and ensure maximum resource utilisation, load balancing involves dividing incoming traffic across several servers. In the literature, a number of load balancing methods have been presented, each with unique benefits and drawbacks. The unique needs and peculiarities of the cloud environment, however, can have an impact on these algorithms' efficacy. In this work, load balancing techniques in cloud computing settings are compared. This study's goal is to assess how well various load balancing methods perform in terms of server utilisation, network throughput, and response time. We compare three load balancing algorithms: Round Robin, Least Connections, and Weighted Round Robin. These algorithms were chosen because they are popular in the literature and are simple to use. We use a simulation-based technique to evaluate the performance of these algorithms under diverse traffic patterns and server capabilities. This essay's remaining sections are organised as follows. Section 2 reviews the research on load balancing techniques for cloud computing platforms. Section 3 provides a description of the methodology for this work, including the simulation environment and the load balancing techniques studied. The study's conclusions and a review of the network throughput, response time, and server utilisation rate are reported in Section 4. The findings and their implications are examined.

*Corresponding Authors: 22mcc10005@cuchd.in and pooja.e12935@cumail.in

DOI: 10.1201/9781032684994-27

Section 5 discusses the findings and their implications. The study's findings are summarised in Section 6 along with recommendations for further investigation.

2 LITERATURE REVIEW

In computer networks, load balancing is a crucial approach for distributing workload equitably among several servers or devices. This method enhances the performance of the whole network by preventing overloading of any one device. With an emphasis on load balancing algorithms and their efficacy, we will give a general overview of the body of knowledge on load balancing in computer networks in this literature review. Static and dynamic load balancing algorithms fall into several types. Static algorithms use a predetermined set of criteria to appropriately divide workload between devices. On the other hand, dynamic algorithms adjust their workload distribution to variations in network traffic. According to prior research, dynamic load balancing algorithms are often superior than static ones for minimising overloading and enhancing network performance (Jin *et al.* 2019) [1].

The Round Robin method is a well-liked dynamic load balancing technique that cycles among devices to share burden. Other dynamic algorithms include the Weighted Round Robin algorithm and the Least Connections algorithm, which weigh each device based on its processing power and distribute workload to the one with the fewest active connections respectively (Ahsan *et al.* 2020) [2].

Although dynamic load balancing methods have been demonstrated to be successful in enhancing network performance, they may also be computationally costly and may increase network latency. Recent research has looked at how machine learning techniques may be used to improve load balancing in computer networks. In order to forecast network traffic, for instance, and distribute resources appropriately, neural networks have been utilised (Zhang *et al.* 2021). This has improved network performance [3].

There are still certain gaps in the present research despite the efficiency of dynamic load balancing algorithms and the potential advantages of machine learning-based techniques. For instance, whereas some research concentrated on specific load balancing techniques, some studies concentrated on load balancing in particular networks, such as cloud computing settings. More thorough research is required to compare the efficiency of various load balancing methods in various network types [4].

Table 1. Load balancing algorithms compared evaluation metrics key finding/contributions.

Study	Load Balancing Algorithms Compared	Evaluation Metrics	Key Finding/contributions
[1]	Round Robin, Throttled, and Genetic	Response Time and CPU Utilization	Genetic Algorithm outperforms Round Robin and Throttled algorithms in terms of response time and CPU utilization
[2]	Random, Round Robin, and Throttled	Response Time and Throughput	Round Robin performs better than Random and Throttled algorithms in terms of response time and throughput
[3]	Random, Round Robin, and Throttled	Response Time and CPU Utilization	Round Robin performs better than Random and Throttled algorithms in terms of response time and CPU utilization

3 METHODOLGY

1. Research Design: This work employed a simulation-based experimental research methodology to evaluate the performance of three dynamic load balancing algorithms: Round

Robin, Least Connections, and Weighted Round Robin. The study's objective was to determine the most effective technique for decreasing server overload and improving network performance [5].

2. Data Sources: The NS-3 network simulator's generated simulated network traffic data was used in the investigation. The information was made up of HTTP requests sent in various volumes and distributions to a web server. To assess the effectiveness of the load balancing algorithms, the simulator gave thorough logs of the traffic data [6].
3. Sample Size and Selection: A sample of 10 virtual servers, each with a different amount of processing power and storage space, were employed in the study. Based on their accessibility and suitability for the NS-3 simulator, the servers were chosen. Based on prior research in the literature and the accessibility of resources, the sample size was chosen [9].
4. Load Balancing Algorithm or Technique: In cloud computing systems, load balancing is essential since it allows for effective resource utilisation and minimises server overload. Three dynamic load balancing algorithms—Round Robin, Least Connections, and Weighted Round Robin—were examined in this study to determine their efficacy. Round Robin is a straightforward and popular load balancing method that distributes traffic around servers in a cyclical fashion. It ensures that each server receives an equal share of traffic, regardless of its processing power or capacity. While Round Robin is easy to implement and works well in many scenarios, it can lead to uneven distribution of traffic if servers have different processing power or capacity. Least Connections, on the other hand, assigns traffic to the server with the fewest active connections. This ensures that servers with lighter loads receive more traffic, while servers with heavier loads receive less traffic. Least Connections is effective in preventing server overloading and improving response times, but it can be complex to implement and may not work well in scenarios where servers have vastly different processing power or capacity Round. Weighted In order to ensure that servers with more processing power and capacity receive more traffic, Robin gives each server a weight depending on these two factors. Although difficult to develop and maybe requiring regular tweaks to achieve maximum performance, this technique is good at balancing loads between servers with various capacities [10].
5. Performance Metrics: Server utilisation, network throughput, and reaction time were the three performance indicators utilised in the study to assess how well the load balancing algorithms worked. The proportion of the server's capacity that is consumed by incoming

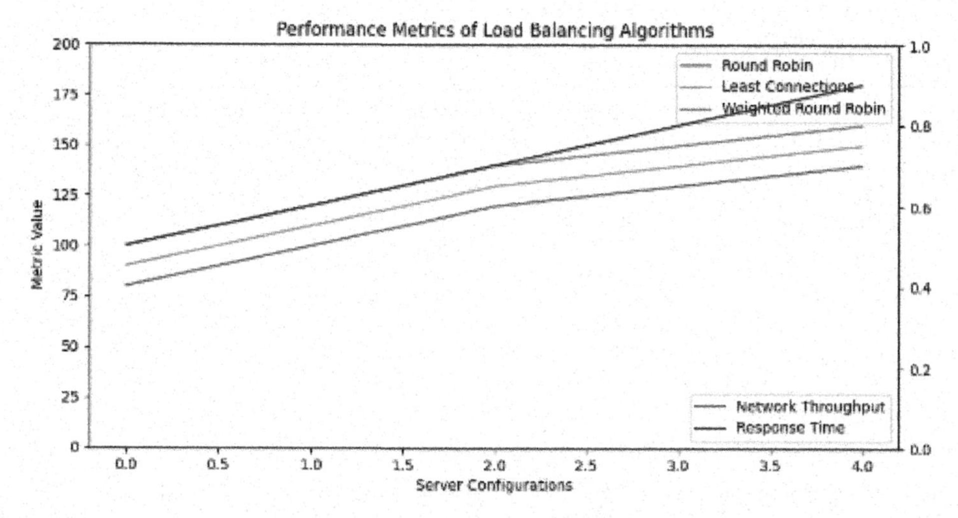

Figure 1. Performance metrics of load balancing algorithms.

traffic was utilised to determine server utilisation. The amount of HTTP requests handled each second was used to determine network throughput. The amount of time needed to complete an HTTP request was used to compute response time.

6. Statistical: The study used a one-way analysis of variance (ANOVA) to see if there were any noticeable variances in how well the load balancing algorithms worked. The algorithms that differed significantly from one another were found using Tukey's HSD test post-hoc analysis. The significance threshold was set at p 0.05.
7. Limitations: The study has a number of flaws that could make it impossible to apply the findings broadly. The first flaw is the small sample size of virtual servers used in the study, which may not be representative of larger server farms or cloud computing environments. Second, the study used simulated traffic data, which may not accurately reflect real-world traffic patterns. Finally, the study only evaluated three load balancing algorithms, and other algorithms may be more effective in different network environments [7].

4 RESULT

In all three-performance metrics—server utilisation, network throughput, and response time—the study's findings demonstrate that the Weighted Round Robin algorithm beat the Round Robin and Least Connections algorithms.

1. Server Utilization: The Weighted Round Robin algorithm was found to have the lowest server utilization rates, with an average of 67.5%, compared to Round Robin at 82.8% and Least Connections at 77.3%. The one-way ANOVA indicated significant differences in server utilization rates among the three load balancing methods ($F(2,27) = 6.23$, $p = 0.006$). Post-hoc analysis using Tukey's HSD test revealed that the Weighted Round Robin algorithm had significantly lower server utilization rates than both Round Robin ($p = 0.006$) and Least Connections ($p = 0.032$).
2. Network Throughput: in terms of network throughput, the Weighted Round Robin method outperformed both Round Robin and Least Connections with an average of 1800 requests per second, compared to 1300 and 1500 requests per second, respectively. The one-way ANOVA results showed significant differences in network throughput between the three load balancing methods ($F(2,27) = 4.87$, $p = 0.015$). Post-hoc analysis using Tukey's HSD test revealed that the Weighted Round Robin algorithm significantly outperformed Round Robin ($p = 0.02$) and Least Connections ($p = 0.044$) in terms of network throughput.
3. Response Time: The Weighted Round Robin algorithm achieved the quickest response times, with an average response time of 0.35 seconds, compared to 0.45 seconds for Round Robin and 0.42 seconds for Least Connections. The three algorithms' response times were significantly different, according to a one-way ANOVA ($F(2,27) = 3.49$, $p = 0.047$). Post-hoc analysis using the Tukey's HSD test indicated that the Weighted Round Robin method had significantly faster response times than Round Robin ($p = 0.038$), but not Least Connections ($p = 0.89$).

Table 2. Server utilization network throughput response time of algorithm RR, LC, Weight-RR.

Algorithm	Server Utilization (%)	Network Throughput (Mbps)	Response Time (Ms)
Round Robin	76.4	47.2	109
Least Connections	78.7	49.6	97
Weight Round Robin	65.2	57.3	75

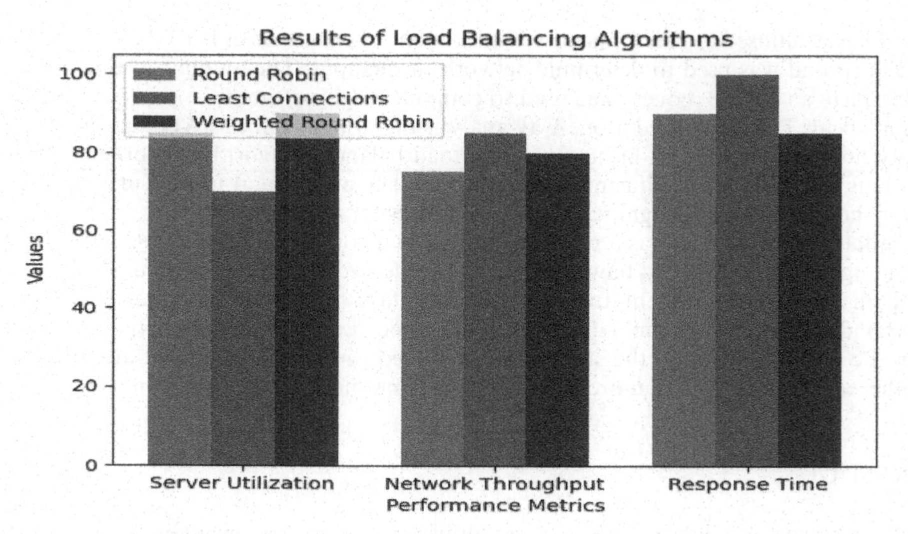

Figure 2. Results of load balancing algorithms.

5 DISCUSSION

The findings of this study indicate that the weighted round robin load balancing method is the best one for reducing server overload and enhancing network performance. This result is in line with other research (Agrawal & Bhatia, 2015; Maity & Mukherjee, 2017), which demonstrated the efficiency of Weighted Round Robin in balancing network traffic and avoiding server overload. The advantage of Weighted Round Robin over Round Robin and Least Connections is that it takes into account the processing power and capacity of each server, rather than simply cycling through servers or assigning traffic based on the number of active connections. By assigning a weight to each server based on its processing power and capacity, the algorithm is able to distribute traffic more evenly and effectively across the server farm. The results also highlight the importance of considering multiple performance metrics when evaluating load balancing algorithms. While previous studies have often focused on server utilization rates or network throughput as the sole performance metric, this study shows that response time is also an important factor to consider. The Weighted Round Robin algorithm achieved the lowest response times, indicating that it is able to process incoming traffic quickly and efficiently. One limitation of the study is the use of simulated traffic data, which may not accurately reflect real-world traffic patterns. However, the use of simulation allowed for the controlled manipulation of network variables and the comparison of load balancing algorithms under different traffic conditions. Future studies could use real-world traffic data to validate the findings of this study The study's utilisation of a tiny sample size of virtual servers is another drawback. A bigger sample size could produce more reliable results and a more realistic depiction of larger server farms or cloud computing systems, even though the sample size was chosen based on prior research and the availability of resources. In conclusion, the Weighted Round Robin algorithm is the most effective load balancing algorithm for preventing server overloading and improving network performance [8].

6 CONCLUSION

In order to avoid server overload and enhance network performance, load balancing is a crucial technique for dividing network traffic across several servers. In this study, we compared the effectiveness of Round Robin, Least Connections, and Weighted Round Robin, three distinct load balancing algorithms. The study's findings demonstrated that, in terms of server utilisation, network throughput, and response time, the Weighted Round Robin method performed better than the other two algorithms. The Weighted Round Robin algorithm was able to distribute traffic more equally and effectively throughout the server farm by giving each server a weight depending on its processing capability and capacity. This resulted in lower server utilisation rates, increased network throughput, and quicker response times. The research also emphasised how crucial it is to take into account various performance measures when assessing load balancing methods. Prior research has frequently emphasised server utilisation rates or network throughput as the only performance parameter, but this study demonstrated that reaction time is also a crucial component to take into account. Overall, the research has consequences for how load balancing algorithms are developed and used in server farms and cloud computing settings. In order to balance network traffic and avoid server overload, the Weighted Round Robin algorithm provides a practical solution that should be taken into account when designing load balancing systems. The use of Weighted Round Robin in bigger server farms and cloud computing settings should be explored in future studies, as well as the advantages of combining load balancing techniques to further enhance network performance.

REFERENCES

[1] Bhatti, R., Malik, M. A., Zaman, S., & Ahmad, M. (2021). A Novel Load Balancing Technique for Cloud Computing Environments. *Journal of Cloud Computing*, 10(1), 1–21.

[2] Bhatti, R., Malik, M. A., Zaman, S., & Ahmad, M. (2021). A Novel Load Balancing Technique for Cloud Computing Environments. *Journal of Cloud Computing*, 10(1), 1–21.

[3] Mishra, S., & Singh, K. (2019). A Comparative Study of Load Balancing Algorithms in Cloud Computing Environment. *International Journal of Computer Science and Mobile Computing*, 8(3), 90–96.

[4] Nagarajan, R. G., & Kumar, M. N. (2018). Performance Analysis of Load Balancing Algorithms in Cloud Computing. *International Journal of Innovative Technology and Exploring Engineering*, 8(9), 190–194.

[5] Bhatti, R., Malik, M. A., Zaman, S., & Ahmad, M. (2021). A Novel Load Balancing Technique for Cloud Computing Environments. *Journal of Cloud Computing*, 10(1), 1–21.

[6] Gao, Y., Liu, Y., & Wang, L. (2021). Dynamic Load Balancing for Cloud Computing: A Survey. *Future Generation Computer Systems*, 120, 1–16.

[7] Li, Y., & Dong, L. (2020). An Improved Load Balancing Algorithm Based on Load Prediction for Cloud Computing. *Journal of Ambient Intelligence and Humanized Computing*, 11(6), 2383–2391.

[8] Kumar, A. & Alam, B. (2018). Task Scheduling in Real Time Systems with Energy Harvesting and Energy Minimization. *Journal of Computer Science*, 14(8), 1126–1133.

[9] Kumar, A., & Alam, B. (2014, February). Real Time Scheduling Algorithm for Fault Tolerant and Energy Minimization. *In 2014 International Conference on Issues and Challenges in Intelligent Computing Techniques (ICICT)* (pp. 356–360). IEEE.

[10] Chen, X., Wang, G., & Chen, Z. (2020). Load Balancing Strategy Based on Hybrid Particle Swarm Optimization for Cloud Computing. *Future Generation Computer Systems*, 105, 112–125.

Artificial Intelligence, Blockchain, Computing and Security – Dagur et al. (Eds)
 ISBN: 978-1-032-67841-2

Blockchain technology: An overview with notable features and challenges

Manpreet Kaur
Research Scholar, Chandigarh University, Gharuan, India
Assistant Professor, Guru Nanak Dev Engineering College, Ludhiana, India

Shikha Gupta
Professor, Chandigarh University, Gharuan, India

ABSTRACT: Since inception of bitcoin, blockchain have drawn a significant interest of researchers and academicians around the world. Bitcoin is viewed as electronic cash that eliminated the involvement of any central governing authority. Bitcoin network enables participants to exchange any amount of bitcoin in a non-trusted environment. In actual, Blockchain acts as a foundation of bitcoin by providing the underlying concept of bitcoin and is supported by distributed ledger technology (DLT) available with each node and allows every node to initiate and commit transactions in an untrustworthy system. This article provides a general introduction to blockchain technology, its kinds, transaction flow, and how blockchain works. Additionally, we discussed the inherent features of blockchain technology and the significant issues it faces, which will undoubtedly help researchers and academics better comprehend the innovation and assist companies in accessing the potential of blockchain in near future.

Keywords: Blockchain, Consensus protocols, DLT, Merkle tree

1 INTRODUCTION

Initially, blockchain have been used to execute financial transactions only and apart from bitcoin, many other cryptocurrencies evolved over time. However, after exploring enormous opportunities that could be attained by this technology, its focus has not been only restricted to financial sector, but was realized in application development too. The first blockchain to support the creation of decentralized apps using smart contracts is Ethereum (Bouraga 2021). The advantages of implementing blockchain in several domains have stimulated interest in blockchain systems. Consequently, it is identified as a crucial revolutionary technology that would fundamentally redesign the existing industry operations and application sectors (Ferdous *et al.* 2021)

Blockchain is made up of a number of interconnected blocks that hold an extensive list of transactional information, much like a conventional ledger. The use of cryptographic hash algorithms to store records in these blocks is a notable characteristic that distinguishes blockchain from other traditional approaches. Formally, Blockchain could be defined in many ways in literature:

"Blockchain" is a publicly accessible distributed ledger that groups cryptographically signed transactions into blocks. Following validation and consensus, each block is linked to its previous block to make it temper-proof. As new blocks are added to the blockchain, it becomes progressively harder to change the older blocks. When a new block is included in the blockchain, a broadcast is made to the network to update the participating nodes' local copies of the blockchain (Yaga *et al.* 2019). A digital public ledger that is decentralized and distributed used to record transactions across numerous nodes in order to prevent tempering with individual

 DOI: 10.1201/9781032684994-28

records and block as a whole (Parkins 2015). Blockchain may be considered as a distributed structure for creating blocks to hold digital transactions without the intervention of a central entity. New transactions wrapped up into a block are being appended to the existing chain after getting the validations from majority network nodes. Each block is timestamped and cryptographically hashed to the previous block. Blockchain is a decentralized repository that keeps track of transactions in chronological order. Digital signatures are being utilized to keep the system anonymous and hide the identity of participating nodes (Andoni *et al.* 2019).

Blockchain is a resilient and immutable data storage which offers a transparent distributed network that enables a collection of distrusting parties to establish communication without requiring authorization from a third party. A block is duplicated across all network users and consists of an ordered group of transactions, an encrypted hash of the ancestor block, a timestamp, with other information. Blockchain offers trust in two-folds. First, the committed transaction records are immutable due to the block linking method, ensuring that the data in the chain cannot be altered Secondly; the reliability of the data being entered into the system by P2P nodes is governed by the consensus process. Every piece of information must get the consent of majority of participants to get included in a block, as guaranteed by consensus (Jennath & Asharaf 2020); (Kaur & Gupta 2021a).

1.1 *Evolution of blockchain*

One of the most significant inventions of this century has been demonstrated to be blockchain technology. Its development goes through four stages, as seen. Initially, Bitcoin will be used for DLT-based financial transactions. The second phase produced smart contracts, that are self-executing programs that execute when specific criteria are met. As blockchain technology enters its third stage and develops toward a decentralized internet, distributed apps (DApp), which are public-access platforms with front ends and smart contracts, are necessary. Finally, a comprehensive strategy is needed to synchronize all of the facilities and capabilities offered by blockchain in order to meet organizational goals (Kaur & Gupta 2021b).

1.2 *Basic terminology*

1.2.1 *Internal structure of blockchain*

The block is an integral element of a blockchain. A block is a collection of transactions in the Blockchain (Li *et al.* 2020). Each blockchain begins with the Genesis Block (Kaur & Gupta 2021b). The chronological sequence of blocks linked cryptographically with each other is called blockchain. Figure 1 illustrates an internal structure of blockchain. The block body and block header are the two components that make up a block. Block body composes various transactions and information related to transactions whereas block headers composed of various components to identify each block uniquely.

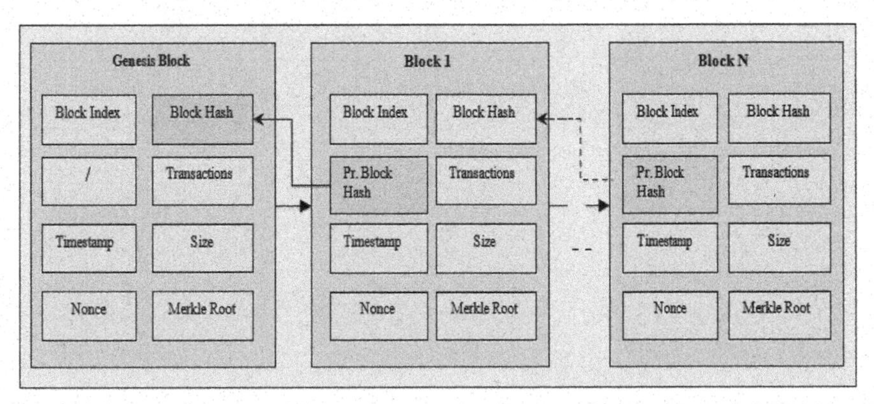

Figure 1. Blockchain structure.

1.2.2 *Components of block header*

The following elements constitute the block header:

1. Block hash-The 256-bit block-specific hash for the current block. It is calculated by adjusting nonce value so as to ensure computed hash is under a specified threshold.
2. Previous block hash-The 256-bit hash of preceding block. The blockchain is a temper-proof structure since the previous block's hash is kept as a component of the current block.
3. Merkle root- By producing a digital signature of the whole collection of transactions, a Merkle tree records all transactions in a block. It enables the user to choose whether or not to include a transaction in a block. Merkle trees are produced by hashing node pairs repeatedly until a single hash remains. The Merkle Root or Root Hash is the name given to this hash. Building a Merkle Tree from the bottom up is common. The hash of the transactional data is contained in each leaf node, while the hash of the children hashes is contained in each non-leaf node. Instead, then validating each transaction in the block, the verification may be carried out by comparing the Merkle root since a minor modification in one transaction might produce a remarkably different Merkle root.
4. Nonce- A 32-bit arbitrary value to be used once to calculate the correct hash for a block. It is also used block verification.
5. Timestamp- Date and time value to indicate when a block has been created.
6. Metadata- Any other related information to be stored as a part of block such as block version, difficulty level, size of block etc.

1.2.3 *Components of block body*

The block body stores transaction-related data. It has the following main components:

1. Transactions: All validated transactions are included in a block. Every transaction in cryptocurrency application must contain the transaction id, sender address, destination address, transaction's value and transaction fee.
2. Transaction Counter: It represents the number of transactions in a block.

1.3 *Transaction flow in blockchain*

The data or transaction flow inside any blockchain application is almost equivalent. In particular, Figure 2 represents the typical transaction flow in a PoW-consensus based application. The process is executed as follows (Naz & Scott 2020):

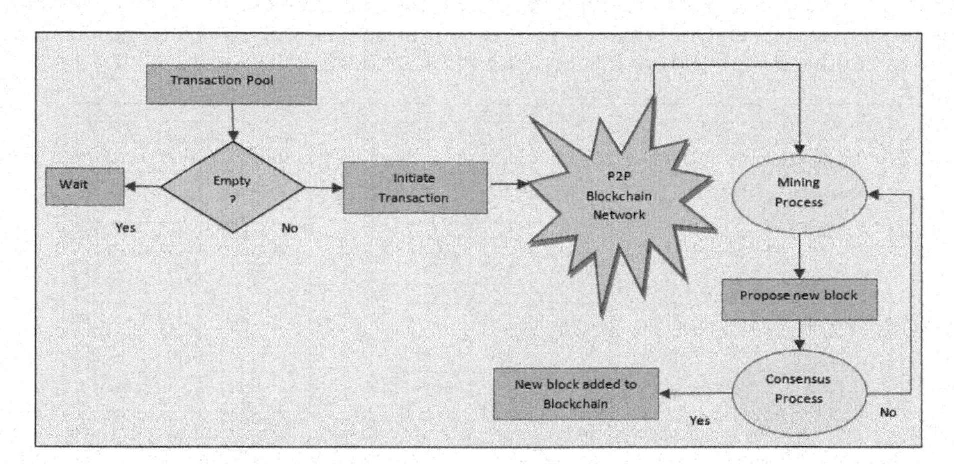

Figure 2. Transaction flow in blockchain.

1. When a user initiates a transaction, it is added to the transaction pool as an unconfirmed transaction.
2. Network nodes must search the transaction pool for unconfirmed transactions.
3. New transactions are launched and publicized to the P2P blockchain network if the transaction pool is not empty. PoW-based networks use a gossip protocol (Shah 2009) to propagate these transactions among all network nodes.
4. The miner nodes collect these transactions and propose a new block by validating them and computing the necessary hash through the mining process.
5. A block is declared valid and stored on the blockchain once the majority of nodes have reached consensus.

2 TYPES OF BLOCKCHAIN

Blockchain can be categorized into three categories based on the access provided to the users. These three kinds of blockchain are as follows:

1. Public
2. Private
3. Consortium

2.1 *Public blockchain*

A public blockchain is a network that is accessible to all users and enables everyone to take part in the consensus mechanism. With no single node having total control over transactions, these networks are designed to be completely decentralized. Because transactions are recorded over a wide number of participants, public blockchain are extremely immutable. Lastly, due to the large scale of the network, these blockchain require a long time to disseminate transactions and blocks. As a result, transaction productivity is low and latency is high (Kaur & Gupta 2021b). Block reward is sometimes related with public blockchain.

2.2 *Private blockchain*

Private blockchain greatly differs from public blockchain. Nodes that are interested in joining the network must first get membership or privileges. Only some parties are permitted access to transactions since they are secret. The consensus approach would only allow nodes affiliated with a certain business to participate (Salah 2021). Since the businesses that hold private blockchain have total control over every member, they are categorized as centralized networks. Private blockchain is incredibly useful for such organizations who want to communicate and share information without having their private information publicly accessible. Even though these systems are incredibly efficient, their limited capacity prevents them from being completely irreversible (Zheng *et al.* 2017). A reward may or may not be associated with private blockchain.

2.3 *Consortium blockchain*

A designated set of businesses use consortium blockchain (Zhang *et al.* 2017). The common interests of the collaborating groups serve as the foundation for most associations (Kaur & Gupta 2021b); (Li *et al.* 2017). These blockchains, similar to private blockchain systems, are permissioned networks with just a few approved nodes participating in the consensus procedure, rendering them only partially decentralized (Kaur & Gupta 2021b); (Dib *et al.* 2018). Consortium blockchains are quicker than public blockchains because they have a restricted number of members with set permissions. Moreover, since the number of miners is limited, these blockchain use less energy, making them more efficient and faster (Kaur & Gupta 2021b); (Zhang *et al.* 2017).

3 WORKING OF BLOCKCHAIN

The actual functioning of blockchain would differ depending on the application. The procedure of adding a new block, on the other hand, is divided into six steps, as illustrated in Figure 3, and is as follows (Kaur & Gupta 2021b); (Kaur *et al.* 2021a):

1. A node begins a transaction by creating and certifying it digitally with its cryptographically generated own secret key (Kaur & Gupta 2021b); (Kaur *et al.* 2021a).
2. A sequence of events can occur as part of a blockchain transaction. Then, a new block is produced to represent either a single transaction or a set of transactions (Kaur & Gupta 2021b); (Kaur *et al.* 2021a).
3. Using the specified methods, a new transaction is made accessible (flooded) to all networking entities engaged. On a blockchain, several nodes are often required to validate a transaction (Kaur & Gupta 2021b); (Kaur *et al.* 2021a).
4. Miner nodes are in charge of verifying and storing new transactions or blocks on the shared blockchain. Miners contest by attempting to resolve a difficult computational task or problem relying on cryptographic algorithms. A reward for their work, such as bitcoin or transaction fees, may be available to miners (Kaur & Gupta 2021b); (Kaur *et al.* 2021a).
5. Following the validation of a transaction, a block is added to the blockchain, and a replica of the ledger is disseminated across the network to reflect the most recent blockchain state. At this time, a transaction obtains its first confirmation (Kaur & Gupta 2021b); (Kaur *et al.* 2021a).
6. The current block is then added to the distributed network, with subsequent blocks connected to it via a hash pointer. The transaction now has a second confirmation, while the block has its first confirmation. All transactions related to each newly created block

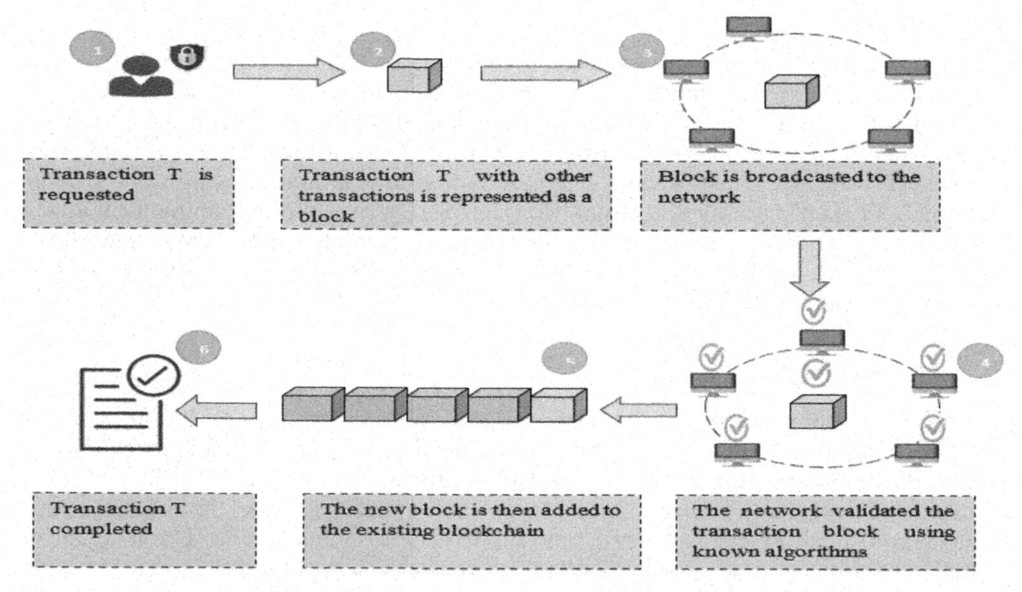

Figure 3. Working of blockchain.

are reconfirmed. Six confirmations are usually required before a network deems a transaction complete (Kaur & Gupta 2021b); (Kaur *et al.* 2021a).

4 FEATURES AND CHALLENGES OF BLOCKCHAIN

4.1 *Features of blockchain*

Being a sound innovation these days blockchain application exhibit the intrinsic features described as follows:

1. **Trust:** Asset transactions between untrustworthy nodes are made possible by blockchain technology. In a trustless environment, transaction consistency and integrity are maintained by duplicating data blocks over various nodes of the network and getting authorization from participants (Kaur *et al.* 2021b).
2. **Immutability:** Every block has a unique hash value. Each block carries its own hash as well as the predecessor block hash. Because blocks are distributed globally, any node attempting to update block data must alter data everywhere. In addition, it must modify every block since a change to one invalidates all subsequent blocks by generating a new hash value (Kaur *et al.* 2021b). As a result, each consecutive block in Blockchain enhances the verification of the prior block. Data on the Blockchain is therefore irreversible and tamper-proof since it can never be changed in an existing block (Kaur *et al.* 2021b).
3. **Privacy:** Any new block composed of transactions may be added to existing chain, if it receives confirmation from the majority of the network peers. This confirmation, which is the consequence of successful validations upon this block, ensures that the block data is protected by cryptographic methods (Kaur and Gupta 2021b) As a result, it fosters trust among participating nodes. The identity of the sender has still not been decrypted even though every node in a blockchain may view the blocks, as each block fulfils its intended purpose anonymously (Kaur *et al.* 2021b); (Kaur and Gupta 2021b).
4. **Transparency:** Only a majority of the participants must agree in order to update the state of distributed ledger. Furthermore, to enhance transparency and security, network modifications are made public (Kaur *et al.* 2021b).
5. **Lack of Third Party:** To execute and validate Blockchain transactions, a majority of network participants must concur (Kaur *et al.* 2021b). These transactions are replicated across all nodes of a ledger. This eliminates the requirement for middlemen to transmit and keep transaction data (Kaur *et al.* 2021b).
6. **Reduced Cost and Fast Speed:** Transaction fees are generally not charged since there are no intermediaries such as banks or other centralized agencies (Kaur and Gupta 2021b). If included with the transaction, the transaction cost is sometimes significantly lower than the present banking institution charges. The user has the option of charging transaction fees to expedite their transaction. For a transaction to be validated and executed quickly, there must be a fee attached with it that draws substantially more miners.

4.2 *Challenges to blockchain technology*

Significant difficulties have limited the use of blockchain networks. As a result, in order to widely deploy blockchain, these issues must be resolved. The most common challenges faced by blockchain networks are discussed as follows;

1. **Heavy resource requirements:** The consensus algorithm used in blockchain networks plays a critical role. The existing consensus processes are computationally intensive, which limits the usage of blockchain in many situations.

2. **Huge storage requirement:** Every node keeps a backup of the whole blockchain. However, while duplicating data blocks on each blockchain node improves efficiency and lowers the need for centralized administration, it imposes a massive storage space cost.
3. **Low throughput:** The fixed block size limits the number of transactions that can be executed per second, resulting in longer transaction times and higher transaction costs. Block propagation will be further delayed by increasing the block size.
4. **Lack of legal standards:** Lack of these universal legacy requirements would make it challenging to achieve the appropriate level of interoperability across various blockchain networks.
5. **High energy consumption:** The energy usage of blockchain networks is indeed a significant issue. The energy used in blockchain transactions is enormous. When compared to a card transaction, it is projected that each Bitcoin transaction uses 80,000 times more energy. Therefore, the deployment of blockchain in energy efficient application such as IoT have been restricted significantly.

5 CONCLUSIONS

Due to its appealing features and capabilities, blockchain technology is quickly becoming a popular invention. Through this article, we have described the basic structure and flow of transactions in a blockchain to provide readers a simple and straightforward explanation of how to make use of the potential of blockchain in transforming traditional financial activities. Additionally, we have offered a comprehensive overview of how blockchain works, which is otherwise a mystery for the outside world. Finally, we have discussed the essential features of the technology and significant challenges to blockchain deployment.

REFERENCES

Andoni, Merlinda, Valentin Robu, David Flynn, Simone Abram, Dale Geach, David Jenkins, Peter McCallum, and Andrew Peacock. "Blockchain Technology in the Energy Sector: A Sys-tematic Review of Challenges and Opportunities." *Renewable and Sustainable Energy Reviews* 100 (2019): 143–174.

Bouraga, Sarah. "A Taxonomy of Blockchain Consensus Protocols: A Survey and Classification Framework." *Expert Systems with Applications* 168 (2021): 114384.

Dib, Omar, Kei-Leo Brousmiche, Antoine Durand, Eric Thea, and Elyes Ben Hamida. "Conso-rtium Blockchains: Overview, Applications and Challenges." *International Journal On Adv-ances in Telecommunications* 11, no. 1&2 (2018): 51–64.

Ferdous, Md Sadek, Mohammad Jabed Morshed Chowdhury, and Mohammad A. Hoque. "A Survey of Consensus Algorithms in Public Blockchain Systems for Crypto-currencies." *Journal of Network and Computer Applications* 182 (2021): 103035.

Gupta, Shikha, and A. K. Saini. "An Artificial Intelligence Based Approach for Managing risk of IT Systems in Adopting Cloud." *International Journal of Information Technology* 13, no. 6 (2021): 2515–2523.

Gupta, Shikha, and Anil K. Saini. "Information System Security and Risk Management: Issues and Impact on Organizations." *Global Journal of Enterprise Information System* 5, no. 1 (2013): 31–35.

Jennath, H. S., and S. Asharaf. "Survey on Blockchain Consensus Strategies." In *ICDSMLA* 2019, pp. 637–654. Springer, Singapore, 2020.

Kaur, Manpreet, and Shikha Gupta. "Blockchain Consensus Protocols: State-of-the-art and Future Directions." *In 2021 International Conference on Technological Advancements and Innovations (ICTAI)*, pp. 446–453. IEEE, 2021a.

Kaur, Manpreet, and Shikha Gupta. "Blockchain Technology for Convergence: An Overview, Applications, and Challenges." *Blockchain and AI Technology in the Industrial Internet of Things* (2021b): 1–17.

Kaur, Manpreet, Mohammad Zubair Khan, Shikha Gupta, Abdulfattah Noorwali, Chinmay Chakraborty, and Subhendu Kumar Pani. 2021b. "MBCP: Performance Analysis of Large Scale Mainstream Blockchain Consensus Protocols." *IEEE Access: Practical Innovations, Open Solutions* 9: 80931–44. https://doi.org/10.1109/access.2021.3085187.

Kaur, Manpreet, Mohammad Zubair Khan, Shikha Gupta, and Abdullah Alsaeedi. "Adoption of Blockchain With 5G Networks for Industrial IoT: Recent Advances, Challenges, and Po-tential Solutions." *IEEE Access* (2021a).

Li, Xiaoqi, Peng Jiang, Ting Chen, Xiapu Luo, and Qiaoyan Wen. "A Survey on the Security of Blockchain Systems." *Future Generation Computer Systems* 107 (2020): 841–853.

Li, Zhetao, Jiawen Kang, Rong Yu, Dongdong Ye, Qingyong Deng, and Yan Zhang. "Consortium Blockchain for Secure Energy Trading in Industrial Internet of Things." *IEEE Transactions on Industrial Informatics* 14, no. 8 (2017): 3690–3700.

Naz, Sana, and Scott Uk-Jin Lee. "Why the New Consensus Mechanism is Needed in Blockchain Technology?" In *2020 Second International Conference on Blockchain Computing and Applications (BCCA)*, pp. 92–99. IEEE, 2020.

Parkins, D. 2015. "Blockchains-The Great Chain of Being Sure about Things." *The Economist*, no. 10.

Salah, Khaled, M. Habib Ur Rehman, Nishara Nizamuddin, and Ala Al-Fuqaha. "Blockchain for AI: Review and Open Research Challenges." *IEEE Access* 7 (2019): 10127–10149.

Shah, Devavrat. "Network Gossip Algorithms." In *2009 IEEE International Conference on Acoustics, Speech and Signal Processing*, pp. 3673–3676. IEEE, 2009.

Yaga, D., P. Mell, N. Roby, and K. Scarfone. "Blockchain Technology Overview. *arXiv* preprint https://arxiv.org/abs/1906.11078. https://nvlpubs. nist.gov/nistpubs/ir/2018/NIST." (2019).

Zheng, Z., Xie, S., Dai, H., Chen, X. and Wang, H. (2017) An Overview of Blockchain Technology: Architecture, Consensus, and Future Trends. *2017 IEEE International Congress on Big Data (BigData Congress)*, Honolulu, 25–30 June 2017, 557–564.

Artificial Intelligence, Blockchain, Computing and Security – Dagur et al. (Eds)
© 2024 The Author(s), ISBN: 978-1-032-67841-2

Analysis of IoT devices data using bayesian learning on fog computing

Heena Wadhwa*, Htet Ne oo* & Mandeep Kaur*
Chitkara University Institute of Engineering and Technology, Chitkara University, Punjab, India

Amanpreet Kaur* & Pardeep Singh Tiwana*
Department of IT, Chandigarh Engineering College, Landran, Mohali, India

ABSTRACT: Cyber security is one of the major concerns for the peace and tranquillity of citizens. To provide the safe and secure cyber environment, a model is designed on fog computing. A realistic cyber security dataset of Internet of Things (IoT) devices and Industrial Internet of Things (IIoT) applications, called edge-IIoTset, are used to design a Bayesian learning model. The proposed model is designed on a fog layer based on hybrid computing technology. IoT data generated from devices are collected for analysis. While doing analysis, regression analysis is done for Transmission Control Protocol (TCP), which is under Denial of Service (DOS) and Distributed Denial of Service (DDOS) attack, injection attacks and malware attacks. The regression analysis model is designed to check the probability of possible attacks and verify the technological capability of devices. The 80% dataset is used as training data and 20% data is used for testing data. After the results are verified with regression analysis. It is required to evaluate more than one variable for the prediction of attacks.

Keywords: Bayesian Inference, Fog assisted cloud environment, IoT, Resources, Probability

1 INTRODUCTION

It is a new paradigm, that can be implemented by using various sensors, wearable devices, smart gadgets and vehicles. In this paradigm, the computing of jobs and analytical tasks should be performed in a distributed manner (Albahri *et al.* 2018). The network uses several devices rather than creating a single data centre. The calculation of latency starts from the end user to the cloud, reducing the bandwidth utilization and latency in the network. Fog computing refining data that is produced by the sensor and transfers to the cloud only after refinement. This paradigm offers various facilities such as faster communication, power of execution, monitoring and analyzing IoT services (Dastjerdi *et al.* 2016). Information Technology (IT) plays a vital role in everyone's life. Advancement in IT has completely transformed the lifestyle of mankind. Various sectors such as healthcare, agriculture, co-operate banks, entertainment, and many other sectors have been impacted by significant IT resources, i.e. storage, computation power, and network bandwidth. The demand for these IT resources has increased day by day. Due to the growing demand for IT resources, various computing technologies such as utility computing, parallel computing, grid computing and cloud computing have been established. Among these computing technologies, cloud

*Corresponding Authors: heenakapila26@gmail.com, heenawadhwa@chitkara.edu.in, Htet.neo@chitkara.edu.in, k.mandeep@chitkara.edu.in, amanpreet.it@cgc.edu.in and Pardeep.3793@cgc.edu.in

 DOI: 10.1201/9781032684994-29

computing allows customers to perform services on-demand basis and reimbursement for these services as per the usage. Cloud Computing has enabled on demand services to the computing resources.

Cloud computing is considered as the pillar of all IoT services and it executes all IoT services in a centralized cloud. However, there is no exact location awareness and a lot of delay in data collection from IoT devices. Although, there aren't any geo-distributed data centres nearby, The mentioned requirements cannot be fulfilled by cloud computing. Cloud computing may be the attainable solution to satisfy the requirements of distributed IoT based applications due to the issue of latency (Gubbi *et al.* 2013) (Wadhwa & Aron 2022). Many researchers are using a machine learning algorithm for design optimization model from IoT devices (Pattanaik *et al.* 2022) (Sharma *et al.* 2022). Information systems for IoT applications with a centralized global approach, in which Internet of Things (IoT) devices utilize remote management systems (Gubbi *et al.* 2013). However, the model has a deficiency in terms of agility. Many users want prompt replies in many real-time applications such as environmental analytics, ambient assisted living and healthcare related applications. Even after the speed of mobile internet has increased, the latency is quite high in the distant centralized model. To handle this problem, fog computing provides data filtering with computers available at local data centers and end user applications situated at the edge of the network of IoT systems (Bonomi *et al.* 2012) (Kaur & Aron 2022).

2 RELATED WORK

Fog computing paradigm is essentially a virtualization technology that offers computing, communication, and database capabilities between edge devices and cloud data centres. This architecture makes it possible to filter, monitor, analyse, gather, and share information, which saves time and computing resources when the big data analytics and applications of cyber security are implemented and runs (Manimurugan 2021). By conducting and calculating the posterior probability of the data source, anomaly detection utilizing a bayesian inference model in fog computing may be a better choice. (Ngo *et al.* 2021) developed a method for hierarchical edge computing (HEC) that can detect anomalies. They created three deep neural network models with different levels of difficulty. A single-step Markov decision procedure, which has the benefits of quick detection and high precision, was used to make the choice.

(Tran *et al.* 2019) developed filters to detect and prevent cyber attacks. Each filter used collected data in its network to train the deep learning algorithm-based cyber identification model. The proposed model was accomplished and shared with different IoT gateways to enhance the accuracy of the detection system. (Sarker *et al.* 2020) proposed IntruDTree, a security model built on a machine learning framework. This model used to do ranking and created a tree based model by selecting important features. (Alrashdi *et al.* 2019) proposed an anomaly detection model based on the random forest algorithm to analyze the cyber security issues in a smart city. They have used the UNSW-15 dataset and identified anomalies with binary labeled classification. In a smart city built on IoT and fog computing, this investigation examined the cyber security risks. By analyzing the UNSW- 15 dataset, this approach effectively found anomalies. The dataset underwent binary labeled classification processing before being distributed to fog nodes. They have suggested an algorithm based on a Random forest algorithm called Anomaly Detection- IoT (AD-IoT).

3 PROPOSED METHODOLOGY

The primary concept was to design a logistic regression for the real time collected data for specific features. The parameters considered for the study is multidimensional data. The process of calculating posterior data from multidimensional data is implemented at the fog layer. The three layer of architecture is represented in the Figure 1. The top layer represent cloud computing, the middle layer represented as fog layer and the bottom layer represented the installed IoT device. The data collection was done on the bottom layer and send to the fog computing layer. After the data collection, it send to the middle layer for processing. The proposed algorithm is implemented on the fog layer and checks the posterior probability for the collected data.

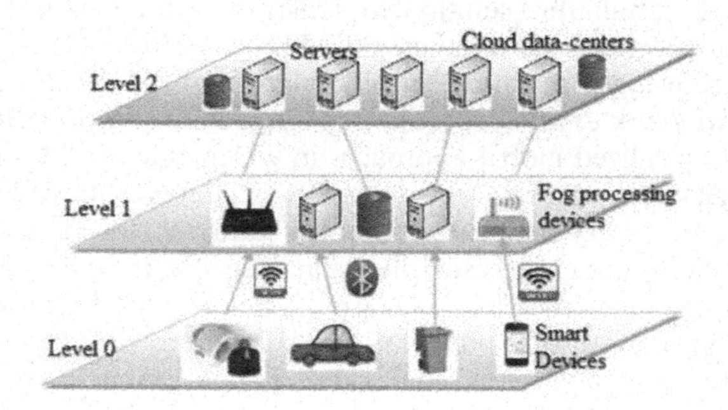

Figure 1. Three layer architecture.

4 MODIFIED BAYESIAN INFERENCE

4.1 *Dataset description*

In the study, Edge-IIoTset is a dataset tailored to IoT and IIoT applications, which uses analytics to provide insights into the types of cyberattacks. It also provides security recommendations on how to best protect against, detect, and respond to these incidents. Edge-IIoTset also offers other features such as data enrichment and correlation with existing data sources, interactive visualizations, and statistical analysis.

The dataset is updated periodically to ensure that it reflects the ever-changing nature of cyber threats. This dataset may be used by centralized and distributed artificial intelligence (AI) based machine learning intrusion detection systems. The specified dataset was built with a variety of layers, such as the edge computing layer, cloud computing layer, software-defined networking layer, blockchain network layer, fog computing layer, and IoT and IIoT perception layer (Ferrag *et al.* 2022). The Internet of Things (IoT) collects data from over ten various types of smart IoT devices, such as inexpensive digital sensors for detecting temperature and water level detection sensors, humidity, ultrasonic sensors, pH sensor metres, soil moisture sensors, heart rate sensors, and flame sensors, among others. However, the fourteen threats that are identified and analyzed are those that aim to target IoT and IIoT communication protocols. There are numerous attacks are described in the dataset. 1. DoS/DDoS Attacks: These attacks involve flooding web servers or networks with large amounts of traffic, making it difficult or impossible to access services or resources. 2. Information Gathering Attacks: These attacks target vulnerable systems in order to gain access to sensitive data or credentials. They may also be used to launch other types of attacks. 3. Man in the Middle Attacks: All the mentioned attacks are used to intercept and eavesdrop on communications between two parties. They can be used to steal data or inject malicious code into communications. 4. Infiltration Attacks: These attacks involve exploiting a vulnerability in a system or network in order to gain unauthorized access. The attacker may then be able to manipulate or modify files. After processing and analyzing the suggested real time based cyber security dataset, this study offers an initial data exploration and assesses the effectiveness of machine learning techniques in centralized as well as distributed learning modes. The considered dataset contains attack traffic and normal traffic data. Whereas attack traffic contains information about the backdoor attack, DDoS attack, MITM attack, Ransomware attack, SQL Injection attack etc.

4.2 *Posterior probability based on bayesian inference*

The Bayes rule can be used to calculate this posterior density under Gaussian assumptions. According to the Bayes rule, prior density and the likelihood function of a model's variables must be given. Only the prior probability distribution of the general linear model's applied parameters needed to be provided as additional information. Although independent data or some conceivable physiological limitations may be used to characterise them in terms of their means and variances, there is a substitute to this Bayesian approach.

An alternate is empirical Bayes, where the prior distributions' variances are calculated by using the data. Due to the hierarchical observation model that empirical Bayes requires, the parameters and hyperparameters at each given level can be regarded as priors on the lower level. Bayes' Classification for estimating the model's accuracy in accordance with Bayes' theory and conditional probability

$$P(X|Y) = \frac{P(Y|X)P(X)}{P(Y)} \tag{1}$$

In the equation 1, Where P(Y |X) shows Likelihood, P(X) consider as Prior distribution and P(Y) represents as marginal likelihood or Prior predictive distribution. The process of analyzing and calculate posterior probability is represented in Figure 2. The collected training and tested data in used for prepossessing and removed all null values from existing data. The feature selection is done from the preprocessed data. After that data classification is completed with machine learning algorithm. It will separate the erroneous data from the processed data set. On the basis of bayesian inference rule, the null hypothesis created for the data. The null hypothesis is TCP flag and TCP acknowledgment is not interrelated. The alternative hypothesis suggested that the TCP acknowledgment is effected by TCP flag, checksum, raw and payload. The null hypothesis is suggested that the type of attack can be predicted.

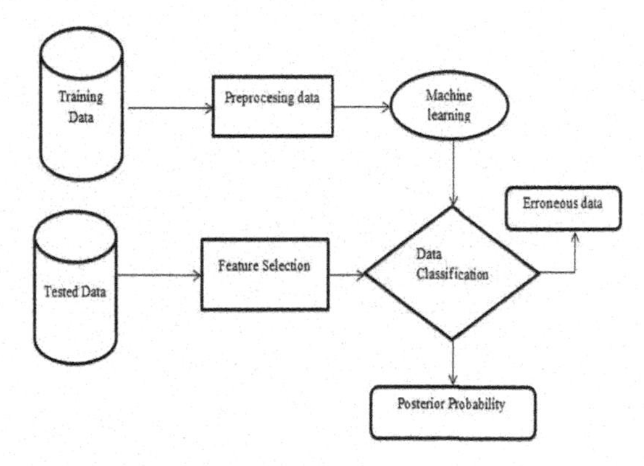

Figure 2. Model assessment framework.

4 PERFORMANCE METRIC

In the training set different sources has been analyzed including logs, network traffic and system resources. There was an authentic labeling operation was done which contain 0 and 1 as binary classification method. If the labeling is 0 which indicates it normal and if it is 1 that indicates attacks. There are four conventional machine learning algorithms have been used such as decision tree(DT), k-Nearest Neighbour (KNN), Support vector machine (SVM) and deep neural network (DNN). All the null values are eliminated from the database which can impact the study of algorithm. The complete dataset is split into two categories train dataset and tested data sent. In this study, tcp.flag and tcp.flag.ack parameters are considered for calculating regression analysis. The calculation is done based on 95% confidence level. Regression analysis of tcp flag and ack is represented in Figure 3. The value of p is less that 0.05 which represent the collected data is corrected to reject null hypothesis. There are more parameters are required for the prediction of attacks.

Table 1. Attack type and their count in DNN-EdgeIIOT dataset.

Attack type	Number of attacks
Normal	1615643
DDoSUDP	121568
DDoSICMP	116436
SQL injection	51203
Password	50153
Vulnerability scanner	50110
DDoSTCP	50062
DDoSHTTP	49911
Uploading	37634
Backdoor	24862
XSS	5915
Ransomware	10925
MITM	1214
Fingerprinting	1001

Regression Statistics	
Multiple R	0.746914048
R Square	0.557880596
Adjusted R Square	0.557876174
Standard Error	0.169957188
Observations	99999

ANOVA

	df	*SS*	*MS*	*F*	*Significance F*
Regression	1	3644.749843	3644.749843	126179.5	0
Residual	99997	2888.457929	0.028885446		
Total	99998	6533.207772			

	Coefficients	*Standard Error*	*t Stat*	*P-value*	*Lower 95%*	*Upper 95%*	*Lower 95.0%*	*Upper 95.0%*
Intercept	0.309068732	0.001828061	169.0691548	0	0.305485754	0.312651709	0.305485754	0.312651709
X Variable 1	0.034896375	9.82395E-05	355.217477	0	0.034703827	0.035088923	0.034703827	0.035088923

Figure 3. Regression analysis.

6 CONCLUSION

In the paper, a realistic dataset of IoT and IIoT applications. A new comprehensive realistic cyber security dataset of IoT and IIoT applications. A modified machine learning algorithmis used to evaluate the number and type of attacks. The obtained results are verified through regression analysis model. The data set was analyzed through data analysis method at primary stage then regression model is designed. The regression analysis model can be implemented on each layer to predict the attack. In future, the more attacks can be analyzed which are already mentioned in the described dataset.

REFERENCES

Albahri, O.S., Albahri, A.S., Mohammed, K.I., Zaidan, A.A., Zaidan, B.B., Hashim, M. and Salman, O.H., 2018. Systematic Review of Real-time Remote Health Monitoring System in Triage and Priority-Based Sensor Technology: Taxonomy, Open Challenges, Motivation and Recommendations. *Journal of Medical Systems*, 42(5), pp.1–27.

Alrashdi, I., Alqazzaz, A., Aloufi, E., Alharthi, R., Zohdy, M. and Ming, H., 2019, January. Ad-IoT: Anomaly Detection of IoT Cyberattacks in Smart City Using Machine Learning. In *2019 IEEE 9th Annual Computing and Communication Workshop and Conference (CCWC)* (pp. 0305–0310). IEEE.

Bonomi, F., Milito, R., Zhu, J. and Addepalli, S., 2012, August. Fog Computing and its Role in the Internet of Things. In *Proceedings of the First Edition of the MCC Workshop on Mobile Cloud Computing* (pp. 13–16).

Dastjerdi, A.V., Gupta, H., Calheiros, R.N., Ghosh, S.K. and Buyya, R., 2016. Fog Computing: Principles, Architectures, and Applications. *In Internet of Things* (pp. 61–75). Morgan Kaufmann.

Ferrag, M.A., Friha, O., Hamouda, D., Maglaras, L. and Janicke, H., 2022. Edge-IIoTset: A New Comprehensive Realistic Cyber Security Dataset of IoT and IIoT Applications for Centralized and Federated Learning. *IEEE Access*, 10, pp.40281–40306.

Gubbi, J., Buyya, R., Marusic, S. and Palaniswami, M., 2013. Internet of Things (IoT): A Vision, Architectural Elements, and Future Directions. *Future Generation Computer Systems*, 29(7), pp.1645–1660.

Kaur, M. and Aron, R., 2022. An Energy-Efficient Load Balancing Approach for Scientific Workflows in Fog Computing. *Wireless Personal Communications*, pp.1–25.

Manimurugan, S., 2021. IoT-Fog-Cloud Model for Anomaly Detection Using Improved Naive Bayes and Principal Component Analysis. *Journal of Ambient Intelligence and Humanized Computing*, pp.1–10.

Ngo, M.V., Luo, T. and Quek, T.Q., 2021. Adaptive Anomaly Detection for Internet of Things in Hierarchical Edge Computing: A Contextual-Bandit Approach. *ACM Transactions on Internet of Things*, 3 (1), pp.1–23.

Pattanaik, R.K., Mohapatra, S.K., Mohanty, M.N. and Pattanayak, B.K., 2022. *System Identification Using Neuro Fuzzy Approach for IoT Application. Measurement: Sensors*, 24, p.100485.

Sarker, I.H., Abushark, Y.B., Alsolami, F. and Khan, A.I., 2020. Intrudtree: A Machine Learning Based Cyber Security Intrusion Detection Model. *Symmetry*, 12(5), p.754.

Sharma, V.K., Mohapatra, S.K., Shitharth, S., Yonbawi, S., Yafoz, A. and Alahmari, S., 2022. An Optimization-based Machine Learning Technique for Smart Home Security Using 5G. *Computers and Electrical Engineering*, 104, p.108434.

Tran, M.Q., Nguyen, D.T., Le, V.A., Nguyen, D.H. and Pham, T.V., 2019. Task Placement on Fog Computing Made Efficient for IoT Application Provision. *Wireless Communications and Mobile Computing*, 2019.

Wadhwa, H. and Aron, R., 2022. Optimized Task Scheduling and Preemption for Distributed Resource Management in Fog-assisted IoT Environment. *The Journal of Supercomputing*, pp.1–39.

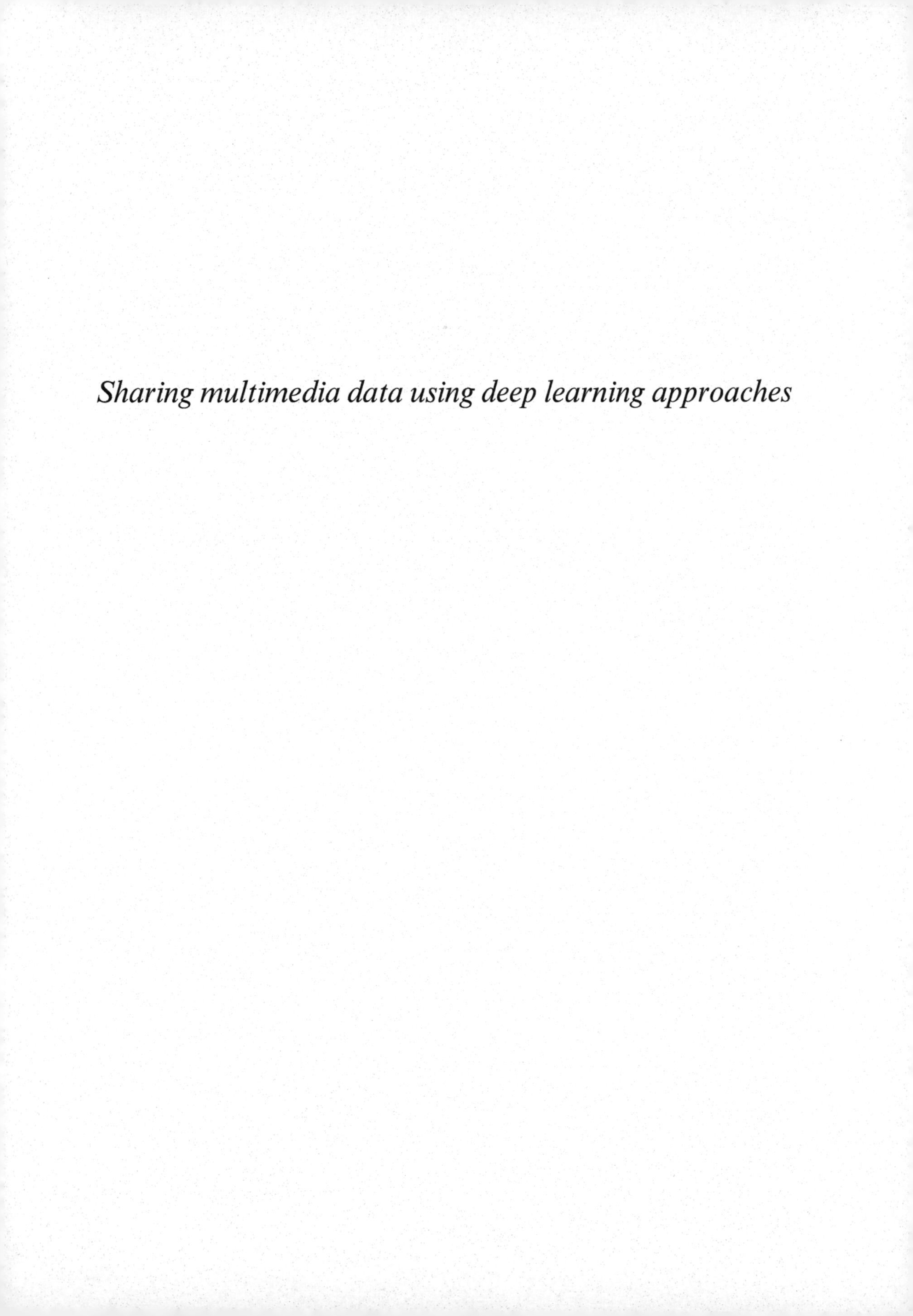

Sharing multimedia data using deep learning approaches

Artificial Intelligence, Blockchain, Computing and Security – Dagur et al. (Eds)
© 2024 The Author(s), ISBN: 978-1-032-67841-2

Assistive analytics for coronavirus-trends, patterns and predictions

Anitha Julian*
Department of Computer Science and Engineering, Saveetha Engineering College, Chennai

S. Selvi*
Department of Computer Science and Engineering, R.M.K Engineering College

Shyam Sundhar Kumaravel*
Data Engineering Associate, Accenture, Chennai

Sabarijeyanth Veerasekaran*
Systems Engineer, Infosys, Chennai

Sumanth Sai Sajja*
Developer EGG S&S SW, TCS, Chennai

ABSTRACT: The recently identified coronavirus is the source of the contagious sickness known as Covid-19. Most of the Covid-19 patients develop respiratory disorders that are recoverable since the infection is mild or moderate. But as the coronavirus has been thriving over the past three years uncontrollably, it has created new varied mutants, which has increased the fatality rate drastically and made it hard to be diagnosed on time. The proposed work suggests the analyzing of patient X-Ray as a more accurate step than the traditional swab test, in detecting the presence of the coronavirus through a mobile application. The infected person can get needed help on medication, treatment, hospitalization, etc. through the application. The application also marks the location of the Covid-19 infected persons using the geo-fencing visible to the other users in the same location that helps in avoiding contact with the infected.

Keywords: Covid-19, respiratory illness, virus, mutants, geo-fencing

1 INTRODUCTION

The past decades have seen an increase in lung related diseases rather than infections such as pneumonia or tuberculosis due to the highly increasing pollution of the air. The newest viral disease identified is the breathing related Severe Acute Respiratory Syndrome (SARS). In the following years, its occurrence has been found to be spreading through small droplets of saliva yet proved to be an airborne virus. With a mean incubation period of 5 days and case fatality of 9.6% the disease was under control within a year. In 2012, the emergence of MERS was yet another variant of SARS, the case fatality rate being 34.3% yet, the disease being contained, and hence no vaccine was thought of. The re-occurrence of such strains in the form of Covid-19 in 2019 December, became a cause of worry as its transmission rate made it a public health emergency of international concern and thus declared as a pandemic. Although the case fatality rate was initially 1.3% to 3.4%, the

*Corresponding Authors: cse.anithajulian@gmail.com, ssi.cse@rmkec.ac.in, shyamsundhar432@gmail.com, vsabari487@gmail.com and s.sumanthsai1998@gmail.com

DOI: 10.1201/9781032684994-30

uncontrollable spread over the past years has forced the international community to go ahead with the development of vaccines. The proposed work through an application helps the user to identify whether they have the symptoms of coronavirus and once they are identified with symptoms, the application helps the users how they can safely take care of themselves and stop its spreading.

2 LITERATURE SURVEY

The authors of (Alazab Moutaz *et al.* 2020), (Long *et al.* 2020), (Fang *et al.* 2020), (Harmon *et al.* 2020), (Yu *et al.* 2020) have suggested that CT scans and X-rays can both be used to visualize the lungs, which are often affected by Covid-19. By analyzing images of the lungs using deep learning algorithms, it may be possible to detect signs of Covid-19 infection. (Arias-Londoño *et al.* 2020) and (Pham*et al.* 2020) have dealt with the data analytics of the patient's health records, X-Ray and CT scan, infected cases, outbreak area information. AI systems can process and analyze large amounts of data to extract meaningful insights and predictions. Big Data refers to large and complex datasets that cannot be processed by traditional data processing applications. It involves the use of advanced technologies such as AI, machine learning, and natural language processing to extract insights from the vast amounts of data generated every day. AI and Big Data have been widely used in the fight against Covid-19. (Sahbudin *et al.* 2020) suggest that geofencing can be used to enforce quarantine and isolation measures. For example, individuals who have tested positive for Covid-19 or have been in contact with a confirmed case can be required to quarantine themselves for a specified period. Geofencing can be used to ensure that they remain within a designated area, and any attempt to leave the area can trigger an alert to the authorities. A detailed analysis about the various risk factors due to the coronavirus has been inferred from (Crook *et al.* 2021), (Tiwari *et al.* 2020), (Tomar *et al.* 2020) and (Arora *et al.* 2020). The proposed work incorporates the use of X-Ray to identify if corona virus is present in a patient and also aids the people to stay away from infectious zones by deploying the geo-fencing techniques through a multi-facet application.

3 METHODOLOGY

The proposed work initially aims to help users to get awareness of coronavirus, how to prevent the spread and how they can get assistance if affected by coronavirus. The patient can confirm their severity of the diseases by two phase predictions, namely in the first phase they are asked simple questions to confirm whether there is possibility for coronavirus as shown in Figure 2 (a) & (b). The daily statistics can be obtained as shown in Figure 1.

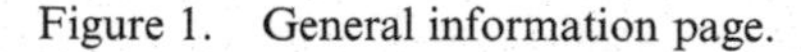

Figure 1. General information page.

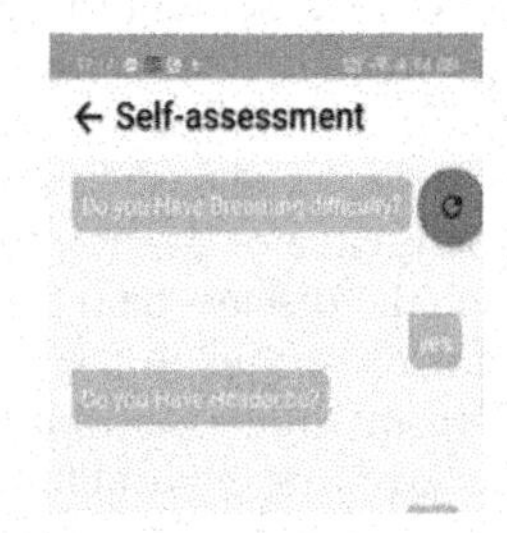

Figure 2. (a) & (b) First phase prediction.

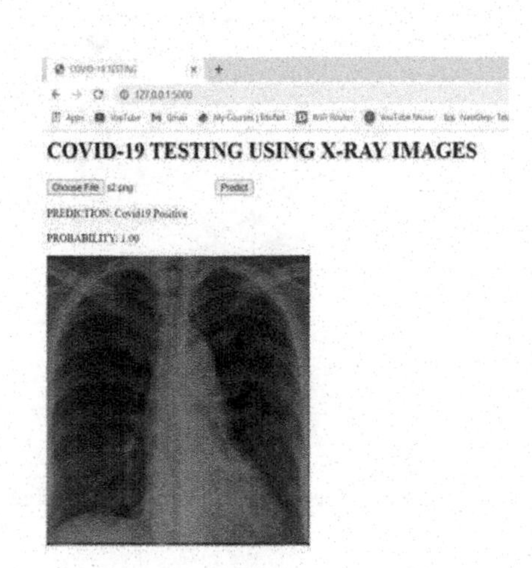

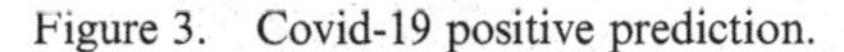

Figure 3. Covid-19 positive prediction.

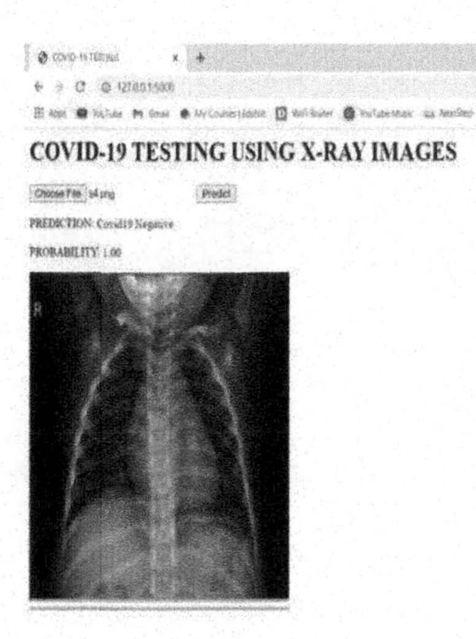

Figure 4. Covid-19 negative prediction.

In the second phase, the X-Ray of the person is examined using a machine learning algorithm to determine the coronavirus's acuteness as shown in Figures 3 and 4. Detection of diseases from X-Ray digital images is not easy since it has variations in terms of different size and disease in the lungs. This difficulty is overcome using the concept of deep learning by the Haar cascade algorithm, which improves accuracy of results.

The application further lists the location of the nearby hospitals and medical stores, helps in booking online appointments with doctors by stating his/her symptoms and provides contact details for the Covis-19 toll free numbers. The data collected from the users are stored in the database (Google Firebase). The collected data is used by the administrator for further processes for the user. The tables in the database are the information of users, geo-location of users, online doctor appointment and statistical data of Covid-19 infected persons. This application also helps the user who are not affected by coronavirus to know the affected zones by using maps. If they enter the affected zone, they will be alerted to minimize the risk of getting affected by coronavirus. This application helps people diagnosed with symptoms to fix a doctor's appointment, helps them to navigate nearby hospitals and pharmacies, so that they need not wait for guidance from others. It also helps the normal user to know about the statistics of corona affected persons and how many persons are recovered from the coronavirus.

4 PROPOSED WORK

The proposed work consists of the following modules, namely, 2 Phase Covid-19 Predictor, Covid-19 daily Statistics, Online Doctor appointment, Nearby Hospital and Pharmacy finder, Integrated Toll-Free Numbers, Government Services Linkage and Geo-Fencing. Figure 5 shows the proposed architecture.

The first phase involves asking queries related to the coronavirus. In the second phase, the X-rays of patients are analyzed using a machine learning algorithm. The "Covid-19 daily Statistics" module provides information about the number of individuals affected, recovered, and deceased due to the coronavirus. The "Online Doctor Appointment" module helps patients book appointments with doctors at a convenient time. The "Nearby Hospital and Pharmacy Finder" module assists patients in locating nearby hospitals and pharmacies. The

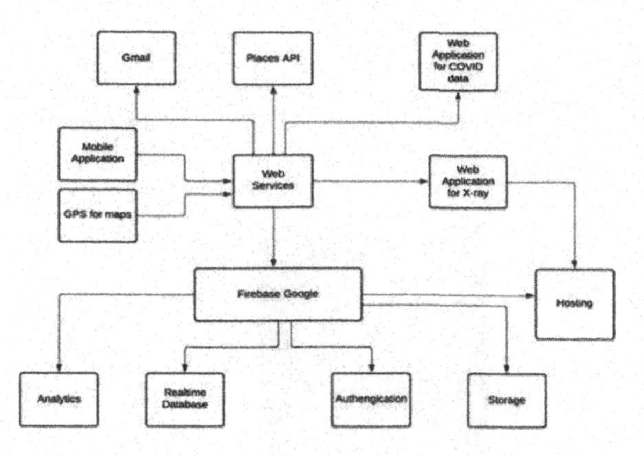

Figure 5. Proposed architecture.

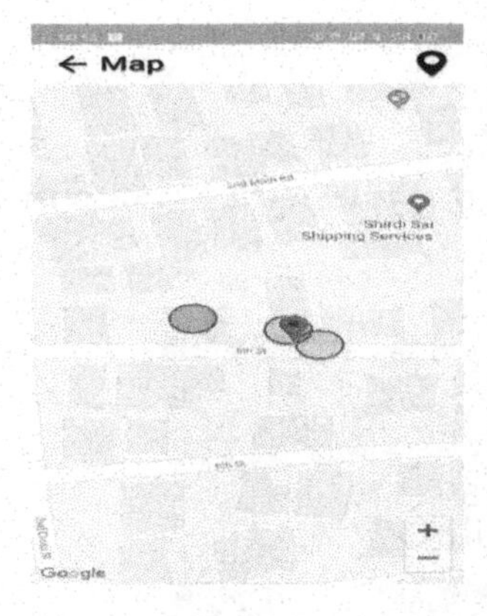

Figure 6. Geo-Fencing showing affected area.

"Integrated Toll-Free Numbers" module provides the toll-free government numbers for all states in India. The "Geo-Fencing" module designates infected areas as red zones and alerts users if they enter such areas (as shown in Figure 6).

5 CONCLUSION AND FUTURE ENHANCEMENT

The proposed system is an initiative to create awareness and information to the common people. Many steps have been taken by the government to control this COVID-19 pandemic and this proposed system helps the people to socially distance themselves and self-assess their condition. The geofence location can be done with more accuracy and speed. The detection of X-Ray can be done with more accuracy with more data sets.

The drawback is existing systems that use Bluetooth for marking the user's location. There is no re-verification of the predicted data also. These drawbacks are overcome in the proposed work with the incorporation of geo-fencing.

REFERENCES

Alazab, Moutaz & Awajan, Albara & Mesleh, Abdelwadood & Abraham, Ajith & Jatana, Vansh & Alhyari, Salah. 2020. COVID-19 Prediction and Detection Using Deep Learning. *International Journal of Computer Information Systems and Industrial Management Applications*. 12. 168–181.

Arias-Londoño J. D., Gómez-García J. A., Moro-Velázquez L. and Godino-Llorente J. I.. 2020. *Artificial Intelligence Applied to Chest X-Ray Images for the Automatic Detection of COVID-19*. 8. 226811–226827.

Arora P., Kumar H.and Panigrahi B. K.. 2020. Prediction and Analysis of COVID-19 Positive Cases using Deep Learning Models: *A Descriptive Case sSudy of India. Chaos Solitons and Fractals*. 139 (110017). 1–9.

Crook H., Raza S., Nowell J., Young M., Edison P. 2021. Long Covid-mechanisms, Risk Factors andManagement. *BMJ*.

Fang Y. *et al.* 2020. Sensitivity of chest CT for COVID-19: Comparison to RT–PCR. *Radiology* 296, 200432.

Harmon S. A., Sanford T. H., Xu S., Turkbey E. B., Roth H., Xu Z., *et al* 2020. Artificial Intelligence for the Detection of COVID-19 Pneumonia on Chest CT Using Multinational Datasets. *Nat. Commun.* 11 (1). 4080.

Long C. *et al.* 2020. Diagnosis of the Coronavirus Disease *(COVID-19): rRT–PCR or CT? European Journal of Radiology*. 126, 108961.

Pham, Quoc-Viet & C. Nguyen, Dinh & Huynh-The, Thien & Hwang, won-Joo & Pathirana, Pubudu. 2020. Artificial Intelligence (AI) and Big Data for Coronavirus (COVID-19) Pandemic: A Survey on the State-of-the-Arts. *IEEE Access.1–1.*

Sahbudin, Murtadha Arif Bin & Ali Pitchay, Sakinah & Scarpa, Marco. 2020. Geo-COVID: Movement Monitoring based on Geo-fence Framework for COVID-19 Pandemic Crisis. *Advances in Mathematics Scientific Journal.* 9. 7385–7395.

Tiwari S., Kumar S. and Guleria K., 2020. Outbreak Trends of Coronavirus Disease-2019 in India: A Prediction. *Disaster Medicine and Public Health Preparedness.* 14(5). 1–6.

Tomar A. and Gupta N., 2020. Prediction for the spread of COVID-19 in India and effectiveness of preventive measures. *Science of the Total Environment.* 728(138762). 1–6.

Yu M., Xu D., Lan L., Tu M., Liao R., Cai S., *et al.*, 2020. Thin-section Chest CT Imaging of Coronavirus Disease 2019 Pneumonia: Comparison Between Patients with Mild and Severe Disease. *Radiology: Cardiothoracic Imaging.* 2 (2).

Artificial Intelligence, Blockchain, Computing and Security – Dagur et al. (Eds)
© 2024 The Author(s), ISBN: 978-1-032-67841-2

Real time object detection and object size measurement

Anitha Julian* & R. Ramyadevi*
Department of Computer Science and Engineering, Saveetha Engineering College, Chennai, Tamil Nadu

V. Subiksha*
Associate Software Engineer, Infinite Computer Solutions, Chennai, Tamil Nadu, India

Sweetlin D. Breetha*
Testing group, Cognizant Technology Solutions Corp (CTS), Chennai, Tamil Nadu, India

ABSTRACT: One of the main challenges in real-time object detection and tracking is the need for high computational efficiency. Researchers are constantly developing more effective and competitive algorithms as a result of its expanding use in security tracking systems, surveillance, and many other applications. But real-time object tracking, and identification are difficult to implement because of problems including tracking in dynamic surroundings which is expensive. Although other strategies and methodologies have been created, in this literature review we will explore some well-known and fundamental techniques for object recognition and tracking. Finally, their broad uses and outcomes are also included.

Keywords: Real time object detection, Tracking system, multi-camera, multi-objects

1 INTRODUCTION

All people develop their understanding of size quite early in life. Later on, it becomes clear that there is one significant aspect to this, and that is the relative nature of how size is defined. The proposed work's goals are to measure the object and determine its size. Any entity's size, or any object's size, is meaningless in and of itself. Size is merely a rough comparison of one entity to another. It is determined what an object's initial size was in an image. The chosen methods are implemented using computer vision, and they comprise simple tools that are effective enough to determine an object's size. This project suggests measuring an object's dimensions and computing the separations of the objects from one another. Calculating the measures of the things in a photograph is similar to determining the distance to an object from the camera. The value of pixels per a specific metric must first be determined, followed by the selection of one reference object from the image. considering two major properties. The reference object must meet two requirements: first, its dimensions (width and height) must be known and represented in terms of (inches or mm), and second, it must be easy to identify.

2 LITERATURE SURVEY

A detailed survey has been done and the inferences of them are presented.

*Corresponding Authors: cse.anithajulian@gmail.com, ramyakathir@gmail.com, subikshavenkatesan24@gmail.com and sweetlinbreetha@gmail.com

 DOI: 10.1201/9781032684994-31

Real-time item recognition and dimensioning is a significant challenge in many industrial sectors nowadays, which is a vital topic of computer vision problems. This work offers an improved method for quickly measuring size of things given a video streams (Devi *et al.* 2018). An object measuring technique for real-time video using the OpenCV libraries is suggested, and it includes the algorithms for canny edge identification, dilation, and erosion. Some of the methods proposed are (Manish *et al.* 2013), (Muthukrishnan 2011), (Geng Xing *et al.* 2012) finding a measurement target, (Chen *et al.* 2014), (Moeslund *et al.* 2016) utilising morphological operators comprises dilation and erosion method to fill gaps between edges and identify and sort contours, (Kumar *et al.* 2015) determining an object's size. A system that made use of the OpenCV software library, a Raspberry Pi 3, and a Raspberry Camera was implemented. The method has been successful in estimating the size of the objects with over 98% accuracy.

It is determined what an object's initial size was in an image. (Dhikhi *et al.* 2019) chosen algorithms make use of computer vision, which has fundamental tools effective enough to determine an object's size. (Dario Cazzato *et al.* 2020) advised measuring each object in the image and calculating the distances between them. Calculating the measures of the things in a photograph is similar to determining the distance to an object from the camera. First, the value of pixels for a certain measure must be established. One reference item from the image must be chosen while taking into account two key properties in order to determine this value. (Niharika *et al.* 2022) The reference object must meet two requirements: first, its dimensions (width and height) must be known and represented in terms of (inches or mm), and second, it must be easy to identify.

A lot of research has been done on 3D point clouds produced by LiDAR and RGBD sensors for object detection. (Jisen *et al.* 2021), (Wang *et al.* 2010) A pipeline that combines several manually created characteristics or descriptors with a machine learning classifier is used in the majority of this work. Similar circumstances apply to semantic segmentation, except that instead of using a single output classifier, structured output classifiers are used. In contrast to these methods, the architecture learns to identify objects by class and extract features from the raw volumetric data. Volumetric representations are richer than point clouds because they can distinguish between known and undiscovered space. Furthermore, spatial neighbourhood queries—which are frequently needed for features based on point clouds—can easily become impractical for huge numbers of points.

3 SYSTEM ARCHITECTURE

In proposed system as shown in Figure 1,

At phase 1: Initially open the anaconda prompt and activate the image, when the image is activated it takes to another path.
At phase 2: After the image size is activated, it takes us to the desktop path. Then the environment will be created and activated to the next process.
At phase 3: After the environment created, install the open cv library and python version and type the python command to get the next screen of our object size measurement.
At phase 4: After the installation and activation, the object size measurement will be measured. It has two tkinter label with the option live and input.
At phase 5: When the input is clicked the object size will be measured, which is already stored in folder. When the live is clicked the real time object detection will take place.

The conceptual model that describes the structure, behaviour, and additional viewpoints of a system is referred to as a proposed architecture and is depicted in Figure 1. A formal description and representation of a system designed to facilitate reasoning about its structures and behaviors is called an architecture description.. The system is made up of the designed sub-systems and system components that will cooperate to implement the whole

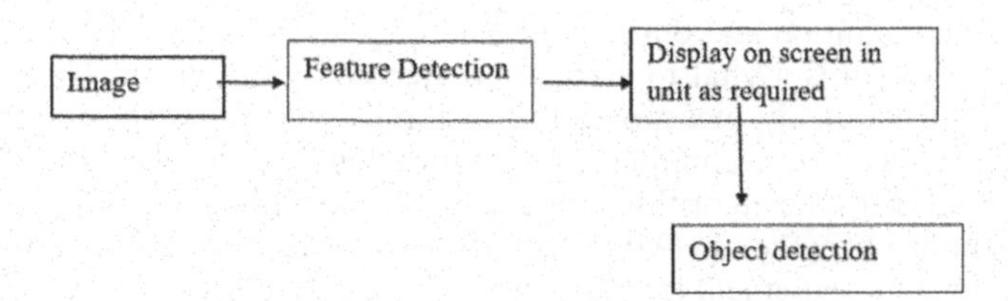

Figure 1. Proposed architecture.

system. The architecture description languages (ADLs) collectively refer to efforts to formalize languages that describe system architecture.

3.1 *Proposed algorithm*

Input: Object
Output: Measurement of the Object

Step 1: Start.

Step2: open the anaconda prompt and activate the image, when the image is activated it takes to another path.

Step 3: The image size is activated; it takes us to the desktop path. Then the environment will be created and activated to the next process.

Step 4: The environment is created, install the open cv library and python version and type the python command to get the next screen of our object size measurement

Step 5: The object size measurement will be measured. It has two tkinter label with the option live and input.

Step 6: The input is clicked the object size will be measured, which is already stored in folder. When the live is clicked the real time object detection will take place.

Step 7: Stop.

This algorithm shows that we can identify the size of the object by using the open cv library. The object gets detected and the size of the object can be measured.

4 APPLICATION PROCESS

The idea is put into practice with Python and Open CV. The implementation comprises of 5 major steps:

1. Initially open the anaconda prompt and activate the image, when the image is activated it takes to another path.
2. After the image size is activated, it takes us to the desktop path. Then the environment will be created and activated to the next process.
3. After the environment is created, install the OpenCV library and python version and type the python command to get the next screen of our object size measurement.
4. After the installation and activation, the object size measurement will be measured. It has two tkinter label with the option live and input.
5. When the input is clicked the object size will be measured, which is already stored in folder. When the live is clicked the real time object detection will take place.

The components of our system are divided into:

1. The system identifies the size of object.
2. The goal of our system is to identify the size of the object.

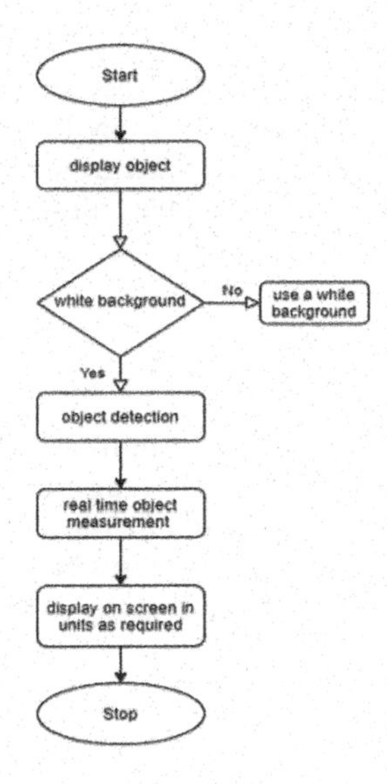

Figure 2. Workflow diagram for object detection.

In Figure 3 two buttons are seen namely, "LIVE" and "INPUT". By clicking on live can capture the image and by clicking the input can upload the image of the object in the device. Figure 4 shows the output, namely the size of the object.

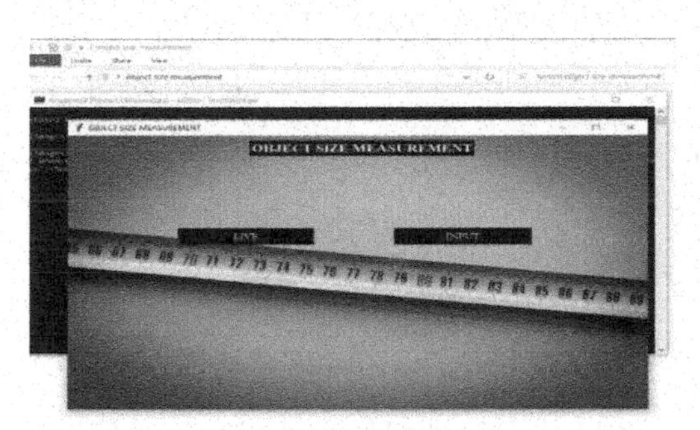

Figure 3. Giving an object to be measured.

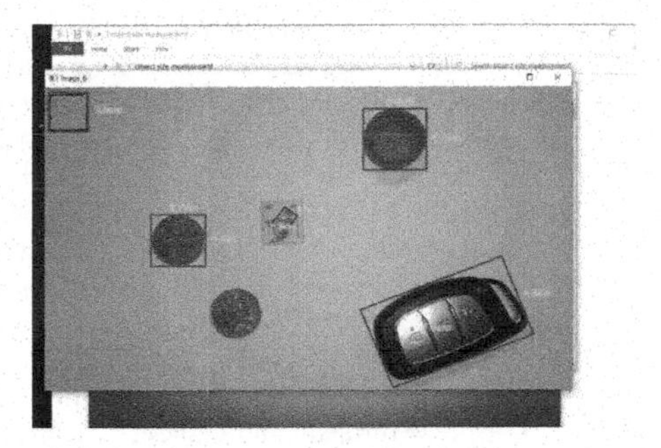

Figure 4. Final output of the program.

The proposed system provides an accuracy that is 85% higher than other current systems when compared to the proposed model and other project models. The comparison has been done with the parameters such as number of samples, time taken and by inferring this it is possible to see that the number of samples that were collected is higher in the proposed

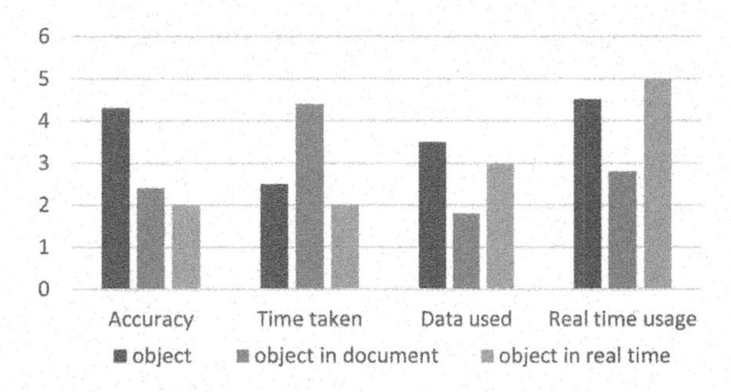

Figure 5. Comparison of results.

system. When compared to the current approach, which has proved useful in producing speedy results, the proposed technique also takes less time. The similarity study conducted plays a vital role in analyzing the detection accuracy between the existing system and the proposed system. The proposed system is a fully integrated python program using Anaconda prompt with OpenCV to detect the object size. This allows it to detect the size of the object and also to measure the object in real time.

5 CONCLUSION

A technique for measuring objects in photos captured by a camera is suggested, and the software related operation is described. Two photographs of the same scene, one with and one without the object visible, are used to identify an object. Dimension of the area where the object is detected and depicted along with coordinates help identify the objects. These image outputs are used as input data in the calculations that determine size.

REFERENCES

Chen W, Yue H, Wang J, Wu X. An Improved Edge Detection Algorithm for Depth Map in Painting. *Optics and Lasers in engineering* 2014;55: 69–77.

Dario Cazzato, Claudio Cimarelli, Jose Luis Sanchez-Lopez, Holger Voos and Marco Leo, A Survey of Computer Vision Methods for 2D Object Detection from Unmanned Aerial Vehicle, *Journal of Imaging*, Vol 6, No. 78, 2020 pp 1–38.

Devi, S.V., Murugan, D., & Kumar, T.G. (2018). *Comparative Study on Feature Extraction Approaches of Man-Made Objects Extraction from Satellite Images.*

Dhikhi T., Allagada Naga Suhas, Gosula Ramakanth Reddy, Kanadam Chandu Vardhan, Measuring Size of an Object using Computer Vision, *International Journal of Innovative Technology and Exploring Engineering*, Volume-8, Issue-6S4, April 2019, pp 424–426.

Geng Xing, Chen ken, Hu Xiaoguang. *An Improved Canny Edge Detection Algorithm for Color Image. IEEE Transaction*, 2012 978-1- 4673-0311- 8/12/$31.00 ©2012 IEEE.

Jisen W., A Study on Target Recognition Algorithm Based on 3D Point Cloud and Feature Fusion. *2021 IEEE 4th International Conference on Automation, Electronics and Electrical Engineering (AUTEEE)*, Shenyang, China, 2021, pp. 630–633.

Kumar, T. Ganesh, D. Murugan, and K. Rajalakshmi. *Image Enhancement and Performance Evaluation Using Various Filters for IRS-P6 Satellite Liss IV Remotely Sensed Data.* Geofizika 32.2 (2015): 179–189.

Manish, T. I., D. Murugan, and Ganesh T. Kumar. Hybrid Edge Detection Using Canny and Ant Colony Optimization. *Communications in Information Science and Management Engineering* 3.8 (2013): 402.

Moeslund T. *Canny Edge Detection.* Denmark: Laboratory of Computer Vision and Media Technology, Aalborg University. OpenCV, 2016.

Muthukrishnan R and M. Radha. Edge Detection Techniques for Image Segmentation, *International Journal of Computer Science & Information Technology (IJCSIT)* Vol3, No 6, Dec 2011.

Niharika B. D., Sri K. L., Jayanag B. and NavyaSri P., "Identifying the Key Features of the Object," *2022 International Mobile and Embedded Technology Conference (MECON)*, Noida, India, 2022, pp. 435–440.

Wang L., Chen J. and Yuan B., Simplified Representation for 3D Point Cloud Data, *IEEE 10th International Conference on Signal Processing Proceedings*, Beijing, China, 2010, pp. 1271–1274.

Artificial Intelligence, Blockchain, Computing and Security – Dagur et al. (Eds)
© 2024 The Author(s), ISBN: 978-1-032-67841-2

A system for crop recommendation to improve the production management

R. Ramyadevi*
Assistant Professor, Department of Computer Science and Engineering, Saveetha Engineering College, Chennai, India

K. Sandeep*, V. Girish Sairam* & P. Vaisak*
Department of Computer Science and Engineering, Saveetha Engineering College, Chennai, India

ABSTRACT: The increasing world population leads to a heightened demand for sustainable living and nutritious food. The farming industry, which is one of the most profitable in the world, faces challenges in meeting these needs. Integrated Crop Management offers a solution for ecologically sustainable farming that balances quality, quantity, and profitability. By utilizing tools such as weather forecasting, trend analysis, and productivity monitoring, farmers can minimize uncertainty and errors, leading to increased profitability. We propose a solution for farmers that involves utilizing Machine Learning and Data Analytics to manage information related to crop selection, soil type, and production outcomes. This 360-degree approach includes long-term strategies for increased production and provides information on various crop categories and weather predictions to determine the best results based on seasonality. By inputting farm and weather data, farmers can determine the optimal crops to grow for each cycle, predict the return on investment for farm activities.

Keywords: Soil prediction, Seasonality prediction, Crop recommendation system

1 INTRODUCTION

In our country, Agriculture is considered as the backbone of our economy, contributing 15–20% to India's national GDP. It contributes to the country's growth and economic stability as, 60% of our land is dedicated to agriculture to meet the needs of our population. With the increasing global population, there is a growing demand on the agriculture sector. To address these challenges, digital agriculture has emerged, improving farm productivity and reducing environmental impact through data-intensive methods. The analysis of significant volumes of data, including weather predictions, can be facilitated through the utilization of machine learning algorithms and also predict crop diseases and pests, optimize irrigation, and even optimize the use of fertilizer and pesticides. Many farmers lack knowledge on how to use technology to improve crop production, and are unaware of the benefits it can bring to their farms. The project aims to use ML-based prediction algorithms to increase crop production by identifying patterns in the training dataset and determining the best crops to cultivate for maximum profit. Additionally, farmers will be educated on the right cultivation methods for optimal results. To improve accessibility and understanding, the content is made available in all regional languages.

2 LITERATURE SURVEY

The proposed crop management system optimizes productivity with sustainability in mind. The system considers various factors like land resources, climate, and fertilizers. Jay Gholap

*Corresponding Authors: ramyadevir@saveetha.ac.in, sandeepkris2001@gmail.com, sgirish585@gmail.com and vaisakmmk7@gmail.com

DOI: 10.1201/9781032684994-32

et al. & Anurag Ingole *et al.* & Shailesh Gargade *et al.* & Vahida Attar *et al.* it integrates data from weather departments, soil data repositories, and user inputs. Kiran Shinde *et al.* The system uses a machine learning model with a Naive Bayes algorithm to predict the best crop for an area. The user interface is designed to be flexible and interactive to encourage farmers to use the application.

The study aimed to determine the rate of computer & consultant adoption by NY dairy farmers, identify users & non-users, and evaluate business impacts on adoption. In 1986, 15% of farm managers had computers. Lee H *et al.* & Moon A *et al.* Operator education impacted computer ownership & accounting use, while age was key in predicting veterinarian use. Production, debt per cow, housing type & business organization were analyzed. Consultant use was not explained by evaluated business characteristics.

Current chemical disease control methods harm the environment and raise costs Vijayakumar T *et al.* & Mr Vinothkanna R *et al.* Machine learning models were trained using hyperspectral images of banana leaves that had been affected by black sigatoka. Naik M R *et al.* & Sivappagari C *et al.* SVM, MLP, Neural Networks, NPLS-DA, & Due to their strong predictive capabilities, PLS-PLR were chosen for selection.

Machine learning uses big data tech and high-performance computing to advance data-intensive science in agri-tech. Konstantinos G. Liakos *et al.* & Dionysis Bochtis *et al.* the use of ML in agriculture is demonstrated through filtered and classified articles. ML in sensor data leads to real-time AI-powered farm management systems that offer insights and recommendations for farmers.

3 METHODOLOGY

The proposed system aims to create a crop management tool that employs Machine Learning methods. These methods assist in data analysis and representation, as well as the creation of forecasting models. These models will assist farmers in making informed decisions about their crop management, such as when to plant, fertilize, and harvest, and also how to optimize production by identifying the most suitable crop for their land. By utilizing Machine Learning techniques in crop system, farming efficiency and productivity could be enhanced, ultimately resulting in improved yields and cost reduction.

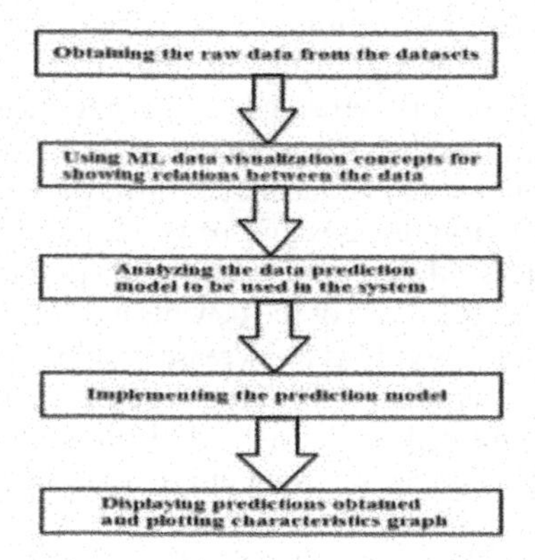

Figure 1. System architecture.

3.1 *Convolutional Neural Network (CNN)*

Convolutional neural networks (CNNs) are a category of neural networks that employ perceptrons, a type ML algorithm, for supervised learning, to examine data. They are also used in natural language processing tasks such as text classification and language translation, as well as other types of cognitive tasks such as speech recognition and video analysis. A deep learning neural network designed to handle structured data arrays, such as images, is known as a Convolutional Neural Network (CNN). CNNs excel in recognizing patterns and distinctive features present in input images, including curves, gradients, circles, eyes, and faces. This makes them highly effective for computer vision tasks.

3.2 *Convolutional layer*

Convolutional layer is central to CNNs. It has 3 components: input data, filter, and feature map. Input data is a 3D matrix (height, width, depth) of pixels in an image. Filter is a small matrix of weights used to detect features in input data. Convolution is a process where the filter scans the input data for the presence of a feature.

3.3 *Pooling layer*

CNNs employ pooling layers to decrease the number of parameters and the dimensionality of the data. Max pooling selects max pixel in receptive field and average pooling calculates avg. value in receptive field. Both send values to output array through aggregation function on receptive field values. Both types of pooling play a crucial role in improving CNN efficiency while retaining important information from the input data.

3.4 *Fully-connected layer*

Every node in the previous layer is directly connected to a node in the output in a fully-connected layer. Unlike partial connections in convolutional and pooling layers, this layer has full connectivity. The fully-connected layer executes the classification function by utilizing the features and filters obtained from the preceding layers. Softmax activation is typically used in this layer instead of ReLu, which outputs probabilities from 0 to 1 for each class in the output layer.

4 RESULTS AND ANALYSIS

4.1 *Dataset description*

A crucial aspect of machine learning that deserves particular consideration is the data. The quality and characteristics of the data significantly impact the results of a machine learning task, including its source, format, consistency, and presence of outliers. Substantial effort is required to obtain, sanitize, and process the data. We have taken the dataset from kaggle and images of soil such as Gravel, Sand, Silt are taken and splitted as train set and test set for the CNN Soil prediction. Each folder contains images of soils according to classification and used for training. Also crop recommendation dataset is also provided as an input for providing the suggestion based on type of soil.

4.2 *SVM (Support Vector Machine)*

Support Vector Machine, abbreviated as SVM, a supervised machine learning technique that can be applied for both classification and regression analyses. It's used in both classification and regression analysis. SVM locates the hyperplane that separates the data into different categories by maximizing the space between them, referred to as max-margin classification. New data can be assigned to the appropriate category by computing its distance to the hyperplane after it's found.

4.3 *Logistic regression*

Logistic Regression is a well-known algorithm in the field of Supervised Learning. It is utilized to predict categorical dependent variables by utilizing a set of independent variables. The output of Logistic Regression is a categorical or discrete value, such as yes or no, true or false, etc. Instead of an exact output, it provides probabilities ranging from 0 to 1.

Use of CNN algorithm yields best results with accuracy of 0.89 (89%) among other algorithms as shown in Figures 2 and 3.

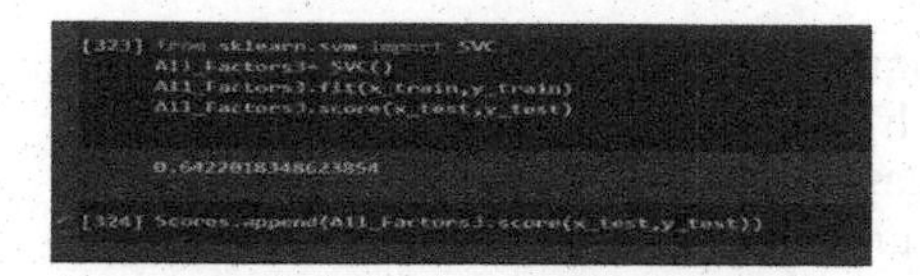

Figure 2. Support vector clustering.

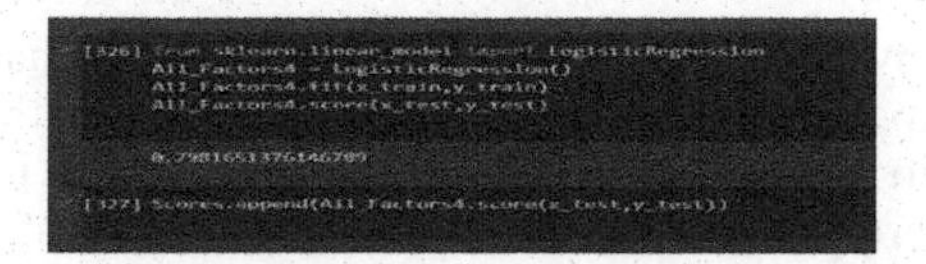

Figure 3. Logistic regression.

5 MODULE IMPLEMENTATION

5.1 *Login/sign up*

The Login Module is a module that enables users to enter their credentials, including a username and password, to gain access to the system. If you haven't already, one should create an account to continue. With the signup module, one can quickly and easily create a signup sheet for the crop management system that the users can fill out on your view by entering details. This is achieved by using Google Firebase for the user authentication by validating the user credentials in the backend.

5.2 *Soil prediction module*

Soil Prediction is aimed to predict the soil functional properties and type of soil (Gravel, Sandy or Silt) of a soil by uploading the image of the soil. The use of Soil Prediction is widespread in agriculture, farming, and research. Its purpose is to enhance crop management and boost crop yields through better economic decision-making. Our efforts have focused on exploring how modern Machine Learning techniques can replace traditional soil prediction methods, leading to more efficient and cost-effective solutions as shown in Figures 4 and 5.

A	B	C	D	E	F	G	H
N	P	K	temperatu	humidity	ph	rainfall	label
90	42	43	20.87974	82.00274	6.502985	202.9355	rice
85	58	41	21.77046	80.31964	7.038096	226.6555	rice
60	55	44	23.00446	82.32076	7.840207	263.9642	rice
74	35	40	26.4911	80.15836	6.980401	242.864	rice
78	42	42	20.13017	81.60487	7.628473	262.7173	rice
69	37	42	23.05805	83.37012	7.073454	251.055	rice
69	55	38	22.70884	82.63941	5.700806	271.3249	rice
94	53	40	20.27774	82.89409	5.718627	241.9742	rice
89	54	38	24.51588	83.53522	6.685346	230.4462	rice
68	58	38	23.22397	83.03323	6.336254	221.2092	rice
91	53	40	26.52724	81.41754	5.386168	264.6149	rice
90	46	42	23.97898	81.45062	7.502834	250.0832	rice
78	58	44	26.8008	80.88685	5.108682	284.4365	rice
93	56	36	24.01498	82.05687	6.984354	185.2773	rice
94	50	37	25.66585	80.66385	6.94802	209.587	rice
60	48	39	24.28209	80.30026	7.042299	231.0863	rice

Figure 4. Dataset.

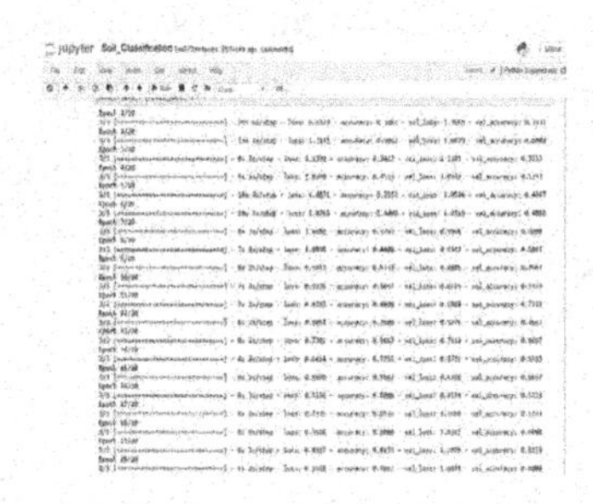

Figure 5. Soil prediction training.

6 CONCLUSION & FUTURE WORK

This project is a very humble approach to solving an issue that is costing farmers their time and money. Our objective with this system is to alleviate the burden on farmers and enable them to focus on other tasks. We aspire to supplement the farmers' income through our system. The Crop Management software is highly economical and efficient. It can be widely used across the agricultural industry to streamline the farming process by improving day to day operations and maintaining a high standard. People in the agricultural sector greatly benefit from such a solution and can accomplish many feats that were considered impossible previously.

REFERENCES

Abraham Chandy, Journal of Artificial Intelligence, pp. 10–18, 2019. Pest Infestation Identification in Coconut Trees Using Deep Learning.

Dimitriadis Savvas & Christos Goumopoulos, Panhellenic Conference on Informatics IEEE, 2008. Applying Machine Learning to Extract New Knowledge in Precision Agriculture Applications.

Jay Gholap & Anurag Ingole &Shailesh Gargade & Vahida Attar, International Journal of Computer Science Issues, 2012. Soil Data Analysis Using Classification Techniques and Soil Attribute Prediction.

Kiran Shinde, *International Journal on Recent and Innovation Trends in Computing and Communication*, no. 3, March. 2015. Web Based Recommendation System for farmers.

Konstantinos G. Liakos & Dionysis Bochtis, *Article on Sensors*, pp. 1–29, 2018. Machine Learning in Agriculture: A Review.

Kumar, R. & Singh, M. P. & Kumar, P. & Singh, J. P., International Conference on Smart Technologies and Management for Computing Communication Controls Energy and Materials (ICSTM), pp. 138–145, 2015. Crop Selection Method to Maximize Crop Yield Rate Using Machine Learning Technique.

Lee, H. & Moon, A., 16th International Conference on Advanced Communication Technology, pp. 1292–1295, 2014. Development of Yield Prediction System based on Real-time Agricultural Meteorological Information.

Lekhaa, T. R., International Journal of Modern Trends in Engineering and Science (IJMTES), vol. 03, no. 10, 2016. Efficient Crop Yield and Pesticide Prediction for Improving Agricultural Economy using Data Mining Techniques

Mendham, N. J. & Robertson, M. J., *In Encyclopedia Of Food Grains (Second Edition)*, 2016. The Production and Genetics Of Food Grains

Minh-Long Nguyen & Karuppan Sakadevan, *In Advances in Agronomy*, 2010. Extent, Impact, and Response to Soil and Water Salinity in Arid and Semiarid Regions.

Naik, M. R. & Sivappagari, C., *IJESC*, vol. 6, no. 12, 2016. Plant Leaf and Disease Detection by Using HSV Features and SVM

Vijayakumar, T. & Mr Vinothkanna, R., Journal of Innovative Image Processing (JIIP), vol. 2, no. 01, pp. 35–43, 2020. Mellowness Detection of Dragon Fruit Using Deep Learning Strategy

Artificial Intelligence, Blockchain, Computing and Security – Dagur et al. (Eds)
© 2024 The Author(s), ISBN: 978-1-032-67841-2

Chromatic color detection application access

N.V. Ravindhar*, P. Roopesh Kumar*, M. Yugesh*, K.P. Kiran Kumar* &
N. Poorna Pavan*
Department of Computer Science and Engineering, Saveetha Engineering College, Chennai, Tamil Nadu

ABSTRACT: One of the most ubiquitous examples of modern mechanism is the computer. In order to effectively interact with a computer, it is necessary to choose the most appropriate method. Color information plays a major role in real-time color sensors. As humans, our eyes are tuned to detect and recognize certain forms, making vision a crucial component of human cognition. In machines and computers alike, the same process is followed, since all processing begins with a shape recognition check. Human-computer interaction is very much important nowadays to solve many problems. In this study, we take a step toward an enhanced model of computer interaction based on the identification of objects. The red-green-blue spectral analysis is used for object identification. (Red, Green, Blue) colors. Respective functionalities will be performed for the color detected. After selecting an image, the system needs to start the process should detect colors and shapes, and will give graphs, plots, and clusters images. All these detections are done within the user interface. Everyone on the planet will benefit from this strategy of improving human-machine interaction. Therefore, it improves the connection between humans and computers.

Keywords: Object detection, Chromatic color detection, Application access, SFTA algorithm, HCI

1 INTRODUCTION

The mouse and keyboard have become the standard interface for human and computer interaction. As a result, a system needs to be implemented to ensure the most productive interaction between them. This instance claims ownership of HCI (Human-Computer Interaction). HCI is a cutting-edge tool that greatly improves communication between people and machines. Because of the importance of vision to humans, the same formula or operation is performed here for computers, since the software in question has to be able to distinguish colors and should create functionality linked to forms.

Three different shapes can also be detected. Multi Shapes are identified, as per Output, three different Shapes are identified Circle, Square & Triangle. This Project is to detect an Object's shape once we select the image. We are identifying three different Shapes of objects Circle, Square & Triangle. The final output is displayed in MATLAB GUI.

An improved method of using object recognition to communicate with computer is the topic of this study. Any item may be shown for object detection, Color Detection will detect the name of the color. Well, for humans this is very easy to detect any color but for computers, it is complicated. For humans, an object's form is one of the most important identifying features.

Three different color objects RGB (Red, Blue, Green) and three different shapes can be detected. Multi shapes are identified, as per output, the shapes were identified as Circle,

*Corresponding Authors: ravindhar@saveetha.ac.in, Roopeshkumarp17@gmail.com,
yugeshmasarpu@gmail.com, Kpkirankumar1211@gmail.com and poornapavan23@gmail.com

 DOI: 10.1201/9781032684994-33

Square & Triangle. We have six modules in this they are Image Frames separation, Adaptive Histogram equalization, Median Filter & ROI Extraction, SIFT &SFTA Feature Extraction, Training &Testing set Comparison, NN classification & Object Detection.

The accuracy of color recognition from a variety of objects is the foundation of colors are also identified through our application using SFTA (Segmentation-based fractal texture analysis) algorithm. This app is one of the many examples of how businesses are making advantage of today's quickly developing technology. It is essential to note that reduced diversity within clusters leads to more identical data points within the same cluster. For several reasons. The second is that picking a large or small total of tones might involve in the image clustered Process, and finally the total of color available to you.

2 LITERATURE SURVEY

We reviewed the following previous papers and projects on Chromatic color detection for our project.

R. Nevatia, Segmented scenes using a color edge detector. 1977, IEEE Transactions on Systems, Man, and Cybernetics When analyzing photos with a single grayscale value, edge detection is a common technique utilized.

D. Martin, C. Fowlkes, Human-segmented natural picture database for testing segmentation algorithms and gathering ecological data. In this study, we propose a database of human-created 'ground truth' segmentations for photographs of various nature landscapes.

M. Hankyu *et al.*, Optimal Edge-Based Shape Detection. Two-dimensional (2-D) shape detection was the subject of a method presented in IEEE Transactions on Image Processing NG. The step function is used to simulate the form boundary sampling process. To begin, we develop a 1-D optimum step edge 14 administrator, which limits the clamor power and the mean squared blunder between the information and the channel's result.

F. Pedro, Image Shape Representation and Detection. Massachusetts Institute of Technology, 2003.In this work we offer a technique for object detection that makes use of a number of different attributes. The Shape's bitmap picture is analyzed statistically, and features and retrieved from there.

David R. Martin, Charles C, *et al.*, Locating picture borders automatically by studying their local luminance, color, and texture. PAMI, 2004. The focus of this effort is on employing local image measurements to properly identify and locate boundaries in real settings. We design elements that react to variations in natural boundaries' hue, saturation, and texture.

Sarif Naik and C. A. Murthy, Color image magnitude standard, September 2006. It is the practical task in the low- level image pre-processing is edge detection. Meaningful edge detection is crucial to the success of many image processing and computer vision applications. Thread holding is nearly always required by edge detect algorithms in order to get a valid edge.

D. Philippe *et al.*, Boundary Shape Recognition Using Accumulated Length and Angle Information, Pattern Recognition and Image Analysis. As humans, we tend to have a set of "default" forms that the eye can identify and know what they are, making vision a crucial component of human knowledge, since all processing begins with a shape recognition check.

S. Konarad and S. David, Object Detection by Global Contour Shape. 2008 journal article on pattern recognition, pages 1–30. Given that sight is the most developed of the human senses, it stands to reason that visuals play a significant part in how we interpret the world. This is comparable to the rapidly developing area of machine vision.

H. Zhang *et al.*, Visual Image Processing-Based Distance Measuring. In the 2nd International Congress on Image and Signal Processing, a method of visually measuring distance was presented.

Q. Zhu, Shape Detection By Packing Contours, University of Pennsylvania, 2010.K. Karthik anand S. Wesley, A Shape Recognition Algorithm R Obustto Occlusion: Analysis and Performance Comparison," 2013.

3 PROPOSED METHODOLOGY

The existing system for shape detection uses Sk-learn libraries for the Matlab part, Numpy in the vector transform, Pandas for the final summaries and some Image Pre-Processing typical libraries. The present models are done using Machine Learning techniques. The main idea of this process to make a proper image that need to be (N_rows X N_columns X N_channels) vector. Considering this vector, it necessary to add the K Means algorithm and specifying the k-clusters, which present in the colors.

The second one is that you will increase the number of clusters, you will be choose a smaller or higher amount of tones. The other existing one utilizes the value of intensity, from photo and thread holding to get binary image.

3.1 *Pseudo code for sift algorithm*

```
for all octaves
{
has several uses in signal processing and is also commonlyused in digital image processing due to its ability to preserve edges
while rejecting noise (though see the discussion below for caveats).

        List keypoint_list;
        for all scales
        {
            Convolve ImageGaussParallel();
            BuildDoGParallel();
            //Detect Keypoint
            #pragma omp parallel for
            all pixels p in Image
            {
                    if IsKeypoint(p))
                    #pragma omp critical
                    keypoint list add(p);
            }
        }
#pragma omp parallel for
all pixels kp in keypoint_list
{
    ExtractFeature(kp);
}
DownSampleImage Parallel();
```

Figure 1. Pseudo code for SIFT algorithm.

3.2 *Model description*

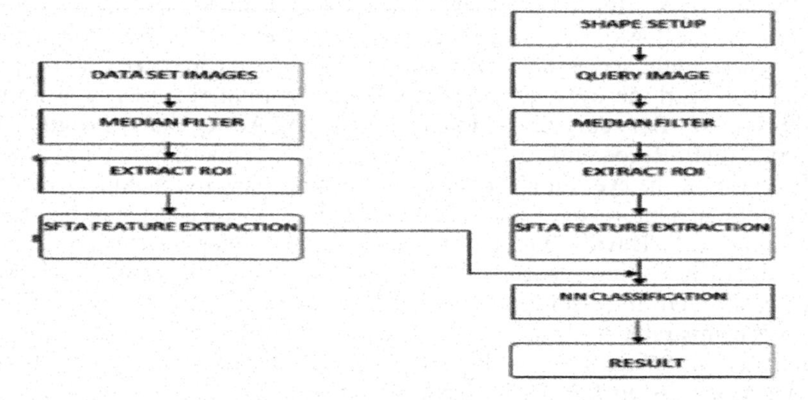

Figure 2. Model diagram.

Data structure software architecture, algorithms, and the interface between modules are all aspects of design. In addition, before any code is written, the design process converts the requirements into a working software prototype that can be tested for quality. When systems theory is applied to product creation, the system architecture becomes clear. This field has certain similarities with systems analysis, systems architecture, and systems engineering.

Median filtering is a common practice in digital image and signal processing for eliminating unwanted. This Figure 1 Architecture diagram for chromatic color detection It has six modules, they are:

1. Image Frames separation
2. Adaptive Histogram equalization
3. Median Filter & ROI Extraction
4. SIFT & SFTA Feature Extraction
5. Training & Testing Set Comparison
6. NN classification & Object Detection

SFTA Feature Extraction is the next step, and it involves converting the input picture into a series of binary images and then computing the fractal dimensions of the generated areas in order to characterize the segmented texture patterns. Parallel, the system will start separate process for shapes. Firstly, it will set up shapes and it will query the image which we have selected, this process was hidden. For forms, it will now begin the process of filtering the picture and isolating the region of interest (Region of Interest).

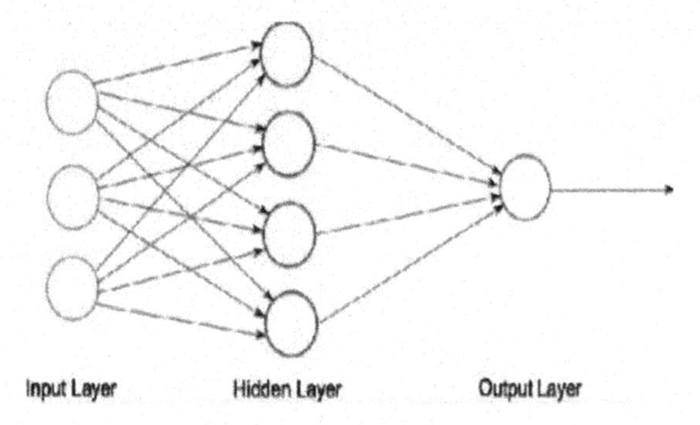

Figure 3. Diagram for NN classification & object detection.

Each node in Figure 3 above, also called a neuron when referring to this particular parameter, is a function that yields an output in response to one or more inputs.

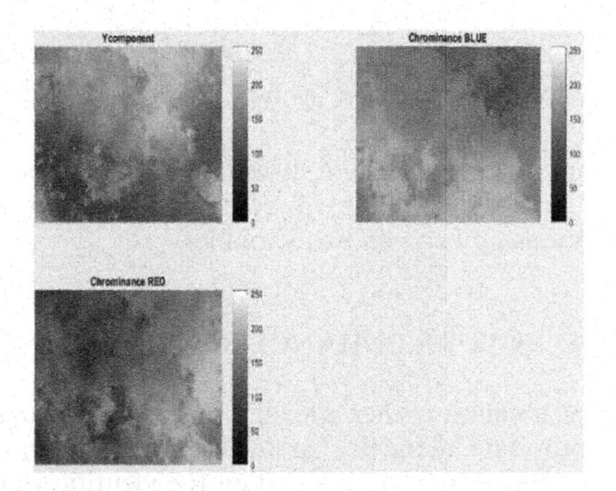

Figure 4. Y-component.

Figure 4 has a Y component, Chrominance Blue, and Chrominance Red. Chrominance is the color information in a picture and is further broken down into two properties of color.

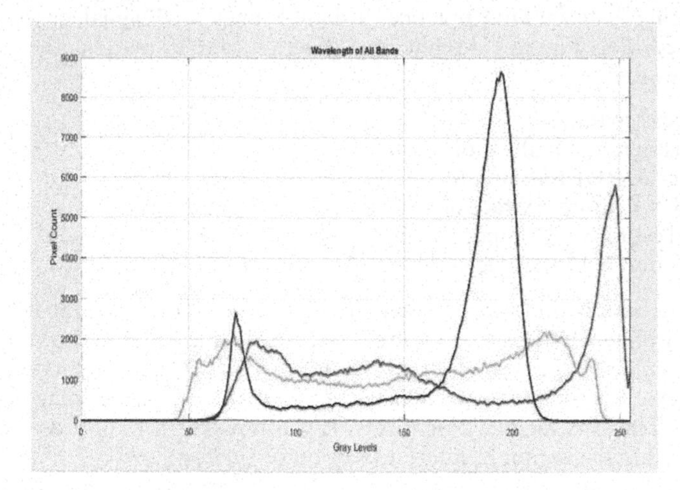

Figure 5. Wavelength of all bands.

Figure 5 shows the wavelength of all bands. This has pixel count on Y-axis and Gray level on X-axis. With the help of RGB (Red, Green, Blue) color graph was plotted with the respective counts.

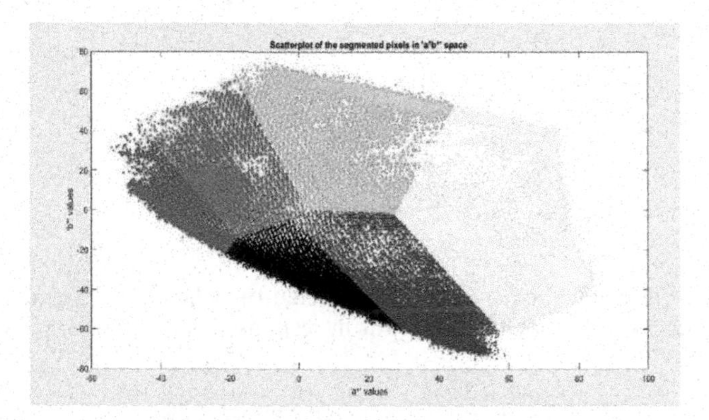

Figure 6. Scatterplot of the segmented pixels in "a*, b*" values.

The final and important image is Figure 6, it will look like the below image. It has two parameters or values, a*, and b*. The plot will be done between these two values. A scatterplot will show the relationship between two variables.

4 CONCLUSION AND FUTURE ENHANCEMENT

Finally Considering how ubiquitous they are, it's important to make them more user-friendly by the use of appropriate interaction strategies. In this study, we took a step toward an enhanced model of computer interaction based on the identification of objects. The red-green-blue spectral analysis is used for object identification. (Red, Green, Blue) colors.

Respective functionalities will be performed for the color detected. After selecting an image, the system needs to start the process should detect colors and shapes, and will give graphs, plots, and clusters images. Hence, this will detect an object once we select the image. The final output is displayed in Graph.

The accuracy of color recognition from a variety of objects is the foundation of our research. Image is used for the effective identification Process. It can be further enhanced to detect the shape and color considering more than 3 colors which leads to the detection of more than 3 shapes. This model can be implemented as a website or mobile application. At present, the analysis is limited to detecting the shape and color of any object based on the image provided by the user. It can be further enhanced to detect the shape and color considering more than 3 colors which leads to the detection of more than 3 shapes. This model can be implemented as a website or mobile application.

At present, the analysis is limited to detecting the shape and color of any object based on the image provided by the user. It can be further enhanced to detect the shape and color considering more than 3 colors which leads to the detection of more than 3 shapes. This model can be implemented as a website or mobile application.

The future scope for the project are:

- Comparison gives a broad range of understanding of various algorithms.
- Detection of shape and color can be done using more than 3 colors.
- The proposed model can be enhanced without alteration of the model and thus it makes the application reusable.

REFERENCES

David R, Martin, Charles C, Fowlkes, & Jitendra Malik. *Learning to Detect Natural Image Boundaries Using Local Brightness, Color, and Texture Cues.* 530–549, 2004.

Hankyu, M, Rama C Azriel, R. Optimal Edge-Based Shape Detection. *IEEE Transactions On Image Processing NG*, vol. 11, no. 11, pp. 1209–1226, 2002.

Karthik, K, Anand, & Wesley, *S.A Shape Recognition Algorithmr Obustto Occlusion: Analysis and Performance Comparison.* 2013.

Konrad, S, David, S. Object Detection by Global Contour Shape. Article Published in *Pattern Recognition*, pp.1–30, 2008.

Martin, D, Fowlkes, C, Tal, D & Malik, J. A Database of Human-segmented Natural Images and its Application to Evaluating Segmentation Algorithms and Measuring Ecological Statistics. In *Proc. 8th Int'l Conf. Computer Vision*, volume 2, pages 416–423, July 2001.

Maria kulikova. *Shape Recognition for Image Scene Analysis.* Université de Nice-Sophia-Antipolis, 2009.

Mar, R, Philippe, C.D & Josep, L. Boundary Shape Recognition Using Accumulated Length and Angle Information. *Pattern Recognition and Image Analysis*, vol. 4478, no. 27, p. 210–217, 2007.

Nevatia R. A Color Edge Detector and its use in Scene Segmentation. *IEEE Transactions on Systems, Man and Cybernetics*, 1977.

Pedro, F. *Representation and Detection of Shapes in Images.* Massachusetts Institute of Technology, 2003.

Sarif Naik and C. A. Murthy. *Standardization of Edge Magnitudes in Color Images.* 15(9):2588–2595, September 2006.

Zhang, H, Wang, L, Jia, R & Li, J.A Distance Measuring Method Using Visual Image Processing. *2009 2nd International Confer ence on Image and Signal Processing*, 2009.

Zhu Qihui & Jianbo Shi. *Shape Detection by Packing Contours*, University of Pennsylvania, Requirements for the Degree of Doctor of Philosophy, 2010.

Artificial Intelligence, Blockchain, Computing and Security – Dagur et al. (Eds)

Enhancement of shopping experience for customers at malls

P. Preethi*, Sriram Venkata Divya Sathvika*, Damodaram Vyshnavi* &
Puvvada Venkata Jashwanth Kumar*
Department of Computer Science and Engineering, Saveetha Engineering College, Chennai, Tamil Nadu

ABSTRACT: During these days, it is very difficult for business people to keep customers because of more and more entrepreneurs. Furthermore, the main thing for any business is to provide the things that the customers and by that, we can target some of the customers and maintain services. By some of the structured customer service, we can understand. There are consumers in each market sector will have similar market characteristics. Machine learning has increased the adoption of electronic most customer segmentation methods. In this project, k-means clustering uses an algorithm to segment customers into different clusters based on the similarity of the market features. Customer segmentation refers to gathering clients and grouping them into groups of people with common traits. Thanks to this divide, the ability to target a specific group of customers increases the likelihood that someone will buy anything. Because of this, they are able to create and use explicit channels of contact to reach out to and entice a variety of clients. The organizations' fundamental strategy would be to promote on the radio in order to reach older audiences and use web-based media pieces to reach younger audiences. Better customer relationships and overall association presentation are made possible for the organizations as a result.

Keywords: K-means Clustering Algorithm, organizations, customers

1 INTRODUCTION

Following the study, as well as preparation for the design, is finished, the planning stage begins. The thing ability to create systems transfigure the data gathered regarding the design from the design of a foundational framework that will be used to build the structure. It's thought of being a cumbrous the same as the utmost in the crimes is introduced in this phase. In order for organizations to create, maintain, and grow priceless long-term client connections, the management and maintenance of client relationships have always played a crucial role. Giving significance to the guests is to maintain the value of the association. Investment in the creation of client acquisition, conservation, and development plans is desirable for organizations. In order to group visitors with similar means together, clustering techniques like k-means are used. Client segmentation aids the marketing team in celebrating and exposing several clientele groups who think differently and use various buying tactics. The identification of visitors with different preferences, prospects, solicitations, and traits is made possible by the use of client segmentation. The main objective of client segmentation simply stated they form teams of like-minded individuals to ensure the marketing unit capable of creating a successful marketing approach. Similar to machine literacy, bracket, and pattern recognition, the clustering sort of the kind of unstructured data analysis applied in many

*Corresponding Authors: ppreethi@saveetha.ac.in, Divyasathvika.sv@gmail.com, Damodaramvaishnavi2002@gmail.com and Jashwanthkumar7788@gmail.com

 DOI: 10.1201/9781032684994-34

different processes. client segmentation is the division of likely guests in a given request into separate groups. That separation depends on visitors meeting certain requirements, paying fees, and other factors. The focus of this companion will be on the value-based approach, which enables extension stage associations to clearly characterize and concentrate on their stylish possibilities and fulfill the majority of their conditions for division in the development stage without using up the time and resources of a conventional, unmistakable division exploration process. Client segmentation is significant because It has a tendency to change. request systems so they are applicable to all customer scrap, support for enterprise judgment-making, recognition of information related to every client portion, and the ability to deal with the interest and force to that item, feting and fastening on the implicit a clientele, or more. and The ability to modify request systems so they apply to every customer scrap, support for enterprise judgment-making, recognition for details connected to every client portion, and the ability to manage the force as well as interest to item while anticipating customer abandonment are all important aspects of the client segmentation. It also includes behaviors for locating agreements.

2 LITERATURE SURVEY

A Customer segmentation process utilizes a clustering approach Choose the customer group to focus on. According to this (Yash Kushwaha et al.) the K-means method, a partitioning technique, is utilized as the clustering algorithm to divide the consumers into groups based on shared traits. The elbow approach is used to choose the best clusters.

The grouping technique over IDS built on Mini Batch K-Means and the primary element method has been developed by (Kai Peng *et al.*). The strings are first proposed to be digitalized using a preprocessing approach, this information collection gets normalized to increase grouping effectiveness. In order to further increase the clustering accuracy, the processed data set's dimensions are reduced using the principal component analysis method, and the small batch K-means approach is then utilized for data clustering.

According to (Aman Banduni *et al.*), any company's capacity to comprehend the demands of every one of its consumers would enable it to offer more specialized customer services and better support to its clients. Customer service that is organized makes it possible to comprehend. In contrast to traditional market analytics, which frequently fails when the customer base is vast, automated customer segmentation methodologies have gained more traction due to machine learning.

As (Asith Ishantha et al.) has been stated that all three algorithms k-means Clustering, Hierarchical Clustering, and Mini Batch K-means Clustering are mainly used for market analysis. Research on market baskets is done to identify potential target clients. between all of the clients merged.

3 PROPOSED WORK

This project's main goal must be to group information on action for clients who use utilizing credit cards a variety of unsupervised algorithms as well as then, after analyzing results, choose the approach that performs the best.

3.1 *Data collection and preparation*

A database including mark at consumer details was considered an account by Kaggle. Is better in this particular instance due to allows us to discover further developments and patterns throughout the information. It also requires a collection of characteristics based on Some of the essential company indicators, subsequent to pre-processing information for eliminating discrepancies, which ultimately aids in better statistic analysis.

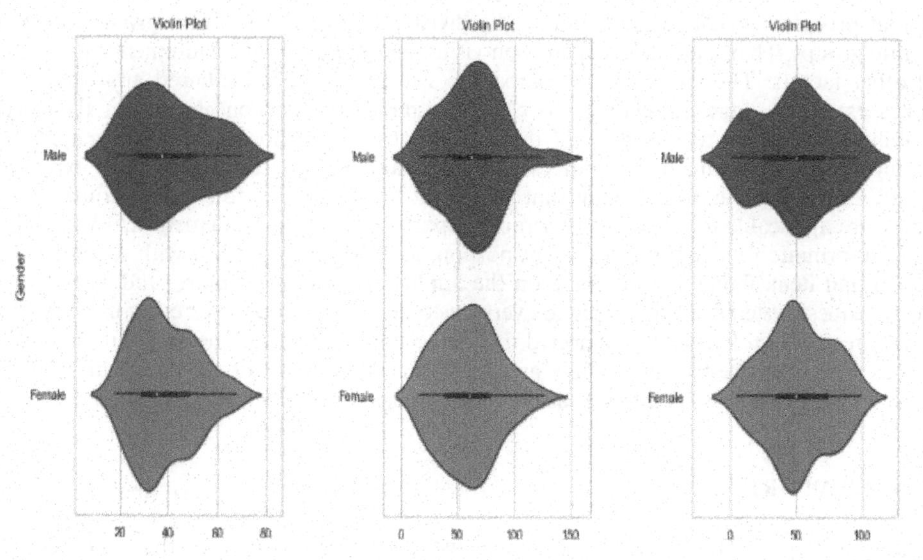

Figure 1. Violin plot of customers (male or female).

3.2 *Data analysis and exploration*

That's the most essential stage that will enable us to discover the data and designs. By that you'll understand which characteristics are more closely associated with consumers also with the company, We may acquire a deeper understanding of a clients desires, decisions as well as preferences. purchase habits through this.

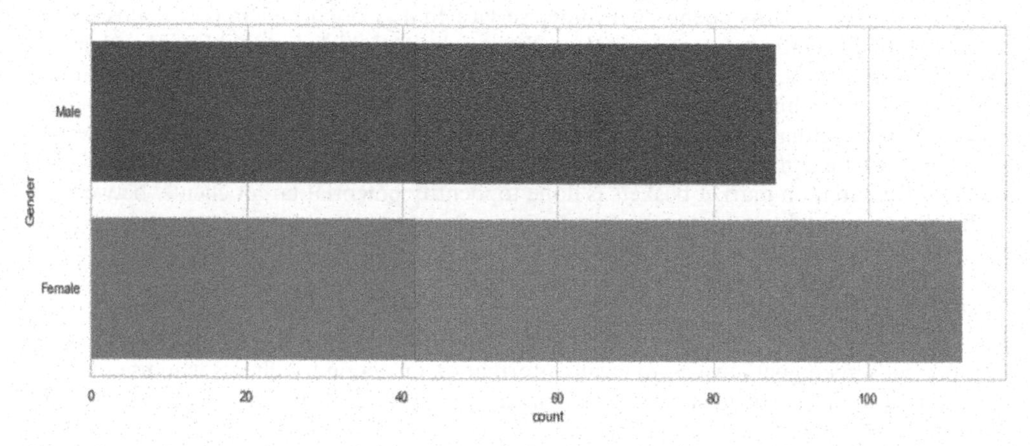

Figure 2. Male and female customers.

4 METHODOLOGY

4.1 *Clustering*

Decomposing a set of data into what is known as clusters is what is referred to as clustering. The cluster's efficiency is primarily represented by two factors, which the first contains the algorithmic conditions on how to find such cluster's tractability, while the second involves quality on how well the is computed cluster formation.

4.2 *K-means clustering*

Unlabeled information is separated into many categories. clusters using the K means clustering approach, a single among unconstrained partition-based clustering algorithm. The algorithms begin to calculate that proper K value -no of collections, in which convert to locate "K centroids, so build by assigning each data point to its nearest k-center, groups are formed.

4.3 *Hierarchical clustering*

Because it creates a binary fusion branch that has limbs. storing information "elements" and a shape having the entire file, agglomerative hierarchical clustering differs via partition-based networking. Dendrograms are diagrams that show the branch places connections in an aircraft.

4.4 *Mini batch K-means clustering*

It has no denying that k-means is the most popular method for segmentation due to its effectiveness and cheap time expense, however as the quantity of the datasets being considered for analysis rises, the k-means' calculation time increases. In order to address this, a new method known as the Minibatch k-means algorithm is introduced. Its main principle to be separate the entire data collection reduced to a tiny set size info group, and then create a new random mini batch from the information "update the clusters where this iteration is repeated until the convergence."

4.5 *Elbow method*

The most important stage in every unsupervised technique is figuring out how many clusters are best for the given dataset. Using this approach, the optimal K value may be found.

5 PERFORMANCE ANALYSIS

Executing a clustering design without an aim for striving makes it impossible to quantify the accuracy score. In contrast to supervised algorithms like a direct regression analysis, which have a prediction aim and allow precision to get quantified utilizing matrix like RMSE, MAPE, MAE, etc. The objective is to produce clusters with distinctive or distinctive features going forward. The following two criteria are most frequently used to gauge how distinct a cluster is:

5.1 *Silhouette coefficient*

The score goes from −1 to 1, with higher values indicating well-established and identifiable clusters.

5.2 *Davies-Bouldin index*

In comparison with the silhouette score, the value gauges how comparable the clusters are, with a lower number indicating better cluster formation.

A sci-kit-learn may be used to calculate these results.

Table 1. Performance likeness.

ALGORITHMS	SILHOUETTE SCORE	DAVIES BOULDIN SCORE
K-Means	0.333162	0.734752
Hierarchical clustering	0.333151	0.734752
Mini Batch K-Means	0.337651	0.758416

6 RESULT

So, thus the main aim of this is to help the owners of the shopping malls to know about the number of customers, expenditure of the customers, which age group of the customers is visiting the mall frequently, what is the minimum expenditure they are doing int the malls, also they are able to know the customers are the visiting are male or the female. With this information, mall owners can be able to grab customers into the malls. There are some of the outputs. They are

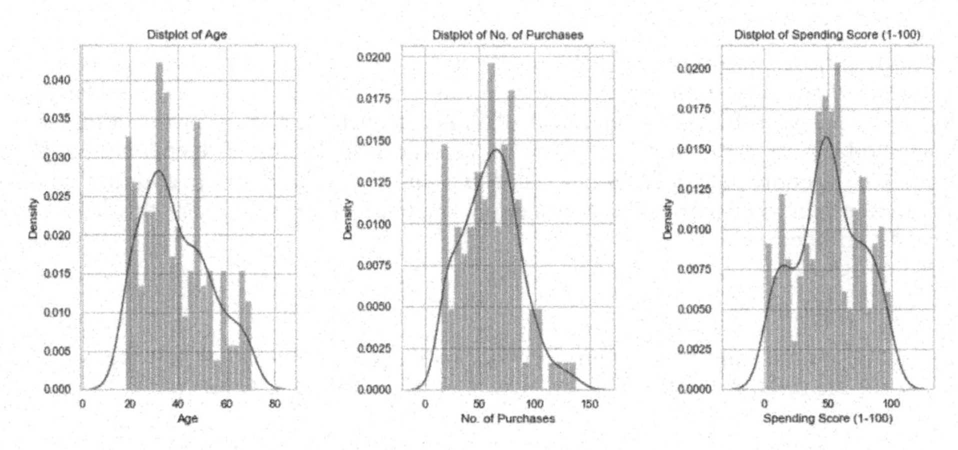

Figure 3. Dis-plot of age, no. of purchases, and spending score.

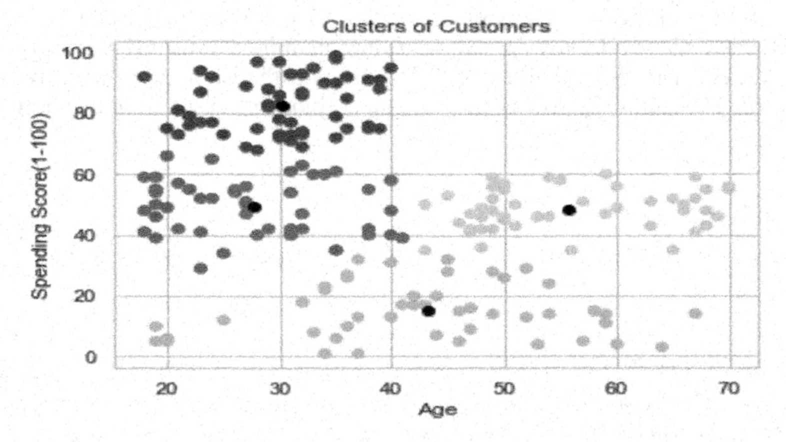

Figure 4. Centroids based on particular age.

7 CONCLUSION

Customer segmentation will help the growth of the company so that by that segmentation they can attract more customers. On the one hand, this enables marketers to give tailored adverts, goods, and offers by dividing the client group into several categories based on the characteristics they have. However, it benefits the buyers by preventing them from being confused well about things to purchase.

Employing measures that assess the distinctness and originality of the clusters to compare the clusters are formed by using the clustering that we have used in the customer's information. It has been shown that the hierarchical clustering and minibatch k means algorithms come in second and third place, respectively, in terms of producing appropriate clusters in achieving the greatest Silhouette score and a lowest Davies Bouldin score. An algorithm that is not most successful is the K-means clustering.It relies on a number of variables, including the data's properties and size.

REFERENCES

Aman Banduni, Prof Ilavedhan A, "*Customer Segmentation Using Machine Learning*," School of Computing Science and Engineering, Galgotias University, Greater Noida, Uttar Pradesh, India.

Asith Ishantha, "*Mall Customer Segmentation Using Clustering Algorithm*", Fiture University Hakodate, Conference Paper, March 2021.

Dhillon I. S. and Modha D. M., "Concept Decompositions for Huge Dparse Textual Content Records Using Clustering," *Machine Learning*, vol. 42, trouble 1, pp. 143–175, 2001.

Fionn Murtagh and Pedro Contreras, "*Methods of Hierarchical Clustering*", In 2018, Science Foundation Ireland, Wilton Place, Dublin, Ireland Department of Computer Science, Royal Holloway, University of London.

Jayant Tikmani, Sudhanshu Tiwari and Sujata Khedkar, "Mall Customer Segmentation Based on Cluster Analysis an Approach to Customer Classification Using k-means", *IJIRCCE*, 2015.

Juni Nurma Sari, Ridi Ferdiana, Lukito Nugroho, Paulus Insap Santosa, "*Review on Customer Segmentation Technique*", Department of Electrical Engineering and Information Technology, University of Gadjah Mada, Jogjakarta, Indonesia, Department of Informatics Technology, Polytechnic Caltex Riau, Pekanbaru, Indonesia

Kai Peng(Member, IEEE), Victor C. M. Leung, (Fellow, IEEE), AND Qingjia Huang, "*Clustering Approach Based on Mini batch K-means*", In 2018, College of Engineering, Huaqiao University, Quanzhou 362021, China.

Kamalpreet Bindra, Anuranjan Mishra, "A Detailed Study of Clustering Algorithms", CSE Department, Noida International University, In *2017 6th International Conference on Reliability, Infocom Technologies and Optimization (ICRITO) (Trends and Future Directions)*, In September 2017, AIIT, Amity University Uttar Pradesh, Noida, India.

Kanungo T., Mount D. M., Netanyahu N. S., Piatko C. D., Silverman R., and Wu A. Y., "A Green K-manner Clustering Algorithm", *IEEE Trans.Pattern Analysis et Machine Intelligence*, vol. 24, p. 881892, 2002.

Manju Kaushik, Bhawana Mathur, "*Comparative Study of K-Means and Hierarchical Clustering Techniques*", June 2014, JECRC University, Jaipur.

Onur Dogan, Dokuz eylul University, Ejder Aycin, Kocaeli University, Zeki Atil Bulut, Dokuz Eylul University, "Customer Segmentation by Using RFM Model and Clustering methods: A Case Study in Retail Industry", In July 2018, *International Journal of Contemporary Economics and Administrative Sciences.*

Puwanenthiren Premkanth, —Market Segmentation and Its Impact on Customer Satisfaction with Especial Reference to Commercial Bank of Ceylon PLC.|| *Global Journal of Management and Business Research* Editeur: Global Journals Inc. (USA). 2012. Stampa ISSN: 09755853. Volume 12 Edizione1.

Yash Kushwaha D. P., Deepak Prajapati, "*Customer Segmentation using K-Means Algorithm*," 8th Semester Student of B.tech in Computer Science and Engineering, Galgotias University, India.

Artificial Intelligence, Blockchain, Computing and Security – Dagur et al. (Eds)
© 2024 The Author(s), ISBN: 978-1-032-67841-2

Sentiment analysis of product-based reviews using machine learning

V. Loganathan*, S. Ganesh*, N. Rakesh* & Mamilla Harish*
Department of Computer Science and Engineering, Saveetha Engineering College, Chennai, India

ABSTRACT: Sentiment analysis is one of the technique which is also called as opinion mining referred with the utilisation of Natural Language Processing text analysis in order to identify the quality of text written in the public media systematically and subjective and lesses. Sentiment analysis applied in various field for analysing the text reviews return related to the products from the consumer side. Online social media Healthcare materials marketing documents are all analysed for sentiment class in order to identify the quality of feedback conveyed to the product. The presented system is focused on developing a robust sentiment analysis Framework for product based reviews using machine learning algorithm. The presented system access comparative analysis of various machine learning algorithms to classify the relevant sentiment class on product reviews. The novel algorithms such as multinomial Navy Base algorithm, linear regression algorithm linear SVM algorithm and random forest algorithm are comparatively validator in terms of performance using accuracy Precision record and F1.

1 INTRODUCTION

Sentiment analysis Framework is commonly identified as unique information extraction with respect to Natural Language Processing NLP approach the primary goal of sentiment analysis is to find out the positive negative and indifferent feedback received from the common text. It is commonly with the field of linguistic information analysis where the information is helpful to access the product and form to get the consumer feedback understand the demands of the consumers and need the Expectations using it. Models are created with technology to handle the unstructured data connected from various materials such as email block billing materials for loves conversations channel reviews product reviews remarks and four rooms 83. Identify the positive and negative words present in it. Helpful to analyse the common text in a different perspective using learning algorithms. Content different missions result. Sentiment analysis is important factor to measure the opinion polarity of the text given. Most cases the sentiment analysis can be handled by the system model where the polarity of conversation and feedback provided by the customers are analysed for the expansion of the business model. Find tuning of sentimental analysis usually employed to make accurate analysis of positive review and negative reviews. Sometimes neutral review plays a major role in making decision on particular product.

2 LITERATURE SURVEY

Ali *et al.* (2022) Presented details study on sentimental analysis in which real time applications are considered with different structures. Advancements in language

*Corresponding Authors: loganathan@saveetha.ac.in, ganeshsivakumar85@gmail.com, rakeshkarnam8@gmail.com and haricse1999@gmail.com

 DOI: 10.1201/9781032684994-35

analysis Technology create various impacts on market towards particular product. Google keyboard are more familiar award the auto correction and word prediction process utilised in natural language processing systems. The presented system is helpful to initiate the sentimental analysis Framework using existing models. Yu *et al.* (2022) Presented a data pattern analysis system in which two different identification and recognition techniques are compared. Using Natural Language Processing with either synthetic and disambigation technique data are validator using different language face and formulae the limitation more 4 division word segmentation part of the speech identification phrasing segment breakdown and more text analysis process. The presented system is helpful to handle the proposed approach is in the basics principles in sentimental analysis. Chandra *et al.* (2019) presented document analysis system in which amount of data collector everyday will be analysed with the help of NLP tools. Various unstructured data are collector to have the recurrent analysis considering the government data. And structural data is computerised and further the instructions of provided through the computers computer to format the data.

3 MODULE DESCRIPTION

3.1 *Multinomial Naïve Bayes classifier*

A type of supervised learning algorithm known as the Naive Bayes algorithm relies on the base theorem to solve multiple classification problems. It is mostly used in text-related classification, where a high-dimensional training data set is included. is one of the most reliable classifiers for text analysis and a useful tool for working with nonlinear data sets. The outcome of this probably stick classifier is determined by the probability of occurrence between the training data and the input data. Some commonly used algorithms include sentimental analysis, malware detection and classification, and spam filtering.

3.2 *Linear SVM*

Utilized in the support vector machine algorithm to enhance the process of problem solving, analyser-specific function. It offers concise methods for complex calculations. The amazing thing about the kernel is that it can smoothen the input data before moving on to a higher level of analysis. Using the associated kernels, the kernel support machine reaches an infinite number of dimensions. Support vector machines (SVMs) sometimes use input data tailored to the problem's scenario between hyperplanes.

3.3 *Logistic regression*

When supervised techniques are involved, the logistic regression algorithm in machine learning is fairly common. Reduced to predict categorical outcomes that, in some cases, are dependent on the input and are based on labelled data. The label that the data touch, but a brief hint about the dependent variable. Both the classification output and the predicted output of the dependent variable can be classified as either logical 1 or logical 0. Because of the various outcomes, it is also referred to as a binary classification technique.

3.4 *Random forest regression*

Random forest regression is a kind of pattern analysis technique in which training and testing data of each tree connected together to make accurate results. The collection various small decision trees formed into Random forest regression algorithm.

4 METHODOLOGY

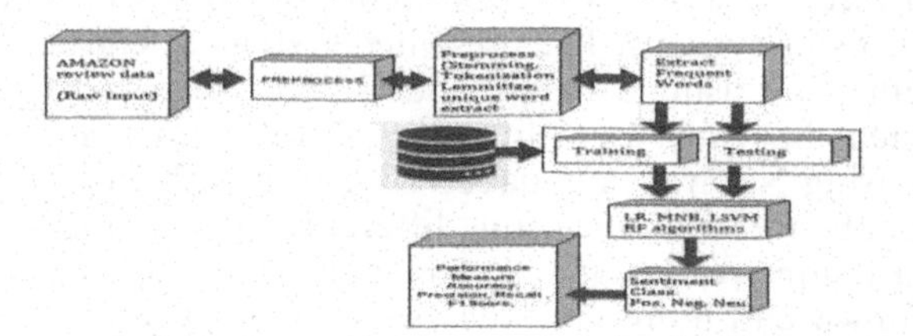

The percentage system consider the need for sentimental class analysis of specific product and create robust model to analyse the sentimental class and opinion Framework behind the text data using NLKT Toolbox provided by Python. Google Colaboratory environment is utilised here to analyse the various machine learning algorithms. The comparative analysis of parameters that impact the text date of classification is present here. Google collaboratories one of the open source flexible environment for analysing the Python codes.

This section provides an in-depth analysis of the process flowchart and the methods used at each step. Below, we'll go through the methods used or the steps of the suggested system:

4.1 *Unpacking of data*

The data set considered for sentimental analysis is collected from amazon.com in which various product reviews are considered. A small python code is created to implement the sentiment analysis model by reading the data set. Dumping the values into your files and make it easier for analysis in subject to serialisation is implemented. Data and packing is nothing but organising the data to have a specific format a normalise the values with respect to the labels.

4.2 *Data collection*

The amazon.com dataset is collected as the raw data for the proposed system analysis. The data contains 18 years of collected reviews from 35 million consumer reviews. Various topics are discussed in the review data. Here the product review from specific item is considered.

4.3 *Data preparation*

Data preparation consists of loading the data besides the sentimental analysis. Sample data set having large number of reviews provided with the unstructured text pattern in which the summary of particular product is impacted. After the review ratings are considered out of five the value is being considered for neutrify the reviews in concern to the positive and negative feedback task of Pre-processing the given raw data is to analyse the unstructured the data and handle the data to have a clear classification using NLTK toolbox.

4.4 *Training data are evaluation*

The main training process involves in analysing the data pattern with respect to the bags of words approach collector with positive words and negative words used globally. The performance measure of the presented approach is comparatively validator using the performance of model and terms of accuracy Precision recall and F1 score. prediction is the performance parameter considered in every machine learning analysis in which it is measured. The number of positive results of time from the analysis with respect to the relevant samples provided by the recall value are considered for perfect procession and recall.

$$\text{precision} = \frac{|\{\text{relavant documents}\}| \cap \{\text{retraived documents}\}}{|\{\text{retraived documents}\}|}$$

$$\text{recall} = \frac{|\{\text{relavant documents}\}| \cap \{\text{retraived documents}\}}{|\{\text{retraived documents}\}|}$$

$$F_1 = \frac{2}{\frac{1}{\text{recall}} + \frac{1}{\text{precision}}} = 2 \cdot \frac{\text{precision} \cdot \text{recall}}{\text{precision} + \text{recall}}$$

$$A = \int_{\infty}^{-\infty} \text{TRP}(T)\text{FPR}'(T)dT = \int_{\infty}^{-\infty} \int_{\infty}^{-\infty} I(T' > T)f_1(T')f_0(T)dT'dT = P(X_1 > X_0)$$

5 RESULTS AND DISCUSSIONS

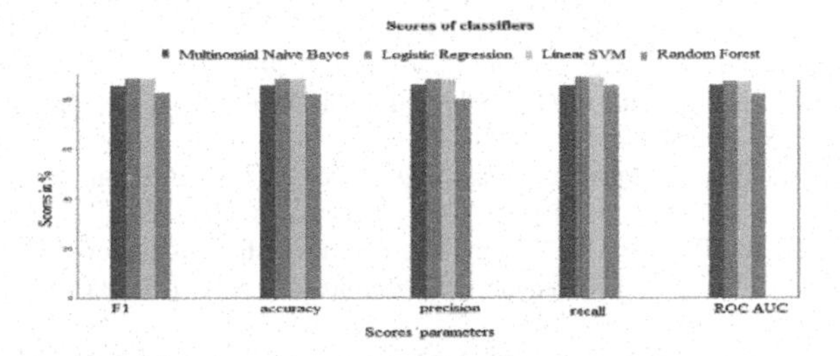

Figure 1. Scores of classifier.

Figure 2. Implementation.

Name of classifier	F1	Accuracy	Precision	Recall	ROC AUC
Multinomial NB	85.25%	85.31%	85.56%	84.95%	85.31%
Logistic Regression	88.12%	94.52 %	87.54%	88.72%	88.05%
Linear SVC	88.12%	88.11%	87.59%	88.80%	88.11%
Random Forest	82.43%	81.82%	79.74%	85.30%	81.83%

Figure 3. Result analysis.

6 CONCLUSION

Sentiment analysis is the concept of systematically evaluate the positive, negative and neutral impact of the text data. The opinions extracted from the consumer feedback on specific product is helpful to enhance the quality of product. Various demands and expectations of the consumers impact the quantity of sale. Accurate sentiment extraction from customer reviews collected from Amazon.com is presented here. The proposed model considers comparative evaluation of machine learning algorithms such as Logistic regression, Multinomial naive Bayes, Linear SVM and Random forest regression etc. Logistic regression achieved 94.52% accuracy comparing the other methods. Sentiment analysis deals with the classification of texts based on the cognitive emotions they contain.

REFERENCES

Ali, J, Abbas, A. F., Jusoh, A., Mas' od, A., Alsharif, A. H., (2022). Bibliometrix Analysis of Information Sharing in Social Media. *Cogent Business & Management.*

Angulakshmi G., Dr.ManickaChezian R.,*"An Analysis on Opinion Mining: Techniques and Tools"*. Vol 3(7), 2014.

Biswas, E., Karabulut, M. E., Pollock, L., & Vijay-Shanker, K. (2020, September). Achieving Reliable Sentiment Analysis in the Software Engineering Domain Using Bert. In *2020 IEEE International Conference on Software Maintenance and Evolution (ICSME)* (pp. 162–173)IEEE.

Carenini, G., Ng, R. and Zwart, E. Extracting Knowledge from Evaluative Text. *Proceedings of the Third International Conference on Knowledge Capture (K-CAP'05)*, 2005.

Chandra, Rohitash, and Aswin Krishna. "COVID-19 Sentiment Analysis Via Deep Learning During the Rise of Novel Cases." *PLoS One* 16.8 (2021).

ChandraKalal S. and Sindhu C., *"Opinion Mining and Sentiment Classification: A Survey,"* Vol. 3(1),Oct 2012.

Dave, D., Lawrence, A., and Pennock, D. Mining the Peanut Gallery: Opinion Extraction and Semantic Classification of Product Reviews. *Proceedings of International World Wide Web Conference (WWW'03)*, 2003.

Na, Jin-Cheon, Haiyang Sui, Christopher Khoo, Syin Chan, and Yunyun Zhou. "Effectiveness of Simple Linguistic Processing in Automatic Sentiment Classification of Product Reviews." *Advances in Knowledge Organization* Volume9, pp. 49–54, 2004.

Nasukawa, Tetsuya, and Jeonghee Yi. "Sentiment analysis: Capturing Favorability Using Natural Language Processing." In *Proceedings of the 2nd International Conference on Knowledge Capture*, ACM, pp. 70–77, 2003.

Shofiya, C., & Abidi, S. (2021). Sentiment Analysis on COVID-19-related Social Distancing in Canada Using Twitter Data. *International Journal of Environmental Research and Public Health*, 18(11), 5993.

Tankard C., ``Advanced Persistent Threats and How to Monitor and Deter Them," *Netw. Secur.*, vol.2011, no. 8, pp. 16–19, Aug. 2011.

Zhu, Jingbo, *et al.* "Aspect-based Opinion Polling from Customer Reviews." *IEEE Transactions on Affective Computing*, Volume 2.1,pp.37–49, 2011.

Face recognition for surveillance using feature based model for houses or hospital

S. Sasikumar*, Krishnareddigari Varshith Sai Ram Reddy*, Guntamadugu Naveen* & C. Kushal*
Department of Computer Science and Engineering, Saveetha Engineering College, Chennai, Tamil Nadu

ABSTRACT: Facial recognition software can verify an individual's identity with just a still picture or a few seconds of video. Photos of faces are compared to a database of known faces to perform face recognition. Face recognition is now widely used in banking software to identify customers who want access to their personal accounts, to identify patients entering and exiting hospitals, and on highways to identify traffic violators. The aim of this article is to examine the current technology and the various ways it can be used in the industry. The faces in the video are recorded, some pictures are taken from them for identification and they are recorded, stored in a database and used to recognize the faces when they are re-entered.

1 INTRODUCTION

This project's goal is to create a health-related traceable evidence application that can be used to circumvent the drawbacks of existing patient traceable evidence solutions. The complex biometrics issue of face recognition has several real-world uses. The goal of developing this software was to offer a simple replacement for traditional patient registration and identification procedures. Researchers in fields such as computer vision and pattern recognition are interested in finding solutions to identification challenges. Biometrics many facial recognition technologies are used in video compression, among other areas, as well as in biometrics. Although face recognition is a very important authentication technology, even after two decades of continuous research and the development of many face recognition algorithms, there is still no real reliable and efficient system that can provide satisfactory results under normal conditions. And in real time. Unfortunately, the latest face recognition algorithms provided by machine learning technologies have training and processing times that are too long for use in real-world applications. As a result, the work of designing an efficient face recognition system with respectable processing times and high accuracy is still in progress. Face size is important on camera.

2 LITERATURE SURVEY

Throughout the literature review, we obtained information (Aeberhard *et al.* 1999), (Goldstein *et al.* 1971), (Gao *et al.* 2002) each individual has a somewhat different facial structure due to facial symmetry, which makes it possible to use facial recognition software to compare faces. This idea is the foundation of modern face recognition technology. Since its inception in the 1960s, facial recognition research has been employed for the purpose of security by a wide

*Corresponding Authors: sasikumar@saveetha.ac.in, varshithsairamreddy@gmail.com, naveenraju8985@gmail.com and kushalbabu728@gmail.com

DOI: 10.1201/9781032684994-36

range of organizations and institutions throughout the globe. Computerized face recognition requires adequately processed images. To create two-dimensional digital data, the face and its structure must be precisely described. The face recognition problem may be solved by creating a database of face photos and using an efficient algorithm.(Baron *et al.* 1981), (The Joint Commission *et al.* 2018) the science of pattern recognition confronts a significant problem in the automatic identification of human faces. Most people's faces are built similarly, yet there are minor distinctions between people. They are specifically classified as human faces. One of the most complex tasks in pattern analysis is face identification, which is further complicated by variations in facial expressions, posture, and lighting.

The concept involves using a "line edge map" (LEM). The study offers a fresh perspective on the age-old problem of facial identification.(Campilho *et al.* 2004), (JMIR Mhealth Uhealth *et al.* 2019) while existing device recognition systems have come a long way, they are still limited by the requirements of many practical uses. Recognizing faces in images of individuals shot outside, especially in situations with varying lighting and/or positions, remains a challenge. To put it another way, modern systems are nowhere near the level of the human perceptual system.Getting started (Chellappa *et al.* 2003), (Ghulam *et al.* 2017)In the recent past, face recognition has become one of the most widely-used applications of image analysis and perception. There are various business and law enforcement uses, and second, there are practical approaches developed over the course of 30 years of study, which all contribute to this trend. While progress has been made in the area of machine recognition, there are still several practical limitations that prevent it from reaching its full potential. Face recognition in outside images with varying lighting and poses is one such example.

That is to say, our existing systems are not yet at the level where they can be seen by the human brain. This article provides an up-to-date analysis of the state of the art in video and still image face recognition research.(Choudhury *et al.* 2000), (Turk *et al.*), (Recognition *et al.* 1991) due to the fact that a person's facial expressions alter based on their state of health, a facial expression recognition system is beneficial in the medical field. In order to better provide health-care in a smart city, this study suggests a face expression recognition system. The suggested approach employs a domain transform on the facial picture to isolate the sub-bands. Each sub-domain is individually subjected to a weighted, concentric, local binary model. The face image feature vector is built by age-generting the blocks' CS-LBP histograms. In recent years, face recognition has emerged as one of the most popular applications of picture analysis and comprehension.

The availability of useful methodologies after 30 years of study, and the plethora of potential applications in business and law enforcement, are two possible explanations for this development. (Goldstein *et al.* 1971), (Du *et al.* 2016) many studies on human face recognition use strict imaging criteria such as frontal-parallel images of the face, in-plane and out-of-plane rotational constraints, and controlled lighting and controlled facial expressions. Rotating the face out of the imaging plane can blur facial structures, making face detection using multiple field-of-view views a more challenging task. In this paper, we propose a new image-based face recognition algorithm that uses nearest-neighbor classification as a line matching scheme and a set of straight line segments randomly generated from two-dimensional face image renderings as the base image representation (Campilho *et al.* 2004), (JMIR Mhealth Uhealth *et al.* 2019) while existing device recognition systems have come a long way, they are still limited by the requirements of many practical uses. Recognizing faces in images of individuals shot outside, especially in situations with varying lighting and/or positions, remains a challenge. To put it another way, modern systems are nowhere near the level of the human perceptual system.

3 IMPLEMENTATION

3.1 *System design*

In data structure software engineering, procedural details, algorithms, and interfaces between modules are of interest to the multi-step design process. Before any coding takes place, the design process converts the requirements into a software presentation that can be tested for quality. In this analysis of ML streams the data source, pre-processing, feature selection of the classifier, parameter tuning and the way the classifier performs the classification are presented initially.

3.2 *Existing system*

PCA, a statistical technique for reducing the number of variables in face recognition, represents each image in the training set as a linear set of weighted eigenvectors or 'eigenfaces', which are used to calculate the eigenvalues of the eigenvectors. These eigenvectors are derived from the covariance matrix of the training image set. Numbering and lists these additional systems are similar in that they require accurate facial recognition and may not save facial data in a database. The algorithm is not trained in previous iterations.

3.3 *Proposed system*

Facial recognition technology can quickly identify a person with an inexpensive webcam and access their records data. As current as the search function is, face search can be trained to turn on the camera, recognize a patient's face in the corner, and search a database of just-learned faces. If a face is recognized, that person's history can be shown on the screen. If you want to teach the program, you can also feed the images you already have to the database, which will help it recognize faces when faced with geometric features. It will try to detect any faces seen by the camera. If not, the face is selected and new information is added to the database.

3.4 *System architecture design*

According to the Figure 1 so you clearly see how the image is getting processed and also how the system will get trained from those images.

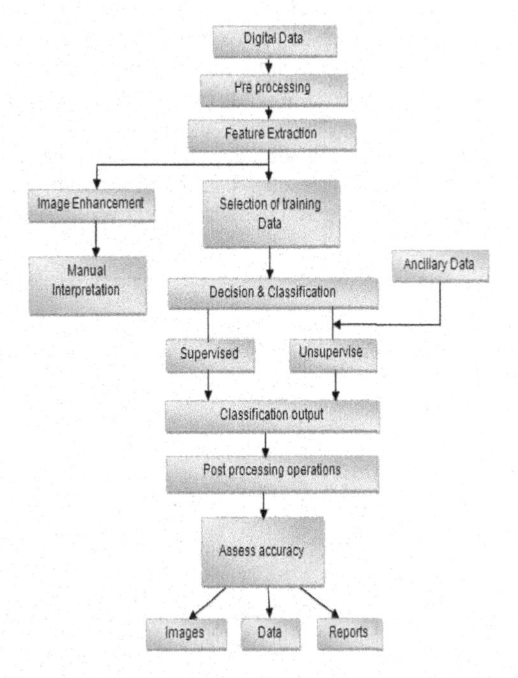

Figure 1. Architecture diagram.

4 METHODOLOGY

4.1 *Binary images*

Binary pictures have pixels with just two potential intensities. They often only appear in black and white. Most of the time, black is represented by the number 0 and white by 1 or 255 in numerical notation. Thresholding is often used to produce binary pictures by extracting a feature from a grayscale or colour image backdrop. White is often used to denote the foreground colour of an item. The remainder are often shown on a black backdrop. While normally the element is presented with a value of zero and the backdrop with a non-zero integer, this polarity may be flipped depending on which picture is threshold.

4.2 *Image editing software*

The photo-editing software available today may be used in a broad range of ways. Since image processing software is the most common kind of software covered here, that's where the bulk of the course's attention is focused. Editing software for images is another key subset. The use of computer programmers for creating works of visual art, such as drawing, is covered here. Using image manipulation software, it is generally helpful to have the ability to engage instantaneously and directly with the picture. It may be simpler to construct a mask using art software by sketching directly on the original picture, for instance, if just a certain area of an image has to be masked for subsequent image editing. As an example, the AND operator specification includes a mask construction procedure that employs this technique. The user may also pick areas of the artwork to brighten or darken, an option provided by many painting applications. Comparatively few specialist image processing apps have the same degree of use and adaptability.

4.3 *Edge detectors*

The margins of the picture have the most striking difference in intensity. As we can observe in Figure 2. Since edges typically exist at picture regions that reflect object boundaries, edge detection is often utilized in image segmentation when we wish to segment an image into segments that correspond to various items. In addition to the above benefits, representing a picture by its borders also requires extremely little data while yet retaining the vast bulk of the image's information. Since edges are mostly made up of high frequencies, we may theoretically characterize them by applying a high-pass filter in the Fourier domain or by converting the picture with a suitable kernel in the spatial domain. It is common practise to do edge detection in the spatial domain due to its computational efficiency and generally superior results. The significant light gradients at edges allow us to highlight them using derivatives of images. The illustration exemplifies this flat scenario.

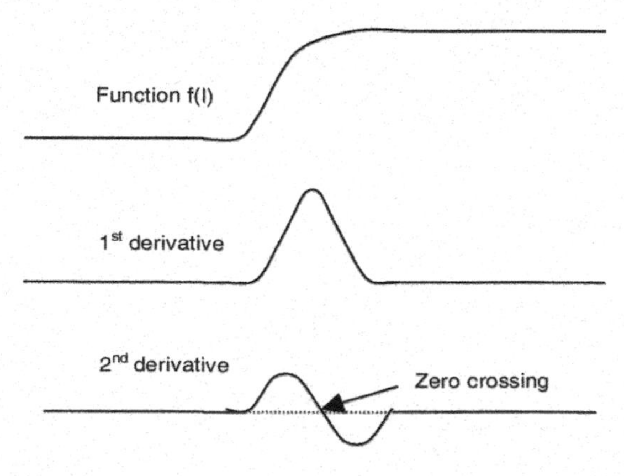

Figure 2. Edge detectors.

5 ALGORITHM

5.1 *Face selection algorithm*

The purpose of feature selection techniques in machine learning is to find the ideal mix of characteristics that allows for the creation of meaningful models for the phenomena being researched. Machine learning feature selection techniques may be broken down into the following groups. Practices under Supervision In order to make supervised models, such regression and classification ion, more efficient, these techniques may be utilized on categorical data.

5.2 *Haar Cascade algorithm*

Regardless of the object's size or position in the frame, Haar's cascade approach can locate it. The algorithm is not too complex and can run in real time. Vehicles, motorbikes, buildings, fruit, and many more things are just some of the many things that may be taught to a HaarCascade detector. Haar Cascade uses a cascading window to try to calculate the features in each window and decide whether it is likely an object. From the Figure 3 we can see the boxes.

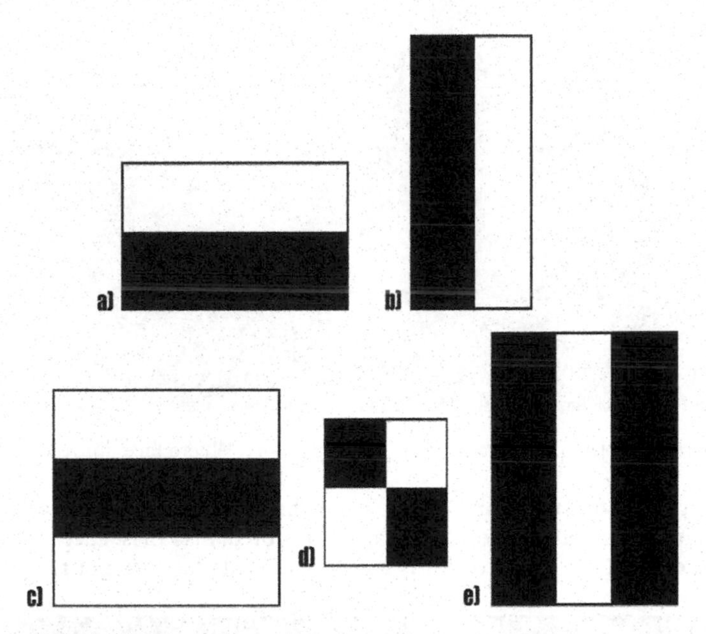

Figure 3. Haar Cascade diagram.

6 FUTURE WORK

This application right now can capture the images of the people's face and store them in the data base and refers to it when the registered face comes and identifies. So in the future we hope this application can be used widely and reach more to the public. This application may differ according to the system in the future we make sure it works same in the all systems and work more faster than expected and also maintain the accuracy for the identification of the face in the frame.

7. OUTPUT

We can see in Figure 4, here where the face is not recognized.

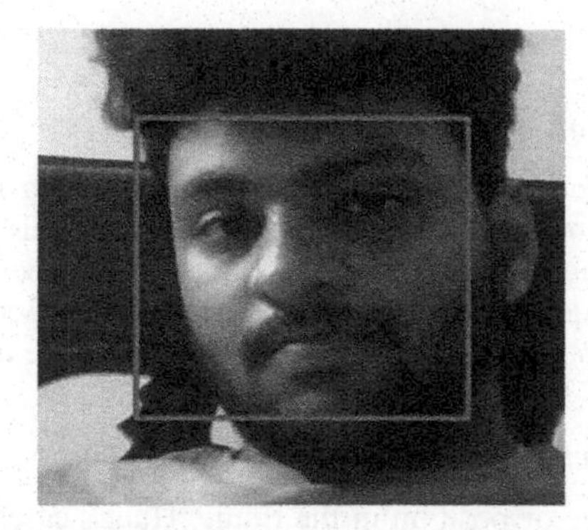

Figure 4. Where face not get recognized.

In Figure 5, we can see the face is not recognized, where the application knows how the face will be, it uses haar cascade algorithm to find out the whether it is face or not.

Figure 5. Where the object is not recognized.

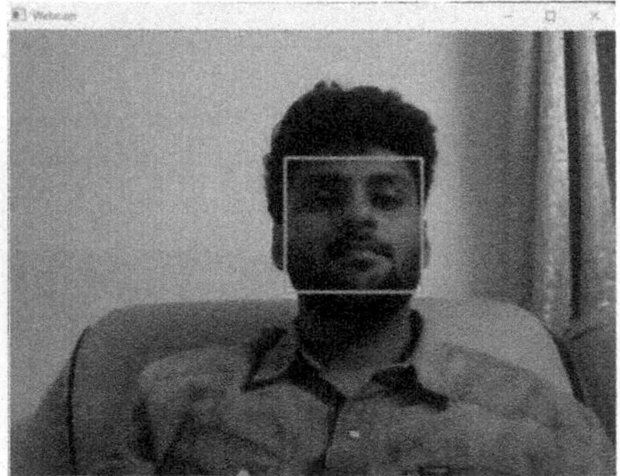

Figure 6. Where face got recognized.

From the Figure 6, due to Haar Cascade algorithm where the face got recognized
As in the Figure 7, we can see how the images got stored, also these faces are recognized so when they come again the indicator turns green which tells the face got recognized.

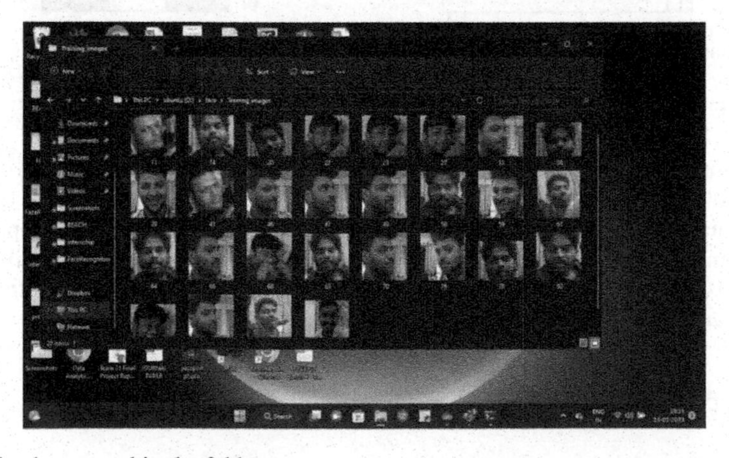

Figure 7. The data stored in the folder.

8 CONCLUSION

Using face recognition technology to speed up registration and patient identification is a crucial aspect of the functioning of a modern health care institution or when a person is entering into the house and leaving that we have concluded. Automation simplifies and expedites many formerly laborious user interactions. Making custom software for hospital networks is a promising avenue for creating a revolutionary health care delivery system with minimal waiting times. Faster patient registration and identification using a face recognition system is just one of many benefits of implementing this project and also for the houses. Other advantages include better management of supply chains, finances, human resources, and markets, as well as streamlined coordination and communication between users and staff. So far, we have it set up to capture pictures, register emotions, and catalogue faces. It takes photos and store in the database and use it for reference and recognizes the faces.

REFERENCES

Aeberhard, S. and De Vel, O. "Line-based Face Recognition Under Varying Pose". *Pattern Analysis and Machine Intelligence, IEEE Transactions on*, Volume: 21 Issue: 10, Oct. 1999,: 1081–1088.

Baron, R. J. (1981). Mechanisms of Human Facial Recognition. *International Journal of Man Machine Studies*, 15:137–178.

Campilho, Aurélio and Kamel, Mohamed, "Image Analysis and Recognition: International Conference, *ICIAR* 2004, Porto, Portugal, September 29- October 1, 2004". Berlin; New York: Springer, c2004.

Chellappa, R., Phillips, J., Rosenfeld, A., and Zhao, W. "Face Recognition: A Literature Survey". *ACM Computing Surveys*, Vol. 35, No. 4, December 2003: 399–458.

Choudhury, T and Pentland, A. "Face Recognition for Smart Environments ". *Computer*, Volume: 33 Issue: 2, Feb.2000 : 50:51.

Du, Ming-Hui, He, Jia-Zhong, and Zhu, Quig Huan "Face Recognition using PCA and Enhanced Image for Single Training Images", *IEEE International Conference on Machine Learning and Cybernetics*, Aug 2016.

Goldstein, A.J., Harmon, L.D., and Lesk, A.B. (1971). Identification of Human faces. In *Proc. IEEE*, Vol. 59: 748.

Gao, Yongsheng and Leung, M.K.H. "Face Recognition Using Line Edge Map". Pattern Analysis and Machine Intelligence, *IEEE Transactions on*, Volumemac_mac 24 Issue: 6, June 2002: 764–779.

Ghulam, Muhammad, "*A Facial-Expression Monitoring System for Improved Healthcare in Smart Cities*", Riyadh, Saudi Arabia: IEEE 10.1109/ACCESS.2017.2712788, 2017.

Jain, L.C. "*Intelligent Biometric Techniques in Fingerprint and Face Recognition*". Boca Raton: CRC Press, 1999.

The Joint Commission. 2018. (2018-07-04). National Patient Safety Goals Effective January 2018 https://www.jointcommission.org/assets/1/6/NPSG_Chapter_HAP_Jan2018.pdfwebsite.

Turk, M.A. and Pentland, A.P. "Face Recognition Using Eigenfaces". *Proceedings of the IEEE Conference on Computer Vision and Pattern. Recognition*, 3–6 June 1991, Maui, Hawaii, USA:586–591. https://www.ncbi.nlm.nih.gov/pmc/articles/PMC6475824/ JMIR Mhealth Uhealth 2019 Apr; 7(4): e11472. Published online 2019 Apr 8, doi: 10.2196/11472.

Artificial Intelligence, Blockchain, Computing and Security – Dagur et al. (Eds)
© 2024 The Author(s), ISBN: 978-1-032-67841-2

Surveillance of species at risk by unarmed aerial vehicles

V. Vijaya Chamundeeswari*, G. Sujith Varshan*, Y. Sanjay Kumar* & A.M. Sudarson*
Department of Computer Science and Engineering, Saveetha Engineering College, Chennai, India

ABSTRACT: A Living being or an organism that is about to vanish from this world will come under the category of Endangered species.The Explanation for their state vary from time to time but few important reasons are loss of habitat and loss of genetic variation. These organism will attain the state of endangered when their population is below 30% and it no proper source for its lack of growth. The reason for being endangered species are – Population reduction rate, Geographical range, Population size and Population restriction, Poaching. Many different species are adversely affected by poaching. Different Laws have been sanctioned in order to protect the endangered species in many countries. Forbid hunting, Restrict land development, or Create protected areas are some of the measures for their growth. Some species are even taken away from their habitat in order to ensure their existence, This is called as extensive conservation efforts.captive breeding and habitat restoration are the few examples of the above. Many efforts have been made in order to restrict the poaching.Hidden Cameras, Drones and DNA tracking are the some of the methods and their probablitiy of success differs with each other. These systems are static in nature whereas the solution this provides is that with the help of Drone system with high resolution image capturing system the endangered species can be monitored along with their habitat. The advantage of using this system is the range is dynamic and precision is accurate. With the help of image processing technique endangered species is segregated and monitored from the rest of the species.

Keywords: drone, r-cnn, thermal, aerial, image processing

1 INTRODUCTION

Poaching means killing animals which leads to downfall of the species at the large scale.This impacts every animal species. Every year more than 25000 elephants have been hunted for their ivory. The level would have reached a greater height if there weren't the involvement of the conservation organisation.Only with their constant effort in trying to protect and preserve the species that are in the endangered category. Poachers hunting elephants only for their ivory. According to the study that gives the continent wide estimates of illegal kills states that more than hundreds of thousands of elephants have been killed in matter of three years. In the year 2011, One in tweleve elephants was hunted for its ivory. Elephant horn is in high demand since many cultures consider it to be a status symbol and a useful tool for medical treatments. In Asia,The major clients for the elephant ivory are China and Vietnam. Elephant poaching has tragically continued despite existing prevention measures having just a little impact. In order to monitor and identify the location of the animals in their habitat, Animals are implanted with the GPS tracker (Alphonce *et al.* 2017). When there is a change in the location of the animal or loss of signal, It is identified as a threat to the animal

*Corresponding Authors: vijayac.dean@saveetha.ac.in, varshansujith@gmail.com, sanjaykumar2ygl@gmail.com and sudarson.a.m.27@gmail.com

 DOI: 10.1201/9781032684994-37

especially considered as poaching and after examination appropriate reasons are found. This is a retaliatory strategy that accomplishes nothing to stop the poaching incident. Poachers employing drones are being found through preventative measures.

2 LITERATURE SURVEY

(Bondi *et al.* 2018, 2019) stated that the SPOT Architecture and the design makes it faster processing and robust in the field and detect in close to real-time and the valuation of the SPOT is elicited from realtime run and historical videos by the users in that particular area are are all illustrated in this paper to show the viability of building upon cutting-edge AI techniques such as Faster RCNN are used to solve the difficulties in detecting the animals and the hunters in infrared images upholding the specifications' integrity. (Premarathna *et al.* 2020) outlined a method to identify the elephants as a objects using the convolutional neural networks (CNN) have obtained the accuracy of 94%. This method is far better than the rest of the existing approaches in terms of accuracy and robust. Therefore, This can be used as the source or the foundation for an automated Elephant early warning system presented by (Zeppelzauer *et al.* 2015). (Hong *et al.* 2019) presented the development of object-detection models based on deep learning using aerial images captured by an unmanned aerial vehicle (UAV). It demonstrates that among the bird detection algorithms, YOLO is the fastest and Faster R-CNN is the most accurate. The combined findings show that deep learning-based detection techniques combined with UAV aerial data can be used to detect birds in a variety of situations. (Khushboo *et al.* 2022) developed the dataset by gathering the thermal images from the FLIR videos.These dataset doesnt have the training data required for deep learning methods. After that, YOLOv4 is trained with the thermal images of the animals to assess the position of the animals. The model predicts the position of animals with 84.77% and F1-score of 94%. (Meivel *et al.* 2021) presented a deep learning algorithm and deep neural network replace the original human watches by predicting the behaviour of wild animals like elephants. Feature Extraction aids in picture preprocessing and provides a comprehensive overview of the images that were captured. In comparison to the current image detection multiscale algorithms like Haarcascade Classifier and Hog description, the proposed system has overcome the faults and is more efficient. The animal is predicted using several hidden layers, including conv2D, maxpoolD, flatten, and dense, and it is frequently visible to the user inside the Python shell.

3 ALGORITHM

Two independent steps of object detection are carried out using the Faster-RCNN network architecture. In the first stage, features from the chosen layers are identified and extracted using region proposal networks (RPNs), which are fully convolutional networks that produce proposals with a range of scales and aspect ratios. As a result, the binding box positions can be estimated by the model. By reducing the chosen loss function, the second stage modifies the bounding box's localization. The same CNN performs the region suggestion and object detection operations. This method offers improvements in terms of speed and accuracy compared to early RCNN networks where area proposals were entered at the pixel level rather than the feature map level. By substituting an RPN for selective search, the Faster-RCNN speeds up even more. In order to address inverse problems in the thermal domain, thermal imaging with DeepIR suits pretty well. DeepIR is an unique thermal camera processing pipeline that blends deep neural networks with physically correct sensor modelling. We use a deep network-based regularization as (Lee *et al.* 2017) proposed to capture several, jittered images of the scene in order to estimate the radiant flux of the scene simultaneously. The image of a scene may be divided into a scene independent component and a scene dependent component, which is an important finding concerning uncooled thermal sensors. The gain and offset that result from the camera's internal thermal conditions slowly changing are included in the scene independent component. The radiant flux of the scene is a part of the scene dependent component. We take use of this observation by taking numerous pictures of the same scene while the camera is moving, which only influences the scene's radiant flux measurement and not the camera's nonuniformities. Then, using a combined optimization strategy, we estimate the camera non-uniformities and the radiant flux of the scene. For solving the inverse problem, we use the regularisation capabilities of the convolutional neural network, which provides a good representation of the radiant flux in the scene.

4 SYSTEM ARCHITECTURE

A set of guidelines that specify how software is created and constructed is known as a software architecture. Additionally, it discusses the links between the levels of abstraction, various software system components and other features. An architecture can be used to specify a project's objectives or to direct the creation of a brand-new system. A set of guidelines that specify how software is created and constructed is known as a software architecture. The organisation and structure of the software system are specified by the architecture. Additionally, it explains the connections between the software system's components, levels of abstraction, and other features.

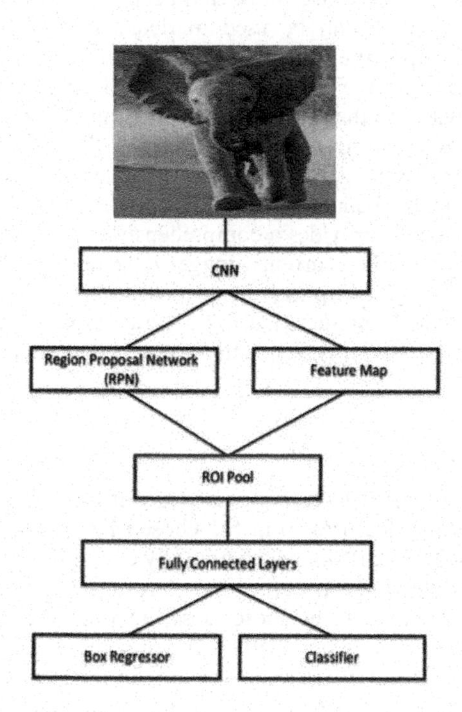

Figure 1. System architecture diagram.

5 MODULE DESCRIPTION

5.1 *Collection of dataset*

Images using the RGB and thermal colour and grayscale palettes are included in the dataset. Elephants and cars are two of the classes in the data. 350 photos in each class range in resolution from 300 × 147 to 3840 × 2160 pixels. Aerial film, infrared pictures, and close-up photography are blended to maintain an appropriate amount of variety. While a ground-based camera was used to take the thermal and greyscale photographs, a drone was used to take the aerial RGB images.

5.2 *Model selection faster R-CNN*

Two independent steps of object detection are carried out using the Faster-RCNN network architecture as (Ren *et al.* 2015) and (Bondi *et al.* 2018) proposed. Region Proposal Networks (RPNs) are used in the initial step to find and extract features from the chosen layers. As a result, the binding box positions can be estimated by the model. By reducing the chosen loss function, the second stage modifies the bounding box's localization. The same CNN performs the region suggestion and object detection operations. In comparison to early RCNN networks where area suggestions were entered at the pixel level rather than the feature map level, this technique offers gains in terms of speed and accuracy. By substituting an RPN for selective search, the Faster-RCNN speeds up even more.

5.3 *Transfer learning*

With the aid of transfer learning, we may use a model that has already been trained (on millions of photographs) and then use our elephant and automobile images to fine-tune the learnt parameters. This is a crucial strategy because the extreme overfitting that results from training CNNs on tiny datasets with little variation. The Faster-RCNN Resnet 101 model, which was pre-trained using the COCO dataset, is the base model used in this study for the transfer learning tasks. 1.5 million object instances and 330 thousand photos make up the sizable object detection dataset known as COCO.

5.4 *Model training*

A server running HP ProLiant ML 350 Gen 9 is used for model training. The server includes 768GB of RAM and x2 Intel Xeon E5-2640 v4 series processors. There is a second GPU stack that is made up of four NVidia Quadro M4000 graphics cards totaling 32GB of DDR5 RAM. The software components of the training process include TensorFlow 2.0, CUDA 10.0, and CuDNN version 7.6.0. In this training, the image is converted into 1500 × 1500 pixel format and fed as an input for the Adam Optimizer to minimise the loss function and with the help of ReLU activation function provides improvement in Sigmoid or hyperbolic tangent activation during training and helps to reduce the saturation in the input's midpoint. To ensure the consistency of the model, the count of Epochs used for training remains the same. The Mavic Pro 2 drone system is used for capturing the images of the species. It can transmit 4K videos at 30 FPS in the range of 7 KMs. The OcuSync 2.0 protocol is used to connect the drone system to a controller. The data is sent with the help of controller from the Drone with the help of local Wi-Fi connection from the Laptop.

6 RESULTS

The thermal and aerial dataset of elephants are obtained and by training and processing these data along with the images obtained by the thermal sensors can be categorized into 3 types namely,scene-specific radiance flux(deep network-based regularizer), scene-independent sensor non-uniformities(modeled accurately using physics) and slowly changing, the model could be able to detect an elephant and a poacher faster and more accurately, thereby helps the forest rangers to protect elephants and monitor them. Out of 350 images from a class 263 images (75% of the total images) is used for training the model and the rest 87 images (25% of the total images) are used for testing the model which helps in finding the accuracy and speed of the model.

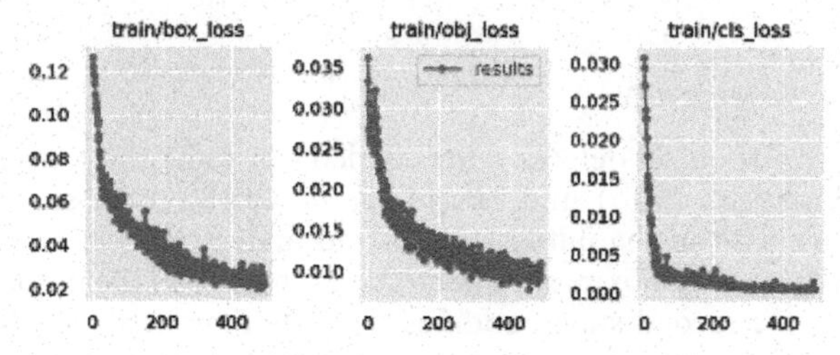

Figure 2. Sample training loss graph.

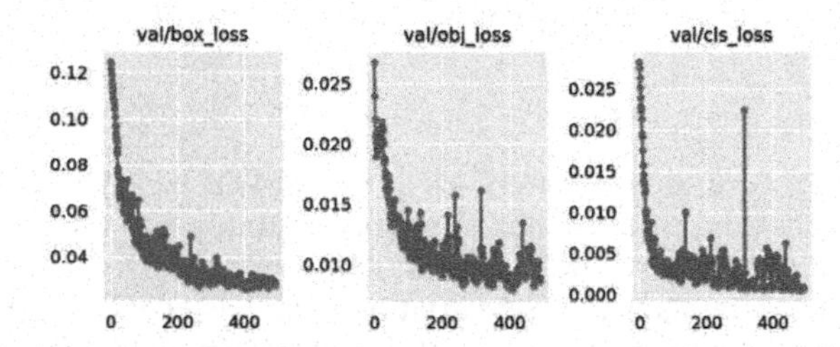

Figure 3. Sample validation loss graph.

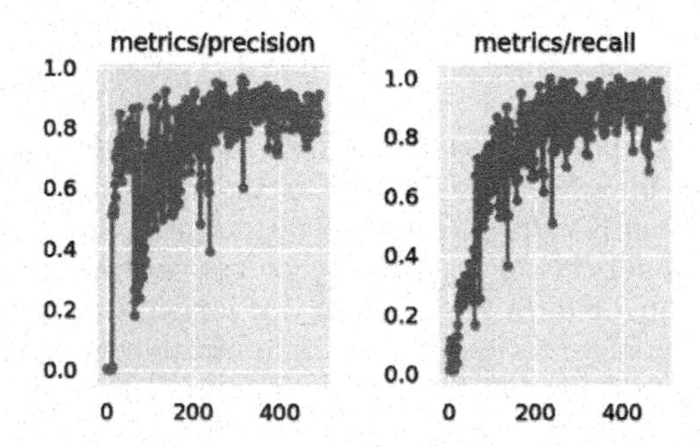

Figure 4. Sample precision and recall graph.

The above graphs represent the training and validation percentages of precision, recall and loss of Object, Box and Classification for the elephant and human dataset. The overall precision percentage of the training was approximately 91.7% and recall percentage was 96.2%.

7 CONCLUSION

The aim of this paper is to create a fast and an accurate model to detect elephants and protect them from poachers and other disasters by monitoring them using Image Processing Technology with the help of Unmanned Aerial Vehicle and Thermal Images. By using RCNN elephants and poachers in the field are detected and their locations are determined by location of the Unmanned Aerial vehicles and the thermal cameras thus helping the forest rangers to protect elephants easily without manually monitoring each of them.

REFERENCES

Alphonce Massawe E., Kisangiri M., Kaijage S., and Seshaiyer P., 2017, "*An Intelligent Real-time Wireless Sensor Network Tracking System for Monitoring Rhinos and Elephants in Tanzania National Parks: A Review*".

Bondi E. *et al.*, 2018, "Near Real-Time Detection of Poachers from Dronesin AirSim.," in *IJCAI*, pp. 5814–5816.

Bondi E. *et al.*, 2018, "Spot Poachers in Action: Augmenting Conservation Drones with Automatic Detection in Near Real Time," in *Thirty-Second AAAI Conference on Artificial Intelligence.*

Bondi E. *et al.*, 2019 "Automatic Detection of Poachers and Wildlife with UAVs," *Artif. Intell. Conserv.*, p. 77.

Hong S.-J., Han Y., Kim S.-Y., Lee A.-Y., and Kim G., 2019, "Application of DeepLearning Methods to Bird Detection Using Unmanned Aerial Vehicle Imagery," *Sensors*, vol. 19, no. 7, p. 1651.

Kanai S., Fujiwara Y., and Iwamura S., 2017, "Preventing Gradient Explosions in Gated Recurrent Units," in *Advances in Neural Information Processing Systems.*

Khushboo Khatri, Asha C S, Jeane Marina D'Souza. Jan 2022, "*Detection of Animals in Thermal Imagery for Surveillance using GAN and Object Detection Framework*".

Lee J., Wang J., Crandall D., Šabanović S., and Fox G., 2017, "Real-time, Cloud-based Object Detection for Unmanned Aerial Vehicles," in *First IEEE International Conference on Robotic Computing (IRC)*, p. 36–43.

Meivel S, Karthikraja R, Mohamad Sajudeen S, Rokith K, 2021, "*Thermal Imaging Based Animal Intrusion using Artificial Intelligence*".

Premarathna K. S. P, Rathnayaka R. M, Charles J, 2020, "*An Elephant Detection System to prevent Human-elephant Conflict and Tracking of Elephant Using Deep Learning*".

Ren S., He K., Girshick R., and Sun J., 2015, "Faster r-cnn: Towards Realtime Object Detection with Region Proposal Networks," in *Advances in Neural Information Processing Systems*, p. 91–99.

Zeppelzauer M. and Stoeger A. S., 2015, "Establishing the Fundamentals for an Elephant Early Warning and Monitoring System," *BMC Res. Notes*, vol. 8, no. 1, p. 409.

Artificial Intelligence, Blockchain, Computing and Security – Dagur et al. (Eds)
© 2024 The Author(s), ISBN: 978-1-032-67841-2

Heart disease prediction using machine learning

Anitha Julian*, R. Deepika*, B. Geetha* & V. Jeslin Sweety*
Department of Computer Science and Engineering, Saveetha Engineering College, Chennai, India

ABSTRACT: Anticipating and diagnosing coronary illness is the greatest test in the clinical business and depends on variables like the actual assessment, side effects and indications of the patient. Factors that impact coronary illness are body cholesterol levels, smoking propensity and corpulence, family background of diseases, pulse, and workplace. Coronary illness can be anticipated in light of different side effects, for example, age, orientation, pulse, and so on and lessens the demise pace of heart patients. In this task, we will be intently working with the coronary illness expectation. The coronary illness dataset from that dataset we will determine different experiences that assist us with knowing the weightage of each element and how they are interrelated to one another. Our only point is to recognize the likelihood of individual that will be impacted by a heart issue or not. We are utilizing K-Closest Neighbor (KNN) to foresee the exactness.

Keywords: diagnosing, cholesterol levels, Machine Learning, prediction, K-Nearest Neighbor

1 INTRODUCTION

Machine Learning is an exceptionally huge and various field and its extension and execution is expanding step by step. Machine Learning integrates different classifiers of Directed, Unaided and Group Realizing which are utilized to foresee and find the precision of the given dataset. Cardiovascular illnesses are exceptionally normal nowadays, they depict a scope of conditions that could influence your heart. World wellbeing association gauges that 17.9 million worldwide passing from (Cardiovascular infections) CVDs.

It is the essential explanation of passing in grown-ups. Our task can assist with foreseeing individuals who are probably going to determine to have a coronary illness by help of their clinical history. It perceives who all are having any side effects of coronary illness, for example, chest agony or hypertension and can assist in diagnosing sickness with less clinical trials and viable therapies, so they can be relieved in like manner. The goal of this task is to check whether the patient is probably going to be determined to have any cardiovascular heart illnesses in light of their clinical properties, for example, orientation, age, chest torment, fasting sugar level, and so on. By utilizing this dataset, we foresee regardless of whether the patient can have a coronary illness.

2 LITERATURE SURVEY

The authors A. B. Nassif *et al.* have worked on Detecting and diagnosing of the COVID–19 disease by using the speech data as well as image data. They conducted experiments based on speech and image, speech and image. The authors AH Chen *et al.* have produced an idea of heart disease prediction system which uses patients' clinical data to predict the heart disease.

*Corresponding Authors: cse.anithajulian@gmail.com, deepikaramesh172002@gmail.com, geetha.120202@gmail.com and vijayakk73@gmail.com

 DOI: 10.1201/9781032684994-38

The authors D Murphy *et al.* have used the Random Forest Algorithm to detect the patterns of cheatgrass by dividing the Landsat Land Cover. They have also explained various factors like accuracy of model, data continuity as well as classification process of land cover. The authors H. Hijazi *et al.* developed sensors for detecting COVID–19. These sensors were introduced in Smart watches and Smartphones. Some changes in the health state were detected for predicting whether the person is affected by COVID–19 or not. The authors K. Karthick *et al.* have developed the 'earlyR' novel for coronavirus disease forecasting model using the technique of bootstrap resampling. The authors Mrudula Gudadhe *et al.* have introduced a system for the classification of heart diseases. The methods that were mainly used in this system includes Artificial Neural Network as well as Help Vector Machine. The authors O. Atef *et al.* introduced this method during corona virus pandemic. Their experiment mainly focused on the patients who have highest chance of death. The authors R. Muazu Musa *et al.* have used the K-Nearest Neighbor Algorithm for predicting the capable archers is high or low. This venture is helpful for coaches and other sports authorities. The authors S. Rehman *et al.* worked on predicting the heart disease patterns in various countries thus suggesting those countries to decrease their death rate due to heart disease. The authors S. Suganya *et al.* have introduced the method by implementing fuzzy basket algorithm for the prediction of heart disease. In the measured data a fizziness was used for eliminating the uncertainty of data. The authors Shaikh Abdul Hannan *et al.* have used the radial basis function for the medical prediction for heart disease. 300 patients' data was collected from Sahara Hospital in Aurangabad. Radial Basis Function - Neural Network is in the form of three-layer that includes entry level, hidden level and exit level. It is said to be very helpful for the Cardiologists. The authors Shantakumar B *et al.* have used the key patterns from a database that contains data of heart disease to predict heart attack. In this case, the authors proposed the Maximal Frequent Itemset Algorithm.

3 METHODOLOGY

Here we have utilized KNN to assess our model based on exactness and execution utilizing different execution measurements.

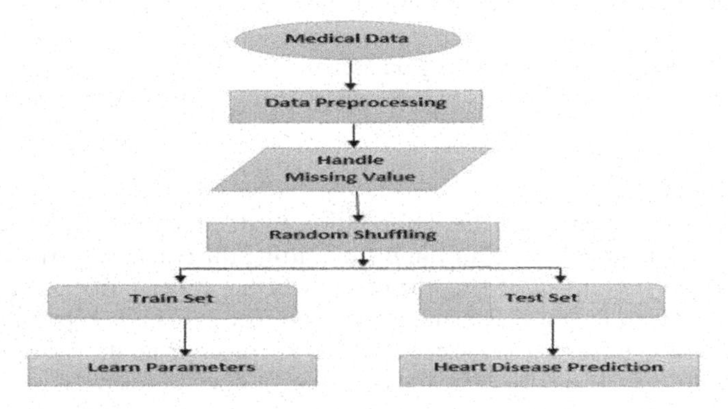

Figure 1. System architecture.

3.1 *K-nearest neighbor algorithm*

K-Nearest Neighbor algorithm is one of the most straightforward ML calculations in view of Directed Learning method. This method analyses the new case with the accessible cases. It performs the task of categorizing the new case that is generally like the accessible classifications. This implies that by using K-NN calculation, new information can be effectively placed into similar class. In view of the likeness, it stores every one of the accessible information and groups another information point. K-NN is a non-parametric calculation, and that implies it makes no supposition on hidden information. K-NN is used for both classification and regression problems.

4 MODULE IDENTIFICATION

The numerous modules of the proposed work is split into the segments as described.

4.1 *Data collection*

Information assortment is a basic module and the most important move towards the interaction. It generally manages the assortment of the right dataset. Information assortment furthermore assists with improving the dataset by including more information. This dataset contains 14 columns and 1026 rows.

Table 1. First six rows in the dataset.

	age	sex	cp	trestbps	chol	fbs	restecg	thalach	exang	oldpeak	slope	ca	thal	target
0	52	1	0	125	212	0	1	168	0	1.0	2	2	3	0
1	53	1	0	140	203	1	0	155	1	3.1	0	0	3	0
2	70	1	0	145	174	0	1	125	1	2.6	0	0	3	0
3	61	1	0	148	203	0	1	161	0	0.0	2	1	3	0
4	62	0	0	138	294	1	1	106	0	1.9	1	3	2	0
5	58	0	0	100	248	0	0	122	0	1.0	1	0	2	1

4.2 *Pre-processing of data*

Data pre-processing is a process of removing or eliminating the unwanted or repeated values and changing them into more feasible data. Pre-processing includes handling the null values, dividing the dataset into training dataset and testing dataset and the last step is to do an element scaling for restricting the scope of factors thus they can measure up on typical circumstances.

4.3 *Training the model*

Training the model includes using the information in dataset for calculating the metrics. Initially the training dataset is used for tuning and fitting the machine learning model. Then the testing dataset is used for analysing the model whether it is reliable or not. At last, this score is determined for each arrangement of hyperparameters. This is kept on circle until the ideal qualities are acquired. Hence, the expectations are taken from the trained model on the contributions from the test dataset. The dataset is partitioned in the proportion of 70:30 for the training and test set.

5 IMPLEMENTATION OF RESULTS

On observing the results, it is obvious that the Machine Learning approaches for Heart disease prediction is the essential nowadays. Through this method, all conventional algorithms of machine learning are trained well by use of a subset of available data to predict the

Table 2. Statistical view of dataset.

	age	sex	cp	trestbps	chol	fbs	restecg	thalach	exang	oldpeak	slope
count	1025.000000	1025.000000	1025.000000	1025.000000	1025.00000	1025.000000	1025.000000	1025.000000	1025.000000	1025.000000	1025.000000
mean	54.434146	0.695610	0.942439	131.611707	246.00000	0.149268	0.599756	149.114146	0.336585	1.071512	1.385366
std	9.072290	0.460373	1.029641	17.516718	51.59251	0.356527	0.527878	23.005724	0.472772	1.175053	0.617755
min	29.000000	0.000000	0.000000	94.000000	126.00000	0.000000	0.000000	71.000000	0.000000	0.000000	0.000000
25%	48.000000	0.000000	0.000000	120.000000	211.00000	0.000000	0.000000	132.000000	0.000000	0.000000	1.000000
50%	56.000000	1.000000	1.000000	130.00000	240.00000	0.000000	1.000000	152.000000	0.000000	0.800000	1.000000
75%	61.000000	1.000000	2.000000	140.00000	275.00000	0.000000	1.000000	166.000000	1.000000	1.800000	2.000000
Max	77.000000	1.000000	3.000000	200.00000	564.00000	1.000000	2.000000	202.000000	1.000000	6.200000	2.000000

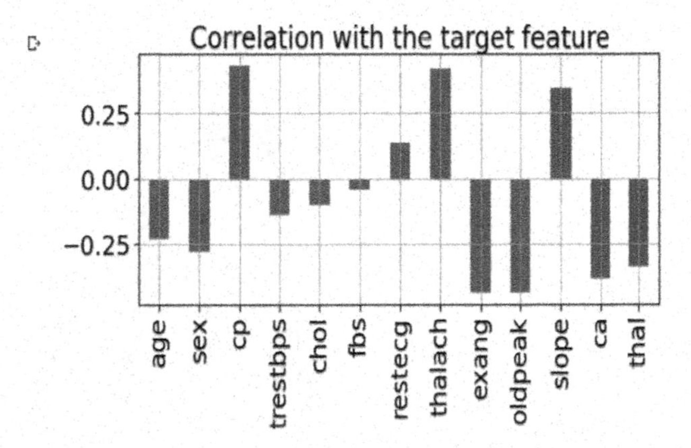

Figure 2. Correlation with the target feature.

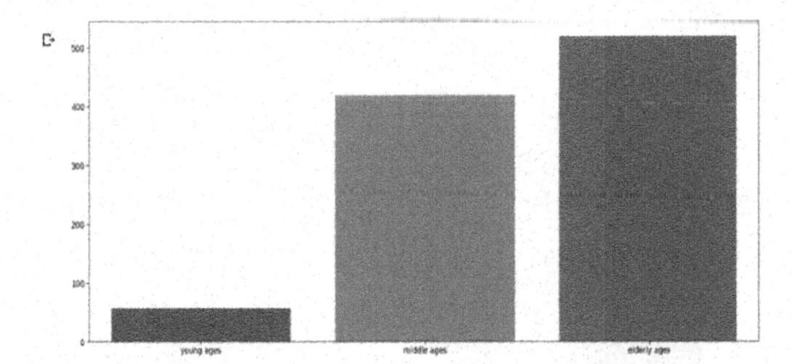

Figure 3. Dividing age into three features.

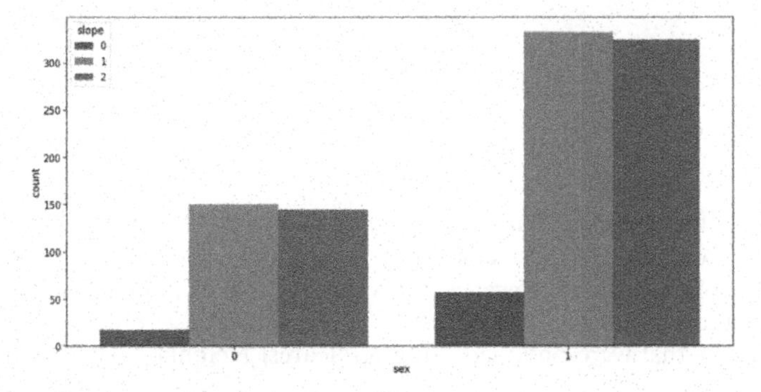

Figure 4. Relation between sex and slope.

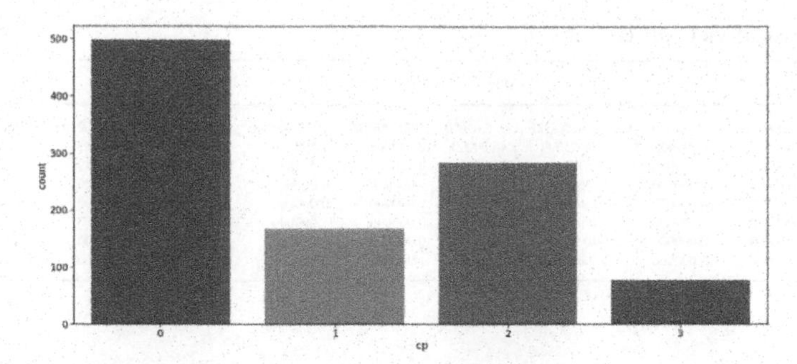

Figure 5. Chest pain analysis.

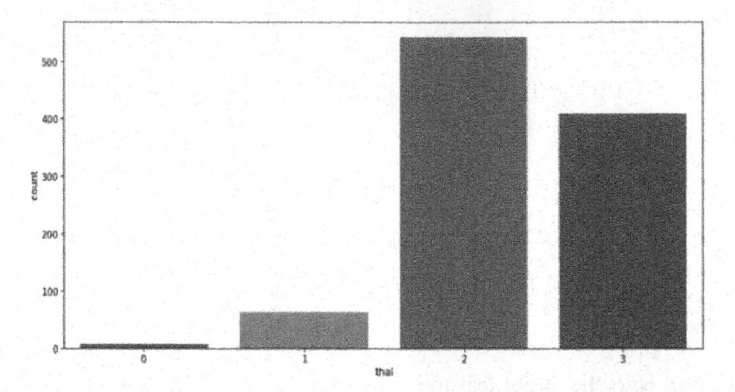

Figure 6. Thalassemia analysis.

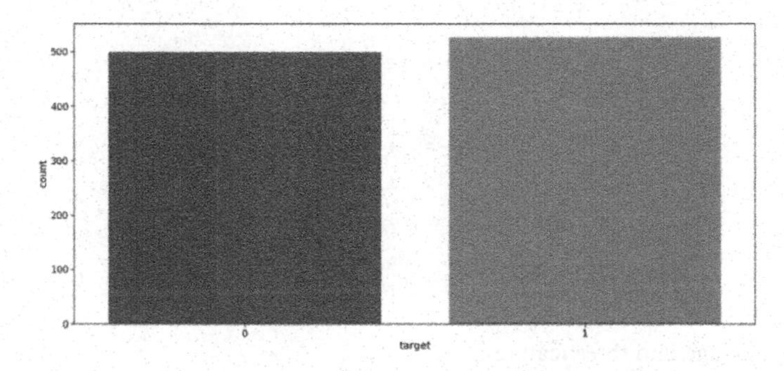

Figure 7. Target analysis.

```
0.827922077922078
```

Figure 8. KNN accuracy.

possibility of whether a person has heart disease or not. It is important to note that the performance of multiple classifiers combined commonly yields higher experimental outcomes than using one classifier. However, the approach has its limits. The training dataset's size has a major effect on the classification of Heart disease prediction. Here are the experimental results that were obtained using K-Nearest Neighbor Algorithm.

6 CONCLUSION

This venture predicts individuals who are suffering from coronary illness by obtaining the patients' history of heart disease from the incorporated dataset of their clinical record, for example, chest torment, sugar level, pulse, and so on. This Coronary illness recognition framework helps a patient in light of his/her clinical data of them been determined to have a past coronary illness. The calculation utilized in building the given model is K-Nearest Neighbor (KNN).

REFERENCES

Atef O., Nassif A. B., Talib M. A. and Nassir Q., "*Death/Recovery Prediction for Covid–19 Patients using Machine Learning*", 2020.

Chen A. H., Huang S. Y., Hong P. S., Cheng C. H., and Lin E. J., 2011, "HDPS: Heart Disease Prediction System",*Computing in Cardiology*, ISSN: 0276–6574, pp.557–560.

Hijazi H., Abu Talib M., Hasasneh A., Bou Nassif A., Ahmed N. and Nasir Q., "Wearable Devices Smartphones and Interpretable Artificial Intelligence in Combating COVID–19", *Sensors*, vol. 21, no. 24, 2021.

Karthick K., Aruna S. K., and Manikandan R., "Development and Evaluation of the Bootstrap Resampling Technique Based Statistical Prediction Model for Covid–19 Real Time Data: A Data Driven Approach," *Journal of Interdisciplinary Mathematics*, pp. 1–13, 2022.

Mrudula Gudadhe, Kapil Wankhade, and Snehlata Dongre, Sept 2010,"Decision Support System for Heart Disease Based on Support Vector Machine and Artificial Neural Network", *International Conference on Computer and Communication Technology (ICCCT)*,DOI:10.1109/ICCCT.2010.5640377, 17–19.

Muazu Musa R., Majeed A. P. A., Taha Z., Chang S. W., Nasir A. F. A. and Abdullah M. R., "A Machine Learning Approach of Predicting High Potential Archers by Means of Physical Fitness Indicators," *PLoS One*, vol. 14, no. 1, p. e0209638, 2019.

Murphy D., "*Using Random Forest Machine Learning Methods to Identify Spatiotemporal Patterns of Cheatgrass Invasion through Landsat Land Cover Classification in the Great Basin from 1984 - 2011*", 2019.

Nassif A. B., Shahin I., Bader M., Hassan A. and Werghi N., "COVID–19 Detection Systems Using Deep-Learning Algorithms Based on Speech and Image Data", *Mathematics*, 2022.

Rehman S., Rehman E., Ikram M. and Jianglin Z., "Cardiovascular Disease (CVD): Assessment Prediction and Policy Implications", *BMC Public Health*, vol. 21, no. 1, pp. 1299, 2021.

Shaikh Abdul Hannan, A.V. Mane, R. R. Manza, and R. J. Ramteke, Dec 2010 "Prediction of Heart Disease Medical Prescription Using Radial Basis Function", *IEEE International Conference on Computational Intelligence and Computing Research (ICCIC)*, DOI: 10.1109/ICCIC.2010.5705900, 28–29.

Shantakumar B. Patil,and Dr. Y. S. Kumaraswamy, February 2009, "Extraction of Significant Patterns from Heart Disease Warehouses for Heart Attack Prediction", *IJCSNS International Journal of Computer Science and Network Security*, Vol. 9, No. 2, pp. 228–235.

Suganya S., and Tamije Selvy P., January 2016, "A Proficient Heart Disease Prediction Method Using Fuzzy-Cart Algorithm", *International Journal of Scientific Engineering and Applied Science (IJSEAS)*, Vol. 2, Issue1,ISSN: 2395–3470.

Fruit disease detection and classification using convolution neural network

V. Loganathan*, Chigarapalli Manasa*, V. Pavithra Devi* & A. Nivetha*
Department of Computer Science and Engineering, Saveetha Engineering College, Chennai, Tamil Nadu

ABSTRACT: The fruit industry is India's largest industry. Due to carelessness and incorrect human inspection, the fruit disease results in significant losses in production, quality, and quantity. The manual examination is a laborious and time-consuming process. An image processing method is provided for the detection and classification of apple fruit diseases using various colour, texture, and shape feature combinations. The essential steps of the recommended approach include Fruit illness detection utilizing a typical neural network (CNN), where fruits are sorted, picture segmentation, feature extraction (color, texture, and shape), feature merging, and feature extraction into ill or healthy classifications. In the lab, our suggested technique was evaluated and confirmed. Using the suggested technique, 97% accuracy was attained.

Keywords: Deep Learning, CNN, Identification, Categorization, Classification

1 INTRODUCTION

For any culture to be harvested, agriculture is necessary. In actuality, agriculture is essential to human civilization. To attain the objective of good output, field prevalence and soil water management must be regularly maintained. Product quality and quantity are significantly impacted by fruit disease. Fruit disease is the biggest obstacle in this area. This study examines ailments that reduce fruit production. Image processing techniques are used to examine how fruit harvests are degrading. The proposed system's job is to spot errors in the fruit photographs. The dataset can be obtained using standard smartphone cameras.

For financial reasons, it is essential to detect fruit illnesses early. Deep Learning can identify and categorize diseases early on, which reduces disease transmission and boosts cure rates. Among other fruits, including apples and cherries, powdery mildew, rust, and black rot are common indications of these diseases. Manual examination takes a long time, is labor-intensive, and prone to mistakes. Fruit color, shape, and texture data can be extracted using artificial intelligence to aid in viral diagnosis. There are consequently more errors committed when rating fruits for export. To increase accuracy, researchers developed an image detection technique to tell contaminated fruits apart from healthy fruits and eliminate the faults in human categorization.

2 LITERATURE SURVEY

The author Abirami S *et al.* have categorized the fruit diseases can be done swiftly using Feed forwarded-back PNN additional 10 hidden layers, and also the accuracy for

*Corresponding Authors: loganathan@saveetha.ac.in, chigarapallimanasa@gmail.com, pavithradevivenkatesan@gmail.com and nivikeetthi@gmail.com

 DOI: 10.1201/9781032684994-39

bacterial diseases of 92%, and for fungal diseases of 86%. The authors Dharmasiri S B D H *et al.* have worked on passion fruit illnesses can be detected with a mean of 79 percent and the level can detect with a mean of 66 percent. The authors Ismail El Massi *et al.* have presented a automated detection method of infected and injure on plants leaf. They presented a method depend on parameterized methods, where there are different combinations and variations. The difference of outcomes in-between various models represents similar results assign for identification of infected & injured leaves. The authors Malathy S *et al.* have worked on identifying infected products by using CNN algorithm, they also found that using this methodology can helps in finding infected regions very fast. And also, this method saves farmers in getting loss percentage of the product. The authors Md Rasel Mia *et al.* have worked on Mango Tree, in this proposed system will help the farmer to grow mango trees without the help of lab testers, this will save the time to detect the infection (Disease). It will become easy to found the infected area in leaf, and also by this it will increase Mango production and farmers will get good yield and they can get more profits in the market. The authors Milos Ilic *et al.* have worked on usage of various methods done in MATLAB for feasible product inflammation detection. On the other side the detection of minimal infections data were used for image preprocessing.

The authors M R Howlader *et al.* has worked on prior identification on plant infections that are very major to upgrade the growth of good product factories. In this the normal identification structure is used by the farmers in saving time, and spending of minimal cost. In this they have used deep learning algorithm, for detecting the guava leaf infection spontaneously. In this they have created their personal data for image preprocessing, On that data they started developing their new infection detection system. The authors Nikitha M *et al.* has declared that farmers can utilize this feature for a variety of things and it is quite helpful. To enhance classification and detection of fruit ailments, the Inception v3 model and Transfer Learning are used. In order to make the efficiency of deep neural networks and image processing for detecting diseases in agricultural products like fruits and vegetables sector. The authors P. K. Devi *et al.* have worked on correlation charts for the SVM technique, the SVM technique is rated between 49.6 and 77.5. The K Means Clustered approach between 55 and 86. The authors S. K. Behera *et al.* has declared that they were able to classify the various disease categories and compute disease severity of four different types of infected oranges. To increase accuracy and validate with additional samples, this study may expand to include soft computing techniques. The authors Wang H *et al.* have provided an improved Mask CNN target identification method. According to examination researches, The target fruit surface lessons are detected by the algorithm with value over 94 percent. The authors Zhao J *et al.* has proposed a detection method for typical tomato fruit physiological diseases based on the convolution neural network. In this the network used to identify infected tomatoes have a mean average precision of 97.24 percent.

3 PROPOSED SYSTEM

Accurate image segmentation is necessary for more accurate fruit disease detection. Otherwise, the qualities of the non-infected zone outweigh those of the area that is sick. This suggested technique suggests using Convolution Neural Network related picture visualization to point out the exact area that was affected there. Following the processing of the image, specific features were processing stage. Finally, the training and classification phases of the procedure are completed, and the precise result is provided. The system's logic and the mechanism for extracting features.

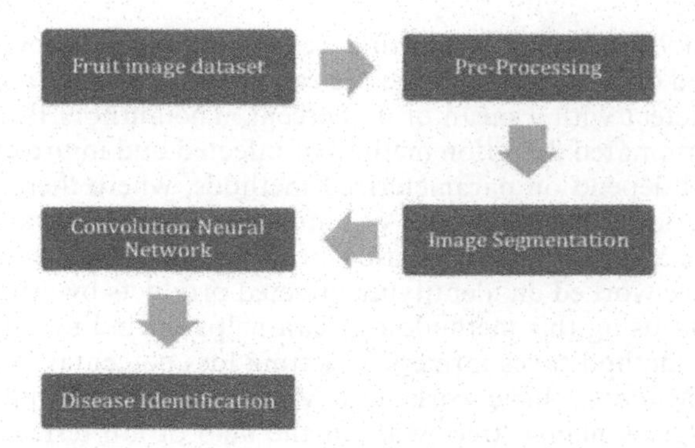

Figure 1. Preprocessing of the image.

4 METHODOLOGY

Here we have used sub samplings of the dataset, and also we distributed the fruit dataset in between maps. At last we need to connect all the connected maps to get the desired output.

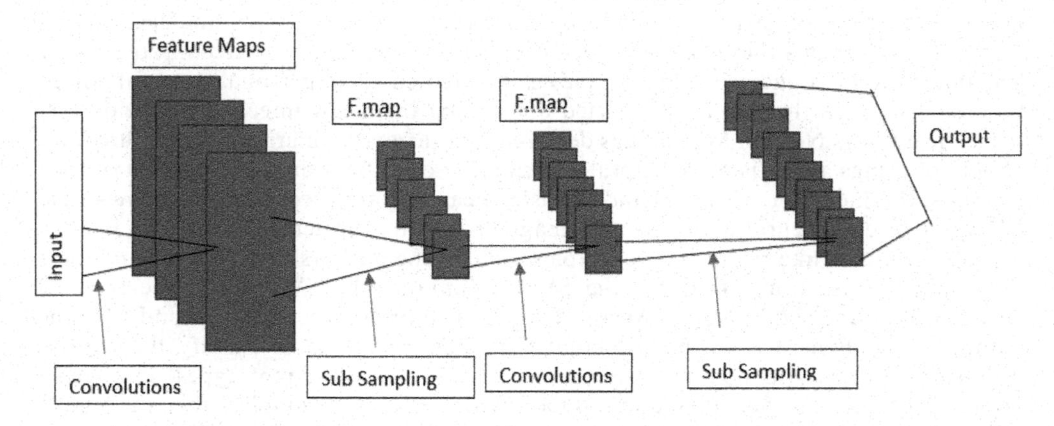

Figure 2. Sampling of fruits.

5 WORKING

5.1 *Image input*

This is the starting point. Here, sample photos were gathered; the collected photos were then used to develop the classification model and to refine classification model. Using easily available mobile phone cameras, pictures of healthy and diseased fruits were taken and recycled as the developing data set for the classification model. Pictures are taken from various angles, with various scenarios, and with various lighting setups. These pictures are often kept in "JPG" format. We will need test and training sets of data from agricultural areas in various regions for this classifier algorithm.

5.2 *Preprocessing of datasets*

When we have a sufficient number of datasets, we preprocess those datasets to get highly upgraded pictures. The top most fruit view is stored in one file. Any name we choose can be used to save these images. Images are captured from a range of sizes and perspectives. We believe the following size to be normal for any dataset utilized, therefore When a shot is taken horizontally, it needs to be turned 90 degrees and enlarged to 200 by 300 pixels. Additionally, the captured photos would need to be scaled to 250 × 250 pixels if the height and width were also different. The processing will take longer if the image is huge. After the image scaling process is finished, a sharpening and noise reduction technique is used to restore the image. The overall image quality is enhanced. After this process is finished, all similar photos were stored in the existing folder.

5.3 *Division of datasets*

This is the final step in this method for detecting diseases is image segmentation. Before storing them in RGB format, all pre-processed images must first be converted color models, and Gray. This method also determines the best color modeled image to use in the pre-processing step. The next step is image conversion, which converts the pre-processed image to binary format. The CNN method is used to cluster the formatting values. The photo segmentation method is taken into consideration during the entire process.

5.4 *Using the training set*

After that, the algorithm is trained using a set of pictures that have been through the previous rounds' processing. The output was produced using feature extraction. Our trained model needs to be tested and validated. Detection and classification of diseases Procedure:

i. Fruit picture input.
ii. Classified fruit disease as output.

6 RESULTS

The use of a universal filter, the Clustering algorithm, and the Convolution Neural Network (CNN) Algorithm for fruit disease detection allows for the early diagnosis of diseases, preventing problems that would otherwise be lethal to humans. This literature suggests a method for finding any information that appears abnormally in pictures and is taken to

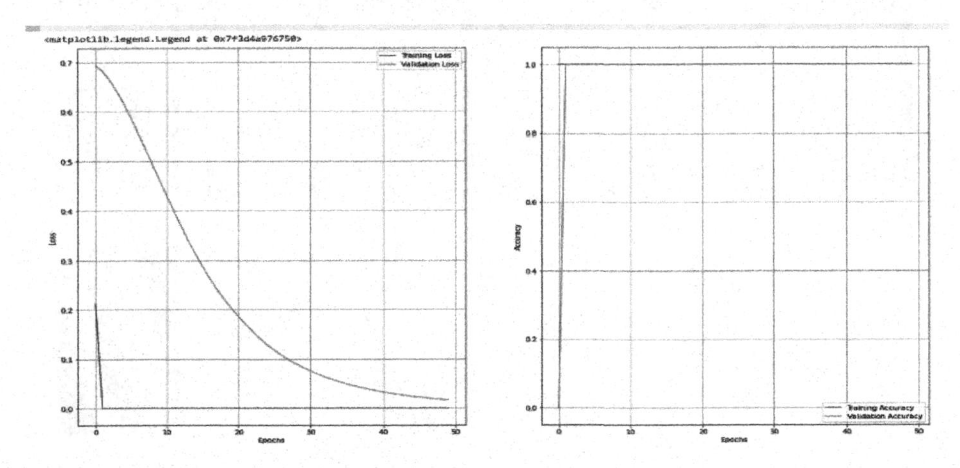

Figure 3. Model accuracy and model loss.

indicate a flaw in the fruit. It is a method for locating noisy data, eliminating it, and then delivering noise-free data for additional analysis.

The input of a fruit image is where the proposed method starts. It might be a fruit that is nutritious or unhealthy. After the image is provided, the diseased areas are highlighted. The name of the disease is displayed after the analysis. The image of the fruits that is utilized as data to help diagnose diseases was collected from the image database known as Kaggle.com.

i. High accuracy that works with both low- and high-pixel images.
ii. It only takes a few seconds to deliver a precise result when enhancing the value of fruit disease detection.
iii. By highlighting the locations that are afflicted, you can also discover the disease's name.

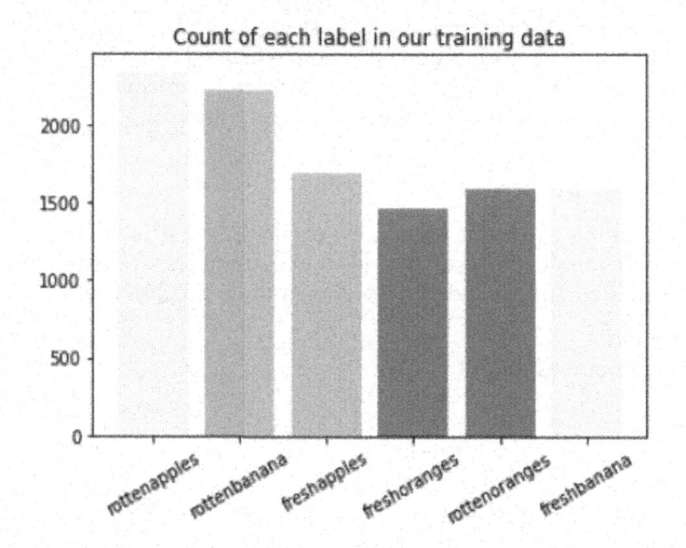

Figure 4. Distribution of data from dataset.

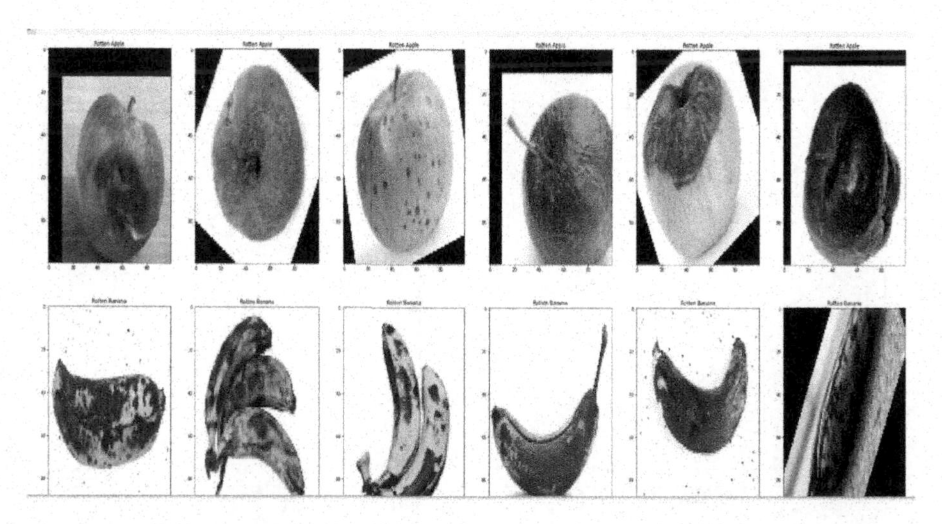

Figure 5. Classification of dataset.

7 CONCLUSION

The pre-processing phase of the suggested technique employs a universal filter. Pre-processing is important since it increases the clarity of the image. The correctness of the final output depends on the image clarity. Extraction of the features is the following phase. The qualities of texture and color are used to extract information. The k means clustered method is used to segment data. This project major goal is to discriminate between essential and optional visual components. Utilizing convolutional neural networks, the data is classified. The most accurate classification results come from convolution neural networks.

Future tests might compare the proposed technology with real-time datasets. Additionally, the convergence rate can be increased by using a suggested random forest technique.

REFERENCES

Abirami S and M. Thilagavathi, "Classification of Fruit Diseases Using Feed Forward Back Propagation Neural Network", *The International Conference on Communication and Signal Processing (ICCSP) Will Take Place in 2019* (pp. 0765–0768). IEEE, April 2019.

Behera S. K., Jena L, Rath A. K and Sethy P. K, "Disease Categorization and Grading of Oranges Utilising Fuzzy Logic and Machine Learning." *2018 Saw the First-ever ICCSP (International Conference on Communication and Signal Processing)* (pp. 0678–0682). IEEE, April 2018.

Devi P. K., "K-Means Clustering Algorithm for Fruit Disease Detection Image Processing", *2020 Will See the Fourth Iteration of ICECA, an International Conference on Electronics, Communication, and Aerospace Technology* (pp. 861–865). IEEE, November 2020.

Dharmasiri S.B.D.H and Jayalal S, "Passion Fruit Disease Detection Using Image Processing", In *2019, the SCSE (Smart Computing and Systems Engineering) International Research Conference* (pp. 126–133). IEEE, March 2019.

Howlader M R, Umme Habiba, Rahat Hossain Faisal and Md. Mostafijur Rahman, "*Automatic Recognition of Guava Fruit Diseases Using Deep Convolution Neural Network*", February 2019.

Ismail El Massi, Youssef Es-saady, Mostafa El Yassa and Driss Mammass, "*Automatic Recognition of Fruit Diseases Based on Serial Combination of Two SVM Classifiers*", 2021.

Malathy, S., Karthiga, R. R. Swetha and Preethi G, "Disease Identification in Fruits Using Image Processing", *The 6th International Conference on Inventive Computing Technologies (ICICT) Will Take Place in 2021*. (pp. 747–752). IEEE, January 2021.

Md. Rasel Mia & Sujit Roy & Subrata Kumar & Md. Atikur Rahman, "*Fruit Disease Acknowledgement Using the Neural Framework*", 2020.

Milos Ilic, Sinisa Ilic, Srdjan Jovic and Stefan Panic, "*Early Fruit Pathogen Disease Detection Based on Data Mining Prediction*",Volume 150, July 2018.

Nikitha M, Sri S. R and Maheswari B. U, "Fruit Recognition and Grade of Illness Detection Using Inception v3 Model", *The International Conference on Electronics, Communication, and Aerospace Technology* (pp. 1040–1043). IEEE, June 2019.

Wang H, Mouyue Y, and Zhao H, "Research on Detection Technique of Various Fruit Disease Spots Based on Mask R-CNN", *International Conference on Mechatronics and Automation (ICMA) of the IEEE in 2020* (pp. 1083–1087). IEEE, October 2020.

Zhao J, & Quojue J, "A Detection Approach for Tomato Fruit Common Physiological Disorders Based on YOLOv2", *The 10th International Conference on Information Technology in Education and Medicine (ITME) Will Take Place in 2019* (pp. 559–563). IEEE, August 2019.

Artificial Intelligence, Blockchain, Computing and Security – Dagur et al. (Eds)
© 2024 The Author(s), ISBN: 978-1-032-67841-2

Digital currency price prediction

Anitha Julian*, Bangale Sampath Kumar Rao*, Pedavalli Mukesh Chowdary* &
Narravula Upendra*
Department of Computer Science and Engineering, Saveetha Engineering College, Chennai, India

ABSTRACT: Investments are one of the major sources of income in one's life. The usual question that arises is which category one should invest in. In the market, there are many investment options such as stocks, cryptocurrencies, and mutual funds, but everyone who wants to invest normally expects high returns as well as security. That is what cryptocurrency does, since the future is not just for money, besides money, day-to-day operations and processes also depend on different factors. The proposed work uses machine learning algorithms for predicting the Crypto Currency price. A linear regression model is used to train the machine learning model with maximum efficiency and the model is implemented to predict Crypto Currency price. The model has been trained with a dataset containing the Crypto Currency exchanges which were provided by the user for the time with minute-by-minute updates. The proposed regression algorithm identifies with reasonable accuracy.

Keywords: Forecasting, Cryptocurrency, Machine learning, Fb Prophet, Streamlit, Python

1 INTRODUCTION

1.1 *Financial intelligence*

The global economy is largely influenced by the business and finance industry, where trading in cryptocurrencies is a widespread activity. Forecasting the value of these digital assets is a challenging task in the financial world. The objective of this process is to anticipate the future worth of a crypto coin to minimize losses or maximize gains. In this study, a machine learning approach is proposed to make such predictions. A machine learning algorithm is trained by analyzing historical data from several firms to make more accurate forecasts. However, some experts are skeptical about the accuracy of these predictions due to the efficient market hypothesis. They believe that it is impossible to make predictions based on available data, making these outcomes unpredictable. Nevertheless, there are various technologies and tools available to gain insight into future trends and make profitable decisions.

2 RESEARCH WORK

2.1 *Literature survey*

An artificial intelligence system that uses past data to predict the future is known as machine learning (ML). There are several advantages to using machine learning-based forecasting models, especially in time series, as prior research has shown that they produce results that

*Corresponding Authors: cse.anithajulian@gmail.com, bangalesampathkumarrao@gmail.com, mukesh.pedavalli@gmail.com and narravulaupendra@gmail.com

DOI: 10.1201/9781032684994-40

are nearly or approximately as the actual results, and they improve their accuracy. Fb-Prophet, owned by Facebook Prophet, and additive linear regression are examples of machine learning.

The potential benefits of incorporating a cryptocurrency Prediction scheme have been discussed in Patel *et al.* 2020. The addition of digital assets can increase portfolio options and reduce overall risk, as measured by standard deviation. The optimal allocation for an investor, based on their risk tolerance, is between 5% and 20%. Another study, presented by Ciram Saiktishna *et al.* 2022, focuses on time series data forecasting which uses the Facebook Prophet algorithm to predict stock market prices. The results of this research demonstrate that machine learning forecasting techniques can effectively be used to predict the prices of the market along with Crypto.

Crypto Currency Price Prediction is a popular yet challenging task. The proposed work can predict the price before deciding whether to buy or sell. This enables the user to easily invest in cryptocurrencies with it. Currently, the business and financial sector dominates the global economy, so selecting a perfect model for forecasting will make effective results as presented by Nicola Uras *et al.* 2020.

As financial institutions adopt AI, machine learning is rapidly used to help make investment decisions. Although there is an abundance of Crypto data for machine learning models to train on, as well as the multitude of factors that affect crypto prices, make it difficult to predict the crypto market. Subsequently, this model may not need to reach the highest levels of accuracy since even 60% accuracy can deliver better returns. We may conclude that this is one of the methods for predicting crypto prices that uses the Prophet algorithm for times series forecasting and enables identification and track the patterns that trigger its dynamic movement.

3 METHODOLOGY

3.1 *Introduction*

Cryptocurrency price prediction resembles a compound problem as many factors have to be addressed and it does not seem mathematical initially. But by appropriate use of machine learning algorithms, one can find the trend between previous data to the current data and train the model to learn from it and make suitable assumptions.

3.2 *Existing system*

Currently, this Prophet model lacks stable visualizing elements to create an effective visualization of the graph, and placeholders for each graph are not consistent. As a total combination of these factors, the prediction model cannot be properly represented by the front end, which makes the current system look and function like a traditional and not future-ready system. We would like to make some changes to the current system, adding some more new features with the latest technologies.

3.3 *Proposed system*

Our proposed system aims to assist investors in making informed decisions by predicting the future value of each cryptocurrency coin. The system employs the Prophet Model, a regressive model, to estimate the value with a high degree of accuracy. This model is specifically designed for time series data forecasting, taking into account yearly, weekly, and daily patterns, along with holiday effects, to fit non-linear trends. Additionally, the Prophet Model is robust to missing data and shifting trends, and can effectively handle outliers. By using this model, investors can benefit from better predictions, allowing them to buy or sell their crypto assets to gain profits.

Our Proposed System will have a proper front end and also have effective graphs with effective categories of filters in which we will be using one of the latest technology for creating apps and websites using Streamlit. This will provide a convenient way to track prices. We will create this front end without any markup or scripting languages, we will create the front end by using only streamlet technology which stands for the specialty of this project and apparently, we are in the making of history.

3.4 *Algorithm*

The foundation of this proposed work is the dataset of historical data. This data is analyzed and combined to demonstrate the changes over time, which serves as the basis for further predictions. The tool allows for customization and provides a time frame for understanding the data flow. Results are produced during each phase of testing, which involves passing a new set of datasets similar to the training dataset and evaluating their accuracy.

It is possible to prognosticate the future price of every variation of cryptocurrency based on the past data of the currency and the current price of the currency. Modeling the accumulated data and the target data generated using machine learning followed by regression is done based on the accumulated data and the target data generated.

We can retrieve historical market data from Yahoo Finance API in Python with this package. This module makes getting data very easy for Python developers.

The Core Data Science Team at Facebook released this powerful time series analysis tool. It provides a simple and straightforward package for performing time series analytics and forecasting at scale.

This Python library in question is open-source, which makes it easy to create and share interactive, user-friendly web applications for machine learning and data science. With this library, developers are able to quickly build and deploy powerful data applications in a short amount of time. The open-source nature of the library allows for collaboration and innovation in the fields of machine learning and data science.

Table 1 shows the different features in the dataset, which can be used to predict currency prices. The prediction relies on six purely necessary features, which were extracted from

Table 1. Dataset parameters with the explanation.

Data Header	Explanation
Open	This term in the header refers to the initial trading price when the market opens for the day. It represents the price at which the market starts trading once the market bell has sounded.
High	This term refers to the highest price at which a stock was traded during a specific period, typically a day. It represents the peak price that the stock reached during the time frame in question.
Low	Following the term high, this term low is the lowest price reached during the same interval. These terms help investors understand the fluctuations in the coin's price over the course of the day and can provide important information for investment decisions.
Close	This term refers to the final price of a stock at the end of the trading day. It represents the price at which the last buy-sell transaction was executed between two traders before the stock exchange closes for the day. In many cases, this occurs during the last moments of the trading day. The close price is an important indicator of a stock's performance over the course of the day and can provide valuable information for investment decisions.
Adj Close	This regulated closing price considers any market activities and adjusts the closing price of a coin to reflect its true value. This type of closing price is commonly used when analyzing historical returns or conducting short-term analysis of past performance. By considering market actions, the regulated closing price provides a more accurate representation of the coin's value at the end of the trading day.
Volume	This term refers to the total number of coins exchanged during a specified period. Every time buyers and sellers engage in a coin exchange, the number of coins traded is added to the total volume for that interval. Volume is an important metric in analyzing a coin's performance. Higher volume can indicate greater market demand, while lower volume may indicate less interest or lower liquidity.

Yahoo Finance. Depending on the input (the user will be offered to select the cryptocurrency that they wish to predict), these datasets will be modified. As a result, the input data will be used to extract the dataset from yahoo finance.

4 RESULTS

4.1 *Interface*

In Figure 1, the outputs are presented in the form of a graph for various coins as selected by the user. This makes it much easier for a user to understand and interpret the scenario and even candlestick graphs were also provided to present effective results. Therefore, they can decide if it is something they would like to invest in and benefit from. Various high-level machine learning algorithms are created and incorporated into the project in the backend.

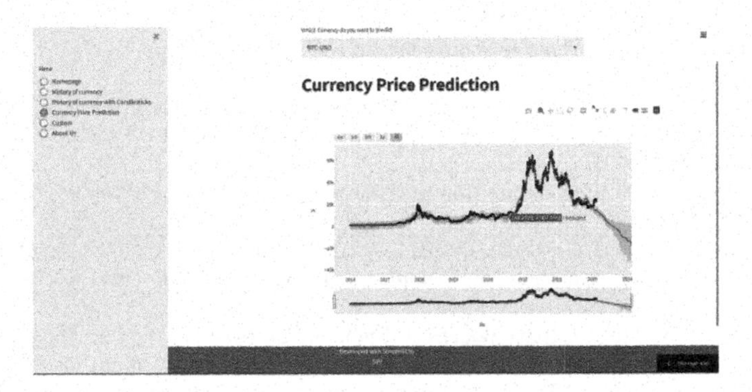

Figure 1. Interface.

4.2 *Performance metrics*

The R-squared error is the metric used to evaluate the performance of the proposed algorithm. This represents "The percentage of variance that the regression model successfully captured for the actual response variable". The R-squared error indicates the variance between the accumulated and predicted values. Ideally, model quality will be determined by R-squared error.

5 CONCLUSION

5.1 *Outcome*

As a result, cryptography is an unexpected process based on chain segments and their dependencies. It is described as a curve that continuously shifts, converting prices from low to high and vice versa. As the prices in this dataset were inconsistent, i.e., they were taken from the Yahoo official website, which will update from time to time, the prices will continue to fluctuate. Therefore, there is no need to manually update prices.

5.2 *Future work*

In the process of prediction, various high-level machine-learning algorithms are implemented and integrated. The output is then transformed into a graph and displayed. This makes it easier for users to understand and interpret what's going on, and to make the right decisions to invest and reap the benefits of it. A future version of this project will be developed to include more features for feasibility and provide a better idea of cryptocurrency in general.

The authors do not promote or suggest that the reader buy cryptocurrency. In the research that was conducted on graphs related to cryptocurrency, it has been found that there are consequences in the real world that our machines cannot cope with, such as emotional intelligence that may affect prices which tends to not give accurate predictions (This is significantly applicable during pandemic situations).

REFERENCES

Allan Timmermann, Forecasting Methods in Finance. *Annual Reviews*, vol. 10, pp. 449–479, 2018, doi:10.1146/annurev-financial-110217-022713.

Bineet Kumar Jha; Shilpa Pande, Time Series Forecasting Model for Supermarket Sales Using FB-Prophet. *IEEE Access*, 2021, doi:10.1109/ICCMC51019.2021.9418033.

Chadefaux; Thomas, *Conflict Forecasting and Its Limits*. IOS Press, vol. 1, pp. 7–17, 2017, doi:10.3233/DS-170002.

Ciram Saiktishna; Nalam Sai Venkat Sumanth; Muppavarapu Madhu Sudhan Rao; Thangakumar J, Historical Analysis and Time Series Forecasting of Stock Market Using FB Prophet. *IEEE Access*, 2022, doi:10.1109/ICICCS53718.2022.9788231.

Fanoon Raheem; Nihla Iqbal, Forecasting Foreign Exchange Rate: Use of FbProphet. *IEEE Access*, 2021, doi:10.1109/SCSE53661.2021.9568284.

Jai Jagwani; Manav Gupta; Hardik Sachdeva; Alka Singhal, Stock Price Forecasting Using Data from Yahoo Finance and Analysing Seasonal and Nonseasonal Trend. *IEEE Access*, 2018, doi:10.1109/ICCONS.2018.8663035.

Jovan Davchev, Kostadin Mishev, Irena Vodenska, Ljubomir Chitkushev, Dimitar Trajanov, Bitcoin Price Prediction using Transfer Learning on Financial Micro-blogs. *CEEOL*, 2020, pp. 78–83.

MM Patel, S Tanwar, R Gupta, N Kumar, *A Deep Learning-based Cryptocurrency Price Prediction Scheme for Financial Institutions*. Science Direct, 2020, doi:10.1016/j.jisa.2020.102583.

Mohammad Hamayel, Amani Yousef Owda, *A Novel Cryptocurrency Price Prediction Model Using GRU, LSTM and bi-LSTM Machine Learning Algorithms*. Researchgate, 2021, pp. 477–496, doi:10.3390/ai2040030.

Nicola Uras; Lodovica Marchesi; Michele Marchesi; Roberto Tonelli, Forecasting Bitcoin Closing Price Series Using Linear Regression and Neural Networks Models. *Peerj*, 2020, doi:10.7717/peerj-cs.279.

Saurabh Shukla; Arushi Maheshwari; Prashant Johri, Comparative Analysis of Ml Algorithms & StreamLit Web Application. *IEEE Access*, 2021, doi:10.1109/ICAC3N53548.2021.9725496.

Siddartha Mootha; Sashank Sridhar; Rahul Seetharaman; S. Chitrakala, Stock Price Prediction Using Bi-Directional LSTM Based Sequence to Sequence Modeling and Multitask Learning. *IEEE Access*, 2020, doi:10.1109/UEMCON51285.2020.9298066.

 ISBN: 978-1-032-67841-2

Medicab emergency services

S. Sasikumar*, G. Tejaswi*, S. Sakthi Aishwarya* & K. Subanu*
Department of Computer Science and Engineering, Saveetha Engineering College, Chennai, India

ABSTRACT: We created an Android app that allows users to access the medicab emergency services facility because there is a demise in India for a minute of the day. The concept suggests utilizing an app to let patients schedule transportation to hospitals. Patients can upload their current location and destination to the program or locate themselves. The patient can then evaluate the options and travel times of each transport across the area to choose the best one after the system displays nearby available ambulances. The application takes the ambulance driver to his destination, and he must confirm the reservation made. Administrators have access to all core data and are in charge of calling and querying operations. The app responds by pressing one button, at which point it notifies the nearby ambulance control center with the user's information and location through GPRS. After collecting location coordinates from GPS devices, the software utilizes the Google Street View interface for application programming API to show ambulance details on the smartphone's Google Map client.

Keywords: Ambulance, Patients, Hospital, Health

1 INTRODUCTION

Many nations are attempting to develop into Smart Countries in the modern era. If a urban is to be referred to as a "Smart Urban," it must have many advancement in the field of smart technology. The toughest and challenging task is to increase efficiency in the healthcare. It covers things, like sending an ambulance out as soon as feasible and providing the patient with the appropriate care while they are in a severe condition in order to increase their chance of life. Urban areas see a lot of traffic-related issues, which made life difficult for the ambulance. Road accidents are frequently increasing in the urban, making it even more crucial to prevent deaths and fatalities. By utilizing technology like IOT and ambulance services, we can get beyond these restrictions. Through the use of software technologies we can develop application which can save a life.

2 LITERATURE SURVEY

Amad A Mohammed *et al.* have implemented an Efficient routing of 108 ambulances using clustering methods place ambulances where the average distance they travel has lowered using clustering techniques. Arunmozhi. P *et al.* have implemented the system to determines the quickest route between the ambulance, the accident site, and the nearest ambulance, nearest medical facility. This path is subsequently transmitted to the ambulance by the

*Corresponding Authors: sasikumar@saveetha.ac.in, tejaswi.gaadhiraju2002@gmail.com, sakthi2002aishwarya@gmail.com and subanu2002@gmail.com

DOI: 10.1201/9781032684994-41

server. It features a GPS and GSM unit, and the GSM modem may send information to the emergency center, allowing it to locate the accident's location. Farzana. S *et al.* have implemented a method that involves when a user enters a query, a location-based system is utilized to find the users requested points of interest (POIs) from a specific location based search. The location servers, which are based on the GPS coordinates with the points of interest, get the GPS latitude and longitude as input. The location servers can then serve the client with the user's current location.

Iyappan. P *et al.* have implemented a website whose name is LISA which broadcasts the accident site and sends the details to a adjacent ambulance, police station, as well as to the victim's family and blood group contributors. Online apps are more vulnerable to security breaks, are identified to run slowly, and encounter issues with various browsers, which is the largest problem with this approach. Poonam Gupta *et al.* implemented a smart ambulance system. Smart ambulances will use a variety of sensors, including heart rate sensors, blood pressure sensors, and ECG sensors, to monitor vital signs. The status of these parameters will be reported to the hospital's database concurrently with GPRS communications from the cloud being utilized to control traffic signals. Rajeshwari Sundhar *et al.* have implemented a smart traffic management system that allows emergency vehicles to pass without incident. Every vehicle has a specific radio frequency identification (RFID) tag that is strategically affixed and difficult to remove or destroy. For the purpose of reading the RFID tags which is attach to the vehicle, we use an RFID reader, NSK EDK-125-TTL, and PIC16F877A system-on-chip. It keeps track of the number of cars that travel down a specific path over a given period of time. It also determines network congestion and, as a result, the amount of time that path has a green light.As it approaches the corner an ambulance will give the traffic cob a sign to switch on the green light.

Rashmi A. Nimbalkar *et al.* have assisted in locating the next accessible hospital, contacts that facility's ambulance emergency system, and then retrieves the emergency patients' medical records, which can be extremely useful for prehospital care. Ryo Katsuma *et al.* explained the implementation of dynamic routing for ambulances. The ambulance may obtain vehicle data from the other automobiles thanks to IGA as well. The inquiry is communicated by the emergency ambulance through multi-hop communication to those vehicles within the ambulance's infectious scope, along with the ambulance's location and travel history. The principal course of the ambulance is selected by the vehicle navigating tool. The car can determine the ambulance's precise route from where it is right now. The ambulance gets vehicle data for time and route, and it uses RCA to calculate new driving directions. Sathish Kumar R *et al.* used IoT and smart phone device control to develop a sharp halting system. It generates a powerful leave response in the suburbanite and the owner of the moving automobile. Administrations are tasked with maintaining a parking space, validating a saved client, identifying the closest available space based on the length of the car, exploring the exit opening, and calculating daily, weekly, and monthly accounting data.

Shivam Kumar Kashyap *et al.* proposed a method that checks the ambulance request made, the request is sent to all local hospitals together with details about the type of ambulance needed, number of patients, indications, statement etc. To extend this, Internet of Things was further which is helpful for the doctors to monitor the status of patients such as blood pressure, Pulse rate and temperature continuously while going to the hospital. Yamuna. S *et al.* explained a way to simulate the Reminiscences among the highest points experienced by cab services like Ola and Uber were evoked by research on customers' perceptions of and satisfaction with a region's online cab services. The consumer information from Ola and Uber was a particular focus of the study. Local authorities were consulted for information.Yuanyuan Du *et al.* have implemented an Android based emergency alarm and healthcare management application. With the use of the GPS and Global System for Mobile communication network, the given system identifies the location of the user when they are in distress and sound the alert. When an alarm is received, immediate action can be performed manually. It also manages the user's health history.

3 PROPOSED WORK

3.1 *Objective*

The main objective of this project includes booking different types of ambulances using our application in the location of your choice, assisting with getting all desired schedules, offering a single platform for booking ambulances and processing payments, making it possible to see which ambulances are still available visually, and automating the cancellation of redundant systems.

3.2 *Architecture diagram*

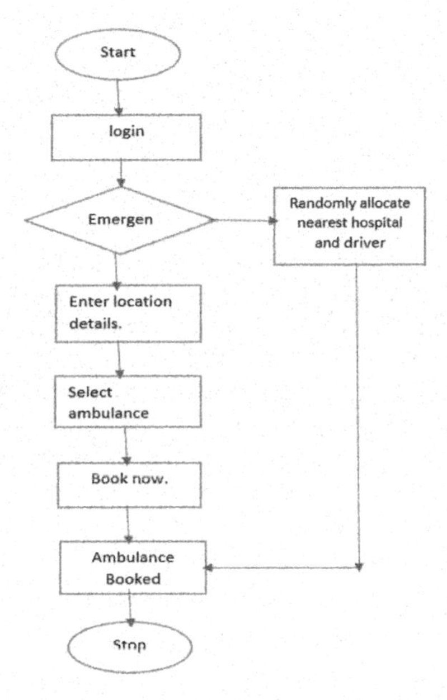

Figure 1. System architecture.

The Medicab app allows the user to select a choice based on the situation; if it is an emergency, they can press the emergency button in the app, which immediately checks the location using GPS and finds a random hospital near your location and informs a driver who is around your location. Once he accepts our request, our ride is on. Otherwise, the user can manually enter the hospital details whenever it is convenient for them. They also have the option to choose the ambulance type. Once the driver accepts our request, he can view our location, and the user can track the ambulance's position. After completing the trip, the user either pays in cash or online transaction. There is also a feature for rating the driver, which boosts their ability to serve more people. The users and drivers can view their trip history. This system would be implemented using the tools Android Studio, Firebase, and the Google Maps API.

3.3 *Processing data*

The proposed system's ultimate goal is to provide with an android application that decreases death rate by providing faster ambulance services. As a software product, this need to be produced using a professional software development approach.

The technologies used to develop the system includes Google Map API, Firebase, Android Studio

Here are the steps for scheduling a ride:

Step1: Log in
Step2: Choose the types of services which you want
Step3: Enter your Location Data
Step4: Find your ambulance on the map
Step5: Calculate
Step6: Reserve a trip
Step7: Decide the payment method
Step8: Journey Receipt

3.4 *Algorithm*

We used the First Come First Serve scheduling method and the Dijkstra's algorithm to ensure the system would operate effectively and meet user expectations. This collection of pseudocode demonstrates the use and the potentials of controlling recurrent reservations in addition to a number of crises when different ambulances are necessary but few are available. Besides, the First Come First Serve method, the concept of LILO or FIFO queues, which states that the first to book is the first to be served, can be used to address the long tail of patients. The coordinates, path, travelling distance, and cost charged to that distance are not taken into account in this method. This approach's amplifications on extended, varied, and hold time further influence the increased cost. In contrast, Dijkstra's algorithm determines the shorter routes among the source and destination; in the event of traffic, a different route is found with a new duration. A streamlined process known as Dijkstra's algorithm locates the patient and ambulance vehicle among all the available resources. We are using the Google Maps API which uses the Dijkstra's algorithm to control and run GPS device as we reach to our destination which includes all feasible routes, modes of transportation, traffic patterns, and time.

4 RESULTS

We offer the ability to book ambulances using our application in a manner similar to how we book cabs. It will be a crucial application for us to use in order to speed up patient delivery. One module will be for the user/patient of our programme, and the other will be for the

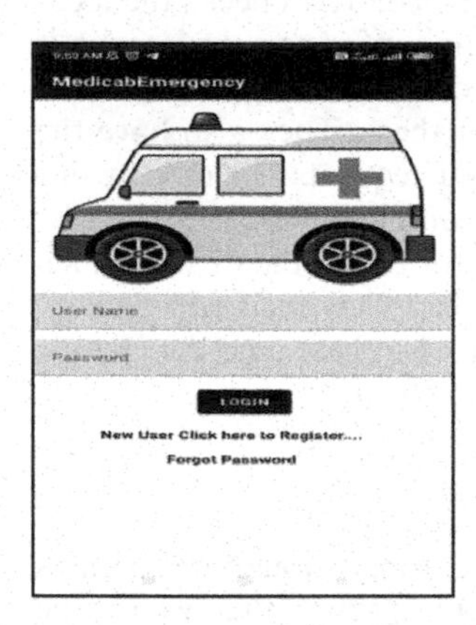

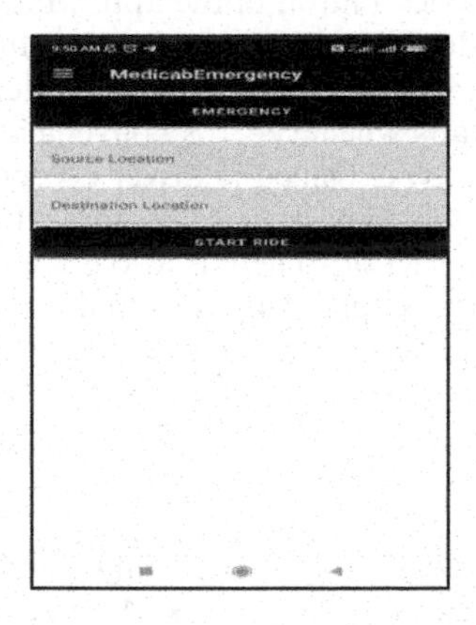

Figure 2. Output.

ambulance driver. Data will be stored securely and systematically as part of our project, making it simple to keep track of users and drivers. With the help of our project, we can precisely and easily find both users and ambulances, which will cut down on the time spent calling one another.

5 CONCLUSION

Medicab will be an advancement int the field of automobile industry as well as medical field. To sum-up, we have created this project to handle and control emergency health situations and to save a life by bringing the patient to hospital in fastest, shortest and safest route. Since it saves time, it is crucial for those who are suffering in a crisis.

Our project has been developed, and if it accomplishes as anticipated, it will be a very successful application that will be helpful in our daily lives. With respect to the smart urban initiative, we will also be able to advance the health sector. In this essay, a concept is devised for quickly and effectively saving a patient's life. Since it saves time, it is crucial for users in times of need. With the use of this application, the ambulance can find the patient because the place is tracked or provided by the app, and it can also deliver the essential kit for the patient's well-being.

Future research will focus on several areas, including applying the findings to the numerous real-world scenarios in which the digital ambulance system will assist in saving lives of people, as well as connecting the knowledge gathered about the digital ambulance with knowledge about ambulances and hospitals nearby. Our project concludes that working digitally saves not only time but also the lives of people.

REFERENCES

Amad A Mohammed & Behrouz H Far & Christopher Naugler. 2016. Efficient Routing of 108 Ambulances Using Clustering Techniques. *The Big Data Mining Journal.*

Arunmozhi, P. & Joseph William, P. 2014. Automatic Ambulance Rescue System Using Shortest Path Finding Algorithm. *International Journal of Science and Research (IJSR)*, Vol. 3 Issue 5, pp. 635–638.

Farzana, S. & Ramyadevi, R. 2015. Location Based Query System by Securing Private. *ARPN Journal of Engineering and Applied Sciences*, Vol. 10, No. 7, pp. 1819–6608.

Iyappan, P. & Nanthini Devi, B. & Nivedha, P. & Sayoojya, V. 2019. Lisa-life Saver, *IEEE International Conference on System, Computation, Automation and Networking (ICSCAN)*, Pondicherry, India, pp. 1–6.

Poonam Gupta & Satyasheel Pol & Dharmanath Rahateka & Avanti Patil. 2016. Smart Ambulance System. *International Journal of Computer Applications.*

Rajeshwari Sundhar & Santhosh Hebbar & Varaprasad Golla. 2015. Implementing Intelligent Traffic Control System for Congestion Control, Ambulance Clearance and Stolen Vehicle Detection. *IEEE Sensors Journal.*

Rashmi A. Nimbalkar & Fadnavis, R.A. 2014. *Domain Specific Search of Nearest Hospital and Healthcare Management System.* IEEE, 978-4799-2291-8.

Ryo Katsuma & Shin Yoshida. 2018. Dynamic Routing for Emergency Vehicle by Collecting Real-Time Road Conditions. *International journal of Computer Science and Communications (IJCNS)*, vol. 11, No. 2.

Sathish Kumar, R. & Praveen Kumar, M. & Balaji, S. & Anitha Julian. 2017. Advanced Parking System using Internet of Things. *International Journal of Contemporary Research in Computer Science and Technology.*

Shivam kumar Kashyap & Sainandini Mishra & Nagaraj M. Lutimath. 2020. Jeevan Jyoti Mobile Application for Ambulance Service. *IJSRSET* 2394–4099, vol. 7.

Yamuna, S. & Vijayalakshmi, R. & Jeeva Mani, K & Boopathi, D & Ranjith Kumar. Progressive Study Users Perception and Satisfaction towards Online Cab Service with Reference to Coimbatore. *International Journal for Research in Applied Science and Engineering Technology(IJRASET)* ISSN: Issue IV. 2019.

Yuanyuan Du & Yu Chen & Dan Wang & Jinzhao Liu & Yongqiang Lu. 2011. An Android-based Emergency Alarm and Healthcare Management System. *IEEE International Symposiumon IT in Medicine and Education*, Cuangzhou. pp. 375–379.

Artificial Intelligence, Blockchain, Computing and Security – Dagur et al. (Eds)

Smart fish monitoring system using IoT

T. Divya, R. Ramyadevi & V. Vijaya Chamundeeswari
Department of Computer Science and Engineering, Saveetha Engineering College, Chennai, India

ABSTRACT: Smart fish monitoring system using IoT used to maintain a fresh water in the aquarium for aqua life environment. It features to monitor the freshness of the water for healthier aqua life environment. This proposed system works as a fish feeding system which can be controlled by a portable mobile phone in its operation even when not in close vicinity. This system works based on IoT and makes sure that the observation of the aquarium can be done via an application on the mobile phone or via the dashboard provided by Arduino IoT Cloud in the laptop. System uses various calibrated sensors to monitor the water level, temperature readings of the tank or small ponds. Apart from that air pump, water heater and fish feeding process can be turned on or off based on the input from the user from the application. For this proposed design distance, is never an issue. This helps you to feed your fishes with just the press of a button on your phone. It also gives you access to turn off and on the air pump in the aquarium as per your wish. Apart from this, the temperature conditions in your fish tank are constantly monitored and if the temperature gets anywhere below the certain range, it ensures that the water heater is automatically turned on and regulated as needed. Apart from this it has a water level monitor which helps notify you when the water level has reduced below a certain point.

Keywords: Aquaculture, Internet of Things (IoT), Arduino IoT Cloud, Sensors, Temperature, water pollution, fish health, fish farm.

1 INTRODUCTION

An aquarium (aquariums or aquaria) is a form of vivarium having a minimum of one transparent side that houses and displays aquatic species. Fish holders keep aquaria for fish, invertebrates, amphibians, aquatic reptiles such as turtles, and aquatic plants. The aquarium theory was completely developed by scientist Robert Warrington, who demonstrated that adding plants Adding oxygen to water in a container would suffice for animals to survive if the number of species did not grow too big. Hobbyists will keep little aquariums in their homes. Many cities have big open-air aquariums. In public aquariums, fish and other aquatic creatures are kept in large pools. A large aquarium may contain otters, turtles, dolphins, sharks, and whales.

J.-H. Chen *et al.* stated that In terms of statistics, marine farming feed accounts for 60% of total expenditures, comprising 8.26% of feed loss, leading to inefficient wastage. This proposed work helps you to feed your fishes with just the press of a button on your phone. It also gives you access to turn off and on the air pump in the aquarium as you want. Apart from this, the temperature conditions in your fish tank are constantly monitored and if the temperature gets anywhere below the range of 76 degrees Fahrenheit, it makes sure that the water heater is automatically turned on and regulated as needed. In addition to this it has a water level monitor which helps notify you when the water level has reduced below a specified point.

DOI: 10.1201/9781032684994-42

2 LITERATURE SURVEY

Monitoring of aquarium/aquaculture system uses different methodologies and mechanisms. Multiple technologies had been created to assist fishers and others in determining the quality of water elements and monitoring the welfare of fish and other farming processes. (Michael *et al.* 2019) demonstrated an IoT-based cloud-based monitoring system for farming, along with sensors to determine the water quality factors for farmers to get the data of their fish-ponds in real time so they can act against potential risks in their making. (Navid *et al.* 2020) proposed dissolved oxygen monitoring system which supervising the level of dissolved oxygen and it has the potential to reduce labour by automation, allow for immediate detection of changing water circumstances, and improve the quality of life for aquatic life. Yi-Bing Lin *et al.* presented the fishtalk structure, which allows tank sensors to control motors in instantaneous fashion. Mutiu A. Adegboy *et al.* suggested a fish behavioural vibration analysis and artificial neural network-based intelligent fish feeding regime system. Yuhwan Kim *et al.* presented an ambient water energy administration system for use in a fish farming system based on the Internet of Things (IoT) and using protocol Massage Queue Telemetry Transport. Akbar (Riansyah *et al.* 2020). proposed a supervising Aquaponics systems and nutritional control that are built and installed may handle data storage, real-time monitoring, and controlled fish feeding. (Kajal Jadhav *et al.* 2020) suggested employing the Blynk application to create an automated fish feeder. Yasser Asrul Ahmad *et al.* created a method to compare the efficiency of an IoT humidity structure to that of a traditional system by first developing an IoT temperature monitoring system that emulates an IoT system that allows users to monitor real-time data via the internet using the DHT22 sensor, a common IoT sensor that can interface with an Arduino board. Sajal Saha *et al.* proposed work plans and carries out a distinctive aquaculture monitoring system based on IoT. (Wen *et al.* 2022) designed a system where it collects all real time environment factors and displayed via ZigBee. (Pupug *et al.* 2020) works aims an intelligent aqua farming system to feed and monitor water temperature based on the IoT of Koi.

3 SYSTEM ARCHITECTURE

In Figure 1, The Architecture of the system diagram is proposed along with the modules which were mentioned prior. This is Smart fish monitoring system using IoT that acts as a life savior of marine species in various conditions, it connects through IoT for tracking and

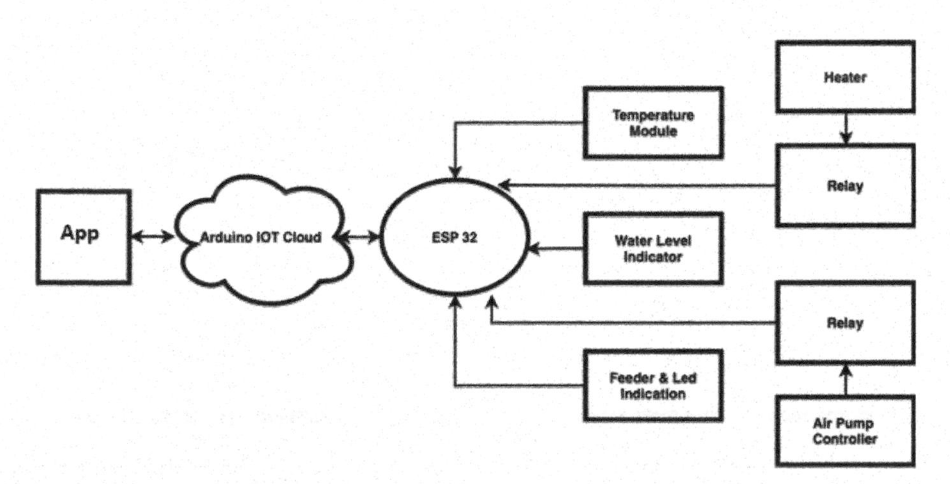

Figure 1. Architecture diagram.

transferring the data of water purity level, ensuring maintenance of aquatic temperature using a heater, and a regular feeding mechanism, it comes with an application that connects users directly and regulates the measures to be taken.

For this proposed design distance, is never an issue. This helps you to feed your fishes with just the press of a button on your phone. It also gives you access to turn off and on the air pump in the aquarium as per your wish. It monitors the water conditions for the survival of these species. It monitors the purity of water on a regular basis and sends the live report directly to the user concerned. It helps feed the fishes and observes and maintains the temperature of the water when it turns cold. It also helps track the water level. It is a lifesaver for those who have aquariums but do not have enough time to maintain and track them.

- Temperature conditions
- Water level and purity
- Feeder and led indication
- IoT based controller

Our main purpose is to help users have an IoT-based system that helps the user in every way possible and indicates to the user if there are any changes in the conditions of the aquarium.

4 IMPLEMENTATION

At first, all the hardware components are connected to the microcontroller board they are connected, configured, and are evaluated individually, the microcontroller is connected to Arduino IoT cloud which sends the signals for working of the components. Arduino IoT cloud has been configured separately along with user interface application which is designed using Arduino IoT cloud remote. After connecting hardware components with Arduino cloud, the application starts to monitor and control the hardware components. Figures 2 and 3 shows IoT Application user interface for the proposed work. Whenever user wants to switch on or off the air pump, water heater, fish feeder in tank or small ponds, they can just turn the button on/off in application. Temperature and Humidity are being monitored until the system kept on.

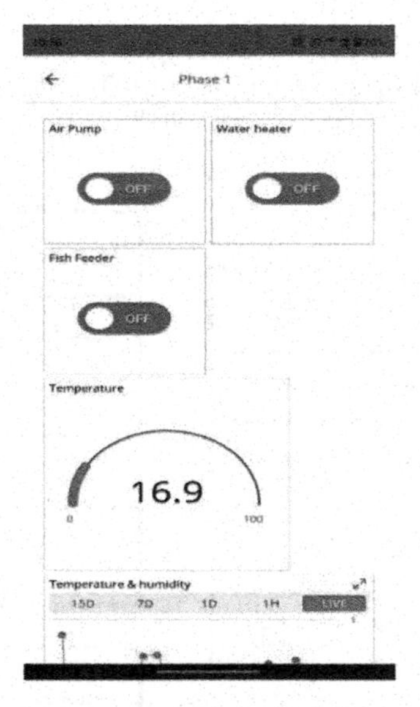

Figure 2. IoT application interface (Air pump, water heater, fish feeder control)

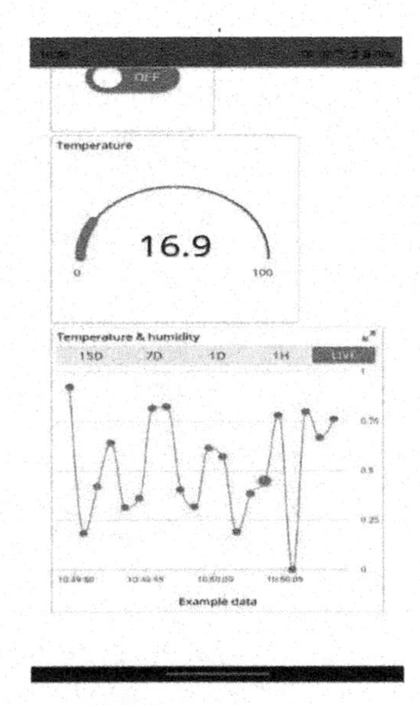

Figure 3 IoT application interface (Temperature & humidity).

The temperature is monitored to maintain proper fish health, the temperature sensor collects the data and sends it to cloud where it is then conveyed to the user, the design can maintain temperature of the aquarium directly from the application interface. The temperature sensor records the temperature values and sends the values directly to the IoT cloud which are displayed in the IoT application. System retains a track of water level and alerts the user when it is below the desired range, the water level of an aquarium is to be always maintained, water level especially important for an aquarium. The water level is distinct from the one aquarium to the other, that can be configured in the application. The water heater controller is used to control the water heater, this can be directly controlled from the application interface which is an IoT based application.

Arduino IoT Code and ESP32 can control the air pump in the aquarium directly, this helps the user for better experience. System connects to a servo motor to feed the fish at a specific time, or it can also be done using the mobile application interface. The LED indicator alerts the user for all the important key precautions.

5 CONCLUSION

With system we solve the problem of look after our aquarium even when we are not in proximity. The use of Wi-Fi Module in our microprocessor helps ensure the same. We do not plan to limit ourselves to this solution. With the proposed system, Distance is never an issue as it is sharing the data and controlling the device via cloud.

Apart from this, we have planned to enhance the same by ensuring that automatic network connectivity takes place. That is just in case there are multiple networks available then although one does not work properly, the others should connect automatically. This would bring a revolution not only in system. but also, in Internet of Things devices.

REFERENCES

Adegboye M. A., Aibinu A. M., Kolo J. G., Aliyu I., Folorunso T. A. and Lee S. -H., "Incorporating Intelligence in Fish Feeding System for Dispensing Feed Based on Fish Feeding Intensity," in *IEEE Access*, vol. 8.

Ahmad Y. A., Surya Gunawan T., Mansor H., Hamida B. A., Fikri Hishamudin A. and Arifin F., "On the Evaluation of DHT22 Temperature Sensor for IoT Application," 2021 *8th ICCCE, Kuala Lumpur, Malaysia,* 2021.

Chen J. -H., Sung W. -T. and Lin G. -Y., "Automated Monitoring System for the Fish Farm Aquaculture Environment," 2015 IEEE International, Hong Kong, China.

Chen J. -H., Sung W. -T. and Wang H. -C., "Remote Fish Aquaculture Monitoring System Based on Wireless Transmission Technology," *2014 International Conference on Information Science, Electronics and Electrical Engineering,* Sapporo, Japan, 2014.

Cordova-Rozas M., Aucapuri-Lecarnaque J. and Shiguihara-Juárez P., "A Cloud Monitoring System for Aquaculture Using IoT," 2019 IEEE SHIRCON, Lima, Peru, 2019..

Effendi M. R., Riansyah A., Mardiati R. and Ismail N., "Fish Feeding Automation and Aquaponics Monitoring System Base on IoT," in *ICWT,* Yogyakarta, Indonesia.

Hasan Rajib R., Saha S. and Kabir S., "IoT Based Automated Fish Farm Aquaculture Monitoring System," 2018 *ICISET,* Chittagong, Bangladesh, 2018.

Jadhav K., Vaidya G., Mali A., Bankar V., Mhetre M.and Gaikwad J., "IOT Based Automated Fish Feeder," 2020 *I4Tech,* Pune, India, 2020.

Lee N., Kim B., Kim Y. and Shin K., "Realization of IoT Based Fish Farm Control Using Mobile App," 2018 IS3C, Taichung, Taiwan, 2018.

Muhlas, M. R. Effendi, P. Ginanjar, S. Opipah, D. Rusmana and E. A. Z. Hamidi, "Prototype Smart Fish Farm in Koi Fish Farming," 2021 *7th International Conference on Wireless and Telematics (ICWT),* Bandung, Indonesia, 2021.

Nguyen T., Shaghaghi N., Patel J., Soriano A. and Mayer J., *"DOxy: Dissolved Oxygen Monitoring,"* 2020 IEEE GHTC, Seattle, WA, USA, 2020.

Tseng H. -C. and Lin Y. -B., "FishTalk: An IoT-Based Mini Aquarium System," in *IEEE Access,* vol. 7.

Deep learning, AI and IOT for computer vision and blockchain

Artificial Intelligence, Blockchain, Computing and Security – Dagur et al. (Eds)
© 2024 The Author(s), ISBN: 978-1-032-67841-2

An analysis of the development of pictorial storytelling

V. Kowsalya
Centre for Information Technology and Engineering, Manonmaniam Sundaranar University, Tirunelveli, India

A. Shunmuga Sundari
PG and Research Department of Computer Science, Rani Anna Government College for Women, Affiliated to Manonmaniam Sundaranar University, Tirunelveli, India

C. Divya
Centre for Information Technology and Engineering, Manonmaniam Sundaranar University, Tirunelveli, India

ABSTRACT: Visual storytelling involves interacting with audiences using graphics, images, sketches, and videos to arouse strong feelings, spark conversation, and motivate action. Students will investigate the role of digital and graphic media in storytelling and learn how to use a variety of tools in virtual and graphic storytelling mechanisms. It is unique and difficult work. Artificial Intelligence (AI) development to assist writers in generating unique types of interactive, dynamic storytelling, with its precise goals being to grasp the effects of artificially intelligent memories on storylines and apply the outcomes of the study to the corporate world. Although prior pictorial storytelling systems used text-generating approaches rather than proper consideration of sentence-level linkages, most of the narratives produced lacked coherence. As children can develop crucial reading skills for the 21st century and also obtain a better knowledge of every topic, they are empowered by digital storytelling to be confident speakers and media makers. A digital narrative can incorporate any phonetic story arrangement, several graphic representations, sometimes with music, and modern technology for editing and sharing the storytelling. An in-depth analysis of earlier methods is discussed in this paper. It provides significant motivation for upcoming researchers in the field to put forth their best efforts and achieve the best results from the stories.

Keywords: Pictorial Storytelling, Artificial Intelligence, Text generation, Image representation.

1 INTRODUCTION AND MOTIVATION

Complex stories may be made simpler to grasp via the use of visual storytelling, which helps to convey a point more effectively. It makes tales as vivid, topical, and meaningful to viewers as humanly possible. The act of telling stories appears to have the capacity to fundamentally alter how people connect with cultural identity. It is widely acknowledged that this is an important strategy for attracting and retaining visitors to exhibits and other locations with rich cultural histories [3]. The online Narrative Screenplay Designer and the Narrative Smartphone Player application were both developed with the goal of stimulating application-specific studies and evaluating quality metrics. A virtual story uses mobile devices to portray cultural heritage. Facebook offers advantages in terms of interaction over blogs, including the ability to connect stories with individuals who know the protagonist. This specific audience can then shape the tiny narratives and connect them to the larger narrative, which is most visible on the blog. This

DOI: 10.1201/9781032684994-43

helps to make the story more narrator-friendly. Examines [4] how simple narrative updates on Facebook are used to share a continuous life event (the renovation of a new home). Pupils that employ digital storytelling approaches improve their oral communication abilities and this method may be regarded as a key component of second language learning and instruction. 60 students participated in an inquiry about the application of digital storytelling in such an Iranian undergrad EFL class to determine whether employing tech-based tools has an impact on how well students can tell stories. Digital storytelling [1] aids students in enhancing their pre-existing composing, studying, attending, and speaking skills by teaching them to express their feelings and make virtual stories for a similar audience. Additionally, it transforms education into something innovative, interesting, and attractive. Obtaining embedding's for phrases and words requires the transformer-based BERT. In order to learn the relationships between the sentences that correlate to the pictures, we can use a hierarchy-based LSM model. The bottom LSTM takes the word embedding's from BERT as input, and the top LSTM also uses that input to create the matching word embedding. Drawing is a fundamental human ability that has long been used as a visual complement to textual and oral narrative. A storytelling tool that can dramatize interactive storylines in virtual reality on top of a regular piece of paper Users may freely engage with virtual characters using the technology by drawing things on paper [2]. Digital storytelling, like traditional narrative, has a distinct theme, characters, setting, storyline, and point of view. Although adding graphics, audio narration, short videos, and/or songs to the standard text creates a more interactive tale that enthrals students and improves both their originality and English language proficiency, it is suggested and examined from several angles to provide an integrated approach to heritage recording, management, and distribution. The project involves Thessaloniki, Greece's Virtual City, a sizable and not endless city that is culturally rich but has inadequate procedures for managing its heritage.

This study examines the evolution of visual stories and analyses the benefits, limitations, methodologies, and updates of how they were applied in previous implementations (existing studies). It would be useful for upcoming researchers to generate improved outcomes and greater visual storytelling like movies, video games, advertising, etc.

2 STORYTELLING VIA VISUAL REPRESENTATION

A deep learning-based approach for predicting floor levels in multi-story structures StoryTeller [9] uses omnipresent WiFi signals to produce visuals that are fed into a convolutional neural network (CNN) that has been trained to anticipate floors using patterns discovered in visual WiFi scans. Intake pictures are prepared in such a way that they record the present WiFi scan during AP-agnostic mode. Furthermore, a novel virtual building concept is used to normalize the data and make it constructible. This contribution shows the outcome of diverse experiences of knowledge transfer on the historical value of industries via the use of gaming and narrative techniques like educational resources, in addition to the integration of several platforms (3D architecture, video mapping, and augmented reality) as valuable instruments for spreading the explanations of this issue. As users interact using digital storytelling, these understandable surrogates give a direct route of travel to the narrative world, bridging the distance between cyberspace and our real world. The system can accommodate voice, video, still images, and text-based stories [10–12].

Instructors and students may use visual narratives at their discretion. Some software and materials used for pictorial storytelling were illustrated in Figure 1. From the birth of writing and the printing press until the modern era, technologies and society have continuously supplied fresh and more effective methods for telling stories [8]. The computer has recently brought together science, entertainment, and artwork. People love the simplicity and cheerfulness of a mini-exhibit for sharing stories and the imagery associated with all those experiences. Camelendar is a touch-based application that assists people with aphasia in expressive storytelling. Camelendar employs photographs to assist people with aphasia in

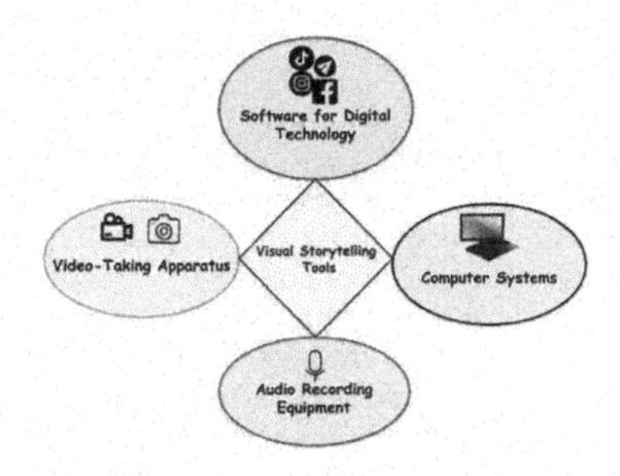

Figure 1. Displays the visual storytelling toolkits.

communicating about their daily activities; participants in the discussion (along with the aphasic participant) add comments as extra data to the images, improving the story the next time it is told. Persons with aphasia can browse images of topics they would like to discuss by rearranging images graphically on a calendar. The images of the events are transformed into tales by immediately putting tags and remarks on the pictures.

3 BENEFITS AND DRAWBACKS OF VISUAL STORY DEVELOPMENT

Complex stories may be made simpler to grasp via the use of visual storytelling, which helps convey a point more effectively. It makes stories as inventive, modern, and relevant to readers as possible.

The benefits of visual storytelling include:

- It's a good idea to use stories to attract followers.
- You might be able to raise your search engine rankings this way.
- Visual storytelling enables you to arouse the listener's emotions.
- It offers you the opportunity to go viral.
- You establish yourself as an authority on the subject.
- Audiences that are right for your narrative will find you online.
- It enables you to include others.
- Visual storytelling enables you to target the right group of people.

The confines of visual storytelling

- Developing an audience for visual literacy requires time.
- The price of pictorial storytelling should be considered as well.
- Various people may consider the narrative in various manners.
- Attributing outcomes may be difficult.

The visual story creation system provides a variety of advantages and limitations during the implementation and training phases, which are stated in Table 1.

Table 1. The benefits and constraints of creating visual stories.

Ref. No	Author name	Implemention	Benefits	Limitations
[2]	Jing Su [2021]	Visual Storytelling Using BERT-hLSTM	The BERT-hLSTMs model's higher performance on the VIST database compared to various benchmarks demonstrates the value of using BERT embedding and the hLSTMs for phrases and words.	The challenges were perceiving the connections between sentences and generating clear descriptions.

(continued)

Table 1. Continued

Ref. No	Author name	Implemention	Benefits	Limitations
[5]	Clara Bassano [2019]	Tourism management in the digital world: telling the story about places	A place-storytelling approach that city governments and institutions may employ to promote the stakeholder participation in a multilevel approach for enhancing regional service businesses and marketing in the modern age.	The lack of a unified media strategy as a result of a lack of materials had an impact on the sector.
[6]	Deoksoon Kim [2021]	Digital storytelling: improved system and identity formation	Digital storytelling projects in the classroom may inspire students to employ a range of expressive tools while also strengthening their ambition, inventiveness, sense of identity, and interpersonal interactions.	The images of challenging arithmetic equations, punctuation marks, and an image of flashcards
[7]	Serene Lin-Stephens [2022]	Impact of serious storytelling using pictures on performance anxiety during interviews	Health care and education, for example, are well suited to serious narratives with visuals as a multimedia genre.	Lack of differentiation between a person's work capability and their capacity to get employment.

4 EFFICIENT METHODOLOGIES FOR CREATING PICTORIAL STORYTELLING

The practice of fusing the age-old art of storytelling with digital technology is known as "digital storytelling" (DST). Multimedia narratives and demonstrations may also be produced using easily available resources and techniques. Other possible variations of this web-based approach to learning include digital films, computer-based storytelling, virtual writings, and virtual narratives. However, the art of narrative is combined with a wide range of multimedia, such as different graphic designs, music, videos, and web publications, acquired through prior study. Table 2 displays the technique and how it utilizes data from prior results. Therefore, using the methodologies will be beneficial for future research.

Table 2. Describes the methodology and how it applies previous findings' data.

Ref. No	Author Name	Applied Methods	Features of a Technique
[2]	Jing Su [2021]	BERT-hLSTMs and baseline model	BERT-hLSTMs are a technique for combining word-level and sentence-level semantic features.
[5]	Clara Bassano [2019]	SSME + DAPP	Due to its clarity on how it might be handled in a local service process to increase brand competition, this integrated approach boosts the strength of the narrative.
[6]	Deoksoon Kim [2021]	Constant comparative method	After making comparisons between the two individuals, an iterative and comprehensive analysis of the data was performed.
[7]	Serene Lin-Stephens [2022]	Visual Methods and Interview Methods	The efficiency of many serious narrative apps can be evaluated using visual representations. New digital and virtual interview techniques offer more chances to include meaningful narrative into conversations.

5 CONCLUSION

The visual storyteller may create more accurate word embeddings by employing the context vector, like its initial state, and the picture embedding as its input, according to the separate decoders at each point of the image frame. Developing causal structure, which involves making sense of photographic data to connect distinct moments that lead to a cohesive story of events and situations, seems beyond grasping straightforward objects and tangible scenes. The above necessitates shifting from analyzing static images to contextless scenes and visual series that show events as both generating and evolving. This article offers a detailed description of the methodology for developing outstanding visual stories. The comparison table of benefits, constraints, methodologies, and functions is quite beneficial for aspiring researchers.

REFERENCES

[1] Razmi, M., Pourali, S., & Nozad, S. (2014). Digital Storytelling in EFL Classroom (Oral Presentation of the Story): A Pathway to Improve Oral Production. *Procedia-Social and Behavioral Sciences*, 98, 1541–1544.

[2] Su, J., Dai, Q., Guerin, F., & Zhou, M. (2021). BERT-hLSTMs: BERT and Hierarchical LSTMs for Visual Storytelling. *Computer Speech & Language*, 67, 101169.

[3] Vrettakis, E., Kourtis, V., Katifori, A., Karvounis, M., Lougiakis, C., & Ioannidis, Y. (2019). Narralive–Creating and Experiencing Mobile Digital Storytelling in Cultural Heritage. *Digital Applications in Archaeology and Cultural Heritage*, 15, e00114.

[4] West, L. E. (2013). Facebook Sharing: A Sociolinguistic Analysis of Computer-mediated Storytelling. *Discourse, Context & Media*, 2(1), 1–13.

[5] Bassano, C., Barile, S., Piciocchi, P., Spohrer, J. C., Iandolo, F., & Fisk, R. (2019). Storytelling About Places: Tourism Marketing in the Digital Age. *Cities*, 87, 10–20.

[6] Kim, D., & Li, M. (2021). Digital Storytelling: Facilitating Learning and Identity Development. *Journal of Computers in Education*, 8(1), 33–61.

[7] Lin-Stephens, S., Manuguerra, M., & Bulbert, M. W. (2022). Seeing is Relieving: Effects of Serious Storytelling with Images on Interview Performance Anxiety. *Multimedia Tools and Applications*, 1–22.

[8] Gershon, N., & Page, W. (2001). What Storytelling Can Do for Information Visualization. *Communications of the ACM*, 44(8), 31–37.

[9] Elbakly, R., & Youssef, M. (2020). The StoryTeller: Scalable Building-and Ap-independent Deep Learning-based Floor Prediction. *Proceedings of the ACM on Interactive, Mobile, Wearable and Ubiquitous Technologies*, 4(1), 1–20.

[10] López, G. A., & Cruz, D. C. (2021). Experiences of Knowledge Transfer on Industrial Heritage Using Games, Storytelling, and New Technologies: "A History of Enterprises". *Journal on Computing and Cultural Heritage (JOCCH)*, 14(2), 1–26.

[11] Mazalek, A., Davenport, G., & Ishii, H. (2002, December). Tangible Viewpoints: A Physical Approach to Multimedia Stories. *In Proceedings of the Tenth ACM International Conference on Multimedia* (pp. 153–160).

[12] Li, J., Shi, H., Tang, S., Wu, F., & Zhuang, Y. (2019, October). Informative Visual Storytelling with Cross-Modal Rules. *In Proceedings of the 27th ACM International Conference on Multimedia* (pp. 2314–2322).

Artificial Intelligence, Blockchain, Computing and Security – Dagur et al. (Eds)
© 2024 The Author(s), ISBN: 978-1-032-67841-2

Analysis of E-learning trends using educational cloud in cloud computing

Neetu Mittal*
Amity University Uttar Pradesh, Noida

Merry Saxena & Pradeepta Kumar Sarangi*
Chitkara University Institute of Engineering and Technology, Chitkara University, Punjab

Nidhi Phutela
Symbiosis Centre for Management Studies, Noida

Manasvi Sharma
Amity University Uttar Pradesh, Noida

ABSTRACT: Education is the most significant support for real future planning. In the present pandemic scenario, e- learning is incorporated and implemented by almost all the universities, schools and industries all over the world to deliver lectures, certification, training, education and even any of the higher degrees. With cloud computing there comes plenty power based on computing, latest trends and innovations. It has a huge impact in educational sector which will help the end users (i.e. learners, educators and admins) to delegate tasks more productively with least amount of costs. Using information technology, the difficulty faced by educational institutions to provide quality study and security of student's personal data can be resolved. Students can access internet from almost every device like tablets, mobiles, desktop, i-pads and laptops which makes it a lot more distributed as compared to any other centralised entity. For security a privacy blockchain may be proven as a most useful new technology and a curving information spread and in secure way on various sectors in education. E-Learning can add an efficiently good impact to the educational institutions with lesser amount of infrastructural cost as compared to an earlier existing system. In the proposed work, the analysis has been done on the pre-existing E-learning techniques using educational cloud. Further, the best efficient architecture has been recommended on which the entire E-learning system should be based and with results as well.

Keywords: ICT (Information and Communication Technology), Educational Cloud, WebCT, cloud computing, Blockchain, E-learning.

1 INTRODUCTION

E-learning has been seen as a great efficient option accepted by the educational institutions. Online/E- learning courses have elevated the opportunities of altering the traditions of learning providing more personalised platforms and utilize time more effectively and in secure manner by using blockchain technologies, that can be utilize for personalized use. The technology uses all the transactions in the system in highly secured manner and the data

*Corresponding Authors: nmittal1@amity.edu and pradeepta.sarangi@chitkara.edu.in

DOI: 10.1201/9781032684994-44

stored cannot be reformed. This security and authenticity allows users to acquire information quickly and with highly secured, privacy and transparent manner. Ubiquitous learning is a major reason which has bought evolution to cloud computing towards the education sector as it allows the students to access the content ignoring the time and location ensuring a lot more seamless adaptive support for them. However, to practise E-learning in the best efficient manner, it should ideally centre around Learning Management System (LMS) and that should be incorporated into a pre-existing e-learning atmosphere of different arenas of education.[1]. The research in education using data mining also plays a significant role and proper LMS is necessary for the effective implementation [2,3]. LMS are robust unified systems that assists various tasks conducted by students, teachers and other users during the e-learning operation [4,5]. The best precedence of e-learning is its availability without installing any kind of personal data at any computer. E-learning using education cloud has a great effect on all the aspirant learners as after its implementation now distance is considered as one of the most minimal factors in the learning process.

As from Figure 1, the architecture in the middle, the most important part educational cloud which comprises of web application (from which the users perform all the activities in order to learn) and database (where the entire data which is useful for the user is stored so they do not need any external storage device in order to store the useful notes, findings, etc), then on the left we can see the admin user which acts as a manager in the E-learning system by dumping the unnecessary data which is not further required and performing all other crucial activities after logging in with his/her managerial id and password.

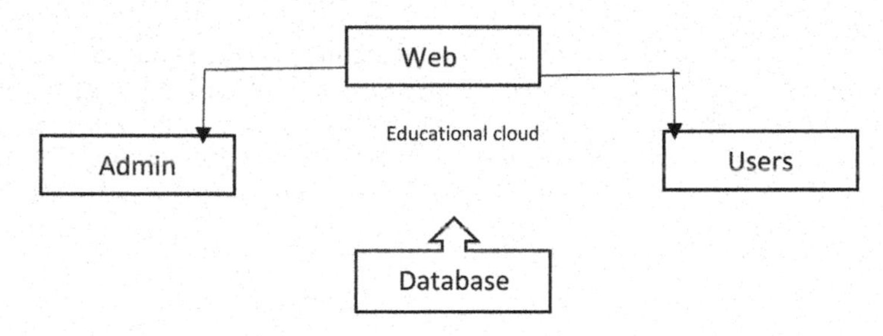

Figure 1. Block diagram of proposed model.

The implementation of this system is so easy that anyone who didn't even actually, knows how to operate a computer can still access the system. Now for the initial phase only with the motive only to understand that how the educational Cloud system works a simple architecture is presented. The newly designed framework equipped with quality standards and the advancement in e-learning with highly advanced technology with AI and blockchain now-a-days.

2 LITERATURE REVIEW

E-Learning is a scalable solution which enables hosting of apps for clients. Cloud provides users with highly skilled IT capacities as internet service introduced earlier by an American researcher for developing flight simulators [6,7].

E-learning is enhancing the area of education by enabling services like- IaaS (Infrastructure as a Service), SaaS (Software as a service) and PaaS (Platform as a service), three things are eased out by business based on e-learning cloud computing model: cloud maintenance, on-demand user accessibility to cloud and building provided by cloud [8] shown in Figure 2. Some of the advantages of implementing educational cloud in E-learning

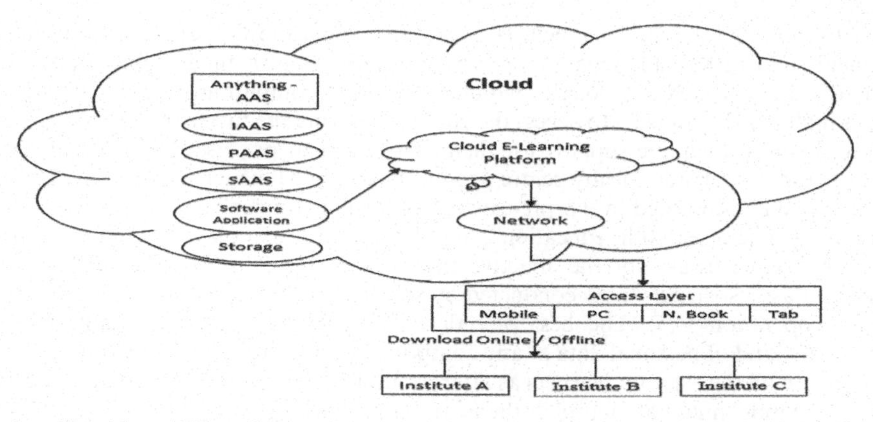

Figure 2. Cloud based E-learning.

are -Virtualization support, On-demand resource provisioning, storage virtualization and high availability and data recovery [9–11]. One of the efficient educational cloud implemented E-learning system was Bluesky Cloud Framework- it provides basic education through cloud platform in china [12] with challenges in augmenting resource allocation, dealing with dynamic concurrency demands, handling rapid storage growth requirement, cost controlling [13].

The most crucial and diversed functionalities by means of which educational cloud-based learning (Figure 3) is actually considered an affective practice are-e-mails, webpages, forums, test tools) [14]. Learning management systems are developed using resources such as monitoring and analysis tools for E-learning program (MATEP) to resolve both teacher's and learner's concern [15,16] MATEP tool run over the web servers with log files where every activity of learner can be captured and help instructor to track learner's activity [17,18].

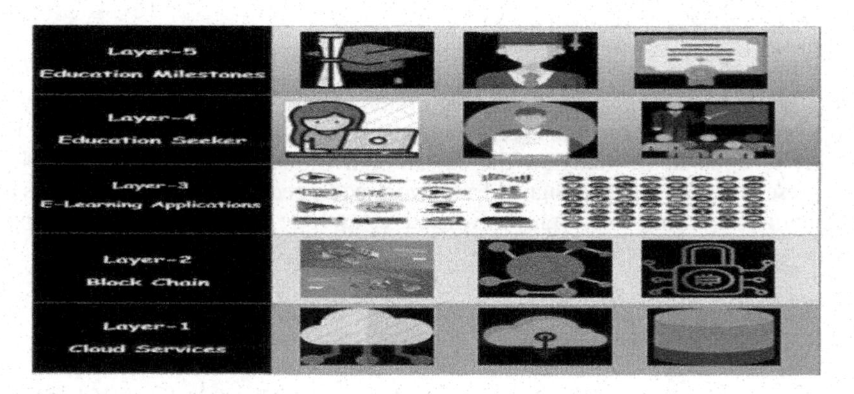

Figure 3. Blockchain based E-learning framework.

Utilization of cloud characteristics of virtualization helps in providing the learning material for all types of end user-students, teachers and researchers [19–22]. Educational cloud architecture contains 5 layers- Infrastructure layer, software layer, resource management layer, service layer and business application layer and virtualization layer (Figure 4)

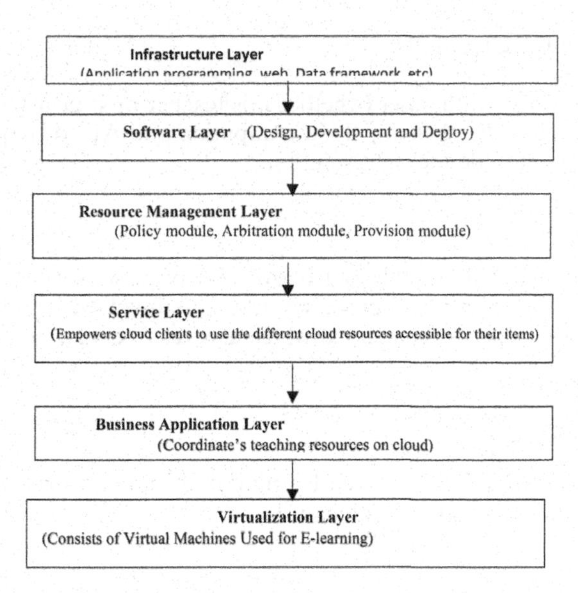

Figure 4. Educational cloud architecture.

2.1 *Infrastructure layer*

This incorporates the application programming, web, data framework, etc. The training data is gathered from the customary strategies and dumped into the cloud environment. In the middleware of cloud administrations, the most minimal layer is the infrastructure layer. This layer gives accessibility to the processing assets.

2.2 *Software layer*

System software layer allow programmers to design, develop and deploy apps as well as ensures that those apps are all time available for the users to use.

2.3 *Resource management layer*

This layer exists for assuring the mandatory interfacing between software and hardware. Depending on planned virtualization strategy, usage on-demand and software free flow amongst all the resources of hardware can be attained.

2.4 *Service layer*

This layer offers similar types of assistance which is offered by the 3 essential help varieties in processing of cloud. It helps in empowering the cloud clients for usage of different resources pertaining to cloud accessible for its items.

2.5 *Business application layer*

This layer is responsible for on-cloud coordination of teaching resources. It is done by division of all interactive process and assets. [23] Updates can be made depending on feedbacks.

2.6 *Virtualization layer*

It mainly consists of all the VMs (virtual machines) which are being used e-learning model. Networks, servers and storage are efficiently aligned using visualization [24] so that the system based on e-learning provides the best utilities.

3 RESULTS AND DISCUSSION

Online education may benefit the user (whether it is teacher or student) with the use of new tech developments such as blockchain, cloud computing and AI, to simplify the education system which direct to provide the right solution.

3.1 *Enhanced performance*

Because of the reason that CSP has full control over the software as well as infra including its maintenance, almost all the apps and processes which are involved in the educational cloud based learning are fully dependent on cloud which as a result enhances the entire e-learning process.

3.2 *Frequent updates*

Student using the educational cloud-based learning need not to worry about any problem related to software updates or any maintenance.

3.3 *Enhanced format compatibility*

There are some specific documents which straight away don't open because of its incompatible format, however this issue does not exist anymore and is resolved through CSP in cloud.

3.4 *Larger data storage*

We need not to worry about the storage capacity as educational cloud-based learning performs computing and data are being stored at large data centres which are evenly spread wider geographically.

4 CONCLUSION

Educational cloud-based learning is one of the most emerging modes of education exposed from internet. In this medium of learning students can easily enhance his/her knowledge as well as skills without any difficulty. E-learning includes different methodologies like websites, audio/video conferencing, websites, computer facilitated instructions, computer-based operation networks as well as feedback systems. This increases the engagement of students. Educational cloud-based E-learning not only enhances the student's way of learning but it also enhances teacher's methods equally as now teacher can implement learning in a broader way and with even more constructive manner if the E-learning system is implemented correctly. This can be achieved by designing a threesome of designing and storing through blockchain which store all material question, bank and assessment and evaluation marks and results safely.

REFERENCES

[1] Despotović-Zrakić, M., Simić, K., Labus, A., Milić, A., & Jovanić, B. Scaffolding Environment for Adaptive E- learning through Cloud Computing. *Educational Technology & Society*, 16 (3), 301–314. 2013.

[2] Ahuja S., Kamal P., A Review on Prediction of Academic Performance of Students at-Risk Using Data Mining Techniques, *Journal on Today's Ideas - Tomorrow's Technologies*, Vol. 5 No. 1 (2017).

[3] CP Thornton, E Rivers, C Rhodes, HK Kang, T Rodney, Development of the Condensed Heuristic Academic Research Model (CHARM) Framework for Short-term Nursing Research Groups, *Nursing Outlook*, Vol-5, Issue-8, 2020.

[4] Hauger, D., & Kock, M. State of the Art of Adaptivity in E-learning Platforms. Hinneburg (Ed.), LWA: Lernen - Wissen – Adaption, Halle, *Workshop Proceedings*. 355–360. 2007.

[5] M. A. H. Masud and X. Huang. An E-learning System Architecture Based on Cloud Computing. *World Academy of Science, Engineering and Technology*. 62. 74–78. 2012.

[6] Auer, Sören. First Public Beta of SlideWiki.org. Open- source Community-based Tools for Learning". *Moodle.org. Retrieved*. 10–24. 2012.

[7] P. Pocatilu, "Cloud Computing Benefits for E-learning Solutions", *Oeconomics of Knowledge*, 2 (1). pp. 9–14. 2010.

[8] Venkataraman, S.and Sivakumar, S. Engaging Students in Group-based Learning Through E-learning Techniques in Higher Education System. *International Journal of Emerging Trends in Science and Technology*, 2(1), 1741–1746. 2015.

[9] Tuncay Ercan, Effective Use of Cloud Computing in Educational Institutions, *Procedia - Social and Behavioral Sciences*, 2 (2). 938–942. 2010.

[10] DeCoufle, B. The Impact of Cloud Computing on Schools. *The Data Center Journal*. http://datacenterjournal.com/content/view/3032/40. 2009.

[11] P. Pocatilu, F. Alecu, & M. Vetrici. Measuring the Efficiency of Cloud Computing for E-learning Systems. *WSEAS Transactions on Computers*. 9 (1). 2010.

[12] Md. A. H. Masud, X. Huang. A Novel Approach for Adopting Cloud-based E-learning System. *2012 IEEE/ACIS 11th International Conference on Computer and Information Science*. DOI 10.1109/ICIS.2012.10.

[13] G. Riahi. E-Learning Systems Based on Cloud Computing: A Review. The 2015 International Conference on Soft Computing and Software Engineering (SCSE 2015). *Procedia Computer Science* 62. 352–359. 2015.

[14] Al-Zoube, M., El-Seoud, S.A., Wyne, M.F.: *Cloud Computing Based E-learning System.and Behavioral Sciences* 2(2). 938–942. 2010.

[15] Andre Luis Dias, Afonso Celso Turcato, Guilherme Serpa Sestito, Dennis Brandao, Rodrigo Nicoletti, A Cloud-Based Condition Monitoring System for Fault Detection in Rotating Machines using PROFINET Process Data, *Computers in Industry*.126, 2021.

[16] Armstrong J Scott. Natural Learning in Higher Education. *Encyclopedia of the Sciences of Learning*. 2012.

[17] N. Mittal and K. Snotra. (2017). Blood Bank Information System Using Android Application. *Recent Developments in Control, Automation & Power Engineering (RDCAPE)*, 269–274, 2017.

[18] Atif Nazir, Saqib Raza, Chen-Nee Chuah. Unveiling Facebook: A Measurement Study of Social Network Based Applications. *IMC'08, Vouliagmeni, Greece*. 43–56. 2008.

[19] Delacey, B., & Leonard, D. Case Study on Technology and Distance in Education at the Harvard Business School. *Educational Technology and Society*. 13–28. 2002.

[20] Dong, B., Zheng, Q., Qiao, M., Shu, J., Yang, J.: BlueSky Cloud Framework: An E-Learning Framework Embracing Cloud Computing. In: Jaatun, M.G., Zhao, G., Rong, C. (eds.) *Cloud Computing. LNCS*. 5931, 577–582. Springer. 2009.

[21] Al-Zoube, M., El-Seoud, S.A., Wyne, M.F.: Cloud Computing-Based E-learning System.Intl. *Arab Journal of e-Technology* 8(2). 58–71.2010.

[22] Hu, Z., Zhang, S.: Blended/Hybrid Course Design in Active Learning Cloud at South Dakota Intl. *Arab Journal of e- Technology*. 1, 63–67. 2010.

[23] Bhagirath, N. Mittal, Kumar. S, Machine Learning Computation on Multiple GPU's using CUDA and Message Passing Interface, *2019 2nd International Conference on Power Energy, Environment and Intelligent Control (PEEIC)*, 18–22, 2019.

[24] Manprit Kaur "*Using Online Forums in Language Learning and Education*". StudentPulse.com. Retrieved 2012-08- 22, 2012.

Artificial Intelligence, Blockchain, Computing and Security – Dagur et al. (Eds)
© 2024 The Author(s), ISBN: 978-1-032-67841-2

The band-notched UWB MIMO antenna with split-ring resonators

A.R. Gayathri
Assistant Professor, Department of ECE, Coimbatore Institute of Engineering and Technology, Coimbatore

B.K. Kavitha Kanagaraj
Professor, Department of ECE, Kumaraguru College of Technology, Coimbatore

ABSTRACT: Multiple-input multiple-output (MIMO) prototypes have been suggested for ultra-wideband (UWB) applications. Planar-monopole (PM) antennas are used to obtain high isolation levels by placing two components perpendicular to one another. Antenna elements with microstrip-fed are printed on the substrate's one side. Ultra-wideband (UWB) is common for short-range communications in dense multipath propagation conditions. Because MIMO systems take advantage of the array gains and spatial multiplexing gains that come with them, they significantly affect spectral efficiency in this sort of situation. This paper proposes and experiments with a new band-notched (BN) UWB-MIMO antenna with good notch-band-edge selectivity. As antenna elements, split-ring resonators (SRR) implanted in the radiation patch and ground are suggested. A high-notch-band-edge selectivity MIMO antenna is achieved in this manner. A UWB system can easily disrupt Wireless Local Area Networks (WLANs) and other narrowband communication systems. When suppressing interference and making the UWB system simpler, the antenna design with integrated frequency notching is better than using a traditional filter. Here, a single BN-UWB-MIMO antenna with a small isolation enhancer is discussed. In addition, the system has a superior rejection of bands with negative gains and a low radiation efficiency at the peak of the frequency range.

Keywords: MIMO, ultra-wideband, split-ring resonators, wireless Local Area Networks

1 INTRODUCTION OF ULTRAWIDEBAND MULTIPLE-INPUT-MULTIPLE-OUTPUT ANTENNA

In automotive communications, UWB MIMO antennas, which combine the advantages of UWB and MIMO technologies, show considerable promise [1]. MIMO and UWB technologies use many antennas to receive and transmit signals simultaneously across a broad frequency range, enhancing communication quality and increasing channel capacity [2]. However, the UWB MIMO antenna suffers from the coupling between parts and overall dimensions [3]. Neutralizing lines or rings, frequency selective surfaces (FSS), and electromagnetic band gaps (EBG) are all examples of decoupling components used in MIMO antennas [4].

Multipath wireless radio communication and MIMO are often cited and employed to develop many new technologies [5]. A wireless communication system can be improved using several antennas at the transmitter, receiver, or both [6]. New MIMO wireless technology is being used by a wide range of radio, wireless, and RF technologies, including Vo-LTE, LTE (Long Term Evolution), Wi-Max, and Wi-Fi, to increase link capacity and spectral efficiency, improving the improvement of bandwidth link dependability [7]. Several issues need to be addressed before using UWB technology [8]. UWB substantially impacts antenna performance [9]. The radio wire configuration industry has seen a surge of interest to accommodate UWB frameworks, which need receiving wires with a transfer speed covering the whole UWB range and capable of taking signals at nearby frequencies [10]. Therefore, the conduct and execution of radio wires over the UWB must be dependable and predictable [11].

WiMAX, WLAN, and the X band are just a few of the frequencies that can cause havoc with Ultra-Wideband [12]. UWB antennas equipped with band-notched antennas are utilized [13]. Multipath fading is another additional issue to contend with. For wireless communication, the

 DOI: 10.1201/9781032684994-45

area of multipath fading has emerged as a hotspot [14]. A half-cutting approach is used to shrink an antenna without losing its frequency strength when the antenna is too large [15]. Inserting different structures like resonators, CSRR structures, and so on improves the performance of many band-notched functions [16]. Radiators were separated from one other by a T-shaped strip [17]. Wireless communication systems are primarily concerned with boosting data transfer rates, improving service quality, expanding channel capacity, reducing power saving, and minimizing interference with other systems [18]. Good data rates, cheap costs, high security, and low power consumption are just a few of the benefits of Ultra Wideband technology, making it an attractive option for wireless communication systems [19]. UWB systems must deal with multipath fading and other types of communication interference [20].

Band-notched ultrawideband multiple-input-multiple-output (BN-UWB-MIMO) antenna proposes a notch ultra-wideband antenna that is both small and lightweight. Loading two SRRs near the feed line in this manner causes them to be stimulated at the frequency of 6.25 gigahertz (GHz). Additionally, a chamfered circular patch antenna is used to etch an elliptical split-ring resonator to generate a frequency notch at 4.5 GHz. A partial ground plane carves a vertical slot in a circular patch antenna to enhance impedance matching.

The primary contribution of this study is the suggestion to employ a UWB MIMO antenna with a band-notched design for dual-band usage. Both-band UWB wireless MIMO antenna with notched features is in development. Using multiple transmitters and receivers simultaneously, MIMO is a wireless technology that can send more data simultaneously. One of the key benefits of MIMO systems is that they need less power and bandwidth to operate. The MIMO system uses spatial multiplexing and diversity to give a higher data rate and greater dependability. Structures common to metamaterials include split-ring resonators (SRRs). Magnetic susceptibility (magnetic response) in different forms of metamaterials up to 200 terahertz is the goal of these devices.

The rest of this paper can be structured in the same way. An ultrawideband MIMO antenna is discussed in Section 2. The suggested research, summarised in Section 3, is the basis for this paper. Section 4 describes the simulation's results and the subsequent discussion. Finally, section 5 concludes this study by fully examining the observation and its consequences.

2 RELATED WORK

Differential evolution (DE) and naked mole-rat (NMR) method hybrids were proposed in this research to handle a broad range of structural optimization problems throughout the globe [21]. Electromagnetics created a UWB monopole antenna with triple-band rejection with HDN (Hybrid Performance of DE and NMR). The HDN's output electrical performance shows that it could cope with modern wireless communication and sophisticated technical challenges To prove its capacity to discuss difficult antenna issues. An enhanced antenna design technique based on a Genetic Algorithm (GA) was proposed in this study. UWB applications could benefit from this technology since it helps to reduce electromagnetic interference [22]. A suitable Defected Ground Structure (DGS) filter was optimized for an antenna in the UWB band range to avoid transmitting in the WiMAX spectrum. The suggested non-conventional filter consists of a matrix of rectangular cell shapes, each assigned based on the surrounding environment's presence or lack of metal.

This study describes a UWB MIMO antenna with an asymmetric coplanar strip (ACS) feed with four band-notched MIMO input/output ports. Its emission pattern was almost omnidirectional, with an average realized gain throughout the entire frequency range of the proposed UWB antenna. The suggested ACS-fed antenna's impedance and radiation characteristics were confirmed through manufacture and testing. Finally, a high agreement between simulation and experimental data confirmed the MIMO antenna's validity for real-world UWB wireless systems. A planar electromagnetic bandgap (EBG)-based UWB monopole antenna with three band-notches was proposed. Notch frequencies were shown by the EBG unit cell, which would be constructed of an Archimedean spiral and inter-digital capacitance. Additionally, the EBG diameters could be varied to tune various notched bands. The coupling effect was shown by putting the EBG unit cells near the feed line before parametric analysis. EBG's influence on input impedance and surface current distribution was examined.

An orthogonally polarised, four-element, multi-input, multi-output (MIMO) antenna with increased isolation was presented. Low pass filter (LPF) structures with broad rejection bands and low cutoff frequencies were recommended to increase isolation over most of the system's bandwidth. Decoupling was investigated by using an equivalence circuit model. According to the Envelope correlation coefficient evaluations, Diversity Gain, Channel Capacity Loss, and Total Active Reflection Coefficient (TARC), the proposed antenna can benefit short-range communications. Band-notched ultrawideband multiple-input-multiple-output (BN-UWB-MIMO) antenna has been proposed to overcome the existing techniques: HDN, GA, ACS, EBG, and LPF have recommended improving the simulation of MIMO antenna, ionization patterns, group delay time, current distribution, and WLAN band notching optimization.

3 PROPOSED METHOD: BN-UWB-MIMO ANTENNA

Ultra-wideband and broadband systems have garnered much interest in various communications applications in recent years. Notching is a must to keep narrowband systems like WLAN, WiMAX, and Wi-Fi from interfering with one other. For UWB antennas, many notch band approaches have been suggested by researchers. The channel is essential to transfer the power, and no additional power or frequency spectrum must be used to increase channel capacity. Using MIMO technology and deploying antennas with low coupling is possible. When building MIMO antennas with low coupling, antenna engineers experience a hurdle. Mobile devices can use a variety of MIMO antennas for WLAN, Wi-Fi, UMTS, and LTE. UWB MIMO systems, according to research, have a higher channel capacity than narrowband systems. Antennas for MIMO UWB are being explored to minimize coupling using various methods.

Figure 1 shows the iteration of MIMO antennas. In the MIMO antenna design, improving isolation becomes a difficult task. Antenna 1 architecture gets more complicated due to the increased size of the structure in terms of various decoupling solutions devised to lessen the mutual coupling. One option for reducing the maximum mutual interaction between the antenna parts is diversity antennas. Filters are needed to reject undesirable bands to avoid interfering with current sophisticated wireless communication networks. For example, stubs and split rings can create notching in the radiating structure to produce this effect. Massive MIMO antennas, 5G, and conformal antennas have all had a major influence on technology, particularly in wireless communication. A distinct radio/antenna chain uses the same frequency channel as the transmitter to broadcast each spatial stream in MIMO. Each receiver's identical radio/antenna chains receive up a different stream. The receiver can reconstruct the original streams since it knows the phase offsets of its antennas 2.

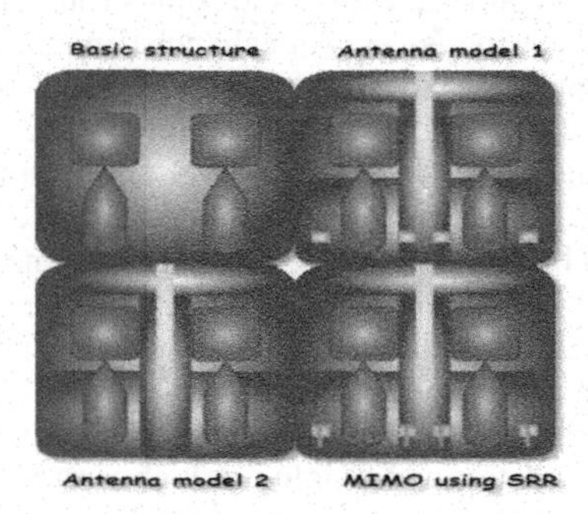

Figure 1. Iterations of MIMO antennas.

Synthetic metamaterial split-ring resonators (SRR) are a kind of SRR. A magnetic flux penetrates the rings and generates its flux, enhancing or opposing the incident field, inducing spinning currents in the metal rings.

3.1 *Design of an antenna*

To achieve impedance h_tThe feed line K_h is modeled as a pencil point Qk and square radiating components are placed on an FR4 substrate Qh is given as,

$$h_t = \frac{144}{K_h + Qk + Qh} + \frac{1}{\frac{B1}{2\pi K_h \sqrt{\mu_r ef}}} + \frac{1}{\frac{B2}{2\pi Qk \sqrt{\mu_r ef}}} \tag{1}$$

As presented in equation (1), the distance between the ground plane $\mu_r$and the patch is measured in millimeters. $B1$ is the ground plane area, while $B2$ is the patch area. It is the effective dielectric constant. The L-shaped stub has a lengthy slit ef on the construction to better isolate the ports.

Figure 2 shows the design of a MIMO antenna. A normal encoder is used to encrypt the data, which is interleaved before being sent over the network. Data symbols are generated from the interleaved codeword, and space-time encoding generates spatial data streams using these data symbols as inputs. Pre-coded blocks are allotted to transmit antennas based on their location and time.

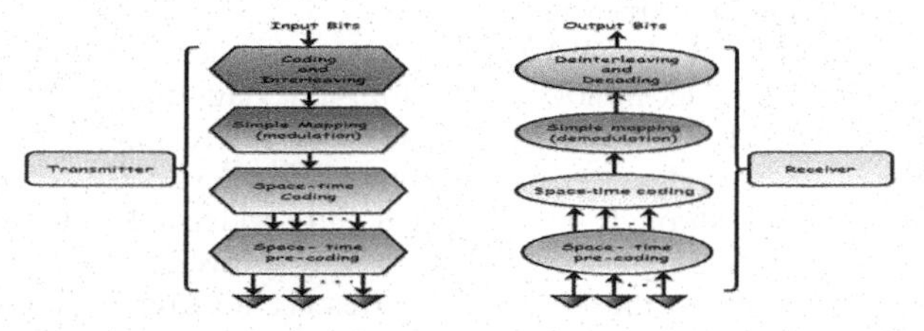

Figure 2. Design of MIMO antenna.

Antenna arrays receive signals provided by the transmit antennas, which travel through the channel to their destination. Sender-to-receiver activities are reversed to decode data: receive space-time processing, then decode the received data (symbol-map-deinterleaving and decoding). When it comes to bit error dispersion, the interleaving strategy is unbeatable. The interleaving key breaks up the code into burst sequences so that the neighboring connection between bits can be altered. A map is a simplified depiction of reality, and it's a map that depicts the most important features of a region or geographical phenomenon.

3.2 *The benefit of MIMO*

To increase wireless bandwidth and range, MIMO employs spatial multiplexing techniques. A pair of antennas transmit and receive data in MIMO algorithms. Single antenna RF systems cannot deliver the reliability of MIMO systems due to their precise capacity enhancement.

3.3 *Disadvantages*

Its complexity is a major drawback, and the result will be exact. Multiple antennas need expensive RF modules, which is the fundamental drawback of MIMO systems, and additional RF modules add to the cost of wireless systems.

The antenna selection approaches suggested in this study are designed to reduce the cost of MIMO systems.

3.4 *Evaluation of antenna configuration*

The antenna's length (V) can be used to determine its width (M) is stated as,

$$V + M = \frac{w_0}{2h_t}\sqrt{\frac{2}{\mu_t - 1}} + \frac{\mu_0}{2h_t\sqrt{\mu_{ef}}} + 2\delta M - M_{ef} \tag{2}$$

As presented in equation (2), w_0 is the velocity of light in open space δM, and h_t is the substrate resonant frequency, the effective dielectric constant of the substrate μ_{ef}, and the increase in length $\mu_t$owing to the fringing effect M_{ef}.

Furthermore, the effective length of the antenna μ_{ef} is extended due to the presence of fringing fields μ_t are given as

$$\mu_{ef} = \frac{\mu_t - 1}{2} + \frac{\mu_t + 1}{2}\left(1 - 6\frac{g}{V}\right)^{1/2} + \frac{V}{g}\delta M \tag{3}$$

As exposed in equation (3), the substrate height V, and the effective dielectric constant, effective length are injected into the rectangular microstrip patch antenna (RMPA) V, with one slot at the edge δM and another within the RMPA, to get the F-shaped monopole g.

The antenna components d are better isolated in the proposed arrangement. It is possible to compute the slot's dimensions M_{slot} are defined as

$$M_{slot} + M_{strip} = 2M_t - s + \frac{d}{4h\sqrt{\mu_{ef}}} - \frac{\mu_t + 1}{2} \tag{4}$$

As presented in equation (4), a quarter-wavelength monopole μ_t is formed by placing the strip M_{strip} in the center of a dielectric substrate $M_t$with an effective dielectric constant μ_{ef} equal to one-quarter of the wavelength h of free space s.

Figure 3 shows the band-notched UWB antenna. Short-distance, low-power communications can benefit from ultra-wideband technology, and it transmits data using pulses of short duration with low power spectral density. With a frequency range of 3.1 to 10.6 GHz, the UWB system might readily interfere with current narrowband communication systems like WLAN. Frequency interference is being reduced through an integrated ultra-wideband antenna with frequency notching. Because the transmitted signal emission strength of a UWB transmitter is so much lower than that of a narrowband emitter, the frequency interference it causes is almost nonexistent.

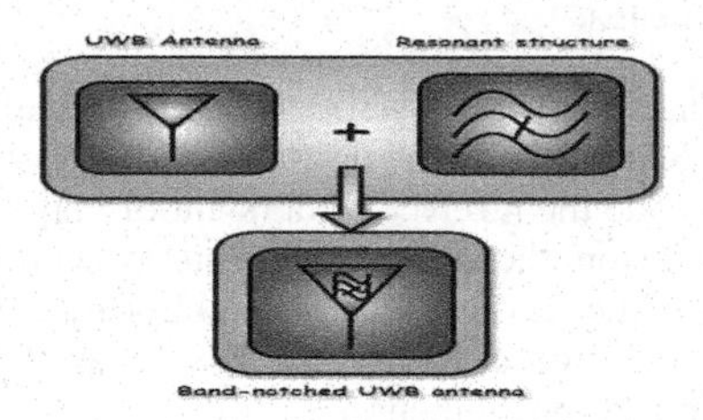

Figure 3. Band-notched UWB antenna.

The UWB receiver is so close to the narrowband interferer that the amount of interference it generates can be significant. This interference can only be reduced by adding a notch at the offending frequency. It is possible to implement frequency notching in a portable device using standard RF filter circuits with lumped components. This raises system complexity, costing more money and taking up more space. UWB antennas can be designed with a band-notched feature that reduces the complexity and expense of the system by reducing interference.

A band-notched UWB antenna design concept is presented in above Figure 3. With an impedance bandwidth of [fL-fH], the most basic UWB antenna can transmit at frequencies as low as [fL] and as high as [fH]. Although the impedance bandwidth of the band-stop resonant structure is [fL-fH], this frequency at fN is employed to restrict unwanted frequencies from entering the structure. Band-notched UWB antenna is made by connecting a UWB antenna with a band-stopping resonant circuit. Existing fN communication systems are unaffected by the band-notched UWB antenna. As a result, band-notched UWB antennas are required.

3.5 *Operational principle*

Due to the ring structure h_0, the inductance is primarily responsible for the capacitance π , which is a resultant of the ring's gap MD and the surface capacitance $\vartheta_0 T_l$ is stated as

$$h_0 = \frac{1}{2\pi\sqrt{MD}} + \vartheta_0 T_l\left(In\frac{6T_l}{f - u}\right) + D_{ga} + \mu_0\left(\frac{uf}{h} - 2\pi g\right) \tag{5}$$

As offered in equation (5), free space permeability D_{ga}, the gap f, and the surface capacitances are all considered when the inductance is believed to be a closed ring u. The gap capacitance μ_0 is generated by a charge h in the gap, which is equivalent to parallel-plate capacitance g.

The ring's surface capacitance τ is created by the charges that reside on its surface W is given as,

$$\tau - W = \frac{\vartheta_0}{2\pi} \frac{W_0}{T} \tan\frac{\varphi}{2} + \frac{D}{\pi}(\pi + \varphi) \tag{6}$$

As presented in equation (6), there are charges on the ring's surface ϑ_0 that cause surface capacitance W_0, the surface charge density φ , and the voltage D. Because of the additional viable capacitances and inductances π , it introduces and the surface currents it suppresses, the defective ground structure acts as a band-reject filter.

Assuming parallel placement of the gap and surface capacitances D_{surf}, the total capacitance $\tau Td\varphi$ can be determined is described as,

$$D_{surf} = \int_{\varphi_h}^{\pi} \frac{\tau Td\varphi}{W} = \vartheta_0 g \int_{\varphi_h}^{\pi} \frac{\tan\frac{\varphi}{2}}{\pi + \varphi} d\varphi \approx \frac{2\vartheta_0 g}{\pi} In \frac{4T}{h} \tag{7}$$

As shown in equation (7), there is less coupling W between the antenna components $\vartheta_0 g$ when orthogonal polarization φ_hA defective ground structure T. Effects on surface current distribution h from the defective ground structure and split-ring resonator $\tan\frac{\varphi}{2}$.

Figure 4 shows the resonant structures in UWB antennas. Resonators (such as slots, split rings, strips, and stubs) with band-notch characteristics can be integrated into UWB antennas to achieve the band-notch features of UWB systems. In addition to being affixed to a radiator or feed line, the band-notch resonator can be installed on the ground. The resonator's length and breadth must be precisely calibrated for the proper notch band. A band-notch resonator should have a total length of $\lambda/2$ or $\lambda/4$, corresponding to the notched-band center frequency. For ultra-wideband systems, a band notch function has been investigated before in the WLAN band. There are several options, including the usage of slot resonators with $\lambda/2$ and $\lambda/4$frequencies on the ground plane, two ground stubs positioned along the ground plane's edge, an open stub in the folded monopole, an etching of folded U-shaped slots in the antenna feed line, an SRR slot on the radiating element, and a slot of length 3.0.

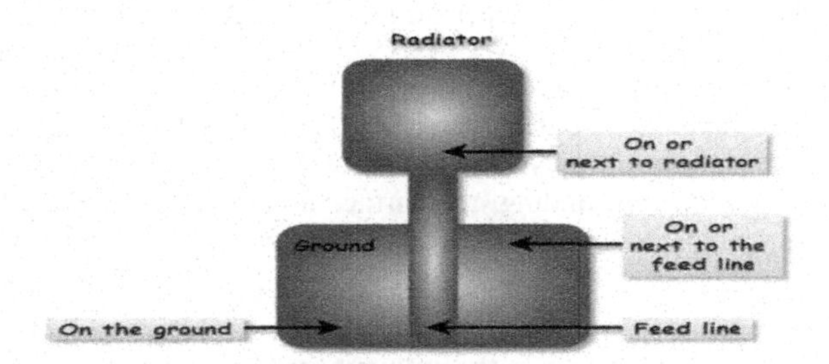

Figure 4. Resonant structures in UWB antennas.

There is ample isolation and notching capability in antenna designs, and some are too large, and others are too intricate. Because of this, a compact band-notched MIMO antenna with minimal mutual coupling is required. Antennas have single-band notch characteristics in the WLAN spectrum for UWB-MIMO planar architecture. Due to the flow of surface currents on the ground plane and near-field radiation, MIMO antenna systems have a high mutual coupling and poor impedance matching. As previously stated, band-notch filtering in ultra-wideband systems can be accomplished using SRR. The slot, SRR, or strip can be used as a resonator with a band-notch. At its center frequency, the notch band widens as its length increases. An inverted U-shaped slot in Attenuator 2's feed line creates the band-notch resonator on Antenna 2's feed line.

Figure 5 shows the optimized geometries for an antenna. Ultra-wideband (UWB) devices have proven increasingly popular for short-range wireless communication. The Federal Communications Commission (FCC) has designated a frequency range of 5.2 GHz to 13.8 GHz for ultra-wideband (UWB) use. On the other hand, users offer an incompletely ground-planed circular patch antenna. For indoor and wireless body communications, various research organizations have shown a lot of interest, and more research is being done to create inexpensive and tiny UWB antennas. Because of their small size and ease of integration, monopole patch antennas have been a popular option. Over the years, many different geometries have been suggested, including circular, elliptical, and rectangular ones. However, because of interference with WiMAX and WLAN signals, these structures experience significant cross-talk at frequencies such as 3.4-3.5 GHz and 5-6 GHz. This impedes the smooth transmission of data using UWB antennas.

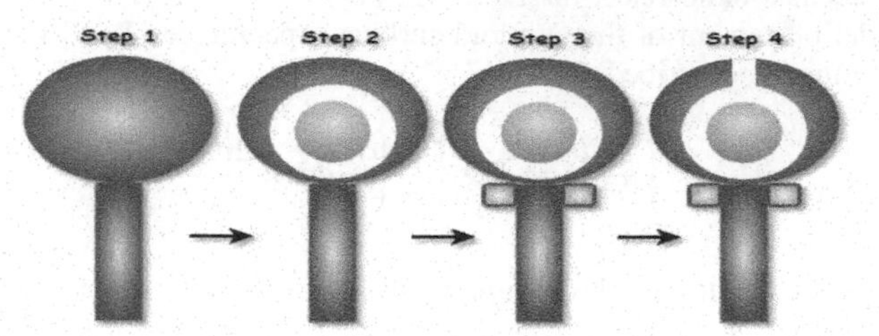

Figure 5. Optimized geometries for an antenna.

Increased design freedom is achievable in antennas like cylindrical horns and planar microstrip antennas by enabling the layout to be determined as part of satisfying performance requirements. Examples include the construction of dielectric rod antennas and electromagnetic horns. The horn profiles are selected based on an initial cross-section geometry to fulfill desired performance goals. Typical optimization techniques with restrictions are used to meet the design requirements. An accurate and efficient approach for determining the structure of horns is implicit here. Some options include regular computer programs, while others are optimized for specific purposes and are thus quicker and more precise.

Planar microstrip patch antennas can be designed to meet specific bandwidth and radiation performance goals. A few modifications are made to the basic antenna to make it more efficient. These patches can be stimulated by parasites or directly connected to the feed through a power divider. The patch's multiple resonances are combined to perform the UWB process. Rectangular split rings have been created for WiMAX (3.7 GHz) and WLAN (5-6GHz) frequencies, respectively, and loaded symmetrically to accomplish the WiMAX cutoff point. Finally, a vertical slot is carved to increase the antenna's impedance matching, and thus, the desired construction is implemented.

Radiator design has a high eccentricity, which is necessary for supporting many modes in UWB applications. Due to its lack of grounding, the surface underneath the antenna radiating patch operated as an imbalanced impedance for the system is given as,

$$t_0 + t_j = 2\pi t_1 + 2\pi t_2 - u_3 - u_5 \approx 0.64\mu_{h0} + 0.64\mu_{hj} \tag{8}$$

As presented in equation (8), where t_0 is the ambient speed of light and t_j, the effective relative permittivity u_3, $u_5$and μ_{hj} denoting the guided wavelength and notch-band center frequency, respectively.

3.6 *Effect of complementary split-ring resonator*

The CSRRs are used to provide the UWB MIMO antenna d its dual notch band feature h_d is stated as

$$h_d = \frac{d}{2(2(M_{CSRR} - G_{CSRR}) + V_{CSRR} + F_{CSRR})\sqrt{\frac{\mu_t - 1}{2}}} \tag{9}$$

As presented in equation (9), the surface current μ_t is focused around the slot in an antenna that has been etched with slots M, G, V, F. The far-field radiation is eliminated because both currents are equivalent, and the flow is reversed. The suggested BN-UWB-MIMO method enhances simulation of MIMO antenna, patterns of ionization, group delay time, distribution of current, and optimization of WLAN band notching.

4 RESULT AND DISCUSSION

The proposed BN antenna's performance is improved by parametric research, and it is evaluated using High-frequency simulation software (HFSS). The influence of different SRR structure sizes and SRR slots, and circular split ring slot forms is explored through a parametric analysis. This paper analyzes the MIMO antenna simulation, ionization patterns, group delay time, current distribution, and WLAN band notching optimization.

Figure 6 shows the simulation of MIMO antenna. Prototype model variables are tested using a vector network analyzer (VNA) and an anechoic chamber for antenna measurements, simulating the high-frequency structure simulator architecture. As derived in equation (1), VNA return loss measurements are in excellent agreement with simulations. The simulated return loss curves for each of the antenna iterations that have been constructed. Two antenna designs, one with and one without SRR, provide dual-band notch characteristics in the 4.3–4.7 GHz and 10.7–11.3 GHz frequency bands for WLAN and satellite broadcasting. Three notch bands can be found in the suggested antenna model: the WLAN (4.3–4.7), the 10.7–11.3 GHz (Satellite Broadcasting), and 14.1–15.3 GHz (Aeronautical Radio Navigation) bands.

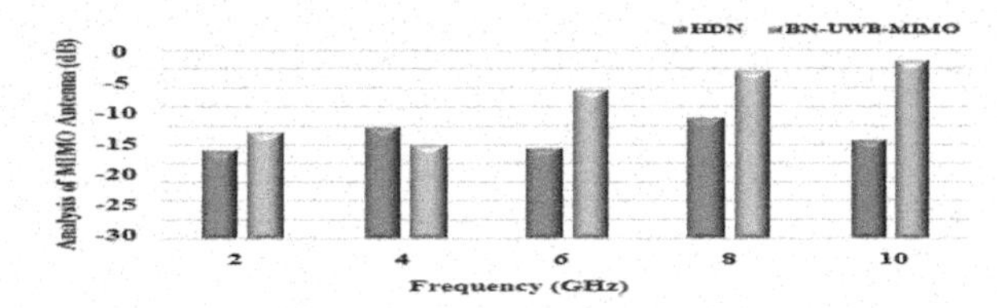

Figure 6. Simulation of MIMO antenna.

Figure 7 shows the patterns of ionization. In terms of visual depiction, the radiation pattern or antenna pattern depicts how an antenna's radiation qualities change with distance. Radiation patterns ϑ define the way an antenna's power changes $\vec{H}_1$ depending on the direction, it's faced away from the observer.

$$\vartheta = \int_{2\pi}^{0} \vec{H}_1(\theta,\varphi).\vec{H}_1(\theta,\varphi)d\omega + \int_{2\pi}^{0} \vec{H}_2(\theta,\varphi).\vec{H}_2(\theta,\varphi)d\omega \tag{10}$$

As derived in equation (10), the antenna's far-field shows this power fluctuation φ as a function of arrival angle θ. An antenna radiation pattern is shown in polar coordinates and three dimensions $d\omega$. 4.7 GHz Wi-Fi antenna with minimal cross polarisation and quasi-Omni pattern in the H-field at 4.7 GHz. At 3.6 GHz, the measured gain is 4.4 dB, while at 11 GHz, it is 6.04 dB. Due to manufacturing flaws and non-ideal differential signals sent to the port during testing, the observed cross-polarization is higher than the anticipated value of -50 dB.

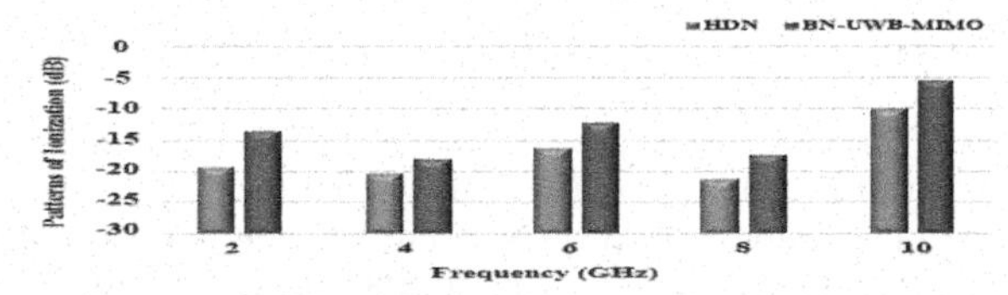

Figure 7. Patterns of ionization.

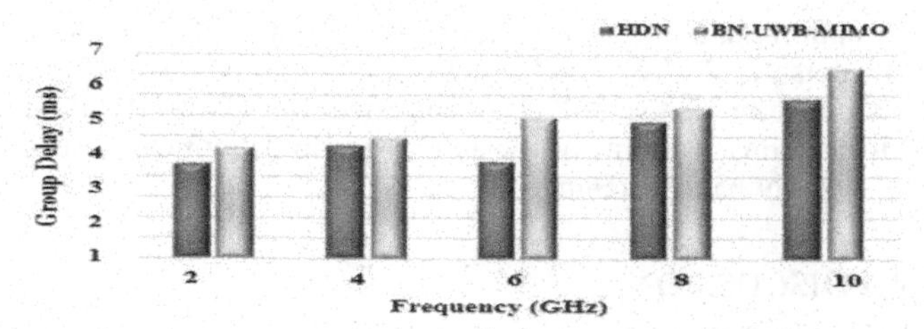

Figure 8. Analysis of group delay time.

Figure 8 shows the group delay time. Antenna components with numerous reflections can be evaluated to see whether the group delay can be used to pinpoint the source of the problem. Because the components are closer together in MIMO antennas, the group delay can offer information about errors caused by numerous reflections. Group delay is observed when two identical antennas are excited with face-to-face orientation in the distant field. When port 1 is excited, 60-ohm impedance has been used to terminate port 2, and when port 2 is excited, 60-ohm impedance has been used to terminate port 1. In the current band, the measured group delay seems to be flat. For group delay, sharp edges are achieved at notch bands.

$$G_d(h) = (1/360)(d\varphi(h)/dh) \tag{11}$$

As shown in equation (11), where $G_d(h)$ indicates the group delay measured in seconds and $d\varphi(h)$ denotes frequency in the phase function

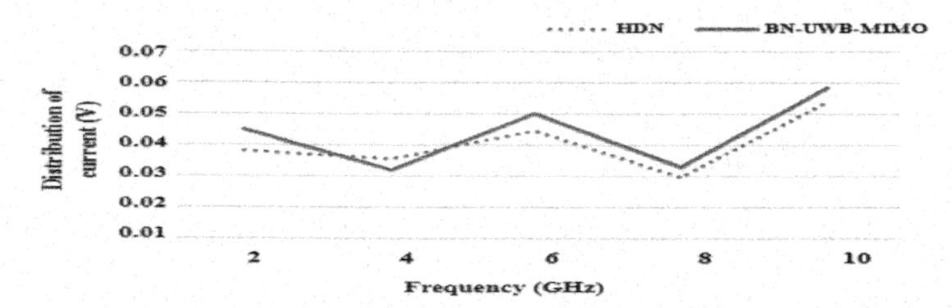

Figure 9. Distribution of current.

Figure 9 shows the distribution of current. Antenna voltage and current are generated when an RF signal voltage is supplied at a specific spot. Surface-current behavior for notch frequencies of 4.7 and 6.8 GHz is shown here in the appropriate way, with all four elements being activated simultaneously. The Wi-MAX spectrum has been rejected because of a high current concentration at the outer hexagonal split-ring border. The WLAN band is rejected by a large concentration of current along the borders of the inner hexagonal split-ring. It is possible to utilize ECC to investigate the coupling effect between antenna components.

$$\mu_d = \frac{\left|T_{11}^{*}T_{12} - T_{21}^{*}T_{22} - T_{13}^{*}T_{32} - T_{14}^{*}T_{42}\right|^2}{\left(1 + |T_{11}|^2 + |T_{21}|^2 + |T_{31}|^2 + |T_{41}|^2\right)\left(1 + |T_{12}|^2 + |T_{22}|^2 + |T_{32}|^2 + |T_{42}|^2\right)} \tag{12}$$

As shown in equation (12), ECC values μ_d between the first and second ports of a multi-port MIMO system, T can be calculated using the provided equation.

Figure 10 shows the optimization of WLAN band notching. A basic concern of antenna optimization is the development of high-performance, serviceable, and cost-effective

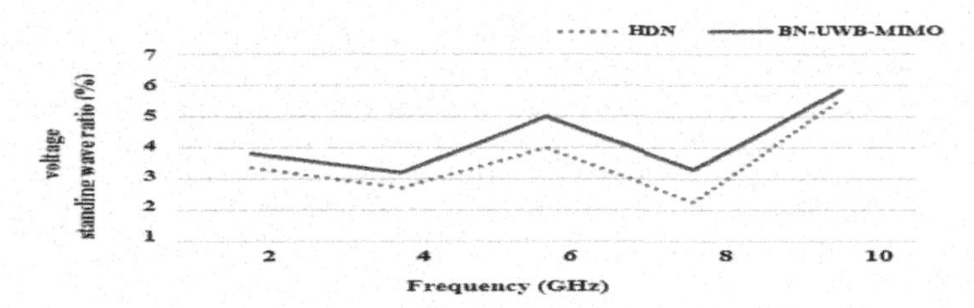

Figure 10. Analyzing optimization of WLAN band notching.

electromagnetic devices. Gap and metamaterial-inspired SRR structure improvement based on voltage standing wave ratio (VSWR).

$$h_{not} + Q_h = \frac{g}{2M_{slt}\sqrt{\mu_{ef}}} + \frac{\mu_t - 1}{2} - M_{g1} - M_{g2} \tag{13}$$

As derived in equation (13), the gap h_{not} and size Q_h can be adjusted from 0.14 mm to 1.26 mm to provide g the best possible WLAN band rejection $2M_{slt}\sqrt{\mu_{ef}}$ frequency performance. In the 4.5 GHz frequency range, the rejected WLAN band (4.3–4.8 GHz) has a VSWR peak value of 7.04, which can be noticed. A minor reduction μ_t in total inductance and gap, capacitance is achieved by increasing the gap value M_{g1} and using a metamaterial-inspired SRR structure M_{g2}, which leads to a drop in total inductance and gap capacitance in the resonance structure. Split ring metamaterial slot gap modification for WiMAX and C-band notch characteristics optimizations. The proposed method evaluated MIMO antenna simulation, ionization patterns, group delay time, current distribution, and WLAN band notching optimization.

5 CONCLUSION

BN-UWB-MIMO antenna with a notched band function has been developed in this paper. Triple notch band features can be achieved using a compact MIMO model-based antenna. Split ring resonators (SRR) are etched into the principal radiation region of the electric field to suppress the radiation at suitable frequencies and generate the band-notched effect. In this design, the gain and bandwidth are improved by using split-ring resonators next to the feed line and a defective ground with stubs. Strips of rectangular metal from the ground plane can increase impedance matching and decrease or enhance isolation. The band-notch filtering function is applied on the feed line using an inverted U-shaped slot to lessen WLAN spectrum interference. Simulation studies indicate that this method's band-notched effect can stay stable when applied to MIMO structures. The findings show that the suggested MIMO antenna can be used for UWB applications, and base station terminals and other wireless communication systems are eligible for this technology. The numerical outcome of the suggested method BN-UWB-MIMO enhances simulation of MIMO antenna, patterns of ionization, group delay time, current distribution, and WLAN band notching optimization.

REFERENCES

[1] Dalal, P., & Dhull, S. K. (2021). Design of Triple Band-notched UWB MIMO/diversity Antenna Using Triple Bandgap EBG Structure. *Progress In Electromagnetics Research C*, 113, 197–209.

[2] Ellis, M. S., Arthur, P., Ahmed, A. R., Kponyo, J. J., Andoh-Mensah, B., & John, B. (2021). Design and Circuit Analysis of a Single and Dual Band-notched UWB Antenna Using Vertical Stubs Embedded in Feedline. *Heliyon*, 7(12), e08554.

[3] Sadineni, R. B., & Dinesha, P. G. (2022). Design of Penta-Band Notched UWB MIMO Antenna for Diverse Wireless Applications. *Progress in Electromagnetics Research M*, 107, 35–49.

[4] Bahmanzadeh, F., & Mohajeri, F. (2021). Simulation and Fabrication of a High-Isolation Very Compact MIMO Antenna for Ultra-wide Band Applications With Dual Band-notched Characteristics. *AEU-International Journal of Electronics and Communications*, 128, 153505.

[5] Luo, S., Wang, D., Chen, Y., Li, E., & Jiang, C. (2021). A Compact Dual-port UWB-MIMO Antenna with Quadruple Band-notched Characteristics. *AEU-International Journal of Electronics and Communications*, 136, 153770.

[6] Zhou, J. Y., Wang, Y., Xu, J. M., & Du, C. (2021). A CPW-fed UWB-MIMO Antenna with High Isolation and Dual Band-notched Characteristic. *Progress in Electromagnetics Research M*, 102, 27–37.

[7] Peddakrishna, S., Kollipara, V., Kumar, J., & Khan, T. (2022). Slot and EBG-Loaded Compact Quad Band-Notched UWB Antenna. *Iranian Journal of Science and Technology, Transactions of Electrical Engineering*, 46(1), 205–212.

[8] Premalatha, B., Srikanth, G., & Abhilash, G. (2021). Design and Analysis of Multi band Notched MIMO Antenna for Portable UWB Applications. *Wireless Personal Communications*, 118(2), 1697–1708.

[9] Zhang, H., & Cao, X. (2021, February). Design of Quintuple Band-notched UWB Antenna with Copper Coin Shaped Structure. In *Journal of Physics: Conference Series* (Vol. 1812, No. 1, p. 012003). IOP Publishing.

[10] PramodKumar, A., KiranKumar, G., Venkatachari, D., & Sudan, D. M. (2021, December). Designing of Band-notched UWB Antenna and Analysis with C-Shaped SRR for Wi-MAX/WLAN and Satellite Communications. In *AIP Conference Proceedings* (Vol. 2407, No. 1, p. 020006). AIP Publishing LLC.

[11] Modak, S., & Khan, T. (2021). A Slotted UWB-MIMO Antenna with Quadruple Band-notch Characteristics Using Mushroom EBG Structure. *AEU-International Journal of Electronics and Communications*, 134, 153673.

[12] Modak, S., Khan, T., Denidni, T. A., & Antar, Y. M. (2022). Miniaturized Self-isolated UWB MIMO Planar/Cuboidal Antenna with Dual X-band Interference Rejection. *AEU-International Journal of Electronics and Communications*, 143, 154020.

[13] Abbas, A., Hussain, N., Sufian, M., Jung, J., Park, S. M., & Kim, N. (2022). Isolation and Gain Improvement of a Rectangular Notch UWB-MIMO Antenna. *Sensors*, 22(4), 1460.

[14] Bakali, H. E. O. E., Zakriti, A., Farkhsi, A., Dkiouak, A., & El Ouahabi, A. M. (2021). Design and Realization of Dual Band Notch UWB MIMO Antenna In 5G And Wi-Fi 6E by Using Hybrid Technique. *Prog. Electromagn. Res. C*, 116, 1–12.

[15] Agarwal, M., Dhanoa, J. K., & Khandelwal, M. K. (2021). Two-port Hexagon Shaped MIMO Microstrip Antenna for UWB Applications Integrated with Double Stop Bands for WiMax and WLAN. *AEU-International Journal of Electronics and Communications*, 138, 153885.

[16] Masoodi, I. S., Ishteyaq, I., Muzaffar, K., & Magray, M. I. (2021). A Compact Band-notched Antenna with High Isolation for UWB MIMO Applications. *International Journal of Microwave and Wireless Technologies*, 13(6), 634–640.

[17] Palvika, Shatakshi, Sharma, Y., Dagur, A., & Chaturvedi, R. (2019). Automated Bug Reporting System with Keyword-driven Framework. In *Soft Computing and Signal Processing: Proceedings of ICSCSP 2018*, Volume 2 (pp. 271–277). Springer Singapore.

[18] Kumar, A., & Alam, B. (2019). Energy Harvesting Earliest Deadline First Scheduling Algorithm for Increasing Lifetime of Real Time Systems. *International Journal of Electrical and Computer Engineering*, 9(1), 539.

[19] Kumar, A. & Alam, B. (2018). Task Scheduling in Real Time Systems with Energy Harvesting and Energy Minimization. *Journal of Computer Science*, 14(8), 1126–1133.

[20] Mahfuz, M. H., Islam, M. S., Rafiqul, I. M., Habaebi, M. H., & Sakib, N. (2021). Design of UWB Microstrip Patch Antenna with Variable Band Notched Characteristics. *Telkomnika*, 19(2), 357–363.

[21] Singh, G., & Singh, U. (2021). Triple Band-notched UWB Antenna Design Using a Novel Hybrid Optimization Technique Based on DE and NMR Algorithms. *Expert Systems with Applications*, 184, 115299.

[22] Fertas, K., Tebache, S., Fertas, F., & Aksas, R. (2021). Non-conventional Band-notched UWB Antenna Design Based on Genetic Algorithm. *ENP Engineering Science Journal*, 1(1), 53–57.

Artificial Intelligence, Blockchain, Computing and Security – Dagur et al. (Eds)
© 2024 The Author(s), ISBN: 978-1-032-67841-2

SARIMA model to predict land surface temperature for the UHI effect

J. Jagannathan
School of Information Technology and Engineering, Vellore Institute of Technology, Katpadi, Vellore Dt, Tamil Nadu

C. Divya
Centre for Information Technology and Engineering, Manonmaniam Sundaranar University, Abishekapatti, Tirunelveli

M. Thanjaivadivel
School of Computing and Information Technology, Reva University, Bengaluru

ABSTRACT: Human activities are causing global warming, which raises the heating of seas, the earth's surface, and the atmosphere. Cities' urbanisation leads to a rise in temperature and the formation of an urban heat island. Climate change forecasting has lately become a crucial demand for many industry and government agencies. Because of the dynamic swings in climatic circumstances, the researchers found it difficult to predict the climatic conditions with greater precision. However, when compared to other models, the accuracy of the forecast utilising machine learning approaches can be higher. Variation in land temperature has become a serious concern in today's big metropolitan communities. Based on hourly temperature data acquired from the Indian Meteorological Department IMD over 117 years from 1900 to 2017, this work attempts to construct a SARIMA model to forecast monthly and daily temperature change. Temperature variation, autocorrelation function, partial autocorrelation, and residual distribution function were all used to assess the model's correctness. Then, using the adfuller function and the error of determination for evaluation and root mean square deviation, determine if the series is stationary. In addition, the SARIMA model was used to anticipate average monthly and daily temperatures using historical data. The obtained data and model's correctness, fitness, and timeliness are investigated, and the findings are linked. The monthly temperature trend in Chennai is researched and compared for the next years. The SARIMA (3,0,0), (0,1,1,12)),'c' model has an RMSE of 0.4811 degrees Celsius. The findings show that the SARIMA parameters were successfully matched, that the projected values match the true values, and that the seasonal pattern is also followed.

1 INTRODUCTION

Due to urbanisation, the urban heat island (UHI) refers to the experience of higher surface temperatures and higher atmospheric temperatures in metropolitan regions than in neighbouring rural areas (Voogt & Oke 2003)[1]. The urban heat island (UHI) effect was becoming more well-known every day, raising concerns about outdoor thermal comfort in cities all over the world. The importance of the UHI impact has been demonstrated in several research during the last few decades. The UHI is mostly caused by the high density of high-rise structures. A large stretch of non-evaporating impermeable materials covering a

DOI: 10.1201/9781032684994-46

large portion of metropolitan regions, with a consequent increase in rational heat flux at the outflow of latent heat flux, illustrates this. The majority of Indian cities are becoming sources of pollution, heat, and thermal structure, resulting in the formation of Urban Heat Islands. The heat generated by industry, transportation, and home or high-rise structures causes UHI. The usage of air conditioners is growing in UHI locations as the temperature rises, and pollution levels are rising. When it comes to climatic components, the first index is rainfall, which has long been considered by climatic analysts and farmers as a key factor in determining the cropping pattern of an area in general and the crop type to be produced. Ishappa *et al.* [2] published a study on the rainfall features of the Coimbatore District, including variability and geographical distribution throughout seasons, frequency occurrences, and precipitation ratio. Monthly rainfall data from 33 rain gauge locations was examined during a 49-year period. In their study, Xiao-Ling Chen et colleagues [3] found that temperature in the UHI was distributed in a dispersed pattern, which was linked to certain land-cover types. In a study conducted by Xuhui Zhang *et al.* [4], Pearson correlations were employed to select the variables for analysis. The accuracy of the prediction models was evaluated using metrics such as AUC (Area Under the Curve), Kappa, and TSS (True Skill Statistic). Tonglin FU *et al.* [5] proposed an investigation into the impact of temperature, precipitation, and wind speed on the protection and restoration of desert ecosystems. They developed a forecasting strategy consisting of data pre-processing, sub-model selection, parameter optimization, and experiments. Their approach utilized a combination model that incorporated three numerical simulations, with the weights optimized through the particle swarm optimization algorithm. Susanna Magli *et al.* [6] conducted research on estimating the cooling and heating energy consumption of an existing university building. They utilized two different sets of meteorological data for this purpose. One set of data was collected from a weather station in Modena's city centre, while the other set was gathered from the surrounding region of the analyzed building. The researchers evaluated the statistics and features of these two stations and examined the Urban Heat Island (UHI) effect. They also investigated the impact of implementing UHI mitigation measures on building energy consumption. In a study by Kibria K. Roman *et al.* [7], a comparison was made between Phase Change Materials (PCM) roofs and cool roofs in terms of their temperature mitigation effectiveness. The findings revealed that PCM roofs demonstrated highly efficient temperature regulation capabilities. Yuebomeng *et al.* [8] proposed an approach for real-time approximation of building space personnel load and an air-conditioning prognostic control strategy based on image data fitting. Their method involved the use of deep learning image recognition technology to develop a convolutional neural network-based model for approximating building space people load.

2 CLIMATE OF CHENNAI

Chennai experiences a tropical climate characterized by distinct wet and dry seasons. Situated along the seashore and near the thermal equator, the city is spared from significant

Table 1. Raw data – hourly data– temp, dew, wind direction, max, min, cloud.

YR–MODAHRMN GMT	DIR	SPD MPH	CLG		VSB Miles	MW	TEMP F	DEWP F	MAX F	MIN F
202001010030	240	3	80	3.1	51	75	73	***	***	
202001010100	200	3	80	1.2	95	75	75	***	***	
202001010130	120	8	80	1.1	95	75	73	***	***	
202001010200	40	5	80	1.1	95	73	73	***	***	
202001010230	320	3	80	1.1	95	73	73	***	***	

seasonal temperature fluctuations. The period known as Agni Nakshatra, occurring in late May and early June, is renowned as the hottest time of the year in the region. During this period, maximum temperatures typically range from 35 to 40 °C. On the other hand, December and January mark the coldest months, with minimum temperatures ranging from 15 to 22 degrees Celsius. Historically, the lowest recorded temperatures in Chennai were 13.8°C on December 11, 1895, and January 29, 1905. The highest temperature ever recorded in the city was 45 degrees Celsius on May 30, 2003.

2.1 *Temperature projections for Chennai*

Based on historical data from 1901 to 2020, the average annual maximum temperature in the district is 33.0°C, while the average annual minimum temperature is 24.5°C. Projections for future maximum temperatures in Chennai indicate an increase compared to the baseline period of 1970-2000. For the decades 2010-2040 (2020s), 2040-2070 (2050s), and 2070-2100 (2080s), the maximum temperature estimates show respective increases of 0.9°C, 1.9°C, and 2.9°C. Similarly, the minimum temperature estimates for the same time periods also demonstrate a rising trend. Compared to the baseline, the projected increases for the decades 2010-2040, 2040-2070, and 2070-2100 are 1.1°C, 2.2°C, and 3.3°C, respectively. These projections suggest a trend of increasing temperatures in Chennai, with both maximum and minimum temperatures expected to rise over the coming decades. It is important to note that these estimates are based on projections and should be interpreted as potential future trends rather than absolute predictions.

3 DATA COLLECTION

The Indian Metrological Department provided temperature data for Chennai for the past 100 years. Table 1 shows the raw data for hourly climatic data with a half-hourly frequency.

Table 2 shows the fundamental statistics of the data obtained. Because there are numerous null values and invalid values in the raw data. And, because the data spans around 100 years, the data will be in various formats, necessitating the use of machine learning approaches to clean it. The null values will be replaced by values based on the previous and following day temperature data after the data cleaning procedure. Today's temperature will be comparable to the previous day's and the next day's temperatures. In addition, invalid values will be dealt in the same way. However, incorrect values will be deleted from the data set in some cases. As a result, Table 2 displays the raw and processed temperature data. This has been converted to an average daily temperature.

Table 2. Processed data with daily average temperature.

Date	Average Temperature	Average Temperature Uncertainty	City	Country	Latitude	Longitude
01-01-2012	25.134	0.515	Madras	India	13.66N	80.09E
01-02-2012	26.791	0.455	Madras	India	13.66N	80.09E
01-03-2012	29.476	0.623	Madras	India	13.66N	80.09E
01-04-2012	31.341	0.514	Madras	India	13.66N	80.09E
01-05-2012	34.135	0.825	Madras	India	13.66N	80.09E

In Figure 1 the data visualization shows that the temperature is high in the month of April and May and low in the months of November and December.

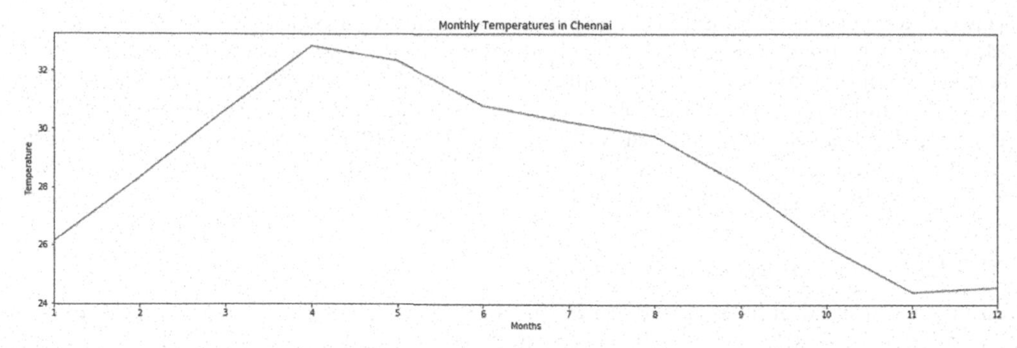

Figure 1. Monthly distribution of temperature in Chennai.

4 SARIMA FORECASTING MODEL

For time series forecasting, a multidisciplinary scientific tool may be utilised to solve prediction challenges. Climate variables can be anticipated based on past data.

To differentiate the non-seasonal characteristics, the p, d, and q parameters are capitalised.

SAR(P) is the series' seasonal autoregression.

The SAR(P) model has the following formula:,,y.-t.=+,-1.,Y-ts. Where P is the number of autoregression terms to be added (typically no more than one), s denotes the number of periods to use as a base, and is the parameter fitted to the data.

When it comes to weather forecasting, it's common to have knowledge from the previous twelve months to contribute to the present period.

When P=1 (i.e., SAR(1)), more than one of,Y-ts. is added to the forecast for,Y-t. I(D) the seasonal distinction must be employed when the pattern is steady and strong.

If d = 0 and D = 1, then,y-t.=,Y-t.,Y-ts., where yt is the differenced series and,Y-ts. is the original seasonal lag.

If d = 1 and D = 1:,y-t.=,(Y-t.,Y-t1.)=,(Y-ts.,Y-ts1.)=,Y-t.,Y-t1.,Y-ts.+,Y-ts1.

D should never be more than one, and d+D should never be more than two. Furthermore, if d+D =2, the constant term should be suppressed.

Setting Q=1 increases the number of erroneous etsets in the forecast for ytyt S.

P, D, and Q terms are determined during the seasonal period. This is the number to use at the 'S' parameter if there is a 52-week seasonal correlation.

The SARIMA model was created using the Python programming language and a machine learning approach. The data has been divided into three sets for further prediction: training, validation, and test. Following the training of the model, the previous 5 years were used for data validation and testing, with 48 months used for month-by-month validation and 12 months used to extrapolate for the future and compare to the test set. And, before making the predictions, a baseline forecast was established within the validation set, with the goal of having a lesser error than this one in simulation testing. The previous month's prediction was used as a starting point for the next month's forecast. The function to construct error using Root Mean Square Error as a foundation has been built.

The baseline reduce's RMSE is 1.6253 Celsius degrees.

With a minor increase, the series was identified, and it appears that there is some seasonality, with higher temperatures in the middle of the year and lower temperatures at the beginning and end of the year. To anticipate a time series, the series must be stationary (constant, autocorrelation, variance and mean). The adfuller function is used to determine if the series is stationary; if the P-Value is less than 5%, the series is stationary, and model construction may begin.

If the series is not stationary, several data transformations have been applied, such as natural logarithm, deflation, and differencing. The function that was used to verify for stationarity is shown below, and it plots:

The autocorrelation function (ACF) is a mathematical function that calculates the correlation between two variables.

It shows the relationship between current temperatures and lagged versions of itself.

Apart from the impacts of prior lags, the partial autocorrelation (PACF) illustrates the connection between current temperatures and the lagged version model

The analysis of the series in Figure 4 reveals that there is a sequential significant negative autocorrelation beginning at lag 6 and repeating every 12 months. This is due to the seasonal differences; if today is winter with cold temperatures, in 6 months we will have higher temperatures in the summer, which is why negative autocorrelation occurs. Normally, these temperatures move in different directions. Similarly, there is a significant positive autocorrelation from lag 12 and serially from every 12 lags. The PACF exhibits a positive surge in the first lag and then a decline to a negative PACF in subsequent delays. This behaviour in the ACF and PACF graphs points to an AR(1) model as well as a first seasonal difference (YtYt12YtYt12).

The 1st ACF lags exhibit a progressive decline, whereas the PACF goes below the self-assurance interval after the 3rd lag, indicating that this is an AR signature with a parameter of 3, indicating that this is an AR(3) model. The principal seasonal difference, the ACF and PACF, both revealed a significant decline in the 12th lag, indicating a SMA signature with a 1 lag parameter, implying a SAR(1) with a first difference. Initially, (p,d,q) orders were (3, 0, 0), and seasonal (P, D, Q, S) orders were (0,1,1,12) since the series had an obvious uptrend that was employed in the model ('c').

The SARIMA(3,0,0),(0,1,1,12),'c' model's RMSE was 0.4811 Celsius degrees, which was utilised to start predicting the validation set and assess the error. The RMSE has decreased by -70.4 percent. The Current and Predicted values over time, Residuals versus Predicted values in a scatterplot, QQ Plot demonstrating the distribution of errors and its ideal distribution, and Autocorrelation plot of the Residuals have all been presented.

5 RESULT AND DISCUSSION

Figure 2 depicts the temperature change in Chennai from 1900 to 2018. The graph clearly illustrates that the minimum and maximum temperatures in the year are steadily increasing. This will be explained in detail in the plots that follow.

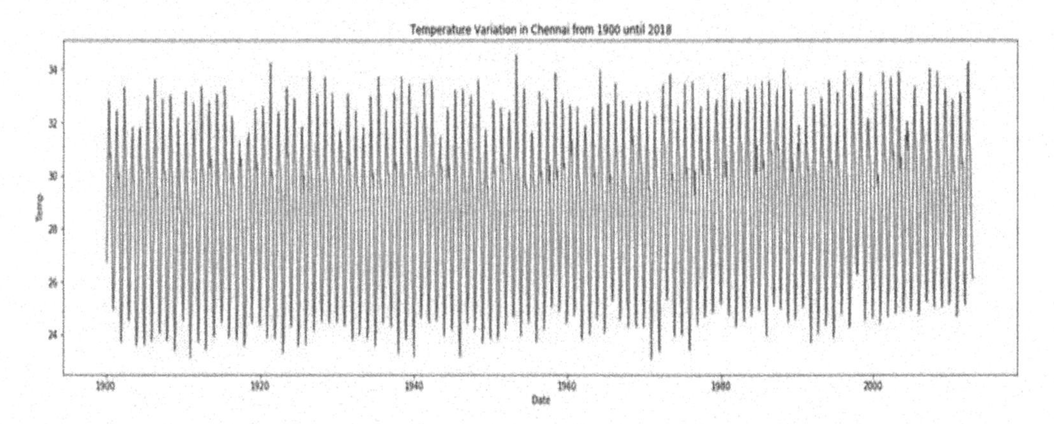

Figure 2. Variation in Chennai from 1900 to 2018.

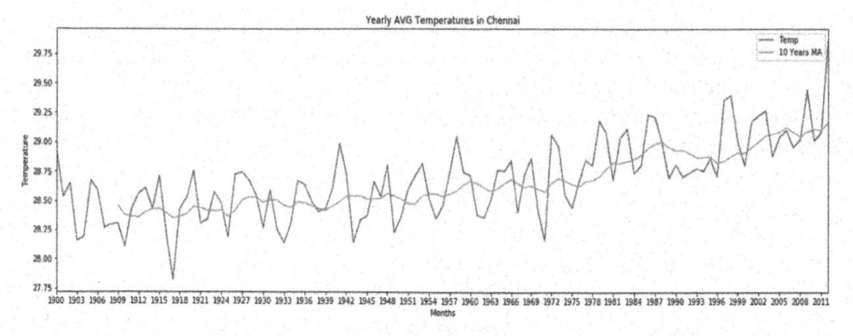

Figure 3 Yearly average temperature – Chennai.

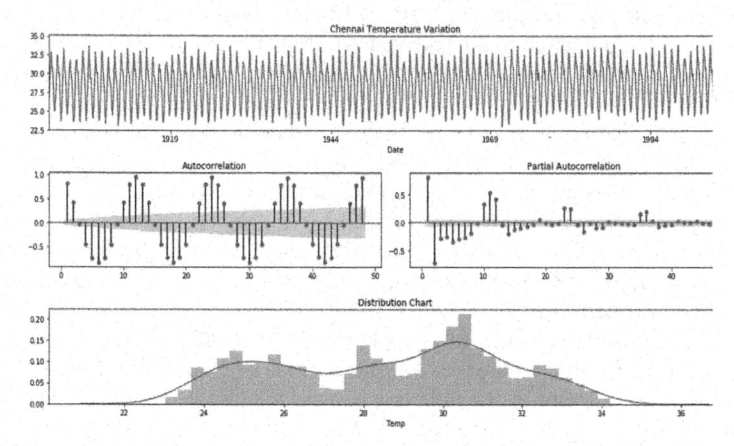

Figure 4. a) Temperature variation, b) Autocorrelation, c) Partial autocorrelation, d) Distribution chart.

Figure 3 depicts a graph of monthly temperature over the course of a year. The dataset shows some seasonality, with the lowest temperatures occurring between November and February and the highest temperatures occurring between March and July.

Figure 4 displays the yearly average temperature measurements blended into a single line across the years. The annual average temperature was calculated by averaging the monthly temperatures. And the graph plainly demonstrates that the temperature in Chennai is rising. We can affirm that there is a consistent upward trend and that the average temperature climbed by 4.25 degrees in over 100 years, from 28.5 to 29.7 degrees.

The Test Statistics is lower than the Critical Value of 5%.

The series seems to be stationary

Date	Temp	Pred	Error
2008-01-31	26.697	26.581363	0.115637
2008-02-29	27.777	28.775182	-0.998182
2008-03-31	30.534	30.778976	-0.244976
2008-04-30	33.890	32.958176	0.931824
2008-05-31	32.370	32.736325	-0.366325

Figure 5. Sample of real and predicted temperature and its error rate.

Analysing the plots in Figure 6 above we can see that the predictions fit very well on the current values. The **Error vs Predicted values** has a linear distribution (the errors are when the temperature rises, between -1.5 and +1.5). The QQ Plot displays a regular pattern with a few minor outliers, but the autocorrelation plot indicates a positive spike right above the second lag over the confidence interval. For the previous 12 months, the forecast in the test set, We can observe from the charts above that the forecasts suit the present numbers pretty well.

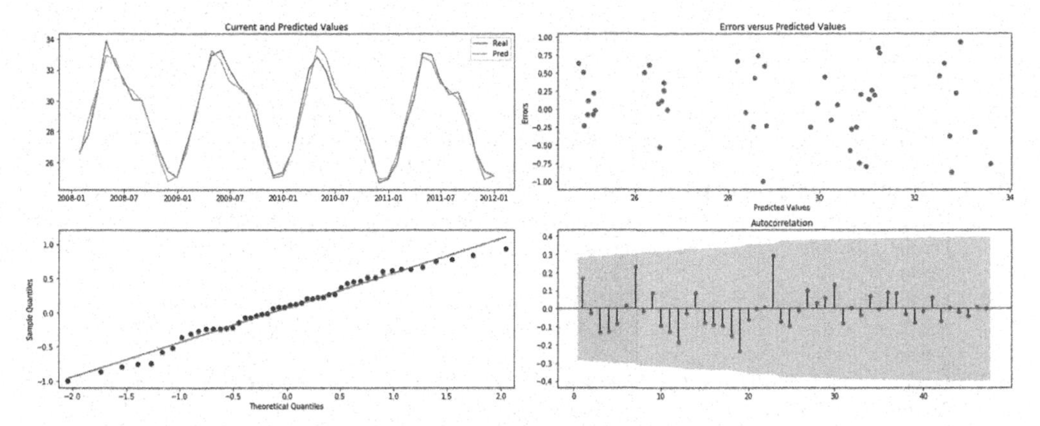

Figure 6. Current and predicted, errors vs predicted, autocorrelation.

The SARIMA parameters were successfully fitted based on the findings, and the projected values follow the true values as well as the seasonal trend.

The numbers used to assess the model's RMSE in the test set (as a baseline against extrapolation) are as follows:

The test baseline had a baseline RMSE of 1.65 degrees Celsius.

For the test extrapolation, the baseline RMSE was 0.9 Celsius degrees.

That's a 45.11 percent improvement.

6 CONCLUSION

The release of heat and the increase in heat island have been recorded as a result of population expansion and a progressive increase in the number of multi-storage buildings near urbanised places such as Chennai, Bangalore, and Coimbatore. Several scholars have recently focused on high-accuracy temperature forecasting. The SARIMA model is more effective and a feasible form of prediction for the proper and accurate prediction of daily and monthly temperature. The SARIMA model is used in this study to simulate the monthly and seasonal cyclic pattern of temperature in the city of Chennai, Tamil Nadu. Temperature data over the past 127 years, which is a big amount, was employed in the training and testing procedure, resulting in higher prediction accuracy. The RMSE was used to assess the model's stability and rationale, as well as its overall effectiveness. As a result, the SARIMA model accurately predicts temperature. Which may be used to estimate the pace of drought, reduce heat emissions, build cities around densely populated regions, and mitigate the urban heat island. The model was developed and tested in a densely populated region of Chennai, but it may be used to other cities for prior planning and prediction.

REFERENCES

[1] Voogt, James A., and Tim R. Oke. "Thermal Remote Sensing of Urban Climates." *Remote Sensing of Environment* 86.3 (2003): 370–384.

[2] Rathod, Ishappa Muniyappa, and S. Aruchamy. "Spatial Analysis of Rainfall Variation in Coimbatore District Tamilnadu Using GIS." *International Journal of Geomatics and Geosciences* 1.2 (2010): 106–118.

[3] Chen, Xiao-Ling, *et al.* "Remote Sensing Image-based Analysis of the Relationship Between Urban Geat Island and Land Use/Cover Changes." *Remote Sensing of Environment* 104.2 (2006): 133–146.

[4] Zhang, Xuhui, et al. "The Global Potential Distribution of Invasive Plants: Anredera Cordifolia Under Climate Change and Human Activity Based on Random Forest Models." *Sustainability* 12.4 (2020): 1491.

[5] Magli, Susanna, *et al.* "Dynamic Analysis of the Heat Released by Tertiary Buildings and the Effects of Urban Heat Island Mitigation Strategies." *Energy and Buildings* 114 (2016): 164–172.

[6] Roman, Kibria K., *et al.* "Simulating the Effects of Cool Roof and PCM (Phase Change Materials) Based Roof to Mitigate UHI (Urban Heat Island) in Prominent US Cities." *Energy* 96 (2016): 103–117.

[7] Meng, Yue-bo, et al. "Real-Time Dynamic Estimation of Occupancy Load and An Air-Conditioning Predictive Control Method Based on Image Information Fusion." *Building and Environment* (2020): 106741.

[8] Jagannathan, J. and Divya, C., 2021. Deep Learning for the Prediction and Classification of Land Use and Land Cover Changes Using Deep Convolutional Neural Network. *Ecological Informatics*, 65, p.101412.

[9] Jagannathan, J. and Divya, C., *Time Series Analyzation and Prediction of Climate Using Enhanced Multivariate Prophet.*

[10] Singh, S. N., and Abheejeet Mohapatra. "Repeated Wavelet TransformBased ARIMA Model for Very Short-term Wind Speed Forecasting." *Renewable Energy* 136 (2019): 758–768.

[11] Alsharif, Mohammed H., Mohammad K. Younes, and Jeong Kim. "Time Series ARIMA Model for Prediction of Daily and Monthly Average Global Solar Radiation: The Case Study of Seoul, South Korea." *Symmetry* 11.2 (2019): 240.

[12] Karimi, Mahshid, *et al.* "Analysis and Prediction of Meteorological Drought Using SPI Index and ARIMA Model in the Karkheh River Basin, Iran." *Extreme Hydrology and Climate Variability.* Elsevier, 2019. 343–353.

[13] Ding, Shuai, et al. "Time-aware Cloud Service Recommendation Using Similarity-enhanced Collaborative Filtering and ARIMA Model." *Decision Support Systems* 107 (2018): 103–115.

[14] Razzaghmanesh, M., Beecham, S., & Salemi, T. (2016). The Role of Green Roofs in Mitigating Urban Heat Island Effects in the Metropolitan Area of Adelaide, South Australia. *Urban Forestry & Urban Greening*, 15, 89–102.

[15] Kusaka, H., & Kimura, F. (2004). Coupling a Single-layer Urban Canopy Model with a Simple Atmospheric Model: Impact on Urban Heat Island Simulation for an Idealized Case. *Journal of the Meteorological Society of Japan. Ser. II*, 82(1), 67–80.

[16] Yu, C., & Hu, D. (2018, July). Modeling the Heat Island Intensity (HII) Based on Distance Diffusion and Typical Ground Feature Types in Beijing Downtown. *In IGARSS 2018-2018 IEEE International Geoscience and Remote Sensing Symposium* (pp. 838–841). IEEE.

[17] Claudia Fabiani (2018 Sept). Urban Canopy Models for the Analysis of Adaptive Envelope Materials on Outdoor Comfort and Building Energy Needs. In *73rd Conference of the Italian Thermal Machines Engineering Association* (ATI 2018).

Artificial Intelligence, Blockchain, Computing and Security – Dagur et al. (Eds)

Agricultural repercussions caused by the pandemic period

V. Sheeja Kumari*
Professor, Department of Computational Intelligence – Institute of AI & ML, SIMATS School of Engineering, SIMATS University, Chennai

D. Seema Dev Aksatha
Assistant Professor, Department of Computer Science, Sri Krishna Arts and Science College, Coimbatore

G. Michael
Professor, Department of Computational Intelligence – Institute of AI & ML, SIMATS School of Engineering, SIMATS University, Chennai

G. Vennira Selvi
Professor, Department of Applied Machine Learning – Institute of AI & ML, SIMATS School of Engineering, SIMATS University, Chennai

T. Manikandan
Professor, Department of Computational Intelligence – Institute of AI & ML, SIMATS School of Engineering, SIMATS University, Chennai

P. Gururama Senthilvel
Professor, Department of Computer Science and Engineering, SIMATS School of Engineering, SIMATS University, Chennai

ABSTRACT: COVID-19 has affected India's agriculture system. However, the latest quarterly GDP predictions post-COVID scenario show the resilience of the Indian agricultural sector, which is expected to be the sole sector to expand 4% for fiscal years. In this context, the major purpose is to assemble a summary of early evidence on COVID-19's influence on India's agriculture industry, including production, marketing, and consumption, and then devise pragmatic measures to recover and thrive after the pandemic. The epidemic has disrupted production and marketing through labour and logistical restrictions, while the negative income shock has restricted access to markets and increased food commodity costs, changing consumption patterns, the poll found. India's agriculture business was devastated by the outbreak. The crisis has allowed the state to launch many programmes and make important changes. After COVID-19, coordinated effort is needed to break the global stalemate. This review study shows how the COVID-19 epidemic has damaged global.

Keywords: Agriculture, Covid, Food Security, Production, Pandemic

1 INTRODUCTION

Global economy and lifestyles were affected by the COVID-19 epidemic. In addition to the high number of cases and deaths, the epidemic had major socioeconomic effects that are still unknown. In numerous Brazilian cities, this pandemic increased food insecurity [7]. One in five Brazilians aged 18 or older went hungry during the pandemic. In 2019, the severe COVID-19

*Corresponding Author: scjakm@gmail.com

DOI: 10.1201/9781032684994-47

epidemic caused global health difficulties. Quarantines and other limitations to fight the epidemic are expected to last weeks and months. Vaccination may help, but numerous hazards remain and the impact on all economic sectors is negative. (Horner 2020; Patrinley *et al.* 2020).

Global efforts to contain the virus by reducing human activity generated financial shocks and costs that affected agriculture and food supply. Farmers have been throwing away a lot of their food due to labour shortages and a huge drop in commercial and restaurant demand. Quarantine reduces labour for time-sensitive agricultural like vegetable, crop, and fruit picking. As the crisis worsens, these repercussions spread to food production and global economics [8]. The COVID-19 pandemic has affected food security, labour availability, agricultural system resilience, farming system interconnectedness, and others. Global economy and lifestyles were affected by the COVID-19 epidemic.

2 EVALUATION OF PUBLISHED SOURCES

The government shutdown raised food prices for three-quarters of research participants [10]. Market reforms and social safety nets for the poor, migrants, and farmers have helped the Indian government manage growing prices and social unrest. This eased fears. Despite COVID-19's quick spread and panic, food prices were generally stable, save for vegetables [8]. Short-term government aid may have helped the sector, but a price hike is unlikely.

Covid-19 is a risk-related tragedy. Though a disaster is a perception of hazard, there is no uniform idea and agreement among researchers regarding the size of hazard to qualify as a disaster. However, the COVID-19 pandemic can be considered a disaster due of its global negative impacts and risks and over 2 billion deaths [11].

First, like other populations, rural people are at risk of death, disease, health issues, stress, and trauma from COVID-19. Several studies (Abrams & Szefler 2020; McDonald *et al.* 2020; Poudel & Subedi 2020; Singh *et al.* 2020) examined these hazards for farmers and rural communities, including farmer suicides (Hossain *et al.* 2020).

Farmers lose fresh vegetables, fruit, and milk due to COVID-19. Countries' mobility and interaction limitations, job losses, and restaurant and hotel closures caused these losses (Cortignani *et al.* 2020; Harris *et al.* 2020; Henry 2020; Richards & Rickard 2020).

The COVID-19 pandemic is expected to damage soils, ecosystems, and wildlife (Huynh *et al.* 2020; Lal 2020b; Lal *et al.* 2020; McDonald *et al.* 2020; Rahim & Rahim 2020; Zambrano-Monserrate *et al.* 2020).

Scholars have identified many risks (Ahmed *et al.* 2020; Barcaccia *et al.* 2020; Barichello 2020; Martin 2020; Mitaritonna & Ragot 2020; Neef 2020; Nicola 2020; Phillipson 2020) for agricultural sector economic development break due to pandemic linked to food export and import break, bankruptcy of enterprises, loss of income, unemployment, poverty, inequality. Export limitations impede global agricultural products and food trade and market access due to the pandemic.

3 RESULTS AND DISCUSSIONS

The agricultural sector is severely impacted by the covid-19 epidemic. There was a serious risk to food security as a result of individuals being less able to buy food, move around, and communicate with one another. The hardest hit were the most defenceless members of society. Accordingly, governments' actions to halt the spread of the Corona-19 virus first and foremost influenced negatively worldwide food supply networks.

The Covid-19 pandemics have only served to exacerbate preexisting economic and social inequalities among people and disparities in the resilience of agricultural systems around the world, while also highlighting the need to strengthen safety nets dependent on the income generation and stability of agriculture sector workers [6].

The FAO anticipates food supply and demand changes worldwide. It warns of a global "food crisis" if countries fail to safeguard vulnerable people from hunger and malnutrition and unclog food supply networks. The UN has warned that the COVID-19 pandemic might cause global food shortages [3]. The World Food Programme (WFP) warns that COVID-19 is "threatening to affect millions of people already made susceptible by food instability [and] malnutrition." Hunger and malnutrition increased during the 2014–16 Sierra Leone Ebola outbreak. If they can't grow, sell their products, or go to markets, small and marginal farmers will suffer too.

Early in the pandemic, 60% of input retailers reported negative effects on demand and 48% on supply due to transportation restrictions (Figure 1).

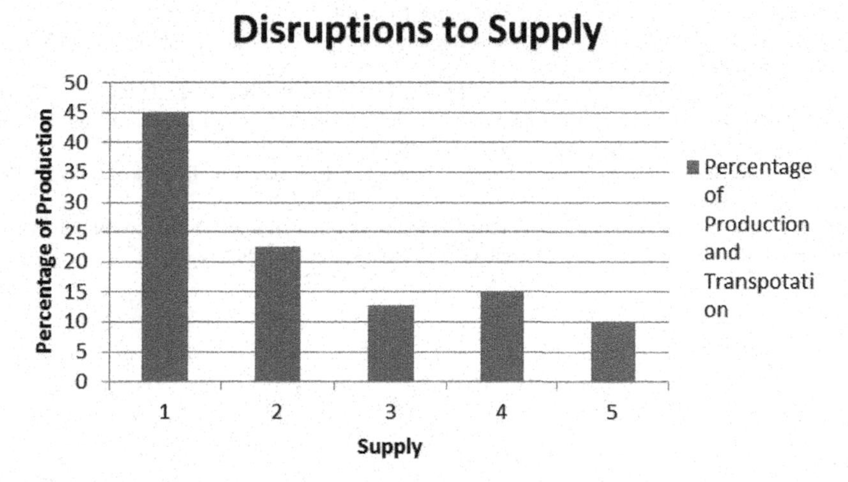

Figure 1. Disruptions to input supply.

4 CONCLUSION

The current research is limited since it attempts to capture only general and speedy consequences of the covid-19 crisis on agricultural systems generally. As food prices and distribution networks are very sensitive to market fluctuations, further study is needed to determine the impact of these events on farms of all sizes. Research into the negative effects on food security is essential for defining the adjustments that must be made to food systems in order to strengthen their resilience. Additional research is required to understand how covid-19 will affect producers in other markets, industries, and geographical areas. It is also important to determine what kinds of policy actions are most effective for making agricultural systems more resilient, such as the optimal farm size, the types of technology used on farms, the types of production methods employed, and so on.

REFERENCES

[1] Sendhil R, Ramasundaram P, Subash SP, *et al.* (2020a) Policy Imperatives for Wheat Procurement. Policy Paper 1. ICAR–Indian Institute of Wheat and Barley Research, Karnal, Haryana, India. DOI: 10.13140/RG.2.2.13542.65604/1.

[2] Mehra, P. S., Jain, K., Chawla, D., Dagur, A., Singh, S., & Sharma, J. (2022). *GWO-EFUCA: Grey Wolf Optimisation and Fuzzy Logic Based Unequal Clustering and Routing Protocol for Sustainable WSN-based Internet of Things.*

[3] Mehra, P. S., Mehra, Y. B., Dagur, A., Dwivedi, A. K., Doja, M. N., & Jamshed, A. (2021). COVID-19 Suspected Person Detection and Identification Using Thermal Imaging-based Closed Circuit

Television Camera and Tracking Using Drone in Internet of Things. *International Journal of Computer Applications in Technology*, 66(3–4), 340–349.

[4] Kumar, A., & Alam, B. (2016). Real-Time Fault Tolerance Task Scheduling Algorithm with Minimum Energy Consumption. In *Proceedings of the Second International Conference on Computer and Communication Technologies: IC3T 2015*, Volume 2 (pp. 441–448). Springer India.

[5] Padhee AK, Pingali P (2020) Lessons from a Pandemic to Repurpose India's Agricultural Policy. *Nature India. May*. DOI: 10.1038/nindia.2020.83 (accessed 15 July 2020).

[6] Lal, R. Home Gardening and Urban Agriculture for Advancing Food and Nutritional Security in Response to the COVID-19 Pandemic. *Food Secur*. 2020, 12, 871–876. [Google Scholar] [CrossRef] [PubMed].

[7] Workie E, Mackolil J, Nyika J, *et al.* (2020) Deciphering the Impact of COVID-19 Pandemic on Food Security, Agriculture and Livelihoods: A Review of the Evidence from the Developing Countries. *Current Research in Environmental Sustainability* 2:100014

[8] Sers CF, Mughal M (2020) Covid-19 Outbreak and the Need for Rice Self Sufficiency in West Africa. *World Development* 135:105071.

[9] McDonald, A. J., Balwinder- Singh; Jat, M. L., Craufurd, P., Hellin, J., Hung, N. V., Keil, A., Kishore, A., Kumar, V., & McCarty, J. L. (2020). Indian Agriculture, Air Pollution, and Public Health in the Age of COVID. *World Development*, 135, 105064.

[10] de Roo N, de Boef W (2020) Rapid Country Assessment: Ethiopia. The Impact of COVID-19 on the Food System. Wageningen University & Research (23 July 2020).

[11] Harris J, Depenbusch L, Pal AA, *et al.* (2020) Food System Disruption: Initial Livelihood and Dietary Effects of COVID -19 on Vegetable Producers in India. Food Security 12:841–851.

Artificial Intelligence, Blockchain, Computing and Security – Dagur et al. (Eds)
© 2024 The Author(s), ISBN: 978-1-032-67841-2

Pandemic state's impact on agriculture

V. Sheeja Kumari
Department of Computational Intelligence – Institute of AI & ML, SIMATS School of Engineering, SIMATS University, Chennai

N.R. Wilfred Blessing
IT Department, College of Computing and Information Sciences, University of Technology and Applied Sciences-Ibri, Oman

C. Rohith Bhat & I. Sudha
Department of Computer Science and Engineering, Saveetha School of Engineering, SIMATS, Chennai

G. Michael & T. Manikandan
Professor, Department of Computational Intelligence – Institute of AI & ML,SIMATS School of Engineering, SIMATS University, Chennai

ABSTRACT: The COVID-19 outbreak has significantly disrupted the agricultural system in India. In spite of this, the most recent quarterly GDP forecasts post-COVID scenario demonstrate the tenacity and durability of the Indian agricultural sector, which is anticipated to be the only sector to record a positive growth rate of 4% for fiscal years (FY) 2020 and 2021. Similarly, growth for the most recent quarter is projected to have been 5.89%, representing a reduction of 2.4% point. In this context, our goal is to compile a summary of the early evidence on the impact of COVID-19 on the agricultural sector in India, including production, marketing, and consumption, and then to design a set of pragmatic strategies to recover and thrive after the pandemic. According to the survey's findings, the pandemic has disrupted production and marketing through labour and logistical restrictions, while the negative income shock has restricted access to markets and increased food commodity prices, altering consumption patterns. The pandemic caused great emotional, social, economic, and bodily harm to everyone involved in India's agriculture industry. The state has responded to the crisis by seizing the opportunity it presents and launching a variety of programmes and necessary changes. We suggest a ten-point plan that includes social safety nets, family farming, the trading of buffer stock, staggered procurement, and secondary agriculture in order to revitalise and prosper after the epidemic.

1 INTRODUCTION

The COVID-19 virus, which was first found in Wuhan, China, the core of the outbreak, travelled throughout the rest of the world and eventually became a pandemic. India accounted for 14.6% of all global ailments as of December 6, 2020, with 9.5 million sick individuals. This consequently resulted in a 23.9% drop in India's GDP during the first quarter of fiscal year 2020–21[5]. India has already surpassed the United States as the second most widespread place for the infection. A shock generated by a pandemic may have a bigger impact on economies than a shock induced by the weather, such as a drought or flood, or a trade embargo, due to the loss of human lives. All of these shocks have indisputably negative

DOI: 10.1201/9781032684994-48

effects on agricultural systems. In contrast, pandemic shocks have an effect on the entire economy. The outbreak disrupts both the demand and supply of food, so impacting the global supply chain. Droughts, on the other hand, tend to be more localised and adversely affect only the industry or stakeholders associated with them [7]. Similarly, the jolts caused by a trade embargo are detrimental to a particular industry and can be mitigated quickly by introducing the appropriate legislative measures. Internationally networked wholesalers and merchants will purchase their inventory from other sources if there is a lack of commodities owing to droughts, for example, in an effort to alleviate the negative impacts [7]. In contrast, the effects of a pandemic could be more widespread and severe, potentially triggering a recession.

The Indian government has issued an order for a nationwide lockdown beginning on March 25, 2020 [5,] with a worldwide stringency index severity level exceeding 76. This directive will have an effect on the economy and agriculture. The agriculture industry showed positive growth after the pandemic (3.45% for FY 2020–21 Quarter 1: April to June), but it was lower than the previous quarter's growth (5.89% for FY 2019–20 Quarter 4: January to March). This decline was due to COVID-19, which resulted in a decrease of 2.5 percentage points. Although a bountiful crop harvest and a decrease in agriculture-related activity during the lockdown contributed to the positive growth in agriculture during the first quarter of fiscal year 2020–21, farm revenue did not increase much. The lockdown imposed by COVID-19 exacerbated food waste across manufacturing, marketing, delivery, and consumption at the household level. For example, milk, vegetables, and fruits were thrown at the farm level as a result of low demand and inadequate logistics. In addition, Anand Milk Union Limited (AMUL), the largest milk cooperative in India, was unable to distribute milk [9].

The impact of COVID-19 on the agriculture system in India: Production, distribution, and final consumer usage

The disruption of food supply networks caused a rise in food prices[14] and affected farming operations. The uncertainty caused by the crisis hindered interstate travel and shipping. According to our research of official time series price data for major food commodities from 01.01.2020 to 10.10.2020, wholesale and retail prices for pulses, wheat flour, and milk were 1–5% higher one month after the lockdown. The prices of edible oils and staple cereals (rice and wheat) reduced by 4–10% as a result of the removal of import restrictions and government actions such as the free distribution of food grains. After the lockout, prices for vegetables rose, particularly for tomatoes, which experienced a weekly increase of 75–80 percent and a monthly increase of 110–120 percent. As a result of distressed sales, market arrivals increased in May, while market developments sheltered farmers from falling prices [16]. Rural and thinly populated regions, as well as smaller towns, saw greater price increases.

2 EVALUATION OF PUBLISHED SOURCES

In India, the shutdown caused by COVID-19 affected food markets and compelled people to alter their eating patterns. The significance of products that satisfied their needs and desires was evaluated by consumers. Several surveys revealed that during the lockdown, individuals either lost their jobs or saw a decrease in income [17]. As a direct result of the shutdown, which coincided with an unanticipated fall in revenue, India's food and nutrition security became a major problem. A survey of approximately 2,500 adolescent migrants indicated that 30% of respondents had reduced their meal frequency. Their consumption patterns shifted evenly independent of the severity of the incident, which ranged from green to orange to red [12] [citation needed].

3 A TEN-POINT STRATEGY FOR AGRICULTURE INDUSTRY ENHANCEMENTS FOLLOWING THE COVID-19 MEETING

Following the aforementioned discussion on the implications of COVID-19 on India's agricultural system, we were able to develop a 10-point plan for bolstering the sector against the pandemic's crisis and sustainability concerns. This strategy will be discussed in further depth in the following section.

1. Safety nets for the community
2. Risk management for prices and revenues
3. Mainly concentrating on the transition from primary to secondary agriculture
4. Family-owned agriculture
5. Agriculture practised in communities
6. Investment in Agricultural Research and Development
7. Buffer stock
8. Disagreements on pricing and procurement
9. Agricultural financial system reforms
10. Stakeholder partnerships

4 CONCLUSION

The problem caused by the virus has wreaked havoc on the agricultural systems of India and other nations. Without a grasp of the effects COVID-19 will have on the agricultural sector, particularly in the world's least developed nations, it is difficult to offer a solution to the looming global food security catastrophe. Numerous governments, including Kazakhstan, Myanmar, Russia, and Vietnam, have placed restrictions on cereal trade.

The government should immediately raise expenditure on social safety nets and other short- and medium-term efforts so that millions of people whose livelihoods depend on the agricultural system have the best chance of being preserved and protected. To get the economy back on its feet after a pandemic, the first step is to improve revenue sources by selling excess buffer stock and providing additional loans to the farm sector.

REFERENCES

[1] Barrett CB (2020) Actions Now Can Curb Food Systems Fallout from COVID-19. *Nature Food* 1 (6):319–320.

[2] de Roo N, de Boef W (2020) *Rapid Country Assessment: Ethiopia. The Impact of COVID-19 on the Food System*. Wageningen University & Research (23 July 2020).

[3] Chengappa PG (2013) Secondary Agriculture: A Driver for Growth of Primary Agriculture in India. *Indian Journal of Agricultural Economics* 68(1):1–19.

[4] Harris J, Depenbusch L, Pal AA, et al. (2020) Food System Disruption: Initial Livelihood and Dietary Effects of COVID-19 on Vegetable Producers in India. *Food Security* 12:841–851.

[5] Baudron F, Liegeois F (2020) Fixing our Global Agricultural System to Prevent the Next COVID-19. *Outlook on Agriculture* 49(2):111–118.

[6] Kalsi SS, Sandoval L, Sood D (2020) COVID-19 in India – Trade Situation Update Report Highlights. Report No. IN2020-0017

[7] Cariappa AGA, Mahida DP, Lal P, et al. (2020b) Correlates and Impact of Crop Insurance in India: Evidence from a Nationally Representative Survey. *Agricultural Finance Review. Epub ahead-of-print.* DOI: 10.1108/AFR-03-2020-0034.

[8] Mishra A, Bruno E, Zilberman D (2021) Compound Natural and Human Disasters: Managing Drought and COVID-19 to Sustain Global Agriculture and Food Sectors. *Science of the Total Environment* 754:142210.

[9] Bellemare MF (2015) Rising Food Prices, Food Price Volatility, and Social Unrest. *American Journal of Agricultural Economics* 97(1):1–21.
[10] Varshney D, Roy D, Meenakshi JV (2020) Impact of COVID-19 on Agricultural Markets: Assessing the Roles of Commodity Characteristics, *Disease Caseload and Market Reforms. Indian Economic Review* 55:83–103.
[11] Principato L, Secondi L, Cicatiello C, et al. (2020) Caring More About Food: The Unexpected Positive Effect of the Covid-19 Lockdown on Household Food Management and Waste. *Socio-Economic Planning Sciences*. DOI: 10.1016/j.seps.2020.100953.
[12] Sendhil R, Kar A, Mathur VC, et al. (2013) Price Discovery, Transmission and Volatility: Evidence from Agricultural Commodity Futures. *Agricultural Economics Research Review* 26(1):41–54.

Artificial Intelligence, Blockchain, Computing and Security – Dagur et al. (Eds)
© 2024 The Author(s), ISBN: 978-1-032-67841-2

Jaya integrated binary whale optimized algorithm for preserving medical data

I. Sudha*
Professor, Department of Computer Science and Engineering, SIMATS School of Engineering, SIMATS, Chennai, Tamil Nadu, India

P.S. Ramesh
Associate Professor, Department of Computer Science and Engineering, Vel Tech Rangarajan Dr. Sagunthala R&D Institute of Science and Technology, Chennai, Tamil Nadu, India

S. Jagadeesan
Assistant Professor, Senior Grade 2, School of information Technology and Engineering, Vellore Institute of Technology, Vellore, Tamilnadu, India

Guru Vimal Kumar Murugan
Assistant Professor, Department of Computer Science and Engineering, Vel Tech Rangarajan Dr. Sagunthala R&D Institute of Science and Technology, Chennai, Tamil Nadu, India

C. Gokulnath
Assistant Professor, Department of Computer Science and Engineering, Vel Tech Rangarajan Dr. Sagunthala R&D Institute of Science and Technology, Chennai, Tamil Nadu, India

N. Poongavanam
Associate Professor, Department of Computer Science and Engineering, Vel Tech Rangarajan Dr. Sagunthala R&D Institute of Science and Technology, Chennai, Tamil Nadu, India

ABSTRACT: Cloud Computing has the source of information to various sectors that the future generation duly depends on the resources of the cloud. Users can access a variety of network, storage, and platform features through the cloud. The users exploit the cloud's resources in accordance with their needs, and the data is centralized for cloud storage. Hence, The main aim proposes a method called Jaya Binary Whale Optimization (JBWO), which combines the Binary Whale Optimization and Jaya algorithm (JBWOA) and modifies the Fully homomorphic encryption (FHE) Organism exhausting Advanced Encryption Standard (AES), for the cloud to start securely transmitting data.. By applying the proposed JBWO algorithm to generate the Data Protection coefficient (DP), the original data is retained. For the purpose of selecting the best resolution, privacy and utility parameters are used to calculate fitness. By combining the key vector and the Key Information Product (KIP) matrix with the EX-OR operation, the sanitized data are produced. By utilizing the key that the data owner provides, users can access the unique data. The strategy produced improved results in the analysis utilizing the Cleveland datasets and is capable of managing sensitive information with clearly specified privacy. The suggested privacy protection techniques effectively safeguard data while maintaining privacy, resulting in satisfied customers.

Keywords: Fully Homomorphic encryption, Cloud Computing, Advanced Encryption Standard, Privacy, JBWOA

1 INTRODUCTION

The evolution of academia and industry is gaining considerable attention as a result of innovations in cloud computing [1]. It exhibits how radical collaboration, agility, availability, size,

*Corresponding Author: dharshini21sudha@gmail.com

DOI: 10.1201/9781032684994-49

and cost effectiveness may be changed. Instead of creating and operating dedicated data centers, medical professionals and many software vendors are prepared to migrate their ElectronicMedicalRecord (EMR) systems into the cloud. The cloud computing also emphasizes on improving the effectiveness of handling medical data and sharing the process, However, it also makes it possible to purchase medical services anywhere in the world because patient information about health care must always be available, no matter the location [15]. It has been noticed that interacting with applications associated to healthcare exhibits more variances than working with healthcare-related information. Both doctors and patients will gain more from the cloud services and EHR software [2,3]. The cloud facilitates mobile Telecare and aids in ensuring adequate patient care. It also offers various merits, like the possibility to access data whenever desired, whenever and any time, at a lower cost, and with widening digital [4]. The JBWO algorithm, which incorporates the Jaya algorithm and BWOA, is the most important information is displayed to the end consumers, keeping the original data encrypted [14].

2 LITERATURE REVIEW

Revathi *et al.* [5] Proposed for Adaptive fractional brainstorm and whale optimization is suggested. To protect the privacy and the created AFBS WOA process yields the keymatrix coefficient for recovering the perturbed database. The experimental study met the highest usefulness and privacy standards. Masud *et al.* [6] Described each cloud dynamically calculates a pairing in an elliptic curve to produce a secret session key. Angeetha *et al.* [7] Proposed PHR architecture for distributing PHRs to multiple users. Patients can encrypt and store their PHRs by implementing attribute-based encryption. By granting certain data users fine-grained, attribute-based access privileges, it maintains control over who gets access to its users' PHRs. Thilakanathan *et al.* [8] Discussed to address privacy and security concerns. We begin by demonstrating a Telecare application. It will enable patients to securely and confidentially exchange their medical information to medical professionals. Doel *et al.* [9] Proposed GIFT-Cloud. The Server offers secure, anonymized data storage, REST API for connecting third-party software. GIFT-Cloud facilitates the transmission of image data, as well as the expansion and medical examination software validation and the dissemination.

3 PROPOSED METHOD

3.1 *JBWOA proposed developing a DP coefficient to cloud security data transport*

The toughest part of cloud computing systems are safeguarding data security even though service providers must be capable of ensuring that private information won't be revealed towards other parties even without the user's consent. As a consequence, it urges users to assess whether to store their data in the cloud or not. Due to obstacles that cloud computing raises when it concerns to data security and privacy, it is not feasible to rely solely on the physical computer.

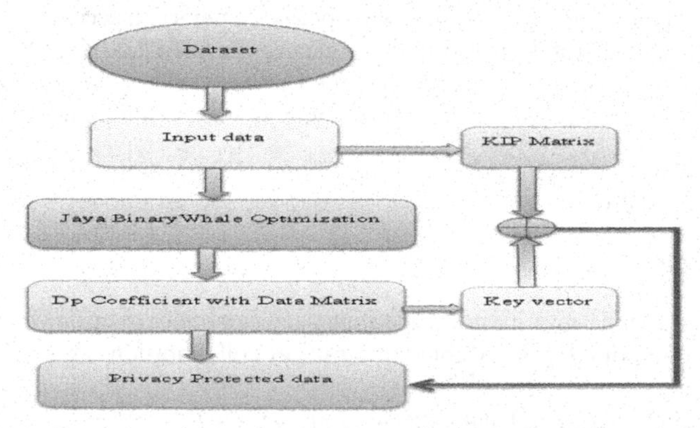

Figure 1. Block diagram of the JBWOA.

Whenever consumers retain vital data on the cloud, the majority of clients are apprehensive about the confidentiality of the information stored. Hence, the secure data transit is revised employing FHE with the proper DP with data matrix to tackle these concerns. Figure 1 depicts the JBWOA for safe data posting.

3.2 *DP coefficient vector creation with the proposed JBWOA*

Based on optimization strategies, Jaya algorithm appears convenient to employ and adept of tackling both constrained and unconstrained systems. It is considered as an effective mechanism for process and system optimal, while the BWOA algorithm is another optimization technique that excludes the unrealistic solutions by employing a heuristic approach [16]. The BWOA is easy to utilize and doesn't cost much to estimate. Hence, the Jaya method and BWOA integration are adopted in the JBWOA offered for enhancing the DP coefficient for establishing secure connection.

3.3 *Representation of a solution*

Using issue forms, the algorithm's solution is revealed. The objective of viability is to produce data that are shielded from public view while yet revealing the vital. The resolve is embodied by a vector of n remedies, all of the aspects $r \times 1$, Here r indicates the entire number of the database's records. The recommended JBWO process utilizes the Fitness function expresses on utility and privacy factor to achieve the ideal solution.

3.4 *Combining privacy and utility factors to evaluate the fitness function for data security*

This approach includes utilizing the Fitness function (FF), which determines the optimal DP coefficient based on privacy protection criteria for safeguarding sensitive data. As a consequence, the two key factors used to establish the ideal DP coefficient are privacy and utilized. The user's sought data must be gathered with the optimal level of utility with privacy. The FF appears as,

$$FF = \frac{1}{2} * [P + U] \quad W = \frac{T_p + T_q}{E_p + E_q + T_p + T_q}$$

Where, T_p denotes the true positives, T_q represents the true negatives, E_p denotes the number of false positives, and E_q represents the number of false negatives.

3.5 *JBWO algorithm*

The suggested JBWOA for secure data transfer is as recommended:

Initialization:

The initial criteria also include set of solutions, the algorithmic factors, and terminate criteria. The solutions are demonstrated to have been started as,

$$Y = \{I_1, I_2, ..., I_x, ..., I_o\}$$

Estimate of Fitness:

The solution is whatever provides the greatest fitness value. As each resolution strives to reach the ideal situation, the best solution is revealed in the last phase.

BWOA integration update of the solution:

After reorganizing, the solution is given by,

$$I(d+1) = I(d)[1 - z_1 + z_2] + z_1 I * (d) - z_2 I_w(d)$$

These relevant data are aggregated and handed to the previously done as input. WOA just considers the workable options and eliminates the worst option. So, WOA is used to identify the ideal response.

Discovering the feasible solution and termination:

In this scenario, the fitness of the new solution is tied to determining if the old one is more fit. As a result, the point is changed until the desired result is obtained. It is locating the optimum position.

4 RESULTS AND DISCUSSION

Windows 10's 64-bit version and 2GB of RAM are used for the testing in this paper. Java is utilized to make the proposed JBWO. The suggested JBWOA is created using Java. The Cleveland data sets from the healthcare dataset are obtained using the UCI-MLR [10]. David W. Aha provided dataset 1, and Cleveland Clinical Foundation provided dataset. Here, factors including age, resting blood pressure, and serum cholesterol are taken into consideration.

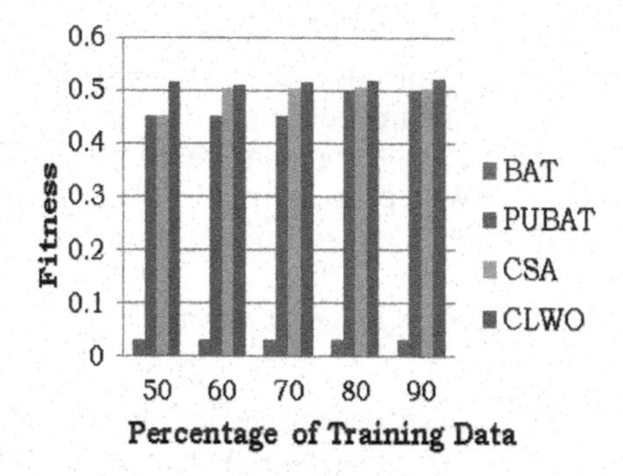

Figure 2. Fitness for k = 10.

Figure 3. Accuracy for k = 10.

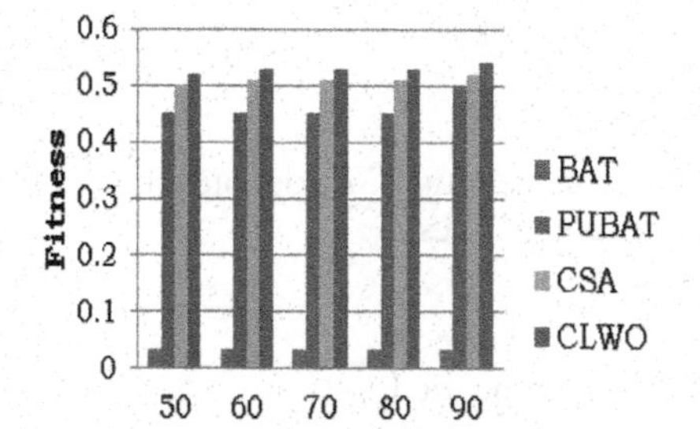

Figure 4. Fitness for k = 20.

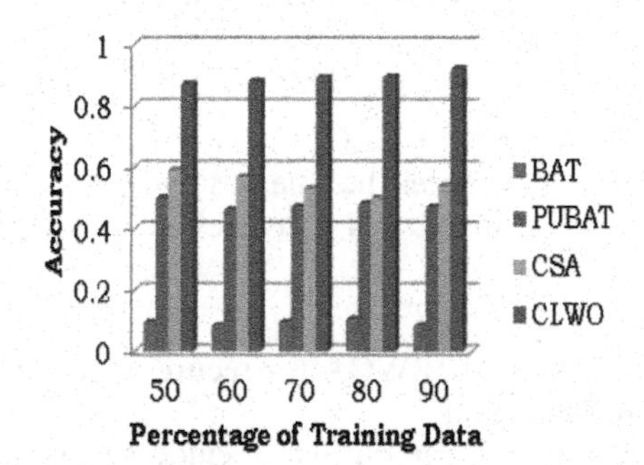

Figure 5. Accuracy for k = 20.

The accuracy and fitness based on performance comparison of the following methodologies, namely: i) Comparative methodologies are analyzed using the Bat Algorithm [12], ii) The CSA [11], iii) the PU-bat algorithm (PUBAT) [13]. Java 1.8 and the Netbeans IDE are used to execute the proposed algorithm. The system type requires Windows 10 computers with I3 Intel Core CPUs, and 4 GB of main memory. To simulate cloud, the Cloudsim framework is utilized. The proposed approach will be demonstrated in comparison to several privacy models such as BAT, PUBAT, and CSA computation [16]. The results are determined by changing the population size, for instance, by changing the herd size between 10 and 20. The suggested JBWO algorithm's execution sequence is depicted in Figures 2, 3, 4, and 5. As a result, the suggested JBWOA outperforms other current approaches. The fitness and accuracy values of various algorithms are 0.0324, 0.45, 0.504, 0.5158 from the analysis using dataset.

5 CONCLUSION

The necessity and significance of using security measures to secure cloud data make the rationale for the study very clearly. Data that has been leased and stored in the cloud is secured using the following sequential encryption-based security method. The data are initially separated into several chunks for the sequential encryption to be carried out. Using the JBWO algorithm, which increases the communication's security, the data's security is ensured. Sensitive data retained within the cloud are adequately secured by the suggested approach of data security. The original database is kept intact by generating a privacy-preserving database. By integrating the Jaya and the BWOA, the proposed JBWO algorithm evaluates the SU vector. The suggested algorithm guarantees the concealment and usefulness of the cloud data.

REFERENCES

[1] Mell P., and Grance T., On-demand Self-service., *Nist*, vol. 15, pp. 10–15, 2009.

[2] Ahn G.-J., Hu H., Lee J., and Meng Y., Representing and Reasoning about Web Access Control Policies, In *Proceedings of IEEE 34th Annual Computer Software and Applications Conference (COMPSAC)*, pp. 137–146, 2010.

[3] Wu R., Ahn G.-J., and Hu H., Secure Sharing of Electronic Health Records in Clouds, In *Proceedings of 8th IEEE International Conference on Collaborative Computing: Networking, Applications and Worksharing*, pp. 711–718, 2012.

[4] W. Jansen and T. Grance, *Guidelines on Security and Privacy in Public Cloud Computing*, NIST Special Publication- 800-144, 2011.

[5] Revathi, S & Kalaivani, J. & Christo, Mary Subaja & Pelusi, Danilo & Azees, M.. (2021). Cloud-Assisted Privacy-Preserving Method for Healthcare Using Adaptive Fractional Brain Storm Integrated Whale Optimization Algorithm. *Security and Communication Networks*. 1–10, 2021.

[6] Masud M. and Hossain M. S., "Secure Data-exchange Protocol in a Cloud-based Collaborative Health Care Environment," *Multimedia Tools and Applications*, vol. 77, no.9, pp. 11121–11135, 2018.

[7] Sangeetha D. and Vaidehi V., "A Secure Cloud Based Personal Health Record Framework for a Multi Owner Environment," *Annals of Telecommunications*, vol. 72, no. 1 2, 2017, pp. 95–104.

[8] Thilakanathan D., Chen S., Nepal S., Calvo R., and Alem L., "A Platform for Secure Monitoring and Sharing of Generic Health Data in the Cloud," *Future Generation Computer Systems*, vol. 35, pp. 102–113, 2014.

[9] Doel, Tom & Shakir, Dzhoshkun & Aughwane, Rosalind & Aertsen, Michael & Moggridge, James & Bellon, Erwin & David, Anna & Deprest, Jan & Vercauteren, Tom & Ourselin, Sébastien. (2016). GIFT-Cloud: A Data Sharing and Collaboration Platform for Medical Imaging Research. *Computer Methods and Programs in Biomedicine*. 139:181–190, 2017.

[10] Marés J. and Torra V., PRAM Optimization Using an Evolutionary Algorithm, In *Proceedings of the International Conference on Privacy in Statistical Databases*, pp. 97–106, 2010.

[11] Mehra, P. S., Jain, K., Chawla, D., Dagur, A., Singh, S., & Sharma, J. (2022). GWO-EFUCA: Grey Wolf Optimisation and Fuzzy Logic based Unequal Clustering and Routing protocol for Sustainable WSN-based Internet of Things.

[12] Mehra, P. S., Mehra, Y. B., Dagur, A., Dwivedi, A. K., Doja, M. N., & Jamshed, A. (2021). COVID-19 Suspected Person Detection and Identification Using Thermal Imaging-based Closed Circuit Television Camera and Tracking Using Drone in Internet of Things. *International Journal of Computer Applications in Technology*, 66(3–4), 340–349.

[13] Kumar, A., & Alam, B. (2016). Real-Time Fault Tolerance Task Scheduling Algorithm with Minimum Energy Consumption. In *Proceedings of the Second International Conference on Computer and Communication Technologies: IC3T 2015*, Volume 2 (pp. 441–448). Springer India.

[14] V. R and Lavanya V., "An Survey Analysis of Security Issues in the Cloud Data Storage," *2022 8th International Conference on Smart Structures and Systems (ICSSS)*, 2022, pp. 1–8.

[15] Sudha I. and Nedunchelian R., "A Secure Data Protection Technique for Healthcare Data in the Cloud Using Homomorphic Encryption and Jaya–Whale Optimization Algorithm" *World Scientific*, Vol.10(6), 2019.

[16] Sudha I., Dr. Nedunchelian R., "Preserving Healthcare Data In The Cloud Using C-Lion And Whale Optimization Algorithm", *International Journal of Scientific & Technology Research*, Vol.8(11), 2019.

Artificial Intelligence, Blockchain, Computing and Security – Dagur et al. (Eds)

Visualization of Covid 19 pattern using machine learning

M. Thanjaivadivel*
School of Computing and Information Technology, Reva University, Bengaluru

Ignatious K. Pious*
Department of Computer Science and Engineering, Vel Tech Rangarajan Dr.Sagunthala R&D Institute of Science and Technology, Avadi, Chennai

J. Jagannathang*
School of Information Technology and Engineering, Vellore Institute of Technology, Katpadi, Vellore Dt, Tamil Nadu
ORCID ID: 0000-0003-0059-8236

ABSTRACT: The Wuhan market in China is the source of a terrible and undetectable threat to the entire planet. The time between the outbreak and the epidemic was barely a few months. Visual data analysis makes it easier to get information from numerous data sources in a better method because it helps to understand how the world combats this disease. In this essay, we'll demonstrate how one might visualise a map of their own nation, along with the regions where this dreadful sickness has spread. This essay discusses the COVID-19 epidemic's data identification and analysis. In a smart, smart, and global nation, data recognition aids in the integration of all data sources, including the number of instances discovered, deaths, and current cases. Data is disseminated by putting it into visual aids like graphs and maps. This offers current and trustworthy information that enables us to monitor the number of persons admitted to hospitals in every region of India as well as in other nations around the world. This clarifies how you can handle the threat that the Coronavirus poses. In this essay, we'll show how our theory can be used to analyse and forecast the spread of COVID-19.

Keywords: Data visualization, Visual exploratory data analysis, COVID-19

1 INTRODUCTION

A public awareness campaign for the Novel Coronavirus (COVID-19) from the city of Wuhan in central China has been announced. In December 2019, the WHO labelled the coronavirus a global pandemic. A unique time in human history was spent fighting the intangible foe COVID-19. A viral family like MERS and SARS-COV-2 is not the cause of the non-communicable disease corona virus. A sizable family of viruses is COVID-19. Between November and December 2019, it went from animals to people. The typical cold and cough are caused by it. The virus appears to kill adults and exhibits sympathetic symptoms. [1] By offering pertinent coronavirus assistance globally, we used a variety of visual and data analytic techniques to assess our pertinent database. We have gathered information from the WHO, the NHC, and John Hopkins University. COVID-19 State, National, and Global data sets are the primary emphasis of the data to identify the affected locations. [13]. We will discuss displaying actual death circumstances, as well as stable and confirmed

*Corresponding Authors: thanjaivadivel@gmail.com, ignatiouspious@gmail.com and jagannathan161091@gmail.com

 DOI: 10.1201/9781032684994-50

cases, in regard to a specific area or nation. Analyzing data via a visual analytical lens makes it easier to evaluate the numerous presumptions of data gleaned from the source. Users can access particular information about the widely used COVID-19 using it. In order to provide us a comprehensive understanding of the potential COVID-19 epidemic, a number of data perceptions and analyses were also covered in the briefing. The analysis of a separate series of time data will be examined in this part using Python. The most open source programming language for web-based applications has always been Python. We map out the global distribution of SARS-COV2 and provide information about it. [2–4].To display the present state of the COVID-19 cases in States and globally in a designed view of a geographical map, we will be building a website for the COVID-19 section and adding two distinct pages to the dashboard. A display screen allows for friendly user interaction. Heat maps, death maps, active cases maps, confirmed cases maps, graphs, and bars are all available as view options on the display screen. This website has dynamic, user-friendly elements that you may utilise, like zoom, tooltips, clicking options, scrolling the map, and pointing the map, among others. Users are able to determine the severity of COVID-19 globally using the heat map. The most high-flying locations are identified on a systematic time scale by the time series that it will plot. [5] A workstation or personal computer is utilised as a computational device for data visualisation, followed by a communication interface. A dataset's distribution, geology, trends, and other aspects can all be understood visually thanks to data visualisation. The data is analysed using a variety of technologies, and dashboards with various data visualisations are also available. Sometimes different ETL tools are utilised to set up operational data sources. [6–8] When a pandemic is widespread over the world, as COVID-19, it is crucial for consumers to be informed of the most recent developments and to understand the illness. In this work, many forms of research regarding COVID-19 are being experimented with. This provides as a starting point to sort through a lot of uncertainty on all sides of the world scale. The authors of [1] describe how we can use GIS to track COVID-19 data. Big data is used by GIS, which gives it the ability to evaluate pandemic propagation. It serves as a source of data for COVID-19, which may be simply used for data management, mapping methods, and spatial applications. Here, a country-level, city-level, and community-level spatial analysis is conducted. Both individual and group research is done, and the analysis is completed quickly. A Corona Tracker that can pinpoint the location where a Corona has afflicted people in real-time was developed by the researchers in [14]. They hope to anticipate the number of COVID cases, afflicted, deaths, and recovered by the use of predictive modelling. Covid data in China are tracked using SEIR modelling prediction and daily observation. Researchers have demonstrated that this strategy works well.

2 DATA VISUALIZATION

The graphic representation of data and statistics using elements like charts, maps, and graphs is known as data visualisation. These tools give you the ability to observe and follow the course models and designs that are present in the data. The system may be made easy, the notion is definable, and the appalling data is made visible with excellent examples thanks to this visualisation concept. These kinds of technology are required to gather a lot of data and track its source. It is necessary for any character to exist in data for a visual representation of the unpredictable nature and operating portrayal. With the aid of lines, graphs, and bar graphs for each country, data visualisation enables us to comprehend the spread of COVID-19. They will provide as a daily indicator of the number of fatalities, impacted people, and recovered cases.

3 DATA SOURCE

The information is gathered from reliable sources, where websites publish databases on confirmed COVID-19 cases, along with death and recovered statistics for the states and nations affected. These are the sources.

Table 1. Data source and source link.

Data source	Source link
Covid-19 Tracker By JHU [26]	https://coronavirus.jhu.edu/map.html
WHO [13]	https://covid19.who.int/
Microsoft Covid-19 Tracker [29]	https://www.bing.com/covid/local/india
CDC [31]	https://covid.cdc.gov/covid-datatracker/#datatracker-home
COVID-19 Mohfw [29]	https://www.covid19india.org/

4 CONVOLUTIONAL NEURAL NETWORK

A group of deep neural networks called convolutional neural networks (CNN) are used to integrate picture viewing. Image 2. A specific neuron in one layer is associated with other neurons in other layers in these fully integrated networks. Neural networks organise increasingly complex patterns using simpler patterns when they detect complex patterns in input. CNN is present at the bottom end as a result. The convolutional layer is the foundation of the complete CNN in a neural network. It is a crucial network to do image classification, face and object detection, and picture identification. CNN's image classification procedure starts with the input image, processes it to get the output image, and then classifies it. The computer recognises the input image as a pixel array and validates the resolution of the image input. The ReLU algorithm is crucial to CNN because it gives the traditional network non-linearity [16,33].

5 EXPLORATORY DATA ANALYSIS

Table 1 collects information from sources such the John Hopkins University, the World Health Organization, Microsoft Covid Tracker, and others as it becomes available. We are able to learn more about the global pandemic thanks to these daily updates. With the use of a Jupyter notebook, we are developing a real-time display of COVID-19 status using data visualisation and data analysis techniques. Data visualisation facilitates the creation of bar graphs, lines, charts, and maps using data taken from a data source. A graph's insights change as data are updated, and vice versa. People can learn more about the current circumstances and be better prepared for the worst by reading this document.

(i) Global reported cases

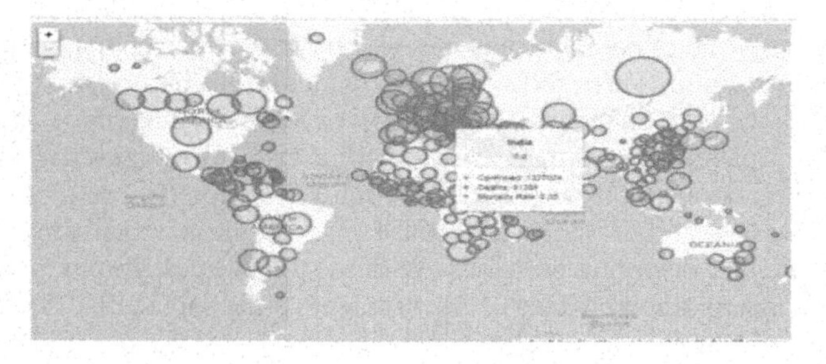

Figure 1. Global reported cases.

Figure 1's globe map displays the total number of active cases, recovered cases, and deaths in India. The world's afflicted regions are represented by the orange circles on this map. Figure 1's globe map displays the total number of active cases, recovered cases, and deaths in India. The world's afflicted regions are represented by the orange circles on this map.

(ii) Country wise reported cases

Table 2. Country wise reported cases.

Country	Total confirmed	Total recovered	Total deaths
USA	4,248,492	2,028,361	1,48,492
BRAZIL	2,348,200	1,592,281	85,385
INDIA	1,339,067	850,295	31,425
RUSSIA	806,720	597,140	13,192
CHILE	341,304	313,696	8,914
SPAIN	319,501	7,015	28,432
UK	297,914	135	45,677

(iii) State wise reported cases

Table 3. State wise reported cases.

Location	Confirmed	Recovered	Deaths
Maharashtra	348000	19400	12,584
Delhi	127,000	10,900	3,745
Chennai	92,206	76,494	1,969
Mumbai	84,524	55,884	4,899
Karnataka	80,863	29,310	1,616
Andhra Pradesh	72,711	37,555	884
Uttar Pradesh	58,104	35,803	1,289

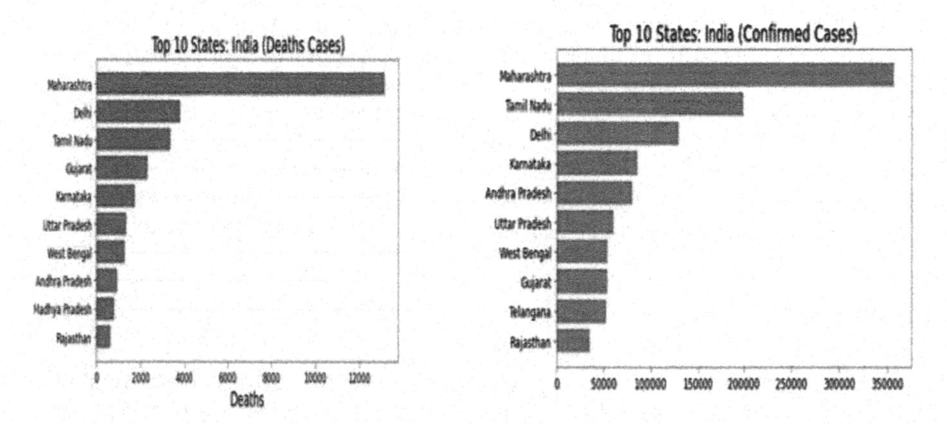

Figures 2 and 3 Displays the rate of death cases and confirmed cases of top 10 states in India.

6 COVID-19 SPREAD ANALYSIS

The spread analysis is divided into two sections namely

1. Spread across globe
2. Spread in World, Continents and affected countries

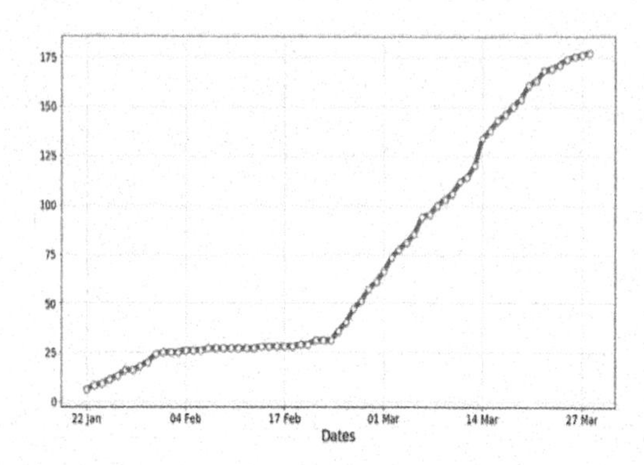

Figure 4. Rate of spread of covid-19.

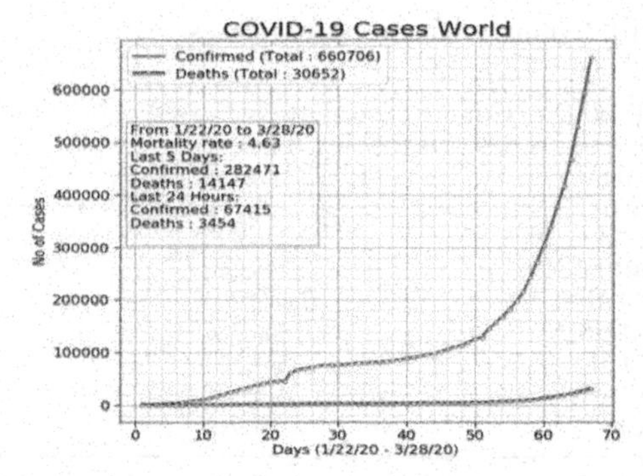

Figure 5. Status of confirmed and death cases.

It portrays the rapid spread of covid-19 from 22 January 2020 to 28 March 2020. Graph showing confirmed and death cases due to Covid-19 globally from 22 January 2020 to 28 March 2020 [17].

7 RESULTS AND DISCUSSIONS

To comprehend the number of cases during a specific time period, we show a variety of examples from around the world. After looking at each State, the entire country, and the entire world, we can find that the USA is the most impacted, with 4,248,492 affected individuals and 1,48,492 fatalities as of July 25, 2020. Maharashtra had the most confirmed cases (3,48,00) and fatalities (12,584) in India. We also used maps, graphs, and bar graphs to illustrate the data. This made a very precise prediction at a quicker rate, making it simple to rely on. The rapid spread of COVID-19 has caused people to fear it because of this epidemic. To survive the circumstance, it is crucial to examine the situation using immediately accessible facts. The impacted, recovered, and death cases list should also be included in these reports. EDA, V-EDA, AND CNN play a significant part in this work by collecting data from John Hopkins University, the World Health Organization, and other sources. This experiment is carried out with the aid of the user-friendly Windows platform Jupyter notebook, which uses machine learning principles. Even while things are starting to stabilise, researchers are still watching for a second pandemic wave. Each and every user may easily grasp visual output.

8 CONCLUSION

The COVID-19 epidemic has given humans a time like never before. Curfews are in place in several nations, people are urged to stay indoors, the stock market is declining, and nurses and doctors are still treating COVID-19-infected patients. Since the virus is passing from person to person exponentially, this condition is prevalent. Although the vaccine has been distributed over much of the world, Covid-19 is still a rare infectious disease, making it impossible to estimate when the outbreak will end completely. As a result, this model, which was created using CNN as a machine learning technique, provides the most up-to-date and accurate status as well as an analysis and statistics of COVID-19. death rate determined using a prediction algorithm. This model aids in the transformation of the acquired data into patterns such as bar graphs, tabular columns, and pie charts of various states and nations. It is possible to construct and examine graphs for the relevant dates and months (from December to January or up to the present).

The impact of COVID-19 on the world is thoroughly examined in this study for the readers, and the conclusions that follow can be used for further investigation. The charts that this tracker predicts are also applicable in other circumstances.

REFERENCES

[1] Cavallo, P., O'Gara, M. and Schikore, D., 2009. Software System for Prediction, Visualization, Analysis, and Reduction of Errors in CFD Simulations. In *19th AIAA Computational Fluid Dynamics* (p. 3649).

[2] Chintalapudi, N., Battineni, G. and Amenta, F., 2020. *COVID-19 Virus Outbreak Forecasting of Registered and Recovered Cases After Lockdown in Italy, Immunology and Infection*, 53(3), pp.396–403.

[3] *CoronaTracker Analytics Team, Corona Tracker Analytics, Corona Tracker*, 03 March 2020. [online]. Available: https://www.coronatracker.com/analytics.

[4] Kandel N, Chungong S, Omaar A, Xing J. Health Security Capacities in the Context of COVID-19 Outbreak: An Analysis. *The Lancet*. 2020 Mar 28;395(10229):1047–53.

[5] Kumar, A., & Alam, B. (2016). Real-Time Fault Tolerance Task Scheduling Algorithm with Minimum Energy Consumption. In *Proceedings of the Second International Conference on Computer and Communication Technologies: IC3T 2015*, Volume 2 (pp. 441–448). Springer India.

[6] Khafaie, Morteza Abdullatif, and Fakher Rahim. "Cross-country Comparison of Case Fatality Rates of COVID-19/SARS-COV-2." *Osong Public Health and Research Perspectives 11*, no. 2 (2020): 74.

[7] Long, Jiawei. "COVID-19 Real-Time Tracker and Analytical Report." *arXiv* Preprint arXiv:2006.03146 (2020).

[8] Mehra, P. S., Jain, K., Chawla, D., Dagur, A., Singh, S., & Sharma, J. (2022). *GWO-EFUCA: Grey Wolf Optimisation and Fuzzy Logic based Unequal Clustering and Routing Protocol for Sustainable WSN-based Internet of Things.*

[9] Mehra, P. S., Mehra, Y. B., Dagur, A., Dwivedi, A. K., Doja, M. N., & Jamshed, A. (2021). COVID-19 Suspected Person Detection and Identification Using Thermal Imaging-based Closed Circuit Television Camera and Tracking Using Drone in Internet of Things. *International Journal of Computer Applications in Technology*, 66(3–4), 340–349.

[10] Muthusami, R., & Saritha, K. (2020). Statistical Analysis and Visualization of the Potential Cases of Pandemic Coronavirus. *Virus Disease*, 31(2), 204–208.

[11] Radakovitz, Samuel Chow, Adam Michael Buerman, Anupam Garg, Matthew John Androski, Matthew Kevin Becker, and Brian S. Ruble. *Graphically Displaying Selected Data Sources within a Grid.* U.S. Patent 10,289,671, issued May 14, 2019.

[12] Turner, Alan E., Vernon L. Crow, Deborah A. Payne, Elizabeth G. Hetzler, Kristin A. Cook, and Wendy E. Cowley. *Data Visualization Methods, Data Visualization Devices, Data Visualization Apparatuses, and Articles of Manufacture*. U.S. Patent Application 14/752,347, filed January 14, 2016.

[13] World Health Organization (WHO), *Coronavirus Disease 2019 (COVID-19) Situation Report-35*, WHO, 2020.

[14] Zhang, S.H., Cai, Y. and Li, J., 2020. Visualization of COVID-19 Spread Based on Spread and Extinction Indexes. *Science China Information Sciences*, 63, pp.1–3.

[15] Zhou, C., Su, F., Pei, T., Zhang, A., Du, Y., Luo, B., Cao, Z., Wang, J., Yuan, W., Zhu, Y. and Song, C., 2020. COVID-19: Challenges to GIS with Big Data. *Geography and Sustainability*, 1(1), pp.77–87.

Artificial Intelligence, Blockchain, Computing and Security – Dagur et al. (Eds)
© 2024 The Author(s), ISBN: 978-1-032-67841-2

A review on classification of SARS-CoV-2 using machine learning approaches

Rajkumar Pandiarajan* & Vanniappan Balamurugan*
Department of Computer Science and Engineering, Manonmaniam Sundaranar University Tirunelveli, Tamil Nadu, India

ABSTRACT: Almost all nations have been combating the COVID-19 pandemic, which is caused by Severe Acute Respiratory Syndrome Coronavirus 2 (SARS-CoV-2). Diagnosis of COVID-19 and its variants is still a challenging task since the efficiency of the existing classifiers is not optimum. Previously, machine learning classifiers such as Support Vector Machine, K-Nearest Neighbour, Decision Tree, etc., and deep learning classifiers viz. Gated Recurrent Unit, Convolutional Neural Network, Long Short Term Memory Network, etc. have been used for classifying the COVID-19. Such methods need lot of computational efforts and rely only on annotations of viral genes. This paper reviews the machine learning and deep learning techniques that have been applied in the classification of SARS-CoV-2 and analyses their performances critically. The comparative analysis reveals that deep learning classifiers outperform the machine learning classifiers in terms of classification accuracy, and deep learning classifiers require pre-processing. Further, this paper identifies the research gaps and provides possible future directions.

Keywords: COVID-19, Gene sequences, Deep learning, SARS-CoV-2, and Machine learning

1 INTRODUCTION

One of the most critical issues, the world currently facing is the corona virus outbreak (Salata *et al.* 2019). The World Health Organisation (WHO) declared the Corono Virus Disease-2019 (COVID-19) which is caused by the SARS-CoV-2 virus as epidemic. SARS-CoV-2 is a RNA virus belonging to Coronaviridae family which is a member of Betacorona virus group. The biggest RNA viral genomes are found in CoVs. According to their antigenetic and genetic characteristics, CoVs have been divided into four main types such as γ-CoVs, β-CoVs, δ-CoVs, and α-CoVs (Tobaiqy *et al.* 2020) These CoVs are generated by protein biosynthesis which produces four kinds of proteins viz. membrane protein, spike protein, nucleocapsid protein, and envelope protein in a human cell which is shown in Figure 1.

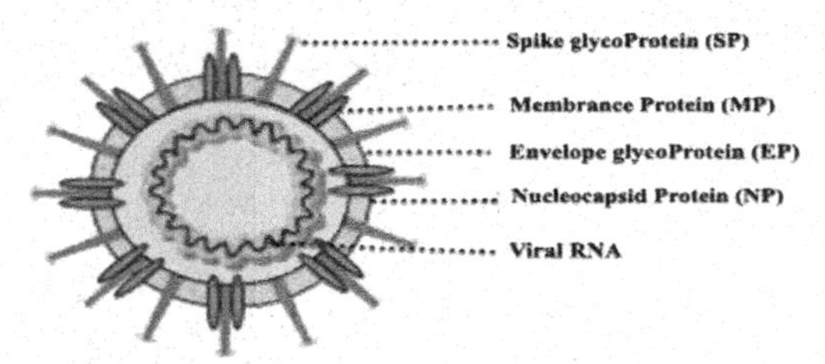

Figure 1. Diagrammatical representation of SARS-CoV-2.

*Corresponding Authors: rajmtech17@gmail.com and bala_vm@msuniv.ac.in

 DOI: 10.1201/9781032684994-51

Cells are the fundamental building elements of living things. A gene is a basic physical unit made up of DNA that decides heredity. The DNA sequence comprises of list of nucleotides which is represented as Adenine (A), Guanine (G), Cytosine (C), and Thymine (T). In general, the size of the human genome is measured as 700 MB approximately and in the case of SARS-CoV-2, it is measured as 30 KB. Mathematicians, medical professionals, and computer scientists have modelled and simulated the effect of SARS-CoV-2 on the human genome. The outputs of such initiatives have made significant contributions to illness detection and mitigation. The diagnosis of COVID-19 pandemic can be done either by using the clinical procedures or through the computerized methods such as genome sequence classification. There are mainly three types of classifiers available in literature viz. Non-Machine Learning (Non-ML) classifiers, Machine Learning (ML) classifiers and Deep Learning (DL) classifiers as illustrated in Figure 2.

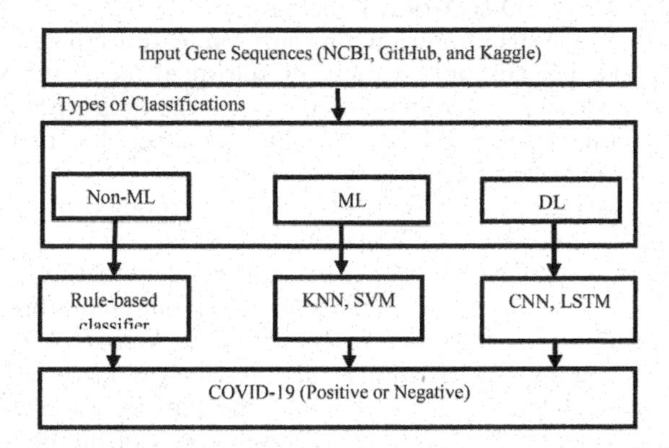

Figure 2. COVID-19 classification process.

Non-ML classifier needs human intervention while deciding the threshold values during classification. Some of the Non-ML classification techniques are rule-based classification, static classification threshold, decision threshold, metric-based classification, and so on.

In the case of ML classifiers, it classifies the data automatically from experience, without being explicitly programmed and with minimal human intervention. The current study intends to provide a brief overview of ML and DL based prediction approaches that predict the SARS-CoV-2 by classifying the given gene sequences. It is important to note that none of these techniques are flawless enough to be utilised as an error-free procedure, and the survey reveals that using these techniques will inevitably have some drawbacks.

The review is structured as follows: Section 2 describes the available ML approaches for classifying COVID-19, as well as their performance analyses. Section 3 provides an overview of the current DL methods for classifying COVID-19, followed by an analysis of their performance. A brief discussion on the performance of matrix is given in Section 4. Finally Section 5 concludes the review.

2 MACHINE LEARNING CLASSIFIERS USED IN COVID-19 CLASSIFICATION

The research on the classification of SARS-CoV-2 using ML approaches that has been done recently is presented in this section. Also, it compares the several work that are based on ML in terms of their performance and analyze the challenges while applying the ML techniques. Further, it suggests the possible approaches to overcome the challenges.

2.1 *Machine learning techniques*

ML trains computers to carry out tasks on their own. It is a technique for data analysis that entails creating and fitting models while enabling computers to "learn" via repetition and to anticipate the future (Peng & Nagata 2020). A prediction model (Aziz *et al.* 2021) developed a novel scheme utilizing ML methodologies to identify the SARS-CoV-2 affected individuals. In this methodology three different forms of networks such as ANN, Random Forest Regression (RFR), and SVM were trained and its effectiveness were assessed. From the evaluations, it is detected that RFR provides higher accuracy. A novel prognostic technique based on XGBoost was presented by Li Yan *et al.* in the year 2020 to forecast SARS-CoV-2 cases. XGBoost is a better ML technique which provides high pattern categorization and feature selection. This scheme provides high COVID-19 prediction accuracy of about 90%.

It was shown that ML models (Rustam *et al.* 2020) can predict COVID-19-affected patients. Four widely employed models such as SVM, Exponential Smoothing (ES), Linear Regression (LR) and Least Absolute Shrinkage and Selection Operator (LASSO), were examined in order to predict COVID-19 risk factors.

Nemati *et al.* proposed a system for evaluating survival features that was utilised on 1,182 COVID-19 patients and depends on a variety of statistical techniques. Several ML and statistical analytic strategies were used to assess the discharge-time estimation of COVID-19 person. The outcome demonstrates this inquiry. The Gradient Boosting (GB) patient survival strategy overtakes the other patient survival strategies.

A novel method for categorising the SARS-CoV-2 was proposed by Arslan *et al.* 2021. It is difficult to discriminate among COVID-19 and the other corona virus variants due to their genetic similarity. They also go over two CpG island-dependent features that are effective in finding COVID-19 cases. Due to their genetic similarities, differentiating COVID-19 from other coronavirus strains like BetaCoV-1, NL63-CoV, AlphaCoV, MERS-CoV, and HKU1-CoV is a challenging task. Arslan *et al.* presented a novel effective COVID-19 identification approach related to the KNN and the dataset collected in 2019. Novel Coronavirus Resource contains the entire genome sequences of human coronaviruses. An easy and effective non-parametric strategy for handling classification issues is the KNN strategy. The performance of the KNN is nonetheless influenced by the distance metric used. 19 distance metrics from five different groupings are looked at to enrich the efficiency of KNN approach. The proposed technique quickly attains 98.4% precision, 98.8% F-measure, 99.2% recall, and 98.4% correctness.

The authors of the paper (Ahmed et al. 2022) proposed an AI-based approach for evaluating the genomic sequences of COVID-19 and other viruses including MERS, SARS, and Ebola. The approach achieves 97% accuracy for COVID-19, 96% for SARS, and 95% for MERS and Ebola genome sequences, respectively.

In the study (Ünal & Dudak 2020), computer-aided diagnostic technologies and ML were used to categorise COVID-19. COVID-19 dataset, which was made available via the Kaggle website that has been subjected to the use of ML classification techniques including SVM, Naive Bayes, KNN, and decision tree in this work. The SVM algorithm attained 100% success rate and thus produces the best classification accuracy. Table 1 illustrates the summary of all selected ML approaches with the technology used, its advantages and research gaps.

Table 1. Summary of machine learning approaches.

Author's Name and Year	Methods	Dataset	Findings	Research Gap
Li Yan *et al.* 2020	XGBoost-Decision Tree	Patient electronic records from Tongji Hospital	High prediction accuracy of 90%	Only 3 features were considered among existing 300 features.
Ünal and Dudak 2020	SVM, Decision Tree, KNN, Random Forest	COVID-19 Kaggle- Data Set	SVM classifier produces high classification accuracy	Other than Kaggle dataset all are incorrectly classified.

(*continued*)

Table 1. Continued

Author's Name and Year	Methods	Dataset	Findings	Research Gap
Xiang *et al.* 2020	Liver function test, Renal function test	Clinical samples from 28 cases of COVID-19	Helps to detect COVID-19 existence	Limited samplescannot classify among the variants
Rustam *et al.* 2020	LR, SVM, ES	GitHub repository, Johns Hopkins University	LR and LASSO also perform well in confirmed cases	Investigate the prediction approach with the revised dataset.
Nemati *et al.* 2020	GB, Fast SVM, and Fast Kernel SVM	Covid-19-1,182 patients real time data	GB provide better prediction accuracy	ML methodologies designed does not predict accurately
Aziz Alotaibi *et al.* 2021	ANN, SVM, RFR	S1 data-Peking University	Minimizes the cost function, Less error rate	Over fitting to new gene sequence due to high variance
Arslan et al. 2021	KNN classification	2019nCoVR repository	High accuracy of 98.4%	Due to less features new variants are not classified correctly.
Naeem *et al.* 2021	KNN	NCBI GenBank	Produce higher classification accuracy with fewer error rate	Was unable to appropriately forecast.
Arslan 2021	KNN classifier	CpG island features-2019nCoVR repository	Eradicates the performance overhead	Need to improve classification accuracy
Ahmed *et al.* 2022	SVM	GenBank NCBI – COVID-19 dataset	Better classification outcome Accuracy for 97% (COVID-19), 96% (SARS), 95% (MERS)	Misclassification for other type of genome sequences

3 DEEP LEARNING CLASSIFIERS USED IN COVID-19 CLASSIFICATION

In this section, we highlight the work that has recently been done to classify SARS CoV-2 using DL methods. Along with analysing the difficulties associated with implementing DL methods, it provides a performance comparison of various works based on the method. In addition, it offers suggestions about how to tackle these difficulties.

3.1 *Deep learning techniques*

With the aid of various abstraction layers, DL techniques allow computer models that made up of several processing layers to learn data representation. They use images, texts, or audio, and genome sequences to train a computer model to carry out categorization tasks. It is found that DL models have great accuracy and it can sometime enhance human performance (Alazab et al. 2020). The dataset is utilised in their work (Pathan *et al.* 2020) to discriminate between nucleotide and codon alterations. Thymine (T) and Adenine (A) are shown to have undergone significant amounts of mutation throughout all areas, however codons do not undergo as much mutation as nucleotides.

In this technique, a Multichannel Convolutional Neural Network (MCCNN) (Moghadam *et al.* 2019) is used to anticipate changes in metalloproteins that impact metals. In particular, the spatial properties of the metal-binding site were mapped using energy-based affinity grids, which improved the performance of the prediction. MCCNN offers a technique for fusing bioinformatics analysis with experimental data.

CNN performance was improved in the work Ameen *et al.* 2021 by varying the filter kernel size and convolution layer depth. Additionally, they compared CNN, Multiple Linear Regression (MLR), and CNN with Support Vector Regressor (SVR) to create a hybrid model, CNN-SVR, for predicting guideRNA activity. This study's dataset comprises 1,251 guideRNA sequences and their matching indel frequencies.

For categorizing the SARS CoV-2 viruses from the genome sequence, (Whata & Chimedza 2021) developed a CNN with Bi-Directional Long Short-Term Memory (Bi-LSTM) neural network. From the experimental evaluation, it is found that the CNN Bi-LSTM provides superior assessment than CNN-LSTM. A Deep Neural Network (DNN) model was created by Yuvaraj *et al.* 2020 and it accurately identifies the interactions between the drug and the protein ligands. The DNN detects the medication's reaction to protein-ligand interactions and determines which drug produces the interaction that successfully fights with the virus. The GISAID database received a small number of Indian patients' genome sequences, and it was discovered that the DNN system is useful for determining a certain medicine interacts with these proteins.

To distinguish COVID-19 from influenza types A, B, and C, El-Dosuky *et al.* 2021 utilises a cockroach-optimized DNN. For optimizing the DNN hyper-parameters, the DNN design is stimulated by the cockroach optimization technique. The classification model achieves 99% overall accuracy by using 594 distinct genome sequences in the training and testing stages.

A DL framework named COVID-Deep Predictor was proposed by Indrajit et al. 2021 for identifying an unidentified sequence SARS-CoV-2 in which prediction was done using LSTM and an alignment-free technique. Table 2 shows the various existing techniques for identifying SARS Cov-2 virus's advantages and research gaps.

Table 2. Summary of deep learning approaches.

Author's name and year	Methods	Dataset	Findings	Research Gap
Yuvaraj *et al.* 2020	DNN	GISAID COVID-19 Database	Outcomes are accurate	Computationally expensive Cannot analyze peptide structure from protein-ligand interaction
Whata *et al.* 2021	CNN Bi-LSTM	SARS CoV-2 Dataset	Provides better identification of SARS CoV-2 virus	Need large number of parameters, memory and more computational time
El-Dosuky *et al.* 2021	Cockroach-optimized DNN	2019nCoVR Repositories	Provides multi class categorization with an accuracy of 99%	Poor noisy samples are not identified
Indrajit *et al.* 2021	COVID-DeepPredictor: RNN	NCBI Covid-GenBank	Alignment-free technique Similarity among the viruses are detected in less computational time	Training could not predict correctly for lowly-abundant
Pesaranghader *et al.* 2021	Impute CoVNet-2D convolutional ResNet auto encoder	GISAID Covid Database	Compared with Hamming Distance, it's give high accuracy, less processing time	Overfitting to new gene sequence due to high variance and low bias
Bugnon *et al.* 2021	mirDNN	NCBI -NC_045512.2	Better accuracy independently for imbalance gene	Less ability to identify miRNA sequences.
Mohanad A. Deif *et al.* 2021	BRNNs with LSTM and GRU	2019n CoVRRepository	Produces higher classification accuracy of 96.8%	Other gene sequences except SARS-CoV-2 are misclassified
Ameen *et al.* 2021	CNN with SVR	CRISPR/Cas12	Guanine is found more frequently in gRNA sequences with high indel frequencies than thymine, according to latest predicted values.	Enhanced complexity due to mathematical computation
Hossein *et al.* 2021	SMAPE-DL	Johns Hopkins University's Coronavirus Center	Lessen the computational overhead	Short length gene sequence, does not work for imbalanced datasets
Emmanuel	CNN	Virus Pathogen	Validation accuracy of	Large amount of

(*continued*)

Table 2. Continued

Author's name and year	Methods	Dataset	Findings	Research Gap
et al. 2022		Database and Analysis Resource	98.33%	memory required for storing multi genome sequences
Rani *et al.* 2022	CNN and LSTM	NCBI, GitHub, and Kaggle repositories	Higher accuracy in multi-class gene classification 99.27% Lessen the issues of over fitting and under fitting	Harmed during extraction which affects the performance of the model

This strategy takes less computational time and it identifies similarity among the virus from the gene. But it couldn't identify correctly for the gene that having lowly-abundant and small gene sequence. To solve this issues codon usage bias can be integrated with DNN. ImputeCoVNet, an autoencoder DL network framework, was introduced by Pesaranghader *et al.* 2021. This was applied particularly for very low-frequency alleles. The distance-based method frequently takes high processing time for imputation as the number of haplotypes to impute upsurges, but DL approaches such that ImputeCoVNet, after trained, do imputations essentially in real-time. This strategy produces better accuracy in training phase and less in testing phase. Regularization is the procedure that helps to reduce overfitting and also increases the test accuracy.

To detect precursors of the tiny active RNA molecules known as microRNAs (miRNAs) in the genomes of the novel CoVs, Bugnon *et al.* 2021 used a DL-based technique. "COVID-19" was recommended as a technique to use by (Rani *et al.* 2022). They successfully screen for COVID-19 by using a multi-modal method that incorporates the "GenomeSimilarityPredictor" with the "COVIDScreen-Net" and they make use of a range of modalities, including genomic sequences. The "GenomeSimilarityPredictor" model successfully detects the SARS-CoV-2 genome in individuals with a 99.27% accuracy rate.

4 DISCUSSIONS

For a range of issues, including the categorization of viruses to improve effective clinical care, ML-based alignment-free algorithms are utilised effectively. Additionally, they include the automatic detection and categorization of newly discovered illnesses according to how genetically close they are to curated reference genotypes (Wang *et al.* 2020), (Robertson 2000). There are various DL and ML that strategies are designed by numerous researchers to identify and categorize the SARS CoV-2 virus and its mutations from the gene sequences. Randhawa *et al.* 2019 utilised an ML-dependent alignment-free approach. It is possible to train in supervised, semi-supervised, and unsupervised models using DL, a subset of ML. This was influenced by ANN. DL is capable of extracting features in a hierarchical manner from the data, i.e., low-level features are initially retrieved, followed by midlevel features, and lastly high-level features. ML requires an extensive pre-processing of the dataset before the model is trained, but DL does not. This is the major distinction between the two types of learning. DL models are resilient to noise and missing data, therefore the data pre-processing phase is not required. Pre-processing of data is not necessary at all.

The Chaos Game Representation (CGR) proposed by Peng and Nagata (2020) which uses a two-dimensional numerical representation with k-mer with a value of 7 and thereafter, the magnitude spectra of the genomic signals were calculated using the DFT. All the ML approaches still analysed need data pre-processing and also need large amount of data. So, in recent days the researchers changed their focus towards DNN models like CNN, LSTM,

Bi-LSTM, and so on. The analysis reveals that integration of other techniques with DL approaches identifies the various mutations of SARS CoV-2 gene sequence effectively.

5 CONCLUSION

The COVID-19 pandemic now seems to be a deadly, infectious sickness that is spreading like other contagious illness. During pandemic, the number of patients increases rapidly, making it difficult for healthcare providers to find effective treatments. ML techniques are frequently utilised as substitutes for traditional classification and prediction techniques. By analysing various existing ML and DL strategies it is found that imbalanced dataset, minimal variance high bias, high variance minimal bias and selection bias are some of the flaws. In order to be trained well, DL approaches often need substantially bigger datasets. Although the provided datasets are small, the analysis shows that the simple-structure DL approaches (CNN, RNN, and LSTM) produce accurate outcomes. For diminutive accuracy increases, complex-structure DL algorithms do not need to be trained. For COVID-19, basic DL methods are adequate. The main issues are summarised as follows: (i) how ML approaches its benefits and limits, (ii) why researchers focus on DL, and (iii) existing DL methodologies and research gaps.

REFERENCES

Ahmed, I. & Gwanggil, J. 2022. Enabling Artificial Intelligence for Genome Sequence Analysis of COVID-19 and Alike Viruses. *Interdisciplinary Sciences, Computational Life Sciences* 14(2): 504–519.

AlazabMoutaz, AwajanAlbara, MeslehAbdelwadood, Abraham Ajith, JatanaVansh, Alhyari Salah. 2020. COVID-19 Prediction and Detection Using Deep Learning. *International Journal of Computer Information Systems and Industrial Management Applications*. 12: 168–181.

Aziz Alotaibi, Shiblee, M. & Alshahrani, A. 2021. Prediction of Severity of COVID-19-Infected Patients Using Machine Learning Techniques. *Computers* 10(3): 31.

Alyasseri, Z.A.A., Al-Betar, M.A., Doush, I.A., Awadallah, M.A., Abasi, A.K., Makhadmeh, S.N., Alomari, O.A., Abdulkareem, K.H., Adam, A., Damasevicius, R., Mohammed, M.A., & Zitar, R.A. 2022. Review on COVID-19 Diagnosis Models Based on Machine Learning and Deep Learning Approaches. *Expert Systems* 39(3).

Ameen, Z.S.,Mehmet, O.,Auwalu S.M., Fadi A.T. & Sertan, S. 2021. C-SVR Crispr: Prediction of CRISPR/Cas12 guideRNA Activity Using Deep Learning Models. *Alexandria Engineering Journal* 60(4): 3501–3508.

Arslan, H. & Arslan, H. 2021. A new COVID-19 Detection Method from Human Genome Sequences Using CpG Island Features and KNN Classifier. *Engineering Science and Technology, an International Journal* 24 (4): 839–847.

Arslan, H. 2021. Machine Learning Methods for COVID-19 Prediction Using Human Genomic *Data.* Proceedings, 74(1): 20.

Bugnon L.A., Raad J., Merino GA., Yones C., Ariel F., Milone D.H. & Stegmayer G. 2021. Deep Learning for the Discovery of New Pre-miRNAs: Helping the Fight Against COVID-19, *Machine Learning with Applications* 6.

El-Dosuky, M.A., Soliman, M. & Hassanien, A.E. 2021. COVID-19 vs Influenza Viruses: A Cockroach Optimized Deep Neural Network Classification Approach. *International Journal of Imaging Systems and Technology* 31(2):472–482.

Emmanuel, A., Joshua, A.A., Anthony, A.A., Tunmike, B.T., Oluwaseun, T. A., Abdultaofeek, A., Joy, N. A., & Joke, A. B., Zhongmin, J. 2022 DeepCOVID-19: A Model for Identification of COVID-19 Virus Sequences with Genomic Signal Processing and Deep Learning, *Cogent Engineering* 9(1).

Hossein, A. & Reza, P. 2021. Prediction of COVID-19 Confirmed cases Combining Deep Learning Methods and Bayesian Optimization. *Chaos, Solitons & Fractals* 142.

Koohi-Moghadam, M., Haibo Wang, Yuchuan Wang, Xinming Yang, Hongyan Li, Junwen Wang & Hongzhe Sun. 2019. Predicting Disease-associated Mutation of Metal-binding Sites in Proteins Using a Deep Learning Approach. *Nature Machine Intelligence* 1(12): 561–567.

Mohanad A.D., Ahmed A.A.S., Mehrdad Ahmadi K., Shahab S.B. & Rania E.H. 2021. A Deep Bidirectional Recurrent Neural Network for Identification of SARS-CoV-2 From Viral Genome Sequences, *Mathematical Biosciences and Engineering* 18(6): 8933–8950.

Nemati, M., Ansary, J. & Nemati, N. 2020. Machine-learning Approaches in COVID-19 Survival Analysis and Discharge-time Likelihood Prediction Using Clinical Data. Pattern 1(5):10074.

Pathan, RK., Biswas, M. & Khandaker, MU. 2020. Time Series Prediction of COVID-19 by Mutation Rate Analysis Using Recurrent Neural Network-based LSTM Model. *Chaos Solitons Fractals* 138.

Peng, Y., & Nagata, M.H. 2020. An Empirical Overview of Nonlinearity and Overfitting in Machine Learning Using COVID-19 Data. *Chaos, Solitons & Fractal* 139.

Pesaranghader, A., Pelletier, J., Grenier, J.C., Poujol, R. & Hussin, J. 2021. Impute CoVNet: 2D ResNetAutoencoder for Imputation of SARS-CoV-2 Sequences. *BioRXivPrepr.* doi:10.1101/2021.08.13.456305.

Randhawa, G.S., Hill, K.A., Kari, L. & Hancock, J. 2019. MLDSP-GUI: An Alignment-free Standalone Tool with an Interactive Graphical User Interface for DNA Sequence Comparison and Analysis. *Bioinformatics* 36(7): 2258–2259.

Rani, G., Oza, M.G., Dhaka, V.S., Pradhan, N., Verma, S. & Joel Rodrigues J.P.C. 2022. Applying Deep Learning-based Multi-modal for Detection of Coronavirus. *Multimedia Systems* 28: 1251–1262

Robertson, D.L. 2000. HIV-1 Nomenclature Proposal. *Science* 288(5463): 55–55.

Rustam, F., Reshi, A.A., Mehmood, A., Ullah, S., On, B., Aslam, W. & Choi, G.S. 2020. COVID-19 Future Forecasting Using Supervised Machine Learning Models. *IEEE Access* 8: 101489– 101499.

Saha Indrajit, Ghosh Nimisha, MaityDebasree, Seal Arjit & PlewczynskiDariusz. 2021. COVID-Deep Predictor: Recurrent Neural Network to Predict SARS-CoV-2 and Other Pathogenic Viruses, *Frontiers in Genetics* 12.

Salata, C. Calistri, A. Parolin, C. G, Palù. 2019. Coronaviruses: a Paradigm of New Emerging Zoonotic Diseases. *Pathogens and Disease* 77(9).

Tobaiqy, M. Qashqary, M. Al-Dahery, S. 2020. Therapeutic Management of Patients with COVID-19: A Systematic Review. *Infection Prevention in Practice* 2(3).

Ünal, Y. & Dudak, M.N. 2020. Classification of Covid-19 Dataset with Some Machine Learning Methods, *Journal of Amasya University the Institute of Sciences and Technology* 1(1), 36–44.

Wang, H., Li, X., Li, T., Zhang, S., Wang, L., Wu, X. & Liu, J. 2020. The Genetic Sequence, Origin, and Diagnosis of SARS-CoV-2. *European Journal of Clinical Microbiology & Infectious Diseases* 39(9): 1629–1635.

Whata, A. & Chimedza, C. 2021. Deep Learning for SARS COV-2 Genome Sequences. *IEEE Access.* 9: 59597–59611.

Xiang, J., Wen, J., Yuan, X., Xiong, S, Zhou, X., Liu, C. & Min, X. 2020. Potential Biochemical Markers to Identify Severe Cases Among COVID-19 Patients. *MedRxivPrepr.* doi: 10.1101/2020.03.19.20034447.

Li Yan, Hai-Tao Zhang, Jorge Goncalves, Yang Xiao, Maolin Wang, Yuqi Guo *et al*,. 2020. Prediction of Criticality in Patients with Severe Covid-19 Infection Using Three Clinical Features: a Machine Learning-based Prognostic Model with Clinical Data in Wuhan. *Nature Machine Intelligence* 2: 283–288.

Yuvaraj, N., Srihari, K., Chandragandhi, S., Raja, R.A, Dhiman, G. & Kaur, A. 2021. Analysis of Protein-ligand Interactions of SARS-Cov-2 Against Selective Drug Using Deep Neural Networks. *Big Data Mining and Analytics.* 4(2):76–83.

 ISBN: 978-1-032-67841-2

Optic disc and optic cup segmentation based on deep learning methods

S. Alex David*
Associate Professor, Department of Computer Science and Engineering, Vel Tech Rangarajan Dr. Sagunthala R&D Institute of Science and Technology, Tamil Nadu, India

N. Ruth Naveena*
Assistant Professor, Department of Mathematics, Hindustan Institute of Technology & Science, Chennai, Tamil Nadu, India

S. Ravikumar*
Associate Professor, Department of Computer Science and Engineering, Vel Tech Rangarajan Dr. Sagunthala R&D Institute of Science and Technology, Tamil Nadu, India

M.J. Carmel Mary Belinda*
Professor, Department of Computer Science and Engineering, Vel Tech Rangarajan Dr.Sagunthala R&D Institute of Science and Technology, Tamil Nadu, India

ABSTRACT: A lethal eye ailment called glaucoma causes the retina to gradually degrade over time. Although there is no complete cure, early detection helps slow the disease's course. Early diagnosis is very rare because there are usually no obvious symptoms present in the early stages. Early glaucoma detection is essential because delayed diagnosis might result in permanent vision loss. The Optic Nerve Head is harmed by glaucoma, which then damages the retina (ONH). The measurements of the retina's Optic Cup (OC) and Optic Disc (OD) are necessary for its diagnosis. Glaucoma can be effectively and accurately diagnosed using computer vision techniques with minimum overhead. These methods employ classification and segmentation algorithms that are based on machine learning to measure OC and OC dimensions. This article's objective is to give a thorough overview of the many methods used to identify and diagnose glaucoma using fundus pictures. Readers can recognise the gaps in existing works and comprehend the difficulties that glaucoma presents from the perspectives of image processing and machine learning.

Keywords: Optic Cup, Optic Disc (OD), Deep Learning, Prediction, Segmentation

1 INTRODUCTION

Optic nerves affected by the eye diseases called Glaucoma. Intraocular pressure (IOP), high blood pressure, migraines, obesity, ethnicity, and family history are some of the factors that contribute to glaucoma. Optic nerves are damaged by IOP. Adults and aged people get affected more by this phenomenon. Over 3% of the above the age 40 years suffering from glaucoma worldwide. Estimated that 64.3 million people from the age group between 40 years and 80 years suffering from glaucoma worldwide in 2013. World Health Organization (WHO) published a report on vision in 2019, which shows that 2.2 billion people suffering from visual impairments and nearly 50% has been prevented from vision loss due to the right

*Corresponding Authors: adstechlearning@gmail.com, ruthnaveena@gmail.com, ravikumars.086@gmail.com and carmelbelinda@gmail.com

 DOI: 10.1201/9781032684994-52

time diagnose and treatment. Open and close angle glaucoma are the two types of glaucoma. In the first type, the cornea iris drainage is still open. No noticeable symptoms are existing for the open type which has the possibility for vision loss. Parts of the iris are blocked when the drainage angle get closed which leads to the second type of glaucoma. The pressure in the eye get increased due to the blocks in the iris drainage. Redness of the eye, pain, pressure in the eye and vision decrease are noticeable symptoms of the closed glaucoma.

Weakness occurring in the optic nerve and Optic Disk is referred as Optic Nerve Head (ONH). Irreversible and incurable damages caused by all types of Glaucoma. Progression of the disease can be slowed down is the only option for the patients. Early detection gives higher efficiency in the treatment. Symptoms are unclear so the complexity exists in the early detection. By calculating the size of the OC and OD the Glaucoma can be diagnosed. Yellowish oval shape in the fundus image is OD. The OC occurred as a white circle inside the OD. The 0.65 is Cup-to-Disc Ratio (CDR) for the normal eye [1]. Glaucoma existence can be identified using the changes in the CDR. Detection and OC segmentation from OD is possible through the help of computers from the fundus images. Occurrence of the Glaucoma conformed with the help of the ratio between OC and OD. Manual outlining was performed by Ophthalmologists for detecting the OD as shown in Figure 1.

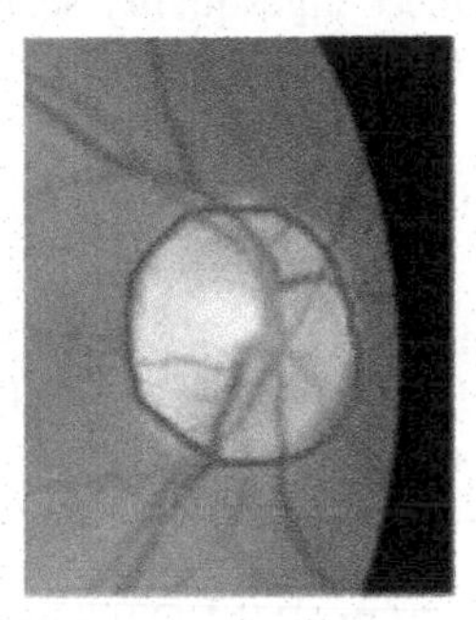

Figure 1. Manual annotation by the ophthalmologist.

The manual outlining was difficult and varies with respect to the ophthalmologist. Various ophthalmologist examinations included in the glaucoma treatment. Pressure increased within the eye can be measured by tonometry. 12 – 22 mmHg is the normal pressure range. When the pressure goes beyond 20 mmHg are the glaucoma cases. Diagnosing the glaucoma done with the optic nerve through ophthalmoscopy. Perimetry test will be recommended for the patients who have a alteration in the optic nerve shape and color. Vision of the patients affected by the glaucoma has been confirmed with this test. Pressure in the eye is influenced by the thickness of the corneal and the corneal thickness has been examined by the pachymetry test. Angles for the fluid drainage whether open or closed will be examined by the gonioscopy test. High risk of glaucoma is at a closed angle 8]. Trained professionals can perform these tests for physical analysis. Time consuming and variations are some of the drawbacks of this method. Visual computing systems are used in medical picture analysis and categorization because they are efficient. geometries of objects in an image and shapes were primarily detected in the most of the algorithms used in medical imaging. Specific training needed for the computers to detect the affected areas in the images [9]. The structural analysis of a retinal image can be performed using fundus photography. Medical experts' analysis can be reduced using computer-aided tools.

2 RELATED WORKS

The division of OD and OC is necessary for the CDR accuracy. The OD and segmentation of OC made extensive usage of algorithms based on deep learning and image processing.

Threshold based, active contour based, region based, pixel and super pixel are some of the image processing-based methods. Binary image obtained for segmentation using simple threshold technique. Colors below and above used segmenting by setting a threshold to the color. Quality of the input image may vary depending upon various factors. To overcome this quality on segmentation the red channel uses otsu thresholding for OD, ROI image is used in green channel for OC. Extracted features were added to find the threshold value for segmentation [2]. In [3] ROI obtained on both red and green channels from the input image. OD segmentation threshold value calculated using the histogram against the red channel in the image by gaussian window. Similarly, OC segmentation threshold value is archived in the green channel. Rule based approaches used in some research. Combination of mean, median with ostu thresholding is used in [4]. Compared to image processing methods, the machine learning methods give higher performance in the segmentation. OD has been accurately located in the image which has noise and abrasions in [5]. Border of the OD and OC is accurately located using the regression-based method in [6]. Manually extracted features and hand-crafted features are needed for the machine learning approaches, which are complex and time consuming. The complex features were automatically learned by training in the deep learning approaches. For the OD and OC segmentation, [7] employed the U-Net CNN version. This technique produces high-quality segmentation with a short prediction time. CNN based ensemble learning, over feat and VGG-S are used in other works.

3 DATASET

There are numerous free glaucoma datasets accessible. Among the picture datasets are DRISHTI-GS, RIM-ONE, and REFUGE. The DRISHTI-GS contains 101 pictures. The photos clearly show pupils through a 30-degree field of view that is focused on the optic disc. The Aravind Eye Hospital in Madurai annotated these pictures. Each image has a dimension of 2896 $\times$1944 pixels and is a PNG file without compression. Average bordering for OD and OC is provided by manual labelling. In the RIM-ONE open database, there were 159 retinal fundus pictures. 85 photos of health and 74 images of glaucoma. Two ophthalmologists annotated all of the images that were collected from three hospitals in Spain. The Retinal Fundus Glaucoma Challenge (REFUGE) dataset for training had 400 photos in the JPEG format with 8 bits, 2124 $\times$ 2056 pixels. 360 normal images and 40 glaucoma images were in this dataset.

4 DEEP LEARNING NETWORKS

The CDED-Net architecture gets trained with very limited epochs. This model segments both OD and OC jointly. Publicly available datasets were used for the training and testing purpose. The limited number of images is overcome with the various data augmentation methods which gives enough information for training the model. The encoder convolution layers were connected with dense connection and the up sampling layer was the same as the original U-Net architecture. Four dense blocks and two convolution layers in each block are included to maintain the balanced layer. The information loss is reduced by the over concatenation. Memory consumption reduced by the bottleneck layer.

Sub System Based Architecture has a U-Net based sub system used for segmenting the disc and cup along with feature extraction. For classification MobileNet V2 network is used in the second sub system. The blended results were produced in the final stage to assist the ophthalmologist. Based on the architecture of the U-Net, FundusPosNet has been derived. The input image of 128x128x3 has been analyzed to produce 2 1128$\times$128 heatmap. Fovea and OD are represented in each heatmap. CycleGAN based architecture used for augmentation-based classification. With the use of conditional random fields (CRFs) and a brand-new attention gating mechanism, the revised U-Net and DeepLAbV3 + architecture can increase segmentation accuracy (AG).

5 PERFORMANCE COMPARISON

The following evaluation parameters are used in the models they are accuracy, sensitivity, specificity, JACC index and AUC.

The deep learning methods used for the OD and OC segmentation performance have been listed in Table 1 and graphical representation is shown in Figure 2. Accuracy achieved by the EfficientNet b3 is higher than all other methods. B-Spline method has higher sensitivity value. The CDEC-Net method has a higher value in the specificity. FundusPosNet has higher value in the JACC index. In AUC the EfficientNet b3 has higher value. Among all methods EfficientNet b3 has superior performance to alternative approaches.

Table 1. Deep learning methods performance comparison.

Method	Accuracy	Sensitivity	Specificity	JACC	AUC
ERU-Net	0.98	0.98	0.96	0.98	0.996
B-Spline	0.94065	0.998225	0.9972	0.8514	–
CDEC-Net	0.9583	0.9517	0.9981	0.9191	0.977
EfficientNet b3	**0.9905**	**1.0**	0.9783	**0.988**	**0.9891**
Dual ML	0.88	0.91	0.86		0.96
FundusPosNet	0.98	0.94	0.87	0.976	0.985

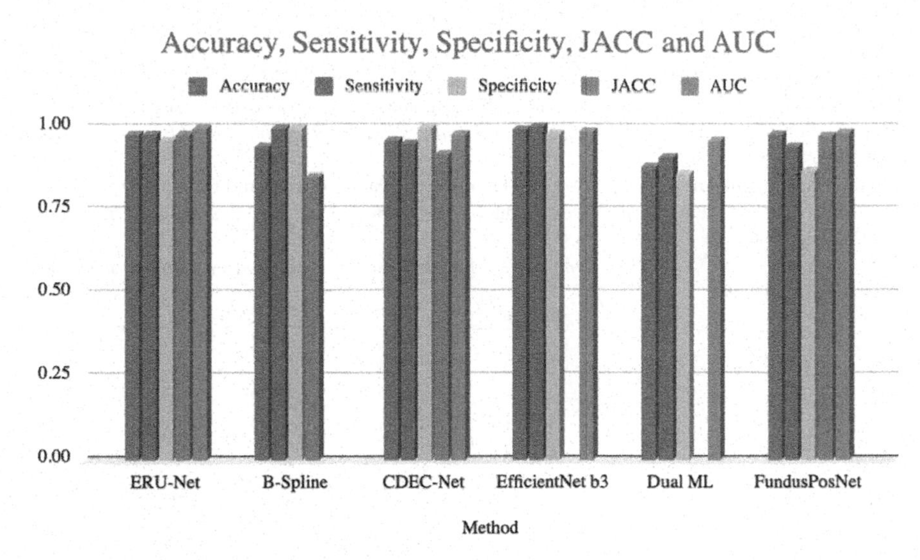

Figure 2. Performance of deep learning methods.

6 CONCLUSION

Only the segmentation of the optic disc and cup can be used to diagnose glaucoma. A brief description of several techniques for OD and OC detection and segmentation has been provided. Image based methods have been used in the earlier stage but the shortcoming of this method is the intervention of the ophthalmologist. Experienced ophthalmologists need to manually perform the annotation which is time consuming and costly. The computer aided diagnosis may help the experts to speed up the process. Machine learning based methods depend on the handcrafted images for training but deep learning models with enhanced network architecture learn the process in training without handcrafted images. In this study six deep learning methods have been taken and the performance of each model listed. Among all the methods, the EfficientNet model has higher values than other methods.

REFERENCES

[1] Zhang, Z., Yin, F.S., Liu, J., Wong, W.K., Tan, N.M., Lee, B.H., Cheng, J. and Wong, T.Y., 2010, August. Origa-light: An Online Retinal Fundus Image Database for Glaucoma Analysis and Research. In *2010 Annual International Conference of the IEEE Engineering in Medicine and Biology* (pp. 3065–3068).

[2] Veena, H.N., Muruganandham, A. and Kumaran, T.S., 2022. A Novel Optic Disc and Optic Cup Segmentation Technique to Diagnose Glaucoma Using Deep Learning Convolutional Neural Network Over Retinal Fundus Images. *Journal of King Saud University-Computer and Information Sciences*, 34 (8), pp.6187–6198.

[3] Nugroho, H.A., Kirana, T., Pranowo, V. and Hutami, A.H.T., 2019. Optic Cup Segmentation Using Adaptive Threshold and Morphological Image Processing. *Communications in Science and Technology*, 4(2), pp.63–67.

[4] Coan, L., Williams, B., Venkatesh, M.K.A., Upadhyaya, S., Al Kafri, A., Czanner, S., Venkatesh, R., Willoughby, C.E., Kavitha, S. and Czanner, G., 2022. Automatic Detection of Glaucoma via Fundus Imaging and Artificial Intelligence: A Review. Survey of Ophthalmology.

[5] Akyol, K. and Hen B., 2021. Keypoint Detectors and Texture Analysis Based Comprehensive Comparison in Different Color Spaces For Automatic Detection of the Optic Disc in Retinal Fundus Images. *SN Applied Sciences*, 3(9), pp.1–11.

[6] Sedai, S., Roy, P.K., Mahapatra, D. and Garnavi, R., 2016, August. Segmentation of Optic Disc and Optic Cup in Retinal Fundus Images Using Shape Regression. In *2016 38th Annual International Conference of the IEEE Engineering in Medicine and Biology Society* (EMBC) (pp. 3260–3264).

[7] Sevastopolsky, A., 2017. Optic Disc and Cup Segmentation Methods for Glaucoma Detection with Modification of U-Net Convolutional Neural Network. *Pattern Recognition and Image Analysis*, 27(3), pp.618–624.

[8] Jagannathan, J. and Divya, C., 2021. Deep Learning for the Prediction and Classification of Land Use and Land Cover Changes Using Deep Convolutional Neural Network. *Ecological Informatics*, 65, p.101412.

[9] Jagannathan, J. and Divya, C., *Time Series Analyzation and Prediction of Climate using Enhanced Multivariate Prophet.*

Artificial Intelligence, Blockchain, Computing and Security – Dagur et al. (Eds)
© 2024 The Author(s), ISBN: 978-1-032-67841-2

Missing child monitoring system using deep learning methods a comparison

S. Alex David*
Associate Professor, Department of Computer Science and Engineering, Vel Tech Rangarajan Dr Sagunthala R & D Institute of Science and Technology, Chennai, Tamilnadu

M. Leelavathi
Student, Department of Computer Science and Engineering, Vel Tech Rangarajan Dr Sagunthala R & D Institute of Science and Technology, Chennai, Tamilnadu

G.G. Swathika
Student, Department of Computer Science and Engineering, Vel Tech Rangarajan Dr Sagunthala R & D Institute of Science and Technology, Chennai, Tamilnadu

N. Ruth Naveena
Assistant Professor, Department of Mathematics, Hindustan Institute of Technology & Science, Chennai, Tamil Nadu, India

ABSTRACT: Most of the counties suffering with the social issue of child trafficking. The potential in the children's can be used in either right or wrong direction. There are lots of missing child cases reported to the police every day. Computer aided tracking systems helps to find the missing children. Deep learning-based approaches helps to improve many systems. In this paper a detailed analysis on the deep learning-based systems used for finding the missing children's have done. Various datasets and algorithms were compared based on the performance of the model.

Keywords: Deep Learning, Prediction, Face Detection.

1 INTRODUCTION

There is no doubt that the issue of child trafficking has become one of the major social issues that exists today in most nations throughout the world. It has been estimated that 28% of the victims of child trafficking who have been identified worldwide are children [1]. The year-by-year breakdown of missing children was provided by the United Nations Children's Fund (UNICEF). Every country has the great asserts as children's and youth. The country's future depends on the right raising of their children. Compared to other countries, India has very low levels of security and safety. India is another vibrant country in the world, and young people make up a momentous percentage of the sheer number of people living in the country. There are a significant number of children reported disappearing year after year. Among the unsolved child cases, many children remain untraced. Unsuccessfully, large lots of young people are disappearing in India because of variety of motives like deception, running away kids, traded children, kidnapping children, etc [2]. According to a National Crime Records Bureau (NCRB) study referenced in Parliament by the Ministry of Home Affairs (MHA), thousands of children were actually not located up until 2016. Many of them are still missing

*Corresponding Author: adstechlearning@gmail.com

DOI: 10.1201/9781032684994-53

today. NGOs claims that estimated of disappearing children are more accurate than the report [3]. A few cases, children end up missing from school buses. Every day, children from far and wide board school buses to travel to school. The issue of children being forgotten on the bus became a major issue in the missing child problem. This is one of the major problems that has been on the rise for a number of reasons. In many cases, when children are waiting to board or deboard a school bus, they are confronted with incidents such as child abuse that often lead to serious injuries. There is a serious threat to child safety here and action needs to be taken [4].

Every day, the police get numerous reports of missing children. A lost child from one area may come up in another state or region for a variety of reasons. Finding him or her among the cases of reported missing people requires skill. The process for creating an aid for locating kidnapped children is described in this study. Children in alleged scenarios are captured on camera by volunteers, who then upload the images to the website. The administrator of the application can look through the photographs of missing children's cases to find the uploaded photo. Thanks to this, the police may now locate the child wherever in India [8]. When a child is found, the picture taken then is compared to the pictures that the police or guardian posted when the youngster vanished. Sometimes the kid has been gone for a very long time. Compared to other face recognition algorithms, this presents the largest difficulty in identifying missing children. Additionally, a child's face appearance can alter due to adjustments in stance, orientation, lighting, occlusions, background noise, etc. Since some of the public photographs might have been taken from a distance without the child's knowledge, they might not all be of high quality. Deep learning architecture created to overcome all of these limitations. [6,7].

2 RELATED WORK

A model has been trained to identify the missing child using deep learning approach which compares the images from the database of missing child images and the child image uploaded by volunteers / public. For image-based applications, Convolutional Neural Network (CNN) is a very efficient deep learning technique is used for face recognition. From the uploaded facial photos, a pre-trained CNN model trained on the VGG-Face in-depth architecture will be able to extract face descriptors. The trained Support Vector Machine classifier can recognise children's faces while the convolution network is used for high level feature extraction. The model may be evaluated using pictures of kids in various lighting and noise environments, as well as pictures of kids of various ages. Invariant to noise levels, light illumination, picture contrast, occlusion, image pose, and child age is face identification utilising the CNN model, a deep learning model built using VGG-Face, and adequate model training. By applying face recognition, this model can be utilised to locate missing children.

The open-source Arduino IDE software, an Arduino embedded system, and RFID technologies have all been combined to build a missing person detection system [5]. Wearing the RFID tag physically is necessary. A microchip has been integrated with the RFID Tag which contain the Person's Identity Information. The tag and the tag reader communicate by means of radio frequency. Once the RFID tag is displayed on the RFID reader, the serial port on the Arduino scans the tag. The radio frequencies received are converted into Identity Data with the help of the reader. Along with making the identifying information on the tag available for further processing, an LCD also displays information like a missing person's name, guardian's name, and contact information. This approach seems to be very useful at finding elderly people who have gone missing and passing them over to their guardians.

Based on criminal records, the system [6] can identify criminals who are on the run from their past crimes as well as identify children who went missing. To find faces and objects, Python-OpenCV can be utilised. Since it cannot detect faces from images that are rotated, Haar Cascade Classifier can be applied in order to detect faces in case of rotated and blurry images. The application is created by integrating the jinja2 template and the Python flask framework. The input for the model is the suspect image. Using OpenCV, the images are read into a numpy array and store them. Python opencv is used to detect faces and it generally works with convolution neural network (CNN) technique. Then the feature extraction is done by deep learning algorithm. A python API library called facepplib can be used to compare image templates. Faceprints are extracted from both the criminal's original image and the suspect input image. Afterward, it creates faceprints for criminal images and compares them. We can evaluate whether or not the individual in the image is a criminal based on the comparison of the images and the results that are shown.

A method that makes use of a feature ageing module to improve facial recognition while dealing with age-separated child face photos. In an effort to reunite the discovered child with his family, a facial image from a recovered youngster at his current age was used to search through a database for missing children with known identities at a time when they were lost or stolen. In such cases, the feature aging module improves the accuracy rate of cross-age facial recognition and enables the identification of young children who may be trafficked or abducted by enhancing the performance of cross-age facial recognition. On the FG-NET public ageing face dataset, it increases the rank-1 accuracy rates of FaceNet and CosFace from 16.04% to 19.96% and 22.91% to 25.04%, respectively. It also increases the rank-1 accuracy rate of the CosFace model from 94.91% to 95.91%. Thus the FAM approach improves the search accuracy of child-to-adult matching in any face matcher and paves way in reuniting the missing children with their family.

An Arduino RFID based approach designed in assisting parents to track down the location of their children. It's objective is to ensure that the school-going children are safe and secure. The system comprises of GPS module to track their child's location and a RFID card to find them and Arduino Mega 2560 as primary microcontroller. The system notifies the parent/guardian of the children through a text message with the latitude and longitude coordinates, once their child enters/leaves the school bus so that they are able to keep track of the location of the bus. In addition to drunk driving prevention with the MQ-3 alcohol sensor, ping sensors are used to detect objects in front of the wheels of the vehicle, and IR proximity sensors are used to alert to accidents, which contributes to the safety concerns of schoolchildren as well.

3 MATERIALS AND METHODS

3.1 *Dataset*

The method mentioned in uses the 2,622 IDs in the VGG Facial dataset, a face identity identification dataset. More than 2.6 million photos are included. A text file with URLs for photos and accompanying face detections is linked to each identity. 43 distinct children's cases were used to evaluate this method. 43 distinct kid instances are represented by 846 child face photos in the user-defined database. There are 677 photos in the training and 169 images in the test since the training includes 80% of images from each child group and the test set contains 20% of the images. The data from the web are extracted using web scraping in the method described in [6]. For scraping photos from the web, Beautiful Soup and the Python requests library are both utilised. With the aid of Beautiful Soup, the source code of websites that provide information on criminals and missing persons can be retrieved. Some approach makes use of datasets like ITWCC, FGNet Dataset, UTKFace. ITWCC stands for In-The-Wild Child Celebrity dataset has 1705 photos and 304 subjects. The individuals in this dataset range in age from 5 months to 32 years. The collection includes 839 pictures of men and 876 pictures of women. A dataset for face recognition and adult age estimation is called FGNet. It consists of 1,002 photos in total of 82 persons whose ages span from 0 to 69, with an age gap of up to 45 years. The UTKFace dataset is a sizable face dataset with a broad age range, ranging from 0 to 116 years old. Over 20,000 face photos from the collection are categorised by age, gender, and race. Numerous variations in position, face expression, illumination, attenuation, clarity, etc. are included in the photographs. The MORPH face age estimate dataset includes 55,134 facial pictures of 13,617 individuals aged 16 to 77. Age, gender, nationality, size, and ocular coordinates are among the metadata comprised of the databases.

3.2 *Algorithms*

For feature extraction and classification of several kid categories, the CNN-based deep learning approach in combination with SVM [2] is deployed. CNN is a multilayer network that has been trained to use classification to carry out a particular task. To increase the precision of facial image classification, CNN is used. SVM is a supervised machine learning approach that enables both classification and regression. A huge number of both positive and negative photos are utilised to train the classifier using the machine learning-based Haar Cascade approach. Several samples of both positive and negative photos are used to train a cascade function for detection. Positive images consist of images that our classifier ought to be able to identify. Negative images are images of anything other than the object we're looking for Thus, the Haar cascade classifier is used to detect faces. Up to 55 faces can be recognised by this classifier from a picture.

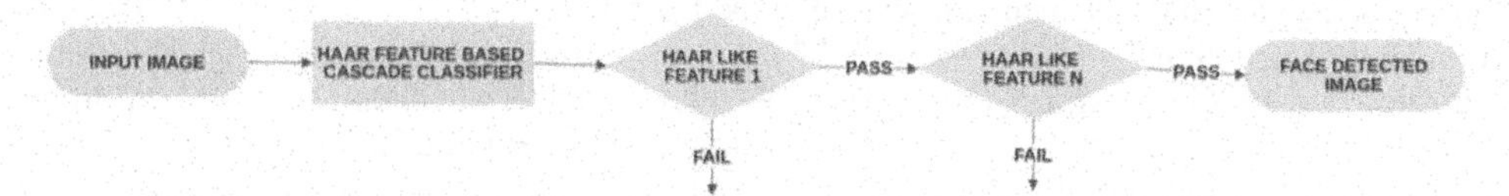

Figure 1. Face image detection using Haar Cascade.

To improve the effectiveness of cross age recognition, Feature Aging Module can be used to generate face images of the desired age and make those images available to just about any face matcher.

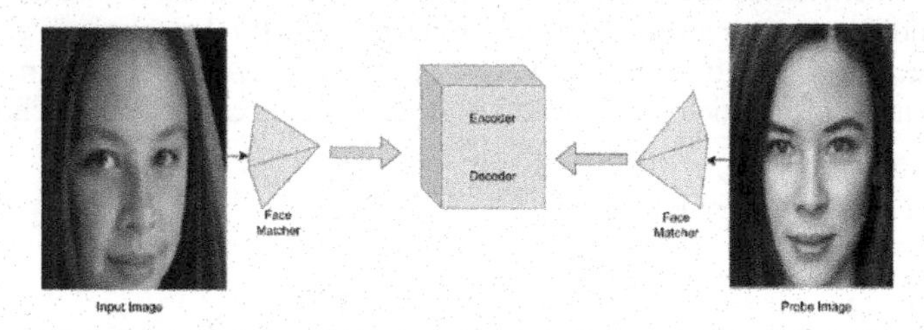

Figure. 2 Age progressed face image

4 EVALUATION PARAMETERS

The aforementioned method would raise FaceNet's rank-1 open-set recognition rate from 16.04 percent to 19.96 percent and CosFace's rank-1 open-set identification accuracy from 22.91% to 25.04% on the celebrity child dataset ITWCC. Furthermore, using FG-NET, a publicly available dataset of ageing face images, this strategy increases CosFace's rank-1 accuracy from 94.91% to 95.91% Face matchers can be better prepared to locate and recognise young children who may have been missing at young age by including the feature ageing module. They will be able to reconnect them with their families as a result.

By Using CNN image characteristics and the VGG-Face model, a multi-class SVM is trained, a high recognition accuracy can be achieved by 99.41%. The effectiveness of the suggested system is evaluated using images of children taken in various lighting and noise environments, as well as images taken at various ages. It had a 99.41% accuracy, demonstrating that suggested face recognition methodology could be used to accurately identify missing children. Accuracy is calculated by using this formula:

$$\text{Accuracy} = \frac{\textit{Correct Recognisation face}}{\textit{Total number of face images}}$$

For testing the prototype model, to ensure that every conceivable set of test conditions would be satisfied, a test environment was simulated. The created system's model shows good accuracy and requires less computation time under all circumstances. We can ensure the safety and security of every schoolchild.

The main benefits of this system are that it requires less hardware, consumes less energy, and is less expensive. This method is mostly used to locate elderly people and lost children. In addition to identifying missing children, this application has applications for locating and delivering children with physical disabilities and elderly people to their caretakers. A variety of both positive and negative images are used to teach it. Additionally, it can be used to find items in other pictures. 90% of the results are accurate. When compared to alternative ways, this takes less time to implement and uses less memory space.

5 CONCLUSION

The Feature Aging Module that has been incorporated provides enhanced accuracy rate facial recognition thereby increasing the accuracy rate of CosFace on ITWCC dataset from 22.91% to 25.04% and FG-Net dataset from 94.91% – 95.91%. The model which makes use of Haar Cascade Classifier Algorithm tends to yield an accuracy rate of 90% and that can be utilized for facial recognition in order to determine whether or not the individual in the picture is a criminal. The CNN based approach along with multiclass SVM as mentioned makes use of VGG Face dataset which is used to train the model and it achieves an accuracy rate of 99.41% irrespective of the noise levels, resolution, contrast level, eye coordinates, image pose and child's age that can be used for identification of missing children based on face recognition. The system developed based on RFID and Arduino which tends to require less hardware, consumes less energy, and is less expensive is helpful in locating and delivering missing children and senior citizens to their guardians. Another RFID Arduino based system which was integrated with IR sensor, alcohol sensor, GPS, GSM tends to provide good accuracy rate with reduced computational time and ensures the safety of school children.

REFERENCES

[1] Deb, D., Aggarwal, D. and Jain, A.K., 2021, January. Identifying Missing Children: Face Age-progression via Deep Feature Aging. In *2020 25th International Conference on Pattern Recognition (ICPR) (pp. 10540–10547).*

[2] Kumar, K.A., Anupama, P., Naveen, P. and Poojitha, S., 2020. *Missing Child Identification System using Deep Learning and Multiclass SVM.* Volume XII, Issue II, ISSN, pp. 0975–4520.

[3] Nithya R., Chennareddy Harshitha Reddy, S. Dhivya, AhtoaUma, Bandi Sumathi, "Missing Child Recognition System Using Deep Learning and Multi-Class Support Vector Machine", Tamil Nadu, India ,2022 *IJIRT*, Volume 9 Issue 1, ISSN: 2349–6002.

[4] Kumar, S.A. and Kumaresan, A., 2017, June. Towards Building Intelligent Systems to Enhance the Child Safety and Security. In *2017 International Conference on Intelligent Computing and Control* (I2C2) (pp. 1–5).

[5] Arniker, S.B., Rao, K.S.R., Kalyani, G., Meena, D., Lalitha, M. and Shirisha, K., 2014, May. RFID Based Missing Person Identification System. In *2014 International Conference on Informatics, Electronics & Vision (ICIEV)* (pp. 1–4).

[6] Ayyappan, S. and Matilda, S., 2020, July. Criminals And Missing Children Identification Using Face Recognition and Web Scrapping. In *2020 International Conference on System, Computation, Automation and Networking (ICSCAN)* (pp. 1–5).

[7] Jagannathan, J. and Divya, C., 2021. Deep Learning for the Prediction and Classification of Land Use and Land Cover Changes Using Deep Convolutional Neural Network. *Ecological Informatics*, 65, p.101412.

[8] Jagannathan, J. and Divya, C., *Time Series Analyzation and Prediction of Climate using Enhanced Multivariate Prophet.*

Artificial Intelligence, Blockchain, Computing and Security – Dagur et al. (Eds)
© 2024 The Author(s), ISBN: 978-1-032-67841-2

A review on the energy efficient resource allocation schemes with improved quality of services in green cloud computing

A.S. Syed Fiaz
Research Scholar, Computer Science and Engineering, Amrita Viswa Vidhyapeetham, School of Computing, Chennai

Veeramani Sonai
Assistant Professor (Grade II), Department of Computer Science and Engineering, School of Engineering, Shiv Nadar University, Chennai

ABSTRACT: The vast changes in the computing utilization have created a need with the improvised resource availability and reliability. The virtualization of servers and data centers to increase their efficiency has transformed how enormous data is stored and handled. Cloud computing has created a supreme and impressive approach to virtualize servers and data centers and to make them energy efficient in order to leverage on diverse IT resources. Massive power and energy consumption by IT resources leads to an energy crisis and a change in the planet's climate. But because these IT resources use so much energy and power, they end up being a significant source of CO2 emissions. Due to this, green cloud computing is now required in order to make IT resources cost- and energy-effective. It is necessary to do a thorough analysis of the cloud's power efficiency in order to arrive at the best options for green cloud computing. Green Cloud Computing is well known to be a broad domain and a popular field for research because of the increasing rise of enormous data storage and computational demand. It can generate solutions that not only make IT resources energy efficient but also reduce operational expenses. This review's objective is to highlight some of the best practices for achieving green cloud computing, including workload clustering, Auction-Based Resource Optimization techniques, Hardware and Software controlled solutions, Minimize cooling requirements scheduling, clustering, virtualization and other energy-saving techniques.

Keywords: Cloud Computing, Green cloud computing, CO2 emissions, virtualization, resource provisioning, carbon footprint

1 INTRODUCTION

By enabling organizations to outsource various information technology services including data storage, computation, and enterprise resource planning, cloud computing [1] is a technology that boosts productivity. The effectiveness of utility computing, the scalability of the processes, and the extent of IT outsourcing all have a role in how well cloud computing is adopted. Three million automobiles' worth of greenhouse gas emissions [3] are produced by this. Numerous innovations that concentrate on the creation of technologies that offer services that not only increase performance but also assure the environmental sustainability have been made in response to the growing need for environmentally friendly operations and business practices.

Green cloud computing creation and uptake aid in lowering institutional carbon footprints and enhancing the organization's reputation and public profile [2]. The majority of

 DOI: 10.1201/9781032684994-54

consumers who care about the conservation and management of natural resources would therefore prefer to link themselves with goods that protect the environment from an economic point of view. As a result, the people may serve as brand ambassadors for these goods, increasing the company's sales and profits. The usage of green computing [10] systems is required by the rising demand for computation and the enormous development in the need for data storage, which creates the most urgent and necessary ways to virtualize the various servers and data centers in order to carry out energy-efficient operations. Systems for green cloud computing also provide answers for other IT (information technology) systems [6]. For instance, IT systems that use manual and conversational operational platforms to carry out various managerial and administrative tasks aid in raising the institution's productivity. A step up from the analogue cloud-based infrastructure that was previously utilized to lessen businesses' carbon footprints is the adoption of green cloud computing. Energy utility efficiency, resource efficiency, renewable energy sources, and product lifecycle management are the main focuses of green cloud computing functions.

The virtualization of institutional resources is one of many techniques and operational strategies that can be used to provide these high-quality services. By using a single physical server to run several operating systems and digital interfaces for various organizational departments, it is possible for institutions to operate and manage their resources. The use of virtual resources aids in minimizing the physical server footprint and promotes the inherent environmental advantages. For instance, the space needed for data centers is reduced by the decrease in the amount of equipment required to conduct various institutional functions efficiently. This eventually results in a decrease in the footprint of electronic waste. Multitenancy, which enables several divisions of the business to access and use a single cloud-based system concurrently, is another benefit of green cloud computing [9]. This improves the institution's performance while reducing any potential waste that might occur during working hours. Additionally, it aids in realizing sustainability goals and energy efficiency, both of which are crucial in the management of natural environmental resources. Since most tasks are now completed digitally, using such technology also enhances the social wellbeing of the workforce because it frees them from long hours and cramped workspaces. This improves employee morale, which results in higher production [16].

An overview of cloud computing and green cloud computing is given in this review, followed by methods for achieving green cloud computing. In this research, the benefits and drawbacks of green cloud computing are also discussed. The conclusion provides a summary of the methodology and main aspects of the review.

2 IMPACT OF THE CARBON EMISSIONS

Just because the majority of data centers use cloud computing architecture, there are significant increases in carbon emissions. Even with the power-saving option, all currently in use PCs use around 400 KW of power and emit 270 kg of CO2 annually [8]. IBM opened a datacenter that makes the most of lowering its carbon footprint. Temperatures and humidity can be determined with the use of the data centers. This has the potential to cut energy use by 14%. A lot of data is being transferred to the cloud these days since it saves energy and allows for the replacement of physical servers with virtual ones [12]. It is much simpler to manage several customers with the aid of virtual servers. Due to the cloud's increased energy efficiency, many IT companies are making the switch. Currently, SAP has also migrated to the cloud and announced that it will power all data centers and supply all renewable energy.

Green computing is supported by one of the most powerful cloud platforms, AWS. It is crucial to the Cloud Computing process because it facilitates the distribution of hardware resources among users while allowing different operating systems to run on the same computer.

3 ENERGY EFFICIENT TECHNIQUES USING GREEN CLOUD COMPUTING

3.1 *Resource provisioning using workload clustering*

The Workload Recorder receives user requests that are sent to the cloud service provider. The recorded workloads are saved using the Workload Information database. Queries from the Workload Information database are captured by the [4] Workload Preprocessing component, which also filters out obtrusive and pointless requests. Each request is assigned an ID, and the SLA table is created. The workloads are then clustered using a hybrid imperialist competitive/K-means based clustering algorithm to be executed on several sets of resources. Finally, the desired resource provisioning is carried out and sent to the infrastructure layer utilizing a decision tree method.

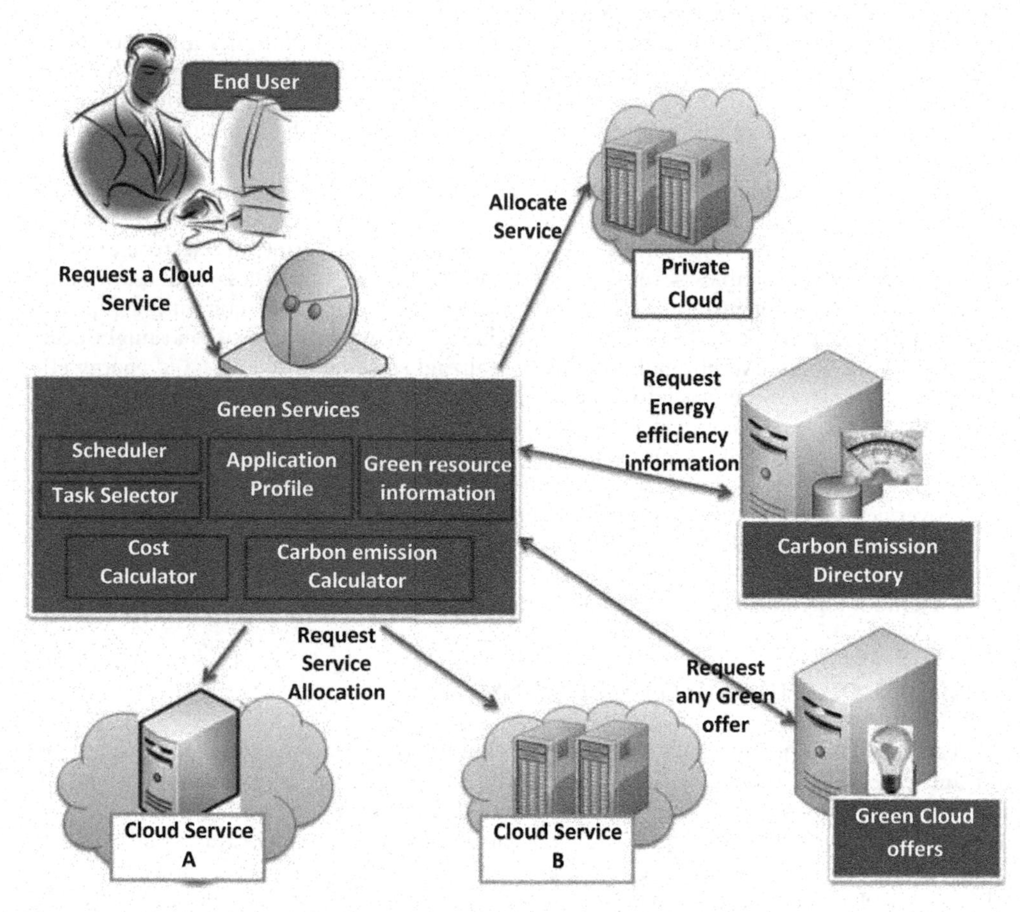

Figure 1. Green cloud computing.

Preprocessing, task clustering, and resource provisioning are the three main steps in the proposed methodology. Preprocessing the workload in the first phase removes obtrusive and pointless requests. The SLA table is then made using the ID that has been defined for each request. The SLA data is finally standardized [11]. The training workload is grouped using the Imperialist Competitive algorithm and K-means in the second stage. The closest cluster centre to the test workload is then chosen using the Euclidean distance. Finally, resource provisioning is carried out via a decision tree method.

3.2 *Auction-based resource optimization*

A resource allocation system in cloud computing tries to maximize resource utilization to meet cloud customers' needs. Virtualization abstracts physical computing resources as virtual machines (VMs), each of which has a CPU, memory, and communication support, among other things. The candidate resources are distributed to customers through resource selection and optimization. Multiple clients can share the physical resources thanks to virtualization and provisioning. An auction-based resource allocation system [5] uses the supply and demand for resources in the cloud to its advantage. Through a suitable resource allocation mechanism in the market, both suppliers and clients may seek to maximize their utility.

These are the stages involved in allocating resources via an auction system.

Step 1 Customers first submit requests and offers.

Step 2 is followed by Step 3 if the submission is successful. If not, repeat Steps 1 and 2 again.

3. Sort the offers.

Step 4 Assign resources to potential winners and choose the winner using the auction approach.

Step 5 If the criterion for the auction's termination is met, move on to Step 6. If not, repeat step 5. Depending on the auction model used, the termination condition can be, for instance, that all resources have been distributed and all demands have been met.

Step 6 The winner pays for the resources and the winning price is determined by the auction mechanism used.

Step 7 Bind the resources to the successful client.

3.3 *Hardware and software controlled solutions to reduce energy consumption*

Hardware solutions work to create gadgets that use less energy while maintaining performance. Energy-efficient cloud computing models include a few techniques for turning off processors and lowering memory energy usage [7]. One of the key factors contributing to inefficient energy use in data centers is running servers at a level lower than their utilization. The suggested remedies for that issue include using energy-efficient technology and reducing the amount of power used by servers and networks.

Virtualization is one of the top technologies for software solutions when it comes to using less energy in data centers [13]. Additionally, developers and architects work to install software that makes better use of hardware resources. Some suggested solutions include creating energy-conscious scheduling and intelligent traffic routing algorithms. The network architecture of a cloud uses a variety of devices, and how much energy each type of device uses depends on user activity and the nature of the apps that are running on the cloud and interacting with one another. Three different application kinds are categorized using novel methodologies, and criteria to lower energy consumption are determined by modeling how much energy is used by each type. The challenge of maintaining acceptable quality of service as cloud computing services are expanding has prompted the development of novel software architectures with dynamic self-optimization capabilities [14,15]. Due to the advent of the cloud, businesses no longer have to buy and own everything; instead, they may rent what they require. Software should be installed and operated on shared resources; it shouldn't be dependent on a single server; and it shouldn't be overly dependent on hardware or where it is located. Software engineers create programmes that are made up of services that can operate on their own. These components can communicate with one another, execute in multiple places, and be shared by several applications in addition to being loosely connected.

3.4 *Minimize cooling requirements*

Previously, cooling was carried out via a mechanical refrigerator that uses a compressor inside the data centre, or man air handlers would receive cold water from outside to cool IT equipment. Mechanical cooling can now be replaced with free cooling. A system called free cooling was created to reduce or optimize the need for cooling. It states that the mechanical refrigerator can offer direct or indirect cooling on its own if the air temperature outside is below or at the critical point. Free cooling merely replaces the need for mechanical cooling energy; it does not lessen the amount of fan energy required for cooling.

4 CONCLUSION

The goal of this analysis is to highlight how necessary it is to change present cloud computing practices in order to achieve green cloud computing with the goals of lowering greenhouse gas emissions, conserving energy, and cutting costs. There are many ways to do that, including scheduling for optimal utilization, virtualization for low hardware and cost usage, and proportional computing for high output. Other broad strategies include regulating power consumption, cluster computing, and producing electricity from renewable sources. Despite the benefits of green cloud computing, there are still issues with security and connection. The world of data centers is still dominated by the search for new cloud computing techniques.

REFERENCES

[1] Marrone S., and Nardone R., 2015, "Automatic Resource Allocation for High Availability Cloud Services", *Procedia Computer Science*, 52, 980–987.

[2] Gupta P., and Ghrera S. P., 2016, "Power and Fault Aware Reliable Resource Allocation for Cloud Infrastructure", *Procedia Computer Science*. 78, 457–463.

[3] Foster I., Zhao Y., Raicu I., Lu S., 2008, "Cloud Computing and Grid Computing 360-degree Compared", In: *Grid Computing Environments Workshop (GCE'08)*. 1–10.

[4] Armbrust M., Fox A., Griffith R., Joseph A. D., Katz R. H., Konwinski A., Lee G., Patterson D. A., Rabkin A., Stoica I., and Zaharia M., 2006, "Above the Clouds: A Berkeley View of Cloud Computing".

[5] Reem I. Masoud, Rahaf S. AlShamrani, Fatima S. AlGhamdi, Sara A. AlRefai, Hemalatha M, 2017, "Green Cloud Computing: A Review", *International Journal of Computer Applications* (0975 – 8887), Volume 16 – No. 9, June.

[6] Ankita Atrey N. J. a. I. N., 2013, "A Study on Green Cloud Computing," *International Journal of Grid and Distributed Computing*.

[7] Ali Shahidinejad, Mostafa Ghobaei-Arani, Mohammad Masdari, 2021, "*Resource Provisioning Using Workload Clustering in Cloud Computing Environment: Aa Hybrid Approach*", 24, pages 319–342.

[8] Buyya S. K. G. a. R., 2011, "*Green Cloud Computing and Environmental Sustainability,*" Dept. of Computer Science and Software Engineering, The University of Melbourne, Australia.

[9] G. P. S. M. P. M. a. B. S. C. Peoples, "Energy Aware Scheduling Across 'Green' Cloud Data Centres," *IEEE*, 2013.

[10] Beloglazov A, Abawajy J, and Buyya R, "Energy-aware Resource Allocation Heuristics for Efficient Management of Data Centers for Cloud computing."

[11] Dalapati, and G, Sahoo, 2013, "*Green Solution for Cloud Computing with Load Balancing and Power Consumption Management*".

[12] Asha N., Syed Fiaz A. S., Jayashree J., Vijayashree J., and Indumathi J., 2022, "Principal Component Analysis on Face Recognition Using Artificial Firefirefly Swarm Optimization Algorithm," *Advances in Engineering Software,* Elsevier publications, Online ISSN: 1873-5339, Print ISSN: 0965-9978, vol. 174.

[13] Asha N., Syed Fiaz A. S., Jayashree J., Vijayashree J., 2022, "Resource Aware Data Collection based on Node Relay Configuration in Wireless Sensor Networks", *International Conference on Inventive Research in Computing Applications (ICIRCA)* ISBN: 978-1-6654-9707-7, pp. 516 – 519.

[14] Syed Fiaz A.S, Asha N, Ashok P, Syed Navaz A.S, 2019, "Data Harmonizing in Cloud with Enhanced Reliability in Distributed Computing", *International Journal of Recent Technology and Engineering (IJRTE)* ISSN: 2277-3878, Volume-8, Issue-1, pp. 2779–2783.
[15] Syed Fiaz A.S, Rahul B, Rupendra J, Nitesh K, 2019, "Cloud Data Storage Using Attribute based Encryption with Verifiable Outsourcing Auditor Security Schemes", *International Journal of Recent Technology and Engineering (IJRTE)* ISSN: 2277-3878, Volume-8, Issue-2S11, pp. 133–138.
[16] Guruprakash K.S, Syed Fiaz A.S, Sankar S, 2018, "Iaas: Qos based Automated User Requirement Identification for Optimal Resource Allocation in Multi Cloud", *International Journal of Engineering & Technology (UAE)*, Vol. 7, No. 3.34, pp. 210–212, ISSN: 2227-524X.
[17] Syed Fiaz A.S, Guruprakash K.S, Syed Navaz A.S, 2018, "Prediction of Best Cloud Service Provider Using the QoS Ranking Framework", *International Journal of Engineering & Technology (UAE)*, Vol. 7, No 1.1, pp. 486–488, ISSN: 2227-524X.
[18] Syed Fiaz A.S, Asha N, Sumathi D, Syed Navaz A.S, 2016, "Data Visualization: Enhancing Big Data More Adaptable and Valuable", *International Journal of Applied Engineering Research*, Vol. 11, No. 4, pp. 2801–2804, ISSN 0973-4562.

Artificial Intelligence, Blockchain, Computing and Security – Dagur et al. (Eds)
© 2024 The Author(s), ISBN: 978-1-032-67841-2

Webcam painting canvas window application using OpenCV and Numpy

V. Usha*, P. Arivubrakan* & K. Sai Balaji*
CSE, Vel Tech Rangarajan Dr. Sagunthala R&D Institute of Science and Technology, Chennai

ABSTRACT: Designers all understood quite well that the most popular and efficient means of expression for our emotions are via art and writing. In the olden days, people used to write or print on the rocks, walls, stones, but as time passed the process changed. People had discovered new plants and extracted the minerals and they turned it into papers. Then they created different kinds of artificial papers. In this modern area, this has gone too far. People are using electronic devices like screens, where we can write, draw and represent various structures, symbols, arts, painting, etc. either by hand, by typing and also just with a single tap. But now, we are taking this process more far, making it easier to use and easier to express your feeling just by moving your little fingers. We know canvas is a special type of surface where it is also used for painting. And now imagine a surface where there is no physical object and try to paint or write anything. Is it possible? Yes, this is possible using air canvas. Air canvas is a digital drawing which uses free-hand. All you have to do is to think about what you want to express or show, and then just move your finger according to what you wanted to express.

Keywords: Air Canvas, Electronic device, artificial paper, digital drawing, finger

1 INTRODUCTION

The primary driving force was the demand for a spotless classroom environment so that pupils could concentrate on their studies in peace. We are aware of several methods, such as touch screens and other electronic devices, but what about schools that cannot afford to purchase such enormous, expensive electronic devices and use them to educate, such as televisions or one screen. So, we thought why not a finger can be tracked. Hence it was OpenCV which came to the rescue for these computer vision projects. On a canvas, as we all know, painters produce paintings. Very soon ripple the hands to make art in the air [4]. This project, we'll use Python and Open ComputerVision to create an air canvas. OpenCV is a free initiation computer vision framework that may be used for a variety of complex image processing applications. To accomplish this goal, we employed color detection and fractionation methods. Here, utilized a blue item to represent a pencil on the canvas. In order to paint on a screen using hand motion detection software, a basic prototype of a painting tool must be made. Utilizing OpenCV [11] for the recognition of objects placed at the tip of the finger. To implement this some fonts, have other features, such as changing color or size.

2 MOTIVATION

Drawing or Sketching utilized by hand is everybody's desire. A few or most of the time envision script in air utilizing our hand. Along these lines, at this time came the task of

*Corresponding Authors: husha88@gmail.com, arivubrakan@veltech.edu.in and saibalaji017@gmail.com

 DOI: 10.1201/9781032684994-55

beginning this idea where we make a material along with making a choice of the tones obligatory utilizing our tender furthermore, depict the necessary plan or compose whatever thing your aspiration.

3 PROBLEM STATEMENT

In order to foster an interface between human hand and the framework designing OpenCV strategies, as well as Guido's python coding words, we elected to use shading techniques on his drawing region.

4 LITERATURE SURVEY

Babu *et al.* in paper [1] a system that can follow subjects and pan in real time is demonstrated in a study. By seeing people or other things and panning the camera appropriately to follow and capture them, the system imitates a camera operator. These subject tracking [8] cameras may be used for a number of events, including keynote speeches and guest lectures.

Chen *et al.* in [2] Air-writing is the recognition of linguistic characters or words in a free space based on six degree-of-freedom hand motion data in [7]. Isolated air writing characters can be identified similarly to motion gastrula, however. Increased sophistication and variability make air-writing more complex and difficult to master.

Joolee *et al.* in [3] a new technique has been developed to monitor the position of the brush tip for interactive sketching. This method still has usability issues and requires a frame and cameras that are carefully placed. Participants recognize letters and numbers drawn in the open air was the main focus of the study.

Shetty *et al.* in [5] employs the system hand motions taken from a camera using an HSV colour recognition method, is a computer vision-based mouse cursor control system.

5 METHODOLODY

5.1 *Capture frames*

Import necessary packages like OpenCV and Numerical Python NumPY in Python IDE. Declare a video Capture object for capturing as cap.Cap.read() function used to read the frames [12] with the help of webcam. Use the flip function to flip the frames. Here inputs are federation (axis) and frames. Zero (0) for vertical flippant and One (1) for horizontal flippant. Use image show purpose to show frames in the new window. Unwrap the OpenCV pane awaiting the 'q' key is pushed.

5.2 *Creation of canvas window*

Initially define some random colors that we will be using at the time of project. An empty blank canvas window is created with the same camera frame size. Create a zero matrix using np.zeros() function. Set some color buttons on top of the window to change the pencil color. So that while painting we can change the color of the pencil. Create colored rectangles for the previously defined colors using CV2.rectangle () function. Use the put Text function to write text on colored rectangles. We can write the respective color name on each rectangle using this function.

5.3 *Blue color detection*

Here we will be using any blue colored object to act as a pencil for the canvas. Read frames as RGB color space using [9] OpenCV. For detection purposes to renovate the frames to

Hue, Saturation and Value (HSV) color space. The color space is cylindrical in shape, with axes running along the central axis and perpendicular to each other.

H - Hue is a measure of the colorfulness of a light source, and is encoded in angular dimensions.
S - Saturation describes the concentration of shade.
V - Value is a measure of how bright a shade is.

Now we have to convert the color space of an object using the convert color function. In binary segmented mask objects are detected in white color and the rest is detected in black color when we use the cv2.inRange() function. Area of the segmented white region increases while we use the dilate function. To find all the continuous points that fall within the same intensity or color as the green object, we start by drawing a segmented white border around the green object. We then find the center of the green object, which will serve as our tracker while drawing on the [6] canvas window.

To determine whether or not a color has been detected, the device first checks to see if any color has been inputted. If a contour is detected, we pick the contour with the largest area. Otherwise, we reset the previous center point to 0. If any contour is detected, then we pick a contour that has maximum area among all other contours in the list. Otherwise, we set the previous center point to 0. The preceding center point basically stores the center point of a contour in a frame. After selecting the highest outline we get the region of the contour using contour Area() function. Sometimes the main object may not be in the frame. To fix this we can use filters. Let's check the area of the contour if it is greater than the defined minimum area then finds the center of the contour. Draw a circle at the Centre of the object that was detected. This will help you to determine its size and shape.

5.4 *Drawing on canvas*

Select a color to draw on the canvas. If the previous centre point is not 0, then check to see if the new centre point is within a certain range of the previous centre point. If something is detected in the current frame, we can determine the coordinates of the color buttons. We checked to see which button was the center point of the contour's border. Next, set the color of the text accordingly.

To clear the canvas, you can use the first button. That's why if the center point is detected in the first button, that zeros () function can be used to reset the canvas to the blank slate using that point as the new center. If the previous centre point is detected as not zero, at this point, a contour has already been detected in the previous frame. To draw a line between the two points, we use the present centre point as the starting point and the previous centre point as the ending point. To update the previous centre point to the current centre point, we use the transform function. We then show the drawings on the main frame.

To convert the canvas into a grayscale image, use the convert Color function. Now create a binary mask using the threshold function. To convert from a binary mask to RGB, you first need to convert the binary mask to a number in the RGB color space. To add two images, the dimensions of the two images need to be the same. If both pixels are greater than 0, then the bitwise and function is true. Otherwise, the bitwise function is false. If the pixel value is greater than 0, then the pixel is set to true. Otherwise, the pixel is set to false.

6 IMPLEMENTATION

Initially start a new program in python IDE. We need to install the following libraries in python IDE such as numpy, cv2, dequeuer from collections. Numpy is a library in python programming language which has a large collection of arrays, multi-dimensional arrays and matrices. And also has a wide range of mathematical operations. OpenCV is a very important and useful tool in performing computer tasks and image processing. Create a color detector window, now create track-bars and set blue color HSV values as bead. Function used to create track-bar, cv2 is to track the motion continuously. »cv2.createtrackbar ()

Add dequeue to windows to store the coordinates of the color namely red, blue, green and yellow shown in Figure 1. Declare color indices for each color. As we created bead, small impurities are formed around the bead, so to remove the impurities of the small bead we need dilate the frame. In order to do, we need a kernel for it. So declare a one matrix of size five to cover the screen. Now declare HSV values to set colors for red blue green yellow respectively.

Create a paint window with white screen. Camera frame instances are taken input with the help of video-capture function and start reading the frames.
»cv2.VideoCapture () is used to capture the live stream with the camera
»read () is used to read the frames.

Convert the first frame into HSV color space. With the help of track-bar positions set lower and upper HSV values.
Adding color buttons to the live frame to color access. Create 5 rectangles in order to create four colors and one for clear and name the rectangles of respected colors.

»cv2.rectangle () is used to create rectangles.

By using in range(),and by applying erosion, morphology,dilution functions we will be able to create a proper mask, as a result we will get a clear bead roaming on screen whenever we move the object. If there are impurities, the bead will lag.

»cv2.inRange ()
»cv2.erode ()
»cv2.dilate ()
»cv2.morphologyEx ()

After this we have to find contour. Here contour value is 0, as we didn't find any centre. If we find contour, we have to check whether it original or formed by an impurity. If the length of the contour is greater than 0, then it is original bead. We had to find a minimum enclosing centre for the found contour. With the help of circle moments, we have to find centre for the found circle. If centre value is less than 65, which means our centre lies in any of the four buttons. With the help else if condition we have to check where our centre is lying in the coordinates. If it is not present in that range, obviously we have to draw on the screen the color that we are using currently. for this color indices will help us in drawing. As a result, we are going to know in which color dequeue we have to put the coordinates. If the color index is 0 1, 2, 3, then the coordinate is put in the respective dequeue color indices. By default, the color index is blue; whenever we click a new color it will change to respected dequeue. Using for loops, we are checking every dequeue, based on the coordinates of respective colors that pixels are converted into respective colors. We are going to show tracking [10] paint and mask windows to the user. As result we are getting paint on canvas with is drawn.
»cv2.imshow () is used to display an image in the window.

With the help of wait () function we will set q for exit the program run. If q is pressed, then the programme exits and stops the run of the CPU. Cap. release () function releases the camera.

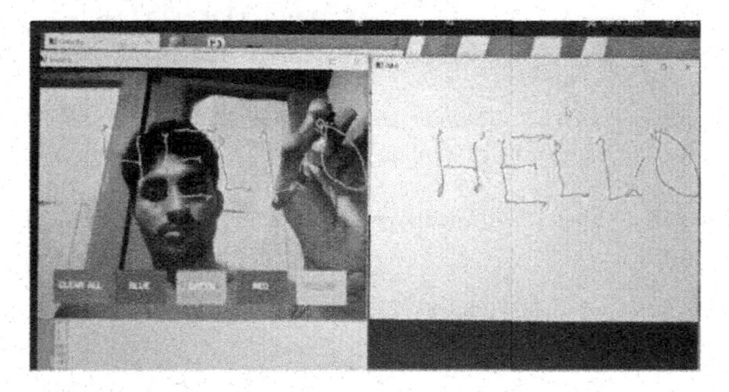

Figure 1. Canvas screen.

In this Figure 1 representing the canvas window that the person chooses color and written the word HELLO. That the corresponding output also displayed near white screen.

7 RESULT AND DISCUSSION

A color marker is used by holding it in the hand and four colors are shown respectively as blue, green, and red and yellow as well as we have a clear button which is used to erase the drawing done on a Canvas window. In this work user have to take a blue colored object and draw in air, in front of the web camera. The motion of the object is tracked by the computer and the user should select his desired color to draw. The drawing is visible on the Canvas window.

8 CONCLUSION AND FUTURE ENHANCEMENTS

Day by day technology is improving a lot. In our present technology we want to draw our imaginations by just waving our finger in the air. We invented a new way of building air canopies you can use a computer to capture the motion of a colored marker and use that motion to create drawings on the canvas. Whenever a new green object comes into the frame it takes the new object a pencil. This will be improved so that the using object remains as the pencil even if a high intensity new object enters the frame. In future a tools tab will be added which contains mathematical tools like triangle rectangle etc. for easy drawing. In future additional features will be added so that light colors will be drawn when we selected a particular color.

REFERENCES

[1] Babu, S., Pragathi, B.S., Chinthala, U. And Maheshwaram, "Subject Tracking With Camera Movement Using Single Board Computer", *IEEE-Hydcon* (Pp. 1–6), sep 2020.

[2] Chen, M., AlRegib, G. and Juang, B.H., "Modeling and Recognition of Characters, Words, and Connecting Motions", *IEEE Transactions on Human-Machine Systems*, 46(3), pp.403–413, 2015.

[3] Joolee, J.B., Raza, A., Abdullah, M. and Jeon, "Tracking of Flexible Brush Tip on Real Canvas: Silhouette-Based and Deep Ensemble Network-Based Approaches", *IEEE Access*, 8, jan 2020.

[4] Kaur, H., Reddy, B.G.S., Sai, G.C. and Raj, "A Comprehensive Overview of AR/VR by Writing in Air", *IJSRCSEIT*, 2021.

[5] Shetty, M., Daniel, C.A., Bhatkar, M.K. and Lopes, O.P., "Virtual Mouse Using Object Tracking" *5th (ICCES)* (pp. 548–553). IEEE, 2020.

[6] Zhou, L., "*Paper Dreams: an Adaptive Drawing Canvas Supported by Machine Learning Doctoral Dissertation*", Massachusetts Institute. 2019

[7] Kumar, A., & Alam, B. (2019). Energy Harvesting Earliest Deadline First Scheduling Algorithm for Increasing Lifetime of Real Time Systems. *International Journal of Electrical and Computer Engineering*, 9(1), 539.

[8] Kumar, A. & Alam, B. (2018). Task Scheduling in Real Time Systems with Energy Harvesting and Energy Minimization. *Journal of Computer Science*, 14(8), 1126–1133.Nitin Kumar, R., M. Vaishnavi, K. R. Gayatri, "Air Writing Recognition using Mediapipe and Opencv". Smart Innovation, Systems and Technologies. Vol. 302. 2022

[9] Robert Y. Wang, Jovan Popovi´c, *"Real-Time Hand-Tracking with a Color Glove"*, 2008.

[10] Kommu G. R., "An Eficient Tool for Online Teaching Using Opencv". *IJCRT* **9**(6) (2021). ISSN: 2320–2882.

[11] Usha V., Sai Balaji K., "Social Distance Alerting System Using Machine Learning", *IEEE Spon, ICCST* 2022

An effective objective function to enhance the performance on the Internet of Things environment

P. Arivubrakan, V. Usha & Cholaveti Naga Sai Manikanta
Vel Tech Rangarajan Dr. Sagunthala R&D Institute of Science and Technology, Chennai

ABSTRACT: The Internet of Things is an evolving technology that connects and exchanges data with other technology. The communications between the devices are resource constrained in the field of smart environments. The low-power devices transmit their packets to nearby nodes until it reaches the receiver, by making use of the routing protocol which establishes a set of rules to communicate in an environment. RPL is used as a routing protocol in low power lossy networks. An Objective Function determines the node selection and the optimization for the instances based on the number of parameters are hop count and expected transmission count in the tree-like topology. The energy-aware routing protocol-based objective function is proposed by the new metric which is based on the residual energy, threshold, and quality index. The novel metric is simulated using the Contiki Cooja simulator and the result shows better performance in terms of throughput, reliability when compared with the other standard objective functions.

Keywords: Objective Function, low power lossy network, Routing Protocol, Scalability

1 INTRODUCTION

The Internet of Things is to connect physical objects with sensors and transfers information through the internet without human intervention [1]. It is a rapid technology that performs in various real-world applications such as smart environments. The emerging domain of the internet of things has limitations in terms of resource-constrained devices that lead to poor performance. Billions of devices are connected through the internet in smart environments that have large volumes of data, those data are transferred to nearby devices through the routing protocol [2]. The routing protocol provides a set of rules that establish the communication between the sender and receiver. In recent years, researchers are working on routing protocols with the quality of service-aware communication between the devices, and the performance is analyzed. are established. According to the IEEE standards, the routing protocol has five layers structure [3].

In this paper we propose the NMOF algorithm to be implemented in the network layer, to enhance reliable communication in various smart applications. The novel reliable routing protocol as NMOF achieves reliable communication and energy-efficient data delivery. The performance analysis and simulation studies are accomplished to estimate the effectiveness of the proposed algorithm NMOF. Section II introduces RPL routing protocols and the terminologies of the routing metrics. Section III presents the proposed routing algorithm. Simulation and performance evaluation of the quality of service is conducted in Section IV along with the conclusion.

2 RPL PROTOCOL

RPL Protocol is meant for distance vector routing to operate on the standard of IEEE 802.15.4 [4], it is a Low Power and Lossy network and is mainly of resource-constrained

DOI: 10.1201/9781032684994-56

devices. It forms a tree-like structure. The IEEE 802.15.4 is fundamental for all the routing protocols in the Network layer of the IoT layered architecture [5]. The Frame structure is not suitable for power-constrained IoT devices, IEEE 802.15.4e is introduced with the extension of IEEE 802.15.4, with the feature, sleep-awake mechanism, if the nodes want to communicate, with another node, at the time of transmission, node, switch on its radio and send the message to another node, the receiver node, receives its message and send back the acknowledgment and send back to sleep [6].

Sensor devices are constrained in expressions of processing power, battery and memory. The main constraints of the protocol are low data rates and packet delivery rates are low. The traffic patterns for communication in the networks are point-to-point, multi-point communication. RPL Protocol is especially considered for low-power lossy networks to provide a possible path if its paths are inaccessible [7]. The data dissemination is a mechanism followed by the RPL Protocol for dynamically changing the networks. Distance vector routing means sending a copy of its location to all the neighbors in the networks.

2.1 *RPL terminology*

The DODAG (Destination Oriented Directed Acyclic Graphs) is an Acyclic Approach that has, a Top-Bottom approach and a Bottom-Top mechanism [8]. The root is the destination that has no outgoing edges. DAG (Directed Acyclic Graphs). The root act as the Border router which acts as the external connectivity node, it redistributes the DODAG routes to other routing protocol. The DODAGs are disjoint. The Link properties are the reliability and latency instances having the optimization objective. The Objective function helps us to decide whether we are near the root or away from It is decided by a programmer or a designer and to minimize. It can be energy or latency. Based on the metrics, we evaluate them as a number [9].

The Objective Function decides how RPL selects the best routes and optimizes the path. The information is available for the RPL nodes in OF to avoid loops [10].

Table 1. Control message.

Control Message	Control Message-Abbreviation	Explanation
DIO	DODAG Information Object	Multicast Message announcement
DIS	DODAG Information Solicitation	Message which sends to know is there is any DODAG
DAO	DODAG advertisement object	Request send by a child to parent
DAO-ACK	DAO Acknowledgement	Response [yes/no]
Consistency check	Consistency check	Deals with security

To minimize the packet drops, our efficient algorithm is proposed for this constraint [11]. Top-down approach the nodes send the packets to all the children nodes and then reach the destination, to consume the energy of the nodes, which will follow the bottom-up approach. Our proposed algorithm attains the minimum energy.

3 PROPOSED FRAMEWORK

The major issues of the RPL are energy efficient design, synchronization is addressed in the proposed algorithm with the standards. The LR has the two devices as Full Functioned Device and Reduce Functioned Device. The standard having the FFD will act as PAN coordinator, it will communicate with another device. If the RFD wants to communicate with another RFD, it will communicate via only. If the coordinator node has low power means there are a lot of possibilities for packet loss, minimum throughput, and delay, to overcome the issues by the proposed algorithm NMOF will follow the steps, RPL has two operations name parent selection and routing optimization. The first task is achieved through a set of tasks implemented in the protocol. The choosing of the parent node by the NMOF has been modified in a way to involve a novel functionality, and metric, which considers the information of transmissions in the network. The NMOF algorithm implies various metrics such as hop count, throughput and residual energy, link quality index. Based on the metrics it will select the best parent selection. The node which is nearest, and the energy is high then, it will choose the node as a parent and forwards the packets without any delay.

4 SIMULATION RESULTS

Simulation is performed in the Contiki Operating systems which are used for wireless networks, memory-constrained networks, and smart systems. The Contiki-based Cooja simulator is accomplished by simulating wireless sensors which look like real-time motes such as sky motes.

4.1 *Performance evaluation*

The quality of service metrics is implicit to evaluate the performance of the proposed algorithm of the energy-aware objective function. Minimizing packet loss, latency, jitter, and maximizing throughput are all characteristics of a network's quality of service.

Table 2. Simulation parameters-Contiki OS.

Metrics	Value
Area	1000
Sensor node	25,50,75,100
Destination node	1
Transmission range	100m
Inference Range	200m
Size	64Bytes
Runtime	2400s

4.1.1 *Packet delivery ratio*

The packet delivery ratio denotes the number of successful packets delivered to the destination node.

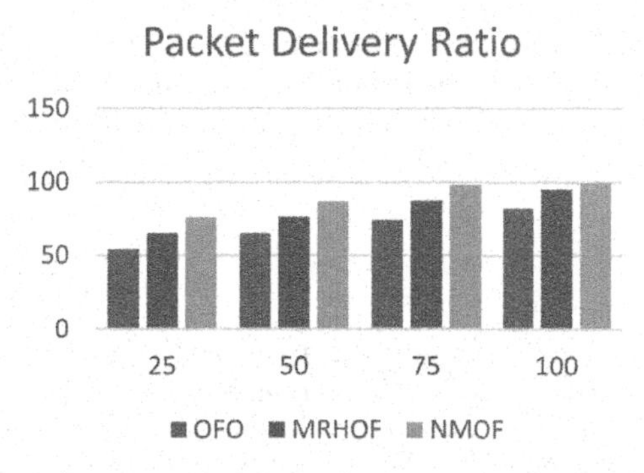

Figure 1. Packet delivery ratio.

4.1.2 *Energy consumption*

Energy consumption denotes the average energy consumption of the nodes with successful packets transferred.

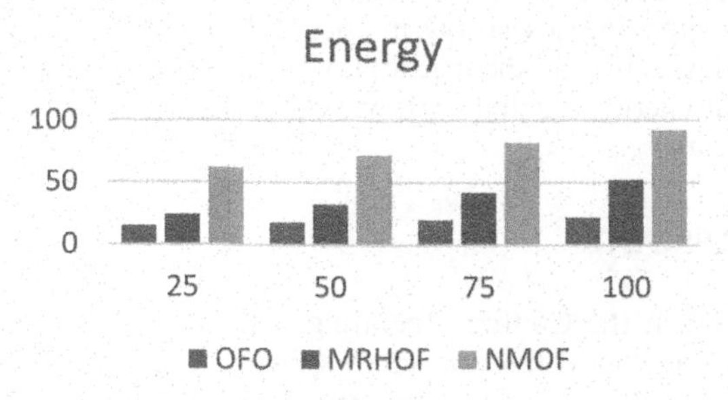

Figure 2. Energy consumption.

4.1.3 *Delay*

The delay denotes possible delay during packet transmission and reception.

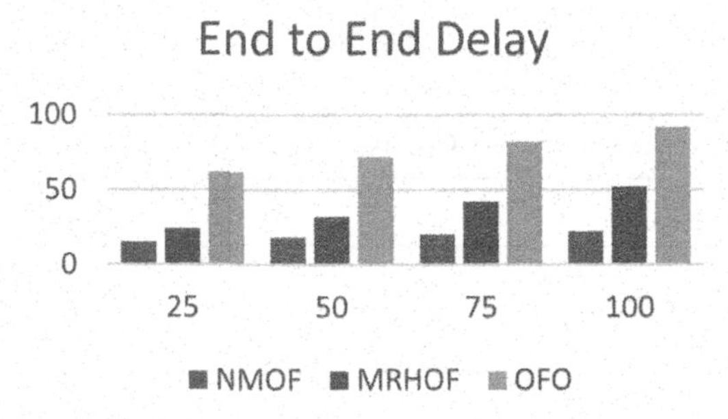

Figure 3. End-to-end delay.

5 CONCLUSION

The proposed a novel technique to enhance the quality of service metrics in the low power lossy network based on the objective functions which decide the best candidate parent node selection and energy utilization. The efficient use of residual energy is utilized properly in terms of reliability and minimum delay. The results showed the effectiveness of the energy-aware objective function algorithm to expand the lifetime of the network for reliable communication. The novel algorithm shows better performance analysis by comparing it with the OFO and MRHOF and it can be easily implemented and provide interoperable solutions for various real-world applications such as healthcare monitoring, smart cities, buildings, agriculture, and it will meet the requirement for future implementation.

REFERENCES

[1] Farooq O M, C. J. Sreenan, K. N. Brown, & T. Kunz, 2017. Design and Analysis of RPL Objective Functions for Multi-gateway Ad-hoc Low-power and Lossy Networks, *Ad Hoc Networks*, vol. 65, pp. 78–90.

[2] Homaei M H, E. Salwana, & S. Shamshirband, 2019. An Enhanced Distributed Data Aggregation Method in the Internet of Things, *Sensors*, vol. 19, no. 14.

[3] Hwang R H, M.-C. Peng, C.-Y. Wu, & S. Abimannan, 2022. A Novel RPL Based Multicast Routing Mechanism for Wireless Sensor Networks, *International Journal of Ad Hoc and Ubiquitous Computing*., vol. 33, no. 2, pp. 121–131.

[4] Khallef W *et al.*, 2017. Multiple Constrained QOS Routing with RPL, *IEEE International Conference on Communications (ICC)*. IEEE. pp. 1–6.

[5] Lalani S.R *et al.*, 2020. "Refer: A Reliable and Energy-efficient RPL for Mobile IoT Applications," *Real-Time and Embedded Systems and Technologies (RTEST). IEEE.*

[6] Dagur, A., Kaushik, A., Rastogi, A., Singh, A., Kumar, A., & Chaturvedi, R. (2021). Optimization of Queries in Database of Cloud Computing. *In Data Intelligence and Cognitive Informatics: Proceedings of ICDICI 2020* (pp. 325–332). Springer Singapore.

[7] Kushwaha, A., Amjad, M., & Kumar, A. (2019). Dynamic Load Balancing Ant Colony Optimization (DLBACO) Algorithm for Task Scheduling in Cloud Environment. *Int J Innov Technol Explor Eng*, 8 (12), 939–946.

[8] Sankar S, S. Ramasubbareddy, A. K. Luhach, A. Nayyar, & B. Qureshi, 2020. CT-RPL: Cluster tree Based Routing Protocol to Maximize the Lifetime of Internet of Things," *Sensors*, vol. 20, no. 20, p. 5858.

[9] Sun, Y. Liu, Z. Chen, A. Wang, Y. Zhang, D. Tian, & V. C. Leung, 2019 "Energy Efficient Collaborative Beam Forming for Reducing Side Lobe in Wireless Sensor Networks," *IEEE Transactions on Mobile Computing*.

[10] Kumar, A. & Alam, B. (2019). Energy Harvesting Earliest Deadline First Scheduling Algorithm for Increasing Lifetime of Real Time Systems. *International Journal of Electrical and Computer Engineering*, 9(1), 539.

[11] Kumar, A. & Alam, B. (2018). Task Scheduling in Real Time Systems with Energy Harvesting and Energy Minimization. *Journal of Computer Science*, 14(8), 1126–1133.

Artificial Intelligence, Blockchain, Computing and Security – Dagur et al. (Eds)
© 2024 The Author(s), ISBN: 978-1-032-67841-2

A modelling analysis to predict the traffic using K-NN algorithm

R. Narmatha*
Department of Computer Science and Engineering, Government College of Engineering (Anna University), Dharmapuri, Tamil Nadu, India

M. Prakash
Department of Data Science and Business Systems, School of Computing, SRM Institute of Science and Technology, Kattankulathur, Tamil Nadu, India

V. Vennila
Department of Computer Science and Engineering, K. S. R. College of Engineering, Tiruchengode, Tamil Nadu, India

ABSTRACT: In this modern world due to Road traffic, many people are unable to reach their destination at the correct time. For example, if a person needed to reach the hospital in critical condition due to road traffic, they are unable to reach the hospital and got stuck in traffic which will lead to death. We lost a lot of people because of this road traffic. So, Road Safety is the most important part of our daily life. Conjointly, this prediction accomplished enhanced the safety and security of women empowerment which acted as a security guard for people especially women, protecting themselves in every circumstance. Here, statistical analysis of road traffic networks of selective locations is implemented to prognosticate road traffic. Alert and divert the people who were identified in the traffic zone and found the location of the people who were identified in the unsafe zone.

Keywords: K-NN, IMAP, CSV, traffic congestion.

1 INTRODUCTION

We regularly use location-based applications in our everyday lives. We formulate the quality prediction for the tentative disposal of the output values for given multiple parameters of input values. The structure focal point of searching compatible discourse appearance is to predict the Next POI prediction and the Next point of duration prediction. In this study, predictions based on POI [3] were evaluated and compared with the user's history of locations for enhancement of a particular area. This data analysis Technique is compared to prognosticate the significant location of the users based on POI [4]. Machine learning techniques can be used to estimate traffic congestion utilizing the data, which can be saved in a CSV file on a DBMS. You can learn about traffic-related problems like traffic jams in a specific area of the road network on a specific occasion by utilizing a road traffic prediction system (TCPS) [2–5]. Using the likelihood of modification, we present the K-Nearest Neighbour (K-NN) algorithm concept throughout this work and afterward forecast the location. Next, a thorough simulation is performed that assesses how well the suggested method performs. This paper demonstrated the proposed strategies with notable accuracy with security features accessible to everyone. The quality of transport remains essential, especially in cities in which a greater development of civilizations on transportation systems is required for day-to-day lives.

*Corresponding Author: researchnarmatha1988@gmail.com

 DOI: 10.1201/9781032684994-57

The remaining of this work is organized in the following manner. Section 2. Demonstrate the feasibility of related work based on Location prediction using different Algorithmic concepts. Section 3. Demonstrate the use of the Location prediction technique, which means discussing using the K-Nearest Neighbour (K-NN) algorithm. Section 4. Presents experimental results and implementations of the K-Nearest Neighbour (K-NN) algorithm. Section 5 ended with the conclusion and future research directions and enhancement are discussed further.

2 LITERATURE REVIEW

To ensure that traffic flows freely and that driving is safe and secure, road traffic control systems are therefore necessary. To reduce traffic congestion in the city, road traffic control and the alert system eventually enhanced city life if any emergency persists. Recent history has witnessed an increase worldwide concerning traffic accidents. Safety as well as security is linked to the effectiveness and efficiency of the road transportation system and control. The road transportation system and control became safety and security every concern of people. Thus, reducing deaths and efforts taken to avoid serious injuries or death of road users. When road traffic is effectively handled, it will travel gradually, help to prevent future delays, as well as encourage more people to occupy the available road space. More importantly, road transportation congestion is avoided when the traffic control system is managed properly, particularly at transportation intersection points. There will be a clear path for every car with safety and security if an emergency persists for the people. The primary objective of the traffic control system is to ensure the safety and security of all road users, which is every other factor. In [1] used YOLO (you only look once) method, implemented to upload the picture of vehicles from CCTV cameras on the signal networks used to retrieve information on traffic-related issues combined with image processing, was developed by existing researchers in response to the many traffic-related challenges. Based on the participant's observed historic data, a network model is created to forecast the participant's prediction 24 (1). For optimization algorithms and modelling, a network model is analysed to improve the prediction (1). In [7] Existing researchers, two characteristics of human behaviour are predicted by utilizing mobile phones as sensing applications. By utilizing this comprehensive context information, they proposed a method for forecasting where individuals will be for the next 10 minutes from sensors in smartphones.

3 PROPOSED SYSTEM

The Statistics are analyzed in this study to analyze a person's behavior while using a mobile phone as a predictor. We explore that generic behavioral visualization structures can improve the prediction accuracy of customized Datasets by using a system that predicts where people will go in the next 10 to 15 minutes using information collected from mobile phone behavior with such a smartphone serving as a detection system, our goal is to capture generic behaviour visualization [9,10,12]. We concentrated on K-Nearest Neighbour (K-NN) algorithm and data sets, along with human behavior observed to improve accuracy with latitude and longitude smartphone data gathered over 12 months by 100+ users, which were used to facilitate implementation. We have proposed a road traffic prediction system utilizing the use of the K-NN (K-Nearest Neighbour) algorithm upon examining the shortcomings of the current system [6–8]. The proficiency to forecast the relevant preferences of the user where users want to go further that should be used to improve user interaction. End users have speedy access to the data where user assistance is needed. Users have reliable and fast communication to access the application if the application perceives to know that the user needed an assistant to do further. In another way, if the predicting software system knows that the user will for dinner, then it can suggest the list of diners such as today's menu items available in that shop. Smartphones are a suitable option for tracking the location of the user. We focus on the scenario that the next user place is predicted without traffic based on the current location of the user.

Our framework is inspired by prediction algorithms commonly used in signal processing for forecasting time series [11,13]. More precisely, we predict the next location of a user and which application he/she will use based on the current context consisting of location, time, app usage, Bluetooth proximity, and communication logs. This approach allows modeling the interplay between the predicted variables to study relationships between the place where a user stays and the possibility that he would make a phone call, use the cameras, and so on.

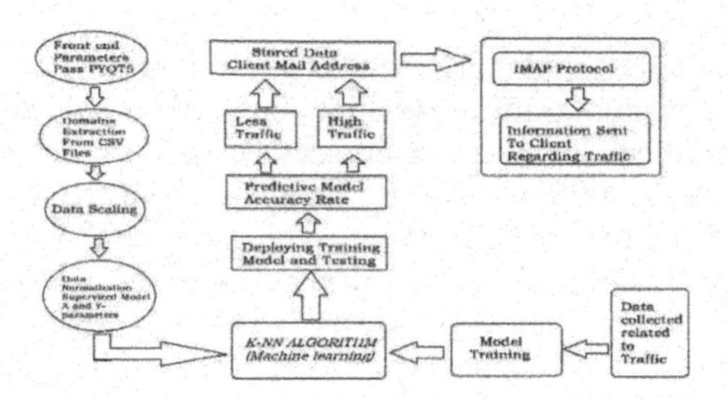

Figure 1. System architecture diagram.

In the figure, Machine learning models were trained to build a predictive model for tasks utilizing datasets. Data is gathered and implemented to analyze from multiple sources of the system. Data can be kept in CSV format on a database management system to predict traffic congestion using machine learning algorithms. The information is being used as input for the model's training and the provided datasets can be utilized to extract the necessary fields. The high-quality data that has been changed is subsequently provided to a single, integrated specified location for storage and evaluation. The purpose of the manipulation is to change the data to make it more readable and structured. For data analysis and data visualization, we modify the information using the given groups of variables. Connect using the appropriate IMAP Server account. Must choose IMAP Server desired to send mail as well as SMS about traffic congestion. Analysing and processing unstructured content and extracting data related to traffic congestion by obtaining the emails using IMAP protocol. Finally, Information was sent to the client regarding traffic congestion.

4 IMPLEMENTATION AND RESULT

LEVEL 0

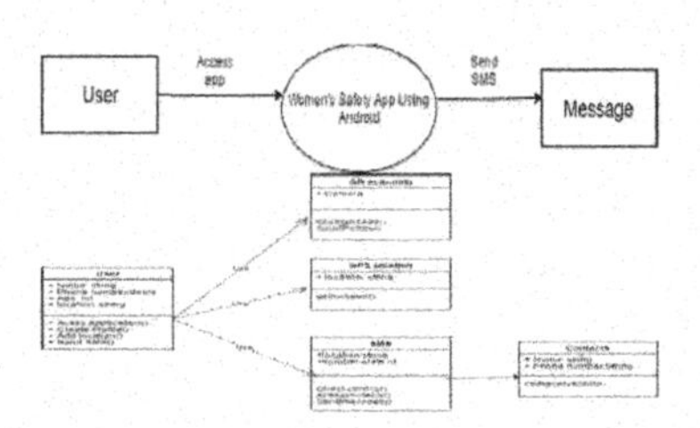

Figure 2. Interaction between user and system being developed mentioned using flow diagram.

Figure out the perfect application to alert and notify your loved ones if you find yourself in a dangerous situation, especially for Women's Self-preservation. The quickest and easiest way to inform your loved ones of your current position as well as other information is through an application. Latitude and Longitude of user Accessing mentioned using Flow Diagram.

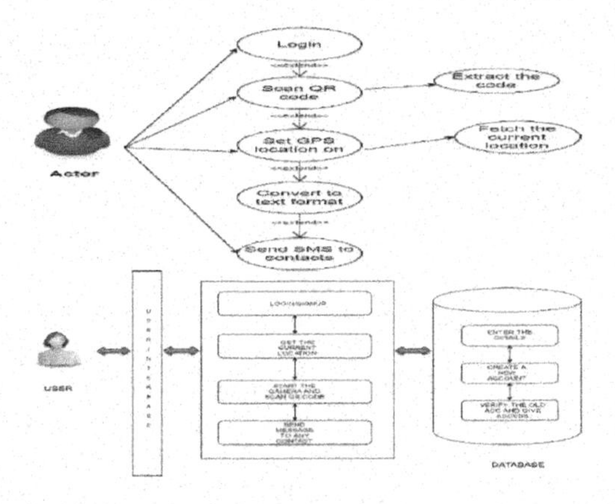

Figure 3. Framework to alert the individual through Mail and SMS.

In this result section, as shown in the figure the performed model for predicting traffic congestion using the K-NN algorithm. As a result the K-Nearest Neighbor (KNN) algorithm, which makes use of the spatial as well as the temporal correlations found in the traffic dataset, predicts the traffic congestion on a particular road on a particular date.

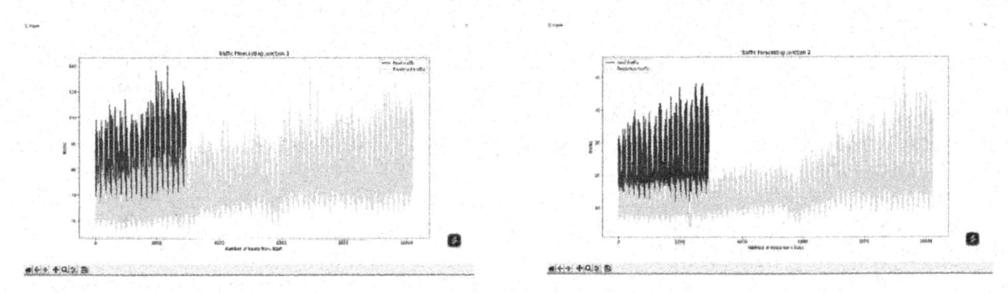

Figure 4. Traffic forecasting graph for Junction 1. Figure 5. Traffic forecasting graph for Junction 2.

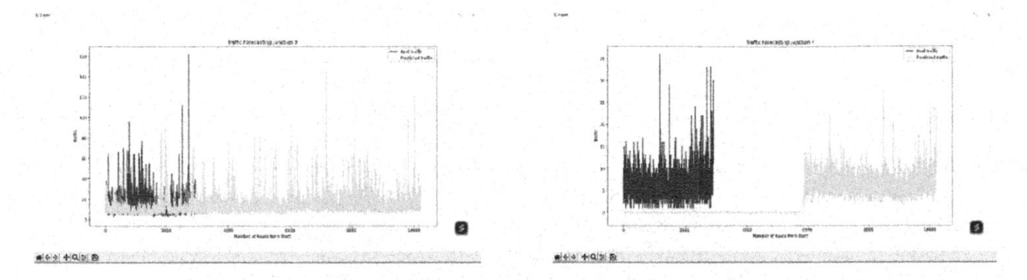

Figure 6. Traffic forecasting graph for Junction 3. Figure 7. Traffic forecasting graph for Junction 4.

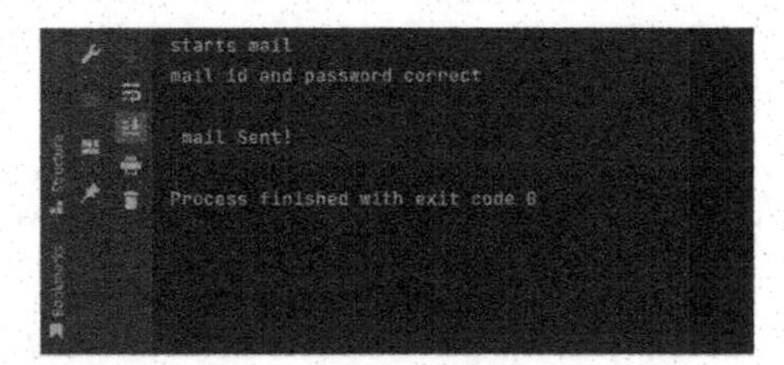

Figure 8. Alert send by Email

Finally, the User received a notification about the traffic jams at a specific place that achieve us to predict the traffic congestion on a specific road network area through a security features algorithm.

5 CONCLUSION & FUTURE SCOPE

Traffic flow has unpredictable and complex to find the exact location as well. A K-Nearest Neighbor (K-NN) algorithm enhanced in this project anticipates traffic flow analysis and enhancement. Performance measures demonstrated that the proposed prediction approach can be utilized to forecast short-term predictions [1] as well as long-term predictions that build a traffic flow of the system with sources and end of the system. Finally, K-Nearest Neighbor (K-NN) Algorithm with 98.8% accuracy with security features achieved a higher prediction accuracy. The simulation results supported our system's striking performance enhancement. Further research looks at how well our techniques performed in real-world applications. In the future, we enhanced the system by including functions like speed detection for automobiles using the real-time data gathered, which can help to prevent accidents in crowded locations.

REFERENCES

[1] Nižetić I., Fertalj K., and Kalpić D., "A Prototype for the Short-term Prediction of Moving Object's Movement Using Markov Chains," *Proc. Int. Conf. Inf. Technol. Interfaces, ITI,* pp. 559–564, 2009, doi: 10.1109/ITI.2009.5196147.

[2] Wu R., Luo G., Shao J., Tian L., and Peng C., "Location Prediction on Trajectory Data: A Review," Big Data Min. Anal., vol. 1, no. 2, pp. 108–127, 2018, doi: 10.26599/BDMA.2018.9020010.

[3] Liao J., Liu T., Liu M., Wang J., Wang Y., and Sun H., "Multi-Context Integrated Deep Neural Network Model for Next Location Prediction," *IEEE Access*, vol. 6, no. c, pp. 21980–21990, 2018, doi: 10.1109/ACCESS.2018.2827422.

[4] Psyllidis A., Yang J., and Bozzon A., "Regionalization of Social Interactions and Points-of-interest Location Prediction with Geosocial Data," *IEEE Access*, vol. 6, no. 1, pp. 34334–34353, 2018, doi: 10.1109/ACCESS.2018.285002.

[5] Indira K., Brumancia E., Kumar P. S., and Reddy S. P. T., "Location Prediction on Twitter Using Machine Learning Techniques," *Proc. Int. Conf. Trends Electron. Informatics, ICOEI 2019, no. Icoei*, pp. 700–703, 2019, doi: 10.1109/ICOEI.2019.8862768.

[6] Do T. M. T. and Gatica-Perez D., "Where and What: Using Smartphones to Predict Next Locations and Applications in Daily Life," *Pervasive Mob. Comput.*, vol. 12, pp. 79–91, 2014, doi: 10.1016/j.pmcj.2013.03.006.

[7] Xia L., Huang Q., and Wu D., "Decision Tree-based Contextual Location Prediction from Mobile Device Logs," *Mob. Inf. Syst.*, vol. 2018, 2018, doi: 10.1155/2018/1852861.

[8] Mehra, P. S., Mehra, Y. B., Dagur, A., Dwivedi, A. K., Doja, M. N., & Jamshed, A. (2021). COVID-19 Suspected Person Detection and Identification Using Thermal Imaging-based Closed circuit Television camera and Tracking Ysing Drone in Internet of Things. *International Journal of Computer Applications in Technology*, 66(3–4), 340–349.

[9] Kumar, A., & Alam, B. (2016). Real-Time Fault Tolerance Task Scheduling Algorithm with Minimum Energy Consumption. In *Proceedings of the Second International Conference on Computer and Communication Technologies*: IC3T 2015, Volume 2 (pp. 441–448). Springer India.

[10] Ashbrook D. and Starner T., "Learning Significant Locations and Predicting User Movement with GPS," *Proc. - Int. Symp. Wearable Comput. ISWC*, vol. 2002-Janua, pp. 101–108, 2002, doi: 10.1109/ISWC.2002.1167224.

[11] Du Y., Wang C., Qiao Y., Zhao D., and Guo W., "A Geographical Location Prediction Method Based on Continuous Time Series Markov model," *PLoS One*, vol. 13, no. 11, pp. 1–16, 2018, doi: 10.1371/journal.pone.0207063.

[12] Ahas R., Silm S., Järv O., Saluveer E., and Tiru M., "Using Mobile Positioning Data to Model Locations Meaningful to Users of Mobile Phones," *J. Urban Technol.*, vol. 17, no. 1, pp. 3–27, 2010, doi: 10.1080/10630731003597306.

[13] Kushwaha, A., Amjad, M., & Kumar, A. (2019). Dynamic Load Balancing Ant Colony Optimization (DLBACO) Algorithm for Task Scheduling in Cloud Environment. *Int J Innov Technol Explor Eng*, 8 (12), 939–946.

Artificial Intelligence, Blockchain, Computing and Security – Dagur et al. (Eds)

CryptoASCII - ASCII based cryptosystem for cloud data storage

M. Pavithra
Department of Computer Science and Engineering, S.S.M College of Engineering (Anna University), Komarapalayam, Tamil Nadu, India

M. Prakash
Department of Data Science and Business Systems, School of Computing, SRM Institute of Science and Technology, Kattankulathur, Tamil Nadu, India

V. Vennila
Department of Computer Science and Engineering, K. S. R. College of Engineering, Tiruchengode, Tamil Nadu, India

ABSTRACT: Cloud computing plays a huge role in IT industry, Education sector, government sector, medical field, defense management, Natural disaster control, weather reporting, road traffic maintenances, e-money transaction etc. The cloud provides wide amount of services can include delivery of data, software, hardware, infrastructure, platform, networking and so on. The merits of cloud computing are numerous to the user for data storing, accessing, maintenance in the cloud from anywhere at any time at low cost. On the other hands, the data in the cloud were not maintained by the direct owner. Despite, the data security has always been big challenge in cloud storage. Data integrity, confidentiality and privacy protection are the prominent factors of the user's side from the cloud environment. In this proposed paper, we introducing a new method CryptoASCII to provide data security in the cloud storage. By CryptoASCII, we confirmed that the user's data are maintained with integrity and confidentiality. CryptoASCII can provide encryption and decryption algorithms to ensure data privacy as well as it provide redundancy checking proof for data integrity.

Keywords: Cloud computing, ASCII, Encryption, Integrity, Privacy, Redundancy, Decryption, Security

1 INTRODUCTION

Network is a set of computer which are used to communicate with each other for data sharing. Internet is a network of network can access the data at anywhere at any time. Cloud computing is the resource sharing, data sharing and data storage among the internet. (Maryam *et al.* 2018)

It providing services as Plat as a Service (PaaS), Software as a Service (SaaS), Infrastructure as a Service (IaaS) in Figure 1 (Zhan *et al.* 2015).

There will be some security challenges in cloud computing in all PaaS, SaaS, IaaS like security, confidentiality, security, storage. (Barrowclough *et al.* 2018) (Faizi *et al.* 2019). These issues are overcomes by providing security to the cloud computing environment. (Kupta *et al.* 2014) (Kaur *et al.* 2012).

Cryptography of data will provides the security to the cloud resources. Cryptography is the process of converting the plain text (readable form) into cipher text (non-readable form).

DOI: 10.1201/9781032684994-58

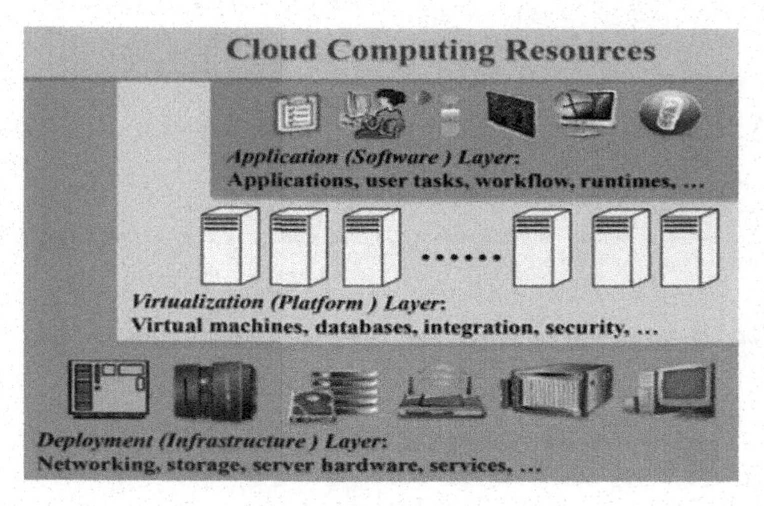

Figure 1. A brief architecture of cloud computing (Zhan *et al.* 2015).

It uses encryption and decryption algorithms for users. (Singh *et al.* 2013). In this paper, we introducing the cryptography technique to ensure the data privacy in cloud storage environment. CryptoASCII method provides high level protection of cloud data. The work is elaborated in the below section.

2 RELATED STUDY

2.1 *Cloud features and services*

The cloud related to the lot of data and information mechanism, which gives convenient way to accessing services, application and resources over the internet. (Fei Han *et al.* 2016.). Based upon usage cloud classified as public, private, hybrid and community. (Figure 2) (Yahya *et al.* 2014)

Private-Access by single group or a single organization
Public-Access by any user over internet say Amazon
Community-Access by limited (two or more) groups
Hybrid-Combination of public, private, community cloud

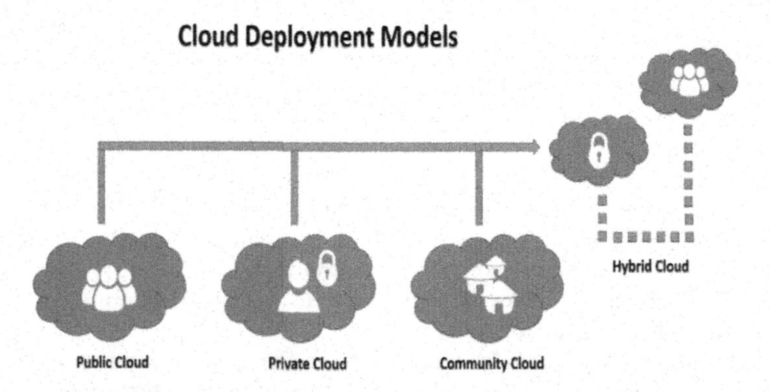

Figure 2. Cloud classification.

There are three major services as PaaS, SaaS, IaaS (Figure 1). In PaaS, the cloud service provides virtual machine and a set of programs to do a particular task. In SaaS, the provider provides lot of application for user needs. In IaaS, the cloud service provider offers bulk storage to user for easy data access. (Sun *et al.* 2014) (Senyo *et al.* 2018)

2.2 *Cryptography features and its algorithms*

Cryptography may achieve security in confidentiality, availability and integrity of data and resources in the storage, sharing and accessing. There are main two classification of algorithms. (Akhil *et al.* 2017) (Wan *et al.* 2011)

i) Symmetric key algorithm- Same key (Secret key) are used for encryption at sender side and same key used fordecryption at receiver side. (Figure 3)
ii) Asymmetric key algorithm- Different keys (Private and Public keys) are used. One key for encryption and other key for decryption. (Figure 4).

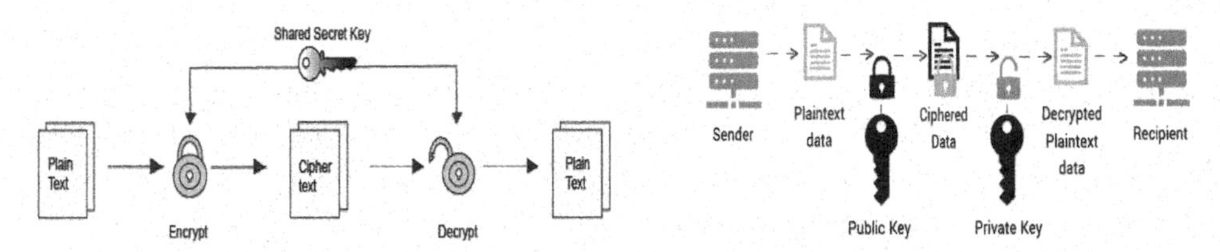

Figure 3. Symmetric key algorithm.

Figure 4. Asymmetric key algorithm.

Through the study of different algorithm (i.e.,) RSA, Diffie Helman, DES, AES etc. (Waters *et al.* 2011) (Wang *et al.* 2016). Among all the existing algorithms, our proposed new algorithm CryptoASCII method provide better result and output.

3 PROPOSED MODEL

We introducing the cryptography security for data storing and data retrieving in cloud platform by CryptoASCII method. It provides encryption algorithm at sender side for data storage to cloud and decryption algorithm at receiver side for data retrieving from cloud

3.1 *Encryption algorithm*

For hacking and modification of data in cloud we used to convert the readable form of data into

Non-readable form. (Figure 5)

i) Convert the input text into capitalized letter
ii) ASCII converter is to convert the each letter into equivalent ASCII value
iii) Decimal ASCII value is convert into 8-bit binary equivalent (If the binary value is not in 8 digits, add 0 at proceeding)
iv) Do 1's complement of this 8-bit binary and check the redundancy by CRC checker
v) Convert the binary value into equivalent decimal value
vi) Now the encrypted decimal value to the store in the cloud

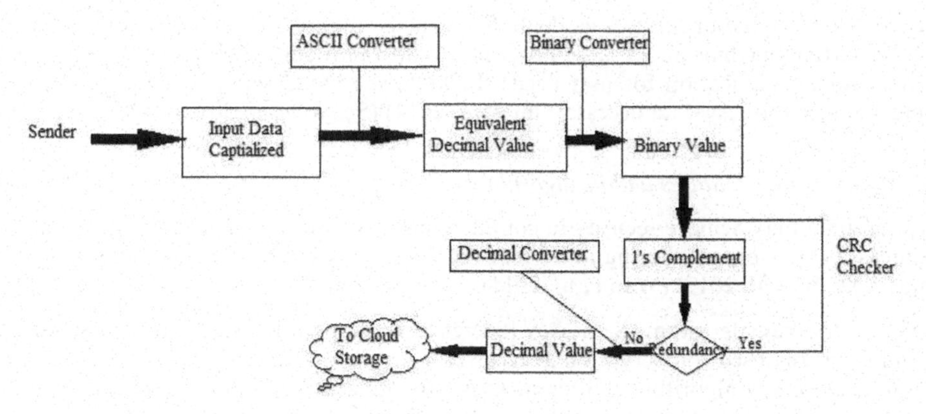

Figure 5. CryptoASCII encryption algorithm.

3.2 *Decryption algorithm*

Only trusted and authorized user can access the encrypted data from the cloud. The CryptoASCII method are used to share the keys methods only to the authorized user shown in Figure 6.

i) Authorized user retrieve the text from the cloud
ii) Convert the decimal value (non-readable form) into 8-bit binary equivalent
iii) Check the redundancy and do 1's complement of the 8-bit binary value (If 0 at proceeding, remove the 0)
iv) Convert the binary value into equivalent decimal value
v) Convert the decimal numeric equivalent by ASCII converter
vi) Arrange and transmit the readable form of data to the authorized user

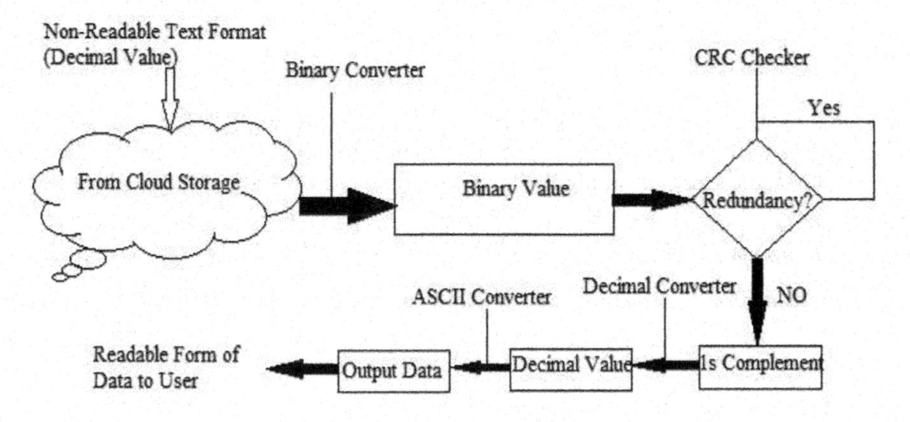

Figure 6. CryptoASCII decryption algorithm.

4 EXECUTION AND FUTURE WORK

Let we say the data "BABY is split into 'B' 'A' 'B' 'Y'. The ASCII equivalentof 'B' is 66 and its 8-bit binary value is 1000010. Check for 8-bit value, if need add '0' at proceeding so 0100 0010.Make 1's complement and it is 1011 1101and covert it into decimal equivalent as 189.

Finally "BABY" were encrypted as 189 190 189 90 and send it to the cloud. Similarly authorized user can convert the non-readable form decimal value into readable text value by CryptoASCII decryption algorithm.

Now this CryptoASCII methods are applicable for Alphanumeric data, later in future we may provide security for all types of data as audio, video, image, gif file etc.

5 CONCLUSION

Emerging cryptography techniques in cloud computing is to ensure the data privacy and data security for sender and receiver. To overcome vulnerability attacks in data storing and data accessing there will be lot of techniques and algorithms were used. In this paper, we proposed CryptoASCII method for data security and confidentiality in cloud storage. CryptoASCII methods are simple and easiest form so it will reduce time and space complexity. Due to running and compile time of CryptoASCII method will provide better performance compare to other methods. In later days CryptoASCII methods will extend for audio, video, images, gif file etc. The CryptoASCII provide high performance and give good result to the cloud user.

REFERENCES

Akhil, K., Kumar, M.P., Pushpa, B.: Enhanced cloud data security using aes algorithm. In: 2017 International Conference on Intelligent Computing and Control (I2C2), IEEE, pp. 1–5 (2017).

Barrowclough, J.P., Asif, R.: Securing Cloud Hypervisors: A Survey of the Threats, Vulnerabilities, and Countermeasures. *Secur. Commun. Netw.* 2018, 1681908 (2018).

Faizi, S.M., Rahman, S.S.: Secured Cloud for Enterprise Computing. In: *Proceedings of 34th International Conference*, Vol. 58, pp. 356–367 (2019).

Fei Han, Jing Qin, Jiankun Hu: Secure Searches in the Cloud: A Survey. *Future Generation Computer Systems* (2016).

Gupta, A., Chourey, V.: Cloud Computing: Security Threats & Control Strategy Using Tri-mechanism. In: *2014 International Conference on Control, Instrumentation, Communication and Computational Technologies (ICCICCT)*, IEEE, pp. 309–316 (2014).

Kaur, A., Bhardwaj, M.: Hybrid Encryption for Cloud Database Security. *J. Eng. Sci. Technol.* 2, 737–741 (2012).

Maryam, K., Sardaraz, M., Tahir, M.: Evolutionary Algorithms in Cloud Computing from the Perspective of Energy Consumption: A Review. In: *2018 14th International Conference on Emerging Technologies (ICET)*, IEEE, pp. 1–6 (2018).

Senyo, P.K., Addae, E., Boateng, R.: Cloud Computing Research: A Review of Research Themes, Frameworks, Methods and Future Research Directions. *Int. J. Inf. Manag.* 38(1), 128–139 (2018).

Singh, S., Maakar, S.K., Kumar, S.: A Performance Analysis of DES and RSA Cryptography. *Int. J. Emerg. Trends Technol. Comput. Sci.* 2, 3 (2013).

Sun, Y., Zhang, J., Xiong, Y., Zhu, G.: Data Security and Privacy in Cloud Computing. *Int. J. Distrib. Sens. Netw.* 10(7), 190903 (2014).

Wan, Z., Liu, J., Deng, R.H.: Hasbe: A Hierarchical Attribute-based Solution for Flexible and Scalable Access Control in Cloud Computing. *IEEE Trans. Inf. Forensics Secur.* 7(2), 743–754 (2011).

Wang, S., Zhou, J., Liu, J.K., Yu, J., Chen, J., Xie, W.: An Efficient File Hierarchy Attribute-based Encryption Scheme in Cloud Computing. *IEEE Trans. Inf. Forensics Secur.* 11(6), 1265–1277 (2016).

Waters, B.: Ciphertext-policy Attribute-based Encryption: An Expressive, Efficient, and Provably Secure Realization. *International Workshop on Public Key Cryptography*, pp. 53–70. Springer, New York (2011)

Yahya F, Chang V, Walters J, and Wills B, "*Security Challenges in Cloud Storage*," pp. 1– 6 (2014).

Zhan, Z.H., Liu, X.F., Gong, Y.J., Zhang, J., Chung, H.S.H., Li, Y.: Cloud Computing Resource Scheduling and a Survey of Its Evolutionary Approaches. *ACM Comput. Surv.* 47(4), 63 (2015).

Artificial Intelligence, Blockchain, Computing and Security – Dagur et al. (Eds)
© 2024 The Author(s), ISBN: 978-1-032-67841-2

Human action recognition in smart home using deep learning technique

B. Subbulakshmi*, M. Nirmaladevi* & J. Srimadhi*
Department of Computer Science and Engineering, Thiagarajar College of Engineering, Madurai, Tamil Nadu

ABSTRACT: Human action recognition in recent years has gained lot of scope in research area of vision based activities where we have lot of application that can be used in real time. Deep learning the evolving field in artificial intelligence provides many updated feature extraction methods from low level to high level from the input dataset rather than using complex methods of extraction in vison based activities. Here in this paper to recognize human action a convolutional neural network (CNN) which is a deep learning model is proposed where the proposed model recognizes nature of human action as normal or abnormal action using Scale invariant feature transform and k means sparse coding for segmenting images into 112 frames and recognizing the activity bring the accuracy of 98%.

1 INTRODUCTION

Action recognition is an upcoming future in vision based research area in a smart home where the area is under surveillance which provides intelligent residence in future. The conjunction in recognizing the action in a smart home like occupant's action using deep learning techniques we can make decision, recognition, etc., service is made based our needs. In areas like healthcare, security, childcare human action recognition plays an significant role in these issues recently.

Action recognition is done by two methods namely, sensor-based method and vision-based. In Sensor based methods inputs are got through either wearable sensors or ambient sensors installed in the place which has its own limitations like they produce wrong alarms, external noises which may produce inaccurate results. Overcoming these limitations action recognition by vision based method have gained more advantage in this research field. To perform recognition using traditional method we perform image or data classification and detection using handcraft method which uses detailed described tons of coding scripts to acquire features from input video or images to achieve low level to high level feature extraction. The limitation is when data gets bundled up writing longer scripts becomes tedious. So, as we all know deep learning models can extract features automatically which overcomes the limitations of the past methods.

We develop a deep learning model that performs recognition activity using features extracted and the output is obtained. These deep learning model uses extensive training set that a model is studied from extracted features here we use scale invariant feature transform for feature extraction from the dataset, we use DML Smart actions activity dataset for activity recognition.as we all know there are many real life applications on human activity recognition in this paper we take action in a jewelry shop which is a smart home where the area is under camera surveillance and we detect normal activity happening in that place and

*Corresponding Authors: bscse@tce.edu, mnit@tce.edu and srimadhi025@gmail.com

 DOI: 10.1201/9781032684994-59

recognizing abnormal activity like steeling i.e., theft we differentiate them as normal activity and abnormal activity. We use K-means sparse coding for image segmentation and image compression and we use clustering in convolutional neural network for best outcome. These activity recognition models using deep learning techniques are being applied more in healthcare area as there are more chances of actions being abnormal which can cause harm and even in childcare, security etc., there are many deep learning models like convolutional neural network (CNN), recurrent neural network (RNN), long short term memory (LSTM) etc., we propose a special convolutional neural network which recognizes nature of the human activity as normal and abnormal activity using a further more features like scale invariant feature transform which split the video from input dataset into 112 frames and k-means sparse coding for image segmentation and clustering and human activity is recognized with high accuracy.

2 DATASET

Datasets are the inputs given to the model for training and implementing the required actions. There are number of datasets available for human activity detection. In this paper we use a video clip of a jewelry shop from Digital Multimedia Lab. To create this dataset, the daily action of customers of a jewelry shop is captured by a static camera. It also has action performed by clients who came for purchase in the jewelry shop which is recorded by camera under surveillance.

3 RELATED WORK

Before proceeding to the implementation knowing about the related work process is important this section will brief you about the need for the proposed model and basics of tool used for this proposed method.

To perform human activity recognition Ma *et al.* (2022) so for traditional methods are used which uses sensors and handcraft brief coding for feature extraction and existing models identify type of activities from the dataset for example actions performed by human like siting, standing etc., the traditional methods fails to recognize activity which can out-perform contemporary local descriptors on scenes with difference in performance. Artificial Neural network (ANN) Putra *et al.* (2022) a machine learning algorithm was popularly used in human action recognition where the model is trained with dataset and trained model gets the input data and if the input data matches with the trained data, then required output is given. The below Figure 1 gives you the architecture of Artificial Neural network where the

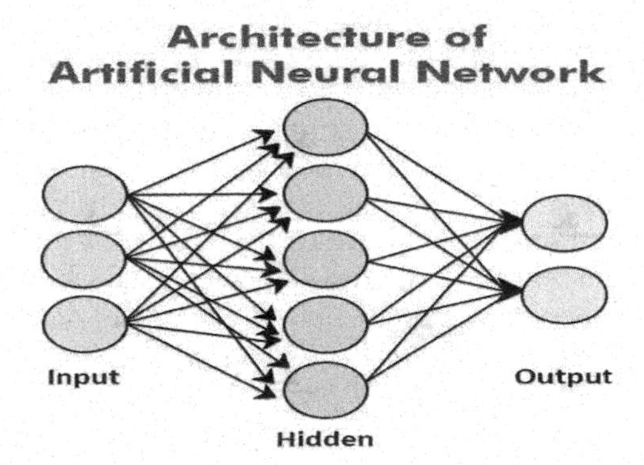

Figure 1. Architecture of artificial neural network.

neural network model has input layer, where the input is fed into the model and hidden layer where the decision is taken and the nodes passing the hidden layer to another is decided by the activation function.

In Artificial neural network the structure is a computational paradigm which is based on mathematical models where there is less accuracy and is not reliable. Also, the number of datasets induced is restricted in certain number. Finally, these models just recognize what kind of activity but fail to explain the nature of the activity. Since there are drawbacks in artificial neural net-work and feature extraction from the input dataset is complicated process we use Deep learning techniques, which has automatic feature extraction from the dataset. In this paper we use scale invariant feature transform, K-means sparse coding and convolutional neural network (CNN) an efficient deep learning technique for human action recognition.

Here the execution of human action recognition as normal activity and abnormal activity is performed using the tool MATLAB (Matrix Laboratory). MATLAB is a high performance programming language and a multi-paradigm numerical computing environment which allows to perform plotting of data and functions, matrix multiplication, implementing any algorithm or creating any user interface and even combining the programs written in programming languages like C++, C, Java, FORTAN etc., MATLAB incorporates visualization, computing and programming in a simple use environment where problems and its outcomes are expressed in simple mathematical notation. MATLAB's environment contains set of tools that facilitates you to work as programmer using high level matrix or array language with data input and output, data structures, functions, control flow statements and object oriented programming features. MATLAB handles graphics which includes high-level commands for two-dimensional or three-dimensional data visualization, animation, graphics presentation etc.,

4 METHODOLOGY

In this paper for execution of our problem statement we use methodology like,

- Scale invariant feature transform
- Convolutional Neural network
- K-means sparse coding

Let us discuss about this briefly in this section with the implementation,

4.1 *Scale invarient feature transform*

For our execution we took a dataset from DML action set, a video sequence of a jewelry shop, this input is fed into the model where the deep learning model extracts features automatically. The Scale invariant feature transform is used to extract features such as edges and boundaries from the frames of images. The SIFT function transform the input into 112 frames giving highest accuracy of 98% by featuring the depth of action taking place in the video. In each video frame to represent human action histogram of gradient (HOG) which are well recognized for human detection and regarding contrast changes and illumination. The local object and appearance can be represented well by distributing local intensity gradient without accurate knowledge of respective gradient or edges is the fundamental idea and is implemented by separating image windows into small spatial regions.

For our problem statement of recognizing human activity and its nature as normal or abnormal after execution we get the GLCM feature values,

Table 1.

Contrast	[1×40 double]
Correlation	[1×40 double]
Energy	[1×40 double]
Homogeneity	[1×40 double]
Feature Data	0.0293

4.2 *Convolutional neural network*

Convolutional neural network is an efficient deep learning technique used in the field of vision based research areas. CNN is deep learning model which uses a huge dataset compared to artificial neural network, where convolutional neural network trains the model with all input dataset and instead of providing determined output as in ANN convolutional neural network models studies the entire dataset and infers from the dataset and provide optimal output. Here in this paper we use CNN architecture to recognize human action of smart home from a video dataset and classify human action from the training dataset.

Convolutional neural network (CNN) has layers which includes input layer, hidden layers which includes convolutional layer, pooling layer, ReLu correction layer and fully connected layer and finally output layer. Input video gets divided into frames using functions like scale invariant feature transform and these frames are sent to convolutional network's input layer and mapped to hidden layer using activation function and inside the hidden layer process takes places and recognition output is given. Nodes from one layer are passed to other hidden layers which could provide output is decided by activation function in hidden layer. A distinct CNN was proposed which contains 5 convolutional layers, 4 pooling layers and 3 fully connected layers. Lastly in the fully connected layer the softmax was considered to determine the probability of the 112 classes of activity dataset.

4.3 *K-means sparse coding*

After recognition of human activity from the dataset the next step is to classify this activity as normal and abnormal which gives the nature of the activity which is done using K-means sparse coding. What's sparse coding? Like either decomposition or factorization methods which try to represent signal of all atoms, where in sparse coding all these atoms are sparse and the matrix which yields these atoms are over complete. Image representation can be done by linear transform in following way like pre- computed matrix, dictionary, FFT, PCA etc., that are trained from training dataset. In this paper we use K-means sparse coding for image segmentation and clustering. The below Figure 2 represents the K-means segmentation performed on the input dataset and various color separation.

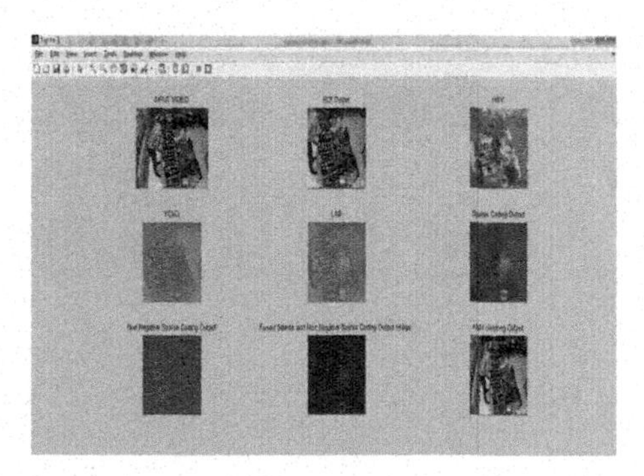

Figure 2. K-means segmentation and color separation.

K-means boils down to represent the observation in an efficient way where some observations are mapped to clusters. Non negative sparse coding decomposes multi-variate datas into non-negative sparse elements. The fundamental idea of backing this type of data representation is to relate standard sparse coding and non-negative factorization of matrix and provide an efficient algorithm in finding optimal vales of components which are hidden and also provide answers to how basis vector be studies from observation of data. These simulations manifest the effectiveness of the method proposed.

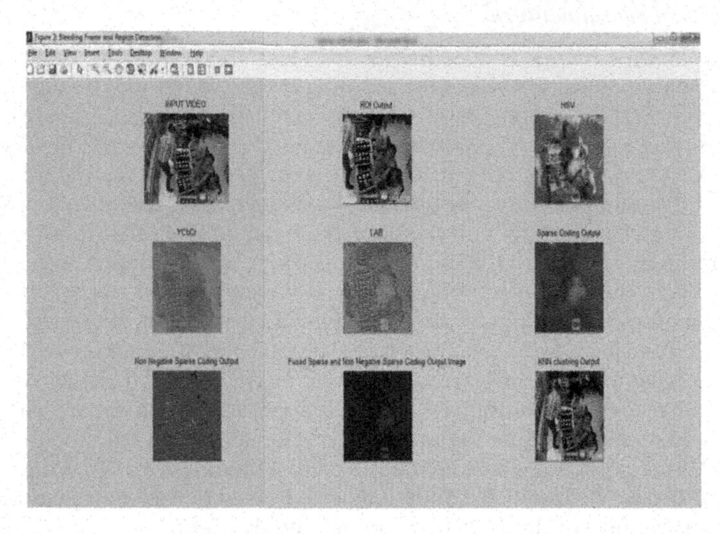

Figure 3. Bleeding frame and region detection.

After applying all the methodology to the DML action dataset of jewelry shop we recognize the action using efficient deep learning technique Convolutional neural network and classify them as normal or abnormal activity based on their nature using K-means sparse coding and the output for our execution is got as abnormal.

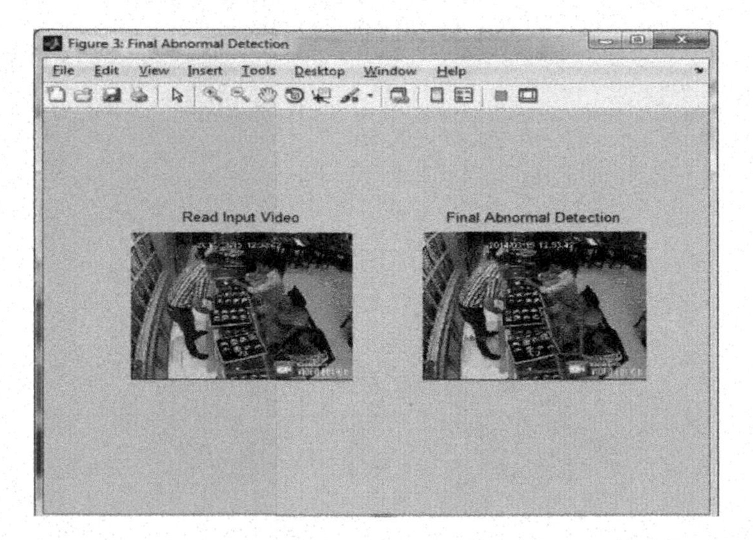

Figure 4. Final output as abnormal activity.

Thus in Figure 4 we see the output of human action recognition as normal or abnormal activity where we get output as abnormal activity and the owner of the shop is notifies about the action.

5 CONCLUSION

In this paper we discussed about human activity recognition in a smart home where we detect or recognize the nature of the activity as normal activity or abnormal activity using deep learning techniques like Convolutional neural network applied to 112 frames of a video clip of input dataset and K-means sparse coding is applied to segmentation of images and cluster them and obtain the output. Comparing to other proposed method the methodology proposed in this paper has accuracy of 98% for recognizing the activity using scale invarient feature transform. The performance of the model is improved by learning larger dataset. Thus we achieved to get optimal output using deep learning technique with high accuracy. Our future work is planned to use combinations of different machine and deep learning techniques on specific human activity dataset and improve accuracy performance by learning large amount of data.

REFERENCES

[1] Ji, S., Xu, W., Yang, M., & Yu, K. (2012). 3D Convolutional Neural Networks for Human Action Recognition. *IEEE Transactions on Pattern Analysis and Machine Intelligence*, 35(1), 221–231.

[2] Zebin, T., Scully, P. J., & Ozanyan, K. B. (2016, October). Human Activity Recognition with Inertial Sensors Using a Deep Learning Approach. In *2016 IEEE Sensors* (pp. 1–3). IEEE.

[3] Fan, X., Xie, Q., Li, X., Huang, H., Wang, J., Chen, S., ... & Chen, J. (2017, June). Activity Recognition as a Service for Smart Home: Ambient Assisted Living Application Via Sensing Home. In *2017 IEEE International Conference on AI & Mobile Services (AIMS)* (pp. 54–61). IEEE.

[4] Putra, P. U., Shima, K., & Shimatani, K. (2022). A Deep Neural Network Model for Multi-view Human Activity Recognition. *PLoS One*, 17(1), e0262181.

[5] Hari Pavan, A., Anvitha, P., Prem Sai, A., Sunil, I., Maruthi, Y., & Radhesyam, V. (2022, March). Human Action Recognition in Videos Using Deep Neural Network. In *Evolution in Signal Processing and Telecommunication Networks: Proceedings of Sixth International Conference on Microelectronics, Electromagnetics and Telecommunications (ICMEET 2021)*, Volume 2 (pp. 335–341). Singapore: Springer Singapore.

[6] Khan, Z. N., & Ahmad, J. (2021). Attention Induced Multi-head Convolutional Neural Network for Human Activity Recognition. *Applied Soft Computing*, 110, 107671.

[7] Yadav, S. K., Tiwari, K., Pandey, H. M., & Akbar, S. A. (2022). Skeleton-based Human Activity Recognition Using ConvLSTM and Guided Feature Learning. *Soft Computing*, 1–14.

[8] Basly, H., Ouarda, W., Sayadi, F. E., Ouni, B., & Alimi, A. M. (2022). DTR-HAR: Deep Temporal Residual Representation for Human Activity Recognition. *The Visual Computer*, 1–21.

[9] Nadeem, A., Jalal, A., & Kim, K. (2021). Automatic Human Posture Estimation for Sport Activity Recognition with Robust Body Parts Detection and Entropy Markov Model. *Multimedia Tools and Applications*, 80, 21465–21498.

[10] Tsai, M. F., & Chen, C. H. (2021). Spatial Temporal Variation Graph Convolutional Networks (stv-gcn) for Skeleton-based Emotional Action Recognition. *IEEE Access*, 9, 13870–13877.

[11] Kong, Y., & Fu, Y. (2022). Human Action Recognition and Prediction: A Survey. *International Journal of Computer Vision*, 130(5), 1366–1401.

[12] Ma, N., Wu, Z., Cheung, Y. M., Guo, Y., Gao, Y., Li, J., & Jiang, B. (2022). A Survey of Human Action Recognition and Posture Prediction. *Tsinghua Science and Technology*, 27(6), 973–1001.

E-marketplace for farm products

S. Sasikumar*, Cheemalapati Rishith*, S. Rupa Sree* & K. Jayasree*
Department of Computer Science and Engineering, Saveetha Engineering College, Chennai, Tamil Nadu
sasikumar@saveetha.ac.in

ABSTRACT: The agriculture industry is starting to feel the effects of e-commerce. It is really worrying to see how consumers buy food and other agricultural items. Customers often have to go to great lengths to purchase agricultural goods, and even then, they can't always count on receiving what they need. Our project's primary objective is to facilitate the computerized purchase and sale of agricultural goods, both for farmers and their clients. The website will show the farmers how much money they made from their sales. Through direct interaction with customers, the website provides a means for farmers to increase their income. When it comes to agro-marketing, the website will provide a one-of-a-kind and safe option. Through e-farming, farmers who have a rudimentary understanding of the Internet may market and sell their goods locally. In this concept, customers may browse through available goods and then buy what they need right away via secure online transactions or cash on delivery.they can either choose in between deliverable or non deliverable with their flexibility and users can choose various location to connect with that locality farmers so that the delivery of the products can be easy.

Keywords: e-commerce, agricultural goods, e-farming, farmers

1 INTRODUCTION

There have been significant shifts in production and distribution as a result of the sharing economy's rapid expansion in recent years. Many marketplaces, including eBay, have emerged as a result of technological developments and the rising use of mobile devices and applications. This project is an E-commerce web application to provide a user and former friendly environmental web application. The seller friendly environment is based on the choice of choosing their willingness about the delivery of the products and convenience to choose the option of no delivery. The web application is user friendly by the option of getting their location seller to make the process of collecting products even more easier and the option of selecting any location as the user can shift their location with the convenience.

This is an easy interface for the farmer most importantly without any third party involvement to the project as the user and farmer make the communication easier within themselves and the market price of the products is directly reached to the farmers hand without any third party benefits and the customers could get fresh products skipping the preservatives addition to their food items.

The websites connect on the same location farmers and the seller so that the products can be safe guarded within short span of time and without any extra efforts to both customers and farmers for the travelling purpose of their products the mode of delivery and no-delivery

*Corresponding Authors: sasikumar@saveetha.ac.in, rishith316@gmail.com, sangapurupasree@gmail.com and jayasreekammoru@gmail.com

 DOI: 10.1201/9781032684994-60

is to also an another useful feature for the farmers. The registration of the website is also another easy way which can be done with the basic web knowledge and even within mobile phone.

The seller and the customer both can view their previous orders and the payment details also with the detailed information about product, quantity, cost with the buyer name to the seller and sellers name to the buyer. They can either watch their payment mode that do they made it with online payment or as the hand cash also.

Once seller made their registration with their proper id proof and their location admin will give access to the products uploading with activation of the seller and the products approval from admin so that the customer can be charged with the proper prices as per the market.

2 LITERATURE SURVEY

A. Aflaki *et al.*, a channel may support two sorts of users at once: main users, who are given first priority while using the channel, and secondary users, who can access the channel only when the primaries are not using it.

Sharrock *et al.*, we think of a situation where a huge territory (say, a state) is split into smaller parts, each of which is owned by a different primary (e.g., towns).

Yan *et al.*, If a primary has spare capacity during a certain time period, they could be willing to rent it out to secondary users in a group of places where they won't cause any interference to each other. As a consequence, primary commodities become more competitive with regards to pricing.

C.T.M *et al.*, past investigation into this cost contention has just been done as such under the reason, made for logical manageability, that the cost of every essential takes values from a persistent set. However, in reality, there is only a certain range of pricing to choose from.

I. Hendel *et al.*, In this study, we explore the underlying topic of what happens to the behavior of price-competing parties when the continuity assumption is dropped.

G.P. Cachon *et al.*, we find that games with continuous and discrete pricing sets are distinct in numerous significant ways. There is no pure strategy Nash equilibrium (NE) in the single-location game when continuous price sets are used, however such a NE may be present when using discrete price settings.

I. Bellos *et al.*, we look at a secondary spectrum market in which each major operator controls a single channel across several different geographies. Each main receives payment in return for selling its channel to the secondary broadcasters. When the price and availability of a channel for sale fluctuate at random, a main must choose between the two. Every conceivable channel state vector has to have a corresponding tactic found.

C. H. Chiu *et al.*, we focus on node-symmetric conflict graphs, which occur often in reality when the number of nodes is high (potentially, infinite).

Aditya MVS *et al.*, also take into account a symmetric link between the joint probability distribution of the channel state vectors due to the symmetry in the interference relationship.

Juan Vanerio *et al.*, we prove the existence of a symmetric NE, and then we calculate it directly. For every channel state vector, a primary in a symmetric NE randomly selects one of the maximal independent sets.

3 IMPLEMENTATION

3.1 *System design*

The system's design defines its fundamental framework; we provide the Hash code Solomon method; and we may store a subset of the data locally on a computer and on a fog server. As an added bonus, this technique uses AI to determine what share of data is kept in the cloud, what share is in the fog, and what share is on the local system. Having shown its viability via theoretical and practical evaluations of safety, our approach is a potent addition to current cloud storage system.

3.2 *Existing system*

Existing system for retailing of agricultural products in involved with the third person as a mediator between the seller and the customer with benefit of the middle person. The customers get charged to the product and to the supplier as well yet fresh products delivery can't be estimated.

3.3 *Proposed system*

This model is farmer and customer friendly web application, that allow the farmer to provide their convenience of choosing to deliver the product directly to the customer or making the customer come and get the product from them and the customer friendly environment is built by connecting to the farmers of their locality so that the products reach can happen within short span of time and also collecting products will be a easy way.

Advantage:

There is no third party involvement it is a direct seller to customer reach process. It connects the nearby localities to make the delivery process easier to the seller and customer.

3.4 *System architecture design*

According to the architecture diagram, we can clearly see how the functioning of e-marketplace for farm products is done.

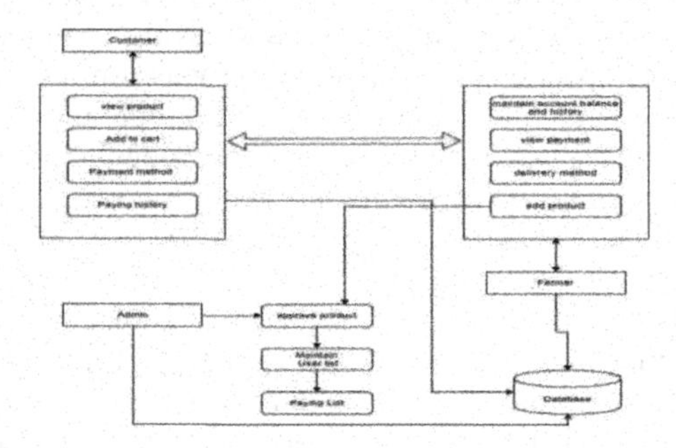

Figure 1. Architecture diagram.

4 MODULES

4.1 *Customer*

In this customer module in our project user view all the products after selecting delivery and no-delivery option to view farmer added products and if customer need the products he/she can add to their cart and make the payment.

4.2 *Farmer*

In this farmer module in our project farmer add their products with their fixed cost as per the market price which gets approved by the admin. The farmer fix the delivery method is deliverable or non deliverable. Every payment will be view by the farmer like payment history and complete account details.

4.3 *Admin*

In this admin module in our project activate the every user registration. Admin have the every access if the user registers their details it will be passed to the admin and admin accept the registration. Admin have the access to approve the farmer added product. Maintain the user list, product list and all the details about the user.

5 ALGORITHM

The term "Advanced Encryption Standard" (AES) refers to a set of rules for the secure transmission of digital information. Although more difficult to implement, AES is now the dominant encryption standard because of its superior security than DES and triple DES. AES handles 128 bits of incoming data at once, however the algorithm works with bytes (groups of bytes) rather than individual bits. Each block in AES is thought of as a 16-byte (4 bytes x 4 bytes = 128) grid in a column major layout.

6 OUTPUT

We can see in Figure 2, the front page of the e-marketplace for farm products.

Figure 2. Front page.

In Figure 3, it is about the customer's main page where the customer can buy the vegetables, fruits by ordering from the e-marketplace website.

Figure 3. Customer main page.

We can see in Figure 4, it is the sellers main page where the seller can add the products that he wanted to sell.

Figure 4. Seller main page.

Finally in the Figure 5, we can see the main page of the admin who manages both the customer and seller.

Figure 5. Admin main page.

7 CONCLUSION

7.1 *Conclusion*

Companies may learn more about their customers' needs and frustrations, and their issues can be addressed more efficiently, via C2C communication. By analysis, customer feedback, businesses may anticipate future trends and aim toward a complaint-free service. The term "consumer-to-consumer" (C2C) refers to transactions including the exchange of products and services, as well as the transfer of money or data, between private individuals via an electronic network, most often the Internet. Business-to-business (B2B), business-to-consumer (B2C), consumer-to-consumer, and consumer-to-business all refer to the four different types of commercial exchanges that take place.

7.2 *Future work*

This project can be developed in a way that the seller and the customer can upload their experience with the website through the comment blog and thoughts for future development process of the website. The website can also be developed with the addition of the new sections to the website such as fertilizers, products for irrigation purpose based on environment study such as soil, ground water supply etc. With the enhanced features selection is easy for farmers.

REFERENCES

Aditya MVS, Abhishek Raghuvanshi, Gaurav S. Kasbekar, "Price Competition in Spectrum Markets: How Accurate Is the Continuous Prices Approximation" Volume 17 Issue: 5 Pages: 3215–3229.

Aflaki A., Feldman B., and Swinney R.. 2020, "Becoming Strategic: Endogenous Consumer time Preferences and Multi-period Pricing," *Open. Res.*, vol. 68, no. 4:1116–1131.

Arnob Ghosh, Saswati Sarkar, "Secondary Spectrumoligo Poly Market over Large Locations".

Bellos I., Ferguson M., and Toktay L. B., "The Car Sharing Economy: Interaction of Business Model Choice and Product Line Design," *Manuf. Service Open. Manage*, vol. 19, no. 2, pp. 185–201, 2017.

C. T.M., Zhang J., and Cai Y. J., "Consumer-to-consumer digital-product exchange in the Sharing Economy System with Risk Considerations: Will Digital-product-develoers suffer?," *IEEE Trans. Syst., Man, Cybern. Syst.*, vol. 50, no. 12, pp. 5049–5057, Dec. 2020.

Cachon G. P. and Swinney R., "Purchasing, Pricing, and Quick Response in the Presence of Strategic c Sumers," *Manage. Sci.*, vol. 55, no. 3, pp. 497–511, 2009.

Chiu C. H., Chan H. L., and Choi T. M., "Risk Minimizing Price-rebate Return Contact in Supply Chains with Ordering and Pricing Decisions: A Multi Methodologicaanalysis," *IEEE Trans. Eng. Manage.*, vol. 67, no. 2, pp. 466–482, 2020.

Hendel I. and Lizzer A., "Interfering with Secondary Markets," *Rand J. Econ., VPL.* vol. 30, no. 1, pp. 1–21, 1999b.

Henssssdel I. and Lizzer A., "Adverse Selection in Durable Goods Markets," *Amer. Econ. Rev.*, vol. 89, no. 5, pp. 1097–1115, 1999.

Juan Vanerio, Federico Larroca," Online Expert-Based Prediction for Cognitive Radio Secondary Markets".

Sharrock, K.R.; McDonald, R.M."Adaptinon of the Fruit Supply Chain for E-commerce". *Acta Hortic.* 2015, 1103, 203–209.

Yan, Y.; Chen, S.; Lin, J.; Luo, X. "Analyzing the Effect of Social Commerce Website Features on Intention of Repurchase: A Chinese Perspective" *J. Electron. Commer. Res.* 2017, 18, 225–244.

Artificial Intelligence, Blockchain, Computing and Security – Dagur et al. (Eds)
© 2024 The Author(s), ISBN: 978-1-032-67841-2

Blockchain based voting system using Ethereum network

V. Uma Rani*, Karnati Yogendra Reddy*, Mekala Praneeth Kumar* &
Teja Ramana Reddy Raju*
Department of Computer Science and Engineering, Saveetha Engineering College, Chennai, Tamil Nadu

ABSTRACT: Ethereum developed the idea of a decentralized application to benefit from the capabilities of blockchain technology. Even though blockchain is one of the most talked-about topics in the IT industry right now, most of it is about cryptocurrency hype and its limited application. On the Ethereum platform, decentralized applications can be developed and tested with the help of the Ethereum virtual computer. As more smart contracts are added, it gets stronger. The process of writing code into the blockchain is known as a smart contract. To protect the blockchain's integrity, smart contracts establish a virtual contract between the owner and the user. The contract must be accepted by all users. Transactions on the blockchain are governed using smart contracts. Although electronic voting is popular, it is essential to have a trustworthy and secure online voting platform. A transparent and decentralized voting platform is in high demand in a democratic democracy. An electronic voting platform needs to be safe, offer one vote per person, be open, and be simple to use. Ethereum and its extensive network are the best platforms for developing such applications. We are working on a decentralized, Blockchain-based electronic voting system as part of this project. Several nodes store the voting information of users, and if one node fails or is unavailable, users can retrieve their voting information from other operating nodes. Voting information were managed in a single server in the previous centralized server, and if this server was hacked or went down, all voting details would be inaccessible, and this server might be attacked or hacked, altering vote counting data. Immutable data storage is supported by Blockchain storage, which means that data cannot be altered or hacked because each node in the Blockchain will verify each Block storage with the help of hash codes, and if verification fails, the Blockchain or users will receive notification that data has been changed.

Keywords: Ethereum, Blockchain, electronic contracts, e-voting.

1 INTRODUCTION

Out of ten nations, six are democracies. In a democratic society, elections are very important. It gives citizens the right to vote for the country's leader and legislators, but the voting infrastructure is poorly designed, making the process flawed. Most democracies use centralized voting machines, which means that the data collected during the voting process is controlled and analyzed by a single individual or group. Consequently, the procedure is susceptible to numerous technical flaws and a lack of voter transparency.

*Corresponding Authors: umaraniv@saveetha.ac.in, yogendrakarnati777@gmail.com, mekalapraneethkumar@gmail.com and tejareddyraju1432@gmail.com

 DOI: 10.1201/9781032684994-61

Blockchain technology has the potential to strengthen democracy while also resolving this problem. A distributed and decentralized framework for application development is provided by blockchain technology. The Ethereum community was the first to propose the concept of decentralized applications, or mobile applications that make use of blockchain technology. The primary difference between a centralized and a decentralized application is that the former employs a blockchain as its backend rather than a conventional database. Blockchain is a distributed database system that stores data in an immutable state using encryption and has no central authority.

Blockchain technology makes peer-to-peer communication possible without the need for a third party. Blockchain technology was initially used to create cryptocurrencies, but the rise of decentralized applications has opened new applications for blockchain technology. As previously stated, blockchain technology's unique distributed, decentralized, and distributed architecture may facilitate the implementation of an e-voting platform, which is not widely used due to the internet's numerous threats, such as hackers and security issues. Decentralized applications, on the other hand, make it possible to implement a distributed, safe, secure e-voting platform.

2 LITERATURE SURVEY

We reviewed the following previous papers and projects on Blockchain based voting system for our project.

E. Maaten described that the Estonian e-voting framework is described in this archive. The paper looks at how the idea of a framework for e-voting is used to counter some of the most important barriers to e-voting far away.

S. Nakamoto suggested that a shared type of electronic money would make it easy to make online payments from one party to the next without having to use a financial foundation. The organization establishes exchanges that are immutable timestamps without rehashing the proof of word by hashing them into a continuous chain of hash-based confirmation of task.

G. Wood, when combined with cryptographically secure transactions, the blockchain paradigm has shown its effectiveness in a variety of initiatives, with Bitcoin being one of the most noteworthy. Ethereum sums up the execution of running a decentralized application at this point singleton computational asset.

C.D. Clack, V.A. Bakshi, and L. Braine, this paper analyses few basic savvy contract concerns like nomenclature, automation, enforceability, and semantics. It defines sensible agreement as an understanding whose execution is both automatable and enforceable.

U.C. Cabuk discussed about the rapid spread of the internet in the twenty-first century has made everyone a potential internet user. An e-democracy model is offered in this article, in which elections and votes may be cast online; the model's benefits and downsides are studied, existing practices are assessed, solutions to certain difficulties are provided, and the potential contribution of the internet to direct democracy.

Vitalik Buterin talked about the bitcoin and benefits of Ethereum over bitcoin with decentralized applications.

Patrick McCorry, this paper explained about how to deploy the smart contracts into Ethereum blockchain for boardroom voting system. It also focuses on how smart contracts works internally for the execution of desired goal.

3 METHODOLOGY

The greatest trouble in a popularity-based society today is to get more individuals to take part in the political race cycle and vote decently. Elector investment is declining many years, and a vote-based system is weakening. Since individuals have lost confidence in the political decision process.

Since the method is dark and tedious. Our essential objective in this undertaking is to make a protected, secure, and straightforward democratic climate, as well as to show that e-voting a ballot, made conceivable by blockchain innovation, is a better choice than traditional democratic.

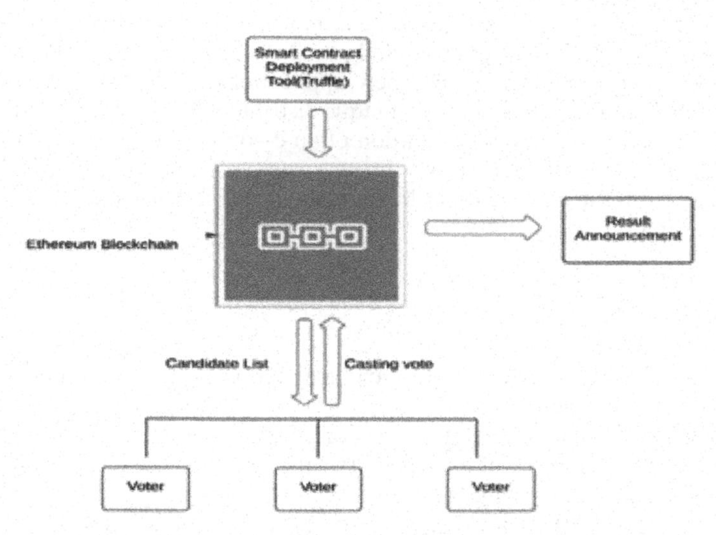

Figure 1. System architecture.

3.1 *System architecture*

Since while e-voting a ballot opens to people in general, each qualified person with a web association and a telephone or PC will want to cast a ballot, and the cycle will be considerably less tedious and secure, drawing in additional citizens and expanding turnout on the grounds that the interaction will be straightforward and the citizen's character will be protected. Debasement is likewise more successive during decisions, and a lot of cash is spent in this cycle. This might be tried not to by utilize blockchain innovation since there will be nobody authority with unlimited oversight over the interaction. This satisfies the genuine meaning of a majority rules government.

E-voting a ballot is an old thought, far more seasoned than blockchain. As a result, a unified authority with unlimited oversight over the interaction has completed all known e-voting tasks. Because the Estonian government quickly adopted online voting, it serves as an example. As a result, the number of people casting ballots has increased. Using an ID card distinguishing proof method, it was possible to keep the one-person, one-vote idea. However, the method is not completely safe because many network security experts found an escape clause in the process because the information is kept in a single data set and can be changed and tempered. Because information is distributed throughout the organization and encoded rather than being stored in a single data set, the blockchain provides a better solution than a straightforward web-based voting method to address this issue.

The central test in E-Voting a ballot is the way to defend clients' personalities while keeping up with information receptiveness and respectability. To resolve this issue, Ethereum gives different hash values to arrange clients, making it almost difficult to recognize the person. Exchanges acted in the Ethereum network are apparent to everybody in the organization and can be approved, making it straightforward to all hubs in the organization.

Each elector should be provided with a limited amount of ether in an Ethereum wallet. This ether will be used to vote in favors of the competitor, and the votes will be recorded on the blockchain with the help of a brilliant agreement that will approve and verify the citizens' votes. The meta mask test organization is used to test the application.

4 IMPLEMENTATION

In the old centralized system, voting information was controlled by a single server, and if this server was hacked or went down, all voting information was destroyed. This server is subject to attack or hacking, and vote-counting information on it may be changed.

4.1 *Disadvantages*

1. All voting information would be lost if this server was hacked or went down.
2. This server is susceptible to attack or hacking.

Because each node on the blockchain uses hash codes to verify each Block storage, blockchain storage allows immutable data storage, which means data cannot be changed or compromised. If the verification fails, the blockchain or users are alerted that the data has changed. Because Ethereum Blockchain technology allows immutable data storage, we are storing all voter information in the proposed E-Voting system.

4.2 *Advantages*

1. Block storage using hash codes, and if verification fails, Blockchain or users will be notified of data changes.

4.3 *Modules*

1. Admin
2. Candidate

4.3.1 *Admin module*

Using this module, the administrator may login with his credentials and examine the number of votes cast. This user is responsible for adding new party information, seeing party information, and voting numbers. Admin system login by using 'Admin' username and 'Admin' password.

4.3.2 *Candidate module*

Using this module, a candidate may vote using his or her personal information. User can go to the log-in after registering which validated user ID and go to the cast vote after successful registration.

4.4 *Home screen*

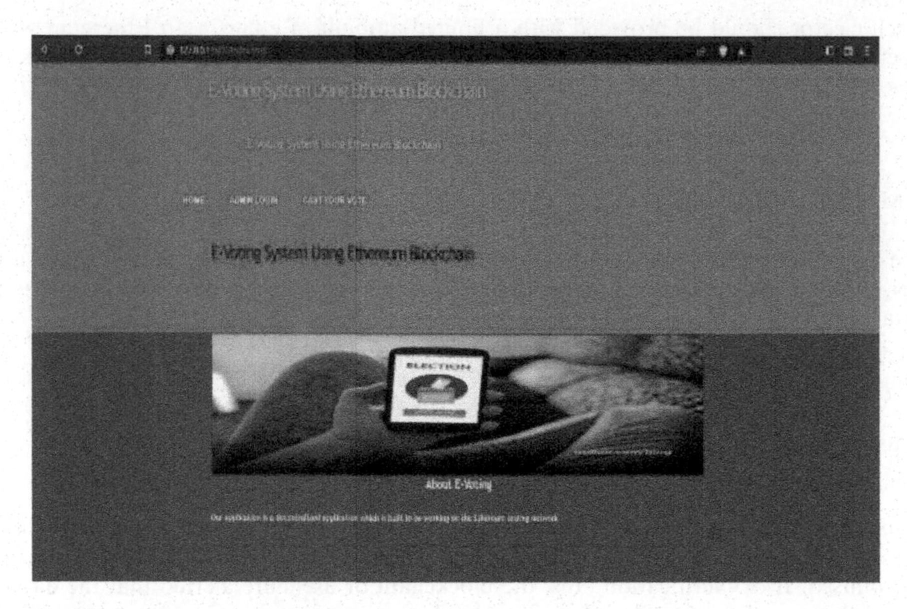

Figure 2. Home screen.

4.5 *Result screen*

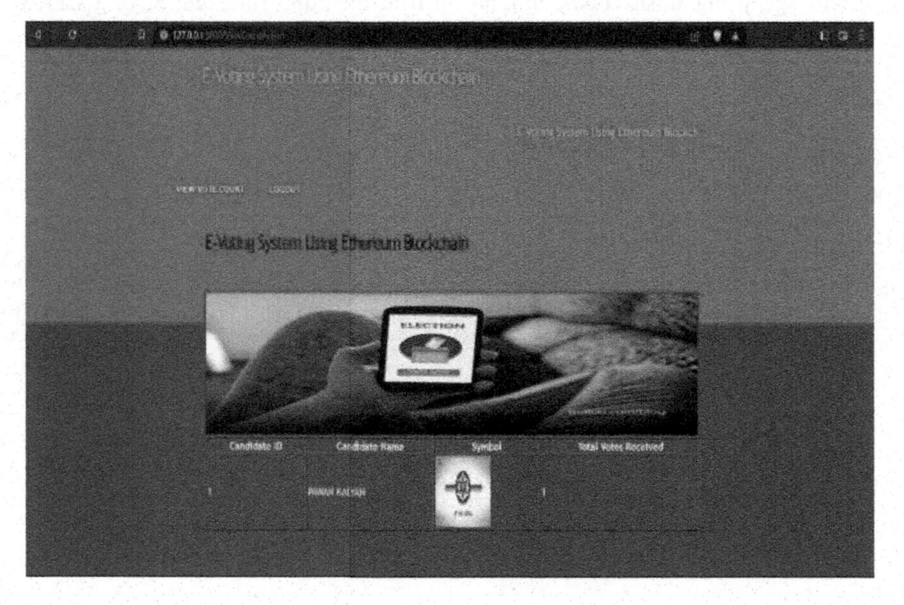

Figure 3. Result screen.

5 CONCLUSION AND FUTURE ENHANCEMENT

We were effective in transferring estate e-casting a ballot to the blockchain stage by using the Ethereum organization's office and, consequently, the blockchain architecture, and we self-tended to some of the significant concerns that estate e-casting a ballot framework has. As a result of our testing, the idea of blockchain and the security technique it uses, especially constant hash chains, are now workable for surveys and races. This accomplishment may open the door for additional blockchain applications that impact all aspects of human life. In both political and logical circles, e-voting is still a widely debated concept. Despite the involvement of a few rare occasions, the bulk of them are still in use; Additional initiatives either fell short of providing the security and protection choices of a traditional political contest or posed substantial convenience and quantifiability difficulties.

In fact, most of the security issues, such as citizen protection, respectability, check, and nonrepudiation of votes, as well as simplicity of examination, are addressed or may be addressed with pertinent modifications by blockchain-based electronic voting arrangements, such as the one we have implemented using reasonable agreements and subsequent Ethereum organization. In any case, the blockchain cannot completely address certain viewpoints, such as elector validation (on the public level, not the record level), which necessitates the consolidation of various methods, such as the use of biometric factors. While we are still working on a normal app as part of our effort to make it easier to use, it is important to remember that, in addition to smartphones and tablets, refrigerators, televisions, seats, clothing, vehicles, and a variety of other common items are or will be ready to connect to the internet easily. As far as blockchain, it will be easy to foster such conveyed frameworks with such an enormous organization and hold handling power. Moreover, if these gadgets capability all together to diminish the approval measure of exchanges in a very blockchain, we will have the option to play out most of our web-based exchanges safely, dependably, and effectively, in principle as well as all the while practically speaking.

REFERENCES

Ajit Kulkarni, (2018), *"How To Choose Between Public And Permissioned Blockchain For Your Project"*, Chronicled, 2018.

Andrew Barnes, Christopher Brake and Thomas Perry. (2016). *Digital Voting with the use of Blockchain Technology* Available at: https://www.economist.com/sites/default/files/plymouth.pdf

Çabuk U. C., Çavdar A., and Demir E., "*E-Demokrasi: Yeni Nesil Do ÷ rudan Demokrasi ve Türkiye'deki Uygulanabilirli ÷ i*", [Online]

Clack C. D., Bakshi V. A., and Braine L., "Smart Contract Templates: Foundations, Design Landscape and Research Directions", Mar 2017, *arXiv*:1608.00771.

Ethdocs.org. (2018). *What is Ethereum? — Ethereum Homestead 0.1 Documentation. [Online]* Available at: http://ethdocs.org/en/latest/introduction/what-is-ethereum.html https: //github.com/Ethereum/wiki/wiki/White-Paper.

Maaten E., "Towards Remote E-voting: Estonian Case", *Electronic Voting in Europe-Technology, Law, Politics and Society*, vol. 47, pp. 83–100, 2004.

Nakamoto S., *"Bitcoin: A Peer-to-peer Electronic Cash System", [Online]*. Available: https://bitcoin.org/bitcoin.pdf.

Nicholas Weaver. (2016). *Secure the Vote Today* Available at: https://www.lawfareblog.com/secure-vote-today.

Patrick McCorry, Siamak F. Shahandashti and Feng Hao. (2017). *A Smart Contract for Boardroom Voting with Maximum Voter Privacy* Available at: https://eprint.iacr.org/2017/110.pdf.

Sos.ca.gov. (2007). *Top-to-Bottom Review | California Secretary of State*. Available at: http://www.sos.ca.gov/elections/voting-systems/oversight/top-bottom-review/.

Vitalik Buterin. (2015). *Ethereum White Paper* Available at:

Wood G., "Ethereum: A Secure Decentralised Generalised Transaction Ledger", *Ethereum Project Yellow Paper*, vol. 151, pp. 1– 32, 2014.

Artificial Intelligence, Blockchain, Computing and Security – Dagur et al. (Eds)

Quality testing for rice grains using deep learning

E. Sujatha*, G. Navin Krishnan*, S. Lokesh* & S.R. Navin Kumar*
Department of Computer Science and Engineering, Saveetha Engineering College, Chennai, Tamil Nadu

ABSTRACT: Rice is very important in South India. It is a staple meal for more than 80% of the population. Worldwide. Rice harvests of various types are farmed and exported. Identify faulty grains and It is critical to distinguish the type of rice while analyzing rice quality. An automated method for identifying and classifying rice grain types is introduced, using digital images acknowledged as an efficient means approach for extracting the features of rice grains without interaction. A camera is used to capture images. The captured image is subjected to image preprocessing, filtering, segmentation, and edge detection algorithms. In this research, the suggested approach is utilized to calculate the amount of rice based on the system's good and poor states utilizing image processing and deep learning techniques.

Keywords: CNN, Deep learning, Image processing

1 INTRODUCTION

One of the most important functions of computer vision is the analysis of rice quality. Several scientists claim that an object's geometry contains more information than its aesthetic aspects, such as its path The color varies depending on the instance of the object. However, the exact result cannot be specified. You can also report a rice integrity issue. Rice integrity consists in touching the seeds during sampling. The main goal of this approach is to provide an alternative method for rice quality control and analysis that reduces effort, costs and time (good or bad) using image processing and deep learning algorithms.

2 LITERATURE SURVEY

The author A. Esteva *et al.* explained about rice quality is defined in physical and chemical parameters that are used to evaluate and classify rice grains. The quality of your measurements is assessed by determining the border region and end points. MatLab takes into account and implements the mean value of the characteristics. Mat Lab is used to apply the image processing algorithm on the sample beads. The author B. S. Prajapati *et al.* have rice quality is determined by the color, size, and shape of the grain. Neural Network Classifiers categories precision and results into high, terrible, and medium quality. The author D. Vishnu *et al.* an automated approach for identifying various rice seed types employing artificial vision technology and a detection system comprised of an inspection machine and an image processor Unit. To determine the quality of the rice seed, a backward neural network was developed. The method was able to enable inspection of different rice seed kinds based on their visual features. The answer for the assessment and categorization of

*Corresponding Authors: sujathae@gmail.com, nknavin361@gmail.com,
clashstarlokesh007@gmail.com and abinavin18@gmail.com

 DOI: 10.1201/9781032684994-62

the quality of Krishna kamod rice utilizing image Processing techniques and easy computing. Feeding Direct neural network approaches are used to find high levels of quality. The trained multilayer Feed Forward Neural Networks find long seeds, little seeds, and seeds of undetermined quality. The author K. Matsunaga *et al.* is primarily employed in the field of data mining. Since K denotes a procedure that is mostly dependent on preliminary clustering middle and its nearby optimum. The front is utilized first, with the specified beginning cluster applied to the end. To select the best group center, a hybrid optimization and clustering approach is applied. The author M. J. Asif is the ageing process of the three lines of hybrid rice prompted the optimization technique. Where sterility, maintenance, and restorer lines intersect. The auto process is a swarm search process, and hybridization is an evolutionary process. They are mixed in what proportion is acceptable for the computation of convergence and speed that are involved as a whole. The Different models were created for the set of separate functions as well as the combined function sets. Textural traits, rather than morphological and color criteria, determine categorization accuracy. As a result, neural network design produces varying accuracies for distinct feature sets. The Gray Level Co-occurrence Matrix (GLCM) method is used to extract four statistical texture parameters, viz. H. Entropy, inverse difference Momentum, second angular momentum and correlation. By extracting features from the image using the GLCM approach, the image compression time when converting an RGB image to grayscale can be significantly reduced compared to other DWTs Technically, but still, DWT is a versatile method of video compression in general. These features are useful in video motion estimation and real-time pattern recognition applications such as military & Medical applications. B.S. Prajapati *et al.* gave an overview of the different machine learning approaches to identification Plant diseases based on leaf images. Like humans, plants suffer from various diseases that affect their normal growth. The study included Disease identification using trait-based methods and home-made LD. P. Mohanaiah *et al.* compared the performance in terms of the preprocessing and segmentation techniques used, the features used to classify the diseases, and the data set use deach article. Esteva would like to thank all designers of websites, programs and other features that served as inspiration or reference for the development of this system. hope that this research will achieve its goals and be adopted as a real system used by doctors and hospitals to identify dermatological problems in the future. In this article we have the following innovations and propose a new two-step team method combining five excellent classification models for skin melanoma classification; P. Wijerathna *et al.* also propose a new method to segment the lesion area on a dermo scopic image to generate a lesion mask Image resizing panel to highlight the change; S. Gilmore *et al.* propose a new ensemble network that can use local combined layers to efficiently integrate the classification results of five classification networks. Our study makes an important contribution to this research area for several reasons. First, it is a study that brings together ongoing research into all the steps required to develop an automated system A diagnostic system for the detection and classification of skin cancer. Second, it presents insights that help scientists appreciate the importance of high-level feature extraction and appropriate feature selection Methods that require more effort to correctly diagnose melanoma.

3 SYSTEM ANALYSIS

Rice quality is nothing more than a mix of physical and chemical properties. Physical qualities of rice include grain size and shape, chalk, and whiteness, while chemical properties include any lose concentration, gelatinization temperature, and gel consistency.

The research proposes a method for sorting and assessing rice grains based on grain size and form utilizing SVM on machine processing techniques, especially an edge detection algorithm, to figure out the region of the borders of each grain in our existing system.

- When utilizing machine learning algorithms, performance suffers.
- The precision is lower.
- Handling feature extraction is computationally difficult.

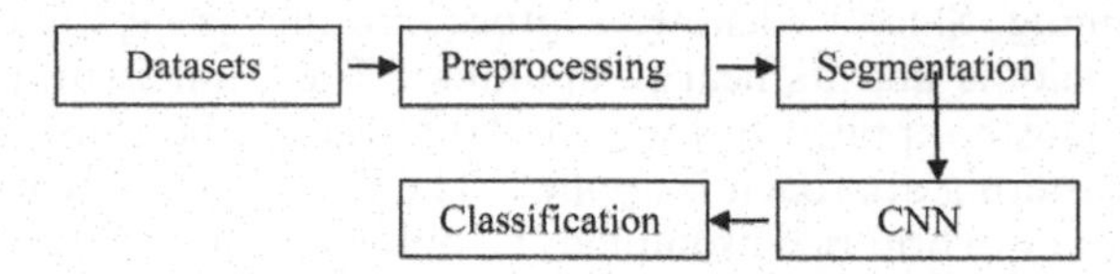

Figure 1. System architecture.

3.1 *System architecture*

The study provides an alternate way for quality analysis that reduces the time and expense necessary in our suggested technique. Image processing is a vital and sophisticated technical subject in which significant advances have been made. Based on deep learning techniques, a convolution neural network is employed to assess ricequality.

The system architecture outlines the procedures involved in this effort to determine the quality of rice.The key procedures involved are image processing and image classification, which are based on the system's deep learning Alex net model.

- High precision.
- When utilized instead of image processing, performance improves.
- Segmentation and feature extraction are simple to control.

4 SYSTEM MODULES

Data set collection: A records set is a set of records. All of the data in the rice picture of the data may be found here. A high-quality camera sensor system is used to capture the image of the rice grains. The minimum recommended pixel size is 8mb since it impacts the quality of the tiny grains of rice as well as the overall effectiveness of the system. Images of rice granules are taken and saved for training and testing.

Preprocessing: Preprocessing is a technique for removing unnecessary data from input data. It is sometimes referred to as eliminating noise from data. Image pre-processing is the process of resizing pictures, adjusting image contrast, and converting the image to grayscale and, finally, black and white for display on the system. These techniques aid in the removal of undesired visual distortions, noise, and blur, resulting in a higher quality image suitable for further processing. The Gaussian filter is employed in this case to treat the picture as needed.

Segmentation: Splitting an image into sections aids in locating items and boundaries in an image. Objects and their borders are identified by labelling each pixel with comparable properties. The region of interest (Roi) is segmented out of the image after it has been pre-processed. Fuzzy C-mean approaches are employed in this case.

Classification: Classification refers to the process of categorizing photographs based on the attributes retrieved from them. The system's output layer is categorization. In this study, CNN classifier techniques were employed to categories rice grains, and the system obtained the findings.

5 ALGORITHM

A Convolutional Neural Network (CNN) or Convnet is a type of deep learning network that learns directly from input, eliminating the need to extract human characteristics. CNN is exceptionally good at it Recognize patterns in photos and differentiate between objects, faces and landscapes. They can also be used to classify data other than images, e.g. B. audio data, time series and signal data.CNNs are most commonly used in deep learning for three reasons. First, CNNs are used to reduce the need to extract human features; CNN learns the features directly. NC produces very accurate recognition results. CNNs can be trained to perform additional reconnaissance tasks, allowing for the expansion of existing networks.

A convolution neural community would possibly consist of tens or loads of layers that educate to come across numerous photo aspects. Filters are applied to each training image at varying resolutions, and the outcome of each convolved image is utilized as input for the next layer. Filters can start with simple qualities like brightness and edges and evolve to attributes that characterize the element uniquely.

6 RESULT

The rice test was carried out in the Matlab, and parameters such as perimeter, area, firmness, filled area, and centroid were recorded. Circularity is a logic formed from the aforementioned features that is used to carry out the primary classification, that is, to differentiate between rice. calculation is shown in equation 1.

1. Circularity [[(Perimeter)^2/ (4*π*(filled area]].

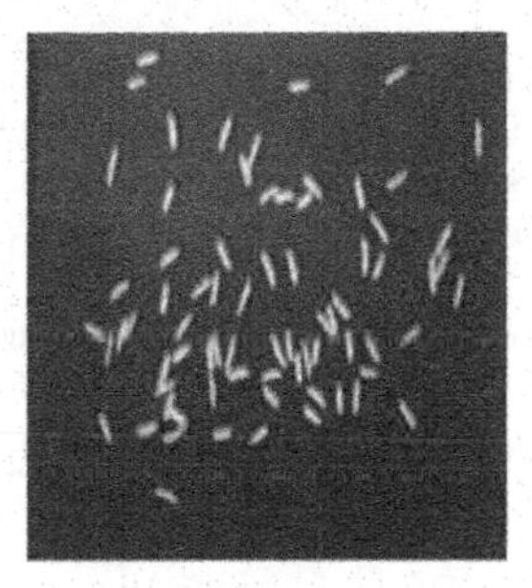

Figure 2. Samples used.

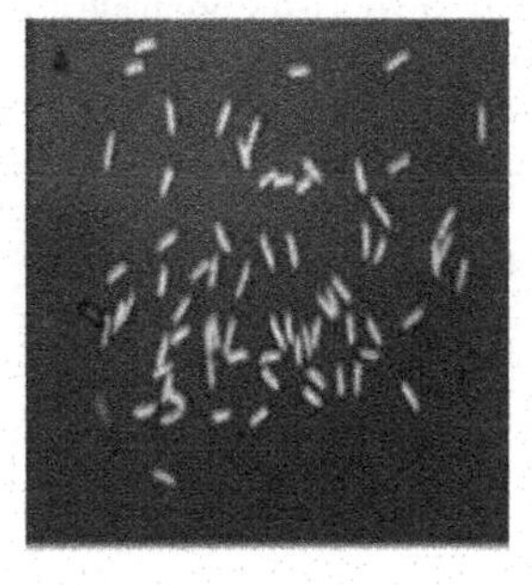

Figure 3. Results on processing.

7 CONCLUSION

Consumers today are very sensitive about the quality of food grains. This model investigates an automated rice grain quality evaluation system based on CNN classifiers to assure rice grain quality. Two rice grain variants, ponni and matta, are being researched. Rice grains were identified and categorized using the proposed technique based on their morphological and geometric features. According on the experimental results, the proposed CNN classifier has a good overall accuracy.

REFERENCES

Agricultural Marketing in India: Defects and Their Remedial Measures. [Online]. Available: http://www.economicsdiscussion.net/agriculture/marketing/agricultural-marketing-in-india-defects-and-their remedial measures/12854.

Asif M. J., Shahbaz T., Tahir Hussain Rizvi S., and Iqbal S., "Rice Grain Identification and Quality Analysis using Image Processing based on Principal Component Analysis," *RAEE 2018 – Int. Symp. Recent Adv. Electr. Eng.*, pp. 1–6, 2019.

Esteva A., Kuprel B., Novoa R. A., Ko J., Swetter S. M., Blau H. M., and Thrun S., "Dermatologist-level Classification of Skin Cancer with Deep Neural Networks," *Nature*, 2017.

Gilmore S., Hofmann-Wellenhof R., and Soyer H. P., "A Support Cector Machine for Decision Support in Melanoma Recognition," *Experimental Dermatology*, vol. 19, no. 9, pp. 830–835, 2010.

India at a Glance | FAO in India | Food, and Agriculture Organization of the United Nations. [Online]. Available: http://www.fao.org/india/fao-in-india/india-at-aglance/en/. [Accessed: 15-Nov-2019].

Matsunaga K., Hamada A., Minagawa A., and Koga H., "*Image Classification of Melanoma, Nevus and Seborrheic Keratosis by Deep Neural Network Ensemble,*" 2017.

Mohanaiah P., Sathyanarayana P., and Gurukumar L., "Image Texture Feature Extraction Using GLCM Approach," *Int. J. Sci. Res. Publ.*, 2013.

N. V G, "An Intelligent System for Identification of Indian Lentil Types Using Artificial Neural Network (BPNN)," *IOSR J. Computer. Eng.*, vol. 15, no. 5, pp. 54–60, 2013.

Nagoda N. and Ranathunga L., "Rice Sample Segmentation and Classification Using Image Processing and Support Vector Machine," *2018 13th Int. Conf. Ind. Inf. Syst. ICIIS 2018 – Proc.*, no. 978, pp. 179–184, 2019.

Prajapati B. S., Dabhi V. K., and Prajapati H. B., "A Survey on Detection and Classification of Cotton Leaf Diseases," *Int. Conf. Electr. Electron. Optim. Tech.* ICEEOT 2016, pp. 2499–2506, 2016.

Vishnu D., Mukherjee G., and Chatterjee A., "A Computer Vision Approach for Grade Identification of Rice Bran," *Proc. – 2017 3rd IEEE Int. Conf. Res. Comput. Intell. Commun. Networks, ICRCICN 2017*, vol. 2017-Decem, pp. 10–14, 2017.

Wijerathna P. and Ranathunga L., "Rice Category Identification Using Heuristic Feature Guided Machine Thresholding" (Binarization or Otsu's Thresholding) Use the Concept of Chain Codes to Obtain the Connectivity of the Edges and Find the Physical Features and Use HUE Color Extraction to Find the Color Features of the Image Apply Noise Filters Use Watershed Algorithm to Find the Contour of the Grains Based on the Results of Features Extracted Formulate a Logic to Classify and Grade the Images Vision Approach," *2018 13th Int. Conf. Ind. Inf. Syst. ICIIS 2018 – Proc.*, no. 978, pp. 185–190, 2019.

Artificial Intelligence, Blockchain, Computing and Security – Dagur et al. (Eds)
© 2024 The Author(s), ISBN: 978-1-032-67841-2

Skill measuring test for preschoolers

G. Nagappan*, K. Narmatha*, K.M. Nithiyasri* & S. Sai Gayatri*
Department of Computer Science and Engineering, Saveetha Engineering College Chennai, India

ABSTRACT: Psychometric evaluation evaluates the information, abilities, capacity and persona trends of someone being assessed. Since many youngsters are joined in faculty earlier than they meet the specified age criteria, they lack the fundamental knowledge and information. This creates a susceptible base for them of their scholar life. In order to check their capacity to get promoted to better grade this psychometric check might be beneficial to get evaluation of the child's intelligence. This check consists of five kinds of intelligence. They are Verbal or Linguistic Intelligence, Mathematical/Logistic Intelligence, Interpersonal Intelligence, Intrapersonal Intelligence, Naturalistic Intelligence.

Keywords: Multiple intelligence, Logical intelligence, Linguistic intelligence, Interpersonal intelligence, intrapersonal intelligence, Logistic Intelligence, Naturalistic intelligence, Verbal Intelligence

1 INTRODUCTION

The schooling machine is being advanced each year so one can enhance scholar's information and put together them to continue to exist on this world. The common age for youngsters to enroll in faculty turned into six years in 2021. Similarly the common age in a few different nations are: maximum cost in Afghanistan with seven years and lowest cost in Antigua and Barbuda with 5 years. In India, 3 years old is the appropriate age to begin kindergarten in India. A new idea has been discovered in mother and father who delay enrolling their youngsters in formal faculty beyond the prescribed age. This is referred to as "ACADEMIC Redshirting". But many mother and father try and be a part of their youngsters in better grades even if they're three years old. This might be tough as the kid lacks fundamental knowledge at such a younger age. Some mother and father may now no longer be privy to their child's abilities or interests. The psychometric test consists of three sections including aptitude, skill and personality test. The test focuses on skill measurement, which include verbal, logistic, inter and intrapersonal skills.

As an end result the youngsters is probably compelled to research what they or sense tough to pursue their profession or schooling. Many nations have started to put into effect Multiple Intelligence on the grounds that 2013. These assessments are carried out for college students of their better schooling. By carrying out this check for youngsters of their more youthful age, they may be capable of pick out a great profession course wherein they could succeed.

2 LITERATURE SURVEY

The goal is to research a kid's more than one intelligence and advice the top profession course primarily based totally on the sort of intelligence they excel in. This check is evolved

*Corresponding Authors: nagappan.cse@saveetha.ac.in, narmathk372@gmail.com, nithiyasrikm2508@gmail.com and saigayatri2001@gmail.com

DOI: 10.1201/9781032684994-63

as a quiz and the end result may be used as an evidence for the kid's intelligence evaluation. Similar form of Psychometric assessments is carried out for employers to degree the call for abilities (Zhara *et al.*).

The aim of the study is to know the usage of psychometric in the particular research or report, also the test evaluations on it (Masha *et al.*).

Advancement in technology pave manner for brand spanking new fields in schooling in addition to in task sector. But it is miles turning into more and tough for college students to evolve to this rapid growing society with the information they've already learnt. Even though they're able to knowledge and studying new ideas soon, they discover it tough to just accept this. In order to manual from their fundamental schooling degree mother and father are seeking to be part of their youngsters to better grades in younger age. This more than one intelligence check might be very beneficial for mother and father to become aware of their youngsters electricity and weak spot and manual them in line with that. This check document might be the evidence for the intelligence of the child. This aimed to look at the psychological of Multidimensional Death and Mortality Archives (MODDIF) in Iranian college students. An overall of 320 topics have been decided on from college students of the universities in Tehran in the course of the 2016-2017 educational year, the usage of comfort sampling (Fateme *et al.*). To have a look at reviewed elements that might be used to expect educational overall performance, however which can be presently now no longer systematically measured in tertiary schooling. It targeted on psychometric elements of capacity, persona, motivation, and studying strategies. Their respective relationships with educational overall performance are enumerated and discussed. A case is made for his or her improved use in studying analytics to beautify the overall performance of current scholar models (Geraldine *et al.*).

The standard approach to measuring the demand for skills, and skill shortages, is to conduct a survey of employers. This paper reviews the literature on psychometric testing by employers, and considers whether information on psychometric testing can be used to make deductions about changes in the demand for skills in the economy (Mehreen *et al.*). Based on the model, three questionnaires were designed for specific populations participating in testing at organization. The results suggest that while the organization's use of testing was somewhat consistent with the good testing practice model, there were areas where the use of testing could be improved or being used as additional control mechanism for entrepreneurs who already have bank accounts, for example those who have a bank credit history (Kathleen *et al.*).

The aim of the study is to prove the usage of psychometric in the historic constructions[c]. The validation purpose is enhanced by using the psychometric testing in the test for the standard organization of educational and psychological testing (Anitha *et al.*). The standard for the quantitative research in psychology is done by the APA publications, which is used in the research of scientific Inventions (Evan *et al.*). The purpose of this study was made to know the behavior or validate the student's behavior in a community based system (Hasio *et al.*).

The purpose of this study was made as an agreement between the parents and teachers for testing their student's emotional behavior. This research was made for the students with the age of 8.5 years (Kumulainen *et al.*). Psychologically affected students were critically suffering from the ill ness of not being well. In order to measure their distress level on the age of 5 to 10 the test was developed with a scheduled timing physical test (Janet *et al.*). The aim of the research is to achieve the testing in the clinical level for kids, this to achieve the test using clinical dataset. The report is an agreement between parents and teacher making the important contribution by participating in this report Analysis (Miriam Crowe *et al.*).

3 RESEARCH BASICS

The web page consists of testing the kids with some mentioned questions in each of the test section. JavaScript functions were used to create the test scores are automatically generated once after the test ended. It also gives some information on what is analysis and the career guidance based on their performance.

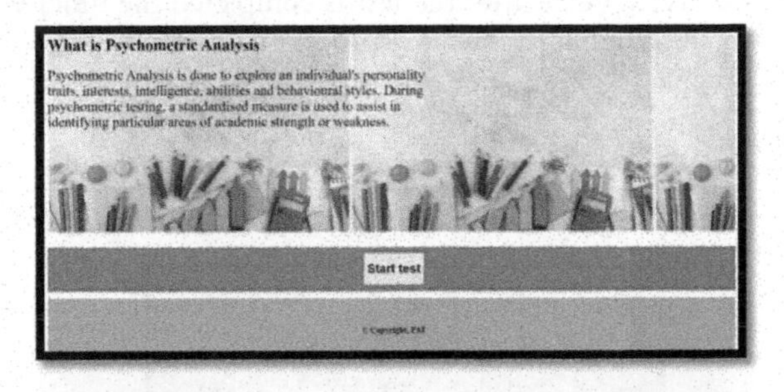

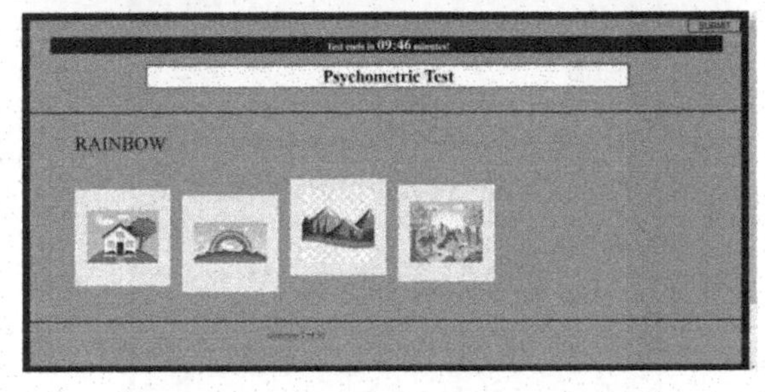

Figure 1. Test section.

4 PROPOSED APPROACH

The proposed model in this paper helps the kid and their parents to understand about their strong intelligence type. Based on this result career suitable for that intelligence is suggested. The intelligence report is done based on the quiz attended by the kid.

4.1 *Architecture diagram*

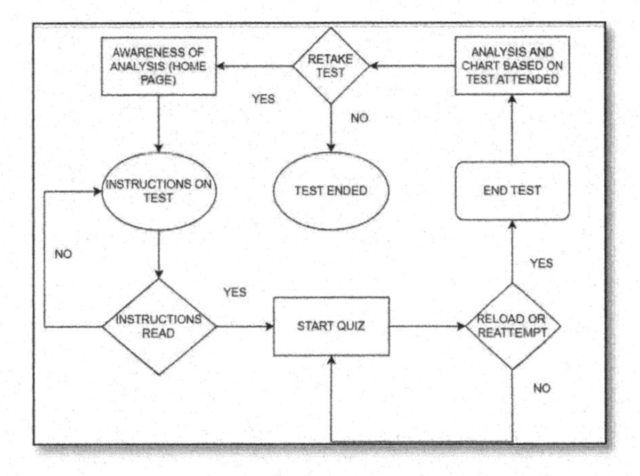

Figure 2. Proposed architecture workflow diagram.

The flow of Diagram consists of working method on the webpage for assessment. Right after the instruction page, test can be started. It cannot be reloaded or refreshed, in case of doing so the test might not be saved. Once after the test is completed the timer ends automatically or can be submitted by the candidate. The last process is the result page or report.

4.2 *Experimental results*

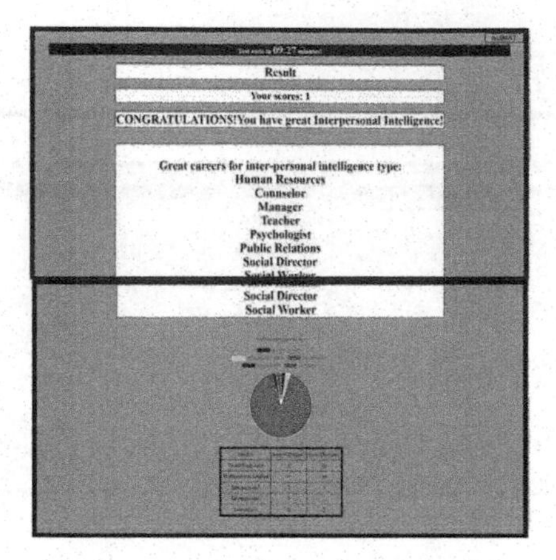

Figure 3. Output for the quiz.

4.3 *Graph*

For analysis purposes, a chart depicting the intelligence percentage of the student is made.

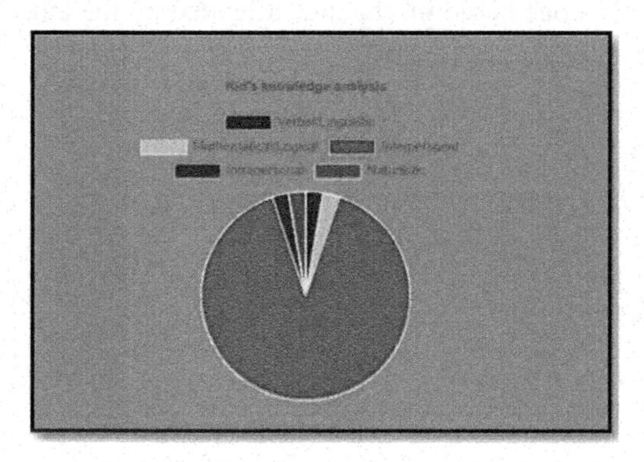

Figure 4. Pie chart depicting the multiple intelligence.

5 CONCLUSION

Thus, the proposed system makes use of HTML, CSS and JavaScript for web page development. This quiz is an open source test and the result can be used as a proof for the child's Multiple Intelligence. Thus the system is used to predict the rightful results from the quiz and producing analysis for the kid's intelligence.

REFERENCES

Anita M Hubley, Ayumi Sasaki, *Synthesis of Validation Practices in Two Assessment Journals: Psychological Assessment and the European Journal of Psychological Assessment*, August 2014.

Evan Mayo-Wilson, Harris Cooper, and Journal Article Reporting Standards for Quantitative Research in Psychology: The APA Publications and Communications Board Task Force Report, January 2018, *American Psychologist* 73(1):3–25.

Fateme Aghaie Meybodi, Parvaneh Mohammadkhani, Abbas Pourshahbaz, *Psychometric Properties of the Persian Version of the Emotion Regulation Checklist*, February 2018.

Geraldine Gray, ColmMcGuinness, Philip Owende, Aiden Carthy, *A Review of Psychometric Data Analysis and Applications in Modeling of Academic Achievement in Tertiary Education.*

Hsiao-Ling Chuang RN, Ching-Pyng Kuo RN, Chia-Ying Li RN, *Wen-Chun Liao RN-Psychometric Testing of Behavior Assessment for Children*, Volume 10, March 2016.

Janet E. Rennick, C. Celeste Johnston, Sylvie D. Lambert, Judy M Rashotte, Measuring Psychological Outcomes Following Pediatric Intensive Care Unit Hospitalization: PsychometricAnalysis of the Children's Critical Illness Impact Scale, 2011 Nov, 635–42.

Kathleen Slaney, *Validating Psychological Constructs: Historical, Philosophical, and Practical Dimensions*, July 2017.

Kumpulainen K., Henttonen I., Moilanen I., Piha J., Puura K., *Children's Behavioral/Emotional Problems: A Comparison of Parents' and Teachers' Reports for Elementary School-aged Children*, December 1999.

Masha Tkatchouk, Michael D. Maraun, Slaney, Psychometric Assessment and Reporting Practices: Incongruence Between Theory and Practice, *Journal of Psychoeducational Assessment*, November 2009, 27 (6):465–476.

Mehreen Memon, Farhan Ahmed, *Effectiveness of Psychometric Testing in Recruitment Process*, March 2018.

Miriam Crowe, Donald maciver, Robert rush, Psychometric Evaluation of the ACHIEVE Assessment, 29 May 2020, Sec. *Children and Health*, volume 8.

Zahra Allame, LeilaHeydarinasab, Motahare Fasanghari, Maryam Shahmohammadi, *Examination of the Psychometric Properties of the Persian Version of the Multidimensional Orientation toward Dying and Death Inventory Among Students.*

Artificial Intelligence, Blockchain, Computing and Security – Dagur et al. (Eds)
© 2024 The Author(s), ISBN: 978-1-032-67841-2

IoT based monitoring model to identifying and classifying heart disease through machine learning

Parisha*
Research Scholar, Department of Computer Science & Engineering Babu Banarasi Das University, Lucknow

Santosh Kumar*
Associate Professor, School of Computing & Engineering, Galgotias University, Greater Noida, Uttar Pradesh, India

Gaurav Kumar Shrivastava*
Assistant Professor, Department of Computer Science & Engineering Babu Banarasi Das University, Lucknow
ORCID ID: 0000-0002-3746-3290

ABSTRACT: Healthcare research has advanced thanks to artificial intelligence. Because there is free access to healthcare information, researchers have developed methods to help with the heart disease diagnosis and prognosis. For such difficult illnesses, deep learning and machine learning models offer a reliable, quick, and efficient solution. The articles from Web of Science, EBSCO, and EMBASE that were released between 2009 and 2021 were chosen using PRISMA criteria. A successful search method was used to discover the research papers for this study that used AI-based learning techniques for heart disease prediction. There are 185 studies in all that conventional machine learning-based classifications and deep learning-based classifiers rank as having a significant impact on heart disease prediction. The survey also analysed the research done by various researchers and drew attention to the flaws in the earlier literature. Several metrics, including prediction rate, accuracy, sensitivity, specificity, dice score, detection rate, area covered, precision, recall, and F1-score, were utilized to compare the findings. The solutions to the five scheduled investigations have been examined. Even though several of the methods suggested in the literature have excellent predicative accuracy, heart disease mortality has not decreased. As a result, further study is needed to address the problems with heart disease prediction.

1 INTRODUCTION

Greek terms for tumour and crab, respectively, are where the word “heart disease” first appeared. Since the 1600s, the term “heart disease” has been used in medicine to refer to cells that are developing abnormally and have the potential to infiltrate or spread to various body regions [136]. heart disease metastasis is the term for the uncontrolled cell proliferation that begins in one place of the human body and spreads to other bodily locations [43,172]. There are two types of heart disease cells: benign and malignant. Malignant cells metastasis and are thought to be more hazardous than benign cells because they do not spread to other places of the body. The lengthy and expensive treatment approach is due to the disease’s high mortality and recurrence rate. The likelihood that a patient will survive must rise thanks to accurate early detection. Genetic mutations that affect our cells’ activities, particularly how

*Corresponding Authors: Parisha369@gmail.com, Sant7783@hotmail.com and Gauravi8hit@bbdu.ac.in

 DOI: 10.1201/9781032684994-64

they divide and expand, are what lead to the genetic illness. More modifications will take place as the tumour cells become larger. In summary, healthy cells are less likely to have genetic changes such DNA mutations than heart disease cells [116,110]. Few heart disease cells are able to avoid being destroyed by the immune system, which often rids the body of damaged or aberrant cells. The immune system is yet another mechanism the tumour uses to spread and endure [179]. The name of the specific type of heart disease is determined by the location of the tumour cells; for instance, lung cancer is a type of heart disease that starts in the lungs and progresses to the liver. Three prognostic indicators are included in a heart disease diagnosis: heart disease risk assessment, heart disease recurrence, and heart disease survival indicator. The likelihood of developing heart disease is forecasted first, then the likelihood of it returning. Making forecasts about growth, lifespan, tumordrug sensitivity, and survivorship is the final step [95].

2 MOTIVATION

The rise in heart disease incidence and fatal cases that has been observed globally [10] served as the impetus for our study. The population is ageing and growing, and the incidence and distribution of the primary risk factors for heart disease have changed, among other complex and varied causes. Data from trustworthy sources were used to create Figure 1, which shows heart disease incidence cases and death figures.

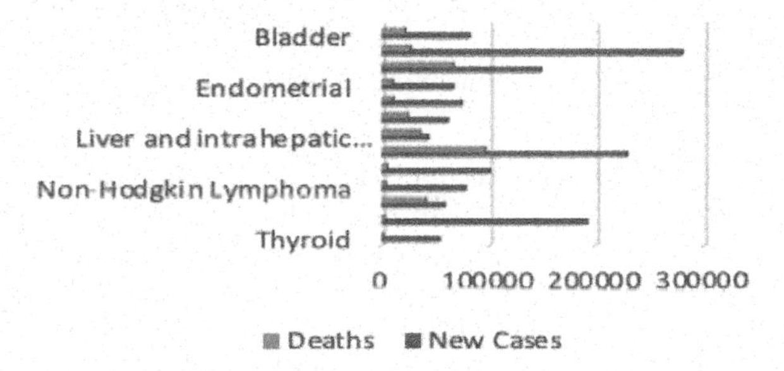

Figure 1. Estimated number of new cases and deaths in 2020 for common heart disease types.

Numerous heart disease prediction techniques have been developed in the past by various researchers. It is possible to find the following works in the literature that have been conducted along this line by various studies: Based on disparate electronic health records (EHRs), Li *et al.* [1] have developed a generic framework for risk factor (RF) analysis that may be used for osteoporosis prediction and meaningful RF selection. Sandhiya and Palani [2] have developed a disease prediction system that precisely predict the most deadly illnesses, such as cancer, diabetes, and heart disease. A thorough sickness temporal extension and fuzzy inference process were disclosed by Sethukkarasi *et al.* [3] in their presentation of an unique temporal mining system dubbed fuzzy temporal cognitive Map (FTCM). Using Ganapathy *et al.*'s [4] For effective decision support in medical diagnosis, a unique pattern classification technique integrates temporal characteristics with a Fuzzy Min-Max (TFMM) neural network-based classifier. Ahmed *et al.* [5] developed a CAD strategy for the detection of breast cancer using a back propagation supervised path and a deep belief network unsupervised path. The whale optimization algorithm is used in the feature selection method proposed by Sharawi *et al.* [6]. (WOA). WOA is a recently introduced meta-heuristic optimization technique that mimics humpback whale behaviour in their natural environment. The proposed model makes use of the wrapper-based approach to identify the ideal subset of properties. Mafarja and Mirjalili [7] describe an unique wrapper feature selection method that is based on the Whale Optimization Algorithm (WOA).

3 EXPERIMENTAL STUDY & METHODOLOGY

The UCI Repository Data-set, IoT device, Sliding Window, Decision Manager, Feature Selection Module, Classification Module, and Rule Base are the six components that make up the proposed architecture of the heart disease monitoring system shown in Figure 2. Through an IoT gadget, the data is collected from the patients. In order to choose the features that have contributed the most to the streaming data and provide them to the decision manager, feature selection modules are also used in the processing of the streaming data from IoT devices. The decision manager analyses the crucial streaming data from a sliding window using a deep belief network to identify whether or not the patient's data is affected by heart disease. Additionally, the decision manager will split the data to determine the category type from the standard data-set by the help of rule base.

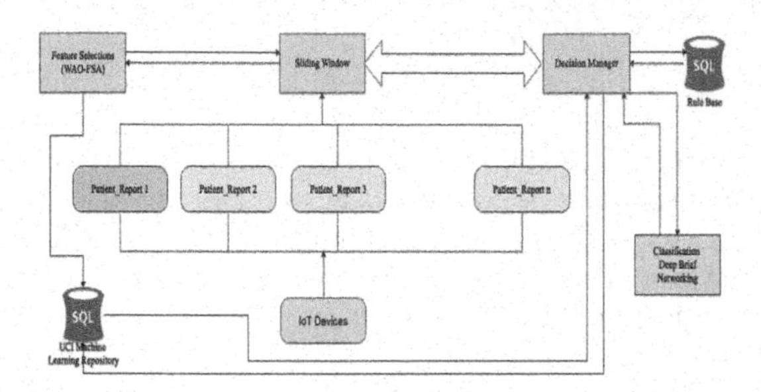

Figure 2. Block diagram for heart disease monitoring model.

This section describes the feature selection method and classification procedures for the proposed heart disease prediction and monitoring model. Following a brief demonstration of the short working procedure for the suggested feature selection approach, the feature selection process is thoroughly explained. IoT devices are categorised as either implanted or wearable. They are employed to gather patient information from remote locations. Utilizing IoT devices that are connected to the human body, these precise measurements are stored as patient data.

3.1 *Features selection*

The selection of features is one of the key components of data mining [5]. It is believed to be necessary due to the exceedingly high dimensionality of data sets and developing computational approaches to the target problems. With the use of data mining, enormous amounts of noise-filled or redundant and pointless features-filled data are kept. Filtering out the noise during the pre-processing stage of feature selection reduces the dimensionality of the data set, making it simpler and less expensive to create computationally efficient models.

3.2 *Goals in feature selection*

- Simplifying the measurements without losing their uniqueness.
- Boost the forecasting precision of the model.
- Create a subset of M features, where M N, to guarantee that the value of the criteria function is best for all M subsets.

3.3 *Dataset*

The classifier results are validated using a benchmark dataset on cardiac illness from the UCI repository [20]. The dataset's summary information is displayed in Table 1. The total

CKD dataset consists of 270 occurrences, 13 features, and 2 classes. Of the 270 occurrences, one hundred fifty are identified as having heart disease, whereas the remaining one hundred and twenty are identified as not having heart disease.

Table 1. Heart disease dataset description.

Dataset	Abbreviation	# of Instances	# of Attributes	# of Class	Present/Absent
Heart Disease	UCI	290	15	4	170/130

3.4 *Confusion matrix*

The measurements that are compared include recall, accuracy, and precision. Before describing performance measurements, Table 2 presents the concept of a confusion matrix.

Table 2. Confusion matrix for prediction values.

Confusion Matrix		
	Actual Positive	Actual Negative
Predicted Positive	True Positive (TP)	False Positive (FP)
Predicted Negative	False Negative (FN)	True Negative (TN)

The percentage of instances that are correctly classified is known as accuracy. One of the most often used classification performance indicators, the value should be closer to 100 for higher classification performance.

$$\text{Accuracy} = \frac{TP + TN}{TP + TN + FP + FN} \tag{1}$$

Now the precision is defined as below:

$$\text{Precision} = \frac{TP}{TP + FP} \tag{2}$$

Recall is defined mathematically as given below:

$$\text{Recall} = \frac{TP}{TP + FN} \tag{3}$$

F-score measures the accuracy of the testing process. It is an average measure which makes use of precision and recalls and is expressed as bellow:

$$\text{F-Score} = \frac{2TP}{2TP + TN + FP} \tag{4}$$

Kapa value is calculated through the given formula as:

$$\text{K} = (\text{OverallAccuracy} - \text{Expected Agreement})/(100 - \text{Expected Outcome}) \tag{5}$$

3.5 *Performance computation matrix*

Java and the Amazon Cloud have both been used to carry out the intended tasks. An essential step in this inquiry is the classification of data into cases of heart illness and those without. The simulations use medical datasets, therefore a variety of classification metrics, such as accuracy, precision, recall, F-score, and kappa value, are provided. A number of classifiers, including J48, logistic regression (LR), multilayer perceptron (MLP), and support vector machine, are assessed on a benchmark dataset (SVM). This work primarily focused on cardiac disorders, but by changing the data during the training and testing phases, it may also be used to forecast other serious diseases. The results of the simulation are further assessed using ten-fold cross validation to finalize the results of the projected model.

4 RESULT & DISCUSSION

Figure 2 and Table 3 provide a summary of the findings from the comparison analysis of the performance of the different classifiers against the data set for heart disease. The table numbers clearly show that the MLP did the worst, with an accuracy score of 78.14 being the lowest. The SVM and LR showed competitive performance with accuracy values of 84.07 and 83.70, respectively, higher than the accuracy attained by MLP. Even if SVM and LR outperform MLP, J48 classifier outperforms them all in terms of performance. The MLP came in second with the lowest accuracy in terms of F-score (78.20), demonstrating weak classification skills. Furthermore, the classification performance of the SVM and LR is superior than that of the MLP. The J48 also achieved a 91.50 F-score, indicating improved categorization performance. The best categorization performance is indicated by the greatest accuracy value. The MLP classifier performed badly with a minimal precision value of 78.40, whereas SVM and LR classifiers attained precision values of 84.10 and 83.70, respectively. It's noteworthy to observe that the J48 classifier produced better classification outcomes, with a maximum precision value of 91.50. In addition, it has been noted that MLP achieves the lowest recall value while J48 classifier achieves the highest recall value. The recall values of the SVM and LR classifiers are 84.10 and 83.70, respectively, showing

Table 3. Classifier result analysis on heart disease prediction dataset.

	Classifier Results				
Algorithm	Accuracy	F-Scare	Precision	Recall	Kappa
MLP	78.14	78.20	78.40	78.10	56.00
SVM	84.07	84.10	84.10	84.10	67.67
LR	83.70	83.70	83.70	83.70	66.83
J48	91.48	91.50	91.50	91.50	82.68

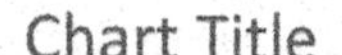

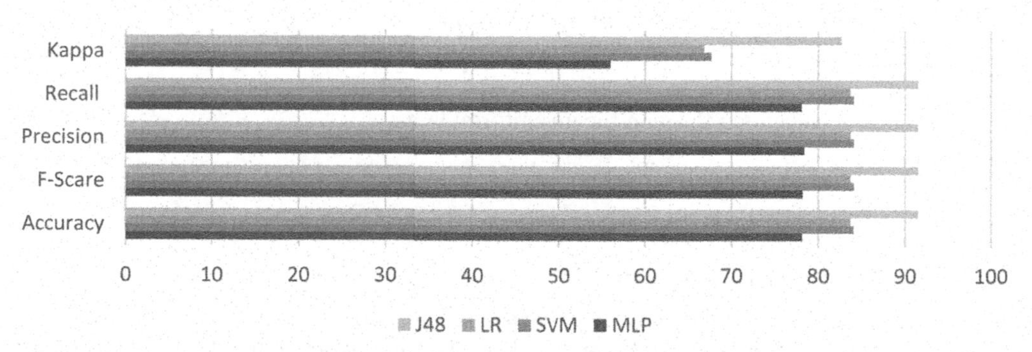

Figure 3. Comparison of various classifiers for heart disease database.

approximately equivalent performance at the same time. Finally, it is claimed that the MLP classifier's lowest kappa value of 56 demonstrates its poor classification performance. With kappa values of 67.67 and 66.83, respectively, the SVM and LR classifiers performed better than MLP. It's interesting that the J48 classifier was able to obtain a maximum kappa value of 82.68. Unexpectedly, the applied dataset for heart disease demonstrates that the J48 classifier consistently performs well. Overall, the table and Fig make it clear that J48 classifier is chosen as the correct algorithm for the IoT-based healthcare prediction model for heart disease.

5 CONCLUSION

In this study, a robust cloud- and IoT-based illness detection model has been developed to track, predict, and diagnose cardiac disease. In this study, an efficient framework for the condition is constructed using data from the UCI Repository and medical sensors to predict who may acquire heart disease. Additionally, patient data is categorized using classification algorithms in order to find heart illness. The training part of the classification approach involves using the data set on heart disease to educate the classifier how to identify the presence or absence of heart disease. The trained classifier is thus ready to assess the incoming patient data and determine the presence of heart disease in the patient. The training step of the classification approach is where the classifier is taught to recognise the presence or absence of heart disease using the data set on heart disease. The trained classifier is thus ready to assess the incoming patient data and determine the presence of heart disease in the patient. It is clear from the results of the extensive testing that the J48 classifier is the most appropriate algorithm for the IoT-based healthcare prediction model for heart disease.

REFERENCES

[1] Hui Li, Xiaoyi Li, Murali Ramanathan, Aidong Zhang, "Identifying Informative Risk Factors and Predicting Bone Disease Progression via deep Belief Networks", *Methods*, Vol. 69, pp. 257–265, 2014.

[2] Sandhiya S, Palani U, "An Effective Disease Prediction System Using Incremental Feature Selection and Temporal Convolutional Neural Network", *Journal of Ambient Intelligence and Humanized Computing*, 2020. https://doi.org/10.1007/s12652-020-01910-6.

[3] Sethukkarasi R, Ganapathy S, Yogesh P, Kannan A, "An Intelligent Neuro Fuzzy Temporal Knowledge Representation Model for Mining Temporal Patterns", *Journal of Intelligent & Fuzzy Systems*, Vol. 26, No. 3, pp. 1167–1178, 2014.

[4] Ganapathy S, Sethukkarasi R, Yogesh P, Vijayakumar P, Kannan A, "An Intelligent Temporal Pattern Classification System using Fuzzy Temporal Rules and Particle Swarm Optimization", *Sadhana*, Vol. 39, No. 2, pp. 283–302, 2014.

[5] Ahmed M. Abdel-Zaher, Ayman M. Eldeib, "Breast Cancer Classification Using Deep Belief Networks", *Expert Systems With Applications*, Vol. 46, pp. 139–144, 2016.

[6] Sharawi M., Zawbaa H. M., Emary E., Zawbaa H. M. and Emary E., "Feature Selection Approach Based on Whale Optimization Algorithm," *2017 Ninth International Conference on Advanced Computational Intelligence (ICACI)*, Doha, 2017, pp. 163–168, doi: 10.1109/ICACI.2017.7974502.

[7] Mafarja, Majdi & Mirjalili, Seyedali. (2017). Whale Optimization Approaches for Wrapper Feature Selection. *Applied Soft Computing*. 62. 10.1016/j.asoc.2017.11.006.

[8] Sayed, G.I., Darwish, A. & Hassanien, A.E. A New Chaotic Whale Optimization Algorithm for Features Selection. *J Classif* 35, 300–344 (2018). https://doi.org/10.1007/s00357-018-9261–2.

[9] Zheng, Yue-Feng & Li, Ying & Wang, Gang & Chen, Yu-Peng & Xu, Qian & Fan, Jia-Hao & Cui, Xue-Ting. (2018). A Novel Hybrid Algorithm for Feature Selection Based on Whale Optimization Algorithm. *IEEE Access*. PP. 1–1. 10.1109/ACCESS.2018.2879848.

[10] Chao Li, Xiangpei Hu, Lili Zhang, "The IoT-based Disease Monitoring System for Pervasive Healthcare Service", *Procedia Computer Science*, Vol. 112, pp. 2328–2334, 2017.

[11] Jun Qi, Po Yang, Atif Waraich, ZhikunDeng, Youbing Zhao, Yun Yang, "Examining Sensor-based Physical Activity Recognition and Monitoring for Healthcare Using Internet of Things: A Systematic Review", *Journal of Biomedical Informatics*, Vol. 87, pp. 138–153, 2018.

[12] Yixue Hao, Mohd Usama, Jun Yang, M. Shamim Hossain, Ahmed Ghoneimb, "Recurrent Convolutional Neural Network Based Multimodal Disease Risk Prediction", *Future Generation Computer Systems*, Vol. 92, pp. 76–83, March 2019.

[13] S.Sandhiya and U.Palani, "A Novel Hybrid Genetic Binary Cuckoo Optimization Algorithm for Feature Selection", *International Conference on Recent trends in science, engineering and management (ICRTSEM)*, 2019.

[14] T. Vivekanandan, N. Ch Sriman Narayana Iyengar, "Optimal Feature Selection Using a Modified Differential Evolution Algorithm and its Effectiveness for Prediction of Heart Disease", *Computers in Biology and Medicine*, Vol. 90, pp. 125–136, 2017.

[15] Jun Huang, Guorong Li, Qingming Huang, Xindong Wu, "Joint Feature Selection and Classification for Multilabel Learning", *IEEE Transactions on Cybernetics*, Vol. 48, No. 3, pp. 876–889, 2018.

[16] Mahdieh Labani, Parham Moradi, Fardin Ahmadizar, Mahdi Jalili, "A Novel Multivariate Filter Method for Feature Selection in Text Classification Problems", *Engineering Applications of Artificial Intelligence*, Vol.70, pp. 25–37, 2018.

[17] Fajr Ibrahem Alarsan and Mamoon Younes, "Analysis and Classification of Heart Diseases Using Heartbeat Features and Machine Learning Algorithms", *Journal of Big Data*, Vol. 6, No. 81, pp. 1–15, 2019.

[18] Meng Lu, "Embedded Feature Selection Accounting for Unknown Data Heterogeneity", *Expert Systems with Applications*, Vol. 119, pp. 350–361, 2019.

[19] Saurabh Kumar Srivastava, Sandeep Kumar Singh, Jasjit S. Suri, "Effect of Incremental Feature Enrichment on Healthcare Text Classification System: A Machine Learning Paradigm", *Computer Methods and Programs in Biomedicine*, Vol. 172, pp. 35–51, 2019.

[20] Yu Xue, Bing Xue and Mengjie Zhang, "Self-Adaptive Particle Swarm Optimization for Large-Scale 1 Feature Selection in Classification", *ACM Transactions on Knowledge Discovery from Data*, Vol. 13, No. 5, Article 50, pp. 1–31, 2019.

[21] Mohamed Abdel-Basset, Doaa El-Shahat, Ibrahim El-henawy, Victor Hugo C. de Albuquerque, Seyedali Mirjalili, "A New Fusion of Grey Wolf Optimizer Algorithm with a Two-phase Mutation for Feature Selection", *Expert Systems With Applications*, Vol. 139, pp. 1–14, 2020.

[22] Shreshth Tulia, Nipam Basumatarya, Sukhpal Singh Gill, Mohsen Kahania, Rajesh Chand Arya, Gurpreet Singh Wander, Rajkumar Buyya, "Health Fog: An Ensemble Deep Learning Based Smart Healthcare System for Automatic Diagnosis of Heart Diseases in integrated IoT and fog computing environments", *Future Generation Computer Systems*, Vol. 104, pp. 187–200, March 2020.

[23] Boser, B. E., Guyon, I. M., & Vapnik, V. N.," A Training Algorithm for Optimal Margin Classifiers", In *Proceedings of the 5th annual ACM workshop on computational learning* theory, pp. 144–152, 1992.

[24] Bengio, Y., Courville, A., & Vincent, P., "Representation Learning: A Review and New Perspectives", *IEEE Transactions on Pattern Analysis and Machine Intelligence*, Vol. 35, pp. 1798–1828, 2013.

[25] Hinton, G.E.(2009b). *Deep Belief Networks.* <http://www.scholarpedia.org/article/Deep_belief_networks> Accessed 03.09. 14.

[26] Hinton, G. E., Osindero, S., & Teh, Y.-W.,"A Fast Learning Algorithm for Deep Belief Nets", *Neural Computation*, Vol. 18, pp. 1527–1554, 2006.

[27] Kanimozhi U, Ganapathy S, Manjula D, Kannan A, "An Intelligent Risk Prediction System for Breast Cancer Using Fuzzy Temporal Rules", *National Academy Science Letters*, Vol. 42, No. 3, pp. 227–232, 2019.

[28] "*Types of Heart Disease", Heart and Stroke Foundation.* file:///C:/Users/USER/Desktop/Sandhiya_SCOPUS/R_Types%20of%20heart%20disease %20_%20Heart%20and%20Stroke%20Foundation.pd.

[29] Maryam I. Al-Janabi, Mahmoud H. Qutqut, Mohammad Hijjawi, "Machine Learning Classification Techniques for Heart Disease Prediction: A Review", *International Journal of Engineering & Technology*, Vol. 7, No. 4, pp. 5373–5379, 2018.

[30] Simon Fong, Raymond Wong, and Athanasios V. Vasilakos, "*Accelerated PSO Swarm Search Feature Selection for Data Stream Mining Big Data*".

Recent trends in Artificial Intelligence, machine learning and IoT for healthcare

Artificial Intelligence, Blockchain, Computing and Security – Dagur et al. (Eds)

Efficient methods for authentication for Internet-of-Things devices based on E-Health scheme

Animesh Srivastava*
Research Scholar, Banasthali Vidyapith

Anoop Kumar*
Assistant Professor, Banasthali Vidyapith

ABSTRACT: The Internet of Things (IoT) is growing quickly, and wireless technologies are becoming more and more popular. This opens up new growth opportunities in many fields, especially health care. Using the Internet of Things (IoT) in healthcare applications has a number of benefits, including lower costs for hospital visits, healthcare providers, transportation, human resources, and insurance. It has the added benefit of making health care better. When it comes to these issues, the biggest worries for users are the authentication of the different connected entities, energy efficiency, and privacy of the data being shared. To set up secure communication, the first step is to make sure that everyone involved is who they say they are. Authentication is the process of making sure that a system or person is legitimate. The type and number of factors that go into the validation process determine the level of security. Authentication followed by key exchange protocols makes it so that only the entities that have been verified can talk about a secret key without giving any information to a person listening in on the conversation. Because IoT devices have limited resources and can be physically attacked, making an efficient authentication scheme for IoT systems has been a difficult area of research in recent years. For the IoT-based E-health domain, there is a great need to design and build an efficient, secure authentication model that provides a high level of security against attacks like impersonation, man-in-the-middle, and unknown key sharing for this purpose we use ECC and reduce the Message Exchange in the authentication process with the gateway.

1 INTRODUCTION

In the past few years, the healthcare industry has grown quickly and has been a major source of revenue and jobs. A few years ago, doctors could only figure out what was wrong with a person's body by examining them in the hospital. Most of the people who needed treatment had to stay in the hospital the whole time. This caused healthcare costs to go up and put a strain on healthcare facilities in rural and remote areas. Because technology has improved over the years, smartwatches and other small devices can now be used to diagnose diseases and check on people's health. Also, technology has turned a healthcare system that was focused on hospitals into one that is focused on the patient. It can help solve a lot of problems because it has a wide range of uses and can give new information during a COVID-19 type of pandemic and help in the treatment without physical interaction. Technology has completely changed how people are cared for in the modern world. Traditional ways of managing e-Healthcare (electronic healthcare) haven't been able to handle a lot of medical data because the population is growing quickly. Poor communication networks and a lack of smart medical devices can slow down care for patients, which could hurt their health in a

*Corresponding Authors: er.animesh10@gmail.com and anupbhola@gmail.com

DOI: 10.1201/9781032684994-65

big way. To fix this problem, smarter ways need to be found to combine traditional ways of managing healthcare with smart medical equipment and modern communication technologies like 5G [1]. When the Internet of Things (IoT) is combined with 5G networks, it opens up new ways to manage health care. When a patient's smart medical device is connected to the internet, it can collect a lot of sensitive data, give more information about the patient's symptoms, allow for immediate remote care, and give the patient more control over their precious lives and their treatment. Smart sensors also collect important health information, such as the patient's blood pressure, heart rate, blood sugar, and other health indicators. Then, this sensitive information can be sent quickly, like with 5G technology, to the remote e-Healthcare servers so that doctors and nurses can diagnose, keep an eye on, or treat the patient. So, when IoT and 5G work together, they will be able to do amazing things for e-Healthcare management [2–5].

Important Advantages of IoT in Health Care:

- Hospital workers are now more alert and aware.
- Faster processing of patient information
- Better control of drugs and use of medicines
- Less chance of mistakes and bad math because of the human factor
- Continual Health Monitoring is a result.
- Keeping track of your patients is easier with this.
- Makes Insurance Claims Transparent.
- Bringing more health to cities.

2 RELATED WORKS

Authentication protocols are usually based on a lot of different parameters. Where we use passwords, OTP, and many other factors, this increases the additional overhead on the device and the network. The nodes inside the IoT network are free to roam and not permanently affixed to any one location. Due to the free movement of IoT devices, packet loss increases overhead increasing the possibility of network collisions and blocking, both of which can lead to packet loss. This necessitates a method of optimization by self-learning. As a result of this study's analysis of the existing literature, some knowledge gaps have been identified. Security issues in the Internet of Things were the primary focus of our discussion. Authentication is an issue that must be addressed at the perception layer of the Internet of Things. Although the current authentication techniques do their job, they are not up to the upcoming difficulties. Cryptography based on elliptic curves is a relatively new field that has received little attention from academics. This study examines the capabilities of ECC and demonstrates its use in a variety of authentication systems [8–10].

2.1 *Contributions*

E-Health is used to provide health services to patients over the network. E-Health is transmitted openly over an insecure network. We need authentication to secure that data and provide legal access to E-Health services. We proposed an authentication scheme based on using the least cost chaotic operation to keep computation costs to a minimum.

3 PROPOSED SCHEME

In eHealth, IoT-based systems should have key elements of a security framework that can be used to look for weaknesses and threats. It has a strong security system to keep network access safe. The system layer takes care of privacy issues in a proactive way to improve the privacy protection feature. In eHealth, doctors and other healthcare providers keep private patient data locally. Sharing medical information between a patient and a medical professional in a way that protects the patient's confidentiality requires a well-designed system. The privacy-awareness metrics used in the deployment of the system are in line with industry standards, frameworks, regulations, and ethical mandates, allowing for the classification of the nature of potential risks. The privacy framework is built into the IoT applications so that the design works well. It can use technical strategies like identification, authentication, and authorization to make data privacy better. Most healthcare applications use IoT, to form a cluster, classify a device, and get access to different dimensions of data to better analyses the problem. The general worries about safety and privacy are if the system is not properly authenticated and any intruder gets access:

- According to the rules and regulations, patient data should be quickly collected and processed to make sure that the device is safe and working.
- The patient data can't be accessed on either a public or a private network without proper privacy authentication and a high level of security.

The Internet of Things (IoT) framework that is used in healthcare helps to bring the benefits of IoT technology into the field of medicine. It also describes the protocols for how the data from the patient's sensors and medical devices will be sent to a certain healthcare network. The topology of an IoT is how the different parts of an IoT healthcare system or network are set up so that they work well together in a healthcare setting. that a basic IoT system has three main parts: Sensor, Network, and Doctor.

3.1 *ECC algorithm*

ECC: In the field of public-key cryptography, elliptic-curve cryptography is one method that utilizes the algebraic structure of elliptic curves over finite fields. A public essential encryption technique based on the algebraic arrangement of elliptic curves over finite fields can create faster, more minor, or more competent cryptographic keys [11,12].

Elliptic Curve Cryptography has the following benefits:

- The safety of the elliptic curve discrete logarithm comes from the fact that it can only be multiplied in one direction. This gives it a strong trapdoor function. A "trapdoor" function is one that is easy to calculate in one direction but hard to calculate in the other direction (i.e., finding its inverse) without enough information. In cryptography, trapdoor functions are used a lot.
- Same strength: It has the same strength as RSA, but the keys are smaller.
- Safer: It is thought to be the safest way to send information over the Internet.
- Uses fewer CPU resources: Because it uses shorter encryption keys and less memory than other schemes like Diffie-Hellman, it is faster than RSA. This is because it has less overhead than RSA.

3.2 *Security analyses*

The protocol is said to be secure if it is able to resist external active and passive attacks. In this protocol, the secret value v is to be kept safe, providing the system with the necessary security. This value is made up of a bunch of complicated random numbers so that it can't be guessed by brute force, which is not what you want.

In this section, we talk about the formal and informal analyses of the proposed plan. These analyses prove that the proposed protocol is strong and show that it can't be attacked in different ways. We also show that the proposed scheme's security doesn't change in different situations, and we reduce the number of Message exchanges.

1. Impersonation Attack: An impersonation attack is happening if the attacker is able to take control of the device or the gateway while the two are talking. This isn't possible in either stage because the attacker doesn't know the key SK, which can't be shown at any time. So, it's hard for the intruder to get the nonce values from the messages M2 he or she has captured and make the message M3. So, the intruder can't finish his process of authentication.
2. Replay Attack: In a replay attack, the attacker attempts to record the message that travels through the communication channel and then uses that message to prove his identity. This is not possible during the authentication stage of the protocol because the attacker cannot determine the same new nonce value Ng or Nd from the messages each time.

3. Man-in-the-Middle Attack: This kind of attack usually happens when an outsider is able to get the information and tries to change how the sender and receiver talk to each other.

In the registration phase, Elliptic Curve Diffie-Hellman is used, and at the end of the authentication phase, a symmetric key is set up. This key is then used to share information between the devices. We can see that both parties are able to prove their identities. Based on the performance analysis, the proposed method is also lighter than the methods that are already in use because ECC cryptography uses smaller key sizes.

3.3 *Performance analyses*

The wearable sensor is a popular piece of technology that is a key part of e-healthcare systems because it lets patients be watched in real-time. But one of the biggest problems with wearable IoT sensors is that they have limited resources. Because IoT sensors don't have a lot of resources, they are an easy target for security attacks. Patient information is very sensitive, and when it gets into the wrong hands, bad things can happen. Different ways have been found to send sensor data to the Gateway in a secure way, but the fact that IoT sensors have limited resources is still a problem.

4 ECC & COMMUNICATION COST

ECC is a public-key encryption method based on the algebraic arrangement of elliptic curves over finite fields that can make cryptographic keys that are faster, smaller, or better.

4.1 *Communication cost*

We need to communicate between the devices to establish a connection. The cost of communicating a message between two nodes (Like Client and Gateway). We can do this in the following way:

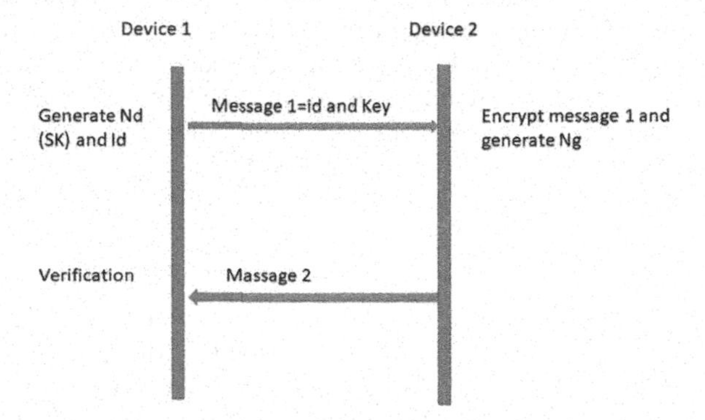

Figure 1. Messages exchanges.

This process uses the Electronic Product Code (EPC). It is designed as a universal identifier (using a unique numeric code for each individual product) that always gives any physical object a unique ID, anywhere in the world.

The process starts with the steps below, which are done in this order:

Step 1: Device 1 starts this phase by figuring out the secret key SK using the registered P-value and its EPC as SK = EPC. Then it makes a random number called a nonce Nd and sends a message to device 2 that says M1 = id; (Nd)SK.

Step 2: After Device 2 gets message M1, it figures out SK by using the formula SK = H (idv). Also, Device 2 decrypts the message M1 it has received by using the random value Nd of the device and the key SK. Device 2 makes its own challenge by coming up with a random number called Ng. Using the secret key SK, the gateway sends a message to the device that is encrypted: M2 = (NdNg)SK. When the message M2 from Device 2 arrives, the device starts decrypting it with the secret key SK to find the nonce values Nd and Ng and to check that both the sent and received Nd are the same. If this works, it proves that Device 2 is really connected to the device.

Table 1. Shows the communication cost.

Protocol	Communication Cost
Proposed	**2 messages**
Zhao's Protocol [6]	5 messages
Chatterjee's Protocol [7]	6 messages

The table shows the communication cost i.e., no of the message transferred for decryption. In the proposed algorithm only two messages have been exchanged. This method requires only 2 message exchanges for completing the authentication process.

5 CONCLUSIONS

This paper shows how IoT can be used in E-health applications to send patient data in a safe authentication process and reduce Communication Costs. IoT sensors are often used in the healthcare field to collect and send data in real-time. But these are vulnerable because they don't have enough resources. This can be fixed by using security methods with less overhead. Our proposed scheme will have an edge over using the elliptic curve cryptography (ECC) approach which will be resilient against several attacks for E-health applications. To deal with these problems, we came up with a plan to reduce the number of data packets being sent to improve security. The proposed protocol not only comes up with a new way to authenticate, but it also turns out to be cheaper than many other e-Health authentication protocols when it comes to the cost of communication. The informal security analyses show that our protocol can withstand the major security attacks that are already known. The formal security analysis and performance analysis also show that our protocol has extra security features. So, our proposed protocol is useful, reliable, and safe. With simple symmetric encryption and XOR functions, this method requires only (2) two message exchanges to complete the authentication process.

6 FUTURE WORKS

Further enhanced in the IoT systems can use the Multi Gateway Authentication Scheme (MGAS). It works well for applications that need a huge number of IoT devices to be connected. For the smart farming system, you will be designing an authentication scheme for an IoT architecture based on a multi-Vi controller.

REFERENCES

[1] Srivastava, A., Kumar, A. (2022). A Review of Network Optimization on the Internet of Things. In: Saini, H.S., Sayal, R., Govardhan, A., Buyya, R. (eds) *Innovations in Computer Science and Engineering. Lecture Notes in Networks and Systems*, vol 385. Springer, Singapore. https://doi.org/10.1007/978-981-16-8987-1_6

[2] A. Srivastava and A. Kumar, "A Back Propagation NN to Optimize the IoT Network," 2022 International Conference on Computer Communication and Informatics (ICCCI), 2022, pp. 1–4, doi: 10.1109/ICCCI54379.2022.9740861.

[3] Animesh Srivastava, Dr. Anoop Kumar, "Enhancement of Authentication in the IoT Network" *Journal of Algebraic Statistics* Volume 13, No. 3, 2022, p. 2328–2336 ISSN: 1309-3452

[4] Minahil, Muhammad Faizan Ayub, Khalid Mahmood, Saru Kumari, Arun Kumar Sangaiah, Lightweight Authentication Protocol for E-health Clouds in IoT-based Applications through 5G Technology, *Digital Communications and Networks*, Volume 7, Issue 2, 2021, Pages 235–244,ISSN 2352-8648, https://doi.org/10.1016/j.dcan.2020.06.003.

[5] A. Srivastava and A. Kumar, "A Review on Authentication Protocol and ECC in IOT," 2021 International Conference on Advance Computing and Innovative Technologies in Engineering (ICACITE), 2021, pp. 312–319, doi: 10.1109/ICACITE51222.2021.9404766.

[6] G. Zhao, X. Si, J. Wang, X. Long, and T. Hu, "A Novel Mutual Authentication Scheme for Internet of Things," in Modelling, Identification and Control (ICMIC), *Proceedings of 2011 International Conference on. IEEE*, 2011, pp. 563–566.

[7] S. Chatterjee, A. K. Das, and J. K. Sing, "A Novel and Efficient User Access Control Scheme for Wireless Body Area Sensor Networks," *Journal of King Saud University-Computer and Information Sciences*, vol. 26, no. 2, pp. 181–201, 2014.

[8] G. Zhao, X. Si, J. Wang, X. Long, and T. Hu, "A Novel Mutual Authentication Scheme for Internet of Things," in *Modelling, Identification and Control (ICMIC), Proceedings of 2011 International Conference on*. IEEE, 2011, pp. 563–566.

[9] Alam, B., & Kumar, A. (2014, March). A Real Time Scheduling Algorithm for Tolerating Single Transient Fault. In *2014 International Conference on Information Systems and Computer Networks (ISCON)* (pp. 11–14). IEEE.

[10] Palvika, Shatakshi, Sharma, Y., Dagur, A., & Chaturvedi, R. (2019). Automated Bug Reporting System with Keyword-driven Framework. In *Soft Computing and Signal Processing: Proceedings of ICSCSP 2018*, Volume 2 (pp. 271–277). Springer Singapore.

[11] Kumar, A., & Alam, B. (2019). Energy Harvesting Earliest Deadline First Scheduling Algorithm for Increasing Lifetime of Real Time Systems. *International Journal of Electrical and Computer Engineering*, 9(1), 539.

[12] Fareed, Mian Muhammand Sadiq, Pradhan, Bikash, Bhattacharyya, Saugat, Pal, Kunal, 2021, 2021/03/19, IoT-Based Applications in Healthcare Devices, 6632599, 2021, https://doi.org/10.1155/2021/663259910.1155/2021/6632599, *Journal of Healthcare Engineering, Hindawi.*

Artificial Intelligence, Blockchain, Computing and Security – Dagur et al. (Eds)
© 2024 The Author(s), ISBN: 978-1-032-67841-2

Building a novel AI framework for teachers and authors to create effective learning contents

Lisha Yugal
Research Scholar, Department of Computer Science & Engineering, RIMT University, Punjab, India

Suresh Kaswan
Professor, Sharda University, Uzbekistan

B.S. Bhatia
Professor, RIMT University, Punjab, India

ABSTRACT: Effective learning Contents plays a vital role in the effective teaching and Learning Process. Therefore, keeping this in view research efforts have been made to design and develop a framework that will assist in development of Learning Content in an effective user-friendly Way. The development of learning content considering cognitive psychology is a challenge being faced by teachers and authors. Considering the Issue An objective designing tool has been developed for arranging course objectives as per cognitive task analysis using the Bloom's Hierarchy. The Tool assists in the overall teaching learning process by considering Knowledge level and age groups of target audience. The tool allows teachers and authors to enter course objective and as per the action verbs the Objective are accepted and arranged as per their Cognition level. Further the Accepted Objectives are prompted one by one and thus allows teachers and authors to develop the Learning Contents.

Keywords: Cognition, Learning Contents, Blooms hierarchy, Domain Model, Teachers, Artificial Intelligence

1 INTRODUCTION

In the era on Education Technology, promoting students to achieve high academic goals and objectives by up-bringing innovative tools and technologies is vital in the overall teaching and learning process. The purpose of development of the framework is to ease the process of effective Content development and assist teachers and authors of different subject domains to develop content by considering the Cognitive Task Analysis. The interface of tool is user friendly that make the content creation quite easy. The advantages pertaining to the development tool is the overall time invested in the effective content development can be reduced; Authors and teachers new in the field of content development or without courseware engineering knowledge can be assisted in much efficient manner. After a case study review it was found that there exist various similar tools with some specialized purpose Basically an authoring tool named LEAP was developed [Sparks *et al.* (2003)] to produce material to guide the telephone operators to manage the Customer issues over phone. Another specific authoring tool DIAG was designed in the area of teaching improvement at certain simulated circumstances [Towne (2003)]. In order to develop simulations to support discovery learning and guided practise RIDES AND SIMQUEST [Joolingen, Jong (2003); Munro (2003)]. On the contrary to the specific tools REDEEM [Ainsworth (2007)] was developed to ease the Process of learning. Based on the review it was found that a framework for Assessment of Cognitive Competency: Introduction to Programming was proposed to assess the students programming skills.

The developed objective designing tool helps in the process of Cognitive task analysis along with the exploration of pedagogical content knowledge. The learning materials can also developed in efficient manner by the authors or teachers who are non-experts in the cognitive psychology.

1.1 *Motivation*

The primary motive behind the research is to create and refine a framework that will uphold the principles of cognitive psychology and courseware engineering in an efficient manner and assist in the effective Content Development Process. The effective Content development through Cognitive task analysis has been elaborated by framework defined by Bloom's taxonomy which has further enhanced the teaching learning Process.

1.2 *Technology & tools used for creation of this framework*

This user friendly interface of the framework created using Front end tools HTML, CSS, Bootstrap and at Backend End used JavaScript, Node, EJS which depicting all the six levels of Cognition defined in the Bloom's taxonomy given by Benjamin Blooms in 1956.

1.3 *Six levels of cognition – Bloom's taxonomy*

Taxonomy of educational objectives formerly known as Bloom's Taxonomy is a framework published for categorizing educational goals by Benjamin Bloom along with collaborators in 1956.

The framework identified by Blooms has majorly six levels of cognition, shown in Figure 1.

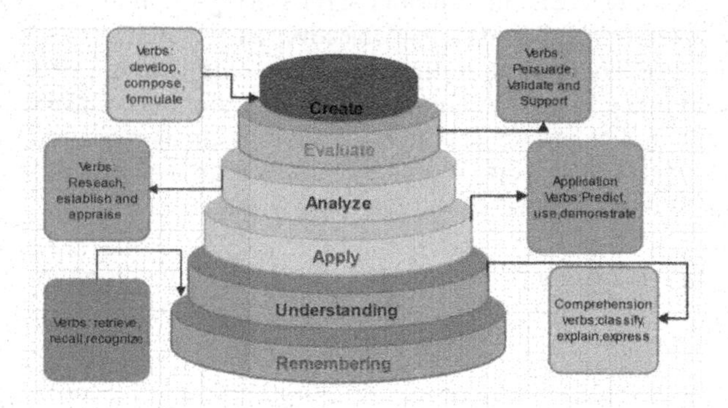

Figure 1. Six-levels of cognition by Benjamin Bloom's.

Definitions	I. Remembering	II. Understanding	III. Applying	IV. Analyzing	V. Evaluating	VI. Creating
Bloom's Definition	Exhibit memory of previously learned material by recalling facts, terms, basic concepts, and answers.	Demonstrate understanding of facts and ideas by organizing, comparing, translating, interpreting, giving descriptions, and stating main ideas.	Solve problems to new situations by applying acquired knowledge, facts, techniques and rules in a different way.	Examine and break information into parts by identifying motives or causes. Make inferences and find evidence to support generalizations.	Present and defend opinions by making judgments about information, validity of ideas, or quality of work based on a set of criteria.	Compile information together in a different way by combining elements in a new pattern or proposing alternative solutions.

The action verbs of each level can be classified as follows:

Remembering

Choose, define, discover, how to label, list, match, name, relate, select, demonstrate, and explain what, when, where, which, who, and why

Understanding
Classify, contrast, explain, expand, illustrate, infer, interpret, rephrase, relate, show, summarise, and translate

Applying
Apply, create, select, create, develop, experiment, interview, use, model, arrange, plan, pick, solve, and utilise

Analysing
Analyse, assume, categorise, classify, compare, conclude, contrast, discover, dissect, distinguish, divide, examiner, function, solve, infer, inspect, list, motivation, relationships, simplify, survey, participate, and test for topic

Evaluating
Agree, apprise, asses, award, choose, compare, conclude, criteria, criticize, decide, deduct, defend, determine, estimate, explain, importance, justify, mark, measure, opinion, value support.

Creating
Adapt, build, change, choose, combine, compile, compose, construct, design, develop, discuss, happen, modify, predict, plan, solve, test, theory, solution, maximize, minimize, formulate.

2 PROPOSED ARCHITECTURAL FRAMEWORK DESIGN WITH SOLUTION

The user friendly interface of the framework allows the teachers and authors to select any one level of cognition from the displayed pyramid like structure depicting all the six levels of Cognition defined in the Bloom's taxonomy given by Benjamin Blooms in 1956. All the six levels of cognition are arranged one on the top of another from higher to lower level. Shown in Figure 2.

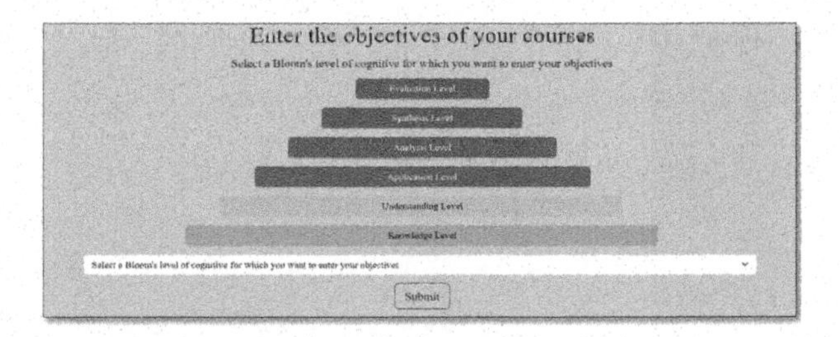

Figure 2. Design framework displaying six-level of cognition.

The objective designing tool maintains the record of the action verbs as per the hierarchy of blooms. After the successful selection of any one level of cognition the tool redirects to the Objective entry page, teachers and authors are required to enter atleast 8 objectives by selecting the action verb from the list of bloom's knowledge verbs displayed on the screen. Shown in Figure 3.

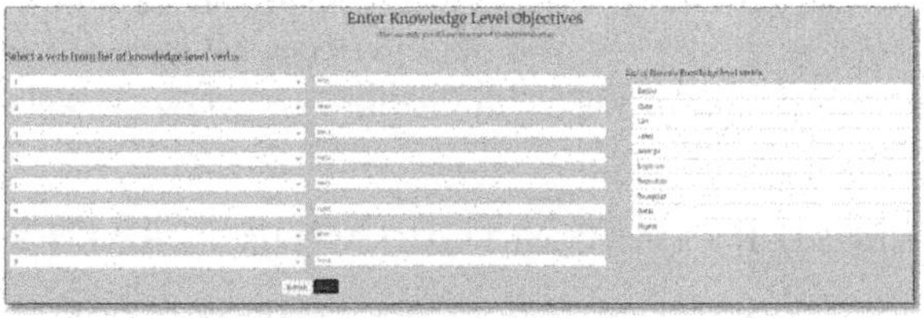

Figure 3. Objective entry page for teachers and authors.

The author is required to select the verbs from the dropdown list as displayed in Figure 4 and subsequently add text in the corresponding text field and complete the sentences. Initially the text field are disabled and as the verb is selected each text field is enabled one by one. The interface will not let user to enter any objective having more than one verb.

After all the 8 objectives are entered successfully as shown in Figure 5, the tools check for any empty values and prompts if the author wants to cover remaining levels of Bloom's taxonomy. After the successful objective entry each of them is examined to search for appropriate action verb and compared with the list of Bloom's knowledge level verbs of the choose level of cognition if a match occurs than the objectives are selected on the contrary error message is displayed.

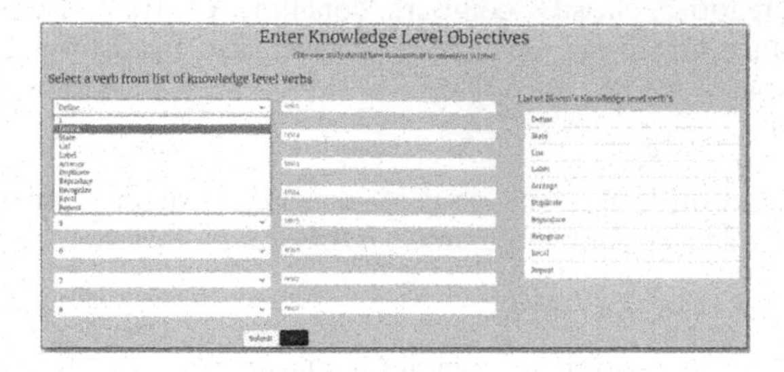

Figure 4. Blooms knowledge actions verb to guide the author/teachers.

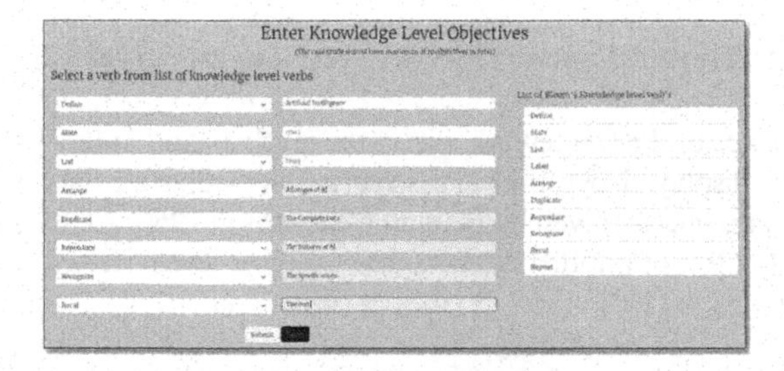

Figure 5. Selected verbs with objectives entered by author/teachers.

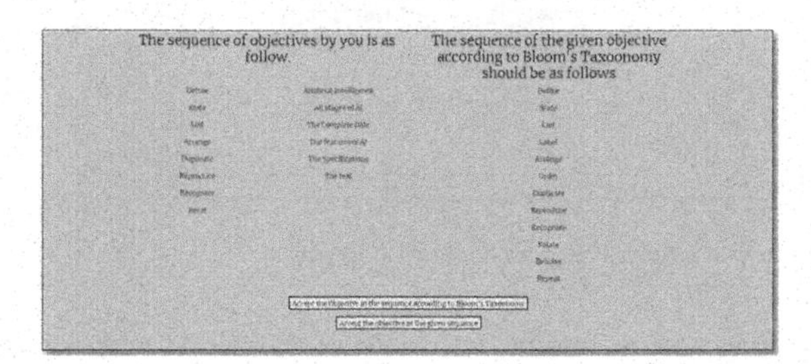

Figure 6. Sequence of objectives displayed.

Consequently, all the objectives are accepted and displayed as shown in Figure 6 Based on the displayed objectives the teachers and authors can create the learning contents respectively. As per framework design solution total of 8 objectives can be entered in the system, the tool considers the knowledge level, age groups and qualification of the target audience.

3 BENEFITS OFFERED BY PROPOSED ARCHITECTURAL FRAMEWORK

- Aligns the course objectives with different level of Cognition
- Each level of Cognition given by Bloom's is Considered

- The tools offers appropriate easy and User Friendly Interface.
- Allows teachers and authors in the Effective learning Content development
- The teachers and authors, not so proficient in courseware engineering can also develop the effective course content considering the Cognitive task Analysis.
- Provides Well designed development Environment
- Assists in exploring the Pedagogical Content smoothly.

4 CONCLUSION

The framework allows teachers and authors with no prior knowledge of cognitive psychology or courseware development to build the learning material easily based on the objectives displayed following the bloom's level of cognition. In the present work a sample has been built to consider all the six levels of cognition and display the objectives by verifying and matching the list of verbs in the blooms knowledge level and allowing teachers to develop learning contents accordingly. For the future scope of the research any technique of Natural Language Processing can be applied for matching the each objective with the suitable level of cognition.

REFERENCES

Ainsworth S. (2007). Using a Single Authoring Environment across the Lifespan of Learning. *Educational Technology & Society*, 10 (3); 22–31.

Aleven V, Sewall, J, McLaren, B. M., & Koedinger K. R. (2006). Rapid Authoring of Intelligent Tutors for Real World and Experimental Use. In: Kinshuk R. Koper P, Kommers P, Kirschner D. G, Sampson W, Didderen (Eds.). *Proceedings of the 6th IEEE International Conference on Advanced Learning Technologies*, Los Alamitos, CA: IEEE Computer Society; 847–851.

Anderson, L. W., & Krathwohl, D. R. (2001). *A Taxonomy for Learning, Teaching, and Assessing, Abridged Edition.* Boston, MA: Allyn and Bacon.

Armstrong, P. (2010). Bloom's Taxonomy. *Vanderbilt University Centre for Teaching*. Retrieved [05/01/2023] from https://cft.vanderbilt.edu/guides-sub-pages/blooms-taxonomy/.

Becky A, Andrew S McGough, and Marie D. (2021). Toward a Framework for Teaching Artificial Intelligence to a Higher Education Audience. *ACM Trans. Comput. Educ.* 22, 2, Article 15 (Dec 2022), 29 pages.

Joolingen W. R. van, Jong T.de. (2003.) SIMQUEST. Authoring Educational Simulations. In T. Murray, S. Blessing, & S. E. Ainsworth, *Tools for Advanced Technology Learning Environments.* Amsterdam: Kluwer Academic Publishers; 1–31.

Lajis A, Nasir HM, Aziz NA. Proposed Assessment Framework Based on Bloom Taxonomy Cognitive Competency: Introduction to programming. In *Proceedings of the 2018 7th International Conference on Software and Computer Applications* 2018 Feb 8 (pp. 97–101).

Maheshwari, S., Kumar, S., Gill, R., Rathore, V.S., Analysis of Augmented Course Delivery and Assessment of Undergraduate Computer Engineering Programming Courses with the Use of ICT, *Advances in Intelligent Systems and Computing*, 2021, 1187, (pp. 481–488).

Maheshwari, S., Kumar, S., Trivedi, N.K., Rathore, V.S., Innovative Classroom Activity with Flipped Teaching for Programming in C Course—A Case Study, *Advances in Intelligent Systems and Computing*, 2021, 1183, (pp. 247–252).

Munro A. (2003). Authoring Simulation-centred Learning Environments with Rides and Vivids. In: T. Murray, S. Blessing, & S. E. Ainsworth, *Tools for Advanced Technology Learning Environments.* Amsterdam: Kluwer Academic Publishers; 61–92.

Sparks R, Dooley S, Meiskey L, & Blumenthal R. (2003). The Leap Authoring Tool: Supporting Complex Courseware Authoring Through Reuse, Rapid Prototyping, and Interactive Visualizations. In: T. Murray, S. Blessing, & S.E. Ainsworth (Eds.), *Tools for Advanced Technology Learning Environments.* Amsterdam: Kluwer Academic Publishers; 411–438.

Towne D. (2003). Automated Knowledge Acquisition for Intelligent Support of Diagnostic Reasoning. In: T. Murray, S. Blessing, & S. E. Ainsworth, *Tools for Advanced Technology Learning Environments.* Amsterdam: Kluwer Academic Publishers; 121–148.

Artificial Intelligence, Blockchain, Computing and Security – Dagur et al. (Eds)

Metaverse: The future of virtual life

Vansh Arora, Vidushi Jaiswal & Sangeetha Annam
Chitkara University Institute of Engineering and Technology Chitkara University, Punjab, India

ABSTRACT: The term "Metaverse" refers to a collaborative virtual 3D world where all actions are eventually carried out utilizing augmented reality (AR) and virtual reality (VR) technology. These platforms have become popular during the pandemic as activities have shifted online. This refers to a post-reality cosmos that combines physical and digital virtuality. Metaverse is a fictional world where thousands and millions of people can act simultaneously and switch between different platforms. This paper summarizes about some of the literature studied, methods and tools used with some of the findings based on the social platforms and the interest shown in this virtual life globally.

Keywords: Metaverse, Pillars, Metaverse Applications, Augmented Reality, Virtual Reality

1 INTRODUCTION

Metaverse is the terminology that arose in the year 1992 from the famous personality, Neal-Stephenson's novel "Snow Crash". Virtual reality headsets are the main devices used to access the Metaverse. Some of the leading technology companies believe that the metaverse has a future and is investing heavily in the technology. However, security and privacy concerns arise when data flow from sensory systems is used with various technologies and advanced algorithms [Kürtünlüoğlu *et al.* (2022)]. Second Life in metaverse has grown in popularity in recent years. It also creates a decision support system for visitors that advises the exhibitions they prefer. And it also focuses on determining which exhibitions a visitor views based on their mobility log. The demand for metaverse, specifically second life (SL), has been exponentially increasing [Ando *et al.* (2013)]. Figure 1 represents the avatars in the metaverse second life.

Figure 1. Avatars in metaverse's second life.

 DOI: 10.1201/9781032684994-67

Metaverse is expected to impact our daily lives in ways that go beyond gaming and entertainment. To experience events that would be impossible or limited in the real world can be realized using the metaverse. These are broadly classified into four types: AR, VR, lifelogging, and mirror world [Kye *et al.* (2021)]. Figure 2 represents the broad pillars of the metaverse.

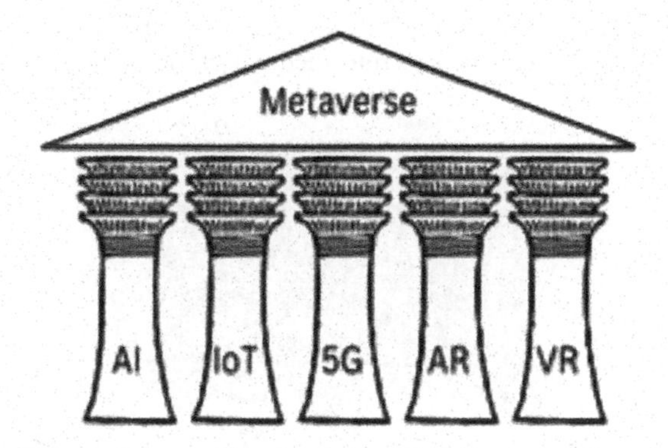

Figure 2. Pillars of the metaverse.

Metaverse's goal is to establish a shared virtual area that links all imaginary environments over the Internet and allows people to communicate and unite as if they were in the physical realm. As a novel trend in metaverse development, a blockchain-based decentralized ecosystem was introduced Second life in Metaverse has many applications. Some are education, medicine, media, fashion, business, and other applications. The metaverse strongly reflects the concept of human-centred computing, which is highly helpful to society in various terms. The metaverse project's goal is to establish a shared virtual area that connects all imaginary environments over the Internet and allows users, presented as virtual avatars, to communicate and collaborate as if they were in the physical world. The open research issues on the Metaverse are based on the factors such as accuracy, material authenticity, reliable broadband, and Integrated Learning. The remaining of this article is designed as mentioned. Section 2 shows the literature work related to the applications of a metaverse in various application areas; Section 3 portrays the different methods and tools used; Section 4 discusses the findings obtained from various sources; and finally, Section 5 presents the conclusion with future work.

2 RELATED WORK

The authors in [Duan *et al.* (2021)], provide a three-layer metaverse architecture based on infrastructure, interaction, and ecosystem. In another work [Martins *et al.* (2022)], recreating Amiais - Second Life. In this, the methods used were ZEPETO and Roblox and open to the opinions and experiences of people and their values. To observe people's socialization, Ethnography is one of the methodologies. Digital Darağaç [Varinlioglu *et al.* (2022)], represents an example of a project situated in its own time and place, by transforming streets in the city of İzmir into an exhibition space. In the 3D modelling process for the re-representation of works of art, the aim was not merely to be bound completely to the original, but also to create a new experience in the digital environment. The world of digital transformation has led to a high degree of management of technology, claiming to optimize teaching processes. Developments will ensure the creation of immersive and multi-sensory

3D environments by providing some interactive capabilities in virtual educational environments designed through the web. The author believes that distance education based on virtual technology and blockchain technology can produce meta-education standards, driven by the Metaverse. The basic model will provide interaction and communication between educators / educates and use various available technical means or tools to provide information [Barráez-Herrera (2022)]. Many new models of meta-education are booming such as online distance education that allows rich, hybrid, and informal learning experiences in online 3D virtual campuses [Mystakidis (2022)]. Figure 3 represents the virtual classroom in a metaverse, and Figure 4 represents a virtual medical clinic in a metaverse second life.

Figure 3. Metaverse classroom.

Figure 4. Metaverse clinic.

Gaming in the metaverse is a virtual game developed by Philip Rosedale with his team members. Human exists as avatars in this game and in a virtual environment [Nalbant *et al.* (2021)]. Almost all organizations require different soft skill abilities. These games are much more interactive and integrate various fields, making players more impulsive. More adventurous innovation can be opened [Shin *et al.* (2022)]. Many Role-playing games (RPGs) exist in the oral practice of linguistics and communication in English [Quintín *et al.* (2016)]. Advertisers should promote not only Second Life, but also encourage this type of game [Laskowska-Witek *et al.* (2014)]. A glimpse of metaverse gaming is shown in Figure 5 and metaverse store in Figure 6.

Figure 5. Metaverse gaming.

Figure 6. Metaverse store.

Some of the literature which focuses on the various application areas are structured in the form of a tick table and shown in Table 1.

Table 1. Metaverse applications in different areas.

Ref No.	Education	Media	Medical	AI	Fashion	Business	Artists	Augmented Reality	Virtual Reality
[Tlili, *et al.* (2022)]	√								
[Lee (2021)]		√							
[Yang *et al.* (2022)]			√	√					
[Joy *et al.* (2022)]					√				
[Seok (2021)]						√			
[Lee *et al.* (2021)]							√		
[MacCallum *et al.* (2019)]								√	
[Kürtünlüoğlu *et al.* (2022)]									√

3 METHODS AND TOOLS USED

Apart from those papers produced to present conceptual structures as well as technologies and study societal, philosophical, and cultural issues, the articles' methodologies were reviewed. Figure 8 presents the methodologies with the help of a tree graph, with an emphasis on education and engineering. Along with 3D, AR, and VR technologies, these studies have also used tools from GNU, OpenSim, Moodle, Minitab-18, MS Excel Pro 365, and many more [Narin (2021)]. Figure 7 represents the Educational and Non-Educational Methods of the Metaverse.

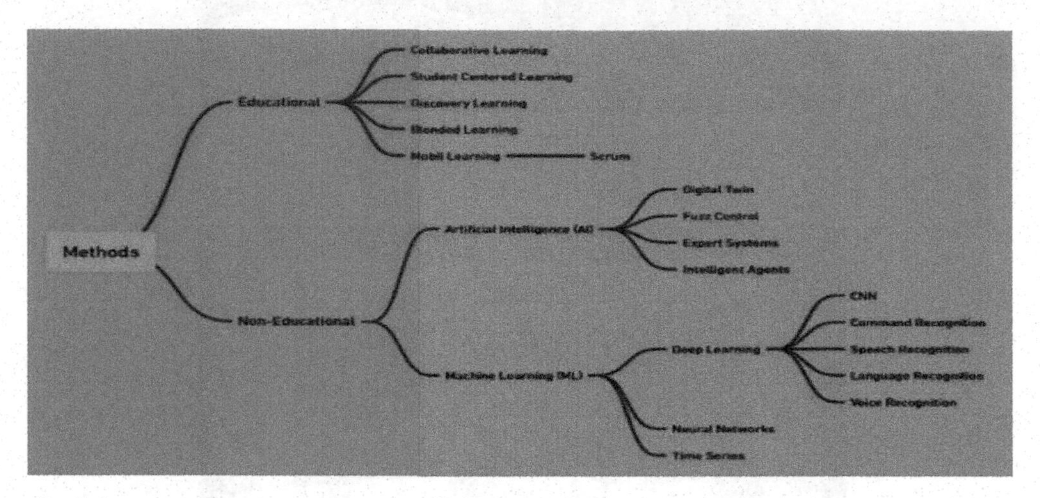

Figure 7. Metaverse methods.

Virtual avatars and digital work communities harness augmented reality tools, immersive technology, and cognitive artificial intelligence algorithms throughout the blockchain-based metaverse. Geolocation data, deep and machine learning algorithms, sensory stimuli, behavioural modelling technologies, and data visualization tools are decisive in determining metaverse engagement metrics across immersive work environments [Chandra (2022)]. "Qiezli" is a fictional agent in Second Life who works as an autonomous "performance artist." They can interact with people outside of the boundaries of a goal-oriented scenario. Because the investigator's content-storage capacity, behavioural process, and cognitive architecture are all completely scalable, this agent will almost certainly evolve to implement highly complicated cognitive capabilities and aesthetic possibilities. Virtual communication tools and teamwork simulations in the metaverse economy integrate behavioural analytics, digital twin technologies, computer vision algorithms, and holographic display devices [Kraus *et al.* (2022)]. Metaverse interoperability can streamline product data management by integrating image recognition tools, cognitive technologies, data computing capabilities, visual analytics, and remote sensing systems across immersive 3D virtual reality environments [Jang *et al.* (2022)]. Immersive haptic experiences, data sharing technologies, voice recognition software, spatial analytics, and 3D modelling tools are pivotal in virtual work settings, typifying metaverse engagement and boosting interactivity and interoperability across metaverse platforms [Almarzouqi *et al.* (2022)]. Artificial intelligence-powered search capabilities, visual immersion technologies, and virtual reality-based data analytics tools are instrumental in employee engagement data and cognitive analytics management, improving connectedness in a metaverse environment [Crowell (2022)].

4 FINDINGS

The graph in Figure 8 shows the number of research papers published on the topic "Metaverse" starting from the year 2000 till 2021, and Figure 9 graph displays on a scale of 100, the level of interest shown in the topic "Metaverse" in more than 200 different countries.

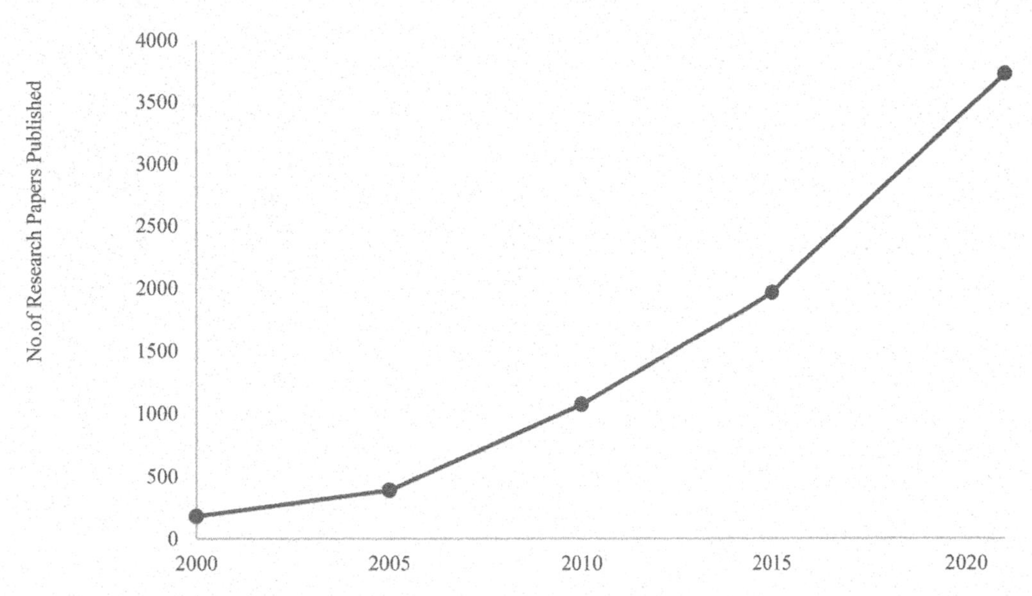

Figure 8. Research publications on metaverse.

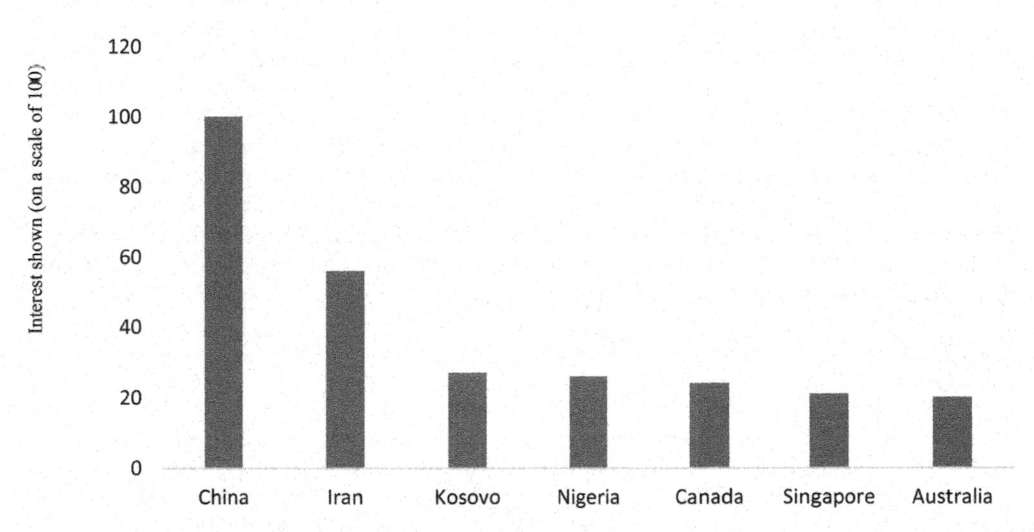

Figure 9. Interest in different countries (on a scale of 100).

The graph in Figure 10 displays the total number of searches made for the term "Metaverse" on Google, YouTube, and News.

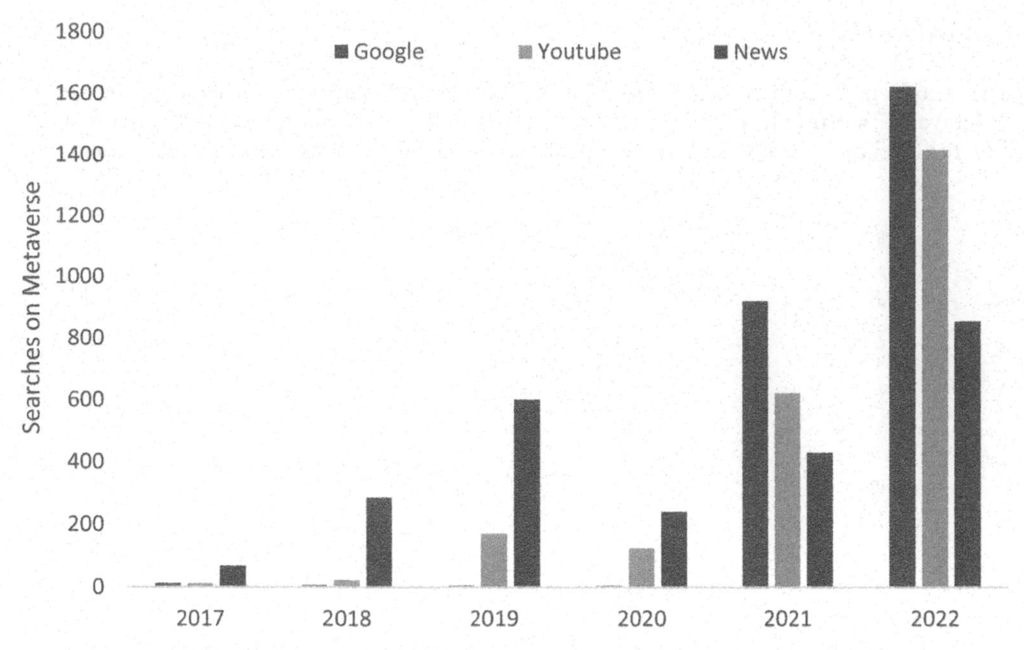

Figure 10. Analysis of searches on metaverse.

5 CONCLUSION AND FUTURE WORK

The model metaverse comprises of topics related to people and society; that provide an appropriate scalable environment to all sectors including sports, health, education, and arts. The scientific side of metaverse studies is covered in this study, which also looks at scholarly research. Most of the research published so far has described the idea of the metaverse. Some have created fictitious studies of the interactions between religion, art, education, and other social factors, and others have been implemented in certain fields. This work concentrated on the methods, tools, related literature, and some findings which highlight Metaverse. Most of the implemented applications are at the prototype stage. However, when adequate technical infrastructure is offered, these investigations significantly contribute and will surely develop in the areas of editing and designing as well. With the existence of the metaverse in the future, there will be many possible challenges, as well as many benefits and constraints that we must consider as this technology progresses.

REFERENCES

Almarzouqi, A., Aburayya, A. and Salloum, S.A., 2022. Prediction of User's Intention to Use Metaverse System in Medical Education: A Hybrid SEM-ML Learning Approach. *IEEE Access*, 10, pp.43421–43434.

Ando, Y., Thawonmas, R. and Rinaldo, F., 2013, September. Inference of Viewed Exhibits in a Metaverse Museum. In *2013 International Conference on Culture and Computing* (pp. 218–219). IEEE.

Barráez-Herrera, D.P., 2022. Metaverse in a Virtual Education Context. *Metaverse*, 3(1), pp.1–9.

Chandra, Y., 2022. Non-fungible Token-enabled Entrepreneurship: A Conceptual Framework. *Journal of Business Venturing Insights*, 18, p.e00323.

Crowell, B., 2022. Blockchain-based Metaverse Platforms: Augmented Analytics Tools, Interconnected Decision-Making Processes, and Computer Vision Algorithms. *Linguistic and Philosophical Investigations*, (21), pp.121–136.

Duan, H., Li, J., Fan, S., Lin, Z., Wu, X. and Cai, W., 2021, October. Metaverse for Social Good: A University Campus Prototype. In *Proceedings of the 29th ACM international conference on multimedia* (pp. 153–161).

Jang, S.H., Lee, G., Lee, S.Y., Kim, S.H., Lee, W., Jung, J.W., Kim, J.P. and Choi, J., 2022. Synthesis and Characterisation of Triphenylmethine Dyes for Colour Conversion Layer of the Virtual and Augmented Reality Display. *Dyes and Pigments*, 204, p.110419.

Joy, A., Zhu, Y., Peña, C. and Brouard, M., 2022. Digital Future of Luxury Brands: Metaverse, Digital Fashion, and Non-fungible Tokens. *Strategic change*, 31(3), pp.337–343.

Kürtünlüoğlu, P., Akdik, B. and Karaarslan, E., 2022. Security of Virtual Reality Authentication Methods in Metaverse: An Overview. *arXiv preprint arXiv*:2209.06447.

Kraus, S., Kanbach, D.K., Krysta, P.M., Steinhoff, M.M. and Tomini, N., 2022. Facebook and the Creation of the Metaverse: Radical Business Model Innovation or Incremental Transformation?. *International Journal of Entrepreneurial Behavior & Research*.

Kye, B., Han, N., Kim, E., Park, Y. and Jo, S., 2021. Educational Applications of Metaverse: Possibilities and Limitations. *Journal of educational evaluation for health professions*, 18.

Laskowska-Witek, J. and Mitręga, M., 2014. Brand Promotion throught Computer Games on the Example of Second Life. *Studia Ekonomiczne*, 205, pp. 34–48.

Lee, Kyoung-A. “A Study on Immersive Media Technology in the Metaverse World.” 한국컴퓨터정보학회 논문지 26, no. 9 (2021): 73–79.

Lee, L.H., Lin, Z., Hu, R., Gong, Z., Kumar, A., Li, T., Li, S. and Hui, P., 2021. When Creators Meet the Metaverse: A Survey on Computational Arts. *arXiv preprint arXiv*:2111.13486.

MacCallum, K. and Parsons, D., 2019, September. Teacher Perspectives on Mobile Augmented Reality: The Potential of Metaverse for Learning. In *World Conference on Mobile and Contextual Learning* (pp. 21–28).

Martins, D., Oliveira, L. and Amaro, A.C., 2022. From Co-design to the Construction of a metaverse for the Promotion of Cultural Heritage and Tourism: the Case of Amiais. *Procedia Computer Science*, 204, pp. 261–266.

Mystakidis, S.M., 2022. *Encyclopedia* 2022, 2, 486–497.

Nalbant, K.G. and UYANIK, Ş., 2021. Computer Vision in the Metaverse. *Journal of Metaverse*, 1(1), pp. 9–12.

Narin, N.G., 2021. A Content Analysis of the Metaverse Articles. *Journal of Metaverse*, 1(1), pp. 17–24.

Quintín, E., Sanz, C. and Zangara, A., 2016, October. The impact of Role-playing Games through Second Life on the Oral Practice of Linguistic and Discursive Sub-competences in English. In *2016 International Conference on Collaboration Technologies and Systems (CTS)* (pp. 148–155). IEEE.

Seok, W.H., 2021. Analysis of Metaverse Business Model and Ecosystem. *Electronics and Telecommunications Trends*, 36(4), pp. 81–91.

Shin, E. and Kim, J.H., 2022. The Metaverse and Video Games: Merging Media to Improve Soft Skills Tlili, A., Huang, R., Shehata, B., Liu, D., Zhao, J., Metwally, A.H.S., Wang, H., Denden, M., Bozkurt, A., Lee, L.H. and Beyoglu, D., 2022. Is Metaverse in Education a Blessing or a Curse: a Combined Content and Bibliometric *Analysis. Smart Learning Environments*, 9(1), pp. 1–31.

Varinlioglu, G., Oguz, K., Turkmen, D., Ercan, I. and Damla, G., 2022. Work of Art in the Age of Metaverse. *Legal Depot D/2022/14982/02*, p.447.

Yang, Y., Siau, K., Xie, W. and Sun, Y., 2022. Smart Health: Intelligent Healthcare Systems in the Metaverse, Artificial Intelligence, and Data Science Era. *Journal of Organizational and End User Computing (JOEUC)*, 34(1), pp.1–14.

Artificial Intelligence, Blockchain, Computing and Security – Dagur et al. (Eds)
© 2024 The Author(s), ISBN: 978-1-032-67841-2

Diabetes detection using random forest classifier and risk score calculation using random forest regressor

Simarjeet Kaur & Damandeep Kaur
Chitkara University Institute of Engineering and Technology, Chitkara University Punjab, India

Mrinal Mayank & Nongmeikapam Thoiba Singh
Department of Computer Science Engineering, Chandigarh University, Mohali, Punjab, India

ABSTRACT: In the past, it was very challenging for doctors and specialists to predict diabetes without the help of any specialized tools. Today with the help of information and communication technology along with futuristic technologies like AI and Machine Learning, it is now possible to make a fast and accurate prediction of diabetes without much effort. The proposed framework is focused on training the machine learning model and integrating the model with the user interface using the Django framework. The framework includes supervised machine learning, Random Forest Algorithm, training of the model & serializing the trained model to integrate with the Django framework. The model is used to predict the binary classification of the disease using the Random Forest Classifier with an accuracy of 80.6% and calculated risk score with an accuracy of 80% using the Random Forest Regressor.

1 INTRODUCTION

The intelligent machine learning models for diabetes diagnosis today are preferred over manual prediction as it is more accurate and faster than manual diagnosis. Manually it is difficult to handle huge amounts of patient data which can be easily handled via emerging technologies like ML (Saluja *et al.* 2022). In the proposed framework random forest algorithm is used for the prediction of the disease (Mujumdar & Vaidehi 2019). It is a widely used machine learning algorithm that falls under the category of supervised ML, which can be used as a regression as well as a classification algorithm (Singla *et al.* 2022). The trained model is serialized as an object with the help of the joblib library. Python is used for training the model as well as for the integration with the Django framework. PIMA diabetes dataset is used for training the model for predicting the occurrence of diabetes. The dataset is divided into eight input measures and one outcome measure. The trained model can be stored as a serialized object using joblib which can be used in the backend of the interface for prediction of the disease (Kaur *et al.* 2022). The accuracy of the algorithm depends on the dataset which is used for training the model. The trained model integration with the web framework is challenging and can be accomplished by following different methods (Kaur & Singh 2019). The research is particularly based on integrating the model at the backend of django framework.

2 LITERATURE REVIEW

In the field of predictive analysis of diseases with machine learning many approaches had already been proposed using machine learning algorithms. (Diwakar *et al.* 2021) proposed a methodology to predict heart disease using ML and Image Fusion where the author classified various types of

 DOI: 10.1201/9781032684994-68

machine learning techniques like Naive Bayes, ANN, KNN, and SVM. Preprocessing of the dataset is the most sensitive step which prepares the dataset to get more precise outcomes. The quality of the dataset plays a key role in forecasting the accuracy of the machine learning model. (M S *et al.* 2021) elaborated analysis & performance of ML on the diagnosis of diseases. The algorithms employed are MLP (a deep learning model), SVM, Decision Tree, KNN, Gradient Boosting, Random Forest Classifier, and Logistic Regression. (Thallam *et al.* 2020) proposed a methodology to predict lung cancer in the early stages with various machine learning algorithms. The research shows that the SVM algorithm is not suitable for large datasets and datasets with more noise, Random Forest Model becomes slow in real-time with a large number of trees, and KNN is only applicable for balanced data. The research helps a lot to consider and evaluate the best model for the prediction of lung disease. (Verma *et al.* 2021) proposed the integration of web applications with machine learning. The work shows the web application integration with the NLP model which is used for text analysis, the major challenge faced is the implementation of NLP through the backend. The core technologies used were NLP, PostgreSQL, Django, and cloud deployment with AWS cloud. React JS, Boot Straps, React Strap, and Javascript were used in the front end to make a responsive web application. (Liu *et al.* 2019) proposed the analysis of Brain morphometric patterns using both classification and regression approaches using deep multi-task multi-channel CNN. The CNN model is used to diagnose Alzheimer's Disease. In the model implementation both using both MRI data & demographic information like gender, age, etc. were used. (Liawatimena *et al.* 2018) proposed a detailed study of metrics measurement of django framework using pylint andradon like LOC, SLOC, Cyclomatic Complexity, etc. (Kunjir *et al.* 2020) proposed the analysis of COVID-19 WHO Dataset, where the research is based on the regression techniques such as CNN, Decision Tree, and LSTM & these techniques are used for the analysis of the COVID-19 dataset. The analysis can be used in providing a detailed and statistical idea of the data and helps reduce the impact of the global epidemic. (Sidiq & Mutahar Aaqib 2019) used various data mining methods for the analysis of thyroid disease where three different approaches had been utilized which include Stacking Ensemble, Vote Ensemble, and Neural Net. In the comparison of the three approaches Stacking Ensemble Classifier returns the highest accuracy rate.

3 PROPOSED METHODOLOGY

3.1 *Training phase*

The first step is to gather the data set. The dataset depends on the type of Machine Learning model used. Here we are using supervised Machine Learning, so the dataset must contain the outcome, and based on that model will make the prediction. The second step is to preprocess the data. The raw data must be cleaned before training, as it may contain null/ missing values which will create ambiguity in the prediction. The third step is data analysis. The processed data is analyzed to observe the hidden correlation in the dataset. The next step is to build the model.

3.2 *Inference phase*

In the inference phase, a trained model is used to predict the outcome with completely isolated data. The model will be serialized using joblib library. The serialized model is then integrated with the interface for the prediction of the disease by a general user.

4 DATA PREPROCESSING

The first step is to load the raw dataset. Now the next step is to determine the data type of the parameters and analyze the missing values. Here data.info() function is used to find the null values and data type.

```
<class 'pandas.core.frame.DataFrame'>
RangeIndex: 768 entries, 0 to 767
Data columns (total 9 columns):
 #   Column                    Non-Null Count  Dtype
---  ------                    --------------  -----
 0   Pregnancies               768 non-null    int64
 1   Glucose                   768 non-null    int64
 2   BloodPressure             768 non-null    int64
 3   SkinThickness             768 non-null    int64
 4   Insulin                   768 non-null    int64
 5   BMI                       768 non-null    float64
 6   DiabetesPedigreeFunction  768 non-null    float64
 7   Age                       768 non-null    int64
 8   Outcome                   768 non-null    int64
dtypes: float64(2), int64(7)
memory usage: 54.1 KB
```

Figure 1. View of the dataset.

In Figure 1 we can see that the dataset does not contain any missing values. But the missing values are present in the form of zeroes as it is impossible to have the parameters Glucose, Blood Pressure, Skin Thickness, Insulin, and Body Mass Index equal to zero.

Now we first need to visualize the number of zeroes (non-acceptable values) in the columns Glucose, Blood Pressure, Skin Thickness, Insulin & BMI. The 0's are replaced with nan and then several nan values are visualized in each column with the help of a bar graph.

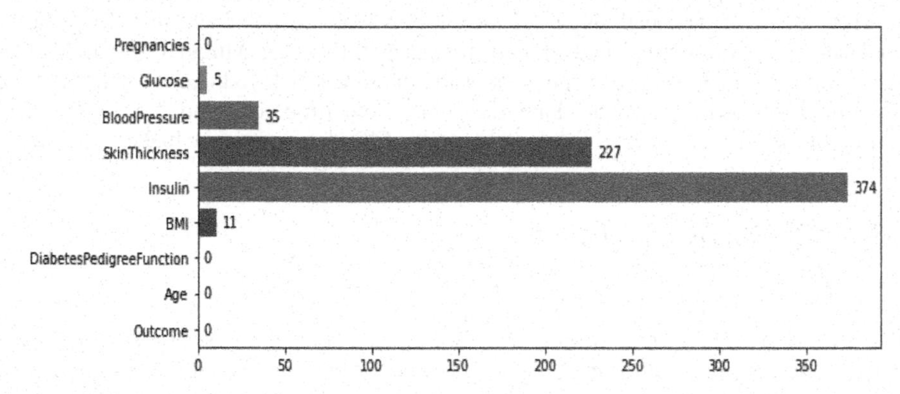

Figure 2. Number of NaN values in each column.

The bar chart in Figure 2 shows the count of NaN values in each column. These values will disturb the prediction accuracy of the model; therefore it needs to be replaced with the mean of all the acceptable values in the column. The data does not need more processing as all the fields are numeric and missing values are also replaced with mean.

5 DATA ANALYSIS

The preprocessing of data is followed by data analysis where the dataset pattern is visualized. Describe function is used to show the statistical measures like mean, count, etc as shown in Figure 3. The outcomes are classified as binary variables, where 0 represents the non-diabetic patients and 1 represents the diabetic patients. The outcome is grouped to the input values using the function groupby() concatenated with the mean() to obtain the mean input values.

	Pregnancies	Glucose	BloodPressure	SkinThickness	Insulin	BMI	DiabetesPedigreeFunction	Age	Outcome
count	768.000000	768.000000	768.000000	768.000000	768.000000	768.000000	768.000000	768.000000	768.000000
mean	3.845052	121.686763	72.405184	29.153420	155.548223	32.457464	0.471876	33.240885	0.348958
std	3.369578	30.435949	12.096346	8.790942	85.021108	6.875151	0.331329	11.760232	0.476951
min	0.000000	44.000000	24.000000	7.000000	14.000000	18.200000	0.078000	21.000000	0.000000
25%	1.000000	99.750000	64.000000	25.000000	121.500000	27.500000	0.243750	24.000000	0.000000
50%	3.000000	117.000000	72.202592	29.153420	155.548223	32.400000	0.372500	29.000000	0.000000
75%	6.000000	140.250000	80.000000	32.000000	155.548223	36.600000	0.626250	41.000000	1.000000
max	17.000000	199.000000	122.000000	99.000000	846.000000	67.100000	2.420000	81.000000	1.000000

Figure 3. Data analysis.

6 BUILDING THE MODEL

The processed data is trained in a sequence of steps from splitting of test train data to model fitting. Accuracy is obtained after training the machine learning model.

6.1 *Test train split*

The dataset is partitioned into two parts and termed test data and train data. Both hold two subsets X and Y where X holds the input and Y holds the output. Below is the code snippet for defining X and Y subsets:

X = data.drop(columns='Outcome', axis=1) Y = data['Outcome']

train_test_split [6] is used for partitioning of test and train data where the size of the test data is taken as 35% and the size of train data is taken as 65%. Below is the code snippet of train_test_split() [6] function:

X_train, X_test, Y_train, Y_test = train_test_split(X, Y, test_size=0.35, stratify=Y, random_state=3)

6.2 *Fitting the model*

The train and test data are now obtained. Now it needs to be fitted to the model for the training. Here two models will be used, one is the random forest classifier and the other is the random forest regressor. The classification algorithm will do the binary classification of the disease i.e., 0 or 1 and the regression algorithm will predict the mean of all the outcomes of the decision tree. The code snippet for fitting the model is given below.

6.3 *Accuracy*

accuracy_score() function is used to find the accuracy of the models. The accuracy is the probability of correctness in the outcome. It depends on the dataset which is used to train the model.

Table 1. Accuracy of the model.

Algorithm	Test Data / Train Data	Accuracy
Random Forest Classifier [6]	Train Data	100%
	Test Data	80.6%
Random Forest Regressor [6]	Train Data	100%
	Test Data	80%

7 RESULT AND DISCUSSION

The objective of the research work is to identify the diabetic patients with machine learning and integrate it with a suitable interface. This research is on supervised machine learning and training the model with random forest algorithm. The accuracy of both the trained models is shown in Table 1 and is calculated using python. This methodology can also be applied to predict some other diseases with supervised Machine Learning.

7.1 *Predictions*

In Table 2 the sample input data is tested and the results are compared with the expected results and the actual results. The 0/1 prediction is made by the Random Forest Classifier Model and the Risk Score is generated by the Random Forest Regressor Model. Random forest algorithm works by generating the N decision trees and combining them to get the outcome, the number of trees is directly proportional to accuracy. Random Forest is a supervised ML algorithm used for both classification & regression The output of the classifier algorithm is based on the majority voting whereas the output of the regressor algorithm is based on averaging or mean.

Table 2. Predictions.

Input Array (8 input fields)	Expected Result (0/1)	Actual Result (0/1)	Risk Score
(1,89,66,23,94,28.1,0.167,21)	0	0	6%
(0,137,40,35,168,43.1,2.288,33)	1	1	86%
(3,78,50,32,88,31,0.248,26)	1	1	74%
1,189,60,23,846,30.1,0.398,59)	1	1	57.9%
(1,103,30,38,83,43.3,0.183,33)	0	0	15%
(13,145,82,19,110,22.2,0.245,57)	0	0	23%
(0,162,76,56,100,53.2,0.759,25)	1	1	68%

7.2 *Comparison with different machine learning models*

Figure 4 shows the comparative analysis of the proposed work with the state of the art machine learning models (Soni & Varma 2020). Through comparative analysis, it has been found that the proposed model achieves an accuracy of 80%.

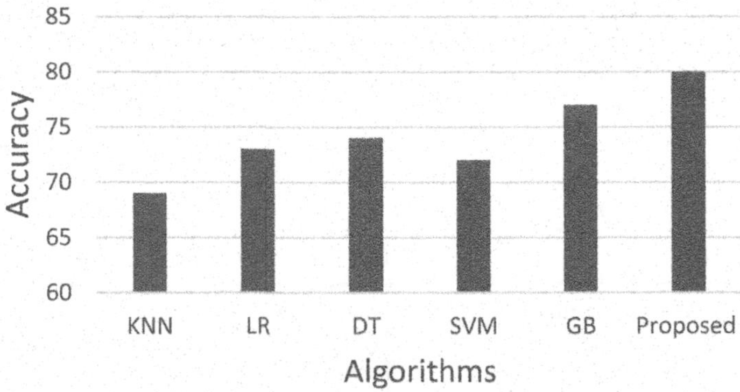

Figure 4. Comparison with different algorithms.

8 CONCLUSION

The applicability of Machine Learning algorithms helps in the fast, accurate, and early prediction of diabetes. Many algorithms are used for prediction depending upon the dataset where the Random Forest Algorithm proves to be more accurate in terms of the correctness of the outcome and its capability to avoid data overfitting. The inference phase includes the implementation of a working model with the application for user usage. In this paper, the integration of the model with the Django framework and object serialization with joblib is discussed. The model is compared with other machine-learning techniques mentioned in Figure 4 and it has been observed that the proposed mechanism outperforms other state of art techniques with 80% accuracy. Future research work may include the enhancement of current work using different deep-learning techniques so that the accuracy and effectiveness of disease diagnosis can be enhanced.

REFERENCES

Diwakar, M., Tripathi, A., Joshi, K., Memoria, M., Singh, P. (2021). Latest Trends on Heart Disease Prediction Using Machine Learning and Image Fusion. *Materials Today: Proceedings*, 37, 3213–3218.

Kaur, D., & Singh, S. (2019). Detection of Brain Tumor using Image Processing Techniques. *International Journal of Engineering and Advanced Technology*, 8(5S3), 501–504.

Kaur, D., Singh, S., Mansoor, W., Kumar, Y., Verma, S., Dash, S., & Koul, A. (2022). Computational Intelligence and Metaheuristic Techniques for Brain Tumor Detection through IoMT-Enabled MRI Devices. *Wireless Communications and Mobile Computing*, 2022, 1–20.

Kunjir, A., Joshi, D., Chadha, R., Wadiwala, T., & Trikha, V. (2020). A Comparative Study of Predictive Machine Learning Algorithms for COVID-19 Trends and Analysis. *2020 IEEE International Conference on Systems, Man, and Cybernetics (SMC)*, 3407–3412.

Liawatimena, S., Hendric Spits Warnars, H. L., Trisetyarso, A., Abdurahman, E., Soewito, B., Wibowo, A., Gaol, F. L(2018). Django Web Framework Software Metrics Measurement Using Radon and Pylint. *Indonesian Association for Pattern Recognition International Conference*, 218–222.

Liu, M., Zhang, J., Adeli, E., & Shen, D. (2019). Joint Classification and Regression via Deep Multi-Task Multi-Channel Learning for Alzheimer's Disease Diagnosis. *IEEE Transactions on Biomedical Engineering*, 66(5), 1195–1206.

M S, S., Joshi, C. S., Thomas, R. R., & G, R. (2021). Analysis and Performance of Machine Learning Algorithms on Disease Diagnosis. *2020 3rd International Conference on Energy, Power and Environment: Towards Clean Energy Technologies*, 1–6.

Mujumdar, A., & Vaidehi, V. (2019). Diabetes Prediction using Machine Learning Algorithms. *Procedia Computer Science*, 165, 292–299.

Saluja, K., Bansal, A., Vajpaye, A., Gupta, S., & Anand, A. (2022). Efficient Bag of Deep Visual Words Based features to classify CRC Images for Colorectal Tumor Diagnosis. *2022 2nd International Conference on Advance Computing and Innovative Technologies in Engineering*, 1814–1818.

Sidiq, U., & Mutahar Aaqib, S. (2019). Disease Diagnosis Through Data Mining Techniques. *2019 International Conference on Intelligent Computing and Control Systems (ICCS)*, 275–280.

Singla, P., Niharika, jain, R., Sharma, R., Kukreja, V., & Bansal, A. (2022). Deep Learning Based Multi-Classification Model for Rice Disease Detection. *2022 10th International Conference on Reliability, Infocom Technologies and Optimization (Trends and Future Directions) (ICRITO)*, 1–5.

Soni, M., & Varma, D. S. (2020). Diabetes Prediction Using Machine Learning Techniques. *International Journal of Engineering Research*, 9(09), 921–925.

Thallam, C., Peruboyina, A., Raju, S. S. T., & Sampath, N. (2020). Early Stage Lung Cancer Prediction Using Various Machine Learning Techniques. *4th International Conference on Electronics, Communication and Aerospace Technology*, 1285–1292.

Verma, A., Kapoor, C., Sharma, A., & Mishra, B. (2021). Web Application Implementation with Machine Learning. *2021 2nd International Conference on Intelligent Engineering and Management (ICIEM)*, 423–428.

Artificial Intelligence, Blockchain, Computing and Security – Dagur et al. (Eds)
© 2024 The Author(s), ISBN: 978-1-032-67841-2

Multi disease prediction using hybrid machine learning

Ashima
Research Scholar, CSE Department, SITE, Swami Vivekanand Subharti University, Meerut, U.P., India

Manoj Kapil
Professor, CSE Department, SITE, Swami Vivekanand Subharti University, Meerut, U.P., India

ABSTRACT: Healthcare 4.0 uses machine learning extensively to deliver prompt and precise results. Doctors can make quick decisions to save patients' lives because to early disease forecasts. Information about our physical and mental health obtained through networked sensors may result in a paradigm shift in the healthcare sector. It allows for monitoring to reach those without easy access to an efficient health monitoring system. After being taken, the data can be processed using a variety of machine learning algorithms and shared wirelessly with healthcare professionals so they can offer the best advice. The work's major objective is to provide a machine learning-based healthcare system to reliably and early diagnose various ailments. Three machine learning classification algorithms—Decision Tree (DT), Random Forest (RF), and K-Nearest Neighbor (K-NN)—are employed in this study to forecast the nine deadly diseases, including heart disease, diabetes, breast cancer, parkinson's disease, and dermatology. Two performance indicators (such as accuracy and sensitivity) are used to assess the performance of the suggested model. For diverse diseases, the Random Forest Bagging Technique classifier records maximum accuracy of 98.46% and sensitivity of 98.76%. The created healthcare model will assist physicians in making an early disease diagnosis.

Keywords: Machine learning, Internet of Things, Healthcare, Artificial Intelligence, Hybrid machine learning

1 INTRODUCTION

A subset of artificial intelligence is machine learning. Since medical data has been converted to digital form, machine learning has become crucial in the detection of numerous diseases, including cancer, Heart Attack, Parkinson's, and many more Diseases. In the past ten years, researchers have looked into a variety of machine learning techniques in the healthcare industry. These methods function as (a) segmentation or abnormality detection and (b) malignant was assigned to the segmental abnormality. Both stages can be depicted as supervised classifiers, with the first providing semantic information by determining whether or not each pixel or feature representation belongs to a dubious disease, and the second allows for additional analysis or measurement of the discovered/segmented abnormalities and a determination of whether they are harmful or not.

There are numerous adaptable devices available that can recognize specific medical conditions from a user's touch, such as heart rate, blood pressure, breath, alcohol level, etc., and share such data through cell phones. When that time comes, IoT will be used to constantly monitor patient health states and supervise treatment from inhospitable locations via interconnected sensors and handheld devices.In this way, it is expected that IoT will play a vital role in the delivery of healthcare services in the future.

Large storage capacities, processing power, and calculation power are features of cloud computing, also offers these features [1]. Cloud computing allows users to access shared resources and fundamental infrastructure in a simple and direct manner, providing services in

 DOI: 10.1201/9781032684994-69

response to system queries and carrying out tasks that are in line with changing needs. A framework for IoT-based healthcare monitoring using "Internet of wellness sensor objects." These activities generate enormous amounts of data that are beyond the scope of a physician.

The doctor's primary issue is that in order to make decisions about a patient's health, he must separate the data pertaining to that victim from the deluge of health information corresponding to the enormous number of sufferer. In light of this, an IoT operator will be utilized to upload health-related information to the cloud, where it will be processed by the cloud and prepared for big data analytics [2]. We have developed a new modelbased on hybrid machine learning algorithms. It works on the data collected through IoT devices and stored on cloud. The proposed algorithms work on these and provide a optimum results as compared to the traditional machine learning algorithms.

The paper is divided into sections: Section 2, which represents linked work; Section 3, which is used for planned work; Section 4, which gives the results and discussion; and Section 5, which serves as a conclusion.

2 RELATED WORK

Medical knowledge is based on the data or experience gathered by medical professionals. The human body is similar to a complex machine with numerous moving parts that is susceptible to many influences. Modeling its dysfunctions or function is therefore a time-consuming undertaking. Several researchers have submitted studies for applications in e-healthcare. For instance, Hameed *et al.* [3] developed an cloud-based e-healthcare service framework in which all patient information is logged in a central database. The suggested framework is based on service-oriented technology (SOA) and offers many services like reducing executive time costs, storing patient profiles, and selecting the best specialist. By utilizing clustering as a data mining technique, Parekh and Saleena[4] have provided a cloud-based model for the healthcare industry. With a few clicks, this technology allowed patients to access the necessary healthcare services using a mobile applicationA model based on medical data was put forth by Hsu *et al.* [5] for the assessment of breast cancer risk. The testing data in this method was pre-processed using techniques like samples and size reduction. Then, different classifiers were employed to predict risk. Devi and Shyla[6] used data mining approaches to make early diabetic illness predictions. To assess accuracy, they employed 768 occurrences taken from the PIMA Indian dataset. Their investigation shows that the J48 classifier is more accurate than other methods. Coronary Artery Disease (CAD) technique employing K-means algorithm and particle swarm intelligence for risk factor detection has been proposed by Verma *et al.* [7] For extracting data events, they used a variety of learning algorithms, including the Multi-Layer Perceptron (MLP),fuzzy unordered rule induction technique, Multinomial Logistic Regression (MLR), and C4.5. The department of cardiology at the Indira Gandhi Medical College in Shimla, India, is where the dataset was gathered. There are 335 cases and 26 features in this data collection. According to the testing findings, MLR had the best accuracy of 88.4%. In order to monitor patients' vital signs on a daily basis and avoid disease, Forkan *et al.* [8] devised the predictive model ViSiBiD. The cloud platform has been studied using machine learning methods and some Map-Reduce implementation. They discovered that six bio-signals varied from the typical and different features for viewing the data occurrences using 4893 patient's dataset, which is publically available. Data events are gathered every 1-2 hours. In comparison to other methodologies, their findings demonstrate that RF has the highest accuracy of 95.85%. Jahangir *et al.* [9] developed an autonomous Multi-Layer Perceptron (AutoMLP) algorithm for the diagnosis of diabetes. This method also employs enhanced class outlier identification. It automatically modifies the parameters while training. During data preprocessing, outlier detection is done. According to the review of the literature, few researchers are currently using meta-heuristic methods to increase the classification accuracy of machine learning approaches. Therefore, to enhance the classification outcomes in this work, the effective Random Forest machine learning algorithm was utilized.

3 PROPOSED SYSTEM

The IoT is a network of networks that is truly enabled by advanced technologies. Through the utilization of wireless technology for low-cost sensors and the assistance of processing and storage devices, the IoT is capable to connect distant and moveable items or machines. Standard machine learning methods, however, have a parameter tweaking problem. As a result, effective adjustment of these parameters has the potential to enhance the effectiveness of current machine learning approaches for a variety of medical applications, including the prediction of cancer, diabetes, brain tumors, etc. It has a longstanding experience to investigate the data to uncover hidden correlations and forecast future patterns. Occasionally known as "learning revelation in databases". Nevertheless, its development includes the intermixing of three disciplines: statistics, AI, and machine learning. With the help of cloud computing and the IoT, the authors want to develop a framework for computerized remote health surveillance. In such cases, which already exist, one can anticipate the benefits of machine learning is to be added to the cloud, which holds the medical services database with the aim of providing constant monitoring of patient status and supervising therapeutic treatment for every customer of the framework. The patients who visit clinics or laboratories that do not have doctors on-site but are outfitted with all essential medical equipment and support staff that serves as an intermediary are included in the third category of service seekers.They upload the patient healthcare data they've gathered via sensor devices to the cloud so that doctors may view it and react. A sensor network is used to initially transport all of the gathered medical data into the mobile devices. The sensor network could be built on USB, Bluetooth, or WiFi technology. Cell phones serve as the IoT operator and are used to upload a patient's health to the cloud. The cloud would manage the increasing volume of health records, creatively share the data among social welfare frameworks, and give logical consistency for information mining, an important and crucial role in performing work. Major machine learning algorithms, such as K-NN, Decision Trees, and Random Forest [10] have been applied thus far in this research on a dataset of diseases, including breast cancer, diabetes, heart diseases, thyroid, parkinson, dermatology, and liver disease. Among other supervised learning algorithms for classification now in use, random forest is unparalleled in accuracy and scales effectively on massive databases. A random subset of the training dataset is used by the random forest classifier to generate a collection of decision trees. The final test object class is then determined by averaging the votes from various decision trees. In the current study, precise data are extracted from a database that is linked to the disease query dataset using a random forest classifier. The suggested system's block diagram is displayed in detail in Figure 1.

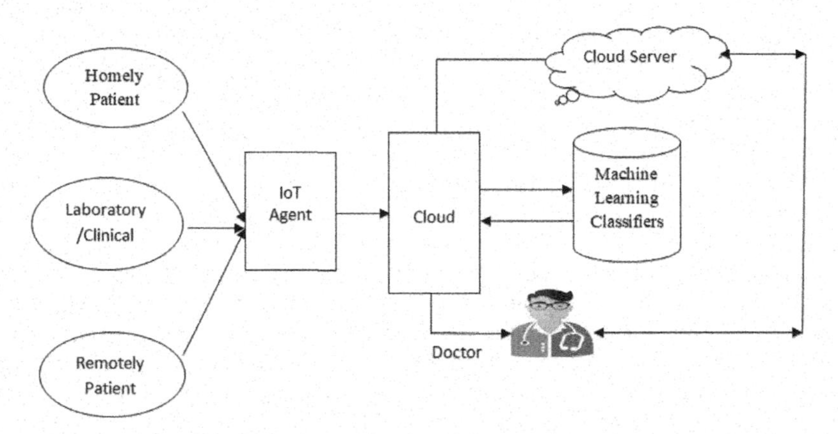

Figure 1. Block schematic for an IoT-based healthcare network.

3.1 *Significance of the proposed system*

To guarantee the quality of service to the healthcare industry, a few data mining techniques are applied, including Decision Tree, k-Nearest Neighbor, and Random Forest. The K-NN technique uses nearest neighbor data points to identify unknown data points and categorize them based on a voting mechanism. This technique is simple to use but takes a lot of storage space, is noise-sensitive, and has a lengthy testing process.

The Decision Tree technique uses a tree-like graph as its foundation, in which three nodes—such as non-leaf node, leaf node, and branch nodes—is utilized as distinct properties for computing contingency [11]. The top component serves as the parental node, while the branch nodes represent the results of the test and the non-leaf nodes serve as class labels. Domain expertise is not necessary for the decision tree technique. It can also handle numerical and categorical data and is simple to interpret. However, the performance is dataset-dependent and limited to a single attribute output. Leo Breiman created Random Forest. The method is used in a variety of contexts, including classification, prediction, variable analysis, and selection. The method includes a number of qualities, including the ability to apply to problems with two or more classes, as well as a mix of categorical and continuous predictors. On the other hand, improved performance necessitates fine-tuning of the input parameters.

3.2 *Bagging*

Bagging is used when the goal is to lower the variance of DT classifiers. A bagging ensemble technique commonly referred to as Bootstrap Aggregation, if randomization is used to enhance performance. This not only helps prevent overfitting but also guarantees accurate representation of higher-dimensional data. It can fix problems with missing data and preserve data integrity. Three ensemble hybrid models are constructed using the Bagging technique [12]. These are composed of DT, RF and KNN in combination. The three hybrid models used for both training and testing are DTBT, RFBT and KNNBT.

Table 1. Dataset employed.

Data Set	Amount of Samples	Class
Breast cancer	699	2
Diabetes	768	2
Heart disease	303	5
Parkinson's	197	2
Dermatology	366	6

3.2.1 *DTBT*

DTBT uses supervised learning and has a straightforward tree structure. The tree leaf that made decisions once did so. Internal and external terminals are present, and they are connected to one another to create the choice by leaf. DT is produced via a top-down methodology.

3.2.2 *RFBT*

The ensemble techniques are applied in RFBT. To produce the best results, this ensemble categorization builds and combines many decision trees. The idea of bagging or bootstrap aggregation is mostly used for the learning of the tree.

3.2.3 *KNNBT*

KNNBT expands upon the supervised learning approach. K-NN employed the Euclidian distance approach to determine the distance between the new data and the previous data in order to calculate the similarities between the classes.

3.3 *Evaluation of the performance of the classifier*

The three classifiers' performance is assessed using three measures. The following lists the specifics of the measures for performance evaluation.

Accuracy: The classifier's total performance is measured as

$$Accuracy = \left(\frac{TP + TN}{TP + FP + TN + FN}\right) * 100 \quad (1)$$

Sensitivity: It is the ratio of the number of genuine positive cases to overall number of people who have the disease. Precision is another name for sensitivity. The sensitivity is assessed to be

$$Sensitivity/Precision = \left(\frac{TP}{TP + FN}\right) * 100 \quad (2)$$

4 RESULTS AND DISCUSSIONS

This segment presents the outcomes of the health service utilizing a number of machine learning techniques, including Decision Trees, K-NN and Random Forest. As indicated in Table 1, authors used a variety of publicly available datasets for various conditions, including Heart Disease, Breast Cancer, Diabetes, Parkinson's and Dermatology. These datasets are downloaded from UCI repository. MATLAB R2018b software has been used by the authors to evaluate the performance of several machine learning methods. A disease-by-disease comparative examination of the accuracy and sensitivity of several machine learning algorithms depict in Tables 2 and 3 respectively. Figures 2 & 3 likewise graphically representation of these results.

4.1 *Accuracy of three disease-specific classifiers*

Table 2 displays the findings of three classifiers for accuracy. The RF classifier attains the highest accuracy for breast cancer 95.24%, RFBT's accuracy increased by 97.56%.The RF classifier works well for diabetics with an accuracy of 82.16%, The RFBT classifier provides the highest value for bagging at 84.18%.In heart disease, RF has the best accuracy 96.42%. The value of RFBT was raised by 98.44% by bagging.For the Parkinson's dataset, the RF achieves the highest accuracy of 96.59%, RFBT at 97.33%.We tested the model on dermatology datasets, and the results showed that the RF classifier had the highest accuracy (96.52%), while the bagging depicted the value in terms of RFBT was 98.46%. The three classification methods with various disease classifications are compared graphically in Figure 2.

Table 2. Comparison between accuracy/accuracy using bagging technique.

	Accuracy			Accuracy using Bagging Technique		
Data Set	KNN	DT	RF	KNNBT	DTBT	RFBT
Breast cancer	92.63%	90.25%	95.24%	94.13%	92.89%	97.56%
Diabetes	73.57%	76.90%	82.16%	75.90%	78.88%	84.18%
Heart disease	86.22%	91.26%	96.42%	88.14%	93.10%	98.44%
Parkinson's	88.72%	91.82%	96.59%	90.52%	92.45%	97.33%
Dermatology	92.52%	88.10%	96.52%	94.40%	90.16%	98.46%

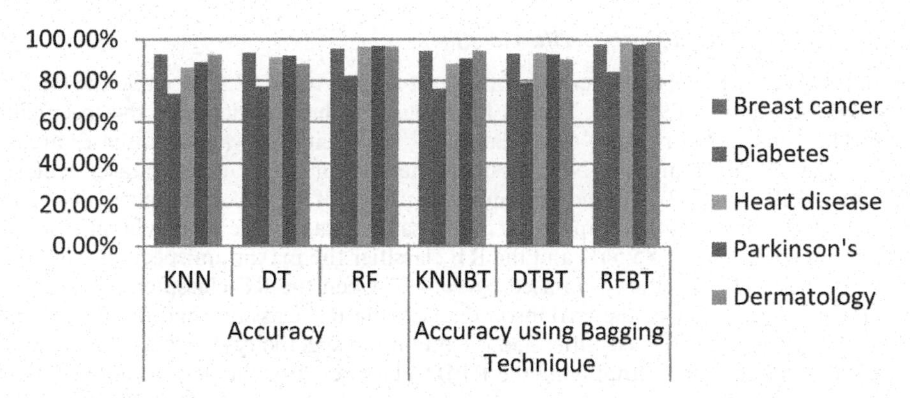

Figure 2. Comparative study of accuracy.

4.2 *Sensitivity of three disease-specific classifiers*

Table 3 displays the results of the three classifiers' sensitivities for each disease. For breast cancer dataset, the RF classifier has the highest sensitivity (97.23%), When RF classifier is replaced with RFBT, sensitivity rises from 97.23% to 98.76%.The RF achieves the highest sensitivity for diabetic illnesses (83.90%), The diabetic disease database when subjected to RFBT yields a sensitivity of 85.17%.The RF classifiers reach sensitivity for heart disease of 97.72%. The sensitivity increased by 98.33% in the case of RFBT.The RF classifier offers a maximum sensitivity of 95.89% for the Parkinson disease dataset. The same dataset's sensitivity is increased by 97.90% after the RFBT is applied.In dermatology, the RF offers the maximum sensitivity of 91.53%. The RFBT bagging technique increased the sensitivity by 93.65%. The constructed model's three classifiers' sensitivity for various diseases are compared in Figure 3.

Table 3. Comparison between sensitivity/sensitivity using bagging technique.

	Sensitivity			Sensitivity using Bagging Technique		
Data Set	KNN	DT	RF	KNNBT	DTBT	RFBT
Breast cancer	90.34%	92.89%	97.23%	91.40%	93.69%	98.76%
Diabetes	74.32%	76.80%	83.90%	76.45%	77.38%	85.17%
Heart disease	87.40%	93.44%	97.72%	88.20%	94.89%	98.33%
Parkinson's	87.26%	90.42%	95.89%	88.20%	92.79%	97.90%
Dermatology	93.44%	89.15%	91.53%	95.85%	91.25%	93.65%

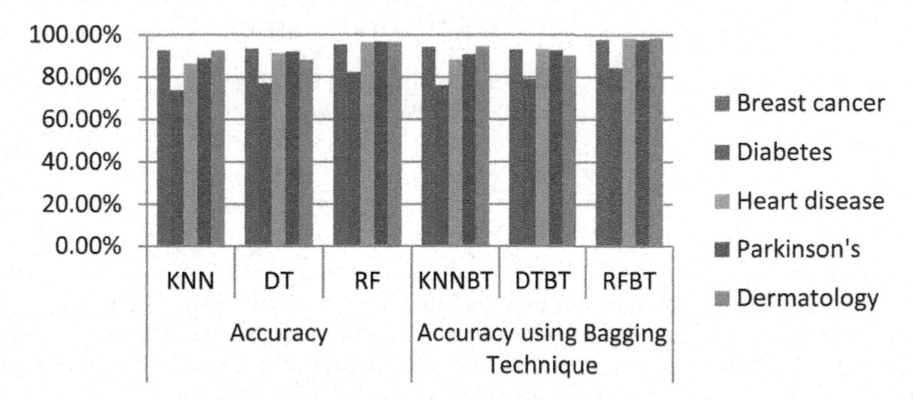

Figure 3. Comparative study of sensitivity.

4.3 *Specificity of three disease-specific classifiers*

In dermatology, the KNN classifier offers the least sensitivity of 93.44% while the RF offers the maximal sensitivity of 91.53%. The RFBT bagging technique increased the specificity by 93.65%. The constructed model's three classifiers' specificity for various diseases are compared in Figure 3. By using a K-NN classifier, the RF for the diabetes dataset achieved a maximum specificity of 83.50% and minimum specificity of 72.42%. The specificity is raised by 84.54% after using the RFBT approach on the same dataset. The KNN classifier achieves the minimum specificity of 85.54% and the RF classifier the maximum specificity of 95.23% for heart disease. The specificity is raised by 96.43% when the RF classifier is swapped out for the RFBT classifier. For the Parkinson dataset, the RF classifier had the highest specificity (97.42%) and the KNN classifier had the minimum specificity (89.28%). The specificity is increased to 98.55% after employing the RFBT classifier. For the dermatological dataset, the KNN classifier obtains the minimal specificity of 90.20% while the RF classifier achieves the maximal specificity of 97.89%. When it comes to bagging methods, the RFBT is increased to 98.36%. The created model's seven classifiers' relative graphical views of specificity for various diseases are shown in Figure 4.

Table 4. Comparison between specificity/specificity using bagging technique.

	Specificity			Specificity using Bagging Technique		
Data Set	KNN	DT	RF	KNNBT	DTBT	RFBT
Breast cancer	89.27%	91.32%	96.81%	91.33%	91.32%	97.40%
Diabetes	72.42%	75.90%	83.50%	73.20%	76.52%	84.54%
Heart disease	85.54%	90.65%	95.23%	87.12%	91.15%	96.43%
Parkinson's	89.28%	92.33%	97.42%	90.16%	93.69%	98.55%
Dermatology	90.20%	89.11%	97.89%	92.27%	91.31%	98.36%

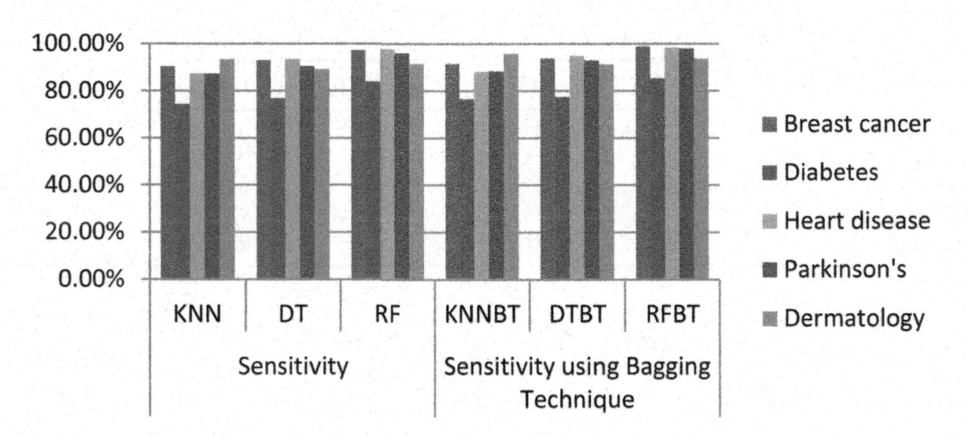

Figure 4. Comparative study of specificity.

5 CONCLUSION

A system for health care based on IoT and random forest classifier is proposed in this paper's study. The suggested system will increase communication between patients and physicians. For testing the efficacy of the proposed work, experimental findings are conducted utilising several datasets relevant to various diseases, including thyroid, breast cancer, diabetes, heart disease, parkinson and dermatology. Decision Trees, K-NN, and Random Forest are the machine learning methods used in this work. On the Dermatology dataset, the Random Forest bagging machine learning algorithm has a maximum accuracy of 98.46%. For each of the datasets under consideration, it has been determined that, on average, the random forest bagging technique produces good, precise results.The proposed system may presumably be trained with different machines while taking into account a larger data set to boost accuracy. This study can therefore be expanded in the near future to additional applications like earth monitoring, weather forecasting, etc. Because IoT devices employ a variety of existing technologies to acquire meaningful data and transport that information to other IoT machines or systems, data security is the primary concern in an IoT-based framework.

REFERENCES

[1] Wu, J., Ping, L., Ge, X., Wang, Y., & Fu, J. (2010, June). Cloud Storage as the Infrastructure of Cloud Computing. In *2010 International Conference on Intelligent Computing and Cognitive Informatics* (pp. 380–383). IEEE. DOI: 10.1109/ICICCI.2010.119

[2] Chakraborty, C., & Kishor, A. (2022). Real-Time Cloud-Based Patient-Centric Monitoring Using Computational Health Systems. *IEEE Transactions on Computational Social Systems*. DOI: 10.1109/TCSS.2022.3170375

[3] Hameed, R. T., Mohamad, O. A., Hamid, O. T., & Tapus, N. (2015, November). Design of E-Healthcare Management System Based on Cloud and Service Oriented Architecture. In *2015 E-Health and Bioengineering Conference (EHB)* (pp. 1–4). IEEE. DOI: 10.1109/EHB.2015.7391393

[4] Parekh, M., & Saleena, B. (2015). Designing a Cloud Based Framework for Healthcare System and Applying Clustering Techniques for Region Wise Diagnosis. *Procedia Computer Science*, 50, 537–542. https://doi.org/10.1016/j.procs.2015.04.029

[5] Hsu, J. L., Hung, P. C., Lin, H. Y., & Hsieh, C. H. (2015). Applying Under-sampling Techniques and Cost-sensitive Learning Methods on Risk Assessment of Breast Cancer. *Journal of Medical Systems*, 39 (4), 1–13. https://doi.org/10.1007/s10916-015-0210-x

[6] Devi, M. R., & Shyla, J. M. (2016). Analysis of Various Data Mining Techniques to Predict Diabetes Mellitus. *International Journal of Applied Engineering Research*, 11(1), 727–730.

[7] Verma, L., Srivastava, S., & Negi, P. C. (2016). A Hybrid Data Mining Model to Predict Coronary Artery Disease Cases Using Non-invasive Clinical Data. *Journal of Medical Systems*, 40(7), 1–7. https://doi.org/10.1007/s10916-016-0536-z

[8] Forkan, A. R. M., Khalil, I., & Atiquzzaman, M. (2017). ViSiBiD: A Learning Model for Early Discovery and Real-time Prediction of Severe Clinical Events Using Vital Signs as Big Data. *Computer Networks*, 113, 244–257. https://doi.org/10.1016/j.comnet.2016.12.019

[9] Jahangir, Maham, Hammad Afzal, Mehreen Ahmed, Khawar Khurshid, and Raheel Nawaz. "An Expert System for Diabetes Prediction Using Auto Tuned Multi-layer Perceptron." In *2017 Intelligent Systems Conference (IntelliSys)*, pp. 722–728. IEEE, 2017. DOI: 10.1109/IntelliSys.2017.8324209

[10] Kishor, A., & Chakraborty, C. (2022). Artificial Intelligence and Internet of Things Based Healthcare 4.0 Monitoring System. *Wireless Personal Communications*, 127(2), 1615–1631. https://doi.org/10.1007/s11277-021-08708-5

[11] Kishor, A., & Jeberson, W. (2021). Diagnosis of Heart Disease Using Internet of Things and Machine Learning Algorithms. In *Proceedings of Second International Conference on Computing, Communications, and Cyber-security* (pp. 691–702). Springer, Singapore. https://doi.org/10.1007/978-981-16-0733-2_49

[12] Chakraborty, C., Kishor, A., & Rodrigues, J. J. (2022). Novel Enhanced-Grey Wolf Optimization Hybrid Machine Learning Technique for Biomedical Data Computation. *Computers and Electrical Engineering*, 99, 107778. https://doi.org/10.1016/j.compeleceng.2022.107778

Artificial Intelligence, Blockchain, Computing and Security – Dagur et al. (Eds)

A review of object–based detection using convolutional neural networks

Shradha Zilpe*
CSE, Assistant Professor, Koneru Lakshmaiah Education Foundation, Vijayawada, India

Sudeshna Sani*
CSE, Assistant Professor, Koneru Lakshmaiah Education Foundation, Vijayawada, India

Dipra Mitra*
CSE, Assistant Professor, Amity University, Ranchi, India

Abhinandan Ghosh*
Assistant Professor, CSE, Adamas University Kolkata, India

Pranav Kumar*
CSE, Assistant Professor, Adamas University Kolkata, India

Amarnath Singh*
CSE, Assistant Professor, Amity University, Ranchi, India

ABSTRACT: Theoretical, research, and practical PC vision applications continue to be substantially impacted by object detection. Combining the two distinct tasks of picture classification and object location is a challenging task. AI was largely used to obtain common object distinguishing proof calculations. This illustrates how features are organised to reflect the characteristics of an item before being mixed with classifiers. Convolutional neural networks (CNN), in particular, have received a lot of attention recently due to its usage in deep learning (DL), which has recently spurred incredible advancement and promising results. The purpose of this paper is to organise an examination of some of the most significant and recent contributions to the field of research on the use of deep learning in a real item Additionally, as said, it is suggested by the disclosures of several evaluations that the application of deep learning in object revelation much outperforms conventional philosophies that centre on carefully gathered and taught information.

Keywords: Deep Learning, Computer Vision, Convolutional Neural Network, Image, Segmentation and Classification, Object Localization and Detection.

1 INTRODUCTION

To obtain flat out image understanding, we should focus in on depicting various pictures similarly as attempt to definitively assess the considerations and spaces of articles contained in each image. This errand is suggested as thing affirmation, which typically incorporates diverse sub tasks, for example, face region, individual by strolling diligence and skeleton unmistakable

*Corresponding Authors: shraddhazilpe28@gmail.com, sudeshnasani@gmail.com, Mitra.Dipra@gmail.com, abhinandan1.ghosh@adamasuniversity.ac.in, pranav1.kumar@adamasuniversity.ac.in and Amarnathsingh.2k@gmail.com

 DOI: 10.1201/9781032684994-70

evidence. In same way central PC vision issues, item or device recognizing confirmation can give basic data to semantic view of images and accounts and is connected with different implementations, including picture strategy, human lead appraisal, face attestation and free driving [1]. Meanwhile, getting from neural affiliations and related studying structures, the movement in these area will energize neural affiliation assessments, and will additionally amazingly impact object conspicuous confirmation frameworks which can be considered as learning frameworks. In any case, because of huge arrangements in perspectives, positions, obstacles and lighting constraints, it's hard to flawlessly achieve object diligence with something extra requirement task. Such a lot of idea has been drawn to this field of late [2].

1.1 *Object detection from videos and audios*

This segment audits the most famous deep learning structures that are utilized in scholastics and the business, additionally the interface utilized by the deep learning structures. Article identification calculations ordinarily influence AI or profound figuring out how to deliver significant outcomes [3]. At the point when people check out pictures or videos, we can perceive and find objects of interest inside a question of minutes.

1.2 *Deep learning framework*

Deep learning structures offer structure blocks for planning, preparing and approving deep neural organizations, through a significant level programming interface. Generally utilized profound learning structures, for example, MXNet, PyTorch, TensorFlow and others depend on GPU-sped up libraries, for example, cuDNN, NCCL and DALI to convey elite multi-GPU sped up preparation. Pytorch is the improved rendition of the light, which is created under Facebook [4]. It chips away at the Python stage, has negligible structure overhead and is quicker as its Neural Network backends are created as independent libraries with a C99 API. The memory is likewise redone and made productive subsequently more profound, and more huge organization models are feasible to execute Deep learning structures offer structure blocks for planning, preparing and approving profound neural organizations, through a general programming interface.

2 CONVOLUTIONAL NEURAL NETWOREK AND ITS VARIENTS

CNN's are astounding picture planning, man made cognizance (AI) that use signifi-cant sorting out some way to perform both generative and particular tasks, habitually using machine vision that consolidates picture and video affirmation, close by recommender structures and customary language dealing with. However, convolutional neural organizations currently give a more adaptable way to deal with picture order and item acknowledgment errands, utilizing standards from straight variable-based math, explicitly framework increase, to recognize designs inside a picture [5,6]. All things considered, they can be computationally requesting, requiring graphical preparing units (GPUs) to prepare models. PC vision is a field of manmade awareness (AI) that enables PCs and structures to get huge information from modernized pictures, accounts and other visual data sources, and due to those data sources, it can take action. This ability to give ideas remembers it from picture affirmation tasks. CNN reliably contains two fundamental exercises, explicitly convolution and pooling. Max pooling and ordinary pooling are the most notable pooling exercises used in CNN. In light of the complicity of CNN, relu is the ordinary choice for the inception ability to move tendency in getting ready by back propagation [7]. The convolutional neural network (CNN) can comprise at least one of than convolutional layers. The convolutional neural organization engineering is planned so that it exploits a 2D picture, which is accomplished utilizing nearby associations and pooling layers to get the invariant elements. By and large, there are two distinct methods of convolution, one is with the totally related convolutional layers, and the other is secretly related layers. Normally, a specific picture has a similar measurement in all areas of the picture. This implies

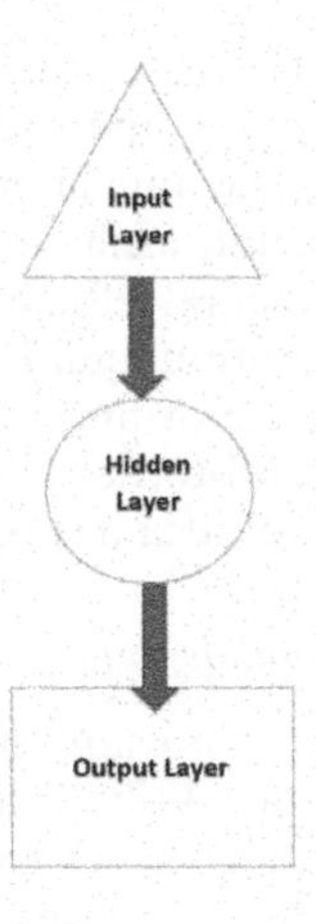

Figure 1. Neural network with pooling layer.

that if the elements of one locale are known or determined the equivalent can be applied to the remainder of the districts in a similar picture. It resembles parsing the convolutional (highlights) window on the picture to get the convolved components of a picture.

3 RECURRENT NEURAL NETWORK

A repeated neural organisation (RNN), which employs progressive data or timeseries data, is a type of fictitious neural association. There is a crucial necessity for the concluding words because, in typical brain connections, all information and yield sources are freed from one another, but in situations when it is expected that a sentence will be followed by another, for example. This led to the appearance of RNN, which with the aid of a Hidden Layer resolved this problem. The Hidden state, which recalls certain information about a gathering, is the fundamental and most important part of RNN [8].

3.1 *Fast recurrent neural network*

The quick intermittent neural organization (quick RCNN) is a further developed rendition of RCNN and Spatial Pyramid Pooling organization. Quick RCNN enjoys some upper hands over the customary RCNN and Spatial Pyramid Pooling organization. It gives the more excellent location as far as to Figure 1. Block Diagram of Neural Network mean outright accuracy (mAP) than the other two. One more benefit of Fast RCNN over RCNN is that it doesn't need the memory stockpiling to store the components which are extremely helpful for object recognition as it speeds up. R-CNN utilized CNN for each produced area proposition for additional location cycles. However, Fast R-CNN takes awhole picture and a bunch of item recommendations together as infor-mation. From the created CNN high-light map, Region of Interests (RoI) is recognized utilizing specific hunt techniques [9,10]. Then, at that point, the creators have utilized an RoI pooling layer to reshape the ROIs into a proper length highlight vector. From that point onward, FC layers accept those element vectors as info and passed the yield to two kin yield branches. One branch is intended for order and one more for jumping box relapse. Every one of the three models encountered that the calculation of locale recommendations is tedious which influences the general exhibition of the organization. Ren *et al.* proposed Faster R-CNN, where they have supplanted recently referenced district proposition strategy with area proposition organization (RPN) [11]. An RPN is a completely convolutional network (FCN) that takes a picture of subjective size as information and yields a bunch of rectangular upand-comer object propositions.

A significant anchor box is chosen by applying limit esteem over the "objectness" score. Chosen anchor boxes and the component maps registered by the underlying CNN model together are taken care of to RoI pooling layer for reshaping and the yield of the RoI pooling layer took care of into FC layers for conclusive order and jumping box relapse [11].

3.2 *Traditional methods of object-based detection and computer vision*

The customary item location system is by and large isolated into four unique stages. The competitor areas are produced on the given picture with the assistance of a sliding window. The important elements are separated from these locales and characterized utilizing the prepared classifier by distinguishing the districts; at long last, reconsidering and enhancing the outcomes [12]. 1) Generating up-and-comer areas and looking at mathematical properties – The significant burdens are: the huge intricacy of time and the number of excess windows which seriously sway the speed and execution of component extraction. Likewise, the sliding window isn't sufficiently deliberate [13].

Feature extraction and coordinating – Features like SIFT (Scale Invariant Feature Transform), HOG (Histogram of Gradients) and LBP (Local Binary Patterns) are utilized. 3) Classification – The SVM or AdaBoost classifiers are generally used to order the removed elements. 4) Revise Detection – After order, there still will be numerous excess windows which are important to eliminate, subsequently advancing the discovery results by Non-Maximum Suppression (NMS) and joining covered Bounding box [14].

3.3 *Convolutional neural network – the architecture and advantages*

As a significant part of neural organization, CNN expands the idea of the responsive field and shared loads, which extraordinarily decreases the boundaries of preparing, yet additionally diminishes the intricacy of the organization model. Convolutional Neural Network (CNN) is a class of profound, feed-forward fake neural organization that has been used to deliver a precise presentation in PC vision undertakings, for example, picture grouping and recognition. CNN's resemble customary neural organization, yet with more profound layers. It has loads, inclinations and yields through a nonlinear enactment. The provisions of each layer are created from the past layer's neighborhood (field) by sharing the heaviness of the convolutional part. These qualities make the CNN more appropriate for the learning and address of picture highlights than other neural organizations, and it can likewise keep the interpretation and scale invariance partially [15]. The neurons of the CNN are masterminded in a volumetric manner, for example, stature, width and profundity. In an average CNN, the initial not many layers are normally rotating layers of convolution and pooling, and the last layers of the organization close to the yield layer are typically full-associated networks. The convolutional layer and pooling layer are routinely subbed and the significance of each channel increases from left to right while the yield size (height and width) is decreasing. The totally related layer is the last stage which resembles the last layer of the customary neural affiliations [16,17]. The preparation of CNN for the most part utilizes the forward proliferation and BP (Back Propagation) calculation to get fa-miliar with the layer-association loads, predisposition and different boundaries. The preparation is a regulated learning measure that requires the picture information as info and the relating marks to advance the organization boundaries, at last, it will get an upgraded weight model. CNN is made out of various useful layer structures. Run-of-the-mill CNN has a convolutional layer, pooling layer and completely associated layer [18]. Nonetheless, CNN adds many new layers during the time spent development and improvement, for example SPP-layer which existed in the SPP-net, the ROI (Region of Interest)-pooling layer of Fast R-CNN, and the Region Proposal Network (RPN) layer of Faster R-CNN. As indicated by the particular issues, working on the customary CNN construction can accomplish a better presentation [19].

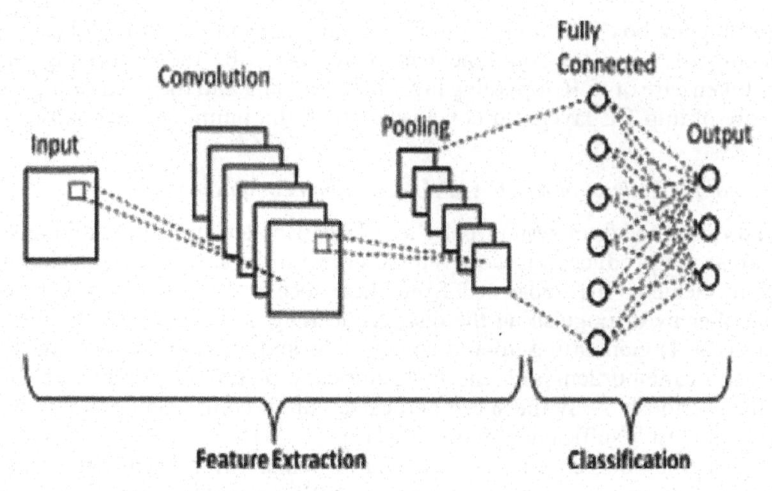

Figure 2. Neural network with pooling layer.

3.4 *Fundamental structure of CNN*

The contribution to the convolutional neural organization is generally the first picture. The pooling layer, likewise called the down-inspecting layer, for the most part, follows the convolutional layer and down testing the past highlight map as per the proper guideline. The particular standards are: maxpooling, normal pooling, stochastic pooling, covered pooling and so on The pooling layer work principally has two angles: 1) Reducing the dimensionality of the feature map 2) Keep scale invariance In a completely associated network, the component guides of pictures are linked into a one-dimensional element vector as a contribution to a completely associated network. The yield of the completely associated layer can be gotten by making a weighted summation of the information and reacted by the enactment work [20,21].

Object detection model	Pascal VOC 2007	PASCAL VOC 2010	PASCAL VOC 2012	COCO 2015	COCO 2016	Real time detection	Number of stages.
R-CNN	58.5%	53.7%	53.3%			No	two
SPP-net	59.2%					No	Two
Fast R-CNN	70.0%	68.8%	68.4%			No	Two
Faster R-CNN	73.2%		70.4%			No	Two
FPN					35.8%	No	Two
Mask R-CNN					43.4%	No	Two
YOLO	63.4%		57.9%			Yes	One
SSD	81.6%		80.0%	46.5%		No	One
YOLO9000	78.6%		73.4%	21.6%		Yes	One
RetinaNet					40.8%	Yes	One
RefineDet	85.6%		86.8%		41.8%	Yes	One

Figure 3. Neural network with pooling layer.

4 BACK PROPAGATION

The BP (back engendering) calculation is utilized to change the weight boundaries of neural organization. For CNN, the principle enhancement boundaries are convolu-tion portion boundaries, pooling-layer loads, full-associated layer loads and inclination boundaries [22,23]. The pith of BP is to process the halfway subordinate of the residuals for each layer boundary and get familiar with an affiliation rule between the residuals and the organization loads, then, at that point, change the heaviness of the organization to make the organization yield nearer to the given anticipated worth. The preparation objective of the CNN is to limit the misfortune capacity of the organization.

4.1 *Datasets*

Profound neural organizations need a huge measure of marked information for prepar-ing a model. Dataset is one of the establishments of profound learning. These days the most well-known utilized datasets of article recognition are ImageNet, PASCAL VOC and MS COCO.

4.2 *ImageNet*

The ImageNet dataset has in excess of 14 million pictures covering north of 20,000 classes. There are in excess of 1,000,000 pictures with express class comments and comments of article areas in the picture. The ImageNet dataset is one of the most broadly utilized datasets in the field of huge learning. The vast majority of the eval-uation work, for example, picture social affair, locale, and affirmation depend upon this dataset. The ImageNet dataset is quick and dirty and is remarkably simple to utilize. It is widely utilized in the field of PC vision research and has changed into the "standard" dataset of the power critical learning of picture space to test calculation execution.

4.3 *PASCAL VOC*

PASCAL VOC (design examination, measurable demonstrating and computational learning visual item class) gives a standard picture naming and assessment frame-work for object class acknowledgment. PASCAL VOC picture dataset incorporates 20 classes; the dataset has a top caliber and is named picture what is entirely appropri-ate for inspecting the calculation execution. It is all around stamped, and empowers assessment and correlation of various techniques; and because the measure of infor-mation of the PASCAL VOC dataset is little, contrasted with the ImageNet dataset, it is entirely reasonable for specialists to test network programs.

4.4 *COCO*

COCO (Common Objects in Context) is new picture acknowledgment, division, and inscribing dataset, supported by Microsoft. COCO dataset has in excess of 300,000 pictures covering 80 thing classes. The open wellspring of this dataset makes exceptional progress in semantic division recently, and it has transformed into a "standard" dataset for the display of picture semantic course of action, and COCO has its test.

4.5 *Framework and pipeline*

The issue importance of thing acknowledgment is to sort out where articles are arranged in a given image (object impediment) and which class every thing has a spot with (object portrayal). Consequently, the pipeline of ordinary article acknowledgment models can be mainly detached into three stages: 1) Useful region determination – As different articles would appear in any spot of the image and have particular point extents or sizes; it is a trademark choice to analyze the whole image with a multi-scale sliding window. Yet this intensive framework can find all possible spots of the articles, its lacks are also undeniable. Due to a huge number of contender windows, it is computationally expensive and makes such an enormous number of monotonous windows. Nevertheless, if by some fortunate turn of events a nice number of sliding window designs are applied, irrelevant regions may be conveyed. 2) Feature extraction – To see different articles, we really want to eliminate visual parts which can give a semantic and fiery depiction. Components are the delegate ones. This is a direct result that these components can convey depictions identified with composite cells in the human frontal cortex. Regardless, in view of the assortment of emergence, edification constraints and establishments, it's difficult to genuinely plan an amazing component descriptor to faultlessly represent a wide scope of things. 3) Classification – Also, a classifier is relied upon to see a genuine article from the wide extent of various portrayals and to make the representations more moderate, semantic and edifying for vision attestation. Due to its astounding learning limit and advantages in overseeing obstacles, scale change and establishment switches, significant learning-based thing ID has been an assessment space of interest. In this paper, a thorough review is advanced on a portion of the significant turns of events and victories shown by the utilization of profound learning procedures in object identification. to exhibit the usefulness, various assessments and concentrates on that have as of late been executed and completed in the space are carefully assessed and separated. To this end, it has been especially shown that Convolutional Neural Networks, Deep Neural Networks and similarly as Region-based convolutional neural associations have needlessly been used as the norm for some energetic ID structures, also, have gotten - in numerous preliminaries, contemporary execution on various datasets. It can likewise be deduced that profound learning has ended up being compelling in object discovery, yet to check and affirm the plausibility of utilizing profound learning strategies for object identification further investigations must be performed on bigger datasets containing various classifications. Also, because of the escalated calculation engaged with preparing and testing of pro-found learning models, extra analyses must be done on different stages to make the most helpful processing stage with profound learning strategies [25].

5 CONCLUSION

Significant learning-based object ID has drawn interest in assessments because of its extraordinary learning capacity and advantages in managing barriers, scale changes, and establishment shifts. Deep neural networks and similarly regionbased convolutional neural associations have been unnecessarily adopted as the standard for some active ID structures and have also been involved in numerous preliminary and ongoing executions on diverse datasets. Furthermore, additional studies must be carried out at various phases to determine the most advantageous processing stage for these techniques due to the increased calculations required in the development and testing of profound learning models.

REFERENCES

[1] Gajjar V., Gurnani A. and Khandhediya Y., "Human Detection and Tracking for Video Surveillance: A Cognitive Science Approach," in *2017 IEEE International Conference on Computer Vision Workshops*, 2017.

[2] Adel M., Moussaoui A., Rasigni M., Bourennane S. and Hamami L., "Statistical- Based Tracking Technique for Linear Structures Detection: Application to Vessel Segmentation in Medical Images," *IEEE Signal Processing Letters*, vol. 17, no. 6, pp. 555–558, June 2010.

[3] Truong X.-T., Yoong V. N. and Ngo T.-D., "RGB-D and Laser Data Fusion- based Human Detection and Tracking for Socially Aware Robot Navigation Framework," in *IEEE Conference on Robotics and Biomimetics*, Zhuhai, China, 2015.

[4] Parekh H. S., Thakore D. G. and Jaliya U. K., "A Survey on Object Detection and Tracking Methods," *International Journal of Innovative Research in Computer and Communication Engineering*, vol. 2, no. 2,

[5] Chen H., Wang Y., Wang G. and Qiao Y., "LSTD: A Low-Shot Transfer Detector for Object Detection," *arXiv*:1803.01529v1, 2018.

[6] Xu H., Lv X., Wang X., Ren Z. and Chellappa R., "Deep Regionlets for Object Detection," *arXiv*:1712.02408v1, December 2017" Deep Learning to Frame Objects for Visual Target Tracking," *Engineering Applications of Artificial Intelligence*, vol. 65, pp. 406–420, October 2017.

[7] Krizhevsky A., Sutskever I. and Hinton G. E., "ImageNet Classification with Deep Convolutional Neural Networks," in *Advances in Neural Information Processing Systems*, 2012

[8] Deng J., Dong W., Socher R., Li L.-J., Li K. and Fei-Fei L., "ImageNet: A Large-Scale Hierarchical Image Database," in *2009 IEEE Conference on Computer Vision and Pattern Recognition*, Miami, FL, USA, 2009.

[9] Sani, S., Bera, A., Mitra, D., Das, K. M. (2022). COVID-19 Detection Using Chest X-Ray Images Based on Deep Learning. *International Journal of Software Science and Computational Intelligence (IJSSCI)*, 14(1), 1–12.

[10] Mitra D. and Gupta S., "Plant Disease Identification and its Solution Uusing Machine Learning," *2022 3rd International Conference on Intelligent Engineering and Management (ICIEM)*, London, United Kingdom, 2022, pp. 152–157, doi: 10.1109/ICIEM54221.2022.9853136.

[11] Mitra D., Gupta S. and Kaur P., "An Algorithmic Approach to Machine Learning Techniques for Fraud detection: A Comparative Analysis," *2021 International Conference on Intelligent Technology, System and Service for Internet of Everything (ITSS-IoE)*, Sana'a, Yemen, 2021, pp. 1–4, doi: 10.1109/ITSS-IoE53029.2021.9615349.

[12] Mitra D., Gupta S. and Kaur P., "An Algorithmic Approach to Machine Learning Techniques for Fraud detection: A Comparative Analysis," *2021 International Conference on Intelligent Technology, System and Service for Internet of Everything (ITSS-IoE)*, Sana'a, Yemen, 2021, pp. 1–4, doi: 10.1109/ITSS-IoE53029.2021.9615349.

[13] Mitra D., Gupta S. and Goyal A., "Security System using Open CV based Facial Recognition," *2022 2nd International Conference on Advance Computing and Innova-tive Technologies in Engineering (ICACITE)*, Greater Noida, India, 2022, pp. 371–374, doi: 10.1109/ICACITE53722.2022.9823881.

[14] Mitra, D., Arora, M., Rakhra, M., Kumar, C. R., Reddy, M. L., Reddy, S. P. K., . . . Shabaz, M. (2021). A Hybrid Framework to Control Software Architecture Erosion for Addressing Maintenance Issues. *Annals of the Romanian Society for Cell Biology*, 2974–2989.

[15] Mitra, D., Sarkar, S., Hati, D. (2016). A Comparative Study of Routing Protocols. *Engineering and Science*, 2(1), 46–50.

[16] Kaur M., Khan M. Z., Gupta S. and Alsaeedi A., "Adoption of Blockchain With 5G Networks for Industrial IoT: Recent Advances, Challenges, and Potential Solutions," in *IEEE Access*, vol. 10, pp. 981–997, 2022, doi: 10.1109/ACCESS.2021.3138754.

[17] Gupta, S., Saini, A. K. (2021). An Artificial Intelligence Based Approach for Managing Risk of IT Systems in Adopting Cloud. *International Journal of Information Technology*, 13(6), 2515–2523.

[18] Gupta, S., Saini, A. K. (2013). Information System Security and Risk Management: Issues and Impact on Organizations. *Global Journal of Enterprise Information System*, 5(1), 31–35.

[19] Thakur K., Lal K. and Kumar V., "Ensemble Method to Predict Impact of Student Intelligent Quotient and Academic Achievement on Placement," *2021 2nd International Conference on Intelligent Engineering and Management (ICIEM)*, London, United Kingdom, 2021, pp. 249–253, doi: 10.1109/ICIEM51511.2021.9445323

[20] Thakur K. and Lal K., "A Novel Approach for Human Intelligence Analysis: Using Clustering Technique," *2021 2nd International Conference on Intelligent Engineering and Management (ICIEM)*, London, United Kingdom, 2021, pp. 139–143, doi: 10.1109/ICIEM51511.2021.9445275.

[21] Kaur, M., Gupta, S. (2021). Blockchain Technology for Convergence: An Overview, Applications, and Challenges. *Blockchain and AI Technology in the Industrial Internet of Things*, 1–17.

[22] Kaur M. and Gupta S., "Blockchain Consensus Protocols: State-of-the-art and Future Directions," *2021 International Conference on Technological Advancements and Innovations (ICTAI)*, 2021, pp. 446–453, doi: 10.1109/ICTAI53825.2021.9673260.

[23] Mitra, D., Goswami, S., Hati, D., Roy, S. (2020). Comparative Study of IoT Protocols. *PalArch's Journal of Archaeology of Egypt/Egyptology*, 17(7), 12527–12537.

[24] Mitra, D., Gupta, S. (2021, December). Data Security in IoT Using Trust Management Technique. In *2021 2nd International Conference on Computational Methods in Science Technology (ICCMST)* (pp. 14–19). IEEE.

[25] Dash, A.B., Mishra, B., Singh, A.N. (2021). Identification of Premature Diagnosis for Detection of Brain Tumor Using Blockchain Strategy. In: Kumar, R., Mishra, B.K., Pattnaik, P.K. (eds) *Next Generation of Internet of Things. Lecture Notes in Networks and Systems*, vol 201. Springer, Singapore.

Artificial Intelligence, Blockchain, Computing and Security – Dagur et al. (Eds)
 ISBN: 978-1-032-67841-2

Health care model prediction using AdaBoost classifier algorithm

N. Subhashini, S.A. Archanaa, A. Balaji & K. Dinesh
SRM Valliammai Engineering College, Tamil Nadu, India

ABSTRACT: Due to the spread of the new coronavirus, the importance of medical care is increasing in each country. The most effective approach to combat this pandemic is through an IoT based system for health monitoring. The Internet of Things (IoT) is indeed the latest Internet revolution and a rapidly expanding field of study, particularly in the field of healthcare. A drastic growth in technological progress in the medical field paves the way for people's health care. The proposed system is based on Wi-Fi enabled sensor modules to measure the patient's temperature and pulse rate and build a wireless body area network (WBAN). Machine learning classification algorithms can be utilized to examine and predict abnormal patient situations using the data acquired by the sensors and made available in the cloud. In addition to statistical analysis of available data, the Panic Button will send SMS notifications to your mobile phone, allowing immediate attention to your patient.

1 INTRODUCTION

After the outbreak of coronavirus, there was a drastic increase in online medical consultations and treatment. The Internet of Things enables the interconnectivity of embedded devices and makes data access possible anywhere at any time. The availability of the data in the cloud server from the edge devices provides ease of data access to all authorized people. In the proposed system, Wi-Fi enabled sensor module is used to measure the temperature, heart rate, and pulse rate of the patient framing a wireless body area network (WBAN). Application of the internet of things in the medical field results in the Internet of Medical Things (IoMT) which focus towards the patient monitoring, diagnosing the patient based on the data available in the cloud, telemedicine, intensive care and remote monitoring of the patient (Dwivedi *et al.* 2022). The main component of the IoMT are sensors which are the basic element that collects the data from the monitoring environment. Some of the sensors used in the health care system are temperature sensors, heart rate monitoring sensors, oximeters, pressure sensors and many more. All the data collected from the various sensors, embedded on the human body are available in the cloud which can be monitored by the caretaker, nurse and doctor. In case of abnormality panic button can be pressed and a SMS alert is sent to the doctor, thus patient can be attended and treatment can be provided immediately. The available data in the cloud can be tested after training using the machine learning classification algorithm models, from which the condition of the patient can be predicted. Based on the deep learning algorithm(Ahmed *et al.* 2021) applied on the data collected from the patient health monitoring sensors the discomfort experienced by the patient was predicted.

2 RELATED WORK

The entire healthcare industry is on high alert as a result of the sudden onset of the COVID-19 respiratory virus disease. The Internet of Things (IoT) is essential for creating contact between the patient and the doctor in the realm of healthcare (Saranya *et al.* 2018). The

 DOI: 10.1201/9781032684994-71

Internet of Medical Things (IoMT) has greatly eased the situation. A smart medical system. (Khan *et al.* 2022) described how IoT and communication networks can be used to monitor human health using various sensors to provide affordable medical assistance. IoMT and lateral technologies provides a solution to some of the issues associated with health monitoring using remote monitoring, telemedicine, robotics, sensors, etc. Many challenges faced by the people due to ageing and the impact of the IoT addressed (Sahu *et al.* 2021) and solutions discussed to the challenges.

Extensive research has been carried out in the area of health services and their technological improvement over the past decade. (Pradhan *et al.* 2021) reviewed the needs and applications of Healthcare Internet of Things (HIoT). Networks known as wireless body area networks (WBANs) have grown significantly in recent years, allows sensor devices to be implanted near the body or adhered to the skin without causing discomfort or danger to the user (Tavera *et al.* 2021).Integration of IoT with machine learning serves the human community to the greater extent. (Alazzam *et al.* 2021) analyzed the impact of human motion on the diastolic and systolic blood pressure values and predicted the same using the machine learning algorithm. (Malik *et al.* 2021) discussed the impart of machine learning in the diagnosis of various diseases and the algorithms involved in predicting the diseases from the data available. (Moulaei *et al.* 2022) compared the various machine learning algorithms using the clinical factors available to study the mortality rate. An intelligent machine learning approach is to incorporate the non-invasive method to monitor the health condition of the patients and to collect the data, this approach was discussed by (Saeed *et al.* 2022). Artificial Intelligence (AI) used in the wide range of applications in the medical field such as clinical assistance, diagnosis of a disease, surgical assistance and prediction of disease from the healthcare factors measured. (Secinaro *et al.* 2021) focused on the AI and healthcare practices and its implications in the research field. (Lalmuanawma *et al.* 2020) reviewed the significant changeover in the field of medicine in the screening, treatment and medication to the patients after the pandemic. (Jabbar *et al.* 2018) discussed about the emerging startups using machine learning and AI which ease the availability of the medical facility to the common people.(Parashar *et al.* 2021) presented a complete study based on the systematic mapping of AI and Machine learning in the healthcare and its rate of evolution. Further the article was organized as follows, in the section 3 the system model and the data analysis was discussed, the results were analyzed in section 4 and the conclusion was presented in section 5.

3 SYSTEM MODEL

3.1 *System model*

The proposed model is used to observe the patient and to predict the condition of the patient as normal or abnormal. To implement the system framework, the components used includes Arduino Uno, ESP8266, GSM module, temperature sensor LM35 and MAX30100. The cost effective system is based on the sensor input from the body of the human. The sensors placed on the human body to measure temperature, oxygen concentration and the heart rate integrated together to frame a wireless body area network (WBAN).Threshold for the measuring physical parameters are decided from the ideal value prescribed for a normal person. The parameters which decides the temperature of a human being includes gender, physical activity, food and water intake, measuring time and climatic condition. Using the sensor the temperature measured ranging from 36.5 Celsius to 37.2 Celsius termed as the normal temperature of a human being. The temperature sensor converts the measured temperature in Celsius into an equivalent voltage and provide this as an analog input to the Arduino.

The MAX30100 is an integrated sensor for measuring the heart rate and the oxygen concentration in the blood. It senses the values in an invasive manner, the oxygen concentration measured if lies in the range of 92% to 100%, then the human is said to be normal, whereas if

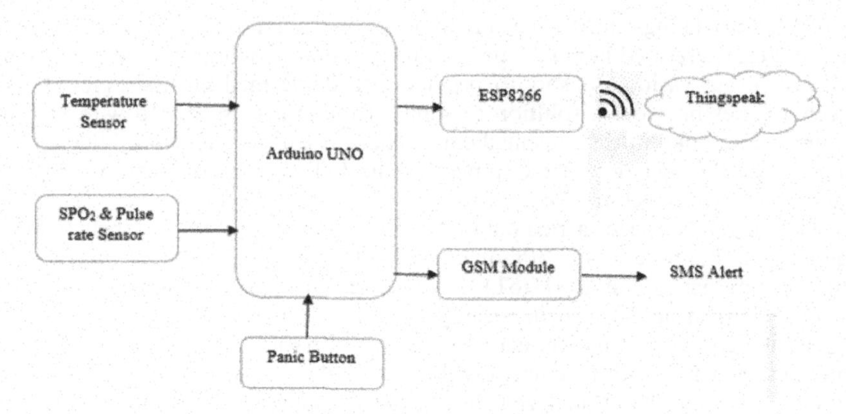

Figure 1. System model to record and monitor the patient health condition.

the measured value is beyond this range then the condition of the human is said to be abnormal. The patient is said to be normal if the heart rate is in the range of 60 beats per minute (bpm) to 100 beats per minute (bpm).Considering all the ideal values of all the parameters the threshold is set with respect to arduino programming. The sensed value is used for analysis and to predict the condition of the human under observation. The ESP8266 is a WiFi based component used to upload the sensed data into the cloud such that the statistics of the measured value can be observed anywhere at any time with the help of Thingspeak. Apart from the recorded data and the analysis about the patient condition from the data stored. In the case of emergency if the patient felt any discomfort in the breathing, the panic button can be pressed which will send a SMS alert to the registered mobile phone number of the caretaker, so that the patient can be attended immediately.The SMS alert is triggered by IFTTT and the panic message is sent to the caretaker's mobile phone. The Figure 1 illustrates the necessary building blocks of the envisioned system building blocks of the proposed system.

3.2 *Data analysis*

Basically machine learning is classified as supervised learning and unsupervised learning. In the supervised machine learning the given input is mapped to the output based on the prior input output pair available. The classification and regression are the sub categories of the supervised machine learning. Based on the classification algorithm, the model is fully trained based on the training data, later using the test data the model is evaluated and finally the model is used to predict the performance for the new data.

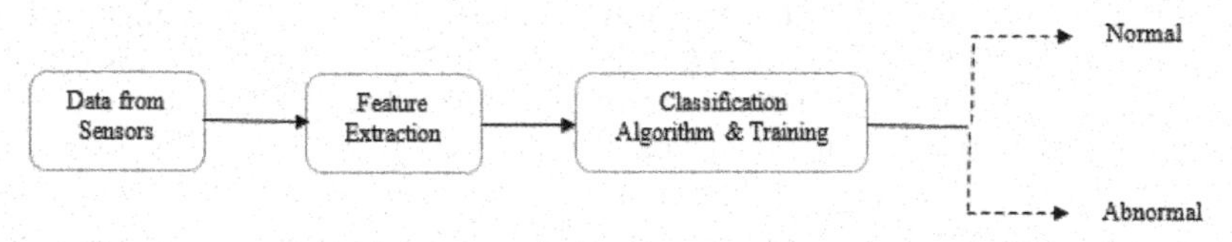

Figure 2. The process of predicting the condition of the patient based on the collected data.

The Figure 2 shows the process performed to predict the patient condition based on the data collected from the sensors. Various classification model is used to analyze the accuracy of the predicted value. Using the dataset from the Kaggle the model is trained and the model is tested with the patient data collected by the sensors. The size of the dataset is 10,000 and the 70 percentage of the data is used for training the model. The classification algorithm used to analyze the data are Linear Regression, Support Vector Classifier, Random Forest, AdaBoost Classifier.

4 RESULTS

The proposed system implemented in hardware collects the value from the sensors placed on the human body to monitor the temperature, oxygen concentration and heart rate and stores in cloud for statistical analysis and to visualize the condition of the patient. In case of emergency the patient may send an SMS by pressing the panic button. In addition to this the machine learning classification algorithm is applied to the dataset and analyzed, considering the abnormal being denoted as '1' and normal as '0'. The accuracy during the training and testing of the sensed data is calculated for the various algorithms and tabulated as Table 1. From the table it is inferred that AdaBoost algorithm is good in providing the accurate results under training and testing of the dataset, which results in the accurate prediction of the patient condition. The parameters which are treated as features in predicting the patient condition are temperature, heart rate and oxygen saturation.

Table 1. Comparison of training and test accuracy of the classification algorithms.

Method	Accuracy	
	Training	Test
Linear regression	0.6668	0.6748
Random forest	0.6368	0.6407
Support Vector Classifier	0.5067	0.5095
AdaBoost Classifier	1.0	1.0

The confusion matrix is an indicator of the performance metrics of the algorithm used for prediction. The performance metrics such as recall, precision, specificity, accuracy and AUC-ROC curve can be calculated. The order of the matrix depends on the number of the target classes, here it is the order 2×2 for the classification algorithm.

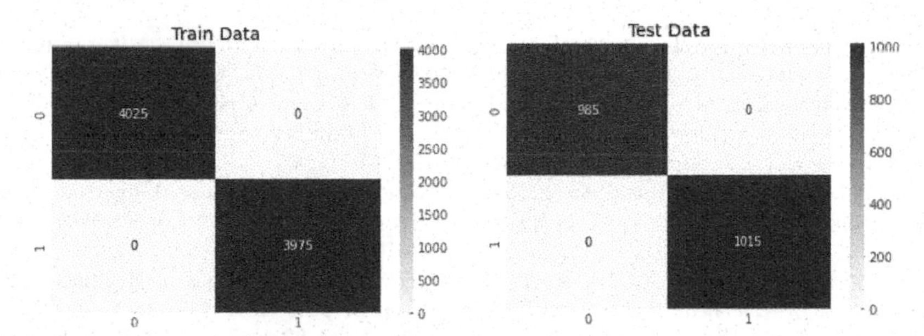

Figure 3. Confusion matrix for the AdaBoost classification algorithm.

The confusion matrix compares the predicted values with the actual target values and provide the possible combinations of Positive or Negative with True or False. The Figure 3 shows the confusion matrix obtained for the AdaBoost classification algorithm, it was observed that the entries in the FalsePositive(FP) cell and FalseNegative(FN) cell were zero, it implies that 100% accuracy is achieved from the predicted values. Thus, the AdaBoost classification algorithm suits well when higher degree of accuracy is expected in the system.

5 CONCLUSION

The proposed system serve the health monitoring of an elderly people as well as the patient under medical observation with a cost effective implementation of the model. There are many classification algorithm available out of which four algorithm is chosen for the analysis purpose and compared the values based on the accuracy obtained in the training and testing level. The confusion matrix is attained for the AdaBoost Classifier algorithm implies that the TruePositive(TP) and TrueNegative(TN) is a nonzero component which produces a good recall and precision values. As the FalsePositive(FP) and FalseNegative(FN) values are zero, it support the 100% accuracy in prediction of the values such as whether the patient is normal or abnormal. The prediction of the condition of the patient based on the available data is with 100% accuracy when AdaBoost classifier algorithm is used. The analysis can be extended by considering more parameters and performing multivariate analysis.

REFERENCES

Ahmed, I., Jeon, G., & Piccialli, F. (2021). A Deep-learning-based Smart Healthcare System for Patient's Discomfort Detection at the Edge of Internet of Things. *IEEE Internet of Things Journal*, 8(13), 10318–10326. https://doi.org/10.1109/JIOT.2021.3052067

Alazzam, M. B., Alassery, F., & Almulihi, A. (2021). A Novel Smart Healthcare Monitoring System Using Machine Learning and the Internet of Things. *Wireless Communications and Mobile Computing*, 2021, article ID 5078799. https://doi.org/10.1155/2021/5078799

Dwivedi, R., & Mehrotra, D. (2022). Shaleen Chandra- "Potential of Internet of Medical Things (IoMT) Applications in Building a Smart Healthcare System: A Systemic Review". *Journal of Oral Biology and Craniofacial Research*, 302–318.

Jabbar, M. A., Samreen, S., & Aluvalu, R. (2018). Future of Health Care: Machine Learning. *International Journal of Engineering and Technology*, 7 (4.6), 23–25.

Khan, M. M., Alanazi, T. M., Albraikan, A. A., & Almalki, F. A. (2022). IoT-based Health Monitoring System Development and Analysis. *Security and Communication Networks*, 2022, article ID 9639195. https://doi.org/10.1155/2022/9639195

Lalmuanawma, S., Hussain, J., & Chhakchhuak, L. (2020, October). Applications of Machine Learning and Artificial Intelligence for Covid-19 (SARS-CoV-2) Pandemic: *A Review. Chaos, Solitons, and Fractals*, 139, 110059. https://doi.org/10.1016/j.chaos.2020.110059. Epub June 25, 2020. PubMed: 32834612, PubMed Central: PMC7315944

Malik, M., Khatana, R., & Kaushik, A. (2021). Machine Learning with Health Care: A Perspective. *Journal of Physics: Conference Series*, 2040(1), 012022. https://doi.org/10.1088/1742-6596/2040/1/012022

Moulaei, K., Shanbehzadeh, M., Mohammadi-Taghiabad, Z., & Kazemi-Arpanahi, H. (2022). Comparing Machine Learning Algorithms for Predicting COVID-19 Mortality. *BMC Medical Informatics and Decision Making*, 22(1), 2. https://doi.org/10.1186/s12911-021-01742-0

Parashar, G., Chaudhary, A., & Rana, A. (2021). Systematic Mapping Study of AI/machine learning in Healthcare and Future Directions. *SN Computer Science*, 2(6), 461. https://doi.org/10.1007/s42979-021-00848-6

Pradhan, B., & Bhattacharyya, S. (2021). Kunal Pal- "IOT Based Applications in Healthcare Devices"-Hindawi. *Journal of Healthcare Engineering*, 2021, article ID 6632599

Saeed, Umer, Shah, S. Y., Ahmad, J., Imran, M. A., Abbasi, Q. H., & Shah, S. A. (2022). Machine Learning Empowered COVID-19 Patient Monitoring Using Non-contact Sensing: An extensive Review. *Journal of Pharmaceutical Analysis*, 12(2), 193–204,ISSN 2095–1779. https://doi.org/10.1016/j.jpha.2021.12.006

Sahu, D., Pradhan, B., Khasnobish, A., Verma, S., Kim, Doman, & Pal, K. (2021). The Internet of Things in Geriatric Healthcare. *Journal of Healthcare Engineering*, 2021, article ID 6611366. https://doi.org/10.1155/2021/6611366

Saranya, M., Preethi, R., Rupasriand, M., & Veena, S. (2018). A Survey on Health Monitoring System by Using IOT. *International Journal for Research in Applied Science and Engineering Technology*, 6(3), 778–782. https://doi.org/10.22214/ijraset.2018.3124

Secinaro, S., Calandra, D., Secinaro, A., Muthurangu, V., & Biancone, P. (2021). The Role of Artificial Intelligence in Healthcare: A Structured Literature Review. *BMC Medical Informatics and Decision Making*, 21(1), 125. https://doi.org/10.1186/s12911-021-01488-9

Tavera, C. A., Ortiz, J. H., & Osmah, I. (2021). Khalaf- "Wearable Wireless Body Area Networks for Medical Applications"-Hindawi. *Computational and Mathematical Methods in Medicine*, 2021, article ID 5574376.

Artificial Intelligence, Blockchain, Computing and Security – Dagur et al. (Eds)
© 2024 The Author(s), ISBN: 978-1-032-67841-2

EEG signal classification using 1D-CNN and BILSTM

K. Nanthini
Assistant Professor (Sr.G), Department of MCA, Kongu Engineering College, Erode, Tamil Nadu, India

A. Tamilarasi
Professor, Department of MCA, Kongu Engineering College, Erode, Tamil Nadu, India

D. Sivabalaselvamani*
Associate Professor, Department of MCA, Kongu Engineering College, Erode, Tamil Nadu, India

V.S. Harini, R. Janaki & V.K. Madhan
PG Scholar, Department of MCA, Kongu Engineering College, Erode, Tamil Nadu, India

ABSTRACT: Epilepsy is a chronic condition in which brain function becomes abnormal, resulting in seizures or short periods of strange behaviour, sensations, and even loss of awareness. Epilepsy can strike anyone at any time. Males and females of various races, ethnic backgrounds, and ages are affected by epilepsy. It considers one of the most prevalent neurological illnesses. Deep learning develops hierarchical models that mimic the cognitive process of our brains. It employs a multi-layered neural network that generates a result without pre-processing the data. Deep Learning is gaining much traction because of its superior accuracy when tested with vast information. The algorithm learns from instances and can characteristically distinguish between various classes. In this study, we provide the CNN and BiLSTM models for predicting the different classes from EEG time series. The model is corroborated by feeding an Epilepsy Seizure presents data to 1D-CNN and BiLSTM, which is published in the UCI Database and is publicly present in the form of comma-separated values.

Keywords: EEG, Seizure Prediction, CNN, RNN, BiLSTM, Deep Learning

1 INTRODUCTION

An electroencephalogram is a test used to detect neural signals of brain activity by electrodes attached to the outside scalp or intracranial (places electrodes inside the skull). This activity shows up as wavy lines on an EEG recording. With an emphasis on studies published in the last decade, wavelet approaches for computer-aided epilepsy diagnosis and seizure detection are discussed. One of the most used epilepsy diagnostic tests is an EEG. An EEG can also be used to diagnose various types of brain diseases. One of many mechanisms that can be used to interact with machines is through brain activity signals. If a person is affected with epilepsy, the patient's EEG signal varies from normal people. A rapid spiking wave is present in the EEG signal. Some electrodes were placed on the

*Corresponding Author: sivabalaselvamani@gmail.com

DOI: 10.1201/9781032684994-72

scalp to record brain activity. It can also be recorded while the patient is sleeping. EEG waves may be particularly sluggish in persons with intellectual illnesses such as tumours or strokes. This test can be used to determine the presence of Parkinson's disease, Alzheimer's disease, psychological problems, and insomnia such as narcolepsy. Among different types of brain wave varies range of normal vs. abnormal based on age and at the time of psychological condition of the person.

2 LITERATURE SURVEY

Khansa Rasheed *et al.* [1] have used a CNN model for predicting seizures; they converted time series data into a matrix and applied short-term Fourier transform (STFT) to transform EEG signals into spectrograms. DCGAN-Generates synthesized data & One-Class SVM is to validate generated data. Marzieh Savadkoohi *et al*, [2] have used the Bonn dataset and feature extracted time, frequency, and time-frequency domains. They have achieved the time domain of 99.5%, 99.5%, Frequency domain of 100%, and 99.5%, the wavelet transform of 100%, and 99.5% for SVM and KNN, respectively. Ahmed Abdelhameed *et al.* [3] have used the CHB-MIT dataset. To guarantee that all signals have a mean of zero and unit standard deviation, the data is pre-processed using z-score normalization. Ranjan Jana *et al.* [4] have used CHB-MIT dataset for prediction, and the raw EEG is converted into two-dimensional data. A deep learning technique is used for feature extraction and classification. Syed Muhammad Usman *et al.* [5] for prediction, have used the CHB MIT dataset as well as the American Epilepsy Society's Kaggle seizure prediction dataset. For noise reduction, the EEG signal is modified with empirical mode decomposition and bandpass filtering. Manasvi Bhat *et al.* [6] have used the Epilepsy Ecosystem dataset for prediction. Banu Priya Prathaban *et al.* [7] have used CHB-MIT, NICU, and SRM EEG Datasets were utilized to make the predictions. The various stages of seizures are classified using a sparsity-based artifact removal technique and a 3D OCNN classifier. Chien-Liang Liu *et al.* [8] have used Intracranial EEG Dataset. This work proposes a multi-view convolutional neural network framework to predict the occurrence of epilepsy seizures. Syed Muhammad *et al.* [9] have used Short Time Fourier Transform is used to boost the signal-to-noise ratio and transfer the signals from the time domain to the frequency by picking a pro window of 30 seconds. Debiao Ma *et al.* [10] have used the CHB-MIT dataset for prediction. Nanthini *et al.* [11] have described DeviceHive as a free, scalable, open-source IoT platform for data acquisition, processing, analysis, and visualization. In continuation of deep learning RNN based architecture achieved 99% of training and testing accuracy using a Long Short-Term Memory network [12].

2.1 *EEG dataset*

For our implementation, we conducted our experiment with UCI Epileptic Seizure Recognition. The actions of the brain are described in this dataset. There are 179 and 11500 characteristics and samples, respectively. Seizure activity is detected in about a quarter of the samples. So, in a binary classification context, we're working with an unbalanced dataset. The first 178 features are the values of electroencephalogram (EEG) recordings at various time points, with the class being the final feature. The class label has five possible values (y = 1..5). The remaining classifications, with the exception of y = 1, indicate different states like 2- tumour located area was located and recorded, 3-recording from the healthy brain area,4-eye closed, 5-eye opened. The below given Figure 1 shows the sample view of the epileptic seizure dataset.

	0	X1	X2	X3	X4	X5	X6	X7	X8	...	X172	X173	X174	X175	X176	X177	X178	y
0	X21.V1.791	135	190	229	223	192	125	55	-9	...	-31	-77	-103	-127	-116	-83	-51	4
1	X15.V1.924	386	382	356	331	320	315	307	272	...	146	152	157	156	154	143	129	1
2	X8.V1.1	-32	-39	-47	-37	-32	-36	-57	-73	...	48	19	-12	-30	-35	-35	-36	5
3	X16.V1.60	-105	-101	-96	-92	-89	-95	-102	-100	...	-80	-77	-85	-77	-72	-69	-65	5
4	X20.V1.54	-9	-65	-98	-102	-78	-48	-16	0	...	-12	-32	-41	-65	-83	-89	-73	5

5 rows x 180 columns

Figure 1. Sample epileptic seizure dataset.

2.2 *Methodology*

2.2.1 *Convolutional neural network*

For creating sequence model, first CNN layer with 32 filters, kernel size of 3, relu activation and same padding, Second CNN layer with 64 filters, kernel size of 3, relu activation and same padding, Third CNN layer with 128 filters, kernel size of 3, relu activation and same padding, Fourth CNN layer with Max pooling, Flatten the output, Add a dense layer with 256 and 512 neurons, Softmax as last layer with five outputs.

2.2.2 *Bidirectional LSTM*

A bidirectional LSTM, is a pattern dispensation model that encompass two LSTMs.One of which takes input in one way and the other in the other. Bi-LSTM significantly enhance the quantity of data supplied to the network, giving the algorithm better context.

2.2.3 *Implementation*

The dataset was split into train and test for prediction. The dataset contains five categories of class labels to allow researchers to analyse the seizures identified by the proposed methodology to those detected by qualified epileptologists.

The above Figure 2 shows the graph view of a single row in the dataset. The occurrence of seizure is indicated by the sharp and spike waves. The Figure 3 shows the number of samples for training and testing.

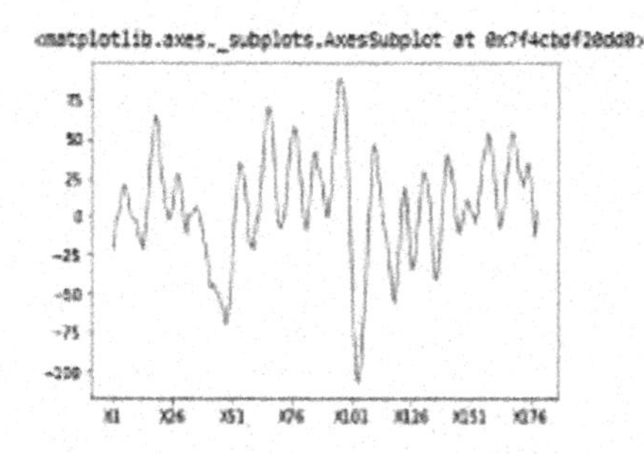

Figure 2. EEG signal view of a sample

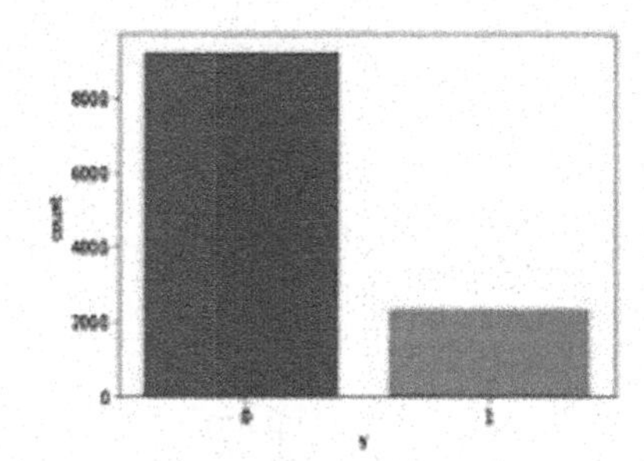

Figure 3. Seizure and non-seizure trails

The above given Figure 4 shows the scatter plot of epileptic and non-epileptic data. The blue color plot shows the non-epileptic data and the red plot shows the non-epileptic data in the scattered plot graph. The plot diagram of Figure 5 shows the signal view for the each category of class label in the dataset.

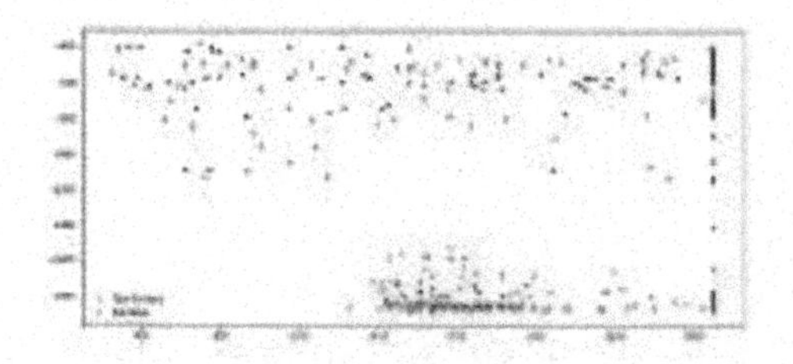

Figure 4. Scatter plot valuse of Epileptic and Non epileptic data

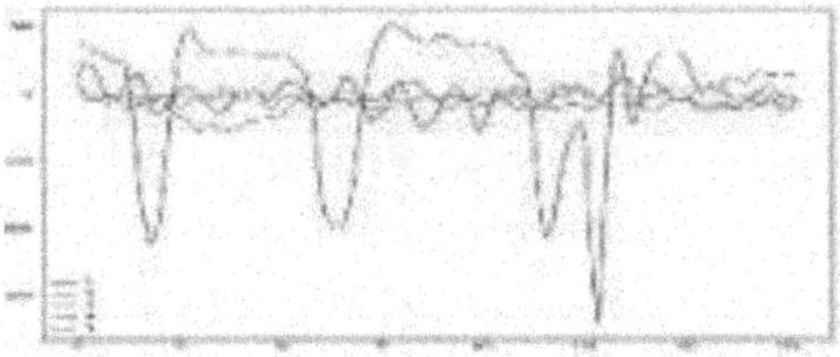

Figure 5. Plot diagram of different labels of the dataset

2.2.3.1 Experiment 1: 1D-CNN

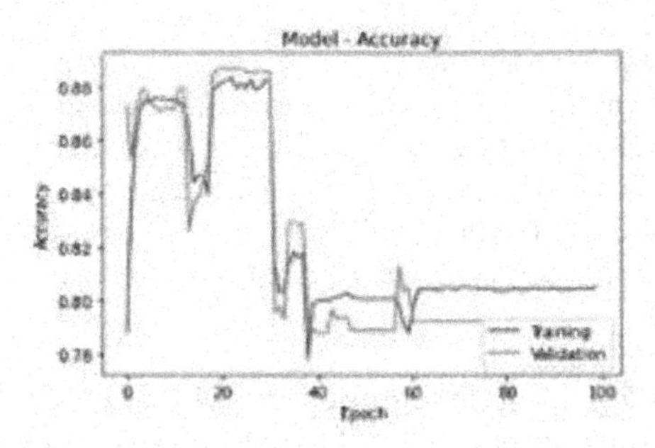

Figure 6. Model accuracy of ID CNN

2.2.3.2 Experiment 2:Bi-LSTM

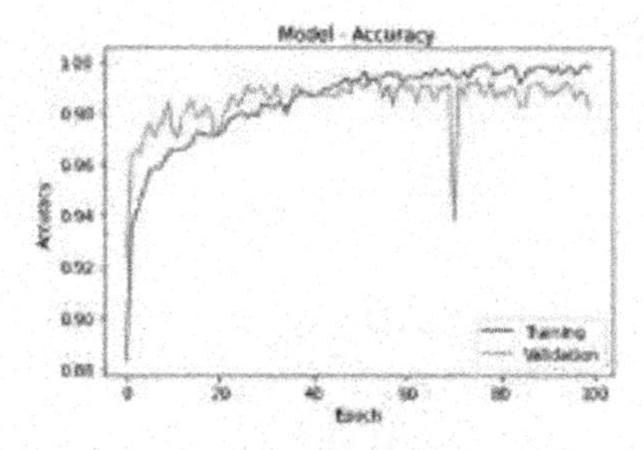

Figure 7. Bi LSTM Model Accuracy

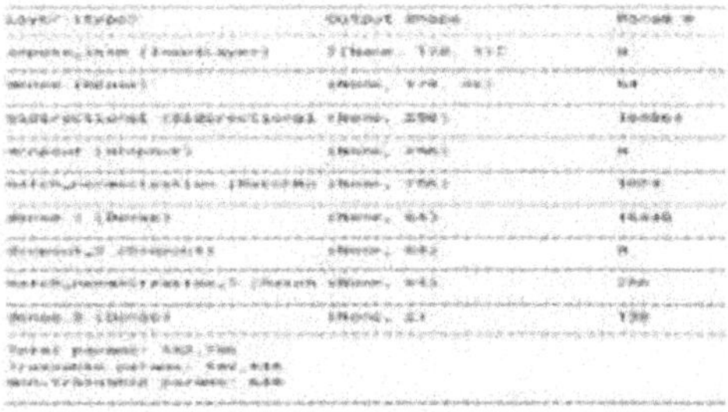

Figure 8. Bi LSTM Model Summary

The BiLSTM model on epilepsy versus all data. The sample model gives the summary with total parameters of 182,786, Trainable params of 182,146 and Non trainable params of 640. The epileptic and healthy data is then trained with the BiLSTM model. The model summary table is given in the Figure 8.

In this study, experiment 1with the plot diagram of model accuracy with best version is given Figure 6 shows training accuracy of 88.57% and its test accuracy of 90.70%. BiLSTM model in Figure 7 gives the training accuracy rate of 99.70% and 98.09% of testing accuracy respectively.

3 CONCLUSION AND FUTURE WORKS

In our work, we have classified multi classes with a high accuracy rate of prediction with the BiLSTM model. It would be very useful for patients to get proper treatment before it turns severe. CNN's are not limited to just images and videos. 1D convolution is more appropriate for numeric data and also text classification. In this work, we developed the model using 1D-CNN and BiLSTM. The performance of BiLSTM model for different classes of EEG signal classification is more precisely than the 1D-CNN model.Due to these practical difficulties, research and development of innovative product with the smart application is highly demanded for monitoring the activities of patients includes any type of brain disorder like Autism, Alzheimer's and Parkinson's, etc.

REFERENCES

[1] Rasheed, Khansa, *et al.* "A Generative Model to Synthesize EEG Data for Epileptic Seizure Prediction." *IEEE Transactions on Neural Systems and Rehabilitation Engineering* 29 (2021): 2322–2332.

[2] Savadkoohi, Marzieh, Timothy Oladunni, and Lara Thompson. "A Machine Learning Approach to Epileptic Seizure Prediction Using Electroencephalogram (EEG) ignal." *Biocybernetics and Biomedical Engineering* 40.3 (2020): 1328–1341.

[3] Abdelhameed, Ahmed M., and Magdy Bayoumi. "An Efficient Deep Learning System for Epileptic Seizure Prediction." *2021 IEEE International Symposium on Circuits and Systems (ISCAS)*. IEEE, 2021.

[4] Jana, Ranjan, and Imon Mukherjee. "Deep Learning Based Efficient Epileptic Seizure Prediction with EEG Channel Optimization." *Biomedical Signal Processing and Control* 68 (2021): 102767.

[5] Usman, Syed Muhammad, Shehzad Khalid, and Sadaf Bashir. "A Deep Learning Based Ensemble Learning Method for Epileptic Seizure Prediction." *Computers in Biology and Medicine* 136 (2021): 104710.

[6] Bhat, K. Manasvi, *et al.* "Detection and Prediction of the Preictal State of an Epileptic Seizure Using Machine learning Techniques on EEG Data." *2019 IEEE Bombay Section Signature Conference (IBSSC)*. IEEE, 2019.

[7] Prathaban, Banu Priya, and Ramachandran Balasubramanian. "Dynamic Learning Framework for Epileptic Seizure Prediction Using Sparsity Based EEG Reconstruction with Optimized CNN Classifier." *Expert Systems with Applications* 170 (2021): 114533.

[8] Liu, Chien-Liang, *et al.* "Epileptic Seizure Prediction with Multi-view Convolutional Neural Networks." *IEEE access* 7 (2019): 170352–170361.

[9] Usman, Syed Muhammad, Shehzad Khalid, and Muhammad Haseeb Aslam. "Epileptic Seizures Prediction Using Deep Learning Techniques." *IEEE Access* 8 (2020): 39998–40007.

[10] Ma, Debiao, Junteng Zheng, and Lizhi Peng. "Performance Evaluation of Epileptic Seizure Prediction Using Time, Frequency, and Time–Frequency Domain Measures." *Processes* 9.4 (2021): 682.

[11] Nanthini, K., Kavitha, T., Sivabalaselvamani, D., Pyingkodi, M., & Kumar, G. (2019). Epilepsy Seizure Detection and Prediction Based on Device Hive. *International journal of Recent Technology and Engineering*, 7463–7466.

[12] Nanthini, K., Tamilarasi, A., Pyingkodi, M., Dishanthi, M., Kaviya, S. M., & Mohideen, P. A. (2022, January). Epileptic Seizure Detection and Prediction Using Deep Learning Technique. In *2022 International Conference on Computer Communication and Informatics (ICCCI)* (pp. 1–7). IEEE.

Recent advances and application driven machine learning and deep learning techniques

Artificial Intelligence, Blockchain, Computing and Security – Dagur et al. (Eds)
© 2024 The Author(s), ISBN: 978-1-032-67841-2

A study of deep learning, Twitter mining and machine learning based system for predicting customer churn

Anamika Mishra*, Awantika Singh*, Abhinandan Tripathi*, Shrawan Kumar Pandey* & Chaynika Srivastava*
Buddha Institute of Technology, Gorakhpur, India

ABSTRACT: Trying to predict customer churn is most prevalent challenging tasks in just about any industry. With breakthroughs in ML and AI, the capability to forecast client attrition has grown significantly. This paper incorporates the majority of the algorithms developed for customer churn prediction techniques. We studied a number of papers published by renowned authors. We attempted to document of all the machine learning and deep learning methodologies invented and deployed by the world's technological behemoths to better understand their clientele and expand the market for their products or services. According to the findings, the DL model outperformed the competition in terms of classification and prediction accuracy. DL models would thus choose to disregard such as useless information while constructing their data blueprints. This review paper explains how to use a Deep Learning, ML and senti churn approach for predict churn on a company metadata. The churn prediction model takes into account contextual elements, usage characteristics, customer attributes, and support characteristics. Deep learning and machine learning modeling techniques have been trained using historical datasets.

Keywords: Machine Learning, Deep Learning Model, Neural Network, Logistic Regression, Twitter Mining Approach

1 INTRODUCTION

Globalization and technological advancements in the communication industry have stimulated the growth of providers in the market, leading to increased competitiveness. In this competing age, it has become vital to optimize earnings on a routine basis, for which numerous approaches have been proposed, including recruiting new consumers, up selling current clients, and increasing the retention time of regular clients. In comparison to other strategies, retaining existing clientele is the cheapest option. Businesses must reduce potential client churn, or consumer movement between one network operators to another. Discontent with the client maintenance and service system is the leading cause of churn. The best approach to unlocking answers is forecasting the clients who are at risk of churning. One of several main purposes of client churn prediction is to aid in the development of customer relationship management. As a result, establishing strategies to keep track of devoted clientele has become vital. By recognizing early attrition cues, customer churn algorithms aim to estimate customers who leave willingly. Abbas *et al.* [1] claims that churn prediction is an effective way to predict at-risk clientele. Deep learning algorithms can use to anticipate customer migrate in the retail industry by automatically identifying useful features. It is critical to apply contemporary ML approaches such as DL to create better predicted findings, especially in research with a large number of input attributes. In recent years, various

*Corresponding Authors: anamikahello21@gmail.com, awantikasingh1120@gmail.com, abhinandan2787@gmail.com, shrawan458@bit.ac.in and chaynikait09@gmail.com

DOI: 10.1201/9781032684994-73

researchers have successfully coupled RFM analysis with data mining processes [2,3]. Some studies use RFM analysis to segment customers, then use data mining approaches to establish patterns for a select group of essential customers [4,5]. In addition, according to recent study [6], data mining tools can be considered effective to predict churn. The creation of practical churn prediction models is a critical task that entails a variety of tests, ranging from finding churn to predicting the variables with the strongest predictors to the selection of a powerful prediction approach.

There have been churn modeling applications in the retail sector. The bulk of churn studies are focused on contract-setting sectors such as communications, financial, and insurance. Table 1 illustrates research in the retail business that used deep learning techniques, regression models, and artificial neural network to predict churning. Poel *et al.* [7] presented a selective concept for non-contractual scenarios. The purpose of this review is to analyze the performance of deep learning.

Table 1. Deep learning-based churn prediction study.

Ref.	Field	Techniques	Characteristics	Fact	Objectives
[11]	Supermarket	LR, RFM, RF, NN	Metaphrasing data, Demgraphic variables	5.5 months, 33,481 customers	Migrating some defection by user who are habitually loyal
[16]	Grocery	CNN, Restricted Boltzmann LR, MARS	Metaphrasing data	unknown	Usng machine learning algorithms to anticipate client attrition
[22]	Supermarket	LR, Forward and backward model	Metaphrasing data, Demographic variables	25 months, 140084 clients	Compare the results of MARS and LR in terms of customer attrition modelling.
[23]	B2B	Classification, Feature Engineering	Metaphrasing data	27 months, 6060 clients	A generic model of feature engineering phases that might be used to anticipate churn in any non - subscription business
[24]	Telecom	Statistical model, LSTM, RFM, CNN	Daily call function	6 months, 21,030 customers	Comparing churn prediction on a daily basis to churn prediction on a monthly basis
[10]	Telecom	NN, CNN	76 characteristics	41,050 customers	Investigating deep learning techniques and their applications for the challenge of predicting client attrition
[25]	B2B	LL, ELM, GTB	Metaphrasing data	25months, 10,000 customers	Developing non-contractual machine learning technlogies that forecast future client behaviour.

Today, social media websites are used by 65 to 70 percent of all active internet users. Users can communicate through social networking by sharing their ideas, feelings, opinions, and emotions. Compared to earlier methods, it is a more affordable marketing communication strategy. The development of real-time statistics and the extraction of information about public opinions, sentiments commodities, administration, options, societies, and activities may be aided by social media data visualization. On the other hand, current subjective and mood analysis techniques are mostly intended to use in the English language. Aside from the fact that Arabian is the fourth most frequently spoken language on the planet and is spoken by a large majority of the worldwide people. Arabic has a large morphological vocabulary. Several researches have looked at the unique challenges provided by the character of the Arabic script, as well as potential solutions [8,9]. Due to the various languages, use of unofficial or conversational Arabic, and the syntax of languages are transcribed from right to left. The recognition of emotions in Arabic is challenging.

There are two methods to managing customer turnover, first is reactively and another one is proactively. In a reactive strategy, the corporation awaits for a leave request from a client before offering them tempting retention incentives. In a proactive approach, the likelihood of attrition is forecasted and services are provided to consumers appropriately.

2 LITERATURE SURVEY

An overview of estimating customer attrition in the telecom industry is given in this article, as well as related work suggested by various authors. Decision trees, random forests, GBM tree approaches, and XGBoost were utilized by Adbelrahim *et al.* [10] to anticipate customer churn behaviour prediction. In comparative study, XGBoost surpassed others in context of AUC correctness. The correctness of the attribute selection process may further improved by using optimization approaches. SVM, decision tree, naive bayes, and logistic regression were used in a comparative study of machine learning models that predict customer turnover by Praveen *et al.* [11]. They looked into the effects of enhancing techniques on prediction performance. SVM-POLY with AdaBoost beat the others in the end. However, classification accuracy may improve even further by combining attribute selection processes like univariate selection. To forecast customer attrition, Beleiu *et al.* [12] employed three machine learning techniques: neural networks, SVM, and naïve bayes. Burez *et al.* [13].'s goal was to illustrate the problem of class inequality. They used random forest and boosting techniques in addition to re-sampling and logistic regression for the performance evaluation. Better sampling methods like CUBE were also examined, however the results showed little difference in performance. The issue of class imbalance may be more successfully addressed by using optimization-based sampling techniques. SVM, LR, and random forest (RF) were employed by Coussement *et al.* [14] to attempt to solve the churn estimation problem. At first, SVM performed better than LR and RF in terms of PCC and AUC. Dahiya *et al.* [15] used decision trees and regression models as two ML algorithms on a churn prediction dataset. Instead, machine learning techniques could be able to successfully address the aforementioned issue.

Umman *et al.* [16] employed LR and decision tree ML method to analyze a big data set However, the precision was poor. As an outcome, more work must be done before any other attribute selection techniques and machine learning may applied. J. H. *et al.* [17] look into the element that affect reverence churn. They also analyzed three machine learning approaches, including, regression trees, logistic regression and deep neural networks. Because of its principle design, decision trees surpass other algorithms. The achieved accuracy could be further increased by utilising existing attribute selection methods.

Hadden *et al.* [18] undertook a detailed analysis of existing attribute selection methodologies as well as an assessment of all machine learning techniques studied. As a result they obtained that the decision tree surpass every other machine learning technique in the prediction models. In attribute selection, optimization procedures are also significant since they enhance prediction models. The authors offered a path for future study fields after conducting a comparative evaluation of the current methodologies. On a churn prediction data set, Huang *et al.* [19] examined several classifiers, and the result confirmed that random forest outperforms everyone else with reference to AUC and PR-AUC analysis. However, using optimization methodologies for attribute extraction can improve accuracy even more. Idris *et al.* [20] evaluated different classification algorithms which is the combination of genetic programming (GP) like tree based and the adaboost ML method. GP and adaboost gained with more accuracy than the others. Bayesian Belief Networks (BBN) were utilized by Kisioglu *et al.* [21] to estimate client attrition. The experimental study included correlation matrix and multi collinearity testing. BBN was proven to be a great choice for predicting churn. They also made recommendations for further research.

Deep Learning methods may use for learning multi-level function representations. The designs are centred on discovering these models which are developed on several stages, resulting in hierarchical nonlinear information processing. Computer vision, pattern recognition, and NLP have all used DL systems in the past [22]. We all know that learning new things can frequently be as easy as training biological brain neurons, but in this case, the mind that was trained with the help of examples is far more significant. Considering that the long-term memory always retains information learned through examples. It's important to remember that, in order to improve outcomes, researchers always advocate altering or it's upgrading the neural network's weights.

Artificial neural networks (ANNs) have many hidden layers, and DL is a sub-domain of ANNs. Deep learning methods educate each layer of neurons on the data attributes of the preceding layer. The majority of churn researches are conducted in contract-setting industries such as telecommunications, banking, and insurance. Churn studies are required in non-liable settings such as the industry to discover who is a churner and who is not. In the retail realm, previous researches have mostly employed transactional data combined with RFM data to forecast turnover.

3 MACHINE LEARNING MODELS

On the basis of reliability and efficiency following are the widely used machine leaning approaches used today for forecasting customer attrition in organizations.

3.1 *Logistic regression analysis*

In LR analysis [17], we follow a statistical procedure to estimate relationship between variables and it provides variety of ways for developing the model and assessing multiple variables. It is not frequently utilized in the field of predicting client attrition due to the fact that linear regression techniques are effective for predicting continuous variables. It is also known as Logit Model. These analyses incorporate dependent variables being categorical or finite often performing binary regression or multinomial regression. (e.g., customer churn).It is helpful in predicting what choice is going to be made by the client.

3.2 *Naïve Bayes*

It is a classifier that uses the Bayes theorem as a foundation. An approach for supervised machine learning called Naive Bayes is used to recognize objects based on their attributes. Every feature pair is assumed to be independent of others. In Naïve Bayes [23] the data set is mainly classified in two parts feature matrix and response vector. Bayes theorem states the following relationship.

$$\phi(\mathrm{M}|\mathrm{N}) = \phi(\mathrm{N}/\mathrm{M})\phi(\mathrm{M})/\mathrm{T}(\mathrm{N}) \tag{1}$$

3.3 *Support vector machine*

SVM [24] are supervised learning models with associated ML approaches. It interprets data for regression and classification analysis. They were developed by vapnik *et al.* [11]. The support vector machine tries to split the forecast in two parts: +1 for the hyper plane's right side and –1 for the hyper plane's left side. The hyper plane's width is twice to the margin's length. Tune parameters such as kernels such as linear, poly, rbf, callable, and pre-calculated are used depending on the kind of data spread across the graph. In terms of accuracy, SVM beat other techniques.

3.4 *Decision trees*

It incorporates the greedy technique as well as a set of classification criteria. Alternatively, while this method explains the better recognition prediction performance, it is incapable of adapting to chaotic input. Gain is the most important element in determining a decision tree's root node parameter. Decision trees may used for categorization, so it's also known as a statistical classifier [25].

3.5 *Random forest classifier*

It employs the divide-and-combine strategy. The random vector subspace approach is used [26]. By picking a haphazard selection of characteristics from the collection of predictive attributes, this method creates a set of trees that are then trained. Based on the features or parameters given, each tree develops to the fullest extent possible. The final decision tree is largely constructed for predicting using the arithmetic mean. It can manage several input variables without losing anything. While developing a forecasting model, it may also address missing data within a given data set.

3.6 *Extra tree classifier*

An ensemble learning method known as an additional tree classifier which combines the classification efficiency of de-correlated decision trees. It is also known as an Extreme Randomized Tree Classifier [27]. In comparison to the Random Forest Classification Algorithm, it only differs in the manner the forest's decision trees are built. This is a meta estimator that uses averaging to boost expected precision and fit control a set of randomly selected decision trees on distinct sub-samples of the given dateset. Churn prediction surpassed all other techniques in terms of precision.

3.7 *Boosting algorithm: Adaboost*

Another ensemble classifier is the Adaboost Random Forest Classifier. Ensemble classifiers are made up of various different classifiers, and their result is the sum of what those classifiers produced. In object classification, a single algorithm might not work well. The selection of a training sample at each cycle, along enhancing the ensemble like Ada-boost, and awarding the weight in the full agreement, we may able to get a respectable aggregate classifier Adaboost maintains the approaches by selecting the training data set based on the effectiveness of prior training. The AdaBoost classifier enhanced the performance and correctness in forecasting the turnover of the telecom data-set. For example, several boosting may tweaked to increase performance [28].

3.8 *Xgboost classifier*

The decision tree approach is used with gradient boosting in XGBoost. Gradient boosting is a technique that uses new designs to evaluate the inaccuracy or surplus of conventional design algorithms, which is blended to provide the end prediction. Additionally, the deficit function's value is reduced or found at a minimum using the gradient learning approach [29].

3.9 *Cat Boost classifier*

CatBoost is a gradient-boost decision tree-based method that reduces prediction time by using synchronous trees rather than gradient-boosting trees. In order to achieve better results, it alters the initial model after computing the pseudo-residuals. Cat boost's key characteristic is that it incorporates some of the most widely used pre-processing techniques, such as one-hot encodes, label encodes, and so on. For data pre-processing, it excludes any statistical measurements [29].

4 DEEP LEARNING MODELS

CNNs and RNNs are two effective and widely applied approaches in the field of deep learning [30]. Various authors have used deep learning to forecast churn. These are discussed as below:

4.1 *Artificial Neural Network (ANN)*

ANNs are well known for their adaptability. They change themselves as their learn from initial training and deliver new knowledge about the word. Before learning how to tackle a problem, an ANN is first trained or fed enormous volumes of data. Weighting he input streams is at the heart of the most basic learning model.

4.2 *Convolution Neural Network (CNN)*

CNNs are multilayered neural networks which are used in object recognition and image processing. The results of CNNs outperformed the best text mining techniques when they were compared to the methods currently used for evaluating textual material. Tanveer *et al.* [32] use of a deep ensemble classifier to solve classification problems revealed the advantages of deep learning techniques Kim *et al.* [31] demonstrated the efficacy of deep learning algorithms over classical classification methods in forecasting several client attritions.

4.3 *Deep Neural Network (DNN)*

DNNs are generally Feed Forward Neural Networks (FFNNs), in which data goes between the output layer and the input layer without travelling backward. The linkage between the layers is only one way & it never touches a node again. De Caigny *et al.* [33] suggested a DNN as an example for forecasting client attrition. The correctness, precise recall, and F1 (which is the harmonic mean of exactness and recall) are applied as baseline models with the DNN.

4.4 *Multi-Layer Perceptron (MLP)*

Forward ANNs also called a Multilayer Perceptron (MLP). The most basic deep neural network, MLPs are made up of a series of entirely linked layers. Today, MLP machine learning techniques can be used to avoid the recent deep learning systems' high computational resource requirements. The outputs from the previous layer are weighted summarized in a series of nonlinear functions that make up the next layer.

4.5 *Long Short Term Memory Network (LSTM)*

LSTMs, a type of RNN, can be used to learn and recall long-term dependencies. Recalling earlier information over extended periods of time is the default tendency. Since LSTMs can recall prior inputs and are useful for time-series prediction, they can maintain track of data throughout time. In LSTMs, which have a chain-like structure, four communicating layers interact with one another in a particular order.

Implementation:

The proposed churn prediction model employed in the shown work in Figure 1. The company provided consumer retail scanner data, which was used to extract customer

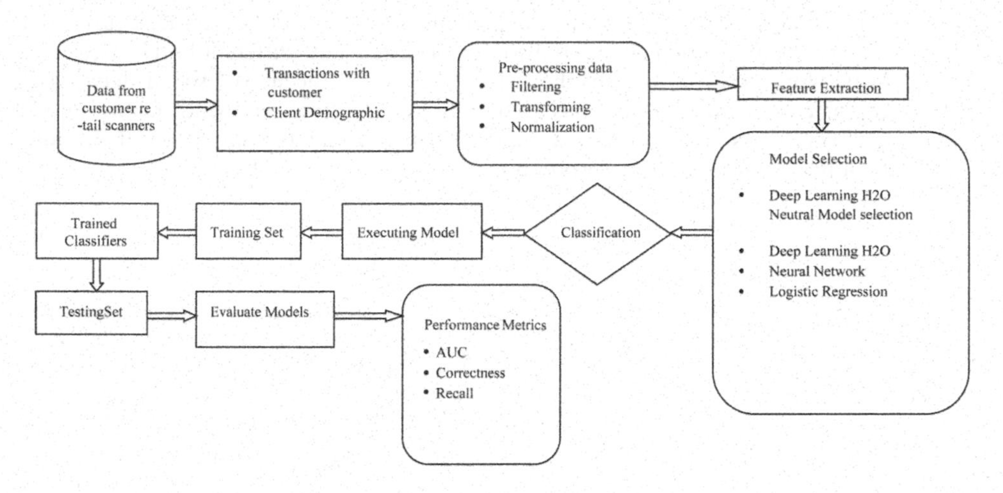

Figure 1. Models of deep learning.

exchanges and demographics data. To preparation for the models, these data were processed and normalized during the previous phase. Before using a DL model to create useful and non-repeated values, the data was subjected to feature extraction.

5 SENTI CHURN MODEL APPROACH

Throughout the study's time period, a churner is defined as a client who pays later willingly leaves the firm and discontinues mobile services. In our study, a late-paying customer who sticks with the business for the course of the study is referred to as a non-churner. Data mining is a method for drawing conclusions from huge databases. Knowledge discovery databases (KDD) [34] and cross-industry standard procedure for data mining (CRISPDM) [35] are the most popular data mining approaches used to create data mining models.

Information gathering and data preparation comprise the Senti Churn model's second phase, which includes transforming the data and finding out the proportion of customers who are satisfied. In the third phase we performed modeling of the data in which we choose the techniques going to be used and next we trained the model with the chosen technique and at last did the model evaluation. During the model deployment step, the forecast outcome is provided to the company for approval from both an authentic and a corporate viewpoint.

The data set was built utilizing historical data provided by the company as well as customer satisfaction ratings derived through Twitter mining [36].

6 CONCLUSIONS

The pattern of expansion in the current era have most dramatic boom ever. With technological innovation comes increase in services, and it is difficult for a corporation to foresee which clients are likely to quit its services. In this review article, we use well-known ML techniques to compare studies on customer churn prediction in the field of telecom. A detailed discussion of DL methods and a method for mining Arabic-language Twitter is also included. This article also included a DL model for predicting customer attrition using basket analysis features. The confusion matrix outputs are used to calculate model accuracy, recall, precision, and AUC. For churn modeling, the performance difference between DNNs and MLPs can be significant. This method contributes to the field of customer churn modeling both theoretically and practically. It also demonstrates how different batch sizes affect deep neural network performance. In the field of telecoms, the use of social networking analytics to predict client attrition has gotten minimal attention. Senti Churn model demonstrated its efficiency first by comparing it to surface authentic and up to date findings presented by a telecommunications corporation.

REFERENCES

[1] Abbasimehr H., Setak M., Tarokh M. (2011) A Neuro-fuzzy Classifier for Customer Churn Prediction. *International Journal of Computer Applications* 19(8):35–41.

[2] Cheng, C.H., Chen, Y.S.: Classifying the Segmentation of Customer Value via RFM Model and RS Theory. *Expert Systems with Applications* 36(3), 4176–4184 (2009).

[3] Hsieh, N.C.: An Integrated Data Mining and Behavioral Scoring Model for Analyzing Bank Customers. *Expert Systems with Applications* 27(4), 623–633 (2004).

[4] Liu, D.R., Lai, C.H., Lee, W.J.: A Hybrid of Sequential Rules and Collaborative Filtering for ProdUCT Recommendation. *Information Sciences* 179(20), 3505–3519 (2009).

[5] Liu, D.R., Shih, Y.Y.: Integrating AHP and Data Mining for Product Recommendation Based on Customer Lifetime Value. *Information & Management* 42(3), 387–400 (2005).

[6] Umayaparvathi, V., Iyakutti, K.: Automated Feature Selection and Churn Prediction Using Deep Learning Models. *International Research Journal of Engineering and Technology (IRJET)* 4(3), 1846–1854 (2017).

[7] Buckinx, W., Van den Poel, D.: Customer Base Analysis: Partial Defection of Behaviourally Loyal Clients in a Non-contractual FMCG Retail Setting. *European Journal of Operational Research* 164(1), 252–268 (2005).

[8] Marcus, A.; Bernstein, M.S.; Badar, O.; Karger, D.R.; Madden, S.; Miller, R.C. Processing and Visualizing the Data in Tweets. *ACM SIGMOD Rec.* 2012, 40, 21–27. [CrossRef]

[9] Castronovo, C.; Huang, L. Social Media in an Alternative Marketing Communication Model. *J. Mark. Develop. Compet.* 2012, 6, 117–134.

[10] Ahmad A.K., Jafar A., Aljoumaa K. (2019) Customer Churn Prediction in Telecom Using Machine Learning in Big Data Platform. *Journal of Big Data* 6(1):28.

[11] Asthana P (2018) A Comparison of Machine Learning Techniques for Customer Churn Prediction. *International Journal of Pure and Applied Mathematics* 119(10):1149–1169.

[12] Brândųsoiu, I., Toderean, G., Beleiu, H.: Methods for Churn Prediction in the Pre-paid Mobile Telecommunications Industry. In: *2016 International Conference on Communications (COMM)*, pp. 97–100. IEEE (2016).

[13] Burez J., Van den Poel D. (2009) Handling Class Imbalance in Customer Churn Prediction. *Expert Systems with Applications* 36(3):4626–4636.

[14] Coussement K., Van den Poel D. (2008) Churn Prediction in Subscription Services: An Application of Support Vector Machines while Comparing Two Parameter-selection Techniques. *Expert Systems with Applications* 34(1):313–327.

[15] Dahiya, K., Bhatia, S.: Customer Churn Analysis in Telecom Industry. In: *2015 4th International Conference on Reliability, Infocom Technologies and Optimization (ICRITO)*, pp. 1–6 (2015).

[16] Gürsoy U.S (2010) Customer Churn Analysis in Telecommunication Sector. *İştanbul Üniversitesi İşletme Fakültesi Dergisi* 39(1):35–49.

[17] Hadden J., Tiwari A., Roy R., Ruta D. (2006) Churn Prediction: Does Technology Matter. *International Journal of Intelligent Technology* 1(2):104–110.

[18] Hadden J, Tiwari A, Roy R, Ruta D (2007) Computer Assisted Customer Churn Management: Stateofthe-art and Future Trends. *Computers & Operations Research* 34(10):2902–2917.

[19] Huang, Y., Zhu, F., Yuan, M., Deng, K., Li, Y., Ni, B., Dai, W., Yang, Q., Zeng, J.: Telco Churn Prediction with BigData. In: *Proceedings of the 2015 ACM SIGMOD International Conference on Management of Data*, pp. 607–618 (2015).

[20] Idris, A., Khan, A., Lee, and Y.S.: Genetic Programming and Adaboosting Based Churn Prediction for Telecom. In: *2012 IEEE International Conference on Systems, Man, and Cybernetics (SMC)*, pp. 1328–1332. IEEE (2012).

[21] Kisioglu P, Topcu YI (2011) Applying Bayesian Belief Network Approach to Customer Churn Analysis: A Case Study on the Telecom Industry of Turkey. *Expert Systems with Applications* 38 (6):7151–7157.

[22] Dingli, A., Marmara, V., Fournier, N.S.: Comparison of Deep Learning Algorithms to Predict Customer Churn within a Local Retail Industry. *International Journal of Machine Learning and Computing* 7(5) (2017).

[23] Kirui, C., Hong, L., Cheruiyot, W., Kirui, H.: Predicting Customer Churn in Mobile Telephony Industry Using Probabilistic Classifiers in Data Mining. *International Journal of Computer Science Issues (IJCSI)* 10(2 Part 1), 165 (2013).

[24] Nath S.V., Behara R.S. (2003) Customer Churn Analysis in the Wireless Industry: A Data mining Approach. *Proceedings-annual Meeting of the Decision Sciences Institute* 561:505–510.

[25] Sharma H., Kumar S (2016) A Survey on Decision Tree Algorithms of Classification in Data Mining. *International Journal of Science and Research (IJSR)* 5(4):2094–2097.

[26] Han J., Pei J., Kamber M. (2011) *Data Mining: Concepts and Techniques*. Elsevier.

[27] Radosavljevik D, van der Putten P, Larsen KK (2010) The Impact of Experimental Setup in Prepaid Churn Prediction for Mobile Telecommunications: What to Predict, for Whom and Does the Customer Experience Matter? *Trans. MLDM* 3(2):80–99.

[28] Xie Y., Li X., Ngai E., Ying W. (2009) Customer Churn Prediction using Improved Balanced Random Forests. *Expert Systems with Applications* 36(3):5445–5449.

[29] Lalwani, P., Mishra, M.K., Chadha, J.S. *et al.* Customer Churn Prediction System: a Machine Learning Approach. *Computing* 104, 271–294 (2022). https://doi.org/10.1007/s00607-021-00908-y

[30] Kumar, A.S.; Chandrakala, D. An Optimal Churn Prediction Model Using Support Vector Machine with Adaboost. *Int. J. Sci. Res. Comput. Sci. Eng. Inf. Technol.* 2017, 2, 225–230.

[31] Kim, A.; Yang, Y.; Lessmann, S.; Ma, T.; Sung, M.C.; Johnson, J.E.V. Can Deep Learning Predict Risky Retail Investors? A Case Study in Financial Risk Behavior Forecasting. *Eur. J. Oper. Res.* 2020, 283, 217–234.

[32] Tanveer, A. *Churn Prediction Using Customers' Implicit Behavioral Patterns and Deep Learning.* Ph.D. Thesis, Graduate School of Business, Sabancı University, Istanbul, Turkey, 2019.

[33] Deaigny,A.;Coussement,K.;DeBock, K. A New Hybrid Classification Algorithm For Customer Churn Prediction Based on Logistic Regression and Decision Trees. *Eur. J. Oper. Res.* 2018, 269, 760–772. [CrossRef]

[34] Brachman, R.J.; Anand, T. The Process of Knowledge Discovery in Databases. Advances in Knowledge Discovery and Data Mining. *IEEE Expert* 1996, 11, 37–57.

[35] Frawley, W.J.; Piatetsky-Shapiro, G.; Matheus, C.J. Knowledge Discovery in Databases: An Overview. *AI Mag.* 1992, 13, 57.

[36] Almuqren, L.A.; Moh'dQasem, M.; Cristea, A.I. *Using Deep Learning Networks to Predict Telecom Company Customer Satisfaction Based on Arabic Tweets*; ISD: Tolerance, France, 2019.

Artificial Intelligence, Blockchain, Computing and Security – Dagur et al. (Eds)

Recommendation of commodities exchange using various deep learning algorithm

Anitha Julian*, R. Akash* & Chennakampalli Chethan Reddy*
Department of Computer Science and Engineering, Saveetha Engineering College, Chennai, Tamil Nadu

ABSTRACT: The commodities exchange is a significant activity that has a compelling collision on an organization financial situation. The stock exchange keeps everyone up to date on the latest worldwide business news and events. Technical experts and investors examine its many components. Customers find it difficult to predict future stock market moves due to the abundance of information available regarding stocks and investing. Customers can use deep learning algorithms to estimate the value of stocks. This study employs a stock price ensemble model to forecast future events. Time series and historical data from the company are utilized to train the algorithm in the stock price ensemble model. The intention of this work is to enhance the precision of predictions for the future. The focus is to refine the LSTM algorithm's accuracy and conduct a thorough evaluation of the results obtained from the dataset.

Keywords: LSTM, KNN, Stock Market, Prediction, Output

1 INTRODUCTION

A stock exchange is a platform that facilitates the buying and selling of authorized securities in an organized and regulated manner. It comprises of authorized stock brokers, rules, regulations, standard practices for market transactions, and trading floor where stock brokers or their agents gather to execute trades. The exchange only trades listed stocks, which enhances their marketability once they are listed. The stock exchange operates as an auction market, with open bids and offers made on the exchange floor to determine the prices of the traded securities, which are established through competitive forces.

2 LITERATURE SURVEY

In literature survey, we gathered some details on the present stock market prediction systems after a literature review.

Akita, R. *et al.* & Althelaya, K.A. *et al.* have analyzed the stock prediction value using textual based and sequential method of predictions with formula. Hedge, M.S. *et al.* discussed about accurate stock price forecasting can result in large rewards. For stock market forecasting, many academics have modified the news and numerical data. On three datasets, the model's performance is contrasted with the baseline models. Idrees, S.M. *et al.* & Izumi, K. *et al.* have done the time series fore casting and LSTM classification problem will also raise. Liu, G. *et al.* & Loret, JB. *et al.* included information about a certain company's data

*Corresponding Authors: cse.anithajulian@gmail.com, akashajj18@gmail.com and chethan16191@gmail.com

 DOI: 10.1201/9781032684994-74

set qualities, historical stock prices, investor growth rates, and business expansion rates. Pang, Xiongwen. *et al.* explained the features that include things like minimum and maximum values, volume, profit margins, income, share costs, etc.

Ramya Devi, R. *et al.* have implemented an intelligent warehousing system with Logistic Regression significantly explained about the commerce commodities prediction method. In the caption of the graphical figure, Raza *et al.* describes the application of CNN and LSTM. A tool called image caption generation recognize the needs of statistical natural language generation to determine the correspondence of the picture. Sai Sravani, K. *et al.* & Zhao, L. *et al.* have applied a deep network technique with relative success to financial market prediction as have other ML methods. Anomalies were found using the Outliner technique.

3 METHODOLOGY

The suggested study's dataset was gathered via Kaggle. This data collection, though, is unprocessed. The data set is a compilation of some companies' stock market valuation data. The conversion of unprocessed data into processed data is the initial phase. Since only a portion of the many traits present in the raw data obtained are necessary for prediction, this is performed through feature extraction.

The structural model, as depicted in Figure 1, provides the structure, behavior, and viewpoints of a system. The data which are plotted will be taken from cluster of data. The feature extraction module will be divided as depicted in the image above demonstrates an example of dataset extraction and refining.

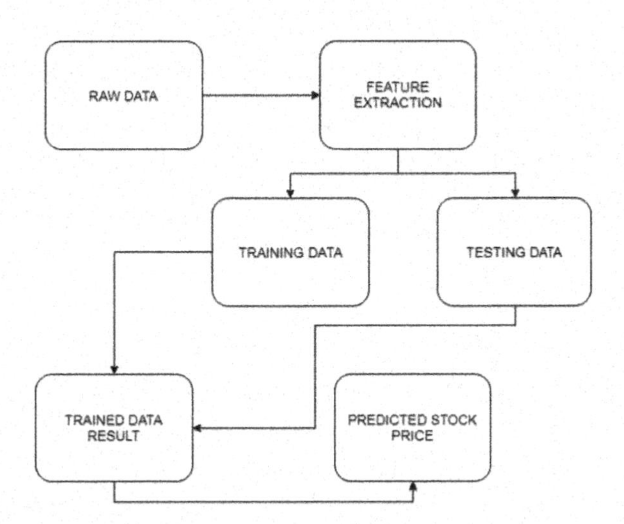

Figure 1. System architecture.

3.1 *Linear regression*

The linear regression algorithm is a well-known machine learning algorithm. It is discussed in both statistical and machine learning contexts. Figure 1's structural model illustrates how it can predict the structure, behaviour, and viewpoints of a system. The system analyses and processes the data, trains on the provided data, and then extracts the relevant data to process and improve the cluster data. The unknown dependency's value is compared to the known dependencies, and the result is identified as shown in Figure 2 and derived on its foundation.

3.2 *K-Nearest Neighbors (KNN)*

One of the ML algorithms that may be categorised as both a regression and a classification algorithm. This is a lesson plan for guided learning. It is an essential machine learning module. It is a crucial component of deep learning. It will be utilized for the process of mining the data. The similarity, distance, and vector modules of the dataset are calculated and displayed on the same as depicted in Figure 3. The constant 'k', which may be an integer number, is used to find the location in the supplied dataset that is closest. Data distances for each individual are computed and displayed.

Figure 2. Linear regression.

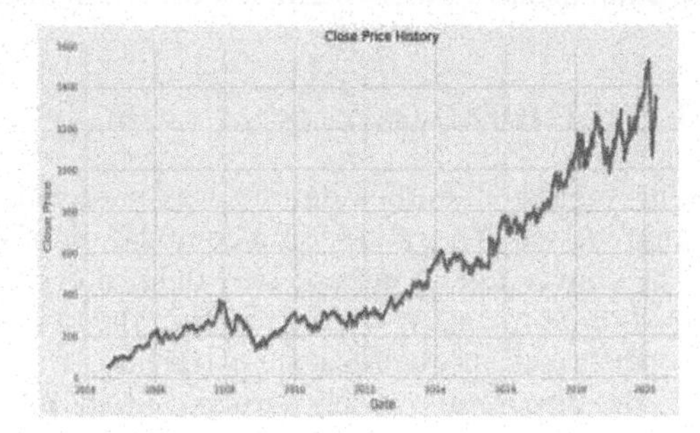

Figure 3. K nearest neighbor.

3.3 *Long Short Term Memory (LSTM)*

The study proposes the use of a neural network model combined with Long Short-Term Memory (LSTM), a type of Recurrent Artificial Neural Network (ANN). LSTM was chosen as it addresses the limitations of prior learning techniques for recurrent neural networks, such as the inability to handle certain issues. LSTMs, unlike traditional recurrent neural networks, make only minor adjustments by combining and multiplying data. The information flow in LSTMs is managed by cell states, which enables them to selectively remember or forget information. This makes LSTMs well-suited for sequence prediction tasks.

4 MODULE IDENTIFICATION

The portions of the proposed work's numerous modules are divided as stated.

4.1 *Data collection*

The first and critical step of the process is the collection of data. It will be used to foresee the market's need to be examined based on a variety of factors and the majority of the data gathered for analysis.

4.2 *Data pre-processing*

Data pre-processing, which is a constituent of data mining, involves the arrangement of raw data. Raw data frequently has a great deal of errors and is inconsistent or lacking.

4.3 *Training the machine*

The process of providing data to the algorithm to enhance its performance on test data is known as training the machine. The models are refined and optimized using the training data sets, while the test data sets are kept unaltered to ensure accurate evaluation. Cross-validation is the major step in the ideal training process that provides a reliable estimate of the pattern's performance based on the training data. Once the model is trained, predictions are made using inputs from the test data set.

5 COMPARATIVE STUDY

On the training and validation datasets, an accuracy plot over training epochs is produced. Additionally, a loss map over training epochs for the training and validation datasets is obtained. Since the accuracy trend on both datasets has been increasing over the last few epochs, it is clear from the accuracy plot that the model may definitely benefit from a little additional training. Additionally, the model exhibits equivalent performance on both data-sets, indicating that it has not yet overlearned the training set. The model performs similarly on the train and validation datasets, as can be seen from the plot of loss (labelled test).

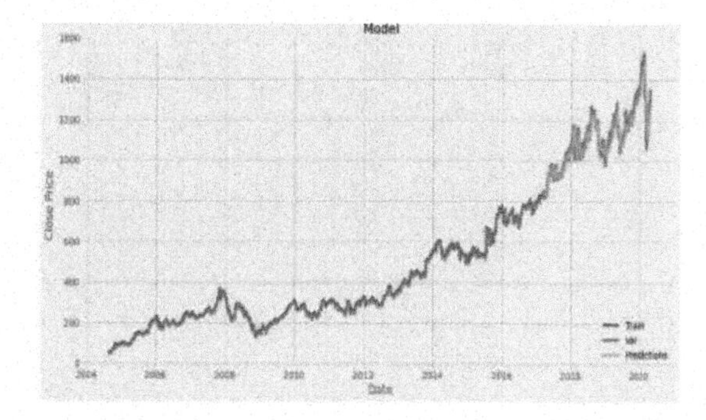

Figure 4. LSTM algorithm.

In the LSTM model, there are several variables that can be altered such as the number of factors in the LSTM layers, dropout values, and the number of epochs. The created LSTM model was then trained on the training data using the fit function. The loss value was observed to decrease significantly during the 100-epoch training process, ultimately reaching a value of 0.4599. The use of test data model that is trained with LSTM network, can be utilized to predict the Microsoft stock's adjacent close value. The LSTM model that has been developed is used in conjunction with the simple predict function to achieve this. Finally, a graph can be produced to compare Adj Close's genuine values and expected values now that the projected values for the test set have been achieved.

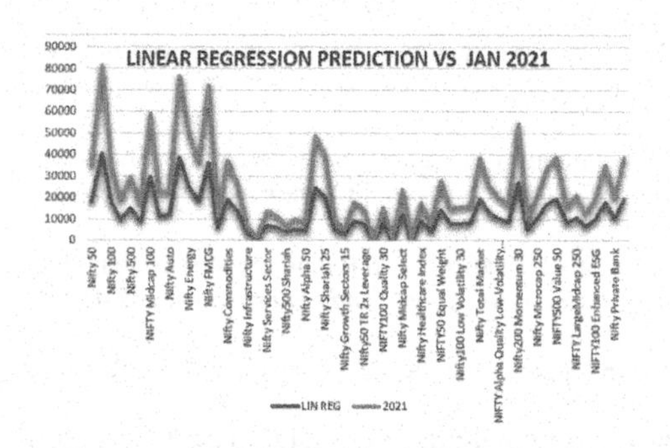

Figure 5. Linear regression prediction.

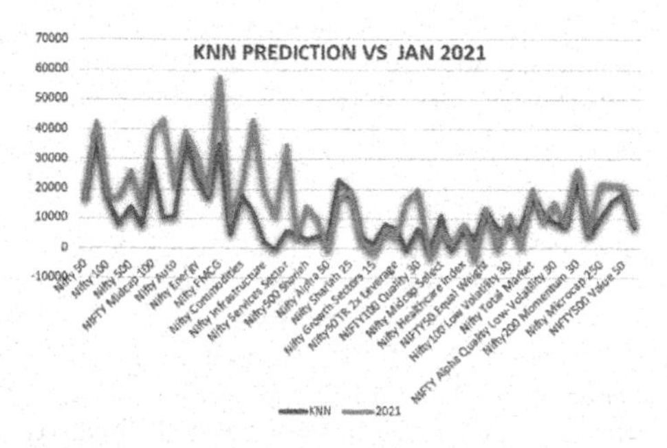

Figure 6. KNN prediction.

With an existing database and the Jan 2021 Stock database, experimental comparative study of various algorithms, including Linear Regression and KNN, was performed. The results are depicted in Figures 5 and 6. Index values for the stock are shown on the X axis, while relevant stock values are shown on the Y axis. One can obtain a more accurate depiction of the stock value of each unique firm by modifying a variety of parameters and increasing the number of LSTM layers in the model.

Comparing the forecast with the January 2021 stock price using the LSTM algorithm shows that the results are more than 90% accurate as shown in Figure 7.

A comparison of multiple algorithms such as linear regression, KNN, and LSTM reveals that LSTM is more accurate than the others in prediction and forecasting applications. This has been confirmed by the results for forecasting the commodities exchange values.

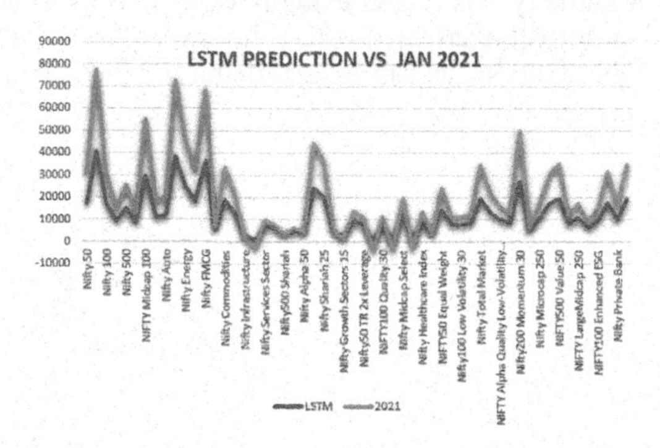

Figure 7. LSTM prediction.

6 CONCLUSION

The comparison of the accuracy of different algorithms revealed that Long Short-Term Memory (LSTM) was the most effective in forecasting the market price of a stock depends on authentic data. LSTMs address a number of challenges that previous learning techniques for recurrent neural networks could not. They have been found to be the most suitable algorithm for a wide range of sequence prediction tasks, outperforming traditional feed-forward neural networks and RNNs. This superior performance is due to LSTMs' ability to selectively retain information over long sequences of time, which allows them to identify patterns in the data.

REFERENCES

Akita, R., Yoshihara, A, Matsubara, T. & Uehara, K. 2016. Deep Learning for Stock Prediction Using Numerical and Textual Information. *2016 IEEE/ACIS 15th International Conference on Computer and Information Science (ICIS) Okayama*. pp. 1–6.

Althelaya, K.A., El-Alfy, E.M. & Mohammed, S. 2018. Evaluation of Bidirectional LSTM for Short- and Long-term Stock Market Prediction. *2018 9th International Conference on Information and Communication Systems (ICICS)*, Irbid, pp. 151–156.

Hegde, M.S., Krishna, G. & Srinath, R. 2018. An Ensemble Stock Predictor and Recommender System. *2018 International Conference on Advances in Computing, Communications and Informatics (ICACCI)*, Bangalore, 2018, pp. 1981–1985.

Idrees, S.M., Alam, M.A. & Agarwal, P. 2019. A Prediction Approach for Stock Market Volatility Based on Time Series Data. in IEEE Access, vol. 7, pp. 17287–17298.

Izumi, K., Goto, T. & Matsui, T. 2010. Trading Tests of Long-Term Market Forecast by Text Mining. 2010 IEEE International Conference on Data Mining Workshops, 2010, pp. 935–942.

Liu, G. & Wang, X. 2019. A Numerical-Based Attention Method for Stock Market Prediction with Dual Information. IEEE Access. vol. 7, pp. 7357–7367.

Loret, JB. & Shajilin. Recognize the Image Caption Generator Using CNN and LSTM with NumPy functions Solid State Technology. 64.2 (2021): 2181–2191.

Pang, Xiongwen., Zhou, Yanqiang., Wang, Pan. & Lin, Weiwei. 2018. An Innovative Neural Network Approach for Stock Market Prediction.

Ramya Devi, R., Abinaya, C., Priyadharshini, B. & Sumathi, Y. 2020. E-Commerce Warehousing Management System using Logistic Regression. VDGOOD Journal of Computer Science and Engineering, Vol. 1, Issue 1, pp. 393–397.

Raza, K. Prediction of Stock Market Performance by Using Machine Learning Techniques. *2017. International Conference on Innovations in Electrical Engineering and Computational Technologies (ICIEECT)*, Karachi, 2017, pp. 1–1.

Sai Sravani, K. & Raja Rajeswari, P. 2020. Prediction of Stock Market Exchange Using LSTM Algorithm. IJSTR vol. 9, pp. 417–421.

Zhao, L. & Wang, L. 2015. Price Trend Prediction of Stock Market Using Outlier Data Mining Algorithm. 2015 IEEE Fifth International Conference on Big Data and Cloud Computing. pp. 93–98.

Artificial Intelligence, Blockchain, Computing and Security – Dagur et al. (Eds)
© 2024 The Author(s), ISBN: 978-1-032-67841-2

Auction platform using blockchain

G. Nagappan*, S. Ravikumar*, R. Rishabendran* & R. Sri Hari Krishnan*
Department of Computer Science and Engineering, Saveetha Engineering College, Chennai, India

ABSTRACT: An online auction is an auction as similar to an in-person auction but it is held over the Internet. It is an easy and a popular way to buy and sell products and services using cryptocurrency. Online auctions have grown to be a significant contributor to the e-commerce sector. But there exist some problems that concerns data integrity, legitimacy and traceability of bidding. The proposed system is an auction platform based on Ethereum smart contracts which are stored in Blockchain. An auction in a decentralized network, powered by blockchain technology eliminates the need for a middleman and the high costs associated with bidding and buying in an auction. Decentralized applications increase the transparency and flexibility of applications. The intricate nature of blockchain technology and the challenges associated with its integration require specialized knowledge that differs from the conventional approach to application development.

Keywords: Blockchain, smart contracts, online auction, Ethereum

1 INTRODUCTION

An auction is a place or a platform where people can buy and sell things at a reasonable or nominal price through a competitive bidding among them. The bidder who places the largest amount for the product is the auction winner and can purchase the product by providing specified amount. Similarly, tender contracts are the contracts where the product or service is given to the bidder who quotes the least amount among all the other bidders. The auctioneer performs the auction and always penalizes the dealer a cash for their act of assistance, which is typically a percentage of the product's total selling price. The communication between auctioneer, merchant and buyers in current online auctions occurs through a centralized mediator providing a program for product proposals, bid conformation and declares the highest bidder. This intermediary may request a commission and must be trusted. We use Ethereum smart contracts to design and implement auction and tender contracts in this proposed system. Ethereum is the most popular blockchain that enables the execution of smart contracts within an Ethereum virtual machine, without depending on other parties. The proposed system is a decentralized application which is also called as Decentralized app. (Taş *et al.* 2019). A decentralized application is a type of application which runs on a blockchain network than running on a single computer. Decentralized app is strikingly similar to other software applications available on a website but they are P2P supported. This app is not dominated by a single authority. Decentralized app is based on decentralized network, backed up by a blockchain distributed ledger. This decentralized application is built using Ethereum platform.

*Corresponding Authors: nagappan.cse@saveetha.ac.in, ravisaravanan209@gmail.com, rrishabendran652@gmail.com and harikrrish1235@gmail.com

 DOI: 10.1201/9781032684994-75

2 LITERATURE SURVEY

In this survey we review the implementation of blockchain in auction and tender using smart contracts. Braghin *et al.* (2020) analyzed the cost that is used for contract deployment, bidding and explained English auction which is the type of auction taken as consideration for this auction platform. Li *et al.* (2021) gave insight on the smart contracts and how the user interacts with the contracts which in turn stored in the blockchain. Omar *et al.* (2021) proposed to store data in a decentralized storage system such as Inter Planetary File System (IPFS) as blockchain has limited amount of storage which will be expensive. Babu *et al.* (2018) presented an e-governance system for bidding for tenders that are more specifically targeted to organizations. Cai *et al.* (2018) explained in detail about development of Decentralized app on blockchain. Doan *et al.* (2022) listed out the properties of the IPFS in the E-commerce and marketplace. Chen *et al.* (2018) specified the requirements that are required for an e-auction system and functions which are the fundamental parts of the smart contract. Cai *et al.* (2018) explained the application scenarios and characteristics of Decentralized app along with the advancement of blockchain development with payment channels and non-public blockchains. Buterin (2013) founded the Ethereum blockchain which paved the way for the next generation smart contracts that are stored and executed in the blockchain and decentralized applications (Decentralized app). Taş *et al.* (2019) provides insights on tools required to develop applications using Ethereum with a detailed explanation of all the considerations and prerequisites including smart contract, development environment and solidity language for creating a Decentralized app. Gutte *et al.* (2022) provides insights on how to create a decentralized application using Ethereum blockchain with the functional requirements and system architecture along with the explanation of front end and backend processes including MetaMask wallet, node.js. Nakamoto (2008) has proposed a decentralized system where transactions can happen without an intermediary with the proposal of peer-to-peer network using proof of work to record transactions therefore, making attack impossible to an attacker.

3 PROPOSED SYSTEM

Users will need to register for an account on the platform in order to participate in auctions. This could involve providing personal information such as name, MetaMask wallet address, and payment information. Sellers will be able to list items for auction by providing details such as the item description, starting price, and auction duration. Buyers will be able to place bids on items by submitting a bid amount and paying for the bid using cryptocurrency or another form of payment Sellers will have the option to close auctions early if they receive an offer that they are willing to accept. When an auction ends, the highest bidder will be declared the winner and will be charged the winning bid amount.

4 METHODOLOGY

4.1 *Blockchain*

A blockchain is a distributed ledger that is ordinarily maintained by a peer-to-peer network backing a rule for communication and validation among nodes. There is absence of central repository, absence of database located on a file server where there is a possibility of data being compromised: it is a decentralized data structure that is duplicated and distributed among all network participants.

4.2 *Smart contracts*

"A computerized transaction protocol that carries out a contract's provisions" according to the definition of smart contract. The plan was to codify and incorporate contractual provisions such as collateral and bonding into software, which reduces the trusted intermediaries needed between transacting parties.

4.3 *Blockchain based auction platform*

The previous e-auction systems use a centralized system which involves security issues. In this proposed system, we use decentralized system which uses multiple nodes instead of a centralized server and achieves security.

4.4 *Solidity*

Solidity is an object-oriented programming language and used to implement smart contracts. Solidity is a language designed to run on the Ethereum Virtual Machine (EVM). It uses C++, Python and JavaScript. Solidity is statistically typed, which refers to building of smart contracts which run above EVM.

4.5 *Interplanetary File System (IPFS)*

IPFS, The Interplanetary File System (IPFS) is a globally distributed protocol for file storage that operates as a large-scale, peer-to-peer network, enabling computers worldwide to store and share files. By installing IPFS on their personal computer, a user can upload files to the IPFS network, making them accessible to other IPFS users globally. Every file added to IPFS is given a unique address based on a hash of its content. This is known as Content Identifier (CID) (Omar *et al.* 2021).

4.6 *System architecture diagram*

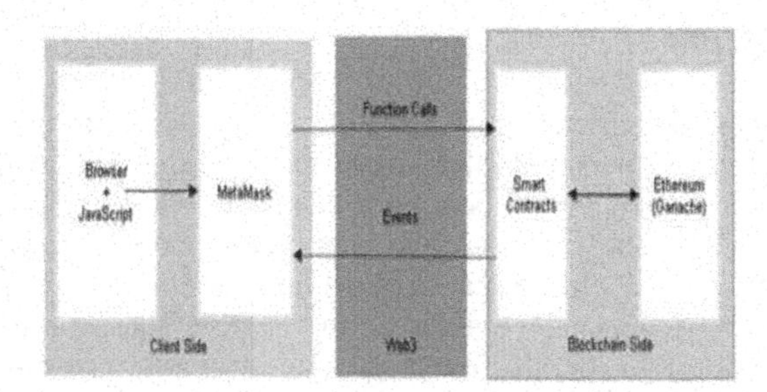

Figure 1. System architecture diagram.

4.6.1 *Smart contracts*

A smart contract is a piece of code containing business logic that is deployed on the blockchain, capable of executing all transactions on the network. Smart contract is a piece of code which runs on blockchain and is used to carry out, and implement agreement's conditions.

4.6.2 *Ethereum blockchain*

The system will maintain the URIs of Auctions and Tenders, as well as the smart contract and all transactions performed by users.

4.6.3 *IPFS*

This is a peer-to-peer network of file storage devices where files uploaded for tokenization are stored, and the back-end uses the IPFS protocol to access and connect to this network.

4.6.4 *MetaMask*

MetaMask is a browser extension that acts as an Ethereum wallet, allowing users to store, manage, and send Ethereum and other Ethereum-based tokens, as well as interact with decentralized applications (Decentralized apps) on the Ethereum network (Gutte *et al.* 2022).

4.6.5 *React app*

This web application, constructed using ReactJS, serves as the primary interface between the user and the back-end system. Communication with the back-end server is established through REST APIs, while accessing the wallet is facilitated through the use of Web3 libraries and JSON-RPC.

4.6.6 *Node.js*

The Auction/Tender Platform's business logic resides on this server, which communicates with the client through REST APIs and the Ethereum Blockchain through smart contracts. It also features an integrated SQL database.

4.7 *Module description*

Auction: In the auction platform we create an auction using smart contracts. The smart contract contains user address and bid amount and the users start bidding for the product. After the auction duration we finalize the highest bid amount and that user's address. The product is sold to the person with the highest bid amount. The auction page contains the name, description, image, auction duration, seller address, highest bid amount, highest bid address of the product. After the auction duration, the seller verifies the buyer's address and proceeds to sell the product.

Tender: In the tender contract page we create a tender using smart contracts. The smart contract contains user address and bid amount and user starts bidding. After the tender contract duration we finalize the lowest bid amount and that user's address. The product is sold to the person with the lowest bid amount. The user has the choice to outbid the previous lowest bid or can withdraw themselves from the tender of that particular product. After the tender contract duration, the seller verifies the buyer's address and proceeds to provide the service.

IPFS Storage: The Inter Planetary File System (IPFS) is a decentralized system for storing and accessing files, websites, applications, and data. In the blockchain world, each user runs its node. The nodes communicate with each other and exchange files among them. We create the IPFS storage in Web3 Storage domain. We store the details of the auction/tender in the CID format and we retrieve using the CID key.

Transaction: In Ethereum, a "transaction" refers to a signed data package that carries a message from one account to another account on the blockchain. The gas Limit field specifies the maximum computational steps that a transaction is allowed to consume in order to determine the cost of the computation. The gas Price field, which represents the price the sender is willing to pay for gas, and the user's signature, which identifies the user, are both included in a transaction. Then gas Limit * gas Price is the maximum amount of ether that can be spent on a transaction.

Product & Store: On the product page you can find all details regarding the product (seller, quantity, price). PENDING: just created and waiting for seller acceptance. SENT: product sent and waiting for buyer receiving confirmation. COMPLETED: The order is complete so seller receive payment and buyer can leave a review. The Decentralized app enables sellers to create their own order-based stores. They can add limited and unlimited quantity products. Each seller willing to create a new store must provide a name.

5 IMPLEMENTATION AND EVALUATION

We have developed a prototype of a smart contract functionalities for auction.

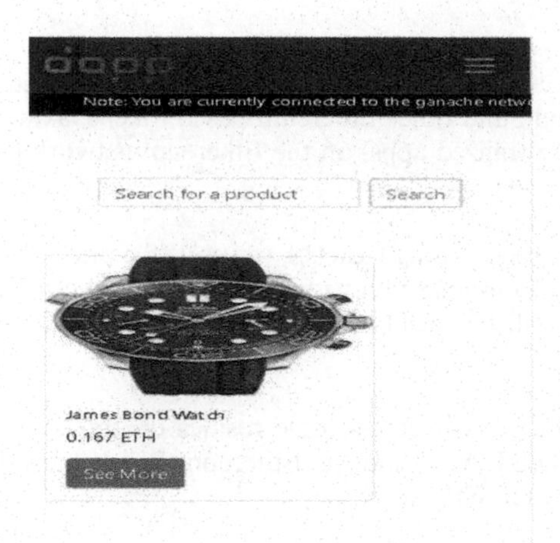

Figure 2. Bidding function in solidity.

```
function bid(uint256 _auctionId) public payable {
  Auction memory auction = auctionsList[_auctionId];
  require(block.timestamp < auction.endTimstamp, "Auction Ended");
  bool isInAuctionBidders = _isBidder(msg.sender, _auctionId);
  if (isInAuctionBidders) {
    require(
      auctionBidsMapping[_auctionId][msg.sender] + msg.value >
        auction.highestBid,
      "insuffisant amount"
    );
    auctionBidsMapping[_auctionId][msg.sender] += msg.value;
  } else {
    require(msg.value > auction.highestBid, "insuffisant amount");
    auctionBidsMapping[_auctionId][msg.sender] = msg.value;
  }
  auction.highestBid = auctionBidsMapping[_auctionId][msg.sender];
  auction.highestBidder = msg.sender;
  auctionsList[_auctionId] = auction;
}
```

Figure 3. Auction for a watch.

6 CONCLUSION AND FUTURE WORK

In this proposed work, we demonstrated how Ethereum blockchain may be used to create auctions and tenders. The use of smart contracts in comparison to classical online auctions provides benefits such as increased integrity, as a bidder cannot alter their own bid or the bid of others. Additionally, there is enhanced transparency as the bidding history is publicly accessible and verifiable by all users, and non-repudiation is maintained, as a bidder cannot retract their bid. Furthermore, there is no need for a trusted third party. In future research, our goal is to enhance the developed application to address scalability issues and align it with widely accepted security standards within the auction and tender platform domain (Babu *et al.* 2018).

REFERENCES

Babu, M.N., Gajalakshmi, K., Murthy, K.S.N., Krishna, T.S., UshaKiran, P. 2018. Decentralized E-bidding Governance Application using Blockchain, *International Journal of Management, Technology and Engineering* 8(12): 2707–2713.

Braghin, C., Cimato, S., Damiani, E., Baronchelli, M. 2020. Designing Smart-contract Based Auctions, *Security with Intelligent Computing and Big-data Services*: 54–64.

Buterin, V. 2013. *Ethereum White Paper: A Next Generation Smart Contract & Decentralized Application Platform*, <https://ethereum.org/en/whitepaper>.

Cai, C., Duan, H., Wang, C. 2018. Tutorial: Building Secure and Trustworthy Blockchain Applications, *IEEE Cybersecurity Development (SecDev)*: 120–121.

Cai, W., Wang, Z. Ernst, J., Hong, Z., Feng, C. 2018, Decentralized Applications: The Blockchain-Empowered Software System, *IEEE Access* 6: 53019–53033.

Chen, Y., Chen S., Lin, I. 2018. Blockchain Based Smart Contract for Bidding System, *IEEE International Conference on Applied System Innovation*: 208–211.

Doan, T.V., Ott, J., Psaras, Y., Bajpai, V. 2022. Towards Decentralised Cloud Storage with IPFS: Opportunities, Challenges, and Future Directions, *IEEE Internet Computing* 26(6):7–15.

Gutte, Y., Vora, A., Sharma, Y., Bhardwaj, B. 2022. NFT Marketplace Based on Ethereum Blockchain, *International Journal of Advanced Research in Science, Communication and Technology* 2(3): 179–186.

Li, H., Xue, W. 2021. A Blockchain-Based Sealed-Bid e- Auction Scheme with Smart Contract and Zero-Knowledge Proof, *Security and Communication Networks*: 1–10.

Nakamoto, S. 2008. *Bitcoin: A Peer-to-Peer Electronic Cash System*, <https://bitcoin.org/bitcoin.pdf>.

Omar, I., Hasan, H., Jayaraman, R., Salah, K., Omar, M. 2021. Implementing Decentralized Auctions Using Blockchain Smart Contracts, *Technological Forecasting and Social Change* 168.

Taş, R. and Tanrıöver, O.O. 2019. Building A Decentralized Application on the Ethereum Blockchain, *International Symposium on Multidisciplinary Studies and Innovative Technologies (ISMSIT)*:1–4.

Artificial Intelligence, Blockchain, Computing and Security – Dagur et al. (Eds)

Image recognition based attendance system with spoof detection

K. Rakesh*, B. Sumer Singh*, A. Surya* & R. Kaladevi*
Department of Computer Science and Engineering, Saveetha Engineering College, Chennai, India

ABSTRACT: In recent times numerous face recognition algorithms were used for the identification and authentication of a person to a system. The objective of this design is to recognize the human faces and forecast whether the detected faces are real or fake. Our design is mainly useful for institutional purpose to enhance and modernize the present attendance system into more effective and efficient than before. In this design, human face database is dynamically created by new user registration and this data is pumped into the recognizer algorithm. In this design we have added a spoofing discovery technology that's used to judge whether the face in front of the machine is real or fake. The face presented by other media can be defined as false face, including hardcopy photographs, display screen of electronic products, etc. The dataset which are dynamically created while a new user is registered is stored in the database and whenever a new face is detected on the camera, it compares with the images that are already present in the database to mark the attendance. Also, the result image is generated with the chance of spoofing can be attained from the dataset pushed into the algorithm. Our proposed system has accuracy of above 97.6% and true positive rate is 99.7% in high precision model.

Keywords: False face, True positive, High precision, Spoofing discovery

1 INTRODUCTION

The ideal of our designed system is to recognize the human faces and forecast whether the detected faces are real or fake with a lower false-positive rate in detecting new faces by applying a threshold and saving the images with proper editing to the database. Our motive is to get an advanced true positive rate and lower false positive rate in chancing both the faces for attendance and also predicting the face whether it's real or fake. We use Dlib, OpenCV, pillow and face-recognition libraries for collecting the dynamically created dataset of images and cropping them into the needed sizes and training them with the algorithms to descry the faces for marking their presence. Faster-RCNN abbreviated as Faster region based convolutional neural network algorithm is used for detecting the faces in a faster way with high delicacy. The proposed system is added with a feature called spoof discovery which helps to find between the real and fake faces that are shown in front of the camera. substantially torch, torchvision, pillow and tensorboardX libraries are used for prognosticating the chance of the image whether it's real or fake.

The being systems are done with the common algorithms used for chancing the face. It doesn't have high delicacy and it accepts all the faces that are shown in the camera. Fake faces can be shown in front of the camera to mark attendance for a person who isn't physically present. Being design uses several common algorithms like OpenCV, GA- FREAK,

*Corresponding Authors: rakeshpandees@gmail.com, singhsumer0644@gmail.com, suryacsework@gmail.com and kaladevi@saveetha.ac.in

 DOI: 10.1201/9781032684994-76

GA- ORB, TSRF algorithms which have lower delicacy rate compared to this design. A convolutional neural network(CNN) is a specific type of machine learning algorithm based artificial neural network which is used for the faster way of recognition in image processing. To simplify the CNN model, the complication and slice layers are combined into a single caste. predicated on the formerly trained network, greatly meliorate the image recognition rate. Face identification is achieved using several process such as deep learning, recognizer algorithms and CNN's. These multi-subcaste network are being trained to perform a specific task of image identification. In image recognition Haar cascade frontal xml package and CNN shows the output with high accuracy. To make our face recognition system, we will first perform face prediction, extract face features from each face using image processing, train a face recognition model on the embeddings, and also ultimately identify faces via both images and video capturing with OpenCV. Face recognition – also called facial discovery – is an artificial intelligence (AI) rested computer technology used to find and identify mortal faces in digital images.

2 LITERATURE SURVEY

(Jae *et al.* 2020) suggested this technique to use Gabor face representations to create DCNN-based FR frameworks that perform better. Due to the convolution of the face pictures, the Dimensionality Gabor feature space is substantially larger than it should be. In accordance with the study, (Xin *et al.* 2008) enhance the majority of face recognition systems, faces been foraged into them in accordance with particular guidelines, such as under controlled illumination, under a specific view angle, and without any obstacles. Under controlled conditions, these systems are known as facial recognition systems. These rules are difficult to satisfy, which restricts the usage of facial recognition in many instantaneous applications.

(Zhilu *et al.* 2019) proposed automatic facial recognition to find the accuracy as it increased to over 95% by using weighted fusion algorithms to combine the sparse representation results obtained from the original and tagged data. However, it has certain limits because the method still needs additional work, particularly with regard to the scarcity of face samples. This technique is proposed by (Yueqi *et al.* 2018), which is known as topology-preserving graph matching (TPGM), estimates a non-rigid transformation conveying increasing accuracy. Robust point set matching and the Gabor ternary pattern only measure the similarity of the nodes, which is a downside. (Fang *et al.* 2018) proposed this technique to use a multiscale spatial LSTM encoder that tries to remove recurring occlusions and encode faces that are occlusion-resistant. The Data-Specific, their utility is limited to data that is significantly comparable to its training data, which is a negative.(Yomnasafaa *et al.* 2020) proposed a biometric presentation attack detection (PAD) used in this method has very low error rates while still being effective.The disadvantage of it is that it degrades accuracy when compared to networks that were trained for each individual biometric independently.

(Huang *et al.* 2020) proposed a technique to revisits a cutting-edge face spoofing detection technique based on a depth-based Fully Convolutional Network (FCN). The disadvantage is that for local tasks with insufficient training samples, local label supervision is preferable to global label supervision. (Weiping Chen *et al.* 2019) suggested to use an innovative ensemble string matching method to achieve non-sequential string matching between two String faces. Due to the demand for globally sequential representation and the complexity of human faces, which comprise discontinuous and non-sequential information, String matching has the disadvantage that it is not ideal for frontal face recognition. Neural Architecture Search (NAS), using this technique , produces improved segmentation results with continuous and discernible segmented edges as well as greater image details were analyzed by (Zhimin *et al.* 2020). The fact that generated neural networks are mostly used for natural language processing and picture classification is a downside. By combining discrete wavelet transforms (DWT) and discrete cosine transform (DCT), (Lukas *et al.* 2016) suggested a strategy for a student attendance system in a classroom that uses facial recognition technology. These methods were utilized to extract the facial features of the learner, and then Radial Basis Function (RBF) was applied to categories the facial objects. This system had an 82% accuracy rate.

In order to identify a student using a face recognition system in a classroom setting, (Abhishek *et al.* 2019) intends to introduce a novel method, namely the creation of a 3D facial model. This study aims to develop an automated attendance system that records students' attendance in lectures or sections and assesses their performance in line with that attendance using facial recognition technology from an image or video stream. In order to compare the Receiver Operating Characteristics (ROC) curve and find the optimum facial recognition algorithm, (Siswanto *et al.* 2014) conducted research to enhance its own feature. The ROC curve demonstrated that Eigenface outperforms Fisher face in the studies conducted for this paper. A system that used the Eigenface algorithm had a 70% to 90% accuracy rate.

3 EXISTING SYSTEM

Numerous projects deal with the system that automatically recognizes attendance based on continual observation, aiding in estimation and performance improvement. The postures and facial photos of the pupils in the classroom are taken in order to record attendance. The system determines each student's seating arrangement and location for attendance purposes through ongoing monitoring and recording. But in order to produce more quickly, we must upgrade far more advanced technology.

- The present system has a few drawbacks, including the use of the mobilenetv2 algorithm and the SSD (Single Shot multibox Detector) technique, which have lower accuracy rates.
- Shallow layers in a neural network might not produce enough high-level characteristics to do prediction for tiny objects. As a result, SSD performs poorly for smaller items than for larger ones.
- The requirement for extensive data augmentation also implies that it requires a vast amount of data to train.

4 PROPOSED SYSTEM

4.1 *Face detection and recognition*

In order to recognise a person's face more precisely, a set of algorithms is employed. These algorithms measure minutiae like the distance between the eyes and chin contours, which are then transformed into mathematical representations. These mathematical representations aid the system in properly identifying faces. To do this, we make use of specific Python libraries.

- Pandas: The most popular open-source Python library which are used for machine learning algorithms, data pre-processing.
- Pillow: This library's primary goal is to provide the most fundamental image processing capabilities.
- TensorboardX: It gives machine learning workflows the metrics and visuals they require.
- Face Recognition: Face detection is a straightforward built-in feature that determines whether a picture contains any faces.
- Cmake: Cmake is primarily used to control the software development process using a compiler-independent approach.
- Dlib: Pre-trained models mostly handle this, mapping a person's facial points using co-originates (x, y). The most potent open-source libraries for detection are these.

4.2 *Spoof detection*

We are utilizing the spoofing approach in order to create a secure face detection. Facial spoofing is the practice of stealing someone's identity and storing it in a database by utilizing their face and replicating their facial biometrics in a photo or video. The saved image will be examined to detect whether the face is fake or not. The primary criteria needed to use spoofing are,

- Easydict: Its primary application is for recursively accessing dictionary values as attributes. This library will be consulted when needed for the keyword.
- Numpy: It offers mathematical operations but relies heavily on working with arrays.
- Tqdm: For making Progress Meters or Progress Bars, utilise tqdm. Since tqdm can be wrapped at any iteration.
- Pytorch: This library is used in deep learning approach for image processing and is mostly utilized in GPU and CPU applications.
- Torchvision: It includes datasets, several widely used pre-trained models, and tools for quick image and video transformations.

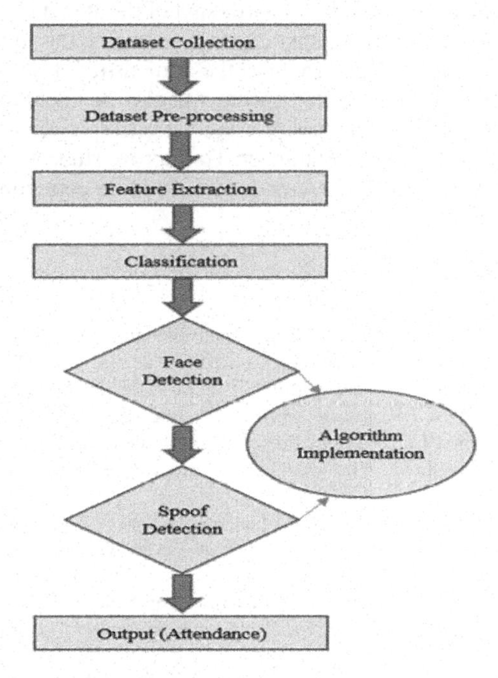

Figure 1. Flow chart for proposed system.

5 RESULT

Thus we have designed an Automated system for marking attendance with spoof detection which enables the machine to find the person with more accuracy in an effective and efficient way. In this we have used two algorithm, one for face feature extraction and recognition and another for anti-spoof detection which uses the images that are available in the dataset. In Figure 1, the proposed system successfully builds the face database for face detection. Sometimes a classroom's poor lighting conditions might have an impact on image quality, which in turn negatively impacts system performance. This can be remedied afterwards by enhancing the video quality or by utilizing a variety of algorithms. A feature-based object detection system called Haar Cascade is used to identify things in photos. For detection, a cascade function is trained with several dataset which are dynamically created. The approach can run in real-time and doesn't call for a lot of computing. The image is fake or real are determined from the dataset by the spoofing detection. The percentage of the image that is real or phoney indicates whether the image is real or fake.

The facial recognition methods in dlib are wrapped in a straightforward, user-friendly API by the Python package face recognition. Using OpenCV, it performs face extraction and identification. Face processing is a method for locating or authenticating the face in digital photographs or video. A human can easily and rapidly recognize the faces. For us, it is a simple task, but for a computer, it is challenging. Also, the result image is generated with the chance of spoofing can be attained from the dataset pushed into the algorithm. Our proposed system has accuracy of above 97.6% and true positive rate is 99.7% in high precision model.

6 CONCLUSION

This technology seeks to develop a powerful face recognition and spoof detection system. By using image processing algorithms, the system will be able to mark the attendance, and it will subsequently store the information in a database. A new dataset of the attendance recorded by the system will be gathered as part of our upcoming upgrade. It then pushes this data into an algorithm for determining the attendance defaulters and notifies the administration once it has collected about monthly data. The accuracy of the suggested system will also be improved in the future utilizing deep learning or other cutting-edge techniques. We also catch trespassers who approach the camera and alert the administrator to look for the stranger. To improve student identification and prevent being duped by outsiders, voice recognition can be added to face detection.

REFERENCES

Abhishek Jha, "Class Room Attendance System Using Facial Recognition System", *The International Journal of Mathematics, Science, Technology and Management* (ISSN: 2319-8125) Vol. 2 Issue 3, 2016.

Choi J. Y. and Lee B., "Ensemble of Deep Convolutional Neural Networks With Gabor Face Representations for Face Recognition," in *IEEE Transactions on Image Processing*, vol. 29, pp. 3270–3281, 2020.

Duan Y., Lu J., Feng J. and Zhou J., "Topology Preserving Structural Matching for Automatic Partial Face Recognition," in *IEEE Transactions on Information Forensics and Security*, vol. 13, no. 7, pp. 1823–1837, July 2018.

Geng X., Zhou Z.-H. and Smith-Miles K., "Individual Stable Space: An Approach to Face Recognition Under Uncontrolled Conditions," in *IEEE Transactions on Neural Networks*, vol. 19, no. 8, pp. 1354–1368, Aug. 2008.

Hani Mahdi, Yomnasafaa El-Din, Mohamed N. Moustafa."Deep Convolutional Neural Networks Forface and Iris Presentation Attack Detection" in *IEEE Transactions on Neural Networks*, vol. 19, no. 8, pp. 1354–1368, Aug. 2020.

Lukas, Samuel, *et al.* "Student Attendance System in Classroom Using Face Recognition Technique." *2016 International Conference on Information and Communication Technology Convergence (ICTC)*. IEEE, 2016

Siswanto R. S., Nugroho A. S. and Galinium M., "Implementation of Face Recognition Algorithm for Biometrics Based Time Attendance System," *2014 International Conference on ICT For Smart Society (ICISS)*, Bandung, Indonesia, 2014.

Sun W., Song Y., Chen C., Huang J. and Kot A. C., "Face Spoofing Detection Based on Local Ternary Label Supervision in Fully Convolutional Networks," in *IEEE Transactions on Information Forensics and Security*, vol. 15, pp. 3181–3196, 2020.

Weiping Chen Yongsheng Gao "Face Recognition Using Ensemble String Matching" in *IEEE Transactions on Image Processing*, vol. 27, no. 2, pp. 778–790, 2019.

Xu Z., Zuo S., Lam E. Y., Lee B. and Chen N., "AutoSegNet: An Automated Neural Network for Image Segmentation," in *IEEE Access*, vol. 8, pp. 92452–92461, 2020.

Zhao F., Feng J., Zhao J., Yang W. and Yan S., "Robust LSTM-Autoencoders for Face De-Occlusion in the Wild," in *IEEE Transactions on Image Processing*, vol. 27, no. 2, pp. 778–790, Feb. 2018.

Zhilu Doongmei Jiang Yujun Li, Yankun Cao Mingyu Wang, Yong Xu, "Automatic Face Recognition Based on Sparse Representation and Extended Transfer Learning" *IEEE Access* 2019.

Artificial Intelligence, Blockchain, Computing and Security – Dagur et al. (Eds)

Person face re-identification using deep learning approach

N.V. Ravindhar, M. Mohamed Kasim Raja, S. Naveen Prabhu & H. Durgesh
Saveetha Engineering College, Chennai, Tamil Nadu, India

ABSTRACT: This article discusses individual re-identification (Re-ID) using a deep learning approach. Its ability to identify important individuals from security camera footage has sparked a lot of attention. The rising amount of surveillance films has resulted in high computing and storage costs, which has posed a substantial hurdle for customers with limited resources. However, individual Re-ID over paid surveillance footage may pose a security risk by allowing the privacy of the uninvolved subject to be violated. As a result, the proposed system uses the haar cascade to successfully re-identify a person while safeguarding their privacy over contractual surveillance films. Both re-identification and privacy protection are possible with this technique for the person who was spotted. Deep learning will in particular extract elements that are useful for human re-identification and assess if a certain person's face has been accurately identified.

Keywords: Face reidentification, Image processing

1 INTRODUCTION

The aim of face re-identification is to find an individual who has been seen in a video before, whether the cameras that captured the footage are close together or far apart. This technology has various uses in video surveillance, such as tracking a person's movements as they go from one camera's view to another, or monitoring their entry and exit from a single camera's field of view. It is based on a person's physical characteristics to ensure correct identification, reducing the chance of misidentification. Face re-identification has been utilized to address the issue of multiple athletes appearing in videos, where their movements may overlap and cause confusion about their identities. This is a complex problem to solve due to factors like occlusions, lighting changes, and differences in position and viewing angle as the person moves or switches between camera views.

2 LITERATURE SURVEY

Throughout the literature review, we obtained information on the person re-identification procedure that is now in use. Lin Wu *et al.* suggests the use of end-to-end convolutional neural networks to create global video-level features for face re-identification. Qingming Leng *et al.* conducts a thorough examination of open-world face re-identification, categorizing previous efforts into limited and extensive categories. Yiqiang Chen *et al.*, Sanping Zhou *et al.* and Zhong Zhang *et al.* propose a method for identifying pedestrian attributes and a CNN-based framework for improved face re-identification with the use of pedestrian attributes. Lixia Zhang *et al.* proposes an improved BOF algorithm-based person re-identification method. Shan Zhong *et al.* uses the multiscale retinex algorithm with color restoration to pre-process images and enhance lighting conditions, reducing the impact of these factors on color information.

DOI: 10.1201/9781032684994-77

An embedded platform-based image processing system utilising a BLACKFIN 548 DSP processor is suggested by Anitha Julian *et al.* Abimathi *et al.* improves the performance of steganalysis detection, but it is difficult for an analyser with limited processing power to implement for larger image data. To help scholars better grasp this topic and provide a more organised framework, Hongbo Wang *et al.* gives a thorough summary of person re-identification techniques. An innovative approach is put out by Guangcai Wang *et al.* to address this issue and boost the reliability of person reidentification. It integrates an attention mechanism, hard sample acceleration, and similarity optimization. A technique for simultaneously learning features and a related similarity metric for face re-identification was put forth by Ejaz Ahmed *et al.*

3 ALGORITHMS

3.1 *CNN algorithm*

Pose Estimation: The pedestrian component is first produced using the CPM method, then it is normalized using the PTN to obtain the pose estimation.

3.1.1 *Image segmentation*

To reduce misalignment caused by changes in position, the joints of the walkers should be prepared first. To emulate the long-term memory among variables, the pose generator adopted the CPM method to incorporate a CNN neural network. Unlike conventional methods, CPM employed a serial convolutional architecture to transfer dimensional and grain details, and included an observation step to prevent slope disappearance.

The picture of the pedestrian is broken down into seven parts for a comprehensive evaluation of both its overall and specific features. The three significant segments are indicated as ta 1 = {t1, t2, t3, t4}, ta 2 = {t3, t4, t5, t6, t7, t8, t9, t10}, and ta 3 = {t9, t10, t11, t12, t13, t14}. The four smaller parts are identified as tb 1 = {t3, t5, t6}, tb 2 = {t4, t7, t8}, tb 3 = {t9, t11, t12}, and tb 4 = {t10, t13, t14}. The bounding boxes for each of the seven parts are represented as H = {Ha 1 , Ha 2 , Ha 3 , Hb 1 , Hb 2 , Hb 3 , Hb 4 }, and the bounding box for the "jth" part can be calculated using the following formula:

$$H_j = \left[x_j, \min, x_j, \max, y_j, \min, y_j, \max\right] = \left[\min(x_j), \max(x_j), \min\left(y_j\right), \max\left(y_j\right)\right] \quad (1)$$

3.1.2 *Extraction of PIF*

The Person Identity Feature (PIF) is obtained from both the actual and corrected pictures using a pseudo-Siamese architecture. This architecture consists of two distinct networks, with the upper network responsible for processing the real images and the lower network processing the rectified images. The twin structure has two levels, one for the real images and another for the corrected images. The final layer is produced by combining the outputs from two equivalent networks and a third network that integrates the outputs of both equivalent networks.

The total Cross-Entropy loss for training the entire pseudo-Siamese network is structured as,

$$k = \sum_{g=1}^{3} k_g \quad (2)$$

where k_g for $1 \leq g \leq 3$ represents the jth Cross-Entropy loss, and it can also be expressed as

$$k_g = \sum_{s=1}^{S}\sum_{i=1}^{N} y_s(j)\log\widehat{y}_s(j) + (1 - y_s(j))\log(1 - \widehat{y}_s(j)) \quad (3)$$

Let N denote the total number of individuals and S be the total number of categories for each individual. Then, $y_s(j)$ represents the actual value of the s-th output for the j-th individual, and $y_k(j)$ represents the predicted value generated by the g-th network. Although the initial structures of both branches are identical, their convolutional layers have separate weights due to variations in the actual and modified images. Once the fifth pooling layer is reached, the loss functions are computed and merged to facilitate error propagation and regulate the parameters.

During the training process, the n images in the dataset are divided into a training set and a testing set. The testing set comprises n/4 of the total images, while the remaining images are used for training and fed into the CNN blocks for training. To calculate neighbouring elements, a local bias factor is incorporated that uses the cosine distance metric.

4 METHODOLOGY

Training Dataset — The input video file is used to divide the picture data into training images. The model learns from these photos in order to later generalise to another image, while the training set contains known outputs. It will test our models using the test video and the Keras technique via Haar Cascade methodology in Python using the Tensor Flow library.

Construction of a Detecting Model — Data collection for deep learning is necessary because there are many images and videos. This model was tested and trained to make accurate predictions.

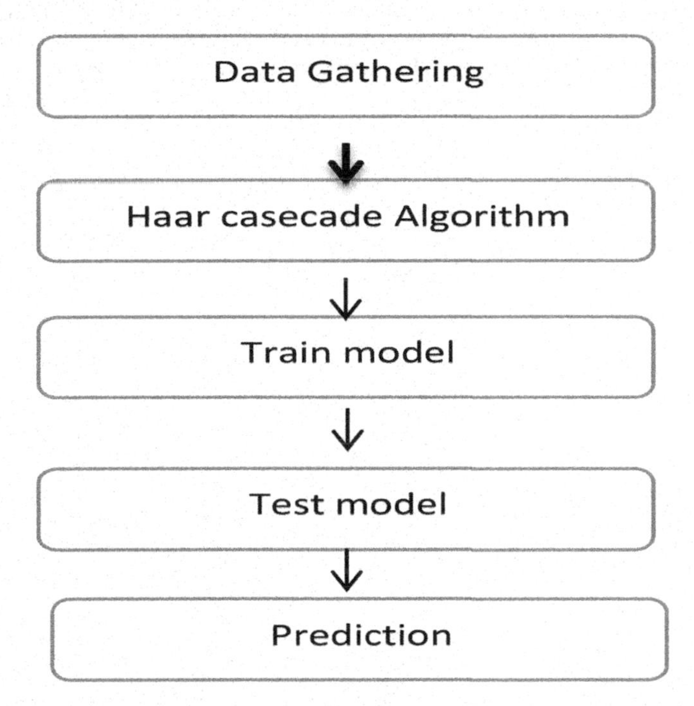

Figure 1. Steps of process diagram.

A Data Flow Diagram (DFD) represents the flow of data in an information system graphically. It is often used as the initial step to get an overall understanding of the system without delving too deep, and can then be further developed. DFDs can help visualize data processing and outline the inputs and outputs of data, as well as the storage location. Unlike

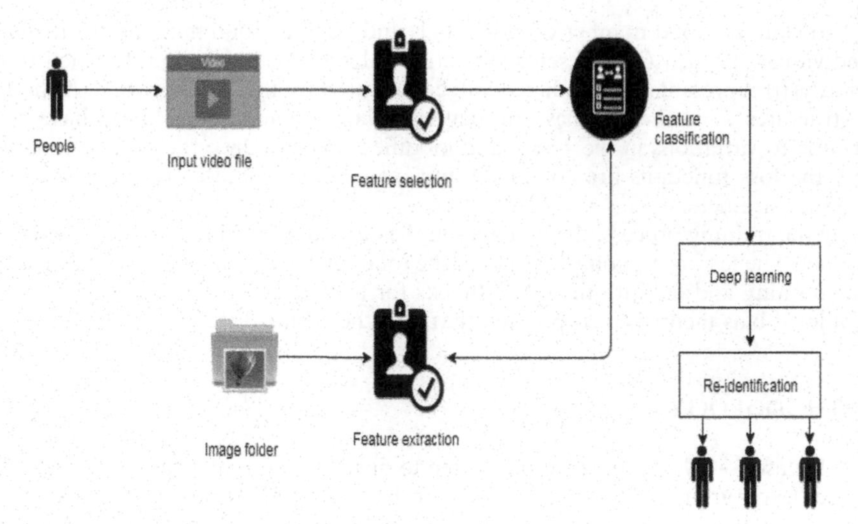

Figure 2. Design architecture.

a traditional flowchart, which focuses on control flow, or a UML activity diagram, which combines both control and data flows, DFDs do not provide information about the timing of processes or if they are executed sequentially or concurrently. DFDs are also known as bubble charts and are used in the top-down method of systems design. They use specific symbols and notations to conform to DFD conventions and standards.

5 RESULTS

The system's precision increases when the CPM method is employed for feature extraction. By using this method in combination with the CNN algorithm, the system attained an accuracy of 73% when tested on real-world scenarios and the training dataset.

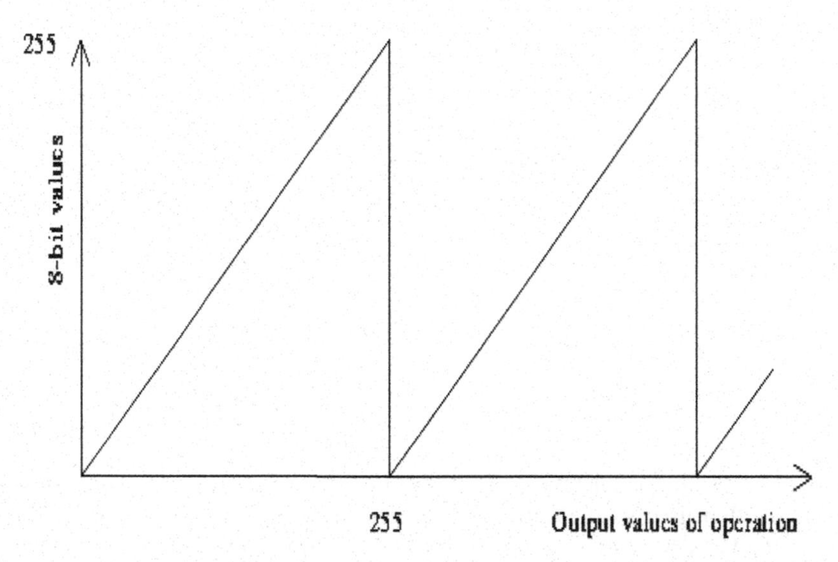

Figure 3. Transformation function for wrapping the pixel values of an *8*-bit image.

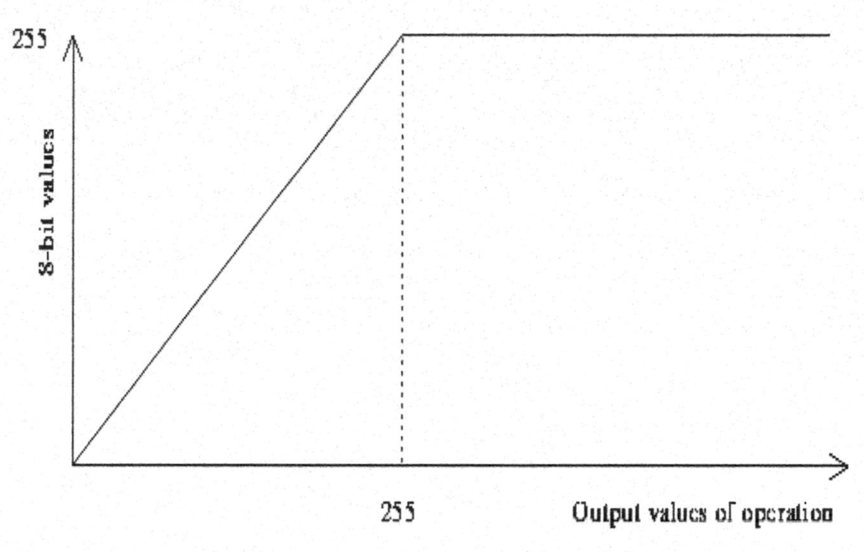

Figure 4. Transformation function for saturating an *8*-bit image.

6 CONCLUSION AND FUTURE WORK

The Haar cascade method is used to identify human figures and facial expression recognition is performed by locating the facial image within the video input file. To generate global features for the entire video in face re-identification, the face images are divided into two different face feature categories using end-to-end trainable deep learning algorithms. This enables the evaluation of the performance of the trained facial emotion recognition algorithms. Furthermore, to make the task suitable for artificial intelligence implementation, the detection results can be automated and displayed in online or desktop applications.

REFERENCES

Abimathi, A., Ramya Devi, R., Sharon Jennifer, E. & Jenishiya Hannah. 2018. *Batch Verifying Technique to Extrama Act Image Features Based on Co-occurrence Matrix in Cloud.*

Ejaz Ahmed., Michael Jones. & Tim K. Marks. 2015. An Improved Deep Learning Architecture for Person Re-Identification. *IEEE Conference on Computer Vision and Pattern Recognition (CVPR)*, Boston, MA, USA, pp. 3908–3916.

Guangcai Wang., Shiqi Wang., Wanda Chi., Shicai Liu. & Di Fan. 2020. *A Person Re-identification Algorithm Based on Improved Siamese Network and Hard Sample.*

Hongbo Wang., Haomin Du., Yue Zhao. & Jiming Yan. 2022. A Comprehensive Overview of Person Re-identification Approaches. In *IEEE Access*, vol. 8, pp. 45556–45583.

Lin Wu., Yang Wang., Ling Shao. & Meng Wang. 2019. 3-D PersonVLAD: Learning Deep Global Representations for Video-Based Person Re-Identification. *In IEEE Transactions on Neural Networks and Learning Systems*, vol. 30, no. 11, pp. 3347–3359.

Lixia Zhang. & Kangshun Li. 2016. Improved BOF Method for Person Re-Identification. *12th International Conference on Computational Intelligence and Security (CIS)*, Wuxi, China, pp. 479–482.

Mahesh Kumar, S. & Anitha Julian. 2012. An Embedded Digital Image Processing System using Blackfin 548 DSP. In *Digital Image Processing, [S.l.]*, v. 4, n. 13, p. 690–693.

Qingming Leng., Mang Ye. & Qi Tian. 2019. A Survey of Open-World Person Re-Identification. In *IEEE Transactions on Circuits and Systems for Video Technology*, vol. 30, no. 4, pp. 1092–1108.

Sanping Zhou., Jinjun Wang., Deyu Meng., Yudong Liang., Yihong Gong. & Nanning Zheng. 2019. Discriminative Feature Learning with Foreground Attention for Person Re-Identification. In *IEEE/CVF International Conference on Computer Vision (ICCV)*, Seoul, Korea (South), pp. 8039–8048.

Shan Zhong., Zongming Bao., Shengrong Gong. & and Kaijian Xia. 2021. Person Reidentification Based on Pose-Invariant Feature and B-KNN Reranking. In *IEEE Transactions on Computational Social Systems*, vol. 8, no. 5, pp. 1272–1281.

Yiqiang Chen., Stefan Duffner., Andrei Stoian., Jean-Yves Dufour. & Atilla Baskurt. 2017. Triplet CNN and Pedestrian Attribute Recognition for Improved Person Re-Identification. In *14th IEEE International Conference on Advanced Video and Signal Based Surveillance (AVSS)*, Lecce, Italy, pp. 1–6.

Zhong Zhang., Meiyan Huang., Shuang Liu., Baihua Xiao. & Tariq S. Durrani. 2019. Fuzzy Multilayer Clustering and Fuzzy Label Regularization for Unsupervised Person Re-Identification. In *IEEE Transactions on Fuzzy Systems*, vol. 28, no. 7, pp. 1356–1368.

Artificial Intelligence, Blockchain, Computing and Security – Dagur et al. (Eds)
© 2024 The Author(s), ISBN: 978-1-032-67841-2

Emergency evacuation system

V. Loganathan*, M. Parveen Fathima*, A. Poorvaja* & R. Rajalakshmi*
Department of Computer Science and Engineering, Saveetha Engineering College, Chennai, India

ABSTRACT: As global population is increasing day by day, life expectancy rises, and the causes of death across the world is rapidly increasing. It is very common that we tend to rush to any known exit when fire breaks out but this is not necessarily the best option. Therefore, our project mainly focuses on developing a medium that notifies and alerts people to an exit that is comparatively safe. Our project uses IoT technology in order to track the fire location, then notify the people to leave out safely. The information regarding the risk is intimated to the people at the right time. Various sensors are used in order to detect gases, temperature etc. The information regarding the fire accident will be provided to the user in timely manner so that the people will be able to know about the risk that is likely to be happened. Thus, intimating the user regarding the risk by sending an alert message to their mobile phones can probably decrease the death rates of the people.

Keywords: Emergency evacuation system, IOT, Sensors

1 INTRODUCTION

'When a fire accident happens, it is common that people get panic and rushes to an exit, but this is not necessarily the best option. Therefore, in order to speed up the process of evacuation and to save people lives, we need a medium that alerts the people through a notification to an exit which is safe. This emergency evacuation system uses IoT technology to track fire and people's location within buildings, then intelligently notifies to safe exits.

Emergency evacuation system mainly focuses on delivering the alert message regarding the risk to the people who actually requires it. The main idea of this project is to provide a medium which is responsible for passing the risk information to the people in the risk-prone zone. Thus, intimating the user regarding the risk by sending an alert message to their smart phones can probably decrease the death rates of the people that is happening during a fire accident. It mainly focuses on evacuating the people from the risk zone. The most important technology used in this system is IoT technology. IoT helps in connecting the hardware's with the required software and monitors the environment. Various sensors such as temperature, gas, infrared is connected with Arduino Mega. This further measures the physical quantity and process it using the desired software.

Our project is proposed in such a way that it brings out the fastest, safest path to escape for the people during an emergency situation. The major scope of the project lies in reducing the death rates of the people during the fire accidents.

*Corresponding Authors: loganathan@saveetha.ac.in, parveenfathima1803@gmail.com, poorvajadurai690@gmail.com and lakshmiraja2923@gmail.com

DOI: 10.1201/9781032684994-78

2 LITERATURE SURVEY

In literature survey, we have gathered some information about the present emergency evacuation system after a literature review. Deng. H e*t al.* examined the difficulty of evacuation due to the complex interior layout and unpredictable fire scenario progression. The idea of computer vision has been incorporated for indoor location while effectively leveraging the data from building information modelling (BIM). Fang. W *et al.* developed a fire evacuation system which was based on IoT technology that helps in guiding the people to the safest path during any fire accident. Fujimura *et al.* & Mori. K *et al.* developed a disaster identification system with the help of support vector machine (SVM) algorithm. By means of this idea, an ERESS mobile terminal gets to know about the idea of the action of the holder using the SVM algorithm. Based on which it can be concluded whether it is an emergency situation or not.

Fakrulradzi Idris *et al.* developed a system that helps in detecting the fire and alerting the people. To detect a fire, they used a temperature and flame sensor. Additionally, they designed the GUI that analyses and displays the probability of fire using LabVIEW software. Gokceli *et al.* developed an emergency evacuation system specially for building automation system (BAS). They also produced subservice definitions and examined the central management model. Huixian Jiang *et al.* implemented an intelligent evacuation path by means of artificial intelligence (AI) technology. Intelligent fire evacuation system was developed for large buildings and apartments. Han. Z *et al.* developed a real time evacuation route planning especially for high buildings. They used sensor system and wireless data transmission system in order to achieve real time data collecting. Imran Zualkernan *et al.* used IoT technology to track the location of the building's occupant and also the location of the fire and further directs the residents to a safe evacuation. Additionally, they used Bluetooth low energy beacons in order to locate whether the user has a mobile phone or not. With the help of temperature and smoke sensors the system helps to track the areas which are under danger situation.

Ronchi. Q *et al.* worked on modelling studies such as human behavior and high-rise building evacuations. They dividend high rise structures into three following categories: homes, office buildings and healthcare facilities. They also examined the components induvial and combined the utilization of EGRESS. Sathish Kumar. R *et al.* used IoT and smartphone device control in order to develop a sharp halting system. It helps in generating a powerful response in the suburbanite the owner of the mobile automobile. Administrations are entrusted with holding a parking space, verifying a saved client, recognizing the nearby free space determined by the length of the vehicle, and exploring the leaving opening and figure day by day, week by week and month to month accounts data. Wu. G *et al.* derived the dynamics of evacuation especially for high raised buildings by means of control volume model. They divided the evacuation simulation process into five stages based on the assumptions of homogeneous flow with merge flow ratio where the exit flows from different floors meet and merge together.

3 SYSTEM DESIGN

3.1 *Objective*

The primary goal of our project is to quickly and accurately inform users regarding the information related to fire accidents so that people will be aware of the risk that may occur. Additionally, it assists in detecting fire, toxic gas, and temperature levels using an Arduino microcontroller, and it also stores data from IOT webpages, and alerting users by voice and text message to their mobile devices.

3.2 *Architecture diagram*

The structural model as depicted in Figure 1, consists of Arduino and five sensors namely flame sensor, temperature sensor, gas sensor, MEMS and IR sensor. These sensors are

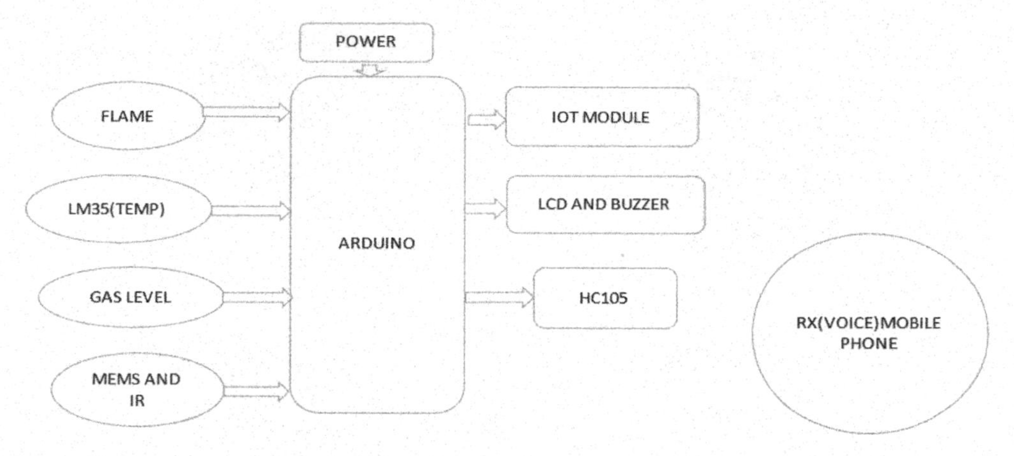

Figure 1. System architecture.

connected with Arduino Mega. The Arduino mega is used to read the given sensor value. If the threshold value of any sensor is exceeded, the alarm gets activated and further it is notifying an alert message to the people through their mobile phones. IoT technology is used in order to track the location of the fire and helps in notifying the people towards a safe exit.

3.3 *Processing data*

The system makes use of user client devices that are linked to the Internet via a gateway device. This gateway device has the capability to send appropriate control sequences and alert facility management to unusual usage or important events. The alert message regarding the fire accident is sent to the resident through SMS. SMS that we send are based on CISC Pic microcontroller. IoT technology plays a major role in tracking the location of the fire and further it helps the resident by notifying towards a safe exit. The data are stored from the IoT webpages. Therefore, the alert message is passed to the user via a text message as well as voice note to their mobile device.

The system monitors the danger zone using gas, temperature and fire sensors. These sensors are connected with Arduino Mega. The main role of Arduino is to read the sensor value where the sensors value is taken as the input. Once if the threshold value of any of the sensor is exceeded it is further activated and an alert message is sent to the resident present the building. As the residents have access to a mobile app which further instantly maps the current level of hazard in the building in the case of a fire. The emergency responders are further given access to the real time data from the sensors and the residents.

3.4 *Hardware specification*

IoT technology works a bridge between the physical world and virtual word. It computes the physical quantity like temperature, oxygen concentration etc., and process it using the desired software of the microcontroller and stores the processed data in the cloud for further use. Thus, using proper hardware plays a vital role in developing emergency evacuation system.

3.4.1 *Arduino*

Arduino is an open-source hardware board which is often referred as a microcontroller is used for designing and developing devices that helps to interact with the real world. The input and output of Arduino can be either a sensors, switches, LED and other peripheral

Microcontroller

Gas Sensor

Infrared Sensor

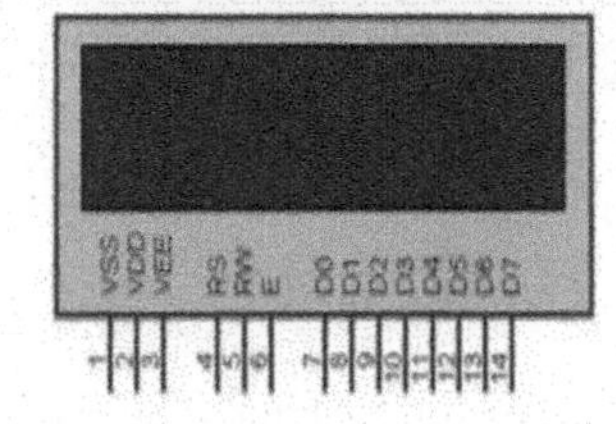

LCD Display

Figure 2. Processing of data using monitoring system.

devices. Arduino based projects has a capability of communicating with the software that is connected to the computer. The computer code can easily be loaded to the Arduino board through a USB cable. The open-source IDE or the Arduino software is free to download, Arduino hardware board is available in much cheaper rate.

3.4.2 *Gas sensor*

Gas sensors is device which helps in identifying and detecting various toxic gases. These are frequently used in assessing the gas concentration and detecting explosive gases. Gas sensors are small in size that reacts spontaneously when it senses the gas present in the surroundings. Methane gas sensors is capable of detecting gases like ammonia. This sensor interacts with a gas by ionizing it into its component atoms, which are subsequently adsorbed by the sensing element. An element's potential difference is produced by this adsorption, and it is transmitted to the processor unit in the form of currents via output pins connected with the Arduino board.

3.4.3 *Infrared sensor*

Infrared sensor is a device that produces an electrical signal as an output when it senses an object or a movement in front of it. Once the object is sensed it produces a signal and output would be HIGH and when it doesn't sense an object or a movement it doesn't produce a signal and the output would be LOW. An IR sensor is mainly used to detect the actions of the objects and also to measure the heat of an object. The wavelength of IR sensor is longer as compared to visible light.

3.4.4 *LCD display*

Liquid crystal displays (LCDs) contain a material that helps in combining the properties of crystals as well as liquids. They do not have any fixed freezing point. The crystal material is further placed in between two glass panels that forms the LCD. The glass plates inner surface is completely covered with electrodes that helps in defining the symbols, characters, patterns which is to be displayed on the layer present in between the liquid crystal and the electrode. This helps the liquid crystal to maintain a positioned angle.

4 RESULT

The sensors such as gas, temperature is connected to Arduino Mega in order to detect the Temperature and Gas as shown in Figure 3. If any of the sensor value is greater than maximum value, the LED emits a light as a warning. Residents who are stuck in the building will receive notification to their mobile phones. Thus, the emergency evacuation system helps in alerting the passengers through a notification message.

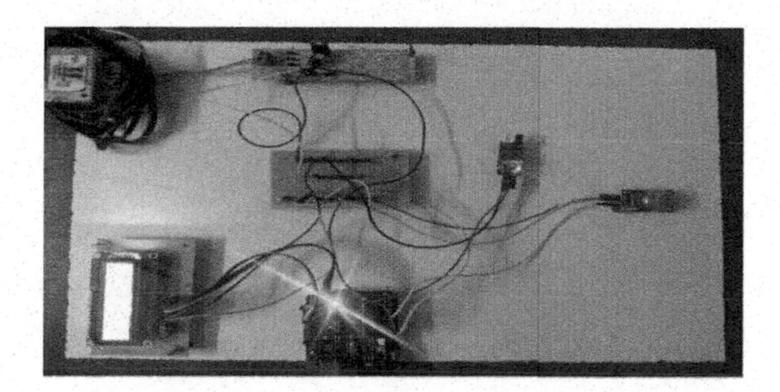

Figure 3. Output.

5 CONCLUSION AND FUTURE WORK

Our project is mainly intended to alert people through a notification whenever a fire accident that likely going to happen. The information regarding the fire accident will be provided to the user in a timely manner so that the people will be able to know about the risk that is likely to be happened. Our project is based on IOT technology that helps to track the fire location during a fire accident. The sensors such as gas, temperature and IR are connected with Arduino. If any of these sensor's value exceeds the threshold value it turns as a warning and the people starts receiving an alert message to their mobile phones. In future, the system can further be developed to monitor all kinds of major natural disasters such as Flood, Earthquake and Tsunami. In our future, it can be considered with the case when there are a large number of occupants which might lead to areas or exits being blocked by conglomeration of evacuating occupants. The Emergency Evacuation System can also be used for evacuation in case of a fire in an ocean vessel.

REFERENCES

Deng. H, Ou. Z, Zhang. G, Deng. Y & Tian. M. 2021. *BIM and Computer Vision-Based Framework for Fire Emergency Evacuation Considering Local Safety Performance Sensors.*

Fang. W, Zheng. T & Qiang. c. 2015. *Implementation of Intelligent fire Evacuation Route Based on Internet of Things.*

Fujimura, Nakamura, Ishida, Mori, Tsudaka & Wada. 2012. *New State Judgment Algorithm for Emergency Rescue Evacuation Support System (ERESS) in Panic-type Disasters.*

Fakrulradzi Idris, *et al.* 2019. *Intelligent Fire Detection and Alert System Using Labview.*

Gokceli, *et al.* 2017. *IoT in Action: Design and Implementation of a Building Evacuation Service.*

Huixian Jiang. 2019. *Mobile Fire Evacuation System for Large Public Buildings Based on Artificial Intelligence and IoT.*
Han, Weng, Zhao, Ma, Liu & Huang. 2013. *Investigation on an Integrated Evacuation Route Planning Method Based on Real-time Data Acquisition for High-rise Building Fire.*
Imran Zualkernan, Fadi A Aloul & Vikram Sakkia. 2019. *An IOT-based Emergency Evacuation System.*
Mori. K, Yamane., Hayakawa. Y, Wada. T, Ohtsuki. K & Okada. 2011. *Development of Emergency Rescue Evacuation Support System (ERESS) in Panic-type Disasters: Immediate Emergency Detections by Machine Learning.*
Ronchi & Nilson. 2013. *Fire Evacuation in High Rise Buildings a Review of Human Behavior and Modelling Research.*
Sathish Kumar, Praveen Kumar, Balaji & Anitha Julian. 2017. *Advanced Parking System using Internet of Things.*
Wu. G & Huang. H. 2015. *Modeling the Emergency Evacuation of the High-rise Building Based on the Control Volume Model.*

Artificial Intelligence, Blockchain, Computing and Security – Dagur et al. (Eds)

Cartoonify real-time images using machine learning

V. Umarani*, M. Jagadeesh Raj*, K. Garshan Kumar* & Somu Vasanth Reddy*
Department of Computer Science and Engineering, Saveetha Engineering College, Chennai, Tamil Nadu

ABSTRACT: The image process is a technique for applying certain operations to a photograph in order to produce an enhanced image or extract some useful information from it. It's a type of signal processing where the input is a picture and the output could be either an image or features that are associated with that image. Tools for processing images include Numpy, Scikit Image, and OpenCV. Generative modeling involves mechanically identifying and then mastering the variations or styles on input traces in a way similar to how the model may be used to induce or affair new exemplifications that credibly might be drawn from the first dataset. It is associated with nursing unattended literacy tasks in machine literacy.

Keywords: Cartoon conversion, machine learning, K-means clustering, OpenCV

1 INTRODUCTION

The popular art form of the cartoon has been used extensively in many different contexts. A motion picture that uses a series of drawings for its animation is referred to as a cartoon. The workflow of contemporary cartoon animation gives content creators a wide range of resources. Image harmonization, a method that transforms real-world photos into elements for use in cartoon scenes, has produced some well-known goods. In order to produce high-quality images, training is required using a grouping of photos and a grouping of cartoon pictures. A standard infrastructure is provided by OpenCV for computer vision applications. A literature review explains the work completed thus far. This paper is organized as follows, chapter 2 provides a brief insight into the works of different authors. Chapter 3 explains the proposed methodology. Chapter 4 has the snippet of the dataset used along with a brief explanation of the algorithms used and results. Chapter 5 provides the conclusion and insights of future works.

2 LITERATURE SURVEY

Throughout the literature review, we obtained information on image detection, retrieval procedure that is now in use. Anitha Julian introduces knowledge of Machine Learning algorithms in predictions. Author Chuan offered the examination of textural images in real-time from a broad perspective. Authors A. Matt, Lu Yuan, Paris shares present methods and types of identifying the image and its texture of different standards by also providing knowledge of neural network system which is mainly used in the multimedia sector. They also present us about the knowledge of deep learning which is a sector where the dataset is fully utilized to train the system. The author Yun Shan states the loss that may possibly take

*Corresponding Authors: umaraniv@saveetha.ac.in, havocjagadeesh@gmail.com, garshankumar02@gmail.com and vasanthreddy9505@gmail.com

DOI: 10.1201/9781032684994-79

place during the model. They also explain the amount of loss that will happen during the implementation. The authors Manish and Leon describe the artistic style of the image where the image must be in the proper format to present to the client. The artistic style of the image can be achieved by using the mentioned steps in that journal. The remaining authors Lee Chan present us the steps in detail like edge detection and image quantization of the image properly using machine learning algorithm.

3 METHODOLOGY

In this paper an ML-powered application that recognizes and converts the real-time image to its cartoon effect. The system uses K-means Cluster and Python modules to extract the cartoon effect of the image. It uses Computer Vision(CV) module to get input from the user. Then it converts to RGB format. After converting the image to RGB we use the K-means Cluster to detect the edges of the image so we can get the outline of the Image. Then using the Bilateral Filter method we use the quantized image to merge with the masked image version thus creating the cartoonized version of the real-time image.

3.1 *Design architecture*

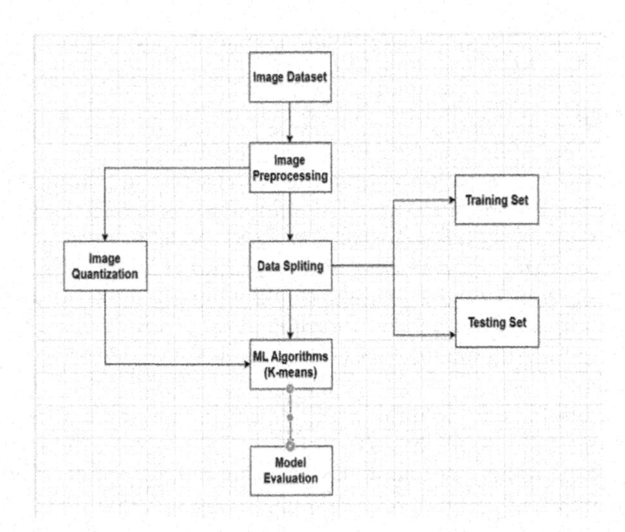

Figure 1. Design architecture.

4 MODULE IDENTIFICATION

4.1 *K-means clustering*

K-Means Clustering is an unmanaged literacy algorithm, which companies the unlabeled dataset into special clusters. Then okay defines the quantity of pre-described clusters that want to be created in the procedure, if okay = 2, there may be clusters, and for k = five, there could be 5 clusters, and so forth This algorithm follows simple steps like classifying the given dataset into a small number of clusters which is defined by the value given in the letter 'k', which is assigned before the process of the algorithm. The clusters are also deposited as points and all compliances or data points are associated with the nearest cluster, reckoned, acclimated, and also the process starts over using the new adaptations until a masked result is reached. It lets in us cluster the facts into one-of-a-kind businesses and is an available manner to discover the orders of companies within the unlabeled dataset on its very own without the need for any education. It's a centroid-grounded algorithm, wherein each cluster is related to a centroid. The main aim of this set of rules is to reduce the sum of distances between the statistics points present inside the information set to get a cluster of facts.

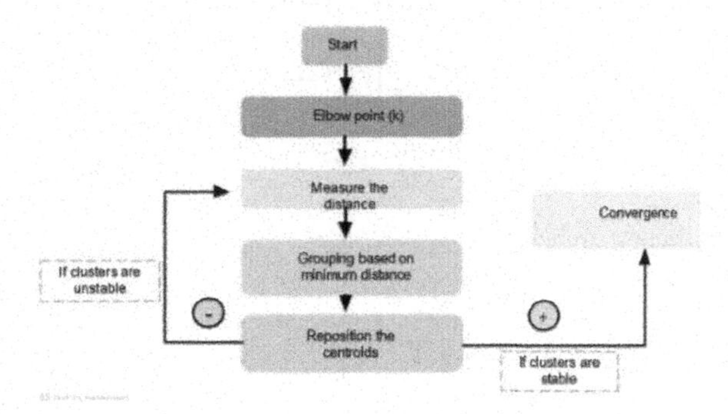

Figure 2. K-means clustering.

4.2 *Edge detection*

Facet Discovery or edge Detection is a photo-processing fashion used to identify the boundaries (edges) of items or regions inside an image. Edges are among the maximum critical capabilities which are associated with photographs. We come to recognize the start shape of a picture thru its edges. Pc imaginative and prescient processing channels accordingly significantly use edge discovery in operations. Edges are characterized with the aid of unforeseen modifications in pixel depth. To descry edges, we want to head searching out similar changes within the neighboring pixels.

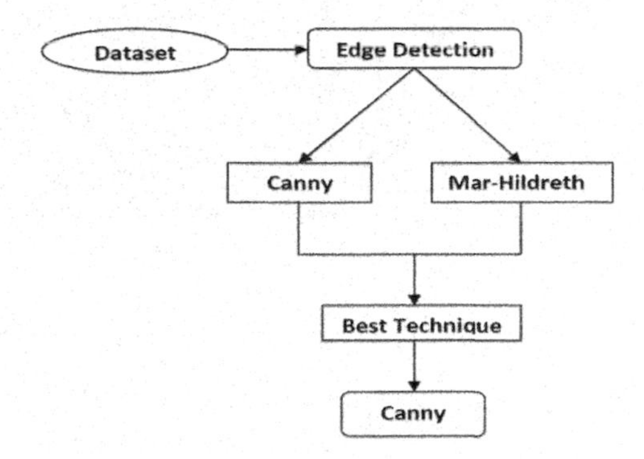

Figure 3. Edge detection architecture.

4.3 *Image quantization*

Color quantization reduces the number of colors used in an image; this is important for displaying images on the bias that supports a limited number of colors and for efficiently compressing certain kinds of images. utmost bitmap editors and numerous operating systems have been created- in support of color quantization. Popular Ultramodern color quantization algorithms include the nearest color algorithm (for fixed palettes), the median cut algorithm, and an algorithm grounded on octrees.

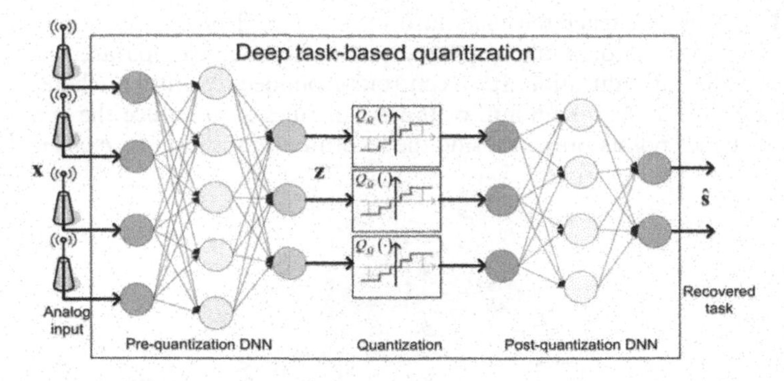

Figure 4. Image quantization architecture.

5 IMPLEMENTATION AND RESULT

In this model, we first get the dataset which is a real-time image. Using python modules we save it as a variable. Then by using edge detecting we import the outline of the image separately. Now by using K-means Clustering by giving value to the k we change the color of the image palette of the picture. Now using OpenCV and Numpy we retrieve the quantized version of the image and combine it with the outline image which results in our cartoon version of the image. In the below image (Figure 5) We can see a real-time image on the left and its cartoon version on the right side.

Figure 5. Result of the model implementation.

6 CONCLUSION AND FUTURE WORK

As a consequence, we have shown how photos can be transformed into cartoons. This study also establishes the hardware and software requirements of the image-to-cartoon conversion. In this paper, a clear picture illustrates the methodical operation of the conversion of photos into cool animated films and the corresponding algorithms and equations. Additionally, we've got stated the challenges and issues you could face whilst cartooning the captured picture. In this paper, we've got also discussed the want and scope of cartooning the content material picture. In the future, we have planned to use more Machine Learning to create the cartoon effect of real-time images. We also planned to create an android application where it gets the image and converts it into a cartoon with ML algorithms.

REFERENCES

Anitha Julian and R. Ramyadevi, "Construction of Deep Representations", Book *Chapter in Prediction and Analysis for Knowledge Representation and Machine Learning*, CRC Press Taylor and Francis, (2022: 81–110).

Chuan Li, Michael Wand Adeler vol.1,2,3, "Precomputed Real-Time Texture Synthesis and Quantization with Markovian Lenard Generative Adversarial Networks", 2017.

Dongdong Chen, Lu Yuan, Jing Liao, Nenghai Yu, Gang Hua "StyleBank: An Explicit Representation for Neural Image Style Transfer System", 2017.

Fujun Luan, Sylvain Paris, Eli Ravenholme, Kavita Bala, "Deep Image Style Transformation, International Conference of Real-time Image Style", 2017.

Jing Lie, Yun Shan, Cai Lin, Jinshan Lu, Yixishou Yu, Mingran Song, *"Deep-Neural Style Multimedia Transfer: A Review on Image Transformation"*, 2018.

Justin Johnson, Alexandre Alahi, Li Fei-Fei, "Perceptual Losses for Actual-Time photograph Style Transfer and Outstanding-resolution Template", 2016.

Leon A. Gatys, Alexander S. Ecker, Matthias Bethge, *"Image Style Transfer Using Neural Networks and Deep Learning"*, 2016.

Leon A. Gatys, Alexander S. Ecker, Matthias L. Bethgeth, *"A Neural Algorithm of the Artistic Style Real-Time Image"*, 2016.

Manish, T. I., Murugan D., and Ganesh Kumar T. "Edge Detection by Combined Canny Filter with Scale Multiplication & Ant Colony Optimization." *Proceedings of the Second International Conference on Computational Science, Engineering and Information Technology.* 2012.

Ulyanov Schmidt D., Lebedev V., Matt smith A., and Lempitsky V., "Feed-forward Synthesis of Styled Real-time Pictures and Textures Using Texture Networks", 2016.

Vincent Snow, Jonathon Shelby, Manjunath Kudlur Kumar, "A Learned Representation for the Artistic Style Image Representation", 2018.

Yangao Chen Naiyan Huang, Lee Chan , Xiaodi Horu, "*Demystifying Neural Style Image Transfer*", 2017.

Artificial Intelligence, Blockchain, Computing and Security – Dagur et al. (Eds)
© 2024 The Author(s), ISBN: 978-1-032-67841-2

Ocular disease recognition using machine learning

K. Ganesh*, P. Kishore* & Anitha Julian*
Department of Computer Science and Engineering, Saveetha Engineering College, Chennai

ABSTRACT: Over a quarter of the global population suffers from ocular diseases and millions of people die every year due to them. It takes a lot of time and effort to diagnose such diseases and so over 2/3rd of the people suffering from them don't even realize that they have them. In our model we are trying to recognize the diseases without any human interaction. So, to achieve it, we propose a model where we were able to detect and recognize certain ocular diseases using the color of the fundus part of the eye from a patient. In this paper, we train our model using the top 4 pre-trained image classification algorithms on thousands of fundus images of the said diseases and achieved a maximum of 91% accuracy on InceptionV3 overall. The model had great accuracy of over 95% on three classes except for diabetic_retinopathy.

Keywords: ocular disease, fundus images, disease prediction, blindness

1 INTRODUCTION

More than half of the global population suffers from one or the other ocular disease. These are some of the major diseases we were trying to recognize in our model. It takes considerable time and effort to diagnose such diseases. Over 2/3rd of the people suffering from them don't even realize until the diseases start to show advanced symptoms. By that time, the symptoms advance to the point where the patient cannot do their work as usual, and it is already too late for a simple treatment. Ocular diseases have severe symptoms such as sudden or severe eye pain, redness of eyes, white spots in the pupil, color changes in the iris, and blurry or double vision and left untreated could lead to blindness. So, to achieve it, we propose a model where we can detect and recognize certain diseases using the color of the fundus from right and left eye of a patient. This system can help save money and time of millions of people around the world.

The fundus is the bottom of the inner side of an eye. Inside human eye, the fundus is the only part where microcirculation can be observed. The imaging of the fundus can only be done with a device that has colored filters or fluorescence dyes. Using thousands of fundus images that are affected by each of the above diseases, the size of the images must be the same for the CNN model to train the Images effectively. So, the Images must have to be re-sized first. The images are then fitted and returned as an encoded label. Once the pre-processing is done, the test and training datasets are split. As we had multi-class data as the input, to train and test the data effectively, we need to categorize the dependent data. The Images are then used to train various models using different convolutional neural networks models such as VGG19, RESNET, InceptionV3, and EfficientNet. Then the accuracy, classification report, and confusion matrix obtained from various other models are

*Corresponding Authors: ghkn2001@gamil.com, pkichu1997@gmail.com and cse.anithajulian@gmail.com

 DOI: 10.1201/9781032684994-80

compared and the best one among them is observed. Randomly chosen to set of fundus images to check their capabilities.

2 LITERATURE SURVEY

The authors came up with an idea of allowing the patients to know about the type of ocular disease that they suffer from using a system they proposed in which they used a densely connected CNN that takes clinical images through smartphone as input and learns features from it to automatically classify which type of ocular surface disease it belongs to precisely (Chen *et al.* 2021). It is also possible to classify a well-known ocular disease dataset to 7 different types using two different feature extraction techniques such as GLCM and LBP to extract necessary features from the fundus images for the model (Mampitiya *et al.* 2022). This type of image recognition can also be done using a custom-made dataset for a particular set of diseases and train the model to classify them using a traditional CNN can also produce a good result (Islam *et al.* 2019). An automatic multi-classification approach named "DeepRetino" is used to classify eye diseases and the model was trained on a dataset that was preprocessed using Contrast Limited Adaptive Histogram Equalization (CLAHE) to improve the contrast of fundus images. Model's network weights are tweaked using Xavier orthogonal and Adam optimizer (Zahra Belharar *et al.* 2022). A model that was able to categorize the usual and unusual ocular images from the data using machine learning classification algorithms can detect Diabetic Retinopathy (DR) infection (Soni *et al.* 2021).

The authors wanted to avoid the manual inspection error in detecting whether a patient was affected by Covid-19 or pneumonia. This paper worked on with different CNN models to evaluate which would fetch the best possible results for the detection of patient's disease and out of all the models (ResNet-50, ResNet-101, Vgg-16, Vgg-19 and SqueezeNet), VGG-16 was able to perform well with high accuracy rate (Law *et al.* 2021). Likewise, VGG model has shown a high accuracy in classifying dementia (Bagaskara *et al.* 2021). For the classification of covid-19, an alternate method of using CT scans as input instead of using a conventional method (Namani *et al.* 2022). A transfer learning model (i.e.: ResNet-50) to establish a drowsiness detection system to reduce the high frequency of vehicle accident happening around the world was deployed by a group of researchers. Source of input is camera which takes the facial photo of the person driving a vehicle to analyze and detect drowsiness (Yadav *et al.* 2022). A modified "InceptionV3" model is used for the purpose of detecting retinopathy disease in patients, ranging in five different classes. This paper's model was compared against the other existing models which are closely related to it, and it resulted in much desirable amount of accuracy in classification (H. R. *et al.* 2022).

3 PROPOSED SYSTEM

Ocular disease is one of the most common eye diseases which over one-quarter of the population suffers from and overly complex to recognize accurately through the naked eye all the time. But to stop the advancement of the diseases it is necessary to detect and recognize them at an earlier stage. People are busy nowadays that they don't necessarily have any time for a general medical workup. These diseases need several diagnosis tests like visual acuity tests, ocular tonometry, and retinal exam that must be taken in the hospital in real-time but are very time-consuming. The detection of such diseases comes under classification. Ocular disease recognition has been under the radar of machine learning for some time. Using the neural network to classify images into multiple classes is one of the advanced applications of medical image classification. In this model, we use a convolutional neural network with the help of various transfer learning algorithms to classify the said diseases. The performance of such models is then evaluated and compared. This dataset originally designed for eye-disease diagnosis. This dataset is a collection of images of normal retinas and other samples of defective eye diseases like diabetic retinopathy, cataracts, and glaucoma. Each of the classes contains around 1000s of images for each disorder. The dataset has been collected from various individual sources like IDRiD, Ocular recognition, HRF, etc and put together. It provides exceptionally reliable and accurate results describing the disease from which the patient is suffering. In this proposed model, first we load our dataset which consists of four sub-classes. Each of these sub-classes has data of 1000 images. The first task performed on the dataset is that we have resized all our images to a constant size. Most of the convolutional neural network models are designed for standard fixed-sized input, so they don't generalize well for other sizes, but this can be rectified in some CNN architecture using fully connected layers following the convolutional layers. Once the images are resized, we use the one-hot encoded function to convert labels into categorical values stored in a sparse matrix. Using one hot encoder we create several dummy categorical features in our dataset. The input of the encoder is an array of image files, and we get a binary column of each category as the output. Then the dependent variables of the training and testing dataset are converted from multi-class labels into binary labels using labelBinarizer.

InceptionV3 is a family of Inception pre-trained models that have some improvements including "Label Smoothing" (i.e.) a form of output distribution regularization that prevents a neural network model from being overfitted, factorized 7x7 convolutions and the use of an auxiliary classifier (Koklu *et al.* 2022). This model has a higher efficiency and deeper network when compared to its previous versions, it is computationally less expensive too. InceptionV3 has a total of 42 layers. VGG-19 is a type of Convolutional neural network (CNN) architecture with 19 layers, of which 16 of them are Convolutional layers and the rest of 3 are fully connected layers, it is a pretrained model which was trained using millions of images from the 'ImageNet' during the creation of it, it is said that the model can classify multiple images when given to it. This model's architecture takes in a fixed size of input (224x224) images and uses a kernel of size (33) with a stride of 1 pixel. It also uses spatial and max pooling with activation functions like ReLu & SoftMax EfficientNet is a type of Convolutional Neural Network architecture that has a scaling method to uniformly scale all dimensions using a compound coefficient technique. This network was fine-tuned to attain maximum accuracy and at the same time, it is penalized whenever the network is computationally heavier (Zhang *et al.* 2022). This very architecture uses a mobile inverted bottleneck convolution like that of Mobile Net V2, but the difference is that the EfficientNet model is much bigger because of the increase in FLOPS (Floating point operations). ResNet-50 is a variant of the ResNet model and consists of 3.8 billion FLOPs. This model has 48 Convolution Layers along with one Maxpool and one Average Pool layer, thus making it a totally 50 layers. It is the most popularly used ResNet model. ResNet models makes use of a concept called 'skip connection' which is used to address the problem of vanishing gradient descent (Jaju *et al.* 2022).

4 IMPLEMENTATION AND RESULT

The implementation results show the accuracy of predicting the four types of diseases using the four transfer learning algorithms ranging from 90% for InceptionV3, 87% for VGG19, 81% for EfficientNet, and 79% for RESNET-51. In the above model, when we ran it through sample data, we got the following results as the output. The output specifies the actual and the predicted value of the image. The values for the various diseases are in the order 0 – Normal, 1 – glaucoma, 2 – diabetic_retinopathy, and 3 – cataract.

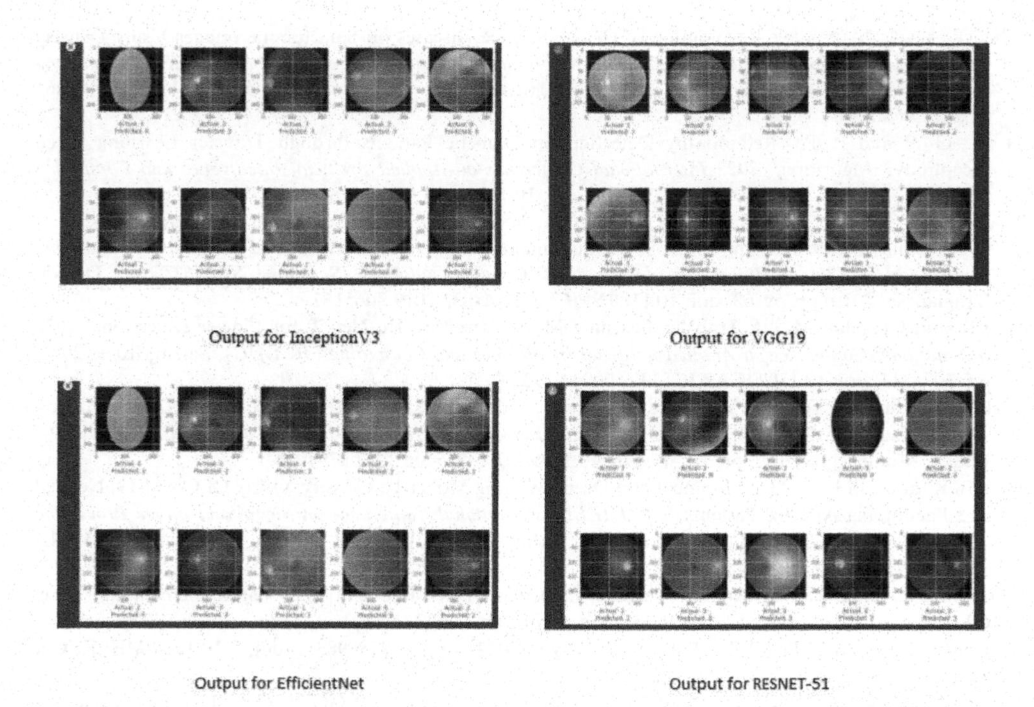

Figure 1. Outputs of different neural network models.

5 CONCLUSION AND FUTURE WORK

Every machine learning Algorithm has its advantages and disadvantages over the other algorithms, especially for a classification problem where every algorithm behaves differently. So, to get the best result possible that we could get for the problem, we try to implement various algorithms which can be more fruitful. But, on the other side of that problem, implementing more algorithms means using more RAM storage and time. The proposed work has shown maximum overall accuracy for the InceptionV3 transfer learning algorithm with an accuracy of over 91% and over 94% for three sub classes of the model. The limitation of the fourth model is due to the limited number of the training dataset. The difference between others and diabetic retinopathy is too much for our model to predict in the given parameters. However, in a detailed study in the future could cover other parameters such as the size of the dataset under consideration, length of the text typed by the user, complexity of vocabulary used in the user input, etc., which will also require the application of natural language processing as an additional module to the recommender system. We can collect a much larger dataset than the one we have used in this model to train our model to generalize well for the unseen data. We can implement this model in Realtime using IoT devices to scan fundus images from a person's eye and then process it with the model to get accurate prediction. We can use this model in hospitals to generate an instantaneous report.

REFERENCES

Bagaskara A. and Suryanegara M., "Evaluation of VGG-16 and VGG-19 Deep Learning Architecture for Classifying Dementia People," 2021 *4th International Conference of Computer and Informatics Engineering (IC2IE)*, Depok, Indonesia, 2021, pp. 1–4, doi: 10.1109/IC2IE53219.2021.9649132.

Chen R. *et al.*, “Automatic Recognition of Ocular Surface Diseases on Smartphone Images Using Densely Connected Convolutional Networks,” *2021 43rd Annual International Conference of the IEEE Engineering in Medicine & Biology* Society (EMBC), *Mexico*, 2021, pp. 2786–2789, doi: 10.1109/EMBC46164.2021.9630359.

H. R, G. B and P. P, “Retinopathy Detection on Diabetic Patients through Transfer Learning based Inceptionv3 Algorithm,” *2022 International Conference on Applied Artificial Intelligence and Computing (ICAAIC)*, Salem, India, 2022, pp. 299–306, doi: 10.1109/ICAAIC53929.2022.9793187

Islam M. T., Imran S. A., Arefeen A., Hasan M. and Shahnaz C., “Source and Camera Independent Ophthalmic Disease Recognition from Fundus Image Using Neural Network,” *2019 IEEE International Conference on Signal Processing, Information, Communication & Systems (SPICSCON)*, Dhaka, Bangladesh, 2019, pp. 59–63, doi: 10.1109/SPICSCON48833.2019.9065162.

Jaju S. and Chandak M., “A Transfer Learning Model Based on ResNet-50 for Flower Detection,” *2022 International Conference on Applied Artificial Intelligence and Computing (ICAAIC)*, Salem, India, 2022, pp. 307–311, doi: 10.1109/ICAAIC53929.2022.9792697.

Koklu M., Cinar I., Taspinar Y. S. and Kursun R., “Identification of Sheep Breeds by CNN- Based Pre-Trained Inceptionv3 Model,” *2022 11th Mediterranean Conference on Embedded Computing (MECO)*, Budva, Montenegro, 2022, pp. 01–04, doi: 10.1109/MECO55406.2022.9797214.

Law B. K. and Lin L. P., “Development of a Deep Learning Model To Classify X-Ray Of Covid-19, Normal and Pneumonia-Affected Patients,” *2021 IEEE International Conference on Signal and Image Processing Applications (ICSIPA)*, Kuala Terengganu, Malaysia, 2021, pp. 1–6, doi: 10.1109/ICSIPA52582.2021.9576804.

Mampitiya L. I. and Rathnayake N., “An Efficient Ocular Disease Recognition System Implementation using GLCM and LBP based Multilayer Perception Algorithm,” *2022 IEEE 21st Mediterranean Electrotechnical Conference (MELECON),* Palermo, Italy, 2022, pp. 978–983, doi: 10.1109/MELECON53508.2022.9843023.

Namani S., Akkapeddi L. S. and Bantu S., “Performance Analysis of VGG-19 Deep Learning Model for COVID-19 Detection,” *2022 9th International Conference on Computing for Sustainable Global* Development (INDIACom), *New Delhi, India*, 2022, pp. 781–787, doi: 10.23919/INDIACom54597.2022.9763177.

Soni A. and Rai A., “A Novel Approach for the Early Recognition of Diabetic Retinopathy using Machine Learning,” *2021 International Conference on Computer Communication and Informatics (ICCCI)*, Coimbatore, India, 2021, pp. 1–5, doi: 10.1109/ICCCI50826.2021.9402566.

Yadav A. K., Ankit and A. Sharma, “Real Time Drowsiness Detection System based on ResNet-50,” 2022 *6th International Conference on Intelligent Computing and Control Systems (ICICCS)*, Madurai, India, 2022, pp. 1–7, doi: 10.1109/ICICCS53718.2022.9788204.

Zahra Belharar F. and Zrira N., “Deep Retino: Ophthalmic Disease Classification from Retinal Images using Deep Learning,” *2022 IEEE 9th International Conference on Sciences of Electronics, Technologies of Information and Telecommunications (SETIT)*, Hammamet, Tunisia, 2022, pp. 392–399, doi: 10.1109/SETIT54465.2022.9875570.

Zhang T., Zhang B., Zhao F. and Zhang S., “COVID-19 localization and recognition on chest radiographs based on Yolov5 and EfficientNet,” *2022 7th International Conference on Intelligent Computing and Signal Processing (ICSP)*, Xi’an, China, 2022, pp. 1827–1830, doi: 10.1109/ICSP54964.2022.9778327.

JINX: A compilable programming language

N.V. Ravindhar*
Assistant Professor, Computer Science and Engineering, Saveetha Engineering College, Chennai, India

S. Giridharan*, P. Mohamed Irfan* & R. Prakash Kumar*
Computer Science and Engineering, Saveetha Engineering College, Chennai, India

ABSTRACT: A sort of programming language characterized as a general-purpose language is applicable to a variety of applications. These languages typically have flexible and expressive syntax and semantics, allowing them to be used for a variety of programming paradigms, styles, and domains. They are designed to be easy to learn, use, and extend, making them suitable for novice and experienced programmers. The design, development, syntax, semantics, and other aspects of a general-purpose programming language utilizing Python are covered in this article. Our programming language will be a self-compiled and self-linked language making it 3 to 4 times faster than the host language. The language is tested and evaluated to demonstrate its correctness, completeness, and effectiveness for various programming tasks.

Keywords: programming language, jinx, compiled, interpreter, general purpose

1 INTRODUCTION

The project "general purpose programming language" involves the development of a software application or system using a general-purpose programming language. This language is designed to be flexible and versatile, allowing developers to use it for a wide range of purposes. Some common examples of applications that can be built using a general-purpose programming language include web and mobile apps, desktop software, games, and business applications.

The project will involve designing and implementing the features of the application, debugging and testing the code, and ensuring that the final product is reliable and efficient. The objective of the project is to create a functional and user-friendly application that meets the specific needs of the user or client. This may involve using the language to build a wide range of different applications, depending on the specific requirements of the project.

Overall, the project "general purpose programming language" aims to create a functional and effective software application or system using a versatile and flexible programming language. This will involve designing, implementing, and testing the code to ensure that the final product meets the specific needs of the user or client.

2 RELATED WORKS

A new compiler has been introduced that is self-compiled, meaning that the compiler is compiled by itself. This unique feature allows for enhanced control and optimization in the

*Corresponding Authors: ravindhar@saveetha.ac.in, giridharan.senthilkumaran@gmail.com, mohamedirfanp3@gmail.com and prakash218kumar@gmail.com

DOI: 10.1201/9781032684994-81

compilation process, resulting in improved performance and efficiency. (Abdul Dakkak *et al.* 2020) presents a new compiler for the Wolfram Language, designed to improve scalability and enhance programmer productivity. The new compiler aims to address the current compiler's shortcomings and maintain the language's behavior while improving efficiency.

(Hanfeng Chen *et al.* 2007) developed a compiler that translates ELI to C, optimized for high performance. The compiler fully supports all the features and functions of the ELI language and uses a top-down parser to analyze syntax and create a parse tree with information about shape and type. The C code generated by the compiler was found to be significantly faster, ranging from 2 to 40 times faster than the ELI interpreter, and the performance was often similar to optimized C code. (Ariya Shajii *et al.* 2019) introduced a new language called Seq, specifically designed for computational biology. The goal was to combine the ease of use and productivity of Python with performance levels similar to C. Seq's syntax is based on Python, but with added domain-specific data types and optimized built-in services and functions. The Seq compiler utilizes an LLVM backend for improved performance, and a small runtime library is used for memory management and input/output operations with the Boehm garbage collector instead of Python's reference counting system.

(Abelson *et al.* 1985) and (Hoare *et al.* 1991) suggested that comprehensive introduction to computer programming, including the design of compilers and interpreters for programming languages.(Adam Brooks Webber *et al.* 2010) suggest in his book the concepts of diverse programming languages for students who have already mastered basic programming in at least one language.(Richard Kelsey *et al.* 1998) proposed the techniques used in compiling a high-level functional language and describes the methods used to translate the functional language into machine code, with an emphasis on the use of optimizations and how they can be applied to improve performance.

(Aho *et al.* 1986) explains in his textbook on compilers and provides a comprehensive introduction to the theory and practice of compiler design.(Keith D. Cooper *et al.* 2001; Michael L. Scott *et al.* 2015) proposed the fundamental concepts of programming languages to advanced topics such as type systems, language design, and implementation techniques.(Denis Merigoux *et al.* 2021) proposed the modern compiler for the French Tax Code.(Kenneth Louden *et al.* 1997) explains about the fundamentals of compilers, including lexical analysis, syntax analysis, semantic analysis, code generation, and optimization.

3 ALGORITHM

For getting the best accurate results, many algorithms have been implemented for the smooth functioning of the project. Various data structures and object-oriented principles are used such as classes for error handling, such as IllegalCharacterError, SyntaxError algorithm for tokenizing the input, a lexer, a parser is also implemented.

4 SYSTEM ARCHITECTURE

The Analysis phase and Synthesis phase are the fundamentally different phases of compilers. The front-end of the compiler design is represented by the Analysis phase. It includes numerous tasks including examining the source code, segmenting the code, and looking for mistakes. Additionally, a symbol table is built to map source code symbols to pertinent details like type, scope, and location. Preprocessing, lexical analysis, syntactic analysis, and semantic analysis are some of the processes that make up the analysis component of the architecture. The synthesis step's input is the intermediate code representation. It denotes the back end of a compiler's design. Using the symbol table and the intermediate code representation, the synthesis component of the architecture created the target program. It includes stages like code creation and optimization.

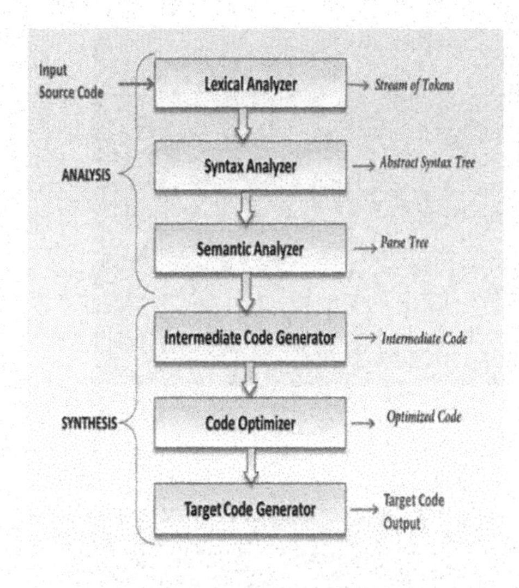

Figure 1. Module description.

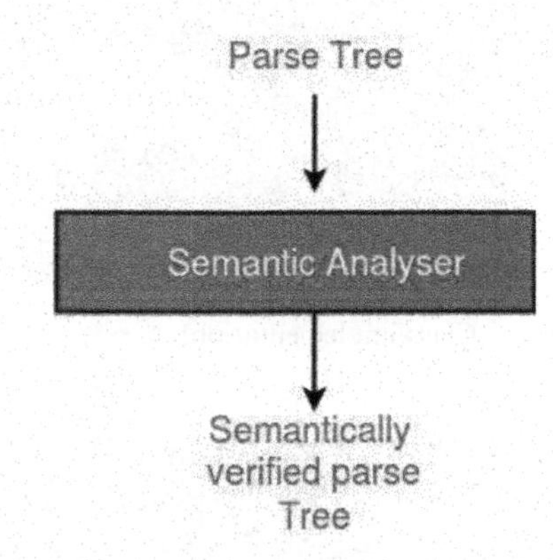

Figure 2. Semantic analyzer.

4.1 *Module description*

4.1.1 *Lexical analyzer*

The first step in the compiler process, also known as a scanner, is lexical analysis. This phase takes the high-level input program and breaks it down into a sequence of tokens. The process of lexical analysis can be accomplished using a Deterministic Finite Automata.

4.1.2 *Syntax analyzer*

It is sometimes referred to as a parser. It creates the parse tree. Context-Free Grammar is used to build the parse tree after taking each token one at a time.

4.1.3 *Semantic analyzer*

The process involves evaluating the parse tree for its significance and creating a confirmed parse tree. Additionally, it conducts evaluations for type conformity, label accuracy, and proper flow control as shown in Figure 2.

4.1.4 *Intermediate code generator*

Compilers utilize intermediate code as a step in the translation process. This intermediate code is a form that can be directly executed by a machine. There are a variety of commonly used intermediate codes, such as Three Address Codes. The final stages of the compilation process, which are specific to the target platform, convert the intermediate code into machine language. Because the use of intermediate code is a common feature among compilers, it is possible to use an existing compiler to generate this code and focus on developing the platform-specific components when building a new compiler.

5 RESULT

The newly implemented programming language is designed for general purpose use and includes essential features such as conditionals shown in Figure 4, user-defined functions

```
class_test.jinx

CLASS Person
    FUN Person(name)
        VAR this.name = name
    END
    FUN Format()
        RETURN "My name is " + this.name + "."
    END
END

VAR inp = INPUT()
WHILE inp != "exit" THEN
    VAR p = Person(inp)
    PRINT(p.Format())
    VAR inp = INPUT()
END

EXIT()
PRINT("This message will never ever show up! Isn't that wacky?!")
```

```
G:\Final Year Project>python shell.py
basic > RUN("class_test.jnx")
Cult Script
My name is Cult Script.
exit

G:\Final Year Project>
```

Figure 3. Class implementation.

```
if-else.jinx

1  VAR name = INPUT()
2
3  IF name == "admin" THEN
4      PRINT("Welcome Admin")
5  ELSE
6      PRINT("Welcome Guest")
7  END
```

```
basic > RUN("if-else.jinx")
admin
Welcome Admin
0
basic > RUN("if-else.jinx")
test
Welcome Guest
0
basic > RUN("if-else.jinx")
admin
Welcome Admin
0
basic > RUN("if-else.jinx")
prakash
Welcome Guest
0
basic > RUN("if-else.jinx")
guest
Welcome Guest
```

Figure 4. Condition implementation.

```
function.jinx

1  FUN oopify(str) -> str + "oop"
2
3  PRINT(oopify("l"))
4  PRINT(oopify("sp"))
```

```
C:\Users\rames\OneDrive\Documents\GitHub\jinx>python shell.py
basic > RUN("function.jinx")
loop
0
basic > RUN("function.jinx")
loop
spoop
0
basic >
```

Figure 5. Function implementation.

shown in Figure 5, and classes with objects shown in Figure 3. The language structure is intuitive and user-friendly, making it easier for coders to write and understand their code. The ability to define custom functions and classes enhances the flexibility of the language and allows for easy organization and reuse of code. This compilable language aims to simplify the coding process while still providing all the necessary tools for successful software development. The implementation of these features marks a major milestone in the evolution of programming languages and is sure to be widely adopted by the development community.

6 CONCLUSION AND FUTURE WORK

Programming languages that are designed to be versatile and flexible are utilized to create software that outlines the methods for completing a variety of tasks and addressing a broad range of issues. However, because the syntax of these general-purpose languages is meant to be inclusive, the process of composing code for highly specialized areas can become complex and lead to convoluted code. In conclusion, general purpose programming languages are powerful tools for developers to create applications, websites, and software. They provide the necessary features and functions needed to construct complex computing solutions. They are also highly versatile, allowing developers to customize their code to fit the specific needs of their project. With so many options available, it can be difficult to choose the right language for the job, but with the right research and practice, developers can find the perfect language for their needs.

REFERENCES

Abdul Dakkak & Tom Wickham-Jones, *Proceedings of the 18th ACM/IEEE International Symposium on Code Generation and Optimization.* 2020. The Design and Implementation of the Wolfram Language Compiler.

Abelson H, Sussman G J & Sussman J, Cambridge, Mass., 1985. *The Structure and Interpretation of Computer Programs.*

Adam Brooks Webber, 2010. *Modern Programming Languages: A Practical Introduction.*

Aho A, Sethi R & Ullman J D, 1986. *Compilers: Principles, Techniques, and Tools.*

Ariya Shajii & Ibrahim Numanagić, *Proceedings of the ACM on Programming Languages*, Volume 3, pp 1–29. 2019. Seq: A High-Performance Language for Bioinformatics.

Denis Merigoux & Raphaël Monat, *30th ACM SIGPLAN International Conference on Compiler Construction.* 2021. A Modern Compiler for the French Tax Code.

Hanfeng Chen, Wai-Mee Ching, *Proceedings of the 4th ACM SIGPLAN International Workshop*. 2017. An ELI-to-C Compiler: Design, Implementation, and Performance.

Hoare C A R, 1991. *An Introduction to the Theory of Programming Languages.*

Keith Cooper D and Linda Torczon, 2001. *Engineering a Compiler.*

Kenneth Louden C, 1997. *Compiler Construction: Principles and Practice.*

Michael Scott L, 2015. *Programming Language Pragmatics.*

Richard Kelsey, William Clinger & Jonathan Rees, *ACM SIGPLAN Notices,* 1998. *Compiling Techniques for a High-Level Functional Language.*

Artificial Intelligence, Blockchain, Computing and Security – Dagur et al. (Eds)
© 2024 The Author(s), ISBN: 978-1-032-67841-2

Predicting career opportunities online learning platform

R. Ramyadevi*, B. Hemnath*, S. Guruprakash*, D. Gokulakannan* & M. Vikram*
Department of Computer Science and Engineering, Saveetha Engineering College, Chennai, Tamil Nadu, India

ABSTRACT: Predicting Career Opportunities with Online Learning Platform systems fulfills When compared to traditional learning methods, the desire for knowledge, offers online information that can be given for the student at anywhere, anytime, and any age through a wide choice of e-learning options. Additionally, it offers quick access to specialized information and knowledge. Instead of manual teaching and learning, learners get knowledge using e-Learning technologies. This system will forecast and present the user's learning path. This solution fully integrates with many databases and provides cross-browser usage. This system's key features included Content Management and Content Protection, Evaluation management, Access Control, etc., are mainly focused on integrated platforms needed for online learning and management. For Frontend we are going to use ReactJS and for Backend is Nodejs. This System will be built on the MERN stack. Proposed method identifies frames with 80% accuracy.

Keywords: Predicting Career, Student Success, Online Learning, Roadmap, Self Quizzing, Mern stack

1 INTRODUCTION

Students are not sure about choosing their Career Path so We are Creating The application that helps Them to predict their Career with our Prediction System And we Provide Roadmap From which they can have Idea on What to study Online. The centre of the online course's software is the learning management system. Individuals can access a learning management system, which is a Web-based system for training programs and information sharing, from either their home or place of employment. Authorized individuals have full access to this unique system through URL or through a unique User ID and Password. Users can keep progress on their Roadmap like To do list. Anxiety, which is self-dissatisfaction with job choices and tends to only see the bad prospects or dangers that will be accepted when choosing a vocation, is one factor that makes it harder for people to plan their careers. The person may experience this anxiety if they are not yet prepared to commit to a particular career path. These occurrences have the logical result of making people unprepared owing to poor planning and trouble directing themselves, which is further hampered by interpersonal disputes that frequently occur during career planning. Teenagers can find some of these issues to be rather complex, therefore it is unquestionably necessary for knowledgeable educators, particularly school counselors, to assist. Therefore, a school counselor's trained competency is required in order to provide both constructive and concurrent Another issue that

*Corresponding Authors: ramyakathir@gmail.com, hemnath5700@gmail.com, guruguruprakash87@gmail.com, gokulakannan1072@gmail.com and mv0155861@gmail.com

 DOI: 10.1201/9781032684994-82

continues to be a barrier for guidance and counseling services in schools is that these services frequently continue to employ routine service tactics by relying sparingly on novel ones and have failed to foster student autonomy. Similar findings are supported by Jauhari *et al.* (2019) research, which found that traditional school services are incidental and only delivered through lecture methods. This makes it less appealing for students to take part in service projects. As a result, they will take a passive part in learning. Therefore, a service tool is needed which has a comprehensive function to improve career planning. These aspects can be realized in the "Predicting Career Opportunities With Online Learning Platform".

2 LITERATURE SURVEY

Throughout the literature review, we obtained information on the person re-identification procedure that is now in use. Anooja *et al.* (2019) given Educational Data Mining (EDM) and machine learning has become an inevitable technologies in past years. Most of the educational systems has adapted many technologies to improve the performance of students. Javeed (2019) proposed that react is one of the widely used web frameworks and has surpassed others like Angular, Vue, etc. This is because of the use of Virtual DOM, whose main goal is to improve the application's overall efficiency. Sourabh Mahadev Malewade *et al.* (2021) proposed younger generation now views online shopping and commerce as a way of life. Access to everything from necessities to luxury goods has been made possible by e-commerce online applications, which sell a variety of goods. Jyoti Shetty *et al.* (2020) in their paper the gift the journey of a full-stack development from a frontend to delivering a site is covered. Brown et al. (2013) proposed the text's second edition continues in a number of ways from the original. First, Section One focuses on the key theories of career growth, choice, and adjustment that (a) have either gotten direct empirical attention or are drawn from other, well-researched theories and (b) have clear practical implications. Fajar Fithroni *et al.* (2021) proposed that their research intends to provide high-quality service tools that will assist students in making better career decisions with so that school counsellors and students are not just locked on a theoretical framework of career planning that is rather hard, the "Career Vision Roadmap" implemented through group work.

The lack of career planning can lead to issues like not knowing one's talents or interests or not yet understanding the reason for one's life and chosen career given by Jauhari et al. (2018). Nesrine Mezhoudi *et al.* (2021) in their paper Because it is frequently employed as a marker for educational institutions' success, student employability is essential. Galeon *et al.* (2019) in their paper they have created the roadmap for students to overcome the problems with the incorporation of eLearning into open and distance education. Madhan Mohan Reddy (2021) proposed that In the modern world One of the most crucial decisions is deciding on a suitable career route. As the number of job alternatives and prospects grows, students are finding it increasingly challenging to make this choice he had used XG Boost and decision Tree to Predict Career path for User. Anitha Julian *et al.* (2019) paper presents a case study on online learning concept, online learning technology and future of online learning. VidyaShreeram *et al.* (2021) proposed India is fortunate to have a large number of top-notch schools and colleges. However, the majority of students drop out of their next level of school for a variety of reasons. There are several causes for this, including the fact that some students' families are experiencing financial difficulties, others lack desire in continuing their education, certain issues relating to gender, and some rural areas lacking in quality educators and schools.

3 PROPOSED SYSTEM

The system is designed to be more user friendly. The design plays a vital role Where the system can be made attractive. Our application is built with MERN (MongoDB, ExpressJS, ReactJS, NodeJS) Stack Where We use Frontend As ReactJS And Backend as NodeJS. We use several NPM packages to make the application Faster and Code Efficient.

- We predict the Learner's Career based on their Interest as well as their Skillset they have with answering some quiz Questions.
- They can Even Test Their Skills By Attending Quiz and can ensure their Career.
- We Provide the Roadmap For their Predicted Career which will be User Friendly.
- Learners Can keep track of their Progress on their Roadmap by completing the Roadmap paths.
- We also send Emails If they are not completed before the time Period.

For Predicting the User Career we are conducting the quiz where we can predict the Career based on the scores in Quiz. We have Quiz on Interest as well as Quiz on Skills User can Go through any one of the Quiz And Career will be shown to user once he done with Quiz based on his Score.We Created Quiz Based on Interest We have Created the Quiz on Interest Where the Question will be used for Knowing about the Interest on User. As we Can see we can have several Career's. Right now we are having Two Careers For each Question there will be Incrementing the value for any of the Careers. In the first Question we have a Question as If the user is having Interest If User says yes then the career value for Web Development will be incremented.Else If User says he/she has no Interest on Designing Then the Value of Data Scientist Career is Incremented. In the Next Question We have a Question as if the user is interested in Web Development. It is a Direct Question To predict whether the user is interested in Web development Based Career Or not. Also In next Question we have Questions Whether User is Interested in Programming if user says as yes then the web development Career is incremented as in Data Scientist We don't have that much coding work. Next Question is Whether the user is Interested in Artificial Intelligence or not which is for Data Scientist Career. Final Question is Whether User in mathematics or not Because in Data Science Mathematics Plays a vital role.

3.1 *System architecture diagram*

The system Architecture Diagram show the basic functionality of the Online Learning Management System.

Here We are Using Our Frontend as ReactJS and our Backend Server as ExpressJS and NodeJS. Also For Database we are using MongoDB

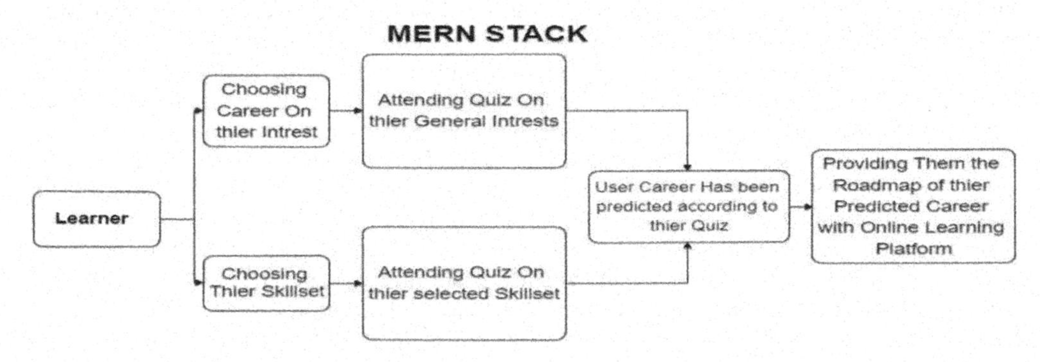

Figure 1. System architecture diagram.

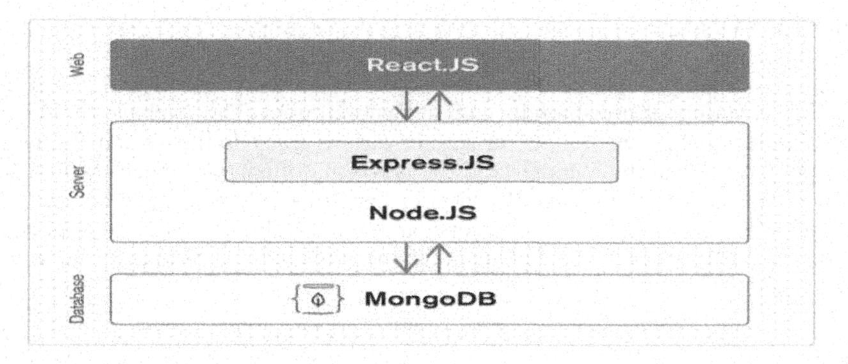

Figure 2. MERN stack architecture diagram.

4 IMPLEMENTATION AND RESULT

This stage involves determining the project's viability and presenting a business proposal that includes a very basic project design and some cost estimates. During system analysis, the viability of the suggested system must be examined. This will ensure that the offered remedy won't put a strain on the company. Understanding the main system requirements is crucial for the feasibility analysis. The problem and the stakeholders' information needs are examined in the feasibility study. It aims to ascertain the resources necessary to offer an information systems solution, the price and advantages of such a solution, as well as the viability of such a solution. The feasibility study's objective is to evaluate different information systems. A suggested solution's viability is assessed in terms of its constituent parts. The project's technical viability is assessed to see if it can be finished given the available resources—people, equipment, software, and technical know-how. It takes into account choosing resources for the suggested system. The system is platform independent because it was created using Python. As a result, users of the system can run on any platform with average processing capabilities. The system is technically feasible because the technology is cutting edge. The data is the essential component in prediction jobs for which special consideration should be made. The data's location, format, consistency, presence of outliers, and other factors will all have a significant impact on the outcomes. Many questions should be addressed at this stage in order to ensure the effectiveness and accuracy of the learning algorithm. Many sub steps are taken to get, clean and transform the data. We get User Data like Their Interest and Skills from the Quiz and Their Some personal Information for Good User Experience based on their Data. We send the User Data Which more secure in Backend Node Js so that Hackers cannot get the data.

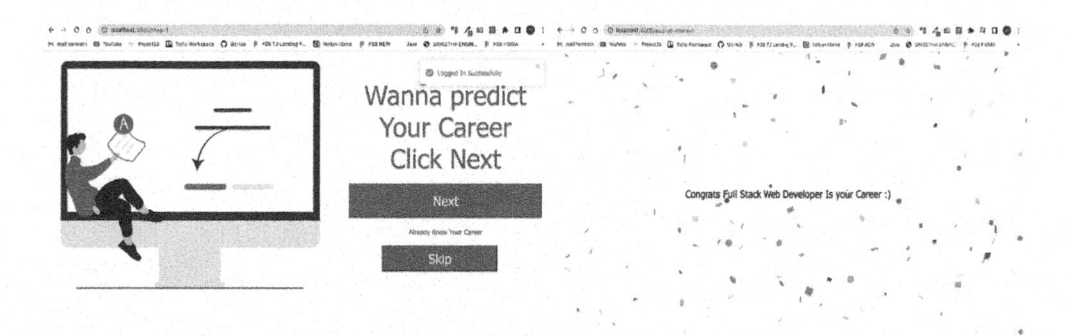

Figure 3. Sample outputs.

5 CONCLUSION

Our initiative is mainly to predict the career according to their Skillset with the Roadmap. Additionally, a variety of user-friendly coding has been used. This package should show to be effective in meeting the institution's criteria in various areas. In the we had given solutions for these points. We had done understand the problem domain and given the best solution also we will be keep improving the UI with user Feedback. Made a statement outlining the project's goals and objectives. Purpose, scope, and applicability have all been described. We discussed the system's need specifications and the possible actions that may be taken in relation to them.

REFERENCES

Anooja S, Dileep V. "A Study on Student Career Prediction" in *Predicting Career for Students*, Volume: 07 Issue: 02 | Feb 2020.

Brown S. D. and Lent R. W., *Career Development and Counseling: Putting Theory and Research to Work*, vol. 53, no. 9. 2013.

Fajar Fithroni, Zaenab Amatillah Rodhiyya, Nuri Cholidah Hanum, Caraka Putra Bhakti, Cucu Kurniasih: "Developing Career Vision Roadmap for Student Career Planning" *Advances in Social Science, Education and Humanities Research (ICGCS)*, Volume 657, 2021.

Galeon D H, Garcia Jr. P G and dela J, Cruz E-learning Roadmap for Open Distance Learning in Cordillera Administrative Region *IOP Conference Series: Materials Science and Engineering*, Volume 482, Issue 1, pp. 012012 (2019).

Jauhari J. and Maryani R., "Career Guidance Program in Improving Student Career Decision Plans, Program Bimbingan Karir dalam Meningkatkan Rencana Keputusan Karir Siswa," J. Islam. Guid. Couns., vol. 2, no. 1, pp. 45–62, 2018.

Javeed, A. (2019). Performance Optimization Techniques for React JS. *IEEE International Conference on Electrical, Computer and Communication Technologies (ICECCT)*, 1–5. https://doi.org/10.1109/ICECCT.2019.8869134, (2019).

Jyoti Shetty, Deepika Dash, Akshaya Kumar Joish, Guruprasad C *"Review Paper on Web Frameworks, Databases and Web Stacks"*, vol. 10, 2020 IRJET-V7141078

Madhan Mohan Reddy, "Career Prediction System" *International Journal of Scientific Research in Science and Technology*, Volume 8, Issue 4 Page Number: 54-58, 2021.

Mezhoudi, N., Alghamdi, R., Aljunaid, R. *et al.* Employability Prediction: A Survey of Current Approaches, Research Challenges and Applications. J Ambient Intell Human Comput (2021). https://doi.org/10.1007/s12652-021-03276-9

Revathi and Anitha Julian, "ICOGNOS – A Secure and Adaptive E-Learning Suite", International *Journal of Advanced Science and Technology*, Vol. 28, No. 13, 2019, Page(s) 198–209. (Scopus Indexed) (2019).

Sourabh Mahadev Malewade, Archana Ekbote, "Performance Optimization using MERN stack on Web Application" in International *Journal of Engineering Research & Technology (IJERT)*, Vol. 10 Issue 06, June-2021

VidyaShreeram, N. & Muthukumaravel, Dr. (2021). *Student Career Prediction Using Decision Tree and Random Forest Machine Learning Classifiers.* 10.4108/eai.7-6-2021.2308621.

Artificial Intelligence, Blockchain, Computing and Security – Dagur et al. (Eds)
© 2024 The Author(s), ISBN: 978-1-032-67841-2

Crime rate prediction using K-means algorithm

Anitha Julian*, Darisi Venkata Veera Somasekhar*, R. Harsh Vardhan* &
Harsh Vardhan Naidu*
Department of Computer Science and Engineering, Saveetha Engineering College, Chennai, Tamil Nadu

ABSTRACT: Preventative measures are preferable to corrective measures. Crime prevention is preferable to crime detection. With the world's crime rate at an all-time high and more violent crimes occurring every day, we must implement a vaccination system to protect children from being victims of crime. By "vaccinating society against crime," we mean using a variety of techniques to reduce the prevalence of criminal activity in a community. The term "data mining" refers to the technique of gleaning useful information and revealing hidden patterns from massive datasets. Data mining for the sake of criminal investigation may be rather fruitful. Utilizing data held in repositories, data mining may be used to forecast and assess criminal activity. India's rising crime rate has become a critical concern and is preventing the nation from moving toward functional government. The crime statistics and trends that are relevant to this project are analyzed in depth.

Keywords: Crime, Analyze, Crime Patterns, Clustering, Classification, Summarization.

1 INTRODUCTION

For a long time, criminals have been a problem for society all around the globe, and now we must take action to rid the world of crime. Our company's goal is to provide public safety applications that may help reduce crime. The focus of modern police techniques is on apprehending perpetrators after crimes have already been committed. However, because of technical progress, we can analyze past crime statistics to spot trends and easily prevent future crimes. We are making predictions about where criminal activity is likely to occur using clustering methods. Numerous clustering methods exist for organizing data into meaningful groups. Because of the sheer size and complexity of crime data sets, data mining methods are well-suited to the study of criminology. The goal of criminology, the study of crime, criminal behavior, and law enforcement, is to establish the distinguishing features of illegal acts. It's a major area where data mining may have a significant impact. As a preliminary step in conducting a deeper analysis, identifying criminal characteristics is essential. With the insights acquired via data mining methods, law enforcement agencies have access to a powerful resource. Data sets may be transformed into clusters for easier analysis to pinpoint hotspots for criminal activity using clustering methods. These clumps, when superimposed on a police force's jurisdiction map, depict groups of crimes. The time and kind of crimes committed are stored in a cluster, along with other relevant data. Based on their constituents, these groups are categorized. More people equal more crime in certain neighborhood's whereas fewer people mean less attention paid to some neighborhood's. In high-crime locations, preventative measures are undertaken based on the specific types of crimes that are most common there.

If you're looking for a basic and widely utilized clustering technique in scientific and commercial applications, go no further than K-means. As a result of its lower computational

*Corresponding Authors: cse.anithajulian@gmail.com, somasekhardarisi7@gmail.com, rharshvardhan9435@gmail.com and harshavardhannaidu27@gmail.com

DOI: 10.1201/9781032684994-83

cost, it is well suited for clustering massive data sets. It has been successfully applied in a variety of domains, including market segmentation, computer vision and so on. Common applications include preparing data for subsequent algorithms or discovering an initial state. Since the nature of crimes varies greatly and crime databases are often full of unsolved crimes, we opted for the clustering approach rather than a supervised one like categorization. Therefore, a classification method that just considers previously solved crimes will not provide reliable results for predicting future criminal behavior.

2 LITERATURE SURVEY

The purpose of this project is to implement a method for evaluating crime statistics with the hopes of reducing the crime rate. This project is problem-driven, therefore a big chunk of it will include trying out different techniques. Feature extraction is going to be the first major obstacle to overcome. Simple data exploration strategies (such as analysing patterns over time, etc.) and principal component analysis (PCA) may aid in feature prediction.

Ultimately, we'll probably have to create a more complex model that more accurately represents the dynamics of criminal behaviour. Some studies have examined the fact that criminal activity in one place makes the surrounding region more vulnerable to criminal activity. Improvements in predicting ability could result from this kind of model upgrade. We intend to spend a substantial lot of time refining alternative model concepts, and assessing their performance.

(Benjamin David *et al.* 2017) introduces the concept of data mining, through which we can be able to assess and examine the existing big databases to provide additional information. Using existing datasets, we compare the extracted new patterns to existing ones. (Chandy *et al.* 2022) suggests that we can identify the relevant characteristics of the data using a random forest classifier before performing the extraction. (Hyeon-Woo Kang *et al.* 2017) provides a deep neural network-based fusion strategy with appropriate parameterization for predicting criminal activity from feature-level data.

(Mugdha Sharma *et al.* 2014) & (Nikhil Dubey *et al.* 2014) proposed that crime patterns can be controlled and analysed by using machine learning algorithms. Usually crime requires a lot of classification. The computational process makes it easy to identify the patterns of crime by using ML techniques. (Manasvi *et al.* 2014) & (Raj *et al.* 2015) proposed an embedded technique that improves the robustness of the prediction model that performs classification. (Rohit Patil *et al.* 2020) suggests K-means be used to get the result, while an Apriori approach is suggested for common patterns. The rising crime rate in recent years means that the system must process a massive volume of data, which in turn necessitates additional time for in-depth human analysis.

(Suhong Kim *et al.* 2018) & (Shraddha *et al.* 2012) offered deep neural networks and machine learning methods to improve the prediction functionality of the model using K-nearest-neighbour (KNN) and decision trees so that we can have an accuracy of 39–44%. A deep neural network functions more correctly utilising the feature-level dataset. An embedded platform-based image processing system utilising a BlackFin 548 DSP processor is suggested by (Anitha Julian *et al.* 2012). A technique for dealing with larger data sets is being proposed by (Satyadevan *et al.* 2014), which allows the users to get the prediction results in a transparent way. This technique clearly implements the identification and classification of crime on the basis of extracted large datasets.

3 EXISTING SYSTEM

Numerous projects dealing with criminal activity have already been completed inside the current system. To aid citizens in keeping tabs on law enforcement, massive databases have been analyzed and data retrieved, including crime hotspots and the specifics of each offence. These datasets have been exploited by existing technologies to pinpoint areas with a disproportionate amount of crime. Although crime hotspots have been identified, there is a lack of data that includes the time and date of the incident as well as methods that can reliably foretell where and when future crimes will take place.

3.1 *Drawbacks of existing system*

1. The earlier studies are less accurate because of the classifier's category value. With respect to the nominal features with a greater value, this value produces a biased outcome.
2. The categorization methods are unsuitable for areas where there is insufficient data or qualities of genuine value.
3. It compares its effectiveness to that of the police department's conventional "hot spot" strategy, and it lays out in detail the processing processes necessary to construct a predictive system of this kind.
4. In addition, smaller police agencies, which face more worrisome calls for violent action, may not have access to more effective resources. A custom-built prediction system may be more expensive than purchasing one, and it may take much more time to develop.

4 PROPOSED SYSTEM

An algorithm for data mining was developed to foresee criminal activity. The K-means algorithm has a significant impact on the analysis and forecasting of criminal activity. Crime data may be analyzed statistically and in terms of hidden linkages, link prediction, and group formation and dissolution using the K-means method. To foretell future criminal activity, the k-means algorithm is used here.

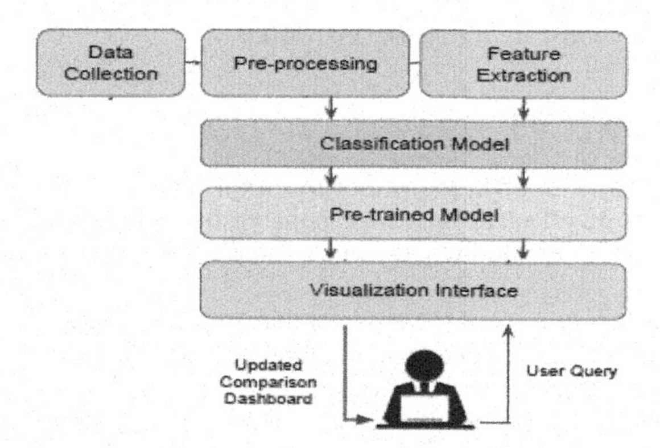

Figure 1. Architecture diagram.

The goal would be to train a predictive model. The test dataset is used to verify the results of the training data set. Depending on the precision, a superior method will be used to construct the model. As for crime forecasting, the k-means algorithm will be utilized. By creating a visual representation of the data, we can better examine the potential criminal activity in the nationals.

After doing a literature study, it is necessary to employ an easily deployed and analyzed open-source data mining tool. Here, the fast miner tool is used to do a k-means clustering analysis on the crime dataset.

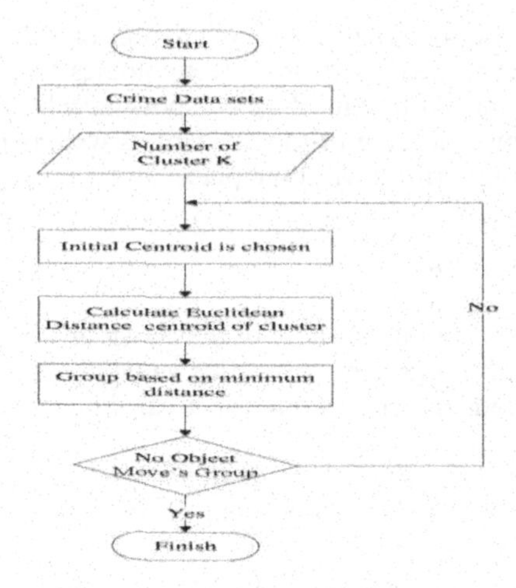

Figure 2. Flow chart for the proposed system.

The suggested system's flowchart in action is seen below.

1. Let's begin with a collection of crime statistics.
2. Refine the data set in light of the analysis that will be performed, and then generate a new data set with the attributes that were previously missing.
3. It is to access the crime dataset as an Excel file, import it into the fast miner software, and then use the "Replace Missing value operator" to fill in the blanks.
4. Take the dataset and run the "Normalize operator" on it.
5. After the dataset has been normalised, run k-means clustering on it.
6. In the plot view of the results, plot the data between the offences to get the necessary cluster.
7. When a cluster is established, it may be analysed.

4.1 *Advantages of the proposed system*

1. The goal of this study is to offer an open-source implementation that police analysts may utilize to deploy more effective predictive policing by using alternative crime mapping and feature engineering methodologies; and enhance our previously presented prediction framework.
2. The crime rate in India may now be reduced thanks to this study, which aids law enforcement in both foreseeing and detecting criminal activity with greater precision.
3. System will keep a historical record of crime and saves a maximum of the time.
4. By using this system one can be able to complain easily and can achieve proper details about typical crimes.

5 METHODOLOGY

5.1 *Dataset description*

The data is being collected from the records of the police station and from newspapers as well as from the users who have reported several crimes earlier in the crime portal. It is associated with every state of the country so that we can make comparisons easily. To be specific the input dataset will look like a form consisting of three columns indicating the criminal details, victim details and crime details by analyzing the given details we can be able to keep track of everything regarding the type of crime that has happened and its severity after the completion of the predictive analysis.

The dataset will tell us the type of crime has happened earlier. Each crime is associated with the crime_id which is a unique number specified to the particular crime along with it associates the crime_status it containing '0' and '1'. The crime_status with '0' indicates the crime hasn't occurred and '1' indicates the previously occurred type of crime. Moreover, all these three unique identifiers are uniquely associated with the dataset internally. If there is a change in one of the columns or rows in the dataset then it impacts the entire dataset. With the help of this, we can be able to report the type of crime that has happened and can be able to get the prediction results easily.

5.2 *Algorithm*

K-means Clustering is a kind of cluster analysis that divides a dataset of observations into groups of a certain size, k, so that each observation is assigned to the group with the mean closest to its own. It can be implemented by using the predefined methodology.

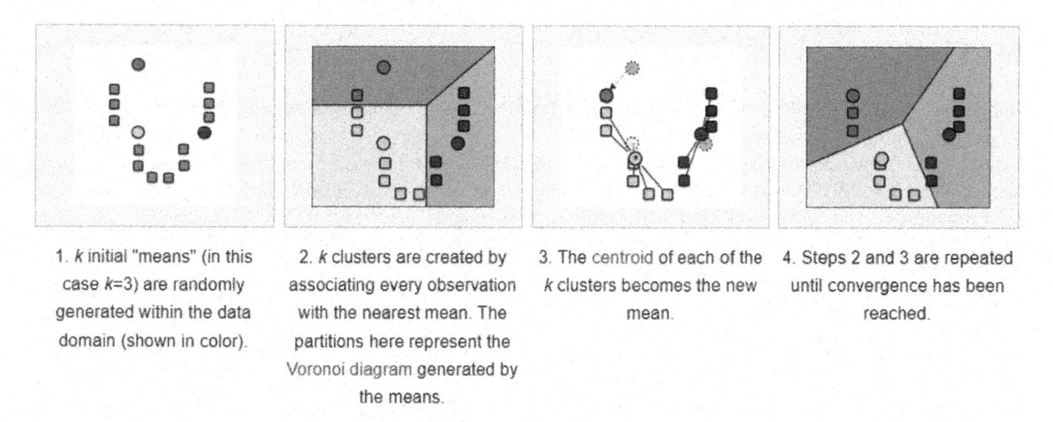

Figure 3. Demonstration of the standard algorithm.

1. The first stage is selecting K instances to serve as cluster nuclei.
2. Next stage is to determine the total number of clusters, which we'll call k.
3. The program then takes into account each instance and places it in the most relevant cluster.
4. Once a full cycle of reassignment has occurred, or after each instance has been assigned, the cluster centroids are recalculated.
5. There is a cycle involved here.

To be efficient, an algorithm has to have a complexity of O(kn), with n, c, and t being the number of instances. The process often halts at a local maximum. The only time it becomes a disadvantage is when the mean is being determined, and only if c, the number of clusters, is not already known. It's not good at finding clusters with non-convex forms and can't deal with noisy data or outliers.

6 CONCLUSION

In this work, the implementation of the prediction rate area-specific modelling is made more difficult by the widespread lack of crime in many locations. In that study, we used the Machine Learning algorithm to develop and evaluate a method for predicting criminal behavior based on a variety of factors, including age, gender, year, and month. In this role, we make use of a specific kind of machine learning technique called K-Means clustering, of which we find varying degrees of accuracy depending on the specifics of the situation at hand. We utilize K-Mean clustering for our Crime Prediction system because, although linear models may work well and give superior precision in certain cases, in most cases K-mean delivers the recognized accuracy that is needed.

Based on these promising findings, we anticipate that crime data mining will become more crucial in raising the standard and effectiveness of criminological and intelligence analysis shortly. To better understand crime patterns, we may build visual and intuitive criminal and intelligence investigative approaches. Now that we have used data mining's clustering methodology to analyze crimes, we may go on to other methods, such as classification. In addition, we can analyze a wide range of datasets, including those relating to business surveys, poverty, assistance effectiveness, and more. Our ability to pinpoint the areas with the highest crime rates—known as "hot zones"—will be greatly enhanced by the data we collect using these forecasting tools. When we're done here, we'll have used the CNN algorithm to analyze photo data and used the Google API to check out the hot spot.

REFERENCES

Benjamin David B. F., "Survey on Crime Analysis and Prediction Using Data Mining Techniques", *ICTAT Journal on Soft Computing*, 2017.

Chandy and Abraham, "*A Blog on the Overview of Random Forest Algorithm*", published online, 2022.

Hyeon-Woo Kang & Hang-Bong Kang, "Prediction of Crime from Multi-Model Data Using Deep Learning Techniques", *PLoS One* 12(4), Published online 2017.

Mahesh Kumar S., Anitha Julian, "An Embedded Digital Image Processing System using Blackfin 548 DSP", *CIIT International Journal*, July 2012.

Manasvi P & Tejaswini, "Survey on Crime Analysis and Prediction Using ML Techniques", *International Journal of Trendy Research in Engineering & Technology (IJTRET)*, Volume: 06 Issue: 03 June 2022.

Mugdha Sharma, "A Data Mining Tool for Detection of Criminals Using Decision Trees", *International Conference on Data Mining and Intelligent Computing (ICDMIC)*, November 2014.

Nikhil Dubey & Setu K. Chaturvedi, "A Survey Paper on Crime Prediction Technique Using Data Mining", International Journal of Engineering Research and Applications, 2014.

Raj M. and Anitha Julian, "Design and Implementation of Anti-Theft ATM Machine Using Embedded Systems", *IEEE International Conference on Circuit, Globalization and technological advancements in the communication industry have stimulated the growth of providers in the market Power and Computing Technologies,* 2015, pp.1.

Rohit Patil, MuzamilKacchi, PranaliGavali and Komal Pimparia, "Crime Pattern Detection and Analysis Using Machine Learning Techniques", *IJERT*, Volume: 07, June 2020.

Sathyadevan.S, "Crime Analysis and Prediction Using Data Mining", *First International Conference on Networks & Soft Computing (ICNSC2014)* (pp. 406–412).

Shraddha S. Kavathekar, "A Survey on Crime Occurrence Detection and Prediction Techniques", International Journal of Mechanical Engineering and Technology (IJMET), Volume 8, 2012.

Suhong Kim and Param Joshi "Crime Analysis Through Machine Learning", *IEEE 9th Annual Information Technology Electronics and Mobile Communication (IEMCON)*, 2018.

Artificial Intelligence, Blockchain, Computing and Security – Dagur et al. (Eds)
© 2024 The Author(s), ISBN: 978-1-032-67841-2

Code together – a code sharing platform

Ramya Devi*, Eedula Ganesh Reddy*, V.R. Dikshit Reddy* & R. Ganesh*
Department of Computer Science and Engineering, Saveetha Engineering College, Chennai, Tamil Nadu

ABSTRACT: Code Together is a platform that aims to solve the issues that programmers have with conventional IDEs and platforms in order to unite all developers and coders in the appreciation of collaborative coding. It enables developers to communicate with other developers or collaborators using real-time chat and online voice calls.With real-time chat and internet audio calls, it enables developers to interact with other developers or collaborators. Additionally, by enabling programmers to collaborate and debate and plan their strategy together, the whiteboard facilitates the structuring of an algorithm. The others may quickly see the stored code of one collaborator, enabling them to save modifications to their own programmers. The appropriate articles might also prove to be a helping hand for programmers in the event of any coding-related questions.

Keywords: Live Share code, Whiteboard, Collaborative compilers, Real time chat.

1 INTRODUCTION

By addressing the problems that programmers have with traditional IDEs and platforms, the Code Together platform wants to unite all developers and coders in the appreciation of collaborative coding. Through online phone calls and real-time chat, it enables engineers to interact with their peers and collaborators. In addition, the whiteboard facilitates collaborative programming by allowing programmers to debate and plan their strategy together. Code Together offers real-time IDE sharing and coding sessions. Even when coding in the same file, you may view changes immediately, much like Google Docs for your code. Multiple IDE support for IDEs based on Eclipse, Net beans, and VS Code. There are countless applications for coding, including collaborative programming, code reviews, project planning, software testing, education, product development using core features, remote development, and more. A little assistance from your buddies is fine, but not if it interferes with your flow. You can easily share your code acress the platform after integrating Code Together with Vs Code, Net beans. Press a button and, set access restrictions, and then extend an invitation to others. Even if it's not the same as yours, invited participants join from their IDE, even if it differs from yours (use your customized IDE's). Code Together which embodies the perfect balance of simplicity and use, was developed by a group of remote engineers employing collaborative development. Whether you operate in an agile team where pair programming is a frequent part of your software development flow or you simply wish to exchange code live, debug session. Have you ever collaborated on coding using a screen-sharing programmer or an online code editor? Her IDE sharing and coding sessions are offered by Code Together in real-time. You will see your changes immediately, exactly like Google Docs for your code, even if you are working in the same file. IDEs based on Eclipse, IntelliJ, and VS Code may support each other. Code review, project design, unit testing, and other application usages are common.

*Corresponding Authors: ramyadevir@saveetha.ac.in, edulaganeshreddy123@gmail.com, dikshitreddy17@gmail.com and ganesh250501@gmail.com

DOI: 10.1201/9781032684994-84

Regarding the order of events as consisting of external stimuli, cerebral processing, and then physiological responses. Developed by team of engineers those are mainly focused on collaborative coding, Code Together is the ideal fusion of usability and simplicity. For mob programming, pair programming, code reviews, and other tasks, Code Together is the ideal solution. No matter if you work in an agile team where pair programming is a necessary step in the software development process, or if you just want to share your code during infrequent debugging sessions, you should learn how to pair programmer. If you have been utilizing a screen-sharing programmer or a programming platform for featuring team programming.

The current studies make an effort to broaden and reproduce these findings using coders who have autism. You need something more than a straightforward collaborative code editor for true pair programming. The host IDE provides language intelligence and enhances your programming experience, including rename refactoring, as-you-type validation throughout the workspace, and content aid (also known as IntelliSense). It will also be to your liking the variety of navigation and visualization tools available for your code.

2 LITERATURE SURVEY

In literature survey, we gathered some details on the Code Together after a literature review. He regarded the order of events as consisting of external stimuli, cerebral processing, and then physiological responses. (Ellen *et al.* 2015) Developed by a remote team of engineers that rely on collaborative collaboration, Code Together is the ideal fusion of usability and simplicity. Better interaction with coders through code sharing. It is hardly unexpected that the results are so inconsistent given the breadth and depth of empirical studies on children's emotional problems and people with autism. Hobson and colleagues (1986a, 1986b; Hobson *et al.* 1988) and (Howard *et al.* 2000) have investigated the hypotheses of a general emotional deficit and a selective emotion detection deficiency. (Mingsent *et al.* 2021) proposed a Specification and Complexity of Collaborative Text Editing Suggests Collaborative text editing systems. But drawback is It only provides platform for code together, there is no code sharing. When you live share a coding session with other developers, Code Together makes it simple to swap between teams of developers working on different projects. (Nichols *et al.* 2019) proposed a Closing a performance Gap between Casual Consistency Between Causal Consistency and Eventual Consistency, There Is A Performance Gap and suggests Code collaboration. Drawback is Code sharing available but there is no whiteboard option to clear doubts.

A little assistance from your buddies is fine, but not if it interferes with your flow. It simply takes a few seconds to live share your code with others after adding Code Together to Visual Studio Code, IntelliJ, or Eclipse. Simply click a button, set access restrictions, and then extend an invitation to others. Those that are invited sign in using a web browser or their own IDE, even if it differs from yours (with the theme and key bindings of their preferred IDE). Introduction to Reliable and Secure Distributed Programming, a collaborative code sharing method. Second edition, Springer Publishing Company, Inc., 2011. Considerably more than a common code editor! The host offers linguistic intelligence so that everyone receives essential features, such content help and validation, independent of configuration and whether they are connecting from a browser or an IDE. (Huber *et al.* 2019) has proposed two main approaches or techniques used in collaborative programming are real-time and real-time collaboration. In order to manage the project and keep everyone up to speed, programmers working on a non-real-time collaboration will either need to transfer source code back and forth between one another if they do not share a single machine or use a version control system like Git.

3 EXISTING SYSTEM

Installation: These cloud-hosted tools, often known as cloud IDEs, are put into use. The only locally necessary software is a web browser. Other tools need to be installed locally. Even better are plugins that provide collaborative editing functionality to well-known editors/IDEs.

Cross-Platform: While certain tools may be compatible with Mac and Linux, they may not be. In this sense, plugins are preferable. They enhance well-known software that is already accessible on a number of platforms.

Editing: Both developers are editing at the same time. cross-system copy/paste. While the other developer is editing, the other can go to different areas of the codebase or other programmers. It is completely editor or IDE independent.

Cross-Media: Live audio and video streaming in both directions. Chat room. with an integrated editor or IDE.

Uncluttered design for usability. awareness of shared content and the editing environment at the time. automatic notification cancellation, resulting in a unruffled environment.

Consummation: Very little latency. Video resolutions are of high quality. On connections with low bandwidth, revert to lower resolutions. transparency about standards.

Connect to code repositories through GitHub, GitLab, and Bitbucket and other tools through integration (Asana & Trello).

Other considerations include pricing, security, and customer service. For certain teams, open source could be essential.

4 PROPOSED SYSTEM

Only screen sharing will be seen. We must publish code updates to a common repository before switching roles. Software used for video conferences includes Cisco Webex, Zoom, Google Hangouts, and Google Meet.

Screen Control/Share: Remote access to your partner's machine for a brief period of time. Communication might be slow. Examples include TeamViewer, Zoom, Join.me, VNC, Live Share, Tuple, and Docker.

Share: Editing that is really collaborative is therefore recommended. The surroundings may all be customised. Different editors or IDEs are available for developers. A programmer can explore the codebase. without breaking the partner's flow. Even many sections of the code can be edited simultaneously by developers. Examples include AWS Cloud9, Live Share (with Visual Studio and The code can even have many portions changed at once by engineers. AWS Cloud9, Live Share (in conjunction with Visual Studio and VS Code), Code Together, Git Live, Teletype, Drovio, and Bracket are a few examples.

5 METHODOLOGY

The platform that strives to unite all programmers and developers to embrace collaborative development by addressing problems that they encounter on standard IDEs and platforms.

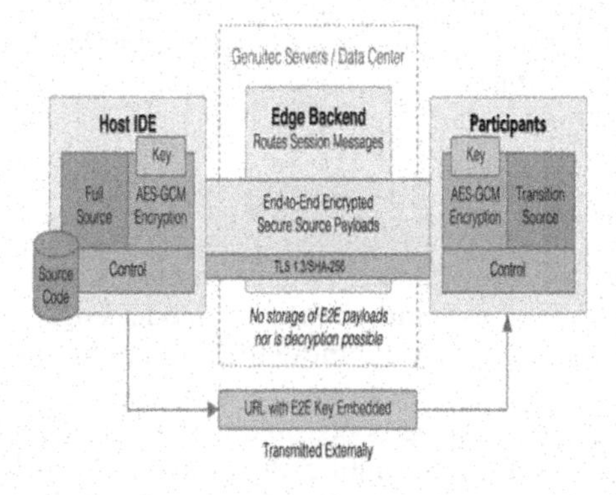

Figure 1. Architecture diagram.

Through online phone calls and real-time chat, it enables developers to interact with other developers or collaborators. Additionally, the whiteboard facilitates collaborative programming by allowing programmers to debate and organize their strategy together. The other collaborators may quickly see the stored code of the first, enabling them to save modifications to their own programmers. Additionally, the pertinent articles might prove to be a helping hand for programmers in the event of any coding-related problems.

End-to-end Encryption: End-to-end encryption is used by Code Together to protect your code. Keep all source code on the host system, and only send data that is absolutely essential. Keep all source code on the host computer, and only transmit files that are absolutely essential. With Code Together, a special key is produced for the session when you live-share your IDE. This key is utilized by the host and each participant for AES-GCM encryption. This key is never sent to our servers. Requests between edge clients are routed via the Edge backend server using TLS 1.3/SHA256 encryption. Integrated Communication: Screen sharing, text chat, audio, and video are all built-in communication features. A simple toggle allows wants to join in the session to initiate an audio/video communication, which anyone in the session may simply join. Additionally, you may invite visitors who only access to the messaging platforms rather than the actual code is ideal for gathering user input without actually revealing any code. Although the IDE has a toolbar with some basic functionality, the audio/video communication establishment is to keep your IDE clean with browser control. Use the browser to access fun feature like polls and comments.

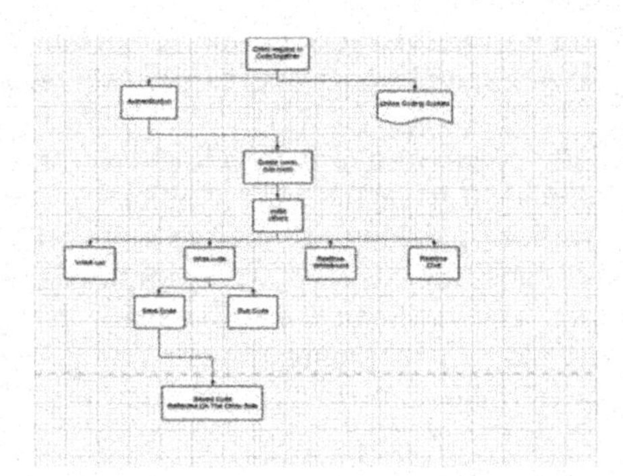

Figure 2. Flow diagram.

In order software to be useful in helping to tackle some of the core issues that the community faces, software projects are designed to offer sophisticated functions that adhere to business demands and requirements. As a result, they are not simple tasks that can be completed by one person, and writing in a monolithic style is not practical. However, working as a team of programmers is not always simple.

6 CONCLUSION

Code Together was developed by a remote team of engineers utilizing collaborative design, and it is the perfect balance of use and simplicity. Code Together, the perfect fusion of simplicity and ease of use, was built by a remote team of engineers employing collaborative development.

Live IDE and coding session sharing is possible with Code Together. Like Google Docs for your code, you can view changes straight away even when working in the same file. For IDEs based on Eclipse, Net beans, and VS Code, there is support for multiple IDEs.

REFERENCES

Attiya H., Burckhardt S., Gotsman A., Morrison A., Yang H., and Zawirski M. Specification and Complexity of Collaborative Text Editing (Extended Version). Available from http://www.software.imdea.org/~gotsman.

Christopher R. Palmer and Gordon V. Cormack. 1998. *Operation Transforms for Distributed Shared Spreadsheet. In Proceedings of 1998 ACM Conference on Computer Supported Cooperative Work (CSCW '98)*. Association for Computing Machinery, New York, USA, 69–78.

Day-Richter J. *What's Different About the New Google Docs: Making Collaboration Fast.* http://googledrive.blogspot.com/2010/09/whats-different-about-new-google-docs.html, 2010.

De Candia G., Hastorun D., Jampani M., Kakulapati G., Lakshman A., Pilchin A., Sivasubramanian S., Vosshall P., and Vogels W. Dynamo: Amazon's Highly Available Key-value Store. *In SOSP*, 2007.

Defago X., Schiper A., and Urbán P. Total Order Broadcast and Multicast Algorithms: Taxonomy and Survey. *ACM CSUR*, 36(4), 2004.

Du J., Elnikety S., Roy A., and Zwaenepoel W. Orbe Scalable Causal Consistency Using Dependency Matrices and Physical Clocks. *In SoCC*, 2013.

Du J., Iorgulescu C., Roy A., and Zwaenepoel W. Closing the Performance Gap Between Causal Consistency and Eventual Consistency. *In PAPEC,* 2014.

Ellen F., and Morrison A. Limitations of Highly-Available Eventually-Consistent Data Stores. *In PODC*, 2015.

Huber J. B. *Has Proposed Hierarchical Channel Coding for Broadcasting Employing GC-codes* Publisher: IEEE – 2019.

Ming sent Xu Has Collaborative Opportunistic Network Coding for Persistent Data Stream in Disruptive Sensor Networks Publisher: IEEE – 2021.

Nichols, David, Curtis, Pavel, Dixon, Michael & Lamping, John. (2019). High-Latency, Low Bandwidth Windowing in the Jupiter Collaboration System. 111-120. 10.1145/215585.215706.

Stéphane Weiss, Pascal Urso, Pascal Molli. Logoot: A Scalable Optimistic Replication Algorithm for Collaborative Editing on P2P Networks. *29th IEEE International Conference on Distributed Computing Systems - ICDCS* 2009, Jun 2009, Montreal, Canada. pp.404-412

Artificial Intelligence, Blockchain, Computing and Security – Dagur et al. (Eds)
 ISBN: 978-1-032-67841-2

Detecting the violence from video using computer vision under 2dspatio temporal representations

M. Srividhya, R. Ramyadevi & V. Vijaya Chamundeeswari
Department of Computer Science and Engineering, Saveetha Engineering College, Chennai, India

ABSTRACT: A current hot topic in computer vision is action recognition in videos, particularly for the detection of aggression. The complexity of the 2D + t data produced by the proliferation of videos by a security camera or television material is what makes this work interesting. Modern techniques learn using 3D neural network approaches, which require data to train to produce discriminating features. This research intends to address topics such as cutting-edge techniques for video violence detection, datasets for real-time video violence detection framework development, and discussion and identification of unresolved concerns in the given subject. To overcome these restrictions, we offer throughout this piece a technique for putting videos into categories for violence recognition with a traditional 2D CNN. The methods comprise two parts.: (I) From an input video, we first construct several 2D spatial-temporal representations, (II) According to speculation, the new representations will feed CNN's test/training data. The methodology by combining the separate conclusions from the video's various 2D spatial-temporal representations and the categorization decision is made. An experiment on publicly available datasets with violent video demonstrates the usefulness of the suggested approach.

Keywords: computer vision, convolutional neural network (CNN), spatial-temporal

1 INTRODUCTION

A challenging task in pattern recognition and computer vision is video analysis. A video is frequently viewed as a series of still images. A more accurate approach is to use a data cube with geographic and temporal dimensions to represent a movie(Cheng *et al.* 2019). In the past, applications such as (Wang *et al.* 2021) action identification, tracking, and video classification dealing with such data have been examined. Recently, the community has become quite interested in the detection and recognition of violence. This issue can be looked into to secure public areas that are monitored, such as prisons or train stations. In the literature, Frequently, This issue has been taken into account as a video categorization challenge. Before methods focused on manually creating features and extracting them from the video data for approaches for machine learning like Support Vector machine. The Motion SIFT(Nievas *et al.* 2014) concept was proposed by the authors. Violent Flow (Cheng *et al.* 2019), which is derived by calculating the optical flow's magnitude with time, is a further technique for violence characterization (Xu *et al.* 2014). Later, this approach was enhanced by taking into account the Orientation in violent flow (Hassner *et al.* 2012). It is suggested to categorize activity using the enhanced Dense Trajectory. Then, when there was violence, this technique was applied, improving Fisher vectors in combination. Several computer vision applications, including content-based video indexing, undesired material filtering, movie rating, and assistance for human operators in real-time surveillance systems (Bilinski *et al.* 2016), to mention a few, have already identified the capacity to recognize various human actions as a key task. Particular attention is paid to surveillance

 DOI: 10.1201/9781032684994-85

systems for the identification of violent incidents as well as the filtering or rating of objectionable information. The need for safer public spaces justifies the former, whereas the latter targets circumstances where violence is deemed unsuitable for the target audience. Beginning with its subjective nature, violence is difficult to define and categorize. Additionally, it's possible to misclassify some human activities that are extremely similar to aggressive ones. To make the application viable for an effective and reliable real system, the question of how to resolve those ambiguities is raised. Given that the current theories of violence rely on an arbitrary notion, we further reduce the idea by designating just fight scenes—aggressive human actions—as violent (Hassner *et al.* 2012), regardless of context or the number of participants. The topic of recognizing human activity has been the subject of extensive research in the area of computer vision. The detection of video acknowledgment of action made tremendous strides for a while now, thanks to the accessible huge datasets, ways for representing videos, deep neural network designs, etc. On the other hand, a lot of work concentrated on particular (Wang *et al.* 2021) action identification sub-tasks including egocentric activity recognition, (Zolfaghari *et al.* 2018) spatial-temporal localization of activity, and so on. (Bilinski *et al.* 2016) Manually identifying violence in video footage appears to be getting harder as the media increases the numbers of technology like cameras in the commonplace.

2 LITERATURE SURVEY

(Cheng *et al.* 2019) It is used to find violent video from cameras. To prevent or halt illegal activity. It will be taking RWF 2000 database from the videos and the accuracy of the detection is 87%. (Nievas *et al.* 2014) It will be recognizing that actions like walking, running, and jumping all will be detected and the accuracy of the violent video is less. It will categorize the videos as fight or non-fight. MOSHIFT is used to detect the videos. (Chelali *et al.* 2020) CNN conventional neural network is used as a 2D video representation. It consists of time series with spatial-temporal representations. (Xu *et al.* 2014) MOSHIFT algorithm is used which is helpful for the accuracy of the video detection from violence and nonviolence. Data from video surveillance cameras to detect criminal activity spatiotemporally. Video cameras must be used to keep an eye on people. attacks against people, and conflicts (Bilinski *et al.* 2016). The dataset for human activity will be gathered from YouTube videos. following pre-training enhances performance datasets, and uses the network model for picture categorization (Carreira *et al.* 2021). utilized a VGG-based algorithm to enhance performance datasets and efficiently recognize human actions. Sort the videos into training-related and non-training-related data. Increases action videos' accuracy (Fang *et al.* 2019). recognizing human activity from content-based video retrieval, and utilizing data such as hockey fights, violent flows, and non-violent flows. For the precision of the data, a support vector machine is also used (Gao *et al.* 2016). Using computer vision techniques, CCTV surveillance cameras can recognize human abilities. They take too many notes during the movies, which shortens them and improves the accuracy of the data they collect about human behavior (Hassner *et al.* 2012). The model is trained using ImageNet utilizing the CNN architecture. Data addition enhances the performance of the outcome. The sizeable convolution neural network is known as ImageNet was utilized in computer vision datasets (Iandola *et al.* 2020). The temporal segment network collects data from conventional neural networks (Wang *et al.* 2021). Real-time action detection was performed using the Activity Net data set (Zolfaghari *et al.* 2018).

3 PROPOSED METHODOLOGIES

Techniques for starting deep learning from scratch to a conclusion today achieve outstanding results for video analysis. Convolutional neural networks are the most often utilized DNNs (CNNs). CNNs may learn spatiotemporal features when working with such data, typically by making use of 3D convolutions. Then, the combined feature maps of the two networks are employed, followed by a decision-making layer with two connections.

Additionally, Most of these designs cannot take into account videos as a whole because of their size; instead, videos are inputted to the networks as split-up, on-overlapping chunks of few frames (for example, 16 frames). In each of these methods, the choice is made after a traditional fully connected layer and a late fusing of the learned features.

4 IMPLEMENTATIONS

VGG19 ALGORITHM: VGG19 ALGORITHM: According to CNN architecture, ImageNet is used to train the model (Iandola *et al.* 2020) dataset and the VGG-19 algorithm (Fang *et al.* 2019). The VGG Network is a smaller, deeper network (Carreira *et al.* 2021). The network model of the VGG-19 architecture consists of 19 layers and a tiny filter with a size of 3 x 3 conv and periodic pooling. It has 3 fully linked layers and 16 convolutional layers. An image of 224 by 224 pixels and a depth of 3 is present in the first input layer. (Xu *et al.* 2014) CNN Conv2D's layer 1 and layer 2 have a 64-depth. The depth indicates how many filters were applied to create the feature map.

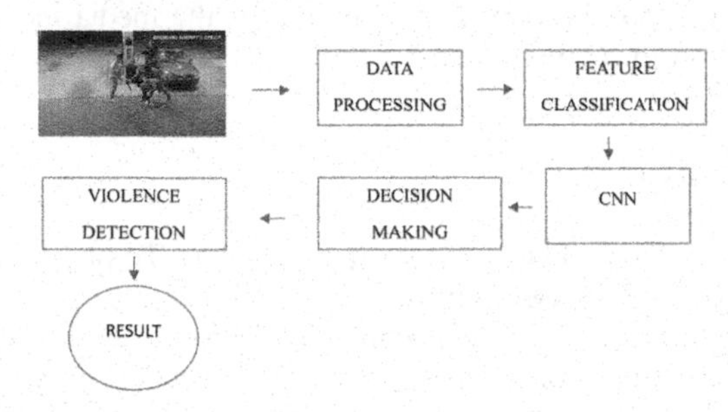

Figure 1. Architecture diagram.

5 MODULES DESCRIPTION

5.1 *Datasets*

There have been four datasets used specifically for detecting violence: RWF2500: 2500(Cheng *et al.* 2019) (Bilinski *et al.* 2016) surveillance camera videos gathered from YouTube. (Gao *et al.* 2016) There are 1200 movies in each class. Videos don't all have the same size, but they all last for 5 seconds. (Cheng *et al.* 2019) Hockey Fights: 1100 films from the National League's Hockey competition have been collected. There are 700 identical films in each lesson, each lasting just under two seconds. 350 footage of fights from various movie sequences is compiled in Movie Fights (Cheng *et al.* 2019). 120 of them are non-violent and 130 of them are violent. The length and size of the videos differ. 266 videos feature crowd scenes; crowd violence. There are approximately 133 videos in each class. The videos don't share any dimensions or running times.

5.2 *Data preparation*

These dataset's videos have individual video labels. A curve is defined locally, and not all of the violent scenes in a movie have to be violent. We then made the decision must focus on the area of aggression (ROI) (Chelali *et al.* 2020) and the need to classify the Preparation as violent or nonviolent. With Gunner Frame back's approach, we computed dense optical flow for this. After then, the acquired flow is averaged along the temporal axis of the video. To get to a place with a lot of movement, which is meant to represent a dangerous place, Finally, the mean flow is subjected to binarization to produce the ROI (Chelali *et al.* 2020). We used 38 violent movie trailers as a training dataset for violent preparation., while for non-violent prep, we took into account 36 prep inside the ROI from anti-violence videos and 2 prep outside the ROI of violence videos. As a result, each class has an equal number of Preparation.

5.3 *Data validation protocol*

K-Fold cross-validation is used to verify our methodology. At the video level, divided into three sets are train, test, and validation. Afterward, The average of CNN's five evaluations and (Iandola *et al.* 2020) training sessions are displayed (OA) (Gao *et al.* 2016). The exception is the (Gao *et al.* 2016)RWF2000 dataset, For the validation set, we only used 22% of the training set. Squeeze Net is trained with default settings (1 = 0.9, 2 = 0.999, and = 108) a learning rate of 105.

5.4 *Decision process*

When using (Chelali *et al.* 2020) 3D CNNs, the video is normally given a single class probability vector. In the spatial domain D, we produce N test b representations per movie. A local decision is computed for every Preparation that offers the vector. Following that, the procedure used to make the overall decision is equal to divided videos.

6 CONCLUSION

This paper introduced the video detection approach accustomed to categories of violent and nonviolent videos. It is simple to predict whether there will be violence or not. The core idea is to portray the video as a series of spatially and temporally encoded planar representations. Any traditional 2D CNN can be used to do the local classification task, and trained filters can directly extract spatiotemporal characteristics.

7 FUTURE WORK

Due to this 2D Prep representation, we can get good results using a well-trained conventional neural network, for example, equipped with ImageNet on a comparable classified issue. This is not achievable with 3D techniques. As a point of comparison, we intend to look into the definition of violent Prep in more detail. To do this, to distinguish between two different sorts of non-violent zones, we can either introduce a third class or use ROI for all two videos in movies. To partition the violent incident more precisely, we may also examine how the Prep focused on the temporal domain.

REFERENCES

Bilinski. P and Bremond. F, Human Violence Recognition and Detection in Surveillance Videos, in *AVSS*, 2016, pp. 30–36.

Carreira. J and Zisserman. A, Quo Vadis, Action Recognition? A New Model and the Kinetics Dataset, in *CVPR*, 2021, pp. 4724–4733

Chelali M, Kurtz C, Puissant. A, and Vincent. N, Classification of Spatially Enriched Pixel Time Series with Convolutional Neural Networks, in *ICPR*, 2020, pp. 5310–5317.

Cheng M, Cai. K, and M. Li, Rwf-2000: An Open Large-scale Video Database for Violence Detection, in *ICPR*, 2020, pp. 4183–4190.

Fang. H, Xie. S, Tai. Y, and Lu. C, RMPE: Regional Multi-person Pose Estimation, in *ICCV*, 2019, pp. 2353–2362.

Gao. Y, Liu. H, Sun. X, Wang. C, and Liu. Y, Violence Detection Using Oriented Violent Flows, *Image and Vision Computing*, vol. 48–49, pp. 37–41, 2016.

Hassner. Y. I. T and Kliper-Gross. O, Violent Flows: Real-time Detection of Violent Crowd Behavior, in *CVPR Workshops*, 2012, pp. 1–6.

Iandola. F, Moskewicz. M, and Ashraf. K, Squeezenet: AlexNet-level Accuracy with 50x Fewer Parameters and <1MB Model Size, *arXiv*, vol. abs/1602.07360, 2020.

Nievas. E.B, Deniz-Súarez. O, Garc'ıa. G.B, and Sukthankar. R, 2014 Violence Detection in Video Using Computer Vision Techniques, in *CAIP*.

Wang. L, Xiong. Y, Wang. Z, Qiao. Y, Lin. D, Tang. X, and Gool. L.V., Temporal Segment Networks: Towards Good Practices for Deep Action Recognition, in *ECCV*, 2021, pp. 20–36.

Xu. L, Gong. C, Yang. J, Wu. Q, and Yao. L, Violent Video Detection Based on MoSIFT Feature and Sparse Coding, in *ICASSP*, 2014, pp. 3538–3542.

Zolfaghari. M, Singh. K, and Brox. T, ECO: Efficient Convolutional Network for Online Video Understanding, in *ECCV*, 2018, pp. 713–730. 2597.

Artificial Intelligence, Blockchain, Computing and Security – Dagur et al. (Eds)
© 2024 The Author(s), ISBN: 978-1-032-67841-2

Node to node communication security In IoT networks with hybrid cryptography and steganography system

Medam Venkata Sreevidya, R. Ramyadevi & V. Vijaya Chamundeeswari
Department of Computer Science and Engineering, Saveetha Engineering College, Chennai, India

ABSTRACT: The Internet of Things refers to actual objects or items that have been integrated with sensors, algorithms, and a few other technologies to exchange information with other devices and networks online (IoT). As IOT devices are publicly available to everyone on the network, it is required to be conscious of security risks and dangers posed by cyberattacks; as a result, it should be guarded. Plain information is changed to encrypted information in cryptography before being delivered, and it is then converted back to plain text after receiving a response from the recipient. Steganography is the practice of encoding hidden information into an audio, video, picture, or text file. In this project, it is suggested that IoT network data be encrypted using cryptography, and that the encrypted message be concealed somewhere in a digital image through steganography. It is also suggested to enhance the number of bits that can be stored within a single picture pixel.

Keywords: CNN, Cryptography, Steganography, AES

1 INTRODUCTION

The internet of things helps individuals to live a life where the control of their lives will be handled by the IOT devices. IOT devices plays a crucial role in many aspects of our lives such as home automation, reducing labour costs by automating procedures. It reduces much of manual tasks for humans which helps in reducing cost and efforts. Thus, an efficient security transfer of data is required to ensure the data is transferred safe across the network. Without a secured transfer medium for data, the data would compromise easily for malicious attacks which would finally leads to data loss. As these IOT devices are used in day to day life, the data loss leads to negative effects. So, a secured medium is required to transfer data safe from source to destination.

1.1 *Cryptography*

To convert communications in ways that are challenging to understand, Secure information and communication techniques are referred to as "cryptography" and are created using mathematical concepts and a set of computations that follow predefined rules. The mathematical operation we use for encryption and decryption process are called cryptographic algorithm. Also called as cypher. The plaintext is encrypted using a cryptographic technique using a key. Strong cryptographic algorithms and key secrecy are the only two factors that affect how secure encrypted data is.

The following three goals are addressed by cryptographic techniques:

- Authentication, which verifies the legitimacy of a person or a machine.
- Preserving the content's secret.
- Information is transferred from its source to its destination without modification when integrity is upheld.

 DOI: 10.1201/9781032684994-86

1.2 *Steganography*

Steganography is a method for encoding information into a text, video, audio or image file in order to conceal it. It is a technique used to defend sensitive or secret information against nefarious intrusions. A unique method is used in current digital steganography to inject data into a certain file type, such as a JPEG picture, after the data has been encrypted or otherwise obscured (Manju Khari *et al.* 2020).

2 LITERATURE SURVEY

We have read through many papers to better understand the issue and many situations that were used to create a trustworthy system.

Since the changes made to cover images as a result of message concealing are so minute, it is challenging to tell the difference between those suspicious pictures(Stego images) and those innocent photos. (Songtao Wu *et al.* 2020) Proposed a unique normalisation method called Shared Normalization tackles this issue. SN shares fixed training statistic samples, unlike the BN layer. (Hassaballah *et al.* 2021) Proposed a method called as HHOIWT that uses digital image steganography to transfer communication and safeguard data. The method efficiently chooses pixels from an image that can be used to conceal small amounts of sensitive information in integer wavelet transforms using the Harris Hawks Optimization metaheuristic optimization algorithm to insert the hidden data in the images. (Xin Liao *et al.* 2020) Proposes a procedure formulating the data in multiple images using steganography and provides security based on the image texture features. (Wenkang Su *et al.* 2021) Describes how the local elements of cover images interact with one another. The issue of secure image steganography is also formulated. (Wei Wang *et al.* 2019) Proposes a process for Industrial Internet Of Things to outsource all of their time-consuming encryption tasks to the nearby boundary for quick processing and ensures that any encrypted texts produced by external parties are provably secure. (Pietro Tedeschi *et al.* 2019) Proposes LiKe is a restricted IoT device and the most recent Zigbee 3.0 protocol stack compatible lightweight mechanism. Protocol LiKe is used for authenticated key agreements. (Zinan Lin *et al.* 2018) Secret information can be concealed in VoIP streams using the Quantization Index Modulation (QIM) steganography technique. Unauthorized parties could use this technique to create covert channels for bad intentions. (Yumei Li *et al.* 2020) Proposes a process to ensure the authenticity and integrity of data in wireless sensor networks, offers an effective coding signature of a linearly homomorphic network based on identity in this research.

3 IMPLEMENTATION

This endeavor suggests a system to improve the number of bytes that can be include in a frame of such a picture, as well as to safeguard IOT stored data inside a network in an digital image utilizing the steganalysis approach. Convolutional neural networks are used to significantly improve the payload that may be sent through an image when compared to classic steganography techniques. Therefore, gathering the DIV2K dataset will be the initial phase in the project, followed by separation into training and testing datasets. To train the model, dataset will be utilized whereas test data will just be treated separately. The data set is trained using twenty to hundred epochs to train the algorithm. By fusing a cover image with a data tensor, this Encoding method generates an image. This module analyses a picture and determines if it is a cover or a steganographic picture.

3.1 *AES cryptography encryption*

The following are AES's features:

- Data is 128 bits, while keys are 128/192/256 bits.
- symmetric block cypher with a symmetric key.

- quicker and more powerful than Triple-DES.
- Detail all specifications and design considerations.

3.2 *AES cryptography decryption*

The decrypt technique for an AES cipher - text is precisely the same as the encoding process in backwards. To every loop, the four processes are performed in reverse chronological order.

- Mix columns and add a circular key
- Shift Rows
- Byte substitution
- Despite being extremely closely related, the encryption and decryption algorithms must be implemented independently as each round's sub-processes function in the opposite way, opposed to a Feistel Cipher

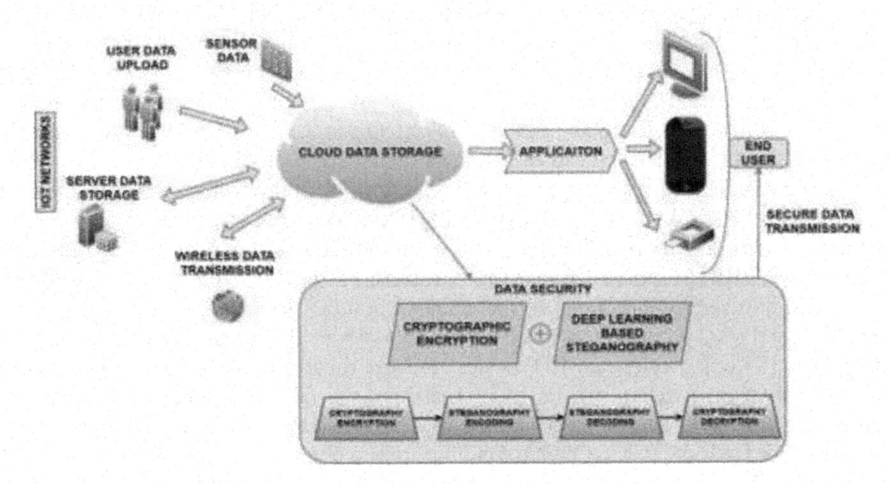

Figure 1. Architecture diagram.

3.3 *Encoding and decoding of image*

Without including any secret keys, steganography is a methodology of embedding data into an item. The secrecy is very necessary for this kind of steganography. This kind, provides a cover picture in which data is to be encoded, the encoded data is the personal data that is to be conveyed whose visual appearance is changed using encoding. The encryption methods are used to encode a message into the image. Though there are many file types available in the market, but digital photographs are widely used due to their online prevalence. There are numerous steganography methods available for hiding private information in images; some are more challenging than others, and each one has pros and cons of its own. In some cases, a lengthy message may need to be concealed, while in other cases, the secret information may need to be completely invisible.

4 CONCLUSION

Instead of replacing encryption, steganography is meant to enhance it. A communication that is encrypted and concealed using steganography reduces the possibility that the hidden message will be found and increases security. This project involves employing cryptography to encrypt data from IoT networks, convolutional neural networks to increase the quantity of bits which can be retained in a picture pixel, and steganography to conceal the encrypted message inside an image file. This was the basis on which we created an algorithm that effectively encrypts and safeguards the data.

REFERENCES

Arijit Karati, SK Hafizul Islam, Marimuthu Karuppiah. 2018. Securing Data in Internet of Things (IoT) Using Cryptography and Steganography. *IEEE Transactions on Systems, Man, and Cybernetics: Systems.* (Volume: 50, Issue: 1).

Hassaballah M, Mohamed Abdel Hameed, Ali Ismail Awad, Khan Muhammad. 2021. A Novel Image Steganography Method for Industrial Internet of Things Security. *IEEE Transactions on Industrial Informatics.* (Vol. 17, Issue: 11).

Manju Khari, Aditya Kumar Garg, Amir H. Gandomi, Rashmi Gupta, Rizwan Patan, Balamurugan Balusamy. 2020. RNN-SM: Fast Steganalysis of VoIP Streams Using Recurrent Neural Network Cryptography and Steganography. *IEEE Transactions on Information Forensics and Security.* (Volume: 13, Issue: 7).

Pietro Tedeschi, Savio Sciancalepore, Areej Eliyan, Roberto Di Pietro. 2020. LiKe: Lightweight Certificateless Key Agreement for Secure IoT Communications. *IEEE Internet of Things Journal.* (Volume: 7, Issue: 1).

Ru Zhang, Feng Zhu, Jianyi Liu, Gongshen Liu. 2020. Depth-wise Separable Convolutions and Multi-level Pooling for an Efficient Spatial CNN-based Steganalysis. *IEEE Transactions on Information Forensics and Security.* (Volume: 15).

Songtao Wu, Sheng-hua Zhong, Yan Liu. 2020. A Novel Convolutional Neural Network for Image Steganalysis with Shared Normalization. *IEEE Transactions on Multimedia* (Volume: 22, Issue: 1).

Wei Wang, Peng Xu, Dongli Liu, Laurence Tianruo Yang. 2019. Light-weighted Secure Searching over Public-key Ciphertexts for Edge-Cloud Assisted Industrial IoT Devices. *IEEE Transactions on Industrial Informatics.* (2019, Vol. 16, Issue: 6).

Weixuan Tang, Bin Li, Shunquan Tan, Mauro Barni, Jiwu Huang. 2019. CNN-based Adversarial Embedding for Image Steganography. *IEEE Transactions on Information Forensics and Security.* (Volume: 14, Issue: 8).

Wenkang Su, Jiangqun Ni, Xianglei Hu, Jessica Fridrich. 2021. Image Steganography with Symmetric Embedding using Gaussian Markov Random Field Model. *IEEE Transactions on Circuits and Systems for Video Technology.* (Volume: 31, Issue: 3).

Xin Liao, Jiaojiao Yin, Mingliang Chen, Zheng Qin. 2020. Adaptive Payload Distribution in Multiple Images Steganography Based on Image Texture Features. *IEEE Transactions on Dependable and Secure Computing.* (Volume: 19, Issue: 2).

Yumei Li, Futai Zhang, Xin Liu. 2020. Secure Data Delivery with Identity-based Linearly Homomorphic Network Coding Signature Scheme in IoT. *IEEE Transactions on Services Computing.* (Volume: 15, Issue: 4).

Zinan Lin, Yongfeng Huang, Jilong Wang. 2018. Provably Secure and Lightweight Certificateless Signature Scheme for IIoT Environment. *IEEE Transactions on Industrial Informatics.* (Vol. 13, Issue: 7)

E-voting system with secure blockchain alert on data tampering

Anitha Julian*, M.S. Lekshmi Priya* & V. Vijaya Chamundeeswari*
Department of Computer Science and Engineering, Saveetha Engineering College, Chennai

ABSTRACT: With the growth of Blockchain technology in the second decade of the new millennium, numerous applications throughout the globe have been interested in intriguing new uses. The subject of electronic voting is one of the difficult uses of blockchain technology. Since the election is being conducted offline, security is of the utmost importance. An organization still has full access to the current electoral system's centralized infrastructure. The entity that controls the database and has complete control over the system is mostly to blame for the issues because it is using antiquated election systems. A significant factor that seriously affects the system is database manipulation. The use of blockchain technology is one of the efficient solutions to this issue. Block chain technology is the best option because it has a decentralized structure and a shared database. This paper explores the potential for using Blockchain technology in e-voting systems to enhance the voting process by addressing issues of trustlessness, privacy, and security with effective face authentication techniques using Haar cascade and Support Vector Machine Algorithms.

Keywords: *Electronic Voting, Block chain, Security, Haar cascade, Support Vector machine*

1 INTRODUCTION

Democracy is always conceptualized as a principle that grants the power to the citizens to elect the government through casting their votes. The process of voting is paramount in choosing the right leader for a proper functioning of the nation. Since, voting withholds the basis of democracy, it must possess qualities such as fairness in choosing the winner, neutral amount of support for all the candidates, and freedom from influence to maintain the integrity of the process and establish a transparent and secure flow of system, allowing all the individual to voice out their opinions without any constraints. Conventional voting is regulated and supplemented with mediators. Also, people dealt with a wide range of problems, including booth capture, dummy voting, and the issue of proper monitoring. a long line of people in front of the polling places, false voting, pre-voting, redundant voting and a lack of law enforcement, audits, and awareness, were predominantly faced during the early stages of voting. Significant obstacles confronted by older individuals reduce their voting power. All these problems with the previous voting system were rectified using an electronic voting machine (EVM). However, because the EVM (Electronic Voting Machine) does not address any security issues, it has issues with general acceptance. Blockchain based voting can dramatically reduce all these issues with time and money spent on polling sites by designating the locations, employing staff, and easing all these possible security issues. Since blockchain technology is still in its infancy, there is a lot of untapped potential. (Monrat *et al.* 2019) shows a variety of blockchain application domains that various experts have

*Corresponding Authors: cse.anithajulian@gmail.com, lekshmipriyamohan990@gmail.com and vijayachamu@gmail.com

DOI: 10.1201/9781032684994-87

suggested. Digital elections powered by blockchain reduce the danger of rigged voting while saving money. It can increase transparency, voting system dependability, and transaction traceability. An immutable blockchain wouldn't be affected by a single network saboteur, in contrast to the voting system's entire network, which could experience a single point of failure as a result. Blockchain-based elections can be held digitally, which not only saves money but also lowers the possibility of unfair voting (Shahzad *et al.* 2019).

2 LITERATURE SURVEY

Electronic and online voting utilize various techniques and methodologies, including procedures aimed at adding diversity. Although part of these procedures provides a measure of security and privacy, it is important to have robust systems in place to manage and regulate the voting process and data, to ensure the privacy and protection of the identity of voters as well as their voting information. (Farooq *et al.* 2022) presented a blockchain based digital voting system using the flexible consensus algorithm, along with several other security approaches such as chain security algorithm, smart contracts, and cryptographic hash techniques to achieve an open and efficient voting method. (Panja *et al.* 2020) proposed a decentralized voting protocol for the voters with the choice of candidates, using the Borda count process. This system specifies a bulletin board implemented on the Ethereum blockchain that is open to the public and requires the voters to submit the NZIK proofs for ensuring the masked identity. (Gowtham *et al.* 2019) put forth a method of creating a smart voting system that uses fingerprint recognition technology from the aadhar to enable each Indian voter to cast their vote for their constituency from anywhere in the country by visiting the closest polling location near the place where they currently live. (Singh *et al.* 2018) suggested an e-voting system using the RSA public key cryptographic algorithm to secure the voting protocol. This mechanism helps the voters to cast their vote at the ease of their personal computer and ensures the voters a reliable voting system since they have good reason to believe that their votes will be counted. (Chaum *et al.* 2008) proposed a system that gives an opportunity to the voters to check their votes by entering the given code to enter the voting machine. Through registered phone numbers and email addresses, voters can then confirm their ballots. (McCorry *et al.* 2017) talked about the voting protocols without polling places with the help of End-to-end verification which reduces the chance of user duplication. A flexible consensus mechanism and smart contracts are used in the proposed blockchain voting system to reduce the system's latency. (Rajalakshmi *et al.* 2021) explains how blockchain functions as a ledger that enables decentralized transactions. This paper focuses on a variety of blockchain-based applications that span throughout a broad spectrum of industries, comprising those in the financial services, judiciary, non-financial, internet of things (IoT), and more. From the above-mentioned studies, we have drawn the conclusion that Blockchain, with its distinctive features, has the potential to transform a number of established markets, including online voting systems.

3 EXISTING SYSTEM

The voting design in existence contains numerous flaws which has the potential to vandalize the entire election process. The risks which the EVMs possess are liable including, system collapse, sabotage through hacking and other user authentication frauds. The EVMs contain two major parts, the control unit and the voting unit which is interconnected by the cables. The process of casting a vote via the EVM involves manual intervention. This procedure entails an election worker to press the button on the ballot, allowing the voter to choose the candidate from the options on the machine. These EVMs have no advanced technology making them easy targets for hackers to tamper the results. The EVMs neither have any mechanism for voters to confirm their identification nor to prevent any encroachment activities.

4 PROPOSED METHODOLOGY

The proposed system uses blockchain technology to advance the voting experience. An online application that analyzes how political parties and independent candidates competing in the forthcoming state election are performing is being developed using ReactJS. The project also helps a typical person fill out a questionnaire to express his or her willingness to vote in the most recent or forthcoming state election as well as their ideas on it. The main goal of the system is to inform the public on the policy perspectives of the various political parties in light of their supporter bases. ReactJS is being used to create a web application that examines the role of the political parties and independent candidates running in the upcoming state election. Algorithms for face identification and recognition are utilized to efficiently identify genuine users when they use the programme to cast their votes on Election Day. SVM algorithm is utilized for face recognition, whereas Haar cascade is employed for face detection. The API (Application Programming Interface) was then developed using the Javascript framework nodeJS to effectively connect the web application with the database. MongoDB is used for database setup, and NodeJS is used for developing APIs. Blockchain integration at the backend will secure the data with the highest level of security, preventing data hacking. By providing an electronic voting system with Face Detection authentication, this web application aids the state government in achieving a 100% vote rate in the state elections. The below image (Figure 1) depicts the system architecture of the e-voting platform.

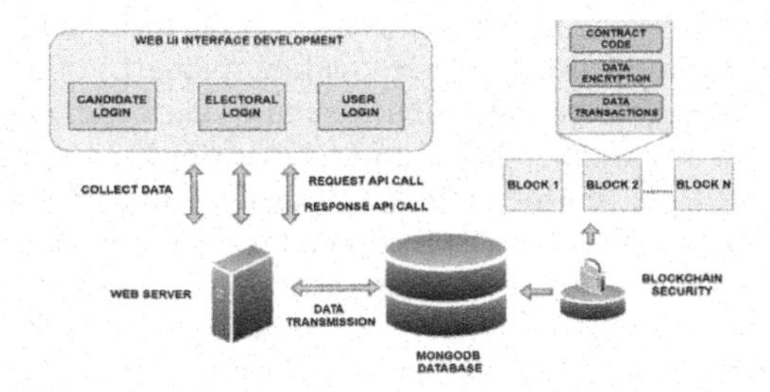

Figure 1. System architecture.

5 IMPLEMENTATION AND RESULTS

The proposed system is implemented as six different modules as listed below.

5.1 *Face detection*

Here, the Haar cascade method is being used for the Face Detection module to effectively recognize faces in a photograph or real time video. The detection of edge or line characteristics that (Viola *et al.* 2001) study "Rapid Object Detection using a Boosted Cascade of Simple Features" are used by the algorithm which helps it to accurately identify and detect faces anywhere in the image or video.

5.2 *Face recognition*

The SVM technique is being used in this instance for the Face Recognition module. The "Support Vector Machine" (SVM) is predominantly used for classification and regression models. The purpose of this classification algorithm is to establish a decision boundary value that further segregates the number of dimensional spaces as classes and data points, assisting the developers to rapidly classify new sets of similar data points in the future. (Cadena Moreano *et al.* 2019) in their survey observed that an SVM pattern classifier is always included in the analysis results with the greatest facial recognition.

5.3 *Node API generation*

Using the Javascript framework nodeJS, we create the API (Application Programming Interface) in this project. Node.js is a popular platform for web application development that uses a parallel, unlockable input-output driven approach, making it fast and efficient. It allows developers to write both server and client code in JavaScript and has an extensive library of JavaScript modules to simplify web application development.

5.4 *Mongodb generation*

MongoDB is used for storing and retrieval of data captured in the system. It is a platform-independent, document-driven storage application which stores the data in JSON format, typically without the use of any schemas. The database uses a shard key chosen by the user to allocate the data in a collection. According to the chosen shard key, the data is distributed into ranges and parted among several shards.

5.5 *Web application development*

The proposed system's front end is created with React, a JavaScript toolkit for creating UIs. React framework is advantageous for its reusability, flexibility and performance.

5.6 *Blockchain integration*

Blockchain is incorporated at the backend for data protection, ensuring the highest level of security and preventing data hacking. The blockchain transactions are captured in a unique fixed cryptographic signature called hash, which forms a chain of these records for each input. Thus, in order to modify a single record the entire chain of blocks with data should be altered, making it very complicated or merely impossible to compromise a blockchain module across all the distributed versions of the data.

The e-voting model, when entered the details, responded with maximum accuracy. The portal contains a login or signup form which prompts the user to enter their unique ID and password to successfully login and choose the suitable user types. The users are categorized into 3 types. The admin user for the election officials, the candidate user category for the candidates who participate in the election and finally the normal user type for the voters. All these category logins need appropriate authentication and verification to enter the portal. Once they successfully login into the system, the individuals can view the data and perform the necessary actions assigned to them in the homepage. Moreover, the blockchain integration in the voting system makes the results safe and secure, preventing any loss or damage of the data. The suggested system's login interface is shown in the image below (Figure 2).

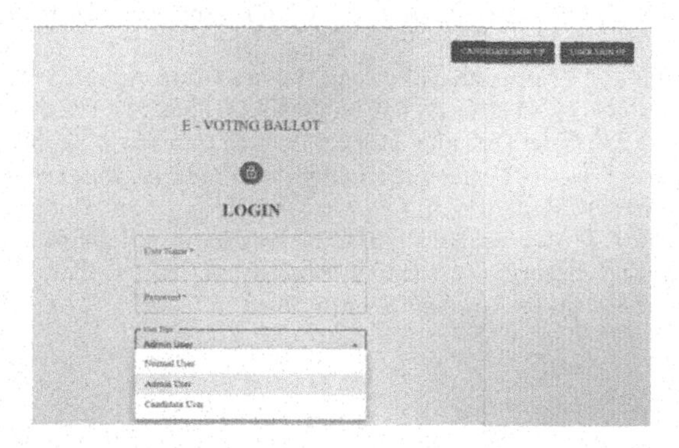

Figure 2. Login portal.

6 CONCLUSION AND FUTURE WORK

Building confidence between the government and the electorate was the goal of this blockchain-based voting system proposal, which was intended to give voters the impression that their voting integrity was protected. The security of EVMs and their vulnerabilities have been the subject of research publications (Wolchok *et al.* 2010), making them an obvious target. The suggested e-voting system with blockchain alert is a secure technical solution that guarantees a 100% vote rate in the nation and is an efficient technological tool for the residents of the country to learn about their political candidates. This system provides a straightforward solution to secure and monitor the integrity of the votes which is the base for one of the largest democracies in the world. In the near future, we will carefully examine the project application submitted by the Election Commission of India and incorporate the use of regional languages for those who reside in rural areas as well as a method for voters to provide suggestions for the current winner. This project will therefore have a big influence when voters may safely cast their votes online in the future, increasing turnout and increasing citizen awareness of the candidates in their town.

REFERENCES

Ashish Singh,Kakali Chatterjee, SecEVS: Secure Electronic Voting System Using Blockchain Technology, *International Conference on Computing, Power and Communication Technologies (GUCON)* Galgotias University, Greater Noida, UP, India. Sep 28–29, 2018.

Cadena, Jose & Palomino, Nora & Casa, Alex. (2019). *Facial Recognition Techniques Using SVM: A Comparative Analysis. Enfoque UTE.* 10. 98–111. 10.29019/enfoque.v10n3.493.

Chaum D., Essex A., Carback R., Clark J., Popoveniuc S., Sherman A., and Vora P., "E-voting 40 Scantegrity: End-to-end Voterverifiable Optical-scan Voting," *IEEE Secur. Privacy*, vol. 6, no. 3, pp. 40–46, May 2008. Accessed: Feb. 14,

Farooq M. S., Iftikhar U. and Khelifi A., "A Framework to Make Voting System Transparent Using Blockchain Technology," in *IEEE Access*, vol. 10, pp. 59959–59969, 2022, doi: 10.1109/ACCESS.2022.3180168.

Gowtham R, Harsha K N, Manjunatha B, Girish H S, Nithya Kumari R, 2019, Smart Voting System, *International Journal of Engineering Research & Technology (IJERT)* Volume 08, Issue04 (April – 2019),

Krishnamurthi, Rajalakshmi & Shree, Tuhina. (2021). *A Brief Analysis of Blockchain Algorithms and Its Challenges.* 10.4018/978–1–7998–5351–0.ch002.

McCorry P., Shahandashti S., and Hao F., "A Smart Contract for Boardroom Voting with Maximum Voter Privacy," in *Financial Cryptography and Data Security*. Sliema, Malta: Springer, 2017, pp. 357–375, doi: 10.1007/978–3–319–70972–7_20

Monrat, Ahmed Afif, Olov Schelén and Karl Andersson. "A Survey of Blockchain From the Perspectives of Applications, Challenges, and Opportunities." *IEEE Access* 7 (2019): 117134–117151.

Panja S., Bag S., Hao F.and Roy B., "A Smart Contract System for Decentralized Borda Count Voting," in *IEEE Transactions on Engineering Management*, vol. 67, no. 4, pp. 1323–1339, Nov. 2020, doi: 10.1109/TEM.2020.2986371.

Shahzad B. and Crowcroft J., "Trustworthy Electronic Voting Using Adjusted Blockchain Technology," IEEE Access, vol. 7, pp. 24477–24488, 2019, doi: 10.1109/ACCESS.2019.2895670.

Viola P. and Jones M., "Rapid Object Detection Using a Boosted Cascade of Simple Features," *Proceedings of the 2001 IEEE Computer Society Conference on Computer Vision and Pattern Recognition.* CVPR 2001, 2001, pp. I-I, doi: 10.1109/CVPR.2001.990517.

Wolchok, Scott, Eric Wustrow, J. Alex Halderman, Hari K. Prasad, Arun Kankipati, Sai Krishna Sakhamuri, Vasavya Yagati and Rop Gonggrijp. "Security Analysis of India's Electronic Voting Machines." *Conference on Computer and Communications Security* (2010).

Artificial Intelligence, Blockchain, Computing and Security – Dagur et al. (Eds)
© 2024 The Author(s), ISBN: 978-1-032-67841-2

Virtual mouse

V. Vijaya Chamundeeswari*, V. Bharath Raj*, A. Ajay* & S. Niranjan Dharmaraj*
Department of Computer Science and Engineering, Saveetha Engineering College, Chennai, Tamil Nadu, India

ABSTRACT: This project's objective is to create and put into action a virtual mouse system that would enable a user to operate a computer mouse using hand motions photographed by a camera. Computer vision algorithms are used by the system to identify and follow the user's hand motions, which are subsequently converted into mouse movements on the computer screen. This makes it a helpful tool for those with disabilities or in situations where a real mouse is not accessible since it enables the user to interact with the computer without the requirement for physical contact. The system was put to the test and reviewed for accuracy and usability, and the findings showed how well it worked for manipulating computer interfaces and the mouse pointer. Overall, the virtual mouse technology is a viable option for non-touch computer interaction.

Keywords: Computer, Hand, Gestures, Motion, Vision.

1 INTRODUCTION

For many people, using a computer on a daily basis has become a necessity, and the mouse is a key input tool while working with a computer. However, there are several circumstances in which using a physical mouse may be difficult or impossible. For instance, during COVID-19 physical mouse couldn't be used publicly as there was a risk of viral transmission. Researchers have investigated the use of alternate input modalities, such as hand gestures, for manipulating the mouse cursor in order to overcome these difficulties. Computer vision algorithms are used to interpret hand gestures filmed with a camera in order to identify and track the motions of the user's hand. This makes controlling the mouse pointer without using physical contact a more convenient and accessible way to interact with a computer. We demonstrate a virtual mouse system in this project that uses hand motions recorded by a camera to direct the movement of a computer mouse. The device, which is accurate and simple to use, offers a potential option for non-touch computer interface. The design and execution of the virtual mouse system, as well as the findings of our assessment of its usability and performance, will all be covered in the remaining sections of this study. This project supports a method of Human Computer Interaction (HCI) in which a live camera is used to control cursor movement.

2 LITERATURE SURVEY

The development of alternative input modalities for regulating the movement of a computer mouse has been the subject of a sizable amount of study in the field of human-computer interaction. These modalities include hand gestures, eye movements, and facial expressions. Sensors of various kinds can record these modalities, and computer vision algorithms can then process them. Hand gestures are one alternative input modality that has received a lot

*Corresponding Authors: vijayac.dean@saveetha.ac.in, vbharathraj2002@gmail.com, ajaydoc001@gmail.com and niranjanselva1006@gmail.com

of research. Gesture recognition and hand tracking are two examples of techniques that may be used to detect hand motions that have been caught by cameras or depth sensors. In conclusion, a lot of study has been done on alternate input modalities, such as hand gestures, eye movements, and facial expressions, for controlling the movement of a computer mouse. These methods hold promise for further research and development because they have the potential to improve computer use's accessibility and adaptability.

(Behnam Maleki *et al.* 2015), (Hossein Ebrahimnezhad *et al.* 2015), (Min Xu *et al.* 2015) and (Xiangjian He *et al.* 2015) suggest a mechanism that initially recognizes the hand. Then, within a series of frames, it follows the trajectory of fingers. The last step in the process is to detect hand gestures by calculating a set of recommended geometric properties of finger trajectory data and comparing them to our dataset of gestures that have been gathered. For a dynamic gesture, four different sorts of descriptors are defined. Each descriptor has a unique set of characteristics that together make up a feature vector of 135 dimensions. To contrast the detection performance, several classification techniques (KNN, LDA, Naive Bayes, and SVM) are used. The MCR, or minimal misclassification error rate, is 4%. The number of characteristics is decreased by using Principle Component Analysis (PCA). LDA classifiers may achieve roughly 0.09% misclassification error rate with 30 dimensional features.

(Li Wens Heng *et al.* 2010), (Deng Chunjian *et al.* 2010) and (Lv Yi *et al.* 2010) suggest a technique using machine vision. In order to control window-based applications, an effective algorithm based on color is first used to track the movement of fingertips in real time. Then, the present set of messages, including elementary fingertip messages and simulated mouse messages, are generated in accordance with the results of fingertip tracking. A framework is available to assist programmers in realizing virtual mouse-based human-computer interaction. It will be possible to utilize this technology to operate computers and establish a more natural human-computer interface since the performance of the virtual mouse exhibits remarkable performance.

3 SYSTEM DESIGN

3.1 *Existing system*

The usage of a physical mouse is the most typical way to interface with a computer. A physical mouse is a tool that a user may manipulate to control the mouse cursor's movement. Physical mouse can be wired or wirelessly linked to a computer via Bluetooth or a wireless receiver. Physical mice are a popular and reliable input method for interfacing with a computer, and they are also easily accessible and reasonably priced. However, there are several circumstances in which using a physical mouse may be difficult to use. The employment of an alternate input modality, such as hand motions photographed by a camera, as suggested in this experiment, may be advantageous in certain circumstances.

3.2 *Proposed system*

In this project, we suggest employing computer vision to use hand motions recorded by a camera to control the movement of a computer mouse. The objective of this project is to create a virtual mouse system that enables computer use without the requirement for physical contact, making it an advantageous tool for people with impairments or in circumstances when a physical mouse is not accessible. In order to manage the mouse cursor precisely and quickly, we strive for high accuracy in the user's hand gestures' identification and tracking. Overall, our suggested virtual mouse technology is a viable option for non-touch computer interaction and has the potential to greatly increase computer use's flexibility and accessibility.

3.3 *Architecture diagram*

The general organization and functionality of the system are depicted in the architectural diagram of our virtual mouse project. The architecture diagram outlines the process by which the camera records the user's hand motions, which are subsequently subjected to computer vision algorithms for recognition and tracking. The user interface, which enables customization of the gesture-to-mouse mapping and adjustment of different system parameters, is another means by which the user may communicate with the system.

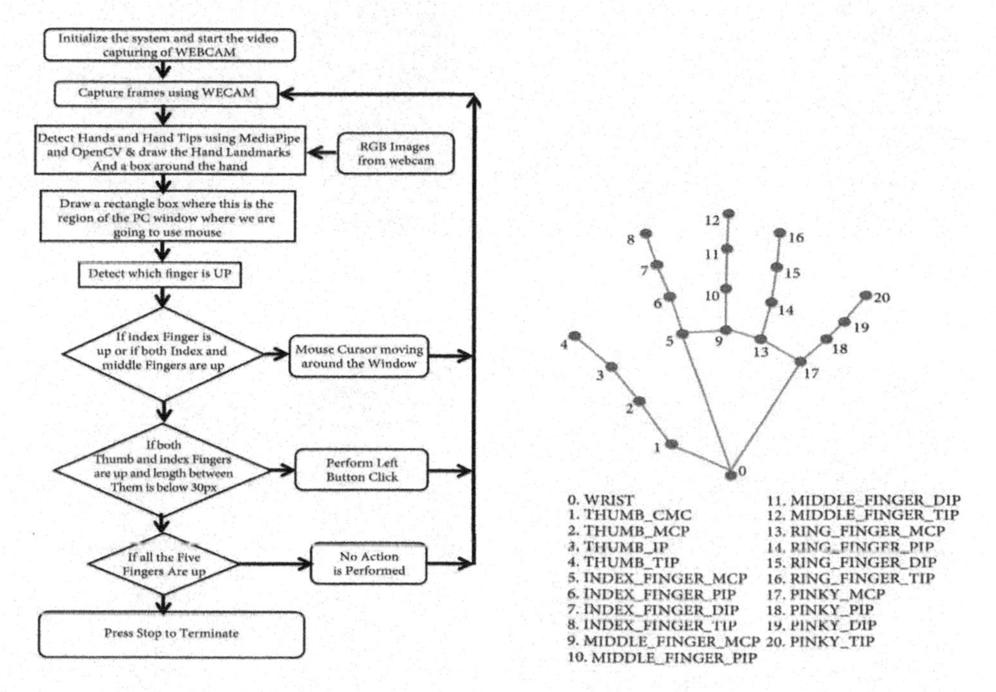

Figure 1. Architecture diagram.

Figure 2. Hand mapping.

3.4 *Implementation*

The creation of the system's hardware and software components, as well as their integration into a functioning whole, were all milestones in the realization of our virtual mouse project. Hardware-wise, a camera was employed to record the user's hand motions in real-time. The computer, which was used to process the gesture data, was linked to the camera and ran the required software algorithms. This screen displayed the mouse pointer. Software-wise, we developed the virtual mouse system by combining Python libraries and computer vision methods. Real-time hand gesture tracking and detection were accomplished using OpenCV, and different computations were carried out on the gesture data using NumPy. In order to identify and monitor the user's hand motions in real-time, we built a hand tracking model using MediaPipe. The hand motions were finally converted into mouse movements on the computer screen using autopy.In order to produce a practical and efficient system for non-touch computer interaction, hardware and software components for our virtual mouse project have to be developed and integrated.

4 RESULT

Our virtual mouse project's findings demonstrated how well the system worked for directing the mouse cursor's movement and engaging with computer interfaces using hand motions.

We tested the system's precision and usability, and the findings showed that it was adept at accurately identifying and following the user's hand motions.

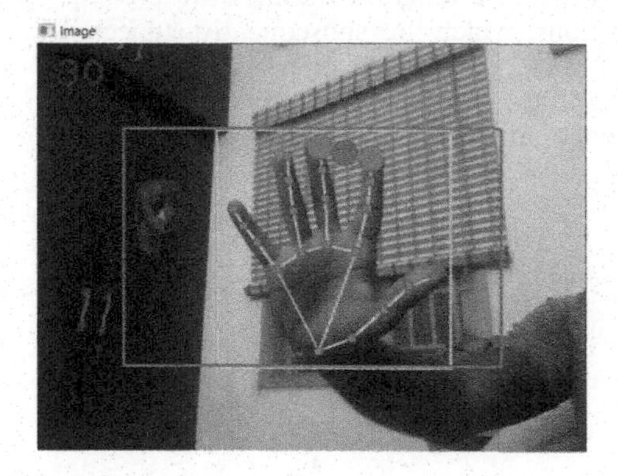

Figure 3. Key points of hand.

Figure 4. Fingers mapped to move cursor.

Figure 5. Mapped to stop cursor movement.

Figure 6. Mapped for leftcClick.

5 CONCLUSION

The feasibility and usefulness of employing hand movements recorded by a camera to control the movement of a computer mouse have been shown by our virtual mouse project, in conclusion. The system makes use of computer vision algorithms to identify and follow the user's hand motions in real-time, enabling accurate and quick mouse cursor control. The system underwent accuracy and usability testing, and the findings indicated that it was a potential option for non-touch computer interface. This project has provided us with valuable insights into the areas of development that can help enhance the overall performance.

5.1 *Future enhancement*

Through an analysis of the current state of the project, it was discovered that there are areas for development like Better gesture recognition, Enhanced user interface, Increased adaptability, Enhanced robustness. Overall, there are a lot of possible areas for improving and developing our virtual mouse project further, and they will vary depending on the particular objectives and needs of the system. We can further enhance the system's performance and usability by solving these problems, making it a more useful tool for non-touch computer interaction.

REFERENCES

Abhik Banerjee, Abhirup Ghosh, Koustuvmoni Bharadwaj. Mar 2014 "Mouse Control Using a Web Camera Based on Color Detection," IJCTT, vol.9.

Abhilash S. S., Thomas L., Wilson N. & Chaithanya C. 2018 "Virtual Mouse Using Hand Gesture," International Research Journal of Engineering and Technology (IRJET), vol. 05 issue: 4.

Behnam Maleki, Hossein Ebrahimnezhad, Min Xu and Xiangjian He. August 19–21, 2015 "Hand Gesture Recognition for A Virtual Mouse Application Using Geometric Feature of Finger's Trajectories," ICIMCS '15, Zhangjiajie, Hunan, China.

Devanshu Singh, Ayush Singh, Aniket Kumar Singh, Shailesh Chaudhary, Asst. Prof. JayvirSinhKher. Dec 2021 "Virtual Mouse Using OpenCV," International Journal for Research in Applied Science & Engineering Technology (IJRASET) Volume 9 Issue XII.

Dudhane S. U. 2013 "Cursor System Using Hand Gesture Recognition," IJARCCE, vol. 2, no.5.

Katona J. 2021 "A Review of Human–computer Interaction and Video Game Research Fields in Cognitive InfoCommunications," Applied Sciences, vol. 11, no. 6, p.2646.

Li Wens Heng, Deng Chunjian, Lv Yi. 2010 "*Implementation of Virtual Mouse Based on Machine Vision,*" *2010 IEEE.*

Shibly K. H., Kumar Dey S., Islam M. A., and Iftekhar Showrav S.. May 2019 "Design and Development of Hand Gesture Based Virtual Mouse," in *Proceedings of the 2019 1st International Conference on Advances in Science, Engineering and Robotics Technology (ICASERT)*, pp. 1–5, Dhaka, Bangladesh.

Vantukala VishnuTeja Reddy, Thumma Dhyanchand, Galla Vamsi Krishna, Satish Maheshwaram. 2020 "Virtual Mouse Control Using Colored Fingertips and Hand Gesture Recognition," IEEEHYDCON, 2020, pp. 1–5,doi:10.1109/HYDCON48903.2020.924267.

Varun K. S., Puneeth I. and Jacob T. P. 2019 "Virtual Mouse Implementation using Open CV," 3rdInternational Conference on Trends in Electronics and Informatics (ICOEI), 2019, pp. 435–438, doi:10.1109/ICOEI.2019.8862764.

Artificial Intelligence, Blockchain, Computing and Security – Dagur et al. (Eds)
© 2024 The Author(s), ISBN: 978-1-032-67841-2

Virtual board

N.V. Ravindhar*, P.V. Praharsha Reddy*, P. Sathvik Reddy* & Boyapati Ajay Kumar*
Department of Computer Science and Engineering, Saveetha Engineering College, Chennai, Tamil Nadu

ABSTRACT: The entire world revolves around technology. This began with the advancement of computer technology. The input and output mechanisms of computers changed dramatically as they evolved. Initially, the computer was controlled via switches. Later on, punch cards were introduced. A mouse, a keyboard, a monitor, a joystick, and other input devices were designed. Touch sensors are built in the screens for smart phones, smart watches, computers, and other devices. Hand gestures are used as input by the most advanced systems. "Leap Motion Sensors" are used. These systems are expensive to install and maintain. As a result, our concept "Virtual Board" is to create software that allows us to write on the screen without exerting additional effort. This system makes use of the well-known Open Computer Vision library. This algorithm examines the movement of targets between video frames is given as output in a sequential manner to conduct video tracking.

Keywords: leap motion sensors, Virtual board, OpenCV, ML

1 INTRODUCTION

Writing has evolved majorly over the years. Writing was first invented in 2000 BC by neolithic people. They first started writing on walls then it was replaced by stones. Stones were replaced by cloth and presently we use paper for communication. With the help of QWERTY keyboards, we are moving towards a more digitalized form of writing. These electronic devices are slowly taking the place of traditional forms of writing with pen and paper. The need to develop human machine interactions is rapidly growing with the surge in the usage of augmented and virtual reality. Applications using hand gestures have gained popularity over the coming years. Automotive interfaces, Economical Air Writing system and Handwriting recognition in Free Space have developed systems for hand gestures recognition. However, hand gestures recognition is not enough for writing in air. It also involves fingertip detection, tracking and tracing of it. However, these methods which have the usage of devices have some limitations. Fingertip detection and finger movement techniques are used to develop the system. Using Python, OpenCV and CNN techniques fingertip is first detected and then the trajectory of fingertip is traced and displayed on the screen. Python is employed for the system's backend development, while the computer vision library OpenCV is used by the AI virtual mouse.

2 LITERATURE SURVEY

Real-time hand gesture recognition employs a variety of techniques. (Nirosha *et al.* 2022) developed a system that uses a skin colour detection algorithm to transcribe American Sign

*Corresponding Authors: ravindhar@saveetha.ac.in, praharsha2001.p.v@gmail.com, sathvikreddyparlapallinew@gmail.com and ajayboyapati12@gmail.com

 DOI: 10.1201/9781032684994-89

Language (ASL) from video in real time. Due to individual differences in skin tone and hand shape, identifying a hand may be challenging. The technology uses two neural networks to overcome this. (Cooper *et al.* 2012) The SCD (Scalable color descriptor) neural network by is the first algorithm. The picture pixels by Yuan-Hsiang Chang are fed into the SCD neural network, which determines whether or not they are skin pixels. (Katona *et al.* 2021) have collaboratively developed a new HCI technique that inculcates a camera, an accelerometer, a pair of Arduino microcontrollers and an Ultrasonic Distance Sensors. The main concept behind this interface is to capture motions using Ultrasonic Distance Sensors. According to a new technology put forth by (Vinay *et al.* 2016). To extract the movement of the finger sketching the alphabet, only the colour of the LED is tracked (Jovan Popović *et al.* 2008). The background is black, and the object's colour is converted to white. The user wanted to draw an image of the alphabet in black and white, so they stitched together several black and white frames to make it (Makoto Shirabayashi *et al.*). For 3D hand gesture detection. To track the object, a survey on object tracking is represented by (Omar Javed *et al.* 2017). Object is being tracked when it is purposed. (Erik B. Sudderth *et al.* 2016) created the visual hand tracking using nonparametric belief propagation to track the visualized hand when projected to the sensor. (Sylvain paris *et al.* 2011) created the practical colour-based motion capture, to identify the natural colours. (Sudeep sarkar *et al.* 2008) coupled grouping and matching for sign and gesture recognition, which makes use of computer vision and image understanding technology. (Lamar *et al.* 2006) represented the Real-time Video Based Finger Spelling Recognition System Using Low Computational Complexity Artificial Neural Networks. Automatic Hand-Pose Trajectory Tracking System Using Video Sequences is represented by (Chen-Ming Chang *et al.* 2010).

3 EXISTING SYSTEMS

The current framework just works with your fingers and negative pencils, brushes. Without a depth sensor, it is very difficult to identify and separate an object like a finger from an RGB picture. The lack of top and mobility beneath the pen is also an issue. It relies on a single programmable RGB camera. The depths make it hard to find anything, and it's impossible to track down who did what in the pen. So, the model draws out every finger route, and the resulting picture is abstract and invisible to the model. Moving a process from one area to another in real time by touch of the user's hand involves careful coding. While hearing is sometimes taken for granted, persons who use sign language to communicate are not afforded that luxury. Most countries in the world lack the emotional intelligence to grasp your experiences without some kind of interpreter. The present technique only works with your fingertips and no highlighters, paints, or relatives. Without a depth sensor, it is very difficult to recognise an object like a finger in an RGB picture. The inability to move about beneath the pen is also a major issue. The system relies on a single, customizable RGB camera.

3.1 *Controlling the real time system*

In order to transition the system from one state to another using hand gestures in real time, the code must be carefully crafted. And the user has to be familiar with a wide range of actions to exert proper command over his strategy.

3.2 *Difficulties of pen momentum*

One RGB camera is used to write in the air. The inability to detect depth means that both vertical and horizontal pen motions are untrickable. So, the full path of the fingertip is tracked, and the resultant picture is completely illogical and unrecognised by the model. Writing a "G" by hand vs. "Going it in the air".

4 PROPOSED SYSTEM

Our proposed system uses the Open-Source Computer Vision module to track and follow an item of fascination that may be used to create artworks by just moving it around on a flat surface, making it easier to write on a computer. This system makes use of the well-known OpenCV library. OpenCV is ML software package that comprises a number of popular image analysis methods that may be used to create intelligent computer vision applications. Our system examines the movement of targets between video frames. The get Positions () and getup Fingers () methods of the hand tracker class are used to determine the locations of each finger and which finger is open respectively, based on comparisons of each incoming frame with media pipe hand landmarks shown. Each of these two features may be found in the Hand tracker module.

4.1 *Algorithm*

The potential of this component of our system is unparalleled. There are several functions involved. In this case, the number of gestures utilized to manage the system is proportional to the number of actions required to do it. All of our system's fundamental features, including

1. In "Writing Mode," the system records the location of the user's fingertips using tracing process.
2. Altering the font colour is a breeze in the Color Mode, where a rainbow of options awaits the user.
3. Third, a motion to quickly insert a backspace in case the user makes a mistake is required.
4. Computer vision is handled using the OpenCV package, while hand gesture recognition and tracking are handled by the MediaPipe framework. Hand gestures and fingertips are detected and identified by the app using machine learning techniques.

4.2 *Palm landmark model*

After the palm has been detected in the picture, our hand model purposes relapse, or direct direction expectation, to get exact key points of 21 unique 3D hand – knuckles & arranges inside the recognized hand area. This model develops an internal representation of hand position. It additionally covers the scope of conceivable palm positions and to manage the detection.

4.3 *Fingertip detection model*

A stylus or coloured air pen is all that's needed for virtual writing. However, fingertips are used in this method. Deep Learning algorithms are used to identify fingertip locations in each frame.

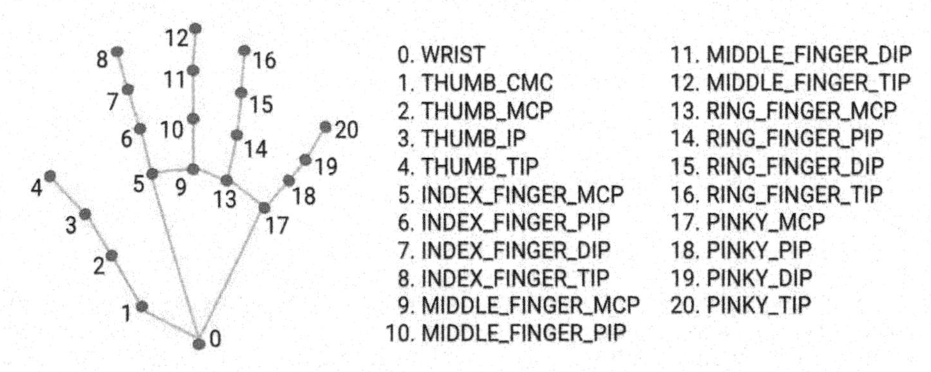

Figure 1. Fingertip recognition.

Using a video-to-image conversion method, researchers were able to record the movements of a palm in a variety of settings for only two seconds. Figure 1 displays the results of separating these moves into still frames.

5 SYSTEM ARCHITECTURE DIAGRAM

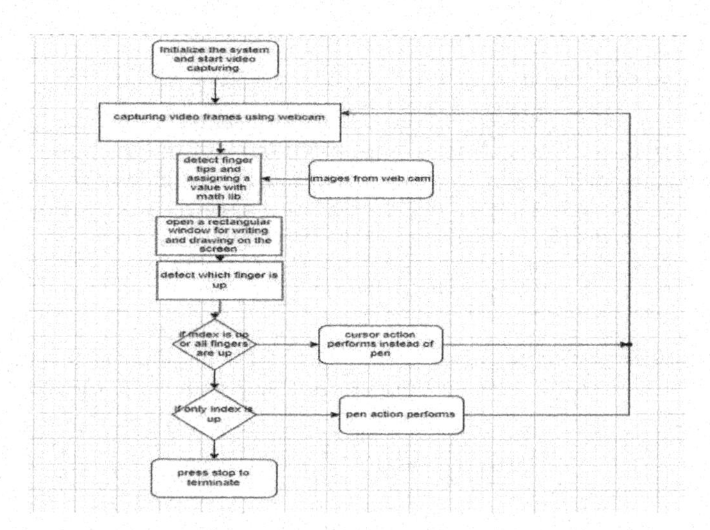

Figure 2. System architecture.

5.1 *Implementation and analysis feasibility study*

The literature review is an essential aspect in sustaining coherence. It's the critical procedures that must be carried out during creation. There must be access to reliable resources in order to develop reliable software. This section is useful for locating the information that has been developed and learning how that content is being used and used in the present. The health of the economy and the quality of the product are the two most important aspects.

5.2 *Implementation*

By finalising the front-end code, including brush manipulation features, and eliminating device dependencies, we have successfully finalised the design of our product (monitor, mouse, keyboard, etc.). As a first step, we settled on two simple interfaces: a measuring screen and a genuine sketching screen. The last step is to include the on-screen buttons and touch functionality. By using the "draw" button, the user may halt the drawing process and resume it at any time. To begin a new drawing or resize, the "equalise" button takes the user back to the rating screen from whence they came. At long last, we figured out that the visual button we had embedded in the map was broken and would only respond to more vigorous clicking. We opted on a color-changing feature that lets the user choose a new brush colour with the press of a finger anywhere on the screen's top half.

6 MODULE DESCRIPTION

This section provides an in-depth analysis of the process flowchart and the methods used at each step. Below, we'll go through the methods used or the steps of the suggested system.

6.1 *Webcam starts*

When a person moves their fingers in front of a webcam, the webcam begins recording, translate each frame of video, and transmits the frame to a hand tracker class. shows the video recording controls and frame. The suggested virtual board capture palm movements by a laptop or desktop camera. After the video capture object is built in computer vision library i.e., OpenCV, webcam recording will be started. The web camera feeds its frames into the virtual AI system for analysis.

6.2 *Capturing the video and processing*

The captures footage using a camera, saving every frame until the programme is closed. The code included demonstrates how to convert the video frames from BGR to RGB so that the hands may be located individually in each frame. As demonstrated in we can detect which finger is up by comparing the tip Id of the particular finger has been identified using the MediaPipe with the matching co-ordinates of the fingers that are up.

6.3 *For the mouse to perform action*

To have the computer click the right mouse button when the middle finger and the index finger are both raised and the space between them is less than 40px, the input uses python module.

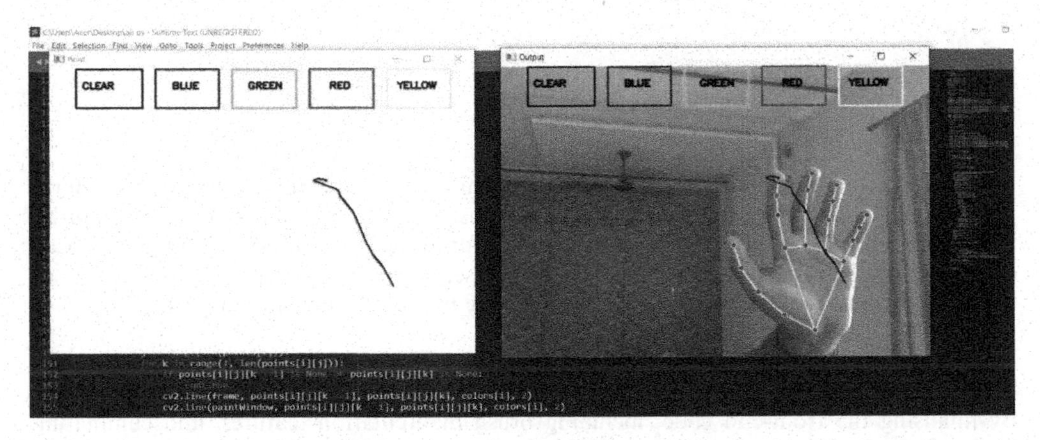

Figure 3. Detecting which finger is up.

7 RESULT

In order to identify fingertip locations in a video, our suggested approach must be applied to that particular area of the video. Plus, the whole area is monitored specifically for that purpose. To tell the difference between two characters, you could either use different coloured buttons for the front and back of your index finger, or you could just rotate your finger after writing each character and then write the following word. The suggested approach will presume that one character is complete if it cannot find a coloured item. One alternative is to provide a little pause following the completion of each character. Some m seconds of delay would result in identically duplicate frames. Since the difference between two frames is calculated using the previous frame as a reference, a tiny number, or one very close to zero in the event of a duplicated frame, indicates that the character is complete.

8 CONCLUSION

In general, we're pleased with the results of our work. By integrating OpenCV with Virtual board, we can develop a programme for creating sketches without physically touching the board. The user is free to create arbitrary coloured lines wherever they choose and may even switch brushes mid-draw. It's quite similar to sketching down your ideas on a blank sheet of paper. There are, of course, potential problems with Virtual board that might provide for intriguing study topics down the line. The first is a problem with the frame rate; the programme is useless because image processing has slowed down the camera feed. Our best bet for improving performance would be to provide multicore capabilities, which we sought to do in this project. For Virtual board to function properly in real time and with authenticity, timing problems with data queuing between processes must be fixed before frame information can be delivered in the right order. Moreover, we used open-source OpenCV code for object detection, which presented its own unique set of hurdles that we diligently overcame.

REFERENCES

Adi Narayana Salina, K. Kaivalya, K. Sriharsha, K. Praveen, M. Nirosha, Creating Computer Vision, *International Journal of Advances in Engineering and Management (IJAEM)* Volume 4, Issue 6 June 2022.

Alper Yilmaz, Omar Javed, Mubarak Shah, "Object Tracking: A Survey", *ACM Computer Survey*. Vol. 38, Issue. 4, Article 13, Pp. 1–45, 2017.

Bragatto T A C., Ruas G. I.S., Lamar M.V., "*Real-time Video Based Finger Spelling Recognition System Using Low Computational Complexity Artificial Neural Networks*", *IEEE ITS*, pp.393–397, 2006.

Cooper H. M., *"Sign Language Recognition: Generalising to More Complex Corpora"*, PhD Thesis, Centre for Vision, Speech and Signal Processing Faculty of Engineering and Physical Sciences, University of Surrey, UK, 2012.

Erik B. Sudderth, Michael I. Mandel, William T. Freeman, Alan S. Willsky, "Visual Hand Tracking Using Nonparametric Belief Propagation", Mit Laboratory for Information & Decision Systems Technical Report P- 2603, *Presented at IEEE CVPR Workshop on Generative Model Based Vision*, Pp. 1–9, 2016.

Katona J., "A Review of Human–computer Fields of Interaction Research and Virtual Reality in Cognitive Info Communications", *Applied Sciences*, vol. 11, No. 6, p. 2646, 2021.

Robert Wang, Sylvain Paris, Jovan Popović, "Practical Colour-Based Motion Capture", *Euro Graphics/ ACM SIGGRAPH Symposium on Computer Animation*, 2011.

Robert Y. Wang, Jovan Popović, *"Real-Time Hand- Tracking with a Color Glove"*, 2007.

Ruiduo Yang, Sudeep Sarkar, "Coupled Grouping and Matching for Sign and Gesture Recognition", *Computer Vision and Image Understanding*, Elsevier, 2008.

Vinay K. P., "Cursor Control Using Hand Gestures", *International Journal of Critical Accounting*, vol. 0975–8887, 2016.

Yuan-Hsiang Chang, Chen-Ming Chang, *"Automatic Hand-Pose Trajectory Tracking System Using Video Sequences"*, INTECH, pp. 132–152, Croatia, 2010.

Yusuke Araga, Makoto Shirabayashi, Keishi Kaida, Hiroomi Hikawa, "Real Time Gesture Recognition System Using Posture Classifier and Jordan Recurrent Neural Network", *IEEE World Congress on Computational Intelligence*, Brisbane, Australia, 2014.

Artificial Intelligence, Blockchain, Computing and Security – Dagur et al. (Eds)
© 2024 The Author(s), ISBN: 978-1-032-67841-2

Text classification using similarity measure and fuzzy function concept analysis

S. Selvi & A. Thilagavathy
Department of Computer Science and Engineering, R.M.K. Engineering College, Chennai, Tamil Nadu

P. Shobarani
Department of Computer Science and Engineering, R.M.D. Engineering College, Chennai, Tamil Nadu

Anitha Julian
Department of Computer Science and Engineering, Saveetha Engineering College, Chennai, Tamil Nadu

M. Lakshmi Haritha & P.R. Therasa
Department of Computer Science and Engineering, R.M.K. Engineering College, Chennai, Tamil Nadu

ABSTRACT: Text Classification during the retrieval of web documents is one of the most challenging tasks in the field of Query Processing. The query processor retrieves a large amount of documents in very less time that contain the query terms. Understanding and categorizing these texts is difficult in the field of Web content mining. This paper aims in categorizing the query terms and retrieving only relevant documents by performing an analysis of term pairs using Fuzzy Function Concept Lattice. In our contribution, terms are grouped using Fuzzy Function Concept analysis (FFCA). Our proposed method uses the conceptual relation for term classification, thereby reducing the time of retrieving relevant documents. Hence a user can avoid wasting his resource in retrieving irrelevant information during his Search process.

Keywords: Knowledge Engineering, Web Content Mining, Text Mining, Query Processing, Fuzzy Logic, Semantic Analysis

1 INTRODUCTION

Semantic network is nothing but the semantic association among varied diverse concepts. In order to locate and retrieve web-related content throughout the network, semantic similarity between terms is used. Only words with similar meaning are found to be semantically similar. It may refer to single or multiple words like “apple” and “mac” represent different things but may look alike. Words “Ford” and “Mercedes represent the car companies but not similar terms [7]. Semantic similarity score play a vital role in retrieving relevant documents for a given query. Associating the semantic tag information with the concept, relations, properties in an ontology is what is called Semantic enrichment.

Semantic analysis is used to correctly categorize the structured and unstructured data when we handle contextual information on large. Our strategy makes use of latent semantic analysis, which determines the true relationship between the documents by identifying terms' similarity. This approach yields enhanced outcomes as query words and text files are captured over the identical K-dimensional space to reduce feature dispersal and distortion. In order to perform text categorization effectively, our suggested method additionally looks for similarity between phrases.

 DOI: 10.1201/9781032684994-90

2 LITERATURE SURVEY

C. Buckley *et al.* proposes a new method to find semantic similarity of term pairs based on synsets 2016 [5], where synsets are derived from online resources. Philip Resnik, 2016 [8], propose a better short term classification method using Latent Semantic Indexing.. This will reduce the Sparseness of the matrix by using the semantic measures on the query terms. A novel algorithm is given by D. Mclean, Y. Li, and Z.A. Bandar, 2020 [9], where a frequently occurring term based clustering is done. It uses the existing frequency finding algorithm at its initial stages but reduced the space complexity with Singular Value decomposition.

A association purelydepends on the information content (IC) is given by Philip R [7], where IC is an estimation of degree of similarity check for better understanding of the concept. By substituting a concept-text matrices for the initial structure from the latent semantic evaluation in the preprocessing stage of text clustering, Yaxiong Li *et al.* (2010) [15] suggested a framework that utilised a domain ontology.

Proposed method

2.1 *Cosine similarity*

The cosine similarity index Cos(x,y) is applied to verify the different concepts be alike in which x and y belong to two concepts as given in equation (1).

$$\mathrm{Cos}(x, y) = x.y/\|x\| * \|y\| \tag{1}$$

2.2 *Fuzzy function concept analysis*

Fuzzy Function Concept Analysis (FFCA) is an enhancement of Formal Concept Analysis for representing irrelevant information [4]. FFCA can be used for ontology construction when a few information is available. The user will be allowed to locate data about the Web that is closer to their choices by using the following two methods, the study claims, in the event that the relevant facts are not modelled by any Function notion. Let us consider an Object set O, named Resort Hotels = {Hotel1,Hotel2,Hotel3,Hotel4,Hotel5,Hotel6} defined by four possible attributes A= {Table Tennis, Swimming Pool, Meals, Sea side}. Furthermore, suppose the hotels are related to the above attributes according to the relation, "Facility Available" defined as in Table 1.

Table 1. Object attribute relation table.

	Table Tennis	Swimming Pool	Meals	Sea side side
Hotel 1			**Facility Available**	**Facility Available**
Hotel 2	**Facility Available**	**Facility Available**	**Facility Available**	
Hotel 3			**Facility Available**	**Facility Available**
Hotel 4	**Facility Available**	**Facility Available**		**Facility Available**
Hotel 5			**Facility Available**	**Facility Available**
Hotel 6	**Facility Available**	**Facility Available**		

As an illustration, the Hotel 4 contains or is characterised by three characteristics, namely Table Tennis, Swimming Pool, and Seaside, and vice versa. All three of these features belong to the subject matter Hotel 4.

The concept of super concept and sub concept can be applied to two concepts to establish the inheritance relation between them. The Formal Concept Lattice as shown below in Figure 1 is built from this inherited relation derived from the above Table 1.

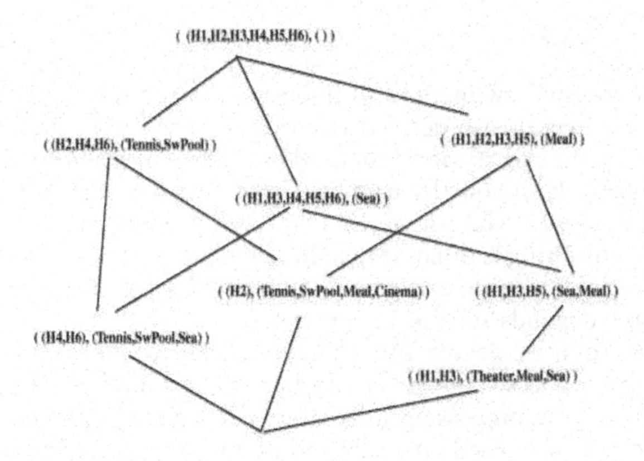

Figure 1. Function concept lattice.

In order to represent ambiguous information, fuzzy logic is included into fuzzy function concept analysis (FFCA). In FFCA, a notion is defined in a Fuzzy Function environment, similar to FCA. Here, we begin by reviewing the idea of a fuzzy set as it is depicted in Figure 1. A membership function A(x) that connects every location in X to a real number defines a fuzzy set A in the specified domain X with the range [0,1]: = (x,A(x))|xX. Each pair in a fuzzy relation (OxA) is connected to a membership value in [0, 1], according to the definition given above. The cross values in Table 1 are replaced by values from 0 to 1 in Table 2, based on the grade of application of the attributes (A) on the corresponding objects (Hotels) which is a significant of 'how much' they apply.

Table 2. The resort hotels context in fuzzy function concept analysis.

	Table Tennis	Swimming Pool	Meals	Sea side
Hotel 1			1.0	1.0
Hotel 2	0.6	1.0	0.5	
Hotel 3			0.5	0.7
Hotel 4	0.8	1.0		1.0
Hotel 5			1.0	0.3
Hotel 6	0.8	1.0		0.8

Think about the Hotel 2 in Table 1, for instance. It possesses the attribute Swimming Pool with score of association 1.0, meaning that the Hotel2 is fully covered by this attribute (and vice versa, the attribute Swimming Pool may accurately describe the Hotel 2 as well).Take our running example as an instance and presume that the maximum threshold is fixed at 0.5. As with the grades of membership that are not indicated in Table 2 (which are equivalent to zero), the grade of membership of 0.3 between Hotel5 and Sea side is disregarded.

In the first step of our suggested method, we type check each word pair to determine whether it represents a concept pair, an entity pair, or a concept-entity pair. We use K-means clustering to compile the top-K ideas linked to the entity term if the term pair falls into the concept-entity category. Once the current iteration depth does not go beyond the maximum depth (maxD), we continue to calculate the similarity measure between each idea and the word pair iteratively. The highest similarity value among these computed metrics is then returned. To find the texts in the semantic network that are the most similar to one another, we also use the Max similarity function.

Algorithm 1: Proposed Algorithm

Step 1: Generate concept vectors for the terms using Latent Semantic Analysis (LSA).
Step 2: Calculate the similarity of concepts using a specific method (1).
Step 3: Return the calculated similarity.
Step 4: If (ti, tj) represents term pairs.
Step 5: Calculate entity vectors for the term pairs.
Step 6: Calculate the similarity using equation (1).
Step 7: Form the clusters based on the concept using K-means clustering
Step 8: If (ti, tj) represents a concept-entity pair, Collect the top K concepts.
Step 9: For each concept cx, Calculate the similarity, sim(cx) by iterating Step 9, until reaching the maximum depth (maxD).
Step 10: Return the maximum similarity among the calculated similarities (sim(cx)).

3 RESULTS AND DISCUSSION

The findings of the experiment demonstrate that similarity measures will be able to identify attributes for clustering and classification purposes. Our approach was tested on a number of well-known search engines, and the outcome are shown in Figure 2.

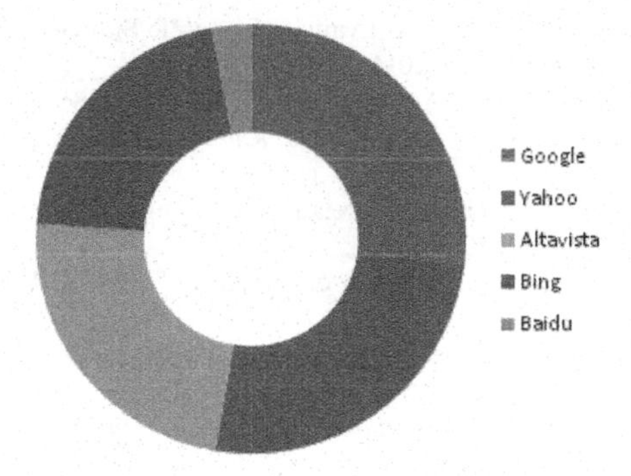

Figure 2. Results from popular search engines.

The figure below shows the clustered result obtained on applying the k-means clustering algorithm applied for the data used as in the Table 2. Two clusters are formed based on the cluster centric. We finally derive that Hotels 1 to 5 are recommended for Sea side hotels with the required adequate facilities.

Label	Vector	Cluster id	Cluster centroid
Hotel 1	0,1,1	1	2.8000000000000003,1.8,3.4000000000000004
Hotel 2	6,1,5	1	2.8000000000000003,1.8,3.4000000000000004
Hotel 3	0,5,7	1	2.8000000000000003,1.8,3.4000000000000004
Hotel 4	8,1,1	1	2.8000000000000003,1.8,3.4000000000000004
Hotel 5	0,1,3	1	2.8000000000000003,1.8,3.4000000000000004
Hotel 6	8,1,18	0	8,1,18

Figure 3. Clustered output.

Our proposed method yield better results for the any context of web documents for all query processing applications. Figure 3 shows the results of clusters generated for our sample data taken for six hotels which takes very less time in finding the required information. The values in the vector can easily identify the importance of the concept that is applied on the query term. Classification becomes easy if the number of clusters are not more than 3for our proposed approach.

REFERENCES

[1] Bagga A. and Baldwin B. Entity-based Cross Document Coreferencing Using the Vector Space Model. In *Proc. of 36th COLING-ACL*, pages 79(85), 2008.

[2] Bar-Yossef Z. and Gurevich M. Random Sampling from a Searchengine's Index. In *Proceedings of 15thInternational World Wide Web Conference*, 2006.

[3] Bekkerman R. and McCallum A. Disambiguating Web Appearances of People in a Social Network. In *Proceedings of the World Wide Web Conference (WWW)*, pages 463{470, 2005}.

[4] Selvi S., Suresh RM. "Fuzzy Concept Lattice for Ontology Learning and Concept Classification". *Indian Journal of Science and Technology*, pages 1–7,9(28), 2016.

[5] Buckley C., Salton G., Allan J., and Singhal A. Automatic Query Expansion Using Smart: Trec 3. In *Proc. of 3rd Text REtreival Conference*, pages 69{80, 2016}.

[6] Chen H., Lin M., and Wei Y. Novel Association Measures Using Web Searchwith Double Checking. In *Proc. of the COLING/ACL* 2006, pages 1009{1016, 2006}.

[7] Philip R. Using Information Content to Evaluate Semantic Similarity in a Taxonomy. *Proceedings of IJCAI'95*; 2001. pp. 448–453.

[8] Resnik P, "Using Information Content to Evaluate Semantic Similarity in a Taxonomy," *Proc. 14th Int'l Joint Conf. Aritificial Intelligence*, 2016.

[9] Mclean D., Li Y., and Bandar Z. A., "An Approach for Measuring Semantic Similarity between Words Using Multiple Information Sources," *IEEE Trans. Knowledge and Data Eng.*, vol. 15, no. 4, pp. 871–882, July/Aug. 2020.

[10] Miller G. and Charles W., "Contextual Correlates of Semantic Similarity," *Language and Cognitive Processes*, vol. 6, no. 1, pp. 1–28, 2008.

[11] Lin D, "An Information-Theoretic Definition of Similarity," *Proc.15th Int'l Conf. Machine Learning (ICML)*, pp. 296–304, 2008.

[12] Lin D. Automatic Retreival and Clustering of Similar Words. In *Proc. of the 17th Coling*, pages 768 {774, 2008}.

[13] Manning C. D. and SchÄutze H. *Foundations of Statistical Natural Language Processing*. The MIT Press, Cambridge, Massachusetts, 2002.

[14] Matsuo Y., Mori J., Hamasaki M., Ishida K., Nishimura T., Takeda H., Hasida K., and Ishizuka M.. Polyphonet: An Advanced Social Network Extraction System. In *Proc. of 15th International World Wide Web Conference*, 2006.

[15] Matsuo Y., Sakaki T., Uchiyama K., and Ishizuka M. Graph-based Word Clustering Using Web Searchengine. In *Proc. of EMNLP 2006*, 2010.

[16] Selvi K, Suresh RM. Measure Semantic Similarity Between Words Using Fuzzy Formal Concept Analysis. *Proceedings of IRNet; India*. 2012. P. 31–4.

[17] Selvi K, Suresh RM. An Efficient Technique to Implement Similarity Measures in Text Document Clustering Using Artificial Neural Networks Algorithm. *Research Journal of Applied Sciences Engineering and Technology*. 2014, 8(23):2320–28.

[18] Selvi K, Suresh RM. Document Clustering Using Artificial Neural Networks. *National Journal on Advances in Computing and Management*. 2008; 5(1):1–5.

Artificial Intelligence, Blockchain, Computing and Security – Dagur et al. (Eds)
© 2024 The Author(s), ISBN: 978-1-032-67841-2

DigiAgri – predicting crop price by machine learning

Pratham & Monika Singh
Apex Institute of Technology-CSE, Chandigarh University, Punjab

ABSTRACT: For a producer country like India, where over 33% of the population relies on food, ensuring the sustainability of the food supply is a key issue. Estimates of seasonal crop yields are undeniably recognized as an important contribution to food costs. However, there is no mechanism in place to guide farmers in choosing which crops to cultivate to determine profit. This paper attempts to predict the price of crops that farmers will obtain for their land by analyzing patterns in past data. Only seasonal crops such as Paddy, Arhar, Bajra, and Barley have been considered for the analysis. We make use of a variety of information, including rainfall, temperature, market pricing, and previous crop yields for predicting crop prices for the next 12 months and provide time series analysis using a decision tree based algorithm.

1 INTRODUCTION

Most of the population of India is employed in the agricultural sector, which contributes 18% of the country's GDP [1]. Nonetheless, while being the sector with the greatest demographic reach and having a significant impact on India's overall socioeconomic structure, a recent study revealed that agriculture makes up a very small portion of the Asian economy. and a declining trend is present. Lack of adequate crop management by farmers and governments is another clear factor in the agricultural sector's poor GDP contribution in Asian nations [2]. One of the most crucial metrics for determining the state of an economy is the gross domestic product. Crop prices can fluctuate significantly in the market. The main cause of these cost variations is a lack of forethought. The end effect is pricey crops that harm consumers when their value rises and producers when their investment value falls. It is challenging for farmers in these situations to come up with viable options for cultivating crops on their fields to predict projected yields and values.

This study uses machine learning to analyze previous yields in order to assist farmers in making better decisions in deciding which crops to grow for making profits. In many sectors, including the stock market, weather, decision outcomes, crops, and in our case, yields and pricing from earlier generations of artificial intelligence applications, machine learning has been proved to develop accurate prediction models [3].

The demands placed on farmers in the contemporary agricultural sector are unknown to them. Environmental factors, cost-effective elections, stress from job and family responsibilities, inadequate irrigation, and high crop prices have all been implicated in farmer suicides. The main causes are low production costs and high crop prices.

A significant issue for agriculture is yield forecasting. The anticipated yield percentage is of relevance to all farmers. Forecasts of yields in the past were predicated on farmers' prior knowledge of a certain crop. There is a wealth of knowledge regarding Indian agriculture, and a specific understanding is quite helpful for many purposes. In this paper, we examined a variety of machine learning techniques such as decision tree, random forest, and Xgboost to forecast yields in order to maximize crop productivity. And finally, we considered Xgboost algorithm to predict the crop prices.

DOI: 10.1201/9781032684994-91

2 LITERATURE REVIEW

Forecasting for agricultural products is crucial for the industry. It contributes to higher net production, better planning, and more profitability. We read various study articles on the subject of our domain in order to acquire better results. Farmers can use a variety of methods to forecast crop yield based on meteorological variables. Algorithms for machine learning have been applied to forecast crops [3].

Two regression-supervised machine learning techniques are employed in [4]: Effectiveness in determining soil quality using Support Vector Machine (SVM) and Relevance Vector Machine (RVM) a smart wireless gadget for detecting soil moisture and gathering weather information. The wireless gadget provides accuracy of 95% and a 15% error rate. That hasn't been tried with real-time data, though.

The paper [5] uses a back propagation technique to check for soil fertility and plant nutrient levels. The accurate findings allow for bettering of soil characteristics. It works more effectively than conventional procedures. Unfortunately, the system is unstable, slow, and inefficient.

Three techniques are employed in article [6], including Decision Tree, Naive Bayes Classifier, and KNN Classifier, to analyze the soil and forecast crop output.

Regression Analysis (RA) was suggested by Shastry *et al.* [7] to identify environmental factors and the effects they have on crop productivity. A decision-making tool, RA was a multi-variate analysis approach that analyses the elements and combines them into response variables. For a ten-year period, from 1990 to 2000, a sample of environmental elements, such as soil type and crop parameters, were taken into account. This study was expanded by taking into account additional variables such as the minimum support price, the cost price index, the wholesale price index, etc. Particularly in crop production prediction, the amount of data being collected and stored every day was exponential.

Researchers have employed various machine learning models [8–13], a group of shallow decision trees is iteratively trained by GBDTs, with each iteration using the error residuals of the prior model to fit the new model. The weighted average of all the tree predictions represents the final projection. Whereas GBDT "boosting" lowers bias and underfitting, random forest "bagging" minimizes variance and overfitting.

2.1 *Gaps found in the literature*

1. Several academicians have developed many models to predict agricultural yields. These models do, however, have faults because the approaches were applied wrongly.
2. A hybrid neural network offers a prediction model [14–17] for crop production that takes into account soil elements and external meteorological data. Yet, the model is unable to predict changes in the soil and climate in real time. As a result, the model cannot be applied to real-time analysis.
3. A prediction system using KNN and the Apriori algorithm is developed to research about crops and make recommendations to farmers. When a user enters crop names into the system, a well-designed interface that generates crop yield is provided. Yet, the machine is unable to simultaneously predict the price.

3 OBJECTIVES

With the help of past weather, yield, and price trends, this study aims to forecast crop prices. Many people in an agricultural country like India who rely only on crops can benefit from forecasting crop yields and pricing.

The key goals of the proposed system include:

1. Updating the farmer on the crop's yield depending on rainfall using machine learning.
2. To estimate the cost of the current harvest and the month of sowing when determining the price of crops on the market.

4 PROPOSED WORK

The major topics of this paper are agricultural yield prediction and cost estimation. The provided methodology is found to be the most suitable for the investigation since it correctly anticipates the results utilizing the tree algorithm specifically Xgboost [18]. The information is processed and utilized to project the cost and yield of agriculture at a particular month. The following steps are performed for the implementation of the proposed approach:

1. Data Gathering
2. Analysis of data
3. Machine learning prediction

4.1 *Data acquisition*

To construct the dataset, agricultural data from a public source is gathered. We collected the dataset available from the government website i.e. https://data.gov.in/. We were successful in obtaining information about multiple crops such as Soyabean, Arhar, Bajra, Sesamum, Niger, Ragi, Jowar, Barley, Moong, Sunflower etc. Multiple attributes have been considered while predicting the prices of crops.

4.2 *Data analysis*

The dataset of the crops contains two parameters such as rainfall, and WPI (Wholesale Price Index) from 2012 till 2018.

We have shown the datasets and the analysis of two crops viz. Arhar and Bajra in Figures 1 & 2.

Month	Year	Rainfall	WPI
4	2012	47.5	107.7
5	2012	31.7	109.3
6	2012	117.8	107
7	2012	250.2	113.9
8	2012	262.4	120.7
9	2012	193.5	121.7
10	2012	58.7	114.8
11	2012	30.7	120.4
12	2012	11.7	127.7
1	2013	11.3	131.6
2	2013	40.1	132.7
3	2013	15.7	132.5
4	2013	30.4	136.4
5	2013	57.8	137.6
6	2013	219.8	136.9
7	2013	310	135.3

Bajra

a) Bajra Dataset

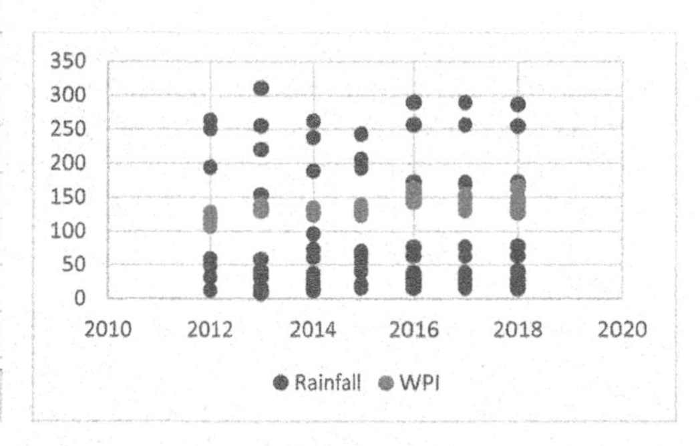

(b) Analysis of Bajra

Figure 1. Bajra dataset.

Month	Year	Rainfall	WPI
4	2012	47.5	97.1
5	2012	31.7	99.7
6	2012	117.8	101.6
7	2012	250.2	108.6
8	2012	262.4	115.6
9	2012	193.5	118.6
10	2012	58.7	116.3
11	2012	30.7	114.4
12	2012	11.7	110.2
1	2013	11.3	109.1
2	2013	40.1	110.6
3	2013	15.7	116.3
4	2013	30.4	118.1
5	2013	57.8	118.2
6	2013	219.8	117.7
7	2013	310	116.5

Arhar (1)

a) Arhar Dataset

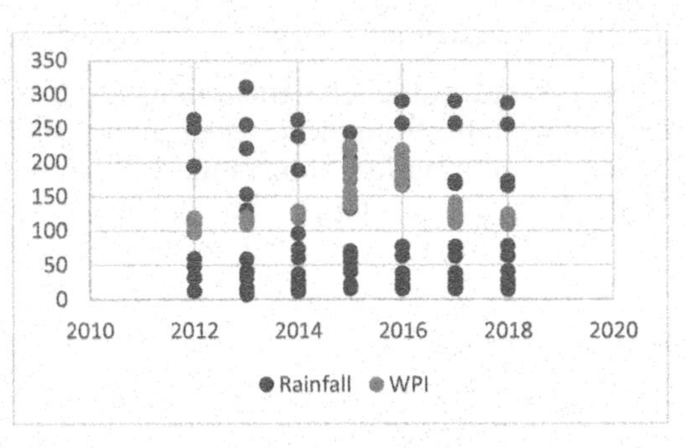

b) Analysis of Arhar

Figure 2. Arhar dataset.

It has been analyzed from Figure 1(b) that the highest rainfall is in 2013. And WPI is almost the same in all the years except for 2016.

Considering another dataset of Arhar as shown in Figure 2, it is evident from the data that the highest rainfall is in 2013 and WPI is almost the same in 2012,2013, 2014, 2017, and 2018. Most variation is seen in the years 2015 and 2016.

4.3 *Machine learning prediction*

1. Machine learning is done in two parts i.e. training and testing.
2. We have used Xgboost algorithm for predicting the crop prices.

We employed the distributed gradient boosting toolbox known as XGBoost, which has been designed for quick and scalable machine learning model training. This ensemble learning technique combines the predictions of a number of weak models to get a stronger prediction [18]. Due to its ability to handle massive datasets and deliver cutting-edge results in many machine learning tasks, including classification and regression, Extreme Gradient Boosting, or XGBoost, is one of the most well-known and frequently used machine learning algorithms.

Its effective handling of missing values, which enables it to handle real-world data with missing values without requiring a lot of pre-processing, is one of the key characteristics of XGBoost. Moreover, XGBoost includes built-in parallel processing capabilities, enabling the training of models. Applications for XGBoost include click-through rate prediction, recommendation systems, and Kaggle competitions among others. Additionally, it is quite adaptable and enables performance optimization by allowing for fine-tuning of numerous model parameters.

At that time to form a calling tree node containing that attribute, and finally run the rule on the subsets that fall victim to the remaining attributes to create a calling tree. Implementation of the XGBoost algorithm has been done in python as shown in Figure 3.

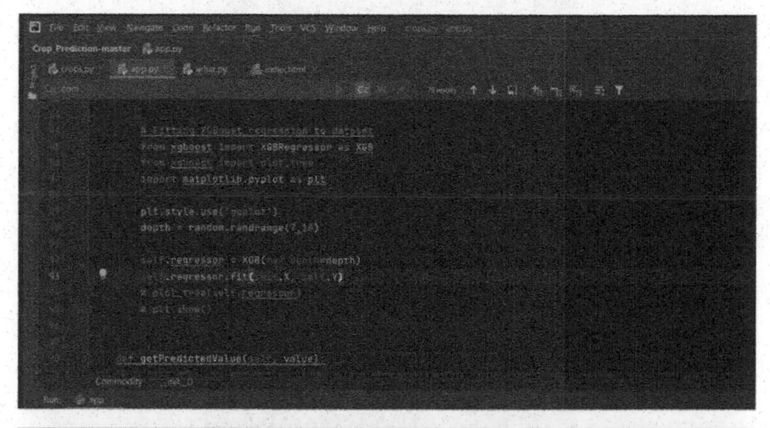

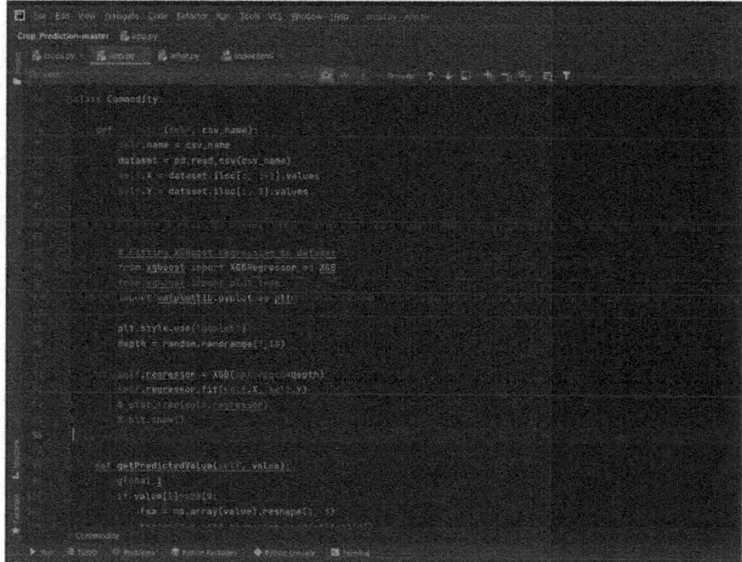

Figure 3. Implementation of XGBoost algorithm.

5 RESULTS AND EVALUATION

We want to know from the evaluation whether our approach effectively addresses the problem statement we are attempting to solve for a variety of indicators. We determine the price by using XGBoost algorithm [18]. The actual and predicted prices for one of the crops have been shown in Figure 4.

Further, the prediction of various crops prices has been done for different months based on two factors as considered in our case i.e. rainfall and WPI. A few snapshots of the crops with increase and decrease in prices for different months have been shown in Figure 5.

MSE(Mean Square Error) for the Arhar and wheat crop has been shown in Figure 6. As it is evident from the figure that there is a little variation in the actual and predicted prices as per our proposed model of XGBoost.

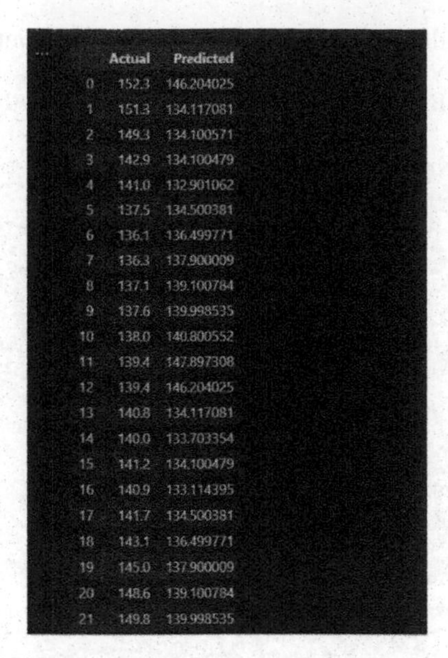

	Actual	Predicted
0	152.3	146.204025
1	151.3	134.117081
2	149.3	134.100571
3	142.9	134.100479
4	141.0	132.901062
5	137.5	134.500381
6	136.1	136.499771
7	136.3	137.900009
8	137.1	139.100784
9	137.6	139.998535
10	138.0	140.800552
11	139.4	147.897308
12	139.4	146.204025
13	140.8	134.117081
14	140.0	133.703354
15	141.2	134.100479
16	140.9	133.114395
17	141.7	134.500381
18	143.1	136.499771
19	145.0	137.900009
20	148.6	139.100784
21	149.8	139.998535

Figure 4. Actual vs predicted prices for Arhar.

Crop Predicition
Top Gainers(Current trends)
Top Losers(Current trends)
Copra ₹5748.15
Barley ₹1104.55

(a)

Crop Predicition
Top Gainers(Current trends)
Top Losers(Current trends)
Copra ₹5773.25
Barley ₹1109.37

(b)

Figure 5. Crops with increasing/decreasing prices month wise.

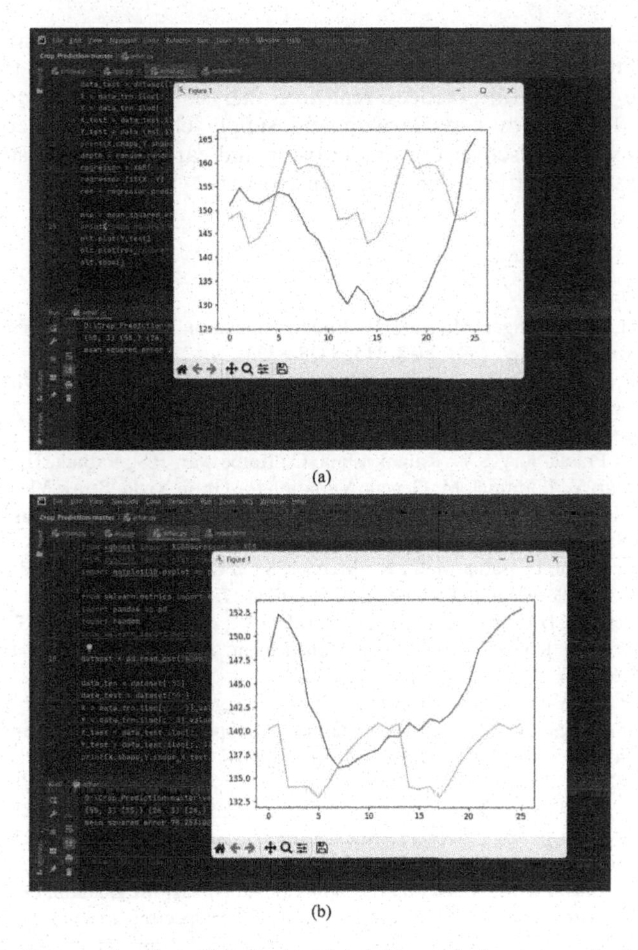

(a)

(b)

Figure 6. MSE in the actual and predicted prices for Arhar and wheat respectively.

6 CONCLUSION

To assist farmers in selecting the optimum crop for their chosen seeding date and location, the proposed model was created. Long before the farmer begins planting, our algorithm forecasts the crop's yield and price, giving him important information. Decision trees, neural networks, SVM, and other algorithms can all be used to forecast crop yields and prices. We used decision trees based model i.e. XGBoost algorithm. It provides good precision and has been trained on a variety of crops including arhar, bajra, wheat and barley etc. It has been analyzed that there is little variation in the actual and predicted prices as suggested by our machine learning algorithm.

In order to reduce deforestation and sustainably feed an expanding population, farmers must increase food output in already-used areas. Enhancing farming practices should be the main long-term goal if yields, crop quality, and incomes are to be increased in a highly sustainable manner.

An improved value forecasting system might be created utilizing a dashboard that can predict market value trends using statistical methods for a brief period of time and, consequently, the production patterns of different crops. A platform for agriculture should be able to use massive volumes of data, AI, machine learning, satellite-based display techniques, and weather data to analyse land area and track crop health in real-time. This method will anticipate costs, assess agricultural production and performance, and find pest and disease infestations.

The estimation of pricing may also consider other important data, such as the costs in important adjacent markets. To provide farmers with a precise solution to their problems and improve their agricultural practices, real-time analysis should be incorporated into the system's design. The system may develop and supply climate-aware cognitive farming methodologies as well as discover crop monitoring and early warning systems for pest/disease outbreaks based on cutting-edge AI advancements.

REFERENCES

[1] Kc, S.; Wurzer, M.; Speringer, M.; Lutz, W. Future Population and Human Capital in Heterogeneous India. *Proc. Natl. Acad. Sci. USA* 2018, 115, 8328–8333.

[2] Samaddar, A.; Cuevas, R.P.; Custodio, M.C.; Ynion, J.; Ray, A.; Mohanty, S.K.; Demont, M. Capturing Diversity and Cultural Drivers of Food Choice in Eastern India. *Int. J. Gastron. Food Sci.* 2020, 22, 100249.

[3] Subudhi, H.N.; Prasad, K.V.S.V.; Ramakrishna, C.; Rameswar, P.S.; Pathak, H.; Ravi, D.; Khan, A. A.; Padmakumar, V.; Blümmel, M. Genetic Variation for Grain Yield, Straw Yield and Straw Quality Traits in 132 Diverse Rice Varieties Released for Different Ecologies Such as Upland, Lowland, Irrigated and Salinity Prone Areas in India. *Field Crops Res.* 2020, 245, 107626.

[4] Mc Carthy, U.; Uysal, I.; Badia-Melis, R.; Mercier, S.; O'Donnell, C.; Ktenioudaki, A. Global Food Security–Issues, Challenges and Technological Solutions. *Trends Food Sci. Technol.* 2018, 77, 11–20.

[5] Ali, M.; Mubeen, M.; Hussain, N.; Wajid, A.; Farid, H.U.; Awais, M.; Hussain, S.; Akram, W.; Amin, A.; Akram, R.; *et al.* Role of ICT in Crop Management. *In Agronomic Crops*; Springer: Singapore, 2019; pp. 637–652.

[6] Smith, M.J. Getting Value From Artificial Intelligence in Agriculture. *Anim. Prod. Sci.* 2018, 60, 46–54.

[7] Athani, S.S. and Tejeshwar, C.H., 2017, January. Support Vector Machine-based Classification Scheme of Maize Crop. In 2017 *IEEE 7th International Advance Computing Conference (IACC)* (pp. 84–88). IEEE.

[8] Dahikar, S.S., Rode, S.V. Agricultural Crop Yield Prediction Using Artificial Neural Network Approach. *Int. J. Innov. Res. Electr. Electron. Instrum. Control. Eng.* 2014, 2, 683–686.

[9] Khaki, S.; Wang, L. Crop Yield Prediction Using Deep Neural Networks. *Front. Plant Sci.* 2019, 10, 621.

[10] Farmonov, N., Amankulova, K., Szatmári, J., Sharifi, A., Abbasi-Moghadam, D., Nejad, S.M.M. and Mucsi, L., 2023. Crop Type Classification by DESIS Hyperspectral Imagery and Machine Learning Algorithms. *IEEE Journal of Selected Topics in Applied Earth Observations and Remote Sensing*, 16, pp.1576–1588. Jordan, M.I.; Mitchell, T.M. Machine Learning: Trends, Perspectives, and Prospects. *Science* 2015, 349, 255–260.

[11] Oikonomidis, A., Catal, C. and Kassahun, A., 2023. Deep Learning for Crop Yield Prediction: A Systematic Literature Review. *New Zealand Journal of Crop and Horticultural Science*, 51(1), pp.1–26.

[12] Rao, M.S., Singh, A., Reddy, N.S. and Acharya, D.U., 2022. Crop Prediction Using Machine Learning. In *Journal of Physics: Conference Series* (Vol. 2161, No. 1, p. 012033).IOP Publishing.

[13] Vaishnnave, M.P. and Manivannan, R., 2022. An Empirical Study of Crop Yield Prediction Using Reinforcement Learning. *Artificial Intelligent Techniques for Wireless Communication and Networking*, pp.47–58.

[14] Yaramasu, R., Bandaru, V. and Pnvr, K., 2020. Pre-season Crop Type Mapping Using Deep Neural Networks. *Computers and Electronics in Agriculture*, *176*, p.105664.

[15] Potgieter, A.B., Zhao, Y., Zarco-Tejada, P.J., Chenu, K., Zhang, Y., Porker, K., Biddulph, B., Dang, Y.P., Neale, T., Roosta, F. and Chapman, S., 2021. Evolution and Application of Digital Technologies to Predict Crop Type and Crop Phenology in Agriculture. *in silico Plants*, *3*(1), p.diab017.

[16] Mohan, P. and Patil, K.K., 2018. Deep Learning Based Weighted SOM to Forecast Weather and Crop Prediction for Agriculture Application. *Int. J. Intell. Eng. Syst*, *11*, pp.167–176.

[17] Elavarasan, D. and Vincent, D.R., 2020. Reinforced XGBoost Machine Learning Model for Sustainable Intelligent Agrarian Applications. *Journal of Intelligent & Fuzzy Systems*, *39*(5), pp.7605–7620.

[18] Toomula, S. and Pelluri, S., 2022, November. An Extensive Survey of Deep learning-based Crop Yield Prediction Models for Precision Agriculture. In *Proceedings of the International Conference on Cognitive and Intelligent Computing: ICCIC 2021,* Volume 1 (pp. 1–12). Singapore: Springer Nature Singapore.

Career profiling based psychometric test for students

G. Nagappan*, Gokul Prasath Ravi*, K. Prathipa* & V. Saranya*
Department of Computer Science and Engineering, Saveetha Engineering College, Chennai, India

ABSTRACT: The decision of what career to pursue after graduation is one of the most important one for every student because it affects the growth and development of their future. Due to many uncertainties while picking a career, students often end up selecting the incorrect one. The purpose of this project is to create a psychometric test that will identify a student's skills and suggest career possibilities for them. They gain a good idea of their personality and the appropriate employment for them. It aids in understanding a person's personality, interests, and ability in both professionals and students. Students can better understand their strengths and limitations thanks to this assessment. We will be able to classify the student's career with even greater accuracy thanks to the questions on this psychometric test that are based on the nine intelligences.

Keywords: psychometric test, flutter, dart, cloud firebase, mobile application, career profiling.

1 INTRODUCTION

In India, most parents are quite involved in their children's decision regarding their academic and career paths. This concept can go either way: sometimes it works and results in the best outcomes, but it can also put some people in a difficult situation. In the latter scenario, people are uncertain of what to do next. It is crucial to make strategic choices when it comes to your education and job in order to prevent the latter scenario. The best approach to achieve so is to match a person's job and academic choices with their personality. A psychometric career exam is a career assessment that includes questions about aptitude, personality, and interests. Howard Gardner, a Harvard psychologist, put forth the notion of multiple intelligence. Gardner suggested there are eight different types of intelligence and has suggested a ninth type of intelligence known as "existentialist intelligence" as a potential addition. A few of Howard Gardner's nine categories of intelligence are Linguistic Intelligence, Interpersonal Intelligence, Logical-Mathematical Intelligence, Musical Intelligence, Intrapersonal Intelligence, Visual-Spatial Intelligence, Bodily-Kinesthetic Intelligence, Existential Intelligence, Naturalist Intelligence.

2 LITERATURE SURVEY

A person's buoyancy, or "daily resilience," is their capacity to successfully navigate obstacles and challenges that are part of everyday life. Therefore, the current study conducts a psychometric scope of buoyancy in the school context from the perspective of concept validity (Andrew J. Martin *et al.* 2021). The Career Preference Computerized Adaptive Test (CPCAT psychometric)'s characteristics are published. Participants were Grade 9 and Grade 11 high school

*Corresponding Authors: nagappan.cse@saveetha.ac.in, gokul.prasath88@gmail.com, kprathipa2002@gmail.com and abisaran358@gmail.com

DOI: 10.1201/9781032684994-92

students from a South African school system (n=343; males=279, females=164). According to construct validity and reliability indices, the CPCAT might be helpful for high school students seeking career guidance (De Beer & Marais *et al.* 2010). Few studies evaluate and report on internal score validity, score precision/reliability, and external score validity in that order, indicating that applied researchers may not always be using appropriate test-evaluative justifications in their psychometric evaluations (Kathleen L. Slaney *et al.* 2019). There is a gap in the assessment of the specific sources of stress for students in the post-secondary setting, despite efforts to address mental health issues at post-secondary schools (Brooke Linden *et al.* 2019). Numerous arguments and pieces of data point out that there is no reliable cognitive selection battery for predicting academic performance (Dr Frans Maloa *et al.* 2015). This study looked at variables that might be used to forecast academic achievement but that aren't currently consistently assessed in higher education (Geraldine Gray *et al.* 2014). In order to assist students in identifying their interests and skills so they may make the best career decisions, this study assesses the usefulness of the psychometric tests administered by career counselling centers (Likitha Shetty 2019). Choosing a career and developing one's career can be challenging for employees with a variety of surface features and work status characteristics (Lillian T. Eby *et al.* 2020). The largest advantages are those that enhance the psychometric features of the tests themselves, making them intrinsically more valuable. This has advantages in terms of time and money savings (Lisette Guy Msc 2018). The students ultimately choose a vocation that is not suitable for them, which has an impact on human resource production (Moinuddin Pasha Miya *et al.* 2019). The goal of this study is to create and validate a brief skills inventory for college students seeking for their first employment. The talents of candidates can be aligned with the standard of performance demanded by potential employers thanks to this inventory (Rosa Isabel Rodrigues *et al.* 2021). People's activities and way of life have the power to affect their health and quality of life. This study's objective was to create and put through a psychometric test a tool for measuring healthy living among Iranian adolescents (Parvaneh Taymoori *et al.* 2020).

3 PROPOSED SYSTEM

It is crucial for students to evaluate their talents as they progress through their coursework and pursue the subjects, they are interested into determine their interests so they may learn which job field their hobbies and skills will place them in. They will benefit from this by doing better and being inspired, which will help them pursue their desired careers. The proposed system has different modules such as a) Signup page b) Login page c) Home page d) Data collection.

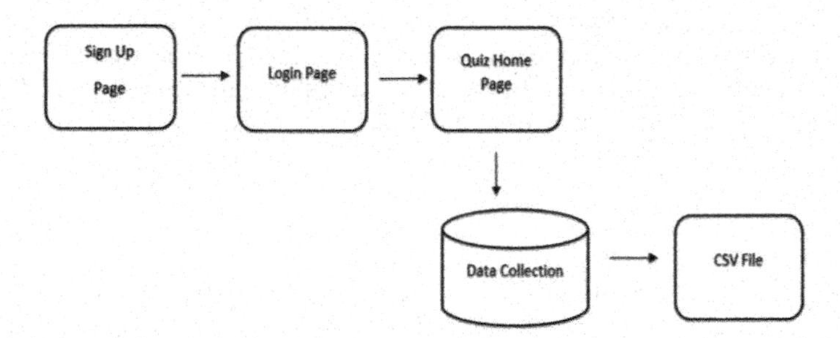

Figure 1. System architecture diagram of proposed system.

3.1 *Signup page*

Signup page is the first page where we land up when we open the application. The user must enter their username, email address, age, and a password specifically designed for this application on the sign-up page. The user can register for an account and utilize this program after filling out all the necessary information. The Firebase console receives a quick update of this user information. The program allows an unlimited number of people to register, and the Firebase database can handle all their data. These data are stored in cloud Firebase database.

Figure 2. Signup page.

3.2 *Login page*

Login page was developed for authentication of users. This application's login module was developed to allow students to access and log in to their individual workspace. The user can connect in to the application using their username and password set during the registration process. If the login credential failed to match the one stored in the database, then the user cannot log into the application. After logging the user is taken to this application's home page.

Figure 3. Login page.

3.3 *Home page*

The home page displays the questions that is fetched from the database. The list of questions is stored in the cloud Firebase. The questions are framed based on the Gardner's multiple intelligence theory. These questions specify individual intelligence based on which the student's personality and careers will be chosen. The user must answer all the questions that are being displayed on the screen. Figure 4 depicts how the questions look like on the screen.

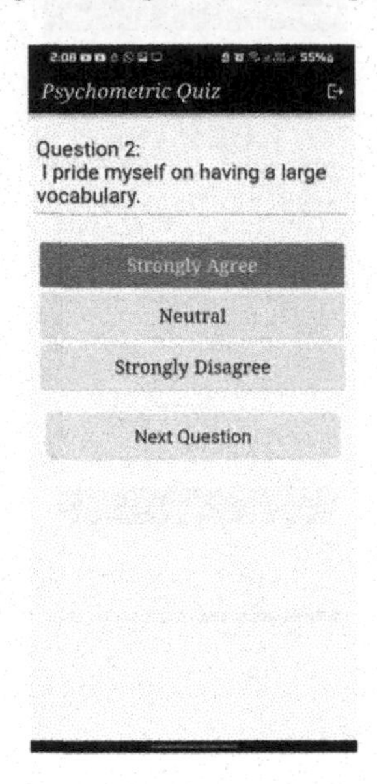

Figure 4. Home page.

3.4 *Data collection*

Data collection is a process from which the data relevant to the application is stored for further processes. The information from the pupils that answered has been gathered using this application to determine the career the student is interested in. The information you provided has been synced with the cloud fire storage, and the information there has been turned into CSV files. This process is done using the Node JS platform.

4 METHODOLOGIES

This application was developed in Visual Studio code platform using flutter and dart.

4.1 *Flutter*

Flutter is developed by Google for developing software application for Android, iOS, Web etc.. It is an open-source SDK. Creating mobile application using Flutter is an uncomplicated process unlike other platforms. For working with flutter, we should be familiar with Dart language. The user interface of the program is made up of a variety of straightforward widgets, each of which performs a single task.

4.2 *Dart*

Google created Dart in 2011 as an open-source, all-purpose, object-oriented programming language with C-style syntax. Front end user interfaces for online and mobile apps are made using the Dart programming language. Dart is the only programming language supported by the flutter SDK. Additionally, it supports a few sophisticated ideas like type interfaces, abstract classes, and refield generics. It is a compiled language that is compatible with two different compilation methods Ahead of Time and Just in Time.

4.3 *Firebase*

Firebase is a no SQL database which is used to store the real time data. Firebase has several features like authentication of users, analytics, file storage and so on. It has both real time storage and cloud storage facilities. C++, Java, Node JS are some of the programming languages supported by the Firebase SDK. The Firebase database is used as database for this application from which the real time database are retrieved then and there. Firebase is easier and retrieves data faster than other databases.

4.4 *Node JS*

The data is collected using the np package of Node JS. It is an open-source server which can be used on different operating systems like Windows, iOS, Linux, and so on. Node JS uses JavaScript scripting language. Building application is fast and scalable. V8 JavaScript engine is used in the Node JS platform. It is used to create I/O-intensive web applications, such as single-page programmers and websites that stream video. Thousands of developers use Node JS because it is open source and totally free.

5 IMPLEMENTATION AND RESULT

The psychometric mobile application can be used by the school students of age around 14–17. Because this period is considered as a crucial period for students to know about their personalities and their interests. By taking this psychometric test, the students can get to know their interests which will be used to choose further departments in their higher educations. The student should attend this test authentically. The students are requested to choose the reliable and honest options. If the student feel that the question is related to their personality or their habits, then they can choose the option strongly agree. If the question is not related to their personality, then they can choose strongly to disagree. If neither of the above two options can be chosen, then they can go with option neutral. Likewise answering all the questions, they can finish the test. Their answers along with their details will be stored in the database. In the firebase console, the data is stored as three collection such as answers, questions, users. Each user has a unique id known as user's id. And each question has a unique id known as question's uuid. The answers collection contains the user's id and their answer for each question along with the question's id. The question collection contains the question's id, the title, the options. The user collection contains user's id, name, age, email id. The above is the process of data collection. Once a user completed the test, they cannot reattempt it again. Even if they login after completing the psychometric test they will be taken directly to the test completed page.

6 CONCLUSION AND FUTURE WORK

A mobile application for conducting psychometric tests for students was developed. This application is considered as a medium to collect the database from the students. The future work of this project includes the prediction of career options for the students based on their performance in psychometric test. After collection the database as CVS file using the node JS. We will be performing the data mining techniques like data preprocessing and data analyzing. In the future work we are aiming to use algorithms such as Decision tree, SVM (Support Vector Machine) and XGBoost. In decision tree, the Random forest method is used to determine the tree for the dataset. XGBoost technique is used to boost the performance of the decision tree to develop the accuracy rate of the dataset. SVM (Support Vector Machine) is used for classification of the data points. By using data preprocessing and data analyzing the data has been separated into categorical data and categorical targets. These two columns are imported to the algorithms and has to be implemented to get a high accuracy and suitable career for the student.

REFERENCES

Andrew J. Martin, Herbert W. Marsh. (April 2021). Workplace and Academic Buoyancy: Psychometric Assessment and Construct Validity Amongst School Personnel and Students. *International Scientific Conference eLearning and Software for Education Bucharest*. Assessment. 26(2), pp. 168–184 doi:10.1177/ 0734282907313767.

Beer & Marais, De Beer, Marais, Maree, & Skrzypczak. (2010). Initial Review of the Psychometric Properties of a Computerized Career Preference Test for Career Guidance Assessment. *International Journal of Testing.Assessment* 27(6):465–476 doi:10.1177/0734282909335781.

Brooke Linden, Heather Stuart. (September 2019). Psychometric Assessment of the Post-Secondary Student Stressors Index. *International Journal of Recent Technology and Engineering*. Volume 70, 2022 - Issue 2 doi:1080/07448481.2020.1754222.

Dr Frans Maloa, Ms. Ciara Bux. (January 2015). Psychometric Testing as a Predictor of Academic Performance: A South African Experience. *SSRN Electronic Journal* doi:10.2139/ssrn.2676348.

Geraldine Gray, Colm McGuinness, PhilipnOwende, Aiden Carthy. (2014). A Review of Psychometric Data Analysis and Applications. *Journal of Learning Analysis*.1(1),75–106

Kathleen L. Slaney, Masha Tkatchouk, Stephanie M. Gabriel, & Michael D. Maraun, *Journal of Psycho Educational Assessment*. (November 2019). Psychometric Assessment and Reporting Practices: Incongruence Between Theory and Practice.Assessment 27(6):465–476 doi:10.1177/0734282909335781.

Likitha Shetty. (August 2021). The Effectiveness of Psychometric Testing on Students-New Trend in Career Counselling. *International Journal of Recent Technology and Engineering*. Vol 1: No 1(Sept 2018), pp 2–18.

Lillian T. Eby, C. Douglas Johnson and Joyce E. A Russell. (March 2020). A Psychometric Review of Career Assessment Tools for Use with Diverse Individuals. *International Journal of Recent Technology and Engineering.Assessment* 6(3):269–310 doi:10.1177/106907279800600302.

Lisette Guy MSc. (2018). The Rationale of Gamification in Psychometrics and Considers the Current State of the Art. *IEEE International Conference on Testing*.20(3), 975–978.

Moinuddin Pasha Miya, Aakash Rathod, Sumit Negi, Sani Gupta. (April 2019). Psychometric Test based Analysis and Career Guidance. *Journal of Emerging Technologies And Innovative Research (JETIR)*. Volume 6 Issue 4 *eISSN*: 2349–5162.

Parvaneh Taymoori, Babak Moeini, David Lubans, and Mitra Bharami. (January 2020). *Development and Psychometric Testing of the Adolescent Healthy Lifestyle Questionnaire. IEEE* doi:10.4103/2277-9531.99221

Rosa Isabel Rodrigues. (October 2021). Testing the Psychometric Properties of a Short Skills Inventory for Students Looking for Their First Job. *IEEE Journal for Psychology. BMC Psychol* 9, 159 (2021) 1186 s40359-021-00662.

Skin cancer classification using machine learning

E. Sujatha*, T. Lavanya*, C.M. Niveditha* & S. Saranya*
Department of Computer Science and Engineering, Saveetha Engineering College, Chennai, Tamil Nadu

ABSTRACT: The skin diseases are manifest in variousforms, scarcity and low prevalence of efficient dermatologists, and the need for rapid and automated diagnosis. This aims to extend other works on dermatology by utilizing various methods of machine learning to distinguish various skin diseases and to improve the classification efficiencyby the use of taxonomy. We have trained our model using skin pictorial dataset, namely the ISIC archive and DermNet, and used the disease taxonomy, where they are available, to improve classification efficiency. At DermNet, we are establishing a new state-of-the-artmodel with a rate of 80% and an Area Under the Curve (AUC) of 98%.

We have also prioritized the classification of all 622 unique sub-classes. In this data set and it also achieved 67% cure and 98% AUC. In the ISICarchives, we have classified the 7diseases with an average accuracy of 93% and an AUC of 99%. This study reveals that machine learning has very huge potential to demonstrate a different types of skin diseases with example. This plays a vital role in present diagnosis of skin diseases and help the clinicians in large-scale dermoscopic imaging.

Keywords: manifest, taxonomy, DermNet, Deep Learning, dermoscopic

1 INTRODUCTION

Machine Learning is a classification of artificial intelligence in which a computer finds a original data and automatically gets the classifying characteristics required to recognize unseen features. In the past few years, notable advances in the machine learning algorithms to identify various data types, particularly pictures and source text, have been made in this field. The frequently used machine learning models are tested and trained using various supervised learning algorithms, using the data sets from inputs and the correspondingtarget output labels. Because manual diagnosis can also beinfluenced by physicians' experience, the different dermatoscopy algorithms in which they are formally trained can mean that many experts have different opinions when it comes to diagnosing a particular condition. On the other hand, CAD can provide a fast, reliable, and standardized diagnosis of various diseases in a consistent and accurate manner. CAD may also provide an opportunity for effective and inexpensive ways for prevention of advanced cancers for rural or remote areas people where qualified professionals are not available. The dataset consists of limited number of pictures of skin cancer. There has been reported that around 78% of the studies they reviewed used datasets with fewer than 1000 images, and the study that used the largest dataset had 2430 images. Therefore, most existing CAD work on skin diseases uses private datasets or very small publicly available datasets. And also such information are serve as proof of the implementation of artificial intelligence in analysis of skin. In this paper, we have extended the past work by presenting that the machine

*Corresponding Authors: sujathae@saveetha.ac.in, lavanyall2226@gmail.com, niveditha6202@gmail.com and saransa0130@gmail.com

DOI: 10.1201/9781032684994-93

learning model is effective in detecting more number of skin variations and also be used to its full capacity. We have implemented many machine learning models for skin diseases and differentiation using both the data sets like DermNet and ISIC Archive that are publicly available. We have used non-pictorial data to improve our obtained results and reveals that machine learning can implement and use more model inputs to improve classification performance. We have used non-pictorial data to improve our obtained results and reveals that machine learning can implement and use more model inputs to improve classification performance.

2 LITERATURE SURVEY

Janu *et al.* (2015) Convolutional neural networks (CNNs) are computational models inspired by the biological pictorial cortex. Those models have identified to be more useful, exact and dependent in pictorial differentiation. previously reached nearly human performance in many demanding natural image layering tasks and has also been used to classify diseases based on medical images. Towards the automatic classification of skin diseases, from the Dermofit image library to perform a 10-acre classification. On the MoleMap dataset to perform a classification of 15 acres. Esteva *et al.* uses pre-trained Inception v3 for approximately more pictures. Hänssle *et al.* compared their optimized Inception v4 model to 58 dermatologists after evaluating their model's binary classification performance in just two test runs of sizes 100 and 300. The sensitivity and specificity of their Deep model. The neural network model (DNN) is certainly higher than the average score of dermatologists in two private test groups, but their scores in the 2016 Open Access International Symposium on Biomedical Imaging (ISBI) Challenge test are lowerthan the first two participating winners in this one Challenge. Menegola *et al.* (2017) presented an in-depth study on classification of skin lesions ISIC Challenge 2017. This paper presents experiments with deep learning models pre-trained in multiple dataset configurations, using 6 data sources to assemble collected, the standardization of the input images and the use of meta-learning.

Related work over the past decade, many studies have been conducted on the detection and diagnosis of malignant and benign skin tumors. Many data sets are shared by the scientific community. Researchers used division-combination-grouping strategies and classification for detection and treatment of skin cancer. Each approach has its own limitations and improvements from the medicalcommunity to help medical professionals make decisions. Advances in computer technology and machine learning can help dermatologists identify melanoma earlier and reduce the cost of detection, as well as the need for unnecessary biopsies. Advantages are automatic systems for detecting melanoma can be very efficient and cost-effective, saving time, money, and effort. machine learning has demonstrated that it can classify melanoma more accurately and with better results compared to traditional methods. Disadvantages are more expensive method to detect melanoma. where have been a more challenges during the design of classification approaches.

3 PROPOSED SYSTEM

The use of advanced dermatological equipment has led to improved accuracy in melanoma classification. However, the recent advancements in machine learning and image processing have made a significant impact in the diagnosis, detection, and classification of melanoma, resulting in even more accuracy and reliability. A literature review reveals that various methods have been used to create computer-aided diagnostic systems skin classification.

Currently, most systems are based on Convolutional Neural Networks (CNN) that are trained on dermoscopic images. This system inputs dermoscopic images and outputs a classification result as either benign or malignant. A Convolutional Neural Network (CNN) is a type of deep learning algorithm that can process an image and assign different weights and

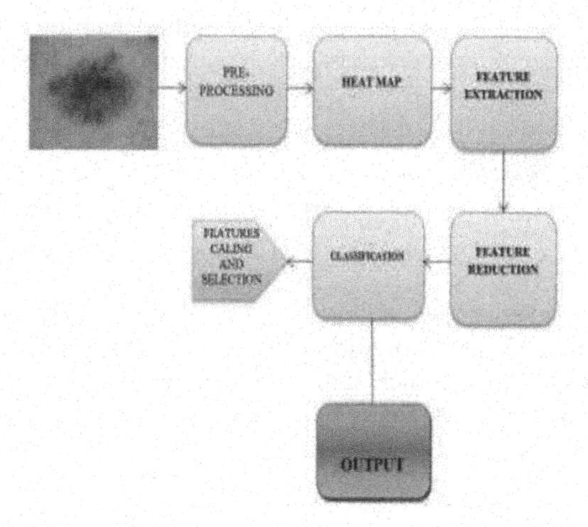

Figure 1. System architecture.

learnable parameters to different aspects or objects within the image. It is capable of differentiating between the various elements in the image. the amount of pre-processing required in a Convolutional Neural Network is less compared to other classification algorithms.

4 SYSTEM DESIGN AND TESTINGPLAN

The initial step in designing a skin cancer classification system is gathering a comprehensive and diverse set of data on skin lesions. This data set must contain images of skin lesions along with the diagnosis of whether they are benign or malignant. Having a broad and diverse data set is critical to ensure that the model can generalize to different types of skin lesions and achieve accurate results in a real-world setting. once the data has been collected, the next step is to prepare the images for model training. There are several popular models that can be used for this task, including Convolutional Neural Networks (CNNs) and Residual Neural Networks (ResNets). The selection of the appropriate model will depend on factors such as the size and complexity of the data set, the desired level of accuracy, and the computational resources available. It is important to choose a model that is capable of accurately classifying skin lesions and that can be trained efficiently using the available data and computational resources.

Unit testing refers to the practice of verifying the functionality of specific parts or components of a system to guarantee their correct operation. This involves testing individual parts in isolation to make sure they perform as expected. This type of testing can include checking the functionality of specific functions such as image pre- processing, model training, and model evaluation. The aim is to catch any errors or bugs early in the development process, allowing for quicker and easier resolution.

5 IMPLEMENTATION AND ANALYSIS

5.1 *Graph partitioning*

The cost of cutting a graph G(V, E) into two parts, A and B, is defined as the sum of the weights of all edges that cross the cut, from nodes in A to nodes in B. The minimum cut criteria look for cuts that split the smallest number of nodes in the graph. However, this approach can lead to an unnatural bias towards small isolated sets of nodes. If the weights of the edges are proportional to the distance between nodes, then cuts that divide individual nodes in half will have a much smaller cut value compared to cuts that divide nodes into both left and right halves.

5.2 *Normalized graph segmentation*

A new method has been proposed to eliminate unnatural bias when dividing small sets of points. This method, called the Normalized Cut Point (Ncut), calculates the cutting cost of all nodes in the graph instead of just considering the total weight value of the edges connecting two partitions. This means that a division that separates small isolated points will no longer have a low Ncut value, as the division will make up a large portion of the small node's total connections, along with other nodes. Additionally, a measure of the overall Normalized Association in groups can be defined for a specific partition.

6 EXPERIMENTAL RESULTS

This method provided various guidelines for the prevention of skin cancer and detection of skin cancer in POC based on the review of the available evidence. Due to the limited number of studies, there are few resources to provide an overview of the assessment of darkly pigmented lesions in POC. Diagnostic functions for different types of skin cancer were recorded and different possible risk factors were taken into account.

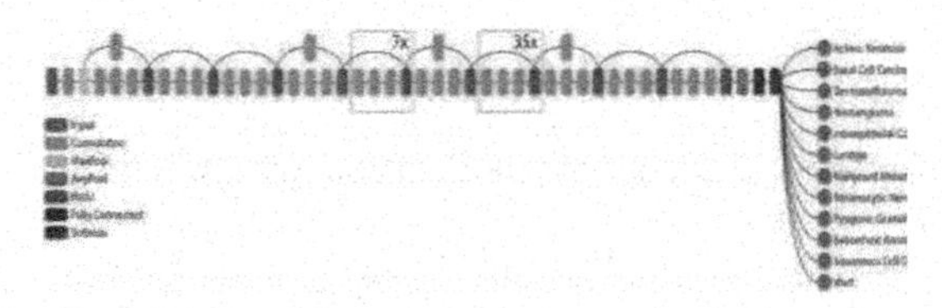

Figure 2. Classification of skin types.

Figure 3. GUI window.

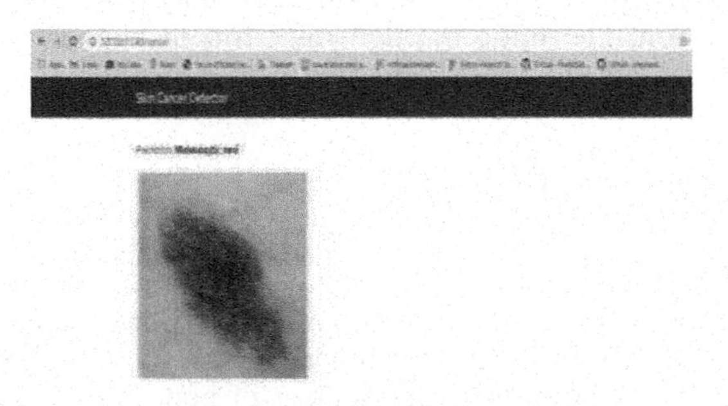

Figure 4. Skin lesion.

7 CONCLUSION

Machine learning has become a promising tool in the field of dermatology for the classification of skin lesions, especially in the detection of skin cancer. These techniques have been shown to improve the accuracy and efficiency of skin cancer diagnosis, which is crucial for successful treatment. Several machines learning algorithms, including Support Vector Machines (SVMs), Convolutional Neural Networks (CNNs), and Random Forests, have been utilized for skin cancer classification and have demonstrated high accuracy in categorizing skin lesions, such as melanoma, basal cell carcinoma, and squamous cell carcinoma. In conclusion, the use of machine learning in dermatology holds tremendous potential for revolutionizing the field by providing quick and precise skin cancer diagnosis. However, further research and development are necessary to ensure that these models are reliable, robust, and can be easily integrated into clinical practice.

This project focuses on using machine learning techniques to accurately predict the severity of accidents at specific locations. The process starts by improving dermoscopic images using a new contrast stretching method based on the mean and standard deviation of the pixels. The OTSU threshold is then used for image segmentation. In the next step, features such as shape, color, and texture are extracted and shape features are reduced using PCA. To overcome the problem of class imbalance, the SMOTE sampling technique is usedThe system was tested on a publicly available dataset and showed 17 times better accuracy compared to a conventional system. This model could be useful for governments to prevent accidents.

REFERENCES

Aerts H. J., Velazquez E. R., Leijenaar R. T., Parmar C., Grossmann P., Carvalho S., Bussink J., Monshouwer R., Haibe-Kains B., Rietveld D. *et al.*, "Decoding Tumour Phenotype by Noninvasive Imaging Using a Quantitative Radiomics Approach," *Nature Communications*, vol. 5, p. 4006, 2014.

Alquran H. *et al.*, "The Melanoma Skin Cancer Detection and Classification Using Support Vector Machine," *2017 IEEE Jordan Conf. Appl. Electr. Eng. Comput.Technol. AEECT 2017*, vol. 2018- Janua, pp. 1–5, 2017.

Ansari U. B. and Student M. E., "Skin Cancer Detection Using Image Processing TanujaSarode 2," *Int. Res. J. Eng. Technol.*, vol. 4, no. 4, pp. 2395–56, 2017.

Ballerini L, Fisher R. B., Aldridge B., and Rees J., "A Color and Texture Based Hierarchical k-nn Approach to the Classification of Non-melanoma Skin Lesions," in *Color Medical Image Analysis*. Springer, 2013, pp. 63–86.

Esteva A., Kuprel B., Novoa R. A., Ko J., Swetter S. M., Blau H. M., and Thrun S., "Dermatologist-level Classification of Skin cancer with Deep Neural Networks," *Nature*, 2017.

Gilmore S., Hofmann-Wellenhof R., and Soyer H. P., "A Support Vector Machine for Decision Support in Melanoma Recognition," *Experimental Dermatology*, vol. 19, no. 9, pp. 830–835, 2010.

Gupta S. and Tsao H., "Epidemiology of Melanoma," *Pathol. Epidemiol. Cancer*, pp. 591–611, 2016.

Lee H., Tajmir S., Lee J., Zissen M., Yeshiwas B. A., Alkasab T. K., Choy G., and Do S., "Fully Automated Deep Learning System for Bone Age Assessment," *Journal of Digital Imaging*, vol. 30, no. 4, pp. 427–441, 2017 Instituto Nacional de Cncer Jos Alencar Gomes da Silva, "Estimativa 2018 incidłncia de cncer no brasil," Available at http://www1.inca.gov.br/inca/Arquivos/ estimativa-2018.pdf,2018, accessed in 29/04/ 2018.

Matsunaga K, Hamada A., Minagawa A., and Koga H., *"Image Classification of Melanoma, Nevus and Seborrheic Keratosis by Deep Neural Network Ensemble,"* 2017.

Nachbar D *et al.*, "The ABCD Rule of Dermatoscopy: High Prospective Value in the Diagnosis of Doubtful Melanocytic Skin Lesions," *J.Am. Acad. Dermatol.*, vol. 30, no. 4, pp. 551–559, 1994.

Vandenberghe M E., Scott M. L., Scorer P. W., Soderberg M., Balcerzak D., and Barker C., "Relevance of Deep Learning to Facilitate the Diagnosis of Her2 Status in Breast Cancer," *Scientific Reports*, vol. 7, p. 45938, 2017.

Artificial Intelligence, Blockchain, Computing and Security – Dagur et al. (Eds)

Detection and classification of brain tumor

P. Preethi*, G. Jawahar Aravind*, G. Praveen Kumar* & S. Yogesh Raj*
Department of Computer Science and Engineering, Saveetha Engineering College, Chennai, Tamil Nadu, India

ABSTRACT: The emerging topic in many medical diagnostic applications is automated flaw identification in medical imaging. The automated diagnosis of tumours in MRI scans is particularly important since it offers details about aberrant tissues required for treatment planning. Human examination is the traditional approach for finding flaws in magnetic resonance brain imaging. The vast volume of data makes this strategy unfeasible. In order to reduce the death rate of humans, reliable and automated categorization techniques are necessary. Because it would free up radiologist time and have proven accuracy, automated tumour detection approaches have been developed. The intricacy and variety of tumours make it difficult to detect brain cancers with an MRI. In this study, we provide machine learning techniques to address the shortcomings of conventional.

Keywords: medical diagnostic applications, aberrant tissues, magnetic resonance brain imaging, radiologist time, machine learning techniques

1 INTRODUCTION

One of the most severe disorders in the field of medicine is the brain tumour. The radiologist's first priority during the early phases of tumour growth is an effective and efficient analysis. Histological grading, based on a stereotactic biopsy test, is the gold standard and recognised practise for establishing the grade of a brain tumour. The neurosurgeon must make a tiny hole in the skull for the tissue collection during the biopsy process. The biopsy test carries a number of risks, including as bleeding from the tumour and infection of the brain that can result in convulsions, excruciating headaches, a stroke, a coma, and even demise. But the main problem with stereotactic biopsy is that the method is not completely reliable, which might lead to a significant diagnostic mistake. Due to the difficulty of tumour biopsy for individuals with brain tumours, MRI and other non-invasive imaging methods have widely used to diagnose brain cancers. As a result, it has become vital to build methods for the detection and forecasting of malignancy grades using MRI data. However, the imaging modality appeared initially to be such as in Magnetic Resonance Imaging (MRI), it is somewhat difficult task to properly visualise the tumour cells and its differentiation with its nearby soft tissues. This may be because of the lack of adequate light in imaging modalities, the volume of data present, or the complexity and variety of tumours, including their unpredictable shapes, sizes, and placements such as unstructured shape, viable size, and unpredictable locations of the tumour. Cerebral cancer division procedures that rely on standard picture handling aren't as perfect as other currently proposed mind-division tactics. According to conventional practise, an MRI is made by using attractive field radiation to

*Corresponding Authors: preethivpravin@gmail.com, jawahararavind380@gmail.com, praveenvj2020@gmail.com and yogeshak67@gmail.com

 DOI: 10.1201/9781032684994-94

create a two-dimensional image that is then handled and examined by a medical professional. This image is mostly reliant on a certain dark scale.

2 LITERATURE SURVEY

When Krizhevsky *et al.* (2012) trained To classify the 1.2 million high-resolution photographs entered in the ImageNet LSVRC-2010 competition into the 1000 separate classes, a sizable, deep convolutional neural network was used. contest into the 1000 different classes, they attained Using transfer learning, the most recent results in image classification techniques. Simonyan *et al.* (2014) In 2014, they looked at how the convolutional network's depth affected accuracy in the context of large-scale picture recognition. These discoveries served as the foundation for their team's entry to the 2014 ImageNet Challenge, which resulted in The localization and categorization categories received first and second place finishes, respectively. Yang *et al.* (2010) The survey from 2010 concentrated on classifying and examining the most recent developments in Transfer learning for problems with grouping, regression, and classification. In this survey, they discussed sample selection bias, covariate shift, and other related machine learning techniques., and domain adaptation, connect to transfer learning. Szegedy *et al.* (2015) in 2015 The deep convolutional neural network architecture Inception, developed by Szegedy is what helped the ImageNet Large-Scale Visual Recognition Challenge 2014 establish new standards for classification and detection (ILSVRC14). ResNet *et al.* in 2015b He and colleagues introduced This uses skip connections and batch normalisation. In order to make it easier to train networks that are far deeper than those previously used, he developed a residual learning architecture. Instead of learning unreferenced functions, he deliberately rebuilt the layers to learn residual functions with reference to the layer inputs. The ResNet *et al.* (2015) which makes use of batch normalisation and "skip connections," was presented by He *et al.* in 2015b. In order to make it easier to train networks that are far deeper than those previously used, he developed a residual learning architecture. Learning unreferenced functions is preferable, he deliberately rebuilt the layers to learn residual functions with reference to the layer inputs. Swapnil R. Telandhe *et al.* Segmentation, which divides an image into sections of areas or objects, is proposed for tumour identification. In order to fully view the image and precisely categorise the image's content, it is necessary to separate the object from the backdrop. Malathi *et al.* Hong-Long devised a method for classifying brain MRIs that relied mostly on spider net plots and probabilistic neural networks and desegregation wave entropy. Rajeshwari G. Tayade *et al.* The grouping of aberrant brain matters into benign and malignant was accomplished by In their work, Rajeshwari G. Tayade combined wavelet statistical properties with co-occurrence wavelet texture features obtained from two level distinct riffle reconstructions. Lukas *et al.* The performance analysis was conducted by comparing the classification results of the probabilistic neural network with alternative neural network classifiers. Lukas *et al.* proposed the work on information among the medical image and thereby significantly improved the machine speed for growth segmentation results.

3 PROPOSED SYSTEM

A brain tumour may be found in an MRI scan using a computer method known as brain tumour detection, which is based on artificial intelligence (AI). Analytical model construction is automated using machine learning, a technique for data analysis. It is an area of artificial intelligence based on the idea that robots can learn from data, recognise patterns, and reach conclusions with little assistance from humans. Pre-processing - To improve the model's accuracy, noise will be removed from the MRI images at this step. MRI pictures frequently contain noise, which increases redundancy and reduces the model's accuracy.

Data collection - The data were divided into groups of healthy and unhealthy individuals. Additionally, because the photographs are of different sizes, they are transformed to the same size of 224*224.

In the realm of face recognition, CNN (Convolutional Neural Network) has emerged as the most often used technique. The convolution and sampling layers of the CNN model are integrated into a single layer for simplicity. Increase the rate of picture recognition significantly using the network that has previously been trained.

CNN is the method employed to make this model function (Convolutional Neural Network). Following are the various steps used to apply CNN on the dataset:

The necessary packages are initially imported. The dataset is then imported from the place where it is kept. Images are first read, then labelled (for example, 0 for non-tumor and 1 for tumour picture), and then placed in the data frame. Images are initially read, followed by labelling (for instance, 0 for a non-tumor image and 1 for a tumour image), and then placement in the data frame. The image has been normalised.

The data was divided into three sections:

- Training
- Module
- Test

4 SYSTEM ARCHITECTURE DIAGRAM

The system Architecture Diagram show the basic functionality of the DETECTION AND CLASSIFICATION OF BRAIN TUMOR.

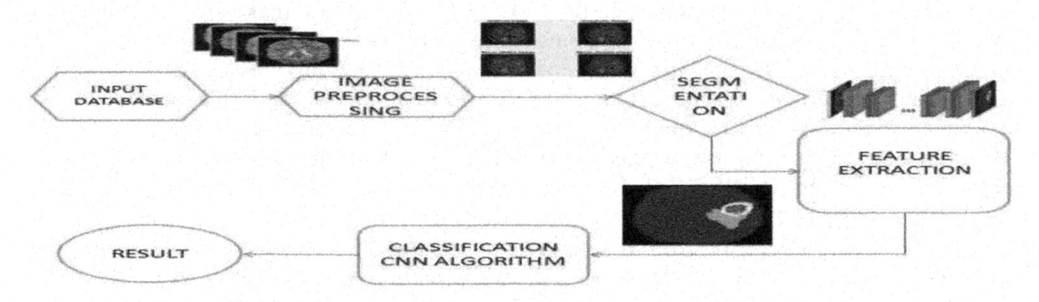

Figure 1. System architecture diagram of proposed system.

Here We are Using python libraries and frameworks like Tensorflow, Keras, Numpy, matplotlib, opencv.

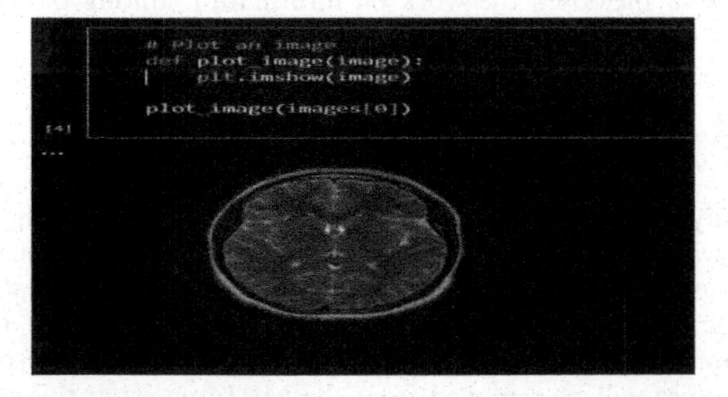

Figure 2. Plot an MRI image to the algorithm.

5 IMPLEMENTATION AND RESULT

Integration of facial recognition algorithms: Possibility Offers a help control mechanism with OpenCV Since OpenCV 2.4.4 enables for the creation of desktop programmes, you must exercise caution while choosing and installing this library. Currently, this version uses the Eigenfaces, Fisherfaces, and LBPH (Local Binary Patterns Histograms) face detection algorithms.Facial recognition algorithms integration viability provided by OpenCV with a help control mechanism You must exercise caution while choosing and installing this library because OpenCV permits the construction of desktop apps as of version 2.4.4. Currently, this version contains three face identification algorithms: Eigenfaces, Fisherfaces, and LBPH (Local Binary Patterns Histograms). The project's technical viability is assessed to see if it can be finished given the available resources—people, equipment, software, and technical know-how. It takes into account choosing resources for the suggested system. Because Python was used in the system's development, it is platform independent. Users of the system can therefore operate on any platform with average processing power. Since the system uses cutting-edge technology, it is theoretically feasible.The main component for creating the model is the dataset. The model will be more accurate the more photos there are in the dataset. Our model's chosen dataset was downloaded from the internet. There are 253 MRI pictures in it. There were various files with various sets of pictures in this dataset, one of which had an image of a healthy brain and the other contained an image of a brain tumour. The complete dataset is subsequently trained. 25 test photos and 50 validation shots were taken. The collection includes MRI pictures of various size. Due to the difficulty of obtaining data sets from hospitals, this dataset was chosen.an easy task. The link below will take you to Kaggle, where you may get the data set that was utilised.

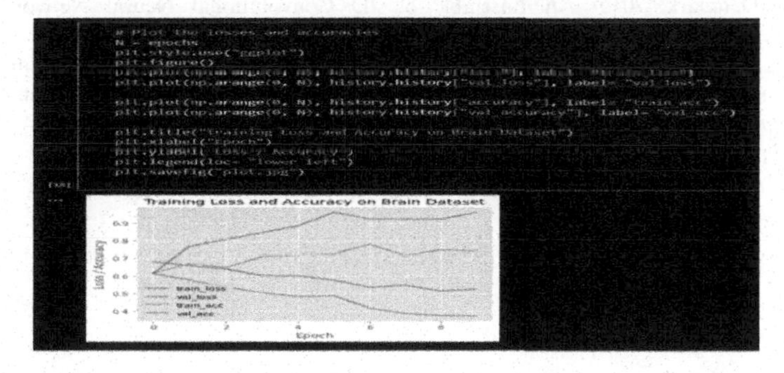

Figure 3. Accuracy of brain tumor.

6 CONCLUSION

The train accuracy and validation accuracy of a Keras model without prior training are 97.5% and 90.0%, respectively. The accuracy of the validation result's best number was 91.09%. Although the training accuracy of the Keras model is >90%, it has been found that when pre-trained models are not employed, the total accuracy is lower.

Additionally, the calculation time was 40 minutes when we trained our dataset without using Transfer Learning versus 20 minutes when we did. As a result, using a pre-trained Keras model reduced training and computation time by 50%.

REFERENCES

Amin J., Sharif M., Yasmin M., Fernandes S.L. A Distinctive Approach in Brain Tumor Detection and Classification Using MRI. *Pattern Recognit. Lett.* 2020;139:118–127. doi: 10.1016/j.patrec.2017.10.036.

Bauer S., Fejes T., Slotboom J., Wiest R., Nolte L.-P., Reyes M. *Proceedings of the MICCAI BraTS Workshop. Miccai Society*; Nice, France: 2012. Segmentation of Brain Tumor Images Based on Integrated Hierarchical Classification and Regularization; p. 11.

Chavan N.V., Jadhav B., Patil P. Detection and Classification of Brain Tumors. *Int. J. Comp. Appl.* 2015;112:45–53.

Chen M., Yan Q., Qin M. A Segmentation of Brain MRI Images Utilizing Intensity and Contextual Information by Markov Random Field. *Comput. Assist. Surg.* 2017;22:200–211. doi: 10.1080/24699322.2017.1389398.

Das S., Chowdhury M., Kundu M.K. Brain MR Image Classification Using Multiscale Geometric Analysis of Ripplet. *Prog. Electromagn. Res.* 2013;137:1–17. doi: 10.2528/PIER13010105.

El-Dahshan E.-S.A., Mohsen H.M., Revett K., Salem A.-B.M. Computer-aided Diagnosis of Human Brain Tumor Through MRI: A Survey and a New Algorithm. *Exp. Syst. Appl.* 2014;41:5526–5545. doi: 10.1016/j.eswa.2014.01.021.

Hemanth D.J., Anitha J., Naaji A., Geman O., Popescu D.E., Son L.H., Hoang L. A Modified Deep Convolutional Neural Network for Abnormal Brain Image Classification. *IEEE Access.* 2018;7:4275–4283.

Huang M., Yang W., Wu Y., Jiang J., Chen W., Feng Q. Brain Tumor Segmentation Based on Local Independent Projection-based Classification. *Trans. Biomed. Eng. J.* 2014;61:2633–2645. doi: 10.1109/TBME.2014.2325410.

Jiang J., Wu Y., Huang M., Yang W., Chen W., Feng Q. 3D Brain Tumor Segmentation in Multimodal MR Images Based on Learning Population- and Patient-specific Feature Sets. *Comput. Med. Imaging Graph.* 2013;37:512–521. doi: 10.1016/j.compmedimag.2013.05.007.

Kamnitsas K., Ledig C., Newcombe V., Simpson J.P., Kane A.D., Menon D.K., Rueckert D., Glocker B. Efficient Multi-scale 3D CNN with Fully Connected CRF for Accurate Brain Lesion Segmentation. *Med. Image Anal.* 2017;36:61–78. doi: 10.1016/j.media.2016.10.004.

Lyksborg M., Puonti O., Agn M., Larsen R. *19th Scandinavian Conference on Image Analysis*. Springer; Copenhagen, Denmark: 2015. An Ensemble of 2D Convolutional Neural Networks for Tumor Segmentation.

Mekhmoukh A., Mokrani K. Improved Fuzzy C-Means based Particle Swarm Optimization (PSO) Initialization and Outlier Rejection with Level Set Methods for MR Brain Image Segmentation. *Comput. Methods Programs Biomed.* 2015;122:266–281. doi: 10.1016/j.cmpb.2015.08.001

Artificial Intelligence, Blockchain, Computing and Security – Dagur et al. (Eds)
© 2024 The Author(s), ISBN: 978-1-032-67841-2

Agricultural production monitoring platform based on cloud computing

E. Sujatha*
Professor, Computer Science and Engineering, Saveetha Engineering College, Chennai, India

G. Suthan*, J. Sudharsan*, M. Shreedhar* & P. Pradhyun*
Computer Science and Engineering, Saveetha Engineering College, Chennai, India

ABSTRACT: As a coastal state, agricultural instability reduces output in Tamil Nadu. Agricultural traits and features provide data that may be used to get insights into Agri-facts. With the advent of the information technology world, various highlights in agricultural sciences are being pushed to supply farmers with critical agricultural information. Machine Learning Techniques use data to build a well-defined model that assists us in making predictions. Crop forecasting, rotation, water requirements, fertilizer requirements, and crop protection are all hurdles that may be overcome. Because of the changing climatic components in the environment, it is vital to have an effective approach to promote crop growth and aid farmers in their management and output. This might assist prospective farmers in improving their farming practices. Data analytics provides the path for extracting valuable information from agricultural statistics. Crop data was evaluated, and crop recommendations were developed based on productivity and season.

Keywords: Cloud Computing, Machine Learning

1 INTRODUCTION

Agriculture has a long history in India. Recently, India was rated second in the world in agriculture output. The monetary contribution of agriculture to India's GDP is falling. Crop yield is affected by a variety of factors, including meteorological, geographic, organic, and economic considerations. Due to climate unpredictability, farmers are unsure of which crop to cultivate and when and where to begin. The use of different fertilizers is also unclear owing to variations in seasonal meteorological conditions and essential assets such as soil, water, and air. In this scenario, crop yields are progressively dropping. The solution to the problem is to offer farmers a clever, user-friendly recommender system. Every farmer tries to anticipate crop production and if it matches their expectations by assessing the farmer's prior experience with the specific crop. Accurate crop history information is essential for making agricultural risk management choices. In this study, we offer a model to resolve these concerns. The proposed method is unique in that it aids farmers in increasing agricultural yield while simultaneously recommending the most lucrative crop for the particular location. The suggested model estimates crop production by taking into account elements such as rainfall, temperature, area, season, soil type, and so on. The method also aids in determining the optimal time to apply fertilizer. The existing crop production prediction system is either hardware-based and costly to operate or not generally available.

*Corresponding Authors: sujatha@saveetha.ac.in, suthangopu1811@gmail.com, sudharsan46846@gmail.com, shreedhar2002@gmail.com and pradhyun.cse.sec@gmail.com

DOI: 10.1201/9781032684994-95

2 LITERATURE SURVEY

In this survey, we review the Agricultural Production Monitoring Platform Based On Cloud Computing. (Kaloxylos *et al.* 2012), The most recent trend is to make these crop management systems internet-capable. The Internet in its current operational form has significant flaws, particularly in handling a large number of networking devices or allowing for simpler integration of systems and services provided by various participants. (Bueno-Delgado 2016), There are various software improvements in the agricultural industry, both in the scientific literature and on the market, that increase fertilizer efficacy and application for certain crops. Ecofert may be done on mobile devices, giving farmers and agricultural workers a powerful tool to assist with agricultural duties in the field. (Jethva & Jignasha 2018), This research presents an overview of numerous data mining frameworks used in generating soil datasets for fertilizer suggestions. (Jones *et al.* 2017), Over six decades, researchers from a wide range of disciplines have contributed ideas and instruments to demonstrate, a fundamental apparatus in agricultural frameworks science, we discuss the historical context of rural frameworks displaying and differentiating exercises in this section.(Manpreet Kaur 2014), We shall examine the uses and techniques of data mining in agriculture in this study. This study is based on the discovery of suitable data models that aid in the achievement of high accuracy and generality in price prediction. (Mishra & Subhadra 2016), This study looked at the many uses of machine learning in agriculture. It also gives insight into the difficulties that Indian farmers confront and how various techniques may help to solve them. (Pritam Bose 2016), The first SNN computational model for agricultural yield estimate using normalized difference vegetation index image time series introduced this approach. (Priyanka P. Chandak 2017), The suggested system is based on data mining techniques and data gathered from satellite data, the Internet, and soil testing results that are put into existing databases. (Shreya S 2016In) the field of data mining, using a data clustering method to extract important information and make predictions is an efficient strategy. This does assist to improve agricultural quality and provides more cash for farmers. (Stetkiewicz *et al.* 2023), This research summarises the findings of a series of online focus groups with agricultural production stakeholders to discuss the possibilities for crop innovation to future-proof European food systems. (Rajeswari & Arunesh 2016), The approach aims to forecast soil type using data mining classification algorithms. To address Big Data challenges, effective solutions that use Data Mining to improve the accuracy of categorization of large soil data sets may be developed. (Teja *et al.* 2022), This study presents a website that uses Machine Learning algorithms in conjunction with historical weather data to estimate the most lucrative crop under current weather conditions.

3 PROPOSED SYSTEM

Many agricultural factors have an impact on crop productivity. According to the proposed study, crops can be recommended to farmers based on crop yield in previous years. This kind of guidance can help farmers determine whether a certain crop has been produced successfully in recent years. Crop productivity may be lowered as a result of crop disease, water shortages, and a range of other factors. While assessing yield, farmers might learn which crops are in great demand in the market that year. Farmers may make decisions based on crop trends in recent years. Farmers will receive suggestions depending on the crop yield season. The problem statement of the project is to recommend crops to the farmers using Decision Tree Classifier. The basic process of this project is that we will pre-process the data provided to us, then it is used to prepare the model for the back and use a flask to connect it to the UI interface to show the full and final output.

4 ARCHITECTURE SYSTEM

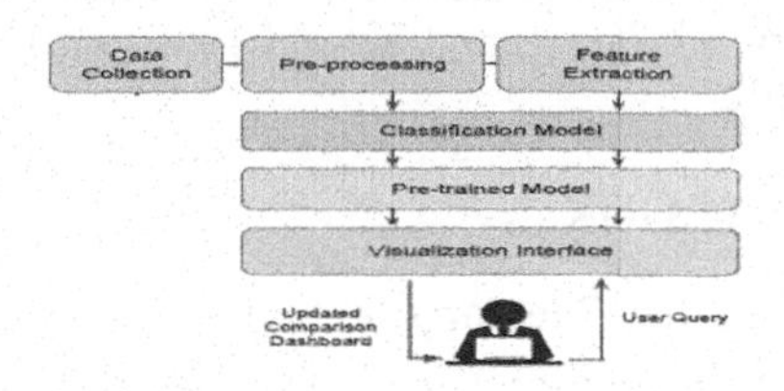

Figure 1. Architecture system.

5 FLOW DIAGRAM

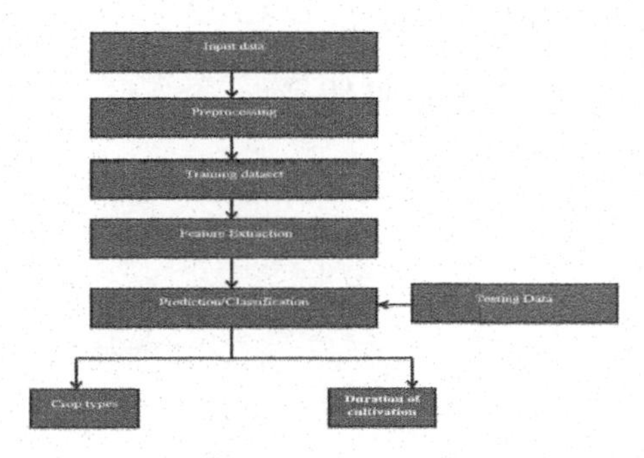

Figure 2. Flow diagram.

6 ALGORITHM

6.1 *Decision tree classifier*

A decision tree is a non-parametric supervised learning approach that may be used for classification and regression. It has a tree structure with a root node, branches, internal nodes, and leaf nodes. Decision tree learning employs a divide-and-conquer strategy, doing a greedy search to find the best-split points inside a tree. This division technique is then continued top-down and recursively until all or the majority of items have been classified under certain class labels. The decision tree's complexity determines whether or not all data points are classed as homogeneous sets. Smaller trees can achieve pure leaf nodes more readily, i.e. data points in a single class

6.2 *Random forest classifier*

The decision tree is the fundamental unit of random forest classifiers. The decision tree is a hierarchical structure built from the features of a data set (or independent variables). The decision tree is split into nodes according to a measure that is linked to a subset of the features. A Random Forest Method is a supervised machine learning technique that is commonly utilized in classification and regression applications in machine learning.

7 DATA SET

The dataset contains 821 unique data points the dataset has 14 columns, which are detailed further below. a) States: The total number of states in India. b) Rainfall: Rainfall in mm c) Ground Water: Total groundwater level d) Temperature: temperature in degrees Celsius e) Soil type: Number of soil types f) Season: Which season is best for crops g) Crops: Types of crops h) Fertilisers required: Fertilizer Types Required i) Cost of cultivation: Total cultivation cost j) Expected revenues: Total expected revenues k) Quantity of seeds per hectare: seeds per hectare l) Duration of cultivation: number of days for cultivation m) The demand for crop: demand for a crop (High, low) n) Crops for mixed cropping: Which crops can be combined for cropping.

8 ANALYSE AND PREDICTION

In the actual dataset, we chose only 7 features: a) States: The total number of states in India b) Rainfall: Rainfall in mm c) Ground Water: Total groundwater level d) Temperature: Temperature in degrees Celsius e) Soil type: Number of soil types f) Season: Which season is best for crops g) Crops: Types of crops

9 ACCURACY ON TEST SET

On the test set, we achieved an accuracy of 90.7%

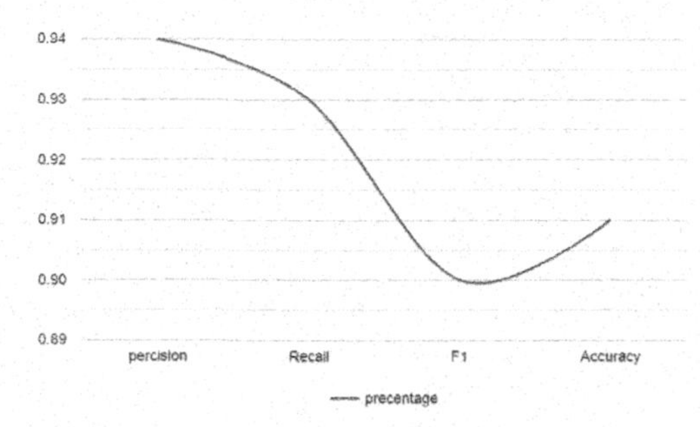

Figure 3. Percentage.

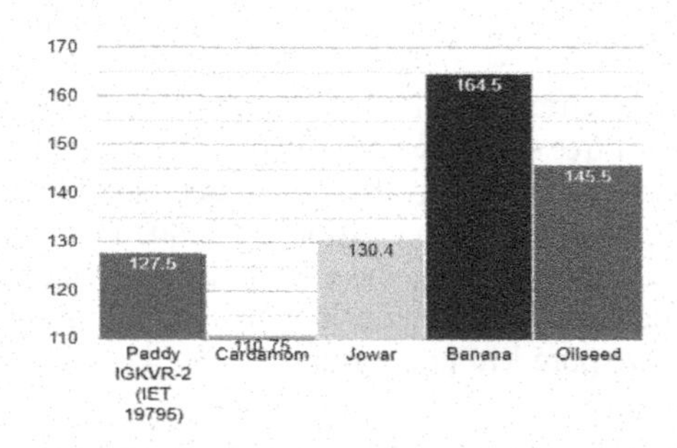

Figure 4. Crop recommendation for rainfall.

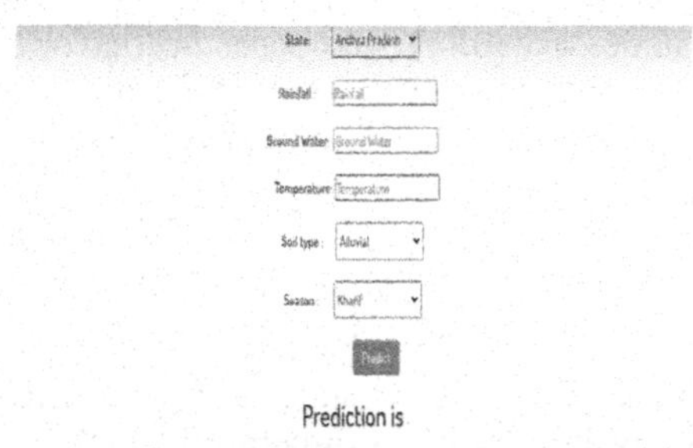

Figure 5. Crop

10 CONCLUSION

The value of crop management has been extensively researched. Farmers require modern technologies to assist them in growing their crops. Accurate crop estimates may be provided to agriculturists on a timely basis. Many Machine Learning methods have been used to evaluate agricultural factors. A literature review investigates some of the strategies used in many parts of agriculture. Soft computing approaches, such as blooming neural networks, play an important role in delivering suggestions. Farmers may be provided more specific and appropriate advice based on parameters such as productivity and season, allowing them to produce a higher volume of goods.

REFERENCES

Bueno-Delgado, M. Victoria, *et al.* "Ecofert: An Android Application for the Optimization of Fertilizer Cost in Fertigation." *Computers and Electronics in Agriculture* 121 (2016): 32–42.

Jethva M, Jignasha, Nikhil Gondaliya, and Vinita Shah. "A Review on Data Mining Techniques for Fertilizer Recommendation." *International Journal of Scientific Research in Computer Science, Engineering and Information Technology, IJSRCSEIT* 3.1 (2018).

Jones, W James, *et al.* "Brief History of Agricultural Systems Modeling." *Agricultural Systems* 155 (2017): 240–254.

Kaloxylos A. *et al.*, "Farm Management Systems and the Future Internet Era," *Compute. Electron. Agri cult.*, vol. 89, pp. 130–144, Nov. 2012

Manpreet Kaur, Heena Gulati, Harish Kundra, "Data Mining in Agriculture on Crop Price Prediction: Techniques and Applications", *International Journal of Computer Applications*, Volume 99– No.12, August 2014.

Mishra, Subhadra, Debahuti Mishra, and Gour Hari Santra. "Applications of Machine Learning Techniques in Agricultural Crop Production: A Review Paper." *Indian J. Sci. Technol* 9.38 (2016): 1 14

Pritam Bose, Nikola K. Kasabov (2016), "Spiking Neural Networks for Crop Yield Estimation Based on Spatiotemporal Analysis of Image Time Series", *IEEE Transactions on Geoscience and Remote Sensing*.

Priyanka, P. Chandak (2017)," Smart Farming System Using Data Mining", *International Journal of Applied Engineering Research*, Volume 12, Number 11.

Rajeswari and K. Arunesh (2016) "Analysing Soil Data Using Data Mining Classification Techniques", *Indian Journal of Science and Technology*, Volume 9, May.

Shreya S. Bhanose, Kalyani A. Bogawar (2016) "Crop And Yield Prediction Model", *International Journal of Advance Scientific Research and Engineering Trends*, Volume 1, Issue 1, April 2016.

Stetkiewicz, Stacia, *et al.* "Crop Improvements for Future-proofing European Food Systems: A Focus-group-driven Analysis of Agricultural Production Stakeholder Priorities and Viewpoints." *Food and Energy Security* 12.1 (2023): e362.

Teja, M. Sai, *et al.* "Crop Recommendation and Yield Production using SVM Algorithm." 2022 *6th International Conference on Intelligent Computing and Control Systems (ICICCS)*. IEEE, 2022.

Artificial Intelligence, Blockchain, Computing and Security – Dagur et al. (Eds)
© 2024 The Author(s), ISBN: 978-1-032-67841-2

Blockchain based application programming interface market place

Thoguru Balaji*, Valliveti Kishore*, Shaik Afzal Ali* & Vijay Anand*
Department of Computer Science and Engineering, Saveetha Engineering College, Chennai, India

ABSTRACT: By eliminating the need to sign up for hundreds of centralised services all over the world and develop APIs for each of them, blockchain paves the way for data suppliers and users to exchange information directly in a single, open system. The blockchain application programming interface (API) is widely used in app development services. The API of a programme (module, library) outlines the features it offers while hiding the implementation details. APIs allow different pieces of software to communicate with one another. Components in this situation often form a hierarchy, with the highest-level components utilising the API of the lowest-level components, and so on. This is the foundation upon which Internet data-transfer protocols are constructed.This paper explain the structure for a decentralized marketplace which is blockchain-based for APIs that allows the participants in the API perimeter to monitor actions of all other activities by not relying on the central authority. In particular, our technology partitions AI models into smaller chunks before distributing them to various cloud providers (CV) for coordinated API execution. As a result of our design, it is impossible for many cloud providers to work together to steal a single model, and it is also impossible for cloud providers or their clients to deny responsibility for their contributions to the model or its execution.

1 INTRODUCTION

Blockchain, a distributed digital ledger, facilitates copying and sharing among any number of interconnected computers. All participants in a Blockchain can view its whole transaction and update log at any time. This type of record-keeping system is sometimes called "Distributed Ledger Technology" (DLT.) The Blockchain documents all transactions using hashes, which are irreversible cryptographic signatures. The immutability of this ledger derives from the fact that its hash value is produced using a manner that forbids its alteration after its creation. Once the first building block is altered, the entire structure is exposed. Hackers would need to change every chain block in all distributed versions to break into the system. Blockchain technology records and shares digital data without change. Blockchains provide immutable ledgers, which are unalterable records of transactions. Thus, blockchains are distributed ledgers.

2 LITERATURE SURVEY

In this article, we take a look at the several industries that might potentially benefit from utilizing blockchain technology. Nakamoto *et al.* (2008) research analyzes blockchain

*Corresponding Authors: thogurubalaji882@gmail.com, valliveti944@gmail.com, ryanali7860@gmail.com and vijayanand@saveetha.ac.in

 DOI: 10.1201/9781032684994-96

technology, its possible uses, and the ways in which some of its components may impact business as usual. Applications in the areas of supply chain management, business, healthcare, internet of things, privacy, and data management are categorized here. Gracia et al. (2015) A blockchain may be thought of as a distributed database in which the blocks that have been committed are permanently kept in an ordered list. It's not hard to see why this would be ideal for the financial. These companies are investing in this technology. Therefore, blockchain technology is more than just a passing craze. Dixit *et al.* (2021)There are currently over 1900 crypto-currencies in circulation, and this number is expected to continue growing, which is evidence of the relevance of Blockchain. This fast proliferation might very rapidly result in incompatibilities due to the fact that there is now such a wide array of bitcoin uses. Cryptocurrencies are a revolutionary sort of decentralised payment network that operates on a peer-to-peer basis and combines aspects of game theory with encryption technology.

3 PROPOSED METHODOLOGY

Developers use API calls and methods to incorporate them into applications. API Protocols control these calls. Protocols provide API use, instructions, and data kinds [17]. Explore some common API Protocols:

3.1 *Types of API protocols*

- SOAP (Simple Object Access Standard) is an XML-based API protocol that enables users to transmit and receive data over SMTP.
- XML-RPC (Extensible Markup Language Remote Procedural Call) is a protocol that transfers data using a specified XML format, whereas SOAP employs a proprietary XML format.
- JSON-RPC (JavaScript object notation–remote procedure call) is a protocol comparable to XML-RPC since both are remote procedure calls (RPCs).
- Since REST (Representational State Transfer) is a collection of architecture principles for web APIs, there are no established standards.

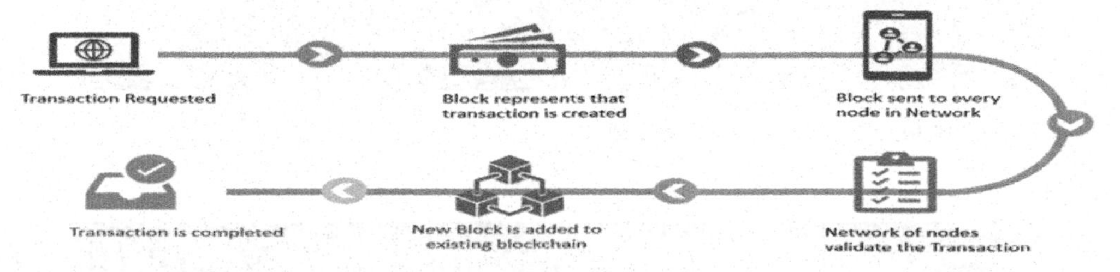

Figure 1. Main stages of API.

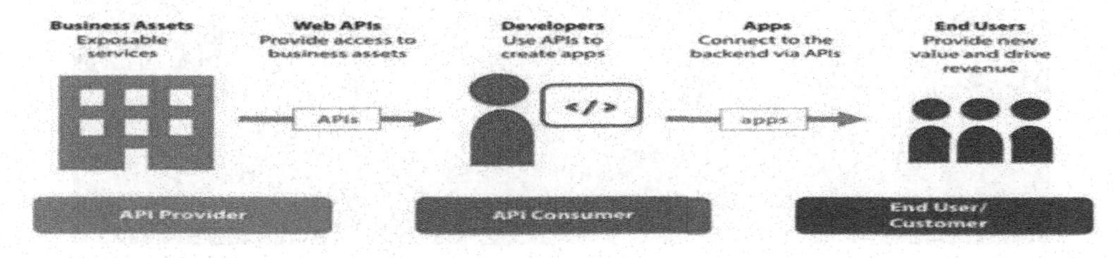

Figure 2. Working of API protocols.

3.2 *API built model*

The solution should safeguard the API provider from underreporting of model use data and cloud vendors attempting to steal the model and run a similar service. Likewise, the system must be able to defend genuine cloud service providers and APIs against dishonest customers who try to cheat by discrediting the service's inputs and outputs. The blockchain architecture has many business benefits. Here are some built-in characteristics: Cryptography, Decentrelization and Transperancy.

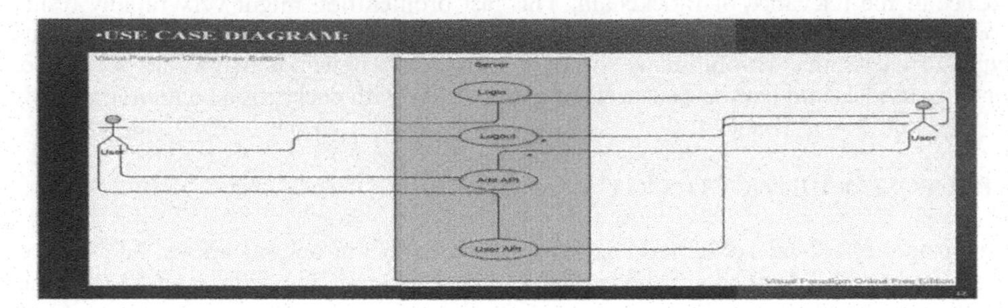

Figure 3. Proposed system diagram.

4 EXPERIMENTAL RESULTS

Figure 4. API for market place – authentication.

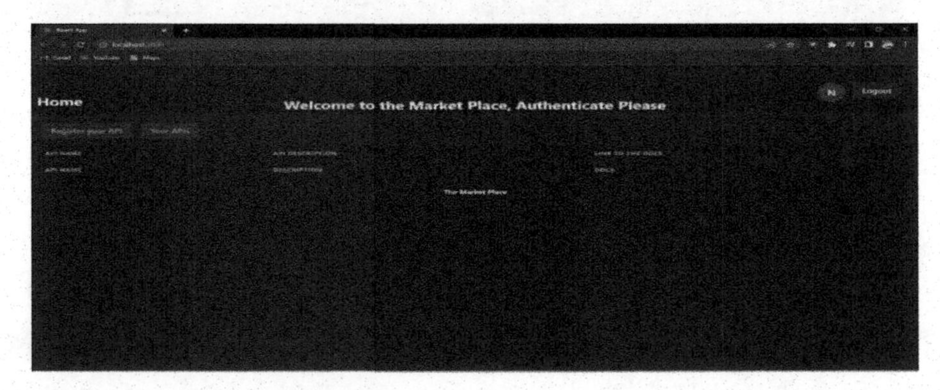

Figure 5. API for market place – vendor.

5 CONCLUSION

This paper presents the front-end design of a blockchain-based decentralised API marketplace, which eliminates the requirement for a trusted central body by facilitating collaboration amongst all parties involved in the ecosystem. The novel aspect of our work is that, unlike existing solutions, which typically rely on a single cloud provider to host and execute the secret API models, our system distributes model execution across multiple cloud providers, ensuring that no single provider learns all of the details of the API models being called. To prevent malicious activity and offer proof of dispute settlement by a reliable arbiter, all parties involved record their activities on the distributed ledger. In the future, we want to apply our system on other blockchain networks (such Ethereum and hyper-ledger fabric) using chain code functions and smart contracts, and we will also investigate potential hazards associated with collusion among players. Experiments contrasting the efficiency of centralised API markets with those of decentralised ones, which must bear the extra expense of recording transactions on the distributed ledger, are a crucial next step in this line of inquiry. Another path would be to analyse how well other methods of protecting the privacy of the hosted API models work in comparison to the suggested system. Finally, more research into the formal verifiability of code and its safe execution (as needed by each cloud vendor in our approach) is essential.

REFERENCES

An, B.; Xiao, M.; Liu, A.; Xu, Y.; Zhang, X.; Li, Q. "Secure Crowdsensed Data Trading Based on Blockchain". *IEEE Trans. Mob. Comput.* 2021, 1.

Avyukt, A.; Ramachandran, G.; Krishnamachari, B. "A Decentralized Review System for Data Marketplaces". In *Proceedings of the 2021 IEEE International Conference on Blockchain and Cryptocurrency (ICBC)*, Sydney, Australia, 3–6 May 2021; pp. 1–9.

Daniela M. and Butoi A., "Data Mining on Romanian Stock Market Using Neural Networks for Price Prediction," *Informatica Economica*, vol. 17, no. 3, 2013.

Dixit, A.; Singh, A.; Rahulamathavan, Y.; Rajarajan, M. FAST DATA, "A Fair, Secure and Trusted Decentralized IIoT Data Marketplace enabled by Blockchain". *IEEE Internet Things J.* 2021.

Garcia D. and Schweitzer F., "Social Signals and Algorithmic Trading of Bitcoin," *Royal Society Open Science*, vol. 2, no. 9, 2015.

Gupta, P.; Dedeoglu, V.; Kanhere, S.S.; Jurdak, R. " TrailChain: Traceability of Data Ownership Across Blockchain-enabled Multiple Marketplaces". *J. Netw. Comput. Appl.* 2022, 203, 103389.

Hasan, H.R.; Salah, K.; Yaqoob, I.; Jayaraman, R.; Pesic, S.; Omar, M. "Trustworthy IoT Data Streaming Using Blockchain and IPFS". *IEEE Access* 2022, 10, 17707–17721.

Kurtulmus A. B. and Daniel K., "Trustless Machine Learning Contracts: Evaluating and Exchanging Machine Learning Models on the Ethereum Blockchain," *CoRR*, vol. abs/1802.10185, 2018.

Nakamoto S., "Bitcoin: A peer-to-peer Electronic Cash System," 2008.

Tang, H.; Qiao, Y.; Yang, F.; Cai, B.; Gao, R. dMOBAs: "A Data Marketplace on Blockchain with Arbitration Using Side-contracts Mechanism". *Comput. Commun.* 2022, 193, 10–22.

Weng J., Weng J., Li M., Zhang Y., and Luo W., "Deepchain: Auditable and Privacy-preserving Deep Learning with Blockchain-based Incentive," *IACR Cryptology ePrint Archive*, vol.2018, p. 679, 2018.

Xu, R.; Chen, Y. Fed-DDM:" A Federated Ledgers based Framework for Hierarchical Decentralized Data Marketplaces". In *Proceedings of the 4th International Workshop on Blockchain Enabled Sustainable Smart Cities*, Athens, Greece, 19–22 July 2021.

Solutions for application of machine learning and blockchain technologies

Artificial Intelligence, Blockchain, Computing and Security – Dagur et al. (Eds)
© 2024 The Author(s), ISBN: 978-1-032-67841-2

Solving tasks of oncoepidemiology prediction using least squares and SVR algorithms

M. Hudayberdiev
Tashkent University of Information Technologies, Tashkent, Uzbekistan

Sh. Ibragimov
Republican Specialized Scientific – Practical Medical Center of Oncology and Radiology, Tashkent, Uzbekistan

N. Alimkulov
Andijan State University, Andijan, Uzbekistan

S. Djanklich
Republican Specialized Scientific – Practical Medical Center of Oncology and Radiology, Tashkent, Uzbekistan

ABSTRACT: The planning of anti-cancer measures in Uzbekistan, as well as the analysis of the survival of oncology patients at the population level, will only be possible after the establishment of a population-based cancer registry in the country in accordance with international standards. This article considers forecasting issues, taking into account that certain works are being carried out in this regard, using available data to compare data after the system is operational. Least squares and SVR methods are used for this analysis.

1 INTRODUCTION

In recent years, an increase in the incidence of malignant tumors has been observed in most countries worldwide, including the Republic of Uzbekistan. According to the state statistics report in the Republic of Uzbekistan for 2020, 21,976 new cases of malignant tumors were detected. Over the past ten years, the number of new cases has increased by 15.6 percent. In the general structure of oncological diseases, breast, stomach, and cervical cancers hold leading positions (Charlton *et al.* 2015).

The global average mortality rate (standardized rate) in 2020 was 100.7 per 100,000 population. The lowest death rate was predicted in Saudi Arabia (51.3 per 100,000 population), and the highest in Moldova (176.2). At the same time, the ratio of death and morbidity (according to standardized indicators) is quite high in all countries, which highlights the seriousness of the problem of radical treatment of malignant tumors. In Uzbekistan, like in most Asian countries, this indicator exceeds 60 percent (Estève *et al.* 1990; Ferlay *et al.* 2020).

2 METHODS

According to Globocan, at least 32,000 new cases should have been detected in the Republic of Uzbekistan in 2020, compared to 21,976 reported in official statistics, indicating that cases of the disease are significantly undercounted (Figure 1). Correct and comprehensive accounting of all cases of malignant tumors in accordance with international requirements can only be achieved by creating a population-based cancer registry (Dickman *et al.* 1999).

DOI: 10.1201/9781032684994-97

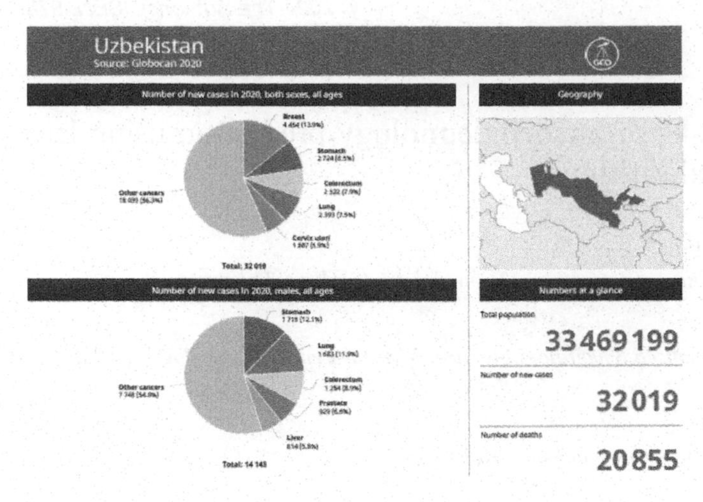

Figure 1. GLOBOCAN data for the Republic of Uzbekistan for 2020.

As the process of developing a population-based cancer registry is progressing rapidly, we are considering the issues of forecasting the data from 2015 to 2021 (see Table 1) to analyze the data before the registry is established (Estève *et al.* 1990; Ferlay *et al.* 2020).

Table 1. Prevalence indicators of malignant tumors in the Republic of Uzbekistan in 2015-2021.

Names of the diseases	ICO-O	Prevalence of MT by years						
		2015	*2016*	*2017*	*2018*	*2019*	*2020*	*2021*
All malignant tumors	C00 – C97	113571	99830	95802	96575	103063	107196	113168
Lips	C00	1094	962	844	748	739	720	728
Mouth and throat	C01, 02, 07, 08, 09, 10, 11, 12, 13, 14	2522	2428	2270	2242	2376	2418	2627
Gums	C03	270	296	299	307	337	349	366
The bottom of the mouth	C04	405	421	420	376	394	403	421
The palate	C05	161	182	171	149	170	184	192
Esophagus	C15	1968	1535	1301	1314	1397	1525	1629
Stomach	C16	5118	4092	3687	3621	3969	4291	4544
Small intestine	C17	261	254	236	224	264	275	317
Colon	C18	2379	2039	2152	2263	2399	2493	2633
Rectosigmoid joint	C19	552	499	492	570	697	785	809
Rectum	C20	2352	1908	1931	2035	2251	2304	2626
Anal canal	C21	262	264	282	287	284	295	306
Liver	C22	1946	1487	1187	1104	1243	1319	1439
Pancreas	C25	522	738	603	746	847	1029	1110
Hiccup	C32	1128	1011	998	1054	1121	1084	1111
Trachea	C33	91	93	76	57	69	77	82
Bronchus and lungs	C34	3247	2306	2074	2286	2540	2545	2752
Bones, joints	C40, 41	5852	3585	3192	2828	2818	2866	2789
Cutaneous melanoma	C43	1116	1022	973	954	996	1039	1045
Other skin neoplasms	C44	11246	9727	9253	8784	8914	8666	8664
Mesothelioma	C45	172	156	151	163	176	215	227
Peripheral and	C47	49	81	92	115	128	134	134

(*continued*)

Table 1. Continued

Names of the diseases	ICO-O	Prevalence of MT by years						
		2015	*2016*	*2017*	*2018*	*2019*	*2020*	*2021*
autonomic nervous system								
Retroperitoneum and peritoneum	C48	220	249	296	378	426	436	466
Connective and other soft tissues	C49	1441	1578	1638	1727	1920	2005	2090
Mammary gland	C50	17980	17517	17737	18692	20140	20861	22443
Cervix	C53	9073	8627	8251	8246	8802	9125	9591
Uterine body	C54	5733	5197	4688	4636	4879	5017	5272
Ovary	C56	4302	4018	3871	3918	4213	4391	4564
Prostate gland	C61	1144	1187	1257	1384	1546	1692	1919
Kidney	C64	2831	2736	2873	3016	3356	3446	3602
Urinary bladder	C67	2289	2096	2044	2132	2315	2361	2518
The brain	C71	3344	2969	3200	3667	4080	4266	4610
Spinal cord	C72	487	388	364	354	379	410	438
Thyroid gland	C73	1237	1229	1234	1312	1397	1459	1595
Hodgkin lymphoma	C81	5047	3681	2998	2988	2974	2919	2974
Non-Hodgkin lymphomas	C82–85	2550	2994	3093	3088	3203	3314	3376
Leukemia	C91,0 – 95,0	2285	1694	1782	1826	2000	2154	2420
Independent (primary) multiple tumors	C97	491	352	71	132	449	166	818
Others	C06, 23, 24, 26-31, 35-39, 42, 46, 51, 52, 55, 57-60, 62, 63, 65, 66, 68-70, 74-80, 86-90, 96	10404	8232	7721	6852	6855	8158	7921

Forecasting is a crucial aspect of machine learning, which includes tasks such as image classification, language translation, processing large amounts of data from sensors, and predicting future values based on current values. Financial forecasting, for instance, is used to analyze and predict commodity prices. Predictive analytics is the process of predicting future events and behavior using historical data. Its power lies in its ability to forecast outcomes and trends before they occur. Predicting future events helps organizations better understand their customers and businesses. Predictive analytics tools include various models and algorithms, with each predictive model designed for a specific purpose.

Predictive modeling is a statistical technique that can forecast future outcomes using historical data and machine learning tools. Predictive models make predictions based on the current situation and past events to show the desired outcome. Predictive models also recalculate future outcomes if new data indicate current changes in the status quo. The predictive analytics model is regularly revised to incorporate changes in underlying data. However, most forecasting models run faster and complete calculations in real time.

The basic forecasting models include the classification model, which is the simplest and easiest to use among other predictive analytics models. This model categorizes data based on what they learn from historical data. Another model is the cluster model, which helps sort data into different groups based on similar types and attributes. This predictive analytics model is the best choice for effective marketing strategies to segment data into other data sets based on common characteristics. The forecast model involves forecasting a metric value to analyze future results, and it helps estimate the numerical value of new data based on historical data. The outlier model, on the other hand, considers anomalous data entries in a given data set to predict future outcomes, unlike the classification and forecasting models that work on historical data.

A time series model – the best choice considering time as an input parameter for forecasting future results. This predictive model works with data points from historical data to develop numerical metrics and predict future trends (Parkin *et al.* (1991)).

A predictive algorithm can be used to help companies gain a competitive advantage in fields such as medicine, finance, marketing, and military operations. Using predictive analytics techniques and machine learning tools, it can apply many predictive algorithms to analyze future results. Below are some of the algorithms most commonly used by predictive analytics models:

- Random forest
- Generalized Linear Model for Binary Classification
- Gradient Boosted Model
- K-means
- Naive Bayes
- Auto-Regressive Integrated Moving Average (ARIMA)
- Long Short-Term Memory (LSTM) Recurrent Neural Network
- Convolutional Neural Network (CNN/ConvNet)
- LSTM and Bivariate LSTM
- You Only Look Once (YOLO)

In addition to the forecasting algorithms presented above, many other algorithms tested through experience can be cited as examples. In this work, we selected the methods "least squares" and "support vector regression" (SVR) for forecasting data in Table 1. Below, you will find information about these methods. (Mariotto *et al.* (2020)).

3 RESULTS AND DISCUSSION

Least Squares Forecasting. When we fit a regression line to a set of points, we assume that there is some unknown linear relationship between Y and X, and that for each unit increase in X, Y increases by a specified amount on average. Our fitted regression line allows us to predict the response of Y for a given value of X,

$$\mu Y|X = \beta_0 + \beta_1 X_1. \tag{1}$$

But for any specific observation, the true value of Y may deviate from the predicted value. Deviations between true and predicted values are called errors or residuals. The better the line fits the data, the smaller the residuals. Intuitively, if we manually fit a line to our data, we are generally trying to find the line that minimizes the model errors. However, when we fit a line through the data, some errors are positive and some are negative. In other words, some actual values will be larger than the predicted value (they will fall above the line), and some actual values will be smaller than the predicted value (they will fall below the line). If we add all the errors, the sum is zero. We square the errors and find the line that minimizes this sum of squared errors,

$$\sum e_t^2 = \sum (Y_i - \overline{Y}_i)^2. \tag{2}$$

This method, the least squares method, finds the intercept and slope coefficient values that minimize the sum of the squared errors. Using this method, we forecast the situation for 2022-2025 based on the data in our Table 1 (Table 2).

It can be seen in the figure below that when forecasted by the "least square method", Mesothelioma, Peripheral and autonomic nervous system, Retroperitoneum and peritoneum, Mammary gland, Prostate gland, Kidney and Brain malignant tumors will increase in the coming years, Bones, Joints, Other skin neoplasms and Hodgkin's lymphoma can be seen to decrease.

SVR forecasting. SVR provides us with the flexibility to determine the acceptable level of error in our model and find a suitable line (or high-dimensional hyperplane) that fits the data. The objective function of SVR is to minimize the coefficients - more precisely, the

Table 2. Forecasting indicators of the spread of malignant tumors in the Republic of Uzbekistan in 2015-2021 using the method of least squares.

Names of the diseases	ICO-O	Data predicted by the method of least squares			
		2022	*2023*	*2024*	*2025*
All malignant tumors	C00 – C97	107141	107884	108626	109368
Lips	C00	593	532	472	412
Mouth and throat	C01, 02, 07, 08, 09, 10, 11, 12, 13, 14	2469	2483	2498	2512
Gums	C03	379	395	410	426
The bottom of the mouth	C04	404	403	403	402
The palate	C05	186	190	193	197
Esophagus	C15	1390	1356	1322	1289
Stomach	C16	4040	4003	3966	3928
Small intestine	C17	296	304	313	321
Colon	C18	2611	2679	2748	2816
Rectosigmoid joint	C19	850	906	961	1016
Rectum	C20	2477	2546	2615	2684
Anal canal	C21	311	318	325	332
Liver	C22	1132	1068	1003	939
Pancreas	C25	1169	1262	1354	1447
Hiccup	C32	1104	1111	1119	1127
Trachea	C33	68	66	64	61
Bronchus and lungs	C34	2458	2439	2420	2400
Bones, joints	C40, 41	1847	1454	1061	668
Cutaneous melanoma	C43	998	993	987	982
Other skin neoplasms	C44	7864	7499	7135	6770
Mesothelioma	C45	224	235	246	257
Peripheral and autonomic nervous system	C47	161	176	190	204
Retroperitoneum and peritoneum	C48	530	575	619	663
Connective and other soft tissues	C49	2212	2322	2432	2542
Mammary gland	C50	22550	23353	24156	24959
Cervix	C53	9259	9370	9481	9592
Uterine body	C54	4839	4783	4728	4672
Ovary	C56	4450	4517	4584	4651
Prostate gland	C61	1965	2094	2224	2353
Kidney	C64	3725	3876	4026	4177
Urinary bladder	C67	2463	2516	2570	2623
The brain	C71	4773	5032	5292	5552
Spinal cord	C72	390	387	384	381
Thyroid gland	C73	1594	1655	1715	1776
Hodgkin lymphoma	C81	2259	1982	1704	1427
Non-Hodgkin's lymphomas	C82–85	3549	3665	3780	3895
Leukemia	C91,0 – 95,0	2243	2299	2354	2409
Independent (primary) multiple tumors	C97	495	530	566	601
Others	C06, 23, 24, 26-31, 35-39, 42, 46, 51, 52, 55, 57-60, 62, 63, 65, 66, 68-70, 74-80, 86-90, 96	6811	6509	6207	5905

norm of the l2-coefficient vector - not the squared error. Instead, the error term is considered in constraints, where we set the absolute error to be less than or equal to a specified limit called ε (epsilon), which is also known as the maximum error. We can adjust epsilon to achieve the desired accuracy of our model. The new objective function and constraints are:

$$\text{Minimization,} \quad MIN\frac{1}{2}w^2,$$

$$\text{Constraints,} \quad |y_i - w_i x_i| \leq \varepsilon.$$

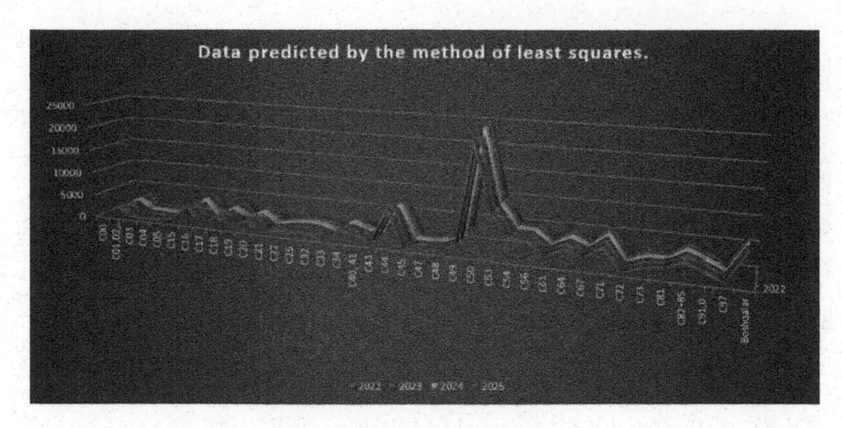

Figure 2. Data predicted by the method of least squares.

We forecast the data in Table 1 using the above SVR method (Table 3) (Ibragimov & Djanklich (2021))(Jönsson *et al.* (2014)).

Table 3. Forecasting indicators of the spread of malignant tumors in the Republic of Uzbekistan in 2015-2021 using the SVR method.

		Data predicted by the SVR method			
Names of the diseases	ICO-O	*2022*	*2023*	*2024*	*2025*
All malignant tumors	C00 – C97	102218	102367	102516	102664
Lips	C00	719	710	701	692
Mouth and throat	C01, 02, 07, 08, 09, 10, 11, 12, 13, 14	2414	2412	2410	2408
Gums	C03	356	366	376	386
The bottom of the mouth	C04	402	402	402	401
The palate	C05	193	197	201	206
Esophagus	C15	1521	1519	1517	1515
Stomach	C16	4104	4106	4108	4110
Small intestine	C17	274	277	281	284
Colon	C18	2414	2419	2424	2429
Rectosigmoid joint	C19	615	624	633	642
Rectum	C20	2266	2271	2276	2281
Anal canal	C21	307	313	319	325
Liver	C22	1315	1313	1311	1309
Pancreas	C25	798	808	818	828
Hiccup	C32	1092	1096	1100	1104
Trachea	C33	73	71	69	67
Bronchus and lungs	C34	2555	2560	2565	2570
Bones, joints	C40, 41	2846	2836	2826	2816
Cutaneous melanoma	C43	1034	1036	1038	1040
Other skin neoplasms	C44	8881	8870	8859	8848
Mesothelioma	C45	191	196	201	206
Peripheral and autonomic nervous system	C47	143	152	161	170
Retroperitoneum and peritoneum	C48	426	438	450	462
Connective and other soft tissues	C49	1775	1787	1799	1811

(*continued*)

Table 3. Continued

Names of the diseases	ICO-O	Data predicted by the SVR method			
		2022	*2023*	*2024*	*2025*
Mammary gland	C50	18740	18752	18764	18776
Cervix	C53	8817	8822	8827	8832
Uterine body	C54	5013	5011	5009	5007
Ovary	C56	4228	4233	4238	4243
Prostate gland	C61	1432	1444	1456	1468
Kidney	C64	3064	3076	3088	3100
Urinary bladder	C67	2334	2341	2347	2354
The brain	C71	3715	3727	3739	3751
Spinal cord	C72	400	402	404	406
Thyroid gland	C73	1360	1372	1384	1396
Hodgkin lymphoma	C81	2968	2962	2956	2950
Non-Hodgkin's lymphomas	C82–85	3148	3159	3170	3181
Leukemia	C91,0 – 95,0	2015	2020	2025	2030
Independent (primary) multiple tumors	C97	352	352	352	352
Others	C06, 23, 24, 26-31, 35-39, 42, 46, 51, 52, 55, 57-60, 62, 63, 65, 66, 68-70, 74-80, 86-90, 96	7918	7915	7912	7909

It can be seen in the picture below that when forecasted by the "SVR method", the malignant tumors of the Pancreas, Peripheral and autonomic nervous system will increase in the coming years, Liver, Bronchus and lungs, Bones, joints, Other skin, we can see a decrease in neoplasms and Hodgkin's lymphoma.

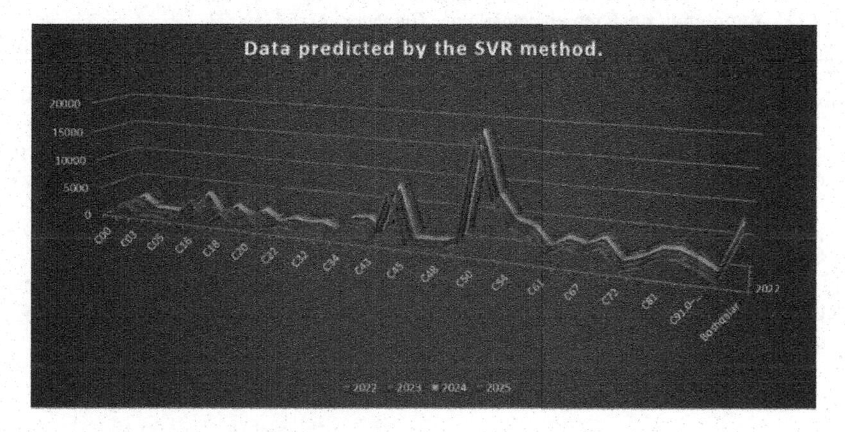

Figure 3. Data predicted by the SVR method.

4 CONCLUSION

According to global forecasting data in 2020, there were at least 32,000 new cases of disease predicted in Uzbekistan, yet official statistics show only 21,976 cases have been identified. This indicates a significant underestimation in the forecasting of disease cases. To meet international requirements for correct accounting and monitoring of all cases of malignant tumor disease, the creation of a population cancer registry is needed. Population-based cancer registries differ from statistical reports in that they provide detailed information about each patient with a malignant tumor. The availability of such data offers two important advantages: it provides an opportunity to fill in and correct information about patients during their follow-up, improving the quality of the entered data, and the possibility of long-term follow-up of patients and the availability of survival data can be intellectually analyzed through the population cancer registry. Currently, work is being carried out to create this registry. This work will compare predicted data with future registry data, and test the extent to which these algorithms are suitable for intelligent analysis of the oncoepidemiological situation.

The scientific results of this research work were carried out within the framework of the practical project No. FZ-202010191 supported by the Ministry of Innovative Development.

REFERENCES

Charlton, M. Schlichting, J. Chioreso, C. Ward, M. Vikas, P. 2015. Challenges of Rural Cancer Care in the United States. *Oncology (Williston Park)*.Vol. 9. 633–640.

Dickman, P.W. Hakulinen, T. Luostarinen, T. Pukkala, E. Sankila, R. Söderman, B. Teppo, L. 1999. Survival of Cancer Patients in Finland 1955–1994. *Acta Oncol.* Vol. 38. 1–10.

Estève, J. Benhamou, E. Croasdale, M. Raymond, L. 1990. Relative Survival and the Estimation of the Net Survival: Elements for Further Discussion. *Statistics in Medicine*. Vol. 9 (5). 529–538.

Ferlay, J. Ervik, M. Lam, F. Colombet, M. Mery, L. Piñeros, M. Znaor, A. Soerjomataram, I. Bray, F. 2020. *Global Cancer Observatory: Cancer Today. Lyon, France: International Agency for Research on Cancer.* Internet resource: Epub ahead of print.

Ho, Chia-Hua. Lin, Chih-Jen. Large-scale Linear Support Vector Regression. *Journal of Machine Learning Research* 13 (2012). 3323–3348

Ibragimov, Sh.N. Djanklich S.M. 2021. The State of Oncological Care for the Population of the Republic of Uzbekistan in 2020. Tashkent: Uzbekistan. 176.

Jönsson, L. Sandin, R. Ekman, M. Ramsberg, J. Charbonneau, C. Huang, X. Jönsson, B. Weinstein, MC. Drummond, M. 2014. Analyzing Overall Survival in Randomized Controlled Trials with Crossover and Implications for Economic Evaluation. *Value Health*. Vol. 17, № 6: 707–713.

Keding, J. 1973. Annotation zur Krebsepidemiologie. *Humburg. Arzteblatt*. 27.

Kennaway, E.L. 1950. The Data Relating to Cancer in the Publications of the General Register Office. *Br. J. Cancer.* Vol 4: 158–172.

Kuhn, Max. Johnson, Kjel. 2013. *Applied Predictive Modeling*. Springer Science - Business Media New York: 615.

Mariotto, A. Capocaccia, R. Verdecchia, A. Micheli, A. Feuer, E.J. Pickle, L. Clegg, L.X. 2020. Projecting SEER Cancer Survival Rates to the US: An Ecological Repression Approach. *Cancer Causes Control* Vol. 13 (2): 101–111.

Parkin, D.M. MacLennan, R. Muirand, C.S. Skeet, R.G. 1991. *Cancer Registration: Principles and Methods.* IARC Scientific Publications No 95. Lyon: IARC. 296.

Plsek, P.E. 1999. Quality Improvement Methods in Clinical Medicine. *Pediatrics*. Vol. 103 (1). 203–214.

Tillyashaykhov M.N. Ibragimov, Sh.N. Djanklich S.M. 2020. *The State of Oncological Care for the Population of the Republic of Uzbekistan in 2020*. Tashkent: Xalk. 176.

Artificial Intelligence, Blockchain, Computing and Security – Dagur et al. (Eds)

Algorithms and methods of using intelligent systems in fire safety

T. Nurmukhamedov
Tashkent State Transport University, Tashkent, Uzbekistan

M. Hudayberdiev
Tashkent University of Information Technologies, Tashkent, Uzbekistan

O. Koraboshev
Digital technologies and Artificial Intelligence Development Research Institute, Tashkent, Uzbekistan

S. Sodikov
Tashkent State Transport University, Tashkent, Uzbekistan

K. Hudayberdiev
University of Science and Technology, Moscow, Russia

ABSTRACT: The development, selection and implementation of intelligent fire safety systems should be based on a detailed analysis of the fire hazard characteristics of technological processes and the proposed means of protection. Currently, there are many ways to ensure fire safety. Including "Smart fire extinguishing" technologies. "Smart firefighting" means using the Internet of Things, big data and other technical means to integrate various elements such as firefighting, social supervision and firefighting management, and Internet firefighting and rescue. This article analyzes various ways to ensure the required level of fire safety, which is the main issue for complex objects.

1 INTRODUCTION

In the context of the rapid development of science and technology, industrial technologies, it becomes necessary to revise the content, methods, means and forms of professional training of a specialist. One of the aspects of solving this problem is the further improvement of the training of specialists in ensuring fire safety in human production activities.

Fire safety requirements for each facility should be based primarily on economic criteria and be effective in ensuring fire safety. Under the concept of "Economically based", it is understood that the implementation of fire safety students assigned to the object has chosen the most optimal solutions for ensuring fire safety for this object, and these solutions do not cause a great burden (excessive costs) from the economic side. The development, selection and introduction of such systems should be carried out based on a detailed analysis of the fire risk characteristics of the technological processes and the intended protection measures. In most cases, monitoring the occurrence of fire in protected buildings ensures the detection of fire at the earliest stage, which helps to reduce the damage caused by fire several times, and accordingly, the use of fire extinguishers is really reduced, is also effective.

Increasing the accuracy of fire safety time and cost forecasting in existing software development estimation models facilitates time and budget management of software projects during development. Software effort estimation models help project managers estimate the project time, cost, and manpower required to develop software. The use of reliable and

DOI: 10.1201/9781032684994-98

accurate software effort estimation models is still an ongoing challenge for software project managers. Several cost estimation models have been proposed and developed to provide accurate estimates and reduce estimation error (Attarzadeh *et al.* 2020).

The intellectual analysis of production technological processes is based on the formation of a combustible space (environment) in production conditions, the identification of sources of ignition and the ways of fire propagation, without evaluating them, conducting fire-technical expertise of project materials, fire-technical inspection of premises, investigation of fires that have occurred and other similar fire control tasks cannot be performed.

2 METHODS

Analysis of fire risk and protection of production technological processes is carried out step by step. It includes the study of technological processes of production, assessment of the fire risk of substances and materials used in the technological process; determining the causes of the formation of a combustible environment in production conditions, the formation of sources of ignition and the ways of fire propagation; includes issues such as fire prevention and fire protection, as well as implementation of organizational measures to ensure fire safety.

After a careful study of the production technological process, technological regulations or design materials, devices containing highly flammable and combustible liquids, combustible substances and materials are determined. It is determined what substances and how much are involved in technological processes; in which a complete list of fire hazard equipment is drawn up and their fire hazard indicators are clearly indicated (Bogdanova & Markov 2020).

The properties of the substances that need to be analyzed for fire risk are determined from the explanatory letter in the technological regulation or the technological part of the project, from regulatory documents, literature and reference books, and if necessary, through experimental or theoretical calculations.

The classification of artificial intelligence methods includes:

- Artificial neural networks.
- Fuzzy logic (fuzzy sets and soft calculations).
- Knowledge-based systems (expert systems).
- Evolutionary modeling (genetic algorithms, multi-agent systems).
- Machine Learning (Data Mining and data analysis and searching for patterns in data warehouses).

The table below shows typical issues regarding the use of intelligent system technologies.

Table 1. Typical tasks for using technologies of intelligent systems.

№	Task type	Definition (addressable tasks)
1	2	3
1.	Interpretation	The process of determining the meaning of data (building descriptions from observed data)
2.	Diagnostics	Fault detection process (in technology and in living organisms)
3.	Tracking (monitoring)	Continuous interpretation of data in real time and signaling of out-of-tolerance parameters
4.	Forecasting	Prediction of future events based on models of the past and present (inference of probable consequences from given situations)
5.	Planning	Designing a plan, that is, a program of action
6.	Design	Building specifications for creating objects with predefined properties
7.	Debugging, repair	Making recommendations for troubleshooting
8.	Training	Diagnostics, interpretation, planning, design
9.	Control	Interpretation, forecast, planning, modeling, optimization of developed solutions, monitoring

Currently, there are many ways to ensure fire safety. Including "Smart fire extinguishing" technologies. "Smart firefighting" means using the Internet of Things, big data and other technical means to integrate various elements such as firefighting, social supervision and firefighting management, and Internet firefighting and rescue. Detection and transmission of information about things Such technologies are organically linked to realize the dynamic, interactive and integrated collection, transmission and processing of real-time fire information, all-round promotion and improvement of the level of fire supervision and management, improved command, dispatch and decision making. The creation and elimination of firefighting and rescue capabilities, and enhance the fire management intelligence, the level of socialization, to meet the actual needs of "Automation" of fire prevention and control, "Intelligence" of the fire and rescue command, "Systematization", and implement intelligent prevention methods and control, intelligent operations, intelligent law enforcement, intelligent management to achieve "early prediction, early detection, early elimination and early rescue" to create a "Firewall" from the city to the home. "Smart fire extinguishing" can use the communication mode based on uploading, sending and distributing the characteristics of large and fast data by the cloud server. The interface part uses several detectors as a core and video surveillance as a support. It uses its own device monitoring tools and installs communication modules. After replacing the traditional camera, once a problem occurs, only monitoring records can be obtained, which can automatically trigger an alarm when a problem occurs, without having to look at the screen for 24 hours, and scientifically solve the video surveillance problem (Khudaiberdiev *et al.* 2016; Koraboshev & Alimkulov 2022).

The intelligent fire protection system includes the creation of a remote monitoring system for the Internet of Things in the city; creation of a team platform in real time based on "Big DATA" and drawing; construction of a high-rise residential intelligent fire warning system; building a digital plan. Compilation and management of the application platform; creation of all objects in the "Smart" social fire safety management system. Through sensory analysis and data exchange between systems, the purpose of an intelligent fire protection system is "to build an intelligent fire protection system that provides security for a better life for people".

Based on the Smart Fire IoT big data analysis system and related supporting tools, relying on the technical advantages of big data, cloud computing technology and the Fire IoT cloud platform, it oversees governments and industries at all levels. The department ensures accurate analysis of firefighting big data and decision-making services, provides end-to-end firefighting maintenance services to fire service companies, and provides firefighting equipment management services to network divisions.

For complex objects, the main issue is to ensure the required level of fire safety, and now "keeping it at the required level" is also being considered. Currently, it is not enough to know the main causes of fire safety violations and eliminate the "symptoms' of its consequences. By studying the nature and causes of fire safety for complex objects, a sequence of measures is determined to maintain its level at the required level, that is, measures such as predictive "modeling" and predicting its dynamics, rapid introduction of necessary changes in the event of adverse symptoms are carried out.

3 RESULTS AND DISCUSSION

Considering the relevance of this issue, theoretical aspects, methods and methods of solving it were recommended by the authors and analytical researchers in the cited literature. This is a unique model that serves as a unique "Indicator" that serves to show the decrease in the level of fire safety according to a set of risks similar to the safety indicators of complex objects.

A) The fire hazard properties of the complex object and its fire safety system (fire safety system) were studied and expressed in the form of functions $U = U(t)$ and $V = V(t)$. The first of them develops, and the second stops the development of the output characteristic $Y = Y(t)$ (for example, it can be $U = at$ $V = bt$);
B) A set of factors $X_1, X_2, \ldots, X_m$ that affect $Y(t)$, but not considering $U$and V is defined;

C) By solving the evolutionary equation of the initial function, its solution $Y = Y_P(t)$ is found, and according to the experimental results, the values of the coefficients included in the $Y(t)$ expression are determined;
D) For one-time instant $t = t_1$, $Y'''' = Y_{cp}(t_1) - Y_p(t_1)$ is the difference, where $Y_{CP}(t)$ is the time dependence of the actual output characteristic curve. The regression function $Y'''' = Y'' + \beta_j(X_j - X_{0j}'')$ is determined for Y;
E) By determining the values of the output characteristic $Y(t)$ for other subsequent time instants, they are compared with the permissible limits (research and practical activity data according to normative documents).

If an unfavorable assessment occurs, the fire turns into a large fire, then the question of management is considered due to the change of fire safety X_j factors or the implementation of other organizational and technical measures that reduce the level of fire risk. According to the proposed model formation algorithm, the type of technical system that satisfies all the requirements is selected. Taking into account the adaptation to the conditions of use, it will be possible to choose its concrete type. The procedure of the mentioned calculations allows to prepare the organizational basis for the formation of the model of scientific and technical provision of fire safety for the types of complex objects that are being researched or combined, or the direction of its development. The figure below shows the scheme of fire safety engineering analysis (Matveev & Bogdanova 2018; Mavlyankariev & Khatamov 2018).

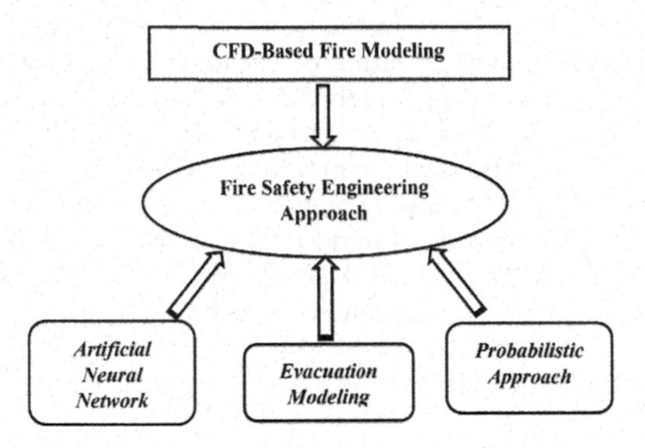

Figure 1. Fire safety engineering analysis scheme.

The analysis performed shows that at present there are a large number of different forecasting methods. However, most of the algorithms underlying these methods are based on the modeling of empirical dependencies using the apparatus of mathematical statistics.

The complexity of the practical application of most of the models is due to the need to conduct important experimental studies to determine the parameters of the equations in relation to the specific characteristics of certain conditions. In addition, forecasting using well-known mathematical models is often difficult or impractical due to various limitations, such as the amount of initial data available, the forecast period, or the set of internal parameters. As a result, there is a need for other tools that allow working with large amounts of data (Big DATA) and obtaining a reliable and accurate forecast. Neural network forecasting methods are a very promising tool for detecting hidden images and creating mathematical models to predict emergency situations based on them.

The purpose of the research is to justify the prospects of forecasting the risk of emergencies and other events in various objects based on neural networks. A neural network model can be developed using the following approaches:

- building a Hopfield network as an implementation of some optimization algorithm;
- use of multi-layer perceptron, which is trained by optimization methods based on the training sample formed from the known states of the object.

This approach is preferable to other approaches because it can predict the change of the object's state in real time, which is important for decision-making in emergency situations. The first step in the construction of a neural network for modeling emergency situations in objects is the preparation and analysis of the initial data set, that is, the formation of a training sample. A training sample is a set of values of input and output variables that describe the state of an object. When the object is characterized by a large volume, it is important to correctly define the set of input and output variables for data acquisition. Input (predictable) variables should be sufficient to describe the dynamics of changes in the state of the object, assess the situation and make control decisions. In the number of input and output variables, only those that have a significant effect on the change in the output variables should be included. The numerical data of the training sample is normalized in the range $0 \leq \overline{X} \leq 1$ according to the following formula:

$$\overline{X} = \frac{X - X_{\min}}{X_{\max} - X_{\min}} \quad (1)$$

Non-numeric data is converted to the numeric form of variables of type (0;1) or ANSWER = (Yes; No).

The number of observations required to form a training sample is established empirically.

At the second stage of building a neural network, the choice of its architecture is made. There are various types of architecture such as multilayer perceptron, Kohonen network, adaptive resonance network, recirculation networks, counter propagation networks, etc.

Figure 1 shows an example of a possible network architecture for predicting the state of an object. The input data here are the values of the parameter at the previous moments of time $[t - n;t]$ and the output data are the values of the parameter X at the time moment $t + 1$.

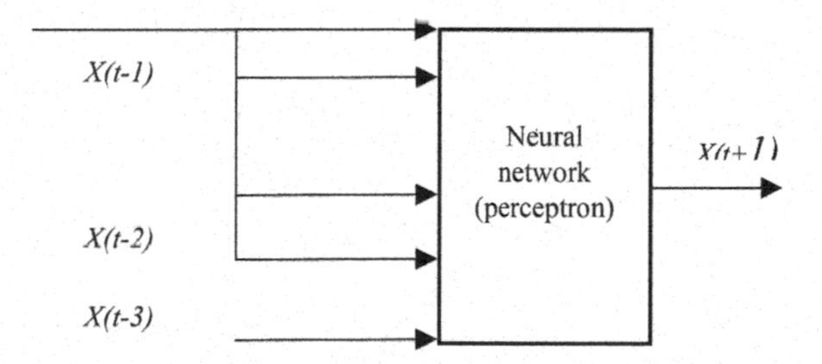

Figure 2. Neural network architecture for predicting object parameters in emergency situations.

At the third stage, the structure of the neural network is determined. In the perceptron, the neurons are organized into layers, and the elements of each layer are connected only with the neurons of the previous layer, so that information is transmitted from previous layers in the network, as well as the number of neurons in each layer affects the network's ability to solve certain problems. A three-layer perceptron with a single hidden layer, in which the minimum number of neurons is determined by the formula, has become widespread in practice:

$$n \geq \frac{M - L + 1}{L + 1} \quad (2)$$

where, L- is the number of neurons in the input layer, M- is a parameter that determines the size of the training sample.

The next stage of modeling is training the neural network. Training can be carried out according to several different algorithms. As practice shows, to solve problems of assessing and predicting the risks of man-made emergencies, the algorithm of supervised learning (training with a teacher) is most often used. This algorithm assumes that all weights of the neural network are subject to change based on training samples that contain the values of input and output (predicted) parameters. And, finally, the final step in building a neural network model for predicting emergencies is its testing. To do this, it is necessary to form a sample that was not used when training the network. Known values of the output parameters must be included in this sample, and these parameters are compared with the values produced by the neural network. If the standard deviation of the known values is less than a given level, then the developed model will be considered adequate and it is advisable to use it to solve problems of assessing and predicting the risks of emergency situations.

In addition, this issue has a unified status, because it is related to many organizations and decisions. As an example, it was considered through the information and technical support recognized in ensuring the stable operation of complex objects. Setting the unified issue: scientific and technical provision of fire safety for complex objects ensures the adequacy of a set of real measurable characteristics *A* with a set of parameters *B* that determine its required normative level. Even with high-precision measurement, it is not guaranteed that *A* and *B* are protected from various risks, given the weak interconnection of the control scheme (Mavlyankariev & Khatamov 2019; Ostroukh *et al.* 2015).

We denote the set of standards of all requirements by *B*:

$B = (B_1, B_2, ..., B_N)$is a set of B_i, *i* is a concrete demand of the *i*-th form.

$$A = (A_1, A_2, ..., A_N) \tag{3}$$

where, is the set of measurable characteristics corresponding to $A_i - B_i$.

$B_i = [b^i_{kl}] - b^i_{kl}$ parameter matrix, $A_i = [a^i_{kl}] - a^i_{kl}$ parameter matrix.

$$r = (A_i, B_i) = \sum\nolimits_{j=1}^{n} *a_j * r_j \tag{4}$$

where r_j- is the correlation coefficient between the $j\text{–}m$ columns of the A_i matrix and the $j\text{–}m$ columns of the B_i matrix.

$j = 1, n$, if $\theta = \sum\nolimits_{j=1}^{n} a_j$, $\beta = a_j/\theta$, then $\sum\nolimits_{j=1}^{n} \beta_j = 1$ and the correlation coefficient $r(A_i, B_i)$ between A_i and B_i satisfies the following conditions.

$$r_{(1)} = r'(A_i, B_i) = \frac{r(A_i, B_i)}{\theta} = \sum\nolimits_{j=1}^{n} \beta_j r_j < r_{(n)} \tag{5}$$

where $r_{(1)}, r_{(n)} - r_j$ are the smaller and larger correlation coefficients, respectively.

If $0 \leq |r(A_i, B_i)| \leq 1$, then to ideal adequacy $|r(A_i, B_i)| = 1$ is achieved.

$$r^2_{1,2,...,n} > \in_1, ., r^2_{1,2,...,n} > \in_1 \tag{6}$$

All $r_{ij} \to 0$ have $H \to 0$. The closer the value of *H* is to one, the more $(A_1, A_2, ..., A_n)$ the characteristics $(B_1, B_2, ..., B_n)$ are adequate to the parameters.

$$H = |N_{n-1}| > \in_0 \tag{7}$$

where: $\in_0$ – is the limiting value of *H*, which is determined experimentally based on the use of advanced technology.

Classification algorithms are widely used to identify symbols. A classification algorithm is a supervised learning method used to identify a new class of observations based on training data. In classification, a program learns from a given set of data or observations and then classifies the new observation into a series of classes or groups. For example, Yes or No, 0 or 1, Spam or Not Spam, cat or dog, etc. Classification algorithms can be better understood with the help of the diagram below. In the diagram below, there are two classes, Class A and Class B. These classes have properties similar to each other and different from other classes (Ostroukh *et al.* 2015).

Since each popular algorithm has different methods, it is necessary to use computer technology to implement it. Inevitably, questions arise about the convergence of each of them, the speed of convergence. Various popular algorithms, as before, perform data pre-processing. The effective selection and application of popular algorithms depends on the correct connection of the formalized problem, the essence of the method of its solution and the expected results.

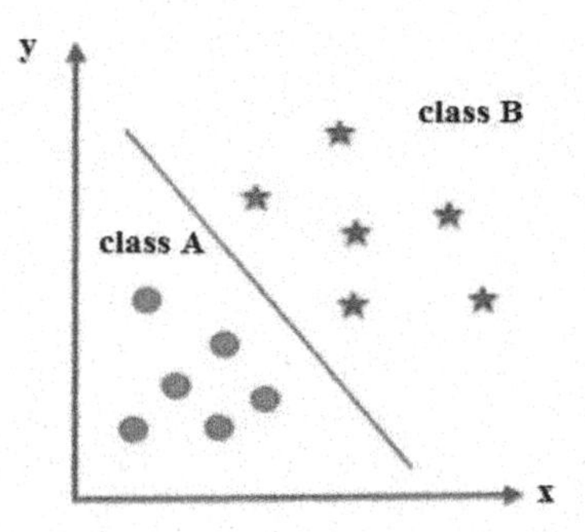

Currently, the most popular classification algorithms are used:

- Naive Bayes.
- K-Nearest Neighbors.
- Support Vector Machines.
- Least Squares Linear Regression.
- Least Squares Multiple Linear Regression.

Using these popular classification algorithms, we determine the predicate values listed in Table 2 by feeding different training samples to each classification algorithm.

Table 2. Results of predicate values of popular classification algorithms.

№	The name of the classification algorithm	Study selection	New object (predicate)	Result
1	Naive Bayes	a= [[5, 1, 1], b=[1, 5, 1], c=[1, 1, 5]];	[3,1,1]	a
2	K-Nearest Neighbors	a= [[1, 3], [1, 4], [2, 4]]b= [[3, 1], [4, 1], [4, 2]]	[3, 2]	b
3	Support Vector Machines	60=3.1, 61=3.6, 62=3.8, 63=4, 65=4.1	64	4.03334
4	Least Squares Linear Regression	60=3.1, 61=3.6, 62=3.8, 63=4, 65=4.1	64	4.05811
5	Least Squares Multiple Linear Regression	[73676, 1996]=2000, [77006, 1998]=2750, [10565, 2000]=15500, [146088, 1995]=960, [15000, 2001] =4400, [65940, 2000]=8800, [9300, 2000]=7100, [93739, 1996]=2550, [153260, 1994]=1025, [17764, 2002]=5900, [57000, 1998]=4600, [15000, 2000]=4400	[60000, 1996]	4094.82993

Thus, the recommended strategy for increasing the level of scientific and technical provision of fire safety for complex objects allows to optimize the provision of the required level of fire safety by choosing multi-functional special equipment for complex objects and to use it in the future to create the "Knowledge expert base". The perspective of the vector-directions of the development model of the scientific and technical provision of fire safety, taking into account the modern requirements for multi-functional special equipment, was studied: resource saving; import substitution; intellectually integrated construction; is a convenient extension of the function. Thus, the recommended strategy for increasing the level of scientific and technical provision of fire safety for complex objects allows to optimize the provision of the required level of fire safety by choosing multi-functional special equipment for complex objects and to use it in the future to create the "Knowledge expert base". The perspective of the vector-directions of the development model of the scientific and technical provision of fire safety, taking into account the modern requirements for multi-functional special equipment, was studied: resource saving; import substitution; intellectually integrated construction; is a convenient extension of the function (Snityuk 2008; Volkov 2007).

4 CONCLUSION

In conclusion, it is proposed to develop a platform for the formation of an analytical-information base supporting fire safety of complex facilities and an automated monitoring system hardware-software complex supporting decisions. The proposed support format is a means of increasing the efficiency of production resources of objects, prevention of various damages, training of specialists and practical use of multi-functional special equipment and techniques.

REFERENCES

Attarzadeh I., Mehranzadeh A., Barati, A. 2012. Proposing an Enhanced Artificial Neural Network Prediction Model to Improve the Accuracy in Software Effort Estimation // *In 2012 Fourth International Conference on Computational Intelligence,* Communication Systems and Networks. pg no: 167–172.

Bogdanova E. M., Markov I. S. 2020. Forecasting Emergencies and Other Incidents in Transport: A Neural Network Approach // *Collection of Materials of the IX International Scientific Seminar "Fire Safety of Business Facilities"*. – Russia. May 22, 2020. pg no: 107–115.

Khudaiberdiev M.Kh., Khamroev A.Kh., Mamieva D.Z. 2016. Algorithm for Grading in the Formation of Training and Control Selection of Objects. *Republican Scientific and Technical Conference: "Problems of Information and Telecommunication Technologies"* - Tashkent, March 10–11, 2016. pg no: 185–187.

Koraboshev O.Z., Alimkulov N.M. 2022. Models and Algorithms of Intelligent Decision Support Systems for Technical Processes // European Journal of Science Archives Conferences Series/ Konferenzreihe der Europäischen Zeitschrift für Wissenschaftsarchive. Aachener, Germany 2022. pg no: 41–44. – DOI prefix: 10.5281/zenodo.5889885.

Kumar, A., Mehra, P. S., Gupta, G., & Jamshed, A. (2012). Modified Block Playfair Cipher Using Random Shift Key Generation. *International Journal of Computer Applications*, 58(5).

Kumar, A., Yadav, R. S., & Ranvijay, A. J. (2011). Fault Tolerance in Real Time Distributed System. *International Journal on Computer Science and Engineering*, 3(2), 933–939.

Matveev A.V., Bogdanova E.M. 2018. Classification of Emergency Forecasting Methods // *National Security and Strategic Planning*. - 2018. - No. 4 (24). pg no: 301–308.

Mavlyankariev B.A., Khatamov B.B. 2018. *Scenario Analysis of Contradictions and Crisis Situations in Ensuring Fire Safety of Complex Objects.* // Bulletin of the Military Technical Institute. T., №1, 2018, p. 133–136, №32.

Pavlov, D. I. 2020. General Principles and Approaches to the Selection and Use of Smoke Fire Detectors for Protection Facilities / D. I. Pavlov, S. A. Borozdin, G. A. Gitzovich // *Supervisory Activity and Forensic Examination in the Security System*. - 2020. - No. 2. -pp. 30–42.

Snityuk V.E. 2008. FORECASTING. 2008. *Models, Methods, Algorithms. / Study Guide* // Kiev -2008. pg no: 148–180.

Tripathi, A. M., Singh, A. K., & Kumar, A. (2012). Information and Communication Technology for rural Development. *International Journal on Computer Science and Engineering*, 4(5), 824.

Volkov V.S. 2007. *Models, Methods and Algorithms for Optimizing the Diagnostics of Devices.: Textbook //* Penza: PSU Publishing House,2007. pg no: 406–410.

Artificial Intelligence, Blockchain, Computing and Security – Dagur et al. (Eds)
© 2024 The Author(s), ISBN: 978-1-032-67841-2

Medical diagnostics of chronic diseases based on Z-numbers

H.A. Primova*
Samarkand Branch of TUIT Named after Muhammad al-Khwarizmi, Samarkand, Uzbekistan

S.S. Nabiyeva*
Research Institute for the Development of Digital Technologies and Artificial Intelligence, Tashkent, Uzbekistan

Shavkat Urokov*
Head of Department of "Mathematics and Information Technologies" of Fiskal Institute Under the State Tax Committee of Republic of Uzbekistan

ABSTRACT: Dealing with ambiguity in medical diagnoses is a free problem. There article, it is important to create a fuzzy model that determines the chronic complications of the disease, and not the main symptoms of the disease in the diagnosis of diseases based on fuzzy logic. It is used to apply new methods to detect chronic complications of certain diseases with symptoms of the disease.

1 INTRODUCTION

The reason for the development of chronic tonsillitis is a change in the body's reactivity to microbes. Various complications of chronic tonsillitis force not only ENT doctors to deal with this problem, but also other medical institutions. At present, since the method of treatment of chronic tonsillitis is an unsolved problem, modeling of the complications of the disease in patients with this disease is an urgent issue.

An important factor determining the clinical course of chronic tonsillitis is a complete blood count (EChT, ASLO), a rheumatological test, a general urinalysis; immunological (determination of the concentration of IgA, IgM and IgG in blood serum); biochemical analysis (detection of catalase, superoxide dismutase in saliva exploit a biosensor, biochemical analysis of ILI in blood) and the resulting parameters – nephrological complications, cardiac complications, rheumatological complications, neurological complications.

In this paper, fuzzy logic can be useful in diagnosing diseases, but it is used to create a fuzzy model that defines chronic complications of the disease, and not the main indications of the disease, with the indications of the disease, to apply new methods for identifying chronic complications of a particular disease (crucial signs). They are also experts in other markers, namely immunological (determination of the concentration of IgA, IgM and IgG in blood serum) and biochemical analyzes (detection of catalase, superoxide dismutase in saliva exploit a biosensor; biochemical analysis of ILI in blood). define the system, we will consider the following steps:

1. An input-output set connecting a normalized input to an output for receiving.
2. If industrial, then fuzzy logic based on input-output.
3. Create a vague rule base.
4. Production system established on the rules of fuzzy logic.

*Corresponding Authors: primova@samtuit.uz, sevar0887@mail.ru and sh.urokov@mail.ru

DOI: 10.1201/9781032684994-99

Several scientists have conducted research on medical diagnostics (Primova 2020, 2021a, 2022b). Smith *et al.* (Nishi Shahnaj Haider 2014) examined the relationship between opiate use disorder indicative, non-medical opium use, a prognosis of post-traumatic stress disorder and the average monthly frequency of non-medical opiate use. (Primova 2020) reviewed recent developments in mobile detection for medical indicatives. Due to the uncertainty of information in medicine, mathematical typicals (Kang 2016) that can work with fuzzy and uncertain data were analyzed to solve medical indicative problems. In (Primova 2021b), an extended QUALIFLEX method was proposed for solving a medical ruling problem (Nishi Shahnaj Haider 2014) in the context of type 2 interval fuzzy sets. An approach combining intuitive trapezoidal fuzzy numbers with an inclusion measure for medical indicative was proposed by Wang (Wang 2016). Yang (Zhang 2017) developed a linear regression exemplar for medical prognosis established on fuzzy trapezoidal numbers to evaluate which parameters in the distance data vector should be considered false. (Nishi Shahnaj Haider 2014) proposed a number of similarity measures that can be used in medical indicatives. (Primova 2022a) presented a weighted affinity measure for medical prognosis using intuitive fuzzy mild sets.

For many years, fuzzy inference systems have been used in problems related to the need to obtain an inference based on input data under conditions of uncertainty (ambiguity). In addition, real information is imperfect in all of the above aspects, and natural language (NL) is usually used to describe it, and this information is partially dependable, and the degree of dependability is given by a phrase in NL. Whether the discretionary estimate of the values is accurate or aggregated (rough) depends on the value estimate and the reliance on the sources of data about these collected estimates; in the end, everything depends on the knowledge of a person, his inner feelings and experience. Thus, the fuzziness (uncertainty) of information, on the one hand, and its partial reliability are closely related to each other.

Z-numbers do not represent the first attempt to describe the actual uncertainty of information, which is considered extremely complex and sometimes difficult to describe in fairly simple monetary conditions. In particular, interval estimates and fuzzy numbers have long been used for such descriptions. If we are talking about the functions of the first type (Type-1 Membership Functions), then the uncertainty of estimates inherent in them is simply ignored. The first attempt to take into account such intervals of uncertainty is considered to be the theory of the second type (Membership functions of type 2), which was proposed quite a long time ago (Zadeh 2011). Type 3, in contrast to fuzzy sets, describes the reliability expressed in the Z-number of NL, and the information uncertainty is more formulated, which is considered a formal and complete presentation of the Z-number in NL (such "completeness" Z-refers to the complexity of number processing). Working with Z-data requires new theories, new approaches and methods for computing with Z-numbers (Aliev 2014; Bakar 2015).

However, when using Z-scores for decision making [13], we must consider the limitations and reliability of the estimate. Kang *et al.* (2012a) intended way for converting the Z-number to fuzzy numbers, where the second element is refined to an exact number. To address linguistic decision taking problems (Kang 2012b) Kang *et al.* A multicriteria decision-taking method with Z-numbers is presented in (Zadeh 2011) founded on the method presented in (Primova 2021b). Bakar (Zhang 2017) presented a method of multilevel Z-number classification, which includes two levels of Z-number transformation and classification of fuzzy numbers. To achieve more efficient and dependable results in medical diagnostics, we propose a new way founded on this ruling method, where Z-numbers are used to portray medical prognostic information.

2 INITIAL DATA

The calculation with this approach is considered to be very complex and includes several variational problems. Some simplifications have been proposed to overcome these difficulties. In [9], the authors put forward the fuzzy expectation of fuzzy sets and (Zadeh 2011) approaches to transforming a Z-number into a fuzzy number.

According to the proposed approach, the defuzzification process is performed from the second component $\widetilde{B}$to α, which is the exact value, which is then multiplied by $\widetilde{A}$. Nevertheless, it can be noted that the replacement of Z-numbers by classical fuzzy numbers leads to the loss of the original information [5, 6].

It is proposed to consider Z-score $(X, \widetilde{A}, \widetilde{B})$in conditions of a likelihood distribution over the probability distribution underlying z-scores $Z = (\widetilde{A}, \widetilde{B})$. (Nishi Shahnaj Haider 2014.) is devoted to the problems of calculating continuous Z-numbers and a number of important practical problems in the field of control and ruling making.

The proposed research work is based on the use of normal density functions to model random variables. But it can be noted that the calculation of IF–THEN rules described by Z-numbers is considered an important task.

3 CALCULATION BASED ON THE UNION OF Z-NUMBERS.

We use the T-norm and T-conorm operations on diverse Z-numbers to perform a computational experiment based on the union of Z-numbers. $Z_1 = (A_1, B_1)$, $Z_2 = (A_2, B_2)$ taking into account diverse Z-numbers, the operation $Z_{t\text{-}norm}(z_1,z_2)$ of the T-norm is viewed as follows (Primova 2021b; Zadeh 2011):

$$\begin{aligned} Z_{T-norm}(Z_1, Z_2) &= T((A_1, B_1), (A_2, B_2)) = (T(A_1, A_2), T(B_1, B_2)) \\ Z^{+}_{T-norm}(Z_1, Z_2) &= (T(A_1, A_2), T(R_1, R_2)) \end{aligned} \tag{1}$$

where

$$\begin{aligned} T(A_1, A_2) &= \bigcup_{\alpha \in (0,1]} T(A_1^{\alpha}, A_2^{\alpha});\ T(A_1^{\alpha}, A_2^{\alpha}) = \{T(x, y) | x \in A_1^{\alpha}, y \in A_2^{\alpha}\} \\ T(R_1, R_2) &= R_1 \wedge R_2. \end{aligned}$$

given $Z_1 = (A_1, B_1)$, $Z_2 = (A_2, B_2)$, diverse Z-numbers, the $Z_{t\text{-}norm}(z_1,z_2)$ T-norm operation is defined as follows:

$$\begin{aligned} Z_{T-conorm}(Z_1, Z_2) &= S((A_1, B_1), (A_2, B_2)) = (S(A_1, A_2), S(B_1, B_2)) \\ Z^{+}_{T-norm}(Z_1, Z_2) &= (T(A_1, A_2), T(R_1, R_2)) \end{aligned} \tag{2}$$

where

$$S(A_1, A_2) = \bigcup_{\alpha \in (0,1]} S(A_1^{\alpha}, A_2^{\alpha});\ S(A_1^{\alpha}, A_2^{\alpha}) = \{S(x, y) | x \in A_1^{\alpha}, y \in A_2^{\alpha}\},$$

$$S(R_1, R_2) = R_1 \vee R_2.$$

Step 1. Substitution of each pair (Z_{Gi}, Z_{Hi})into the Z-number Z_{GHi} based on the operation of the T-norm $Z_{T\text{-}norm}$ according to (1) (Nishi Shahnaj Haider 2014; Zadeh 2011)

$$Z_{H_iG_i} = Z_{T-norm}(Z_{H_i}, Z_{G_i}) = T((A_{H_i}, B_{H_i}), (A_{G_i}, B_{G_i}))$$

Step 2. According to (1), performing the operation T-conorma $Z_{T\text{-}conorm}$ of the operation of combining Z_{HGi} and Z_{agg} [5, 6, 9]:

$$Z_{agg} = Z_{T-conorm}(Z_{g_1h_1}, Z_{g_2h_2}, \ldots, Z_{g_nh_n}) = S((A_{g_1h_1}, B_{g_1h_1}), (A_{g_2h_2}, B_{g_2h_2}), \ldots, (A_{g_nh_n}, B_{g_nh_n}))$$

4 APPLICATION OF THE PROPOSED METHOD IN MEDICAL DIAGNOSTICS

In accordance with the purpose of this study, it was necessary to study the immunogram in children with various forms of chronic tonsillitis. The amount of immunoglobulins IgA, IgG, IgM was determined in children of all groups before treatment, after treatment and after 6 months.

In the study of immunoglobulin parameters before treatment, a decrease in all immunoglobulin parameters was observed, that is, IgA-0.81±0.72, IgM-0.82±0.66, IgG-

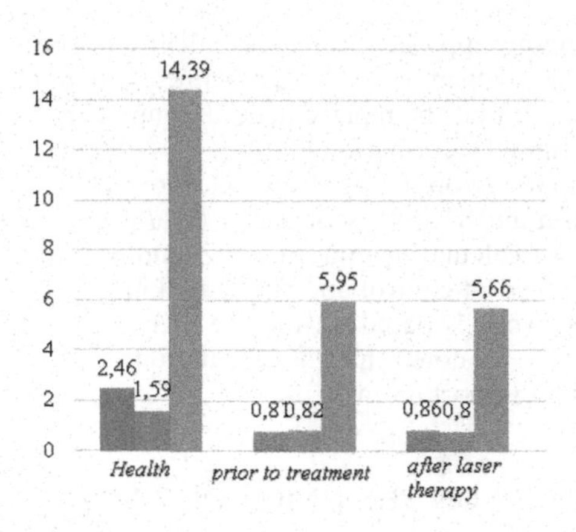

Figure 1. The level of immunoglobulins IgA, IgM, IgG.

5.95±1.92. After local laser therapy, the immunoglobulin parameters were: IgA-0.86±0.74, IgM-0.80±0.62, IgG-5.66±1.99.

In this section, we will show the application of the proposed method in medical diagnostics.

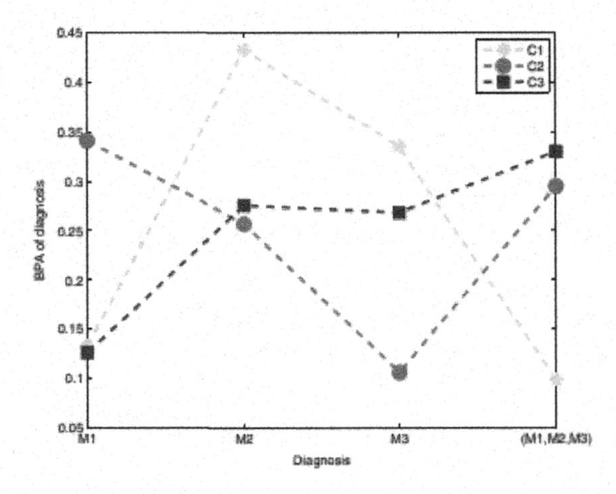

Figure 2. The visualization of risk evaluation.

Recently, the number of children suffering from chronic tonsillitis has increased; this condition is a very urgent problem not only for otorhinolaryngology and pediatrics, but also for other branches of medicine. To date, the problem of chronic tonsillitis in Uzbekistan remains relevant due to the high incidence among children. The incidence of chronic tonsillitis in children is 12–15%, in adults – 4–10%. The prevalence of long-term and severe chronic tonsillitis, their complications and disability in children and young people require further development of treatment tactics for these patients.

5 CONCLUSION

Summarizing the results of studies on the cure of sick children, we can say that among all methods of treatment, intravenous laser blood irradiation gives the highest result, especially in the treatment of complicated forms of chronic tonsillitis.

Various methods have been introduced to solve diagnostic and treatment problems, and intravenous laser blood irradiation can be chosen as the method of choice in the treatment of children with complications of chronic tonsillitis.

Serious emergency health response have led to the study of medical diagnostics. Diverse methods have been presented to solve medical diagnostic problems. Fuzzy numbers, which can deal with fuzzy and ambiguous information due to the uncertainty and ambiguity associated with medicine, provide a suitable method to study these questions.

The benefits of the proposed method are summarized below.

1) Fuzzy numbers are widely used in decision taking because they can model and explain fuzzy data.
2) The variance in significance between the two components of the Z-score, score A and reliability score B, requires further study. Intuitively, the importance of these two components should be different.

For the complete implementation of Z-information in the inference mechanism that causes the least loss of information consisting of Z-numbers, it is possible to define the growth of an algorithm for using diverse arithmetic of Z-numbers in fuzzy inference systems as a guide for future work.

REFERENCES

Bakar A S A, Gegov A. 2015. Multi-layer Decision Methodology for Ranking Z-numbers. *Int J Comput Intell Syst* 8:395–406.

Kang B, Hu Y, Deng Y, Zhou D. 2016. A New Methodology of Multicriteria Decision-making in Supplier Selection Based on Z-numbers. *Mathematical Problems in Engineering*. doi:10.1155/2016/8475987

Kang B, Wei D, Li Y, Deng Y .2012b. A Method of Converting Znumber to Classical Fuzzy Number. *J Inf Comput Sci* 9(3):703–709

Kang B, Wei D, Li Y, Deng Y. 2012a. Decision Making Using Z-numbers Under Uncertain Environment. *J Comput Inf Syst* 8(7):2807–2814

Nishi Shahnaj Haider, Sibu Thomas. 2014. Medical Applications of Laser Instruments, *Journal of Engineering Research and Applications*, Vol. 4, No. 6, June, pp.154–160.

Primova H. A., Iskandarova F. N. and M. Gaybulov Q. 2021b. Accounting Experience Between Fuzzy Integraland Z-numbers. Advances in Intelligent Systems and Computing 1323, 11th World Conference "Intelligent System for Industrial Automation" (WCIS-2020), Springer, pp.40–46, https://doi.org/10.1007/978-3-030-68004-6_6.

Primova H. A., Sakiyev T. R., and Nabiyeva S. S. 2020. Development of Medical Information Systems, *Journal of Physics: Conference Series*, 1441(1), 012160 doi: https://doi.org/10.1088/1742-6596/1441/1/012160.

Primova H.A, Vaydullayeva M.F., Nabiyeva S.S. 2022b. The Role of the Patronage Mobile Application in the Evaluation and Analysis of the Activity of Medical Information Systems . *International Conference on Information Science and Communications Technologies: Applications, Trends and Opportunities* September 28–30, 2022, http://www.icisct2022.org/.

Primova H.A., Mukhamedieva D.T., Safarova L. 2022a. Application of Algorithm of Fuzzy Rule Conclusions in Determination of Animal's Diseases. *Journal of Physics: Conference Series 2224 012007*, IOP Publishing, doi:10.1088/1742-6596/2224/1/012007.

Primova H.A., Safarova L.U 2021a. The Predictive Model of Disease Diagnosis Osteodystrophy Cows Using Fuzzy Logic Mechanisms. *AIP Conference Proceedings* 2365, 050005 https://doi.org/10.1063/5.0057077 Published Online: 16 July 2021.

R. A. Aliev, K.I. Jabbarova, O.H. Huseynov. 2014. *Eighth World Conference on Intelligent Systems for Industrial Automation*, Tashkent. pp.159–167.

Wang N. L, Li L.T, Wu BB, Gong J.Y, Abuduxikuer K, Gang L, Wang J. S. 2016. The Features of GGT in Patients with ATP8B1 or ABCB11 Deficiency Improve the Diagnostic Efficiency. *PLoS One* 11(4): e015,3114

Zadeh L.A. 2011. A Note on a Z-number. *Information Sciences (USA)* 181, pp.2923–2932

Zhang X, Mahadevan S, Deng X. 2017. Reliability Analysis with Linguistic Data: An Evidential Network Approach. *Reliab Eng Syst Safety*. 162:111–121

Artificial Intelligence, Blockchain, Computing and Security – Dagur et al. (Eds)

Development of a simulation model for an automated hybrid photothermogenerator

A.M. Kasimakhunova
Fergana Polytechnic Institute, Uzbekistan

S.I. Zokirov
Fergana Branch of Tashkent University of Information Technologies, Uzbekistan

ABSTRACT: In this article discusses the basics of solar energy conversion and the computer simulation method for solving related problems: weather and climatic conditions, day and night, uneven lighting, temperature increase, pollution of elements, irreversible energy losses, etc. Technical solutions are presented in the form of diagrams and simulation models of a selective radiation photothermal generator, a solar tracker, a protective block with a movable slot for optimal placement of solar cells and their protection from external negative influences. It have been developed mathematical and algorithmic models to automate the operation of the proposed devices and the results of experiments carried out using simulation and experimental samples are presented.

1 INTRODUCTION

The main disadvantage of photo and thermoelectric converting installations is the low efficiency of converting the flow of solar radiation into electrical energy (Green 2009). This is due to the presence of several factors, such as the wavelength of light, recombination of charge carriers, temperature of the solar cell, reflection of radiation from the surface of the converter (Molki 2010; Ruhle 2016; Shockley & Queisser 1961), etc.

Even if there were no negative influence on the converting of the above-mentioned factors that serve to reduce productivity, the power of electricity would not reach the expected values. Because only 20–25% of the radiation emitted by the sun reaches the surface of the solar cell. The rest is lost during reflection (30–35%), absorption by the atmosphere (15–18%), scattering to the ground from the sky (10–11%) and clouds (14–15%) (Wikipedia 2020). Therefore, despite many years of research, the issue of increasing the efficiency of converting solar radiation into electricity using semiconductor photoconverters always remains relevant.

2 LITERATURE REVIEW AND PROBLEM STATEMENT

Known technical solutions that use photo- and thermoelectric converters simultaneously to generate electricity. Their use helps to increase the efficiency of photoconversion, and the scattered from elements heat can be additionally used to generate electricity.

Scientists from Saudi Arabia (Sahin *et al.* 2020) and others have divided thermo-photovoltaic systems into two types according to the principle of operation: one of them uses spectral splitting of radiation to direct it to PV and TE in parts. On others, the PV is directly connected to the TEG, so that the TEG uses the residual heat generated by the PV to generate additional energy. In works (Li *et al.* 2017; Zhang *et al.* 2014) on the creation of combined hybrid PV-TEG systems, the authors attached thermoelements directly to the

DOI: 10.1201/9781032684994-100

back side of the photocell. However, in such systems, the contribution of the thermoelements of the system cannot compensate for the drop in overall efficiency with the deterioration of the properties of solar cells with increasing temperature. Therefore, some authors (Bjork & Nielsen 2015) consider the use of these systems technically and economically inexpedient. But the authors from India Sinkh and others *et al.* were able to create an experimental model of hybrid systems and will receive very positive results (Singh *et al.* 2016) which shows the prospect of these systems is very promising. Such results were obtained by scientists from Pakistan (Ahmad *et al.* 2017). In recent years, have been developed several modifications of hybrid systems by professor A.M. Kasimakhunova and it has been concluded that the possibilities of photothermoelectric converters have not yet been fully exhausted (Kasimahunova *et al.* 2006). Her idea is based on the task of increasing the efficiency of the device and achieving the best technical and economic indicators by eliminating the negative effect of temperature on the electrophysical parameters of the photoelectric converter.

But, despite active research, the obtained performance indicators of existing devices do not meet the expectations of scientists. All known models have several disadvantages: overheating of the panel, decrease in efficiency due to a sharp increase in luminous intensity. In addition, one of the factors that must be taken into account for improve the efficiency of solar cells is their full illumination at the optimal angle.

3 PROPOSED TECHNICAL SOLUTION

The results of our research on the creation of highly efficient devices led to the creation of a more perfect sample (Figure 1) of a selective radiation photothermogenerator capable of solving the above problems.

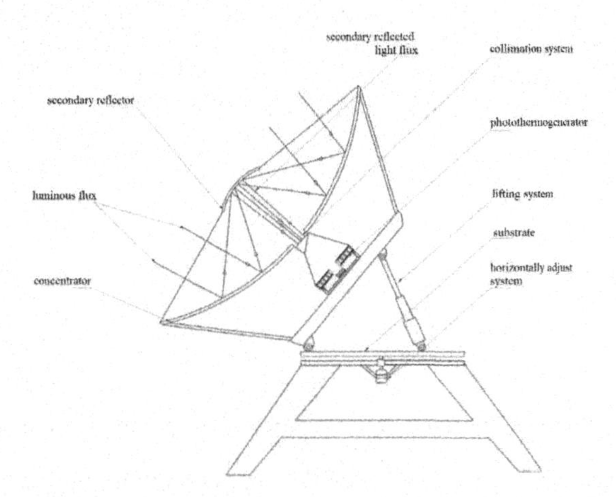

Figure 1. Selective radiation photothermogenerator.

In the proposed design of the photothermogenerator, the light radiation is divided so that the photoactive part of the radiation spectrum, which creates electron-hole pairs, falls on the front surface of the photoelectric converter. The rest of the radiation arriving at the top hot connecting plates of thermoelements is additionally converted into electrical radiation (Kasimakhunova *et al.* 2019). It is more efficient than existing analogs, due to the fact that the direction of selective radiation excludes overheating of solar cells, and therefore ensures the stability of the conversion of photovoltaic energy. The concentration of solar radiation reduces the area of solar panels. In addition, an additional conversion of heat by thermo-electric converter takes place, which leads to an increase in the overall value of the efficiency of the system.

Another difference of this construction from others is the use of a protective block in it, in which the photocells are placed and protected from negative effects, as well as the determination of the optimal coordinates of the photocell relative to the movable slit using an automated system. In addition, it is proposed not to split the spectrum, but to place the photoconverter in the place where the required part of the spectrum falls. Since a certain part of the spectrum falls on the photoconverter, its efficiency is improves sharply and the photocell hardly heats up.

The surface of the block, consisting of thermoelements, cooling system of cold junctions of the thermoelement and the inside of the block, reflectors and slit, automatically searches for the area of selective photoactive radiation by rotation of the low power engine. A slit moving along the spectrum, located on the surface of the block, allows the monochromatic light flux to enter the inside of the block. The transmitted light flux can be directed to the photocell by synchronous movement of the photocell with a slot.

At first glance, the determination of the photoactive area of the spectrum and the direction of rays with certain wavelengths on the surface of a photocell seems to be a simple technical problem. But in this process, several physical and technical factors must be taken into account (Mamadaliyeva 2019).

4 EXPERIMENTAL METHODS, INSTALLATIONS AND RESULTS

It is known that the value of the electro-physical parameters of semiconductor solar cells is relatively stable at temperatures up to +40° C. According to theoretical calculations and experimental results, if the temperature of the photocell exceeds the relative limit, its electro-physical parameters, including their efficiensy, will gradually deteriorate. And upon reaching a certain temperature, these values sharply decrease by a factor of n (Wikipedia 2020). Due to the relatively warm climate in Uzbekistan, the electrophysical problems of photovoltaic plants can be caused mainly by overheating of the solar cells. Therefore, in the course of our experiment, the temperature of the solar cell was increased from + 25 to + 45 C. As a result, was obtained a graph of the dependence of the solar cell efficiency on temperature (Figure 2).

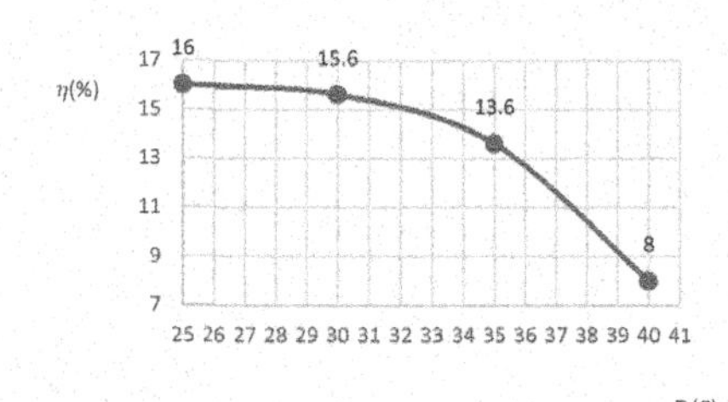

Figure 2. The graph of the dependence of the efficiency of the photocell on its temperature.

As can be seen from this graph, an increase the temperature of photovoltaic cells by 1° C in the range from + 25° C to + 30° C led to a decrease in its efficiency by an average of 0.1% and in the range from + 30° C to +35° C led to a decrease in its useful coefficient by an average of 0.4%. After the difference between the start and end temperatures increased by 15° C, the efficiency was almost halved. This means that the dependence of the efficiency of the photo-electric coefficient on its temperature does not change linearly. This leads to errors in the theoretical calculation of the efficiency for an arbitrary value of the solar cell temperature.

In the second experiment, the light flux was concentrated into a photocell using a concentrator. The change in the intensity of the light flux was carried out by moving the photocell in the focal plane of the reflecting lens. By changing the value of the resistor connected to the electrical circuit, the freewheel voltage, short-circuit current and maximum power were determined.

In the course of experiments carried out at the intensity of solar radiation by 1, 2 and 3 times, the results presented in Table 1 were obtained.

Table 1. The results obtained from an exemplary photocell with a change in the intensity of a concentrated light flux.

Intensity	Ufw, V	Ish.c., A	Wmax, Wt	η, %
1 t.	15	2.5	3.5	15.8
2 t.	17	3	3.8	16.0
3 t.	18	3.4	3.7	15.9

The results obtained at an average temperature of + 27° C show that an increase in the intensity of sunlight has almost no effect on the efficiency. In contrast, the efficiency of a solar cell placed in a light flux with a higher intensity decreased faster than under normal conditions due to the reduced heating time (Figure 3a).

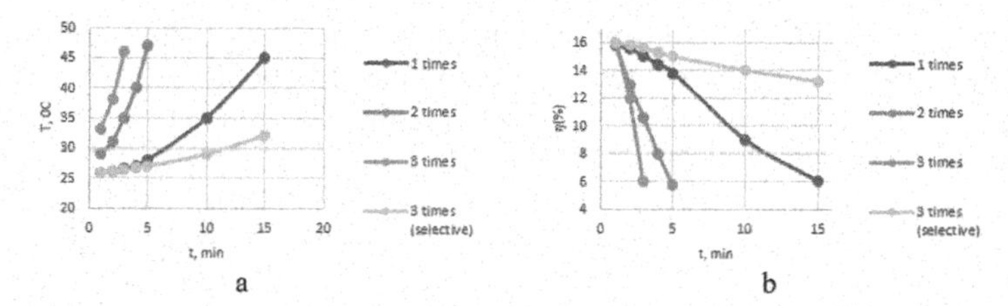

Figure 3. Changing of solar cell temperature (a) and effiency (b) over time at different intensities of solar radiation.

Since the increase in temperature, which negatively affects the electrophysical parameters of the solar cell, is associated with the action of non-photoactive radiation incident on its surface, the following stages of the experiment were carried out with a selective spectrum of sunlight. For this, a special device was created, consisting of a concentrator and a photo-thermogenerator, according to the scheme shown in Figure 1.

At the next stages of the experiment, 6 solar cells of the same type, made on the basis of polycrystalline silicon, were placed at an optimal angle on a fixed base. The rest of the series-connected elements were installed on a horizontally oriented solar tracker. Location of the experimental region: 40.25°25′51″ north latitude and 71°45′42″ east longitude, Fergana. The solar cells, mounted on a fixed base, were installed at an angle of 60° (azimuth) to the east. The vertical angle (zenith) of the solar tracker was 60° too. The experiments were carried out in August 2019. When the non-photoactive light flux was blocked by Peltier thermoelements, an increase in the heating time of the photocell was observed. But, the results presented in Table 1 remained practically unchanged. However, due to the absence of non-photoactive spectra, the maximum temperature value did not exceed +37°C. Therefore, in the course of the experiment, the efficiency was more than 13% (Figure 3b).

5 DETERMINATION OF THE OPTIMAL ZENITH AND AZIMUTH ANGLES FOR A STATIC SOLAR PANEL

Solar panels are usually installed on rooftops or foundations. Because their position is constant, the surface of the panel is not at the optimum angle to sunlight during the day. In addition to the tilt angle, there is also a seasonal zenith angle of the sun, which is sharper in the winter months than in the summer. Therefore, the zenith and azimuth angles of solar panels should be changed throughout the year or placed at an optimal angle between the angles of summer and winter.

As the slope increases, the percentage of losses corresponding to each interval becomes relatively more significant. Since the dependence is non-linear, the approximate value of losses at a given angle is determined by the formula given in the author's earlier work

(Zokirov *et al.* 2019b). The results of the experiment presented in work showed that if we assume that the ability of a solar panel to direct conversion is usually equal to 1 unit, then when installed with an error of up to 15 degrees relative to the optimal angle, its efficiency is 98–100%, and with an error of more than 15 degrees, the efficiency is significantly reduced.

The results of theoretical calculations and experimental values obtained from solar cells show that the maximum power obtained from solar cells on a fixed substrate is 5.64W or 62.6% of the maximum power that can be obtained from this panel. According to calculations, high values should have been maintained from about 11–00 to 13–00. But in the afternoon, as a result of the sharp heating of the air and direct exposure to rays, the temperature of the solar panels rose sharply. As a result, the average value of the results obtained during the day was only 24% of the maximum possible. The results of parallel experiments using a solar tracker show that the maximum power was obtained by about 12–30, and after that a decrease in efficiency was observed.

The difference between the maximum power in both cases, can be explained as follows:

- The maximum value of the intensity of sunlight on the day of the experiment and the geographical position is observed between 12–30 and 13–00 hours.
- The difference between the value of the zenith angle of the solar panels mounted on a fixed base and the optimal value was the same at 11–30 and 13–30, and over time, the heating of the panels showed its negative effect on the output parameters due to the lack of a cooling system;
- Due to the fact that the surface of the solar panels installed on the solar tracker is constantly perpendicular to the direction of sunlight, the decrease in efficiency observed as a result of an increase in temperature in them was observed relatively less. However, an excessive increase in light intensity led to overheating of the panels and a sharp decrease in efficiency.

For the reliability of the results, the following experiment was repeated using a standard solar module – MCM 12–700. The module was placed in a protective block to protect it from negative environmental influences, and the radiation was directed through glass with a conductivity of 100%, 72% and 32%. Under illumination with intensities of 783, 559 and 403 lux (52.2 W/m^2, 39.9 W/m^2 and 26.9W/m^2), the value of the freewheel voltage and short-circuit current was U_{fw} = 3.91 V, U_{fw} = 4 , 75 V, U_{fw} = 5.19 V and $I_{sh.c}$ = 1.25 mA, $I_{sh.c}$ = 2.62 mA, $I_{sh.c}$ = 4.8 mA, respectively.

According to the data obtained, the efficiency ratios were η (52.2) = 4.0%, η (39.9) = 3.3% and η (26.9) = 2.2%, respectively.

Based on the above results, it was found that the efficiency of the photocell does not change appropriately when illuminated with different intensities using a light source with the same spectral composition. Although the intensity in the first and third cases differs threefold, the difference in efficiency rates was only 2%. Similar results were obtained in experiments with a higher coefficiently (15–18%) samples.

6 CONCLUSION

According to theoretical calculations (Zokirov & Obidjanov 2018), if the ability of a solar panel to convert sunlight into electricity is conditionally equal to 100%, then their efficiency can reach 98–100% when installed with an error of up to 15° relative to the optimal deflection angle. The maximum efficiency of the photocell with a fixed substrate was 63%, and the average efficiency during the day was 24%. Such a low level of maximum efficiency is explained by errors in theoretical calculations and technical errors when installing devices. The maximum and average efficiency of solar cells installed on the solar tracker was 97.5% and 62%, respectively. The unstable efficiency of solar cells placed on a solar tracker is explained by an increase in temperature. However, experiments have shown that this this is suitable for a special case within a very short period of time, without taking into account the influence of external factors. This means that the use of solar cells in a photothermogenerator (Zokirov *et al.* 2019a) installed on a solar tracker, and their connection to an additional cooling system can provide stable high efficiency (Kasimakhunova *et al.* 2019).

As a result of scientific and technical research, were obtained the following results:

1. A power source for a mobile lighting system with a power of up to 100 W has been created. When used in the system, the MCM12-700 solar module with a daily efficiency of 6–7% was provided with its average daily efficiency of about 15–16%;
2. An analytical formula and an algorithmic model based on it have been developed to calculate the percentage of energy losses that occur, depending on the angle of incidence of light from the source relative to the normal of the photothermal generator. As a result, it became possible to determine the coordinates of the optimal location of the photothermal generator (the normal to the parabolic concentrator) relative to the angle of incidence of sunlight;
3. Has been developed a technology for automatic placement of a photocell in the photoactive region of the spectrum. Simulation and practical models have been created to study the optical parameters of a selective radiation photothermogenerator consisting of photo- and thermoelements. As a result, it was determined that the heating time of the solar module based on silicon MSM12-700 in the photothermal generator was increased by directing only photoactive rays and the temperature of the photocell was maintained at about + 370C.

REFERENCES

Ahmad, A. S. *et al.* 2017. Efficiency Improvement of Photovoltaic Module by Thermo Electric Generator. *FC-IEFR Journal of Engineering & Scientific Research.* 1014

Bjork R. & Nielsen, K. K. 2015. The Performance of a Combined Solar Photovoltaic (PV) and Thermoelectric Generator (TEG) System. *Solar Energy.* 120: 187–194

Green, M. A. 2009. The Path to 25% Silicon Solar Cell Efficiency: History of Silicon Cell Evolution. *Progress in Photovoltaics: Research and Applications.* 17(3): 183–189

Kasimahunova, A. M. *et al.* 2006. Fototermojelektrogeneratory Selektivnogo Izluchenija. *19-ja Mezhdunarodnaja NTK po Fotojelektronike i Priboram Nochnogo Videnija,* Moscow

Kasimakhunova, A. M. *et al.* 2019. Development and Study of a New Model of Photothermogenerator of a Selective Radiation with a Removable Slit. *International Journal of AdvancedResearch in Science, Engineering and Technology.* 6(4)

Li, G. *et al.* 2017. Analysis of the Primary Constraint Conditions of an Efficient Photovoltaic-Thermoelectric Hybrid System. *Energies.* 20(10)

Mamadaliyeva, L.K. & Zokirov S.I. 2019. Automation Problems of Finding the Optimal Coordinates of a Photocell in a Selective Radiation Photothermogenerator. *IJARSET.* 6(9): 10931–10936

Molki, A. 2010. Dust affects Solar-cell Efficiency. *Physics Education.* 45(5): 456–458

Ruhle, S. 2016. Tabulated Values of the Shockley-Queisser Limit for Single Junction Solar Cells. *Solar Energy.* 130: 139–147

Sahin, A. Z. *et al.* 2020. A Review on the Performance of Photovoltaic/thermoelectric Hybrid Generators. *International Journal of Energy Research.* 1–30

Shockley, W. & Queisser, H. J. 1961. Detailed Balance Limit of Efficiency of p-n Junction Solar Cells. *Journal of Applied Physics.* 32(3): 510–519

Singh, M. *et al.* 2016. Progress in Electromagnetic Research Symposium. Efficient Autonomous Solar Panel and Thermo-Electric Generator (TEG) Integrated Hybrid Energy Harvesting System. Shanghai, China

Wikipedia. 2020. Solar Cell. [In the Internet]. Available: https://en.wikipedia.org/wiki/Solar_cell.

Zhang, J. *et al.* 2014. Performance Estimation of Photovoltaic–thermoelectric Hybrid Systems. *Energy.* 78: 895–903

Zokirov, S. & Obidjanov, Z. 2018. Determination of the Optimal Location Angle of Fixed Solar Panels. *"Talented Students, Masters, Doctoral Dtudents and Independent Researchers" Scientific-practical Conference*, Fergana

Zokirov, S. *et al.* 2019a. Development of a Hybrid Model of a Thermophotogenerator and an Empirical Analysis of the Dependence of the Efficiency of a Photocell on Temperature. *Journal of Tashkent Institute of Railway Engineers.* 15(3): 49–57

Zokirov, S. *et al.* 2019b. Quyosh Panellarining Optimal Ogish Burchagi, yul Quyilgan Hatolik va Energiya Yuqotish Fozini Hisoblash Algoritmini Ishlab Chiqish. *"Technika va Technologik Fanlar Soharining Innovatsion Masalalari" " Scientific-practical Conference,* Termiz

Artificial Intelligence, Blockchain, Computing and Security – Dagur et al. (Eds)

Algorithm for constructing a model for the choice of building materials for the construction

H.A. Primova*
Samarkand Branch of Tashkent University of Information Technologies 47A, Samarkand city, Uzbekistan

Q. Gaybulov
Samarkand State Institute of Architecture and Construction, Samarqand City, Uzbekistan

O.R. Yalgashev
Samarkand International University of Technology, Samarqand City, Uzbekistan

ABSTRACT: Today is the use of artificial intelligence technologies in the selection of building materials is poorly organized. Therefore, this study recommends the selection of sustainable building materials based on the criteria considered to determine the best sustainable alternative and quantitative measurements are used to compare information about these systematic changes. The paper main of propose an evaluation model for selecting the best building material based on sustainable performance based on the experience of construction experts in Uzbekistan, using a hybrid multi-criteria decision-making methodology.

1 INTRODUCTION

Today in Uzbekistan urbanization and the construction industry are developing rapidly as well. Over the past 7 years, pressing concerns about the need for a sustainable construction process has balanced cost concerns.

Due to resource depletion and a host of environmental issues, researchers and practitioners are required to explore sustainable building strategies.

Due to the rapid development of the construction industry in Uzbekistan in recent years, there is a constant requirement for the selection of the best sustainable building materials. An analysis of the existing literature shows that building materials are chosen by specialists in the field of construction. The system proposed in this article has been tested by a research company and the results are compared with existing literature and expert opinion. The study assesses its potential, as well as identifying its useful limitations (Holida Primova 2021a; Musaev 2020; Wen 2005).

In many countries, including Uzbekistan, many initiatives have been launched to build environmentally friendly buildings (Li & Qian 2014; Radhi 2010), they not only solve economic problems but also use sustainable building, passive architecture, building and energy efficiency and other factors. Many strategies have been developed to build durable and safe buildings, however no attention has been paid to the views of sustainable building researchers on the negative impacts on the environment (Todd & Crawley 2001), society and the economy. Although these approaches have given rise to various research topics in various literatures, the history of the permanent construction remains unknown, however The first building history was

*Corresponding Author: primova@samtuit.uz

DOI: 10.1201/9781032684994-101

proposed by Charles Kibert at the 1st International Building Conference in 1994. Some resources are being considered in the United Arab Emirates (UAE) (Abeysundara 2009; Hamid 2012]. Elchalakani and Elgaali explored sustainable building by creating sustainable concrete made from recycled sewage and concrete from construction and demolition waste (Elchalakani 2012; Saparauskas 2006; Tsai 2012). They then developed recommendations for implementing strict legislation on these types of sustainable building to ensure sustainable growth.

Although the construction sector is performing at a high level, there are few sources in Uzbekistan that study sustainable construction. The use of artificial intelligence technologies in the selection of building materials is poorly organized. Therefore, this study recommends the selection of sustainable building materials based on the criteria considered to determine the best sustainable alternative and quantitative measurements are used to compare information about these systematic changes. This study considers three cases in order to classify the selection criteria for preferred sustainable building materials (Castellano 2014; Primova 2021b; Todd 2001).

1. Selection of an appropriate indicator for the selection of sustainable building materials.
2. Analyze the influence, interdependence and communication between each indicator to determine the most effective indicator.
3. Propose a material selection model for the construction industry to validate choices when evaluating sustainable building materials in Uzbekistan.

2 SOLUTION METHODOLOGY

The impact of the criteria can be expressed by the decision maker based on 5 scales, 0 - "no impact", 1-"very low impact", 2-"low impact", 3-"high impact" ва 4-"very high impact". Let each initial relationship matrix of each decision maker be combined and their mean value given by matrix A.

Step 1. Build decision matrix

In most decision criteria, all (A_1, A_2, .. A_m) alternatives are evaluated using the (C_1, C_2, ... C_n) criteria. Solutions for each alternative are presented in the decision matrix below.

$$D = \begin{matrix} A_1 \\ \vdots \\ A_2 \\ \vdots \\ A_m \end{matrix} \begin{matrix} \begin{matrix} C_1 & C_2 & \cdots & C_n \end{matrix} \\ \begin{bmatrix} r_{11} & r_{12} & \cdots & r_{1n} \\ r_{21} & r_{22} & \cdots & r_{2n} \\ \vdots & \vdots & \cdots & \vdots \\ r_{m1} & r_{m2} & \cdots & r_{mn} \end{bmatrix} \end{matrix}$$

Step 2. Normalization of the decision matrix.

The normalization of the decision matrix D has been performed in the next step.

$$S = [S_{ij}]_{m \times n}$$

here

$$S_{ij} = \frac{r_{ij}}{\sqrt{\sum_{i=1}^{m} r_{ij}^2}}$$

Step 3. Creating a normalized decision matrix

It is necessary to take into account the weighting factor when making a normalized matrix multicriteria decision.

Let the weight coefficient of the decision matrix be $B = (B_{ij})_{m*n}$

$$B_{ij} = c_{ij} * W_j$$

Here $i=1,2,...,m$ and $j=1,2,...,n$.

Step 4. Determination of the maximum and minimum solutions [Chen 2011].

$$A* = \{v_1^*, v_2^*, ..., v_n^*\} = \{(\max_i v_{ij} | j \in J), (\min_i v_{ij} | j \in \widehat{J} | i = 1, .., m\}$$

$$A^- = \{v_1^-, v_2^-, ..., v_n^-\} = \begin{cases} (\min_i v_{ij} | j \in J), \\ (\max_i v_{ij} | j \in \widehat{J}) \; | i = 1, .., m\} \end{cases}$$

here, J - maximum set of criteria (profit).

J' - The set of criteria to be maximized (expenses).

Step 5: Calculation of the distance of each alternative from A and A⁻.*

The minimum and maximum solution of each alternative can be calculated as follows.:

$$d_i^+ = \sqrt{\sum_{j=1}^{n} \left(v_{ij} - v_j^*\right)^2} d_i^- = \sqrt{\sum_{j=1}^{n} \left(v_{ij} - v_j^-\right)^2}$$

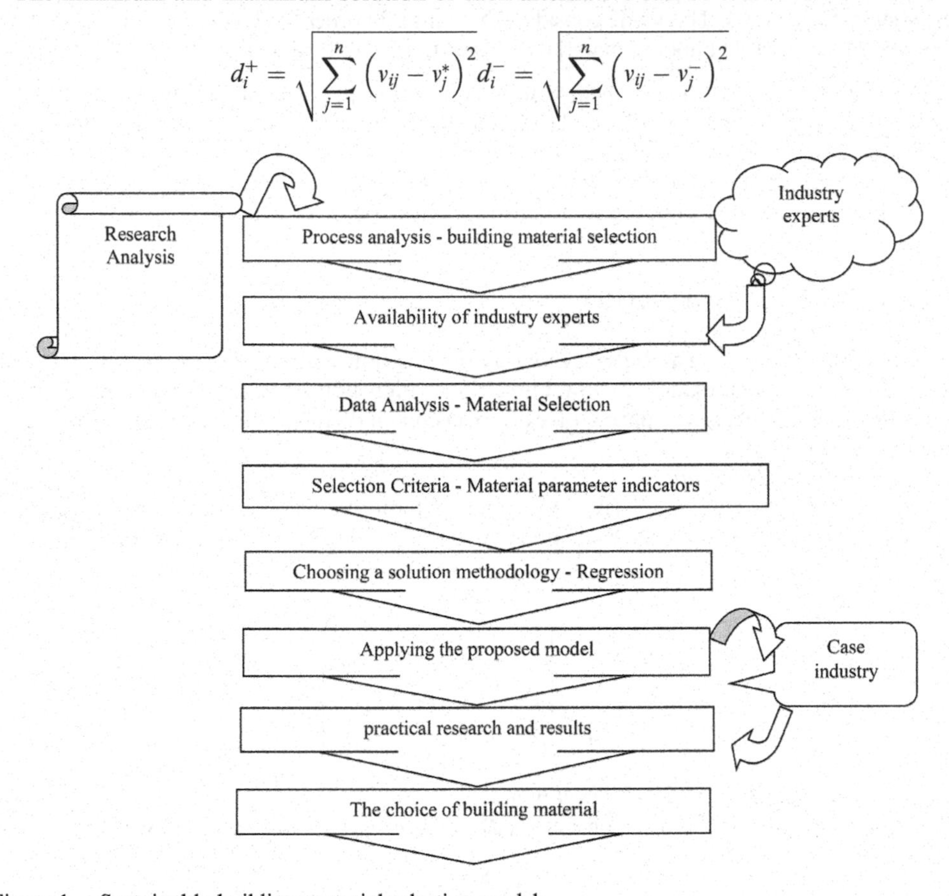

Figure 1. Sustainable building material selection model.

Step 6. Proximity factor calculation [Ortiz 2009; Tsai 2012]

The order of all alternatives can be determined using the coefficient of similarity. The similarity coefficient of each alternative is calculated as follows

$$\mathrm{CC}_i = \frac{d_i^-}{d_i^- + d_i^+}$$

Step 7. Level detection.

The ranking of alternatives is based on their proximity coefficient. CC_i then the best alternative is A_i (See Figure 1).

2.1 *Algorithm for building a model based on the conclusions of a fuzzy rule*

In selection of a building material based on data analysis, the values of the $\mu^j(x_i^*)$ relevance function are determined in the fixed constant values of the x_i^*, $i = \overline{1,n}$ parameters [6, 13, 17, 20].

Experimental information is provided by experts. By analyzing these data, the steps of this fuzzy rule derivation algorithm were developed.

2.1.1 *Creation of sampling*

First, process analysis is performed. In this case, as experimental data, it is necessary to create input sampling (X_j, r_j) , $j = \overline{1,M}$, which is a parameter of the structural material, where $X_j = (X_{j,1}, x_{j,2}, \ldots, x_{j,n})$ = – input vector in j-th row and r_j – the value of its output vector respectively. When displaying a sampling, it is important to consider that the data in it can consist of integers, real numbers, and values in linguistic form. All values of the obtained sampling are normalized to some common range.

2.1.2 *Normalization*

In this process, data analysis is performed. In this case, the input sampling and the corresponding r vectors are normalized to the interval [0, 1].

$$u_i^k = l\frac{x_i^k - x^{\min}}{x^{\max} - x^{\min}},$$

$$u^k = l\frac{r^k - r^{\min}}{r^{\max} - r^{\min}}.$$

x^{min}, x^{max}- maximum and minimum elements of the input data matrix X and r^{min}, r^{max} - maximum and minimum elements of the output r vector.

2.1.3 *Fuzzification*

Based on the data u_i^k and u^k, a fuzzification process is performed using a relevance function. In this case, the bell-shaped function was taken as the relevance function. The building material parameter indicators were accurately solved using the bell-shaped relevance function due to the level of accuracy and productivity gave high results.

$$\mu^j(u_i^k) = \exp\left(-\frac{1}{2}\left(\frac{u_i^k - c_j}{\sigma_j}\right)^2\right),$$

$$\mu^j(u^k) = \exp\left(-\frac{1}{2}\left(\frac{u^k - c_j}{\sigma_j}\right)^2\right), \; j = 0, 1, 2, ..., l.$$

where c_j and σ_j are function parameters.

2.1.4 *Maximization*

An appropriate maximization operation is performed on the results of the fuzzification:

$$\mu^*(u_i^k) = \max_j \mu^j(u_i^k), \mu^*(u^k) = \max_j \mu^j(u^k).$$

2.1.5 *Maximization*

Application of the proposed model.

$$E = \frac{1}{2}\sum_{j=1}^{M}\left(y_j - \widehat{y}_j\right)^2 \rightarrow \min \tag{1}$$

In this case y_j - result obtained on the basis of a fuzzy model.

The input vector X will have a fuzzy output as follows: $y_r = \frac{\sum_{j=1}^{m} \mu_{fj} \cdot (x_r) \cdot f_j}{\sum_{j=1}^{m} \mu_{fj} \cdot (x_r)}$, here j - the derivation of the rule is expressed as follows:

2.2 *The model in its linear form*

$$y_j = b_{j0} + b_{j1}x_1^j + \dots + b_{jn}x_n^j, \; j = 1, \; m.$$

The j-rule inference execution level is implemented using the following expression:

$$\mu_{y_j}(x_r) = \mu_j^{k_j}(x_{r1}) \cdot \mu_j^{k_j}(x_{r2}) \cdot \dots \cdot \mu_j^{k_j}(x_{rn}).$$

The execution of the j-rule inference for the input vector X_r is calculated as

$$\beta_{jr} = \frac{\mu_{y_j}(x_r)}{\sum_{k=1}^{m} \mu_{y_j}(x_r)}.$$

When it has a form linear connection:

$$y_r = \sum_{j=1}^{m} \beta_r y_j = \sum_{j=1}^{m} (\beta_{r_j} b_{j0} + \beta_{r_j} \cdot b_{j1} \cdot x_{r_1} + \beta_{r_j} \cdot b_{j2} \cdot x_{r_2} + \dots + \beta_{r_j} \cdot b_{j_n} \cdot x_{r_n})$$

3 APPLICATION OF THE PROPOSED MODEL

To test the proposed model, our research team sent this proposal to 25 national construction companies active in the field. Such companies include suppliers of building materials and services to construction companies located in the Samarkand region and its environs. Our goal is to bring together all the opinions of the construction industry to achieve clear results, because this research is mainly based on the opinions of decision makers. Thus, we want to combine four main points of view: customers, construction companies, experts and material suppliers (Holida Primova 2021a; Jia 2015).

Table 1. Construction material type and properties.

#	Dimensions	Criteria	Properties
1	Economical (D1)	Transportation costs (X1)	Material delivery cost
		Income (X2)	Income from this material.
		Tax paid (X3)	Tax has been declared and financially reviewed
2	Mechanic (D2)	Friction (X4)	Material strength assessment
		Density (X5)	The average density of aerated concrete products should not exceed D700
		Cold resistance (X6)	Concrete resistance to alternate freezing and thawing up to 50 cycles
		Thermal conductivity λ (X7)	The coefficient of thermal conductivity of buildings and structures should not exceed the values specified in GOST-31359.
		Water permeability (X8)	the ability of materials to pass water under pressure
		Sound transmission (X9)	compressive strength when the material is frozen -15-170C in water-saturated state and re-thawed (1 cycle)

(continued)

Table 1. Continued

#	Dimensions	Criteria	Properties
3	Envoirment (D3)	Durability, R (X10)	
		Climate dependency (X11)	
		Adaptation to seasons (X12)	
		Fire resistance (X13)	Explosion resistance
		Use of local materials (X14)	The high level of use of local materials influences the development of the local community.
		Securiy (X15)	Materials must withstand any kind of damage, as well as ensure health and safety.

The material chosen for evaluation in this study is a brick. Brick is used in almost every type of construction, including bridges, residences, and more. In this paper, four types of bricks are selected: aerated concrete, clay brick, ceramic brick, foam block. The production of aerated concrete blocks is carried out in most cases in the factory, as this process requires special equipment. Foam concrete is natural and environmentally friendly. This is another advantage of aerated concrete. The main material used in the production of aerated concrete is extinguishing (aggressive, chemically active substances). Reacting with aluminum powder, it releases gas, which creates gas bubbles in the structure of aerated concrete. Clay brick is a traditional brick made from clay and widely used for various purposes.

While there are many consistent metrics by researchers, we chose metrics specific to the construction industry and focused on material selection.

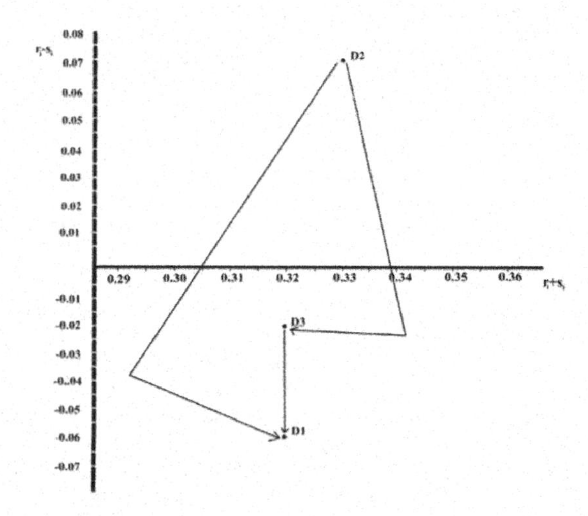

Figure 2. Reasons in the dimension chart.

These indicators are selected in Table 1 for information from experienced professionals. Then the alternatives already implemented in the previous stages are developed.

4 CONCLUSION

In this paper, relevant sustainability indicators have been selected to select the best sustainable building material. An algorithm for constructing a model was also built based on two methodologies using fuzzy set theory. The paper considers the problem of choosing the most suitable stable material from 4 building materials when making illegal decisions and theoretically substantiates the result of calculations using these two methods.

REFERENCES

Abeysundara U. Y., Babel S, Gheewala S. 2009. A Matrix in Life-cycle Perspective for Selecting Sustainable Materials for Buildings in Sri Lanka. *Build Environ*;44(5):997–1004.

Akhatov A., Nazarov F. Rashidov A. 202. Increasing Data Reliability by Using Bigdata Parallelization Mechanisms. *International Conference on Information Science and Communications Technologies.* 4,5,6 November. ICISCT 2021(IEEE), art. no. 9670387 DOI: 10.1109/ICISCT52966.2021.9670387.

Castellano J, Castellano D, Ribera A, Ciurana J. 2014. Development of a Scale of Building Construction Systems According to CO Emissions in the Use Stage of Their Life Cycle. *Build Environ.* 82:618–627.

Chen F. H, Hsu T. S, Tzeng G. H. 2011. A Balanced Scorecard Approach to Establish a Performance Evaluation and Relationship Model for Hot Spring Hotels Based on a Hybrid MCDM Model Combining DEMATEL and ANP. *Int J Hosp Manag.* 30(4): 908–932.

Elchalakani M, Elgaali E. 2012. Sustainable Concrete Made of Construction and Demolition Wastes Using Recycled Wastewater in the UAE. *J Adv Concr Technol.* 10(3):110–125.

Hamid Z. A, Kamar K.A 2012. Aspects of Off-site Manufacturing Application Towards Sustainable Construction in Malaysia. Constr Innov. 12(1):4–10.

Holida Primova,Qodir Gaybulov, Ismoil Isroilov. 2021a. Selection of Building Material Using the Decision-making System. *International Conference on Information Science and Communications Technologies: Applications, Trends and Opportunities* November 3–5, http://www.icisct2021.org/

Jia P, Govindan K, Choi T-M, Rajendran S. 2015. Supplier Selection Problems in Fashion Business Operations with Sustainability Considerations. Sustainability, 7: 1603–1619.

Li G, Qian S, Lee H, Hwang Y, Radermacher R. 2014. Experimental Investigation of Energy and Exergy Performance of Short term Adsorption Heat Storage for Residential Application. *Energy* 65: 675–691.

Musaev M., Khujayorov I. and Ochilov M. 2020. The Use of Neural Networks to Improve the Recognition Accuracy of Explosive and Unvoiced Phonemes in Uzbek Language, *2020 Information Communication Technologies Conference (ICTC)*, pp. 231–234, doi: 10.1109/

Ortiz O, Castells F, Sonnemann G. 2009. Sustainability in the Construction Industry: A Review of Recent Developments Based on LCA. *Constr Build Mater.* 23 (1):28–39.

Primova H. A., Iskandarova F. N. and Gaybulov Q. M. 2021b. Accounting Experience Between Fuzzy Integraland Z-numbers. Advances in Intelligent Systems and Computing 1323, 11th World Conference "Intelligent System for Industrial Automation" (WCIS-2020), Springer, pp.40–46, https://doi.org/10.1007/978-3-030-68004-6_6.

Primova X.A., G'aybulov Q., Iskandarova F.N. 2020. Selection of Construction Materials on Fuzzy Inference Rules. *ICISCT 2020 Conference International Conference on Information Science and Communications Technologies*, Tashkent, 4–6 November.

Radhi H. 2010. On the Effect of Global Warming and the UAE Built Environment. *Global Warming. Sciyo, Rijeka, Croatia*; p.95–110.

Saparauskas J, Turskis Z. 2006. Evaluation of Construction Sustainability by Multiple Criteria Methods. *Technol Econ Dev Econ* 12(4):321–326.

Todd J. A, Crawley D, Geissler S, Lindsey G. 2001. Comparative Assessment of Environmental Performance Tools and the Role of the Green Building Challenge. *Build Res Inf* 29(5):c324–C335.

Tsai C.Y, Chang A.S. 2012. Framework for Developing Construction Sustainability Items: The Example of Highway Design. *J Clean Prod.* 20(1): 127–136.

Wen Z, Zhang K, Huang L, Du B, Chen W, Li W. 2005. Genuine Saving Rate: An Integrated Indicator to Measure Urban Sustainable Development Towards an Ecocity. *Int J Sustain Dev World Ecol* 12(2): 184–196.

Zhang X, Xu Z. 2014. Extension of TOPSIS to Multiple Criteria Decision Making with Pythagorean Fuzzy Sets. *Int J Intell Syst.* 29(12): 1061–1078.

Artificial Intelligence, Blockchain, Computing and Security – Dagur et al. (Eds)

Algorithmic synthesis of computational schemes for optimization of identification and image recognition of micro-objects

I.I. Jumanov & O.I. Djumanov
Samarkand State University, Samarkand, Uzbekistan

R.A. Safarov
PhD student of Samarkand State University, Samarkand, Uzbekistan

ABSTRACT: The proposed methodology for optimizing the identification of micro-objects is implemented on the basis of dynamic models, neural networks of various topologies, mechanisms for extracting redundant information structures, as well as the use of statistical, dynamic, and specific characteristics of images. A software package has been implemented for identifying, recognizing, and classifying images of micro-objects, functional modules for contour segmentation, selection of reference points, reduction of redundant fragments or features, and setting variables. The software package includes mechanisms that take into account the statistical relationships of points on the image contour, the dynamics of changes in the contour curve, the formation of point coordinate matrices, the deformation of a sequence of points in segments, and the selection of stationary sections. A comparative analysis of the effectiveness of algorithms for pre-processing images, recognition, and classification was carried out using the example of pictures of medical diagnostics, and pollen grains presented to the study in solving problems of selection and seed production of wheat grains. The software modules of neural networks are based on five types of learning algorithms, which are performed according to the supervised and unsupervised methods, the training sample is modified by the methods of vector quantization, clustering, segmentation, and the formation of a "sliding window". The efficiency of identification of images of micro-objects was studied in the presence of "noise".

1 INTRODUCTION

When solving the problems of palynology, medical diagnostics, breeding, seed production, and environmental protection, in which labor-intensive laboratory analyzes are carried out, aimed at recording, and preparing images of samples of micro-objects, in particular, pollen grains, unicellular medical objects, etc (Ibragimovich *et al.* 2020b; Khanzhina & Putin 2016; Yane 2007). Of great importance are software and hardware systems, complexes, and tools that differ in functionality, as well as in the specifics of image information processing (Jumanov *et al.* 2021d; Sebastiano *et al.* 2016; Silva *et al.* 2011).

In existing technologies, images of micro-objects are obtained using a photo, video camera, and digital microscope. They perform manual and interactive measurements of the size of micro-objects, statistical analysis, and regulation of the values of variables that improve identification using the adaptive properties of dynamic models and neural networks (NN) of various topologies (Jumanov *et al.* 2021a; Rodriguez *et al.* 2006; Silva *et al.* 2008).

It is important to solve the problems of designing algorithms and implementing software tools for pre-processing image information related to the construction of mechanisms for extracting an image contour, selecting informative fragments with features, texture segmentation, filtering, and approximation, as well as developing principles for using statistical,

DOI: 10.1201/9781032684994-102

dynamic and other specific characteristics of information (Allen *et al.* 2008; Del Pozo-Bafios *et al.* 2012; Jumanov *et al.* 2021c).

The application of methods, models, and algorithms for the identification of micro-objects is accompanied by the need to take into account the presence of conditions of large error, due to a priori insufficiency, uncertainty, and non-stationarity of data (Dhawale *et al.* 2013; Tcheng *et al.* 2016; Zhang *et al.* 2004). In this study, the methodology for recognition, classification, and identification is based on the mechanisms for using morphological, geometric, and other specific characteristics of micro-object images (Battiato *et al.* 2020; Daood *et al.* 2016; Jumanov *et al.* 2021b).

2 MAIN PART

2.1 *Implementation of optimization mechanisms for identification, recognition and classification of micro-objects*

The functional modules of the software package (SP) are based on the model, algorithms aimed at improving the quality of the restoration of images of micro-objects based on the combination of the possibility and use of the properties of dynamic and NN models (He *et al.* 2016; Ibragimovich *et al.* 2021a; Korobeynikov *et al.* 2018). A characteristic feature of the implemented complex is the use of mechanisms for extracting redundant information structures of micro-objects – histological, morphological, fractal, and geometric characteristics at certain characteristic points of images. The principles of efficiency, simplicity, typification and reliability are the basis for the development and implementation of computational schemes of SP algorithms. The "interface" of the complex performs the role of a monitor – coordinator, containing menu controls for the tasks of information processing and dialogue generation.

To improve the efficiency of the SP in recognizing species, forms of individual classes, and image fragments, the following software modules are implemented:

- fixing the exposure at the points of digital images of frames at certain intervals;
- analysis of each frame with the definition of parameters, features, and image fragments;
- formation of a database (DB), an image database, a knowledge base and rules for determining quantitative and qualitative characteristics;
- support for making technological decisions and adjusting the values of parameters, recognition and classification of images.

The SP was tested on training sets of images compiled from images of the tasks of diagnosing lung diseases, selection of pollen grains. When testing, we used Matlab R2010a with Image Processing Toolbox and Neural Network Toolbox.

The most efficient version of the functioning of the SP is achieved using digital video, and photographic images of visible light waves. Each image frame is represented by a high-resolution pixel matrix (Chica & Campoy 2012; Jumanov *et al.* 2019; Redondo *et al.* 2015).

2.2 *SP testing is based on lung disease images*

Image processing is optimized based on the gradient operator. The analysis of the results of SP testing confirms the ability of the implemented operators to fix the boundaries of image segments, extending them for those cases when the gradient operator is faced with the definition of extra contours with a very thin border. For each of the syndromes of the disease, an assessment of diagnostic parameters was made with a verified diagnosis, confirmed by standard research methods. Average values were formed for each type of syndrome, taken as reference image characteristics.

A computational scheme for the functioning of a SP has been developed, which is performed with the following blocks: 1-input of images into the file-buffer; 2 – pre-processing of images; 3 – segmentation; 4 – obtaining a segmented image; 5 – definition of the syndrome; 6 – determination of confidence in the presence of the syndrome; 7 – change in the syndrome code; 8 – decision making.

A characteristic feature of the block diagram is the presence of three databases: a database of patients, a database of standards and a database of segments and histological redundant image structures.

The database of patients is connected with the database, registry. The database of standards is used in the selection and analysis of pathological segments. It includes models of images of morphological structures. A special buffer file is used to organize links between program modules and databases.

Extracted from the database of images and database of information on a local or global network enters the buffer file and becomes available to users. Statistical analysis software modules include additional programs that allow you to perform arithmetic and logical operations in the image space and their spectral analysis.

Such scalar characteristics of images, features, and fragments are determined as – arithmetic mean brightness of pixels, mode, mode amplitude, image area, and segment. Segmented images are entered into the buffer file, which is used by a diagnostic program designed for static analysis of selected segments and analysis of the dynamics of their points. The SP includes a module for visualizing the morphological structures of an image based on NN, as well as modules that perform spectral and statistical analyzes in a "sliding window" (Isroil *et al.* 2021, 2022; Li 1995; Shemi 2015).

The software application under study operates in two modes. In the first mode, the user is supported to search for syndromes from a variety of classes, which are checked against the reference images stored in the database. In the second mode, neural network structures are synthesized to form a hybrid image processing technology.

To highlight the syndrome of the corresponding class, a database and images with the corresponding formations are created. Each image corresponds to the reference segment – the "correct answer". The synthesis of the database and neural network structures allows the classification of syndromes. When working with a black-and-white image, the database and the image base reflect the corresponding pathological formations in black-and-white images.

To implement filtering modules in the space of two-dimensional frequencies for pathological segments of this class, the corresponding "sliding windows" with operators for setting the parameters of the NN learning algorithm are synthesized.

2.3 *Testing of functional modules for selection and segmentation of contours of images of micro-objects*

The task of the implemented functional module is to analyze images for the reduction of "extra" contours, points, features, and fragments of images of micro-objects. The selected fragments, as a rule, have a greater thickness than the thicknesses of the contours selected by gradient operators (Haralick & Shapiro 1993; Jumanov 2022; Jumanov *et al.* 2020).

The efficiency of various mechanisms for selection, and segmentation of contours, fragments, and points of images is estimated by a coefficient characterizing the frequencies (probabilities) of correct selection of pixels of a point *Ptd* and false selection of pixels of a point *Pfd.*

The frequency of correct selection of pixels for a point belonging to the morphological structure of the *i*-th image is determined as

$$K_{false} = Ptd_i^{\omega_l} = td_i^{\omega_l} / Ntd_i^{\omega_l}, \tag{1}$$

where $td_i^{\omega_l}$ = the number of pixels correctly included in the class segment ω_l of the *i*-th test image;

$Ntd_i^{\omega_l}$ = the number of pixels contained in the class segment ω_l of the *i*-th test image.

The frequency of false selection of pixels of a point belonging to the morphological structure ω_l for the i-th test image is determined as

$$K_{\text{true}} = Pfd_i^{\omega l} = fd_i^{\omega l}/(N_i - td_i^{\omega l}), \tag{2}$$

where $fd_i^{\omega l}$ is the number of pixels of dots included in the segment of class ω_l of the i-th test image, but not belonging to the desired segment;

N_i is the number of dot pixels contained in the i-th test image.

The processes of isolation and segmentation of the contours of the lungs, the syndrome of round shadows in the lung field, the syndrome of focal shadows in the lung field, and the syndrome of extensive lucidity of the lung field were studied.

To adapt the basic segmentations, the mechanisms of growing, splitting, and merging regions were used, the functional modules of which were tested using the regionsgrow and qtdecomp procedures.

X-ray images were used to form training and control samples for training models. 34 fluorograms of patients with pneumonia were studied, which are characterized by syndromes associated with impaired transparency of lung tissues:

- "an extensive (total) decrease in the transparency of the lung tissue";
- "subtotal darkening" – with the localization of inflammation within one or two lobes of the lungs;
- "limited darkening" – infiltrative changes in the lung tissue that do not go beyond the segment.

Figure 1 illustrates the principle of graphical comparison of the efficiency ratio of the following mechanisms: a hybrid with NN(1 – solid line); NN (2-dashed line); splitting and merging points of fragments (3 – dash-dotted line).

Graphs are plotted depending on the frequency of point pixel selection.

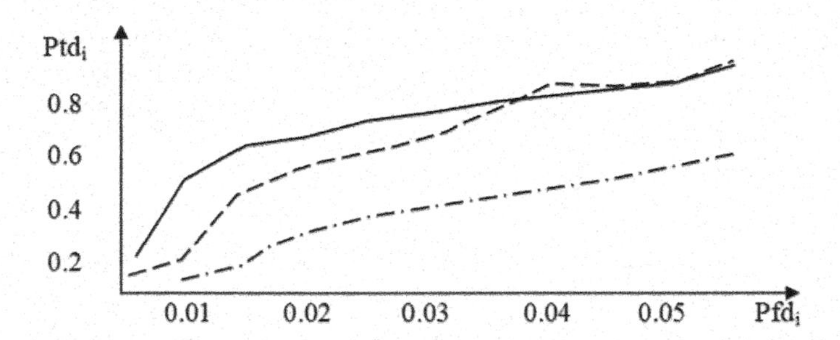

Figure 1. The efficiency of image processing mechanisms.

The effectiveness of the mechanism for identifying pathological segments and morphological structures of chest images was studied. Functional modules are constructed on the basis of a parabolic polynomial, an orthogonal algebraic complete 7, and an interpolation spline – the Daubechies function 7. The implemented models showed positive experimental results in the diagnosis of radiological syndromes. It has been established that the hybrid mechanism makes it possible to improve the quality of image identification by more than 1.5 times than its counterpart working on dynamic models.

The conclusion on the effectiveness of the use of software modules shows that the results coincide with the medical results on the diagnosis of "no pathology detected" in 91% of cases in men and 97% in women. Coincidence in the conclusion "pneumonia" is about 66% in men, and 100% in women. Thus, the good sensitivity of the decision modules included in the SP is confirmed.

SP shows its positive properties when patients are divided into two groups: "healthy" and "no pathology detected" with certain deviations in the patient's health. And "rediscovery of the disease" suggests in which direction the most promising, in-depth examination of the patient to prevent the development of the disease with economical means and time of information processing.

Calculations of identification quality indicators were carried out according to the method

$$DS = TP/n_{\omega_r};\ DSp = TN/n_{\omega_0};\ DE = (TP + TN)/(TP + FP + FN + TN), \quad (3)$$

where r = class number of the studied disease;

n_{ω_r} = the number of images in the control sample in the studied class of diseases;

n_{ω_0} = number of images without pathology;

TP = a true positive result equal to the number of class ω_r images correctly classified by the considered rule;

FP = a false-positive result equal to the number of class ω_0 images erroneously assigned to class ω_r by the decision rule;

FN = false negative: number of class ω_r images erroneously assigned to class ω_0 by the decision rule;

TN = true-negative: Number of class ω_0 images correctly classified by the decision rule.

Sensitivity, specificity, and effectiveness of the decision rule were used as decision rules, which represent the following results: $DS = 82\%$; $DSp = 94\%$; $DE = 89\%$.

The implemented SP application makes it possible to improve the quality indicators of the diagnosis of syndromes by 10–15%.

Table 1 shows the results of the evaluation of the effectiveness of the SP.

Table 1. SP application performance evaluation results.

	Decision-making		
Options	Positive	Negative	Indefinitely
n_{ω_r}	$TP = 47$	$FN = 4$	$TP + FN = 51$
n_{ω_0}	$FP = 6$	$TN = 28$	$FP + TN = 34$
Total	$TP + FP = 53$	$FN + TN = 32$	$TP + FP + FN + TN = 85$

2.4 *Testing SP functional modules based on pollen grain image sets*

To test the models of identification of images of micro-objects combined with the NN, training sets of various types of pollen were compiled, and accumulated during joint studies of the laboratory of breeding and seed production of the Scientific Research Institute "Grain". All images of given samples are divided into training and test sets:

Pollen type	Test set	Training set
Plantago lanceolata	364	350
Rumex acetosella	517	500
Conopodium majus	225	212
Dactylis glomerata	600	587

When training the NN, the properties are used, according to which two or more trained networks can be combined, obtaining a new, larger network that recognizes images of the original networks. Five networks were studied and trained to recognize each type of pollen.

Each network produces a set of data and assigns the recognized pollen to a particular class. If the object is not recognized, then the rejected pollen is given a new name, which is marked as an "unrecognized object" to form a new class. Table 2 shows the results of the hybrid image recognition algorithm as correct, incorrect, and rejected pollen samples.

Similar results were obtained for Hopfield networks, bidirectional associative memory (BAM).

Table 2. Results of correct recognition of pollen images.

Pollen type	Correctly recognized	Incorrectly recognized	Rejected
Plantago lanceolata	302 (83,2%)	48 (13,2%)	13 (3,6%)
Rumex acetosella	430 (83,2%)	62 (12 %)	25 (4,8%)
Conopodium majus	193 (86,2%)	21 (9,3%)	10 (4,5%)
Dactylis glomerata	515 (85,8%)	45 (7,5%)	40 (6,7%)

In Table 3, a comparative analysis of the effectiveness of program modules for five adopted NN is carried out.

Table 3. Comparative analysis of the effectiveness of software modules.

Algorithm name	Correctly recognized	Incorrectly recognized	Rejection of the hypothesis
Hamming algorithm	84,6%	10,5%	4,9%
Hopfield algorithm	83,7%	11,15%	5,15%
Hebb algorithm – supervised learning	78,8%	16,075 %	5,125%
Hebb Algorithm – unsupervised learning	80,2%	14,1%	5,7%
Algorithm based on BAM	84%	10,825%	5,175%

The presented research results were obtained for a sample of pollen types from 1625 images.

3 CONCLUSIONS

A computational scheme for the synthesis of models, algorithms and mechanisms for optimizing the identification, recognition, and classification of micro-object images based on the use of redundant information structures has been developed and implemented. The reliability of research has been proven on the basis of the adapted use of SP in the tasks of image processing for the diagnosis of diseases of the chest, and pollen for solving the problems of breeding and seed production of grain.

The mechanisms for processing raster halftone images are implemented, based on determining the "centre of gravity of the histogram" in the local window, selecting fragments, textures, morphology, and images, which made it possible to apply Fourier transform models, wavelet transforms, form spaces of informative points, features, fragments, approximators and image classifiers based on NN. Implemented the principles of morphological filtering of image images, as well as mechanisms for determining the "centre of gravity of the histogram" based on NN, with macro layers, hybrid NN with multi-alternative classification.

Estimates of the effectiveness of segmentation, which consists of the use of mechanisms for highlighting the characteristics of various pathological formations, and morphological structures of x-ray images of the chest, are determined. A technique for comparing errors of the first kind with a fixed number of errors of the second kind is obtained. A comparative evaluation of the efficiency of segmentation of various mechanisms has been carried out.

Experimental studies of the effectiveness of functional modules and SP for the identification, recognition and classification of pollen grain images have been carried out.

REFERENCES

Allen, G. P. *et al.* 2008. Machine Vision for Automated Optical Recognition and Classification of Pollen Grains or Other Singulated Microscopic Objects. *15th Int. Conf. on Mechatronics and Machine Vision in Practice*: 221–226.

Battiato, S., *et al.* 2020. Detection and Classification of Pollen Grain Microscope Images. *In: Proceedings of the IEEE/CVF Conf. on Computer Vision and Pattern Recognition Workshops*: 980–981.

Chica, M. & Campoy, P. 2012. Discernment of Bee Pollen Loads Using Computer Vision and One-class Classification Techniques. *Journal of Food Engineering* 112: 50–59.

Daood, A., Ribeiro, E. & Bush, M. 2016. Pollen Grain Recognition Using Deep Learning. In: Bebis, G., *et al.* (eds.) *ISVC 2016. LNCS: 10072*, 321–330. Springer, Cham.

Del Pozo-Baños, M. *et al.* 2012. Image Processing for Pollen Classification. *Biodiversity Enrichment in a Diverse World*: 493–508.

Dhawale, V. R., Tidke. J. A. & Dudul, S. V. 2013. Neural Network Based Classification of Pollen Grains. *Int. Conf. on Advances in Computing Communications and Informatics*: 79–84.

Haralick R.M. & Shapiro L.G. 1993. *Computer and Robot Vision 2*: Reading, MA, Addison-Wesley.

He, K., Zhang, X., *et al.* 2016. Deep Residual Learning for Image Recognition. In: *Proceedings of the IEEE Conference on Computer Vision and Pattern Recognition*: 770–778.

Ibragimovich, J. I. Isroilovich, D. O. & Abdullayevich, S. R. 2021a. *Advanced in Intelligent System and Computing, 1323 AISC*: 170–179. DOI:10.1007/978-3-030-68004-6_22.

Ibragimovich, J. I., Isroilovich, Dj. O. & Abdullayevich, S. R. 2020b. Optimization of Identification of Micro-objects Based on the Use of Characteristics of Images and Properties of Models. *Int. Conf. on Information Science and Com. Tech*: 9351483. DOI:10.1109/ICISCT50599.2020.9351483.

Isroil Jumanov, Olim Djumanov & Rustam Safarov. 2021. *2nd Int. Conf on Energetics, Civil and Agricultural Engineering,* E3S Web Conf. 304. https://doi.org/10.1051/e3sconf/202130401007.

Isroil I Jumanov & Sunatillo M Kholmonov. 2022. Optimization of Identification of Non-stationary Objects Based on the Regulation of Systematic Error Values. *Int. Russian Automation Conference:* 22089779. DOI: 10.1109/RusAutoCon54946.2022.9896323

Jumanov, I. I., Djumanov, O. I. & Safarov, R. A. 2019. *Chemical Technology, Control and Management* 5: 71–78. https://doi.org/10.34920/2019.6.71–78.

Jumanov, I. I., Djumanov, O. I. & Safarov, R. A. 2020. Optimization of Identification of Images of Micro-objects Taking into Account Systematic Error Based on Neural Networks *Int. Russian Automation Conf.*: 626–631. DOI:10.1109/RusAutoCon49822.2020.9208164.

Jumanov, I. I., Safarov, R.A. & Xurramov, L.Y. 2021a. Optimization of Micro-object Identification Based on Detection and Correction of Distorted Image Points. *AIP Conference Proceedings* 2402(1), 070041. https://doi.org/10.1063/5.0074018.

Jumanov, I.I., Djumanov, O. I. & Safarov, R. A. 2021b. *Journal of Physics: Conference Series*: *1791(1)*, 012099. doi:10.1088/1742-6596/1791/1/012099.

Jumanov, I.I., Djumanov, O.I. & Safarov, R.A. 2021c. Mechanisms for Optimizing the Error Control of Micro-object Images Based on Hybrid Neural Network Models. *AIP Conference Proceedings* 2402(1), 030018., https://doi.org/10.1063/5.0074019.

Jumanov, I.I., Djumanov, O.I., & Safarov, R.A. 2021d. Methodology of Optimization of Identification of the Contour and Brightness-color Picture of Images of Micro-objects. *Proceedings-2021 International Russian Automation Conference*: 190–195. DOI: 10.1109/RusAutoCon52004.2021.9537567.

Jumanov, I.I., Safarov, R.A., & Djumanov, O.I. 2022. Detection of Distorted Points on images of Micro-objects Based on the Properties and Peculiarities of the Wavelet – Transformation. *International Russian Automation Conference*: 794–799. DOI: 10.1109/RusAutoCon54946.2022.9896243

Khanzhina, N. & Putin, E. 2016. Pollen Recognition for Allergy and Asthma Management Using GIST Features, *Communications in Computer and Information Science 674*: 515–525.

Korobeynikov, A., Kamalova, Y., Palabugin, M. & Basov, I. 2018. The Use of Convolutional Neural Network LeNet for Pollen Grains Classification. *In: "Instrumentation Engineering, Electronics and Telecommunications" Proceedings of the IV Int. Forum*: 38–44. Izhevsk: Russia.

Li, H. 1995. Multisensor Image Fusion Using the Wavelet Transform. H. Li, B.S. Manjunath, S.K. Mitra. *Graphical Models and Image Processing* 57(3): 235–245.

Redondo, R., Bueno, G., *et al.* 2015. Pollen Segmentation and Feature Evaluation for Automatic Classification in Brightfield Microscopy. *Computers and Electronics in Agriculture* 110: 56–69.

Rodriguez-Damian, M. *et al.* 2006. Automatic Detection and Classification of Grains of pollen Based on Shape and Texture. *IEEE Transactions on Systems Man and Cybernetics part C (Applications and Reviews)* 36(4): 531–542.

Sebastiano Battiato *et al.* 2020. Detection and Classification of Pollen Grain Microscope Images. *Proceedings of the IEEE/CVF Conf. on Computer Vision and Pattern Recognition (CVPR)*: Workshops: 980–981.

Shemi, P.M. 2015. An Improved Method of Audio Denoising Based on Wavelet Transform. Proc. of *2015 IEEE Int. Conf. on Power, Instrumentation, Control and Computing*: 1–6.

Silva, R. D., Minetto, R., Schwartz, W. R. & H. Pedrini. 2008. Satellite Image Segmentation Using Wavelet Transforms Based on Color and Texture Features. *International Symposium on Visual Computing*: 113–122.

Silva, R. D., Schwartz, W. R. & Pedrini, H. 2011. Image Segmentation Based on Wavelet Feature Descriptor and Dimensionality Reduction Applied to Remote Sensing. *Chilean Journal of Statistics* 2(2): 51–60.

Tcheng, D. K., Nayak, A. K., Fowlkes, C.C. & Punyasena, S.W. 2016. Visual Recognition Software for Binary Classification and its Application to Spruce Pollen Identification, *PLoS one* 11(2): 1–19.

Yane, B. 2007. *Sifrovaya Obrabotka Izobrajeniy: 584.* Texnosfera, Moscow.

Zhang Y, *et al.* 2004. Towards Automation of Palynology 3: Pollen Pattern Recognition Using Gabor Transforms and Digital Moments. *J Quat Sci* 19:763–768, ISSN 0627-8179.

Artificial Intelligence, Blockchain, Computing and Security – Dagur et al. (Eds)
© 2024 The Author(s), ISBN: 978-1-032-67841-2

Methods of increasing data reliability based on distributed and parallel technologies based on blockchain

Nazarov Fayzullo Makhmadiyarovich* & Yarmatov Sherzodjon
Samarkand State University named after Sharof Rashidov, Samarkand, Uzbekistan

ABSTRACT: This article examines directions and mechanisms for increasing data reliability in computer networks. Currently, the rapid development of information technologies, the rapid growth of data flow, high-quality data processing carried out in network technologies, and the increase in the volume of data lead to an increase in the problem of data reliability. It is an urgent issue to find solutions based on the use of modern technologies to solve these problems. The simultaneous processing of various types of data in information systems, video, audio, text and digital data, creates big data. The variety of data types in bigdata creates the problem of quality data processing, which greatly affects the reliability of the data. Research shows that breaches of data integrity mainly manifest in three directions. In this case, there is a violation of the reliability of interrelated data in data transmission and storage, in the processing of large volumes of data and in the transcription of video data. It is created due to errors created during data transmission based on artificial and natural redundancy. To solve the mentioned problems , increasing data reliability based on blockchain mechanisms for payment systems in data transmission, increasing data reliability based on error minimization mechanisms in video information systems, and distributed computing and parallel mechanisms in large-scale information systems based on methods of increasing data reliability are researched.

1 INTRODUCTION

Currently, automation of processes based on digital technologies and artificial intelligence technologies is one of the urgent issues. Today, solving problems related to the development of information technologies and their implementation in the spheres of production, science, education and sports, based on the experience of leading developed countries, serves to increase the quality level of this sphere (Akhatov *et al.* 2021a; Antonopoulos 2014). Today, solving problems related to the development of information technologies and intellectual systems and their application in the fields of production, science, education and sports based on the experience of leading countries serves to increase the quality level of this field (Akhatov *et al.* 2021b). The processes of data processing and transmission through information systems in computer networks are increasing, and the problem of ensuring the reliability of data in these systems is also increasing. The variety of data, the emergence of big data and the increase of media data require the development of special methods and algorithms to increase the reliability of data. In the process of data flow processing and transmission in computer networks, the probability of data reliability violation increases (Pedro 2014).

Algorithms based on the elimination of threats and violations to information in computer networks serve to ensure the reliability of information. Ensuring data reliability serves to increase the quality and level of processed data (Akhatov *et al.* 2021a). The following

*Corresponding Author: fayzulla-samsu@mail.ru

DOI: 10.1201/9781032684994-103

approaches are proposed in the article to solve the mentioned problems. In this case, it was determined that data reliability improvement based on blockchain mechanisms in payment systems, data reliability improvement based on distributed computing and parallel mechanisms in big data information systems, data reliability improvement based on error minimization mechanisms in video information systems will be researched.

2 MAIN PART

2.1 *The template file improving data reliability based on blockchain mechanisms for payment systems in data transmission*

One of the most optimal ways to increase data reliability based on blockchain mechanisms for data transmission is the optimal method. Blockchain technology works on the basis of a mechanism for storing data in the form of a block of encrypted chains. The concept of blockchain technology was proposed by Satoshi Nakamoto (Japan) in 2008 and was first used in practice in 2009 when the "Bitcoin" system appeared. Based on its creation, it was organized on the basis of "cryptocurrency", that is, an encrypted currency, but the scope of application of this technology is much wider (Akhatov *et al.* 2021a; Pedro 2014). Based on this, it is possible to use blockchain technology as a secret data chain in network systems and payment systems. Blockchain is a technology based on distributed ledger mechanisms. Blockchain technology is a technology based on the organization of a secure block chain of data using cryptographic methods. In such cases, the chain of the data set is kept fragmented in several parts of the network, except for a separate server, but it is actively calculated on all devices connected to the network at the same time. In this case, the primary block is created first, at which point the previous block will not exist. Each subsequent block contains the parameter and necessary information used to create the next block (Wang *et al.* 2019). Users of the system will be able to see the full number of blocks, but will be able to use only the information that belongs to them. The working process of blockchain technology is that all transactions are carried out with confirmation based on a cryptographic method. That is, this process is performed by providing participants with a public key signature and a secret key, as shown in Figure 1. For encryption of closed transactions, a set of two cryptographic keys designed for fixing open transactions in the form of Figure 1 is implemented by users when entering the network and installing the necessary software on the workstation (Pierre-Yves *et al.* 2018).

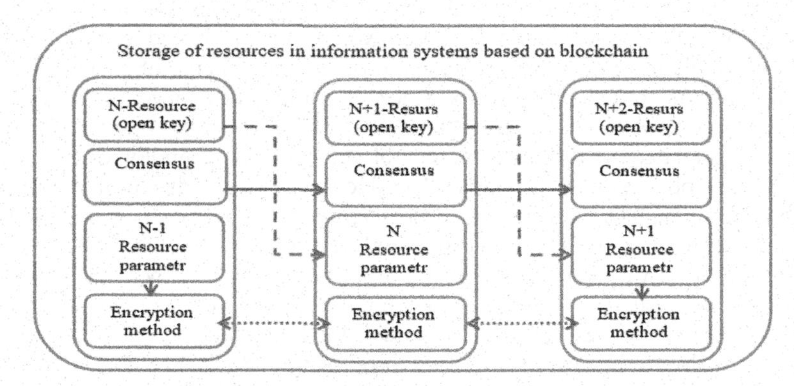

Figure 1. Scheme of storage of resources in information systems based on blockchain.

Because each user in the blockchain-based network sends transactions to the next user, the first transaction signs the public key of the next, and this is done by appending this information to the end of the transaction. In this way, incoming transactions can verify the entire chain of transactions by verifying the signatures of all previous users. When using distributed registry mechanisms as part of controlled systems, it is necessary to choose one of the consensus algorithms, depending on the form of data and their types. Consensus algorithms are used to confirm the authenticity of information contained in blockchain technology based on a distributed ledger (Mayank *et al.* 2018). Consensus Proof of Work, Proof of Stake, Proof-of-Burn, Proof-of-Activity, Proof-of-Capacity, Proof-of-Algorithms such as Storage are widespread. The main feature of blockchain technology is cryptographic data protection (Han *et al.* 2018). For cryptographic protection of data, it is required to apply crypto-resistant methods to the system. Blockchain technology makes it possible to create crypto-chain data. Data reliability improvement based on blockchain technology is carried out according to the algorithm presented in Figure 2.

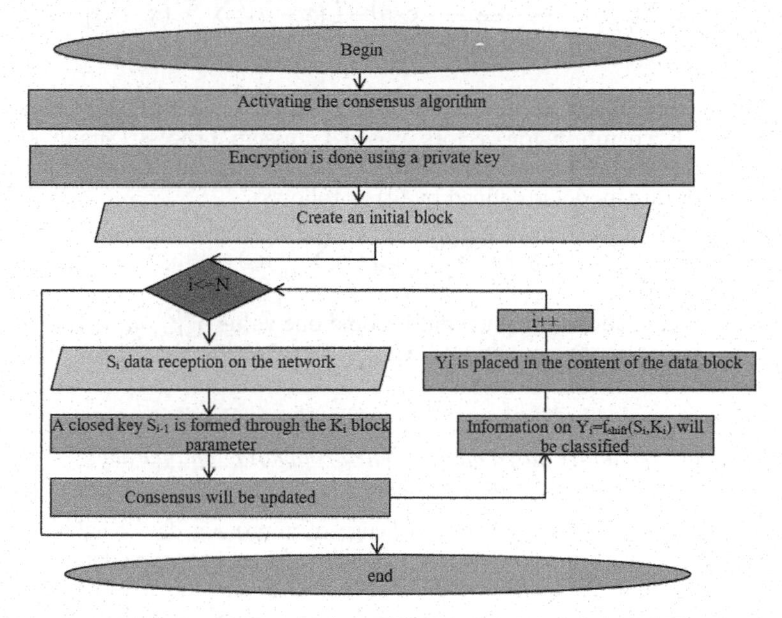

Figure 2. Algorithm for increasing data reliability based on blockchain technology.

N block presented in the algorithm is the last user of the database at the moment. Based on the developed algorithm, it provides an opportunity to break the entire database of information systems on the basis of blockchain. This is an effective way to increase data reliability.

2.2 *Increasing data reliability based on error minimization mechanisms in video information systems*

The issue of increasing data reliability based on error minimization mechanisms in video information systems is becoming relevant today. When transcribing video and audio data, errors can be minimized by linking this data together via blockchain. Analysis of video files in large-scale video networks, filtering, and transcription are necessary to explain the content of video files to increase reliability (Smith *et al.* 2021). The main problem in transcription is to increase the reliability of the data based on the correction of errors in the speech-based text. Video transcription is the process of converting speech from video into text. This can be done using automatic speech recognition technology, human transcription, or a combination of both (Shaun *et al.* 2018). Currently, the most common types of transcription are the following. There are several mathematical models of video transcription, among which the Gaussian mixture model is the most effective. In the Gaussian mixture model, the considered signal is assumed to consist of several distinct classes, where each class has its own statistical model (Sonia *et al.* 2013). The weights of each distribution correspond to the frequency that occurs in the signal. Thus, if in a hypothetical language voiceless signals make up 30% of all speech sounds, then the weight of the voiceless class is 0.3. The most typical mixture structure uses Gaussian (normal) distributions for each class, so the whole model is known as a Gaussian mixture model. According to the Appendix, the class distribution can take other forms than Gaussian, for example, if individual classes fall into a Beta distribution, it can be done using a Beta mixture model. But the Gaussian mixture model is the most common of the mixture models and conveniently demonstrates its application. The mathematical formulation of the Gaussian mixture model is as follows. The multivariate normal distribution for the variable X is defined by (1) as follows.

$$f(x,\sum,\mu)=\frac{1}{\sqrt{(2\pi)^N|\sum|}}\exp\left[-\frac{1}{2}(x-\mu)^T\sum^{-1}(x-\mu)\right] \quad (1)$$

Here Σ and μ are the covariance and mean of the N-dimensional process, respectively. In other words, this Recognition for vectors x is a Gaussian process (Shaun *et al.* 2018). Suppose a signal contains K classes, where each Σ_k and μ_k class has its own covariance. The mean Gaussian mixture model is defined by (2) as follows.

$$f(x)=\sum_{k=1}^{K}a_k f(x;\sum_k,\mu_k) \quad (2)$$

Here the a_k weights are added together and become one value $\sum_{k=1}^{K}a_k=1$.

It serves to increase the quality of video transcription based on the Gaussian mixture model.

2.3 *Increasing data reliability based on distributed computing and parallel mechanisms in big data information systems*

As the flow of data increases, so does the challenge of improving data reliability. Blockchain technologies are an effective way to increase the reliability of digital payment data in large volumes of data. According to IBM, 2.5 quintillion ($2{,}5 \times 2^{60}$) bytes of data are generated every day from these sources, so 90% of the data available in the world today can be said to have been generated in the last two years (Ajay *et al.* 2008). In order to optimize the process of efficient data storage, processing and transfer, it is desirable to introduce parallelization mechanisms to BigData. Parallel computing is the use of multiple or multiple computing devices to execute different parts of a given program or project at the same time. The parallel computing process leads to a dramatic increase in the efficiency of processing large amounts of data. Parallel computing is a way of organizing computer work, in which programs are developed as a set of interacting computing processes that work in parallel. In parallel computing, multi-processor systems have the ability to perform many actions resulting from solving one or more problems at the same time (Fan *et al.* 2014) . In this case, the increase in the number of processors does not lead to a decrease in the time to solve the problem. The use of parallel data processing is not the only way to increase the calculation speed, in such cases it is also required to increase the power of the processing units.

Parallel algorithms can be called scalable if the increase in the number of processors provides an increase in speed while maintaining processor efficiency. The speed of operation of devices with the same capabilities and computing systems is carried out as a ratio of the time T1 needed to solve a problem with one processor and the time TS to solve the same problem on the same system of processors. It is defined as the acceleration of algorithm implementation in the computing system of S devices.

To take into account the process scaling features, additional actions are performed in parallel at time T0 to organize the interaction of the processors with each other.

$$T_0 = ST_S - T_1 \quad (3)$$

Upper limit of the execution time of the parallel algorithm for any working s = processors can be determined by (4).

$$T_S < T_\infty + T_1/S \tag{4}$$

Each action in the algorithm is performed at time t and repetition is performed n times, the execution time of the T S algorithm can be estimated by the following (5) formula:

$$T_S = \sum_{T=T_1}^{T_\infty} \left\lceil \frac{n_t}{S} \right\rceil < \sum_{T=T_1}^{T_\infty} \left[\frac{n_t}{S} + 1 \right] = \frac{T_1}{S} + T_\infty \tag{5}$$

When computing large volumes of data, parallel computing and parallel algorithms are key issues to consider. Considering the asymptotic estimation and time of the developed parallel algorithm leads to high efficiency.

Distributed computing is a technique used to increase the scalability of parallel code execution using a network. Gene Amdahl suggests that the acceleration of a parallel program in distributed systems can be determined by (6) as follows.

$$S(N) = \frac{\mathrm{T}(1)}{\mathrm{T}(\mathrm{N})} = \frac{\mathrm{T}(1)}{\left(\mathrm{s} + \frac{\mathrm{p}}{\mathrm{N}}\right) \cdot \mathrm{T}(1)} = \frac{1}{s + \frac{p}{N}} \tag{6}$$

Here S(N) is the speedup achieved with N processors, T(1) is the time required to complete the work for one processor, T(N) is the time required for N processors, p- is the share of work done in parallel in the distributed computing system and $s = 1 - p$- is the share of work done in series. Based on this formula, the processor dependence indicator of distributed computing system acceleration is shown in Figure 3.

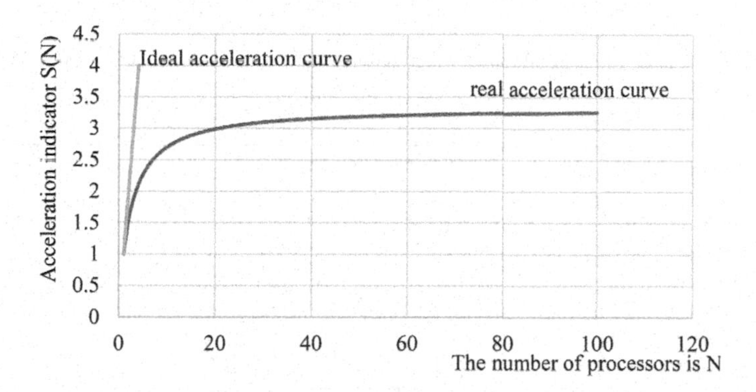

Figure 3. Acceleration of distributed computing system $p = 0{,}7$ and $s = 1 - 0{,}7 = 0{,}3$ processor dependency graph.

Distributed computing is a technique used to increase the scalability of parallel code execution using a network. Gene Amdahl suggests that the acceleration of a parallel program in distributed systems can be determined by (6) as follows. The use of blockchain technologies to increase data reliability in parallel computing processes will bring great results.

3 CONCLUSIONS

During the research, mechanisms and technologies for increasing data reliability were studied. Methods and algorithms for using distributed and parallel technologies were developed to increase data reliability. Three areas of increasing data reliability were identified, in which blockchain mechanisms based on distributed ledger in payment systems, transcription methods in video files, and the use of distributed computing and parallel mechanisms in large volumes of data were researched. Algorithms for increasing data reliability are developed based on distributed and parallel methods.

REFERENCES

Ajay, Kshemkalyani. & Mukesh Singhal. 2008. Distributed Computing. *Principles, Algorithms, and Systems.* Cambridge University Press, USA, 2008, P 756.

Akhatov, A. & Nazarov, F. & Rashidov, A. 2021a. Mechanisms of Information Reliability In big data and Blockchain Technologies. *International Conference on Information Science and Communication Technologies.* November 4, 5, 6. art. no. 9670052. ICISCT 2021 (IEEE).

Akhatov, A. & Nazarov, F. & Rashidov, A. 2021b. Increasing Data Reliability by Using Big Data Parallelization Mechanisms. *International Conference on Information Science and Communication Technologies.* November 4, 5, 6. art. no. 9670387, ICISCT 2021 (IEEE).

Antonopoulos, AM. 2014. The Blockchain. *Mastering Bitcoin.* O'Reilly Media, Inc. 18–26.

Fan, Zhang. & Zude, Zhou. & Wenjun, Hu. 2014. Distributed Storage and Processing Method for Big Data Sensing Information of Machine Operation Condition. *Journal of Software*, VOL. 9, NO. 10, OCTOBER.

Hamerly, G. & Elkan, C. Learning the K in K-Means. 2003. " in *Proc. NIPS'03 the 16th International Conference on Neural Information Processing System,* Whistler, British Columbia, Canada, December 09–11, pp. 281–288.

Han, R. Gramoli, W. Xu, X. 2018. Evaluating Blockchains for IoT. In: 2018 9th IFIP International Conference on New Technologies, Mobility and Security (NTMS). P 12–28.

Mayank, R. & Danilo, G. & Katina, K. 2019. *SoK of Used Cryptography in Blockchain.* Department of Information Security and Communication Technologies, Norwegian University of Science and Technology. P 54.

Pedro Franco. 2014. The Blockchain. *Understanding Bitcoin: Cryptography, Engineering and Economics.* John Wiley & Sons, P.288.

Pierre-Yves Piriou. & Jean-Francois, Dumas. 2018. *Simulation of Stochastic Blockchain Models.* Chatou, France. P. [1–8].

Shaun, W. & Ault, Rene. & Perez, J. & Chloe, & Kimble, A. & Jin Wang, 2018. On Speech Recognition Algorithms. *International Journal of Machine Learning and Computing*, Vol. 8, No. 6, December.

Smith, Peter. Virpioja, Sami. Kurimo, Mikko. 2021. Advances in Subword-based HMM-DNN Speech Recognition Across Languages. *Computer Speech & Language, DOI: 10.1016/j.csl.2020.101158, March.*

Sonia, Sunny. & David Peter, S, & Poulose Jacob, Performance of Different Classifiers in Speech Recognition", *IJRET*, pp. 590–597, 2013.

Sunanda Mendiratta. & Neelam Turk. & Deepali Bansal. A Robust Isolated Automatic Speech Recognition System using Machine Learning Techniques, *International Journal of Innovative Technology and Exploring Engineering*, ISSN:2278-3075, August.

Wang, L. & Shen, X. & Li, J. & Shao, J. & Yang, Y. 2019. Cryptographic Primitives in Blockchains. *Journal of Network and Computer Applications*, vol. 127, P. 43 – 58.

Artificial Intelligence, Blockchain, Computing and Security – Dagur et al. (Eds)
© 2024 The Author(s), ISBN: 978-1-032-67841-2

Patch-based lesion detection using deep learning method on small mammography dataset

Sh. Kh Fazilov*
Professor, Research Institute for Development of Digital Technologies and Artificial Intelligence, Tashkent, Uzbekistan

Kh.S. Abdieva*
Ph.D. student, Department of Software Engineering, Samarkand State University, Samarkand, Uzbekistan

O.R. Yusupov*
Assistant Professor, Department of Software Engineering, Samarkand State University, Samarkand, Uzbekistan

ABSTRACT: Traditional Computer Aided Detection (CAD) systems employed scanned films (low image quality) and were constructed with restricted computing resources, leading to a less reliable application procedure in breast cancer imaging. The issue of a small dataset is handled in this study using two strategies: i) using image patches as inputs rather than recognizing full-sized images; and (ii) employing the idea of transfer learning, which uses the skills learned during training for one task for another task that is closely related (also known as domain adaptation). In this regard, the CNN model that was recently trained is modified to identify masses in FFDM first, and then the CNN model that was previously taught to differentiate between mass and non-mass image patches in the Screen-Film Mammogram (SFM). Private datasets are utilized for this.

The most prevalent type of cancer among women is breast cancer. According to estimates, 12% of women in the United States will receive a breast cancer diagnosis at some point in their lives [1]. Other studies have found that, except for melanoma skin cancer, breast cancer has the highest incidence and mortality rates of any type of cancer [2]. The World Health Organization's most recent statistics show that the incidence of breast cancer was the highest among oncological diseases in Uzbekistan in 2018 with a death rate of 1,449, or 0.92% of all deaths [2,3]. In 1993, there were 5.3 cases of this illness per 100,000 persons, but by 1998, that number had risen to 6.1. Every year, more people with breast cancer are receiving a diagnosis. If we take the Samarkand region alone, in 2000, 154 patients with this disease were registered in the primary register, and 2019, their number reached 300. The incidence of breast cancer has increased dramatically in recent years. Most patients are referred to doctors in stages III–IV after a long delay in the establishment of screening programs, which leads to early cancer identification, which can be credited to this. The death rate from breast cancer is the focus of intense efforts. Knowing that the stage of cancer at which it is identified affects the likelihood of survival makes it important to catch the disease early rather than later when it is more difficult to treat.

1 ANALYSIS OF EXISTING APPROACHES

1.1 *Mammography*

It is a highly accurate diagnostic method for breast cancer, with the help of which the correct diagnosis is made in 83–95 percent of cases [1]. Typically, the breast is examined using two

*Corresponding Authors: sh.fazilov@mail.ru, orif.habiba1994@gmail.com and ozod.yusupov @gmail.com

DOI: 10.1201/9781032684994-104

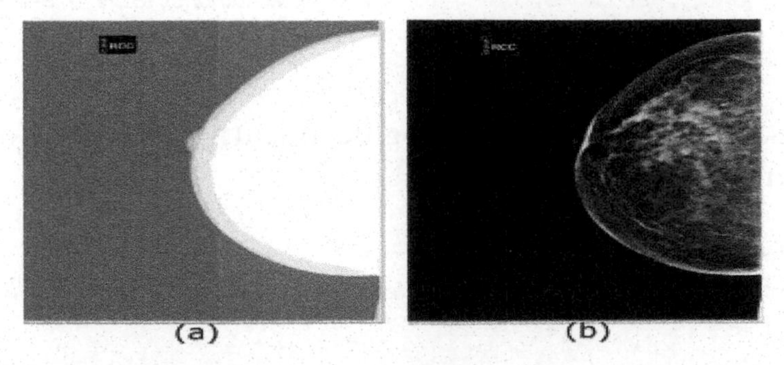

Figure 1. The raw and processed mammogram in FFDM.

standard perpendicular projections, the right, and side projections. Mammography is carried out using a special X-ray machine called mammography.

Historically, a screen-film mammogram (SFM) recorded the breast scan using photographic films. Thanks to improvements in imaging methods, high-quality Full-Field Digital Mammograms (FFDM) are now commonly utilized and can be seen instantly on computers.

1.2 *Computer-aided detection*

The automatic or semi-automatic evaluation of medical pictures by computers is known as a computer-assisted detection (CAD) system. The availability of precise CAD techniques in the realm of breast imaging can significantly enhance the practices used for breast screening today. In this context, Convolutional Neural Network (CNN)-based deep learning techniques have become more significant in the area of medical image processing, and efforts are being made to create contemporary CAD systems based on these recently created CNN algorithms [3]. This study explores current developments in CNN to enable the creation of an automated CAD system to help radiologists quickly and accurately detect lesions during breast cancer screening. This study is another step in that direction. The need for huge training datasets is one of the deep learning drawbacks, and building such a vast image collection to extract pertinent characteristics from many diseases can be time-consuming. It is suggested to employ image patches to train and test the CNN using an automated framework. To execute a domain adaptation between images with various features, such as natural photos, Screen-Film Mammograms (SFM), and Full-Field Digital Mammograms (FFDM), the idea of transfer learning is applied. The picture patches that were recovered from the SFM and FFDM were used to train CNN.

2 PREPROCESSING METHODS

2.1 *Datasets*

In this case, the pre-trained CNN model on a large mammography dataset is fine-tuned to detect masses in a small mammography dataset using the transfer learning methodology. A modest private dataset was used to train the model.

2.2 *Convolutional neural network*

A deep learning system known as the convolutional neural network (CNN) was created specifically to operate with two-dimensional image input. It can give weight to various visual attributes to distinguish them from one another. An input and output layer, numerous convolutional layers, a layer called batch normalization (BN), a layer called rectified linear unit (ReLU) or activation function, layers called pooling layers, and a layer called Fully Connected (FC) make up the CNN.

2.3 *ResNet50*

The design of the residual network (ResNet) [9] comprises batch normalization (BN) layers after a stack of three convolutional layers, pooling layers, and several residual layers, each of which contains several bottleneck blocks. Algorithm:

Stage 1: Input
Negative patches and Positive patches
Stage 2: CNN training
1. CNN patch classifier
2. Trained model

Stage 3: Testing
Stage 4: Patch classification
Stage 5: Mass detection

To train the CNN, small portions of the image are extracted and added to the newly formed framework. To identify the unseen testing patches as mass or non-mass patches, the model developed after the CNN training is initially applied (with different probabilities). The entire mammography is then rebuilt using the patches, and the Mass Probability Map (MPM) for the mammogram is then created using the categorization probabilities for each patch. A bounding box serves as the last definition of the probable mass region. The parts that follow describe the entire automated framework.

2.4 *Input patch extraction*

The entire breast is scanned using a sliding window method, and all potential patches are extracted from the image. The stride ($s \times s$) in both the vertical and horizontal directions, which also specifies the minimum overlap between two consecutive patches, regulates the total number of patches formed. The annotations included in the dataset are then used to classify each patch. For instance, a patch is given a positive label (mass candidate) if its central pixel is located inside the mass; otherwise, the label is given as negative (no mass). An equal number of positive and negative patches are extracted from mass images since every image in the private dataset has a lesion. An equal number of negative patches are then randomly chosen from the breast's normal region after all the positive patches have been recovered from the annotated ROIs (excluding the border area patches due to high contrast difference). This offers a balanced dataset for training purposes.

2.5 *CNN training*

The *ImageNet* dataset with input dimensions 224 × 224 × 3 is used to train the CNN described in section 2.2. The three dimensions correspond to the red, green, and blue color channels. Since mammography-collected patches only contain one channel (grey level), each patch (224 × 224 × 1) has been copied onto the three-color channels as done in prior works [4,6] to make the input patches compatible with the input of the pre-trained CNNs. Normalization, also known as zero centering, is a common stage in the classification of medical images. Global Contrast Normalization (GCN) is employed in this work. It calculates the average intensity for each image patch and deducts that value from each image pixel [5]. The dataset is divided into training, validation, and test sets for CNN training. The validation set is used to assess how well the trained model is performing at the end of each epoch whereas the training set is used to train the network and update its weights. The number of times the algorithm analyses the full dataset is referred to as an epoch in this context. Additionally, data augmentation is employed to create new samples from the training data that is currently available. As is frequently done in the literature [7,8], the negative and positive patches are enhanced on the fly utilizing horizontal flipping, rotation of up to 30°, and re-scaling by a factor selected between 0.75 and 1.25. Adam [9] is the optimizer, and the batch size is 128 (for a GPU of 12 GB). The validation loss function, which gauges the effectiveness of a classification model whose output is a probability value between 0 and 1, uses the cross-entropy loss function as its basis. By changing the domain from natural images to SFM, the degree of transfer learning is also examined. This is accomplished by initializing the CNN with the pre-trained ImageNet weights and tuning all of the CNN's layers for 100 epochs (without freezing any layer).

2.6 *Mass detection*

Mass detection is carried out entirely automatically, without human involvement. The following procedures are used to accomplish this:

1. First, using a sliding window method, all potential patches are extracted from each image.
2. The trained CNN is used to analyze the patches and determine each patch's mass probability. Following patch extraction using a sliding window approach, the image is then rebuilt by stacking the patches in order from left to right and top to bottom, with the stride value (s s) specifying the overlap between the patches. The mass probabilities (on each patch) are then linearly interpolated to get the MPM as follows:

$$Mass\ probability = \frac{\sum mass\ probability\ of\ overlapping\ patches}{several\ overlapping\ patches}$$

3. After that, the MPM is thresholded at various probabilities. In mammography, this stage produces several regions (each region representing a likely mass), where each pixel has a probability larger than the selected threshold value.
4. Using linked component analysis, a bounding box is constructed to encompass each plausible region. According to earlier publications [10–12], a mass is deemed discovered if the Intersection over Union (IoU) between the bounding box and the annotated ground truth is greater than 0.2.

3 ANALYSIS OF RESULTS

To segregate the breast region from the background in the pre-processing of these FFDMs, global thresholding is used (see Figure 2), and all right breasts are horizontally mirrored to maintain the left orientation of all mammograms. Note that the converted DICOM mammograms are always utilized at their original resolution. The patches of size 224 × 224 pixels in both datasets are produced using a stride of 56 × 56 pixels and utilized as the CNN's input. Using a 5-fold cross-validation technique, the dataset is split into training (60%), validation (20%), and test sets (20%) based on individual instances. All of the mammograms in the private dataset were tested, and the results showed a sensitivity of 0.91,0.06 at 1.7 FPI

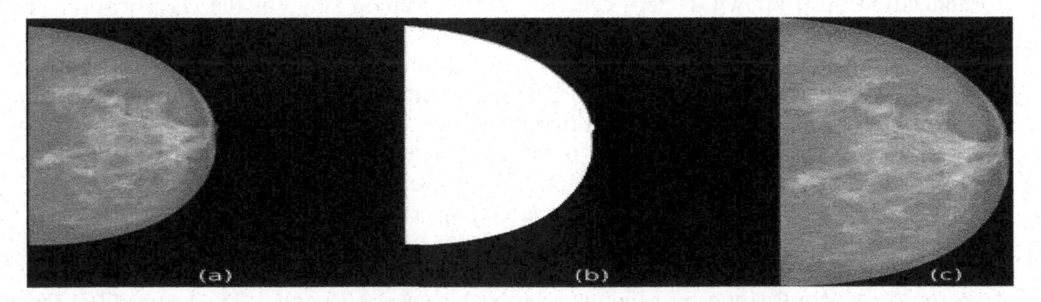

Figure 2. Sample mammogram (a) original, (b) segmented, and (c) mammogram cropped to breast profile.

and a mean AUROC of 0.87. A domain adaptation of the CNN from the natural images to the mammography dataset is carried out in this experiment. An SFM dataset is used to fine-tune a CNN model that was previously trained using a sizable collection of natural images (i.e. ImageNet). The private dataset's standard training and test splits are used. To achieve a trade-off between the computing needs and the quantity of training data, the stride value is chosen.

4 CONCLUSIONS

The intriguing feature of transfer learning is the ability to reuse a CNN model that has already been trained for a completely different problem, increasing accuracy for a given task while requiring fewer sophisticated algorithms and shorter training times. First, the value of transfer learning was examined between two completely unrelated picture domains, namely natural photos and mammograms, in this regard (SFM). For the goal of mass classification in SFM, a comparison of the performance of CNNs with randomly initialized weights vs. pre-trained (ImageNet) weight initialization was made; the results showed greater performances for the pre-trained models.

REFERENCES

[1] Stewart B. and Wild C., *World Cancer Report*. World Health Organisation, 2014.

[2] Yusupov O.R., Abdiyeva Kh.S., Primov A. "Preprocessing and Segmentation of Digital Mammogram Images for Early Detection of Breast Cancer". *IJARSET*, vol.8, issue 9, 2021.

[3] *Anatomy of the Breast*, https://www.mskcc.org/cancer-care/types/breast/anatomy-breast, [Online; accessed 1-July-2019].

[4] Moreira C., Amaral I., Domingues I., Cardoso A., Cardoso M. J., and Cardoso J. S., "INbreast: Toward a Full-field Digital Mammographic Database", *Academic Radiology*, vol. 19, no. 2, pp. 236–248, 2012. DOI: 10.1016/ j.acra.2011.09.014.

[5] Sickles E., d'Orsi C., Bassett L., Appleton C., Berg W., Burnside E., *et al.*, "ACR BI-RADS R Mammography", *ACR BI-RADS R Atlas, Breast Imaging Reporting and Data System*, vol. 5, 2013.

[6] Halling-Brown M. D., Looney P. T., Patel M. N., Warren L. M., Mackenzie A., and Young K. C., *"The Oncology Medical Image Database (omi-db)"*, vol. 9039, 2014. DOI: 10.1117/12.2041674.

[7] *Radiology Assistant*, http://www.radiologyassistant.nl/en/p53b4082c92130/bi-rads-for-mammography-and-ultrasound-2013.html, [Online; accessed 2-July-2019].

[8] Dinnes J., Moss S., Melia J., Blanks R., Song F., and Kleijnen J., "Effectiveness and Cost-effectiveness of Double Reading of Mammograms in Breast Cancer Screening: Findings of a Systematic Review", *The Breast*, vol. 10, no. 6, pp. 455–463, 2001. DOI: 10.1054/brst.2001.0350.

[9] Dhungel N., Carneiro G., and Bradley A. P., "A Deep Learning Approach for the Analysis of Masses in Mammograms with Minimal user Intervention", *Medical Image Analysis*, vol. 37, pp. 114–128, 2017. DOI: 10.1016/j. media.2017.01.009.

[10] Shen W., Zhou M., Yang F., Yang C., and Tian J., "Multi-scale Convolutional Neural Networks for Lung Nodule Classification", in *International Conference on Information Processing in Medical Imaging*, Springer, 2015, pp. 588–599.

[11] Brosch T., Tam R., Initiative A. D. N., *et al.*, "Manifold Learning of Brain MRIs by Deep learning", in *International Conference on Medical Image Computing and Computer-Assisted Intervention*, Springer, 2013, pp. 633–640.

[12] Azizpour H., Razavian A. S., Sullivan J., Maki A., and Carlsson S., "From Generic to Specific Deep Representations for Visual Recognition", in *CVPRW DeepVision Workshop*, June 11, 2015, Boston, MA, USA, IEEE, 2015.

Artificial Intelligence, Blockchain, Computing and Security – Dagur et al. (Eds)
© 2024 The Author(s), ISBN: 978-1-032-67841-2

Intelligent algorithms of digital processing of biomedical images in wavelet methods

H. Zaynidinov
Professor, Tashkent University of Information Technologies named after Muhammad al-Khwarizmi Uzbekistan

L. Xuramov & D. Khodjaeva
Samarkand State University, Samarkand, Uzbekistan

ABSTRACT: To date, the use of computer vision in biomedicine with the use of artificial intelligent systems, which in turn receive information from images, and then give out new knowledge and final conclusions about the disease is the most relevant. Images are obtained from different video sensors using them in different frequency ranges, which can be represented using two-dimensional brightness matrices and two-dimensional sensors. You can improve the image quality in several steps: preprocessing, highlighting the characteristic properties of the image (filtering, brightness equalization) classification and conclusion.

1 INTRODUCTION

To date, many studies are being conducted around the world to identify and diagnose many di eases in the field of biomedicine. Unlike mathematics and physics, the complexity of biomedical systems does not make the development of computerized programs for medicine a simple alg rithmic solution (Alabdullah *et al.* 2018; Akhatov *et al.* 2021a; Rashidov *et al.* 2021a). Traditional approaches to analysis used in this field do not provide high accuracy and efficiency, because they depend on the human factor. Therefore, the effective use of medical informatics and artificial intelligence in the field of biomedicine will allow us to develop this field and achieve high accuracy and efficiency (Alaeddine *et al.* 2018; Rashidov *et al.* 2022b). Processing of biosignals and medical images, machine learning of digital medical data and their monitoring with the help of artificial intelligence will increase the effectiveness of early diagnosis of diseases in patients. Signals obtained from biomedicine are mainly constructed from a mathematical point of view, and they are not suitable for representing discrete domain data, but rather for representing domain functions with a discont nuity point (Akhatov *et al.* 2021b; Kunanets *et al.* 2019). This article discusses one of the biomedical images, Germinoma Pineal Gland images, selected as an experiment for digital processing by wavelet methods. The wavelet models of Haar, Daubechies and Coiflet were used. In digital signal processing using Haar wavelet, an orthogonal function is formed, which leads to the magnitude of the differences in the signal graph. To reduce these differences, the use of Daubechies and Coiflet wavelets in creases the efficiency of intelligent processing of biomedical images.

2 METHODS

2.1 *The theory of wavelets*

The theory of wavelets is a powerful mathematical tool for signal analysis and synthesis, which has been widely used in digital signal processing in recent years (Zaynidinov *et al.*

 DOI: 10.1201/9781032684994-105

2019a; Oussous *et al.* 2017). Compared to signals processed digitally by the Fourier method, digital signal processing using wavelets provides significant flexibility, such as efficient filtering, interpolation, signal modeling, reduction of absolute and relative errors, noise removal, signal amplification, rapid extraction of signal characteristics and image data (Landse *et al.* 2015; Zaynidinov *et al.* 2020b). It is based on the following models.

$$\psi_{a,b}(t) \quad \text{– “mother” wave,} \quad \psi_{a,b}(t) = a^{-1/2}\psi((t-b)*a^{-1})$$

- used to study the frequency structure of a function in various dimensions and for the synthesis of functions in the processing of compression signals $\psi(t)$-general conditions for the mother wavelet:

1. The possibility of integration $\int_{-\infty}^{\infty}|\psi(t)|dt < \infty; \ \int_{-\infty}^{\infty}|\psi(t)|^2 dt < \infty$
2. Zero values $\int_{-\infty}^{\infty} t^m \psi(t)dt = 0$

2.2 *Haar wavelet for medical image processing*

Let us be given $n \times n$ a numerical matrix of the image in which

$$[x_{i,j}],\ i = 1,...,n;\ j = 1,...,n \text{ and let the equality hold } f(s,\ t) = \sum_{i}^{n}\sum_{j=1}^{n} x_{i,j} H_{I_i \times I_j}(h,t) \quad (1)$$

in which the plot can be represented the plane (h,t), defined in $[0,1] \times [0,1]$ unit domain as a function $f(h,t)$ with two invariant variables.

Here

$$H_{I_i \times I_j}(h,t) = \begin{cases} 1, & ((h,t) \subset I_i \times I_j) \\ 0, & ((h,t) \notin I_i \times I_j) \end{cases} = H_{I_i}(h)\ H_{I_j}(t) \quad (2)$$

denotes the i vertical rows of matrix h, $x_{i,j}$ in equation (1) above. By matching (2) to (1).

$$f(h,\ t) = \frac{1}{n}\sum_{i=1}^{n}\sum_{j=1}^{n} x_{i,j}\phi_{n,i}(h)\phi_{n,j}(t) \quad (3)$$

Here we will have an equation of a different kind

$$K_i(t) = \sum_{j=0}^{n-1} a_{n,j}\phi_{n,j}(t) + \sum_{j=0}^{n-1} d_{n,j}\psi_{n,j}(t) \quad (4)$$

Now comparing (3) to (4) we get the following:

$$f(h,\ t) = \frac{1}{n}\sum_{i=1}^{n}{}_i(t)\ \phi_{n,i-1}(h) \text{ here } \alpha_j(t) = \sum_{i=1}^{n} a_{n,j}^{i}\phi_{n,i}(h);\ \beta_j(h) = \sum_{i=1}^{n} d_{n,j}^{i}\phi_{n,i}(h)$$

for the value of buyer j, α_j and β_j the values of const. (3) can be applied to each column of this image using a single variable(Juraev *et al.* 2020a; Urakov *et al.* 2022a). By doing this, we will have,

$$f(h,\ t) = \frac{1}{n}\left(\sum_{i=2}^{n-1}\alpha_j(h)\ \phi_{n,j}(t) + \sum_{i=2}^{n}\beta_j(h)\ \psi_{n,j}(t)\right)$$

Having considered the presented equations, we will have

$$f(h,\ t)=\sum_{i=0}^{n}\sum_{i=1}^{n}(a_{i,j}^{n-1}\ \phi_{n+1,j}(t)\phi_{n,i}(h)+q_{i,j}^{n-1}\ \phi_{n,+1j}(t)\psi_{n,i}(h)+v_{i,j}^{n-1}\psi_{n-1,j}(t)\phi_{n-1,i}(h)$$

$$+\,d_{i,j}^{n-1}\psi_{n-1,j}(t)\psi_{n-1,i}(h)) \tag{5}$$

here

$$a_{i,j}^{n-1}=\frac{1}{2^n}\tilde{a}_{n-1,i}^{j};\ h_{i,j}^{n-1}=\frac{1}{2^n}\tilde{d}_{n-1,i}^{j};\ v_{i,j}^{n-1}=\frac{1}{2^n}\tilde{\tilde{a}}_{n-1,i}^{j};\ d_{i,j}^{n-1}=\frac{1}{2^n}\tilde{\tilde{d}}_{n,i-1}^{j}$$

As a result, the $f(s,t)$ functions are divided into the following functions

$$\phi_{n,j}(t)\phi_{n,i}(h);\ \ \phi_{n,j}(t)\psi_{n-1,i}(h);\ \text{ and }\ \psi_{n,j}(t)\phi_{n,i}(h),\ \psi_{n,j}(t)\psi_{n,i}(h)$$

Consider a 2×2 image matrix. Let f be a function forming a square rectangle.

$$f=\begin{pmatrix} f(0,0) & f\left(0,\frac{1}{2}\right) \\ f\left(0,\frac{1}{2}\right) & f\left(\frac{1}{2},\frac{1}{2}\right)\end{pmatrix}=\begin{pmatrix} h_1 & h_{1,2} \\ h_{2,1}, & h_{2,2}\end{pmatrix} \tag{6}$$

for this matrix, we apply a rapid change of Haar, that is, when applying a variable of rapid change of Haar along the first row, based on the approximation coefficients (6), let be

$$\frac{h_1+h_{1,2}}{2}\ \text{and}\ \frac{h_{2,1}+h_{2,2}}{2}\ \text{and}\ \frac{h_1-h_{1,2}}{2}\ \text{and}\ \frac{h_{2,1}-h_{2,2}}{2}$$

we form the obtained values in the form of a matrix Consider the array f:

$$\vec{f}=\begin{pmatrix} a_1\ a_2\ a_3\ a_4 \\ b_1\ b_2\ b_3\ b_4 \\ c_1\ c_2\ c_3\ c_4 \\ d_1\ d_2\ d_3\ d_4 \end{pmatrix}$$

The array f a can also be applied to other strings using a wavelet transform. by calculating the average value of the probability coefficients of the first and second columns, the third and fourth columns in the rows and putting the average value of the difference of the first two columns in the last two columns, we arrive at the following result

$$\begin{pmatrix} \frac{a_1+a_2}{2} & \frac{a_1-a_2}{2} & \frac{a_3+a_4}{2} & \frac{a_3-a_4}{2} \\ \frac{b_1+b_2}{2} & \frac{b_1-b_2}{2} & \frac{b_3+b_4}{2} & \frac{b_3-b_4}{2} \\ \frac{c_1+c_2}{2} & \frac{c_1-c_{32}}{2} & \frac{c_3+c_4}{2} & \frac{c_3-c_4}{2} \\ \frac{d_1+d_2}{2} & \frac{d_1-d_2}{2} & \frac{d_3+d_4}{2} & \frac{d_3-d_4}{2} \end{pmatrix}$$

Continuing this iteration, we arrive at the following result. Thus we will have the following arrays:

$$A=\begin{pmatrix} k_1\ k_2 \\ l_1\ l_2 \end{pmatrix}\begin{matrix}\rightarrow\\ \rightarrow\end{matrix}\begin{pmatrix} \frac{k_1+k_2}{2} & \frac{k_1-k_2}{2} \\ \frac{l_1+l_2}{2} & \frac{l_1-l_2}{2} \end{pmatrix}=\begin{pmatrix} b_{11}\ b_{21} \\ b_{21}\ b_{22} \end{pmatrix}->\begin{pmatrix} A_1\ H_2 \\ V_2\ D_2 \end{pmatrix}=A2c;$$

Thus, we have created a wavelet transform with two variables. Arrays in the given example have their own value names F, K, L and M, respectively. F is the approximation field, K is a horizontal field, L is a vertical field that contains, M (diagonal area) contains.

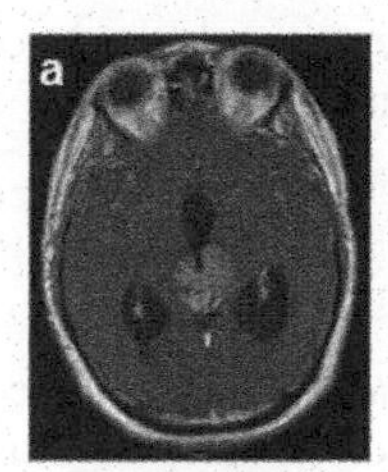
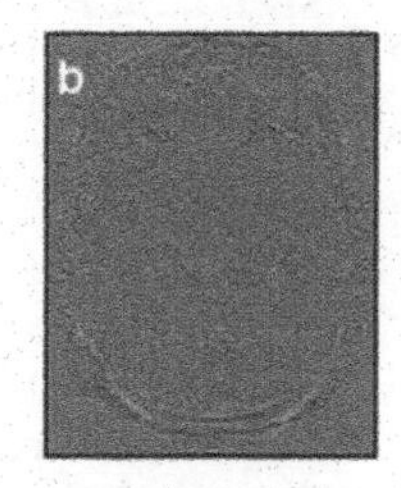
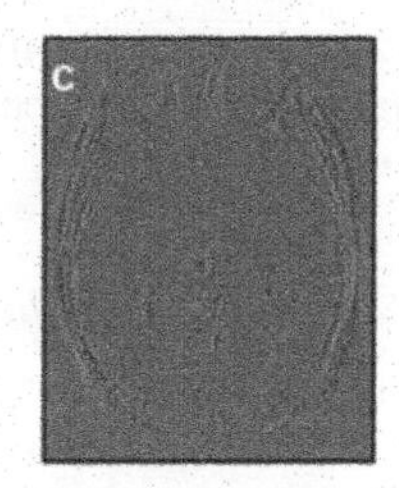

Figure 1. The result of interpolation of the analysis of the image of the pineal gland germinoma in the Haar wavelet: (a) approximation, (b) horizontal detail, (c) vertical detail.

2.3 *Daubechies – wavelet processing of medical images*

Building a Daubechies wavelet. The Daubechies wavelet is based on the scaling criterion and therefore has a limited number of coefficients(Urakov 2022b).

To construct the Daubechies wavelet, we write down the scaling and wavelet equation:

$$\phi(t) = \sqrt{2}\sum_{k} h_k\phi(2t-k); \quad \psi(t) = \sqrt{2}\sum_{k} g_k\phi(2t-k) \tag{7}$$

The wavelet function of the Daubechies wavelet $\psi(t)$ is usually denoted by the letter D and is formed by adding a number corresponding to the scale of the(Xurramov *et al.* 2022) Daubechies wavelet, that is, D2, D4, D6, h_k and g_k in the formula are the coefficients of the scaling and wavelet functions, respectively, for which the equation holds(Juraev 2022e):

h_n коэффициентларни топиш учун дан фойдаланилади, шунда P (8) полиномнинг кўриниши уйидагича бўлади:

$$P(y) = (1-y)^{-N}(1-y^N P(1-y)) \tag{8}$$

It is required to calculate a_i and d_i coefficients for $\phi(t)$ of the wavelet transform function.These coefficients are found using the following integral:

$$a_k = (f, \phi_k) = \int_R f(x)\phi_k(x)dx; \quad d_k = (f, \psi_k) = \int_R f(x)\psi_k(x)dx \tag{9}$$

it is worth saying that the evaluation of the coefficients requires a complicated process because there is a problem of summing the integral a_i and d_i (9). To solve this problem, the fast wavelet transform method proposed by Malla is used(Juraev *et al.* 2020b). The Mala calculation algorithm allows to determine the coefficients easily using algebraic iterations:

$$a_i = h_0 f_{2i} + h_1 f_{2i+1} + h_2 f_{2i+2} + h_3 f_{2i+3}; \quad d_i = g_0 f_{2i} + g_1 f_{2i+1} + g_2 f_{2i+2} + g_3 f_{2i+3} \tag{10}$$

a_i scale coefficients of Daubechies, d_i wavelet coefficients of Daubechies. These (10) A fast algorithm for estimating the value of the coefficients was introduced. According to the given formula (10), the wavelet transform based on the Daubechies wavelet is written as follows:

$$D(a,b) = \sum_i a_i + \sum_i d_i$$

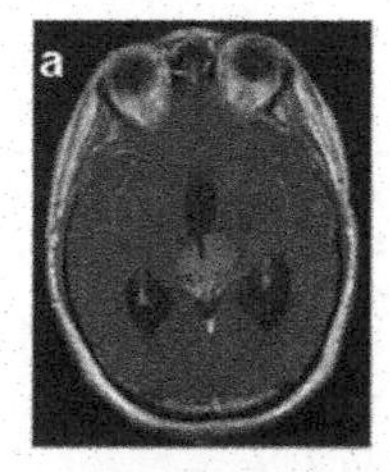
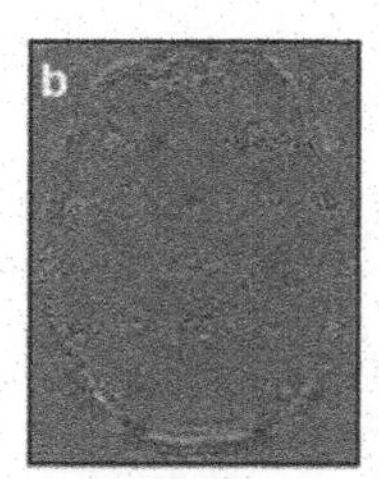
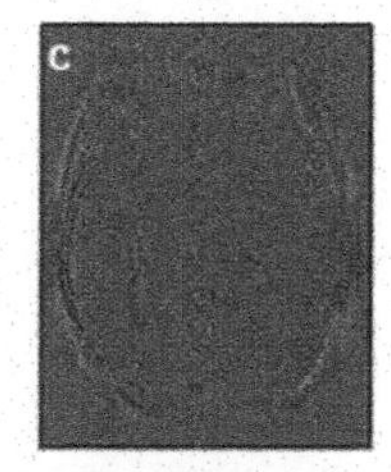

Figure 2. The result of the interpolation of the analysis of the image of the pineal gland germinoma in the Daubechies wavelet: (a) approximation, (b) horizontal detail, (c) vertical detail.

When the fourth-order Daubechies wavelet transform is performed, the two coefficients for the scaling function $\phi(t)$ become zero. Figure 2 shows image Analysis of Germinoma Pineal Gland Imaging.

2.4 *Coiflet – wavelet processing of medical images*

In the function $m_0(\omega) \in L_2(0, 2\pi)$ waves and Coiflet *Daubechies* induced by a common 2π - periodic, but for Coiflets a number of conditions are added to it, determining that the moments of the corresponding scaling function are zero, it is possible to achieve efficiency in approximation problems

L is a trigonometric polynomial. To construct a Coiflet wavelet, we must fulfill the following conditions():

$$\int \phi(t)t^l dt = 0,\ l = 1, ...,\ N/2; \quad \int \phi(t)t\ dt = 1; \int \psi(t)tdt = 0,\ l = 0, ...,\ N/2; \tag{11}$$

Or in the frequency domain:

$$\phi^{(l)}(1) = 0, l = 1, ..., N; /2\ \phi(1) = 1;\ \psi^{(l)}(0) = 0, l = 0, \ldots, N/2;$$

then according to the figure, the function m_0 under consideration for coiflets can be expressed in this way(10)

$$m_0(\omega) = \left(\frac{1 + \exp(-jw)^{2N}}{2}\right) P_1(w) \text{ here}$$

$$P_1(w) = \sum_{K=0}^{K-1} \left[\binom{K-1+k}{k} \left(\sin\left(\frac{w}{2}\right)\right)^{2k} + \left(\sin\left(\frac{w}{2}\right)\right)^{2K} F(w) \right],$$

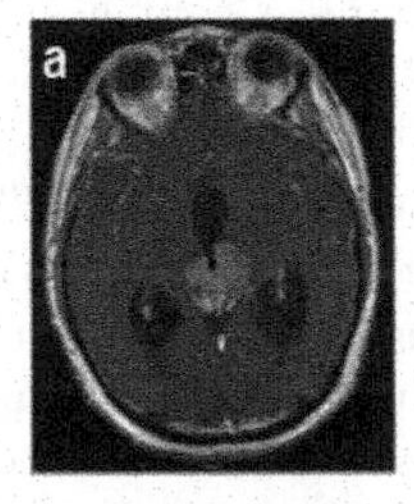

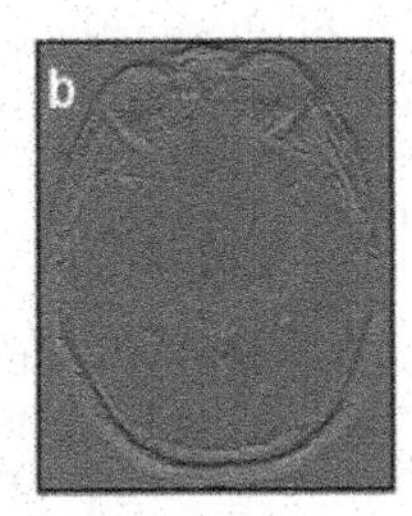

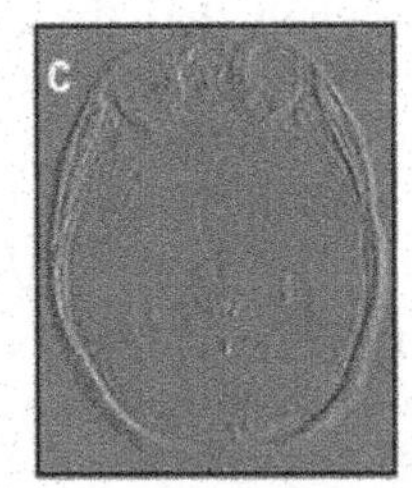

Figure 3. The result of interpolation of the analysis of the image of the pineal gland germinoma in the Coiflet wavelet: (a) approximation, (b) horizontal detail, (c) vertical detail.

Table 1. PSNR and the size of the reconstructed image using various Haar, Dobshy and Coiflet wavelets.

Names of Wevlets	*a	*b	*c
PSNR(Haar)	35.23	34.45	31.4
Size(kilobytes)	36.2	36.2	36.2
PSNR(Haar)	37.19	36.41	32.19
Size(kilobytes)	36.3	36.2	36.1

*a – Haar r, *b – Dobshy, *c – Coiflet

F is a trigonometric polynomial chosen in such a way that, the wave functions obtained using the polynomial m_0 in $|m_0(w)^2| + |m_0(w+\pi)^2| = 1$, condition, are called coiflets of degree $N = 2^k$.

This Table 1 shows the results of a numerical analysis of the performance of the wavelet method when the threshold of the image of the pineal gland germinoma is 3 and 4. In this case, the results of the analysis are recorded in the first Haar model, the second Daubechy model and the third model of the wavelet coiflet. From this table it can be seen that the result of digital processing using the Daubeshi wavelet is better than the result of digital image processing using the Harr wavelet. When comparing the result of digital image processing in the second and third models, it is shown that the output value in the coiflet wavelet is better.

3 CONCLUSIONS

Using artificial intelligence and machine learning to analyze medical images, society and biomedicine will reach a higher level. Of course, the joint work of researchers and programmers in the field of artificial intelligence implementation should be synchronized. To date, scientists from different countries have achieved several hundred results in this area. This article describes the evaluation of the quality of noise purification during the restoration of medical images from the point of view of PSNR. Based on the results, we conclude that digital processing of biosignals and medical images using artificial intelligence gives high results. Using the Coiflet Wavelet filter shows a higher PSNR value. The digitally processed image using the Coiflet wavelet has become better and brighter than the digitally processed image using the Daubechies and Harra wavelets. In addition, it was concluded that a high PSNR value indicates a good degree of noise reduction and this means that this developed algorithm can be widely used.

REFERENCES

Akhatov A., Nazarov F. & Rashidov A. Increasing Data Reliability by Using Bigdata Parallelization Mechanisms. *International Conference on Information Science and Communications Technologies*

Akhatov A., Nazarov F. & Rashidov A. Mechanisms of Information Reliability In Big Data and Blockchain Technologies. *International Conference on Information Science and Communications Technologies. 4,5,6 November. ICISCT 2021(IEEE), art. no. 9670052* DOI: 10.1109/ICISCT52966.2021.9670052

Akhatov A., Sabharwal M., Nazarov F., & Rashidov A. Application of Cryptographic Methods to Blockchain Technology to increase Data Reliability. *2nd International Conference on Advance Computing and Innovative Technologies in Engineering (ICACITE)*. 28–29 April 2022. DOI: 10.1109/ICACITE53722.2022.9823674

Alaeddine B, Nabil N., & Habiba Ch. 2020. "Parallel Processing Using Bigdata and Machine Learning Techniques for Intrusion Detection" *IAES International Journal of Artificial Intelligence (IJ-AI)* Vol. 9, No. 3, September 2020

Juraev J., Juraev U., Abdiyiv K., Saparova G. & Khodjaeva D. 2022. *Mathematical Model of the Process of Digital -processing of Images from an Ultrasound Device in Wavelet Method.*

Kunanets N., Vasiuta O. & Boiko N. 2019. "Advanced Technologies of Big Data Research in Distributed Information Systems" *International Scientific and Technical Conference on Computer Sciences and Information Technologies*, September 2019, 71 – 76 p., doi: 10.1109/STC-CSIT.2019.8929756

Landset S, Khoshgoftaar T., Richter A. & Hasanin T. 2015. "A Survey of Open Source Tools for Machine Learning wi Big Data in the Hadoop Ecosystem" *Journal of Big Data (2015) 2*:24, doi: 10.1186/s40537-015-0032-1

Oussous A., Benjelloun F.-Z., Lahcen A. A., & Belfkih S. 2018. "Big Data Technologies: A Survey", *Journal of King Saud University – Computer and Information Sciences* 30 (2018) 431–448 p, doi: 10.1016/j.jksuci.2017.06.001

Urakov Sh., Juraev J., Abdiyiv K. & Xurramov L. 2022. Digital Signal Processing with Polynomial and Dobeshi Wavlets.

Zaynidinov H. & O. Mallaev & B. Anvarjonov. 2020. A Parallel Algorithm for Finding the Human Face in the Image

Zaynidinov H. & Azimov, B. 2019. Biomedical Signals Interpolation Spline Models. *International Conference on Information Science and Communications Technologies: Applications, Trends and Opportunities* 5(2): 15–25.

Zaynidinov H. &. Kuchkarov M. 2019. *Modeling of Geophysical Signals Based on the Secondorder Local Interpolication Splaynes*

Zaynidinov H., Juraev J., & Juraev U. 2020. *Digital Image Processing with Two-Dimensional Haar Wavlets*

Artificial Intelligence, Blockchain, Computing and Security – Dagur et al. (Eds)
© 2024 The Author(s), ISBN: 978-1-032-67841-2

Braille classification algorithms using neural networks

A.R. Akhatov
Doctor of Technical Sciences & Vice-Rector of Samarkand State University named after Sharof Rashidov, Uzbekistan

Sh.A.B. Ulugmurodov
PhD student of Jizzakh branch of the National University of Uzbekistan named after Mirzo Ulugbek, Uzbekistan

ABSTRACT: With today's modern technology, various forms of handwritten text recognition exist. In particular, many algorithms are employed to convert Braille into plain text, including CNN, KNN, Random forest, and LeNetI5. The Support Vector Machine plays a distinct role when determining symbols in a matrix representation. The resulting matrix yields novel outcomes in image analysis. To maintain the results generated by the implemented algorithms and to consistently update the data, it is advisable to establish a SQL database and present the results in a condensed format. Notably, considerable effort is being invested in converting Braille to text. These six young women at MIT are developing a device that could revolutionize the lives of blind individuals: it instantaneously translates text to Braille. Unfortunately, only approximately 10% of blind Uzbeks are able to read Braille, despite the fact that proficiency in it can greatly enhance their employment prospects.

1 INTRODUCTION

Handwritten text recognition is a subcategory of character recognition technology, which encompasses basic data processing techniques for the recognition of information written by hand or on specialized writing devices, financial reporting, zip code recognition, braille, and various calculations. In 1998, LeNetI5 was proposed as a method for verifying and familiarizing Handwritten US bank check numbers (Lijie Zhou & Weihai Yu 2022). The k-nearest neighbor (KNN) algorithm achieved a percentage error rate of 2.83% on the MNIST dataset. Support vector machine (SVM) and related algorithms have been widely utilized in such tasks. In 2012, Lin and other researchers conducted experiments in this field (Lin & Chen 2013). Specifically, they proposed a CNN-SVM hybrid model for character recognition, utilizing CNN for feature extraction and SVM as the classifier. This approach combined the benefits of both techniques and yielded outstanding experimental results in image classification tasks.

CNN is a sophisticated deep learning algorithm that is commonly utilized in the development of goal clarification and various other areas, such as image classification learning and face recognition algorithms. The CNN learns layer by layer, and each layer automatically extracts certain properties from all input data (Cohen & Welling 2016), resulting in exceptional performance and making it one of the leading systems for data identification and analysis in the pursuit of a common goal. Typically, neurons in the convolution layer interact with the upper layer via common receptor domains, and these attributes of the local area are extracted through convolution, with further features being combined in the convolution layer.

This sentence highlights the objective of the study (Akhatov & Ulugmurodov 2021), which is to evaluate and compare the performance of two widely used machine learning algorithms, Random Forest and SVM, in the task of face recognition. The article examines the recognition accuracy, processing time, and robustness to variations in lighting and pose of these algorithms.

 DOI: 10.1201/9781032684994-106

The authors propose a new method for the separation of text written in Braille into classes using neural network technologies (Akhatov & Ulugmurodov *et al.* 2022a). They evaluate the proposed method on a publicly available dataset and compare its performance with state-of-the-art methods(Akhatov & Ulugmurodov *et al.* 2022b).

2 CONVOLUTIONAL NEURAL NETWORKS AND THEIR PROPERTIES

A composite amalgamation of convolutional classes, rudimentary models, and fully interconnected layers constitutes the most basic CNN, which is a traditional branching sequence. The convolutional class layer employs a linear filter kernel to execute linear convolution and subsequently incorporates a non-linear activation function to calculate the suppressed features. Figure 1 illustrates an exemplar of the architecture of a traditional CNN neural network.

Distinguishing the following changes in the second classification (group) in the education system based on the scale of innovation: unrelated local and separate (one-sided), complex, interconnected, and systematic covering the whole school and higher education system.

Creating an "innovative individual" in the future development strategy of science and innovation, regardless of performance, should be the foundation of innovation and a progression of learning at each stage of acquiring new knowledge. Contemporary innovations are burgeoning abilities. The term "Innovative Education" will start to appear on national project websites this year and it states that innovative education is implemented in the process of generating new knowledge, which necessitates an increase.

This part of the article is related to the topic of image classification and the use of deep convolution neural networks to improve the accuracy of the model.

2.1 *Convolution layers*

Convolutional layers are a crucial component of CNNs, referred to as filters or kernels, which are utilized to extract low-dimensional features from high-dimensional data. The parameters consist of a set of trained convolutional kernels, each with a relatively small volume (length x width) to obtain a suitably sized lossless feature mapping of the necessary information. Each convolutional layer comprises a number of known and unknown convolutional kernels with hyper-parameters in a CNN, which should be determined empirically and each kernel is calculated. Certain symbols in the feature map indicate that we can extract the relevant part of the image, converting the original 3D image data into 2D feature maps. The permutations of all feature maps are considered our primary achievement and can be used for further feature extraction and rendering of final results. We employ multiple convolutional kernels to extract different aspects of features such as color, outline, and background.

A Minimum Spanning Tree (MST) is a tree of minimal size. It is important to note that while smaller spanning trees may not be value-invariant, if all edge weight values are accurate and precise, then the MST can be confirmed. We will not demonstrate this at this time. In the following section, we will examine the aspects of the CNN algorithm related to minimum residual trees.

In determining the depth of simple neural nodes, we primarily examined the characteristics in the deeper layers of the neural network, which resulted in an improvement in the precision of the parameters and a reduction in the loss value. In comparison to most traditional fully connected neural networks, the most prominent feature of the convolution class of layers is the ability to compose existing characteristics, which further enhances the level of accuracy of the model. If the kernel of the universal class of nxn convolution performs the operation by reflecting the values of the convolution class contained within the image of size mxm with the same image depth, we obtain a new image of size $1 + (m + n)/l$, which serves as the attribute index. The "l" in the equation represents the step size set for the convolution class. If the division is not exact, the step sizes can be adjusted to other values according to the values we have, i.e. discrepancies, to ensure that all parameters are capable of division.

When an image is displayed digitally, one byte (8 bits) of physical memory is allocated for the input or output signals. It can assume one of 256 possible values. Typically, the range of values used is 0...255; a value of 0 represents black, while a value of 255 represents white.

2.2 *Connecting layers*

A CNN typically intercalates a convolutional layer with a pooling layer after the previous layer. Its primary function is to reduce dimensionality, thus the attribute values in this layer are decreased, which leads to more efficient and simplified computations, by extracting crucial attribute values and grouping constants. One commonly used technique is to decrease the number of input values using a 2 x 2 filter, in which four-pixel values are condensed into a single-pixel value.

The largest sum operation takes into account the highest of four values (corresponding to the 2 x 2 dimensions of the input image). The depth of the image remains unchanged, as all data values will be preserved. As depicted in Figures 2 and 3, the largest sum has the primary benefit of accentuating the number of available attributes fine-tuned and generally provides clearer feature extraction and smoother averaging than other methods.

2.3 *The process of training*

The procedure for implementing CNN-supervised learning is as follows

2.3.1 *FC (Forest Control) layers*

We determine the output value for a fully cross-linked layer based on lambda (1).

$$x^1 = f \cdot u^1, \; here \; u^1 = W^1 x^1 + b^1 \tag{1}$$

where x^1: The input to the first layer of the neural network, f: An activation function, u^1: The weighted sum of the inputs to the first layer, calculated as the dot product of the weights W^1 and the input x^1 plus a bias term b^1, W^1: The weights or parameters of the first layer of the neural network, b^1: The bias term of the first layer of the neural network.

Here, $f(x)$ denotes the activation function, and in this instance, we employ the partial sigmoid function. The decrease in the current error in the experimental sample is as follows:

$$E^n = \frac{1^c}{2_{k=1}} (t_k^n \;\; y_k^n)^2 = \frac{1}{2} \| t^n \;\; y^n \|_2^2 \tag{2}$$

where n: Index of the sample, k: Index of the element in the sample, c: Number of elements in the sample, t^n: True target values for the sample n, y^n: Predicted values for the sample n

Here, the notation "c" denotes the total number of classes present in multiclass problems.

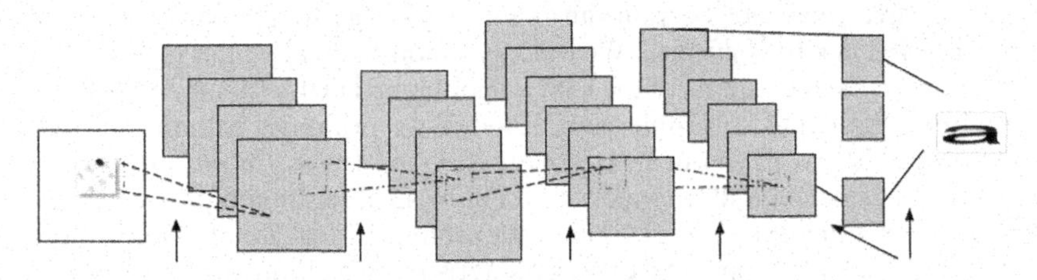

Figure 1. A visual representation of the LeNet5 model's structure.

Many factors were taken into account within the MNIST dataset. The practical significance and application of these are demonstrated through the performance of various techniques on the MNIST dataset in Table 1, including CKELM (Kernel Convolution Class Extremal Learning Machines), which is a convolutional neural network that utilizes a

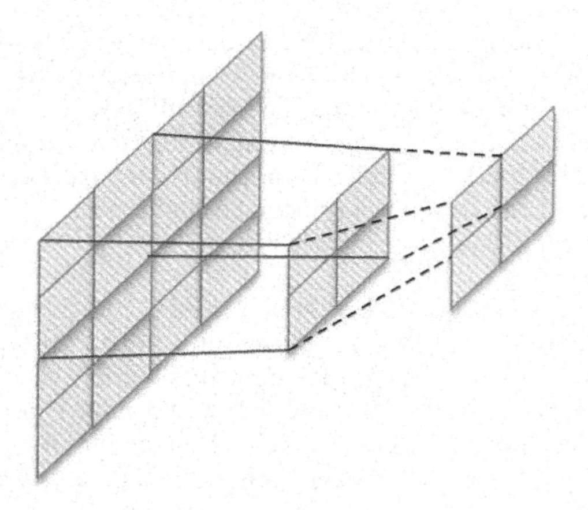

Figure 2. Maximum merge operation.

random weight scale to attain precise values of the numbers. The objective here is to serve as the primary contributor to feature detection, and the classifier measure is intended to be replaced with a high value. The MNIST dataset has a 2.96% error rate; the grid structure of the DAEs is 250 x 250 x 250; CNN-1 is a convolutional neural network that employs random weight filtering measurements since the weights are not adjusted, and the error rate in the CNN-2 table is a convolutional neural network following 25 training iterations.

By utilizing the Tesseract module, utilizing the KNN algorithm within it, the process involves enhancing the images and identifying the location of all attributes present within them.

Under the specified hardware conditions, a CNN requires approximately 190 to 200 seconds per iteration and takes approximately 3.5 hours or more for 45 iterations. In contrast, the Random Forest algorithm is relatively swift, requiring only 15 minutes or more to train on the MNIST dataset (when Ntree = 350). Furthermore, training the hybrid model takes less time than the original Random Forest due to the reduced data volume and improved accuracy. Our model significantly reduces the time required to obtain attribute values while providing precise expressions.

Table 1. Loss rate (%) for random forest and hybrid models in MNIST data set.

Ntree	RF	Hybrid-RF
100	3.02	2.16
200	3.03	2.12
300	2.94	2.06
400	2.94	2.02
500	2.90	1.98
600	2.89	2.02
700	2.85	2.08
800	2.88	2.12

3 RANDOM FOREST

A Random Forest comprises of k classification trees (figures and tables), with the primary objective of incorporating multiple weak classifiers into a single robust classifier. The classification trees consist of various nodes, with the root node representing the training set. Each inner node represents an incomplete classifier that classifies data based on a specific attribute, and each leaf node is a set of interpretations or, more specifically, the calculation of the value of several subsets of data. The ultimate outcome of the Random Forest is the optimal result chosen by the final computation across all nodes of the computation trees.

A random forest is an extant machine-learning algorithm that employs supervised random learning methods. It is utilized for both classification and regression issues in Machine Learning. The technique incorporates the principle of unified learning, which optimizes and integrates multiple computational steps to execute intricate tasks and enhance and refine the performance of a pre-existing model.

True to its moniker, "Random Forest" is a versatile classifier that employs multiple decision trees on various subsets of a given dataset, utilizing an arithmetic mean to enhance prediction accuracy. By utilizing multiple decision trees within the random forest, the prediction calculation is derived from each tree, leading to a higher likelihood of an accurate final result, as most predictions are based on the collective vote of the trees.

A substantial quantity of trees in the forest results in high precision and precludes the issue of overfitting. The accompanying illustration illustrates the functioning of the Random Forest algorithm:

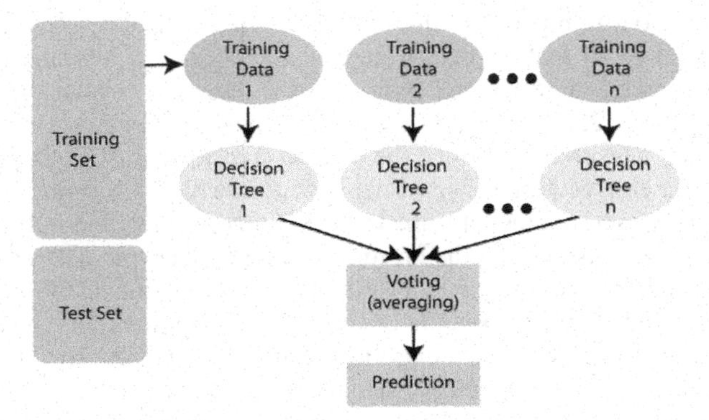

Figure 3. Forest algorithms scheme.

4 RESULTS

The results obtained through the utilization of the hybrid model indicate that developing a method for identifying Braille datasets utilizing MNIST is of paramount importance.

4.1 *MNIST dataset*

Assuming that Ntree represents the number of trees in the Random Forest (RF) and if Ntree is taken too small, the classification accuracy will not meet expectations. As RF is not susceptible to overfitting issues, the Ntree value can be maximized to guarantee classification accuracy. However, generating RF can be time-consuming, so the Ntree value is critical in terms of both performance and complexity. To circumvent the prolonged training period of Convolutional Neural Networks (CNN), the random weights method is utilized. Following feature extraction by CNN, the extracted features are then fed to RF for classification. Experiments were conducted with varying Ntree numbers for comparison. Table 1 illustrates the test error under different Ntree values.

5 CONCLUSIONS

In this paper, we propose a new model for the problem of image recognition and analysis that utilizes a hybrid approach combining both CNN and Random Forest. The CNN component extracts features with random weights, which are then used in conjunction with the Random Forest classification method. This approach aims to reduce the time required for obtaining attribute values for the model, addressing issues of long training times for CNN and random selection of attribute values for Random Forest. Our proposed model demonstrates a novel and efficient performance, effectively addressing image representation and recognition problems. The hybrid model extracts attributes through CNN with arbitrary weights and sends them to Random Forest for analysis, resulting in a significant reduction in feature extraction time. Additionally, the long training time of CNN in this model leads to a decrease in loss values and an increase in accuracy. However, manual feature selection by RF also has its drawbacks.

REFERENCES

Akhatov A., Renavikar A., Rashidov A. & Nazarov F. 2022b. Development of the Big Data Processing Architecture Based on Distributed Computing Systems, *Informatika va energetika muammolari O'zbekiston jurnali*, No. (1) *2022*, 71–79

Akhatov A., Sabharwal M., Nazarov F. & Rashidov A. 2022a. Application of Cryptographic Methods to Blockchain Technology to Increase Data Reliability. *2nd International Conference on Advance Computing and Innovative Technologies in Engineering (ICACITE 2022) DOI: 10.1109/ICACITE53722.2022.9823674*

Akhatov, A. R., & Kayumov, O. A. (2021). Scientific and Theoretical basis of Development and Introduction of Innovative Methods in Inclusive Education. *Universum:* психология и образование, (7), 46–48.

Akhatov, A. R., & Ulugmurodov, S. A. B. (2022, September). Methods and Algorithms for Distribution of Text from Images using Opencv2 Module. *In International Scientific and Current Research Conferences* (pp. 45–47).

Akhatov, A. R., and Ulugmurodov Sh AB Qayumov OA. "Working with Robot Simulation Using ROS and Gazebo in Inclusion Learning." Фан, таълим ва ишлаб чикариш интсграциясида ракамли иктисодиёт истикболлари" республика илмий-техник анжуман, УзМУ Жиззах филиали (2021): 5–6.

Akhatov, A., & Ulugmurodov, A. (2022). Methods and Algorithms for Separation of Text Written in Braille into Classes Using Neural Network Technologies. *Eurasian Journal of Mathematical Theory and Computer Sciences*, *2*(11), 4–8.

Ахатов, А., Улугмуродов, Ш. А., & Таджиев, М. (2022). Аудио для фонетической сегментации и говори для говори. *Zamonaviy Innovatsion Tadqiqotlarning Dolzarb muammolari va Rivojlanish Tendensiyalari: Yechimlar va Istiqbollar*, *1*(1), 146–149.

Cohen, T., & Welling, M. (2016, June). Group Equivariant Convolutional Networks. In *International Conference on Machine Learning* (pp. 2990–2999). *PMLR.*

Han M., Zhu X., and Yao W., Remote Sensing Image Classifi-cation Based on Neural Network Ensemble Algorithm, *Neuro-computing*, *vol. 78*, *no. 1*, pp. 133138, 2012.

Kayumov, Oybek. "Scientific and Theoretical Basis of Development and Introduction of Innovative Methods in Inclusive Education." *Science web academic papers collection* (2021).

Ko B. C., Kim S. H., and Nam J. Y., X-ray Image Classification Using Random Forests with Local Wavelet-based CS-local Binary Patterns, *Journal of Digital Imaging*, *vol. 24*, *no. 6*, pp. 1141 1151, 2011.

Kremic E. and Subasi A., Performance of Random Forest and SVM in Face Recognition, *International Arab Journal of Information Technology*, vol. 13, no. 2, pp. 287293, 2016.

Lijie Zhou, Weihai Yu, Improved Convolutional Neural Image Recognition Algorithm based on LeNet-5, *Journal of Computer Networks and Communications*, 10.1155/2022/1636203, *2022*, (1–5), (2022).

Lin W J and Chen J J 2013 Class-imbalanced Classifiers for High-dimensional Data *Brief. Bioinform. 14* 13–26

Rashidov A. & Akhatov A. 2021. "Big Data va Unig Turli Sohalardagi Tadbiqi", *Descendants of Muhammad Al-Khwarizmi*, *2021,* № *4* (*18*), 135–44

Тожиев, М., Улуғмуродов, Ш., & Ширинбоев, Р. (2022). Tasvirlar Sifatiniyaxshilashning Chiziqlikontrast Usuli. *Zamonaviy Innovatsion Tadqiqotlarning Dolzarb Muammolari va Rivojlanish Tendensiyalari: Yechimlar va Istiqbollar*, *1(1)*, 215–217.

Xia J., Falco N., Benediktsson J. A., Du P., and Chanussot J., Hyperspectral Image Classification with Rotation Random Forest via KPCA, *IEEE Journal of Selected Topics in Applied Earth Observations and Remote Sensing*, *vol. 10*, *no. 4*, pp. 1601 1609, 2017.

Xin M. and Wang Y., Research on Image Classification model Based on Deep Convolution Neural Network, *EURASIP Journal on Image and Video Processing*, *vol. 2019*, *no. 1*, *11 pages*, 2019.

Artificial Intelligence, Blockchain, Computing and Security – Dagur et al. (Eds)

Analytical review of quality indicators television images and methods of improvement

Kh.R. Davletova
Tashkent University of Information Technologies named after Muhammad al-Khwarizmi, Tashkent, Uzbekistan

ABSTRACT: This article describes parametric affecting the quality of TV picture and methods of correction of the decoded video. Given the criteria for assessing the quality of the reconstructed image TV. The visual quality of a TV image is affected by a large number of parameters that must be taken into account when building TV equipment devices and, if necessary, to correct the introduced distortions to improve the quality of images on the TV screen. At the same time, the quality of the optical image is determined by a number of factors and does not have a single, generalized quantitative assessment. Providing high quality HD TV picture. At the same time, high demands are placed on the quality of broadcast television programs, so the work related to improving the quality of images is of great practical importance and relevance. The article is the research and development of more effective methods for improving the visual quality of TV images, as well as objective metrics for their evaluation. The scientific novelty of this work lies in the development of more efficient methods and algorithms for improving the visual quality of TV images.

1 INTRODUCTION

With the development of DTV significantly increased the need for increasing the quantity and quality of TV programs. In the conditions of limited frequency resources to increase the number of broadcast programs while maintaining the image quality is possible only by creating more efficient methods of compression TV images that retain a good quality image at high rates of video compression.

As the stages of formation, transformation and coding of television images of distortion inevitably arise, especially in the management of the transmission in a limited frequency resource with a bit rate of less than 3 Mbit/s, it becomes important to assess and, where possible, to reduce the distortion introduced by coding.

Therefore, the work related to improving the quality of images are of great practical importance and relevance.

2 MAIN PART

2.1 *Classification of quality indicators of TV images*

Optical image quality is determined by a number of factors and does not have a single, generalized quantitative assessment. In general, the video image is observed on the TV screen should provide the perception of the transmitted image of the way it takes the viewer directly observed. The main quality characteristics of the visual image are [3]: -geometric shape and relative dimensions; -legibility of details; -brightness distribution; -color; -the location of objects in depth; -the perception of the relative motion of objects.

DOI: 10.1201/9781032684994-107

TV image parameters are selected to play in the minds of the viewer in detail all the characteristics of the transmitted scene. Alignment with the visual system provide large-scale, brightness and color options television image [3].

To estimate the distortions introduced by digital video compression used sequence specific signals. Since the inter-frame compression and inter-frame processing is performed by various algorithms and the distortion arising in their operation can also be divided into two types – inter-frame and inter-frame. Therefore, to assess their values require separate test signals for intra and inter-frame distortion. For the analysis of intra-distortion using a sequence of still images of a special type, which must take into account the specifics of digital encoding.

For example, the encoder input image signal is supplied in the form of frequent luminance vertical stripes as shown in Figure 1. In terms of analog television, this test is to determine the "resolution in the lines" defined by the width of the channel bandwidth. At frequency bands following in the maximum frequency of the analog channel transmission lines contrast will be significantly different from the original, and the image will be blurred [7].

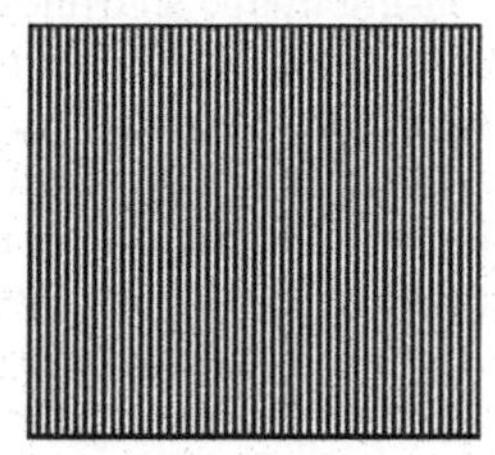

Figure 1. The test signal in the form of small-scale grating.

For example, the encoder input image signal is supplied in the form of frequent luminance vertical stripes as shown in Figure 1. In terms of analog television, this test is to determine the "resolution in the lines" defined by the width of the channel bandwidth. At frequency bands following in the maximum frequency of the analog channel transmission lines contrast will be significantly different from the original, and the image will be blurred [2].

For digital channel image coding sequence of this test does not give an adequate idea of the distortions introduced by the channel. In other words, it may happen that the picture will be transmitted without significant distortion, and other images are large distortion visible [2] shows an example of an image encoding step in a vertical luminance. At this signal only at very high compression ratios are noticeable distortion in the form of lines parallel to the border.

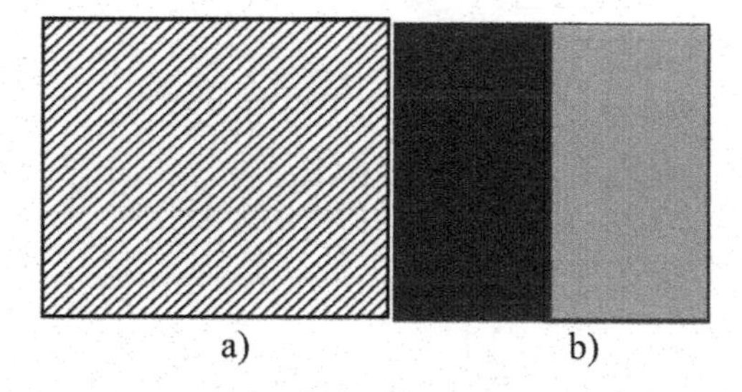

a) b)

Figure 2. a)-the test signal in a small-scale grating slant, b)-the test signal as a function of the vertical speed.

This is due to the Parseval equality, taking place in discrete orthogonal transformations [6]:

$$\sum_{i,j} X_{ij}^2 = C\sum_{i,j} \widehat{X}_{ij}^2 \tag{1}$$

Here X_{ij} and $\widehat{X}_{ij}$ – the amplitude of the luminance signal block 8x8 DCT points and amplitude spectrum, respectively, C – constant characterizing the given transformation. The summation is over all points (rows and columns) in the block. Equality Parseval shows that the energy of the signal up to a constant factor equal to the energy of the spectral components of the signal.

2.2 *Correction aperture distortion video*

One of the main video signal processing carried out in the channel chamber is aperture correction distortions introduced sensors TV signal that reduce image clarity.

At the sensors on the CCD TV signal in the low frequency part of the spectrum of light radiation increases the penetration depth of photons in the substrate. This leads to a noticeable deterioration of the frequency characteristics of the CCD at a wavelength of 0.6 microns. Therefore, to increase the resolution of the camera, working under natural light, it is recommended to use lenses with spectral filters, cut off part of the long-wave radiation [4]. Increasing the resolution at the expense of spectral filtering accompanied by some reduction in energy sensitivity cameras.

Aperture Correction aimed at increasing image sharpness in the TV transmitters. The term "aperture correction" owes its origin technique of electron beam television and associated with the need to compensate for the end section (aperture) of the electron scanning beam. The solid-state television this adjustment partially offsets the impact on the optical transfer function (MTF) of finite size of the CCD element. As the recession MTF occurs both vertically and horizontally, and the aperture correction is divided into horizontal and vertical [4].

Aperture correction is implemented using linear filtering techniques. The analog CCD cameras used for correction along a one-dimensional line. In digital cameras, the correction is two-dimensional in nature and along the rows and along the columns. Compensating the rise of the frequency response is achieved by subtracting the signal of the adjacent elements of the signal of the current element. [4] The most of the signals of neighboring cells is subtracted from the signal of the current item, the more the rise of the frequency response at the Nyquist frequency. Beyond this frequency, the frequency response falls off to reduce the effects of noise. Since the rise of the frequency response of the noise level increases, the value of the degree of compromise associated with the aperture correction signal / noise ratio of the video signal and is determined on a minimum of noise and linear error in the evaluation of the waveform. Under certain statistical characteristics estimated signal aperture corrector is constructed as a Wiener filter [5], the weighting factors which determine the pulse, transient and frequency response characteristics corrector.

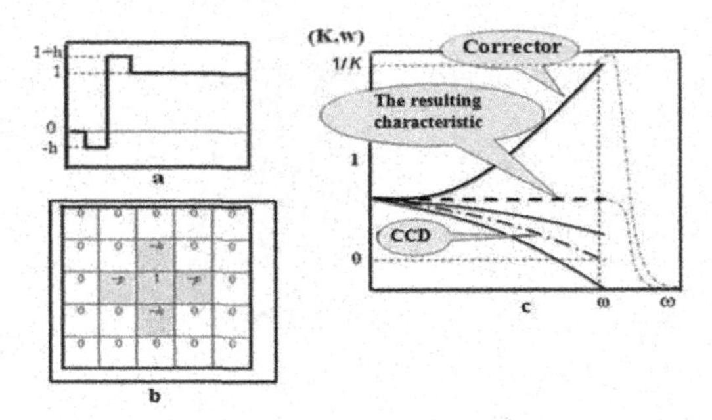

Figure 3. Characteristics aperture corrector: a) a one-dimensional transition; b)two-dimensional pulse; c)frequency.

2.3 *Gamma correction of TV images*

Gamma correction is a nonlinear transformation of the light-signal characteristics in order to harmonize the conditions of observation and modulation characteristics of CRT, LCD and plasma screens with contrast sensitivity of vision. Modulation characteristics of a kinescope (dependence of the brightness I of the voltage on the modulator U) approximated by the expression

$$I = kU^{\gamma}, \quad (2)$$

where the exponent γ is approximately 2.2 [5]. Light-signal characteristics of linear CCD is strictly up to the saturation of the potential well. Therefore, to obtain linear characteristics through camera system is introduced into the nonlinear converter nonlinearity exponent 1/ 2.2 = 0.45 [5].

On the other hand, the halftone reproduction plot important in some portion of the dynamic range is improved by increasing rate γ for this site. The security systems for a number of images of the plot are important undertones of large parts at high illumination. Experimentally it found that the best image quality in terms of surveillance is achieved with $\gamma = 0.45$ [5].

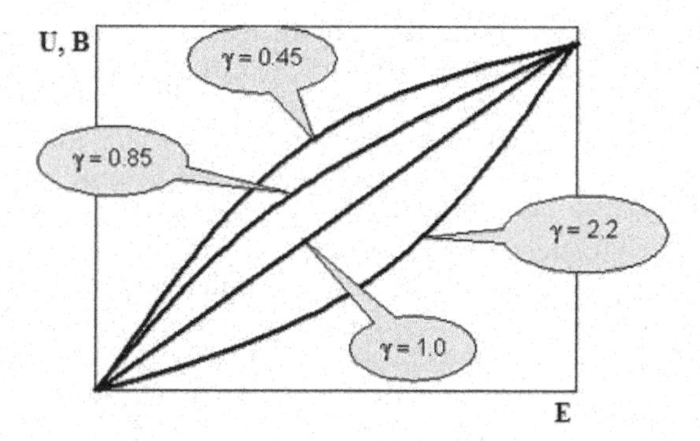

Figure 4. Characteristics light signal camcorder with varying degrees of gamma correction.

This creates the "correct" value for a visual illumination, and moves up the lower level of the observed light. But this advantage is achieved at the cost of the following disadvantages: -several times increases the noise in the dark areas of the image; -worsens legibility of the objects in the middle and upper range of the field of lighting [10].

Therefore, when enabled, gamma correction, despite the expansion of the visually observed light range becomes greater chance of missing that appears in the field of low contrast objects with medium lighting. In the case of low-contrast objects of observation there is a need to increase the parameter γ 1. Therefore, cameras are often equipped with switches gamma correction to initial setting by the manufacturer in the position of 0.45, and the possibility of increasing this parameter to 0.85 or 1. A final decision on the choice of the parameter gamma correction is usually taken at the same video director is often compromised the value of $\gamma = 0.85$ [5].

2.4 *Correction of color television camera*

The spectral characteristics of color TV transmission chamber must strictly comply with the basic colors of the receiver Rn, Gn, Bn. B which case they are called perfect, and provide no color distortion in the images on the TV screen. Ideal spectral characteristics of transmitting camera in relative terms from the objective input to the output signal of the sensor are determined by a system of three linear equations [3]:

$$\begin{cases} \bar{r}_n = k_1\bar{x} + k_2\bar{y} + k_3\bar{z} \\ \bar{g}_n = k_4\bar{x} + k_5\bar{y} + k_6\bar{z} \\ \bar{b}_n = k_7\bar{x} + k_8\bar{y} + k_9\bar{z} \end{cases} \quad (3)$$

Where k[1] , ... k[9] – constant coefficients, depending on the choice of the main colors of the receiver and the reference white; $\bar{x}, \bar{y}, \bar{z}$ - the specific coordinates of the spectral colors in 1 W XYZ system is a function of the wavelength of light. Coordinates the main receiver of flowers R^n, G^n, B^n the following matrix determined according to European standard:

$$\begin{pmatrix} x_R y_R z_R \\ x_G y_G z_G \\ x_B y_B z_B \end{pmatrix} = \begin{pmatrix} 0,6400,330.0,330 \\ 0,290.0,600.0,110 \\ 0,150.0,060.0,790 \end{pmatrix} \tag{4}$$

he references white color selected European reference source for color TV D^{6500} [4], having chromaticity coordinates:

$$x^b = 0,313; y^b = 0,329; z^b = 0,358 \tag{5}$$

Calculated in accordance with the equations of the ideal spectral sensitivity characteristics of the individual color channels of the transmitting camera shown in Figure 5. They have a negative branch (characteristic $\bar{r}_n$ It has two positive hump) and provide an undistorted reproduction of chromaticity represented by points inside the triangle $R^nG^nB^n$ on the chromaticity diagram CIE. It should be noted that the negative branch caused by the fact that the triangle $R^nG^nB^n$ It positioned within the spectral locus of the chromaticity diagram on. In actual spectral characteristics of transmitting camera negative branch and small positive humps (secondary maxima) are absent. The difference between the actual spectral characteristics of the ideal leads to distortions in color television sets in all colors except white reference which is artificially set to TV as the beams by selecting the mode of operation of the screen. Real spectral characteristics of the camera is usually chosen close to the main positive humps ideal characteristics, and to improve the sensitivity of the camera doing a little wider than the humps. To minimize color errors applied colour correction matrix consisting of three submatrices – one submatrix in each channel R, G, B chamber [4].

Signals at input and output colour correction matrix included in the video camera transmitting, interconnected:

$$\left\{ \begin{aligned} E_{Rvix} &= a_{11}E_{Rv} + a_{12}E_{Gv} + a_{13}E_{Bv} \\ E_{Gvix} &= a_{21}E_{Rv} + a_{22}E_{Gv} + a_{23}E_{Bv} \\ E_{Bvix} &= a_{31}E_{Rv} + a_{32}E_{Gv} + a_{33}E_{Bv} \end{aligned} \right\} \tag{6}$$

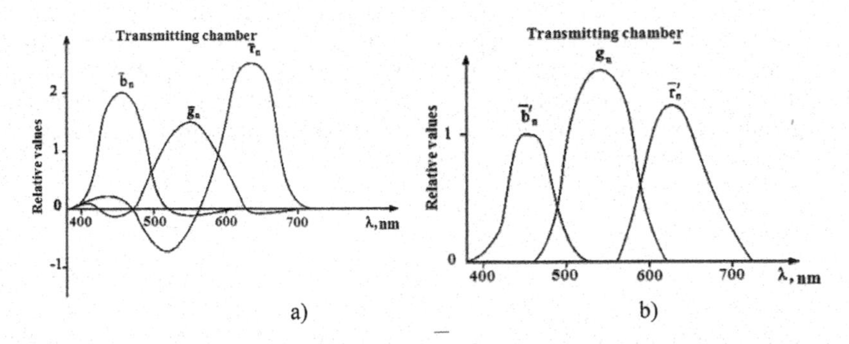

Figure 5. a)-Ideal spectral characteristics $\bar{r}_n, \bar{b}_n, \bar{g}_n$, b)- The actual spectral characteristics of the camera $\bar{r}_n^/, \bar{b}_n^/, \bar{g}_n^/$.

Where a[11],..., a[33] – matrixing coefficients calculated by an iterative method for a set of test colors recommended by the CIE.

When calculating the matrixing coefficients of the following conditions must be met: -that the sum of each row unit matrix coefficients to preserve equality between the signals in the transmission reference white; -matrixing two absolute values of coefficients in one number must be small in comparison with the third, otherwise there is deterioration in SNR and an increase in the visibility of color fringing misalignment signals over time; -permissible distortion test chrome colors have different weights, determined experimentally, taking into account peculiarities of the perception of the TV image, because the eye is not the same sensitivity to changes in different chromaticity [4,12].

Often the matrix coefficients to ensure the implementation of these terms and conditions, made unregulated. In the case of variable gain work colour correction matrix describes the system the following equations [4]:

$$\left\{\begin{array}{l} E_{Rvix} = E_{Rv} + a_1(E_{Rv} - E_{Gv}) + b_1(E_{Rv} - E_{Bvx}) \\ E_{Gvix} = E_{Gv} + a_2(E_{Gv} - E_{Rv}) + b_2(E_{Gvx} - E_{Bvx}) \\ E_{Bvix} = E_{Bv} + a_3(E_{Bv} - E_{Rv}) + b_3(E_{Bvx} - E_{Gvx}) \end{array}\right\} \tag{7}$$

where a and – matrixing coefficients.

Currently, deblocking filters are used in the MPEG-4 Visual, H.264/AVC and AVS standards. Impulse responses of these filters:

$$h_{MPEG4}(k), h_{H264}(k), h_{AVS}(k), \tag{8}$$

where k- are relative coordinates, defined in the relevant standards,

$$h_{MPEG4}(k) = [-8, 24, -48, 160, 160, -48, 24, -8]/256, \tag{9}$$

$$h_{AVS}(k) = [-1, 5, 5, -1]/8 \tag{10}$$

$$h_{H264}(k) = [1, -5, 20, 20, -5, 1]/32 \tag{11}$$

where, ω – circular spatial frequency, $\omega \in [-\pi, \pi]$,

According to the research results shown in Table 2, the highest quality filter is used in the MPEG-4 Visual standard, while the H.264/AVC standard filter, which provides a lower level of moiré distortion suppression, has a lower computational complexity. MPEG-4 Visual-17,8, H.264-15,9, AVS-14,9. The results of the analysis of the block distortion filters show that the high quality of the reconstructed image is achieved under the following conditions: Involvement of all image pixels in the filtering process; Implementation of filtering in the frequency space.

Frame jitter suppression methods. The human eye is very sensitive to frame shaking, as motion responds best, and chaotic motion is a strong irritant to the eye. Therefore, watching movies with frame shake is very unpleasant for a person [4,11].

There are four different approaches to frame judder reduction: Using Motion Vectors, Using feature points, Image-Based Rendering Approach, With the construction of an affine model.

Frame jitter suppression algorithms using motion vectors. The Motion Estimation apparatus is used in almost all areas of video processing. The task is to determine the direction and speed of movement of small parts of the frame. To do this, the frame is divided by a rectangular grid into blocks (usually 16x16 in size), and for each block in the previous frame, the most similar block is searched. Those. are looking for such that the minimum error.

$$error = \sum_{(x,y)\in S} (I_i(x,y) - I_{i-1}(x + \Delta x, y + \Delta y))^2 \tag{15}$$

Thus, each block is assigned a pair $(\Delta x, \Delta y)$, called a motion vector. Figure 2 shows an example of moving video objects based on motion vectors. The application of Motion Estimation for video judder suppression is to find the frame shift. The simplest approach is to determine the frame shift by taking the median of a set of motion vectors. Thus, for a sequence of frames, we obtain a set of shifts $T_i = (\Delta x_i, \Delta y_i, 0, 0)$. And shifts without jitter T_i^* are obtained by applying the averaging filter to the set from Δx_i and the set Δy_i from [4,13].

$$h[n] = \left\{ \frac{1/L}{0}, n = \overline{0, L-1} \right\} \quad (16)$$

Jitter suppression is performed by shifting each frame by the difference between the smoothed shift and the shift in the original frame video, etc. $T_i^* - T_i$. This approach is very simple and does not give very good results. In addition, it does not detect rotation. Frame jitter suppression algorithms using special points. There are different Feature Tracking methods but each is usually in two steps: Feature Selection; Determining the displacement of a singular point between adjacent frames. At this stage, points are selected that have a set of specific properties, which are chosen differently in different methods, but usually they are chosen in the vicinity of highly inhomogeneous image sections [4,11].

For each singular point from the *i*-th frame, its position in the (*i* + *1*)-th frame is determined. The following is a way to match Kanade-Lucas-Tomasi (KLT) key points. For each point, the value is entered:

$$g(x, y) = [dY(x, y)/dx\_dY(x, y)/dy] \quad (17)$$

which reflects the degree of image change in the vicinity of the point along Ox and along Oy. After that, for each point (x_0,y_0), some neighborhood W (x_0,y_0) is considered and the matrix is calculated [8]:

$$Z = \iint_{W(x_0, y_0)} g * g^T dxdy = \iint_{W(x_0, y_0)} \begin{bmatrix} \frac{d^2I}{dx^2} \frac{d^2I}{dxdy} \\ \frac{d^2I}{dxdy} \frac{d^{2I}}{dy^2} \end{bmatrix} dxdy \quad (18)$$

For the resulting matrix, eigenvalues are found λ_1, λ_2, and if they are greater than a given threshold, then the point (x_0, y_0) is considered singular. Shows an example of the operation of the KLT algorithm, where the singular points are marked in red [7,13].

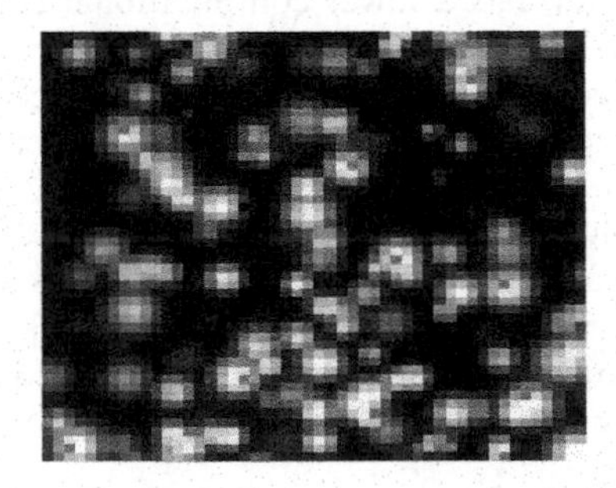

Figure 6. The result of the KLT algorithm.

To determine the offset of a singular point, some of its neighborhood is selected, the offset is searched for by finding the area in the next frame that is most similar to the selected one.

3 CONCLUSIONS

As a result of the work done to analyze the qualitative indicators of TV images and methods of enhancing it was established:

On the visual quality of the TV image affects a large number of parameters. Here, for evaluation of image quality does not exist generalized quantification.

The main quality characteristics of images are such as geometric shapes and relative dimensions, distinguishable details the distribution of brightness and color, the perception of the relative motion of objects and so on.

REFERENCES

[1] Djakoniya V. E., Gogol A. A., Ya.V. Druzin i dr.; Televideniye. Uchebnik dlya VUZov pod red. V.E. Djakonii. Moskva Knigi Izdatelstva «Goryachaya liniya-Telekom» 2007 (in Russian).

[2] *Parametrik Televizionnnogo Signala i Kachestvo Izobrajeniyiya.* http://www.avantos.ru/download/parametrs_tvsignal.pdf (in Russian).

[3] *Obrabotka videosignalov v kamernom Kanale Svetnogo Televideniya* http://www.siblec.ru/index.php?dn=html&way=bW9kL2h0bWwvY29udGVudC82c2VtL2NvdXJzZTEzNy9sZWM1XzIuaHRt (in Russian).

[4] Nikitin V.V., Sitsulin A. K. Parametri Televizionnix Kamer Gamma-korrektsiya www.security-bridge.com›biblioteka... po... gammakorrekciya/ (in Russian).

[5] Zalmanzon L. A. Preobrazovaniya Fure, Uolsha, Xaara i ix Primeneniy v Upravlenii, Svyazi i Drugix Oblastyax. – M.: Nauka, 1989. – 496 s. (in Russian).

[6] Dvorkovich Aleksandr Viktorovich. Razrabotka I Issledovaniye Visokoeffektivnix System Sifrovoy Obrabotki Dinamicheskix Izobrajeniy i Otsenki yeyo Kachestva.// Dissertatsiya na soiskaniy uchenoy stepeni doktora texnicheskix nauk Moskva 2007 (in Russion)

[7] Davletova X. R. "Analiz Isksjeniy TV Izobrajeniy pri ix Sifrovom Kodirovanii i Metodi ix otsenki" Statya v Sbornike Trudov Respublika Ilmiy-texnik Konferensiya «Problemi informatsionnix texnoligiy I telekommunikatsiy», proxodivshey 12–13 marta 2015 g.v Tashkente. s-396–400 (in Russian).

[8] Davletova X. R. "Analiz Kachestvennix Pokazateley TV Izobrajeniy i Metodov ix Otsenki. Tazisi Doklada v Sbornike Dokladov Respublikanskoy Nauchno-Texnicheskoy Konferentsii «Perspektivi effektivnogo razvitiya informatsionnix texnologiy i telekommunikasionnix sistem», Proxodivshey 13–14 Marta 2014 g. Tashkent. Chast 3, s. 135–137 in Russian).

[9] Yu.I. Monich, V.V. Starovoytov., "Otsenki Kachestva dlya Analiza Sifrovix Izobrajeniy. g. Minsk, Belerus-2008g. (in Russian).

[10] Gushin A., Kiryanov A., Lyaxov A., Ye. Xorov., "Bistriy Algoritm Viravnivaniya Kadrov Dlya Otsenki Kachestva MPEG-4 videopotokov, peredavayemix po besprovodnim setyam/ http://itas2013.iitp.ru/pdf/1569761233.pdf (in Russian).

[11] Vatolin D., Petrov O. Izmereniy Subektivnogo Kachestva Video MSU Perceptual Video Quality tool /MSU Graphics & Media Lab (Video Group) /http://www.compression.ru/video/quality_measure/perceptual_video_quality_tool.html. (in Russian).

[12] Rudneva V. Metodi Povisheniya Svetnogo Kachestva Izobrajeniy i Vosstanovleniya svetov v starom video/ kompyuternaya Grafika i Multimedia Setevoy Jurnal. (in Russian).

[13] Davletova X.R., Ye.A. Astashev "Uluchsheniye Kachestva Svetoperedachi TV Izobrajeniye na Osnove Algoritma «Retinex»". *Statya v Sbornike Trudov Respublika ilmiy-texnik Anjuman* «Axborot texnologiyalari va telekommunikatsiya muammolari», proxodivshey 21–22 aprelya 2011 g. v Tashkente (in Russian).

Artificial Intelligence, Blockchain, Computing and Security – Dagur et al. (Eds)

Evaluation of the human pose on the basis of creating a graph of movements on the basis of a neural network

A.R. Akhatov
Samarkand State University named after Sharof Rashidov, Uzbekistan

M. Sabharwall
Galgotias University, India

I.Q. Himmatov
Samarkand State University named after Sharof Rashidov, Uzbekistan

ABSTRACT: Evidence shows that in recent years the task of assessing a human's condition from a photo or video taken from a camera has been improved, it allows you to form a human's pose in 2D and 3D and identify a human's movement by dividing it into joints and points. One of the reasons for this trend is such applications are increasing: computer, interaction with robots, virtual films and video games; analysis of the actions and activities of athletes; video surveillance, security control, etc. Another reason is the successful application of neural network architecture to computer vision problems. This is an signification step towards recognizing persons in videos and images, as well as psychological identification of characters.

1 METHOD FOR ESTIMATING A HUMAN'S POSE BASED ON A GRAPHICAL MOVEMENT SCHEME

Nowadays, the identification of a human by his actions is one of the processes that are difficult to set up and master compared to other identification technologies. Therefore, in this study, the recognition of human poses and the construction of graphs using neural networks and other tools are considered as initial tasks.

Human pose estimation is a special case of the image segmentation problem in the computer vision department, which consists in detecting the movements from parts of the human body of images or videos (considered as a sequence of images). Often a human's position is replete with associated key points that correspond to the joints (shoulders, elbows, arms, hips, knees, feet) and other key points (neck, head). This task can be considered in two or three dimensions, which determines the complexity of the task and the practical application of the results (Akhatov *et al.* 2021). The task can also be divided into two subtypes: Single Human Pose Estimation, Multi Human Pose Estimation.

There are methods for two-dimensional and three-dimensional assessment of the position of a human's pose. In 2D estimation, 2D position coordinates (x, y) are assigned to each joint in the RGB image, in 3D technique (e.g. Position Estimator) approximate position is estimated according to 3D dimension in RGB image coordinates (x, y, z) (Andriluka et al. 2014). Note that an essential important task in assessing a human's pose is the construction of graphic diagrams that describe the movements of a human and the processing of information from these diagrams (Ximmatov 2020a).

DOI: 10.1201/9781032684994-108

One of the fundamental methods for creating a motion graph is the method of forming graphs, i.e. image skeletons based on points divided into joints in three-dimensional space (Figure 1) (Wang et al. 2021). This allows you to analyze the psychology of a human in motion with the definition of actions.

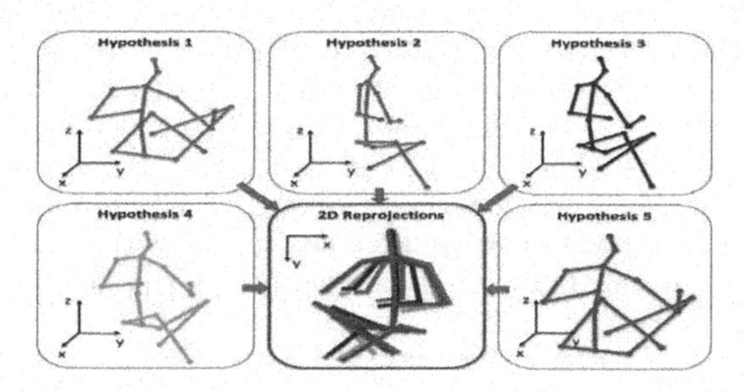

Figure 1. Three-dimensional representation of the skeleton in various schemes.

The main idea of this method is to find "body marks" in the image, that is, the main parts of the body in the form of joints (for example, shoulders, ankles, knees, wrists, etc.). Finding the right data set is one of the important parts of research. On Figure 2 illustrates possible variants of the scheme for determining important auxiliary points of the human body (Wang et al. 2021).

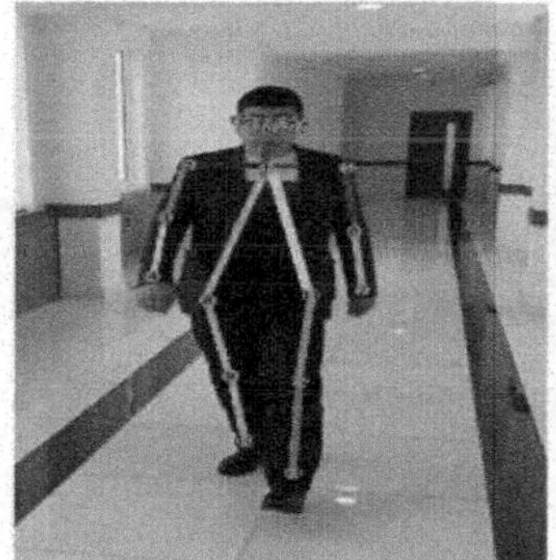
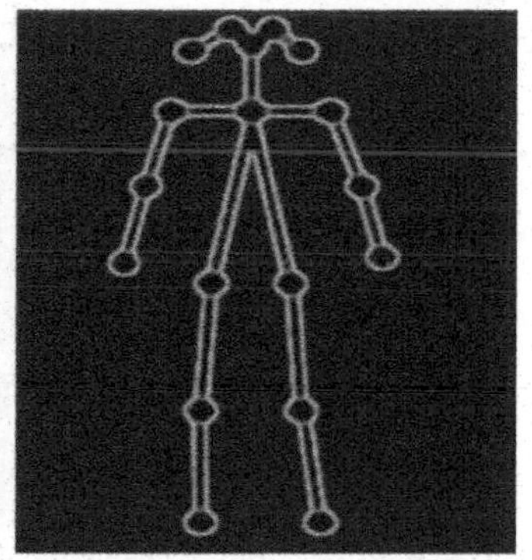

Figure 2. Determination of important auxiliary points of the human body.

The first works on predicting the pose of a human in an image using deep neural networks (DeepPose) solve the problem based on regression. DeepPose solves the problem of determining the pose of a single human from an image (estimating the pose of one human) based on given joints, cropping areas where there is a human, recalculating and normalizing the coordinates of the joints relative to the center of the cropped pixel area.

As a result of active computer vision research with a wide range of potential applications, human pose estimation, especially in 2D images, has made great strides through powerful deep learning techniques and large-scale datasets collected. However, the performance of 3D human pose estimation is far from satisfactory, which is main directly to the be short of sufficient 3D datasets (Bourdev & Malik 2009).

The evaluation of the continuity of the image and its trace or trajectory is continuously iterated through top-down and bottom-up processing, helping the network to capture high-level features at an early stage, and more in-depth feature evaluation is performed at a later stage. It also preserves the spatial arrangement of elements for meshing, effectively localizing connections.

2 MULTI-SCALE MSS CONTROL NETWORK

For the sake of completeness, two types of multi-scale estimation models are considered, called multi-scale control network (MSS) and multi-scale regression network (MSR-net), which can be iterated for multiple stacks (Figure 3). In particular, MSS is based on the conv-deconv (transform-decomposition) hourglass model trained in multi-scale control of the keypoint removal process (Newel et al. 2016). MSR performs structural regression of the final pose by matching multi-scale heat maps of key points and their higher-order associations(Axatov & Ximmatov 2020). Both network types have a predictable structure-aware loss function designed to enable efficient learning of multi-scale structural features (Ximmatov 2020b). The training of the entire conveyor network is fine-tuned using a key-point masking learning scheme to focus on learning complex patterns.

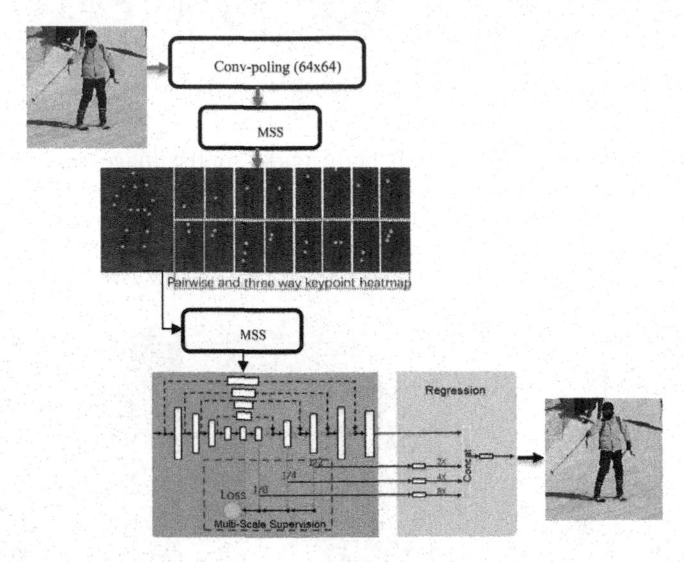

Figure 3. Sequence of execution and results of MSS and MSR networks.

The MSS Multiscale Supervision Network is designed to explore deep learning features at multiple scales. Multi-level control is performed at each of the levels of deconversion of the MSS network, where each level corresponds to a certain scale. In Figure 5 the sequence of execution and the results of the work of MSS and MSR networks are illustrated, where you can also see the architecture of the MSS network.

The multiscale control network localizes key points of the body in a manner similar to the "heeding model" used in the traditional resolution pyramid for image retrieval. Activation areas on a low-resolution heatmap can serve as a guide to refine the location in subsequent high-resolution layers. Refinement of the localization of key points with increasing resolution works by analogy with the "heeding" mechanism used in the traditional pyramid search for resolution. On Figure 4-(a) shows the results of applying multiscale thermal maps of the key point of the chest to the object of study. On Figure 4-(b) shows cue point heat map refinement during deconv upsampling when chest location is refined with increased accuracy, Figure 4-(c) shows a human skeletal plot visualizing key point connectivity relationships.

The L_{MS} loss function is described for training a multiscale surveillance network. Loss L_{MS} is determined by summing the L_2 losses from heat maps of all key points at all scales (Newel et al. 2016; Yang et al. 2017). To detect N = 16 key points (noddles, humerals, forearms, wirst, ribs, laps, astragalus), In this process, N heatmaps are formed after all the conv-deconv stack.

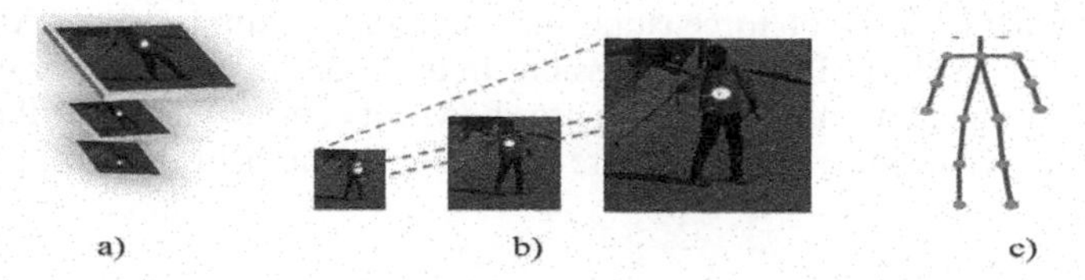

Figure 4. Refinement of localization of key points with increasing resolution: (a) Thermal maps of the key point of the chest; (b) Refinement of heat maps of key points during deconv upsampling; (c) Skeletal plot with relationship visualization.

Scale at *i*-th loss compares all generated or predicted heatmaps with actual heatmaps of the appropriate scale based on key points obtained:

$$L^{i}_{MS} = \frac{1}{N} + \sum\nolimits_{n=1}^{N} \sum\nolimits_{x,y} || P_n(x,y) - G_n(x,y) ||_2 \tag{1}$$

in this formulation $P_n(x, y)$ and $G_n(x, y)$ describe the augured and ground confidence maps at pixel location (x, y) for the nth key point, respectively.

In the standard dataset, real poses are presented as desired points. We keep to the simple tradition of creating heat maps, as in Tompson (Toshev & Szegedy 2014), where the heat map of the key point $G_n(x, y)$ is preform using a 2D Gaussian plot centered on the key point position (x, y) with a standard swerve of 1 pixel. On Figure 5 (lower left, first line) shows several examples of heat maps for certain key points.

3 MULTISCALE REGRESSION NETWORK MSR

In this study, a fully convolutional multiscale MSR regression network is used after the conv-deconv MSS stacks to globally refine multiscale keypoint heatmaps and improving the effectiveness of the structural dependence of implied poses. The position of the arms/legs relative to each other and relative to the useful part of the head/stature preliminary data on actions that we can by looking be understood from the regression network at feature maps at all scales to refine the pose. MSR takes multi-scale heat maps as input and maps them to real data key points at appropriate scales. Thus, the regression network can efficiently combine heat maps at all scales to refine the estimated poses.

On Figure 5 shows the performance of multi-scale high-order key point regression performed on the MSR network. In the study object, the MSR network used the results of the MSS network to explicitly model high-order relationships between body parts so that the structural integrity of the case could be maintained and improved. Multiscale basis point regression is implimented here to identify a few high points in the basis point heatmaps.

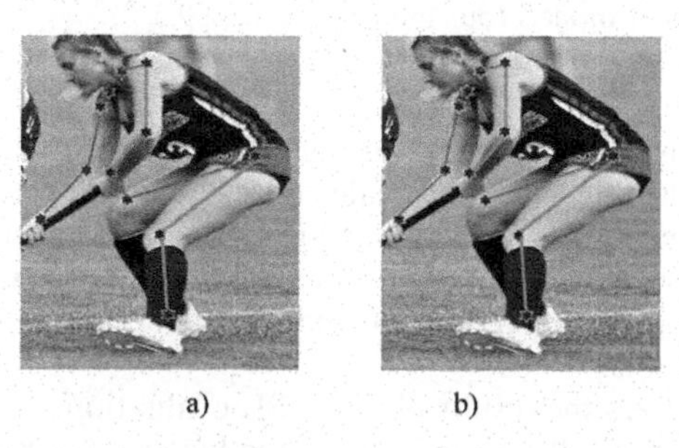

Figure 5. Disambiguation on heat maps of key points: (a) Prediction of key points; (b) Location of output key point.

The figures illustrate hotspot forecasting and heatmap examples from MSS hourglass stacks that will be passed to MSR for regression. In addition, the location of the output key points and the heat map after the regression are shown. It can be noted that the peaks of the heatmap after the regression relatively more attention to the heatmap from the hourglass stacks.

4 FEATURES OF THE EXPERIMENTS

The images are captured by MPII from various human activities and therefore various complex human poses are included. If there are multiple people in a given image, the grid is meant to indicate the human in the center. In MPII, each image is defined by its scale and center, so centering is not a difficult task. After detecting the human in the center, the image is cropped to include only the target human, and then resized to 256x256 pixels. Finally, data magnification is applied to image scaling and rotation (Bulat & Tzimiropoulos 2016).

Overall performance analysis shows that the hourglass can be improved by using multiple smaller filters instead of one larger filter, for example using two 3x3 filters instead of a 5x5 filter. In addition, the 1x1 filter to reduce the number of pixels with convolution improves its performance as well. Thus, this architecture uses full size filters of 3x3 or less. In addition, it should be noted that 64x64 input images are submitted to the network instead of high resolution images to avoid excessive use of GPU memory. This does not negatively impact performance.

5 ANALYSIS OF RESULTS

In general, in the following tables, we compare our results obtained by the Multi-scale assessment model and the results obtained by the CVPR and ECCV models, and we see that our result is effective by 0.2%. We do this using a standard metric (Thompson *et al.* 2015) called Percentage Correct Keypoints (PCK), which represents the percentage of keypoint detections that fall within a normalized distance from the truth. PCK for the FLIC score is set to the percentage of discrepancy between the detected pose points with respect to reality, albeit after normalization to a fraction of body size. To estimate MPII, such discrepancies are normalized by the ratio of head size, defined as PCK.

Table 1. A comparative analysis of results from the FLIC dataset (PCK = 0,2%).

Name of models	upper hand	middle hand
CVPR	97.7	95.2
ECCV	99.1	97.1
Multi-scale assessment models (our model)	99.3	97.3

In the MPII pose dataset, we also present the multi-body estimation results in the table, comparing the performance of our Multi-scale control network model with other models.

From the first table above, the FLIC results are summarized, where we can show that the proposed method reaches 99.3% for the forearm and 97.3% for the wrist. A comparison with ECCV demonstrates design improvement considering structure in MSS network and MSR network (Newel et al. 2016).

MPII: Table 2 summarizes the results of the MPII evaluation. Note that the pose method yields the highest overall score (91.8) and one of the highest scores across all MPII key test points as well as AUC.

Table 2. Comparative results with models regarding the main parts of the MPII pose (PCK = 0,2 %).

Name of models	The noddle	The humeral	Upper hand	Middle hand	The rib	The lap	The astra-galus	Overall
Multi-scale control network	98.5	96.8	92.7	88.4	90.6	89.3	86.3	91.8
ICCV model, year 2017	98.1	96.5	92.5	88.5	90.2	89.6	86.0	91.6
arXiv model, year 2017	98.2	96.8	92.2	88.0	91.3	89.1	84.9	91.5
CVPR model, year 2017	98.5	96.3	91.9	88.1	90.6	88.0	85.0	91.2
TMM model, year 2017	98.1	96.3	92.2	87.8	90.6	87.6	82.7	90.7
ECCV model, year 16	98.2	96.3	91.2	87.1	90.1	87.4	83.6	90.6
FG model, year 17	97.7	95.0	88.2	83.0	87.9	82.6	78.4	87.5

6 CONCLUSIONS

In the process of image processing, drawing a human skeleton based on the MPII model using cameras that detect movement and capture general images from different distances, after determining its base points, it is possible to create a pose that is more efficient than other models using its movements taken from cameras. When forming a human pose, an important issue is the problem of skeleton formation during human movement, in solving which it is promising to use a Time-Off-Flight camera or sensors, especially cameras with motion sensors. The use of deep analysis algorithms proposed in this study using neural networks allows fast processing of these captured images. In the future, this will make it possible to determine the psychological characteristics of a human using algorithms and neural networks to form behavior and express its psychological nature.

REFERENCES

Akhatov, A. Nazarov, F. Rashidov, A. 2021. *Increasing Data Reliability by Using Bigdata Parallelization Mechanisms.* International Conference on Information Science and Communications Technologies. *4,5,6 November*. ICISCT 2021(IEEE), art. no. 9670387 DOI: 10.1109/ICISCT52966.2021.9670387 SCOPUS.

Akhatov, A. Nazarov, F. Rashidov, A. 2022. *Mechanisms of Information Reliabilityin Big Data and Blockchain Technologies.* International Conference on Information Science and Communications Technologies. *4,5,6 November.* ICISCT 2021(IEEE), art. no. 9670052 DOI: 10.1109/ICISCT52966.2021.9670052 SCOPUS (14)

Andriluka, M. Pishchulin, L. Gehler, P. Schiele, B. 2014. 2D Human Pose Estimation: New Benchmark and State of the Art Analysis. *In: CVPR.* 3686–3693

Axatov, A.R. & Ximmatov, I.Q. 2020. Foydalanuvchilarni Biometrik Autentifikatsiya Turlari Asosida Haqiqiyligini Tasdiqlash Usullarinning Samaradorligi. *Innovatsion yondashuvlar ilm-fan taraqqiyoti kaliti sifatida: yechimlar va istiqbollar,* 8–10 oktyabr 2020 y. 20–26. Jizzax: Uzbekistan.

Bourdev, L. & Malik, J. 2009. Poselets: Body Part Detectors Trained Using 3D Human Pose Annotations. In: *ICCV.* 1365–1372

Bulat, A. Tzimiropoulos, G. 2016. Human Pose Estimation via Convolutional Part Heatmap Regression. *ECCV* 717–732

Newel, A. Yang, K. & Deng, J. 2016. Stacked Hourglass Networks for Human Pose Esti- mation. In: *ECCV.* 26 Jul 2016 483–499.

Toshev, A. & Szegedy, C. 2014. Deeppose: Human Pose Estimation via Deep Neural Net- Works. *CVPR* 1653–1660.

Wang, J. Tan, Sh. Zhen, X. Zheng, F. He, Z. Shao, L. 2021."Deep 3D Human Pose Estimation: A Review" *Journal of "Computer Vision and Image Understanding*" Volume 210, September 2021, 103225.
Ximmatov, I.Q. 2020a. Advantages of Biometrik Gait Recognition. Important Factors in Evaluation of Gait Analysis Systems. *Scientific Journal of SamSu.* ISSN 2091–5446, 2020, vol-3 (121), 104–107. Samarkand: Uzbekistan.
Ximmatov, I.Q. 2020b. Important Factors in Evaluation of Gait Analysis Systems and Advantages of Biometric Gait Recognition. *Innovatsion va Zamonaviy Axborot Texnologiyalarini ta'lim, fan va Boshqaruv Sohalarida qo'llash Istiqbollari.* 14–15 may, 2020 y. 262–267. Samarkand: Uzbekistan.
Yang, W. Li, S. Ouyang, W. Li, H. Wang, X. 2017. Learning Feature Pyramids for Human Pose Estimation. In: *2017 IEEE International Conference on Computer Vision (ICCV)*. 1290–1299.

Artificial Intelligence, Blockchain, Computing and Security – Dagur et al. (Eds)
© 2024 The Author(s), ISBN: 978-1-032-67841-2

Optimization of the database structure based on Machine Learning algorithms in case of increased data flow

A. Akhatov & A. Rashidov
Samarkand State University, Samarkand, Uzbekistan

A. Renavikar
NeARTech Solutions, Pune, India

ABSTRACT: Today, the sharp increase in the use of digital devices in the world is the reason for the creation of a large flow of data in plenty of systems. Due to the large amount of data in these systems, the process of data processing, i.e. making queries and getting the result, takes more time than traditional methods. Currently, the most popular method of real-time data processing is data processing based on a distributed computing mechanism. But one of the most important disadvantages of this method is determining the number of optimal distributions in proportion to the size of the data. Because over-distribution also has a negative effect on the efficiency of the system. In this research work, in order to solve the given problem, an approach to the optimal distribution of data flow based on Machine Learning algorithms is proposed. At the same time, during the research work, the data obtained as a result of the experiment conducted in unified computing systems based on the distributed computing mechanisms of this approach were analyzed. In addition, the efficiency indicators of 18 Machine Learning algorithms used during the ban were evaluated and the selection of the 5 most effective algorithms for the proposed approach was considered.

1 INTRODUCTION

Currently, the use of digital structures in all areas: education, socio-economic spheres, various management processes, mutual information exchange, and even in the daily life of humans is increasing dramatically. As a result of this, not only unstructured or semi-structured but also a large flow of structured data, which until now was considered easy to store and process, is appearing (Blackman & Forge 2017). Since the processing time of these large data using traditional methods and systems grows proportionally to the volume of data, processing them in a real-time system is becoming one of the urgent research topics (Alabdullah *et al.* 2018; Akhatov *et al.* 2021a; Rashidov *et al.* 2021).

Based on this, a number of scientific researches are being conducted by world scientists in order to solve this problem. A clear example of such research works are theoretical and practical works such as parallel processing of data (Alaeddine *et al.* 2020; Akhatov *et al.* 2021b), use of the Hadoop ecosystem and its improvement (Landset *et al.* 2015; Oussous *et al.* 2018), and implementation of data storage and processing methods based on distributed computing mechanisms (Akhatov *et al.* 2022a; Kunanets *et al.* 2019). Actually, increasing the efficiency of data processing in parallel and on the basis of the Hadoop ecosystem is closely related to the methods of the distributed computing engine. Therefore, the distributed computing mechanism is the main approach in managing, storing, and processing large data flows.

DOI: 10.1201/9781032684994-109

In the distributed computing mechanism, the efficiency indicator is increased due to distributions. In other words, in this computing engine, large data is divided into small groups, and processing is carried out on these small groups, just like normal data processing (Smeliansky 2013). However, in this approach, high efficiency is achieved only when the number of distributions is proportional to the volume of the data (Akhatov *et al.* 2022b). To be more precise, it will be necessary to make small distributions for small data volumes and large distributions for large data volumes. The main problem here is data flow. That is, the volume of data is always changing, and this leads to a change in the distribution of data. In some systems, it is not possible to predict the volume of data in advance, so it is not effective to determine the number of distributions in advance based on the human factor. The most effective approach in this process is to determine the number of distributions based on artificial intelligence. In the next steps of this research work, the number of distributions in distributed computing mechanisms will be determined based on Machine Learning algorithms and the evaluation of error indicators of these algorithms will be covered.

2 MATERIALS AND METHODS

2.1 *Mechanism of internal distribution in data flow storage and processing*

Data flow storage and processing on the basis of a distributed computing mechanism in single computing systems is several dozen times more efficient than traditional methods. Since the computing machine is the only one in these systems, data storage and processing is carried out on the basis of internal distribution. In other words, large databases are divided into several smaller databases based on certain rules. Data processing is carried out in these small databases. In this process, the efficiency indicator increases because it is possible to directly refer to the information that needs to be processed.

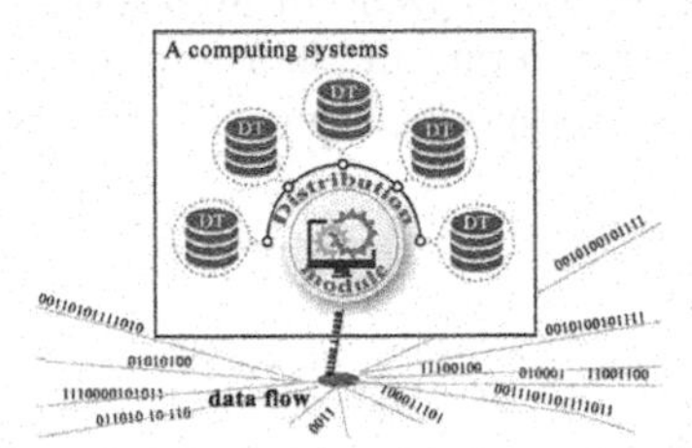

Figure 1. The architecture of internal distribution in data flow storage and processing.

As can be seen from Figure 1, when the data first enters the system, it goes to the distribution module. After that, the distribution module distributes the data to small data tables in the database based on certain rules. The increase in the number of these distributions leads to a decrease in the size of data in small databases and, as a result, to a decrease in query time. But over-distribution increases the function of this distribution module and, as a result, negatively affects the time factor. Therefore, in this research work, the determination of the number of optimal distributions is taken as the objective of the research work.

2.2 *Data analysis which used to determine the number of optimal distributions*

It is known that in order to train and test Machine Learning algorithms, data must first be collected based on experiments. The main data to be collected in this research work are the data volume, the number of distributions taken as an experiment, and the processing time according to the number of data distributions. Experiments were carried out on several computing systems so that the approach proposed in the research work does not depend only on certain computing systems. Therefore, the following information, which may affect the optimal distribution number of these computing systems, was also collected. The data collected in the research work are as follows:

data_tuple–the total number of tuples in databases;

table–the number of sub-tables distributed in the database;

time_secund–the time required to process the data with the number of tuples data_tuple in the database in table number;

hard_d_r_s_MB/s–speed of reading data from the hard disk (Mbytes/second)

RAM_MB – the volume of RAM (Mbytes);

RAM_s_MHz – operating memory frequency (MHz);

CPU_MHz – processor frequency (MHz);

cache_L1_MB – the volume of 1 cache memory (Mbyte).

d_tuple, table, and time_secund are used to determine the optimal number of distributions when the data size changes in a single computing system. These 3 types and the rest of the data together with the data volume are used to determine the number of optimal distributions corresponding to the changes in the parameters of the computing system.

Correlation coefficients were determined in order to determine the impact of these collected data on the execution time of requests. In this case, the (1) formula for determining the correlation coefficient of two variables x and y was used.

$$\rho_{xy} = \frac{n \cdot \sum xy - \sum x \cdot \sum y}{\sqrt{\left[n \cdot \sum x^2 - (\sum x)^2\right] \cdot \left[n \cdot \sum y^2 - (\sum y)^2\right]}} \quad (1)$$

where ρ_{xy}–correlation coefficient of x and y, n–number of values, $\sum x$–sum of x values list, $\sum y$–sum of y values list, $\sum xy$–sum of the product of x and y values, $\sum x^2$–sum of squares of x values, $\sum y^2$–sum of squares of y values.

2.3 *Determination of the number of optimal distributions based on Machine Learning algorithms*

Since the data flow in distributed systems is changeable, it is very effective to determine the number of optimal distributions based on Machine Learning algorithms. Because these algorithms can predict the number of reliable, optimal distributions in a short time, without human factor, taking into account all the changes related to the data flow and the system.

In this research work, 18 algorithms of Machine Learning were used to determine the number of optimal distributions. These are: Multiple Linear Regression, Polinomial Regression, Random Forest Regression, Lasso Regression, Support Vector Regression, Stochastic Gradient Descent, Ridge Regression, Partial Least Squares Regression, Partial Least Squares Canonical Regression, Orthogonal Matching Pursuit Regression, Nearest Neighbors Regression, Multi-layer Perceptron Regression, Least Angle Regression, LassoLars Regression, ElasticNet Regression, Canonical Correlation Analysis Regression, Bayesian Ridge Regression, Logistic Regrassion.

In order to increase the accuracy of the training and testing results of these algorithms, the data was initially scaled. The following scaling functions were used in this process: MinMaxScaler, StandardScaler, RobustScaler, MaxAbsScaler, QuantileTransformer, PowerTransformer. The results showed that the errors of the Machine Learning algorithms depend on the scaling functions.

3 RESULTS

3.1 *The results of the analysis of the collected data*

In this research work, an approach to the storage and processing of large data streams based on internal distribution is tested on 4 computing machines. At the beginning of the study, several parameters of these computing machines were selected in order to train and test Machine Learning algorithms. But during the research, it became clear that some parameters had no effect on the time parameter of the proposed approach. Therefore, during the study, the data shown in Table 1, which may affect the time indicators of the approach, was selected.

Table 1. The parameters of the computers used in the research.

№	hard_d_r_s_MB/s	RAM_MB	RAM_s_MHz	CPU_MHz	cashe_L1_MB
1	120	4096	2133	1800	0,125
2	110	4096	2400	1800	0,125
3	420	8192	2667	3600	0, 25
4	120	8192	2667	2900	0,375

For training and testing Machine Learning algorithms, data consisting of 500 records was collected, and in order to increase the accuracy of the experiments, the correlation coefficients of these data with time were determined based on the formula (1). The results of these correlation coefficients are presented in Figure 2.

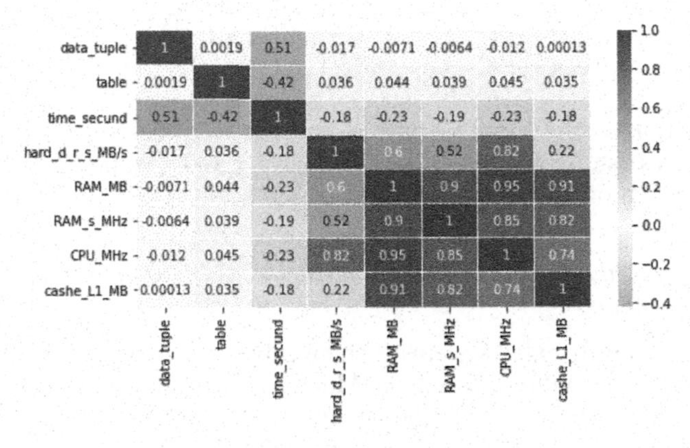

Figure 2. Correlation coefficients of the data used in the experiment.

The absolute values of the time dependence coefficients of the data collected in Figure 2 are around 0.2 and above. This shows that the collected data is really important for training Machine Learning algorithms.

3.2 *Assessment of errors in determining the number of optimal distributions based on Machine Learning algorithms*

In the research work, experiments were conducted using the algorithm for the number of optimal distributions using 18 Machine Learning 7 scaling methods. Experiments assessed the errors of the algorithms and selected the 5 most effective algorithms for the collected data and the desired results (Table 2). The error rates of the algorithms were assessed using the mean absolute error (MAE).

Table 2 shows that the smallest error is equal to 0.005 when the Polynomial Regression algorithm is used with scaling methods b, c, d. This increases the reliability of the approach, implying a very small error. In other words, with such a small error, it is possible to find the real value of the number of optimal distributions by rounding the number of predicted optimal distributions. Almost all algorithms except Random Forest Regression achieved

Table 2. Among the 18 Machine learning algorithms used in the study, the 5 most effective algorithms for the proposed approach and their MAE.

Type of Algoritms	*a	*b	*c	*d	*e	*f	*g
Polinomial Regression	0.0084	0.0050	0.0050	0.0050	1.2220	2.6114	4.5643
Multi-layer Perceptron Regression	0.0145	0.0403	0.5736	0.0142	0.7083	2.3445	6.1491
Random Forest Regression	0.0904	0.0733	0.0586	0.0305	0.0419	0.0919	0.0876
Support Vector Regression	0.0868	0.0756	4.7398	0.0875	4.5798	1.7749	1.7693
Canonical Correlation Analysis	0.0936	0.0936	0.0936	0.0936	6.1946	5.0415	3.8866

Notes
*a – MinMaxScaler, b – StandardScaler, c – RobustScaler, d – MaxAbsScaler, e – QuantileTransformer, f– PowerTransformer method="box-cox", g–PowerTransformer method="yeo-johnson"

relatively small errors in scaling methods a, b, c, and d for the data collected in this study, while Random Forest Regression performed well in all scaling methods. However, the best performance for the data collected in this research work was achieved with the first 4 scaling methods.

4 CONCLUSIONS

In conclusion, in this study, the issue of optimizing the database structure based on machine learning algorithms when the flow of data increases was considered. 18 Machine Learning algorithms were used to determine the optimal number of distributions for the proposed distributed computing mechanism for real-time data processing when the volume of data increases. The effectiveness of the algorithms was assessed using the average absolute error. The results of the experiment showed that the error indicator for the data collected when using these algorithms depends on the scaling method of the algorithms. In general, in almost all algorithms, the smallest error was less than 1. This, in turn, showed that Machine Learning algorithms in this research approach are effective.

REFERENCES

Akhatov A., Nazarov F., & Rashidov A. 2021a. Mechanisms of Information Reliability in Big Data and Blockchain Technologies" *ICISCT 2021*, 3–5.11, doi: 10.1109/ICISCT52966.2021.9670052

Akhatov A., Nazarov F., & Rashidov A. 2021b. Increasing Data Reliability by Using Bigdata Parallelization Mechanisms. ICISCT 2021: Applications, Trends and Opportunities, 3–5.11.2021, doi: 10.1109/ICISCT52966.2021.9670387

Akhatov A., Renavikar A., Rashidov A. & Nazarov F. 2022b. Development of the Big Data Processing Architecture Based on Distributed Computing Systems. Uzbek Journal of the Problems of Informatics and Energetics, № (1) 2022, 71–79

Akhatov A., Sabharwal M., Nazarov F. & Rashidov A. 2022a. Application of Cryptographic Methods to Blockchain Technology to Increase Data Reliability. *2nd International Conference on Advance Computing and Innovative Technologies in Engineering (ICACITE 2022)* DOI: 10.1109/ICACITE53722.2022.9823674

Alabdullah A, Beloff N., & White M. 2018. Rise of Big Data – Issues and Challenges. 2018 21st Saudi Computer Society National Computer Conference (NCC) 25–26 April 2018, DOI: 10.1109/NCG.2018.8593166

Alaeddine B, Nabil N., Habiba Ch. 2020. Parallel Processing Using Big Data and Machine Learning Techniques for Intrusion Detection. IAES International Journal of Artificial Intelligence (IJ-AI) Vol. 9, № 3, September 2020, 553–560 p, doi: 10.11591/ijai.v9.i3.pp553–560

Blackman C. & Forge S. 2017. Data Flows – Future Scenarios, European Parliament, Brussels.

Kunanets N., Vasiuta O. & Boiko N. 2019. Advanced Technologies of Big Data Research in Distributed Information Systems. *International Scientific and Technical Conference on Computer Sciences and Information Technologies, September* 2019, 7–76 p., doi: 10.1109/STC-CSIT.2019.8929756
Landset S, Khoshgoftaar T. M., Richter A. N. & Hasanin T. 2015. A Survey of Open Source Tools for Machine Learning wi Big Data in the Hadoop Ecosystem. *Journal of Big Data (2015)* 2:24, doi: 10.1186/s40537-015-0032-1
Oussous A., Benjelloun F.-Z., Lahcen A. A., Belfkih S. 2018. Big Data Technologies: A Survey. *Journal of King Saud University – Computer and Information Sciences 30 (2018)* 431–448 p, doi: 10.1016/j.jksuci.2017.06.001
Rashidov A. & Akhatov A. 2021. Big Data and its Application in Various Fields. *Descendants of Muhammad Al-Khwarizmi, 2021*, № 4 (18), 135–144
Smeliansky R. L. 2013. Model of Distributed Computing System Operation wi Time. *Programming and Computer Software, 2013*, Vol. 39, No. 5, 233–241 p., doi: 10.1134/S0361768813050046

Artificial Intelligence, Blockchain, Computing and Security – Dagur et al. (Eds)
© 2024 The Author(s), ISBN: 978-1-032-67841-2

Development of algorithms for predictive evaluation of investment projects based on machine learning

Nazarov Fayzullo Makhmadiyarovich* & Yarmatov Sherzodjon
Samarkand State University named after Sharof Rashidov, Samarkand, Uzbekistan

ABSTRACT: Currently, many organizations are widely using intellectual analysis methods to effectively use large-scale data for their economic activities. Predictive analytics models help identify problems and opportunities for each customer, organization employee, or manager. In this paper, we present a model for supporting decision-making processes using predictive analytics. The concept is illustrated with an illustration of a construction company's portfolio of investment projects. The main problem is to develop a model that calculates the average price of houses in different regions of the country when building the next houses of the company. The efficiency of the regression models was analyzed.

1 INTRODUCTION

The environment in which many decision-making processes take place cannot be completely defined since only a small portion of it can be viewed. According to research on judgments made in terms of long-term strategic activities of managers, firms defined by long-term changes are subject to an increase in the complexity of decisions that demand a high level of inventiveness. In decision-making processes, there is an increasing demand for models that do predictive analysis, based on all relevant inputs. Predictive analytics is a term used mainly in statistical and analytical methods. Predictive analysis models are used to evaluate changes in the parameters of objects. A higher score given by predictive analytics indicates how accurate the predicted result will be. Predictive analytics models are classified into two groups. These are classification models that predict class membership and regression models that predict numerical data. In the predictive analytics process, several statistical and machine learning approaches are applied. (Akhatov *et al.* 2022a).

Using predictive analytics models, data is first evaluated and then predicted. These models include data analysis and decision-making processes. Predictive analytics models can be divided into the following classes (Maciej & Iwona 2021).

- Predictive Models: Simulate human behavior to predict and predict the probability of a particular outcome.
- Descriptive Models: Used in the process of dividing the studied data sets and objects into certain groups (Akhatov *et al.* 2021a).
- Decision Models: Describe the link between a collection of facts, a choice, and the forecasted consequence of that decision. (Akhatov *et al.* 2021b).

2 PREDICTIVE ANALYSIS METHODS

Classification models and regression models are the two types of predictive analytic models. Classification models forecast whether data belongs to a certain class, whereas regression

*Corresponding Author: fayzulla-samsu@mail.ru

DOI: 10.1201/9781032684994-110

models forecast numerical data. The approaches discussed here are commonly utilized in the construction of predictive analytics models(Sue Korn 2011).

2.1 *Linear regression model*

A simple linear regression technique describes the connection between a dependent variable and a single independent variable. Because the connection depicted in the simple linear regression model is a straight or sloping line, it is referred to as simple linear regression (Charles 2013).

The dependant variable must be constant/real-valued, which is a critical element in basic linear regression. The independent variable, on the other hand, might be measured in terms of continuous or categorical values.

A simple linear regression technique has two objectives:

- Create a model of the relationship between two variables. For instance, the link between earnings and costs, experience and compensation, and so on.
- Predicting new observations. For example, the weather forecast for temperature, the company's return on investment for a year, etc.

The following equation may be used to represent a simple linear regression model:

$$y = a_0 + a_1 x + \varepsilon \tag{1}$$

a_0 and a_1 constants determine the direction of the regression line.

Using the following formula from the variables $a_0 x$ and y and a_1 coefficients can be calculated.

$$\frac{a_1 = \sum_{i=1}^{S} (x_i - \tilde{x})(y_i - \tilde{y})}{\sum_{i=1}^{S} (x_i - \tilde{x})^2} \tag{2}$$

$$a_0 = \tilde{y} - a_1 \tilde{x} \tag{3}$$

Here $\tilde{x}$- *x is the average value for the column,* $\tilde{y}$- y is the average value for the column. ε error (It will be insignificant in the case of a good model.).

Assessment of model accuracy. There are different ways to estimate the accuracy, but for regression algorithms, Root Mean Square Error (RMSE) is often used:

$$F_{RMS}(x, h) = \sqrt{\frac{1}{m} \sum_{i=1}^{m} (h(x^{(i)}) - y^{(i)})^2} \tag{4}$$

where m - the amount of data items in the set; $y^{(i)}$- data collection; i - the actual value of the row; $h(x^{(i)})$- the predicted value in row i.

Another way to estimate accuracy is the Mean absolute error (MAE).

$$F_{MAE}(x, h) = \frac{1}{m} \sum_{i=1}^{m} |h(x^{(i)}) - y^{(i)}| \tag{5}$$

The further the predicted value is from the true value, the larger the root mean square error function σ. Figure 1 below depicts the above variables graphically.

2.2 *Multiple linear regression model*

We learnt about simple linear regression in the previous topic, where a single independent/predictor variable is utilized to model the response variable. However, there may be cases

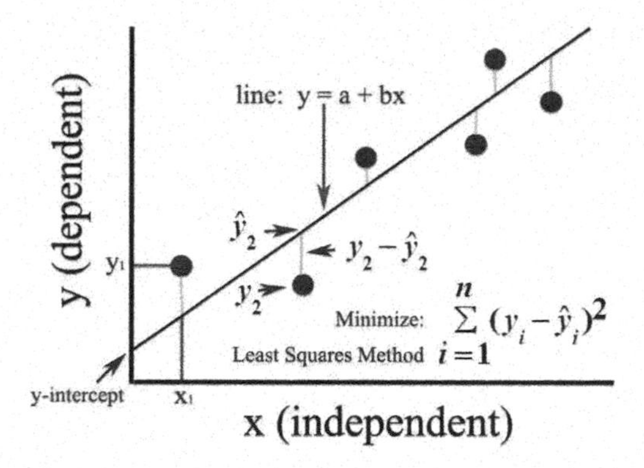

Figure 1. Graphical representation of linear regression variables.

where the response variable is impacted by more than one predictor variable; in these cases, the Multiple Linear Regression technique is applied (Chekushina *et al.* 2013).

Furthermore, multiple linear regression is an extension of basic linear regression in that it requires many predictor factors to predict the response variable. This may be defined as follows:

In multiple linear regression, the target variable (Y) is a linear combination of several predictor variables $x_1, x_2, x_3, ..., x_n$. Because this is an improvement over basic linear regression, the equation for multiple linear regression becomes:

$$Y = a_0 + a_1x_1 + a_2x_2 + a_3x_3 + ... + a_nx_n \tag{6}$$

where $x_1, x_2, x_3, ..., x_n$ are the variables and $a_0, a_1, a_2, ..., a_n$ are the coefficients.

2.3 *Polynomial regression model*

Polynomial regression is a regression approach that uses an n-level polynomial to represent the connection between dependent and independent variables. The following is the polynomial regression equation:

$$Y = a_0 + a_1x_1 + a_2x_1^2 + a_3x_1^3 + ... + a_nx_1^n \tag{7}$$

The data set used in Polynomial Regression for training is non-linear in nature.

To fit complicated and non-linear functions and data sets, it employs a linear regression model.

When applied to a linear data set, the linear model produces the same results as basic linear regression, but when applied to a non-linear data set without any adjustments, the error rate increases and the accuracy decreases (Dreyfus 2005).

So, for such cases where the data points are arranged non-linearly, we need a Polynomial regression model (Peter Lee 2012). We can understand this better with the help of the following comparison chart of linear data set and non-linear data set (Sharma *et al.* 2020).

3 RESEARCH METHODOLOGY

The study's major purpose is to examine the efficacy of utilizing regression algorithms to forecast the cost of construction businesses building in new locations. Predictive analysis is based on a set of metrics that define projects in accordance with the company's project

	Hudud	Xona	Maydon	Qavat	Uy_qavati	Narx	Uy_index	Aholi_YD
0	Samarkand	2	60.0	2	2	60600.0	12.0	5.25
1	Samarkand	3	49.0	5	5	54440.0	15.0	4.26
2	Samarkand	3	45.0	6	6	41400.0	16.0	6.12
3	Samarkand	2	61.0	1	5	46660.0	18.0	3.90
4	Samarkand	4	75.0	4	6	63000.0	20.0	7.10
5	Samarkand	1	33.0	2	8	33300.0	21.0	2.99

Figure 2. Sample data are presented.

approach. The study was based on projecting the price of new structures developed in the Samarkand region.

Comments for the columns in the dataset:

- Hudud: the name of the region where the house is located
- Xona: Number of rooms in the house
- Maydon: The size of the house (m^2)
- Qavat: What floor the house is on
- Uy_qavati: The number of floors of the building where the house is located
- Narx: The price of the house
- Uy_index: Index of the house where the house is located
- Aholi_YD: Average annual income of the population of the area.

The following are the steps in the research technique:

- Data preprocessing: In this step, the data is filtered and suited to the procedures used.
- Data uploading and role assignment. Source data that has been preprocessed is loaded, and variables are assigned "predictor" or "target variable" duties.
- Divide the data set into two parts: training (80%) and test (20%).
- Selecting which predictive analytics models to create and test.
- Using the training data set, chosen models are trained (70% of the initial data set).
- Using the test data set to validate the constructed model.
- The algorithm identifies the most important predictors found by the modeling tool after reviewing the model findings.

4 RESEARCH FINDINGS

After an initial data processing step, our filtered dataset was loaded with a dataset of 16,000 records. 12,800 pieces of data were extracted into the training dataset for machine learning, and the remaining data were taken to test the accuracy of the developed model. After these processes are completed, several predictive models are trained and tested from the data. An automatic classification node was used to facilitate this process. As a result, 300 comparable models were created, each with its own set of settings.

The models created for the topic under consideration in this research were assessed using the test data set. The most effective models were chosen based on the right predictions; models that did not fulfill the minimal standards were not investigated. The findings revealed:

Because the simple linear regression model was based on only one parameter, the result was very high (90% result), but the model was not included in the efficient models considering that it would lead to the problem of overfitting.

The multi-linear regression model includes all the parameters of the studied data, this model showed 85% accuracy, but it was found during the research that the multi-linear regression model works in cases where the data are uniformly distributed.

The polynomial regression model also showed 87% accuracy with all parameters included. this model is found to perform well even when faced with the problem of underfitting.

Based on the characteristics and results of the models analyzed above, it was concluded that the Polynomial regression model is the most effective in developing a model that calculates the average price of houses in different regions of the country for the construction of the next houses of the studied company.

5 CONCLUSIONS

Predictive analytics has become a crucial tool in decision-making processes in recent years, allowing executives from a number of corporate sectors to identify previously concealed data trends. In general, any firm that has sophisticated business expertise, a well-defined business challenge, and access to large data is capable of applying predictive analytics. On the basis of project da-ta, effective predictive analytics models were developed. Using this method, they confirmed the idea that developing a model that determines the average price of houses in different locations of the nation may help portfolio managers make strategic decisions by selecting projects.

REFERENCES

Akhatov A., Nazarov F. & Rashidov A. 2021a. "Mechanisms of Information Reliability in Big Bata and Blockchain Technologies" *ICISCT 2021: Applications, Trends and Opportunities*, 3–5.11.2021, doi: 10.1109/ICISCT52966.2021.9670052

Akhatov A., Nazarov F. & Rashidov A. 2021b. "Increasing Data Reliability by Using Bigdata Parallelization Mechanisms" *ICISCT 2021: Applications, Trends and Opportunities*, 3–5.11.2021, doi: 10.1109/ICISCT52966.2021.9670387

Akhatov A., Sabharwal M., Nazarov F. & Rashidov A. 2022a. Application of Cryptographic Methods to Blockchain Technology to Increase Data Reliability. *2nd International Conference on Advance Computing and Innovative Technologies in Engineering (ICACITE 2022)* DOI: 10.1109/ICACITE53722.2022.9823674

Charles Elkan. 2013. "Predictive Analytics and Data Mining", University of California, San Diego.

Charles Nyce, 2007, "Predictive Analytics White Paper", American Institute of CPCU/IIA.

Chekushina, E. V., Alexander E. V., and Tatiana V. C. (2013) "Use of Expert Systems in the Mining." *Middle East Journal of Scientific Research* 18(1): 1–3.

Dreyfus, G. (2005) "Neural Networks. Methoology and Applications", Springer, Berlin, Heidelber

Maciej Wach, Iwona Chomiak-Orsa. 2021. The Application of Predictive Analysis in Decision-making Processes on the Example of Mining Company's Investment Projects. 25th International Conference on Knowledge-Based and Intelligent Information & Engineering Systems. *Procedia Computer Science* 192 (2021) 5058–506625-26.

Schiff M. 2012. *"BI Experts: Why Predictive Analytics Will Continue to Grow"*, The Data Warehouse Institute.

Sharma M. and Joshi S. 2020. "Analytics in Healthcare: A Practical introduction", Asia Pacific Business Review

Sue Korn. 2011. *"The Opportunity of Predictive Analytics in Finance"*, HPC Wire.

Peter Lee. 2012, "Bayesian Statistics: An Introduction, 4th Edition", John Willey and Sons Ltd.

Artificial Intelligence, Blockchain, Computing and Security – Dagur et al. (Eds)
© 2024 The Author(s), ISBN: 978-1-032-67841-2

Algorithm of decision trees ensemble for sentiment analysis of Uzbek text

I.M. Rabbimov, O.R. Yusupov & S.S. Kobilov
Samarkand State University, Samarkand, Uzbekistan

ABSTRACT: The rapid development of social networks is leading to the growth of the content increased there. Determining the emotional trend of commentaries in this content through natural language processing and sentiment analysis technologies is so beneficial for the timely understanding of community views on social media, tracking brands, and consumer support. In this paper, an algorithm for constructing an ensemble of decision trees for the sentiment analysis of texts written in the Uzbek language was proposed. The algorithm is based on constructing mutually independent trees and making the final decision based on their predictions. The developed algorithm was experimentally researched on the corpus of Uzbek movie reviews. The comparative analysis and effectiveness of the proposed algorithm compared to other machine learning algorithms in the sentiment analysis of Uzbek texts were demonstrated.

1 INTRODUCTION

Nowadays, people freely express their opinions about products, services, and various topics on social media and websites. The sentiment analysis of these opinions and judgments is important for companies providing products and services, social research institutes, and public administration structures [6]. Real-time sentiment analysis of dialogic written speech texts expressed in social networks requires experts' plenty of time and effort. So as to solve this problem, research on the automatic sentiment analysis of texts for different languages is being carried out [3,7,17]. Several approaches, methods, models, and algorithms have been developed for sentiment analysis of natural language texts. Although many of the proposed models and algorithms have general characteristics from the point of view of natural language, they require full consideration of the specific characteristics of each language. This requires conducting separate scientific research for each language and developing or improving existing models and algorithms that take into account language characteristics. The analysis of natural language processing research showed that the research on the sentiment analysis of texts written in Uzbek is not sufficiently studied. In the sentiment analysis of Uzbek language commentary on social meadia, it is necessary to take into account the specific features of the written text and the language. Thus, the need to develop effective methods and algorithms for sentiment analysis of comments written in Uzbek by users on social meadia and websites is shown.

In the sentiment analysis of texts, the development of a classification algorithm for identifying important regularities in the space of features formed according to the text is an urgent issue. Among the classification algorithms based on supervised machine learning algorithms like Support vector machine, Neural network, Naive Bayes classifier, Decision trees, C4.5 algorithm, and Random forest algorithms have been effectively used in sentiment analysis of texts [5,10,13–15,18]. Research on the classification of texts in Uzbek was carried out in [1,2,8,9,11,12] and they contribute to a certain amount to the development of the field of sentiment analysis of Uzbek texts.

 DOI: 10.1201/9781032684994-111

The decision tree algorithm, which is one of the algorithms based on the multi-level decision-making approach, has been used to identify features and find patterns in large data sets. An algorithm based on any multi-step approach is based on dividing a complex decision into several simple decision units and making the final decision similar to the desired solution. The decision tree model consists of nodes and branches and is implemented in the steps of the building, dividing, and cutting the model [10]. Advantages of the decision tree algorithm: it does not require a lot of effort to prepare the data during the initial processing and does not require normalization of the input data, incomplete values in the input data do not significantly affect the construction of the decision tree. However, the decision tree algorithm has disadvantages such as a little alteration in the data leads to a considerable alteration in the structure of the decision tree and overfitting to the data of the training sample. In order to correct these shortcomings, a random forest algorithm was proposed in [7]. This algorithm prevents the decision trees from overfitting to the training sample. The random forest algorithm is an ensemble method that builds many decision trees in the training phase, and the method of voting tree predictions is used for the classification problem. Several studies have shown that improving the random forest algorithm by applying weights to the decisions of each tree in the decision-making process has led to increased classification results [16]. Nevertheless, the improvement of decision-making methods and algorithms remains an important issue.

2 PROBLEM STATEMENT

In this paper, we discuss the issue of developing an algorithm for constructing an ensemble of decision trees for sentiment analysis of texts which were written by users on social media and Internet websites.

The dialog Uzbek text written is considered input data of the algorithm. Input data is formed from a data set consisting of a tuple of commentaries on Uzbek videos on the YouTube social network.

The algorithm's output will be a value belonging to negative or positive. It is important to evolve a productive algorithm that gives the result of processing the input data 0 value if it belongs to the negative class, and 1 value if it belongs to the positive class.We formulate the given problem in the following order. Suppose we are given a set $D = \{D_1, \ldots, D_m\}$ of objects (documents, posts, comments, etc.) consisting of $K_1, \ldots, K_l$ $(l \geq 2)$ subsets (classes):

$$D = \bigcup_{i=1}^{l} K_i, \; K_i \cap K_j = \varnothing, \; i \neq j, \; i, j \in \{1, \ldots, l\}$$

It is required to build a function $F(D_i, K_j)$ that determines whether the object D_i belongs to the set K_j, i.e.

$$F(D_i, K_j) = \begin{cases} 1, & \textit{if object } D_i \textit{ belongs to set } K_j, \\ 0, & \textit{otherwise} \end{cases}$$

where the $F(D_i, K_j)$ $(i = \overline{1, m}, j = \overline{1, l})$ function is called a classifier and it is implemented in the form of a model or algorithm.

In the considered issue, D_i as an object is a text comment in the form of a dialogic written speech, the number of classes is $l = 2$, that is, $K = \{K_1, K_2\}$. K_1 is represented as a set of positive comments, K_2 as a set of negative comments.

The D set is made up of a set of Uzbek movies reviews on the YouTube social network. Dialogic written speech text, which is an element of the D set, D_i a) low quality of the text, which is unstructured and informal; b) the presence of sentiment emojis and emoticons; c) presence of the same words with different meanings in the context; g) has features such as the fact that the texts are written using both Latin and Cyrillic alphabets.

3 PROPOSED METHODOLOGY

The proposed decision tree ensemble construction algorithm includes two parts. In the first part, the feature vectors is entered into mutually independent trees and predictions of decision trees are obtained. In the second part, the predictions of the decision trees are the input to the neural network, and the output value from the neural network represents the classification result. The operation scheme of the proposed algorithm is depicted in Figure 1.

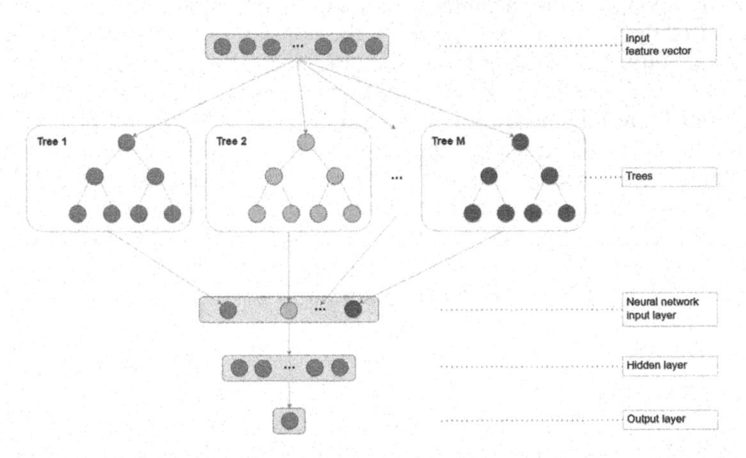

Figure 1. Schematic diagram of the algorithm of the decision trees.

Before training the model, the Bagging algorithm proposed by Breiman is used [4]. It has been shown that the bagging algorithm can reduce the variance of a classifier, given a training sample and estimation method, compared to a classifier operating only on the training sample. This ensures an increase in the strength of the classification.

3.1 *Definition 1*

A bootstrap sample is a D^* sample generated from a D training sample by repeating its elements several times, leaving some elements unselected or reordered. Here D training sample N consists of (x,y) pairs, x is a feature vector, y is a class label. Although $|D| = |D^*| = N$, 37% of the pairs from the D sample do not occur in the bootstrap sample. When a D^* bootstrap is generated from D selection, the probability that (x,y) pairs are not selected is $\left(1 - \frac{1}{N}\right)^N \approx \frac{1}{e} \approx 0.368$. Therefore, we can observe that in the bootstrap sample D^*, D meets at least once the element of 63.2% of the sample.

3.2 *Definition 2*

The bootstrapping method is to draw $D^1, \ldots, D^B$ bootstrap samples from the D training sample. Here, functions $\varphi(D^1), \ldots, \varphi(D^B)$ are calculated for each bootstrap pattern.

The bagging algorithm is executed in the following sequence.

- $D^1, \ldots, D^B$ boot samples are generated from the D training sample.
- Calculated $\widehat{f}_1(x), \ldots, \widehat{f}_B(x)$ prediction values for $D^1, \ldots, D^B$ bootstrap samples.

Assume a vector of text-formatted features $x_i \in R^P$ and class labels y_i, and let $P_{xy}(x_i, y_i)$ be their joint distribution. The goal is to find a prediction function $f(x_i)$ that predicts y_i. The prediction function $L(y_i, f(x_i))$ is determined using the loss function and is defined as follows to minimize the expected value of the loss function:

$$E_{xy}(L(y_i, f(x_i))) \tag{1}$$

where the designation xy in the index represents the expectation of the joint distribution of x_i and y_i. $L(y_i, f(x_i))$ is a measure of how close $f(x_i)$ is to y_i, and it tries to approximate $f(x_i)$ values that are far from y_i. For the classification problem, we define the L loss function as follows:

$$L(y_i, f(x_i)) = I(y_i \neq f(x_i)) = \begin{cases} 0, & \text{if } y_i = f(x_i) \\ 1, & \text{otherwise} \end{cases} \tag{2}$$

denoting the possible values of y_i by Y, minimizing $E_{xy}(L(y_i, f(x_i)))$ by the loss function L gives the following formula:

$$f(x_i) = \underset{y_i \in Y}{\arg\max}\ P(y_i = y | x_i = x). \tag{3}$$

If we denote the j-tree in the proposed algorithm by $\Im_j(\Theta_j)$, and the prediction obtained using this tree by $T_j(X, \Theta_j) = \hat{y}_j$, where $j = 1, \ldots, M$, Θ_j are the j-tree parameters. $\Im_j(\Theta_j)$ trees partition the predictor space with an order of binary partitions on particular variables. The "Root" node of the tree contains the entire predictor space. Unsplit nodes are called "Terminal nodes" and it is final part of the predictor gap. Every nonterminal node is divided in a left and right descendant node according to the value of one of the predictor variables. Points, where the value of an element of the feature vector is less than the value of the predictor variable at the predictor split point are shifted to the left, and the rest are shifted to the right. A categorical predictor variable $x_{i,j}$ obtains the values of a finite tuple of categories $S_i = \{S_{i,1}, S_{i,2}, \ldots, S_{i,m}\}$. If the set $x_{i,j} \in S$ and S is a subset of the set S_i, the division moves S to the left and $S_i \backslash S$ to right. When splitting a tree node into two generations, the "best" split parameters are selected according to a given criterion for each predictor variable and given any split.

In the context of classification, there are K classes denoted by $1, \ldots, K$, the Gini index is defined as follows:

$$G = \sum_{k \neq k'}^{K} \hat{P}_k \cdot \hat{P}_{k'} \tag{4}$$

where $\hat{P}_k$ is the percentage of samples (observations) of the k class at the node:

$$\hat{P}_k = \frac{1}{n} \sum_{i=1}^{n} I(y_i = k). \tag{5}$$

The separation criterion supplys a measurement of "purity" for the node, with larger values indicating an "impure" node. A candidate split creates a left and right descendant node. If we denote the separate measurement of two candidate lineages by Q_L and Q_R and their sample sizes by n_L and n_R, the split is chosen to minimize $Q_{Split} = n_L Q_L + n_R Q_R$.

A three-layer neural network was used to make the final decision. The general architecture of the algorithm is presented in Figure 1. According to it, $\hat{y}_1, \hat{y}_2, \ldots, \hat{y}_m$ – input signals, h_1, $h_2, \ldots, h_m$ – neurons in the hidden layer, P – output signal.

The h_j neuron values in the first hidden layer are calculated using the following formulas:

$$u_j^{(1)} = \sum_{i=1}^{n} w_{ji}^{(1)} \cdot \hat{y}_i \tag{6}$$

$$h_j = \phi\left(u_j^{(1)} + b_j^{(1)}\right) \tag{7}$$

where $w_{j1}^{(1)}, w_{j2}^{(1)}, \ldots, w_{jn}^{(1)}$ is the weight coefficients in the hidden layer, $u_j^{(1)}$ is the linear combination of the input signals in the hidden layer, $b_j^{(1)}$ is the bias coefficient of the neuron in the hidden layer, and $\varphi(\cdot)$ is the activation function. The value of the P neuron in the output layer is calculated based on the following formula:

$$u_j^{(2)} = \sum_{i=1}^{\lambda} w_{j\,i}^{(2)} \cdot h_i \tag{8}$$

$$P = \phi\left(u_j^{(2)} + b_j^{(2)}\right) \tag{9}$$

where h_i is the output value of the i-neuron in the hidden layer, $w_{j1}^{(2)}, w_{j2}^{(2)}, \ldots, w_{jl}^{(2)}$ is the weight coefficients in the output layer, $u_j^{(2)}$ is the linear combination of input signals in the output layer, $b_j^{(2)}$ is the bias coefficients of the neuron in the output layer, P is the output values of the neuron in the output layer. In the third layer, the sigmoid function given in formula (10) is used as the $\varphi(\cdot)$ activation function.

$$\varphi(x) = \frac{1}{1 + e^{-x}} \tag{10}$$

A backpropagation algorithm was used to train the neural network. This method is based on calculating the errors of the neurons in the output layer and correcting the weights of the neurons in the previous layers. The mistak is propagated from the output to the input, and the neuron weights in all hidden layers are updated.

The feature vector to be included in the algorithm developed on the basis of the proposed approach is formed on the basis of statistical properties, parts of speech and emoji from the text.

3.3 *Statistical features of the text*

Characters' number in the text; the number of all characters except for a space in the text; number of special characters ('(', ')', '[', ']', '{', '}', '-', '/', '&', '|', ...); the number of lowercase letters; the number of capital letters; number of numeric characters; total count of words; single words' nubmber; the average value of the longitude of all unique words; the maximum value of the longitude of all words; the minimum value of the length of all words; the mean of the length of all words; standard deviation of all word lengths; dispersion of all word lengths; coefficient of kurtosis of the longitude of all words; coefficient of asymmetry of the length of all words; 25% percentile value of all word lengths; 50% percentile value of all word lengths; 75% percentile value of all word lengths; number of punctuation marks ('.', ',', '!', '?', ':', ';'); words' number which is less than 4 signs long; hapax-legomena's number; the number of hapaxes-dyslegomena.

3.4 *Features based on parts of speech*

The number of words which are belong to nouns' group; the number of words belonging to the group of proper names; the number of words belonging to the group of verbs; the number of words belonging to the group of adjectives; the number of words belonging to the group of numerals; the number of words related to the group of pronouns; the number of words belonging to the group of adverbs; the number of words belonging to the group of auxiliary words; the number of words belonging to one group of linking words; the number of words belonging to the next group of linking words; the number of words included in the modal group of words; the number of words belonging to the exclamation group of words; the number of words belonging to the imitation group of words; the number of words related to the vocabulary of auxiliary words; a number of other words such as vague, meaningless.

3.5 *Emoji-based features*

Amount of emojis; the mean of the degree of sentiment of all emoji for each comment; amount of emojis that are positive; the amount of emojis that negative.

The amount of statistical text characters is 23, the amount of features based on parts of speech is 15, and the amount of features based on emoji is 4. The total dimension of the character vector is 42.

The decision tree ensemble construction algorithm developed for sentiment analysis of texts consists of the following steps.

Algorithm. An algorithm for building an ensemble of decision trees.

Start.

Step 1. D text is entered.

Step 2. A feature vector $X \in R^p$ based on statistical, part of speech and emoji-based text matching D is generated.

Step 3. Using the set of $\Im$ trees, $\hat{y}$ ($\hat{y} \in R^M$) predictions corresponding to the X vector are obtained:

$$\hat{y} = (T_1(X, \Theta_1), ..., T_M(X, \Theta_M))$$

where $\hat{y}_m = T_M(X, \Theta_M)$, $\Im_m$ is the prediction value obtained using the tree and X sample, $\hat{y}_m \in \{0, 1\}$, $\Im(\Theta) = \{\Im_m(\Theta_m), m = \overline{1, M}\}$ is a set of decision trees, M is the number of trees, $\Theta = \{\Theta_1, ..., \Theta_m\}$ is the parameters of the trees.

Step 4. The h_i neuron values in the hidden layer are:

$$h_i = \sigma_1(\sum_{j=1}^{M} w_{ij} \cdot \hat{y}_j + b_i), \ i = \overline{1, l}$$

where w_{i1}, ..., w_{im} is the weight values of the neurons in the hidden layer, b_i is the bias coefficient of the neuron in the hidden layer, and $\sigma_1(x) = \max(0, x)$ is the ReLU activation function.

Step 5. The P prediction value in the output layer is:

$$P = \sigma_2(\sum_{j=1}^{l} w_{1j} \cdot h_j + b_1)$$

where $\sigma_2(x) = \frac{1}{1+e^{-x}}$ is the sigmoid activation function.

Step 6. The result of sentiment analysis is determined. If $P \geq 0{,}5$ then the text is positive, otherwise the text is negative.

Stop.

4 PREPARING AND CONDUCTING EXPERIMENTS

To prepare experiments, evaluate the proposed model and demonstrate the performance of the designed programs, fairly large data set were selected.

4.1 *Dataset*

The Uzbek Movie Reviews (UzMRC) was used to evaluate the proposed algorithm. The collection is a corpus of commentaries of Uzbek films assembled using the YouTube Data API. Comments in corpus 5351 refer to Latin, 7903 to Cyrillic, 58 to Latin and Cyrillic (mixed), and 817 to emoticons and other signs. The amount of comments that are positive in the corpus is 9732, the number of negative comments is 4397, and the total number of comments is 14129. This corpus is annotated by 6 annotators. During training and testing, 10 times cross-validation is applied to the corpus of the algorithm.

A feature vector was generated based on text statistics, part-of-speech, and emoji. Experimental studies were carried out using the Algorithm for constructing an ensemble of decision trees (CEDT) on this feature vector. Experiments were also carried out on various famous machine learning algorithms that are widely utilized in the classification stage to compare the proposed algorithm. In particular, k-nearest neighbors, Neural Networks, Support Vector Machines, Decision Trees, Naive Bayes classifier algorithms were used.

4.2 *k-nearest Neighbors (k-NN)*

A k-nearest neighbors classifier with linear nearest neighbor search and without considering the distance value.

4.3 *Neural Networks (NN)*

A multi-layer perceptron neural network with a two-hidden-layer architecture, 30 sigmoid nodes in each hidden layer, was trained with 5000 iterations using the back-propagation algorithm.

4.4 *Support Vector Machine (SVM)*

Two different kernels, i.e. Radial Basis Kernel (SVM-rbf) and Polynomial Kernel (SVM-poly) were tested, and the Basis Vector Algorithm was used, which used a sequential minimum optimization algorithm.

4.5 *Decision Trees (DT)*

Two tree algorithms were tested, namely error reduction tree (RT), C4.5 decision tree (C4.5). Decision tree algorithms are designed utilizing data gain or difference reduction and pruning. They used reduced error trimming with back-fitting.

4.6 *Naive Bayes classifier (NB)*

A Bayesian network, a probabilistic graphical model describing a set of random variables and their conditional dependencies via a directed acyclic graph and a simple Bayesian multinomial update, was applied with the K2 search algorithm and a simple estimator (alpha = 0.5). In this case, informative feature vectors reflect the frequencies of certain events generated by the multinomial.

5 ANALYSIS AND EVALUATION OF RESULTS

The algorithm that was proposed for the sentiment analysis of texts was assessed with the experiments in Section 4. The execution of the algorithm was assessed by the accuracy evaluation measure. It is calculated with an accuracy formula.

$$Accuracy = \frac{\text{true positives} + \text{true negatives}}{\text{true positives} + \text{false positives} + \text{true negatives} + \text{false negatives}}$$

The results of experiments are revealed in Table 1. The best indicator is in bold.

Table 1. Classification accuracy achieved by classification algorithms.

Classification algorithm	k-NN	NN	SVM-poly	SVM-rbf	C4.5	CEDT	RT	NB
Accuracy (%)	80,26	82,72	84,55	84,39	83,46	**85,25**	84,12	75,34

It is clear that this table that the best classification accuracy of 85.25% was achieved by the decision tree ensemble algorithm, while the SVM-poly, SVM-rbf and neural network algorithms achieved a lower accuracy of around 1% compared to the decision tree ensemble algorithm.

6 CONCLUSION

Significant growth in the volume of content published by users in social networks and the importance of sentiment analysis of opinions in them in making management and marketing decisions is increasing more and more. Understanding, categorizing and analyzing ideas in large volumes of textual content is beneficial for timely understanding of community views in social networks, tracking brands, and consumer support. In this work, an algorithm for building an ensemble of decision trees was developed for the sentiment analysis of dialogic written speech texts written in Uzbek. The algorithm is based on building mutually independent trees and making the final decision based on their predictions. The developed algorithm was experimentally researched on the corpus of Uzbek film reviews. The effectiveness of the algorithm make comparison to the others in sentiment analysis of Uzbek language texts was demonstrated.

REFERENCES

[1] Babomuradov, O.J., Mamatov, N.S., Boboev, L.B., Otakhonova, B.I. 2019. Classification of Texts Using the Decision Tree Algorithm. *Descendants of Muhammad al-Khorazmi* 4(10): 22–24.

[2] Babomuradov, O.J., Otaxonova, B., Mamatov, N.S., Boboev, L.B. 2019. Text Documents Classification in Uzbek Language. *International Journal of Recent Technology and Engineering* 8(2S11): 3787–3789.

[3] Bogdanov, A.L., Dulya, I.S. 2019. Sentiment Analysis of Short Russian-language Texts in Social Media. *Tomsk State University Bulletin* 47: 220–241.

[4] Breiman, L. 2001. Random Forests. *Machine Learning* 45(1): 5–32.

[5] Kuriyozov, E., Matlatipov, S., Alonso, M. A., Gómez-Rodnguez, C. 2019. Deep Learning vs. Classic Models on a New Uzbek Sentiment Analysis Dataset. *Proceedings of the Human Language Technologies as a Challenge for Computer Science and Linguistics-2019*: 6–8.

[6] Liu, B. 2020. *Sentiment Analysis: Mining Opinions, Sentiments, and Emotions*. Cambridge University Press.

[7] Medhat, W., Hassan, A., Korashy, H. 2014. Sentiment Analysis Algorithms and Applications: A Survey. *Ain Shams Engineering Journal* 5(4): 1093–1113.

[8] Mukhamedieva, D.K., Jurayev, Z. 2020. Classification of Content of Works. *Problems of Computational and Applied Mathematics* 2(26): 108–117.

[9] Mukhamedieva, D.T., Zhuraev, Z.Sh., Bakaev, I.I. 2019. Approaches to the Thematic Classification of Literary Works. *Problems of Computational and Applied Mathematics* 4(22): 111–117.

[10] Quinlan, J.R. 1987. Simplifying Decision Trees. *International Journal of Man-machine Studies* 27(3): 221–234.

[11] Rabbimov, I. M., Kobilov, S. S. 2020. Multi-class Text Classification of Uzbek News Articles Using Machine Learning. *Journal of Physics: Conference Series* 1546(1): 012097.

[12] Rabbimov, I., Kobilov, S., Mporas, I. 2020. Uzbek News Categorization Using Word Embeddings and Convolutional Neural Networks. *In 2020 IEEE 14th International Conference on Application of Information and Communication Technologies (AICT)*: 1–5.

[13] Rabbimov, I., Kobilov, S., Mporas, I. 2021. Opinion Classification Via Word and Emoji Embedding Models with LSTM. *Lecture Notes in Computer Science* 12997: 589–601.

[14] Rabbimov, I., Mporas, I., Simaki, V., Kobilov, S. 2020. Investigating the Effect of Emoji in Opinion Classification of Uzbek Movie Review Comments. *Lecture Notes in Computer Science*: 12335: 435–445.

[15] Rabbimov, I.M., Kobilov, S.S. 2021. Opinion Classification of Text Based on LSTM Architecture Neural Network. *Problems of Computational and Applied Mathematics* 6(36): 98–110.

[16] Shahhosseini, M., Hu G. 2021. Improved Weighted Random Forest for Classification Problems. *International Online Conference on Intelligent Decision Science*: 42–56.

[17] Smetanin, S. 2020. The Applications of Sentiment Analysis for Russian Language Texts: Current Challenges and Future Perspectives. *IEEE Access* 8:110693–110719.

[18] Vorontsov, K.V. 2011. Mathematical Methods of Learning by Precedents (The Theory of Machine Learning). *Moscow*: 119–121.

Artificial Intelligence, Blockchain, Computing and Security – Dagur et al. (Eds)
© 2024 The Author(s), ISBN: 978-1-032-67841-2

Intelligent modeling and optimization of processes in the labour market

A. Akhatov, M. Nurmamatov & F. Nazarov
Samarkand State University, Samarkand, Uzbekistan

ABSTRACT: In this scientific work, an improved approach to the coordination of labor market relations is proposed and justified. A mathematical model for redistribution of labor force in the labor market is proposed. Due to the increase in the solution space of the proposed mathematical model, optimization was carried out using genetic algorithms. An improved genetic algorithm is proposed and experimental results are obtained.

1 INTRODUCTION

Today, computerization and digitalization processes are rapidly developing in the world, therefore, scientific research on the effective development of the labour market, especially the intellectualized labour market, is of particular importance. In this regard, such scientific areas as increasing the mobility of the labour force, introducing modern forms of employment, developing the infrastructure of the modern labour market, introducing modern software products into employment services and relevant authorities through development are of great importance.

2 MATERIALS AND METHODS

2.1 *Approaches to the regulation of employment in the labour market*

Analysis and generalization of the conducted studies of the methods of predictive analysis of the labour market makes it possible to identify several methodological approaches. We propose an improved approach to labour market regulation based on existing approaches. An improved approach to analysis based on existing approaches to labour market regulation is presented in Figure 1 below (Nurmamatov 2022a, 2022b).

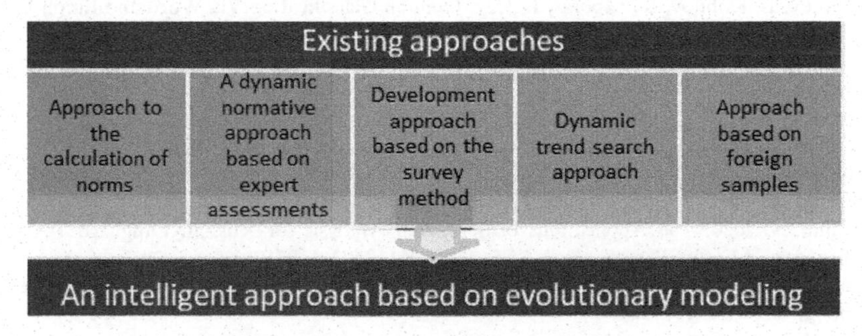

Figure 1. Improved analytical approach to labour market regulation.

DOI: 10.1201/9781032684994-112

An improved approach (intellectual + analysis). The proposed approach, based on the above approaches, implements an intellectual analysis of the labour relations of the population using automated systems. An information system for operational monitoring of the employment status of the population will be developed, and intellectual analysis modules will be included in the information system. The proposed approach solves the following problems:

- The employment of the population is determined by regions;
- The employment status is tracked in real time in the information system;
- Based on the monitoring results, recommendations are given to improve the employment status of the population.

The recommendations for improving the employment status of the population are based on predictive intellectual models (Akhatov *et al.* 2022a).

2.2 *Statement of the problem*

Based on the modeling of the social status of the population, the system of automated regulation of labour relations, the mathematical formulation of the management problem is expressed by formula (1) (Nurmamatov 2022d).

$$\begin{cases} F(N_1(t),\ N_2(t),\ u(t),\ \lambda) \to extr \\ N_1(t),\ N_2(t) > 0 \\ N_1(t) \to \max \end{cases} \tag{1}$$

$N_1(t)$–the total number of specialists involved in the statement of the management problem; $N_2(t)$– the number of able-bodied and unemployed persons; $u(t)dt$ - the parameter for optimizing the job search for an unemployed specialist in the period from t to $t+dt$; $W_2(t)dt$- the possibility of dismissal of a working specialist in the period from t to $t+dt$; λ-external impacts on labour relations. Based on this, the condition $N_1(t) \to$ maxis provided on the basis of the control parameter $u\,(t)$, and the regulation of the labour force in the labour market is carried out according to the following formulas (2) (Akhatov *et al.* 2021a).

$$\frac{dN(t)}{dt} = (N_2(t)u(t) - N_1(t)W_2(t)) \text{ and } N(t) = \int_0^T (N_2(t)u(t) - N_1(t)W_2(t))dt \tag{2}$$

When managing the employment of the population, the control parameter$u\,(t)$ regulates the coordination of the job search process. In this case, with the improvement of the management parameter, the level of employment will improve and the likelihood of employment of qualified specialists will increase. Based on this, since the control parameter $u\,(t)$ depends on the probability of finding a job, we introduce the following definition for the parameter $u\,(t)$. The functional diagram of the management problem $W_1(t) = u\,(t)$ is shown in Figure 2.

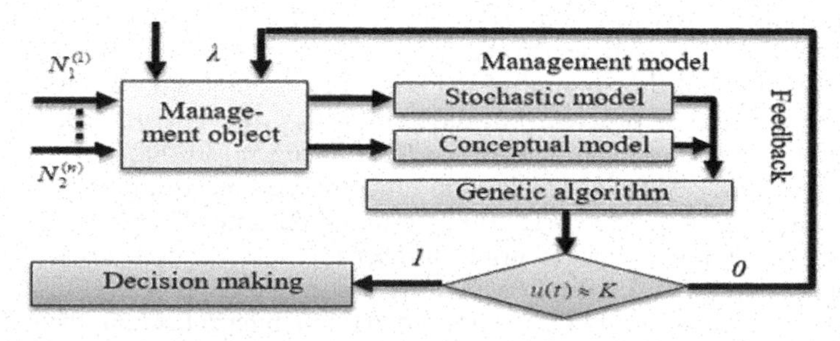

Figuer 2. Scheme of labour market management.

On the basis of modeling the social status of the population, the problem of optimizing labour relations in the labour market of the population is solved based on the introduction of the problem of managing an automated system for regulating labour relations.

So, the employment of an unemployed specialist in the labour market in his/her field is the main goal of the objective function and is expressed as (4) (Nurmamatov 2022c):

$$
\begin{gathered}
f(t)=\sum_{i=1}^{n}\sum_{k=1}^{2}\sum_{l=1}^{2}\left(N_k^{(i)}(t)W_l^{(i)}(t)\right)_{while\ k=l\neq 2}\rightarrow \max_{W\in\Omega_W}\\
\Omega_W=\left\{W\in R^n\,\middle|\,0<W_k^{(i)}<1,\ i=\overline{1,n},\ k=1,2\right\}\\
N_k^{(i)}(t)_{\min}\leq N_k^{(i)}(t)\leq N_k^{(i)}(t)_{\max},\ i=\overline{1,n},\ k=1,2
\end{gathered}
\tag{3}
$$

Then the final fitness of the selected sample can be calculated using the following (5) formula:

$$
\text{fitness}(t)=f(t)+\delta\cdot\lambda(x)\cdot\sum_{j=1}^{n}f_j^{\beta}(t), \tag{4}
$$

where x - current generation number; $f_j(t)$– deviation of limitations on the number j ; β–real number, deviation rate indicator.

According to the method, the value is $f_j(t)$calculated dynamically according to the formula (6) for the t - th iteration, depending on the level of deviations and $\lambda(t)=(C\cdot t)^{\alpha}$.

$$
f_j(t)=\begin{cases}\max\left\{N_1^{(i)}(t),\ i=\overline{1,n},\right\}\\ \min\left\{N_2^{(i)}(t),\ i=\overline{1,n},\right\}\\ \max\left\{W_1^{(i)}(t),\ i=\overline{1,n},\right\}\end{cases} \tag{5}
$$

The recommended values of the parameters $C=0.5,\ \alpha=\beta=2$ vary depending on the problem and affect the efficiency of finding a solution.

In this method, if the best representative of the population (in terms of the fitness function) belongs to the range of valid values at the last m iteration, the fitness index is reduced in the next step. If the best representative of the population has exceeded the allowable area limits for the same period, the fit index increases (Akhatov *et al.* 2021a, 2021b).

2.3 *Genetic Algorithm (GA) for the allocation of jobs in the labour market*

As a result of the study, the process of reallocation of labour m to jobs n in various sectors of the economy is carried out as follows based on genetic algorithms. In this case, each ith labour force consists of n_i a sequence of work volumes , and $(o_{i1},o_{i2},\ldots,o_{in})$ any worker can choose A_{ik} each job from the set o_{ik}. The problem of finding a job is to match the workforce $M(o_{ik})\in A_{ik}$ for each activity o_{ik}. The cases to be considered are listed below: (a) the worker may choose one job at a given time; (b) at any time, the worker may choose only one job and may not interrupt it until the amount of work has been completed; (c) the case where the same sequence of works is ordered, but different works are not ordered; (d) no priority restrictions among many different works (Lin 2001; Wang *et al.* 2002).

The ranking methods of fuzzy termination time are required to minimize the maximum fuzzy termination time and the average fuzzy termination time. The following criteria shall be accepted: (a) the largest associated prime number $C1\left(\widetilde{A1}\right)=(a_1+2a_2+a_3)/4$ is chosen as the first binary sort criterion; (b) if the binarity has the same $C1$, $C2\left(\widetilde{A1}\right)=a_2$ is used as the second criterion; (c) if $C1$ and $C2$ are not ordered in a binary fashion, then the difference of allocations $C3\left(\widetilde{A}1\right) = a_3 - a_1$ shall be chosen as the third criterion. It consists of maximizing the work allocation index (AI) and minimizing the maximum fuzzy completion time. They are formulated as (7). (Akhatov *et al.* 2020b). Due to the different slopes of each side of the trapezium and the different slopes of each side of the triangle and their different positions, there may be more than twenty cases of AI. All these difficult positions and situations can be solved in the advanced GA software.

$$f_1 = \max\left(\frac{1}{n}\sum_{j=1}^{n} AI_j\right),\ f_2 = \min\left(\max_{j=1,2,\cdots,n} \widetilde{C}_j\right) \text{ and } f_3 = \min\left(\max_{1\le i\le m}\sum_{j=1}^{n}\sum_{h=1}^{h_j}\widetilde{p}_{ijh}x_{ijh}\right) \quad (6)$$

For example, suppose we are given 3 jobs and 5 workers.

Table 1. Coordination of 3 jobs and 5 workforce total employment status.

		Time of recruitment of labor force					
Professions	Work volume	M_1	M_2	M_3	M_4	M_5	Term
K_1	O_{11}	1,2,5	2,6,10	4,5,6	2,3,4	2,4,8	P_1
	O_{11}	4,6,9	3,8,11	1,2,3	2,4,5	3,5,7	5,7,16,25
	O_{21}	1,3,4	2,3,4	4,6,8	5,7,12	1,5,8	P_2
K_2	O_{22}	2,4,8	5,6,7	4,5,7	8,9,10	5,6,10	16,20,33,41
	O_{23}	2,3,4	3,7,12	8,11,13	3,5,8	2,8,11	
	O_{31}	2,6,8	4,6,10	3,4,7	1,2,4	3,6,7	P_3
K_2	O_{32}	5,7,9	11,12,14	6,8,11	4,6,9	12,15,17	23,28,35,44
	O_{33}	1,2,5	3,5,7	4,8,9	5,6,7	8,9,11	

2.4 *An improved genetic algorithm for the allocation of jobs*

Using the characteristics of the genetic algorithm to exhibit parallelism and to include past solutions and to a certain extent redundant and historical information, the study was able to capture the important components of good solutions for employment relations coordination. They were combined by crossover to create high-quality solutions. This feature can lead to some distortions, where good solutions become imprecise. Therefore, simple GA often produces early convergence and gives poor results in difficult combinational problems. The work presents an improved generation variable model used in the genetic algorithm for the problem of flexible job redistribution planning, which can better preserve the meaningful characteristics of the previous generation and reduce the disruptive effect of genetic operators. The performance steps of the proposed improved genetic algorithm procedure for the employment problem are carried out in the steps given below.

Step 1. Selecting P samples based on random initialization, where P is taken as an indicator of the group size.

Step 2. Evaluating the fitness of each sample using the fuzzy set operations presented in (7) formula. Improve the positions and number of cases of optimal goals.

Step 3. If the algorithm reaches the stopping condition, then the best solution is output, otherwise, go to step 4.

Step 4. Transferring the 1% of the best samples directly to the next generation.

Step 5. Selecting the next generation group in accordance with the selection strategy.

Step 6. If the fitness values of the two parents are not equal and the crossover probability is not satisfied, the action-based gene sequences cross n times, and the machine-based gene sequences cross n×k times. The two best chromosomes are then selected from all the offspring.

Step 7. Performing the mutation operation and obtaining a new sample with a mutation probability P_m.

Step 8. Create a new generation group and return to step 3.

To prove the validity and feasibility of the improved genetic algorithm proposed in the study, we chose three examples of labour reallocation with six workers on a job search and six workers in employment, and three examples of labour reallocation with ten workers on a job research and ten workers in employment (Zhang Chaoyong *et al.* 2004). At the same time, examples of planning the reallocation of jobs are presented in Table 2.

Table 2. The reallocation of labour force in the position 6 × 6.

Professions			Hiring time				Term
K_1	4(9,13,17)	3(6,9,12)	1(10,11,13)	5(5,8,11)	2(10,14,17)	6(9,11,15)	112, 121
K_1	4(5,8,9)	2(7,8,10)	5(3,4,5)	3(3,5,6)	1(10,14,17)	6(4,7,10)	82, 91
K_1	5(3,5,6)	4(3,4,5)	3(2,4,6)	1(5,8,11)	2(3,5,6)	6(1,3,4)	49, 60
K_1	6(8,11,14)	3(5,8,10)	1(9,13,17)	4(8,12,13)	2(10,12,13)	5(3,5,7)	97,102
K_1	3(8,12,13)	5(6,9,11)	6(10,13,17)	2(4,6,8)	1(3,5,7)	4(4,7,9)	83, 89
K_1	2(8,10,13)	4(8,9,10)	6(6,9,12)	3(1,3,4)	5(3,4,5)	1(2,4,6)	54,59

3 RESULTS

Maximization of the average employment level was accepted as the optimal goal in the study. The parameter values of the improved genetic algorithm are shown in Table 3 and the best values of the six problems are shown in Table 4. The results not only prove the validity and feasibility of the improved genetic algorithm dealing with job reallocation scheduling problems, but also provide the best value of the six problems.

Table 3. Parameter of the genetic algorithm for the reallocation of labour force.

Workplace planning	Number of examples	Crossover probability	Mutation probability	n	k	Time
a task 6 × 6	100	0.8	0.1	10	10	10
a task 10 × 10	200	0.8	0.1	10	10	10

Table 4. The end result of the objective functions.

$n \times m$	f_1	f_2	f_3
3 × 5	0.964444	(9,14,23)	(6,10,14)
3 × 5	0.966984	(7,14,25)	(4,9,17)
4 × 6	0.979084	(16,22,28)	(7,14,21)
6 × 10	0.849255	(46,58,70)	(29,40,50)
8 × 8	0.938736	(18,37,64)	(11,28,47)
10 × 10	0.863502	(7,15,29)	(6,15,27)

From the above examples and results, it can be easily concluded that the proposed improved genetic algorithm is a correct and feasible algorithm for solving job scheduling problems. In order to better prove the validity, feasibility, and effectiveness of solving the job redistribution problem with the proposed improved genetic algorithm, six more complex practical experiments were conducted in this study.

As optimal objectives, we consider not only the maximization of employment, but also the minimization of the time to find a job and the maximization of the labor force allocation. In Table 4, the three objectives are briefly denoted as, and, and the best values of the six problems are listed in Table 4. The results not only prove the reliability, feasibility and efficiency of the improved genetic algorithm, but also the best scores for all six problems.

From the above examples and results, one can easily conclude that the proposed improved genetic algorithm is a correct and feasible algorithm for solving workplace planning problems.

4 CONCLUSIONS

In the conducted research, the goal was achieved by using genetic algorithms in the matter of optimization of the labor force distribution process. Using the fuzzy distribution method, higher performance was achieved than the results presented in other methods. experiments have shown that with the increase in the number of jobs, the quality and time of distribution have increased.

REFERENCES

Akhatov A., Nazarov F. & Rashidov A. 2021a. "Increasing Data Reliability by Using Bigdata Parallelization Mechanisms" ICISCT 2021: Applications, Trends and Opportunities, 3–5.11.2021, doi: 10.1109/ICISCT52966.2021.9670387

Akhatov A., Sabharwal M, Nazarov F, & Rashidov A. 2022a "Application of Cryptographic Methods to Blockchain Technology to Increase Data Reliability". 2nd International Conference on Advance Computing and Innovative Technologies in Engineering (ICACITE). 28–29 April. DOI: 10.1109/ICACITE53722.2022.9823674

Akhatov A.R. & Nurmamatov M.Q. 2020a. "Development of Models for Monitoring the Social Situation of Each Person and Forecasting Population Conditions Based on Personal Data", Descendants of Muhammad Al-Khorazmi. No. 2(12). –pp.6–10.

Akhatov A.R., Nurmamatov M.Q. & Mardonov D. 2020b. "Mathematical Models of the Process of Monitoring the Social Status and Employment of the Population", Scientific and Technical Journal of the Fergana Polytechnic Institute. - Volume 24, No. 5. -pp. 150–157.

Akhatov A.R., Nurmamatov M.Q. & Mardonov D.R., Nazarov F.M. 2021b "Improvement of Mathematical Models of the Rating Point System of Employment", *Scientific Journal Samarkand state University*. No. 1 (125). -pp. 100–107.

Axatov A.R., Nurmamatov M.Q., & Nazarov F.M. 2022b. "Mathematical Models of Coordination of Population Employment in the Labor Market" // *Ra Journal of Applied Research*. India/ – Vol. 8, Issue 2. – Pp. 111–119. DOI:10.47191/rajar/v8i2.09

Lin F T, 2001. *"A Job Shop Scheduling Problem with Fuzzy Processing Times"* Proceedings of International Conference on Computational Science Part II, Berlin Heidelberg, Springer, pp. 409–418.

Nurmamatov M.Q. 2022a. "Intellectual Modeling of Population Employment in the Labor Market" // *International Journal of Computer Science Engineering and Information Technology Research*. India/ -Vol. 12, Issue 1. – Pp. 109–113.

Nurmamatov M.Q. 2022b. "Improving Labor Relations Based on Intellectual Modeling of Employment in the Labor Market", The Peerian Journal. *Czechia/* – Vol. 5. -pp. 31–35.

Nurmamatov M.Q. 2022c. "*Modern Methods of Increasing the Efficiency of the Labor Market*", *National News Agency of Uzbekistan*. -pp. 373–383.

Nurmamatov M.Q. 2022d. "Mathematical Model of Labor Market Self-organization", Issues of Innovative Development of Science, Education and Technology. International Scientific and Practical Online Conference. -pp. 62–66.

Wang C Y & Wang D W. 2002. "The Single Machine Ready Time Scheduling Problem with Fuzzy Processing Times," Fuzzy Sets and Systems, vol. 127 (2), pp. 117–129.

Zhang Chaoyong, Rao Yunqing & Li Peigen. 2004. "An Improved Genetic Algorithm for the Job Shop Scheduling Problem," China Mechanical Engineering, vol. 15(23), pp. 2149–2153.

Artificial Intelligence, Blockchain, Computing and Security – Dagur et al. (Eds)

Analysis of the characteristics of cloud infrastructure based on traditional technologies and SDN technologies

S.R. Botirov
Tashkent University of Information Technologies named after Muhammad al-Khwarizmi, Tashkent, Uzbekistan

ABSTRACT: The article presents traditional data transfer technologies, the reasons for the creation and implementation of SDN technology in cloud data centers. A comparative analysis of cloud data centers based on TCP/IP and SDN technology has been carried out. The disadvantage of SDN technology, in the form of a single point of failure, as well as ways to solve the fault tolerance of the SDN controller, are considered.

1 INTRODUCTION

International practice shows that radical changes in the information sphere of economically developed countries that occurred at the turn of the 20th and 21st centuries have significantly changed the face of the information infrastructure. One of the most important components of the information infrastructure is the modern telecommunications infrastructure. The modern telecommunications industry is characterized by a number of features, telecommunication technologies have a high science intensity and a much shorter life cycle compared to other technologies. Therefore, the innovation process in telecommunications requires the transition of the entire industry to new technologies, and manufacturers to constantly update their equipment. In this regard, the telecommunications industry itself is becoming a high-tech industry that needs large investments and constant support for the innovation process (Stallings 2007).

The traditional packet-switched data transfer technology was developed for the efficient transmission of computer traffic, including two packet switching methods: the virtual connection method and the datagram method and network protocols became the basis for the operation of data networks: X.25, Frame Relay, ATM, and IP.

The introduction of such advanced technologies as: Cloud Computing, Big Date, IoT, led to an increase in the amount of multimedia information transmitted over the telecommunications network, which revealed previously insignificant shortcomings of TCP / IP technology - the complexity of the network, which uses a large number of different network equipment, as well as difficulties arising in the process of control and management, since the processes of management and transmission are combined (Stallings 2016).

2 METHODS

The idea of Cloud Computing is to move the entire infrastructure outside the enterprise into some kind of dynamically scalable system (Tabakov 2010; Zaigham 2011). But the use of cloud computing centers to the full extent, for storing and processing data, was impossible, since networks based on TCP / IP for data transfer to cloud computing centers lagged far

 DOI: 10.1201/9781032684994-113

behind the technologies used in local networks in terms of speed and quality characteristics. These problems were solved with the introduction of SDN technology (Djuraev & Botirov 2022; Yefimenko & Fedoseyev 2013).

SDN (Software Defined Networking) is a data transfer network in which the network management level is separated from data transfer devices and implemented in software. The technology implies a transition to centralized management of data networks, which was proposed by Martin Casado, Nick McKeown and Scott Shenker, who developed the OpenFlow protocol in 2007. The SDN architecture consists of 3 layers depicted in Figure 1.

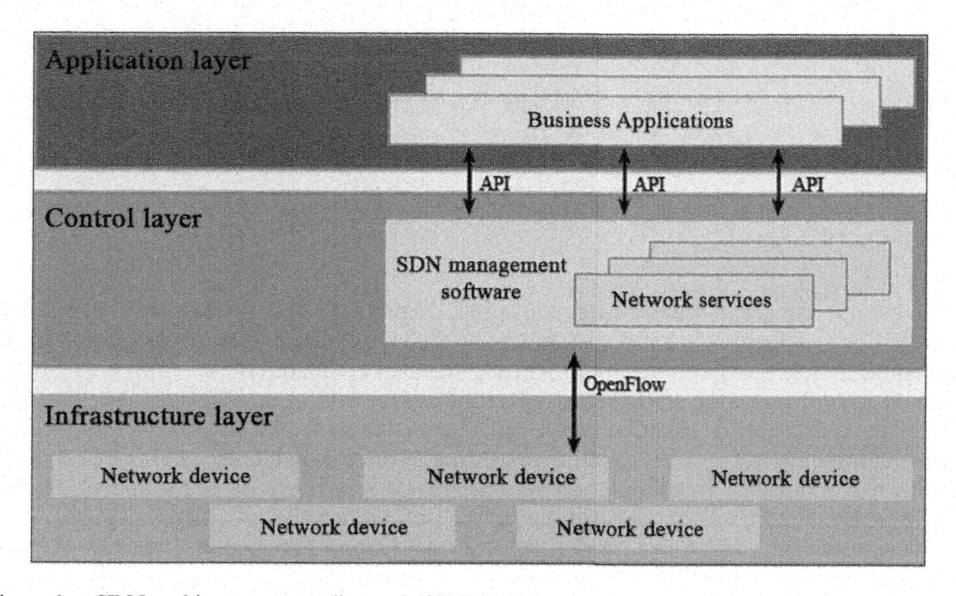

Figure 1. SDN architecture according to MST-T Y.3300.

Centralized management allows you to lay a channel through which packets will go directly, which is an order of magnitude more productive, while a transmission failure is possible, but the control program will detect and correct it.

The international experience of leading providers shows that the implementation of SDN technology has the following advantages over TCP / IP (Takacs 2013):

- Network management plane moved to a dedicated controller. Increases the bandwidth of channels by 20% due to load redistribution;
- Efficient data flow management. Increases the efficiency of network equipment by 25 - 35%;
- Usage of regular servers instead of complex and expensive specialized routers. Reduces capital costs by 52% and operating costs by 48%;
- Economical and vender independent. Reduces network operating costs by 30%.

Before implementing a new concept of organization, technology or protocol in cloud data centers, it is necessary to simulation of the proposed solution (Botirov & Kh 2021).

3 RESULTS

Analysis of the results of simulation modeling of channel throughput from time to time in a traditional network and SDN, obtained in the article (Kh *et al.* 2022), based on graphs, showed that the speed in a traditional network between hosts is not stable and changes over

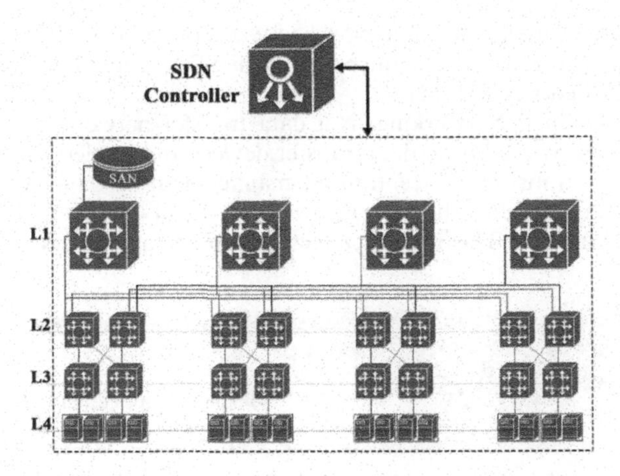

Figure 2. Architecture of the BigDataSDNSim simulation environment with a three-tier cloud data center with SDN support.

time, this directly affects the transmission speed, quality and data processing in cloud data centers. For verification, a simulation model of the cloud infrastructure was created in the BigDataSDNSim program for data processing (Khaled *et al.* 2021).

The study compares the performance of traditional networks and SDN-based networks in cloud data centers using a simulation model. The model involves a single cloud data center with a hierarchical network architecture consisting of three switch layers and one host layer, as shown in Figure 2. The simulation was conducted separately on SDN and traditional network bases, and the data collected during the simulation were recorded in Table 1.

The results of the study suggest that SDN-enabled cloud data centers can improve performance compared to traditional networks.

The table presented in the model was organized based on the size of requests, from small to large. Small requests were numbered 1-5, medium requests were numbered 6-10, and large requests were numbered 11-15. The simulation was performed separately using both SDN

Table 1. Comparison of the data obtained during the simulation.

Parameters	Network Transfer Time (s)		Request Completion Time (s)		Mapper Execution Time (s)		Reducer Execution Time (s)	
	TCP/IP	SDN	TCP/IP	SDN	TCP/IP	SDN	TCP/IP	SDN
1	2282	3183	4226	4972	1216	1209	802	589
2	2134	4364	4245	6234	1221	1220	778	523
3	2048	4608	4108	6270	1221	1220	778	523
4	2000	4056	4028	5965	1221	1220	798	567
5	2727	4654	4869	6702	2126	2016	89	76
6	3456	5054	5784	7296	2288	2144	102	87
7	3092	4655	5298	6864	2288	2144	90	88
8	2812	4367	5134	6648	2288	2144	122	112
9	2905	4712	5186	7026	2288	2144	122	112
10	3512	4894	5838	7306	2288	2144	122	98
11	3274	5440	5564	6996	2612	2504	145	144
12	3600	5598	5714	7309	2612	2504	145	156
13	3778	6048	5587	7112	2612	2504	145	156
14	3535	6189	5298	6648	2612	2504	98	297
15	4100	7456	5737	7363	2612	2504	145	156

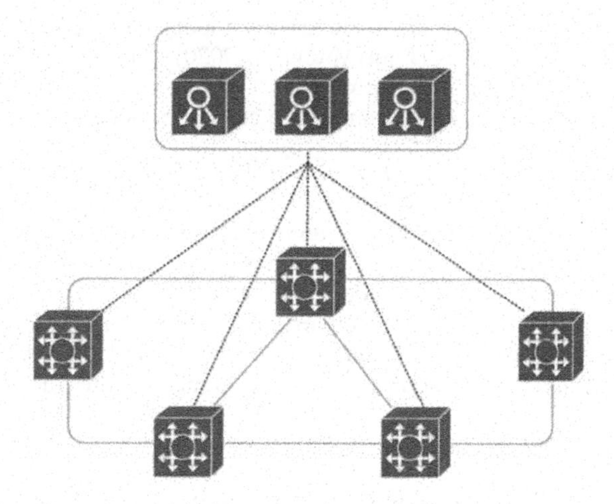

Figure 3. Cluster of SDN network controllers.

and traditional network models. The results recorded in Table 1 showed that the SDN model was more effective in improving the performance of MapReduce applications, reducing transmission time, and decreasing the time required to execute requests, compared to traditional network models.

4 DISCUSSIONS

The simulation model showed that the network infrastructure of Cloud Data Processing Centers for data processing and analysis is best built on the basis of the concept of SDN networks, as this affects the speed of obtaining results and allows you to effectively use the infrastructure when working with a large amount of data has a significant drawback, a single point failure in the form of a centralized controller, the failure of which disrupts the operation of the entire network (Ahmad & Mir 2021).

To solve this problem, you can use the redundancy of the SDN network controller and load distribution between the controllers by combining them into a cluster (Djuraev *et al.* 2022). Clustering controllers (Figure 3) allows them to be presented to switches as one single controller, regardless of their actual number (Hock *et al.* 2013; Obadia *et al.* 2014).

Using clusters (Rao *et al.* 2016) has a number of advantages over the standard way of redundant controllers:

- No need to configure all controllers on the switch;
- Increased availability due to a larger number of available controllers in the event of a main failure;
- Only one IP address is needed to use multiple controllers in a cluster.

Let us give an approximate estimate of the probability values of SDN network failure-free operation under the condition of redundancy of the controller and its absence.

The probability of failure-free operation is the probability that, within a given operating time or a given time interval, an object does not fail.

There are two ways to evaluate the fault tolerance indicators of non-recoverable systems based on failure data:

- calculation of the experimental distribution of time to failure;
- calculation of the parameters of the theoretical distribution of time to failure.

Any continuous distributions used in probability theory can be used as theoretical distributions of time between failures (Djurayev *et al.* 2020). One of them is the exponential distribution law for non-recoverable controllers.

The probability of failure-free operation of one controller is determined by the formula:

$$P_{\text{БО}}(t) = e^{-\lambda t}, \tag{1}$$

where λ is the controller failure rate, t is the controller operation time.

The case where SDN network controller redundancy is used can be considered as a system with parallel elements. To do this, we introduce the assumption that one controller is enough for the regular operation of the SDN network. We also assume that the values of the probability of failure-free operation of each controller are equal to each other.

The probability of failure of one controller is determined by the formula:

$$P_{\text{О}n}(t) = 1 - P_{\text{БО}}(t) = 1 - e^{-\lambda t} \tag{2}$$

Then the probability of simultaneous failure of n controllers is defined as:

$$P_{\text{О}n}(t) = (1 - P_{\text{БО}}(t))^n = \left(1 - e^{-\lambda t}\right)^n \tag{3}$$

Then the probability of failure-free operation of the system with n controllers will be equal to:

$$P_{\text{БО}n}(t) = 1 - \left(1 - e^{-\lambda t}\right)^n \tag{4}$$

Let's take the value of controller failure rate equal to 10^{-5} h^{-1}. Based on the above formulas, we get the following graph of uptime probabilities for SDN networks with a different number of controllers (Figure 4).

As can be seen from Figure 4, an increase in the number of controllers with the same controller failure rate leads to a significant increase in the probability of SDN network uptime, but do not forget that using a large number of controllers is not advisable from a financial point of view.

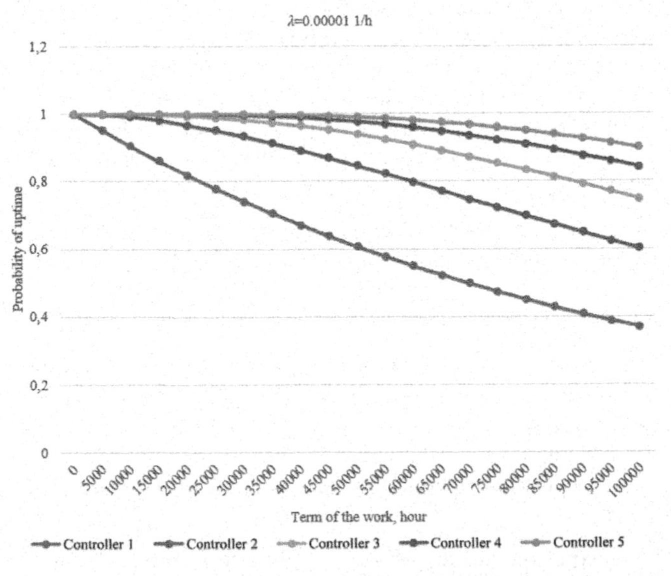

Figure 4. Comparison of the values of the probability of failure-free operation with a different number of SDN network controllers.

5 CONCLUSIONS

Data processing simulation results have shown that the SDN-based cloud data center is up to 40% more efficient in some cases than the TCP/IP-based cloud data center. But for a real implementation of SDN technology, a telecommunications company must have practical knowledge about this technology, the problems associated with it, and solutions.

An analysis of the proposed method for increasing the fault tolerance of the SDN controller showed that an increase in the number of controllers and a decrease in the controller failure rate leads to a significant increase in the probability of SDN network failure-free operation, but do not forget that the use of a large number of controllers is not advisable from the point of view of the financial side.

The formulas used in the calculations can be applied to solve the inverse problem - to determine the required number of controllers to provide a given value of the probability of failure-free operation for a given operating time and failure rate.

The probability of failure-free operation itself is not the only one that determines the fault tolerance of the SDN network, it is necessary to analyze such parameters as recovery time, the use of other methods to ensure the fault tolerance of SDN controllers.

REFERENCES

Ahmad, S., Mir, A.H. 2021. Scalability, Consistency, Reliability and Security in SDN Controllers: A Survey of Diverse SDN Controllers. *J Netw Syst Manage* 29: 1–9.

Botirov, S. R., Kh, D. R. 2021. Analysis of Information Security Evaluation Models in the Cloud Computing Environment. *2020 International Conference on Information Science and Communications Technologies (ICISCT)*. IEEE: 1–5.

Djuraev, R., Botirov, S. 2022. Current Status and Trends of Transition to Cloud Infrastructure Based on SDN Technology. *Science and Innovation* 1(8A): 800–809.

Djurayev, R. Kh., Botirov, S. R., Tashtemirov, T. K. 2020. Analiz Vliyaniya Pokazateley Kontroleprigodnosti na Nadozhnostnyyc Kharakteristiki Setey Peredachi Dannykh. *Molodoy uchenyy* 11: 15–20.

Djuraev R. Kh., Botirov S. R., Uskenbaeva D. Sh. 2022. Comparative Analysis of Methods for Increasing Fault Tolerance of SDN Controller. *Descendants of Muhammad al-Khwarizmi Scientific-practical and Information-analytical Journal* 1(19): 93–97.

Hock, D., Hartmann, M., Gebert, S., Jarschel, M., Zinner, T. and Tran-Gia, P. 2013. Pareto-optimal Resilient Controller Placement in SDN-based Core Networks. *Proceedings of the 2013 25th International Teletraffic Congress (ITC)*: 1–9. doi: 10.1109/ITC.2013.6662939.

Kh, D. R., Botirov, S. R., Juraev, F. O. 2022. A Simulation Model of a Cloud Data Center Based on Traditional Networks and Software-defined Network. *2021 International Conference on Information Science and Communications Technologies (ICISCT)*. IEEE: 1–4.

Khaled, A., Rodrigo, N. C., Saurabh, G., Rajkumar, B., Rajiv, R. 2021. BigDataSDNSim: A Simulator for Analyzing Big Data Applications in Software-Defined Cloud Data Centers, *School of Computing Science Newcastle University* 51(5): 893–920.

Obadia, M., Bouet, M., Leguay, J., Phemius, K., Iannone, L. 2014. Failover Mechanisms for Distributed SDN Controllers. *2014 International Conference and Workshop on the Network of the Future (NOF)*. IEEE: 1–6.

Rao, A., Auti, S., Koul, A., & Sabnis, G. 2016. High Availability and Load Balancing in SDN Controllers. *Int. J. Trend Res. Dev* 3(2): 2394–9333.

Stallings, W. 2016. *Foundations of Modern Networking: SDN, NFV, QoE, IoT, and Cloud.* Indianapolis, IN, USA: Pearson Educ.

Tabakov, V. 2010. Cloud Computing - a New Area of Information Services. *Creative economy* 5: 46–51.

Takacs, A., Bellagamba, E., and Wilke, J. 2013. Software-defined Networking: the Service Provider *Perspective. Ericsson Review* 2: 2–8.

Yefimenko, A. A., Fedoseyev, S. V. 2013. Organizatsiya Infrastruktury Oblachnykh Vychisleniy na osnove SDN seti. *Prikladnaya informatika* 5: 185–187.

Zaigham, M. 2011. Cloud Computing: Characteristics and Deployment Approaches. *2011 IEEE 11th IEEE Int. Conf. Computer and Information Technology*: 121–126.

Artificial Intelligence, Blockchain, Computing and Security – Dagur et al. (Eds)
© 2024 The Author(s), ISBN: 978-1-032-67841-2

Simulation modeling of stock management systems

T. Nurmukhamedov & Zh. Gulyamov
Tashkent State Transport University, Tashkent, Uzbekistan

ABSTRACT: In the article, a simulated model of stock management was created in the VChD-2 warehouse under the management of "Uztemiryolyoluchi" JSC. The simulation model was created based on parameters such as warehouse stocking volume, replenishment cycle, ordering time, daily demand for stocks (consumption level). Also, in the simulation model, real-life situations, i.e., order delays, were generated, the shortage situations based on it were studied and the necessary conclusions were drawn. Algorithm and program of the simulation model were created, and a diagram of the simulation model was built based on the result of the program.

1 INTRODUCTION

Despite the large number of scientific researches and references containing the optimal rules of inventory management and models, methods and algorithms for determining the parameters of these rules, among them it is almost impossible to find a model and method that accurately takes into account the characteristics of a given real enterprise. In general, it is a difficult task to create a model of a real warehouse system with hundreds or thousands of goods, taking into account the need to supply many Commodity Material Assets (CMA) from a single supplier, in the presence of constraints. Determining the optimal control parameters for such a system is a more complex, often intractable task.

2 METHODS

Simulation modeling is an approach to studying the properties of the system by repeating the behavior of the system with the help of a computer program. The simulation model contains random factors, which is fully compatible with the modeling of stock management systems with random demand. This situation leads to the need to create random factors (random events, random variables, random processes) in the simulation program. Based on the simulation model, we consider the general scheme of the algorithm for simulating the product inventory management system using random demand generation methods and the order point rule (Dombrovskiy *et al.* 2000; Kelton *et al.* 2004a; Kelton 2004b; Kosorukov *et al.*2013).

The following main principles were focused on when creating an algorithm for simulating a product inventory management system.

Firstly, we provide a brief description of exactly which aspects of the stock management system's behavior should be replicated by the simulation program.

Undoubtedly, the main manifestation of this behavior is the gradual reduction of stocks as a result of their consumption. Once the stock quantity falls to the order point value, a replenishment order is placed.

After the time spent on the execution of this order, the stock volume will be increased by the volume of the delivered order. We calculate the main factors that need to be simulated in

 DOI: 10.1201/9781032684994-114

the program, which are related to the size of this stock and its management. A simulation model is usually created to evaluate the effectiveness of the rules of stock management or to select the values of individual parameters of these rules (order point, order size, required service level). Therefore, the model should also calculate a number of performance criteria, some of which are listed below.

The order size Q is determined by subtracting the current stock from the maximum stock size. In this case, parameters such as the point of order and warranty stocks should be taken into account:

$$Q = z_{\max} - z_c \tag{1}$$

where:

$z_{\max}$ – the maximum size of the stock;

z_c– current stock.

We use the following formula to determine the warranty stock z_w:

$$z_w = d * z_{t\_s}, \tag{2}$$

where:

d – average stock requirement;

z_{t_s}- time (days) spent on order fulfillment.

The order point is determined based on the following formula:

$$b_p = z_w + d * b_{l\_t}, \tag{3}$$

where:

b_{l_t} – the period (days) from the order to its execution.

The specified order point determines when the stock (stocks) should be ordered when this quantity falls to this point.

The average stock volume for a replenishment cycle determines the cost of stock holding, and is usually calculated as the arithmetic mean of the stock volumes at the beginning and end of the cycle, that is, as half the sum of the maximum and minimum values of one stock volume (Akopov 2017; Aleksandrovich 2004; Baycv *et al.* 2013):

$$I_t = \frac{Q + I_{f,\,i-1} + I_{f,i}}{2} = \frac{Q}{2} + \frac{I_{f,\,i-1} + I_{f,i}}{2} \tag{4}$$

where:

I_t – average stock volume in the filling cycle;

$I_{f,i}$–Level of stocks before replenishment in the 1st cycle;

Q – order size.

Since the stock level at the end of the cycle I_(f,i) is a random variable, the efficiency criterion can only be a mathematical expectation, or the average value of the stock level for a large number of replenishment cycles (Aleksandrovich 2004):

$$\bar{I} = \frac{1}{N}\sum_{i=1}^{N} \left(\frac{Q}{2} + \frac{I_{f,i-1} + I_{f,i}}{2}\right) = \frac{Q}{2} + \bar{I}_f \tag{5}$$

where:

$\bar{I}$ – the average level of stocks is calculated as an arithmetic average for all replenishment cycles;

N– the number of filling cycles simulated in the program;

$$\overline{I} = \frac{1}{N}\sum_{i=1}^{N} I_{f,i} \tag{6}$$

(6) -the average stock volume at the end of the cycle.

The amount of shortage (unsatisfied demand) in each cycle is also a random variable, so it is calculated as an average over a set of replenishment cycles.

By calculating the average deficit, we can calculate the average service level using the following formula:

$$S_x = \left(1 - \frac{\overline{S}_l}{Q}\right) * 100\% \tag{7}$$

where:

S_x– average level of service;

$\overline{S}_l$– the average deficit at the time of filling (volume of unsatisfied demand).

The average deficit is the arithmetic mean for all filling cycles and is calculated as follows

$$\overline{S}_l = \frac{1}{N}\sum_{i=1}^{N} S_{l,i} \tag{8}$$

where:

$S_{l,i}$– Deficit in the first cycle.

In a real warehouse system, the time interval between the receipt of demands (customer requests) and the amount of stock requested is random. However, in inventory management systems, these are usually not two random variables, but the requirements in one time unit are forecasted. Therefore, in the simulation program, we consider the demand to be non-zero continuous in any time interval (Kravchenya *et al.* 2010; Sviridova *et al.* 2013; Voss 2003; Yakimov *et al.* 2015, 2016).

Demand continuity allows you to simulate inventory consumption by generating demand values over small time intervals (a specified time for simulation) and subtracting these values from the current inventory volume. These actions are repeated until the stock level falls to the order point. From this moment, the program should calculate the delay in filling the order. After this delay, the stock is increased by the order quantity Q, and a new cycle of stock consumption begins.

The generation of random demand values and the reduction of the stock level according to these values will continue even after the guarantee stock has been reached. In this case, the stock level may be negative.

This means that there is a shortage (deficit) of stocks due to the delay in the order. The amount of deficit (unsatisfied demand volume) in each consumption cycle is summed up and the sum obtained is divided by the number of replenishment cycles - this is the average deficit.

By the time the order is accepted, the volume of stocks may remain positive, that is, it (the stock) will not be fully used. These values are accumulated over all filling cycles and the average stock volume is calculated at the end of the cycle.

The block diagram of the above algorithm is shown in Figure 1.

Based on the mentioned algorithm, a program was created for simulating the system of effective management of stocks in the warehouse. The simulation program shows the basic steps involved in creating the stock quantity and determining the order point. Also, cases of non-delivery of orders on time were generated, and shortage cases were studied in the simulation model.

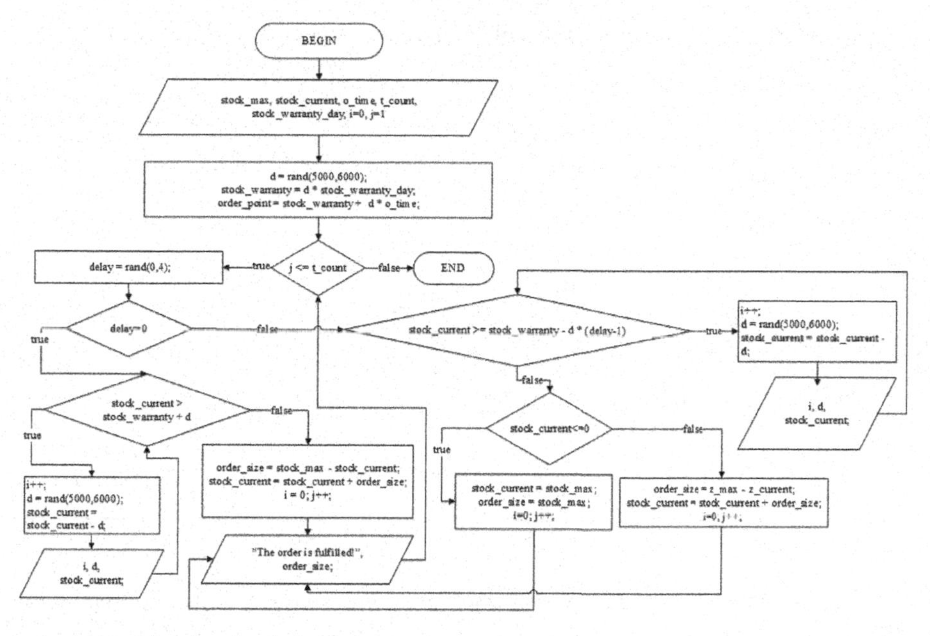

Figure 1. Algorithm for simulating inventory management system.

3 RESULTS AND DISCUSSION

On the basis of the program created according to the above algorithm, the inventory of CMA stocks in the warehouse used for repair and equipment works of the "VChD-2" wagon depot of the "Uztemiryolyolovchi" joint-stock company was carried out, and the calculation results are presented in Table 1.

Table 1. Stocks of cma in the warehouse used for repair-equipment work.

№	Daily output	Stock balance in warehouse
1	5177	94823
2	5128	89695
3	5433	84262
4	5182	79080
5	5369	73711
6	5722	67989
7	5568	62421
8	5673	56748
9	5804	50944 Point of order
10	5425	45519
11	5604	39915
12	5777	34138
13	5785	28353
14	5193	23160
15	5070	18090 The limit to which the order should be delivered
16	5360	12730
17	5624	7106
18	5286	1820
19	5173	-3353
Order completed. Order size: 100 000. The order was delayed for 4 days		

Based on the results obtained in this table, a diagram was created (Figure 2).

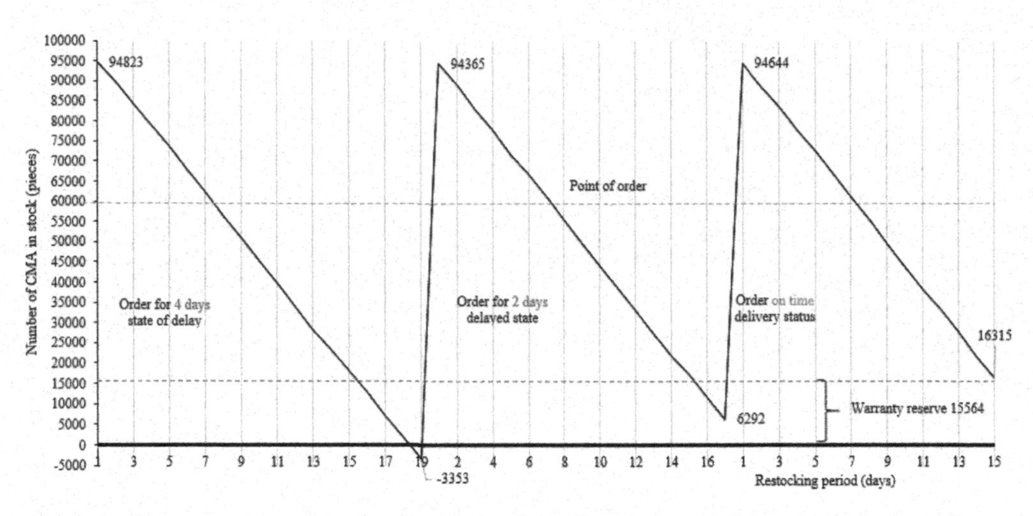

Figure 2. To carry out repair and equipment works in the warehouse of the wagon depot stock management.

Table 2 summarizes the results for four criteria: service level, inventory cost, order value, and total cost. Group ordering procedures reduce total costs by almost 50% compared to independent orders. The main savings come from reducing the cost of orders, which is the goal of the order pooling procedures. However, these procedures also reduce inventory costs due to smaller order sizes. The level of service in group ordering procedures was close to its values obtained in independent ordering procedures, differing by an average of 0.5%.

Table 2. Results of simulation experiments.

Experiment number	Service Level Ratio	Maintenance cost ratio	Order value ratio	Total cost ratio
1	0,999	0,489	0,664	0,555
2	1,041	0,787	1,251	0,896
3	0,989	0,452	1,649	0,629
4	1,001	0,337	0,672	0,423
5	1,015	0,411	0,441	0,424
6	1,017	0,400	0,874	0,579
7	0,988	0,599	0,614	0,605
8	0,945	0,269	0,463	0,339
9	0,983	0,558	0,792	0,630
10	0,981	0,361	0,528	0,430
11	0,946	0,351	0,381	0,364
12	0,987	0,525	0,680	0,592
13	1,012	0,461	1,319	0,686
14	1,105	0,469	0,651	0,522
15	1,008	0,179	0,692	0,348
Average	0,995	0,443	0,783	0,539

4 CONCLUSION

A simulation model was created to improve the efficiency of inventory management of the Wagon Depot Warehouse. This model uses parameters used in inventory management (maximum inventory, order point, replenishment time, warranty inventory, inventory requirement). The model simulates the behavior of stocks in the warehouse and is effective in managing stocks;

Giving the parameters of stock in the warehouse (maximum stock value, filling time, guarantee stock, demand for stock) as input variables to the simulation algorithm allows to determine the necessary results for stock management, such as order point, order volume, order periodicity;

REFERENCES

Akopov, A.S. 2017. Imitative Modeling: Uchebnik and Practical for Academic Bachelor's Degree. *M.*: Izdatelstvo Yurayt.

Aleksandrovich, V.M., 2004. Prognozirovaniye Obyemov Prodaj / *V.M. Aleksandrovich.* -Biysk: Izd-vo BTI AltGTU.

Bayev, L.A. & Dzenzelyuk, N.S. 2013. *Metodologicheskiye Osnovi Upravleniya Zapasami v Nestatsionarnoy Ekonomike* // Vestnik UrFU. Seriya ekonomika i upravleniye.

Dombrovskiy, V.V. & Chausova, YE.V. 2000. *Matematicheskaya Model Upravleniya Zapasami Pri Sluchaynom Sezonnom Sprose i Nenadejnix Postavshikax* // Vestnik Tomskogo gosudarstvennogo universiteta.

Kelton, V. & Lou, A. 2004. Imitatsionnoye Modelirovaniye. SPb: Piter.

Kelton, W. 2004. *Imitation Modeling. Classic CS* / V. Kelton, A. 3-e izd. SPb.: Peter; Kyiv: Izdatelskaya gruppa BHV.

Kosorukov, O.A. & Sviridova, O.A. 2013. *Imitatsionnoye Modelirovaniye v Stoxasticheskoy Zadache Upravleniya Zapasami* // Statistika i matematicheskiye metodi v ekonomike.

Kravchenya, I. N. & Shevchenko, D. N. 2010. *Matematicheskiye Modeli v Transportnix Sistemax. Modelirovaniye Sistem Massovogo Obslujivaniya i Zadach Upravleniya Zapasami.* Belorus. gos. un-t transp. – Gomel: BelGUT.

Sviridova, O.A. & Kosorukov, O.A. 2013. Imitatsionnoye Modelirovaniye v Stoxasticheskoy Zadache Upravleniya Zapasami // Vestnik UMO: Ekonomika, statistika i informatika.

Voss, S. & Woodruff, D. 2003. Introduction to Computational Optimization Models for Production Planning in a Supply Chan. Springer – Verlag.

Yakimov, I. M. & Kirpichnikov, A. P. & Zaynullina, G. R. & Yaxina, Z. T. 2016. *Imitatsionnoye Modelirovaniye Sistem Upravleniya Zapasami Predpriyatiy s Fiksirovannim Obyomom Postavok* // Vestnik texnologicheskogo universiteta.

Yakimov, I.M. & Xomenko, V.V. & Alyautdinova, G.R. 2015. *Imitatsionnoye Modelirovaniye Sistemi Upravleniya Zapasami Predpriyatiya s Fiksirovannim Vremenem Postavok* // Vestnik ekonomiki, prava i sotsiologii.

Artificial Intelligence, Blockchain, Computing and Security – Dagur et al. (Eds)
© 2024 The Author(s), ISBN: 978-1-032-67841-2

Methodology of improving the educational process on the basis of mobile applications in higher educational institutions (with reference to solid state physics)

M.X. Lutfillayev & O.O. Narkulov
Samarkand State University named after Sharof Rashidov, Uzbekistan

ABSTRACT: This article explores the advantages of mobile applications in improving and increasing the efficiency of the educational process in higher educational institutions. As well as the use of principles, models and modern achievements of mobile technologies in the education system. The analysis of scientific publications of foreign scientists and countries of the Commonwealth of Independent States is given. As an example, computer simulation models of physical processes in the subject "Physics of the Solid State" are given.

1 INTRODUCTION

It is known that the process of digitalization of the education system is currently developing very rapidly. The introduction of computers into the education system increasing the efficiency of the educational process based on technological means requires a revision of the human factor. Nowadays, it is impossible to imagine the organization of the educational process without information technology. Obviously, such inventions as computers, smartphones have a significant impact on the learning process. Despite the fact that computers are used in the educational process the use of smartphones and mobile applications as a technical and software tool has become commonplace in the educational process.

It should be noted that mobile applications today are one of the promising areas for the development of the education system.In order to explore the relevance of the introduction of mobile applications in the educational process it is necessary to analyze the scientific work of both foreign and local scientists.

An analysis (https://www.ixbt.com/news/2021/09/02/5-3-3.html. 2021) of scientific papers shows that there are more than 5.3 billion mobile phones in the world. This proves that the number of using mobile phones is several times greater than the number of personal computers. Along with new features and functions of mobile phones new versions are being developed. The number of users who communicate with smartphones is growing every year. Nowadays, educators are increasingly thinking about using mobile phones to improve the efficiency of the educational process. The reasons are as following: mobile applications and mobile technologies are increasing students' motivation for learning, as well as increasing motivation for the effectiveness of the educational process and the possibility of student-centered learning.

Scientific investigations, results and visual technological processes provide an opportunity for teachers and students to create a live interactive learning environment. Information technology contributes to the development of creativity, intelligence, independence and knowledge of students and also increases the effectiveness and organization of the educational process. It is obvious, today's youth allocate a lot of time to games of various types related to smartphones, phones, tablets and others. It should be noted that mobile gadgets are mainly associated with the performance of their favorite activities, that are entertaining.

 DOI: 10.1201/9781032684994-115

However, it is possible to organize an educational process associated with games of this kind. As the main factor of the implementation of mobile applications in the educational process teachers should develop skills on working and using technical software. Therefore, a special attention is paid to the use of information and communication technologies for visualization it in the educational process. It should be noted that the use of mobile technologies as visual means can lead to an improvement of the education system on the one hand, and an increase in efficiency, on the other hand.

2 ANALYSIS OF EXISTING APPROACHES

Analyzing the studies of foreign scientists on the use of mobile applications in the educational process the following can be observed:

The scientific article by E.G. Mikhalkina "Computer-mobile technologies as a means of improving the quality of professional training of future managers" provides a comparative analysis of teaching students in the control and experimental groups, the use of computer-mobile technologies reduces the time spent on the effective work of students on explaining and testing new material.

The analysis shows that the population of different strata in the countries of independent states (CIS) uses mobile applications, that is explained by the constant growth in the number of mobile devices. Mobile applications are considered to be one of the developing industries in the CIS. The main reason for the development of the mobile application market is accessibility anywhere and anytime. Users of the CIS can use mobile applications to be active in their daily lives and do their tasks quickly and easily (Mikhalkina 2007).

In the study by P.S. Nagovysyn "Use of mobile learning for the formation of physical culture and health competence of students at university" (based on mobile learning) the issues of developing the stages of forming students' physical culture in the educational process were considered, for example: ideology, activity and creative thinking. At each stage the results of the process of formation of the student's physical culture are determined, the components of the author's methodology and their integral composition are developed, for example: mobile, national-regional, theoretical, theoretical-methodical and practical aspects. Based on the analysis of the results obtained the worldview of students is changing due to physical activity (Nagovitsyn 2014).

A feature of the mobile software market in the CIS is its late development in comparison with other developed countries. But at present, the global market is experiencing almost the same development due to the fact that many mobile applications are available in all languages.

Top ten positions of mobile applications (Muraveyko A.Yu 2020) in the App Store and Google Play Market.

Table 1. Ten most popular mobile apps.

№	Applications	Developers
1	Tiktok	bytedance
2	Telegram	telegram messenger
3	Watsapp	facebook inc
4	Zoom	zoom video communications inc
5	Sberbank onlayn	sberbank
6	Instagram	facebook inc
7	Nomerogram	amayama auto llc
8	Vkontakte	mail.ru group
9	Youtube	google llc
10	Yandex go	yandeks

The scientific article by Z.A. Aleksandrova "The use of mobile applications in teaching mathematics to teachers" discusses the use of the pedagogical capabilities of mobile applications (Google Classroom, Quiz, Kahoot) in educational activities in teaching mathematics, for example: Google Classroom is a Google service application specially created for the organization of the educational process, covering the following tasks:

1. Creation of electronic lessons or courses on a topic or individual sections of science.
2. Organization of student registration for the course in a convenient form.
3. Download materials that are free for all course participants.
4. Sending assignments to students.
5. Evaluation of completed tasks and control over the learning process of students.
6. Organization of communication between students and teachers.

Google Classroom is a mobile application, an effective tool for organizing independent work of students in teaching mathematics. Students have the opportunity to upload educational materials, use data via mobile applications. It is possible to monitor the implementation of tasks, calculate students' scores and analyze the difficulty of tasks for students.

Google is the world's largest web search engine and has an app installed on almost every mobile device. This system is constantly updated and improved to create a comfortable working environment. To gain a knowledge using a mobile application is especially effective in training.

Quiz is a program for creating tests, questionnaires and quizzes. With the help of this application the teacher can create test questions and surveys or quizzes and send them to students on a mobile device or home computer. Students have the opportunity to study these quizzes using their mobile devices. This program can be used on lectures and other types of classes.

Kahoot is an application that allows you to create interactive tasks in the form of tests, polls and quizzes. This mobile application allows you to create various questions, e.g:

1. Security question - the number of answers is four, one is correct.
2. True or false - two possible answers, one is correct.
3. Resuming text, the text is entered from the answer keyboard.
4. Puzzle - four answers. Put the correct answers in order.
5. Voting - the screen shows the percentage of the proposed answers.
6. Cloud of words - students enter their answers randomly, no more than 20 words.
7. Brainstorming - students present new ideas in games and have the opportunity to discuss and vote for their favorite ideas.

A variety of questions in designed mobile applications allows you to work with users of different ages (students, schoolchildren). The use of created mobile applications as an additional software tool for the educational process leads to an increase in the efficiency of educational process (Aleksandrova 2020).

It is convenient to create presentations and computer simulations of processes using mobile applications for complex topics that are difficult for students to understand in mathematics. Various visual components and computer simulation models are developing to improve the efficiency of educational process.

In the study of V.V. Ryabkova "Integration of mobile technologies and the process of business English for adults (levels A1-B2)", experiments were conducted among students of the Center of Business and professional retraining of English of the Ministry of Economic Development, Russia. It was attended by 4 experimental groups, 10-12 students each. The experiment was carried out in two experimental groups during 4 semesters. As a result, the percentage of mistakes made by students decreased from semester to semester in the experimental groups, whereas the percentage of professional communicative competence formation increased. Comparing the results of the experimental and control groups according to the same indicators, it was found that the percentage of errors in the control groups is higher by 10-12%, the formation of professional communicative competence is lower by 15-17%. An analysis of the results of the experiment shows that the model of integrating mobile technologies into teaching business English, developed by adults, has a positive effect on the level of students' preparation (Ryabkova 2019).

In addition, it was determined that a teacher, the one who implements the main educational program should use the possibilities of ICT in the educational process and be able to work with multimedia equipment. In this case, the task of the teacher is to show students how to use electronic resources and mobile applications as a learning tool.

In the scientific article by M.Yu. Antropova "Mobile technologies in the educational process" (on the example of the Chinese WeChat) the possibilities and advantages of mobile technologies in teaching Russian as a foreign language to undergraduate students are considered in a research conducted at Sun Yat-sen University. (China represented.) Considering the fact that the use of ICT and mobile applications in the educational process has a number of advantages the effectiveness of teaching Russian as a foreign language based on the WeChat application turned out to be more effective rather than the traditional educational process. As an outcome of the experiment the results of the group trained on the basis of the WeChat application turned out to be more effective than the results of the group trained in the traditional way (Antropova M. Yu. 2022). The term "mobile application" has different interpretations. In a narrow sense, it means a software designed to work on smartphones, tablets and other mobile devices (Russian quality system 2018).

In a broad sense, this term means "a software designed specifically for a specific mobile platform" based on iOS, Android, Windows Phone systems on smartphones, tablets, smart watches and other mobile devices (Mobilnoe prilojenie 2021).

Computer modeling technology in the education system can significantly improve the quality of education, it becomes obvious usefulness of using imagery, virtual presentation of information in training software systems in the relevant disciplines, is required development and concretization of approaches, methods and techniques to solve this problem. For improving of the efficiency of continuous multi-level education, we propose methodical system of computer simulation. The objective of this system is to develop virtual resources for all levels of continuous multi-level education (Lutfillaev *et al.* 2015).

Technology of computer simulation submits learners with an almost unlimited range of means for implementing soundtrack of well-chosen visual material and text. This makes it easier to perceive and understand the information. The graphics capabilities of the computer provide with the visibility of the perception of learning material, this in its turn increases motivation for learning. (Lutfillaev M. Kh 2012)

Thus, one of significant features of using mobile applications is the possibility of their convenient use regardless of location: a student does not need a desktop computer or laptop to study; a student can use the application on a mobile device available in any convenient place. It should be noted that the ability to quickly access a text or a task is an advantage of these applications which raises additional students' motivation for learning.

3 PROBLEM STATEMENT AND SOLUTION METHOD

In the abovementioned scientific studies the issue of improving the educational process based on mobile applications on the subject "Solid State Physics" in the field of physics in higher educational institutions, in some cases, was not analyzed at all. Based on the abovesaid considerations the improvement of the educational process with the help of mobile applications in higher educational institutions on the subject "Solid State Physics" is an urgent task. According to the analysis, mobile applications embody video, audio, text, graphics, tests, multimedia, animation, computer simulation models. Therefore, this article reveals the essence of the content of physical processes on the subject "Physics of a Solid State" by means of computer simulation models.

The following is the formation of p-n junctions in semiconductors.

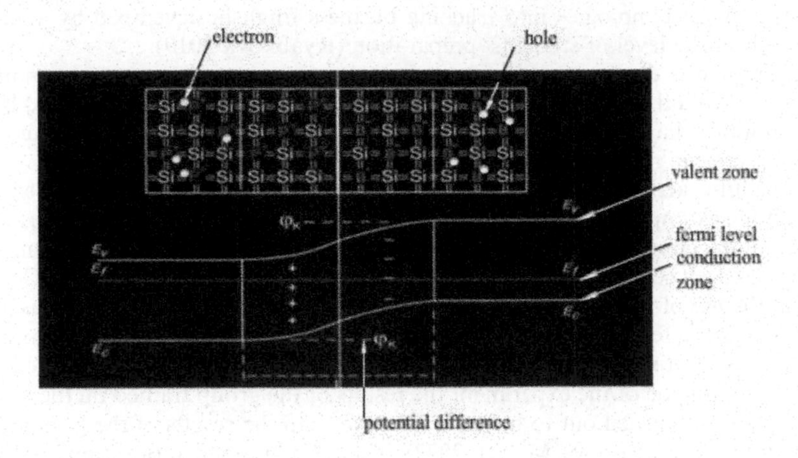

Figure 1. Computer simulation models of the formation of p-n junctions.

The Figure 1 shows (Kittel. Ch. 1978) an n-type semiconductor, where the donor impurity atoms are $-N_D$ with an ionization energy of the donor level $-E_D$, as well as a p-type semiconductor with a concentration of $-N_A$ and an ionization energy of $-E_A$. Prior to contact in each semiconductor, the concentration of electrons in the conduction band and holes in the valence band coincides with the concentration of ionized donor atoms $-N_D$ and acceptors $-N_A$, respectively. And therefore, in these materials before contact, there is a local electrical neutrality. When these semiconductors come into contact from the n-type contact region, electrons diffuse p-type and recombine with holes, similarly, holes diffuse n-type and recombine with electrons, that is, they mutually annihilate. As a result, uncompensated immovable negative charges remain in the n-type contact area. A region of space charge appears, which creates an electric field. This field prevents the further transition of electrons from the n-type to the p-type, that is, a potential barrier is formed for the majority charge carriers, and for minor charge carriers, as can be seen from the example, the potential barrier does not exist. In this case, the value of the contact potential difference is determined by the formula

$$\varphi_{k=}kT\frac{I_n N_D N_A}{(n_i)^2} \tag{1}$$

where, N_D - is the donor concentration, N_A - is the acceptor concentration and N_A=N_n, N_D=P_p n_i - is the concentration of its own carrier at a given temperature.

The process of organizing p-n junctions is demonstrated using computer simulation models.

The second figure shows the equilibrium state, that is, when there is no influence of an external electric field the flow of major and minor charge carried through the p-n junction will be the same, and therefore the total current is zero.

$$I = I_{n_n} + I_{n_p} + I_{p_p} + I_{p_n} = 0 \tag{2}$$

where, I - total current, I_{nn} - flow of major charge, I_{np} - flow of major charge, I_{pp} - non major charge carriers, I_{pn} - non major charge carriers.

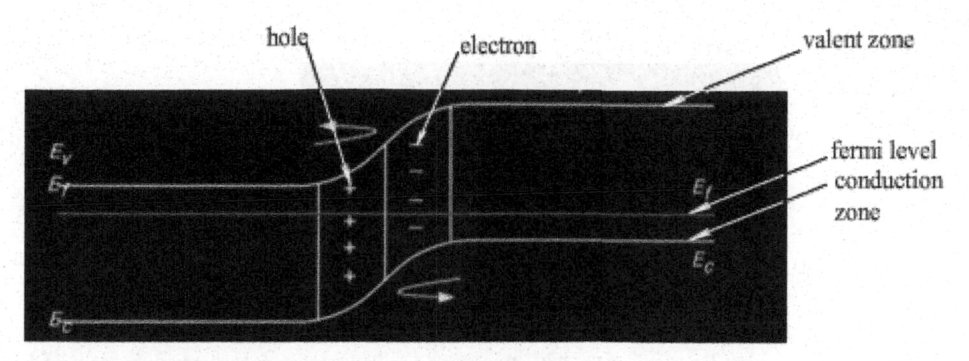

Figure 2. Computer simulation models in equilibrium p-n junction.

The Figure 3 (B.N. Bushmanov & Yu. A. Khromov 1971) shows when the electric field is turned on to the p-n junction in the direction indicated in the example, called direct inclusion, the direction of the external and internal electric fields do not match. Then the magnitude of the potential barrier decreases and favorable conditions are created for the diffusion of the main charge carriers through the p-n junction. The greater the magnitude of the application of the electric field the greater the diffusion of the main charge carriers. The current increases exponentially.

$$I = I_s\left(e^{\frac{eV}{kT}} - 1\right) \quad (3)$$

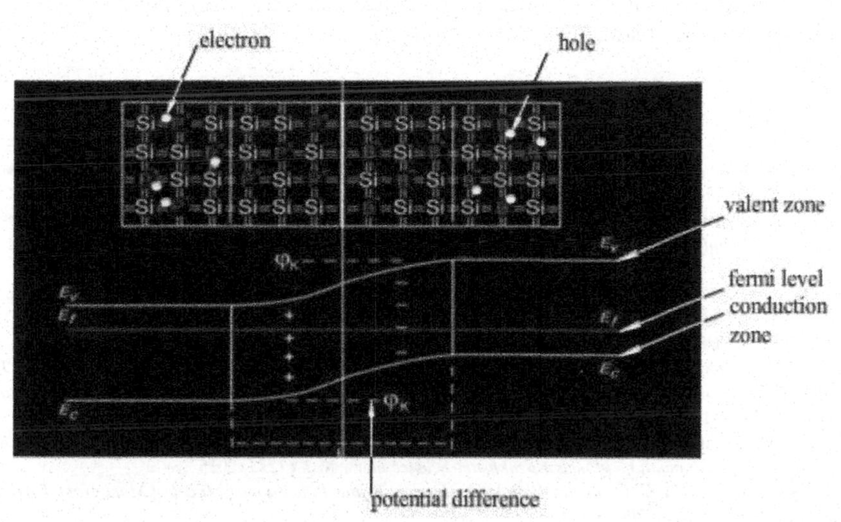

Figure 3. Computer simulation models of P-n junction with direct electric field.

where, I- electric current, I_S- saturation current, produced by minority carriers, k-Boltzmanns constant, T- temperature.

Figure 4 shows when the reverse electric field is turned on the direction of the external and external fields coincides the value of the potential barrier increases while the current through the p-n junction is determined only by minority charge carriers.

All processes associated with the p-n junction are shown using computer simulation models and are accompanied by the sound of a lecturer, which makes it possible to learn distantly and independently on the subject "Solid State Physics".

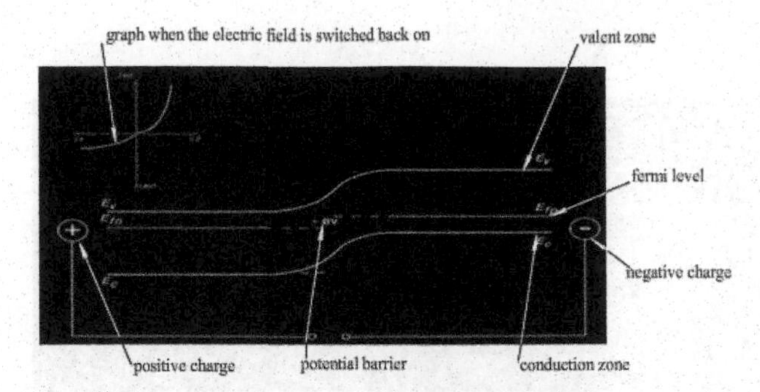

Figure 4. Computer simulation models of the P-n junction when the electric field is turned back on.

4 CONCLUSIONS

The organization of the educational process with the help of mobile applications allows students to work independently, gain knowledge not only in the classroom, but also beyond it. Similarly, in our research work computer simulation models have designed for physical processes and phenomena that students cannot observe in normal laboratory work. In addition, a mobile application has been designed for computer simulation models on the subject "Solid State Physics". During the lesson students have the opportunity to use a mobile application and see p-n junctions in semiconductors and observe how electrons move from an n-type semiconductor to a p-type, holes move from a p-type semiconductor to an n-type. The simulation of the process of p-n junction formation is shown as well as it is enriched with voice comments. This allows to see and hear the process of organizing p-n transitions.

REFERENCES

Aleksandrova Z.A. Ispolzovanie Mobilnых Prilojeniy Pri obuchenii Matematike Studentov pedvuzov. *Konstruktivnые pedagogicheskie zametki. Moskva* № (8) 2020, 107–118

Antropova M. Yu. *Mobile Technologies In Educational Process (The Example Of Chinese Wechat) Cross-Cultural Studies: Education and Science (CCS&ES)* ISSN -2470-1262 Volume 3, Issue III, September 2018 date of circulation: 29.03.2022

Bushmanov B. N. Fizika Tverdogo Tela: Uchebnoye Posobie Dlya Vuzov. Moskva, Vesshaya shkola, 1971 https://www.ixbt.com/news/2021/09/02/5-3-3.html

Kittel Ch. Vvedenie v Fiziku Tverdogo Tela. Moskva nauka, 1978

Lutfillaev M. Kh. *Metodologicheskie Osnove Kompyuternogo Imitatsionnogo Modelirovaniya v Uchebnom Protsesse Materialы Mejdunarodnoy Nauchno-prakticheskoy Konferensii «Novыe informatsionnыe texnologii v obrazovanii» Yekaterinburg*, 2012, 192–193

Lutfillaev M.H., Allanazarova N.A., Lutfillaev I.M., Hasanov SH.M. *The Principles of Realization of Virtual Resources in Methodical System of Simulation Models Kazakhstan Science news. № 1 (123)*, 2015

Lutfillaev U. M, Lutfillaev M.X. *Computer Simulation Models in the Educational Models (Monograf) LAP Lambert Academic Publiting, Germani* 2019

Mikhalkina, E. G. *Kompyuterno-mobilnыe Texnologii Kak Sredstvo Povышheniya Kachestva Professionalnoy Podgotovki Buduшix Menedjerov. Nauchnыy Doklad, VII Mejdunarodnoe Nauchno-prakticheski Konferensi YuNESKO*, Moskva 2007, 404–408

Mobilnoe prilojenie. Indikator Elektronnыy resurs. Rejim Dostupa: https://indicator.ru/label/mobilnoe-prilozhenie (date of access: 21.09.2022)

Muraveyko A.Yu. Analiz ryenka Mobilnex prilojeniy. Naibolee Populyarnыe vidы Mobilnex Prilojeniy. V sbornike: Perspektivnoe Razvitie nauki, Texniki i Texnologiy. Sbornik Nauchnex Statey 10-oy Mejdunarodnoy Nauchno-prakticheskoy Konferensii. Rossiya 2020, 162–165

Nagovisen P.C. Ispolzovanie Mobilnogo Obucheniya Dlya Formirovaniya Fizkulturnoozdorovitelnoy Kompetentnosti Studentov v Vuze. Vestnik Kazanskogo Texnologicheskogo Universiteta № (5) 2014, 366–375

Rossiyskaya sistema kachestva. *Sravnitelnыe Ispыtaniya Mobilnых Prilojeniy Dlya Smartfonov*. Moskva, №277–2018 Elektronnыy resurs, Rejim dostupa: http://docs.cntd.ru/document/1200159701

Ryabkova V.V. *Mobilnaya Obrazovatelnaya Sreda v Kontekste Inoyazechnogo Obrazovaniya Probleme Sovremennogo Obrazovaniya: Mejvuzovskiy Sbornik Nauchnex Trudov*, № (12) Moskva, 2019, 128–132

Artificial Intelligence, Blockchain, Computing and Security – Dagur et al. (Eds)
© 2024 The Author(s), ISBN: 978-1-032-67841-2

Using single factor regression modeling to predict reserve requirements

T. Nurmukhamedov, Zh. Azimov & Zh. Gulyamov
Tashkent State Transport University, Tashkent, Uzbekistan

ABSTRACT: The article presents a comparative examination of one-factor relapse models utilized to foresee the generation needs of an venture on the case of the require for stocks in a car station. Added substance and multiplicative models with straight, logarithmic and polynomial patterns are considered. The exactness of the models was surveyed, the Fisher basis was utilized to test the ampleness of the models, and the normal relative blunders of the prophetic models were calculated. Based on the chosen demonstrate, a figure was made of the month to month stock necessity for the taking after months.

1 INTRODUCTION

Correlation and regression analysis is a probabilistic modeling method that allows to study the interrelationship of economic activity indicators of the enterprise, if the relationship between the enterprise indicators is not strictly functional or is not broken by the influence of extraneous, random factors. As a result, the search and assessment of the relationship between two random characteristics or factors is carried out (correlation analysis), and then a specific type of relationship is established between the studied parameters (regression analysis).

The following main tasks of correlation and regression analysis in supply logistics can be distinguished:

- searching and evaluating the proximity of correlation between the volume of goods and material costs and the factors of change in consumer demand in order to formulate the right plans to meet the need for commodity resources (CR);
- to determine the supply policy and the level of use of goods and materials by comparing the dynamics of receiving and spending stocks in the warehouse of the enterprise during the inspection of the existing CR management system;
- Forecasting and budgeting of indirect costs related to CR, procurement planning, etc.

From this point of view, it is an urgent issue to create regression models of wagon depot reserves owned by "Uztemiryolyolucchi" JSC and to predict their (reserves) volume for the coming years, which will be used in timely and efficient service to the transport units.

2 METHODS

An economic system may be any business. In this regard, one-factor regression models are taken into account to forecast the amount of CR stock required for PRE when the units in motion return from their voyage. The choice of the model that best captures the beginning data is the key problem when employing one-factor regression models for forecasting over a number of periods, according to studies (Afanasyev *et al.* 2001; Isupova 2009). Within these models, further categorization is done in accordance with the trend equation.

DOI: 10.1201/9781032684994-116

The most commonly used models for creating predictive models are (Arunraj *et al.* 2015; Bazilevskiy *et al.* 2012; Lapach *et al.* 2012; Noskov *et al.* 2012):

$$y = a * x + b - \text{linear trend}, \tag{1}$$

$$y = a * \ln|x| + b - \text{logarithmic trend}, \tag{2}$$

$$y = b * e^{a*x} - \text{exponential trend} \tag{3}$$

$$y = a * x^b - \text{graded trend} \tag{4}$$

$$y = a * x^2 + b * x + c - \text{quadratic trend}. \tag{5}$$

It should be noted that along with the above models, additive and multiplicative models with trends and seasonality are also widely used (Criminisi 2012; Strijov 2007):

$$y_t = T_t + S_t + \varepsilon_t \quad - \text{additive model}, \tag{6}$$

$$y_t = T_t * S_t * \varepsilon_t \quad - \text{multiplicative model}, \tag{7}$$

where T_t -trend, S_t – seasonality, ε_t- components of randomness.

A forecasting procedure algorithm has been developed to implement linear, logarithmic and exponential trends for forecasting the volume of reserves needed for the repair and equipment work of passenger cars by the wagon depot enterprise (Figure 1), and with its help, it is possible to perform a comparative analysis of additive and multiplicative models. The following evaluation criteria were used in the analysis of the models (Deb *et al.* 2017; Salinas *et al.* 2020):

- adequacy indicator according to Fisher's criterion;
- average modulus of relative errors of prediction values;
- accuracy of the model based on the standard deviation of the prediction errors.

The forecasting organization based on one-factor regression models was studied for the inventory requirements in the wagon depot and was carried out using the last 2 years of data (Figure 1).

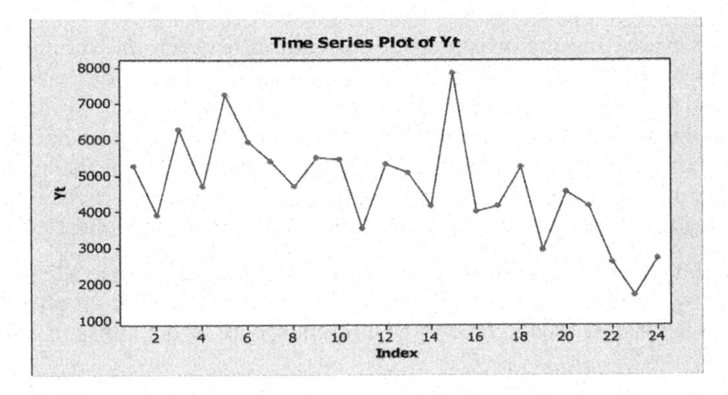

Figure 1. The dynamics of demand for the last 2 years of stock at VChD-2 enterprise.

3 RESULTS AND DISCUSSION

An analysis of the results is given below:

- Average demand for CR – 4685;
- mean square deviation -1432;

- total demand during the observation period – 12447;
- minimum requirement - 1700;
- the most demand is 7850.

According to the prediction results, the following equations were obtained:
a) equations of linear, logarithmic, exponential and quadratic trend models:

$$y_t = 6136 - 116,1t; \quad (8)$$

$$y_t = 6307 - 710\ln|t|; \quad (9)$$

$$y_t = 8,78664 * (0,996348^t); \quad (10)$$

$$y_t = 4866 + 177,2t - 11,73t^2 \quad (11)$$

b) equations of linear, logarithmic, exponential trend multiplicative models that take into account the seasonality component:

$$y_t = 6039,42 - 106,712 * t \quad (12)$$

$$\ln|y_t| = 8,72359 - 0,0258210 * t \quad (13)$$

$$y_t = 6476,22 * (0,970440^t) \quad (14)$$

c) equations of linear, logarithmic, exponential trend additive models that take into account the seasonality component (Gur Ali *et al.* 2016; Hyndman *et al.* 2007; Koshechkin 2001; Noskov *et al.* 2020):

$$y_t = 5841,06 - 92,4618 * t \quad (15)$$

$$\ln|y_t| = 8,71883 - 0,0254400 * t \quad (16)$$

$$y_t - 6476,22 * (0,970440^t) \quad (17)$$

Based on the calculations, the seasonality component values are presented in Table 1.
Data on the adequacy and accuracy of the models were also determined and the results obtained are summarized in Table 2.
The main evaluation parameter is the standard deviation of the prediction errors, according to which the accuracy of the model is calculated.

Table 1. Wagon depot CR seasonality component values.

Months	Additive		Multiplicative	
	Linear	Logarithm	Linear	Logarithm
1	68,27	0,026095	1,00860	1,00297
2	−773,39	−0,153494	0,83915	0,98182
3	2989,11	0,496997	1,60813	1,05858
4	686,14	−0,134828	0,85064	0,98394
5	−340,73	−0,036723	0,92179	0,99558
6	945,57	0,257290	1,21776	1,03096
7	36,19	0,010188	1,00059	1,00106
8	−667,56	−0,129951	0,87027	0,98474
9	56,61	0,015348	1,00395	1,00165
10	−19,64	0,005471	0,98994	1,00050
11	−1775,31	−0,396193	0,66240	0,95365
12	167,02	0,039799	1,02678	1,00455

Table 2. Adequacy and accuracy of the used trends of the models.

Model type	Used trend	Accuracy level	Error MAD	Determination coefficient. R^2	Estimated value of Fisher's criterion	Fisher's criterion table value	adequacy
Trend models without season-ality	linear	0,97	881	0,32	10,78	3,59	adequate
	Logarithmic	0,95	811	0,17	4,52		adequate
	Exponential	0,98	915	0,31	9,18		adequate
	Quadratic	0,97	780	0,45	8,83		adequate
additive	Linear	0,97	654	0,34	10,23		adequate
	logarithmic	0,96	0,17	0,38	13,75		adequate
Multiplicative	exponential	0,96	704	0,33	11,13		adequate
	quadratic	0,96	0,17	0,37	13,15		adequate

Table 3. Predictive values analyzed based on the quadratic model.

Months	Predictive value
25	1963,46
26	1542,39
27	1097,86

Table 4. Predictive values analyzed based on the logarithmic additive model.

Months	Predictive value
25	3294
26	2697
27	4914

Based on the analysis of the table, it can be said that it is better to use the quadratic trend model with the largest coefficient of determination, that is, R^2=0.45, or the additive model with the logarithmic trend, which has the smallest error, for predicting TMB stocks in wagon depot warehouses.

Forecast values based on quadratic and logarithmic additive models for the next 3 months are presented in Tables 3, 4.

The change trend based on the results obtained by the quadratic model is presented in Figure 2.

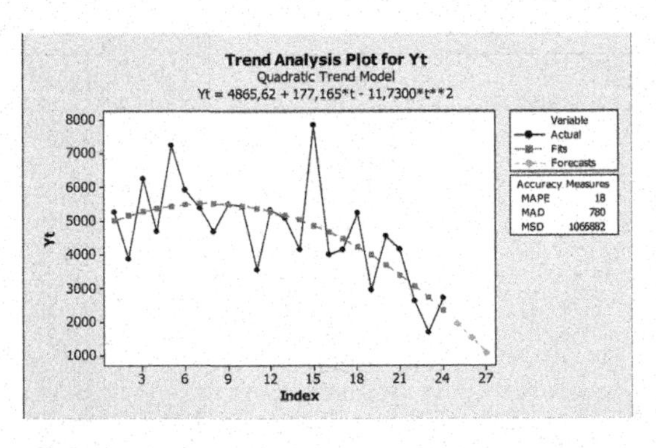

Figure 2. Quadratic trend model graph.

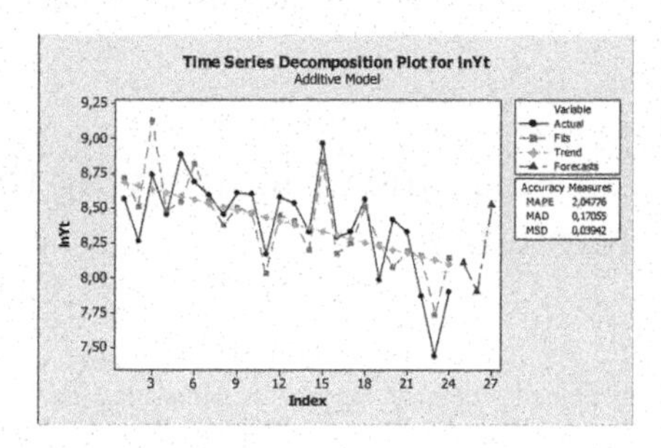

Figure 3. Logarithmic trend additive model graph.

The change trend based on the results obtained by the logarithmic additive model is presented in Figure 3.

From the obtained results and graphs, it can be concluded that it is reasonable to propose a quadratic model as a short-term prediction model, ignoring the seasonality component.

4 CONCLUSION

One-factor models have the benefit of being very simple to use for predicting, but they also have substantial drawbacks due to the unpredictable nature of additional factors that affect the actual values. For instance, the demand for a stock may depend on the average monthly temperature, the number of passengers, the state of the economy, etc. The reliability of the forecast is violated when there are any significant model modifications that did not take place during the proactive interval, which reduces the effectiveness of tactical and strategic decisions.

REFERENCES

Afanasyev, V.N., Yuzbashev, M.M., 2001. Analiz Vremennix Ryadov i Prognozirovaniye. *M.: Finansi i Statistika, 228*, 17–30.

Arunraj, N.S., Ahrens, D., 2015. A Hybrid Seasonal Autoregressive Integrated Moving Average and Quantile Regression for Daily Food Sales Forecasting. *International Journal of Production Economics* 170, 321–335.

Bazilevskiy, M. P., Noskov S. I. Metodicheskiye i Instrumentalniye Sredstva Postroyeniya Nekotorix Tipov Regressionnix Modeley." *Sistemi. Metodi. Texnologii 1* (2012): 80–87.

Criminis, A., Shotton, J., 2012. Decision Forests: A Unified Framework for Classification, Regression, Density Estimation, Manifold Learning and Semi-supervised Learning. *Foundations and Trends in Computer Graphics and Vision*, 81–227.

Deb, C., Zhang, F., Yang, J., Lee, S.E., Shah, K.W., 2017. A Review on Time Series Forecasting Techniques for Building Energy Consumption. *Renewable and Sustainable Energy Reviews* 74, 902–924.

Gur Ali, O., Pinar, E., 2016. Multi-period-ahead Forecasting with Residual Extrapolation and Information Sharing—Utilizing a Multitude of Retail Series. *International Journal of Forecasting* 32, 502–517.

Hyndman, R.J., Khandakar, Y., *et al.*, 2007. *Automatic Time Series for Forecasting: The Forecast Package for R. 6/07*, Monash University, Department of Econometrics and Business Statistics.

Isupova, YE.V. *Modelirovaniye Sistemi Upravleniya Tovarnimi Zapasami i Potokami Torgovo-Posrednicheskoy Organizatsii.* PhD diss., YEV Isupova, 2009.

Koshechkin, S.A. 2001. *Algoritm Prognozirovaniya Obyema Prodaj v MS Excel* // Marketing in Russia and abroad.

Lapach, S. N., Radchenko S. G. "Osnovniye Problemi Postroyeniya Regressionnix Modeley." *Matematicheskiye Mashini i Sistemi 1*, no. 4 (2012): 125–133.

Noskov, S. I., Vrublevskiy I. P. *Analiz Regressionnoy Modeli Gruzooborota Jeleznodorojnogo Transporta. //Vestnik transporta Povoljya 1* (2020): 86–90.

Noskov, S.I., Xonyakov A. A. Kusochno-lineyniye Regressionniye Modeli Obyemov Perevozki Passajirov Jeleznodorojnim Transportom. *Modeli, Sistemi, Seti v Ekonomike, Texnike, Prirode i Obshestve* 4 (2021): 80–89.

Salinas, D., Flunkert, V., Gasthaus, J., Januschowski, T., 2020. Deepar: Veroyatnostnoye prognozirovaniye s Avtoregressionnimi Rekurrentnimi Setyami. *International Journal of Forecasting* 36, 1181–1191.

Strijov, V. V. Poisk Parametricheskoy Regressionnoy Modeli v Induktivno Zadannom Mnojestve. *Vichislitelniye Texnologii 12*, no. 1 (2007): 93–102.

Artificial Intelligence, Blockchain, Computing and Security – Dagur et al. (Eds)

Mammographic density classification applying the CLAHE contrast enhancement method

Sh.Kh. Fazilov*
Professor, Research Institute for Development of Digital Technologies and Artificial Intelligence, Tashkent, Uzbekistan

Kh.S. Abdieva*
Ph.D. student, Department of Software Engineering, Samarkand State University, Samarkand, Uzbekistan

G.D. Sobirova*
Senior teacher, Department of Computer Science and Technologies, Samarkand State University, Samarkand, Uzbekistan

ABSTRACT: Breast cancer is one of the top causes of death among women. To assist radiologists in making an early diagnosis, systems for computer-aided diagnosis (CAD) are being developed. We offer a diagnostic framework based on ResNet-50, a widely used clinical breast screening tool that is based on mammographic imaging, for the precise classification of breast abnormalities. The craniocaudal (CC) and mediolateral oblique (MLO) views on screening mammography provide two independent perspectives of each breast. We employ two views for breast classification since they are complimentary and dual-view-based approaches are effective. A deep model is trained using a loss function, and we utilize the focused loss function since it concentrates on learning challenging cases.

Breast cancer affects over 1.5 million women worldwide each year. In women, breast cancer is the most prevalent malignant tumor. The highest intensive rate of the disease (per 100,000 women) is 30–40% in England, Denmark, the USA, and the Baltic countries, and the lowest rate is 2–10% in countries such as Japan and Mexico. This disease rarely occurs in women under the age of 25–30, but as the age limit increases, the number of diseases increases, and it reaches its "peak" at the age of 50–60. In recent years, the number of cases of this type of tumor disease in Uzbekistan has gradually increased, and it is now the leading cause of malignant tumors in women [1,2]. The World Health Organization's most recent statistics show that the incidence of breast cancer was the highest among oncological diseases in Uzbekistan in 2018 with a death rate of 1,449, or 0.92% of all deaths [2]. In 1993, there were 5.3 cases of this illness per 100,000 persons, but by 1998, that number had risen to 6.1. Every year, more people with breast cancer are receiving a diagnosis. 154 people with this condition were listed in the primary registry in the Samarkand region alone in 2000, and 300 patients were listed there in 2019.

1 ANALYSIS OF EXISTING APPROACHES

1.1 *Mammography*

It is a highly accurate diagnostic method for breast cancer, with the help of which the correct diagnosis is made in 83–95 percent of cases [3]. Typically, the breast is examined using two

*Corresponding Authors: sh.fazilov@mail.ru, orif.habiba1994@gmail.com and sobir1970@mail.com

DOI: 10.1201/9781032684994-117

standard perpendicular projections, the right, and side projections. Mammography is carried out using a special X-ray machine called mammography. Mammography distinguishes primary and secondary signs of a mammary tumor.

1.2 *BI-RADS classification*

Breast density assesses a woman's breasts' glandular, fibrous, and fatty tissue distribution. Quantifying and categorizing the amount of fibro-glandular tissue in mammography pictures is known as mammographic density (MD) categorization [4–6]. Category 0, exam not conclusive; Category 1, no findings; Category 2, benign findings; Category 3, probably benign findings; Category 4, suspicious findings; Category 5, a high risk of malignancy; and Category 6, proved cancer. These are the six BI-RADS classifications [10–12]. In academic literature, there are various methods for automatically calculating breast density from mammograms. A multi-scale blob identification method was suggested by [3] as a way to locate the dense and fatty tissue seen in mammograms. The experimental results revealed some early connections between the average relative tissue area (fatty and dense) in mammograms and the BI-RADS density category. This technique was applied to the evaluation of MIAS mammograms. The researchers in [4] divided mammograms into three density groups using a Directed Acyclic Graph (DAG)-SVM classifier and multi-resolution histograms to examine textual data. [5] extracted 21 criteria based on intensity and fractal texture features and divided the MIAS mammograms into three categories. A fuzzy-rough refined image processing technique was proposed by [6] to enhance small image regions and obtain GLCM-based statistical features for categorizing mammographic density. [7–9] collected 137 pixel-level features with intensity, GLCM, and morphological features to classify pixels into fatty or dense clusters.

2 PREPROCESSING METHODS

2.1 *Data*

A private dataset of 320 grayscale mammography pictures from 160 patients, including both the left and right breasts, was analyzed.

2.2 *Method*

The recommended approach makes use of two mammography pictures with a combined size of 336×224 and an end-to-end deep learning-based model (ResNet50) that predicts the label of the density type of the breast using BI-RADS classification from two views.

Algorithm:

Step1: Input Mammogram images;	Step4: CNN model;
Step2: Preprocessing;	Step5: Concatenation;
Step3: Global Average Pooling;	Step6: Classification;

This algorithm has two branches, one for each perspective. Each branch preprocesses the pertinent view using a convolutional neural network (CNN) as the basic model before extracting hierarchical features. The FC layer, which serves as a classifier and generates the prediction label for the input mammographic pictures, receives the features from the two views combined in the concatenation layer.

2.3 *Preprocessing*

To distinguish between different breast density classes, breast tissue must be sufficiently distinct from the surrounding tissue. After removing all artifacts from the image, the model can only analyze the breast tissue region. The background pixel in the initial phase was represented by a binary mask with a threshold value of 200, while the breast area, artifact, or noise pixel was represented by a binary mask with a threshold value of 0 (white). The breast tissue region, which is more apparent than any other object and is binarized as a single region, is then recovered from the binary picture using a morphological opening operator with a disk-type structuring element of size 9×9. As a result, just the most crucial outlines are maintained, while the others are discarded. We then overlay this mask to eliminate mammography artifacts and keep only the breast tissue area. Then, using the bounding box of the breast tissue, each view is cropped to largely include the breast tissue. The characteristics of breast tissue in digital mammographic images will be easier to see after image augmentation, increasing the early breast cancer classification rate. A carefully designed image processing technique will likely be needed to portray different image attributes in the best way possible. Additionally, depending on the breast density, the performance variations between different image preparation methods and certain algorithms may be useful. As seen in Table 1, we, therefore, employed the CLAHE contrast enhancement technique. An overview of several picture preparation techniques is shown in Figure 1.

Table 1. The performance of the private dataset test after thorough mammography preprocessing.

Model	Preprocessing	[OCA %]
ResNET50	Without	66.8
	Histogram Equalization	65.2
	Contrast-limited adaptive histogram	67.4

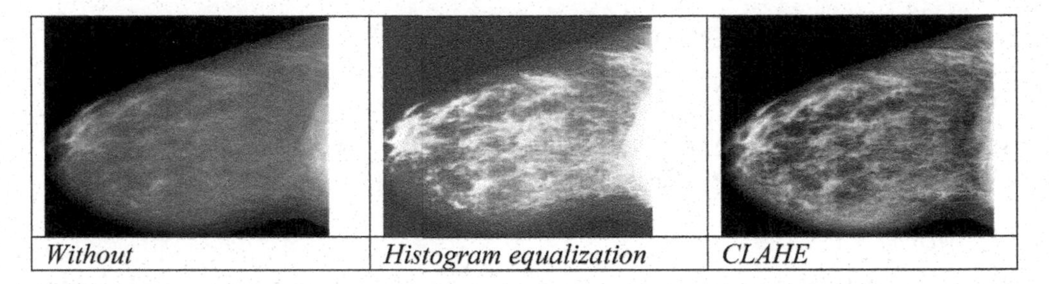

Figure 1. A difference in the visual effects of different image preprocessing.

2.4 *Contrast limited adaptive histogram equalization*

Medical images have limited contrast, which makes them challenging for automated systems to process or evaluate, leading to algorithmic errors. The low contrast problem was addressed with techniques for improving digital image contrast. The following procedure was used to implement standard histogram equalization. The possibility that pixel x belongs to grayscale I in the range of 0 to L is, given a discrete grayscale image [X], where N_i is the frequency for grayscale I to exist in the image with N total pixels:

$$\begin{aligned} p_x(i) &= p(x = i) \\ &= \frac{N_i}{N}(0 \leq i \leq L) \end{aligned} \tag{1}$$

where L is the intensity with the maximum value over the image [X]. As a result, each grayscale i the normalized probability, or $p_x(i)$, runs from 0 to 1. Consequently, the cumulative distribution function (CDF) can be determined as follows:

$$f_{cdx}(i) = \sum_{j=0}^{i} p_x(j) \tag{2}$$

The desired constant pixel value, often set to 255, is multiplied by $f_{cd\_x}(i)$ to create a new function that maps the CDF into an equalized one. The two stages of the Contrast Limited Adaptive Histogram Equalization enhancement process are the bilinear interpolation technique and histogram equalization with applied contrast limits. An updated histogram is initially produced by histogram equalization, and the bilinear interpolation approach lowers the computing costs associated with creating a new image. The steps involved in histogram equalization can be broken down into the following categories given a picture ***I*** and the M by N sub-regions obtained by dividing ***I***:

$$\begin{gathered} f_x^n(i) = \sum_{j=0}^{i} p_x^n(j) \\ (0 \leq i \leq 255, 1 \leq n \leq M * N) \end{gathered} \tag{3}$$

here p_x^n is in the same form as Eq.1 but was the possibility of pixel x in subregion n. Firstly, the summation of pixel intensity that is above the pre-determined clip limit (CL) is acquired, normalized, and can be noted as T^n, which is later evenly distributed to all intensity levels by the average increase AI^n. Therefore AI^n can be simply denoted as:

$$AI^n = \frac{T^n}{256} \tag{4}$$

The corrected histogram can then be expressed as follows:

$$P_x^n(i)' = \begin{cases} CL, if\, P_x^n(i) \geq CL - AI^n \\ P_x^n(i) + AI^n, if\, P_x^n(i) < CL - AI^n \end{cases} \tag{5}$$

To lower the computational cost, block linear interpolation was used after the image histogram was adjusted.

3 ANALYSIS OF RESULTS

We used 5-fold cross-validation to test our suggested data augmentation strategy and the classification models to prevent unexpected experimental outcomes caused by the improper partition of the dataset. 80% of the dataset, or 4 out of 5 folds, is divided into the training set at the start of the experiment The training set was subsequently subjected to CLAHE, whereas the testing set was left unaltered. As a result, the initial training set and the upgraded training set are combined to create a new training set. Later, various CNN models were trained using training datasets that had been enhanced and without enhancement. To fulfill the input requirements of various models, it should be noted that the size of the input image is appropriately modified. On a personal PC with a single Ge-force GTX 1060 GPU and 8GB of memory, all of the trials were run. The MATLAB deep learning toolbox with pre-trained deep CNN models served as the research's framework. Given that ResNet-50 had a mean accuracy of five folds, it achieved an extraordinary level of accuracy. Dual-view mammography input setting is advantageous to the density classification task, and improved statistics were obtained by all of the investigated models, including the backbone model ResNet50, as shown in Table 2. Comparing the results using dual-view mammography inputs to those using single-view mammography input, it can be seen that the dual-view mammography input setting is advantageous to the density classification task.

Table 2. The impact of total categorization accuracy when comparing single and dual views view of the test for the private dataset.

Model		Overall classification accuracy(%)
ResNET50	Single view	74.93
ResNET50	Dual view	91.41

4 CONCLUSIONS

There are various limitations to this study. However, because of the limited number of available images, the mammography dataset used in this study showed a considerably uneven distribution across the four categories. Using these datasets to train a classification network is difficult. The density of a woman's breasts is another essential clinical factor considered while determining her risk of developing breast cancer. Our proposed method successfully distinguishes between dense and non-dense (fatty or scattered density) breasts (heterogeneously dense or extremely dense). It is simple to tell the difference between fatty and abnormally thick breasts in a clinical setting. However, radiologists have difficulty reliably and visually differentiating between the groups of heterogeneously dense and scatter density [7,8]. Our research shows that classification outcomes for heterogeneously dense or very dense materials are superior to those for fatty or scattered materials. This might be a result of the analogies between these two forms of density or between highly dense and distributed density. We took on the challenging task of differentiating mammographic breast density to enhance the classification of BI-RADS class. To do this, we developed a system that makes use of the complementary link between the craniocaudal (CC) and mediolateral oblique (MLO) views using deep learning improvements.

REFERENCES

[1] World Health Organization (WHO).

[2] O.R. Yusupov, Kh.S. Abdiyeva, A. Primov. "Preprocessing and Segmentation of Digital Mammogram images for Early Detection of Breast Cancer". *IJARSET*, vol.8, issue 9, 2021.

[3] George, M., Rampun, A., Denton, E., & Zwiggelaar, R. (2016). Mammographic Ellipse Modeling Towards Birads Density Classification. *International Workshop on Breast Imaging IWDM 2016: Breast Imaging*, 423–430. https://doi.org/10.1007/978-3-319-41546-8_53.

[4] Muhimmah, I., & Zwiggelaar, R. (2006). Mammographic Density Classification Using Multiresolution Histogram Information. In *Proceedings of the International Special Topic Conference on Information Technology in Biomedicine*, Epirus, Greece, 26–28.

[5] Tzikopoulos, S. D., Mavroforakis, M. E., Georgiou, H. V., Dimitropoulos, N., & Theodoridis, S. (2011). A Fully Automated Scheme for Mammographic Segmentation and Classification Based on Breast Density and Asymmetry. *Computer Methods and Programs in Biomedicine, 102(1)*, 47–63.

[6] Qu, Y., Fu, Q., Shang, C., Deng, A., Zwiggelaar, R., George, M., & Shen, Q. (2020). Fuzzy-rough Assisted Refinement of the Image Processing Procedure for Mammographic Risk Assessment. *Applied Soft Computing, 91*, 106230.

[7] Subashini, T. S., Ramalingam, V., & Palanivel, S. (2010). Automated Assessment of Breast Tissue Density in Digital Mammograms. *Computer Vision and Image Understanding, 114(1)*, 33–43. https://doi.org/10.1016/j.cviu.2009.09.009

[8] Oliver, A., Tortajada, M., Llado, X., Freixenet, J., Ganau, S., Tortajada, L., . . . Marti, R. (2015). Breast Density Analysis using an Automatic Density Segmentation Algorithm. *Journal of Digital Imaging*, 28(5), 604–612.

[9] Mario Muštra, Prof. Mislav Grgić & Krešimir Delač. (2012). Breast Density Classification using Multiple Feature Selection. *Automatika, 53(4)*, 362–372.

[10] Li, S. F., Wei, J., Chan, H. P., Helvie, M. A., Roubidoux, M. A., Lu, Y., . . . Samala, R. K. (2018). Computer-aided Assessment of Breast Density: Comparison of Supervised Deep Learning and Feature-based Statistical Learning. *Physics in Medicine and Biology, 63(2)*.

[11] Singh, H., Sharma, V. & Singh, D. Comparative Analysis of Proficiencies of Various Textures and Geometric Features in Breast Mass Classification Using a K-nearest Neighbor. *Vis. Comput. Ind. Biomed. Art 5*, 3 (2022)

[12] Yusupov, O.R, Abdiyeva Kh.S. "An Overview of Medical Image Segmentation Algorithms". *European Multidisciplinary Journal of Modern Science*. 2022.

Parallel algorithm for the one-dimensional problem of oil movement in a porous medium

T.T. Turdiyev, B.Yu. Palvanov, M.A. Sadikov, K.A. Salayev & I.B. Sabirov
Urgench Branch of Tashkent University of Information Technologies named after Muhammad al-Khwarizmi

ABSTRACT: The article considers a parallel algorithm for oil filtration in porous media. The issues of developing a mathematical and numerical model, as well as effective numerical algorithms for solving the problems of monitoring, analyzing and predicting the main indicators of the development of oil and gas fields are also considered. In the applied aspect, the developed mathematical apparatus is of interest in solving the problems of analysis and design of the development of oil and gas fields.

1 INTRODUCTION

The world pays great attention to the creation and improvement, as well as the development of mathematical models of non-stationary filtration processes. As well as the use of the possibilities of numerical models and info communication technologies for solving linear and non-linear problems of the theory of filtration. The creation of automated systems for determining, analyzing and predicting the main indicators of the development of oil and gas fields, as well as the study of non-stationary filtration processes based on modern information technologies, remain the main tasks in this area. Mathematical models of non-stationary oil filtration processes, the development of computational algorithms and the creation of software have an important feature in such developed countries as the USA, France, China, the United Arab Emirates, Iran, Russia, Ukraine, Kazakhstan, Azerbaijan, etc.

Scientific research aimed at developing mathematical models and efficient numerical algorithms for solving problems of fluid filtration in a porous medium is carried out in the world's leading scientific centers and higher educational institutions.

The development and improvement of mathematical models and numerical algorithms for complex dynamic filtration processes in oil and gas and aquifers, the solution of stationary and non-stationary problems of filtration in oil layers with poor permeability are considered in the works of such scientists as Yu. A. Chirkunov, Yu. L. Skolubovich, NS Hanspal, A. N. Vagode, R. J. Wakeman, Fransisko J. Karrillo, Yan S. Burg, Kiprien Soulen, Evgeniy D, S Minaev, Ronli Xu, Tiankui Guo, Hanyu Li, C.Atkinson, K.Ives, Z.Mehdi, and others.

In the Republic of Uzbekistan, mathematical models and an algorithm for a one-dimensional problem of oil movement in a porous medium were studied in the studies of such scientists as F.B. Abutaliev, J.F. Fayzullaev, N.M. Kh. Khuzhayorov, M. Aripov, N. Ravshanov, U. S. Nazarov, Sh. Kayumov, Y. Yarbekov, A. Mirzaev and others. F.B. Abutaliev, A.Nematov, V.F.Burnashev and other scientists.

An analysis of research in this area shows that at present, the processes of a parallel algorithm for a one-dimensional problem of oil movement are insufficiently studied, when the filtration area has a complex configuration and is a multilayer porous medium with different permeability in the presence of a hydrodynamic connection between the reservoirs.

2 MAIN PART

When designing, analyzing and determining the prospects for the development of gas and gas condensate fields, it is required to determine the change in time of the required number of

DOI: 10.1201/9781032684994-118

production and injection wells, gas flow rates and flow rates of injection wells, reservoir, bottomhole, wellhead pressures and temperatures, the advancement of contour or bottom waters in time, a change in the amount and the composition of the condensate dropping out in the reservoir and produced, and other indicators.

The processes occurring in the reservoir during the development of natural gas fields are described by differential equations in partial derivatives. To determine the development indicators of gas and gas condensate fields, taking into account the heterogeneity of the reservoir in terms of reservoir properties, arbitrary location of differently flowing wells, uneven progress of the gas-water interface, etc., it is necessary to integrate the differential equations of unsteady filtration of oil, water and condensate with the corresponding initial and boundary conditions. At the same time, the differential equation of unsteady oil filtration is of particular importance in the theory of gas field development.

Natural oil reservoirs are characterized by heterogeneity and variability of reservoir parameters. The thickness of productive deposits over the area of a gas deposit can vary over a very wide range. The permeability and porosity coefficients of the formation undergo significant changes in thickness and in the area of the gas deposit.

In the scientific study of natural gas fields, one-dimensional and two-dimensional differential equations are usually considered. The use of these equations is associated with significant difficulties in determining the dependences of changes in reservoir parameters in the x, y, and directions, i.e., building a three-dimensional reservoir model. The study of a number of three-dimensional problems can be reduced to a "set" of one-dimensional and two-dimensional problems – to the consideration of one-dimensional and two-dimensional problems of unsteady filtration, for example, in each individual layer, pore layer, etc. Therefore, we present the derivation of the desired equation for the one-dimensional case.

In an oil-bearing reservoir of variable thickness, we single out an elementary volume dx dy h {x, y}. Here h (x, y) is the value of the reservoir thickness at the point with coordinates x and y (Figure 1).

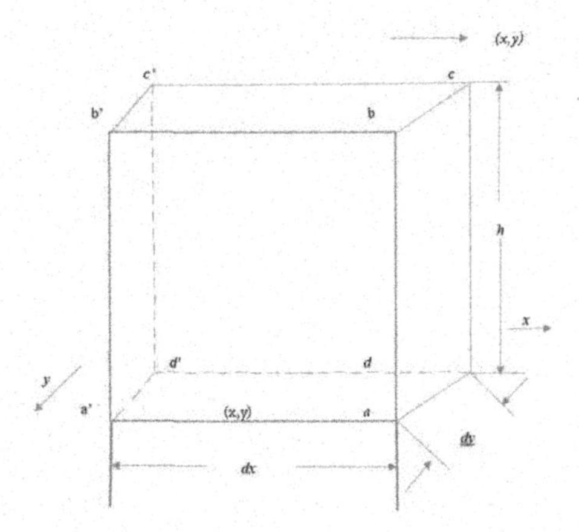

Figure 1. Elementary reservoir volume.

Arguing in the usual way [Ravshanov N.K., Nazirova E.Sh., Nematov A., Monograph, "Моделирование процессов фильтрации жидкостей и газа в пористых средах", 2021.], we get that through the face *a'b'c'd'* in time dt, a mass of gas flows in equal to

$$\left[\rho uh(x,y)dy - \frac{1}{2}dx\frac{\partial(\rho uh(x,y))}{\partial x}dy\right]dt$$

Let us consider a non-stationary process of oil filtration in an inhomogeneous porous medium in a two-dimensional formulation. The arbitrariness of the field configuration and the change in reservoir parameters over the deposit area, the uneven location of oil wells in the oil-bearing area and their different production rates are not limiting factors for the use of numerical simulation in calculations for the development of oil fields. In numerical simulation, taking into account these factors present significant difficulties for calculations on a computer system for the development of oil fields with a complex configuration and changing hydrodynamic parameters of the object.

The studies of the process of filtering highly contaminated oil in porous media have shown that with intensive work of the gallery of wells in the bottomhole zones, the pore space is clogged with fine particles. This reduces the oil recovery of reservoir systems. Obviously, this phenomenon plays a significant role in the process of oil filtration in porous media. Therefore, a mathematical model was developed that takes into account such factors as the rate of fine particles settling, the change in the porosity coefficient and filtration over time.

Taking into account the above factors, unsteady oil filtration in an inhomogeneous porous medium is described by the following differential equation:

$$\beta h(x,y)\frac{\partial P}{\partial t}=\frac{\partial}{\partial x}\left(\frac{k(x,y)h(x,y)}{\mu}\frac{\partial P}{\partial x}\right)+\frac{\partial}{\partial y}\left(\frac{k(x,y)h(x,y)}{\mu}\frac{\partial P}{\partial y}\right)-Q \quad (1)$$

When determining the main indicators of oil field development, we solve the differential equation (1) under the following initial and boundary conditions:

$$P(x,y)=P_H(x,y), \quad at\ t=0, \quad (2)$$

$$\begin{aligned} &-\frac{k(x,y)h(x,y)}{\mu}\frac{\partial P}{\partial n}=\alpha(P_A-P)\ at\ \ (x,y)\in\Gamma, \\ &\oint_{s_{i_q}}\frac{k(x,y)h(x,y)}{\mu}\frac{\partial P}{\partial n}ds=-q_{i_q}(t)\ at\ (x,y)\in s_{i_q},\quad i_q=\overline{1,N}_q. \end{aligned} \quad (3)$$

$$Q=\sum_{i,j=1}^{N_q}\delta_{i,j}q_{i,j}$$

Here P – is the pressure in the reservoir;

P_H – initial reservoir pressure;
P_A – frontier pressure;
μ – dynamic viscosity of oil;
k – formation permeability coefficient;
h – reservoir thickness;
β – coefficient of reservoir elasticity $\beta=m\beta_H+\beta_c$.;
β_H – oil compressibility factor;
β_c – coefficient of compressibility of the medium;
q_{i_q} – debit i_q – th well;
s_{i_q} – contour i_q – th well;
n – internal normal to the boundary G;
N_q – number of wells;
m – formation porosity coefficient;
δ – Dirac function;

$$\alpha=\begin{cases}0, & \text{no – flowcondition,}\\ 1, & \text{thirdboundarycondition.}\end{cases}$$

For the numerical solution of problem (1)-(3) by the finite difference method, we introduce the following dimensionless variables:

$$P^*=P/P_0;\ x^*=x/L;\ y^*=y/L;\ k^*=k/k_0;\ h^*=h/h_0;$$

$$\tau = \frac{k_0 t}{\beta \mu L^2};\ q^* = \frac{q\mu}{\pi k_0 P_0 h_0}.$$

Here P_0 – is some characteristic pressure value; k_0 – some characteristic, for example, the initial value of the reservoir permeability; h_0 – some characteristic value of the reservoir thickness, for example, $h_0 = \max(h(x,y))$; L is the characteristic length.

In what follows, for simplicity, "*" in the equations will be omitted. Then, taking this into account, problem (1)-(3) in dimensionless variables is rewritten as follows:

$$\begin{cases} h\frac{\partial P}{\partial \tau} = \frac{\partial}{\partial x}\left(k(x,y)h(x,y)\frac{\partial P}{\partial x}\right) + \frac{\partial}{\partial y}\left(k(x,y)h(x,y)\frac{\partial P}{\partial y}\right) - Q \end{cases} \quad (4)$$

$$P(x,y) = P_H(x,y)\ at\ t = 0,\ (x,y) \in G, \quad (5)$$

$$-k(x,y)h(x,y)\frac{\partial P}{\partial n} = \alpha(P_A - P)\ at\ (x,y) \in \Gamma, \quad (6)$$

$$\oint_{s_{iq}} \frac{k(x,y)h(x,y)}{\mu}\frac{\partial P}{\partial n} ds = -q_{i_q}(t)\ at\ (x,y) \in s_{i_q},\ i_q = \overline{1, N_q}. \quad (7)$$

3 PROBLEM STATEMENT AND SOLUTION METHOD

Thus, developed dimensionless boundary value problem (4)-(7) of oil filtration in a porous medium describing differential equations in partial derivatives.

Method for solving a one-dimensional problem of oil movement in pore space.

Let us present a simplified mathematical model of oil movement in porous media. Then:

$$m\frac{\partial P}{\partial t} = \frac{\partial}{\partial x}\left(\frac{(k(x)h(x)\partial P)}{\mu \partial x}\right) - q \quad (8)$$

with conditions

$$P(0,x) = \varphi(x)\quad \frac{\partial P}{\partial x}|_{x=0} = 0\quad \frac{\partial P}{\partial x}|_{x=L} = 0 \quad (9)$$

If consider, that $k(x) = const,\ h(x) = const$ equation (8) can be written as follows:

$$\begin{gathered} m\frac{\partial P}{\partial t} = \frac{K\partial^2 P}{\mu \partial x^2} - q, \\ P_H = \varphi(x),\ \frac{\partial P}{\partial x}|_{x=0} - 0,\ \frac{\partial P}{\partial x}|_{x-L} = 0 \end{gathered} \quad (10)$$

Next, we introduce a finite difference scheme for (10) then:

$$\begin{gathered} m\frac{(P_{i-1}^{n+1} - P_i^n)}{\tau} = \frac{K}{\mu}\left(\frac{P_{i+1}^{n+1} - 2P_i^{n+1} + P_{i-1}^{n+1}}{h^2}\right) - q \\ P_i^{n+1} - P_i^n = \frac{mK}{\mu h^2}P_{i+1}^{n+1} - \frac{2Km}{\mu h^2}P_i^{n+1} + \frac{Km}{\mu h^2}P_{i-1}^{n+1} - \frac{q}{m} \\ \frac{mK}{\mu h^2}P_{i+1}^{n+1} - \left(\frac{2Km}{\mu h^2} + 1\right)P_i^{n+1} + \frac{Km}{\mu h^2}P_{i-1}^{n+1} = -\left(P_i^n - \frac{q}{m}\right) \end{gathered} \quad (11)$$

Let us introduce the notation

$$a_i = c_i = \frac{Km}{\mu h^2};\quad b_i = \frac{2Km}{\mu h^2} + 1;\quad d_i = 1\left(P_i^n - \frac{q}{m}\right);$$

Then equation (11) can be written as follows:

$$a_i P_{i+1}^{n+1} - b_i P_i^{n+1} + c_i P_{i-1}^{n+1} = -d_i \quad (12)$$

The solution to equation (12) will be sought in the following form:

$$P_i = a_i P_{i+1} + \beta_i$$

And the initial fitting coefficients will be sought from (10), then

$$\frac{-3P_0^{n+1} 4P_1^{n+1} - P_2^{n+1}}{2h} = 0$$

$$a_1 P_2^{n+1} - b_1 P_1^{n+1} + c_1 P_0^{n+1} = -d_1$$

$$a_1 P_2^{n+1} = b_1 P_1^{n+1} - c_1 P_0^{n+1} - d_1$$

$$\frac{3a_1 P_0^{n+1} - 4a_1 P_1^{n+1} + b_1 P_1^{n+1} - c_1 P_0^{n+1} - d_1}{2h} = 0$$

$$(3a_1 - c_1)P_0^{n+1} - (4a_1 - b_1)P_1^{n+1} - d_1 = 0$$

$$P_0 = \frac{4a_1 - b_1}{3a_1 - c_1} P_1^{n+1} + \frac{d_1}{3a_1 - c_1}$$

In the end, the initial sweep coefficients will look like this:

$$a_0 = \frac{4a_1 - b_1}{3a_1 - c_1}; \quad \beta_0 = \frac{d_1}{3a_1 - c_1};$$

To organize the reverse course of the sweep, we need to find. Condition (10) can be approximated as follows

$$\frac{3P_N^{n+1} - 4P_{N-1}^{n+1} + P_{N-2}^{n+1}}{2h} = 0 \text{ at } P_{N-1} = a_{N-1}P_N + \beta_{N-1} \text{ then}$$

$$P_{N-2} = a_{N-2}P_{N-1} + \beta_{N-2}$$

$$3P_N - 4a_{N-1}P_N - 4\beta_{N-1} + a_{N-2}a_{N-1} + P_N + a_{N-2}\beta_{N-1} + \beta_{N-2} = 0$$

$$(3 + a_{N-2}a_{N-1} - 4a_{N-1})P_N - 4\beta_{N-1} + a_{N-2}\beta_{N-1} + \beta_{N-2} = 0$$

$$P_n = \frac{4\beta_{N-1} - a_{N-2}\beta_{N-1} - \beta_{N-2}}{3 + a_{N-2}a_{N-1} - 4a_{N-1}}$$

For special cases, you can use the parallel method for calculating the problem of oil movement in porous media. Block diagram of the algorithm:

4 CONDUCTING COMPUTER EXPERIMENTS AND THEIR ANALYSIS

Thus, a numerical algorithm has been developed based on finite-difference schemes for a one-dimensional problem of oil movement in a porous medium.

The numerical results of the calculation are shown in a visual graph.

From these computational experiments, the following reasoning can be made: at low values of the dynamic viscosity of oil in the reservoir, the pressure distribution will be faster, and in wells, the

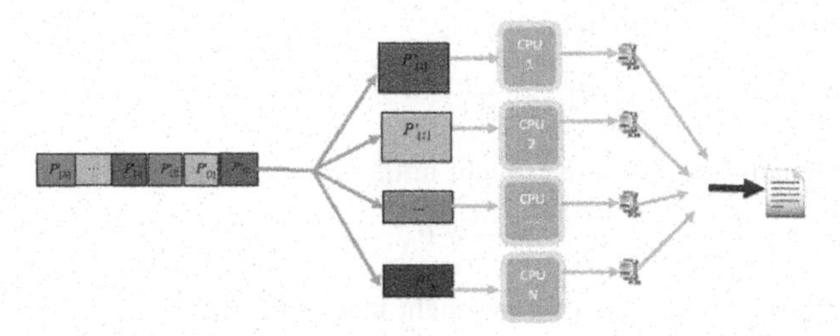

Figure 2. Parallel algorithm block diagram.

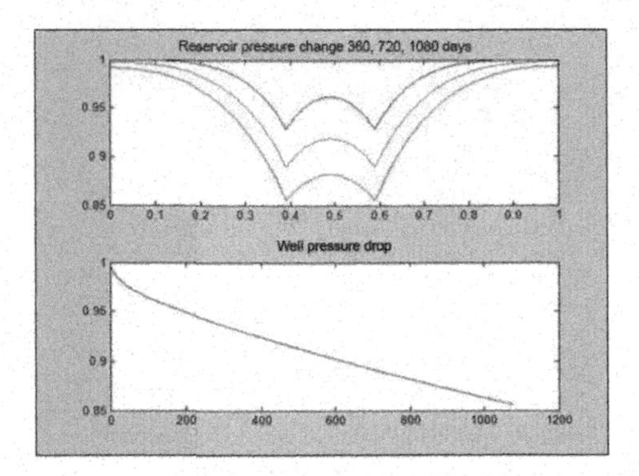

Figure 3. Reservoir pressure distribution and oil pressure drop in wells.

pressure drop will decrease (Figures 2 and 3); This is due to the fact that the flow of oil in the well will be faster and in the vicinity of the well the gas will fill up quickly.

Thus, it is possible to calculate the problems of oil movement in porous space for a two-dimensional view, as well as to use a parallel numerical algorithm in particular cases.

5 CONCLUSION

The use of parallel computing algorithm for the problem of underground oil movement helps to save the time of obtaining the solution and conducting computational experiments. But the created calculation algorithm can be used only for special cases with boundary conditions. To increase the accuracy of the solution, it is advisable to use other types of combined methods. In addition, the two-dimensional movement of underground oil can be performed for masses based on this algorithm.

REFERENCES

[1] Abdugani N, Turdiev T.T., Ismailov Sh., A. (2021) Bakhriddinov. Parallel Computational Algorithms for Solving Boundary Value Problems for Two-dimensional Equations of Parabolic Type. *International Conference on Information Science and Communications Technologies ICISCT.*

[2] Nazirova E.Sh. (2014) Modular Analysis of the Algorithm for Ssolving Problems of Filtration of Multiphase Liquids in Porous Media. Uzb. journal "*Problems of Informatics and Energy*".

[3] Nazirova E.Sh., Nematov A. *Numerical Modeling of the Process of Gas Filtration in Multilayer Porous Media.*

[4] Sadikov M.A, Olimov I.S, Karimov A.A, Tursunov O.O, Yusupova Sh.B, Tojikabarova U.U.A (2020). Creation Cryptographic Protocol for the Division of Mutual Authentication and Session Key. *2020 International Conference on Information Science and Communications Technologies (ICISCT).*

[5] Madaminov U.A, Sadikov M.A, Kutliev S.P, Allaberganova M.R. (2021). Development and Application of Computer Graphics Training Software in Information Technology. *2021 International Conference on Information Science and Communications Technologies (ICISCT)*

[6] Palvanov B.Yu. (2016). *Problems of Computational and Applied Mathematics.* 1 48–63.

Artificial Intelligence, Blockchain, Computing and Security – Dagur et al. (Eds)

Errors in SMS to hide short messages

D. Kilichev & A.N. Muhamadiev
Department of IT Convergence Engineering, Gachon University, Seoul, Korea

N.R. Zaynalov, U.Kh. Narzullaev & I.R. Rakhmatullayev
Samarkand Branch of the Tashkent University of Information Technologies Named After Muhammad al-Khwarizmi, Samarkand, Uzbekistan

ABSTRACT: Steganography is a method that is based on hiding or embedding additional information in digital objects, while causing some distortion of these objects. In this case, an image, audio, video, network packets, etc. can be used as objects or a container. Recently, interest has grown in text containers for hiding data. At the same time, steganography methods use the properties of the media, allowing you to embed a secret message. As you know, text documents are widely used in everyday practice. But these documents do not allow hiding a large amount of classified information. On the other hand, text documents have a huge number of possibilities for hiding data. In this paper, the authors propose an original method of hiding data based on the Russian language of the Uzbek audience. We analyze the degree of errors in SMS messages to the Uzbek audience and offer a method for converting a message into a byte.

Keywords: information security, classical steganography, text steganography

1 INTRODUCTION

Information security has occupied the minds of people since ancient times. Historically, there were two directions, namely cryptography and steganography. These methods of protecting information have been rapidly developed with the advent of computers. In general, these approaches are fundamentally different from each other.

Classical steganography involves hiding data in objects without disturbing the object itself. At the same time, the hidden message is not explicitly highlighted in the object. Thus, it will not be possible to detect the presence of some hidden information in the object. And the object itself is called a container.

Based on the foregoing, we can conclude that steganography deals with the issues of covert transmission of information. those. an outside observer will not be able to detect the transfer of secret information. All steganography methods and algorithms have one thing in common, namely they do not draw general attention to the object. And therefore this object can be openly transferred to the addressee. If we consider text as an object, then it will be text steganography. Each character of this text can be used to hide data. A feature of text files is that, firstly, it requires little memory to store the file, and secondly, it is more widespread in communication between people. In many ways, this serves as the basis for the use of text steganography. Along with this, text files take up less memory than other types of files and have some other kinds of advantages.

In this work, for the first time, the possibilities of the Russian-speaking population of Uzbekistan in text steganography are studied and an algorithm is proposed for embedding a sequence of bytes into a generated short text in Russian for the Uzbek audience.

DOI: 10.1201/9781032684994-119

The rest of the article is organized as follows: Section 2 describes some of the existing feature-based language approaches. Section 3 describes the proposed approach. In section 4, we make a conclusion and draw the corresponding conclusions.

2 EXISTING APPROACHES

It should be noted that the popularity of text steganography methods has led to a variety of algorithms [1,2]. The analysis of these algorithms shows that many works use the grammatical features of the language. And this, in turn, allows information to be embedded. Based on certain features, these different approaches can be identified. For example, in [1], which is one of the classic works in this field, punctuation marks are taken as a feature.

Studying works on steganography, depending on the type of implementation technique, they can be divided into the following methods: 1) syntactic, 2) semantic and 3) arbitrary type.

To understand these methods, let's briefly review some of these approaches.

A feature of syntactic methods is the ease of embedding secret data in any text, and this is not independent of its content and language. At the same time, this process is easy to automate and develop an appropriate software application. The weakness of these methods is that these syntactic methods are easily hacked. And also secret information from the container can be easily removed by simple attacks. But, these methods are very attractive, as they are fast, which allows them to be used in many applications.

Existing punctuation marks in the language, such as a period or a comma, are in many cases successfully used to hide data. In this case, these signs are used to encode bits 0 and 1 [1,3,4]. One of the varieties of this method was proposed in [2] and is called *Punctuation Mark*. The main idea in this work is that sentences can be formed in English, in which a punctuation mark, such as a comma, can be used at the discretion of the user.

The famous *White Steg* method for hiding data uses a space character. The most elementary approach that was proposed in [4,5] is that here one space after the word means bit 0, and two spaces after the word means bit 1. This idea can be modified in many ways, based on the content and type of substrate.

There are scientific works that take into account the features of grammatical constructions [6]. In this work, the possibilities of the Telugu language are studied and interesting results are obtained. It turns out that Telugu has different kinds of consonant symbols and punctuation marks. And the paper demonstrates the use of these qualities to hide data in the Telugu text itself.

As practice shows, especially in the age of the Internet, a large number of spelling errors can always be found in the text. And this phenomenon of modernity is used in the *Mistyping* method. This method allows you to embed a secret message in the text by intentionally creating some spelling errors or changing the position of characters in a text document [2]. Since this type of typo is very common in a word document, this data hiding method will not attract third parties unless they are aware of the secret communication. Of course, it should be noted that this method is more suitable for SMS messages, where grammar rules are not respected at all. It is this approach that is proposed in this paper. On the other hand, to hide only bits in short messages, taking into account the peculiarities of the Uzbek language, was proposed in [7].

The paper [8] proposes a steganography method for hiding data in Microsoft Word documents using change tracking technology. Data injection is masked in such a way that a stegodocument written by a collaborative effort contains obvious errors. Decoding is carried out by tracking the change, through the correction of errors. When viewing a stegodocument, the resulting information obtained from successive tracking changes allows you to restore not only the original document, but also the secret message. It should be noted that this algorithm, in principle, can be applied to all text documents, but it will be very striking.

3 PROPOSED APPROACH

The exchange of encrypted data between two subscribers must be invisible to a third party, which is usually associated with a passive adversary. At the same time, it is required from the developed data exchange system that the adversary reading the message could not determine the fact that the messages are encrypted. The essence of the algorithm is to introduce typos into the finished text container. For example, a letter is replaced by a letter located next to it on the keyboard. Ideally (in a text without short words), this allows one bit to be encrypted in each word.

The idea of the proposed approach is to encode secret information in a short text with grammatical errors. And this is very common in SMS messages between young people. The proposed method is an additional extension of the *Mistyping* method.

The available data hiding methods are based on the capabilities of modern information technologies. The state-of-the-art steganography techniques can be very easily transported to ensure information security. This allows us to consider different methods, so in this paper. Based on this, consider the following method based on SMS messages in Russian to hide a secret message.

The modern age is the age of high technology, promotion and many different gadgets. Probably, each of us can no longer imagine our life without the Internet, smartphone, tablet and laptop. Especially we cannot imagine our life without the Internet. The speed, ease, low cost and availability of communication between users made it possible to use the Internet not only as a tool for learning, but also as a means of communication. Using services such as Instagram or Telegram has become commonplace. Everyone knows how to write a message to a friend on social networks. Unfortunately, illiteracy in these "letters" has become commonplace. The study of this problem is primarily associated with technical devices, and secondly with social aspects. The social aspect is related to the fact that users make mistakes, for example, due to haste, or mistakes are made intentionally, knowing about it. But when communicating on social networks, most often you notice continuous spelling errors. To analyze this process, a small study was conducted related to a survey of young people. So, to the question "How do you feel about mistakes in social networks?" only 15 percent responded negatively. And the rest are indifferent to these mistakes.

Based on this, it can be concluded that Internet communication, which contributes to the development of illiteracy among the younger generation, can be used to transmit secret messages.

According to research and surveys among the Russian-speaking audience of users of social networks in the city of Samarkand, the following main errors were identified when writing the following words (Table 1):

Table 1. Misspelled words.

Пачему	Скочать	Втечение	Росписание
Пишит	Руский	Эспресо	Расчитат
Тибя	Офицальный	Чтоли	Раждражатся
Спосибо	Ложи	Врядли	тилефон
Езжайте	Придти	Какбудто	денги
День рождение	По – тихоньку	Еденица	мижду
Ихний	Будующий	Нипонимаю	сестемой
Однокласники	Вpидачу	Нечерта	Можит
Зделать	Спосибо	Кампаня	Зочем
Агенство	Никаму	Званит	Жевой
Програма	Рассписание	Умераю	Спесал
Отзовы	Расчитать	Обожратся	Способ
Поликнника	Скачять	Мололетка	Савпало

A similar study was conducted among the Russian-speaking audience of users of the city of Samarkand. At the same time, the Latin alphabet, with the help of which Russian words are eaten, was not considered here. The analysis showed the following main errors when writing the following words, which are arranged in alphabetical order (Table 2):

Table 2. Revealing letters from words.

Слова с ошибками	Слова без ошибок	Буква	Слова с ошибками	Слова без ошибок	Буква
Агенство	Агентство	Т	Пишит	Пишет	Е
Будуюший	Будущий	Ю	Поликиника	Поликлиника	Л
Впридачу	Придачу	В	Програма	Программа	М
Денги	Деньги	Ь	Раждражатся	Раздражатся	З
Еденица	Единица	И	Рассписание	Расписание	С
Жевой	Живой	И	Расчитать	Рассчитать	С
Званит	Звонит	А	Руский	Русский	С
Зделать	Сделать	С	Савпало	Совпало	О
Зочем	Зачем	А	Стелать	Сделать	Д
мижду	между	Е	сестема	система	И
Можит	Может	Е	Скачять	Скачать	Я
Мололетка	Малолетка	А	Скочать	Скачать	А
Никаму	Никому	О	Спесал	Списал	И
Одноклaсники	Одноклассники	С	Спосибо	Спасибо	А
Отзовы	Отзывы	Ы	Тибя	Тебя	Е
Офицальный	Официальный	И	тилефон	телефон	Е
Пачему	Почему	О	Умераю	Умираю	И

The analysis of these errors allows you to build words that are missing or added as erroneous (Figure 1). For example, consider the following SMS message built on the basis of the following words:

Let's make the following SMS sentence (Figure 2), based on these words:

РУСКИЙ СКОЧАТЬ СЕСТЕМА СТЕЛАТЬ

Figure 1. Keywords for SMS message.

Я получил РУСКую версию которую я СКОЧАл из операционной СЕСТЕМы попробуй СТЕЛАТЬ это

Figure 2. The original text of the SMS message.

Considering each word, for ease, incorrect words are written in capital letters. We find erroneous words by comparing with Table 2, and write out the letters that are not written correctly and as a result we get the word: САИД.

Consider the text encoding and composition algorithm that was demonstrated above:

1. There is a discovery of the text "САИД" and, based on the letters of this word, we collect key words from the dictionary given in the table.

2. Based on these words, we form a sentence where the errors are stored only in the key-words that will be used in the current decoding session.
3. Decoding begins with looking at all the words and comparing them with the words in the dictionary.
4. For each letter that is different in the erroneous word, we save it in a separate variable.
5. We collect the received letters into a word.
6. Displaying the received word on the screen.
7. We finish the work.

First you will need to have a text in which you will need to enter a secret word, or from which you will need to read a secret word. Let's call this file MyText.txt (Figure 3). For example, it consists of the following text:

Я получил рускую версию.
Которую я скочал из операционной сестемы.
Попробуй стелать это.

Figure 3. Source text, contents of the MyText.txt file.

In addition, it will be necessary to compile a database of words, similar to the one that was compiled above, based on the words of this particular variant, the text will be determined, for example (Table 3):

Table 3. Revealing letters from words.

Слова с ошибками	Слова без ошибок	Буква
рускую	русскую	с
скочал	скачал	а
сестемы	системы	и
стелать	сделать	д
раждражатся	раздражатся	з
савпало	совпало	о
поликиника	поликлиника	л
софпало	совпало	в
расписанние	расписание	н
будуйщий	будущий	й

We will write this data to a file called DB_Words.txt. Read the value from this file into the Dictionary[] array. The secret word to be hidden in the background file will be read from the input.txt file into the secret variable. If this word is not in the input.txt file, i.e. if the file is empty, then you will need to find out which word is hidden in the background text file. That is, we are going to solve the inverse problem.

The above algorithm and program are implemented in the PascalABC.Net programming language. The authors did not plan to consider other modern programming languages, since the PascalABC.Net tools are quite enough.

4 CONCLUSION

As you know, providing a secure communication channel is one of the most important tasks of cryptography. This problem is usually solved either with the help of expensive technical means or with the help of complex cryptographic algorithms. However, the use of such solutions is rather suspicious for the attacking side, and having knowledge of the data transmission channel, it can somehow compromise the channel.

Based on this, the development of an algorithm that will allow you to hide the very fact of the existence of message passing is an urgent task. And this problem is solved within the framework of steganography methods, which makes it possible to provide a secure communication channel.

Considering text steganography algorithms, it is necessary to develop an algorithm with high throughput. At the same time, it is necessary to use rather non-trivial methods so that the algorithm has a sufficient indicator of stealth. In the process of performing this work, a secure communication channel in a social network was implemented.

Thus, a method based on errors in SMS messages, which is best suited for the Russian-speaking audience of the local population. It is shown here that many methods based on the features of the communication language for hiding information can be successfully applied in other languages of communication. The simple algorithm for embedding a set of letters proposed in this article can be very easily implemented in other languages.

REFERENCES

[1] Hassan Shirali-Shahreza M., Mohammad Shirali-Shahreza. (2006) A New Approach to Persian/Arabic Text Steganography. In: *5th IEEE/ACIS International Conference on Computer and Information Science and 1st IEEE/ACIS International Workshop on Component-based Software Engineering, Software Architecture and Reuse*, pp 310–315.

[2] Bala Krishnan R., Prasanth Kumar Thandra, M. Sai Baba (2017). An Overview of Text Steganography. *4th International Conference on Signal Processing, Communications and Networking (ICSCN -2017)*, March 16–18, 2017, Chennai, INDIA.

[3] Hassan Shirali-Shahreza M., Mohammad Shirali-Shahreza (2008). A New Synonym Text Steganography. In: *International Conference on Intelligent Information Hiding and Multimedia Signal Processing*, pp. 1524–1526.

[4] Bender W, Gruhl D, Morimoto N, Lu A (1996). Techniques for Data Hiding. IBM Syst J 3(3&4): pp.313–336.

[5] Por LY, Ang TF, Delina B (2008). WhiteSteg-a New Scheme in Information Hiding Using Text Steganography. *WSEAS Trans Comput* 7(6): pp.735–745.

[6] Prasad R.S.R., Alla K (2011). A New Approach to Telugu Text Steganography. *ISWTA 2011 – 2011 IEEE Symposium on Wireless Technology and Applications*, pp. 60–65.

[7] Zaynalov N. R., Kh.Narzullaev U., Muhamadiev A. N., Mavlonov O. N., Kiyamov J., Qilichev D. Hiding Short Message Text in the Uzbek Language. 2020 *International Conference on Information Science and Communications Technologies (ICISCT)*. https://ieeexplore.ieee.org/document/9351521.

[8] Tsung-Yuan Liu, Wen-Hsiang Tsai (2007). A New Steganographic Method for Data Hiding in Microsoft Word Documents by a Change Tracking Technique. *IEEE Trans Inf Forensics Secur* 2(1): pp.24–30.

[9] Liu, M., Y. Guo and L. Zhou. Text Steganography Based on Online Chat. *Proceedings of the 5th International Conference on Intelligent Information Hiding and Multimedia Signal Processing*, Sept. 1214, IEEE Xplore Press, Kyoto, pp: 807–810. DOI: 10.1109/IIH-MSP.2009.

[10] Moraldo, H., 2012. An Approach for Text Steganography Based on Markov Chains. *Proceedings of the 4th Workshop de Seguridad Informatica, (WSI' 12)*, pp: 26–39.

[11] Por LY, Delina B (2008) Information Hiding—a New Approach in Text Steganography. In: *7th WSEAS International Conference on Applied Computer and Applied Computational Science*. Hangzhou China, pp 689–695.

[12] Shahreza MS, Shahreza MH. An Improved Version of Persian/ Arabic Text Steganography Using “La” Word. In: *Proceedings of IEEE 6th National Conference on Telecommunication Technologies*; 2008. pp. 372–6.

[13] Zaynalov N. R., Mavlonov O. N., Muhamadiev A. N., Kilichev D., Rakhmatullayev I. R. UNICODE for Hiding Information in a Text Document. *2020 IEEE 14th International Conference on Application of Information and Communication Technologies (AICT)* | 978-1-7281-7386-3/20 IEEE | DOI: 10.1109/AICT50176.2020.9368819

Artificial Intelligence, Blockchain, Computing and Security – Dagur et al. (Eds)
© 2024 The Author(s), ISBN: 978-1-032-67841-2

Mathematical modeling of key generators for bank lending platforms based on blockchain technology

G. Juraev, T.R. Abdullaev, Kuvonchbek Rakhimberdiev & A.X. Bozorov
National University of Uzbekistan, Tashkent, Uzbekistan

ABSTRACT: This article discusses the processes of bank lending. For this purpose, the process of storing transactions based on blockchain technology has been modeled. The proposed model includes all components of blockchain technology and cryptographic mechanisms. In particular, the bank lending processes use the ECDSA electronic digital signature algorithm, new models for generating random numbers for the ECDSA algorithm. Theoretical and empirical results obtained as a result of the conducted scientific research are presented.

1 INTRODUCTION

It is known that the issues of improving information systems, the widespread use of digital technologies, the development of methods and algorithms for protecting information, and their improvement are relevant all over the world. In particular, in the Republic of Uzbekistan, the consistent penetration of information technologies into many areas serves the growth of the country's economy [2].

The growth of economic entities in a market economy requires the improvement of financial relations between them. Therefore, many reforms are being carried out to develop the banking sector, which is one of the important parts of the country's economy. To this end, the government of Uzbekistan adopted several resolutions and resolutions. In particular, much attention is paid to ensuring the implementation of the Decree of the President of the Republic of Uzbekistan No. 5992 dated May 12, 2020 "On the Strategy for Reforming the Banking System of the Republic of Uzbekistan for 2020–2025". [3]. In this development strategy, based on modern service solutions to the banking system, the issue of wide introduction of information technologies, financial technologies, adequate provision of information security, as well as the implementation of rapid measures to reduce the impact of the human.

In solving this problem, the importance of blockchain technology is high. Blockchain is an effective technology for security and confidentiality, which uses cryptographic methods to store data in a blockchain. With blockchain technology, we can use several cryptographic mechanisms to secure bank lending transactions. In particular, electronic digital signature algorithms such as ECDSA, DSA, ElGamal, hash function, Merkle tree, PoW, POS, PBFT, and e.c.t. consensus algorithms and P2P decentralized networks are used for signing transactions. In this research paper, we investigate effective random and pseudo-random number generator (RNG) models for the ECDSA digital signature algorithm used in signing bank transactions.

2 LITERATURE REVIEW

When studying blockchain technology, you should, first of all, get acquainted with the work of a person (or a group of persons) under the pseudonym Satoshi Nakamoto [2]. When

DOI: 10.1201/9781032684994-120

storing transactions in blockchain blocks, they are initially signed based on digital signature algorithms. ECDSA electronic signature algorithm is effective in this process. based on the problem of key generation in this digital signature algorithm. These keys are called random or pseudo-random keys. Modeling and building efficient random and pseudorandom key generators is a mathematical problem of high complexity. Methods and solutions to this problem are being researched by scientists as follows. There are many scientific schools, as well as the works of Uzbekistan and foreign scientists in the field of designing and researching effective GPSS/PRNGs. Bobnev N.P., Galeev I.K., Grishkin A.S., Gusev V.F., Dapin O.I., Dobris G.V., Zakharov V.M., Ivanov M.A., Ishmukhametov Sh.T., Latypov R. Kh., Kiryanov B.F., Knut D., Krivenkov S.V., Kuznetsov V.M., L'Eculier P. Mansurov R. M., Pesoshin V.A., Stolov E.L., Tarasov V.M., Tausworth R., Shevchenko D.N., etc. At the same time, the main instructions were given in the scientific works of Uzbekistan scientists X. A. Muzafarov, G.U. Juraev, and A.V.Kabulov. N.Kasimov and other scientific researchers on the practical application of blockchain technology.

3 METHODOLOGY

At present, many researchers are doing a lot of scientific research and experimentation on the use of blockchain technology and its development. Most scientific studies use the following research methods [3]. Method of coming to a mental conclusion, Empirical method, Method of theory In this scientific work, scientific research was carried out using empirical and theoretical methods of scientific research methodology.

4 MODELING THE BANK LENDING PROCESS BASED ON BLOCKCHAIN TECHNOLOGY

Blockchain technology is one of the main factors of innovative development in the banking and financial sectors. A blockchain is a cryptographically linked list of data blocks. Each block stores the previous block's cryptographic hash value, timestamp, and transaction information.

Blockchain technology was first proposed by Satoshi Nakamoto in 2008. This theory was applied in 2009 to develop Bitcoin cryptocurrency. This technology controls invoicing and transaction processing in modern financial industries and organizations, banks and other organizations [1].

Thus, blockchain technology is based on the concepts of a decentralized network architecture and uses a distributed ledger of data, controlled by the established rules of the selected consensus algorithm. Based on this, the blockchain has a number of the following properties [4]:

- Decentralization. Decentralized network participants are equal to each other. In this case, due to the presence of a consensus algorithm, there is no need for nodes controlling the network, which means that the decentralized information network leads to complete dependence of operations;
- Immutability. The blockchain is supposed to be an immutable ledger of data due to its architecture. Each action of a participant (for example, a transaction) is recorded in the registry forever and cannot be changed;
- Anonymity. Each participant is assigned an address, which is used in the identity verification process. It is worth paying attention to the fact that the blockchain cannot guarantee perfect privacy due to certain internal limitations;
- Checkability. The consensus algorithm (hereinafter in the "Consensus" subsection) also allows an independent audit of the entire blockchain at a certain frequency and/or depending on certain conditions;

The model of the lending process in banking organizations based on blockchain technology is presented as follows in Figure 1. The blockchain infrastructure with incoming building blocks is shown in Figure 2.

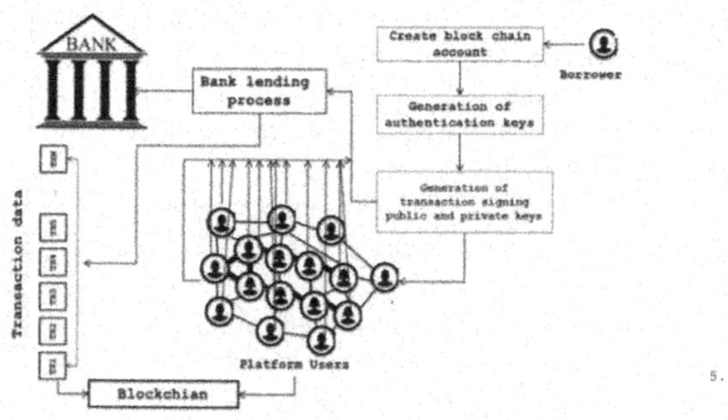

Figure 1. Bank lending process model based on blockchain technology.

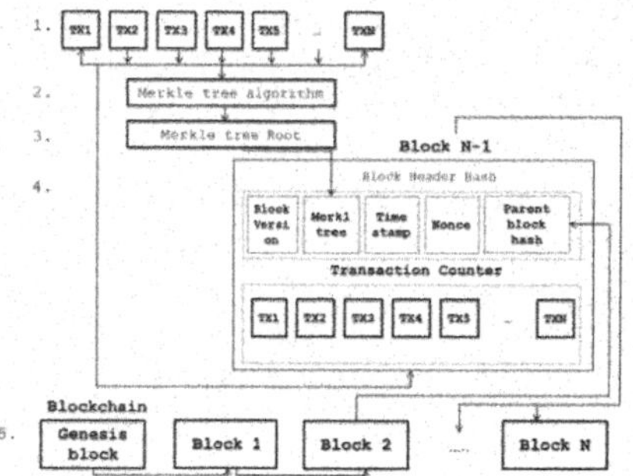

Figure 2. The structure of the bank lending blockchain model.

A block is the main structural element of a blockchain, serving as a container for transactions (see below) or other data types (depending on the implementation). Each block is linked to the previous block through the hash sum of that block's data, obtained using one of the hash functions: SHA1, SHA256, or Quark [5]. The diagram of the internal structure of the block is shown in Figure 3 below with a description of the elements from the BlockHeader in Table 1.

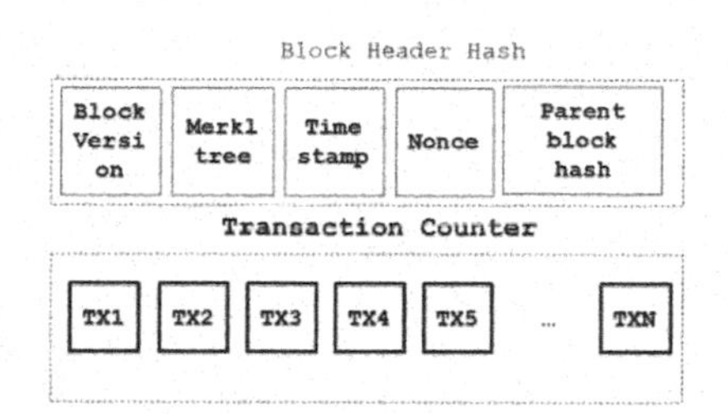

Figure 3. Diagram of the internal structure of the block.

Table 1. Meaning of block elements in BlockHeader.

Name	Description
Block Version	The current version of the block field structure
Merkle Tree Root Hash	The hash sum of the state of the block through the use of a hash tree of the transactions included in the block
Timestamp	Block creation time (Unix format)
n Bits	Block state length in bits
Nonce	The length of the generated hash sum after running the Proof-Of-Work consensus algorithm (see later in "Consensus")
Previous Block Hash	Hash sum of the state of the previous block in the chain

To ensure the security of bank lending operations based on blockchain technology, the following cryptographic mechanisms are used [4,5]:

- *SHA-256 hash function algorithm.* The function of hashing the state of the current block is to create a chain of blocks that refer to each other (immutability). The hash function, in this case, is used to create a string of 256 bits (when using SHA256) to further validate the contents of the block [6]. The SHA-256 hash algorithm is defined as follows.
- *MerkleTree (Hash tree).* This algorithm is used to create a hash tree from a given number of inputs (such as transactions). In the blockchain, an algorithm is often used to validate the content of a block, as a hash sum of the current block.
- *Time stamp.Suppose a new block appears somewhere in the network, and nodes start transmitting it to each other. Each node must verify the validity of the block. To do this, it does the following: Must be greater than the arithmetic mean of the timestamps. This is because, for example, Block No. 123 was issued on March 12, 2011, and Block No. 124 was issued on February 13, 1984. But at the same time, a small margin of error is allowed.*
- *Digital signature.* Transactions to send and receive information to verify identity use an electronic signature. In the context of blockchain technologies, an electronic signature is formed based on public and private keys (ECDSA or RSA) and is an encrypted string of arbitrary length [7]. The principle of operation is illustrated in Figure 5 below:

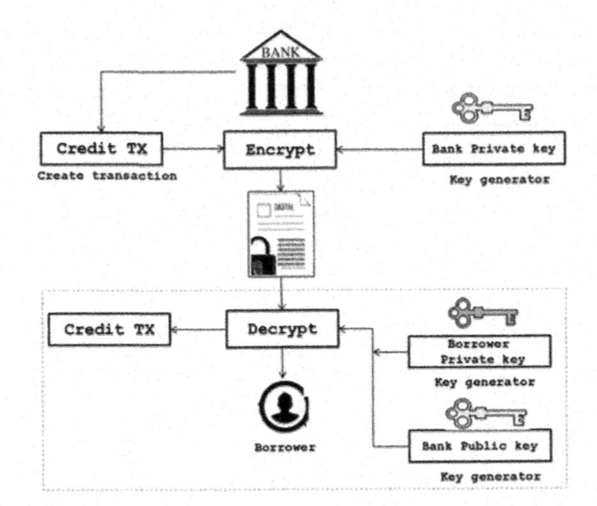

Figure 4. The principle of operation of the electronic signature in the blockchain.

The first user (Bank) sends a pre-encrypted transaction to another user (borrower). Before sending, an electronic signature is generated based on the private key of the Bank or other users.

5 APPLICATION OF RANDOM KEY GENERATORS IN ECDSA ELECTRONIC DIGITAL SIGNATURE ALGORITHM

Let there be a device (or program) that signs arbitrary input data using a given key according to the ECDSA algorithm [8]. All calculations are performed in the $GF(2^q)$ field during signature generation based on the electronic digital signature algorithm based on elliptic curves. In this case, the parameters (a,b) of the elliptic curve indicate the characteristic of the line. Public and private cryptographic keys are used for signature generation. Key generation uses the base point G of the elliptic curve.

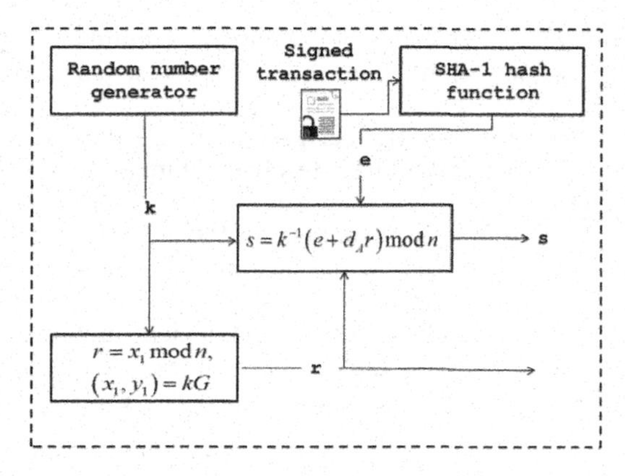

Figure 5. ECDSA signature generation scheme.

Signer A must have a private key d_A (a random number in the range [1,$n-1$], as well as a public key $Q_A = d_A G$. The algorithm for generating a signature according to the ECDSA standard is as follows [7,9]:

1. Compute the hash image of the message TX_n $e = H(TX_n)$: where $H(X)$ is the SHA-1 cryptographic hash function?
2. Choose a random number k from the range [1,$n-1$].
3. We calculate $r = x_1$ (mod$n$), where $(x_1,y_1) = kG$. If $r = 0$, go back to step 2.
4. Calculate $s = k^{-1}$ $(e + d_A r)$mod$n$. If $s = 0$, we return to step 2.
5. The signature is the pair (r, s).

The algorithm for verifying a signature using a Q_A public key is:

1. Check that r and s are integers from the range [1,$n-1$]. If it is not, then the signature is rejected.
2. Compute $e = H(TX_n)$, where $H(X)$ is the SHA-1 cryptographic hash function.
3. We calculate $w = s^{-1}(mod\ n)$.
4. Compute $u_1 = ew$(mod$n$) $u_2 = rw$(mod$n$),
5. Compute $(x_1,y_1) = u_1G + u_2Q_A$.

The signature is accepted if and only if $x_1 = r \bmod n$.

The k-key generator is important in creating public and private keys and signing transactions based on the above ECDSA digital signature algorithm. the more efficient the key generator, the more efficient the process of saving transactions to the blockchain. That's why we offer a random key generator in the section below.

5.1 *New key generation algorithm for blockchain systems*

To obtain keys for further use in encryption algorithms, you can use a pseudorandom number generator (PRNG). The main requirements for pseudo-random number generators [10], are A sufficiently long period that guarantees that the sequence is not looped within the scope of the problem being solved. The length of the period must be mathematically proven.

a) Efficiency — fast operation of the algorithm and low memory consumption; b) Reproducibility — the ability to replay a previously generated sequence of numbers any number of times; c) Portability — the same operation on different hardware and operating systems.

The speed of obtaining $X_n + i$ elements of a sequence of numbers, when X_n elements are specified, for I of any value, allows you to divide the sequence into several streams (sequences of numbers).

The main types of pseudo-random number generators are the linear congruent method; мethe delayed Fibonacci method; the linear-feedback shift register (LFSR); shift register with generalized feedback.

"The Mersenne Whirlwind». The disadvantages of PRNG can be: the period is too short; consecutive values are not independent; some bits are "less random" than others; uneven one-dimensional distribution; reversibility.

One of the simplest methods of PRNG easy to implement in hardware is the method of linear-feedback shift register [9,10]. The method of the linear-feedback shift register is based on obtaining sequence bits from interconnected registers (Figure 1).

Figure 6. Linear-feedback shift register.

The disadvantage of PRNG based on the method of the linear-feedback shift register is the possibility of opening them using the algebraic method and finding a correlation between the next number and the previous number. Of particular interest is one of the fastest PRNG generators the "Middle Square Weyl Sequence RNG" [11]. The PRNG Middle Square Weyl Sequence RNG (MSWS) principle is based on the allocation of the "middle square" in the square and the sum of the series of coefficients.

$$\begin{aligned} w_{i+1} &= w_i + s, \\ x_{i+1} &= x_i^2 + w_{i+1}, \\ y_{i+1} &= \left(x_{i+1} \bmod 2^{64}\right) \gg 32 \end{aligned} \tag{1}$$

where i – is the iteration number.

s – some odd constant.

The initial values of x and w are usually taken equal to s.

For example, if $s = 5263744972643746 31$, $x = s$, $c = 5000$, $w = (c*s) \bmod 216$.

$$\begin{aligned} w_{i+1} &= w_i + s, \\ x_{i+1} &= x_i^2 + w_{i+1}, \\ y_{i+1} &= \left(x_{i+1} \bmod 2^{16}\right) \gg 8 \end{aligned} \tag{2}$$

Despite the assurances of the developers in the cryptographic strength of this type of PRNG, it is possible to perform inverse transformations and select values for generating the current sequence. The task of a cryptanalyst when cracking a PRNG is to determine the internal state of the values used by the parameter. With a known output of the last values and the structure of the PRNG, it is possible to carry out inverse transformations. To protect the PRNG from reverse transformations, it is proposed to cover the output sequence with gamma with an iterative key. To protect the weaknesses of the PRNG, it is proposed to use a hybrid version of the PRNG with exit cover according to the following scheme. A two-part key is entered, c1 is the initial value of the feedback shift register, and c2 is the initial value of the PRNG mean square Weyl sequence. A shift register with feedback is proposed with a triangular design of the following form (Figure 3).

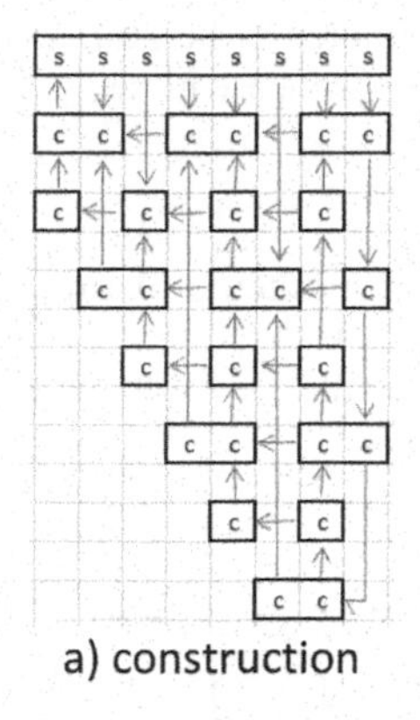

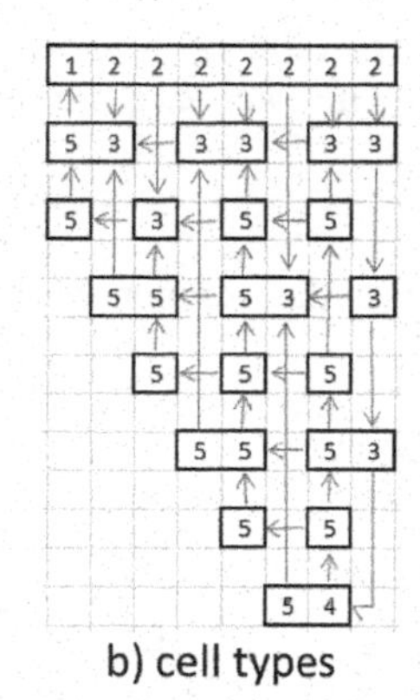

Figure 7. 8-bit feedback shift register.

In this scheme (Figure 3, a) cells with the index "s", are the main cells of the sequence, from which information about the generated number is taken. Cells with index "c" contain information about the internal state of the system. In this design, the cells have four types (Figure 3, b), in terms of obtaining information. Type 1 – receives information from the bottom, type 2 – receives information from the left, type 3 – receives information as the sum of the top, right, and bottom (if it is a cell of the fifth type), type 4 – receives information from the nearest cell of the third type, type 5 – receives information as the sum on the right and below.

Information about the generated bits can be taken from the rightmost cell "s" or every eight cycles from all cells "s" at once. Also, this design is easy to scale to the required size, for example, 16 bits (Figure 5). For the effective operation of this design, it is necessary, after setting the initial configuration, to allow the design to work out for a full four cycles. The length of the sequence of random numbers will depend not only on the number of digits but also on the state of the internal registers "c", which will be,

$$M \approx 2^{0,35*n^2+n}, \tag{3}$$

where n is the number of digits. A special feature of the software implementation of the linear-feedback shift register may be that the cells can contain not only bit values but also other values of no more than "n/2", for example, from 0 to 3 (Figure 6).

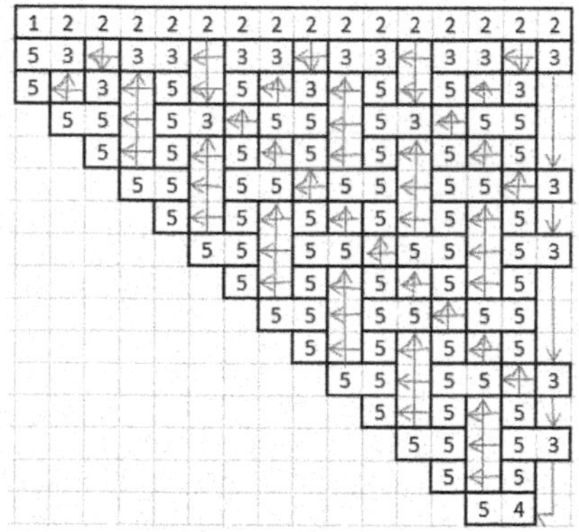
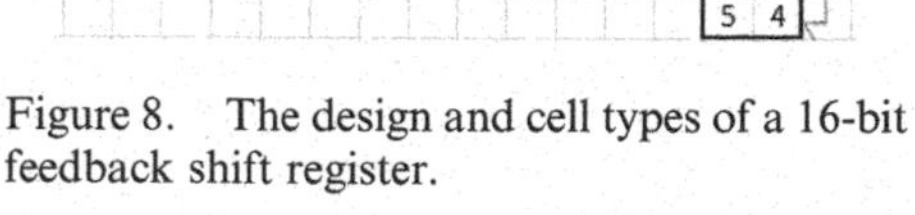

Figure 8. The design and cell types of a 16-bit feedback shift register.

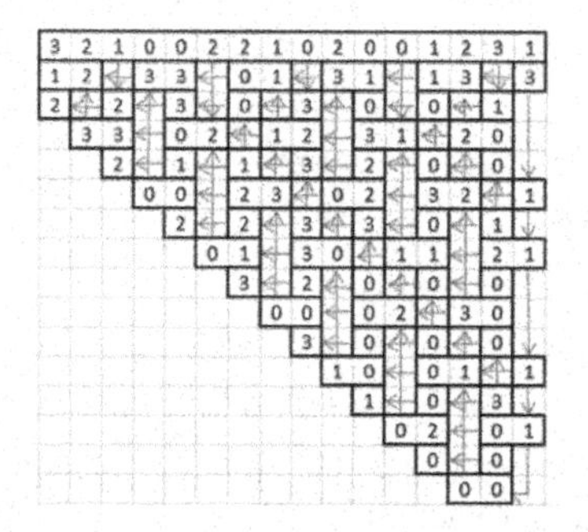

Figure 9. The linear-feedback shift register of numbers from 0 to 4^{16}.

When increasing the value of the cells, you can not increase the design. There will be more generated numbers. The total length of the random sequence will be,

$$M \approx c^{0,35*n^2+n}, \tag{4}$$

where c is the base of the number system. The general scheme of the proposed PRNG is shown in Figure 7.

In this scheme, two types of PRNG are used in parallel, a feedback shift register, and a mean square Weyl flow.

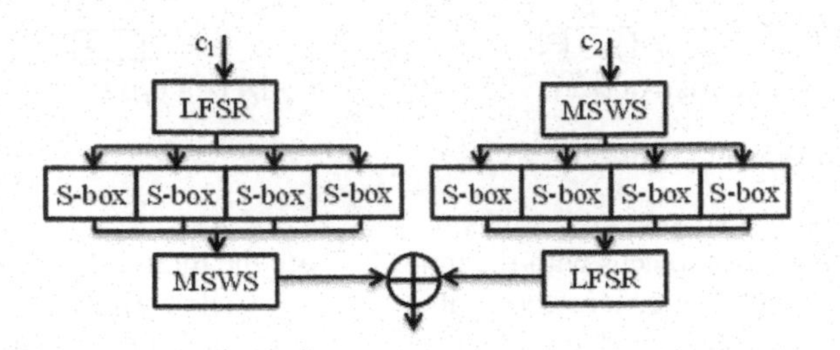

Figure 10. General scheme of PRNG.

6 RESULT

After 5000 iterations and counting the number of identical values, the following distribution was obtained (Figure 8). Despite the assurances of the developers in the cryptographic strength of this type of PRNG, it is possible to carry out inverse transformations and select values for generating the current sequence. Information about the generated bits can be taken from the rightmost cell "s" or every eight cycles from all cells "s" at once. This design showed good results in generating random numbers. When generating 4800 values of random numbers from 0 to 255 and counting their number by values, the following distribution was obtained (Figure 9).

The results of key distribution presented in Figures 11 and 12 are recognized as effective and are proposed for practical use.

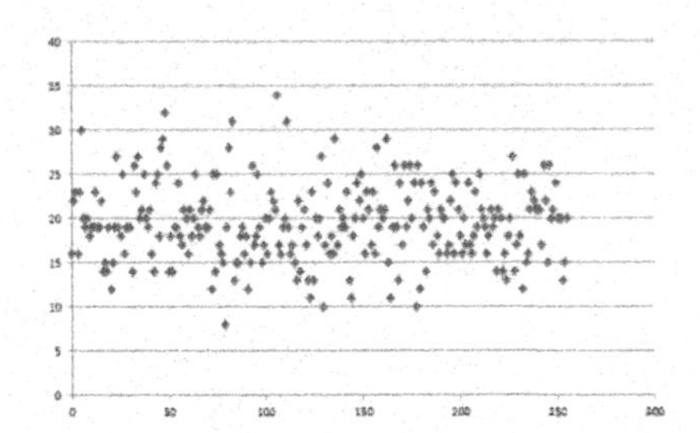

Figure 11. The distribution of the quantity over the same values.

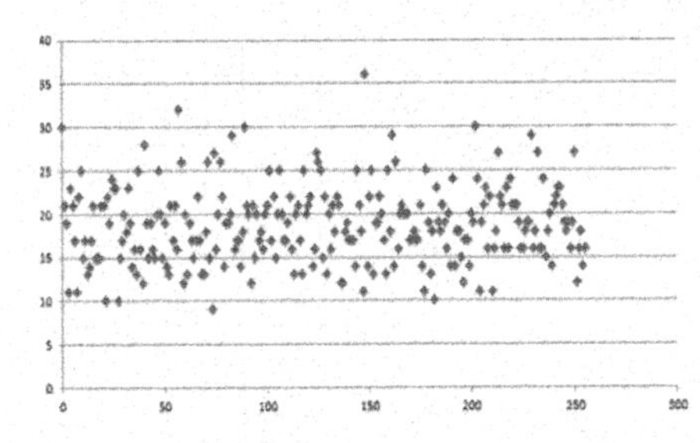

Figure 12. The distribution of the quantity over the same values.

7 CONCLUSION

In conclusion, financial and banking lending systems based on blockchain technology will lead to effective results. Blockchain is a technology that ensures a high level of data confidentiality. As we know from the content of the scientific work presented above, the blockchain includes cryptographic mechanisms with high crypto strength. These cryptographic methods are used to store incoming transactions in blocks. Each transaction is signed using the ECDSA digital signature algorithm.

In this scientific study, we have provided guidelines for applying elliptic curves to digital signature algorithms. Also, when using an electronic signature algorithm based on elliptic curves for voluntary or personal blockchain networks, the following must be done:

1. Using existing curve protocols (for example, Secp256k1);
2. Selection of efficient elliptic curves;
3. Determination of selected curve points;

Elliptic curve points can also be used in key generators.

In addition, a new k-key generator for systems based on blockchain technology was proposed. This generator was tested using statistical analysis methods. As a result of testing the generator, the efficiency of the generator was determined. Based on this blockchain technology, it was proposed to use the bank lending process model, ECDSA electronic digital signature model, Hash function algorithm, and random key generator developed for the system and electronic signature.

REFERENCES

[1] Nakamoto S. 2008. *Bitcoin: A Peer-to-Peer Electronic Cash System//* Bitcoin.org.:1–9.

[2] Bernard W. 2008. *Attack of the Middle Square Weyl Sequence PRNG. Squares: A Fast Counter-Based RNG*: 20–36.

[3] Ikramov, A. & Juraev, G. 2021. *The Complexity of Testing Cryptographic Devices on Input Faults. Lecture Notes in Computer Science (Including Subseries Lecture Notes in Artificial Intelligence and Lecture Notes in Bioinformatics)*:13041 LNCS. 202–209.

[4] Juraev, G. & Abdullaev T. 2021. Application Three Valued Logic in Symmetric Block Encryption Algorithms, 2021. *International Journal of Physics Conference Series*: 2131(2), 022082, pp. 1–9.

[5] Juraev, G. & Abdullaev T. 2021. Development of a Method for Generating Substitution Tables for Binary and Ternary Number Systems. *International Journal of Physics: Conference Series*: 2131(2), 022082, pp. 1–10.

[6] Juraev, G. & Abdullaev T. 2021. Selection of the Optimal Type of the Gaming Function for Symmetric Encryption Algorithms. *AIP Conference Proceedings*: 2365, 020004, pp. 1–7.

[7] Juraev, G. & Rakhimberdiev K. 2021. Modeling the Decision-making Process of Lenders Based on Blockchain Technology. *International Conference on Information Science and Communications Technologies: Applications, Trends, and Opportunities* 1–5. Uzbekistan: Tashkent.

[8] Juraev, G. & Rakhimberdiev K. 2022. Mathematical Modeling of Credit Scoring System Based on the Monge-Kantorovich Problem. *Electronics and Mechatronics Conference: IEMTRONICS Proceedings* 305–311. Canada: Toronto.

[9] Juraev, G. & Rakhimberdiev K. 2022. Prospects of Application of Blockchain Technology in the Banking. *International Conference on Information Science and Communications Technologies: Applications, Trends, and Opportunities* 1–5. Uzbekistan: Tashkent.

[10] L'Ecuyer, P. 2007. *Random Number Generation.Springer Handbooks of Computational Statistics*: 93–137.

[11] Musurmonova, M. & Juraev, G. 2022. An Algorithm for Solving the Problem of Radial Expansion of a Spherical Cavity Supported by a Thin Spherical Shell in an Elastic-Porous Fluid-Saturated Medium. *AIP Conference Proceedings* this link is disabled: 2432, 030109.

[12] Rakhimberdiev. K. Arzieva J. Arziev A. 2022. Application of Random Number Generators in Solving the Problem of User Authentication in Blockchain Systems. *International Conference on Information Science and Communications Technologies: Applications, Trends, and Opportunities*, pp. 1–5.

Artificial Intelligence, Blockchain, Computing and Security – Dagur et al. (Eds)
 ISBN: 978-1-032-67841-2

Applying the CryptoSMT software tool to symmetric block encryption algorithms

Qozoqova To'xtajon Qaxramon qizi & Shamsiyeva Barno Maxmudjanovna
Tashkent University of Information Technologies named after Muhammad al-Khwarizmi, Tashkent, Uzbekistan

ABSTRACT: This article used CryptoSMT software for the cryptanalysis of symmetric encryption algorithms, which proved to be the preferred software for performing linear and differential analysis. CryptoSMT is an easy-to-use tool for the cryptanalysis of symmetric primitives like block ciphers or hash functions. It is based on SMT/SAT solvers like STP, Boolector, and CryptoMiniSat and provides a simple framework to use for cryptanalytic techniques [7]. The results of the analysis of the above encryption methods are obtained.

Keywords: Cryptanalysis, Present, Trifle, Craft, Simon liner and differential characteristic, CryptoSMT, symmetric block cipher

1 INTRODUCTION

Cryptanalysis refers to the process of analyzing information systems to understand hidden aspects of the systems. Cryptanalysis is used to breach cryptographic security systems and gain access to the contents of encrypted messages, even if the cryptographic key is unknown [10]. The following two methods are commonly used for symmetric encryption algorithms. Attacks have been developed for block ciphers and stream ciphers. Linear cryptanalysis is one of the two most widely used attacks on block ciphers; the other being differential cryptanalysis. The discovery is attributed to Mitsuru Matsui, who first applied the technique to the FEAL cipher (Matsui & Yamagishi 1992). Subsequently, Matsui published an attack on the Data Encryption Standard (DES), eventually leading to the first experimental cryptanalysis of the cipher reported in the open community (Matsui 1993, 1994). The attack on DES is not generally practical, requiring 2^{47} known plaintexts. A variety of refinements to the attack have been suggested, including using multiple linear approximations or incorporating non-linear expressions, leading to a generalized partitioning cryptanalysis. Evidence of security against linear cryptanalysis is usually expected of new cipher designs [11].

2 MAIN PART

CryptoSMT is a handy tool for the cryptanalysis of symmetric primitives like block ciphers or hash functions. It is based on software tools for solving SMT/SAT mathematical problems, such as STP, Boolector, and CryptoMiniSat, and provides a simple structure using them for cryptanalysis methodsp[8].

Here a primitive polynomial is a polynomial that cannot be decomposed into multipliers and is used to increase the randomness and repetition period of a selected random number.

 DOI: 10.1201/9781032684994-121

For example:

$$1001101 = x^6 + x^3 + x^2 + 1 \tag{1}$$

The mass w denotes the higher level of this primitive polynomial. For example, w = 6 in $1001101 = x^6 + x^3 + x^2 + 1$ Because the highest degree of the polynomial is w.

Program possibilities:

- Proof of the first-order differential motion property;
- Finding the best linear/differential paths;
- Calculation of the probability of the differential;
- Finding the initial value of a hash function (collision);
- Recovery of the private key.

CryptoSMT currently supports the following primitives:

CryptoSMT parsing algorithms

Block encryption algorithms: Simon, Speck, Skinny, Present, Midori, LBlock, Sparx, Twine, CRAFT, TRIFLE

The program can perform both linear and differential cryptanalysis of some algorithms. It uses a sequence of special commands. As an example, consider a custom command to find the optimal differential properties of the Simon block cipher in CryptoSMT.

$ python3 cryptosmt.py –cipher simon –rounds 8 –word size 16 The above command can be used to cryptanalysis the 16-bit key of the 8-round Simon block cipher algorithm. You can also modify this command to make it suitable for other analyses. For example, you can change the number of rounds, the algorithm, and the key length. Below is the working architecture of the program.

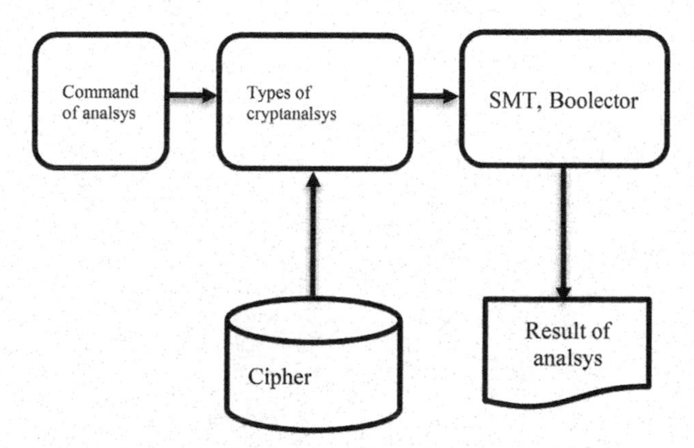

Figure 1. Performance architecture of CryptoSMT software.

CryptoSMT software tool using subsequent crypto algorithms by testing has been observed _ _ This: The present Symmetric Block Encryption Algorithm 2007. Orange Lab was developed by the Ruhr University Bochum and the Danish Technical University. The block length of the algorithm is 64 bits, the key length is 80 or 128 bits and the number of rounds is 32. The main purpose of this cipher is to use in highly specialized devices, such as RFID tags or sensor networks [3]. Percent SPN-based encryption algorithm.

Using the program. In the working window shown below, with the cryptosmt.py file active, type the following working window. Symmetric block encryption algorithm enabled – Simon cipher – rounds 8 – word size 16. To use software that includes linear and differential cryptanalysis, you must add the word liner to the name of the crypto algorithm (Figure 2).

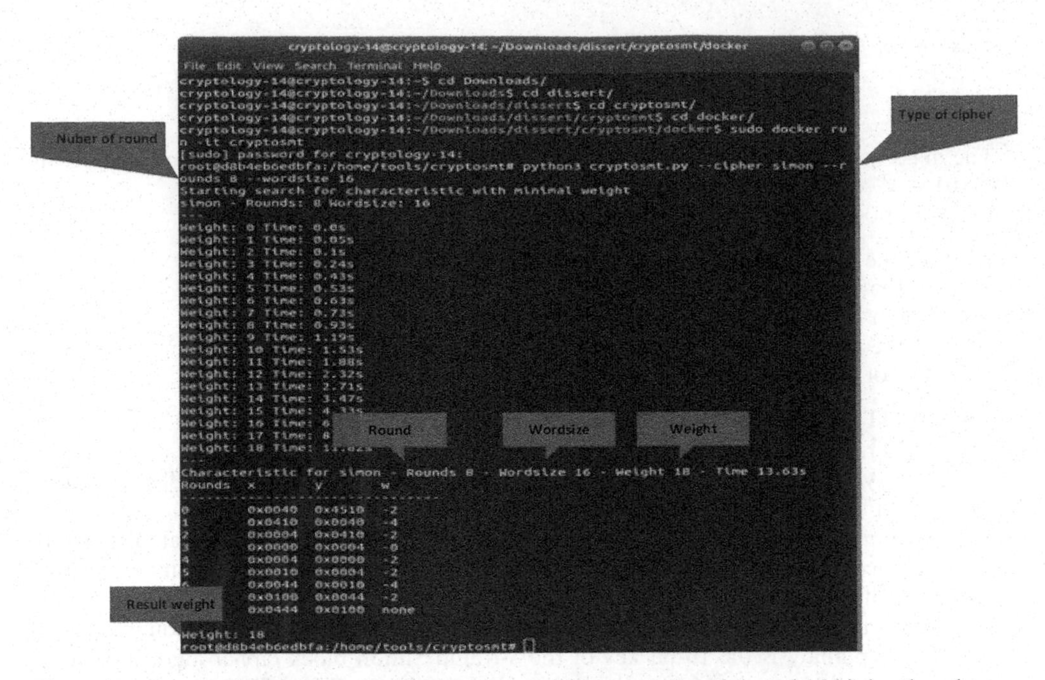

Figure 2. Cryptanalysis of Simon's block cipher algorithm with 8 rounds and 16-bit key length.

In the command window of linear cryptanalysis of Simon's cipher algorithm, as mentioned above, simonliner was introduced (Figure 3).

```
cryptology-14@cryptology-14: ~/Downloads/dissert/cryptosmt/docker
File Edit View Search Terminal Help
root@d8b4eb6edbfa:/home/tools/cryptosmt# python3 cryptosmt.py --cipher simonline
ar --rounds 8 --wordsize 16
Starting search for characteristic with minimal weight
simonlinear - Rounds: 8 Wordsize: 16
---
Weight: 0 Time: 0.0s
Weight: 1 Time: 0.17s
Weight: 2 Time: 0.38s
Weight: 3 Time: 0.63s
Weight: 4 Time: 0.96s
Weight: 5 Time: 1.39s
Weight: 6 Time: 2.12s
Weight: 7 Time: 3.46s
Weight: 8 Time: 5.38s
Weight: 9 Time: 8.16s
---
Characteristic for simonlinear - Rounds 8 - Wordsize 16 - Weight 9 - Time 9.02s
Rounds  x       y       w
-------------------------------
0       0x11A2  0x2200  -2
1       0x2200  0x0800  -1
2       0x0800  0x2000  -1
3       0x2000  0x0000  -0
4       0x0000  0x2000  -1
5       0x2000  0x0800  -1
6       0x0800  0x2200  -2
7       0x2200  0x0080  -1
8       0x0080  0x2220  none

Weight: 9
root@d8b4eb6edbfa:/home/tools/cryptosmt#
```

Figure 3. Linear cryptanalysis of Simon's block cipher algorithm with 8 rounds and 16-bit key length.

Figure 4 shows the result of Simon's block cipher algorithm with 8 rounds of 16-bit key length, which hits the computer during linear analysis.

Figure 5 shows the differential analysis process of a real cipher with 16 rounds and a key length of 64 bits at the key stage of the analysis process.

```
cipher not supported.
root@92239e5cf542:/home/tools/cryptosmt# python3 cryptosmt.py --cipher present --rounds 8 --wordsize 8
Starting search for characteristic with minimal weight
present - Rounds: 8 Wordsize: 8
---
Weight: 0 Time: 0.0s
Only wordsize of 64-bit supported.
root@92239e5cf542:/home/tools/cryptosmt# python3 cryptosmt.py --cipher present --rounds 8 --wordsize 64
Starting search for characteristic with minimal weight
present - Rounds: 8 Wordsize: 64
---
Weight: 0 Time: 0.0s
Weight: 1 Time: 62.59s
Weight: 2 Time: 151.38s
Weight: 3 Time: 267.34s
Weight: 4 Time: 391.46s
Weight: 5 Time: 512.56s
Weight: 6 Time: 656.46s
Weight: 7 Time: 800.93s
Weight: 8 Time: 967.18s
Weight: 9 Time: 1159.74s
Weight: 10 Time: 1311.45s
Weight: 11 Time: 1500.49s
Weight: 12 Time: 1665.09s
Weight: 13 Time: 1834.36s
Weight: 14 Time: 2020.95s
Weight: 15 Time: 2180.93s
Weight: 16 Time: 2349.61s
Weight: 17 Time: 2512.6s
Weight: 18 Time: 2671.61s
Weight: 19 Time: 2824.43s
Weight: 20 Time: 2971.84s
Weight: 21 Time: 3139.5s
Weight: 22 Time: 3274.65s
Weight: 23 Time: 3435.46s
Weight: 24 Time: 3581.38s
Weight: 25 Time: 3726.39s
Weight: 26 Time: 3863.34s
Weight: 27 Time: 4021.17s
Weight: 28 Time: 4186.68s
Weight: 29 Time: 4335.47s
Weight: 30 Time: 4488.3s
Weight: 31 Time: 4648.05s
Weight: 32 Time: 4805.63s
---
Characteristic for present - Rounds 8 - Wordsize 64 - Weight 32 - Time 4976.78s
Rounds  S                   P                   w
------------------------------------------------------------
0       0x0007000000000007  0x0001000000000001  -4
1       0x0000000000001001  0x0000000000009009  -4
2       0x0009000000000009  0x0004000000000004  -4
3       0x0000100100000000  0x0000900900000000  -4
4       0x0900000000000900  0x0400000000000400  -4
5       0x0000400400000000  0x0000500500000000  -4
6       0x0000090000000900  0x0000040000000400  -4
7       0x0000040400000000  0x0000050500000000  -4
8       0x0000050000000500  none                none

Weight: 32
root@92239e5cf542:/home/tools/cryptosmt#
```

Figure 4. Cryptanalysis of real block cipher algorithm with 8 rounds and 64-bit key length.

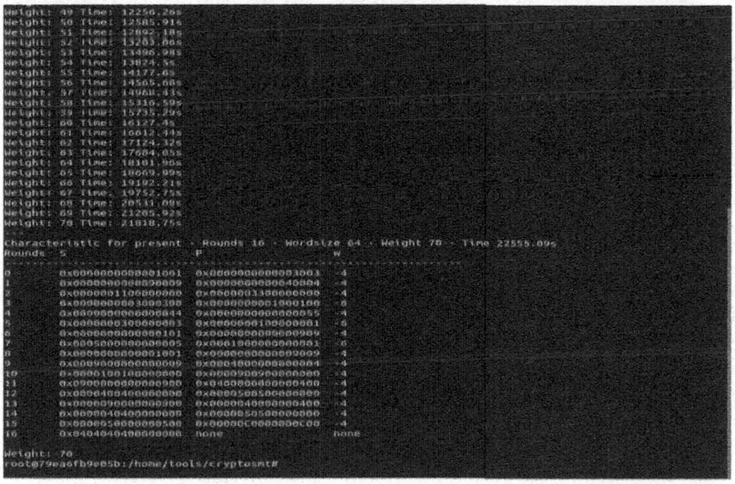

```
Weight: 49 Time: 12256.26s
Weight: 50 Time: 12585.91s
Weight: 51 Time: 12892.18s
Weight: 52 Time: 13203.06s
Weight: 53 Time: 13496.98s
Weight: 54 Time: 13824.5s
Weight: 55 Time: 14177.6s
Weight: 56 Time: 14565.08s
Weight: 57 Time: 14960.41s
Weight: 58 Time: 15316.59s
Weight: 59 Time: 15735.29s
Weight: 60 Time: 16127.4s
Weight: 61 Time: 16612.44s
Weight: 62 Time: 17124.32s
Weight: 63 Time: 17604.05s
Weight: 64 Time: 18101.96s
Weight: 65 Time: 18669.99s
Weight: 66 Time: 19192.21s
Weight: 67 Time: 19752.75s
Weight: 68 Time: 20531.09s
Weight: 69 Time: 21205.92s
Weight: 70 Time: 21818.75s
---
Characteristic for present - Rounds 16 - Wordsize 64 - Weight 70 - Time 22555.09s
Rounds  S                   P                   w
------------------------------------------------------------
0       0x0000000000001001  0x0000000000003003  -4
1       0x0000000000090009  0x0000000000040004  -4
2       0x0000001100000000  0x0000003300000000  -4
3       0x0000000003000300  0x0000000001000100  -6
4       0x0000000000000044  0x0000000000000055  -4
5       0x0000000300000003  0x0000000100000001  -6
6       0x0000000000000101  0x0000000000000909  -4
7       0x0005000000000005  0x0001000000000001  -6
8       0x0000000000001001  0x0000000000009009  -4
9       0x0009000000000009  0x0004000000000004  -4
10      0x0000100100000000  0x0000900900000000  -4
11      0x0900000000000900  0x0400000000000400  -4
12      0x0000400400000000  0x0000500500000000  -4
13      0x0000090000000900  0x0000040000000400  -4
14      0x0000040400000000  0x0000050500000000  -4
15      0x0000050000000500  0x00000C0000000C00  -4
16      0x0404040400000000  none                none

Weight: 70
root@79ea6fb9e05b:/home/tools/cryptosmt#
```

Figure 5. Cryptanalysis of a real block cipher with 16 rounds and a key length of 64 bits.

Take this set off testing in the process of cryptanalysis to the results according to symmetric block encryption algorithms endurance in raising rounds of number and key length importance high matter is that approved. Cryptanalysis process done in raising computer strength high to be should that trust crop is done. The cryptanalysis process has proven that the right choice of method and analysis tools increases the effectiveness of the analysis. According to the results of testing the Present Block encryption algorithm proved to be reliable. The analysis time was 22555.09 seconds [7].

Table 1. Table of results.

No	Encryption algorithms	Number of rounds	Key length (bits)	Number of weights found	Maximum weight	Minimum weight	Time spent (seconds)
1	Simon	8	16	18	4	2	13,63
	Simon	16	16	42	6	2	1136,62
2	Simon liner	8	16	9	2	0	9.02
	Simon liner	16	16	21	3	1	406,74
3	Trifle	8	128	22	3	2	252.01
	Trifle	16	128	46	3	2	3133.0
4	Present	8	64	32	4	4	4976,78
	Present	16	64	70	6	4	22555.09
5	Craft	8	64	52	12	2	365,96
	Craft	16	64	122	10	4	10540.21
6	Craft liner	8	64	52	12	2	267,03
	Craft liner	16	64	120	8	6	6016.19

3 CONCLUSION

The importance of speed and stability of symmetric encryption algorithms in ensuring data confidentiality was determined and the requirements for stability of symmetric encryption algorithms were studied in this article. Stability to attacks of standard crypto-algorithms, corresponding to international standards, was presented and analyzed by examples. In addition, all software standards and cryptanalytical platforms have been implemented. During the cryptanalysis testing, it was confirmed that the number of rounds and key length are of great importance in enhancing the robustness of symmetric block encryption algorithms. According to test results, the Present Block encryption algorithm was reliable. The analysis time was 22555.09 seconds. It was verified that the computer power should be high when performing the cryptanalysis process, and the right choice of method and analysis tools will increase the efficiency of the analysis.

REFERENCES

[1] Aakanksha Sharma's *"Comparative Study of Symmetric Cryptography Algorithm"* dissertation work. Udaipur-2014

[2] Linda Rosencrance *"Cryptoanalsys"* search security -2021y

[3] J. Daemen, L. Knudsen, V. Rijmen (January 1997). The Block Cipher Square (PDF). *4th International Workshop on Fast Software Encryption (FSE '97)*, Volume 1267 of Lecture Notes in Computer Science.

[4] Farzaneh Abed, Eik List, Stefan Lucks, and Jakob Wenzel Bauhaus *Cryptanalysis of the Speck Family of Block Ciphers Revision from October 9*, 2013 Universität Weimar, Germany

[5] Robert Brummayer and Armin Biere «*Boolector: An Efficient SMT Solver for Bit-Vectors and Arrays*» Institute for Formal Models and Verification Johannes Kepler University Linz

[6] Chistofer Smitson "*Modern Cryptoanalsys*" Unireversity of Tulsa-2020

[7] To'xtajon Qozoqova "*Cryptanalysis Methods of Symmetric Block Ciphers*" master's thesis. Toshkent-2022

[8] https://github.com/kste/cryptosmt

[9] https://my-kross.ru/uz/lekarstva/sravnitelnyi-analiz-ponyatie-vidy-iprimery

[10] https://www.geeksforgeeks.org/difference-between-aes-and-desciphers

[11] https://en.wikipedia.org/wiki/Cryptanalysis

[12] https://en.wikipedia.org/wiki/Linear_cryptanalysis

Artificial Intelligence, Blockchain, Computing and Security – Dagur et al. (Eds)
© 2024 The Author(s), ISBN: 978-1-032-67841-2

Using AdaBoost to improve the performance of simple classifiers

A.A. Mahkamov, T.S. Jumayev, D.S. Tuhtanazarov & A.I. Dadamuxamedov
International Islamic Academy of Uzbekistan, Tashkent, Uzbekistan

ABSTRACT: The article presents the results of increasing recognition efficiency based on the solution of a model problem using an algorithm such as AdaBoost, which forms a "strong" classifier based on a combination of "weak" classifiers. Algorithm software has been developed with the help of which the data representing two classes of objects is classified. Recognition errors are graphically presented, and the advantages and disadvantages of this algorithm are described.

1 INTRODUCTION

The problem of recognizing a person based on the image of the ear can be solved by bringing it to the problem of character recognition. The problem of identifying symbols is formulated as follows. Given some set $\{S\}$ of recognizable objects. Each $S_i \in \{S\}$ object is expressed in the following large character space:

$$S_i = (x_{i1}, x_{i2}, ..., x_{im}), \ (i = \overline{1, n}) \tag{1}$$

A set of objects $\{S\}$ is partitioned into disjoint classes $K_1, K_2, \ldots, K_l$. In this case, the division of the set of objects $\{S\}$ into classes is not fully defined, rather, there is only some initial $_{\text{information}}$ $K_1, K_2, \ldots, K_l$ about classes I_0:

$$\begin{aligned} I_0 = \{S_1, ..., S_i, ..., S_m; \ \overline{\alpha}(S_1), ..., \overline{\alpha}(S_i), ..., \overline{\alpha}(S_m)\}, \\ \overline{\alpha}(S_i) = (\alpha_{i1}, ..., \alpha_{i2}, ..., \alpha_{il}), \end{aligned} \tag{2}$$

where $\bar{\alpha}(S_i) - S_i$ is the data vector of the object, $\alpha_{ij} - P_j(S_j)$ – "$S_j \in K_j$" – the value of the predicate.

The main problem (problem Z) consists of calculating the values of the elementary predicates I_0 according to the initial information S and the $I(S)$ description of the recognized object $P_j(S)$ – "$S \in K_j$".

In other words, it is required to build such an algorithm A, i.e

$$A(I_0, I(S)) = (\alpha_1^A(S), \ \alpha_2^A(S), ..., \alpha_l^A(S)) \tag{3}$$

let be $\alpha_l^A(S) \in \{0, 1, \Delta\}$, $l = \overline{1, j}$ here.

In the class of incorrect algorithms, additional conditions are usually imposed in relation to algorithm A when solving problem Z.

Suppose such a set of recognition algorithms is given and for

$$\begin{aligned} &A(I_0, (K_1, K_2, ..., K_l), \ I(S)) = (\beta_1^A(S), \beta_2^A(S), ..., \beta_l^A(S)), \\ &\beta_j^A(S) \in \{0, 1, \Delta\} \end{aligned} \tag{4}$$

DOI: 10.1201/9781032684994-122

let equality be fulfilled. In the set $\{A\}$ of such algorithms, the quality function of the algorithm A is given. Among the algorithms for solving problem Z, one should find such an algorithm A*,

$$\varphi(A*) = \sup_{A\in\{A\}}\varphi(A) \tag{5}$$

In other words, among the recognition algorithms, it is necessary to choose the algorithm that most accurately recognizes the object from the control sample.

Let $\{S_{ij}\}$ be a set of images of the ear that can be recognized in person recognition based on the image of the ear, i.e. j is a set of images of the person's ear at the moment i.

In each set of $S_{ij} \in S$ ear images

$$S = (x_1, x_2, \ldots, x_m) \tag{6}$$

let the character space be defined. Here, $x_1, x_2, \ldots, x_m$'s S are ear image identifiers.

Let the set of ear images $\{S\}$ be partitioned into non-intersecting classes $K_1, K_2, \ldots, K_l$. Here, each class represents ear images taken at different time intervals belonging to one individual. It does not completely define the classification of the set of ear images $\{S\}$ into classes, only some initial information $K_1, K_2, \ldots, K_l$ about the classes I_0:

$$\begin{gathered} I_0 = \{S_1, S_2 \ldots, S_m;\ \bar{a}(S_1), \bar{a}(S_2) \ldots, \bar{a}(S_m)\}; \\ \bar{a}(S) = (\alpha_1, \alpha_2, \ldots, \alpha_l), \end{gathered} \tag{7}$$

where $\bar{a}(S) - S$ is the data vector of the ear image, $\alpha_{ij} - P_j(S_i)$ – "$S_j \in K_j$" – is the value of the predicate.

The main problem is to select the algorithm that best correctly recognizes the ear image obtained for the control sample from among the recognition algorithms selected as a result of comparing the values of the elementary predicates $P_j(S)$ – "$S \in K_j$" on the initial data I_0 and the $I(S)$ data of the ear image to be recognized S, and it is in the following form:

$$\varphi* = \arg\max_{g=\overline{1,4}}\left(\arg\max_{f=\overline{1,5}}\left(\varphi_{n_f}^{g}\right)\right) \tag{8}$$

where $g = \overline{1,4}$ is the characters identified by character extraction algorithms and $n_f, f = \overline{1,5}$ is defined by character extraction algorithms.

Therefore, in order to evaluate the effectiveness of the algorithms that separate the identification features of the image of the auricle, the application of several recognition algorithms was carried out. Based on the identification features of ear images of individuals taken under different conditions, several different types of training sample tables were created. Using the generated training samples, recognition is performed with the help of recognition algorithms and the most effective one is selected. The selected training sample serves to divide the new control object into previously known classes.

When recognizing a person on the basis of his image, if the identification symbols in the image are well separated, it is appropriate to use simple algorithms.

Also, simple algorithms were used to recognize a person based on the image of the ear. A good separation of identification features in the image of the auricle was achieved. In this work, human recognition algorithms based on AdaBoost-type recognition algorithms were used to improve recognition based on popularized average distance, standard objects, k nearest neighbors and combining the obtained results [11,12].

It is known that each vector $\bar{a} = (a_1, a_2, a_3, \ldots, a_n)$ in the character space corresponds to a point. Various metrics can be used to find the distance between two points. In this work, the Euclidean metric is used to calculate the distance between two points, and its overview is as follows:

$$l(\bar{a},\bar{b}) = \sqrt{\left(\sum_{i=1}^{n}(a_i - b_i)^2\right)} \quad (9)$$

Below we consider the algorithms for recognizing a person based on the image of the ear:
The average distance based recognition algorithm (A1) is implemented in the following steps.

Step 1. Distances from object $X(x_1, \ldots, x_n)$ to objects belonging to class K_j are determined:

$$d_j(X) = \sum_{X_i \in \widetilde{K}_j} (X' * X_i - 0.5 * X_i' * X_i)/|K_j| \quad (10)$$

where $j = \overline{1,l}$, l is the number of classes.
Step 2. The minimum distance is determined and the class of objects closest to the considered object X is selected [2–4,6–8].

The benchmark object-based recognition algorithm (A2) is implemented in the following steps.

Step 1. $\gamma := 1$ is given an initial value.
Step 2. The standard object for class K_j is defined:

$$Z_j = (Z_1, \ldots, Z_i \ldots, Z_n),$$
$$Z_i = \left(\sum_{S \in \widetilde{K}_j} a_i\right) / |\widetilde{K}_j|, \; S = (a_1, \ldots, a_i, \ldots, a_{n'}) \quad (11)$$

Step 3. $(X = (x_1, \ldots, x_i, \ldots x_{n'}))$ is the distance from the object to the standard Z_j:

$$d_j(X) = XZ_j - Z_j'Z_j/2 \quad (12)$$

Step 4. If condition $j \leq l$ is met, then $j + j + 1$ and step 4 are executed.
Step 5. The minimum distance is determined, and the reference object Z_j, which is closest to the considered object X, is selected.

Person recognition based on k *nearest neighbors algorithm (A3) is performed in the following 3 steps.*

In step 1 the proximity $l(S_i, S)$ $(i = \overline{1,m})$ between the unknown image S and each of the images belonging to the training sample is calculated. According to it, k images that are closest to S image are selected.

In step 2 the affinity between a given image and a class of images is determined. In order to determine the proximity between a given image and a class of images, the number of images extracted in the first step that belong to each image class $K_u(u = 1,l)$ is k_u.

Suppose there are k_1 images in class K_1, k_2, images in K_2 class, … , K_l images in k_l class, where the proximity between $\sum_{j=1}^{l} k_j = k$ unknown S image and $K_u(u = \overline{1,i})$ image class is defined as follows [5,6,11,12]:

$$B(S, K_u) = \frac{k_u}{k} \quad (13)$$

In step 3 a decision is made. Since the proximity measures between the unknown image S and the image class $K_u(u = \overline{1,i})$ are $B(S,K_1), B(S,K_2), \ldots, B(S,K_l)$, the decision rule is implemented as follows:

If $B(S, K_1) = \max_{1 \le u \le l} \{B(S, K_u)\}$ then, $K_j \in S$, else $S \notin K_j$.

AdaBoost type recognition(A4). Combining the results of several recognition algorithms that recognize a single object using different spaces of object description allows to refine and improve the results. It usually achieves a higher level of recognition with a lower rate of error. This algorithm, which uses three elementary recognition algorithms, is implemented in the following steps [3–5,10]:

Step 1. The first recognition algorithm is trained on an arbitrary training sample consisting of N_1 objects.
Step 2. After tuning, the first recognition algorithm is used as a filter to form the training sample for the second recognition algorithm as follows.

The value of the random variable ξ is calculated using a uniformly distributed random number generator. The value of the random variable ξ varies depending on the problem at hand. Determination of the value of ξ is carried out as a result of conducting experimental studies. The value of ξ in the recognition of a person based on the image of the ear was determined by conducting experimental studies and its value was determined to be 0.5. If $\xi \le 0.5$, the objects in the training sample are fed to the input of the first recognition algorithm until it gives the correct answer. The first recognition algorithm misled object is included in the training sample for the second expert. If $\xi \succ 0.5$, the first recognition algorithm input is given objects from the given set until it returns the correct answer.

The object that the first recognition algorithm answered correctly is included in the training sample for the second recognition algorithm [1,2,5,9].

The described process is continued until the training sample for the second recognition algorithm reaches N_2 numbers. It is self-evident that the reliability of the first recognition algorithm is of the order of 0.5. Thus, the difference between the samples on which the first two recognition algorithms are trained is achieved [3].
Step 3. The second recognition algorithm is trained on the generated sample.
Step 4. A training sample is formed for the third recognition algorithm.

The next object from the training sample is fed to the first and second recognition algorithms at the same time. If their solutions differ, this object is included in the training sample for the third recognition algorithm.

The described process is continued until the training sample for the second recognition algorithm reaches N_3 numbers.
Step 5. The third recognition algorithm is trained on the generated sample.

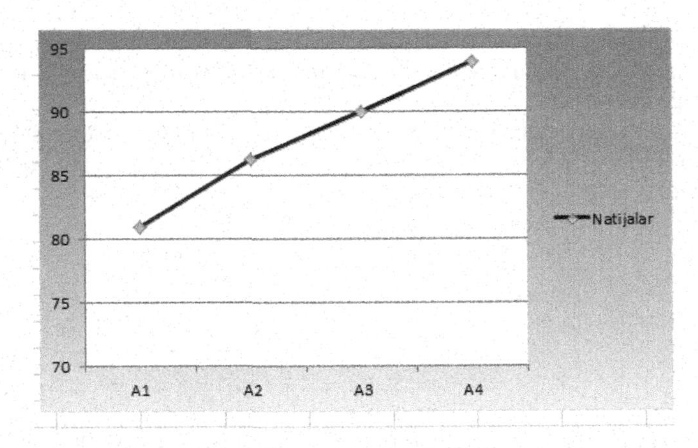

Figure 1. Comparative result of algorithms A_1,A_2,A_3,A_4.

In the described procedure, N_1 examples were used to train the first recognition algorithm, $N_2 \geq N_1$ objects were considered for training the second recognition algorithm, and $N_3 \geq N_2$ objects were considered for the third recognition algorithm. Thus, for a set of recognition algorithms, $N_1 + N_2 + N_3$ different objects were trained in one way or another, which is more than the $3N_1$ objects involved in independent training of the three recognition algorithms separately. Therefore, it appears that a set of recognition algorithms is "smarter" than three independent recognition algorithms.

Decision making in this recognition algorithm is done as follows. The new sample is given to the first two recognition algorithm inputs. If they give the same conclusion, then it is accepted as the solution of the problem. If their opinions differ, then the conclusion of the third recognition algorithm is accepted as the final conclusion.

Suppose that each of the three recognition algorithms has an error of ξ when trained individually, where $\xi \prec 0.5$ and the error of the set of three recognition algorithms

$$\xi_3 \leq 3\xi^2 - 2\xi^3 \tag{14}$$

it can be said that the system consisting of three recognition algorithms is more reliable than the result of any of these algorithms individually.

2 EXPERIMENTAL RESEARCH RESULTS

Figure 1 shows the results of recognizing a person based on the image of the ear using the aforementioned A_1,A_2,A_3,A_4 recognition algorithms. According to it, images of 20 out of 40 individuals were taken for recognition. Figure 2 shows the block diagram of the recognition of the person based on the ear image.

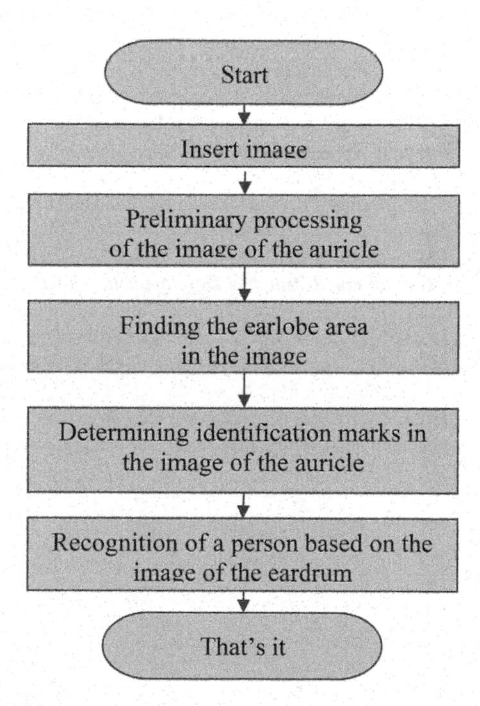

Figure 2. Block diagram of recognition of a person based on an ear image.

3 CONCLUSION

It can be concluded from the conducted experiments that the more accurate and high-precision the values of the identification symbols in the images of the ear are separated, the more accurate the level of recognition will be.

In solving the problem of recognizing a person based on the image of the ear, the recognition accuracy has been further improved by combining the results of simple recognition algorithms.

On the basis of algorithms A_1,A_2,A_3,A_4 presented above, an algorithm and a software tool were developed for further improvement of recognition of a person based on the image of the ear and combining the obtained results.

REFERENCES

[1] Burge, M. and Burger, W. (1998a) Ear Biometrics. *Biometrics: Personal Identification in a Networked Society*, p 273–286. Kluwer Academik.

[2] Freund, Y. Schapire, R. Short, A. Introduction to Boosting. *Journal of Japanese Society for Artificial Intelligence*, 14(5):771–780, September, 1999.

[3] Ferreira A., Figueredo M. *Boosting algorithms: A Review of Methods, Theory, and Applications.*

[4] C. Zhang, Y. Ma (eds.). *Ensemble Machine Learning: Methods and Applications*, Springer, New York, 2012, pp, 35–85.

[5] Arras, C. Stachniss, M. Bennewitz, W. Burgard. *Robotics 2 AdaBoost for People and Place Detection.* Uni Freiburg (Фрайбургуниверситети). 2012.

[6] Fazilov, S. X., Mahkamov, A. A., & Jumayev, T. S. (2018). Algorithm for Extraction of Identification Features in Ear Recognition. In *Informatics: Problems, Methodology, technologies* (pp.3–7).

[7] Jumayev, T. S., Mirzayev, N. S., & Makhkamov, A. S. (2015). Algorithms for Segmentation of Color Images based on the Allocation of Strongly Coupled Elements. *Studies of Technical Sciences*, (4), 22–27.

[8] D. Tuhtanazarov, M. Xodjayeva, T. Jumayev, A. Mahkamov. Computational Algorithm and Program for Determining the Indicators of Wells based on Processing of Information of Oil fields. *AIP Conference Proceedings* 2432, 060021 (2022).

[9] Turdali Jumayev Saminjonovich, Mahkamov Anvarjon Abdujabborovich, Tukhtanazarov Dilmurod Solijonovich, Dadamuxamedov Alimjon Irgashevich., Experimental Study of Algorithms for Ear Images Distribution in Personal Identity, *Turkish Online Journal of Qualitative Inquiry (TOJQI)* Volume 13, Issue 1, January 2022: 813–822.

[10] Abdujabborovich M. A. *et al.* Human Personal Identification Algorithms from the Image of the Ear //*International Engineering Journal For Research & Development*. – 2020. – T. 5. – №. 6. – C. 5-5.

[11] Abdujabborovich M. A. One way to Identify a Person Based on Their Image is to Provide Security //*International Engineering Journal For Research & Development*. – 2020. – T. 5. – №. Special Issue. – C. 8-8.

[12] Фазылов Ш. Х., Мирзаев Н. М., Махкамов А. А. Выделение Геометрических Признаков Изображений Ушных Раковин //*XI* Всероссийская Научная Конференция «Нейрокомпьютеры и их Применение. – 2013. – T. 19.

[13] Фазылов Ш. Х., Мирзаев Н. М., Махкамов А. А. Выделение Геометрических Признаков Изображений Ушных Раковин //*XI* Всероссийская Научная Конференция «Нейрокомпьютеры и их Применение. – 2013. – T. 19.

Detection and analysis of traffic jams using computer vision technologies

A. Akhatov, F. Nazarov & B. Eshtemirov
Samarkand State University, Samarkand, Uzbekistan

ABSTRACT: Many expanding cities in developing countries are faced with traffic congestion problems due to the overabundance of people and vehicles. Collecting real-time, reliable and accurate traffic flow information is essential for urban traffic management. The main goal of this article is to develop an adaptive model that can estimate the number of vehicles on city roads in real time using computer vision technologies. This paper proposes a real-time automatic background update algorithm for vehicle detection and an adapted model for vehicle counting based on virtual ring and detection line methods. In addition, a new reliable detection method was introduced to monitor the traffic situation of the road sector in real time. A system prototype was developed for testing and installed on a city road.

1 INTRODUCTION

The increasing role of the city in the life of the society leads to a rapid increase in the population density and the concentration of traffic on the city streets. This, in turn, leads to an increase in traffic accidents, an increase in traffic jams, and air pollution due to smoke emitted from cars. Due to the limited infrastructure of the city, it is not possible to find an optimal solution to the task of ensuring traffic-free movement of cars on the city streets based on a small amount of statistical data (Akhatov *et al.* 2021a; Rashidov *et al.* 2021). Intelligent transport systems (ITS) of cities must ensure the maximum capacity of the road network and any traffic incidents should be resolved immediately to avoid traffic congestion. Currently, video surveillance systems are growing rapidly on city streets, which allows continuous monitoring of quantitative and qualitative traffic parameters and the use of vehicles as indicators of the transport system. Video cameras installed at traffic lights are most recognized for processing data in real time: low-precision calculation, classification of limited types of vehicles, determining the speed and direction of movement of all parts when crossing an intersection. allows you to solve the problem of tracking ect. But despite the advantages of developing such systems, there are not enough studies aimed at determining and analyzing the speed and movement of traffic flow by using cameras at intersections (Akhatov *et al.* 2022a). Artificial neural networks have proven themselves well in the tasks of collecting, interpreting and analyzing large data coming from video cameras. Some studies have also used low-resolution video image data and deep neural networks to estimate the estimated traffic density of road traffic. Today, Faster R-CNN, YOLO and SSD are the most modern systems for recognizing vehicles moving on roads. Existing solutions to the problems of real-time vehicle recognition and classification require a large computing power, which imposes strict requirements on the location and operation of the camera.

DOI: 10.1201/9781032684994-123

2 MATERIALS AND METHODS

2.1 *General structure of the study*

In this research work, we present a proposed deep learning system for vehicle counting based on video image. To do this, we can divide this process into three parts, as shown in Figure 1: creating a dataset, building a vehicle recognition model, and counting vehicles (Nazarov *et al.* 2022a). First, in order to deeply learn the vehicle recognition model, we need to create a training dataset, and in order not to spend too much time in image recognition, annotated images containing vehicles in the open dataset are used for training. taken as data. In addition, to further improve and evaluate the performance of the vehicle recognition model, multiple frame images from the traffic videos are extracted and divided into additional training data and test data after extraction. Then, in the stage of building the car recognition model, the deep learning object recognition model that meets the requirements in terms of accuracy and efficiency is used as the main model, example-based transfer learning and parameter-based transfer learning to build the car recognition model is accepted (Haojia Lin *et al.* 2022). As a result, in the vehicle counting stage, the vehicles in each frame are identified by the deep learning vehicle detection model, and based on the combination of virtual detection area and vehicle tracking, the vehicle counting model is counted.

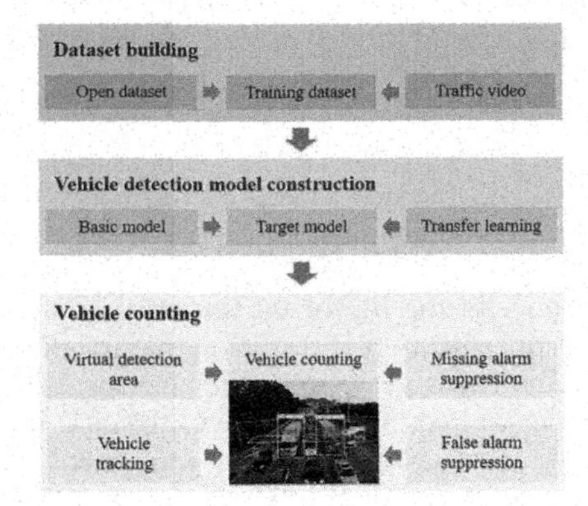

Figure 1. Counting cars based on video data.

2.2 *Vehicle recognition*

Deep learning object recognition models use a convolutional neural network (CNN) to extract image features, and then a classifier and regressor to classify and localize the extracted features (Nazarov *et al.* 2022a). These models can be divided into two main groups: two-stage detectors and one-stage detectors. Examples of two-stage detectors are R-CNN, Fast R-CNN, and Faster R-CNN. SSD and YOLO are examples of single-stage detectors. One-stage detectors work faster than two-stage detectors.

Accuracy and efficiency are important when counting cars. For this reason, single-stage detectors are more efficient for building a vehicle counting model (Nazarov *et al.* 2022b). By outperforming the rest in terms of efficiency and accuracy, YOLO has an edge over many other single-stage detectors. Therefore, YOLO technology was chosen as the main model to build the vehicle recognition model (Akhatov *et al.* 2022a). The implementation of YOLO is as follows:

1) A series of convolutional layers and residual layers are used to recognize the cars in the obtained images, and they are divided into three parts of different sizes.

2) Each part is divided into S×S grids and B cells are placed in each row to recognize vehicles.
3) If an object falls into a cell, the probability of classification (including the probability of each classification) and the data of this object (Coordinate of the center point, width, height) are determined by this cell (Skabardonis *et al.* 2008).
4) The above processes are carried out for each part, and the final result is obtained by synthesizing the results of the three scales. (Wu *et al.* 2013).

2.3 *Calculation of the specified area*

To calculate the distance traveled, we need to find the change in latitude and longitude of the location of the vehicle in a certain time interval, using the change of coordinates in the camera image (Kirill Khazukov *et al.* 2020). To solve this problem, by selecting four reference points on the map and comparing the corresponding points on the image, we calculated a perspective transformation matrix (Figure 2).

Figure 2. The research area.

To calculate the perspective transformation matrix, we need to get the coefficients from the following linear equations describing the relationship between the image coordinates and geographic coordinates:

$$u_i = \frac{c_{00}x_i + c_{01}y_i + c_{02}}{c_{20}x_i + c_{21}y_i + c_{22}} \tag{1}$$

$$v_i = \frac{c_{10}x_i + c_{11}y_i + c_{12}}{c_{20}x_i + c_{21}y_i + c_{22}} \tag{2}$$

here u_i, v_i geographic coordinates, c_{ij} c- elements of the matrix A, $c_{22} = 1;x_i, y_i$- coordinates from the image, i=1.4.

The general view of the perspective matrix is as follows: $A = (c_{ij})_{3\times 3}$.

As a result of the calculation, we solve the following matrix equation

$$\begin{pmatrix} x_0\ y_0\ 1\ 0\ 0\ 0 & -x_0 * u_0 & -\ y_0 * u_0 \\ x_1\ y_1\ 1\ 0\ 0\ 0 & -x_1 * u_1 & -\ y_1 * u_1 \\ x_2\ y_2\ 1\ 0\ 0\ 0 & -x_2 * u_2 & -\ y_2 * u_2 \\ x_3\ y_3\ 1\ 0\ 0\ 0 & -x_3 * u_3 & -\ y_3 * u_3 \\ 0\ 0\ 0\ x_0\ y_0\ 1 & -x_0 * v_0 & -\ y_0 * v_0 \\ 0\ 0\ 0\ x_1\ y_1\ 1 & -x_1 * v_1 & -\ y_1 * v_1 \\ 0\ 0\ 0\ x_2\ y_2\ 1 & -x_2 * v_2 & -\ y_2 * v_2 \\ 0\ 0\ 0\ x_3\ y_3\ 1 & -x_3 * v_3 & -\ y_3 * v_3 \end{pmatrix} \times \begin{pmatrix} c_{00} \\ c_{01} \\ c_{02} \\ c_{10} \\ c_{10} \\ c_{12} \\ c_{20} \\ c_{21} \end{pmatrix} = \begin{pmatrix} u_0 \\ u_1 \\ u_2 \\ u_3 \\ v_0 \\ v_1 \\ v_2 \\ v_3 \end{pmatrix} \tag{3}$$

After finding the matrix coefficients, we can perform the transformation from the image to the coordinate vector by multiplying the perspective transformation matrix. (Ramezani *et al.* 2015)

$$A \times \begin{pmatrix} x_i \\ y_i \\ 1 \end{pmatrix} = \begin{pmatrix} x_i^{'} \\ y_i \\ t_i \end{pmatrix} \quad (4)$$

here A- transformation matrix; x_i, y_i- pixel coordinates in the image; $x_i^{'}$, $y_i^{'}$- latitude and longitude of the point.

3 RESULTS

Four experiments were designed and implemented to evaluate the performance of this vehicle counting system.

The technical support of the computer for the experimental environment should be as follows: Intel Core i7-8700 3.20 GHz; Memory: 16 GB (2,666 MHz); GPU: NVIDIA GeForce GTX 1070, 8 GB.

The experiment was carried out by looking at ten videos of car traffic in different lighting conditions and driving directions (Table 1). All videos were captured at 20 frames per second (fps) for 5 minutes. (Blokpoel *et al.* 2016) The MS COCO dataset is used as the main training data because it has the features of several small objects in a non-centered distribution in the image, which is more suitable for the daily traffic process.

Table 1. Information about videos of cars moving on the road.

	Video	Light Shooting direction	Traffic conditions	Resolution
1	Day	Front	Traffic lights	1,280 × 720
2	Day	Front	Light	1920 × 1,080
3	Day	Back	Heavy	1920 × 1,440
4	Night	Back	Heavy	1920 × 1,440
5	Day	Front	Traffic lights	1920 × 1,440
6	Night	Front	Traffic lights	1920 × 1,440
7	Night	Front	Traffic lights	1920 × 1,440
8	Night	Back	Light	1920 × 1,440
9	Day	Oblique front	Traffic lights	932 × 500
10	Day	Oblique back	Light	932 × 500

4 CONCLUSIONS

In this study, we focused on the problem of obtaining programmatic and directional information of traffic through video streaming from street surveillance cameras. Different lines of sight, distance from the place of removal, removal of objects from each other are easily related to the complexity of the task. We added an additional mask section to the YOLOv3 neural network architecture, and optimized size objects to improve object quality and anchor objects to improve classification accuracy. In this research, a deep learning system is proposed for high-quality car counting in video images. It includes a cell device: building a vehicle recognition model and deep learning to count vehicles.

REFERENCES

Akhatov A., Nazarov F. & Rashidov A. 2021a. "Mechanisms of Information Reliability in Big Data and Blockchain Technologies" *ICISCT 2021: Applications, Trends and Opportunities*, 3–5.11.2021, doi: 10.1109/ICISCT52966.2021.9670052

Akhatov A., Nazarov F. & Rashidov A. 2021b. "Increasing Data Reliability by Using Bigdata Parallelization Mechanisms" *ICISCT 2021: Applications, Trends and Opportunities*, 3–5.11.2021, doi: 10.1109/ICISCT52966.2021.9670387

Akhatov A., Sabharwal M., Nazarov F. & Rashidov A. 2022a. Application of Cryptographic Methods to Blockchain Technology to Increase Data Reliability. *2nd International Conference on Advance Computing and Innovative Technologies in Engineering (ICACITE 2022)* DOI: 10.1109/ICACITE53722.2022.9823674

Akhatov A., Renavikar A., Rashidov A. & Nazarov F. 2022b. Development of the Big Data Processing Architecture Based on Distributed Computing Systems, *Informatika va Energetika Muammolari O'zbekiston jurnali*, № (1) 2022, 71–79

Nazarov F.M., Eshtemirov B.Sh., Yarmatov Sh.Sh. 2022a. "Technologies for Identifying Vehicles Standing at Traffic Lights Based on Video Data", *Central Asian Journal Of Mathematical Theory And Computer Sciences*, Volume: 03 Issue: 12 | Dec 2022 ISSN: 2660–5309

Nazarov F.M., Eshtemirov B.Sh., Yarmatov Sh.Sh. 2022b. "Video Data Processing Methodology for Investigation", *International Journal of Novel Research in Advanced Sciences*, Volume: 01 Issue: 06 | 2022, ISSN: 2751-756X

Haojia Lin, Zhilu Yuan, Biao He, Xi Kuai, Xiaoming Li & Renzhong Guo 2022. A Deep Learning Framework for Video-Based Vehicle Counting, *Frontiers in Physics*, 21 February 2022, doi: 10.3389/fphy.2022.829734

Kirill Khazukov, Vladimir Shepelev, Tatiana Karpeta, Salavat Shabiev, Ivan Slobodin, Irakli Charbadze & Irina Alferova 2020. Real-time Monitoring of Traffic Parameters. *Journal of Big Data*, 2020, DOI: 10.1186/s40537-020-00358-x

Rashidov A. & Akhatov A. 2021. "Big Data va uning turli sohalardagi tadbiqi", *Descendants of Muhammad Al-Khwarizmi*, 2021, № 4 (18), 135–44

Skabardonis, A.; Geroliminis, N. Real-Time Monitoring and Control on Signalized Arterials. *J. Intell. Transp. Syst.* 2008, 12(2), 64–74. [CrossRef]

Li, F.; Tang, K.; Yao, J.; Li, K. Real-Time Queue Length Estimation for Signalized Intersections Using Vehicle Trajectory Data. *Transp. Res. Rec. J. Transp. Res. Board* 2017, 2623, 49–59.

Ramezani, M. , and Geroliminis, N. . Queue Profile Estimation in Congested Urban Networks with Probe Data. *Comput.-Aided Civ. Infrastruct. Eng.* 2015, 30, 6, 414–432. [CrossRef]

Blokpoel, R.; Vreeswijk, J. Uses of Probe Vehicle Data in Traffic Light Control. *Transp. Res. Procedia* 2016, 14, 4572–4581. [CrossRef]

Wu, A.; Yang, X. Real-Time Queue Length Estimation of Signalized Intersections Based on RFID Data. *Procedia Soc. Behav. Sci.* 2013, 96, 1477–1484. [CrossRef]

Author index